EXTERIOR ANALYSIS

Using Applications of Differential Forms

Erdoğan S. Şuhubi

Emeritus Professor, Istanbul Technical University
Istanbul, Turkey

Exterior Analysis

Using Applications of Differential Forms

Erdoğan S. Şuhubi

AMSTERDAM • BOSTON • HEIDELBERG • LONDON • NEW YORK • OXFORD
PARIS • SAN DIEGO • SAN FRANCISCO • SYDNEY • TOKYO

Academic Press is an Imprint of Elsevier

Academic Press is an imprint of Elsevier
225, Wyman Street, Waltham, MA 02451, USA
The Boulevard, Langford Lane, Kidlington, Oxford OX5 1GB, UK
Radarweg 29, PO Box 211, 1000 AE Amsterdam, The Netherlands

Library of Congress Cataloging-in-Publication Data
A catalog record for this book is available from the Library of Congress

British Library Cataloguing in Publication Data
A catalogue record for this book is available from the British Library

ISBN: 978-0-12-415902-0

For information on all Academic Press publications
visit our web site at *store.elsevier.com*

TABLE OF CONTENTS

PREFACE ix

CHAPTER I **EXTERIOR ALGEBRA**

1.1. Scope of the Chapter 1
1.2. Linear Vector Spaces 2
1.3. Multilinear Functionals 19
1.4. Alternating k-Linear Functionals 24
1.5. Exterior Algebra 33
1.6. Rank of an Exterior Form 37
Exercises 46

CHAPTER II **DIFFERENTIABLE MANIFOLDS**

2.1. Scope of the Chapter 51
2.2. Differentiable Manifolds 52
2.3. Differentiable Mappings 93
2.4. Submanifolds 103
2.5. Differentiable Curves 110
2.6. Vectors. Tangent Spaces 112
2.7. Differential of a Map Between Manifolds 120
2.8. Vector Fields. Tangent Bundle 126
2.9. Flows over Manifolds 133
2.10. Lie Derivative 142
2.11. Distributions. The Frobenius Theorem 153
Exercises 168

CHAPTER III **LIE GROUPS**

3.1. Scope of the Chapter 175
3.2. Lie Groups 175

3.3. Lie Algebras 180
3.4. Lie Group Homomorphisms 187
3.5. One-Parameter Subgroups 189
3.6. Adjoint Representation 194
3.7. Lie Transformation Groups 197
Exercises 200

CHAPTER IV TENSOR FIELDS ON MANIFOLDS

4.1. Scope of the Chapter 207
4.2. Cotangent Bundle 208
4.3. Tensor Fields 209
Exercises 216

CHAPTER V EXTERIOR DIFFERENTIAL FORMS

5.1. Scope of the Chapter 219
5.2. Exterior Differential Forms 220
5.3. Some Algebraic Properties 224
5.4. Interior Product 226
5.5. Bases Induced by the Volume Form 236
5.6. Ideals of the Exterior Algebra $\Lambda(M)$ 244
5.7. Exterior Forms under Mappings 253
5.8. Exterior Derivative 260
5.9. Riemannian Manifolds. Hodge Dual 269
5.10. Closed Ideals 286
5.11. Lie Derivatives of Exterior Forms 290
5.12. Isovector Fields of Ideals 301
5.13. Exterior System and Their Solutions 306
5.14. Forms Defined on a Lie Group 317
Exercises 321

CHAPTER VI HOMOTOPY OPERATOR

6.1. Scope of the Chapter 327
6.2. Star-Shaped Regions 327
6.3. Homotopy Operator 330
6.4. Exact and Antiexact Forms 336
6.5. Change of Centre 340
6.6. Canonical Forms of 1-Forms, Closed 2-Forms 341

6.7. An Exterior Differential Equation 352
6.8. A System of Exterior Differential Equations 355
Exercises 363

CHAPTER VII LINEAR CONNECTIONS

7.1. Scope of the Chapter 365
7.2. Connections on Manifolds 366
7.3. Cartan Connection 377
7.4. Levi-Civita Connection 388
7.5. Differential Operators 398
Exercises 401

CHAPTER VIII INTEGRATION OF EXTERIOR FORMS

8.1. Scope of the Chapter 405
8.2. Orientable Manifolds 406
8.3. Integration of Forms in the Euclidean Space 414
8.4. Simplices and Chains 416
8.5. Integration of Forms on Manifolds 423
8.6. The Stokes Theorem 429
8.7. Conservation Laws 446
8.8. The Cohomology of de Rham 459
8.9. Harmonic Forms. Theory of Hodge-de Rham 476
8.10. Poincaré Duality 482
Exercises 484

CHAPTER IX PARTIAL DIFFERENTIAL EQUATIONS

9.1. Scope of the Chapter 487
9.2. Ideals Formed by Differential Equations 488
9.3. Isovector Fields of the Contact Ideal 500
9.4. Isovector Fields of Balance Ideals 520
9.5. Similarity Solutions 563
9.6. The Method of Generalised Characteristics 569
9.7. Horizontal Ideals and Their Solutions 584
9.8. Equivalence Transformations 604
Exercises 655

CHAPTER X CALCULUS OF VARIATIONS

10.1. Scope of the Chapter 657
10.2. Stationary Functionals 657
10.3. Euler-Lagrange Equations 661
10.4. Noetherian Vector Fields 670
10.5. Variational Problem for a General Action Functional 679
Exercises 693

CHAPTER XI SOME PHYSICAL APPLICATIONS

11.1. Scope of the Chapter 695
11.2. Conservative Mechanics 696
11.3. Poisson Bracket of 1-Forms and Smooth Functions 705
11.4. Canonical Transformations 710
11.5. Non-Conservative Mechanics 716
11.6. Electromagnetism 731
11.7. Thermodynamics 742
Exercises 748

REFERENCES 751

INDEX OF SYMBOLS 755

NAME INDEX 757

SUBJECT INDEX 759

PREFACE

Exterior forms come into existence at its inception as elements of the exterior algebra, which is in essence a Grassmann algebra, formed by defining an operation of exterior product over a linear vector space. However in early 20th century, works of Poincaré, and particularly of Cartan make it possible to extend this algebraic structure as to include the exterior differential forms by employing exterior products of differentials of coordinates. Nevertheless the main impetus to the elaboration of the theory of exterior differential forms has been the development of the concept of differentiable manifolds that are topological spaces equipped with an appropriate structure which blends topological properties with some kind of differentiability. It was then possible to define exterior differential form fields on differentiable manifolds that are locally equivalent to Euclidean spaces and to introduce an analysis of forms in which only first order derivatives survive. It was soon realised that this analysis would be one of the most powerful, perhaps indispensable tools of the modern differential geometry and many mathematical properties could be relatively easily revealed by almost algebraic operations. On the other hand, it is perhaps not wrong to claim that the mathematical structure of theoretical physics today is entirely based on the formalism of differential geometry. We also observe that this formalism is increasingly infiltrating into engineering sciences to study some fundamental problems and even in many practical applications. Therefore exterior analysis is no longer in the realm of mathematicians. It seems that it would now be quite beneficial for physicist and engineers to acquire a rather good skill in dealing with exterior forms. The aim of this book to expose the field of exterior analysis to readers of mathematical, physical or engineering origin, to acquaint them with the fundamental concepts and tools of exterior analysis, to help them gain certain competence in using these tools and to emphasise advantages provided by this approach. In doing so it is tried, without sacrificing mathematical rigor, not to expose the reader to very advanced, somewhat esoteric mathematical approaches. It is attempted to design the book as self sufficient as much as possible. An advanced mathematical background is not necessary to follow the exploration of the subject for an attentive reader.

The book comprises 11 chapters. **Chapter I** is a brief introduction to the exterior algebra. First, linear vector spaces over which exterior algebra will be defined are explored, linear and multilinear functionals which map vector spaces and their Cartesian products, respectively, into field of scalars are defined by means of dual spaces. Exterior forms are introduced as alternating multilinear functionals and the exterior algebra, which is essentially a Grassmann algebra, is constructed by appropriately defining a degree augmenting exterior product. **Chapter II** is concerned with differentiable manifolds that are topological spaces which are locally homeomorphically equivalent to Euclidean spaces. This equivalency is provided by charts and atlases covering the manifold. This structure over the manifold enables one to differentiate real valued functions over manifolds and mappings between manifolds, to treat the fibre bundle obtained by joining to each point of the manifold the tangent space at that point as a differentiable manifold. A differentiable function between two manifolds generates a differential of that function which maps linearly tangent spaces one another at corresponding points. Finally, the Lie derivatives which measure variations of one vector field with respect to another and the Lie algebra which they give rise on tangent spaces are defined and its various applications are explored. Special emphasis is put on distributions and their role in forming submanifolds. **Chapter III** is devoted to the study of the Lie groups that are topological spaces endowed with a continuous group operation. Lie algebras generated by left- and right-invariant vector fields are introduced and various properties are considered. Lie groups of transformations mapping a manifold into itself are also investigated. Covariant and contravariant tensors defined previously on vector spaces and their duals are extended in **Chapter IV** to tensor fields on manifolds by making use of tensor products of local basis vectors on tangent spaces and their duals. **Chapter V** is one of the fundamental chapters of the book. It deals with exterior differential forms. Noting that differentials of local coordinates constitute a basis, as was observed by Cartan, for the dual of the tangent space, differential forms are defined as completely antisymmetric covariant tensors of various orders. An exterior algebra over the manifold is built by using a degree augmenting exterior product operation. Some algebraic properties of exterior forms are revealed and a degree decreasing operation called the interior product of a form with a vector field is defined. The non-zero volume form on the manifold is employed to derive a new system of top-down generated basis for the exterior algebra that will prove to be extremely useful is several important applications. Then the ideals of the exterior algebra are defined, the exterior derivative of differential forms is introduced as to satisfy certain requirements. The Riemannian manifolds that are equipped with a metric tensor making it possible to measure distances on the manifold are explored and the Hodge dual of a form is defined. After briefly glancing over closed ideals, the Lie derivative of a form with respect to a vector field is introduced. By utilising this concept isovector fields under which an ideal remains invariant and characteristic vector fields whose interior products with forms within an ideal remains in that ideal are defined. This chapter ends with the study of exterior systems and their solutions. **Chapter VI** deals with the homotopy operator on contractible manifolds. This operator helps us to understand the relationship between closed forms with vanishing exterior derivative and exact forms that are exterior derivatives of some other forms. The forms occupying the kernel of the homotopy operator is named as antiexact forms. Homotopy operator is then effectively employed to obtain solutions of system of exterior equations. **Chapter VII** is concerned with the various types of linear connections on manifolds that helps connect tangent spaces at different points. Through the connection coefficients one can define covariant derivative of tensor fields that are also tensors. Torsion and curvature tensors of the manifold are then introduced and it becomes possible to obtain more concrete forms of various differential operators in Riemannian manifolds. In **Chapter VIII** the integration of forms is examined. The integral of a form on a manifold whose dimension is equal to the degree of the form is actually an appropriately defined Riemann integral. However, in order to make this operation realisable on a manifold that might have a rather complicated structure, such a manifold must acquire some new properties. One approach is to cover a manifold with some geometric objects such as simplices of very simple shapes, chains by forming unions of simplices, cycles that are chains with vanishing boundaries. Another approach is to make use of the partition of unity that is a topological property. The fundamental result in the integration of forms is Stokes' theorem which equates the integral of the exterior derivative of a form on an appropriate domain to the integral of this form on the boundary of that domain. De Rham cohomology group of a manifold is the quotient space of closed forms with respect to exact forms whereas homology group that reflects some topological properties of the manifold are defined as the quotient space of cycles with respect to cycles that are boundaries of some chains. Stokes theorem, together with properties of exact sequences, helps reveal certain very interesting relationships between these two seemingly unrelated groups. The long **Chapter IX** is devoted to the discussion of partial differential equations by employing exterior analysis. To this end, a system of partial differential equation of finite order is enlarged to a great extent by introducing auxiliary dependent variables to a system of first order equations. It is then shown that the solution of that system of differential equations coincides with the solution of an ideal of the exterior algebra on an extended manifold generated by certain exterior forms. An important part of this chapter is devoted to the determination of isovector fields of the ideal. These vector fields will generate groups of symmetry transformations (Lie groups) that leave the system of differential equations invariant, thereby transforming one solution onto another one. The reduced determining equations for isovector components are explicitly obtained for balance equations which play an important part in physical and engineering sciences. Furthermore, the classical method of characteristics is generalised by employing isovector fields so that they yield solutions of the system for particular initial data satisfying certain requirements. In another approach, the ideal in question is enlarged as to include some arbitrariness which can then be somewhat removed by certain assumptions to lead certain particular solutions. Finally equivalence transformations that map a solution of a member of a family of differential equations with certain common properties to the solution of another member of the same family are discussed, the determining equations for isovector fields are given and their explicit solutions are further provided. **Chapter X** contains the treatment of classical variational calculus by use of exterior analysis. The making of a functional defined by an integral stationary requires the vanishing of some exterior forms. This way one obtains the well-known Euler-Lagrange equations. Moreover, variational symmetries are defined and Noetherian vector fields generating these symmetries and resulting conservation laws are discussed. In the final **Chapter XI**, the application of exterior analysis to some physical fields such as analytical mechanics, electromagnetism and thermodynamics is discussed in some detail.

The book contains 115 examples which are hoped to explain and illustrate the main text and 250 exercises which may help the reader to acquire certain competence on the subject.

I gratefully acknowledge the initial support provided by Turkish Academy of Sciences during the preparation of this work.

Finally, I would like to express my sincere thanks to ELSEVIER and especially to the Acquisition Editor Ms. Patricia Osborn, the Math Publisher Ms. Paula Callaghan, the Editorial Project Manager Ms. Jessica Vaughan and the Production Manager Mr. Mohanapriyan Rajendran for their kind and constructive cooperation during the whole process of the publication of this book.

Istanbul, October 2012 Erdoğan S. Şuhubi

CHAPTER I

EXTERIOR ALGEBRA

1.1. SCOPE OF THE CHAPTER

An operation that helps us to extend in some way the notion of vectorial product in the classical vector algebra to vector spaces with dimensions higher than three is called *the exterior product* and a vector space equipped with such an operation assigning a new vector to every pair of vectors in the vector space is called an *exterior algebra*. This operation was introduced in 1844 by German mathematician Hermann Günter Grassmann (1809-1877). Thus the exterior algebra is sometimes known as the ***Grassmann algebra***.

We first define in Sec. 1.2 linear vector spaces axiomatically over which the exterior algebra will be built. Some pertinent attributes of vector spaces to which we will have recourse frequently are briefly discussed there. These are concepts of linear independence and basis, linear operators, the algebraic dual space that is the linear vector space formed by linear functionals over this vector space and some significant properties of dual spaces of finite-dimensional vector spaces and finally exact sequences. Then, the multilinear functionals that are mappings from the finite Cartesian product of vector spaces into the field of scalars that are linear in each of their arguments are considered in Sec. 1.3. It is shown that by properly defining the operation of tensor product it becomes possible to endow the Cartesian products of vector spaces with a structure of a vector space and it is observed that multilinear (k-linear) functionals may be expressible in terms of elements of that space called tensors (contravariant on the vector spaces, covariant on their duals). Afterward we investigate briefly in Sec. 1.4 alternating k-linear functionals that are completely antisymmetric with respect to their arguments and the operation of alternation which help produce completely antisymmetric quantities. The generalised Kronecker deltas and Levi -Civita symbols that facilitate to a great extent the implementation of this operation are also discussed in detail. The exterior product of vectors are then defined by means of the operation of alternation on tensor products. It is then shown that a completely antisymmetric covariant tensor representing

Exterior Analysis, DOI: 10.1016/B978-0-12-415902-0.50001-3

an alternating k-linear functional is expressible by using exterior products. Such a tensor will be called an *exterior form*. Exterior products of exterior forms are defined in such a way that two exterior forms generate another form of different degree. Thus, this enables us to construct in Sec. 1.5 an exterior algebra over a vector space. This chapter ends in Sec. 1.6 with the discussion of the concept of rank of a form that makes it possible sometimes to reduce an exterior form to a simpler structure.

1.2. LINEAR VECTOR SPACES

In order to define a linear vector space abstractly we consider an *Abelian* (*commutative*) *group* $\{G, \#\}$ [after Norwegian mathematician Niels Henrik Abel (1802-1829)] and a *field* $\{\mathbb{F}, +, \times\}$. 1 is the identity element of the field with respect to multiplication. A binary operation $\mathbb{F} \times G \to G$ is denoted by $*$. Hence, this binary operation assigns a member $\alpha * x \in G$ of the group to an arbitrary scalar $\alpha \in \mathbb{F}$ and an arbitrary member $x \in G$ of the group. Furthermore, we shall assume that this binary operation $*$ will obey the following rules for all $\alpha, \beta \in \mathbb{F}$ and $x, y \in G$:

$$(i).\ (\alpha \times \beta) * x = \alpha * (\beta * x).$$
$$(ii).\ (\alpha + \beta) * x = (\alpha * x) \# (\beta * x), \quad \alpha * (x \# y) = (\alpha * x) \# (\alpha * y).$$
$$(iii).\ 1 * x = x.$$

The algebraic system $\mathcal{V} = \{G, \mathbb{F}, \#, +, \times, *\}$ satisfying these conditions is called a ***linear vector space*** over the field $\mathbb{F}$. Members of the group are named as ***vectors*** whereas members of the field as ***scalars***. The operation # is known as *vector addition* and the operation $*$ as *scalar multiplication*.

Sometimes it becomes advantageous to replace the field of scalars by a ring with identity in the system described above. Such an algebraic system is then called a ***module***. We will have opportunities to deal with modules in later parts of this work.

As far as we are concerned, the field of scalars $\mathbb{F}$ will either be the real numbers $\mathbb{R}$ or complex numbers $\mathbb{C}$. Accordingly, we shall consider either *real* or *complex vector spaces*. However, in this work, we shall be mostly interested in real spaces. Moreover, in order to simplify the notation we prefer to use the same symbol $+$ to designate addition operations both in the group and in the field while identity elements with respect to these operations will be represented, respectively, by the symbols $\mathbf{0}$ and 0. Usually, we shall not use any symbol for scalar multiplication as well as for the product of two scalars of the field by adopting the familiar convention employed in the multiplication of real or complex numbers. Although one might think that representation of different operations by the same symbol would cause some complications, we should observe that the real nature of these symbols are unambiguously revealed within the context of expressions in which they are involved. Thus it is unlikely that misinterpretations may ever arise concerning these operations. Nevertheless, a much more detailed definition of a linear vector space can also be given as follows.

I. $+$ *is a binary operation on a set V, whose members are called vectors, having the following properties*:

(i). *$u+v\in V$ for all $u,v\in V$ (closed operation).*
(ii). *$u+v=v+u$ for all $u,v\in V$ (commutative operation).*
(iii). *$(u+v)+w=u+(v+w)$ for all $u,v,w\in V$ (associative operation).*
(iv). *There exists an **identity element** $\mathbf{0}\in V$ such that $u+\mathbf{0}=u$.*
(v). *There exists an **inverse element** $-u\in V$ for each $u\in V$ such that $u+(-u)=\mathbf{0}$.*

*These properties are tantamount to say that the set V is an Abelian group with respect to the operation $+$. The element $\mathbf{0}$ is called the **zero vector** and $u+v$ is called the **vector sum** of vectors u and v. We usually employ the abbreviated notation $u-v$ to denote $u+(-v)$.*

II. *Let $\mathbb{F}$ be a field of scalars. Scalar multiplication over the Abelian group V is so defined that it satisfies the following relations*:

For all $\alpha,\beta\in\mathbb{F}$ and $u,v\in V$ we have

(i). *$\alpha u\in V$ (closed operation).*
(ii). *$(\alpha\beta)u=\alpha(\beta u)$ (associative operation).*
(iii). *$(\alpha+\beta)u=\alpha u+\beta u,\ \alpha(u+v)=\alpha u+\alpha v$ (distributive operation).*
(iv). *$1\cdot u=u$.*

*Here 1 is the identity element of the field of scalars with respect to the multiplication. We call the set V satisfying all axioms in **I** and **II** a **linear vector space over the field $\mathbb{F}$**. The scalar multiplication is represented by the symbol $\cdot$ although we would often prefer to omit it.*

We can deduce some fundamental properties of linear vector spaces from the foregoing axioms:

(a). If we write $u=1\cdot u=(1+0)\cdot u=1\cdot u+0\cdot u=u+0\cdot u$, we immediately obtain

$$0\cdot u=\mathbf{0}$$

for all $u\in V$.

(b). From $\mathbf{0}=0\cdot u=(1-1)\cdot u=1\cdot u+(-1)\cdot u=u+(-1)\cdot u$, it follows that

$$(-1) \cdot u = -u$$

for all $u \in V$.

(c). Since $\alpha u = \alpha(u + \mathbf{0}) = \alpha u + \alpha \cdot \mathbf{0}$, we find that

$$\alpha \cdot \mathbf{0} = \mathbf{0}$$

for all $\alpha \in \mathbb{F}$.

Example 1.2.1. Let us consider the set $\mathbb{F}^n$ where n is a positive integer. $\mathbb{F}^n$ is the Cartesian product $\underbrace{\mathbb{F} \times \mathbb{F} \times \cdots \times \mathbb{F}}_{n}$. An element $u \in \mathbb{F}^n$ is an ordered n-tuple $u = (\alpha_1, \alpha_2, \dots, \alpha_n)$ where $\alpha_1, \alpha_2, \dots, \alpha_n \in \mathbb{F}$. For elements $v = (\beta_1, \beta_2, \dots, \beta_n) \in \mathbb{F}^n$ and $\alpha \in \mathbb{F}$ let us define the vector addition and scalar multiplication by making use of the operations in the field $\mathbb{F}$ as follows

$$u + v = (\alpha_1 + \beta_1, \alpha_2 + \beta_2, \dots, \alpha_n + \beta_n), \ \alpha u = (\alpha\alpha_1, \alpha\alpha_2, \dots, \alpha\alpha_n).$$

It is then straightforward to see that the set $\mathbb{F}^n$ so equipped is a linear vector space. The zero vector $\mathbf{0} \in \mathbb{F}^n$ is the n-tuple $(0, 0, \dots, 0)$ and the inverse of the vector u is $-u = (-\alpha_1, \dots, -\alpha_n)$. With the same rules $\mathbb{R}^n$ becomes a real vector space while $\mathbb{C}^n$ is a complex vector space.

If we increase n indefinitely, the elements of the set $\mathbb{F}^\infty$ are *sequences* of scalars given by

$$u = (\alpha_1, \alpha_2, \dots, \alpha_n, \dots).$$

With the same rules $\mathbb{F}^\infty$ becomes also a linear vector space. ■

Example 1.2.2. Let us consider the set $\mathcal{F}(X, \mathbb{F})$ of all scalar-valued functions $f : X \to \mathbb{F}$ on an abstract set X. We define the sum of two functions in that set and the multiplication of a function with a scalar by the following rules

$$(f_1 + f_2)(x) = f_1(x) + f_2(x), \ (\alpha f)(x) = \alpha f(x).$$

We then see at once that this set acquires the structure of a vector space over the field $\mathbb{F}$. The zero vector $\mathbf{0}$ of this space corresponds naturally to zero function mapping all members of X to 0. ■

Let V be a vector space and $U \subseteq V$ be a subset. If the subset U is a linear vector space relative to operations in V, then the subset U is said to be a ***subspace*** of V. Subspaces are sometimes called *linear manifolds*. It may easily be verified that the necessary and sufficient conditions for a subset $U \subseteq V$ to be a subspace are (i) $u_1 + u_2 \in U$ for all $u_1, u_2 \in U$ and (ii) $\alpha u \in U$ for all $\alpha \in \mathbb{F}$ and $u \in U$. It is clear that we must have $\mathbf{0} \in U$. Every linear vector space has obviously two trivial subspaces: zero subspace $\{\mathbf{0}\}$ and the space itself.

As is well known, an ***equivalence relation*** R on an arbitrary set X is a subset $R \subseteq X^2$ of the Cartesian product $X^2 = X \times X$ which is *reflexive* $(x \in X \Rightarrow (x,x) \in R)$, *symmetric* $\big((x_1,x_2) \in R \Rightarrow (x_2,x_1) \in R\big)$ and *transitive* $\big((x_1,x_2) \in R,\, (x_2,x_3) \in R \Rightarrow (x_1,x_3) \in R\big)$. The set of all elements of X that are related to an element $x \in X$ by the equivalence relation is called an ***equivalence class*** $[x]$. It is readily seen that $\bigcup_{x \in X} [x] = X$ and *equivalence classes are all disjoint sets*. Therefore, equivalence classes constitute a ***partition*** on the set X. The set $X/R = \{[x] : x \in X\}$ is called the ***quotient set*** with respect to the equivalence relation R.

Let U be a subspace of the vector space V. We define a relation $\sim$ on V such that $u \sim v$ implies $u - v \in U$ for $u, v \in V$. Since $u - u = \mathbf{0} \in U$ we have $u \sim u$, i.e., the relation is reflexive. If $u \sim v$, namely if $u - v \in U$ we obtain $v - u = -(u - v) \in U$ and we see that $v \sim u$, i.e., the relation is symmetric. On the other hand, if $u \sim v$, $v \sim w$, namely, both $u - v \in U$ and $v - w \in U$, we then get $u - w = u - v + v - w \in U$. Hence we find that $u \sim w$, i.e., the relation is transitive. We then conclude that the relation so defined is an equivalence relation. Thus, this relation decomposes the vector space V into disjoint equivalence classes. Therefore an equivalence class, or a ***coset***, associated with a vector $v \in V$ is defined as the set

$$[v] = \{v + u : \forall u \in U\}. \tag{1.2.1}$$

Sometimes the notation $[v] = v + U$ is also used. We know that the set of all equivalence classes $V/U = \{[v] : v \in V\}$ is the quotient set. If we can devise appropriate rules for the addition of element of this set and for the scalar multiplication we are then able to endow the quotient set V/U with a vector space structure. To this end, we define vector addition and scalar multiplication on V/U by the following rules

$$[v_1] + [v_2] = [v_1 + v_2], \quad \alpha[v] = [\alpha v] \tag{1.2.2}$$

where the scalar α is an element of the field over which the vector space V is defined. The validity of this definition becomes evident if we note that

$$\begin{aligned}(v_1 + u_1) + (v_2 + u_2) &= (v_1 + v_2) + u_1 + u_2 \in [v_1 + v_2]\\ \alpha(v + u) &= \alpha v + \alpha u \in [\alpha v]\end{aligned}$$

for all $v_1, v_2 \in V$ and $u, u_1, u_2 \in U$. The set V/U equipped with such a structure is called the ***quotient space***, or more accurately, the ***quotient space of V modulo U***. The zero element of this vector space is the coset $U = [\mathbf{0}]$ and the inverse of an element $[v]$ is the coset $[-v]$. Since an equivalence class $[v] \in V/U$ is assigned to each vector $v \in V$, we can say that there exists a surjective mapping $\phi : V \to V/U$. ϕ is often called the ***canonical mapping*** of V onto V/U and we can write $[v] = \phi(v)$. Due to definitions (1.2.2) we immediately deduce that the mapping ϕ must satisfy the relations $\phi(v_1 + v_2) = \phi(v_1) + \phi(v_2)$ and $\phi(\alpha v) = \alpha\phi(v)$. Thus *the canonical mapping is linear* [*see p.* 9]. Obviously, ϕ is not injective in general.

Let U_1 and U_2 be two subspaces of the vector space V. We define the set $U_1 + U_2$ by

$$U_1 + U_2 = \{u = u_1 + u_2 : \forall u_1 \in U_1, \forall u_2 \in U_2\} \subseteq V.$$

It is straightforward to see that this set is a *subspace* of V that is called the ***sum*** of subspaces U_1 and U_2. One must note that the sum of two subspaces is completely different from their *union* $U_1 \cup U_2$ as sets. It is easy to see that $U_1 \cup U_2$ is not in general a subspace. The ***intersection*** of two subspaces U_1 and U_2 is the set of all vectors belonging to both subspaces. It is then properly denoted by $U_1 \cap U_2$. In contrast to the union, one easily observes that the intersection of two subspaces, in fact, the intersection of a family of subspaces, is again a subspace. The intersection of subspaces cannot be empty since all subspaces must contain the zero vector. We say that two subspaces U_1 and U_2 of V are ***disjunct*** if $U_1 \cap U_2 = \{\mathbf{0}\}$.

Let U_1 and U_2 be two subspaces of the vector space V and let the subspace $U = U_1 + U_2 \subseteq V$ be the sum of these subspaces. If there corresponds to each vector $u \in U$ a *uniquely determined* pair of vectors $u_1 \in U_1$ and $u_2 \in U_2$ such that $u = u_1 + u_2$, we then say that the subspace U is the ***direct sum*** of subspaces U_1 and U_2 and we write $U = U_1 \oplus U_2$. It is quite easy to see that *the sum U of two subspaces U_1 and U_2 is a direct sum of these subspaces if and only if U_1 and U_2 are disjunct, that is, if and only if $U_1 \cap U_2 = \{\mathbf{0}\}$.*

Let V be a linear vector space and let V_1 be a subspace of V. If we can find another subspace V_2 of V such that

$$V = V_1 \oplus V_2$$

any such subspace V_2 is said to be ***complementary*** to V_1 in V. It can be shown by employing the celebrated ***Zorn lemma*** [German-American mathematician Max August Zorn (1906-1993)] that *there exists at least one subspace which is complementary to a given subspace of a linear vector space.* However, a complementary subspace is generally not uniquely determined. It is rather straightforward to observe that the restriction $\phi|_{V_2}$ of the canonical mapping $\phi : V \to V/V_1$ is injective, consequently, the function $\phi|_{V_2} : V_2 \to V/V_1$ is bijective. Therefore, $\phi|_{V_2}$ is an isomorphism between the spaces V_2 and V/V_1. We thus conclude that *any subspace of V which is complementary to a subspace V_1 is isomorphic to the quotient space V/V_1.*

This result reflects the fact that *all complementary subspaces of V_1 in V are isomorphic* to one another [*see* p. 10 for the definition of isomorphism].

Let $S_n = \{v_1, v_2, \dots, v_n\}$ be a non-empty set of a *finite*, say $n > 0$, number of elements of a vector space V. The vector l formed by the sum

$$l = \alpha_1 v_1 + \alpha_2 v_2 + \cdots + \alpha_n v_n \in V$$

where $\alpha_1, \alpha_2, \dots, \alpha_n \in \mathbb{F}$ are arbitrary scalars is called a ***linear combination*** of the vectors in S_n. We call the set S_n as ***linearly independent*** if and only if the relation

$$\alpha_1 v_1 + \alpha_2 v_2 + \cdots + \alpha_n v_n = \mathbf{0} \tag{1.2.3}$$

is satisfied when all scalar coefficients vanish, namely, when $\alpha_i = 0$ for all $1 \le i \le n$. On the other hand, if the expression (1.2.3) is satisfied with scalar coefficients not all of which are zero, the set S_n is called as ***linearly dependent***. If *all non-empty finite subsets* of a possibly infinite set $\mathcal{A} \subseteq V$ are linearly independent, we say that the set $\mathcal{A}$ is ***linearly independent***. In such a set $\mathcal{A}$ no element of $\mathcal{A}$ can be expressed as a finite linear combination of some other elements of $\mathcal{A}$. *It is quite clear that a linearly independent set cannot be empty and cannot contain the zero vector.* Let us denote the *subspace* which is the collection of all finite linear combinations of vectors in $\mathcal{A}$ by $[\mathcal{A}]$. This subspace is called the ***linear hull*** of the set $\mathcal{A}$.

Theorem 1.2.1. *A subset $\mathcal{A}$ of a vector space V is linearly independent if and only if each vector in the subspace $[\mathcal{A}]$ can be uniquely represented as a finite linear combination of vectors in the set.*

Let the set $\mathcal{A}$ be linearly independent and let us assume that a vector $v \in [\mathcal{A}]$ is expressible as two different finite linear combinations of vectors in $\mathcal{A}$. But we can of course naturally combine vectors appearing in the first and the second representations into a single finite set such as $v_1, v_2, \dots, v_k$. We can then write

$$v = \sum_{i=1}^{k} \alpha_i v_i = \sum_{i=1}^{k} \beta_i v_i$$

where some of scalar coefficients $\{\alpha_i\}$ and $\{\beta_i\}$ may of course be zero. It then follows from the above expression that

$$\sum_{i=1}^{k} (\alpha_i - \beta_i) v_i = \mathbf{0}$$

which yields $\alpha_i = \beta_i$ for all $1 \le i \le k$ since all of the vectors involved are linearly independent. Hence the vector v has a unique representation. Conversely, let us assume that every vector in the subspace $[\mathcal{A}]$ has a *unique* representation in the form of finite linear combination of vectors in A. Since the set A is also contained in $[\mathcal{A}]$ this uniqueness should also be valid for all vectors in $\mathcal{A}$. This simply means that any element of $\mathcal{A}$ cannot be expressible as a linear combination of other vectors in $\mathcal{A}$. Hence $\mathcal{A}$ is a linearly independent set. □

If the linear hull of a linearly independent subset $\mathcal{B}$ of a vector space V is the entire space V, that is, if $[\mathcal{B}] = V$, then the set $\mathcal{B}$ is called a ***basis*** for the vector space V. In this case, every vector v in the vector space is expressible *in exactly one way* as a *finite linear combinations* of some vectors in B. Therefore, each vector $v \in V$ can be represented by the sum

$$v = \sum_{e_\lambda \in \mathcal{B}} \alpha_\lambda(v) e_\lambda \tag{1.2.4}$$

where scalar coefficients $\alpha_\lambda(v) \in \mathbb{F}$ that are determined uniquely for any given vector v *do not vanish only for a finite number vectors* $e_\lambda \in \mathcal{B}$ and they are called ***components*** of the vector v with respect to the basis $\mathcal{B}$. The basis $\mathcal{B}$ might be an infinite, even uncountably infinite, set but the expression (1.2.4) must involve only a sum of finite number of vectors that may of course be different for each vector $v \in V$. Such a basis, if it exists, is called an ***algebraic basis*** or ***Hamel basis*** because it was first introduced, albeit in a limited framework, by German mathematician Georg Karl Wilhelm Hamel (1877-1954). We can also readily show that a linearly independent set $\mathcal{B}$ of V is a basis *if and only if* it is *maximal* with respect to linear independency. Here the term maximal is used to indicate that every subset of V containing the set $\mathcal{B}$ is *linearly dependent*. One can prove by resorting to the Zorn lemma that every *non-zero* vector space has an algebraic basis. However, like almost every proposition based on Zorn lemma, we have no algorithm at hand to determine such a basis although we definitely know that it exists. Furthermore, we cannot say that there exists a unique basis.

It is now quite clear that a non-zero vector space V might possess several, possibly infinitely many, bases. But it can be shown that all Hamel bases have the same cardinality. This cardinal number is called the ***dimension*** of the vector space V and is denoted by $\dim(V)$. If $V = \{\mathbf{0}\}$ we adopt the convention that its dimension is 0. If the dimension of a vector space is a finite integer, then this space is ***finite-dimensional***, otherwise it is ***infinite-dimensional***. In this work, we shall mostly be dealing with finite-dimensional vector spaces. When we would like to underline this fact we shall usually write, say, $V^{(n)}$.

In a vector space V, the ***line segment*** joining two vectors u and v is defined as the subset $\{\alpha u + (1-\alpha)v : 0 \le \alpha \le 1\} \subset V$. A non-empty subset A of a vector space V is called a ***convex*** subset if it contains every line segments joining any pair of vectors $u, v \in A$. In other words, a set $A \subseteq V$ is a convex set if $\{\alpha u + (1-\alpha)v : 0 \le \alpha \le 1\} \subset A$ for all

$u, v \in A$. If A is a subspace of V, it is clear that it becomes automatically a convex set.

Example 1.2.3. Let us consider the vector space $\mathbb{F}^n$ introduced in Example 1.2.1 and define the vectors $e_1, e_2, \dots, e_n \in \mathbb{F}^n$ as

$$e_1 = (1, 0, \dots, 0),\ e_2 = (0, 1, \dots, 0),\ \dots,\ e_n = (0, 0, \dots, 1).$$

It is obvious that an arbitrary vector $u = (\alpha_1, \alpha_2, \dots, \alpha_n) \in \mathbb{F}^n$ can now be expressed by the following linear combination

$$u = \alpha_1 e_1 + \alpha_2 e_2 + \cdots + \alpha_n e_n.$$

From the definitions of vectors $e_1, e_2, \dots, e_n$ we see at once that the relation

$$\alpha_1 e_1 + \alpha_2 e_2 + \cdots + \alpha_n e_n = (\alpha_1, \alpha_2, \dots, \alpha_n) = \mathbf{0}$$

is satisfied if and only if $\alpha_1 = \alpha_2 = \cdots = \alpha_n = 0$. Hence the set $\mathcal{B} = \{e_1, e_2, \dots, e_n\} \subset \mathbb{F}^n$ is linearly independent and all linear combinations of vectors in $\mathcal{B}$ generate the vector space $\mathbb{F}^n$. Hence $\mathcal{B}$ is an algebraic basis for $\mathbb{F}^n$. Since the cardinal number of the set $\mathcal{B}$ is n, the dimension of the vector space $\mathbb{F}^n$ is n.

On the other hand, if we consider the vector space $\mathbb{F}^\infty$ we can easily verify that the countably infinite set $\{e_1, e_2, \dots, e_n, \dots\} \subset \mathbb{F}^\infty$ where

$$e_1 = (1, 0, \dots, 0, \dots),\ \dots,\ e_n = (0, 0, \dots, 0, 1, 0, \dots), \dots$$

are linearly independent and any vector $u = (\alpha_1, \alpha_2, \dots, \alpha_n, \dots)$ is uniquely represented by

$$u = \sum_{n=1}^{\infty} \alpha_n e_n.$$

However, it is quite evident that each vector $u \in \mathbb{F}^\infty$ cannot be expressed as a *finite* linear combinations of vectors $e_1,\ e_2, \dots, e_n, \dots$. Therefore, the countably infinite subset $\{e_1, e_2, \dots, e_n, \dots\} \subset \mathbb{F}^\infty$ cannot be a Hamel basis for the vector space $\mathbb{F}^\infty$. ■

If a function $A : U \to V$ between vector spaces U and V defined on the same scalar field $\mathbb{F}$ possesses the properties

$$A(u_1 + u_2) = A(u_1) + A(u_2) \in V,\ \ A(\alpha u) = \alpha A(u) \in V$$

for all $u, u_1, u_2 \in U$ and $\alpha \in \mathbb{F}$, then it is called a ***linear operator*** or a ***homomorphism*** since it preserves algebraic operations. It is evident that all linear operators of this kind constitute also a vector space $\mathcal{L}(U, V)$. If the inverse linear operator $A^{-1} : V \to U$ exists, then A is a ***regular linear operator***. The ***null space*** of a linear operator A is the subspace $\mathcal{N}(A) =$

$\{u \in U : Au = \mathbf{0}\} \subseteq U$ and its ***range*** is the subspace $\mathcal{R}(A) = \{v \in V : Au = v, \forall u \in U\} \subseteq V$. Sometimes $\mathcal{N}(A)$ is denoted by Ker (A), ***kernel*** of A, and $\mathcal{R}(A)$ by Im (A), ***image*** of U under A. We see that $\mathcal{N}(A) = \{\mathbf{0}\}$ if A is *injective* and $\mathcal{R}(A) = V$ if it is surjective. The necessary and sufficient condition for a linear operator to be regular is that it has to be bijective, i.e., $\mathcal{N}(A) = \{\mathbf{0}\}$ and $\mathcal{R}(A) = V$. A bijective linear mapping between two vector spaces preserving operations is called ***isomorphism*** and such spaces are said to be ***isomorphic***. It is straightforward to see that *compositions of isomorphisms is also an isomorphism*. It is a simple exercise to show that if $A : U \to V$ is an isomorphism and the set $\mathcal{B} \subseteq U$ is an algebraic basis for U, then the set $A(\mathcal{B})$ is an algebraic basis for V.

The ***rank*** $r(A)$ of a linear operator $A : U \to V$ is the dimension of its range and its ***nullity*** $n(A)$ is the dimension of its null space. Let N_A be a complementary subspace of the null space $\mathcal{N}(A)$ in U so that one writes $U = \mathcal{N}(A) \oplus N_A$. We consider the restriction $A^\dagger = A\big|_{N_A}$ of the linear transformation A to the subspace N_A. Each vector $u \in U$ is now expressed as a unique sum $u = u_1 + u_2$ where $u_1 \in \mathcal{N}(A)$ and $u_2 \in N_A$. We immediately notice that $\mathcal{R}(A^\dagger) = \mathcal{R}(A)$. Next, let us assume that $A^\dagger u_2 = \mathbf{0}$. We thus have $Au = Au_1 + Au_2 = Au_2 = A^\dagger u_2 = \mathbf{0}$. In consequence, we see that $u_2 \in \mathcal{N}(A)$. But, $\mathcal{N}(A) \cap N_A = \{\mathbf{0}\}$, therefore, it follows that $u_2 = \mathbf{0}$ which means that *the restriction of a linear transformation to the complementary subspace of its null space is injective, hence it is an isomorphism of N_A onto $\mathcal{R}(A)$. Consequently, if the set $\mathcal{B}_1$ is a basis for N_A, then $A^\dagger(\mathcal{B}_1)$ is a basis for $\mathcal{R}(A)$*. We thus conclude that if U is an n-dimensional vector space, then $\mathcal{R}(A)$ has to be finite-dimensional so that one gets the simple, but rather useful, relation

$$\dim (U) = n = n(A) + r(A).$$

If linearly independent vectors $e_1, \dots, e_n \in U^{(n)}$ are chosen as a basis for a finite-dimensional vector space, then each vector $v \in U^{(n)}$ is *uniquely* expressible as

$$u = \alpha_1 e_1 + \alpha_2 e_2 + \cdots + \alpha_n e_n, \quad \alpha_i \in \mathbb{F}, \quad 1 \le i \le n.$$

Let us denote $a = (\alpha_1, \alpha_2, \dots, \alpha_n) \in \mathbb{F}^n$. We then see that there exists a mapping $F : U^{(n)} \to \mathbb{F}^n$ determined by the relation $F(u) = a$. We deduce immediately from definition that F is a bijective linear operator. We thus conclude that the spaces $U^{(n)}$ and $\mathbb{F}^n$ are *isomorphic*.

Let U be a vector space defined over a field of scalars $\mathbb{F}$. A linear operator $f : U \to \mathbb{F}$ that assigns a scalar number $f(u)$ to each vector u in U is known as a ***linear functional***. The term functional was coined by French mathematician Jacques Salomon Hadamard (1865-1963) in 1903. The linear vector space $U^* = \mathcal{L}(U, \mathbb{F})$ formed by linear functionals is called the ***dual***, or more appropriately the ***algebraic dual***, of the vector space U.

Consider vector spaces $U^{(m)}$ and $V^{(n)}$ with bases $\{e_i\}$ and $\{f_i\}$ respectively. Let $A : U \to V$ be a linear operator. We can then write

$$v = Au = A\Big(\sum_{i=1}^{m} u^i e_i\Big) = \sum_{i=1}^{m} u^i A e_i = \sum_{i=1}^{m}\sum_{j=1}^{n} u^i a_i^j f_j = \sum_{j=1}^{n} v^j f_j$$

from which it follows that $v^j = \sum_{i=1}^{m} a_i^j u^i, j = 1, \dots, n$. This relation is expressible in the matrix form $\mathbf{v} = \mathbf{A}\mathbf{u}$ where $\mathbf{A}$ is the $m \times n$ matrix $[a_i^j]$ and $\mathbf{u}, \mathbf{v}$ are column matrices $[u^i], [v^j]$. The matrix $\mathbf{A}$ is a representation of the linear operator A with respect to some chosen bases in U and V.

Let us now consider a finite-dimensional vector space $U^{(n)}$. If a basis of this space is $\{e_1, e_2, \dots, e_n\}$, then every vector $u \in U$ is written as $u = \sum_{i=1}^{n} u^i e_i$ where $u^i \in \mathbb{F}$. The value of a linear functional $f \in U^*$ on a vector u can now be evaluated as follows:

$$f(u) = \sum_{i=1}^{n} f(u^i e_i) = \sum_{i=1}^{n} u^i f(e_i) = \sum_{i=1}^{n} u^i \alpha_i \in \mathbb{F} \tag{1.2.5}$$

where the scalar numbers α_i are prescribed by

$$\alpha_i = f(e_i) \in \mathbb{F}, \quad i = 1, \dots, n. \tag{1.2.6}$$

This means that the action of any linear functional on a vector u is completely determined by an ordered n-tuple $a = (\alpha_1, \alpha_2, \dots, \alpha_n) \in \mathbb{F}^n$. Thus, there is a mapping $T : U^* \to \mathbb{F}^n$ such that $T(f) = a$. If $T(f_1) = a_1$, $T(f_2) = a_2$, we then deduce from (1.2.5) that $T(f_1 + f_2) = a_1 + a_2$, $T(\alpha f) = \alpha a$. Hence, T turns out to be a linear operator. Each ordered n-tuple of scalars $(\alpha_1, \alpha_2, \dots, \alpha_n)$ determines a linear functional f. Therefore T is surjective. On other hand if $T(f_1) = T(f_2) = a$ we find $T(f_1 - f_2) = \mathbf{0} = (0, 0, \dots, 0)$ and (1.2.5) leads to the conclusion that $(f_1 - f_2)(u) = 0$ for all $u \in U^{(n)}$. This simply implies that $f_1 - f_2 = 0$ or $f_1 = f_2$. Thus T is injective. Consequently, the linear operator T is bijective. This indicates that the vector space U^* is *isomorphic* to $\mathbb{F}^n$ just like the space $U^{(n)}$. Since isomorphic spaces must have the same dimension, the dimension of the space U^* is also n. Furthermore, U^* and $U^{(n)}$ must be isomorphic to one another because they are isomorphic to the same space $\mathbb{F}^n$. Let us now consider n linearly independent vectors $(0,\dots,\underset{i}{1},\dots,0), i = 1,\dots,n$ in the vector space $\mathbb{F}^n$ such that in the ith vector only its ith entry is 1 and all the others are zero. We can then obtain n linear functionals $f^i \in U^*, i = 1,\dots,n$ corresponding to those vectors in $\mathbb{F}^n$ through the isomorphism $T^{-1} : \mathbb{F}^n \to U^*$.

The definition (1.2.6) leads now to relations

$$f^i(e_j) = \delta^i_j, \quad i,j = 1,2,\dots,n \tag{1.2.7}$$

where δ^i_j denotes the ***Kronecker delta*** [it is so named because it was first introduced by German mathematician Leopold Kronecker (1823-1891)]. It is equal to 1 if $i = j$ and to 0 if $i \neq j$. Hence, it essentially represents the $n \times n$ unit matrix. The set of linear functionals $\{f^i\}$ so obtained is linearly independent. To see this, we consider the zero functional given by

$$c_1 f^1 + c_2 f^2 + \cdots + c_i f^i + \cdots + c_n f^n = 0$$

where $c_1, c_2, \dots, c_n \in \mathbb{F}$. Because the value of this functional on the basis vectors $e_j, j = 1,\dots,n$ of the vector space U must be zero, we obtain

$$\sum_{i=1}^{n} c_i f^i(e_j) = \sum_{i=1}^{n} c_i \delta^i_j = c_j = 0, \quad j = 1,2,\dots,n.$$

This means that all linear functionals $f^1, f^2, \dots, f^n$ are linearly independent and constitute a basis for the dual space U^* since its dimension is n. Hence, an arbitrary linear functional $f \in U^*$ can now be uniquely represented in the following form:

$$f = \sum_{i=1}^{n} \alpha_i f^i, \quad \alpha_i \in \mathbb{F}, \quad i = 1,\dots,n.$$

Let $\{e_1,\dots,e_n\}$ be the basis in U which we have employed to generate the basis $\{f^i\} \subset U^*$. Then the value of a functional $f \subset U^*$ on a vector $u \in U$ can be calculated as follows

$$f(u) = \sum_{i=1}^{n}\sum_{j=1}^{n} \alpha_i u^j f^i(e_j) = \sum_{i=1}^{n}\sum_{j=1}^{n} \alpha_i u^j \delta^i_j = \sum_{i=1}^{n} u^i \alpha_i$$

which is the same as (1.2.5). We easily observe that the relations

$$f^i(u) = u^i, \quad f(e_i) = \alpha_i \tag{1.2.8}$$

are satisfied.

The two foregoing ordered sets of basis vectors $\{f^1, f^2, \ldots, f^n\}$ of U^* and $\{e_1, e_2, \ldots, e_n\}$ of U are called ***dual*** (or ***reciprocal***) bases. In view of the relations (1.2.7), we may also say that they form a set of ***biorthogonal bases***.

Sometimes it proves to be more convenient to use the notation $\langle f, u\rangle$ instead of $f(u)$. This symbolism is known as the ***duality pairing*** and it is clear that it describes a mapping $\langle\,\cdot\,,\,\cdot\,\rangle : U^* \times U \to \mathbb{F}$ which may be called a ***bilinear functional*** or a ***bilinear form*** due to the obvious reason that this functional has the following properties:

$$\begin{aligned}\langle f_1 + f_2, u\rangle &= \langle f_1, u\rangle + \langle f_2, u\rangle, \\ \langle f, u_1 + u_2\rangle &= \langle f, u_1\rangle + \langle f, u_2\rangle, \\ \langle \alpha f, u\rangle &= \langle f, \alpha u\rangle = \alpha\langle f, u\rangle.\end{aligned} \tag{1.2.9}$$

With this notation (1.2.7) can be rewritten as

$$\langle f^i, e_j\rangle = \delta^i_j. \tag{1.2.10}$$

Since one can write $f(u) = \sum_{i=1}^{n} \alpha_i u^i$, *it is obvious that if* $f(u) = 0$ *for all* $u \in U^{(n)}$, *we then get* $\alpha_1 = \cdots = \alpha_n = 0$, *namely,* $f = 0$; *conversely, if* $f(u) = 0$ *for all* $f \in U^{*(n)}$, *we then have to write* $u^1 = \cdots = u^n = 0$ *so that* $u = 0$.

We shall now discuss the change of basis in finite-dimensional vector spaces. We choose first a basis $\{e_i\}$ in a vector space U and consider another basis $\{e'_i\}$. Since both basis are to be linearly independent sets, this operation is obviously carried out by use of a *regular matrix* $\mathbf{A} = [a^i_j]$ *such that* $\det \mathbf{A} \neq 0$ as follows

$$e_j = a^i_j\, e'_i = \sum_{i=1}^{n} a^i_j\, e'_i, \tag{1.2.11}$$

where we have employed the celebrated *summation convention* proposed by the great German physicist Albert Einstein (1879-1955). *Repeated indices (usually superscripts and subscripts), that are sometimes called* ***dummy indices*** *because we can freely rename them without actually affecting the meaning of an expression, will imply a summation over the range of these indices. When we would like to suspend this rule we will underline the relevant indices*. Henceforth, we shall always resort to the Einstein summation convention to simplify the appearance of rather complicated expressions, at least notationally, by dispensing with the symbol Σ. It follows from the relations

$$u = u^j e_j = u^j a^i_j e'_i = u'^i e'_i$$

that the components of a vector u with respect to the new basis in terms of old components are given by

$$u'^i = a^i_j u^j. \tag{1.2.12}$$

Let the reciprocal basis in the dual space U^* to the basis $\{e'_i\}$ be $\{f'^i\}$. Hence the value of $f \in U^*$ on every vector $u \in U$ is found as

$$f(u) = u'^i \alpha'_i = u^j a^i_j \alpha'_i = u^j \alpha_j$$

from which it follows that

$$\alpha'_i = b^j_i \alpha_j, \qquad \{b^j_i\} = \mathbf{B} = \mathbf{A}^{-1}. \tag{1.2.13}$$

On the other hand, when we consider the relations

$$f'^i(u) = u'^i = a^i_j u^j = a^i_j f^j(u)$$

that must be satisfied for all vectors $u \in U$ we are led to the following transformation rules

$$f'^i = a^i_j f^j, \quad f^i = b^i_j f'^j. \tag{1.2.14}$$

Let us now consider a sequence of linear vector spaces $\{V_n\}$ and a sequence of linear operators $A_n : V_n \to V_{n+1}$, that is, homomorphisms, represented diagrammatically as

$$\cdots \longrightarrow V_{n-1} \overset{A_{n-1}}{\longrightarrow} V_n \overset{A_n}{\longrightarrow} V_{n+1} \longrightarrow \cdots. \tag{1.2.15}$$

The sequence $V^\bullet = \{V_n, A_n\}$ is called an ***exact sequence*** if $\mathcal{R}(A_{n-1}) = \mathcal{N}(A_n) \subset V_n$ for all n, This of course requires that $A_n \circ A_{n-1} = 0$. However, we observe easily that this condition alone is not sufficient for the above sequence to be exact. In fact, if $v_{n-1} \in V_{n-1}$, then $A_{n-1}(v_{n-1}) \in \mathcal{R}(A_{n-1}) \subseteq V_n$. Since by definition we assume $A_n\big(A_{n-1}(v_{n-1})\big) = 0$, we can only infer that $A_{n-1}(v_{n-1}) \in \mathcal{N}(A_n)$ implying merely that $\mathcal{R}(A_{n-1}) \subseteq \mathcal{N}(A_n)$. If, at each stage, the image of one homomorphism is contained in the kernel of the next homomorphism, this *increasing* sequence is called a ***cochain complex***. Clearly, an exact sequence is also a cochain complex, but the converse statement is generally not true. Let us consider two cochains $V^\bullet = \{V_n, A_n\}$ and $U^\bullet = \{U_n, B_n\}$. A ***cochain homomorphism*** $C^\bullet : V^\bullet \to U^\bullet$ is a set of homomorphisms $\{C_n : V_n \to U_n\}$ such that the following diagram commutes for all n:

$$\begin{array}{ccccccc} \cdots \longrightarrow & V_n & \overset{A_n}{\longrightarrow} & V_{n+1} & \longrightarrow \cdots \\ & \downarrow C_n & & \downarrow C_{n+1} & \\ \cdots \longrightarrow & U_n & \overset{B_n}{\longrightarrow} & U_{n+1} & \longrightarrow \cdots \end{array}$$

We can thus write $C_{n+1} \circ A_n = B_n \circ C_n$ for all n.

An exact sequence of the form

$$0 \longrightarrow U \overset{A}{\longrightarrow} V \overset{B}{\longrightarrow} W \longrightarrow 0 \qquad (1.2.16)$$

is called a ***short exact sequence***. Obviously A is injective because $\mathcal{N}(A) = \{0\}$ whereas B is surjective since $\mathcal{R}(B) = W$. Hence, A has left inverses and B right inverses so that there are homomorphisms $L : V \to U$ and $R : W \to V$ such that $L \circ A = i_U, B \circ R = i_W$ where i_U and i_W are identity mappings A simple example to a short exact sequence is provided by the quotient space V/U produced by a subspace $U \subseteq V$:

$$0 \longrightarrow U \overset{\mathcal{I}}{\longrightarrow} V \overset{\phi}{\longrightarrow} V/U \longrightarrow 0$$

where $\mathcal{I} : U \to V$ is the inclusion mapping, i.e., $u = \mathcal{I}(u) \in V$ for all $u \in U$ and $\phi : V \to V/U$ is the canonical mapping [*see p.* 6]. We know that $\mathcal{N}(V/U) = U$ so that we may write $\phi \circ \mathcal{I} = 0$.

A salient property of exact sequences is revealed in the following theorem known as ***the five lemma***.

Theorem 1.2.2. *Let $V^\bullet = \{V_n, A_n\}$ and $U^\bullet = \{U_n, B_n\}$ be two exact sequences and $C^\bullet = \{C_n : V_n \to U_n\}$ be a cochain homomorphism. Let us consider the five consecutive elements of these sequences corresponding to $n-2, n-1, n, n+1, n+2$. If $C_{n-2}, C_{n-1}, C_{n+1}, C_{n+2}$ are isomorphisms, then C_n must also be an isomorphism.*

The commutativity of the diagram below with rows of exact sequences

$$\begin{array}{ccccccccccc} \cdots \longrightarrow & V_{n-2} & \overset{A_{n-2}}{\longrightarrow} & V_{n-1} & \overset{A_{n-1}}{\longrightarrow} & V_n & \overset{A_n}{\longrightarrow} & V_{n+1} & \overset{A_{n+1}}{\longrightarrow} & V_{n+2} & \longrightarrow \cdots \\ & \downarrow C_{n-2} & & \downarrow C_{n-1} & & \downarrow C_n & & \downarrow C_{n+1} & & \downarrow C_{n+2} & \\ \cdots \longrightarrow & U_{n-2} & \overset{B_{n-2}}{\longrightarrow} & U_{n-1} & \overset{B_{n-1}}{\longrightarrow} & U_n & \overset{B_n}{\longrightarrow} & U_{n+1} & \overset{B_{n+1}}{\longrightarrow} & U_{n+2} & \longrightarrow \cdots \end{array}$$

requires that $C_{n+1} \circ A_n = B_n \circ C_n : V_n \to U_{n+1}$ for each n.

Let us first show that the homomorphism C_n is injective. Let $v_n \in V_n$ and assume that $C_n(v_n) = 0 \in U_n$. Then $C_{n+1}\big(A_n(v_n)\big) = B_n\big(C_n(v_n)\big) = 0$. Since C_{n+1} is an isomorphism, we obtain $A_n(v_n) = 0$. Therefore,

$v_n \in \mathcal{N}(A_n) = \mathcal{R}(A_{n-1})$ so that there exists $v_{n-1} \in V_{n-1}$ such that $v_n = A_{n-1}(v_{n-1})$. Then $B_{n-1}\big(C_{n-1}(v_{n-1})\big) = C_n\big(A_{n-1}(v_{n-1})\big) = C_n(v_n) = 0$ implying that $C_{n-1}(v_{n-1}) \in \mathcal{N}(B_{n-1}) = \mathcal{R}(B_{n-2})$ so that we may choose $u_{n-2} \in U_{n-2}$ such that $B_{n-2}(u_{n-2}) = C_{n-1}(v_{n-1})$. Since C_{n-2} is an isomorphism, there exists $v_{n-2} \in V_{n-2}$ such that $C_{n-2}(v_{n-2}) = u_{n-2}$. Then, we obtain $C_{n-1}\big(A_{n-2}(v_{n-2})\big) = B_{n-2}\big(C_{n-2}(v_{n-2})\big) = B_{n-2}(u_{n-2}) = C_{n-1}(v_{n-1})$. Because C_{n-1} is an isomorphism, we get $v_{n-1} = A_{n-2}(v_{n-2})$. Since $A_{n-1} \circ A_{n-2} = 0$ because $\{V_n, A_n\}$ is an exact sequence, we thus find $0 = A_{n-1}\big(A_{n-2}(v_{n-2})\big) = A_{n-1}(v_{n-1}) = v_n$. Hence, $v_n = 0$ which amounts to say that C_n is injective.

We shall now show that C_n is surjective. Let $u_n \in U_n$ be an arbitrary vector. We then have $u_{n+1} = B_n(u_n) \in U_{n+1}$. Since C_{n+1} is an isomorphism, there exists a vector $v_{n+1} \in V_{n+1}$ so that $B_n(u_n) = C_{n+1}(v_{n+1})$. We thus have $C_{n+2}\big(A_{n+1}(v_{n+1})\big) = B_{n+1}\big(C_{n+1}(v_{n+1})\big) = B_{n+1}\big(B_n(u_n)\big) = 0$ because $B_{n+1} \circ B_n = 0$ since $\{U_n, B_n\}$ is an exact sequence. We thus find $A_{n+1}(v_{n+1}) = 0$ because C_{n+2} is an isomorphism. Since v_{n+1} belongs to the null space of A_{n+1}, then there exists a vector $v_n \in V_n$ such that $v_{n+1} = A_n(v_n)$ because $\mathcal{N}(A_{n+1}) = \mathcal{R}(A_n)$. Let us now consider the vector $u_n - C_n(v_n) \in U_n$. Recalling that $B_n(u_n) = C_{n+1}(v_{n+1})$, We readily observe that

$$\begin{aligned} B_n\big(u_n - C_n(v_n)\big) &= B_n(u_n) - B_n\big(C_n(v_n)\big) \\ &= B_n(u_n) - C_{n+1}\big(A_n(v_n)\big) \\ &= B_n(u_n) - C_{n+1}(v_{n+1}) = 0. \end{aligned}$$

Since $u_n - C_n(v_n) \in \mathcal{N}(B_n)$, there exists a vector $u_{n-1} \in U_{n-1}$ satisfying the relation $u_n - C_n(v_n) = B_{n-1}(u_{n-1})$ and we have $u_{n-1} = C_{n-1}(v_{n-1})$ for some $v_{n-1} \in V_{n-1}$ because C_{n-1} is an isomorphism. Let now consider the vector $v_n + A_{n-1}(v_{n-1}) \in V_n$. We can then write

$$\begin{aligned} C_n\big(v_n + A_{n-1}(v_{n-1})\big) &= C_n(v_n) + C_n\big(A_{n-1}(v_{n-1})\big) \\ &= C_n(v_n) + B_{n-1}\big(C_{n-1}(v_{n-1})\big) \\ &= C_n(v_n) + B_{n-1}(u_{n-1}) = u_n \end{aligned}$$

implying that C_n is surjective. Since this linear operator is both injective and surjective, then C_n is an isomorphism. □

Let $V^\bullet$ be a cochain given by (1.2.15) such that $\mathcal{R}(A_{n-1}) \subseteq \mathcal{N}(A_n)$. The quotient space of $\mathcal{N}(A_n)$ with respect to its subspace $\mathcal{R}(A_{n-1})$ is the vector space

$$H^n(V^\bullet) = \mathcal{N}(A_n)/\mathcal{R}(A_{n-1}) = \mathrm{Ker}\,(A_n)/\mathrm{Im}\,(A_{n-1}). \qquad (1.2.17)$$

$H^n(V^\bullet)$ is called the ***nth cohomology group*** due to the fact that a vector space is an Abelian group. An element of the vector space $H^n(V^\bullet)$, called a ***cohomology class***, is an equivalence class $[v_n] = \{v_n + A_{n-1}v_{n-1}\}$ involving all vectors $v_{n-1} \in V_{n-1}$ where $A_n v_n = 0$. We shall now demonstrate the following theorem commonly known as ***the zigzag lemma***.

Theorem 1.2.3. *Let us consider the following short exact sequence*

$$0 \longrightarrow U^\bullet \overset{A^\bullet}{\longrightarrow} V^\bullet \overset{B^\bullet}{\longrightarrow} W^\bullet \longrightarrow 0 \tag{1.2.18}$$

where $U^\bullet = (U_n, d), V^\bullet = (V_n, d), W^\bullet = (W_n, d)$ *are cochains so that* $d^2 = 0$ *and* $A^\bullet = \{A_n\}, B^\bullet = \{B_n\}$ *are cochain homomorphisms. Then there exists a homomorphism* $\Gamma : H^n(W^\bullet) \to H^{n+1}(U^\bullet)$ *such that the sequence*

$$\cdots \overset{\Gamma}{\longrightarrow} H^n(U^\bullet) \overset{A_n}{\longrightarrow} H^n(V^\bullet) \overset{B_n}{\longrightarrow} H^n(W^\bullet) \overset{\Gamma}{\longrightarrow} H^{n+1}(U^\bullet) \overset{A_{n+1}}{\longrightarrow} \cdots \tag{1.2.19}$$

is exact.

We consider the following commutative diagram whose rows are short exact sequences and columns are cochains:

$$\begin{array}{ccccccccc}
 & & \vdots & & \vdots & & \vdots & & \\
 & & \downarrow d & & \downarrow d & & \downarrow d & & \\
0 & \longrightarrow & U_n & \overset{A_n}{\longrightarrow} & V_n & \overset{B_n}{\longrightarrow} & W_n & \longrightarrow & 0 \\
 & & \downarrow d & & \downarrow d & & \downarrow d & & \\
0 & \longrightarrow & U_{n+1} & \overset{A_{n+1}}{\longrightarrow} & V_{n+1} & \overset{B_{n+1}}{\longrightarrow} & W_{n+1} & \longrightarrow & 0 \\
 & & \downarrow d & & \downarrow d & & \downarrow d & & \\
0 & \longrightarrow & U_{n+2} & \overset{A_{n+2}}{\longrightarrow} & V_{n+2} & \overset{B_{n+2}}{\longrightarrow} & W_{n+2} & \longrightarrow & 0 \\
 & & \downarrow d & & \downarrow d & & \downarrow d & & \\
 & & \vdots & & \vdots & & \vdots & &
\end{array}$$

We thus infer that for all n, the homomorphism A_n is injective and B_n is surjective and $\mathcal{R}(A_n) = \mathcal{N}(B_n)$. Similarly, we have $\mathcal{R}_n(d) \subseteq \mathcal{N}_{n+1}(d)$ and this gives rise to cohomology groups $H^n(U^\bullet)$, $H^n(V^\bullet)$, $H^n(W^\bullet)$ for all n along columns of cochains. The linear operator $A_n : U_n \to \mathcal{N}(B_n) = \mathcal{R}(A_n) \subseteq V_n$ is evidently bijective so that it is an isomorphism, hence its inverse $A_n^{-1} : \mathcal{N}(B_n) \to U_n$ exists. Equivalence classes in the quotient space $V_n/\mathcal{N}(B_n)$ are given by $[v_n] = \{v_n + A_n u_n : u_n \in U_n\}$. Then the operator B_n interpreted as $B_n : V_n/\mathcal{N}(B_n) \to W_n$ becomes an isomorphism so that one has the inverse $B_n^{-1}w_n = [v_n]$. Therefore, we may define a linear operator Γ by

$$\Gamma = A_{n+1}^{-1} \circ d \circ B_n^{-1} : W_n \to U_{n+1} \tag{1.2.20}$$

which is unique within the precepts of the cohomology. Due to the commutativity of the diagram, we infer from (1.2.20) that

$$d \circ \Gamma = d \circ A_{n+1}^{-1} \circ d \circ B_n^{-1} = A_{n+2}^{-1} \circ d^2 \circ B_n^{-1} = 0.$$

It straightforward to see that we also get the relation $0 = A_{n+1}^{-1} \circ d^2 \circ B_{n+1}^{-1} = \Gamma \circ d$. Let us now consider a representative w_n of the equivalence class $[w_n] \in H^n(W^\bullet)$ so that $dw_n = 0$. We then obtain $d(\Gamma w_n) = 0$. Hence, $\Gamma w_n \in H^{n+1}(U^\bullet)$. Thus, Γ is a homomorphism as follows

$$\Gamma : H^n(W^\bullet) \to H^{n+1}(U^\bullet).$$

Let us take a vector $w_n \in W_n$. Since B_n is surjective, there exists a representative vector $v_n \in V_n$ of an equivalence class $[v_n]$ such that $B_n v_n = w_n$. Because we have to consider the cochain $W^\bullet$, let us assume that $w_n \in \mathcal{N}_n(d) \subseteq W_n$ so that $dw_n = 0$. Due to the commutativity of the above diagram we find that $dB_n v_n = B_{n+1} dv_n = 0$. Thus, $dv_n \in \mathcal{N}(B_{n+1}) = \mathcal{R}(A_{n+1})$. Since A_{n+1} is injective, there is a unique vector u_{n+1} such that $A_{n+1}u_{n+1} = dv_n$. It follows from the commutativity of the above diagram that $A_{n+2}\, du_{n+1} = dA_{n+1}u_{n+1} = d^2 v_n = 0$ so that $du_{n+1} \in \mathcal{N}(A_{n+2})$. Since A_{n+2} is injective, we get $du_{n+1} = 0$. Hence, u_{n+1} belongs to a cohomology class. Obviously, it is expressed as $u_{n+1} = \Gamma w_n$. However, we have to show that this result is independent of the choice of representative of the equivalence class. Let us consider another vector $v_n' \in [v_n]$. We then must write $v_n - v_n' \in \mathcal{N}(B_n)$. Exactness requires that there exists a $u_n \in U_n$ such that $A_n u_n = v_n - v_n'$. Now the commutativity of the diagram implies that

$$A_{n+1} du_n = dA_n u_n = d(v_n - v_n').$$

It then follows from cochain and exact sequence properties that there are $u_{n+1}, u_{n+1}' \in U_{n+1}$ such that $A_{n+1}u_{n+1} = dv_n$ and $A_{n+1}u_{n+1}' = dv_n'$. Since A_{n+1} is injective, the relation $A_{n+1}(u_{n+1} - u_{n+1}' - du_n) = 0$ yields $du_n = u_{n+1} - u_{n+1}'$, hence we get $du_{n+1} = du_{n+1}'$. Consequently, u_{n+1} and u_{n+1}' belong to the same cohomology class.

We now consider an element $w_n = dw_{n-1} \in W_n$ where $w_{n-1} \in W_{n-1}$. Since $dw_n = 0$, we get $w_n \in H^n(W^\bullet)$. In view of the surjectivity of B_{n-1} we can write $B_{n-1}v_{n-1} = w_{n-1}$ for a vector $v_{n-1} \in V_{n-1}$. Let $v_n = dv_{n-1}$ so that $dv_n = d^2 v_{n-1} = 0 \in V_{n+1}$. We have seen above that there exists a unique vector $u_{n+1} \in U_{n+1}$ such that $A_{n+1}u_{n+1} = dv_n = 0$. Since A_{n+1} is injective, we have $u_{n+1} = 0$. This of course implies that all elements in the equivalence class $[w_n] \in H^n(W^\bullet)$ are mapped under the operator Γ onto the same equivalence class $[u_{n+1}] \in H^{n+1}(U^\bullet)$. Hence, Γ is a well defined operator.

Finally, we have to show that the sequence

$$\cdots \xrightarrow{\Gamma} H^n(U^\bullet) \xrightarrow{A_n} H^n(V^\bullet) \xrightarrow{B_n} H^n(W^\bullet) \xrightarrow{\Gamma} H^{n+1}(U^\bullet) \xrightarrow{A_{n+1}} \cdots$$

is exact. To this end, it suffices to prove exactness at $H^n(U^\bullet)$. Because, the sequence is exact at $H^n(V^\bullet)$ since $\mathcal{R}(A_n) = \mathcal{N}(B_n)$ and proof at $H^n(W^\bullet)$may be accomplish in the same fashion. Let $[w_{n-1}] \in H^{n-1}(W^\bullet)$ and take the element $\Gamma[w_{n-1}] \in H^n(U^\bullet) = \mathcal{R}_{n-1}(\Gamma)$ into account. It then immediately follows from (1.2.20) that $A_n\Gamma[w_{n-1}] = [dv_{n-1}] = [0]$ where $[v_{n-1}] = B_{n-1}^{-1}[w_{n-1}]$. Consequently, we obtain $\mathcal{R}_{n-1}(\Gamma) \subseteq \mathcal{N}(A_n)$. Conversely, let us now consider an equivalence class $[u_n] \in \mathcal{N}(A_n)$ of the cohomology group $H^n(U^\bullet)$. Since $A_n[u_n] = [0] \in H^n(V^\bullet)$, we find that $A_n[u_n] = [dv_{n-1}]$. We then define $w_{n-1} = B_{n-1}v_{n-1} \in W_{n-1}$ and the cohomology class $[w_{n-1}] \in H^{n-1}(W^\bullet)$. Since $\Gamma[w_{n-1}] = A_n^{-1}dB_{n-1}^{-1}[w_{n-1}]$ $= A_n^{-1}dB_{n-1}^{-1}B_{n-1}[v_{n-1}] = A_n^{-1}[dv_{n-1}] = A_n^{-1}A_n[u_n] = [u_n] \in \mathcal{N}(A_n)$, we see that $[u_n]$ is the image of an equivalence class $[w_{n-1}]$ under Γ. Thus, we get $\mathcal{N}(A_n) \subseteq \mathcal{R}_{n-1}(\Gamma)$ and we finally find $\mathcal{R}_{n-1}(\Gamma) = \mathcal{N}(A_n)$ Hence, the sequence is exact at $H^n(U^\bullet)$. We shall not repeat the analysis to prove exactness at $H^n(W^\bullet)$. □

Finally, for later applications, we have to emphasise the fact that what we have said so far are equally valid for *modules*.

1.3. MULTILINEAR FUNCTIONALS

Let $(U_1, U_2, \ldots, U_k)$ be ordered k-tuple of linear vector spaces defined over the same field of scalars $\mathbb{F}$. Let us consider a scalar-valued function $\mathcal{T} : U_1 \times U_2 \times \cdots \times U_k \to \mathbb{F}$ on the Cartesian product of these vector spaces. If the function $\mathcal{T}(u_{(1)}, u_{(2)}, \ldots, u_{(k)}) \in \mathbb{F}$, where $u_{(\alpha)} \in U_\alpha$, $\alpha = 1, 2, \ldots, k$, is *linear in each one of its arguments*, that is, if the following relations

$$\begin{aligned} \mathcal{T}(\ldots, u_{(i)} + v_{(i)}, \ldots) &= \mathcal{T}(\ldots, u_{(i)}, \ldots) + \mathcal{T}(\ldots, v_{(i)}, \ldots) \quad (1.3.1)\\ \mathcal{T}(\ldots, \alpha u_{(i)}, \ldots) &= \alpha\mathcal{T}(\ldots, u_{(i)}, \ldots),\ \alpha \in \mathbb{F} \end{aligned}$$

are satisfied for all $1 \le i \le k$, then the function $\mathcal{T}$ is called a ***multilinear functional*** (or a ***k-linear functional***). In finite-dimensional vector spaces whose dimensions and bases are $n_1, \ldots, n_k$ and $\{e_i^{(\alpha)}\} \in U_\alpha$, $i = 1, \ldots, n_\alpha$,

$\alpha = 1, \ldots, k$, we can then write $u_{(\alpha)} = \sum_{i=1}^{n_\alpha} u^i_{(\alpha)} e^{(\alpha)}_i$, without having recourse to the summation convention. Multilinearity then leads to the following value of the functional at vectors $u_{(1)} \in U_1, u_{(2)} \in U_2, \ldots, u_{(k)} \in U_k$

$$\mathcal{T}(u_{(1)}, u_{(2)}, \ldots, u_{(k)}) = \sum_{i_1=1}^{n_1} \sum_{i_2=1}^{n_2} \cdots \sum_{i_k=1}^{n_k} t_{i_1 i_2 \cdots i_k} u^{i_1}_{(1)} u^{i_2}_{(2)} \cdots u^{i_k}_{(k)} \tag{1.3.2}$$

where $n_1 \times n_2 \times \cdots \times n_k$ number of scalar $t_{i_1 i_2 \cdots i_k}$ are defined by

$$t_{i_1 i_2 \cdots i_k} = \mathcal{T}(e^{(1)}_{i_1}, e^{(2)}_{i_2}, \ldots, e^{(k)}_{i_k}) \in \mathbb{F}. \tag{1.3.3}$$

We thus conclude that the set of scalars $\{t_{i_1 i_2 \cdots i_k}\}$ completely determines the action of a k-linear functional on any set of k number of vectors $u_{(1)} \in U_1$, $u_{(2)} \in U_2$, $\ldots, u_{(k)} \in U_k$. We can thus say that they unambiguously characterise a multilinear functional.

Let us now suppose that $U_1 = U_2 = \cdots = U_k = U^{(n)}$. The value of a multilinear functional $\mathcal{T} : U^k \to \mathbb{F}$ on vectors $u_{(1)}, u_{(2)}, \ldots, u_{(k)} \in U$ can now be found from (1.3.2) and (1.3.3) as follows

$$\begin{aligned} \mathcal{T}(u_{(1)}, u_{(2)}, \ldots, u_{(k)}) &= t_{i_1 i_2 \cdots i_k} u^{i_1}_{(1)} u^{i_2}_{(2)} \cdots u^{i_k}_{(k)}, \\ t_{i_1 i_2 \cdots i_k} &= \mathcal{T}(e_{i_1}, e_{i_2}, \ldots, e_{i_k}), \;\; 1 \le i_1, i_2, \cdots, i_k \le n \end{aligned} \tag{1.3.4}$$

where we experience no difficulty in resorting to the summation convention because the range of all indices is the same now, from 1 to n. In this case, we can introduce a more advantageous representation of a multilinear functional as an operator. To this end, we shall first introduce the tensor product of two vector spaces.

Let U and V be two linear vector spaces defined on the same field of scalars $\mathbb{F}$. As is well known, the Cartesian product $U \times V$ of these spaces is formed by ordered pairs (u, v) where $u \in U$ and $v \in V$. There is initially no algebraic structure on this product set. However, by making use of known operations on vector spaces U and V, we may define appropriate operations on the set $U \times V$ so that it may be equipped with a structure of a linear vector space. The resulting vector space will be called the ***tensor product*** of spaces U and V and will be denoted by $W = U \otimes V$. Let us choose operations of vector addition and scalar multiplication on W in such a way that tensor product of vectors $u \otimes v \in U \otimes V$ has to satisfy the following bilinearity conditions:

$$(i).\; u \otimes (v_1 + v_2) = u \otimes v_1 + u \otimes v_2,$$

$$(ii).\ (u_1 + u_2) \otimes v = u_1 \otimes v + u_2 \otimes v,$$
$$(iii).\ (\alpha u) \otimes v = u \otimes (\alpha v) = \alpha(u \otimes v), \quad \alpha \in \mathbb{F}.$$

Let us note that the same symbol $+$ in the foregoing expressions represents, in fact, different addition operations in three different vector spaces U, V and W. We can thus write

$$(u_1 + u_2) \otimes (v_1 + v_2) = u_1 \otimes v_1 + u_1 \otimes v_2 + u_2 \otimes v_1 + u_2 \otimes v_2.$$

The space W is then defined as the collection of all *finite sums* $\sum_i u_i \otimes v_i$ where $u_i \in U$ and $v_i \in V$. If we consider finite-dimensional vector spaces $U^{(m)}$ and $V^{(n)}$ with respective bases $\{e_i\}$ and $\{f_j\}$, a vector $w \in W$ is evidently expressible as $w = w^{ij} e_i \otimes f_j$.Hence, W is an mn-dimensional vector space with a basis $\{e_i \otimes f_j\}$. The tensor product can evidently be extended on Cartesian products of arbitrary number of vector spaces.

Let us now consider the n-dimensional dual space U^* of an n-dimensional vector space U. It is quite clear that an element, or a vector, of the tensor product $\otimes^k U^*$ can now be represented by

$$\mathcal{T} = t_{i_1 i_2 \cdots i_k} f^{i_1} \otimes f^{i_2} \otimes \cdots \otimes f^{i_k} \tag{1.3.5}$$

where $\{f^i\}$ is the reciprocal basis in U^* corresponding to the basis $\{e_i\}$ in U. We define the value of the element $\mathcal{T}$ on an ordered k-tuple of vectors $(u_{(1)}, u_{(2)}, \dots, u_{(k)}) \in U^k$ as

$$\mathcal{T}(u_{(1)}, \dots, u_{(k)}) = t_{i_1 \cdots i_k} u^{j_1}_{(1)} \cdots u^{j_k}_{(k)} f^{i_1}(e_{j_1}) \cdots f^{i_k}(e_{j_k})$$

In view of (1.2.7), we then find that

$$\mathcal{T}(u_{(1)}, u_{(2)}, \dots, u_{(k)}) = t_{i_1 i_2 \cdots i_k} u^{i_1}_{(1)} u^{i_2}_{(2)} \cdots u^{i_k}_{(k)}.$$

We immediately see that the above relation leads to $(1.3.4)_2$ for vectors e_{i_1}, $e_{i_2}, \dots, e_{i_k}$. Hence (1.3.5) does in fact play the part of a k-linear functional on U^k and the tensor product $\otimes^k U^*$ is the vector space in which such k-linear functionals inhabit. We say that the elements of this vector space are ***k*-covariant tensors** and the number k is known as the ***order*** of the tensor. The scalar coefficients $t_{i_1 i_2 \cdots i_k}$ are then called the ***components*** *of such a tensor* with respect to bases $f^{i_1} \otimes \cdots \otimes f^{i_k}$. It is easily observed that the tensor product $f^{i_1} \otimes \cdots \otimes f^{i_k}$ of basis vectors constitutes a basis for the space $\otimes^k U^*$. Indeed the value of the zero element in $\otimes^k U^*$

$$t_{i_1 i_2 \cdots i_k} f^{i_1} \otimes f^{i_2} \otimes \cdots \otimes f^{i_k} = 0$$

on vectors $e_{j_1}, e_{j_2}, \ldots, e_{j_k} \in U$ vanishes naturally so that one obtains

$$t_{i_1 i_2 \cdots i_k} f^{i_1}(e_{j_1}) f^{i_2}(e_{j_2}) \cdots f^{i_k}(e_{j_k}) = t_{j_1 j_2 \cdots j_k} = 0$$

for all coefficients. Hence, the dimension of this vector space is n^k. Obviously, the sum of two tensors of the same kind and multiplication of a tensor by a scalar are again the following tensors of the same kind:

$$\mathcal{T}_1 + \mathcal{T}_2 = (t^{(1)}_{i_1 i_2 \cdots i_k} + t^{(2)}_{i_1 i_2 \cdots i_k}) f^{i_1} \otimes f^{i_2} \otimes \cdots \otimes f^{i_k}$$
$$\alpha \mathcal{T} = (\alpha t_{i_1 i_2 \cdots i_k}) f^{i_1} \otimes f^{i_2} \otimes \cdots \otimes f^{i_k}.$$

This is of course a direct consequence of $\otimes^k U^*$ being a linear vector space. We can now naturally define the tensorial product of a k-covariant tensor and an l-covariant tensor by

$$\mathcal{T}_1 \otimes \mathcal{T}_2 = t^{(1)}_{i_1 \cdots i_k} t^{(2)}_{j_1 \cdots j_l} f^{i_1} \otimes \cdots \otimes f^{i_k} \otimes f^{j_1} \otimes \cdots \otimes f^{j_l}.$$

The result is obviously a $(k+l)$-covariant tensor.

Let us now change the basis $\{e_i\}$ in the vector space U to another basis $\{e'_i\}$ as in (1.2.11). We know that the reciprocal basis $\{f^i\}$ in the dual space U^* changes to a reciprocal basis $\{f'^i\}$ through the relations (1.2.14). Consequently, the same tensor $\mathcal{T}$ is represented with respect to two different bases as follows

$$\begin{aligned}\mathcal{T} &= t_{j_1 j_2 \cdots j_k} f^{j_1} \otimes f^{j_2} \otimes \cdots \otimes f^{j_k} = t'_{i_1 i_2 \cdots i_k} f'^{i_1} \otimes f'^{i_2} \otimes \cdots \otimes f'^{i_k} \\ &= t_{j_1 j_2 \cdots j_k} b^{j_1}_{i_1} b^{j_2}_{i_2} \cdots b^{j_k}_{i_k} f'^{i_1} \otimes f'^{i_2} \otimes \cdots \otimes f'^{i_k}\end{aligned}$$

from which we immediately deduce that the following rule of transformation between components of a k-covariant tensor must be valid:

$$t'_{i_1 i_2 \cdots i_k} = b^{j_1}_{i_1} b^{j_2}_{i_2} \cdots b^{j_k}_{i_k} t_{j_1 j_2 \cdots j_k}. \tag{1.3.6}$$

In a similar fashion we may define a multilinear (k-linear) functional on the dual space U^* of a vector space. Such a functional $\mathcal{T} : (U^*)^k \to \mathbb{F}$ assigns a scalar number $\mathcal{T}(f^{(1)}, f^{(2)}, \ldots, f^{(k)}) \in \mathbb{F}$ to an ordered k-tuple of linear functionals $(f^{(1)}, f^{(2)}, \ldots, f^{(k)}) \in (U^*)^k$ and obeys the rules

$$\mathcal{T}(\ldots, f^{(i)} + g^{(i)}, \ldots) = \mathcal{T}(\ldots, f^{(i)}, \ldots) + \mathcal{T}(\ldots, g^{(i)}, \ldots)$$
$$\mathcal{T}(\ldots, \alpha f^{(i)}, \ldots) = \alpha \mathcal{T}(\ldots, f^{(i)}, \ldots), \alpha \in \mathbb{F}.$$

By resorting to the reciprocal basis $\{f^i\} \in U^*$ corresponding to the basis $\{e_i\} \in U$, we can of course write $f^{(m)} = \alpha^{(m)}_i f^i, \alpha^{(m)}_i \in \mathbb{F}, 1 \le m \le k$ and we obtain

$$\mathcal{T}(f^{(1)}, f^{(2)}, \dots, f^{(k)}) = t^{i_1 i_2 \cdots i_k} \alpha^{(1)}_{i_1} \alpha^{(2)}_{i_2} \cdots \alpha^{(k)}_{i_k}, \tag{1.3.7}$$
$$t^{i_1 i_2 \cdots i_k} = \mathcal{T}(f^{i_1}, f^{i_2}, \dots, f^{i_k}).$$

The ensemble of scalar numbers $t^{i_1 i_2 \cdots i_k}, 1 \le i_1, i_2, \cdots, i_k \le n$ entirely determines the action of a multilinear functional $\mathcal{T}$ on $(U^*)^k$. Let us now define an element in the tensor product $\otimes\, ^k U$ by

$$\mathcal{T} = t^{i_1 i_2 \cdots i_k} e_{i_1} \otimes e_{i_2} \otimes \cdots \otimes e_{i_k}.$$

$\mathcal{T}$ is called a ***k-contravariant tensor***. It is evident that the linearly independents elements $e_{i_1} \otimes e_{i_2} \otimes \cdots \otimes e_{i_k}$ constitute a basis for the vector space $\otimes\, ^k U$. n^k number of scalars $t^{i_1 i_2 \cdots i_k}$ are said to be *components* of this tensor with respect to bases $e_{i_1} \otimes \cdots \otimes e_{i_k}$. Let us define the value of the tensor $\mathcal{T}$ on k linear functionals $f^{(1)}$, $f^{(2)}, \dots, f^{(k)}$ by the relation

$$\mathcal{T}(f^{(1)}, f^{(2)}, \dots, f^{(k)}) = t^{i_1 i_2 \cdots i_k} f^{(1)}(e_{i_1}) f^{(2)}(e_{i_2}) \cdots f^{(k)}(e_{i_k}).$$

In view of (1.2.6) we find that

$$\mathcal{T}(f^{(1)}, f^{(2)}, \dots, f^{(k)}) = t^{i_1 i_2 \cdots i_k} \alpha^{(1)}_{i_1} \alpha^{(2)}_{i_2} \cdots \alpha^{(k)}_{i_k}.$$

It is clear that the product of a k-contravariant tensor and an l-contravariant tensor is a $(k + l)$-contravariant tensor. We now consider a change of basis in the vector space U. We then obtain

$$\mathcal{T} = t^{j_1 j_2 \cdots j_k} e_{j_1} \otimes e_{j_2} \otimes \cdots \otimes e_{j_k} = t'^{i_1 i_2 \cdots i_k} e'_{i_1} \otimes e'_{i_2} \otimes \cdots \otimes e'_{i_k}$$
$$= t^{j_1 j_2 \cdots j_k} a^{i_1}_{j_1} a^{i_2}_{j_2} \cdots a^{i_k}_{j_k} e'_{i_1} \otimes e'_{i_2} \otimes \cdots \otimes e'_{i_k}$$

from which we deduce the following rule of transformation for components of a contravariant tensor

$$t'^{i_1 i_2 \cdots i_k} = a^{i_1}_{j_1} a^{i_2}_{j_2} \cdots a^{i_k}_{j_k} t^{j_1 j_2 \cdots j_k}. \tag{1.3.8}$$

We can also easily define tensors of mixed type. A k-contravariant and l-covariant ***mixed tensor*** is an element of the vector space $\otimes\, ^k U \otimes\, ^l U^*$ and can be written in the form

$$\mathcal{T} = t^{i_1 i_2 \cdots i_k}_{j_1 j_2 \cdots j_l} e_{i_1} \otimes e_{i_2} \otimes \cdots \otimes e_{i_k} \otimes f^{j_1} \otimes f^{j_2} \otimes \cdots \otimes f^{j_l},$$
$$t^{i_1 i_2 \cdots i_k}_{j_1 j_2 \cdots j_l} = \mathcal{T}(f^{i_1}, f^{i_2}, \dots, f^{i_k}, e_{j_1}, e_{j_2}, \dots, e_{j_l}),$$
$$1 \le i_1, i_2, \cdots, i_k \le n, \; 1 \le j_1, j_2, \cdots, j_l \le n.$$

The value of this tensor on linear functionals $f^{(1)}, f^{(2)}, \dots, f^{(k)} \in U^*$ and vectors $u_{(1)},\; u_{(2)}, \dots, u_{(l)} \in U$ is given by

$$\mathcal{T}(f^{(1)},\dots,f^{(k)},u_{(1)},\dots,u_{(l)}) = t^{i_1 i_2\cdots i_k}_{j_1 j_2\cdots j_l}\alpha^{(1)}_{i_1}\alpha^{(2)}_{i_2}\cdots\alpha^{(k)}_{i_k}u^{j_1}_{(1)}u^{j_2}_{(2)}\cdots u^{j_l}_{(l)}.$$

It is quite obvious that we do not have to select the ordering in the tensor products in the foregoing way. We may, of course, consider a different ordering such as $U\otimes U^*\otimes U^*\otimes U\otimes U^*\otimes\cdots$. The indices of components of this type of a tensor occupy accordingly proper upper and lower positions. It is evident that different ordering of spaces in the tensor product will give rise to different types of tensors of the same order.

If, in a mixed tensor of order $k+l$, we remove the tensor product between the functional f^{j_m} and the vector e_{i_n}, then the relation $f^{j_m}(e_{i_n}) = \delta^{j_m}_{i_n}$ between reciprocal basis vectors reduces the order of the tensor. We thus obtain a $(k-1)$-contravariant and $(l-1)$-covariant tensor, in other words, a tensor of order $k+l-2$ defined by the relation

$$\mathcal{T}_c = t^{i_1\cdots i_n\cdots i_k}_{j_1\cdots i_n\cdots j_l}e_{i_1}\otimes\cdots\otimes e_{i_{n-1}}\otimes e_{i_{n+1}}\otimes\cdots\otimes e_{i_k} \\ \otimes f^{j_1}\otimes\cdots\otimes f^{j_{m-1}}\otimes f^{j_{m+1}}\otimes\cdots\otimes f^{j_l}.$$

This operation is called a ***contraction***. The components of the contracted tensor are given as follows:

$$_c t^{i_1\cdots i_{n-1}i_{n+1}\cdots i_k}_{j_1\cdots j_{m-1}j_{m+1}\cdots j_l} = t^{i_1\cdots i_{n-1}i i_{n+1}\cdots i_k}_{j_1\cdots j_{m-1}i j_{m+1}\cdots j_l}.$$

1.4. ALTERNATING k-LINEAR FUNCTIONALS

Let us consider a multilinear functional $\omega : U^k \to \mathbb{F}$ where U is a finite-dimensional vector space so that for vectors $u_i\in U,\ i=1,\dots,k$ we have $\omega(u_1,u_2,\dots,u_k)\in\mathbb{F}$. We know from Sec 1.3 that the multilinear functional ω may be represented by a k-covariant tensor. We say that the multilinear functional ω is an ***alternating k-linear functional*** or a ***k-vector*** or a ***multivector*** if it becomes zero whenever any two of its arguments are equal. It can be shown that such an alternating multilinear functional enjoys the following properties:

1. *An alternating k-linear functional is completely antisymmetric in the sense that its value changes only its sign whenever any two of its arguments are interchanged.*

To understand the effect of interchanging the argument vectors u_i and u_j let us take into account the expansion

$$\omega(u_1,\dots,u_i+u_j,\dots,u_i+u_j,\dots,u_k) = \\ \omega(u_1,\dots,u_i,\dots,u_i,\dots,u_k)$$

$$+\,\omega(u_1,\dots,u_i,\dots,u_j,\dots,u_k)$$
$$+\,\omega(u_1,\dots,u_j,\dots,u_i,\dots,u_k)$$
$$+\,\omega(u_1,\dots,u_j,\dots,u_j,\dots,u_k)=0.$$

If we note that the first and the fourth terms in the above expression is zero by definition, we obtain from the middle lines the following property of *complete antisymmetry* for every pair of arguments:

$$\omega(u_1,\dots,u_i,\dots,u_j,\dots,u_k)=\,-\,\omega(u_1,\dots,u_j,\dots,u_i,\dots,u_k)$$

Thus if $U=U^{(n)}$, then the value of an alternating k-linear functional on vectors $u_1,u_2,\dots,u_k\in U$ are given by

$$\omega(u_1,u_2,\dots,u_k)=\omega_{i_1i_2\cdots i_k}u_1^{i_1}u_2^{i_2}\cdots u_k^{i_k} \tag{1.4.1}$$

where the scalars $\omega_{i_1i_2\cdots i_k}=\omega(e_{i_1},e_{i_2},\dots,e_{i_k})\in\mathbb{F}$ are *completely antisymmetric* with respect to k indices $i_1,i_2,\cdots,i_k$ taking the values from 1 to n. Hence, for every pair of indices the relation

$$\omega_{i_1\cdots i_p\cdots i_q\cdots i_k}=\,-\,\omega_{i_1\cdots i_q\cdots i_p\cdots i_k} \tag{1.4.2}$$

is satisfied. It is then straightforward to see that the number of independent components of such coefficients are given by $\binom{n}{k}=\dfrac{n!}{k!\,(n-k)!}$.

2. *The value of an alternating k-linear functional on linearly dependent vectors is zero.*

Let us assume that at least one of the k vectors is a linear combination of the remaining $k-1$ vectors. When we expand the functional by employing multilinearity, we see that it is expressible as a sum of terms in each of which at least two arguments in the functional are equal. Hence the value of the functional becomes zero. Consequently *if $k>n$ all k-linear functionals on a vector space of dimension n are identically zero.*

3. *Any alternating n-linear functional on a linear vector space $U^{(n)}$ that vanishes on an ordered basis $\{e_1,e_2,\dots,e_n\}$ of $U^{(n)}$ is identically zero.*

If we insert ordered vectors $u_i=u_i^j\,e_j,i=1,\dots,n$ into the functional, expand the resulting expression by making use of multilinearity, equate to zero the terms involving repeated arguments and exploit the property of antisymmetry, we see that the value of the functional is a linear combination of terms in the form $\pm\,\omega(e_1,e_2,\dots,e_n)$. In case $\omega(e_1,e_2,\dots,e_n)=0$, the value of the functional becomes eventually zero on every ordered n-tuple of vectors.

We can generate a completely antisymmetric quantity from a quantity with k indices, say $a_{i_1 i_2 \cdots i_k}$, through the ***alternation mapping***. Let us denote a permutation of indices $i_1, \dots, i_k$ by $\sigma_m(i_1, i_2, \dots, i_k)$. As is well known the total number of all such permutations is $k!$. We now introduce the following quantity through the alternation mapping

$$a_{[i_1 i_2 \cdots i_k]} = \frac{1}{k!} \sum_{m=1}^{k!} (-1)^{\kappa(\sigma_m)} a_{\sigma_m(i_1, i_2, \dots, i_k)} \tag{1.4.3}$$

where $\kappa(\sigma_m) = 0$ if $\sigma_m(i_1, i_2, \dots, i_k)$ is an even permutation whereas $\kappa(\sigma_m) = 1$ if it is odd. We know that a permutation is realised by means of a number of transpositions performed by interchanging successive indices. A specified permutation is called an even permutation if the number of transpositions performed is even and odd if that number is odd. We can immediately verify that the quantity $a_{[i_1 i_2 \cdots i_k]}$ is completely antisymmetric. *Henceforth, the indices inside a square bracket will always represent the completely antisymmetric part.* As an example, let us consider a quantity a_{ijk} with three indices. We then find that

$$a_{[ijk]} = \frac{1}{3!}(a_{ijk} + a_{jki} + a_{kij} - a_{ikj} - a_{kji} - a_{jik}).$$

If $a_{i_1 i_2 \cdots i_k}$ is already completely antisymmetric, then it is clearly understood that $a_{i_1 i_2 \cdots i_k} = a_{[i_1 i_2 \cdots i_k]}$.

Since the coefficients $\omega_{i_1 i_2 \cdots i_k}$ are completely antisymmetric, only the completely antisymmetric parts of terms $u_1^{i_1} u_2^{i_2} \cdots u_k^{i_k}$ in a k-fold sum as in (1.4.1) can contribute to the sum so that we can write

$$\begin{aligned} \omega(u_1, u_2, \dots, u_k) &= \omega_{i_1 i_2 \cdots i_k} u_1^{i_1} u_2^{i_2} \cdots u_k^{i_k} \\ &= \omega_{i_1 i_2 \cdots i_k} u_1^{[i_1} u_2^{i_2} \cdots u_k^{i_k]}. \end{aligned} \tag{1.4.4}$$

The components of a completely antisymmetric quantity $\omega_{i_1 i_2 \cdots i_k}$ whose indices satisfy inequalities $1 \le i_1 < i_2 < \cdots < i_k \le n$ will be called its *essential components*. Because all other components are either zero or determined by essential components, sometimes, only with a change of sign. The expression (1.4.4) can then be written in the following form by using essential components

$$\omega(u_1, u_2, \dots, u_k) = k! \sum_{1 \le i_1 < i_2 < \cdots < i_k \le n} \omega_{i_1 i_2 \cdots i_k} u_1^{[i_1} u_2^{i_2} \cdots u_k^{i_k]}. \tag{1.4.5}$$

As an example, we consider a 2-linear alternating functional $\omega(u_1, u_2) = \omega_{ij} u_1^i u_2^j$ and $n = 3$. Since $\omega_{ij} = -\omega_{ji}$ we obtain at once with $k = 2$

$$
\begin{aligned}
\omega(u_1, u_2) &= \omega_{12} u_1^1 u_2^2 + \omega_{21} u_1^2 u_2^1 + \omega_{13} u_1^1 u_2^3 \\
&\quad + \omega_{31} u_1^3 u_2^1 + \omega_{23} u_1^2 u_2^3 + \omega_{32} u_1^3 u_2^2 \\
&= \omega_{12}(u_1^1 u_2^2 - u_1^2 u_2^1) + \omega_{13}(u_1^1 u_2^3 - u_1^3 u_2^1) + \omega_{23}(u_1^2 u_2^3 - u_1^3 u_2^2) \\
&= 2(\omega_{12} u_1^{[1} u_2^{2]} + \omega_{13} u_1^{[1} u_2^{3]} + \omega_{23} u_1^{[2} u_2^{3]}) \\
&= \omega_{12} u_1^{[1} u_2^{2]} + \omega_{21} u_1^{[2} u_2^{1]} + \omega_{13} u_1^{[1} u_2^{3]} \\
&\qquad + \omega_{31} u_1^{[3} u_2^{1]} + \omega_{23} u_1^{[2} u_2^{3]} + \omega_{32} u_1^{[3} u_2^{2]} \\
&= \omega_{ij} u_1^{[i} u_2^{j]}.
\end{aligned}
$$

The operation of alternation can be performed much more systematically by introducing the ***generalised Kronecker delta***. We shall define in an n-dimensional space the generalised Kronecker delta of order $k \leq n$ by means of the following *symbolic determinant*

$$
\delta^{i_1 i_2 \cdots i_k}_{j_1 j_2 \cdots j_k} = \begin{vmatrix} \delta^{i_1}_{j_1} & \delta^{i_1}_{j_2} & \cdots & \delta^{i_1}_{j_k} \\ \delta^{i_2}_{j_1} & \delta^{i_2}_{j_2} & \cdots & \delta^{i_2}_{j_k} \\ \vdots & \vdots & & \vdots \\ \delta^{i_k}_{j_1} & \delta^{i_k}_{j_2} & \cdots & \delta^{i_k}_{j_k} \end{vmatrix} \tag{1.4.6}
$$

where the range of all indices $i_1, i_2, \cdots, i_k$ and $j_1, j_2, \cdots, j_k$ is, of course, from 1 to n. Since a determinant changes only its sign when we interchange either its two columns or its two rows we immediately notice that n^{2k} number of quantities $\delta^{i_1 i_2 \cdots i_k}_{j_1 j_2 \cdots j_k}$ are completely antisymmetric with respect to its superscripts or its subscripts so that only the sign of the relevant quantity changes when we interchange any two of its upper indices or lower indices and it becomes zero when any two indices in upper or lower positions are equal. If the indices $\{i_1, i_2, \cdots, i_k\}$ and $\{j_1, j_2, \cdots, j_k\}$ are *not* chosen from a same subset of the set $\{1, \ldots, n\}$ involving k distinct numbers, then at least one row of the determinant (1.4.6) is zero owing to the definition of the Kronecker delta. Hence, the corresponding generalised Kronecker delta vanishes. On the other hand, if the upper and lower indices are *both even or odd permutations* of the same distinct k numbers the generalised Kronecker delta becomes $+1$ whereas it becomes -1 if *one is an even and the other is the odd permutations* of these k numbers. To see this, it suffices to note that when we choose upper and lower indices from the same set of distinct indices we can obviously set $i_1 = j_1,\ i_2 = j_2, \ldots, i_k = j_k$ by properly interchanging row and columns in the determinant, in other words, by properly permuting upper and lower indices. In this case the determinant reduces simply to

$$\mp\begin{vmatrix} 1 & 0 & \cdots & 0 \\ 0 & 1 & \cdots & 0 \\ \vdots & \vdots & & \vdots \\ 0 & 0 & \cdots & 1 \end{vmatrix} = \mp 1.$$

It is clear that if it is necessary to make either even or odd permutations in both upper and lower indices then the value of the generalised Kronecker delta would be $+1$. However, if it is required to make even permutation in one set of indices and odd permutation in the other set the value would, of course, be -1. It is clear that if $k > n$, the generalised Kronecker delta becomes identically zero.

Since the generalised Kronecker delta is completely antisymmetric with respect to both upper and lower indices, it follows from the definition (1.4.6) that

$$\delta^{i_1 i_2 \cdots i_k}_{j_1 j_2 \cdots j_k} = k!\, \delta^{i_1}_{[j_1} \delta^{i_2}_{j_2} \cdots \delta^{i_k}_{j_k]} = k!\, \delta^{[i_1}_{j_1} \delta^{i_2}_{j_2} \cdots \delta^{i_k]}_{j_k}. \tag{1.4.7}$$

Indeed, we can readily observe this property in two simple examples below for $k = 2$ and $k = 3$

$$\delta^{ij}_{kl} = \begin{vmatrix} \delta^i_k & \delta^i_l \\ \delta^j_k & \delta^j_l \end{vmatrix} = \delta^i_k \delta^j_l - \delta^i_l \delta^j_k = 2\, \delta^i_{[k} \delta^j_{l]} = 2\, \delta^{[i}_k \delta^{j]}_l$$

$$\begin{aligned} \delta^{ijk}_{lmn} &= \begin{vmatrix} \delta^i_l & \delta^i_m & \delta^i_n \\ \delta^j_l & \delta^j_m & \delta^j_n \\ \delta^k_l & \delta^k_m & \delta^k_n \end{vmatrix} = \delta^i_l \delta^j_m \delta^k_n - \delta^i_l \delta^j_n \delta^k_m + \delta^i_m \delta^j_n \delta^k_l - \delta^i_m \delta^j_l \delta^k_n \\ &\quad + \delta^i_n \delta^j_l \delta^k_m - \delta^i_n \delta^j_m \delta^k_l \\ &= 3!\, \delta^i_{[l} \delta^j_m \delta^k_{n]} = 3!\, \delta^{[i}_l \delta^j_m \delta^{k]}_n. \end{aligned}$$

Consider a quantity $A^{i_1 i_2 \cdots i_k}$ with k indices. It is rather straightforward to see that (1.4.7) leads to the relation

$$\delta^{i_1 i_2 \cdots i_k}_{j_1 j_2 \cdots j_k} A^{j_1 j_2 \cdots j_k} = k!\, A^{[i_1 i_2 \cdots i_k]} \tag{1.4.8}$$

Let us now rewrite the expression (1.4.4) defining an alternating k-linear functional in the form

$$\omega(u_1, u_2, \ldots, u_k) = \omega_{i_1 i_2 \cdots i_k} f^{[i_1}(u_1) f^{i_2}(u_2) \cdots f^{i_k]}(u_k)$$

where, as usual, the vectors, or linear functionals $\{f^i\} \subset U^*$ constitutes the reciprocal basis in the dual space with respect to the basis $\{e_i\} \subset U$. Thus we can represent this alternating k-linear functional acting on an element $(u_1, u_2, \ldots, u_k)$ of the Cartesian product U^k [*see* (1.3.5)] by the following expression

$$\omega = \omega_{i_1 i_2 \cdots i_k} f^{[i_1} \otimes f^{i_2} \otimes \cdots \otimes f^{i_k]} \tag{1.4.9}$$

by employing the tensor product. Resorting to the relation (1.4.8) we can transform the expression (1.4.9) into

$$\omega = \frac{1}{k!}\, \omega_{i_1 i_2 \cdots i_k} \delta^{i_1 i_2 \cdots i_k}_{j_1 j_2 \cdots j_k} f^{j_1} \otimes f^{j_2} \otimes \cdots \otimes f^{j_k}.$$

We now define the ***exterior product***, or ***wedge product***, of k basis vectors in the dual space U^* by the relation

$$\begin{aligned} f^{i_1} \wedge f^{i_2} \wedge \cdots \wedge f^{i_k} &= \delta^{i_1 i_2 \cdots i_k}_{j_1 j_2 \cdots j_k} f^{j_1} \otimes f^{j_2} \otimes \cdots \otimes f^{j_k} \\ &= k!\, f^{[i_1} \otimes f^{i_2} \otimes \cdots \otimes f^{i_k]}. \end{aligned} \tag{1.4.10}$$

We can then represent (1.4.9) in the form

$$\omega = \frac{1}{k!}\, \omega_{i_1 i_2 \cdots i_k} f^{i_1} \wedge f^{i_2} \wedge \cdots \wedge f^{i_k}. \tag{1.4.11}$$

For instance, we find that

$$\begin{aligned} f^i \wedge f^j &= f^i \otimes f^j - f^j \otimes f^i, \\ f^i \wedge f^j \wedge f^k &= f^i \otimes f^j \otimes f^k + f^j \otimes f^k \otimes f^i + f^k \otimes f^i \otimes f^j \\ &\quad - f^i \otimes f^k \otimes f^j - f^k \otimes f^j \otimes f^i - f^j \otimes f^i \otimes f^k. \end{aligned}$$

It is clear that the exterior product introduced by (1.4.10) is completely antisymmetric. In view of the representation (1.4.11), we call an alternating k-linear functional as an ***exterior form of degree k*** or simply a ***k-form***. *Such a form is obviously a completely antisymmetric k-covariant tensor.* The value of a k-form on linearly independent k vectors $u_1, u_2, \ldots, u_k \in U$ is given by (1.4.4). However, if we recall the definition of a determinant we can immediately recognise that a quantity $u_1^{[i_1} u_2^{i_2} \cdots u_k^{i_k]}$ is expressible by a determinant as follows:

$$k!\, u_1^{[i_1} u_2^{i_2} \cdots u_k^{i_k]} = \begin{vmatrix} u_1^{i_1} & u_2^{i_1} & \cdots & u_k^{i_1} \\ u_1^{i_2} & u_2^{i_2} & \cdots & u_k^{i_2} \\ \vdots & \vdots & & \vdots \\ u_1^{i_k} & u_2^{i_k} & \cdots & u_k^{i_k} \end{vmatrix}.$$

We can thus write

$$\omega(u_1,u_2,\dots,u_k)=\omega_{i_1i_2\cdots i_k}u_1^{[i_1}u_2^{i_2}\cdots u_k^{i_k]}=\frac{1}{k!}\,\omega_{i_1i_2\cdots i_k}\begin{vmatrix}u_1^{i_1}&u_2^{i_1}&\cdots&u_k^{i_1}\\u_1^{i_2}&u_2^{i_2}&\cdots&u_k^{i_2}\\\vdots&\vdots&&\vdots\\u_1^{i_k}&u_2^{i_k}&\cdots&u_k^{i_k}\end{vmatrix}$$

$$=\frac{1}{k!}\,\omega_{i_1i_2\cdots i_k}\begin{vmatrix}f^{i_1}(u_1)&f^{i_1}(u_2)&\cdots&f^{i_1}(u_k)\\f^{i_2}(u_1)&f^{i_2}(u_2)&\cdots&f^{i_2}(u_k)\\\vdots&\vdots&&\vdots\\f^{i_k}(u_1)&f^{i_k}(u_2)&\cdots&f^{i_k}(u_k)\end{vmatrix}$$

By employing essential components, we can also transform this expression into the form

$$\omega(u_1,u_2,\dots,u_k)=\sum_{1\le i_1<i_2<\cdots<i_k\le n}\omega_{i_1i_2\cdots i_k}V^{i_1i_2\cdots i_k}(u_1,u_2,\dots,u_k) \tag{1.4.12}$$

Here

$$V^{i_1i_2\cdots i_k}(u_1,u_2,\dots,u_k)=\begin{vmatrix}u_1^{i_1}&u_2^{i_1}&\cdots&u_k^{i_1}\\u_1^{i_2}&u_2^{i_2}&\cdots&u_k^{i_2}\\\vdots&\vdots&&\vdots\\u_1^{i_k}&u_2^{i_k}&\cdots&u_k^{i_k}\end{vmatrix} \tag{1.4.13}$$

may be interpreted as the *k-dimensional volume*[1] of the projection of k-dimensional parallelepiped formed by vectors $u_1,u_2,\dots,u_k$ in n-dimensional vector space on a subspace generated by *axes* $i_1<i_2<\cdots<i_k$. As an example, let us consider $n=3,\ k=2$ and a 2-form

$$\omega=\frac{1}{2}\omega_{ij}\,f^i\wedge f^j$$

whose value on vectors u_1 and u_2 is given by

$$\omega(u_1,u_2)=\omega_{12}V^{12}(u_1,u_2)+\omega_{13}V^{13}(u_1,u_2)+\omega_{23}V^{23}(u_1,u_2)$$

where one identifies the numbers $V^{12}(u_1,u_2)=u_1^1u_2^2-u_1^2u_2^1$, $V^{13}(u_1,u_2)=u_1^1u_2^3-u_1^3u_2^1$ and $V^{23}(u_1,u_2)=u_1^2u_2^3-u_1^3u_2^2$ as *areas* of parallelograms that are projections of the parallelogram formed by vectors u_1 and u_2 in the 3-dimensional space, respectively, on the planes generated by 12-, 13- and

[1]One must notice the fact that this number does not correspond to the real invariant geometric volume. As is easily observed, this number is dependent on the selected basis of the vector space U. But it is non-zero for linearly independent vectors.

23-axes. We can now say that a k-form defined on an n-dimensional vector space U makes it possible for us to evaluate certain linear combinations, with coefficients of that form, of *k-dimensional volumes* projected onto k-dimensional subspaces from a k-dimensional parallelepiped formed by k linearly independent vector in U.

Let us now consider an n-form as follows

$$\omega = \frac{1}{n!}\omega_{i_1 i_2 \cdots i_n} f^{i_1} \wedge f^{i_2} \wedge \cdots \wedge f^{i_n}. \tag{1.4.14}$$

Since the indices have to be permutations of the numbers $1, 2, \ldots, n$, the only essential component is $\omega_{12\cdots n}$. In order to express this situation more systematically we now introduce the ***Levi-Civita symbol*** [after Italian mathematician Tullio Levi-Civita (1873-1941)] with covariant indices as

$$e_{i_1 i_2 \cdots i_n} = \begin{cases} 0, \text{if any two indices are equal,} \\ +1, \text{if indices } (i_1, \cdots, i_n) \text{ is an even permutation of } (1, \ldots, n), \\ -1, \text{if indices } (i_1, \cdots, i_n) \text{ is an odd permutation of } (1, \ldots, n),. \end{cases}$$

The symbol $e^{i_1 i_2 \cdots i_n}$ with contravariant indices is defined in exactly the same fashion. On the other hand, it is easy to see that we have the relation

$$e_{i_1 i_2 \cdots i_n} e^{i_1 i_2 \cdots i_n} = n! \tag{1.4.15}$$

since each term in the above sum will take the value $+1$ for every permutation. We can thus write for an n-form

$$\begin{aligned} \omega &= \frac{1}{n!} e_{i_1 i_2 \cdots i_n} \omega_{12 \cdots n} e^{i_1 i_2 \cdots i_n} f^1 \wedge f^2 \wedge \cdots \wedge f^n \\ &= \omega_{12 \cdots n} f^1 \wedge f^2 \wedge \cdots \wedge f^n \end{aligned}$$

Since $\binom{n}{n} = 1$ there exists indeed only one linearly independent form, for instance, $f^1 \wedge f^2 \wedge \cdots \wedge f^n$. *All other n-forms are scalar multiples of that form.* The value of this form ω on linearly independent n vectors $u_1, u_2, \ldots, u_n \in U$ are given by

$$\begin{aligned} \omega(u_1, u_2, \ldots, u_n) &= \omega_{12\cdots n} \begin{vmatrix} u_1^1 & u_2^1 & \cdots & u_n^1 \\ u_1^2 & u_2^2 & \cdots & u_n^2 \\ \vdots & \vdots & & \vdots \\ u_1^n & u_2^n & \cdots & u_n^n \end{vmatrix} \\ &= \omega_{12\cdots n} V_n(u_1, u_2, \ldots, u_n). \end{aligned}$$

We may interpret the determinant V_n as the volume of an n-dimensional parallelepiped formed n vectors in the space U. If these vectors are linearly independent we know that the above determinant cannot vanish so that we have $V_n(u_1, u_2, \ldots, u_n) \neq 0$. If we rename the basis vectors $e_1, \ldots, e_n$ in U properly we can set $V_n(e_1, \ldots, e_n) = +1$ and we find

$$\omega_{12\cdots n} = \omega(e_1, e_2, \ldots, e_n)$$

as it should be. If $k = n$, the generalised Kronecker deltas are obviously expressible in terms of Levi-Civita symbols in the following way

$$\delta^{i_1 i_2 \cdots i_n}_{j_1 j_2 \cdots j_n} = e^{i_1 i_2 \cdots i_n} e_{j_1 j_2 \cdots j_n}. \tag{1.4.16}$$

The determinant V_n can now be written as

$$V_n = n!\, u^{[1}_1 \cdots u^{n]}_n = e_{i_1 i_2 \cdots i_n} u^{i_1}_1 u^{i_2}_2 \cdots u^{i_n}_n.$$

But this expression is completely antisymmetric with respect to indices 1, $\ldots, n$. Therefore, we can also write

$$V_n = \frac{1}{n!} e_{i_1 \cdots i_n} e^{j_1 \cdots j_n} u^{i_1}_{j_1} \cdots u^{i_n}_{j_n} = \frac{1}{n!} \delta^{j_1 \cdots j_n}_{i_1 \cdots i_n} u^{i_1}_{j_1} \cdots u^{i_n}_{j_n}. \tag{1.4.17}$$

It then readily follows from (1.4.17) that the relation

$$\begin{aligned} e_{k_1 \cdots k_n} V_n &= \frac{1}{n!} \delta^{j_1 \cdots j_n}_{k_1 \cdots k_n} e_{i_1 \cdots i_n} u^{i_1}_{j_1} \cdots u^{i_n}_{j_n} \\ &= e_{i_1 \cdots i_n} u^{i_1}_{[k_1} \cdots u^{i_n}_{k_n]} = e_{i_1 \cdots i_n} u^{i_1}_{k_1} \cdots u^{i_n}_{k_n} \end{aligned} \tag{1.4.18}$$

is valid for determinants.

It is straightforward to realise that the addition of k-forms on a vector space U and their multiplication with scalars are again k-forms. To see this let us consider two k-forms α and β:

$$\alpha = \frac{1}{k!} \alpha_{i_1 \cdots i_k} f^{i_1} \wedge \cdots \wedge f^{i_k}, \quad \beta = \frac{1}{k!} \beta_{i_1 \cdots i_k} f^{i_1} \wedge \cdots \wedge f^{i_k}.$$

The sum $\gamma = \alpha + \beta$ of these forms will naturally be

$$\gamma = \frac{1}{k!} \gamma_{i_1 \cdots i_k} f^{i_1} \wedge \cdots \wedge f^{i_k}, \quad \gamma_{i_1 \cdots i_k} = \alpha_{i_1 \cdots i_k} + \beta_{i_1 \cdots i_k}.$$

Similarly, for an arbitrary scalar λ the form $\eta = \lambda \alpha$ is given by

$$\eta = \frac{1}{k!} \eta_{i_1 \cdots i_k} f^{i_1} \wedge \cdots \wedge f^{i_k}, \quad \eta_{i_1 \cdots i_k} = \lambda \alpha_{i_1 \cdots i_k}.$$

Hence k-forms constitute a linear vector space which will be denoted by $\Lambda^k(U)$. This vector space is well defined for $1 < k \leq n$. Obviously, there are $\binom{n}{k}$ linearly independent k-forms in this space. All forms whose degrees satisfying $k > n$ are identically zero. If we define exterior forms for $k = 1$ by the expression

$$\omega = \omega_i f^i, \ \ \omega_i \in \mathbb{F} \tag{1.4.19}$$

the spaces $\Lambda^k(U)$ will be completely determined for $1 \leq k \leq n$. There are evidently n linearly independent 1-form since $\binom{n}{1} = n$.

1.5. EXTERIOR ALGEBRA

We shall now try to define the product of two exterior forms in such a way that the result will again be an exterior form. Thus, we will be able to construct an exterior algebra. Let us consider the forms $\alpha \in \Lambda^p(U)$ and $\beta \in \Lambda^q(U)$ given below such that $p \leq n$, $q \leq n$ and $p + q \leq n$:

$$\alpha = \frac{1}{p!}\,\alpha_{i_1 i_2 \cdots i_p} f^{i_1} \wedge f^{i_2} \wedge \cdots \wedge f^{i_p},$$
$$\beta = \frac{1}{q!}\beta_{j_1 j_2 \cdots j_q} f^{j_1} \wedge f^{j_2} \wedge \cdots \wedge f^{j_q}.$$

The ***exterior product*** $\alpha \wedge \beta$ of forms α and β will now be defined in the following fashion

$$\alpha \wedge \beta = \frac{1}{p!\,q!}\,\alpha_{i_1 i_2 \cdots i_p} \beta_{j_1 j_2 \cdots j_q} f^{i_1} \wedge f^{i_2} \wedge \cdots \wedge f^{i_p} \wedge f^{j_1} \wedge f^{j_2} \wedge \cdots \wedge f^{j_q}$$

where the exterior product of basis vectors is, of course, determined by

$$f^{i_1} \wedge f^{i_2} \wedge \cdots \wedge f^{i_p} \wedge f^{j_1} \wedge f^{j_2} \wedge \cdots \wedge f^{j_q} = \delta^{i_1 i_2 \cdots i_p j_1 j_2 \cdots j_q}_{k_1 k_2 \cdots k_p l_1 l_2 \cdots l_q} f^{k_1} \otimes f^{k_2} \otimes \cdots \otimes f^{k_p} \otimes f^{l_1} \otimes f^{l_2} \otimes \cdots \otimes f^{l_q}.$$

With this definition we are obviously led to the result $\alpha \wedge \beta \in \Lambda^{p+q}(U)$. The coefficients of the form $\alpha \wedge \beta$ should be completely antisymmetric with respect to $p + q$ indices. But they are already completely antisymmetric with respect to the first p and the last q indices. Therefore, the number of independent components will be $(p + q)!/p!\,q!$ and if we define

$$\gamma_{i_1 i_2 \cdots i_p j_1 j_2 \cdots j_q} = \frac{(p+q)!}{p!q!}\alpha_{[i_1 i_2 \cdots i_p} \beta_{j_1 j_2 \cdots j_q]} \tag{1.5.1}$$

we obtain

$$\begin{aligned}\gamma &= \alpha \wedge \beta \\ &= \frac{1}{(p+q)!}\gamma_{i_1\cdots i_p j_1\cdots j_q} f^{i_1}\wedge\cdots\wedge f^{i_p}\wedge f^{j_1}\wedge\cdots\wedge f^{j_q}.\end{aligned} \tag{1.5.2}$$

If $p+q>n$ we clearly find $\alpha\wedge\beta=0$. As an example, consider

$$\alpha=\alpha_i f^i \in \Lambda^1(U), \quad \beta=\frac{1}{2!}\beta_{jk} f^j\wedge f^k \in \Lambda^2(U)$$

where β_{jk} are antisymmetric. For $n>3$ we obtain

$$\gamma=\alpha\wedge\beta=\frac{1}{2!}\alpha_i\beta_{jk}\, f^i\wedge f^j\wedge f^k=\frac{1}{2!}\alpha_{[i}\beta_{jk]}\, f^i\wedge f^j\wedge f^k.$$

On the other hand, we find that

$$\begin{aligned}\alpha_{[i}\beta_{jk]} &= \frac{1}{3!}(\alpha_i\beta_{jk}+\alpha_j\beta_{ki}+\alpha_k\beta_{ij}-\alpha_i\beta_{kj}-\alpha_k\beta_{ji}-\alpha_j\beta_{ik}) \\ &= \frac{1}{3}(\alpha_i\beta_{jk}+\alpha_j\beta_{ki}+\alpha_k\beta_{ij})=\frac{1}{3}\gamma_{ijk}.\end{aligned}$$

Hence the exterior product $\alpha\wedge\beta$ has the standard structure

$$\gamma=\frac{1}{3!}\gamma_{ijk}\, f^i\wedge f^j\wedge f^k \in \Lambda^3(U).$$

Just from the definition of the exterior product of forms, we conclude that the exterior product is *distributive*, namely

$$\alpha\wedge(\beta+\gamma)=\alpha\wedge\beta+\alpha\wedge\gamma,\ (\alpha+\beta)\wedge\gamma=\alpha\wedge\gamma+\beta\wedge\gamma. \tag{1.5.3}$$

Here we have, naturally, considered the addition of forms of the same degree. It is evident that the exterior product so defined is *associative*:

$$\alpha\wedge(\beta\wedge\gamma)=(\alpha\wedge\beta)\wedge\gamma=\alpha\wedge\beta\wedge\gamma. \tag{1.5.4}$$

However, the exterior product is not generally *commutative*. Let us consider the forms $\alpha\in\Lambda^p(U)$ and $\beta\in\Lambda^q(U)$. We can show that the relation

$$\beta\wedge\alpha=(-1)^{pq}\alpha\wedge\beta \tag{1.5.5}$$

is valid. Indeed, in order to transform the form $\alpha\wedge\beta$ into the form $\beta\wedge\alpha$, we are compelled to interchange the exterior products $f^{i_1}\wedge\cdots\wedge f^{i_p}$ and $f^{j_1}\wedge\cdots\wedge f^{j_q}$ as blocks. To this end, we first put the vector f^{i_p} at the end of the second sequence by successively interchanging it with vectors f^{j_1}, f^{j_2}, $\ldots$, f^{j_q}. Every transposition gives rise to the multiplication by -1. Thus the form is eventually multiplied by $(-1)^q$. Since this operation should be repeated p times for vectors $f^{i_p},\ f^{i_{p-1}},\dots,f^{i_1}$ we obtain the relation (1.5.5). It follows now from (1.5.5) that if $\alpha\in\Lambda^1(U)$, we then of course find $\alpha\wedge\alpha=0$.

The vector space of k-forms $\Lambda^k(U)$ on an n-dimensional vector space U is not an algebra since it is not closed with respect to the exterior product. If we use the notation $\mathbb{R}=\Lambda^0(U)$ to denote the field of real numbers, the sequence of spaces $\Lambda^k(U)$ starts then with $\Lambda^0(U)$ and ends with $\Lambda^n(U)$. Let us now define a vector space $\Lambda(U)$ by the following direct sum:

$$\Lambda(U)=\Lambda^0(U)\oplus\Lambda^1(U)\oplus\Lambda^2(U)\oplus\cdots\oplus\Lambda^n(U). \qquad (1.5.6)$$

It is obvious that the vector space $\Lambda(U)$ now becomes an algebra under the exterior product. In other words, for all forms $\alpha,\beta\in\Lambda(U)$ we find $\alpha\wedge\beta\in\Lambda(U)$. We call the algebra $\Lambda(U)$ as the ***exterior algebra***. However, this vector space is constructed as a direct sum of some linear vector spaces. Therefore, it is called a ***graded algebra***.

We are now going to show that the k-forms $f^{i_1}\wedge f^{i_2}\wedge\cdots\wedge f^{i_k}$, $1\le i_1<i_2<\cdots<i_k\le n$ constitute a basis for the vector space $\Lambda^k(U)$. To this end, it suffices to prove that those forms are linearly independent. With arbitrary scalars $\alpha_{i_1i_2\cdots i_k}$, let us write

$$\sum_{1\le i_1<i_2<\cdots<i_k\le n}\alpha_{i_1i_2\cdots i_k}f^{i_1}\wedge f^{i_2}\wedge\cdots\wedge f^{i_k}=0$$

Let us choose an arbitrary index set of k distinct numbers $\{i'_1,i'_2,\cdots,i'_k\}$ out of the set $\{1,2,\dots,n\}$. Let the index set of $n-k$ natural numbers that is the complement of this subset with respect to the set $\{1,2,\dots,n\}$ be the subset $\{j'_{k+1},\cdots,j'_n\}$. The exterior product of the foregoing expression by the $(n-k)$-form $f^{j'_{k+1}}\wedge\cdots\wedge f^{j'_n}$ will be

$$\sum_{1\le i_1<i_2<\cdots<i_k\le n}\alpha_{i_1i_2\cdots i_k}f^{i_1}\wedge f^{i_2}\wedge\cdots\wedge f^{i_k}\wedge f^{j'_{k+1}}\wedge\cdots\wedge f^{j'_n}=0.$$

However, the set $\{j'_{k+1},\cdots,j'_n\}$ is the complement of the set $\{i'_1,\cdots,i'_k\}$ with respect to the set $\{1,2,\dots,n\}$. Consequently, all terms in the above sum except the one corresponding to those indices vanish because at least two basis vectors (actually 1-forms) would be equal. We thus see that only the term

$$\alpha_{i'_1\cdots i'_k}f^{i'_1}\wedge\cdots\wedge f^{i'_k}\wedge f^{j'_{k+1}}\wedge\cdots\wedge f^{j'_n}=\pm\,\alpha_{i'_1\cdots i'_k}f^1\wedge\cdots\wedge f^n=0$$

survives in that zero form. The value of that form on n linearly independent vectors $u_1,u_2,\dots,u_n\in U$ is given by

$$V^n(u_1, u_2, \dots, u_n)\, \alpha_{i'_1 \cdots i'_k} = 0.$$

Since $V^n(u_1, u_2, \dots, u_n) \neq 0$ we find that $\alpha_{i'_1 \cdots i'_k} = 0$. Since the choice of indices is entirely arbitrary, we conclude that all scalar coefficients must vanish. Hence, the set of forms $\{f^{i_1} \wedge \cdots \wedge f^{i_k} : 1 \leq i_1 < \cdots < i_k \leq n\}$ constitutes a basis for the vector space $\Lambda^k(U)$. The cardinality of this set is $\binom{n}{k}$ implying that the dimension of the vector space $\Lambda^k(U)$ is $\binom{n}{k} = \dfrac{n!}{k!\,(n-k)!}$. The basis of the vector space $\Lambda(U)$, which is defined by the direct sum (1.5.6), is clearly determined by the union of bases of component vector spaces. Since the basis of the vector space $\Lambda^0(U)$ is 1, the basis of $\Lambda(U)$ is prescribed by

$$\{1\} \cup \{f^i\} \cup \cdots \cup \{f^{i_1} \wedge \cdots \wedge f^{i_k} : i_1 < \cdots < i_k\} \cup \cdots \cup \{f^1 \wedge \cdots \wedge f^n\}.$$

Therefore the *dimension of the exterior algebra* $\Lambda(U)$ on a vector space $U^{(n)}$ is given by the integer

$$N = \sum_{k=0}^{n} \binom{n}{k} = 2^n. \tag{1.5.7}$$

We say that a k-form is a ***simple form*** if it is expressible as an exterior product of k linearly independent 1-forms, that is, if a k-form is simple it can be written as follows

$$\omega = \omega^{(1)} \wedge \omega^{(2)} \wedge \cdots \wedge \omega^{(k)}, \ \ \omega^{(i)} \in \Lambda^1(U), \ \omega \in \Lambda^k(U) \tag{1.5.8}$$

where $\omega^{(m)} = \omega_i^{(m)} f^i$, $m = 1, 2, \dots, k$. We thus obtain

$$\omega = \omega^{(1)}_{[i_1} \cdots \omega^{(k)}_{i_k]} f^{i_1} \wedge f^{i_2} \wedge \cdots \wedge f^{i_k} = \frac{1}{k!} \omega_{i_1 i_2 \cdots i_k} f^{i_1} \wedge f^{i_2} \wedge \cdots \wedge f^{i_k}.$$

Here the scalar numbers $\omega_{i_1 i_2 \cdots i_k} = k!\, \omega^{(1)}_{[i_1} \cdots \omega^{(k)}_{i_k]}$ are components of the form ω. The value of a simple k-form on k linearly independent vectors u_1, $u_2, \dots, u_k \in U$ can now be evaluated as follows

$$\omega(u_1, u_2, \dots, u_k) = \omega^{(1)}_{i_1} \cdots \omega^{(k)}_{i_k} \begin{vmatrix} u_1^{i_1} & u_2^{i_1} & \cdots & u_k^{i_1} \\ u_1^{i_2} & u_2^{i_2} & \cdots & u_k^{i_2} \\ \vdots & \vdots & & \vdots \\ u_1^{i_k} & u_2^{i_k} & \cdots & u_k^{i_k} \end{vmatrix}$$

$$
= \begin{vmatrix} \omega^{(1)}_{i_1} u^{i_1}_1 & \omega^{(1)}_{i_1} u^{i_1}_2 & \cdots & \omega^{(1)}_{i_1} u^{i_1}_k \\ \omega^{(2)}_{i_2} u^{i_2}_1 & \omega^{(2)}_{i_2} u^{i_2}_2 & \cdots & \omega^{(2)}_{i_2} u^{i_2}_k \\ \vdots & \vdots & & \vdots \\ \omega^{(k)}_{i_k} u^{i_k}_1 & \omega^{(k)}_{i_k} u^{i_k}_2 & \cdots & \omega^{(k)}_{i_k} u^{i_k}_k \end{vmatrix} = \begin{vmatrix} \omega^{(1)}(u_1) & \omega^{(1)}(u_2) & \cdots & \omega^{(1)}(u_k) \\ \omega^{(2)}(u_1) & \omega^{(2)}(u_2) & \cdots & \omega^{(2)}(u_k) \\ \vdots & \vdots & & \vdots \\ \omega^{(k)}(u_1) & \omega^{(k)}(u_2) & \cdots & \omega^{(k)}(u_k) \end{vmatrix}
$$

1.6. RANK OF AN EXTERIOR FORM

Let us consider a form $\omega \in \Lambda^k(U)$ on an n-dimensional vector space U (unless stated otherwise we shall always consider a finite dimensional vector space):

$$
\omega = \frac{1}{k!}\,\omega_{i_1 i_2 \cdots i_k} f^{i_1} \wedge f^{i_2} \wedge \cdots \wedge f^{i_k}. \tag{1.6.1}
$$

We now choose a certain linear combinations of reciprocal basis vectors in the dual space U^* as follows

$$
g^\alpha = c^\alpha_i f^i, \quad i = 1, 2, \ldots, n; \quad \alpha = 1, 2, \ldots, m. \tag{1.6.2}
$$

c^α_i are some scalar coefficients. We shall assume that the vectors g^α are linearly independent. In other words, the rank of the rectangular matrix $[c^\alpha_i]$ should be m. Therefore, the transformations (1.6.2) will be meaningful if only $m \leq n$. Let us suppose that these transformations reduce the form (1.6.1) into the following k-form

$$
\omega = \frac{1}{k!}\,\Omega_{\alpha_1 \alpha_2 \cdots \alpha_k} g^{\alpha_1} \wedge g^{\alpha_2} \wedge \cdots \wedge g^{\alpha_k}.
$$

The least integer m found in this fashion, that is, $r = \min m$, is called the ***rank*** of the form ω. In order to determine the rank of a form, we have to look for the nontrivial, linearly independent solutions of the following homogeneous equations

$$
\omega_{i_1 i_2 \cdots i_k} h^{i_1} = 0, \quad h^{i_1} \in U^*. \tag{1.6.3}
$$

If we find linearly independent $n - r$ solutions $h^a, a = r+1, r+2, \ldots, n$ we can then write $h^\alpha = \gamma^\alpha_a h^a, \alpha = 1, 2, \ldots, r$. Hence, the rank of the rectangular matrix $[\gamma^\alpha_a]$ must be r for vectors h^α to be *linearly independent among themselves*. We will see that this number denotes also the rank of the form ω. If $r = n$, then we clearly get $h^i = 0,\ i = 1, 2, \ldots, n$. In this case, we cannot reduce the number of basis vectors or forms appearing in (1.6.1).

Let us now assume that the rank of the form is satisfying the condition $r < n$. It follows from equations (1.6.3) that

$$\omega_{\alpha i_2\cdots i_k}h^\alpha + \omega_{a i_2\cdots i_k}h^a = (\omega_{\alpha i_2\cdots i_k}\gamma_a^\alpha + \omega_{a i_2\cdots i_k})h^a = 0.$$

Since we supposed that the vectors h^a are linearly independent, we then see that the relations

$$\omega_{\alpha i_2\cdots i_k}\gamma_a^\alpha + \omega_{a i_2\cdots i_k} = 0, \tag{1.6.4}$$

where $a = r+1,\dots,n; i_m = 1,\dots,n; m \geq 2,$ should be satisfied. Let us now define the linearly independent vectors

$$g^\alpha = f^\alpha - \gamma_a^\alpha f^a, \ \alpha = 1,2,\dots,r \tag{1.6.5}$$

and insert the vectors $f^\alpha = g^\alpha + \gamma_a^\alpha f^a$ into the first factor in the exterior product in (1.6.1). In the first step we obtain

$$\begin{aligned}
\omega &= \frac{1}{k!}\omega_{i_1 i_2\cdots i_k}f^{i_1}\wedge f^{i_2}\wedge\cdots\wedge f^{i_k}\\
&= \frac{1}{k!}(\omega_{\alpha_1 i_2\cdots i_k}f^{\alpha_1} + \omega_{a_1 i_2\cdots i_k}f^{a_1})\wedge f^{i_2}\wedge\cdots\wedge f^{i_k}\\
&= \frac{1}{k!}\big[\omega_{\alpha_1 i_2\cdots i_k}(g^{\alpha_1} + \gamma_{a_1}^{\alpha_1}f^{a_1}) + \omega_{a_1 i_2\cdots i_k}f^{a_1}\big]\wedge f^{i_2}\wedge\cdots\wedge f^{i_k}\\
&= \frac{1}{k!}\big[\omega_{\alpha_1 i_2\cdots i_k}g^{\alpha_1} + (\omega_{\alpha_1 i_2\cdots i_k}\gamma_{a_1}^{\alpha_1} + \omega_{a_1 i_2\cdots i_k})f^{a_1}\big]\wedge f^{i_2}\wedge\cdots\wedge f^{i_k}\\
&= \frac{1}{k!}\omega_{\alpha_1 i_2\cdots i_k}g^{\alpha_1}\wedge f^{i_2}\wedge\cdots\wedge f^{i_k}
\end{aligned}$$

where we made use of the relation (1.6.4) in the fourth line. In the second step, we are led to

$$\begin{aligned}
\omega &= -\frac{1}{k!}\omega_{i_2\alpha_1\cdots i_k}g^{\alpha_1}\wedge f^{i_2}\wedge\cdots\wedge f^{i_k} =\\
&-\frac{1}{k!}g^{\alpha_1}\wedge\big[\omega_{\alpha_2\alpha_1 i_3\cdots i_k}g^{\alpha_2} + (\omega_{\alpha_2\alpha_1 i_3\cdots i_k}\gamma_{a_2}^{\alpha_2} + \omega_{a_2\alpha_1 i_3\cdots i_k})f^{a_2}\big]\wedge f^{i_3}\wedge\cdots\wedge f^{i_k}\\
\mathrm{k} &= \frac{1}{k!}\omega_{\alpha_1\alpha_2 i_3\cdots i_k}g^{\alpha_1}\wedge g^{\alpha_2}\wedge f^{i_3}\wedge\cdots\wedge f^{i_k}.
\end{aligned}$$

Continuing this way, we arrive at the following result in the kth step

$$\omega = \frac{1}{k!}\omega_{\alpha_1\alpha_2\cdots\alpha_k}g^{\alpha_1}\wedge g^{\alpha_2}\wedge\cdots\wedge g^{\alpha_k}. \tag{1.6.6}$$

This clearly means that the k-form ω is now generated by basis forms $\{g^{\alpha_1}\wedge g^{\alpha_2}\wedge\cdots\wedge g^{\alpha_k}\}$. The cardinality $\binom{r}{k}$ of this set is of course less than the cardinality $\binom{n}{k}$ of the original basis set. If $r = k$, i.e., if the rank of the form is equal to its degree, we get $\binom{k}{k} = 1$ and the form ω can be represented by

$$\omega = \omega_{1\cdots k}\, g^1 \wedge g^2 \wedge \cdots \wedge g^k. \tag{1.6.7}$$

In order to see this, it suffices to note that one has

$$\omega_{\alpha_1\cdots\alpha_k} = e_{\alpha_1\cdots\alpha_k}\omega_{1\cdots k}, \quad g^{\alpha_1} \wedge \cdots \wedge g^{\alpha_k} = e^{\alpha_1\cdots\alpha_k} g^1 \wedge \cdots \wedge g^k$$

and $e_{\alpha_1\cdots\alpha_k} e^{\alpha_1\cdots\alpha_k} = k!$. If we write $\widetilde{g}^{\,1} = \omega_{1\cdots k}\, g^1$, (1.6.7) now becomes

$$\omega = \widetilde{g}^{\,1} \wedge g^2 \wedge \cdots \wedge g^k.$$

We thus conclude that *every k-form whose rank is equal to its degree can be reduced to a simple form*. Conversely, if a k-form is simple it can be written in the form (1.5.8) as follows

$$\omega = \omega^1 \wedge \omega^2 \wedge \cdots \wedge \omega^k$$

where 1-forms $\omega^\alpha = \omega_i^\alpha f^i, \alpha = 1, \ldots, k, i = 1, \ldots, n$ are of course linearly independent. Therefore, the rank of the rectangular matrix $[\omega_i^\alpha]$ must be k. Thus we can state the following theorem:

Theorem 1.6.1. *A form $\omega \in \Lambda^k(U), k \leq n$ is a simple form if and only if its rank is equal to its degree.* □

We now apply the general approach which we have developed above to a 2-form owing to its rather simple structure. We know that an arbitrary form $\omega \in \Lambda^2(U^{(n)})$ is expressible as

$$\omega = \frac{1}{2}\,\omega_{ij}\, f^i \wedge f^j \tag{1.6.8}$$

where ω_{ij} constitutes an antisymmetric $n \times n$ matrix of real numbers. In order to find the rank of the form ω we have to determine nontrivial, linearly independent solutions $h^i \in U^*$ of the homogeneous equations

$$\omega_{ij}\, h^j = 0. \tag{1.6.9}$$

Since ω_{ij} is an antisymmetric matrix, its rank is always an even number, say, $r = 2m$ where m is a positive integer. Therefore, the dimension of the null space of the linear operator represented by the matrix ω_{ij}, or the number of linearly independent vectors spanning this subspace would be $n - 2m$. In other words, $2m$ vectors out of n vectors satisfying the equations (1.6.9) are expressible as linear combinations of the remaining

$n - 2m$ vectors. Thus the rank of the form ω becomes $2m$. If n is an even number, it may happen that the rank of the form may be equal to n. In this case, it will not be possible to reduce the form. However if n is an odd number, $2m$ will, of course, always be smaller than n. Consequently, in this case a 2-form is always reducible.

Example 1.6.1. Let us first begin with a relatively simple case of $n = 3$. By using the essential components, we can express a 2-form by the following expression

$$\omega = \omega_{12}\, f^1 \wedge f^2 + \omega_{13}\, f^1 \wedge f^3 + \omega_{23}\, f^2 \wedge f^3.$$

Obviously the rank of this form is 2. Indeed, the equations $\omega_{ij}\, h^j = 0$ are now written in the form

$$\begin{aligned} \omega_{12}\, h^2 + \omega_{13}\, h^3 &= 0 \\ -\,\omega_{12}\, h^1 + \omega_{23}\, h^3 &= 0 \\ -\,\omega_{13}\, h^1 - \omega_{23}\, h^2 &= 0 \end{aligned}$$

whence we deduce by the assumption $\omega_{12} \neq 0$ that

$$h^1 = \frac{\omega_{23}}{\omega_{12}} h^3, \qquad h^2 = -\frac{\omega_{13}}{\omega_{12}} h^3.$$

Let us now define 1-forms

$$g^1 = \omega_{12}\big(f^1 - \frac{\omega_{23}}{\omega_{12}} f^3\big), \; g^2 = f^2 + \frac{\omega_{13}}{\omega_{12}} f^3.$$

We immediately see that the form ω reduces to

$$\omega = g^1 \wedge g^2$$

■

Example 1.6.2. In order to explore a little bit more complicated case, let us now choose $n = 4$. By using the essential antisymmetric components we can express a 2-form as follows

$$\begin{aligned} \omega = {} & \omega_{12}\, f^1 \wedge f^2 + \omega_{13}\, f^1 \wedge f^3 + \omega_{14}\, f^1 \wedge f^4 + \omega_{23}\, f^2 \wedge f^3 \\ & + \omega_{24}\, f^2 \wedge f^4 + \omega_{34}\, f^3 \wedge f^4. \end{aligned}$$

The rank of the form ω can now be either 4 or 2. If the rank is 4, ω is evidently not reducible. Let us consider the equations

$$\begin{aligned} \omega_{12}\, h^2 + \omega_{13} h^3 + \omega_{14}\, h^4 &= 0 \\ -\,\omega_{12}\, h^1 + \omega_{23}\, h^3 + \omega_{24}\, h^4 &= 0 \end{aligned}$$

$$\begin{aligned} & -\omega_{13}\, h^1 - \omega_{23}\, h^2 + \omega_{34}\, h^4 = 0 \\ & -\omega_{14}\, h^1 - \omega_{24}\, h^2 - \omega_{34}\, h^3 = 0. \end{aligned}$$

$\Delta = (\omega_{12}\omega_{34} - \omega_{13}\omega_{24} + \omega_{14}\omega_{23})^2$ is the determinant of the coefficient of these linear equations. If $\Delta \neq 0$, then the rank of the form is 4. If only $\Delta = 0$, then the rank reduces to 2. When $\Delta = 0$ the solution of the above homogeneous equations is given by

$$h^1 = \frac{\omega_{23}}{\omega_{12}} h^3 + \frac{\omega_{24}}{\omega_{12}} h^4, \quad h^2 = -\frac{\omega_{13}}{\omega_{12}} h^3 - \frac{\omega_{14}}{\omega_{12}} h^4$$

with the assumption $\omega_{12} \neq 0$. Hence, the transformations

$$g^1 = \omega_{12}\Big[f^1 - \frac{1}{\omega_{12}}(\omega_{23} f^3 + \omega_{24} f^4)\Big], \; g^2 = f^2 + \frac{1}{\omega_{12}}(\omega_{13} f^3 + \omega_{14} f^4)$$

lead to the expression

$$\omega = g^1 \wedge g^2. \qquad \blacksquare$$

It is possible to introduce a *canonical structure* for 2-forms imposed by their ranks.

Theorem 1.6.2. *Let ω be a 2-form whose rank is $2m$. There exist linearly independent 1-forms $g^1, g^2, \dots, g^{2m}$ such that ω is expressible in the following canonical form*

$$\begin{aligned} \omega &= g^1 \wedge g^{m+1} + g^2 \wedge g^{m+2} + \cdots + g^m \wedge g^{2m} \qquad (1.6.10) \\ &= \sum_{i=1}^{m} g^i \wedge g^{m+i} \end{aligned}$$

We can easily prove this theorem by resorting to mathematical induction. By employing the essential components we can write the form ω in the following manner

$$\begin{aligned} \omega = \omega_{12}\, f^1 \wedge f^2 + \omega_{13}\, f^1 \wedge f^3 + \cdots + \omega_{1n}\, f^1 \wedge f^n \\ + \omega_{23}\, f^2 \wedge f^3 + \omega_{24}\, f^2 \wedge f^4 + \cdots + \omega_{2n}\, f^2 \wedge f^n + \overline{\Phi}_1 \end{aligned}$$

where $\overline{\Phi}_1$ is a quadratic form depending only to basis forms $f^3, f^4, \dots, f^n$. Let us then rewrite it as follows

$$\begin{aligned} \omega = f^1 \wedge (\omega_{12}\, f^2 + \omega_{13}\, f^3 + \cdots + \omega_{1n}\, f^n) \qquad (1.6.11) \\ + f^2 \wedge (\omega_{23}\, f^3 + \omega_{24}\, f^4 + \cdots + \omega_{2n}\, f^n) + \overline{\Phi}_1. \end{aligned}$$

If we assume that $\omega_{12} \neq 0$, we can define 1-forms g^1 and g^{m+1} by

$$g^1 = f^1 - \frac{1}{\omega_{12}}(\omega_{23}\, f^3 + \omega_{24}\, f^4 + \cdots + \omega_{2n}\, f^n) \qquad (1.6.12)$$
$$g^{m+1} = \omega_{12}\, f^2 + \omega_{13}\, f^3 + \cdots + \omega_{1n}\, f^n.$$

When we insert the forms (1.6.12) into the expression (1.6.11) we conclude after some manipulations that

$$\begin{aligned}
\omega = \big[g^1 &+ \frac{1}{\omega_{12}}(\omega_{23}\, f^3 + \omega_{24}\, f^4 \cdots + \omega_{2n}\, f^n)\big] \wedge g^{m+1} + \\
&\frac{1}{\omega_{12}}(g^{m+1} - \omega_{13}\, f^3 - \cdots - \omega_{1n}\, f^n) \wedge (\omega_{23}\, f^3 + \omega_{24}\, f^4 \cdots + \omega_{2n}\, f^n) \\
&+ \overline{\Phi}_1 = g^1 \wedge g^{m+1} - \frac{1}{\omega_{12}}\, g^{m+1} \wedge (\omega_{23}\, f^3 + \cdots + \omega_{2n}\, f^n) \\
&\quad + \frac{1}{\omega_{12}}\, g^{m+1} \wedge (\omega_{23}\, f^3 + \cdots + \omega_{2n}\, f^n) \\
&\quad - \frac{1}{\omega_{12}}(\omega_{13}\, f^3 + \cdots + \omega_{1n}\, f^n) \wedge (\omega_{23}\, f^3 + \cdots + \omega_{2n}\, f^n) + \overline{\Phi}_1
\end{aligned}$$

or

$$\omega = g^1 \wedge g^{m+1} + \Phi_1.$$

The new quadratic form Φ_1 will evidently involve only $n-2$ number of 1-forms $f^3, f^4, \dots, f^n$. Thus its rank will be at most $2m-2$. If this number is not zero, namely, if $\Phi_1 \neq 0$, we then repeat this operation this time for the form Φ_1. After repeating this operation k number of times, we reach to the conclusion

$$\omega = \sum_{i=1}^{k} g^i \wedge g^{m+i} + \Phi_k.$$

The rank of the quadratic form Φ_k depending on $n-2k$ number of 1-forms will now at most $2m-2k$. Therefore, when we repeat this operation a sufficient number of times the form Φ_k will eventually vanish and we shall arrive at the relation (1.6.10). □

Example 1.6.3. We consider a 2-form on a 4-dimensional vector space given by its essential components:

$$\begin{aligned}
\omega = \omega_{12}\, f^1 \wedge f^2 + \omega_{13}\, f^1 \wedge f^3 + \omega_{14}\, f^1 \wedge f^4 + \omega_{23}\, f^2 \wedge f^3 \\
+ \omega_{24}\, f^2 \wedge f^4 + \omega_{34}\, f^3 \wedge f^4.
\end{aligned}$$

The number m can now be at most 2. We define as above

$$g^1 = f^1 - \frac{1}{\omega_{12}}(\omega_{23}\, f^3 + \omega_{24}\, f^4),$$
$$g^3 = \omega_{12}\, f^2 + \omega_{13}\, f^3 + \omega_{14}\, f^4.$$

When $\omega_{12} \neq 0$, we then easily find that

$$\omega = g^1 \wedge g^3 + \frac{\omega_{12}\omega_{34} - \omega_{13}\omega_{24} + \omega_{14}\omega_{23}}{\omega_{12}}\, f^3 \wedge f^4.$$

Let us now write

$$g^2 = \frac{\omega_{12}\omega_{34} - \omega_{13}\omega_{24} + \omega_{14}\omega_{23}}{\omega_{12}}\, f^3,$$
$$g^4 = f^4$$

we obtain

$$\omega = g^1 \wedge g^3 + g^2 \wedge g^4.$$

On the other hand, if the relation $\omega_{12}\omega_{34} - \omega_{13}\omega_{24} + \omega_{14}\omega_{23} = 0$ is satisfied, then the rank of the form ω reduces to 2 and the canonical form becomes

$$\omega = g^1 \wedge g^3.$$ ■

In view of Theorem 1.6.2, it is now understood that any 2-form on a vector space $U^{(n)}$ whose rank is an even number is always expressible in the following canonical form

$$\omega = \sum_{\alpha=1}^{m} g^\alpha \wedge g^{m+\alpha}.$$

$r = 2m$ is the rank of the form and $g^1, g^2, \dots, g^{2m}$ are linearly independent 1-forms. We now define 2-forms

$$\omega_\alpha = g^\alpha \wedge g^{m+\alpha} \in \Lambda^2(U), \quad \alpha = 1, 2, \dots, m. \tag{1.6.13}$$

Due to properties of the exterior product we immediately observe that the relations

$$\omega_{\underline{\alpha}} \wedge \omega_{\underline{\alpha}} = 0, \quad \omega_\alpha \wedge \omega_\beta = \omega_\beta \wedge \omega_\alpha \tag{1.6.14}$$

are satisfied. We can now write

$$\omega = \sum_{\alpha=1}^{m} \omega_\alpha.$$

Let us next consider the form $\omega^k = \underbrace{\omega \wedge \omega \wedge \cdots \wedge \omega}_{k} \in \Lambda^{2k}(U)$. Owing to the commutation rule $(1.6.14)_2$ we readily realise that the well known multinomial expansion

$$\omega^k = \Big(\sum_{\alpha=1}^{m} \omega_\alpha\Big)^k \tag{1.6.15}$$
$$= \sum_{k_1+k_2+\cdots+k_m=k} \frac{k!}{k_1!k_2!\cdots k_m!}\,\omega_1^{k_1} \wedge \omega_2^{k_2} \wedge \cdots \wedge \omega_m^{k_m}$$

would be valid just like in the classical algebra. But, if $k_\alpha > 1$, then we have $\omega_\alpha^{k_\alpha} = 0$, $\alpha = 1, 2, \ldots, m$ due to $(1.6.14)_1$. Hence only the terms corresponding to $k_1 = k_2 = \cdots = k_m = 1$ and involving only the exponents k_α meeting the restriction $k_1 + k_2 + \cdots + k_m = k$ will survive. When we take $k = m$, this expansion will of course yield

$$\omega^m = m!\,\omega_1 \wedge \omega_2 \wedge \cdots \wedge \omega_m$$
$$= m!\,g^1 \wedge g^{m+1} \wedge g^2 \wedge g^{m+2} \wedge \cdots \wedge g^m \wedge g^{2m}.$$

Hence ω^m is a simple form. This result should be anticipated because the rank of the form ω^m is equal to its degree. The relation (1.6.15) implies clearly that $\omega^k = 0$ if $k > m$. This scheme suggests a rather simple method to determine the rank of a quadratic form: If $\omega^m \neq 0$, but $\omega^{m+1} = 0$, then the rank of the quadratic form ω is $r = 2m$. If $k < m$, it then follows from (1.6.15) that

$$\omega^k = k!\sum k\text{-fold exterior products of forms } (\omega_1, \omega_2, \cdots, \omega_m)$$

whence we deduce with a little care that ω^k is represented by

$$\omega^k = k! \sum_{\alpha_1=1}^{m-k+1} \sum_{\alpha_2=\alpha_1+1}^{m-k+2} \cdots \sum_{\alpha_k=\alpha_{k-1}+1}^{m} \omega_{\alpha_1} \wedge \omega_{\alpha_2} \wedge \cdots \wedge \omega_{\alpha_k}$$
$$= k! \sum_{\alpha_1=1}^{m-k+1} \sum_{\alpha_2=\alpha_1+1}^{m-k+2} \cdots \sum_{\alpha_k=\alpha_{k-1}+1}^{m} g^{\alpha_1} \wedge g^{m+\alpha_1} \wedge g^{\alpha_2} \wedge g^{m+\alpha_2} \wedge \cdots \wedge g^{\alpha_k} \wedge g^{m+\alpha_k}.$$

As an application of what we have obtained so far let us try to answer this question: under what conditions a quadratic form

$$\omega = \frac{1}{2}\omega_{ij}\, f^i \wedge f^j$$

is expressible as $\omega = g^1 \wedge g^2$, i.e., as an exterior product of two 1-forms? In order to realise this situation, the rank of the form must be 2, namely, we must have $m = 1$, and consequently $\omega^2 = \omega \wedge \omega = 0$. This relation then gives rise to

$$\omega^2 = \frac{1}{4}\,\omega_{ij}\,\omega_{kl}\, f^i \wedge f^j \wedge\; f^k \wedge f^l = \frac{1}{4}\,\omega_{[ij}\,\omega_{kl]}\, f^i \wedge f^j \wedge\; f^k \wedge f^l = 0$$

or $\omega_{[ij}\,\omega_{kl]} = 0$. By making use of the relation (1.4.8), we should note that one can write

$$\omega_{[ij}\,\omega_{kl]} = \frac{1}{4!}\,\delta^{pqrs}_{ijkl}\,\omega_{pq}\,\omega_{rs}.$$

Moreover, it follows from the definition of the generalised Kronecker delta that we arrive at the expansion

$$\begin{aligned}
\delta^{pqrs}_{ijkl} = \begin{vmatrix} \delta^p_i & \delta^p_j & \delta^p_k & \delta^p_l \\ \delta^q_i & \delta^q_j & \delta^q_k & \delta^q_l \\ \delta^r_i & \delta^r_j & \delta^r_k & \delta^r_l \\ \delta^s_i & \delta^s_j & \delta^s_k & \delta^s_l \end{vmatrix} &= \delta^p_l\delta^q_k\delta^r_j\delta^s_i - \delta^p_k\delta^q_l\delta^r_j\delta^s_i - \delta^p_l\delta^q_j\delta^r_k\delta^s_i \\
&+ \delta^p_j\delta^q_l\delta^r_k\delta^s_i + \delta^p_k\delta^q_j\delta^r_l\delta^s_i - \delta^p_j\delta^q_k\delta^r_l\delta^s_i - \delta^p_l\delta^q_k\delta^r_i\delta^s_j + \delta^p_k\delta^q_l\delta^r_i\delta^s_j \\
&+ \delta^p_l\delta^q_i\delta^r_k\delta^s_j - \delta^p_i\delta^q_l\delta^r_k\delta^s_j - \delta^p_k\delta^q_i\delta^r_l\delta^s_j + \delta^p_i\delta^q_k\delta^r_l\delta^s_j + \delta^p_l\delta^q_j\delta^r_i\delta^s_k \\
&- \delta^p_j\delta^q_l\delta^r_i\delta^s_k - \delta^p_l\delta^q_i\delta^r_j\delta^s_k + \delta^p_i\delta^q_l\delta^r_j\delta^s_k + \delta^p_j\delta^q_i\delta^r_l\delta^s_k - \delta^p_i\delta^q_j\delta^r_l\delta^s_k \\
&- \delta^p_k\delta^q_j\delta^r_i\delta^s_l + \delta^p_j\delta^q_k\delta^r_i\delta^s_l + \delta^p_k\delta^q_i\delta^r_j\delta^s_l - \delta^p_i\delta^q_k\delta^r_j\delta^s_l \\
&- \delta^p_j\delta^q_i\delta^r_k\delta^s_l + \delta^p_i\delta^q_j\delta^r_k\delta^s_l\,.
\end{aligned}$$

Therefore, we obtain

$$\begin{aligned}
\omega_{[ij}\,\omega_{kl]} &= \frac{1}{24}(\omega_{lk}\,\omega_{ji} - \omega_{kl}\,\omega_{ji} - \omega_{lj}\,\omega_{ki} + \omega_{jl}\,\omega_{ki} + \omega_{kj}\,\omega_{li} - \omega_{jk}\,\omega_{li} \\
&\quad - \omega_{lk}\,\omega_{ij} + \omega_{kl}\,\omega_{ij} + \omega_{li}\,\omega_{kj} - \omega_{il}\,\omega_{kj} - \omega_{ki}\,\omega_{lj} + \omega_{ik}\,\omega_{lj} \\
&\quad + \omega_{lj}\,\omega_{ik} - \omega_{jl}\,\omega_{ik} - \omega_{li}\,\omega_{jk} + \omega_{il}\,\omega_{jk} + \omega_{ji}\,\omega_{lk} - \omega_{ij}\,\omega_{lk} \\
&\quad - \omega_{kj}\,\omega_{il} + \omega_{jk}\,\omega_{il} + \omega_{ki}\,\omega_{jl} - \omega_{ik}\,\omega_{jl} - \omega_{ji}\,\omega_{kl} + \omega_{ij}\,\omega_{kl}) \\
&= \frac{1}{3}(\omega_{ij}\,\omega_{kl} - \omega_{ik}\,\omega_{jl} + \omega_{il}\,\omega_{jk}).
\end{aligned}$$

Hence, the conditions which we are looking for turn out to be

$$\omega_{ij}\,\omega_{kl} - \omega_{ik}\,\omega_{jl} + \omega_{il}\,\omega_{jk} = 0.$$

A non-degenerate quadratic form $\omega \in \Lambda^2(U)$ with maximal rank on a linear vector space $U^{(n)}$ is called a ***symplectic form***. The maximality of the rank implies that $r = n$ if the dimension n of the vector space is an even number, and $r = n - 1$ if it is an odd number. Non-degeneracy means that the relation $\omega(u) = 0$ for a vector $u \in U$, or in explicit form, the set of equations

$$\omega_{ij}u^j = 0 \tag{1.6.16}$$

has only the trivial solution $u = 0$. If n is an even number, then the maximal rank will imply the existence of non-degeneracy. However, if n is an odd number, then the maximal rank should be less than n so that equations (1.6.16) will be satisfied by a vector $u \neq 0$. Consequently, the form ω will be degenerate. We thus conclude that *a symplectic form can only be defined on vector spaces with even dimensions.*

Exterior forms have several other algebraic properties. However, we prefer to postpone to treat them on differentiable manifolds later in Chapter V within a much more general context.

I. EXERCISES

1.1. Let U be a linear vector space. U_1 and U_2 are its finite-dimensional subspaces. (a) Show that their sum $U_1 + U_2$ is also a subspace whose dimension is given by

$$\dim(U_1) + \dim(U_2) - \dim(U_1 \cap U_2).$$

(b) Find the basis set of the subspace $U_1 \cap U_2$. (c) Show that the subset $U_1 \cup U_2$ is generally not a subspace. (d) Show that $[U_1 \cup U_2] = U_1 + U_2$. (e) Show that $U_1 \cup U_2 = U_1 + U_2$ if and only if one of the relations $U_1 \subseteq U_2$ or $U_2 \subseteq U_1$ are satisfied.

1.2. U is a vector space and $u_0 \in U$ is a given *fixed* vector. If $\mathbb{F}$ is the field of scalars over which this vector space is defined, then we introduce two new operations # and $*$ that can be interpreted as the vector addition and scalar multiplication as follows

$$u_1 \# u_2 = u_1 + u_2 + u_0, \quad \alpha * u = \alpha u + (\alpha - 1)u_0$$

for all $u_1, u_2, u \in U$ and $\alpha \in \mathbb{F}$. Show that the triple $(U, \#, *)$ is also a linear vector space.

1.3. If n is a positive integer, show that the set $\mathbb{Q}^n$ is a linear vector space over the field of rational numbers $\mathbb{Q}$.

1.4. Construct explicitly three subspaces U_1, U_2 and U_3 of the vector space $\mathbb{R}^3$ such that

$$U_1 \cap U_2 = U_1 \cap U_3 = U_2 \cap U_3 = \{\mathbf{0}\}$$

but $U_1 \cap (U_2 + U_3) \neq \{\mathbf{0}\}$.

1.5. If the subspaces U_1, U_2 and U_3 of a vector space V satisfy the relations $U_1 \cap U_2 = U_1 \cap U_3, U_1 + U_2 = U_1 + U_3$ and $U_2 \subset U_3$, show that we have necessarily $U_2 = U_3$.

1.6. Let us consider the linear vector space of functions differentiable up to the nth order on an open interval $\mathcal{I}$ of $\mathbb{R}$. Show that the necessary and sufficient condition for the set of such functions $\{f_1(x), f_2(x), \dots, f_n(x)\}$ to be linearly independent at the point x is that the following determinant does not vanish at that point

$$W(x) = \begin{vmatrix} f_1(x) & f_2(x) & \cdots & f_n(x) \\ f_1'(x) & f_2'(x) & \cdots & f_n'(x) \\ \vdots & \vdots & \vdots & \vdots \\ f_1^{(n-1)}(x) & f_2^{(n-1)}(x) & \cdots & f_n^{(n-1)}(x) \end{vmatrix} \neq 0.$$

The above determinant $W(x)$ is known as the ***Wronskian*** of the set of functions $\{f_1(x), f_2(x), \dots, f_n(x)\}$ [after Polish-French mathematician Josef-Maria Hoëné Wronski (1778-1853)].

1.7. Are the functions $\{1, \sin x, \cos x, \sin 2x, \cos 2x\}$ linearly independent ?

1.8. The complex numbers $\{\alpha_1, \alpha_2, \dots, \alpha_n\}$ are satisfying the conditions $\alpha_i \neq \alpha_j$ when $i \neq j$. Show that the set of functions $\{e^{\alpha_i x} : i = 1, 2, \dots, n\}$ is linearly independent.

1.9. Show that the set P_n of polynomials with real coefficients whose degrees are less than or equal to n constitute a linear vector space. Is the subset

$$S = \Big\{p(x) \in P_n : \int_0^1 p(x)\,dx = 0\Big\} \subseteq P_n$$

a subspace?

1.10. Show that $m \times n$ real matrices constitute a vector space $M_{mn}(\mathbb{R})$ with dimension mn and determine a basis for this space.

1.11. A square matrix satisfying the relation $\mathbf{A} = \mathbf{A}^{\mathrm{T}}$ is ***symmetric*** whereas if it satisfies the relation $\mathbf{A} = -\mathbf{A}^{\mathrm{T}}$ it is ***antisymmetric***. $\mathbf{A}^{\mathrm{T}}$ is the transpose matrix. Show that a symmetric matrix and an antisymmetric matrix of the same order are linearly independent.

1.12. Let U be a finite-dimensional vector space and let $A : U \to U$ be a linear transformation. Show that the following statements are equivalent:

$$(i).\ \mathcal{N}(A) \cap \mathcal{R}(A) = \{\mathbf{0}\}.$$
$$(ii).\ \mathcal{N}(A^2) \subset \mathcal{N}(A).$$
$$(iii).\ \mathcal{N}(A) \oplus \mathcal{R}(A) = U.$$

1.13. A is a linear transformation which maps the vector space $V^{(n)}$ into itself. For a given basis $\{e_1, \dots, e_n\}$, let us suppose that the transformation A satisfies the relations

$$Ae_i = e_1 + e_2 + \cdots + e_n, \ i = 1, \dots, n$$

What is the value of A at a vector $v = v_1e_1 + \cdots + v_ne_n$? Find the null space and the range of A.

1.14. U and V are vector spaces and $A : U \to V$ is a linear transformation. Let a finite-dimensional subspace of U be U_1. Show that

$$\dim[A(U_1)] = \dim U_1 - \dim[\mathcal{N}(A) \cap U_1]$$

1.15. Let K be a convex subset of a vector space U. For a finite number of vectors $u_1, u_2, \ldots, u_n$ arbitrarily chosen from the set K and for scalars $\alpha_i \geq 0$, $i = 1, \ldots, n$ obeying the condition $\sum_{i=1}^{n} \alpha_i = 1$, show that their linear combination belongs to K, namely, $\alpha_1 u_1 + \alpha_2 u_2 + \cdots + \alpha_n u_n \in K$. If A is a linear transformation from U into a vector space V, prove that A maps convex sets in U onto convex sets in V.

1.16. Let U, V, W be linear vector spaces and let $A : U \to V$ and $B : V \to W$ be linear transformations. Show that

$$r(BA) \leq \min\{r(A), r(B)\}, \quad n(BA) \leq n(A) + n(B).$$

1.17. Show that V is a zero vector space if and only if the sequence $0 \to V \to 0$ is exact.

1.18. Show that the linear operator $A : U \to V$ is an isomorphism if and only if the sequence $0 \to U \overset{A}{\to} V \to 0$ is exact.

1.19. Let $0 \to U \to V \to W \to 0$ be an exact sequence where U, V, W are finite-dimensional vector spaces. Show that $\dim(V) = \dim(U) + \dim(W)$.

1.20. Let $0 \to V_1 \to V_2 \to \cdots \to V_n \to 0$ be an exact sequence where each V_i is finite-dimensional. Show that $\sum_{i=1}^{n}(-1)^i \dim(V_i) = 0$.

1.21. $\mathcal{T} : U \times V \to \mathbb{F}$ is a bilinear functional. One can define two kernels or null spaces for $\mathcal{T}$: the subspace $\mathcal{N}_U(\mathcal{T}) = \{u \in U : \mathcal{T}(u, v) = 0, \ \forall v \in V\} \subseteq U$ and the subspace $\mathcal{N}_V(\mathcal{T}) = \{v \in V : \mathcal{T}(u, v) = 0, \ \forall u \in U\} \subseteq V$. $\mathcal{T}$ is called a *non-degenerate* transformation if $\mathcal{N}_U(\mathcal{T}) = \{\mathbf{0}\}$ and $\mathcal{N}_V(\mathcal{T}) = \{\mathbf{0}\}$. We denote quotient spaces of U and V with respect to these subspaces by $U/\mathcal{N}_U(\mathcal{T})$ and $V/\mathcal{N}_V(\mathcal{T})$, respectively. We define a bilinear functional on the Cartesian product of these spaces, i.e., $\mathcal{S} = U/\mathcal{N}_U(\mathcal{T}) \times V/\mathcal{N}_V(\mathcal{T}) \to \mathbb{F}$ by the relation $\mathcal{S}([u], [v]) = \mathcal{T}(u, v)$. Show that the functional $\mathcal{S}$ is non-degenerate.

1.22. $\mathcal{T} : U \times U \to \mathbb{F}$ is a symmetric bilinear functional, i.e., for each $u_1, u_2 \in U$ one has $\mathcal{T}(u_1, u_2) = \mathcal{T}(u_2, u_1)$. Show that $\mathcal{T}$ satisfies the *polarisation identity*

$$4\mathcal{T}(u_1, u_2) = \mathcal{T}(u_1 + u_2, u_1 + u_2) - \mathcal{T}(u_1 - u_2, u_1 - u_2).$$

A real functional $Q : U \to \mathbb{R}$ is called a *quadratic functional* if it satisfies the relation $Q(\alpha u) = \alpha^2 Q(u)$ for all $\alpha \in \mathbb{R}$ and $u \in U$. A quadratic functional is derivable from a symmetric bilinear functional in the following manner

$$\mathcal{Q}(u) = \mathcal{T}(u, u), \ \ u \in U.$$

Conversely, show that a symmetric bilinear functional can be generated from such a quadratic functional through the relation

$$\mathcal{T}(u_1, u_2) = \mathcal{Q}\Big(\frac{u_1 + u_2}{2}\Big) - \mathcal{Q}\Big(\frac{u_1 - u_2}{2}\Big) = \frac{1}{4}\big[\mathcal{Q}(u_1 + u_2) - \mathcal{Q}(u_1 - u_2)\big].$$

1.23. Let U and V be linear vector spaces. Show that the tensor products $U \otimes V$ and $V \otimes U$ are isomorphic vector spaces.

1.24. If $u_i \in U_i, i = 1, 2, \dots, n$, show that the equality $u_1 \otimes u_2 \otimes \cdots \otimes u_n = 0$ is satisfied if and only if anyone vector is zero, i.e., if $u_i = 0$ for at least one $1 \leq i \leq n$.

1.25. If $u_i, u_i' \in U_i, i = 1, 2, \dots, n$, then verify that the equality $u_1 \otimes u_2 \otimes \cdots \otimes u_n = u_1' \otimes u_2' \otimes \cdots \otimes u_n' \neq 0$ is satisfied if and only if $u_i' = \alpha_{\underline{i}} u_{\underline{i}}$, $\alpha_i \in \mathbb{F}$, $\alpha_i \neq 0$, $i = 1, 2, \dots, n$ such that $\alpha_1 \alpha_2 \cdots \alpha_n = 1$.

1.26. U and V are vector spaces, and $U_1 \subset U$ and $V_1 \subset V$ are subspaces. Verify that the relation $(U_1 \otimes V) \cap (U \otimes V_1) = U_1 \otimes V_1$ is valid.

1.27. U and V are vector spaces, and $U_1, U_2 \subset U$ and $V_1, V_2 \subset V$ are subspaces. Verify that the following relation is valid

$$(U_1 \otimes V_1) \cap (U_2 \otimes V_2) = (U_1 \cap U_2) \otimes (V_1 \cap V_2).$$

1.28. A 2-covariant tensor $\mathcal{T}$ on a vector space $U^{(n)}$ is called a *symmetric tensor* if $\mathcal{T}(u_1, u_2) = \mathcal{T}(u_2, u_1)$ for all $u_1, u_2 \in U$, and an *antisymmetric tensor* if $\mathcal{T}(u_1, u_2) = -\mathcal{T}(u_2, u_1)$. If the set $\{f^i\}$ is a basis for the dual space U^*, we write $\mathcal{T} = t_{ij} f^i \otimes f^j$. Show that the components of the tensor $\mathcal{T}$ must satisfy the conditions $t_{ij} = t_{ji}$ if it is symmetric, and the conditions $t_{ij} = -t_{ji}$ if it is antisymmetric. Show further that these conditions do not depend on the choice of bases in U^*.

1.29. Show that any 2-covariant tensor is expressible *uniquely* as the sum of one symmetric and one antisymmetric tensor.

1.30. Let $\mathcal{T} = t_{ij} f^i \otimes f^j$ and $\mathcal{S} = s_{ij} f^i \otimes f^j$ be symmetric non-zero tensors on a vector space $U^{(n)}$. Show that if components of these tensors satisfy the equality

$$t_{ij} s_{kl} - t_{il} s_{jk} + t_{jk} s_{il} - t_{kl} s_{ij} = 0,$$

then the relation $t_{ij} = (t_{kk}/s_{ll})\, s_{ij}$ is valid. This result is known as *Schouten's theorem* [Dutch mathematician Jan Arnoldus Schouten (1883-1971)].

1.31. The components of a 2-covariant tensor are satisfying the relations

$$\alpha t_{ij} + \beta t_{ji} = 0$$

where $\alpha, \beta \in \mathbb{F}, \alpha, \beta \neq 0$. Show that this tensor must be either symmetric or antisymmetric.

1.32. A 3-tensor $\mathcal{T} = t^i_{kl} e_i \otimes f^k \otimes f^l$ on a vector space $U^{(2)}$ is explicitly given by

$$\mathcal{T} = -e_2 \otimes f^1 \otimes f^2 + 6\, e_1 \otimes f^2 \otimes f^2 - 3\, e_2 \otimes f^2 \otimes f^1.$$

Find all contracted tensors.

1.33. A mixed 3-tensor on a vector space $U^{(2)}$ is given by

$$\mathcal{T} = 3\, e_1 \otimes e_2 \otimes f^1 - e_2 \otimes e_2 \otimes f^2 + e_2 \otimes e_1 \otimes f^2.$$

A new basis for $U^{(2)}$ is determined by transformations

$$e_1' = e_1 - 2e_2,\ e_2' = e_1 + e_2.$$

Find the components of this tensor with respect to the new basis.

1.34. Evaluate the quantities $e_{ijk}e^{lmn}, e_{ijk}e^{imn}, e_{ijk}e^{ijn}, e_{ijk}e^{ijk}$ where the indices take the values $1, 2, 3$.

1.35. Using the definition of generalised Kronecker delta, show that one can write

$$(a).\ \delta^{ij}_{kl}\,\omega_{ij} = \omega_{kl} - \omega_{lk}$$
$$(b).\ \delta^{ijk}_{lmn}\,\omega_{ijk} = \omega_{lmn} - \omega_{lnm} + \omega_{mnl} - \omega_{nml} + \omega_{nlm} - \omega_{mln}.$$

1.36. Find the values of $\delta^{k\,k+1\cdots n\,1\,2\cdots k-1}_{1\,2\cdots n-1\,n}$ and $\delta^{1\,2\,\cdots\,2n\text{-}1\,2n}_{1\,n+1\,2\,n+2\,3\,n+3\,\cdots\,n\,2n}$.

1.37. Let the basis and its reciprocal for a vector space $U^{(n)}$ and its dual $U^{*(n)}$ be $\{e_i, f^i, i = 1,\ldots,n\}$, respectively. Then verify that for $1 \le k \le n$, one finds $f^{i_1} \wedge f^{i_2} \wedge \cdots \wedge f^{i_k}(e_{j_1}, e_{j_2},\ldots,e_{j_k}) = \delta^{i_1 i_2\cdots i_k}_{j_1 j_2\cdots j_k}$.

1.38. We consider the following members of the exterior algebra $\Lambda(U^{(4)})$: $\alpha = \alpha_{13}f^1 \wedge f^3 + \alpha_{24}f^2 \wedge f^4$, $\beta = \beta_1 f^1 + \beta_4 f^4$, $\gamma = \gamma_{14}f^1 \wedge f^4 + \gamma_{23}f^2 \wedge f^3$, $\theta = \theta_{123}f^1 \wedge f^2 \wedge f^3 + \theta_{234}f^2 \wedge f^3 \wedge f^4$ where all coefficients are scalars. Evaluate the forms (a) $\alpha \wedge \beta - \beta \wedge \gamma + \theta$, (b) $\alpha \wedge \alpha + 3\gamma \wedge \gamma - 2\theta \wedge \beta$, (c) $\beta \wedge \theta + \alpha \wedge \gamma$.

1.39. Let us consider the forms $\alpha \in \Lambda^2(U), \beta \in \Lambda^1(U)$. Show that one can write

$$(\alpha \wedge \beta)(u_1,u_2,u_3) = \alpha(u_1,u_2)\beta(u_3) - \alpha(u_1,u_3)\beta(u_2) - \alpha(u_2,u_3)\beta(u_1)$$

for all $u_1, u_2, u_3 \in U$.

1.40. If we choose to omit the factor $1/k!$ in the definition (1.4.11) of an exterior form $\omega \in \Lambda^k(U)$, show that the exterior product of such types of forms turns out to be no longer associative.

1.41. Let us consider an exterior form $\omega \in \Lambda^{n-1}(U^{(n)})$, $\omega \neq 0$ on a vector space $U^{(n)}$. Show that the forms α satisfying the equality $\alpha \wedge \omega = 0$ constitute an $(n-1)$-dimensional subspace of $\Lambda(U^{(n)})$ and there exist 1-forms $\alpha^1, \alpha^2, \ldots,$ α^{n-1} such that ω is expressible as $\omega = \alpha^1 \wedge \alpha^2 \wedge \cdots \wedge \alpha^{n-1}$.

1.42. If U is finite-dimensional, then show that the vector spaces $\left(\Lambda^k(U)\right)^*$ and $\Lambda^k(U^*)$ are isomorphic.

1.43. The exterior form $\omega \in \Lambda^3(U^{(4)})$ is given by

$$\omega = a_1 f^1 \wedge f^2 \wedge f^3 + a_2 f^1 \wedge f^2 \wedge f^4 + a_3 f^1 \wedge f^3 \wedge f^4 + a_4 f^2 \wedge f^3 \wedge f^4.$$

Find its rank and its reduced form .

CHAPTER II

DIFFERENTIABLE MANIFOLDS

2.1. SCOPE OF THE CHAPTER

The concept of manifold is essentially propounded to extend the definition of surfaces in classical differential geometry to higher dimensional spaces. This relatively new concept was first introduced into mathematics by German mathematician Friedrich Bernhard Riemann (1826-1866) who was the first one to do extensive work generalising the idea of a surface in a three-dimensional space to higher dimensions. The term manifold is derived from Riemann's original German term, *Mannigfaltigkeit*. This term is translated into English as *manifoldness* by English mathematician William Kingdon Clifford (1845-1879). Riemann's intuitive notion of a Mannigfaltigkeit evolved into what is formalised today as the concept of manifold. German mathematician Herman Klaus Hugo Weyl (1885-1955) gave an intrinsic definition for differentiable manifolds in his lecture course on Riemann surfaces in 1911–1912 at Göttingen University uniting analysis, geometry and topology. However, it was American mathematician Hassler Whitney (1907-1989) who clarified the foundational aspects of differentiable manifolds during the 1930s. Especially, the Whitney embedding theorems provided a firm connection between manifolds and Euclidean spaces.

In Sec. 2.2 we first briefly review topological spaces to which differentiable manifolds also belong. We define fundamental notions and focus on various relevant properties of topological spaces. We then introduce a metric space as a special topological space and finally the Euclidean space that proves to be very important for our investigation. A manifold, also a differentiable manifold, is defined as a topological space that is *locally* equivalent to the Euclidean space. This amounts to say that each point of the manifold belongs to an open set which is homeomorphic to an open set of the Euclidean space. These open sets covering the manifolds are called charts and an atlas is a collection of charts. Certain operations such as differentiation are not allowed on manifolds as topological spaces. However, the local equivalence with the Euclidean space enables us to perform these

Exterior Analysis, DOI: 10.1016/B978-0-12-415902-0.50002-5

operations on manifolds by means of the Euclidean space on which such operations are carried out quite easily. Although the topological structure of a manifold does not allow us to evaluate directly the derivative of a real-valued function on a manifold we will be able to describe it indirectly in Sec. 2.3 by making use of local charts and well known differentiability in the Euclidean space. We further extend this description to define differentiable mappings between manifolds. In Sec. 2.4 we utilise differentiable mappings to define submersions, immersions and embeddings between manifolds and we discuss various approaches to generate submanifolds via those mappings. Differentiable curves embedded on manifolds are considered in Sec. 2.5. Sec. 2.6 is devoted to the construction of the tangent space of a manifold at a given point as the vector space of all tangent vectors at that point of all differentiable curves through that point which are constructed by employing local images of these curves in the Euclidean space. A more convenient vector space that is isomorphic to the tangent space is introduced as the space of linear operators determined as derivatives of a scalar function in the direction of tangent vectors. In Secs. 2.7 we define the differential of a differentiable mapping between two manifolds as a linear operator mapping a tangent space into another at the corresponding points of manifolds. We show in Sec. 2.8 that the fibre bundle generated by patching all tangent spaces at all points of the manifold can be equipped with a differentiable structure through which we can define a vector field on the manifold. We investigate properties of a mapping called *flow* generated by trajectories of a vector field, namely, by curves tangent to the vector field in Sec. 2.9. The Lie derivative that measures the variation of a vector field on a manifold with respect to another vector field is defined in Sec. 2.10. This derivative, which is also called the Lie product, is utilised to construct a Lie algebra on the tangent space. Finally, in Sec. 2.11 we define a distribution produced by choosing same dimensional subspaces of the tangent spaces at every points of the manifold. It is shown that these elementary fragments of vector subspaces attached to every points of the manifold can be patched together smoothly to form a submanifold if and only if the distribution is involutive, i.e., if its vectors constitute a Lie subalgebra. This is known as the Frobenius theorem.

2.2. DIFFERENTIABLE MANIFOLDS

Let M be a non-empty set. $\mathcal{P}(M)$ denotes the power set of M which is the collection of all subsets of M, the set M itself and the empty set $\emptyset$. Let $\mathfrak{M} \subseteq \mathcal{P}(M)$ be a class of subsets of M. Let us assume that the class $\mathfrak{M}$ satisfies the following *axioms*:

(i). M and $\emptyset$ belong to the class $\mathfrak{M}$.
(ii). The union of *any number* of members of $\mathfrak{M}$ (even uncountably many) belongs to the class $\mathfrak{M}$.
(iii). The intersection of *any finite number* of members of the class $\mathfrak{M}$ belongs to the class $\mathfrak{M}$.

Such a class $\mathfrak{M}$ is called a ***topology*** on the set M. The ordered pair $(M, \mathfrak{M})$ is called a ***topological space***. Unless it causes an ambiguity, a set M endowed with a topology will also be usually called a topological space M. However, we should remark that several topologies may be defined on the same set M generating different topological spaces. We usually name the elements of a topological space as its ***points***. The members of the topology $\mathfrak{M}$ will be called ***open sets*** of M. Therefore a set $U \subseteq M$ is open if and only if $U \in \mathfrak{M}$. If the complement V' of a subset $V \subseteq M$ with respect to M is open, that is, if $V' \in \mathfrak{M}$, then V is called a ***closed set***. Since $M' = \emptyset$ and $\emptyset' = M$, we conclude that the sets M and $\emptyset$ are *both open and closed sets*, simultaneously. Whether the topological space M contains subsets other than those two sets having this property is closely related to the topological concept of *connectedness*. We immediately see that the class of closed set will satisfy the following rules directly obtainable from the familiar de Morgan laws of the set theory: (i) X and $\emptyset$ are closed sets, (ii) the intersection of *any number* of closed sets (even uncountably many) is a closed set, (iii) the union of *any finite number* of closed sets is a closed set.

The ***relative topology*** on a subset $A \subseteq M$ is the class of subsets of A defined by $\mathfrak{M}_A = \{U_A = A \cap U : U \in \mathfrak{M}\}$. It is straightforward to show that $(A, \mathfrak{M}_A)$ is a topological subspace. Indeed, $\emptyset \in \mathfrak{M}$ and $M \in \mathfrak{M}$ implies that $\emptyset = A \cap \emptyset \in \mathfrak{M}_A$ and $A = A \cap M \in \mathfrak{M}_A$. Let us consider a family of subsets $\{V_\lambda \in \mathfrak{M}_A : \lambda \in \Lambda\}$ where Λ is an index set. Then for each $\lambda \in \Lambda$, there exists an open set $U_\lambda \in \mathfrak{M}$ such that $V_\lambda = A \cap U_\lambda$. We thus obtain for the arbitrary union $\bigcup_{\lambda \in \Lambda} V_\lambda = \bigcup_{\lambda \in \Lambda} (A \cap U_\lambda) = A \cap \big(\bigcup_{\lambda \in \Lambda} U_\lambda\big) \in \mathfrak{M}_A$. We now choose a *finite* index set $\{\lambda_1, \lambda_2, \ldots, \lambda_n\} \subseteq \Lambda$. Since $\bigcap_{i=1}^{n} U_{\lambda_i} \in \mathfrak{M}$, we eventually obtain $\bigcap_{i=1}^{n} V_{\lambda_i} = \bigcap_{i=1}^{n} (A \cap U_{\lambda_i}) = A \cap \big(\bigcap_{i=1}^{n} U_{\lambda_i}\big) \in \mathfrak{M}_A$. We thus conclude that the class $\mathfrak{M}_A$ complies with the axioms of topology. *It should be noted that the set $U_A \in \mathfrak{M}_A$ may not in general be an open set of* $\mathfrak{M}$. If only A itself is an open set of X, then open sets of relative topology coincide with the open sets of M. Evidently, the closed sets of the relative topology are of the form $A \cap U'_\lambda$.

A subset N_p of M is called a ***neighbourhood*** of the point p if there exists an open set U_p such that $p \in U_p \subseteq N_p$. An ***open neighbourhood*** of the point p is just an open set of M containing p. Let A be a subset of a topological space M. If a point $a \in A$ belongs to an open set contained in A, i.e., if there is a set $U \subseteq A$, $U \in \mathfrak{M}$ such that $a \in U$, then a is an ***interior point*** of the set A. In other words, *if the set A is a neighbourhood of the point $a \in A$, then a is an interior point of A*. We can thus propose at once that *the set $A \subseteq M$ is open if and only if A is a neighbourhood of each of its points.*

In fact, let us first assume that A is open and $a \in A$. Due to the obvious relation $a \in A \subseteq A$, the set A is a neighbourhood of the point a. Now let us suppose that A is a neighbourhood of each of its points. Therefore, for each $a \in A$, there exists an open set U_a such that $a \in U_a \subseteq A$. We next define the open set $V = \bigcup_{a \in A} U_a$. Since $U_a \subseteq A$ for each $a \in V$, we find that $V \subseteq A$. On the other hand, each point of the set A belongs to a set U_a and consequently to V. This implies that $A \subseteq V$. We thus obtain the result $A = V$. Hence the set A is open. □

Collection of all neighbourhoods of a point is called the system of neighbourhoods of that point. If each neighbourhood of a point p contains at least one member of a family of neighbourhoods $\{N_{p\lambda} : \lambda \in \Lambda\}$, where Λ is an index set, then this family is a ***fundamental system of neighbourhoods*** of p. A topological space is called a ***first countable space*** if each of its points has *a countable fundamental system of open neighbourhoods*. The set of all interior points of a set $A \subseteq M$ is called the ***interior*** of A and is denoted by $\mathring{A}$. It is easy to see that *the largest open set contained in A is its interior $\mathring{A}$*. It is rather straightforward to verify that $(A \cap B)^\circ = \mathring{A} \cap \mathring{B}$.

The ***closure*** of a subset $A \subseteq M$ is the intersection of all closed sets containing A. We denote the closure of a set A by $\overline{A}$. Since the intersection of any number of closed sets is also closed, we deduce that $\overline{A}$ is a closed set. Hence, the closure of a set A is then *the smallest* closed set containing A. We can then show the following proposition:

Let A be any non-empty subset of a topological space M. A point $p \in M$ belongs to the closure $\overline{A}$ if and only if the intersection of each neighbourhood of p with A is not empty.

We first consider a point $p \in \overline{A}$ and assume that there exists a particular open neighbourhood $U_p \in \mathfrak{M}$ of p such that $U_p \cap A = \emptyset$. We thus have $A \subseteq U_p'$. But, since U_p' is closed we conclude that $\overline{A} \subseteq U_p'$. Therefore, we reach to the contradiction that the point p belongs to both U_p and U_p'. Consequently, we ought to take $U_p \cap A \neq \emptyset$. Hence, every open neighbourhood of each point in the closure of the set A must intersect A. Now, conversely, we assume that the intersection of each open neighbourhood of a point $p \in M$ with A is not empty, but p does not belong to $\overline{A}$, that is, for all $U_p \in \mathfrak{M}$ we should have $U_p \cap A \neq \emptyset$, $p \notin \overline{A}$ so that the point p has to belong to the open set $\overline{A}'$. Consequently, there must exist an open neighbourhood U_0 of p such that $U_0 \subseteq \overline{A}'$. This open set U_0 cannot intersect A and this gives rise to a contradiction so that $p \in \overline{A}$. Hence we are led to define the closure of a set A as the set $\overline{A} = \{p \in M : U_p \cap A \neq \emptyset$ for all $U_p \in \mathfrak{M}\}$. □

It can easily be verified that $\overline{A \cup B} = \overline{A} \cup \overline{B}$ and if $A \subseteq B$, then one deduce at once that $\overline{A} \subseteq \overline{B}$.

The ***boundary of a subset*** $A \subseteq M$ is defined by $\partial A = \overline{A} - \mathring{A} = \overline{A} \cap (\mathring{A})'$. The boundary ∂A of a set A is always closed since it is described by the intersection of two closed sets.

Let A and B be two subsets of a topological space M. If $B \subseteq \overline{A}$, then we say that A is a ***dense set*** in B. On the other hand, if $B = \overline{A}$, then A is called an ***everywhere dense set*** in B. When $B = M$, a set A which is dense in M naturally has to satisfy the relation $M = \overline{A}$. Therefore, a set dense in M is always an everywhere dense set in M. A topological space M is called a ***separable space*** if it possesses a *countable* dense subset $A = \{p_1, p_2, \ldots p_n, \ldots\}$ so that one gets $\overline{A} = M$.

A topological space M is called a ***Hausdorff space*** if each pair of its distinct points p_1, p_2 have *disjoint* neighbourhoods, that is, if $p_1, p_2 \in M$ such that $p_1 \neq p_2$, then there exist open sets U_1 and U_2 so that $p_1 \in U_1$, $p_2 \in U_2$ and $U_1 \cap U_2 = \emptyset$ [after German mathematician Felix Hausdorff (1869-1942)].

Let M be a Hausdorff space. If $p \in M$, then the singleton $\{p\}$, i.e., the set of just the single point p is a closed set.

To observe this, let use take any point $q \in \{p\}' = M - \{p\}$. Since $q \neq p$, there are disjoint open sets $U_p, U_q \in \mathfrak{M}$ such that $U_p \cap U_q = \emptyset$. Therefore, the open set U_q does not contain the point p and we get $U_q \subset \{p\}'$ implying that the point q is an interior point of the set $\{p\}'$. Since all points of the set $\{p\}'$ are interior points, it is open and therefore the set $\{p\}$ *is closed.* □

A subclass $\mathfrak{N}$ of the power set $\mathcal{P}(M)$ is a ***basis*** for a topological space $(M, \mathfrak{M})$ (the term ***open basis*** will, in fact, be more appropriate) if every open set in the topology $\mathfrak{M}$ is expressible as a union of some sets in $\mathfrak{N}$. Elements of $\mathfrak{N}$ are called ***basic open sets***. If we are given a class of subsets $\mathfrak{N} \subset \mathcal{P}(M)$ satisfying naturally the condition $M = \cup\{N\}$ where $N \in \mathfrak{N}$, we cannot usually generate a topology on M by considering all unions of subsets in $\mathfrak{N}$ because the intersection axiom of the topology does not hold in general. It is rather straightforward to see that the necessary and sufficient condition for a class of subsets $\mathfrak{N}$ of a set M to constitute a basis for a topology are provided as follows:

A subclass $\mathfrak{N} \subset \mathcal{P}(M)$ *with the condition* $M = \cup\{N : N \in \mathfrak{N}\}$ *is a basis for a topology on* M *if and only if for any two sets* $N_1, N_2 \in \mathfrak{N}$ *and any point* $p \in N_1 \cap N_2$*, there exists a set* $N_3 \in \mathfrak{N}$ *such that* $p \in N_3 \subseteq N_1 \cap N_2$. □

For instance in a topology on $\mathbb{R}$ basic open sets are open intervals. It is shown in real analysis that every open sets in $\mathbb{R}$ is expressible as a *countable* union of open intervals. A topological space M possessing a *countable basis* is called a ***second countable space***. Such a topological space enjoys several pleasant and rather remarkable properties. For instance a *second countable space is a separable space.* This property is quite easy to show. Let $\mathfrak{N}$ be a countable basis for a topological space M. We choose a point $p_N \in N$ in each non-empty set $N \in \mathfrak{N}$ and then introduce the subset $D = \{p_N : N \in \mathfrak{N}\}$ of M. D is obviously a countable set. Since there is a member of the basis, and consequently, a point of D, in every neighbourhood of each point of M, the countable set D would be dense in X.

Compactness. A ***cover*** $\mathcal{A}$ of a set X is a collection of some subsets of X whose union is X, that is, $\mathcal{A} = \{U_\lambda \subseteq X : \lambda \in \Lambda\}$ where Λ is an index set is a cover of X if and only if $X = \bigcup_{\lambda \in \Lambda} U_\lambda$. If a subclass $\mathcal{B}$ of $\mathcal{A}$ is also a cover of X, then $\mathcal{B}$ is a ***subcover*** of X. A cover $\mathcal{A}$ is an ***open cover*** of a topological space M if all members of $\mathcal{A}$ are open sets. *If every open cover* $\{U_\lambda \subseteq M : \lambda \in \Lambda\}$ *of a topological space* M *has a finite subcover*, namely, if one is able to write $M = \bigcup_{i=1}^{n} U_{\lambda_i}, \lambda_i \in \Lambda$ where n is finite integer, then M is a ***compact topological space***. *Compactness of a subspace of* M *is naturally defined with respect to its relative topology.*

We can show that closed subspaces of compact topological spaces are also compact.

Let M be a topological space and $A \subset M$ be a closed subspace. We consider an arbitrary open cover $\{V_\lambda\}_{\lambda\in\Lambda}$ of A. We know that $V_\lambda = U_\lambda \cap A$ where $\{U_\lambda\}_{\lambda\in\Lambda}$ is a class of open sets in M. Since A' is open, the class $\{U_\lambda, A' : \lambda \in \Lambda\}$ is an open cover of the space M. Since M is compact this cover must have a finite subcover $\{U_{\lambda_i}, A' : \lambda_i \in \Lambda, i = 1, 2, \ldots, n\}$ so that one can write $A' \cup U_{\lambda_1} \cup \cdots \cup U_{\lambda_n} = M$. Since $M = A \cup A'$, we conclude that $A \subseteq U_{\lambda_1} \cup \cdots \cup U_{\lambda_n}$ and finally $A = V_{\lambda_1} \cup \cdots \cup V_{\lambda_n}$. This means that A is compact. □

In Hausdorff spaces the converse of the above statement is also valid. *Let* M *be a Hausdorff space and let* $A \subset M$ *be a compact subspace. Then* A *is closed.*

In order to prove this proposition, we have to show that A' is an open set. We take a point $p \in A'$. Since M is a Hausdorff space, for any point $a \in A$, we can find disjoint open sets $U_{p,a}$ and U_a containing the points p and a, respectively. The class $\{U_a \cap A : a \in A\}$ is an open cover of A in relative topology. But A is compact, hence there is a finite set $\{a_1, \ldots, a_n\}$ $\subset A$ such that $A = \bigcup_{i=1}^{n} U_{a_i} \cap A \subseteq \bigcup_{i=1}^{n} U_{a_i}$. It is now clear that the finite intersection $U = \bigcap_{i=1}^{n} U_{p,a_i}$ is an open neighbourhood of the point p and $U \cap A$ $= \emptyset$. We thus obtain $U \subset A'$. Hence the arbitrary point p is an interior point of A', i.e., A' is open and A is closed. We can now easily deduce the following corollary: *if M is a compact Hausdorff space, then a subspace is compact if and only if it is closed.* □

A subspace of a topological space M is called ***relatively compact*** if its closure is compact. A topological space each point of which admits a compact neighbourhood is called a ***locally compact space***. *If M is a locally compact Hausdorff space, we can replace the term "compact neighbourhood" by "relatively compact neighbourhood"*. Indeed, let the point $p \in M$ admit the compact neighbourhood N. Since M is a Hausdorff space, N is closed. On the other hand, the relation $\mathring{N} \subset N$ implies that $\overline{\mathring{N}} \subset N$. $\overline{\mathring{N}}$ is a closed subset of a compact set. Therefore, it is compact. Hence. p has an open neighbourhood with a compact closure.

A useful generalisation of compactness is ***paracompactness***. This concept was introduced in 1944 by French mathematician Jean Alexander Eugène Dieudonné (1906-1992). Let $\mathcal{A} = \{U_\lambda \subseteq M : \lambda \in \Lambda\}$ be a class of subsets of a space M. Another class of subsets $\mathcal{B} = \{V_\gamma \subseteq M : \gamma \in \Gamma\}$ is called a ***refinement*** of class $\mathcal{A}$ if and only if for any $V_\gamma \in \mathcal{B}$ there exists a $U_\lambda \in \mathcal{A}$ such that $V_\gamma \subseteq U_\lambda$. An open cover $\mathcal{A}$ of a topological space M is called ***locally finite*** if every point $p \in M$ has a neighbourhood that intersects only finitely many sets in the cover. In other words $\mathcal{A} = \{U_\lambda \subseteq M\}$ is locally finite if each point $p \in M$ has a neighbourhood $V(p)$ such that the set $\{\lambda \in \Lambda : V(p) \cap U_\lambda \neq \emptyset\}$ is finite. *M is a* ***paracompact space*** *if any open cover of M admits an open refinement that is locally finite.* It is obvious that every compact space is also paracompact.

It can be shown that a locally compact, second countable Hausdorff space M is paracompact.

Let $\{V_i : i \in \mathbb{N}\}$ where $\mathbb{N}$ denotes the set of natural numbers be a countable basis for M. We shall first form a countable basis with compact closure. By our assumption, there exists a relatively compact open set U_p containing a point $p \in M$. Since U_p is expressible as union of some basic open sets, there is a set V_{i_p} such that $p \in V_{i_p}$ and $V_{i_p} \subseteq U_p$ whence we obtain $\overline{V}_{i_p} \subseteq \overline{U}_p$. But $\overline{U}_p$ is compact. Being a closed subset of a compact set, $\overline{V}_{i_p}$

is also compact. Therefore $\{V_{i_p} : p \in M\} \subseteq \{V_i : i \in \mathbb{N}\}$ is a countable relatively compact basis. Let us now suppose that $\{U_i\}$ is such a basis. Next, we construct inductively a sequence of *nested* open sets $\{W_i\}$ with the following properties: (i) $\overline{W}_i$ is compact, (ii) $W_i \subset \overline{W}_i \subset W_{i+1}$, (iii) $M = \bigcup_{i=1}^{\infty} W_i$. We further adopt the convention that $W_0 = \emptyset$. We take $W_1 = U_1$. Hence, $\overline{W}_1 = \overline{U}_1$ is compact. We now introduce the open set

$$W_k = U_1 \cup U_2 \cup \cdots \cup U_{j_k} = \bigcup_{i=1}^{j_k} U_i$$

Since $\overline{W}_k = \overline{\bigcup_{i=1}^{j_k} U_i} = \bigcup_{i=1}^{j_k} \overline{U}_i$ is a finite union of compact sets, it is also compact. So it must be covered by finitely many elements of the open cover $\{U_i\}$. We then take the index j_{k+1} as the least positive integer greater than the index j_k so that one is able to write

$$\overline{W}_k \subseteq \bigcup_{i=1}^{j_{k+1}} U_i.$$

We then define the next member of the sequence as

$$W_{k+1} = \bigcup_{i=1}^{j_{k+1}} U_i.$$

This completes the construction of the sequence $\{W_i\}$. The property (iii) is then satisfied automatically. Let $\{U_\lambda : \lambda \in \Lambda\}$ be an arbitrary open cover of M. The set $K_i = \overline{W}_i - W_{i-1} = \overline{W}_i \cap W'_{i-1}$ is compact since it is a closed subset of the compact set $\overline{W}_i$. We obviously get $K_1 = \overline{W}_1$. On the other hand, properties of the sequence imply that K_i is contained in open set $Z_i = W_{i+1} - \overline{W}_{i-2} = W_{i+1} \cap \overline{W}'_{i-2}$. For $i \geq 3$, we can choose a finite subcover of the open cover $\{U_\lambda \cap Z_i : \lambda \in \Lambda\}$ of the compact set K_i. For the compact set $K_2 = \overline{W}_2 - W_1$, we choose a finite subcover of the open cover $\{U_\lambda \cap W_3 : \lambda \in \Lambda\}$. Similarly, the compact set K_1 will be covered by a finite subcover of $\{U_\lambda \cap W_2 : \lambda \in \Lambda\}$. Because of the relation $W_i \subset \overline{W}_i$, we get $W_i - W_{i-1} \subset \overline{W}_i - W_{i-1} = K_i$. Since the sequence $\{W_i\}$ is nested, we obviously obtain

$$M = \bigcup_{i=1}^{\infty} W_i = \bigcup_{i=1}^{\infty} (W_i - W_{i-1}) \subset \bigcup_{i=1}^{\infty} K_i$$

implying that $M = \bigcup_{i=1}^{\infty} K_i$ where each K_i is covered by finitely many members of the open cover $\{U_\lambda \cap Z_i : \lambda \in \Lambda\}$. It is straightforward to see that this open cover is a locally finite, countable refinement which consists of a countable union of finite unions. Hence, M is a paracompact space. □

Let us consider topological spaces $(M, \mathfrak{M})$ and $(N, \mathfrak{N})$. It is a simple exercise to see that we can endow the Cartesian product $M \times N$ with a topology by choosing its open sets as unions of *elementary open sets* $U \times V$ where $U \in \mathfrak{M}$, $V \in \mathfrak{N}$. Such a topology on $M \times N$ is called the ***product topology***. This definition may be, of course, extended to Cartesian product of any number of topological spaces. For instance, in $\mathbb{R}^n$ the elementary open sets are ***open n-rectangles*** obtained as Cartesian products $(a_1, b_1) \times \cdots \times (a_n, b_n)$ of open intervals in $\mathbb{R}$. It is easy to see that $\mathbb{R}^n$ *is a second countable topological space* because it has a countable basis that is the collection of all Cartesian products $\prod_{i=1}^{n}(a_i, b_i)$ where $(a_i, b_i) \in \mathbb{R}$ is an open interval with *rational* end points.

$(M, \mathfrak{M})$ and $(N, \mathfrak{N})$ are topological spaces. *The function* $f : M \to N$ *is* ***continuous at the point*** $p_0 \in M$ *if for each neighbourhood* V *of the image point* $f(p_0) \in N$*, there exists a neighbourhood* U *of the point* p_0 *such that* $f(U) \subseteq V$. Another completely equivalent definition may be given as follows: *the function* f *is continuous at a point* p_0 *if the inverse image* $f^{-1}(V)$ *of every neighbourhood* V *of the point* $f(p_0)$ *is a neighbourhood of the point* p_0. Indeed, if the set U is a neighbourhood of p_0 satisfying the relation $f(U) \subseteq V$, we immediately get $U \subseteq f^{-1}(f(U)) \subseteq f^{-1}(V)$. Conversely, suppose that the set $f^{-1}(V)$ is a neighbourhood of p_0. If we write $U = f^{-1}(V)$, we find that $f(U) = f(f^{-1}(V)) \subseteq V$.

A function $f : M \to N$ is ***continuous*** on M if it is continuous at every point of its domain. *We can easily show that* f *is a continuous function if and only if the inverse image of every open set in* N *is an open set in* M*, i.e., if* $f^{-1}(V) \in \mathfrak{M}$ *for all* $V \in \mathfrak{N}$.

Let f be a continuous function. Consider an arbitrary open set $V \in \mathfrak{N}$ and define the set $U = f^{-1}(V) \subseteq M$. Let p be a point of U. We obviously have $f(p) \in V$. Since V is an open set, $f(p)$ is an interior point of V. Thus, there exists an open set $V_{f(p)}$ such that $f(p) \in V_{f(p)} \subseteq V$. Due to the continuity of f, the set $f^{-1}(V_{f(p)}) \subseteq U$ is a neighbourhood of p. Hence, there exists an open set $U_p \in \mathfrak{M}$ such that $p \in U_p \subseteq U$. All points of U are, therefore, interior points, that is, U is an open set. Conversely, let us now assume that for all $V \in \mathfrak{N}$, we have $f^{-1}(V) \in \mathfrak{M}$. Consider an arbitrary point p in M and assume that $f(p) \in V \in \mathfrak{N}$. The set $U = f^{-1}(V)$ is an open neighbourhood of the point p. Consequently, f is continuous at all points of M. □

It is not too difficult to demonstrate that the following definitions for the continuity of functions are equivalent:

(a). *The function f is continuous.*
(b). *The inverse image of every open set is open.*
(c). *The inverse image of every closed set is closed.*
(d). *For every subset $B \subseteq N$, the relation $\overline{f^{-1}(B)} \subseteq f^{-1}(\overline{B})$ is satisfied.*
(e). *For every subset $A \subseteq M$, the relation $f(\overline{A}) \subseteq \overline{f(A)}$ is satisfied.*

It is evident from the definition of the continuity that the composition of continuous functions is also a continuous function.

One can easily demonstrate that images of compact sets are also compact under continuous functions. We thus have to prove that *if $f : M \to N$ is a continuous function from a compact space M into a topological space N, then the set $f(M) \subseteq N$ is a compact subspace.*

We assume that the class $\{V_\lambda\}_{\lambda\in\Lambda}$ is an arbitrary open cover of the range $f(M) \subseteq N$ in its relative topology. We know that its members are in the form $V_\lambda = U_\lambda \cap f(M)$ where U_λ are open sets in N. Obviously, the class $\{f^{-1}(V_\lambda)\}_{\lambda\in\Lambda}$ is a cover of M implying that $M = \bigcup_{\lambda\in\Lambda} f^{-1}(V_\lambda) = \bigcup_{\lambda\in\Lambda} f^{-1}\big(U_\lambda \cap f(M)\big) = \bigcup_{\lambda\in\Lambda} f^{-1}(U_\lambda) \cap f^{-1}\big(f(M)\big) = \bigcup_{\lambda\in\Lambda} f^{-1}(U_\lambda) \cap M = \bigcup_{\lambda\in\Lambda} f^{-1}(U_\lambda)$. The continuity of f requires that the class $\{f^{-1}(U_\lambda)\}_{\lambda\in\Lambda}$ is an open cover of M and must have a finite subcover since M is compact. We thus obtain $M = \bigcup_{i=1}^{n} f^{-1}(U_{\lambda_i}), \lambda_i \in \Lambda, i = 1, \dots, n$, and hence, we find that $f(M) = \bigcup_{i=1}^{n} f\big(f^{-1}(U_{\lambda_i})\big) \subseteq \bigcup_{i=1}^{n} U_{\lambda_i}$. The class $\{V_{\lambda_i} = U_{\lambda_i} \cap f(M)\}$ is a finite subcover of $f(M)$ in its relative topology since one can clearly write $f(M) = \bigcup_{i=1}^{n} V_{\lambda_i}$. Therefore, $f(M)$ is a compact subspace of N. □

We can then deduce the following corollary: *if a bijective function $f : M \to N$ from a compact space M into a Hausdorff space N is continuous, then the inverse function $f^{-1} : N \to M$ is also continuous.*

In order to prove that the function f^{-1} is continuous, it would be sufficient to show that the image $f(A)$ in N of an arbitrary closed set A in M is also closed. Since A is closed, it must be a compact subspace of M. Since f is a continuous function $f(A)$ will be a compact subspace of N. Hence $f(A)$ is closed. □

Since topologies are governed by open sets, it is evident that in order to establish a ***topological equivalence*** between two topological spaces, it would be sufficient to be able to transform open sets in one space to open sets in the other. This mapping must be bijective to ensure numerical equivalence. If $h : M \to N$ is a continuous bijective mapping, then the inverse images of open sets in N would be open in M. If the inverse mapping $h^{-1}: Y \to X$ is continuous as well, then the images of open sets in M will be open in N. A bijective mapping $h: M \to N$ between topological spaces $(M, \mathfrak{M})$ and $(N, \mathfrak{N})$ is called a ***homeomorphism*** if both h and h^{-1} are continuous. Such topological spaces M and N are said to be ***homeomorphic***. We thus conclude that two spaces are topologically equivalent if we can show that there exists a homeomorphism between them. If h is a homeomorphism, then we get $h(U) \in \mathfrak{N}$ for all $U \in \mathfrak{M}$ and, conversely, $h^{-1}(V) \in \mathfrak{M}$ for all $V \in \mathfrak{N}$. It can, therefore, be said that a homeomorphism is an ***open***, continuous and bijective mapping. A property which remains invariant under a homeomorphism is called a ***topological property***, namely, a topological property observed in a topological space remains unchanged in all homeomorphic images of this space. For instance, we see at once that Hausdorff property is a topological property. It is quite obvious that the inverse of a homeomorphism or a composition of two homeomorphisms are also homeomorphisms. It is not difficult to observe that the set of all homeomorphisms of a topological space onto itself equipped with a binary operation defined as the composition of two homeomorphisms constitute a group with respect to this operation.

In the light of the above statements we can conclude at once that *if the function $f: M \to N$ from a compact space M onto a Hausdorff space N is continuous and bijective, then the mapping f is a homeomorphism.* In this case, N must clearly be a compact space as well.

Let R be an equivalence relation on a topological space $(M, \mathfrak{M})$ [*see* p. 5]. We know that the set $[p]$ consisting of all points that are related to $p \in M$ through R is an equivalence class. Each point in the set $[p]$ generates the same equivalence class. Thus distinct equivalence classes are disjoint sets. They form a partition of the set M. The set $M/R = \{[p] : p \in M\}$ has already been called the *quotient set*. Therefore, to each point p in the set M there corresponds a unique equivalence class $[p]$ in the set M/R, that is, there is a function $\pi : M \to M/R$ such that $\pi(p) = [p]$. π is called a ***canonical*** or ***natural projection***. It is evident that the canonical mapping π is surjective, but it is also clear that it is not injective. We now define a class of subsets of M/R by

$$\mathfrak{M}_R = \{V \in \mathcal{P}(M/R) : \pi^{-1}(V) \in \mathfrak{M}\}.$$

It is easily seen that this class is a topology on M/R. The relations $\emptyset = \pi^{-1}(\emptyset) \in \mathfrak{M}$ and $M = \pi^{-1}(M/R) \in \mathfrak{M}$ mean that $\emptyset \in \mathfrak{M}_R$ and $M/R \in \mathfrak{M}_R$. Let us now consider a family of sets $\{V_\lambda : \lambda \in \Lambda\} \subseteq \mathfrak{M}_R$ where Λ is an index set. Our definition implies that $U_\lambda = \pi^{-1}(V_\lambda) \in \mathfrak{M}$ so that one can write

$$\pi^{-1}(\bigcup_{\lambda\in\Lambda} V_\lambda) = \bigcup_{\lambda\in\Lambda} \pi^{-1}(V_\lambda) = \bigcup_{\lambda\in\Lambda} U_\lambda \in \mathfrak{M}.$$

Thus one has $\bigcup_{\lambda\in\Lambda} V_\lambda \in \mathfrak{M}_R$. Let $\{V_{\lambda_i} : 1 \le i \le n\} \subseteq \mathfrak{M}_R$ be finite family. Because of the relation $\pi^{-1}(\bigcap_{i=1}^{n} V_i) = \bigcap_{i=1}^{n} \pi^{-1}(V_i) = \bigcap_{i=1}^{n} U_i \in \mathfrak{M}$, we obtain $\bigcap_{i=1}^{n} V_i \in \mathfrak{M}_R$. Hence, $\mathfrak{M}_R$ is a topology and the pair $(M/R, \mathfrak{M}_R)$ is a topological space. We call $\mathfrak{M}_R$ ***the quotient topology*** and $(M/R, \mathfrak{M}_R)$ ***the quotient space***. It is quite clear that through the topology so defined the canonical projection π is rendered continuous.

Certain topological spaces possess quite a useful property called *the partition of unity*.

Partition of Unity. Let M be a topological space and $\{V_i : i \in \mathcal{I}\}$, where $\mathcal{I}$ is an index set, be a locally finite open cover of M. Hence, we have $M = \bigcup_{i\in\mathcal{I}} V_i$ and every point $p \in M$ has an open neighbourhood U_p whose intersection with only finitely many members of the cover is not empty. If a family of continuous functions $f_i : M \to [0,1]$ satisfies the conditions

$$(i).\ \textit{supp}\,(f_i) \subset V_i \text{ for each index } i,$$

$$(ii).\ \sum_{i\in\mathcal{I}} f_i(p) = 1 \text{ for each } p \in M$$

then the family of ordered pair $\{V_i, f_i\}$ is called a partition of unity. Here the ***support of a function*** $f : M \to \mathbb{R}$ is defined as the *closed set*

$$\textit{supp}\,(f) = \overline{\{p \in M : f(p) \neq 0\}} = \overline{f^{-1}(\mathbb{R} - \{0\})} \subseteq M.$$

Since the family $\{V_i\}$ is locally finite there are only finitely many, say N number of non-empty open sets $V_i \cap U_p$ containing a point p. Consequently, $f_i(p) \neq 0$ only for a finite N number of functions f_i so that at any point p the sum $\sum_{i\in\mathcal{I}} f_i(p)$ must contain only finitely many terms and one can write $\sum_{k=1}^{N(p)} f_{i_k}(p) = 1,\ \{i_1, \dots, i_N\} \subset \mathcal{I}$. Naturally the number $N < \infty$ may be dependent on the position points of M.

Let M be a topological space on which there exists a partition of unity $\{V_i, f_i\}$ as defined above and let the family $\{U_\lambda : \lambda \in \Lambda\}$ be an open cover of M. *If for each member V_i of locally finite open cover one can find an open set U_{λ_i} such that supp $(f_i) \subset U_{\lambda_i}$ then we say that the partition of unity $\{V_i, f_i\}$ is subordinate to the open cover $\{U_\lambda : \lambda \in \Lambda\}$.*

As we shall see later in dealing with integration on manifolds, the existence of a partition of unity on a topological space will prove to be very effective in reducing certain global properties to some local properties.

Connectedness. If a topological space $(M, \mathfrak{M})$ cannot be expressible as the union of two non-empty disjoint open sets, that is, if $M \neq U_1 \cup U_2$; $U_1, U_2 \in \mathfrak{M}$, $U_1 \cap U_2 = \emptyset$, we say that it is a ***connected space***. Conversely, if there exist such open sets U_1 and U_2 so that $M = U_1 \cup U_2$, then M is a ***disconnected space***. In a disconnected space we naturally have $U_1' = U_2$ and $U_2' = U_1$. Hence the sets U_1 and U_2 are both open and closed sets in topology $\mathfrak{M}$ whence we conclude that a topological space is connected if it cannot be expressed as the union of two disjoint closed sets. It is straightforward to see that a space M being connected means that only the sets $\emptyset$ and M are both open and closed. Indeed, if M possesses a proper subset A that is both open and closed, then its complement A' ought to be both open and closed. Since $M = A \cup A'$ and $A \cap A' = \emptyset$, M becomes expressible as the union of two disjoint open or closed sets. Hence, M is a disconnected space.

A ***connected subspace*** of a topological space M is a subspace $A \subset M$ that is connected with respect to its relative topology. According to this definition, a subset A is connected if it cannot be contained in the union of two open sets of M whose intersections with A are disjoint and non-empty.

It is almost straightforward to show that the image of a connected space under a continuous function between two topological spaces is also connected.

Another concept of connectedness which is not entirely equivalent to the one described above may be introduced by resorting to a more geometrical approach. Let M be a topological space and $\mathcal{I} = [0, 1] \subset \mathbb{R}$ in which the topology is determined by open intervals. A ***path***, or an ***arc*** on the space M is defined as the continuous mapping $\phi : \mathcal{I} \to M$. We say that ϕ joins the points p_1 and p_2 in M if $\phi(0) = p_1$ and $\phi(1) = p_2$. If $\phi(t) \in M$ for every $t \in [0, 1]$, then *the path ϕ stays in the space M*. If any two points in the space M can be joined by a path staying in M, then M is called a ***path-connected*** or an ***arc-connected*** space. If this property is valid for a subspace of M, then this subspace is path-connected. Such a space is schematically described in Fig. 2.2.1.

If M is path-connected, N is a topological space and $f : M \to N$ is a continuous mapping, then we immediately deduce from the fact that composition of continuous mappings is also continuous, the subspace $f(M)$ is path-connected as well. *If a topological space M is path-connected, then it is also connected.* However, the converse statement is generally not true.

When $\phi(0) = \phi(1) = p_1$, we say that the path is *closed*. If every closed path in the space M can be contracted continuously to a point inside the path, the space M is called ***simply connected***. Equivalently, we say that a connected topological space M is simply connected if a path connecting any two points of M can be continuously deformed into every other curve connecting these two points.

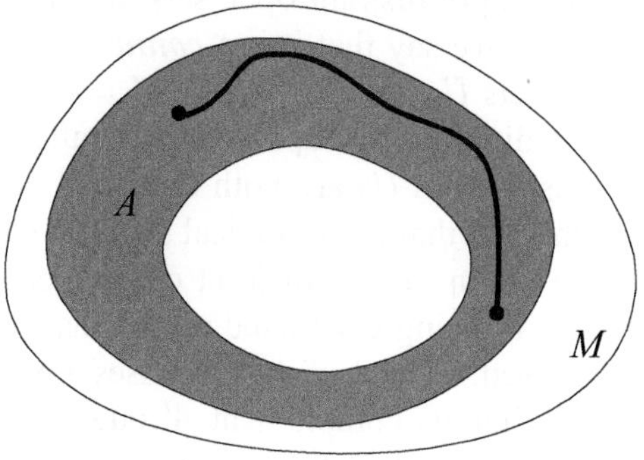

Fig. 2.2.1. A path-connected subspace.

Metric Spaces. A topology on a set M can be defined sometimes by means of a real-valued function. Let M be a non-empty set. Let us suppose that we can define a real-valued, non-negative function $d : M \times M \to \mathbb{R}^+$ on this set. We further impose the following conditions on the function d:

(i). *For each* $p_1, p_2 \in M$ *one has* $d(p_1, p_2) \geq 0$.
(ii). $d(p_1, p_2) = 0$ *if and only if* $p_1 = p_2$.
(iii). *For each* $p_1, p_2 \in M$ *one has* $d(p_1, p_2) = d(p_2, p_1)$.
(iv). *For each* $p_1, p_2, p_3 \in M$ *one has* $d(p_1, p_2) \leq d(p_1, p_3) + d(p_3, p_2)$.

The inequality (iv) above is known as *the triangle inequality*. We call such a function $d(p_1, p_2)$ a ***metric*** on the set M and we interpret its value as the ***distance*** between two points p_1 and p_2 of the set M. In fact, we can easily verify that the metric concept coincides entirely with the familiar distance concept in the Euclidean space. The pair (M, d) is called a ***metric space***. The ***open ball*** of radius r centred at the point $p \in M$ is defined as the set

$$B_r(p) = \{p_1 \in M : d(p, p_1) < r\} \subset M. \tag{2.2.1}$$

We can generate a topology on a metric space called ***metric topology*** *by noting that open balls constitute a basis for a topology*. Consider a class of subsets $\mathfrak{B}_d = \{B_r(p) : p \in M, r \geq 0\}$ of the set M. It is evident that $M = \cup\{B_r(p) : p \in M, r > 0\}$. $\emptyset \in \mathfrak{B}_d$ since $B_0(p) = \emptyset$. In order to show that the class $\mathfrak{B}_d$ is in fact a basis for a topology on M, all we have to do is to demonstrate that any point in the intersection of two open balls belongs to an open ball contained in that intersection. Let us consider two open balls centred at the points p_1 and p_2 with radii r_1 and r_2, respectively. If their intersection is empty, the criterion is automatically satisfied. Hence, we assume that the intersection of these open balls is not empty and take a point p in their intersection $B_{r_1}(p_1) \cap B_{r_2}(p_2)$ into consideration. Hence we can write $d(p_1, p) < r_1$ and $d(p_2, p) < r_2$. Let us now choose

$$r = \min\{r_1 - d(p_1, p), r_2 - d(p_2, p)\} > 0.$$

The open ball $B_r(p)$ is contained both in the sets $B_{r_1}(p_1)$ and $B_{r_2}(p_2)$. For an arbitrary point $q \in B_r(p)$ the triangle inequality yields $d(p_1, q) \leq d(p_1, p) + d(p, q) < r_1 - r + r = r_1$ implying that $q \in B_{r_1}(p_1)$. In the same fashion, we obtain this time $d(p_2, q) \leq d(p_2, p) + d(p, q) < r_2 - r + r = r_2$ and $q \in B_{r_2}(p_2)$. We thus find that $B_r(p) \subset B_{r_1}(p_1) \cap B_{r_2}(p_2)$. Consequently, the class $\mathfrak{B}_d$ constitutes a basis for a topology on M. Each open set of this topology is given by unions of some open balls, that is, if U is an open set, then it is expressible as $U = \bigcup_{p \in U} B_{r(p)}(p)$ for some $r(p)$. The set

$$B_r[p] = \{p_1 \in M : d(p, p_1) \leq r\} \subset M \tag{2.2.2}$$

is called a ***closed ball*** with centre $p \in M$ and radius r. It is easy to verify that $B_r[p]$ is a closed set. It can easily be observed that $\overline{B_r(p)} \subseteq B_r[p]$. Let us consider all open balls centred at a point whose radii are rational numbers. We immediately observe that these open balls constitute a countable fundamental system of neighbourhoods of that point. Therefore, *metric spaces are first countable spaces*.

Metric spaces has quite a distinctive property. *They are all Hausdorff spaces*. Indeed, if we consider two distinct points of a metric space M, we must have $d(p, q) = r_1 > 0$ whenever $p \neq q$. By choosing $r \leq r_1/2$, one easily demonstrates that it is always possible to find two open balls with radius $r > 0$ such that $B_r(p) \cap B_r(q) = \emptyset$.

Let us consider a sequence of points $\{p_n\} \subset M$. This sequence converges to a point $p \in M$, if there exists a natural number $N(\epsilon)$ for each $\epsilon > 0$ such that $d(p_n, p) < \epsilon$ whenever $n \geq N(\epsilon)$. The sequence $\{p_n\}$ is called a ***Cauchy sequence*** [French mathematician Augustin-Louis Cauchy (1789-1857)] if to each $\epsilon > 0$ there corresponds a natural number $N(\epsilon) \in \mathbb{N}$ such that $d(p_m, p_n) < \epsilon$ whenever $m, n \geq N$. If every Cauchy sequence in a metric space is convergent, then we say that this metric space is ***complete***. It can be shown that a *subspace of a complete metric space is complete if and only if it is closed* .

It can also be proven that metric spaces are paracompact spaces though we have to omit its difficult proof because it is beyond our scope.

Let A be a subset of a metric space. *The diameter of* A is defined by the non-negative number $D(A) = \sup_{p_1, p_2 \in A} d(p_1, p_2)$. If $D(A) < \infty$, then A is a *bounded set*. Obviously open and closed balls of radius r are bounded and their diameters are both $2r$.

The standard metric on the set of real numbers is $d(x, y) = |x - y|$. Let us now consider the set $\mathbb{R}^n$. If $x \in \mathbb{R}^n$, then $x = (x^1, x^2, \ldots x^n)$ is an ordered n-tuple of real numbers where $x^i \in \mathbb{R},\ i = 1,\ 2, \cdots, n$. Next, we define the function

$$d(x, y) = \Big(\sum_{i=1}^{n} |x^i - y^i|^2\Big)^{1/2} \tag{2.2.3}$$

for a pair of points $x, y \in \mathbb{R}^n$. It is straightforward to observe that this function is actually a metric on $\mathbb{R}^n$. We name the set $\mathbb{R}^n$ equipped with this *standard metric* as the n-dimensional ***Euclidean space*** E_n. Since $\mathbb{R}^n$ formed by the Cartesian product of the real line n times, the real numbers $\{x^i\}$ determining a point $x \in E_n$ are called *Cartesian coordinates* of that point. The collection of all such numbers constitutes the *coordinate cover* of E_n. *The length* or *the norm* of an element $x \in E_n$ is given by

$$\|x\| = \Big(\sum_{i=1}^{n} |x^i|^2\Big)^{1/2} \tag{2.2.4}$$

so that we can write $d(x, y) = \|x - y\|$.

A ***norm*** on a complex vector space V defined over a field of scalars $\mathbb{F}$ is a real-valued, non-negative function $\|\cdot\| : V \to \mathbb{R}^+$ satisfying the following conditions:

(i). $\|v\| \geq 0$ *for all* $v \in V$ *and* $\|v\| = 0$ *if and only if* $v = \mathbf{0}$.
(ii). $\|v\| = |\alpha|\|v\|$ *for all* $v \in V$ *and* $\alpha \in \mathbb{F}$.
(iii). $\|u + v\| \leq \|u\| + \|v\|$ *for all* $u, v \in V$.

We say that a vector space equipped with a norm, i.e., the ordered pair $(V, \|\cdot\|)$ is a ***normed linear space*** or a ***normed vector space*** or simply a ***normed space***. By taking $\alpha = 0$ and $\alpha = -1$, we obtain $\|\mathbf{0}\| = 0$ and $\|-v\| = \|v\|$, respectively. (iii) is known as the *triangle inequality*. It is then rather easy to establish directly by induction that the following inequality holds for a number of vectors $v_1, v_2, \ldots, v_n \in V$:

$$\|v_1 + v_2 + \cdots + v_n\| \leq \|v_1\| + \|v_2\| + \cdots + \|v_n\|.$$

For any two vectors $u, v \in V$, we can write $\|u - v\| \geq \|u\| - \|v\|$ and $\|u - v\| = \|v - u\| \geq \|v\| - \|u\|$. We thus find

$$\|u - v\| \geq \big|\|u\| - \|v\|\big|$$

for all $u, v \in V$. These properties of the norm amply justify our interpreting the norm of a vector as its ***length***. By means of the norm, we can introduce a function $d : V \times V \to \mathbb{R}^+$ as follows:

$$d(u, v) = \|u - v\|.$$

Evidently, this function satisfies the conditions $d(u, v) \geq 0$; $d(u, v) = 0$ if and only if $u = v$ and $d(u, v) = d(v, u)$. Furthermore, one can write

$$d(u, v) = \|u - w + w - v\| \leq \|u - w\| + \|w - v\| = d(u, w) + d(w, v)$$

so that d holds the triangle inequality. Hence, we realise that the function d so defined is actually a metric on the vector space V. We call this metric generated by the norm, the ***natural metric*** on the normed space V. But, in addition to its commonly known properties, this metric satisfies the following equalities for all $u, v, w \in V$ and $\alpha \in \mathbb{F}$:

$$d(\alpha u, \alpha v) = |\alpha| d(u, v), \quad d(u + w, v + w) = d(u, v).$$

The last relation indicates the fact that the distance between two vectors does not change by their parallel translations.

It is now clear that a normed space is a Hausdorff space equipped with a metric topology induced by its natural metric. In this topology, *open* and *closed balls of radius r centred at a vector v* are of course defined, respectively, by

$$B_r(v) = \{u \in V : \|u - v\| < r\}, \ B_r[v] = \{u \in V : \|u - v\| \leq r\}.$$

The basis for this topology is the class $\{B_r(v) : \text{for all } v \in V \text{ and } r > 0\}$. We obviously have $B_0(v) = \emptyset$, $B_0[v] = \{v\}$. One immediately verifies that an open ball $B_r(v)$ is obtained by just simply translating all vectors in the open ball $B_r(\mathbf{0})$ of radius r centred at the zero vector $\mathbf{0}$ by the vector v. If M is a subset of V, the set $v + M = \{v + u : \text{for all } u \in M\}$ is said to be the ***translation*** of the set M by the vector v. We thus have $B_r(v) = v + B_r(\mathbf{0})$. The same property will also be valid for closed balls. Unlike general metric spaces, it can easily be demonstrated that one always obtains $\overline{B_r(v)} = B_r[v]$ in all normed spaces.

Let us consider a sequence of vectors $\{v_n\} \subset V$. This sequence converges to a vector $v \in V$ if there exists a natural number $N(\epsilon)$ for each $\epsilon > 0$ such that $\|v_n - v\| < \epsilon$ whenever $n \geq N(\epsilon)$. The sequence $\{v_n\}$ is a

Cauchy sequence if there exists a natural number $N(\epsilon)$ for each $\epsilon > 0$ such that $\|v_n - v_m\| < \epsilon$ for all $n, m \geq N(\epsilon)$. *If every Cauchy sequence relative to its natural metric of a normed space V is convergent,* then V is called a *complete* normed space. A complete normed space is named as a ***Banach space*** [after Polish mathematician Stefan Banach (1892-1945)].

An ***inner product*** on a complex vector space V is a scalar-valued function $(\,\cdot\,,\,\cdot\,) : V \times V \to \mathbb{F}$ satisfying the following rules:

(i) $(u, v) = \overline{(v, u)}$ *for all vectors* $u, v \in V$.
(ii) $(\alpha u, v) = \alpha(u, v)$ *for all vectors* $u, v \in V$ *and scalars* $\alpha \in \mathbb{F}$.
(iii) $(u + v, w) = (u, w) + (v, w)$ *for all vectors* $u, v, w \in V$.
(iv) $(u, u) > 0$ *for all non-zero vectors* $u \in V$.

An overbar here denotes the complex conjugate. We can easily extract from this definition some novel results:

(a). $(\mathbf{0}, v) = (0 \cdot u, v) = 0 \cdot (u, v) = 0$ and similarly $(u, \mathbf{0}) = 0$ from which we naturally deduce that $(\mathbf{0}, \mathbf{0}) = 0$.

(b). Since $(u, u) = \overline{(u, u)}$ in compliance with (i), one finds $(u, u) \in \mathbb{R}$ and the property (iv) becomes meaningful. If $(u, u) = 0$, we then obtain that $u = \mathbf{0}$.

(c). The inner product is linear in its first argument because of the properties (ii) and (iii). On the other hand, we can easily observe that

$$(u, v + w) = \overline{(v + w, u)} = \overline{(v, u)} + \overline{(w, u)} = (u, v) + (u, w),$$
$$(u, \alpha v) = \overline{(\alpha v, u)} = \overline{\alpha(v, u)} = \overline{\alpha}\ \overline{(v, u)} = \overline{\alpha}(u, v).$$

Hence the inner product is additive in its second argument but is not homogeneous because of the fact that the conjugate of the scalar multiplier is involved. This situation is known as the ***conjugate linearity***. Thus, the inner product on a complex vector space is a ***sesquilinear*** **($1\frac{1}{2}$- *linear*)** function with respect to its two arguments.

(d). If $(u, w) = (v, w)$ or $(w, u) = (w, v)$ for all $w \in V$, then we find that $u = v$. We can indeed prove this by simply taking $w = u - v \in V$ in the relation $(u - v, w) = 0$.

For a real-valued inner product on a real vector space, the property (i) is reduced to the *symmetry* condition $(u, v) = (v, u)$. A real inner product is linear in its second argument too since $(u, \alpha v) = \alpha(u, v)$ for $\alpha \in \mathbb{R}$. Hence, an inner product on a real vector space is a ***bilinear*** function.

A linear vector space endowed with an inner product is called an ***inner product space***.

Inner product must hold an important relation which is called ***Cauchy-Bunyakowski-Schwarz's inequality*** or briefly the ***Schwarz inequality***

[German mathematician Karl Hermann Amandus Schwarz (1843-1921) and Russian mathematician Viktor Yakovlevich Bunyakowski (1804-1889) who had actually discovered this inequality that had appeared in one of his books published in 1859]. *Let H be an inner product space. The inequality $|(u,v)| \leq \sqrt{(u,u)(v,v)}$ holds for all non-zero vectors $u, v \in H$. The equality is valid if and only if the vectors u and v are linearly dependent.*

If one of the vectors in that inequality is zero, the relation holds trivially as $0 = 0$. For any two vectors $u, v \in H$ with $v \neq \mathbf{0}$ and any scalar number $\alpha \in \mathbb{F}$, we can write

$$0 \leq (u - \alpha v, u - \alpha v) = (u,u) - \alpha\overline{(u,v)} - \overline{\alpha}(u,v) + |\alpha|^2(v,v).$$

The right-hand side vanishes if and only if $u = \alpha v$, namely, if two vectors are linearly dependent. Let us next choose $\alpha = (u,v)/(v,v)$ to cast the above inequality into the form

$$(u,u) - \frac{|(u,v)|^2}{(v,v)} - \frac{|(u,v)|^2}{(v,v)} + \frac{|(u,v)|^2}{(v,v)} = (u,u) - \frac{|(u,v)|^2}{(v,v)} \geq 0$$

or

$$|(u,v)|^2 \leq (u,u)(v,v).$$

The square root of the above inequality yields the Schwarz inequality. □

The Schwarz inequality helps us to show that a norm is derivable from the inner product. *Let us define* $\|u\| = \sqrt{(u,u)}$. We immediately see from the definition that $\|u\| \geq 0$ for all $u \in V$ and $\|u\| = 0 \Leftrightarrow u = \mathbf{0}$. If $\alpha \in \mathbb{F}$, then we readily observe that $\|\alpha u\| = \sqrt{(\alpha u, \alpha u)} = \sqrt{|\alpha|^2(u,u)} = |\alpha|\|u\|$. Moreover, we easily obtain that $\|u+v\|^2 = (u+v, u+v) = (u,u) + (u,v) + \overline{(u,v)} + (v,v) = \|u\|^2 + \|v\|^2 + 2\Re(u,v)$. $\Re(u,v) \leq |\Re(u,v)| \leq |(u,v)|$ yields through Schwarz's inequality $\Re(u,v) \leq \|u\|\|v\|$. We thus obtain $\|u+v\|^2 \leq \|u\|^2 + \|v\|^2 + 2\|u\|\|v\| = (\|u\| + \|v\|)^2$. Hence the triangle inequality

$$\|u+v\| \leq \|u\| + \|v\|$$

follows at once. □

By this definition of the norm, Schwarz's inequality is expressed as

$$|(u,v)| \leq \|u\|\|v\|.$$

The norm generated by an inner product imposes a restriction that any two vectors in an inner product space must obey the ***parallelogram law***. Let $u, v \in H$. We have

$$\|u+v\|^2 = \|u\|^2 + \|v\|^2 + 2\Re(u,v),\ \|u-v\|^2 = \|u\|^2 + \|v\|^2 - 2\Re(u,v).$$

By adding those two expressions, we obtain

$$\|u+v\|^2 + \|u-v\|^2 = 2(\|u\|^2 + \|v\|^2).$$

This relation reflects the fact that a well-known relation of the classical geometry, namely, the sum of the squares of two diagonals of a parallelogram being equal to the sum of the squares of all of its four sides is still valid in an arbitrary inner product space.

The natural norm induced by an inner product generates now a *natural metric* on the vector space V through the function

$$d(u,v) = \|u-v\| = \sqrt{(u-v,u-v)}.$$

An inner product space which is complete relative to its natural metric is called a ***Hilbert space*** [after German mathematician David Hilbert (1862-1943)]. It goes without saying that a Hilbert space is also a Banach space.

A quite a simple generalisation of the classical Heine-Borel theorem [German mathematician Heinrich Eduard Heine (1821-1881) and French mathematician Félix Édouard Justin Émile Borel (1871-1956)] of real analysis leads to the result that every subset of E_n that is closed and bounded is compact. Let us consider an open set of E_n and a point inside this set. Hence, there exists an open ball centred at this point and contained in the open set. On the other hand, there is a closed ball with the same centre inside this open ball that is closed and bounded. Therefore, E_n is a *locally compact* metric space. If there is no ambiguity, we prefer to employ henceforth the notation $\mathbb{R}^n$ instead of E_n to denote the Euclidean space that illustrates the formation of this space more clearly. $\mathbb{R}^n$ is also a complete metric space. It can be shown that the class of open balls generated by the metric (2.2.3) constitutes a *countable basis* for the metric topology on $\mathbb{R}^n$. Thus the *second countable metric space* $\mathbb{R}^n$ *is a paracompact topological space* according to the statement in *p*. 57.

Let (M,d) and (N,ρ) be two metric spaces and consider a function $f: M \to N$. The topological concept of continuity takes now a purely metrical form. We say that the function f is *continuous* at a point $p \in M$ if for each number $\epsilon > 0$ there corresponds a number $\delta(\epsilon;p) > 0$ such that $\rho\big(f(p), f(p_1)\big) < \epsilon$ for all points $p_1 \in M$ satisfying $d(p,p_1) < \delta$. If f is a continuous function and if we can find for each ϵ a number $\delta(\epsilon)$ that is independent of points p, then f is called a *uniformly continuous function.*

Manifold. A ***differentiable manifold*** is essentially a topological space. But it is also equipped with a particular structure that makes it possible to support differentiable mappings, vectors, tensors and exterior differential forms associated with those topological spaces.

Let us first consider a more general definition. An n-dimensional ***topological manifold*** M *is a Hausdorff space every point of which belongs to an open set that is homeomorphic to an open set of the Euclidean space* $\mathbb{R}^n$. *These open sets constitute an open cover of* M. Thus a topological manifold is *locally equivalent* to the Euclidean space $\mathbb{R}^n$.

It proves to be advantageous for a topological manifold to be a paracompact space if we wish to develop a workable theory of integration on manifolds. That is the reason why many authors prefer to assume that the principal ingredient of a topological manifold is a second countable, hence a separable, locally compact Hausdorff space. As we have mentioned earlier, the concept of manifold stems from the desire to make an abstraction of the classical notion of smooth surfaces in the Euclidean space, to endow a topological space with a local structure supporting differentiability and to be able patch together these local structure to cover the entire manifold. The above definition means that when we consider a point $p \in M$, there will be a connected open set $U \in \mathfrak{M}$ containing the point p and a homeomorphism $\varphi : U \to V \subseteq \mathbb{R}^n$. Thus, the function φ is bijective, and φ and φ^{-1} are continuous. Since the metric space $\mathbb{R}^n$ is a Hausdorff space, it is imperative that M has also the Hausdorff property in order this homeomorphism to exist. Obviously the set $V = \varphi(U) \subseteq \mathbb{R}^n$ is open, hence it is expressible as a union of some open balls in the Euclidean space $\mathbb{R}^n$. A ***chart*** on M is the pair (U, φ). n is the *dimension* of this chart. The open set U is the domain of the chart. Let us now write $\varphi(p) = \mathbf{x} = (x^1, x^2, \dots x^n) \in \mathbb{R}^n$ and we choose clearly continuous functions $g^i : \mathbb{R}^n \to \mathbb{R}, i = 1, 2, \ \dots, n$ by the rule $g^i(\mathbf{x}) = x^i$. We say that the real-valued continuous functions $\varphi^i = g^i \circ \varphi : \ U \to \mathbb{R}, \ i = 1, \ \dots, n$ are the ***coordinate functions*** of the chart (U, φ) providing the mapping $\varphi^i(p) = x^i$ whereas the real numbers $(x^1, x^2, \dots x^n)$ will be called the ***coordinates*** of the point $p \in M$ in the chart (U, φ) (Fig. 2.2.2). Thus a chart gives rise to a ***local system of coordinates*** on the manifold. If every point of a topological manifold M has an n-dimensional chart, we say that M is an n-dimensional manifold. When we want to emphasise its dimension we denote this manifold by M^n. The union of local coordinates systems in all charts covering M constitute the ***coordinate cover*** of the manifold M. If there is a point $p_0 \in U$ such that $\varphi(p_0) = \mathbf{0}$, then we say that the local coordinate system is *centred* at p_0.

Let us consider a function $f : M \to \mathbb{R}^n$. The function $f^i = g^i \circ f : M \to \mathbb{R}$ is called the ith *component function* of f.

The inverse mapping $\varphi^{-1} : V \to U$ is called a ***parametrisation*** of the open set U. The coordinates $x^1, x^2, \dots x^n$ are then called ***parameters*** of U.

The coordinate lines on M are the images of Cartesian coordinate lines on $\mathbb{R}^n$ under the mapping $\varphi^{-1} : \varphi(U) = V \to U$ (Fig. 2.2.2).

It is now clear that the manifold M behaves *locally* just like an open set of the Euclidean space $\mathbb{R}^n$ in the *vicinity* of the point $p \in M$. Since the Euclidean space is locally compact and homeomorphism preserves compactness, *a finite-dimensional topological manifold must also be locally compact*. In fact, let us consider a point $p \in M$ contained in a chart (U, φ). The point $\varphi(p)$ will be in an open neighbourhood in $\mathbb{R}^n$. Hence, it belongs to an open ball inside $\varphi(U)$. Since the closure of this open ball is contained in a closed ball that is both closed and bounded, then $\varphi(p)$ has a compact neighbourhood K. Because the function φ^{-1} is continuous, then the point p also must have the compact neighbourhood $\varphi^{-1}(K)$ in the open set U.

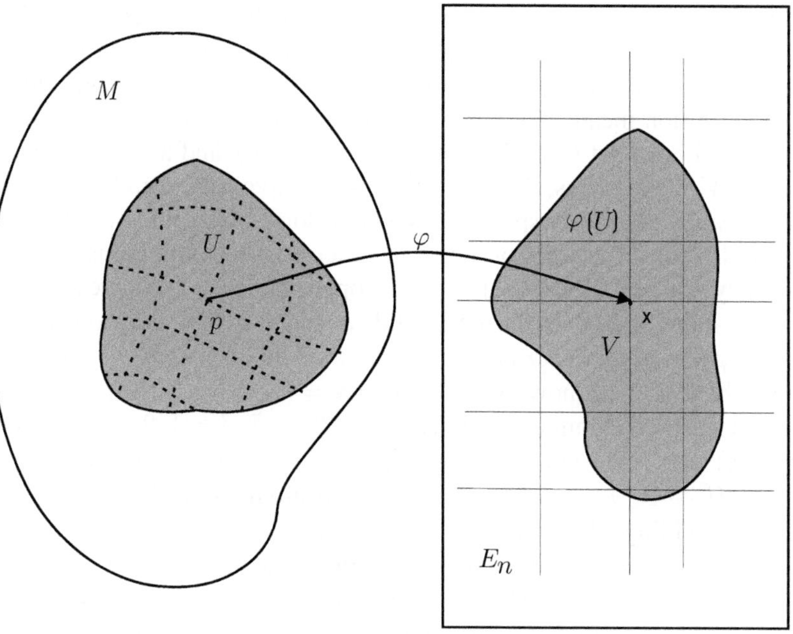

Fig. 2.2.2. A chart on the manifold M.

A ***C^k-atlas*** $\mathcal{A}$ on a topological manifold M is a family of charts $\mathcal{A} = \{(U_\alpha, \varphi_\alpha) : \alpha \in \mathcal{I}\}$ where $\mathcal{I}$ is an index set. Moreover this family must satisfy the following conditions: (i) all charts have the same dimension and the union of their domains constitute an open cover of the manifold, that is, $M = \bigcup_{\alpha \in \mathcal{I}} U_\alpha$, (ii) consider two different charts $(U_\alpha, \varphi_\alpha)$ and (U_β, φ_β) of the atlas. Let us assume that $U_\alpha \cap U_\beta \neq \emptyset$. Images of the open intersection $U_\alpha \cap U_\beta$ under mappings φ_α and φ_β will usually be different open set in $\mathbb{R}^n$

(Fig. 2.2.3). On the overlapping domain $U_\alpha \cap U_\beta$ of the homeomorphisms φ_α and φ_β, we can define the following ***transition functions***:

$$\varphi_{\alpha\beta} = \varphi_\beta \circ \varphi_\alpha^{-1} : \mathbb{R}^n \to \mathbb{R}^n, \qquad (2.2.5)$$
$$\varphi_{\alpha\beta}^{-1} = \varphi_\alpha \circ \varphi_\beta^{-1} : \mathbb{R}^n \to \mathbb{R}^n.$$

$\varphi_{\alpha\beta}$ is also a homeomorphism because it is the composition of two homeomorphisms. A better description of these homeomorphisms may be illustrated more clearly as

$$\varphi_{\alpha\beta} : \varphi_\alpha(U_\alpha \cap U_\beta) \to \varphi_\beta(U_\alpha \cap U_\beta),$$
$$\varphi_{\alpha\beta}^{-1} : \varphi_\beta(U_\alpha \cap U_\beta) \to \varphi_\alpha(U_\alpha \cap U_\beta).$$

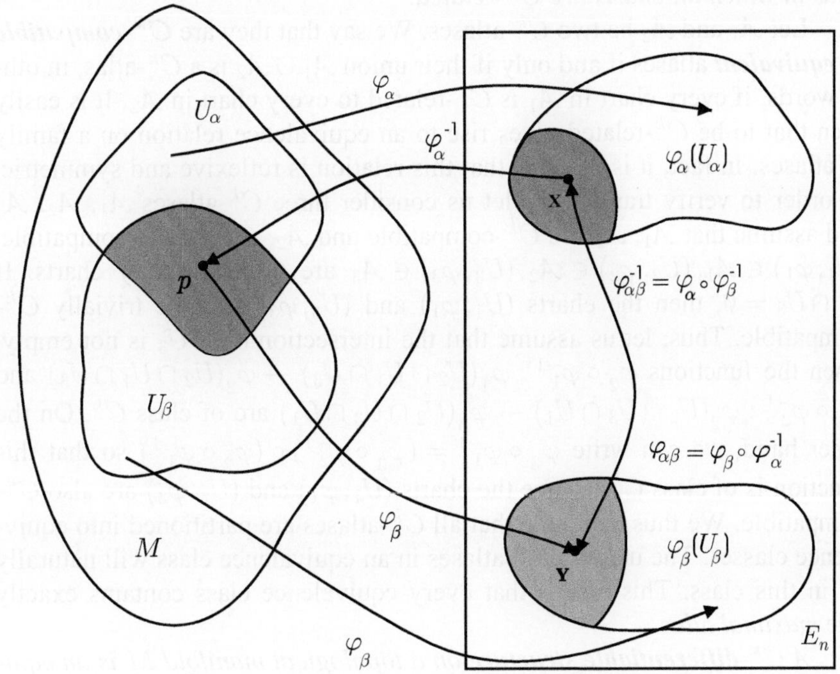

Fig. 2.2.3. Overlapping charts on a manifold M.

Let us denote the coordinates in charts $(U_\alpha, \varphi_\alpha)$ and (U_β, φ_β) by $\{x^i\}$ and $\{y^i\}$, respectively. Then, the transition mapping $\varphi_{\alpha\beta}$ leads to a relation between images $\mathbf{x}$ and $\mathbf{y}$ of the same point $p \in M$ with respect to two overlapping charts in the form $\mathbf{y} = \varphi_{\alpha\beta}(\mathbf{x}) \in \mathbb{R}^n$ that can be expressed as

$$y^i = \phi^i_{\alpha\beta}(x^j),\ i,j = 1,2,\dots n;\ \mathbf{x} \in \varphi_\alpha(U_\alpha \cap U_\beta) \tag{2.2.6}$$

Naturally, the mapping $\varphi^{-1}_{\alpha\beta}$ yields the inverse relation:

$$x^i = \psi^i_{\alpha\beta}(y^j),\ i,j = 1,2,\dots n;\ \mathbf{y} \in \varphi_\beta(U_\alpha \cap U_\beta). \tag{2.2.7}$$

The foregoing relations corresponds clearly to a coordinate transformation on the open set $U_\alpha \cap U_\beta$. We know that partial derivatives are defined on $\mathbb{R}^n$. We say that the charts $(U_\alpha, \varphi_\alpha)$ and (U_β, φ_β) are ***C^k-compatible*** if the functions $\phi^i_{\alpha\beta}$ are continuously differentiable of order k or they are of class C^k. This of course means that they have continuous partial derivatives with respect to all variables x^j up to and including order k. Two charts are ***C^k-related*** if either they are C^k-compatible or $U_\alpha \cap U_\beta = \emptyset$. *A C^k-atlas is an atlas in which all charts are C^k-related.*

Let $\mathcal{A}_1$ and $\mathcal{A}_2$ be two C^k-atlases. We say that they are ***C^k-compatible*** or ***equivalent*** atlases if and only if their union $\mathcal{A}_1 \cup \mathcal{A}_2$ is a C^k-atlas, in other words, if every chart in $\mathcal{A}_1$ is C^k-related to every chart in $\mathcal{A}_2$. It is easily seen that to be C^k-related gives rise to an equivalence relation on a family of atlases. In fact, it is obvious that this relation is reflexive and symmetric. In order to verify transitivity, let us consider three C^k-atlases $\mathcal{A}_1$, $\mathcal{A}_2$, $\mathcal{A}_3$ and assume that $\mathcal{A}_1$, $\mathcal{A}_2$ are C^k-compatible and $\mathcal{A}_2$, $\mathcal{A}_3$ are C^k-compatible. $(U_1, \varphi_1) \in \mathcal{A}_1, (U_2, \varphi_2) \in \mathcal{A}_2, (U_3, \varphi_3) \in \mathcal{A}_3$ are three arbitrary charts. If $U_1 \cap U_3 = \emptyset$, then the charts (U_1, φ_1) and (U_3, φ_3) become trivially C^k-compatible. Thus, let us assume that the intersection $U_1 \cap U_3$ is not empty. Then the functions $\varphi_2 \circ \varphi_1^{-1} : \varphi_1(U_2 \cap U_1 \cap U_3) \to \varphi_2(U_2 \cap U_1 \cap U_3)$ and $\varphi_3 \circ \varphi_2^{-1} : \varphi_2(U_2 \cap U_3 \cap U_1) \to \varphi_3(U_2 \cap U_3 \cap U_1)$ are of class C^k. On the other hand, we can write $\varphi_3 \circ \varphi_1^{-1} = (\varphi_3 \circ \varphi_2^{-1}) \circ (\varphi_2 \circ \varphi_1^{-1})$ so that this function is of class C^k. Hence the charts (U_1, φ_1) and (U_3, φ_3) are also C^k-compatible. We thus conclude that all C^k-atlases are partitioned into equivalence classes. The union of all atlases in an equivalence class will naturally be in this class. This means that every equivalence class contains exactly one *maximal atlas*.

*A C^k-**differentiable structure** on a topological manifold M is an equivalence class of C^k-atlases.* We can also say that *a C^k-**differentiable structure** on a topological manifold M is the choice of a maximal C^k-atlas. A **C^k-differentiable manifold** is a topological manifold equipped with a C^k-differentiable structure.*

If real-valued functions (2.2.6) and (2.2.7) with real variables have continuous derivatives of all orders, we obtain a C^∞-atlas and C^∞-differentiable manifold. A C^∞-differentiable manifold will also be called a ***smooth***

manifold. If the coordinate transformations are real analytical functions, that is, if they are expressible as convergent power series, then we get an ***analytical manifold*** or a C^ω-class manifold. It is evident that every C^ω-function is also a C^∞-function. But we know that the converse statement is generally not true. We can thus write symbolically $1 \leq k \leq m \leq \infty < \omega$. A C^m-differentiable structure $\mathcal{A}$ prescribed on a manifold M determines a unique C^k-differentiable structure on M for $k \leq m$. In order to see this it suffices to enlarge the set of admissible charts by adding all charts which are C^k-related with charts in $\mathcal{A}$ to the structure $\mathcal{A}$. Conversely, we can ask this question: when we are given a C^k-differentiable structure, is it possible to obtain a C^m-atlas for $m \geq k$ by discarding some charts? The answer to this question is provided by the following classical theorem whose proof we avoid to give because it is quite long and rather difficult.

Theorem 2.2.1 (Whitney's theorem). *Every C^k-structure with $k \geq 1$ on a second countable topological manifold is C^k-equivalent to a C^ω-structure.* □

This theorem means that if we locally make a coordinate transformation $y^i = f^i(x^j)$ of class C^k on an n-dimensional second countable topological manifold, there exist such functions $z^i = g^i(y^j)$ of C^k-class that the composition $z^i = g^i(f^j(x^m))$ is of C^ω-class, that is, they are analytical functions.

This theorem had been proven by Whitney[2]. That a C^0-manifold cannot be equivalent to a C^1-manifold can be shown through a more difficult theorem. According to the Whitney theorem we can choose all second countable or separable differentiable manifolds as analytical manifolds without loss of generality. However, it is not very comfortable to work with C^ω-functions as it is with C^∞-functions. Therefore, it will prove to be more advantageous to consider smooth manifolds. Henceforth, unless stated otherwise we take only *smooth manifolds* into consideration.

It is possible to extend above definitions to infinite dimensional manifolds. However, for this purpose we have the replace the Euclidean space by a Banach space, that is, by a complete normed space. In this case a chart $(U_\alpha, \varphi_\alpha)$ implies that the homeomorphism φ_α maps an open set U_α of the manifold M to an open subset V of a Banach space $\mathcal{V}$ such that $\varphi_\alpha(p) = v \in V$ where $p \in M$. The differentiable structure is now defined by Fréchet differentiability of the transition function $\varphi_{\alpha\beta} = \varphi_\beta \circ \varphi_\alpha^{-1} : V \to V$ on the overlapping domain $U_\alpha \cap U_\beta$ of the homeomorphisms φ_α and φ_β. A brief definition of the Fréchet derivative is given below.

Let $\mathcal{U}$ and $\mathcal{V}$ be Banach spaces and let $T : \mathcal{U} \to \mathcal{V}$ be a possibly non-

[2]Whitney, H., Differentiable manifolds, *Ann. of Math.* **37**, 645-680, 1936.

linear operator. Suppose that $\Omega = \mathcal{D}(T) \subseteq \mathcal{U}$ is an open set. If a *continuous linear operator* $T'(u) \in \mathcal{U} \to \mathcal{V}$ exists at a vector $u \in \Omega$ such that

$$\lim_{\|\Delta u\| \to 0} \frac{\|T(u + \Delta u) - T(u) - T'(u)\Delta u\|}{\|\Delta u\|} = 0$$

for all vectors $\Delta u \in \mathcal{U}$, then $T'(u)$ is called the ***Fréchet derivative of the operator*** $\boldsymbol{T}$ ***at a vector*** $\boldsymbol{u}$. $T'(u)$ depends possibly nonlinearly on the vector u. This derivative was introduced by French mathematician Maurice René Fréchet (1878-1973) in 1925. The domain of the operator T' contains naturally all vectors in $\mathcal{U}$ at which the Fréchet derivative of T can be defined. The above definition amounts to say clearly that for each $\epsilon > 0$, there exists a number $\delta(\epsilon) > 0$ such that

$$\frac{\|T(u + \Delta u) - T(u) - T'(u)\Delta u\|}{\|\Delta u\|} < \epsilon$$

or

$$\|T(u + \Delta u) - T(u) - T'(u)\Delta u\| < \epsilon \|\Delta u\|$$

for all $\Delta u \in U$ satisfying the condition $\|\Delta u\| < \delta$. It is then straightforward to see that the following relation is valid:

$$T(u + w) - T(u) = T'(u)(w) + \omega(u; w), \quad \lim_{\|w\| \to 0} \frac{\|\omega(u; w)\|}{\|w\|} = 0.$$

We thus conclude that the existence of the Fréchet derivative at a vector u brings about the possibility of evaluating the vector $T(u + w) - T(u)$ approximately through a continuous linear operator for all vectors w with sufficiently small norms.

It is straightforward to see that the Fréchet derivative may also be expressible in the form

$$T'(u)(w_1) = \lim_{t \to 0} \frac{T(u + tw_1) - T(u)}{t}.$$

By following exactly the same procedure we have employed in evaluating the Fréchet derivative of T, we can of course define the Fréchet derivative of the operator $T'(u)$ as

$$T''(u)(w_1, w_2) = \lim_{t \to 0} \frac{T'(u + tw_2)(w_1) - T'(u)(w_1)}{t}$$

for all $w_1, w_2 \in \mathcal{U}$. If this derivative exists, then the operator $T''(u)$ is called the second Fréchet derivative of T at u. This operator must be linear in each vector w_1 and w_2. This approach permits us to define higher order derivatives as well. Let us suppose that the $(k-1)$th order Fréchet derivative $T^{(k-1)}(u)$ is known. Then the kth order Fréchet derivative can be similarly defined as follows

$$\begin{aligned}T^{(k)}(u)(w_1, w_2, \ldots, w_k)\\ = \lim_{t\to 0} \frac{T^{(k-1)}(u+tw_k)(w_1,\ldots,w_{k-1}) - T^{(k-1)}(u)(w_1,\ldots,w_{k-1})}{t}\end{aligned}$$

for all ordered sets of vectors $w_1, w_2, \ldots, w_k \in \mathcal{U}$. Evidently, the operator $T^{(k)}(u) : \mathcal{U}^k \to \mathcal{V}$ is an k-linear function, that is, it is linear in each vector $w_i \in \mathcal{U}, i = 1, \ldots, k$. We can immediately extract from the definition that the operator $T^{(k)}(u)$ may be formally expressed in the following form

$$T^{(n)}(u)(\boldsymbol{w}) = \frac{\partial^n}{\partial t_1 \partial t_2 \cdots \partial t_n} T(u + t_1 w_1 + t_2 w_2 + \cdots + t_k w_k)\Big|_{t_1=t_2=\cdots=t_k=0}$$

where $\boldsymbol{w} = (w_1, w_2, \ldots, w_k)$.

In this work, we shall always deal with finite-dimensional manifolds.

Open Submanifold. Let U be an *open subset* of a differentiable manifold M with a differentiable structure. We can define a differentiable structure on U by

$$\mathcal{A}_U = \{(U \cap U_\alpha, \varphi_\alpha|_{U\cap U_\alpha}) : (U_\alpha, \varphi_\alpha) \in \mathcal{A}\}.$$

since $U \cap U_\alpha$ are open sets covering U. It is clearly seen that the open set U endowed with this structure becomes itself a differentiable manifold called an ***open submanifold*** of M. Since the same homeomorphism is utilised, *this open submanifold has evidently the same dimension as the manifold M.*

Product Manifolds. Let us consider two differentiable manifolds M of dimension m and N of dimension n. We choose, respectively, atlases $\mathcal{A}_M = \{(U_\alpha, \varphi_\alpha) : \alpha \in \mathcal{I}\}$ and $\mathcal{A}_N = \{(V_\beta, \psi_\beta) : \beta \in \mathcal{J}\}$ from the differentiable structures of these manifolds. The set $M \times N$ of the Cartesian product of these manifolds can now be equipped with a structure of an $(m+n)$-dimensional differentiable manifold by choosing the topology on $M \times N$ as the *product topology* and by introducing an atlas in the form $\mathcal{A}_{MN} = \{(U_\alpha \times V_\beta, \omega_{\alpha\beta}) : (\alpha, \beta) \in \mathcal{I} \times \mathcal{J}\}$. Here, the mapping $\omega_{\alpha\beta}$ is identified by

$$\omega_{\alpha\beta} : U_\alpha \times V_\beta \to \varphi_\alpha(U_\alpha) \times \psi_\beta(V_\beta) \subset \mathbb{R}^m \times \mathbb{R}^n = \mathbb{R}^{m+n}.$$

Thus, if $p \in U_\alpha \subset M$ and $q \in V_\beta \subset N$, then we have to write

$$\omega_{\alpha\beta}(p,q) = (\varphi_\alpha(p), \psi_\beta(q)) = (\mathbf{x}, \mathbf{y}) \in \varphi_\alpha(U_\alpha) \times \psi_\beta(V_\beta) \subset \mathbb{R}^{m+n}$$

where $\mathbf{x} = \varphi_\alpha(p)$ and $\mathbf{y} = \psi_\beta(q)$.

We now consider some samples of manifolds.

Example 2.2.1. Cartesian Spaces. The *standard* manifold structure on the Euclidean space $\mathbb{R}^n$ is prescribed by an atlas including a single chart $(\mathbb{R}^n, i_{\mathbb{R}})$ where $i_{\mathbb{R}} : \mathbb{R}^n \to \mathbb{R}^n$ is the identity mapping. Coordinate functions φ^i, $i = 1, \dots, n$ are just Cartesian coordinates $\{x^i : i = 1, \dots, n\}$ of a point of $\mathbb{R}^n$. As a differentiable manifold $\mathbb{R}^n$ is called the ***affine space***.

The space $\mathbb{R}$ acquires a manifold structure with the single chart $(\mathbb{R}, \varphi_1)$ where $\varphi_1 : \mathbb{R} \to \mathbb{R}$ is given by $\varphi_1(x) = x$. Similarly if we replace φ_1 by $\varphi_2(x) = x^3$, then $\mathbb{R}$ becomes a manifold with the chart $(\mathbb{R}, \varphi_2)$. But these two atlases are not compatible, because the mapping $\varphi_{12} : \mathbb{R} \to \mathbb{R}$ given by $\varphi_{12}(x) = \varphi_1 \circ \varphi_2^{-1}(x) = x^{1/3}$ is not differentiable at the point $x = 0$. ■

We can observe at once that every open subset of $\mathbb{R}^n$ is an n-dimensional manifold. Furthermore, we can easily show that an *n-dimensional connected manifold is equivalent to an open submanifold of $\mathbb{R}^n$ if and only if its atlas contains only a single chart.* Indeed, if the entire manifold M is homeomorphic to a single open set of the space $\mathbb{R}^n$, then its atlas has only one chart. Conversely if the atlas of an n-dimensional manifold M has only one chart, then the entire space M is homeomorphic to a single open submanifold of $\mathbb{R}^n$.

Example 2.2.2. Finite-Dimensional Vector Spaces. Let V be an n-dimensional real vector space equipped with an arbitrary norm. We choose a set of basis vectors by $(e_1, e_2, \dots, e_n)$. Then any vector $v \in V$ is expressed as $v = x^1 e_1 + x^2 e_2 + \cdots + x^n e_n$ where $x^i \in \mathbb{R}$, $i = 1, \dots, n$. If we denote $x = (x^1, x^2, \dots, x^n) \in \mathbb{R}^n$, it becomes obvious that there is an isomorphism and hence a linear homeomorphism $\varphi : V \to \mathbb{R}^n$ such that $x = \varphi(v)$. It then follows that V is also an n-dimensional smooth manifold since $\mathbb{R}^n$ is a smooth manifold. Evidently this property is independent of the choice of the basis in V. As a concrete example to finite-dimensional vector spaces, let us consider the set of $m \times n$ matrices defined on real numbers. According to the rule of matrix addition and scalar multiplication, this set is an mn - dimensional vector space. Indeed, we can write any member of this set in the form $\mathbf{M} = a^{\alpha i}\mathbf{M}_{\alpha i}, \alpha = 1, \dots, m, i = \ , \dots, n$ where the matrix $\mathbf{M}_{\alpha i}$ has 1 in its row α and its column i while all other entries are 0. These mn linearly independent matrices $\mathbf{M}_{\alpha i}$ constitute a basis for this vector space. This vector space is isomorphic to the space $\mathbb{R}^{mn}$ whose points are identified by elements $(a^{11}, \dots a^{1n}, \dots, a^{m1}, \dots, a^{mn})$ of matrices. Hence, such matrices constitute an mn-dimensional smooth manifold.

We denote the set of $n \times n$ real square matrices by $gl(n, \mathbb{R})$. $gl(n, \mathbb{R})$

is a smooth manifold of n^2-dimension. Let us consider the subset $GL(n,\mathbb{R})$ of regular matrices of $gl(n,\mathbb{R})$. This set is called the ***general linear group***. Let $\det : gl(n,\mathbb{R}) \to \mathbb{R}$ be the determinant function. In this case the general linear group is expressed as the following set difference:

$$GL(n,\mathbb{R}) = gl(n,\mathbb{R}) - \det{}^{-1}\{0\}.$$

Since the determinant is a continuous function and the singleton $\{0\} \in \mathbb{R}$ is a closed set, then $\det^{-1}\{0\} \in gl(n,\mathbb{R})$ is a closed set. Thus $GL(n,\mathbb{R})$ is an open set, that is, it is an open submanifold of the manifold $gl(n,\mathbb{R})$.

We can obtain similar results on matrices defined on the field of complex numbers. But, a complex number is given by two real numbers. Consequently, the dimension of the real manifold to which space of matrices is homeomorphic becomes twice as much. For instance, the ***general linear group*** $GL(n,\mathbb{C})$ of regular $n \times n$ complex matrices is a smooth manifold of $2n^2$-dimension. ■

Example 2.2.3. The Sphere in $\mathbb{R}^3$. We consider a spherical surface of radius R in $\mathbb{R}^3$. Any point $P(x,y,z)$ of this 2-dimensional surface $\mathbb{S}^2$ can be written in the form

$$\begin{aligned} x &= R\sin\theta\cos\phi, \quad 0 \le \theta \le \pi, \quad 0 \le \phi \le 2\pi, \\ y &= R\sin\theta\sin\phi, \\ z &= R\cos\theta \end{aligned}$$

by employing spherical coordinates (θ,ϕ). By defining $x^1 = \theta$ and $x^2 = \phi$ we can determine a function $\varphi_1 : \mathbb{S}^2 \to \mathbb{R}^2$ mapping $\mathbb{S}^2$ on the region $[0,\pi] \times [0,2\pi]$ of $\mathbb{R}^2$ (Fig. 2.2.4).

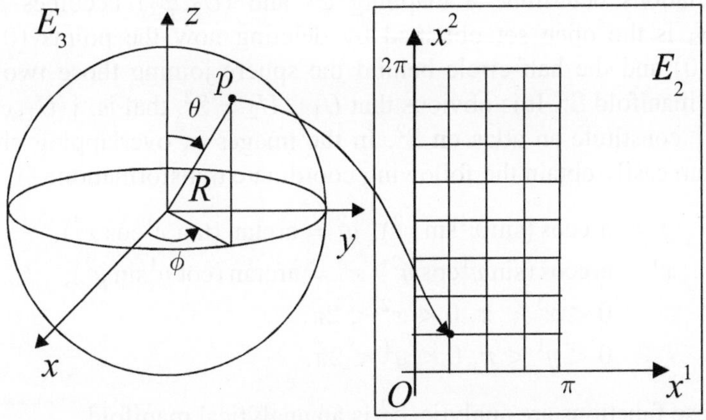

Fig. 2.2.4. 2-dimensional sphere.

It is straightforward to find the inverse function φ_1^{-1}:

$$x^1 = \arctan \frac{\sqrt{x^2 + y^2}}{z}, \quad x^2 = \arctan \frac{y}{x}$$

Unfortunately, φ_1 is not a homeomorphism on the entire sphere. The poles $(0, 0, R)$ and $(0, 0, -R)$ of the sphere ($\theta = 0$ and $\theta = \pi$, respectively) are mapped onto sets $\{0\} \times [0, 2\pi]$ and $\{\pi\} \times [0, 2\pi]$ in E_2. Furthermore, the image of a point on the half-circle $\phi = 0$ are two points on the lines $x^2 = 0$ and $x^2 = 2\pi$. Hence, on this set φ_1 is not even a function. In order to make the mapping φ_1 a homeomorphism, we exclude from the set $\mathbb{S}^2$ the poles $(0, 0, R)$, $(0, 0, -R)$ and the half-circle $\phi = 0$ joining them. Thus we have to choose $\theta \in (0, \pi)$ and $\phi \in (0, 2\pi)$ and to restrict φ_1^{-1} on the open set $(0, \pi) \times (0, 2\pi)$ in $\mathbb{R}^2$, in other words, we have to take $0 < x^1 < \pi$, $0 < x^2 < 2\pi$. It is now evident that the set

$$U_1 = \mathbb{S}^2 - \{(0, 0, R)\} \cup \{(0, 0, -R)\} \cup \{\phi = 0\} \subset \mathbb{S}^2$$

is open since it is the homeomorphic image of the open set $(0, \pi) \times (0, 2\pi)$. Consequently (U_1, φ_1) is a chart but it cannot cover the entire manifold $\mathbb{S}^2$. This result should be expected because the sphere $\mathbb{S}^2$ is a closed and bounded subset of the manifold $\mathbb{R}^3$. In order to find another chart, let us choose now the point $(0, R, 0)$ as a pole of the sphere. As above, we write

$$\begin{aligned} x &= R \sin y^1 \sin y^2, \quad 0 < y^1 < \pi, \quad 0 < y^2 < 2\pi, \\ y &= R \cos y^1, \\ z &= R \sin y^1 \cos y^2. \end{aligned}$$

These relations determine a mapping φ_2 and (U_2, φ_2) becomes a chart where U_2 is the open set obtained by deleting now the points $(0, R, 0)$, $(0, -R, 0)$ and the half-circle behind the sphere joining those two points from the manifold $\mathbb{S}^2$. It is obvious that $U_1 \cup U_2 = \mathbb{S}^2$, that is, $\{(U_1, \varphi_1)$ and $(U_2, \varphi_2)\}$ constitute an atlas on $\mathbb{S}^2$. In the images of overlapping charts in $\mathbb{R}^2$, we can easily obtain the following coordinate transformation:

$$\begin{aligned} y^1 &= \arccos(\sin x^1 \sin x^2), \; y^2 = \arctan(\tan x^1 \cos x^2); \\ x^1 &= \arccos(\sin y^1 \cos y^2), \; x^2 = \arctan(\cot y^1 \sin y^2); \\ &0 < x^1 < \pi, \; 0 < x^2 < 2\pi; \\ &0 < y^1 < \pi, \; 0 < y^2 < 2\pi. \end{aligned}$$

Since these functions are analytic, $\mathbb{S}^2$ is an analytical manifold. ■

Example 2.2.4. The Sphere in $\mathbb{R}^{n+1}$. Let us consider the n-dimensional spherical hypersurface $\mathbb{S}^n$ with radius R in $\mathbb{R}^{n+1}$. If we denote the

Cartesian coordinates in $\mathbb{R}^{n+1}$ by $(x_0, x_1, \ldots, x_n)$ then the set $\mathbb{S}^n$ is determined by the equation

$$x_0^2 + x_i x_i = R^2,\ i = 1, 2, \ldots, n.$$

We choose the pole k of $\mathbb{S}^n$ as the point $(R, 0, \ldots, 0)$. We specify the subspace E_n by the condition $x_0 = 0$. To describe the mapping $\varphi_1 : \mathbb{S}^n \to \mathbb{R}^n$, we impose that the image point $q = \varphi_1(p)$ of a point $p \in \mathbb{S}^n$ in $\mathbb{R}^n$ is the point of intersection of the straight line joining the pole k and the point p with the *hyperplane* $\mathbb{R}^n$ (Fig. 2.2.5). This mapping is known as ***stereographic projection***. If the coordinates of the point p are (x_0, x_i), then the relation $x_0 = \mp\sqrt{R^2 - x_i x_i}$ must be satisfied. Let the *unit* basis vectors in $\mathbb{R}^{n+1}$ be $\{e_0, e_i\}$. The vector e_0 is in the direction $\overrightarrow{Ok}$, while basis vectors in $\mathbb{R}^n$ are $e_i, i = 1, \ldots, n$. Let us denote the coordinates of the point $q \in \mathbb{R}^n$ at which the line joining the points k and p intersects the space $\mathbb{R}^n$ by $\mathbf{y} = (y_1, y_2, \ldots, y_n)$. Thus, we can write

$$Re_0 + \lambda\big[(x_0 - R)e_0 + x_i e_i\big] = y_i e_i$$

where λ is a real parameter. Then, it follows that

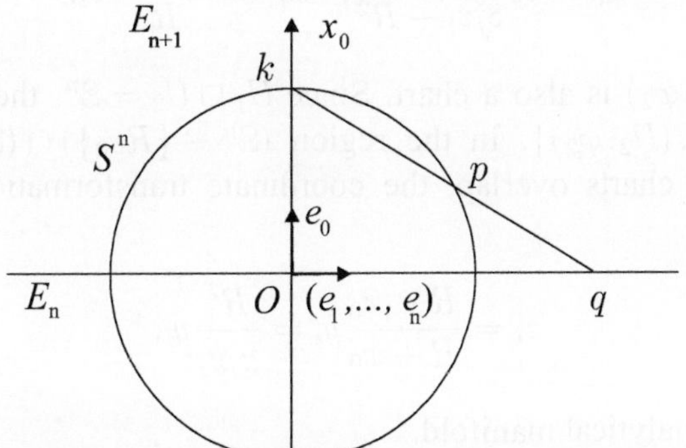

Fig. 2.2.5. Stereographic projection for an n-dimensional sphere.

$$\lambda = \frac{R}{R - x_0}, \quad y_i = \frac{Rx_i}{R - x_0}, \quad y_i y_i = R^2 \frac{R + x_0}{R - x_0}.$$

Hence, the points on $\mathbb{S}^n$ with the same elevation x_0 form now an $(n-1)$-dimensional sphere $\mathbb{S}^{n-1}$ in $\mathbb{R}^n$. The radius of that sphere is of course given by $R\sqrt{(R + x_0)/(R - x_0)}$. It is greater than R if $0 < x_0 < R$ and less than R if $-R < x_0 < 0$. Let $\mathbf{x} = (x_0, x_1, \ldots, x_n) \in \mathbb{S}^n$, then the above relations prescribe a mapping $\varphi_1 : \mathbb{S}^n \to \mathbb{R}^n$ where $\varphi_1(\mathbf{x}) = \mathbf{y} \in \mathbb{R}^n$. The inverse mapping $\varphi_1^{-1} : \mathbb{R}^n \to \mathbb{S}^n$ is easily found to be

$$\mathbf{x} = \varphi_1^{-1}(\mathbf{y}), \quad x_0 = R\frac{y_i y_i - R^2}{y_j y_j + R^2}, \quad x_i = \frac{R - x_0}{R} y_i.$$

We can immediately observe that φ_1 is not a homeomorphism on the entire $\mathbb{S}^n$. Indeed, the pole k determined by $x_0 = R$, $x_i = 0$ is mapped on a "set of infinities" in $\mathbb{R}^n$ under φ_1. We can simply observe that φ_1 becomes a homeomorphism if we delete the single point k from $\mathbb{S}^n$. Thus (U_1, φ_1) is a chart where $U_1 = \mathbb{S}^n - \{Re_0\}$ is an open set. We can next introduce another chart by choosing the point $\{- Re_0\}$ as another pole of $\mathbb{S}^n$ and by defining the function $\varphi_2 : U_2 \to \mathbb{R}$ as follows

$$\mathbf{z} = \varphi_2(\mathbf{x}), \quad z_i = \frac{Rx_i}{R + x_0}, \quad z_i z_i = R^2 \frac{R - x_0}{R + x_0}$$

where $\mathbf{z} = (z_1, \dots, z_n) \in \mathbb{R}^n$ and $U_2 \subset \mathbb{S}^n$ is the open set $\mathbb{S}^n - \{- Re_0\}$. The inverse mapping $\varphi_2^{-1} : \mathbb{R}^n \to U_2$ is easily provided by the following relations

$$x_0 = R\frac{R^2 - z_i z_i}{z_j z_j + R^2}, \quad x_i = \frac{R + x_0}{R} z_i.$$

Obviously (U_2, φ_2) is also a chart. Since $U_1 \cup U_2 = \mathbb{S}^n$, then we have the atlas $\{(U_1, \varphi_1), (U_2, \varphi_2)\}$. In the region $(\mathbb{S}^n - \{Re_0\}) \cap (\mathbb{S}^n - \{-Re_0\})$ where the two charts overlap, the coordinate transformation $\varphi_2 \circ \varphi_1^{-1}$ is found to be

$$z_i = \frac{R - x_0}{R + x_0} y_i = \frac{R^2}{y_j y_j} y_i.$$

Thus $\mathbb{S}^n$ is an analytical manifold. ■

Example 2.2.5. Torus. We denote the surface of a 2-torus in $\mathbb{R}^3$ by $\mathbb{T}^2$. This surface is obtained, for instance, by rotating a circle with radius b whose distance of its centre from z-axis is a about that axis (Fig. 2.2.6). We can thus write $\mathbb{T}^2 = \mathbb{S}^1 \times \mathbb{S}^1 = (\mathbb{S}^1)^2$ as a product manifold. The manifold $\mathbb{S}^1$ represents a 1-dimensional sphere, namely, a circle. Thus, one-dimensional torus $\mathbb{T}^1$ is just the circle. In view of Example 2.2.3, the manifold $\mathbb{S}^1$ has an atlas with two charts homeomorphic to $\mathbb{R}^1$. In this case we expect that the product manifold $\mathbb{T}^2$ will have an atlas with four charts homeomorphic to open subsets of $\mathbb{R}^2$.

On the other hand, a torus may be determined parametrically in $\mathbb{R}^3$ by the following relations

$$x = (a + b \sin\theta)\cos\phi,$$
$$y = (a + b \sin\theta)\sin\phi,$$
$$z = b\cos\theta$$

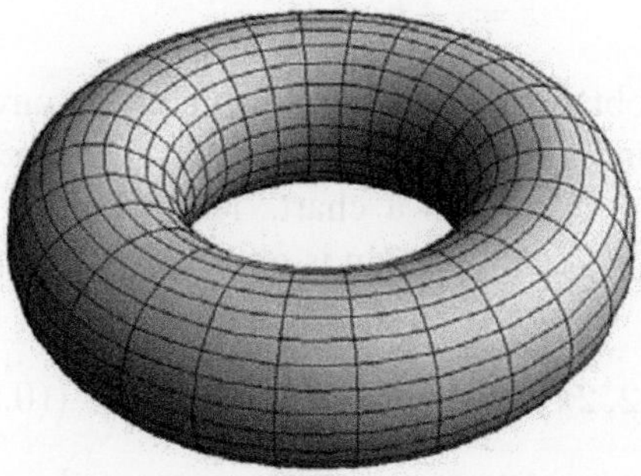

Fig. 2.2.6. 2-dimensional torus.

where the condition $b < a$ should be satisfied. The parameters ϕ and θ measure the angles along small and large circles. If we write $x^1 = \theta$, $x^2 = \phi$ these relations define a mapping $\varphi_1^{-1} : \mathbb{R}^2 \to \mathbb{T}^2$. But to render this mapping injective we have to restrict its domain to an open set in $\mathbb{R}^2$ prescribed by inequalities $0 < x^1 < 2\pi$, $0 < x^2 < 2\pi$. Let U_1 be the open set obtained by deleting from $\mathbb{T}^2$ the circle with radius a at the plane $z = b$ and the circle with radius b at the xz-plane centred at the point $x = a, z = 0$. (U_1, φ_1) then becomes a chart. We define a new mapping φ_2 by

$$x = -(a + b \sin y^1)\sin y^2,$$
$$y = (a + b \sin y^1)\cos y^2,$$
$$z = b\cos y^1.$$

Let U_2 be the open set obtained by deleting from $\mathbb{T}^2$ the circle with radius a at the plane $z = b$ and the circle with radius b at the yz-plane centred at the point $y = a, z = 0$. It is straightforward to see that (U_2, φ_2) is now a chart. The region $U_1 \cap U_2$ in which two charts overlap is the union of two open sets V_1 and V_2 that are *disconnected* where

$$V_1 = \varphi_1^{-1}\big((0, 2\pi) \times (\pi/2, 2\pi)\big),\ V_2 = \varphi_1^{-1}\big((0, 2\pi) \times (0, \pi/2)\big).$$

There are analytical coordinate transformations $y^1 = x^1$, $y^2 = x^2 - \frac{\pi}{2}$ on V_1

and $y^1 = x^1$, $y^2 = x^2 + \frac{3\pi}{2}$ on V_2. Finally, let us consider the mapping φ_3 given by

$$\begin{aligned} x &= (a + b\cos z^1)\cos z^2, \\ y &= (a + b\cos z^1)\sin z^2, \\ z &= \ -b\sin z^1. \end{aligned}$$

The open set U_3 is obtained by deleting from $\mathbb{T}^2$ the circle with radius $a+b$ at the plane $z=0$ and the circle with radius b at the xz-plane centred at the point $x=a, z=0$. (U_3, φ_3) is a chart. The region $U_1 \cap U_3$ in which the charts (U_1, φ_1) and (U_3, φ_3) overlap is obviously the union of two open sets W_1 and W_2 that are *disconnected* where

$$W_1 = \varphi_1^{-1}\big((\pi/2, 2\pi) \times ((0, 2\pi))\big), \ W_2 = \varphi_1^{-1}\big((0, \pi/2) \times (0, 2\pi)\big)$$

There are analytical coordinate transformations $z^1 = x^1 - \frac{\pi}{2}$, $z^2 = x^2$ on W_1 and $z^1 = x^1 + \frac{3\pi}{2}$, $z^2 = x^2$ on W_2. The charts (U_2, φ_2) and (U_3, φ_3) overlap on $U_2 \cap U_3$ which is the union of open sets Z_1, Z_2, Z_3 and Z_4. These sets are given by

$$\begin{aligned} Z_1 &= \varphi_2^{-1}\big((0, \pi/2) \times (0, 3\pi/2)\big), \ Z_2 = \varphi_2^{-1}\big((0, \pi/2) \times (3\pi/2, 2\pi)\big) \\ Z_3 &= \varphi_2^{-1}\big((\pi/2, 2\pi) \times (0, 3\pi/2)\big), Z_4 = \varphi_2^{-1}\big((\pi/2, 2\pi) \times (3\pi/2, 2\pi)\big). \end{aligned}$$

Analytical coordinate transformations on these four sets are determined by the following expressions, respectively

$$\begin{aligned} z^1 = y^1 + \frac{3\pi}{2}, z^2 = y^2 + \frac{\pi}{2};\ \ z^1 = y^1 + \frac{3\pi}{2}, z^2 = y^2 - \frac{3\pi}{2}; \\ z^1 = y^1 - \frac{\pi}{2},\ z^2 = y^2 + \frac{\pi}{2};\ \ z^1 = y^1 - \frac{\pi}{2}, z^2 = y^2 - \frac{3\pi}{2} \end{aligned}$$

Since $U_1 \cup U_2 \cup U_3 = \mathbb{T}^2$, we conclude that 2-torus has an analytical atlas with three charts $\{(U_1, \varphi_1), (U_2, \varphi_2), (U_3, \varphi_3)\}$.

An n-torus may be described in a similar fashion as a product manifold $\mathbb{T}^n = \mathbb{S}^1 \times \mathbb{S}^1 \times \cdots \times \mathbb{S}^1 = (\mathbb{S}^1)^n$. ■

Example 2.2.6. Klein Bottle $\mathbb{K}^2$. The Klein bottle is a 2-dimensional manifold in the space $\mathbb{R}^4$ [It was introduced in 1882 by German mathematician Felix Christian Klein (1849-1925)]. We denote the coordinates in $\mathbb{R}^4$ by (x, y, z, v). $\mathbb{S}^1$ is a circle with radius b at the xz-plane whose centre is the point $(a, 0, 0, 0)$. We assume that $a > b$. Klein bottle is produced by the following process: while turning the centre C of that circle about O in the xy-plane by an angle ϕ, we rotate its plane in 4-dimensional space about the axis OC that remains perpendicular to the zv-plane by an angle $\phi/2$

(Fig. 2.2.7). It can be shown that this operation is tantamount to first forming a cylindrical surface by gluing two mutual edges together of a rectangular strip, then trying to glue one edge of this cylinder to the other after giving a half-twist with respect to the other one. In 3-dimensional space this operation cannot be realised without intersecting the surface. Therefore, Klein bottle can be considered as a manifold only in a 4-dimensional space. It cannot be embedded into $\mathbb{R}^3$ since in such a mapping self-intersections should not be permissible. However, it is possible to immerse this surface into 3-dimensional space if we allow self-intersections [for properties of these sort of mappings see Sec. 2.4]. These immersions are found to be unfortunately not unique. Two different immersions is depicted in Fig. 2.2.8.

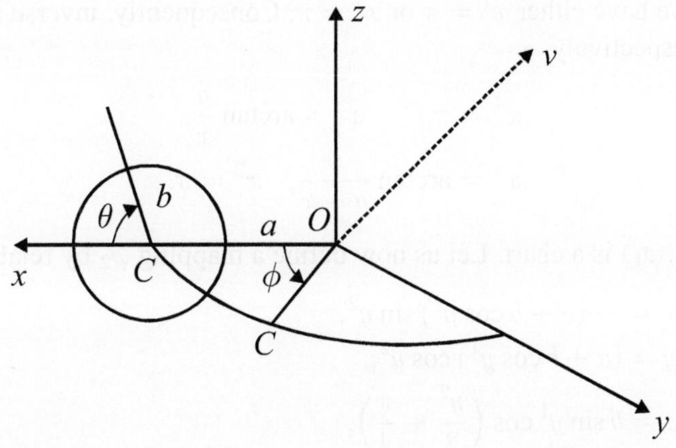

Fig. 2.2.7. Description of Klein bottle in 4-dimensional space.

It is now obvious that a point on Klein bottle is represented parametrically by equations

$$
\begin{aligned}
x &= (a + b\cos\theta)\cos\phi,\\
y &= (a + b\cos\theta)\sin\phi,\\
z &= b\sin\theta\cos(\phi/2),\\
v &= b\sin\theta\sin(\phi/2), \quad 0 \le \theta \le 2\pi,\ 0 \le \phi \le 2\pi
\end{aligned}
$$

When we eliminate these parameters, Klein bottle is given in Cartesian coordinates with the following relations

$$
\begin{aligned}
y(z^2 - v^2) - 2xzv = 0\\
x^2 + y^2 + z^2 + v^2 - 2a\sqrt{x^2 + y^2} + a^2 - b^2 = 0
\end{aligned}
$$

With $x^1 = \theta$, $x^2 = \phi$, these relations determine a mapping $\varphi_1 : \mathbb{K}^2 \to \mathbb{R}^2$. However, in order to render the mapping $\varphi_1^{-1} : \mathbb{R}^2 \to \mathbb{K}^2$ injective, we have to restrict its domain in $\mathbb{R}^2$ to the open set determined by the inequalities $0 < x^1 < 2\pi$, $0 < x^2 < 2\pi$. Hence the domain of φ_1 is the open set U_1 obtained by deleting from $\mathbb{K}^2$ the circles $\phi = 0$ given by $x - a = b\cos\theta$, $z = b\sin\theta$ and $\theta = 0$ given by $x = (a+b)\cos\phi$, $y = (a+b)\sin\phi$. Thus, the inverse mapping φ_1^{-1} is found as follows when $z \neq 0$

$$\sin x^1 = \frac{\sqrt{z^2+v^2}}{b}, \quad \cos x^1 = \frac{\sqrt{x^2+y^2}-a}{b},$$
$$x^2 = 2\arctan\frac{v}{z} = \arctan\frac{y}{x}.$$

If $z = 0$, we have either $x^1 = \pi$ or $x^2 = \pi$. Consequently, inverse mappings become, respectively

$$x^1 = \pi, \qquad x^2 = \arctan\frac{y}{x},$$
$$x^1 = \arctan\frac{v}{a-x}, \quad x^2 = \pi.$$

Hence (U_1, φ_1) is a chart. Let us now define a mapping φ_2 by relations

$$\begin{aligned}
x &= -(a + b\cos y^1)\sin y^2, \\
y &= (a + b\cos y^1)\cos y^2, \\
z &= b\sin y^1 \cos\left(\frac{y^2}{2} + \frac{\pi}{4}\right), \\
v &= b\sin y^1 \sin\left(\frac{y^2}{2} + \frac{\pi}{4}\right), \ 0 < y^1 < 2\pi, \ 0 < y^2 < 2\pi
\end{aligned}$$

where y^2 is representing now the angle in xy-plane measured from y-axis. We can easily observe that the mapping φ_2 is a homeomorphism on the open set U_2 obtained by deleting from $\mathbb{K}^2$ the circle with radius $a+b$ in xy-plane and the circle with radius b centred at $y = a$ and located on the bisecting plane of yz- and yv-planes. Hence, (U_2, φ_2) is a second chart and it contains the set $\{x^2 = 0\}$. We see that $U_1 \cap U_2 = V_1 \cup V_2$ where V_1 and V_2 are open disconnected sets given by

$$V_1 = \varphi_1^{-1}\big((0,2\pi) \times (\pi/2, 2\pi)\big), \quad V_2 = \varphi_1^{-1}\big((0,2\pi) \times (0, \pi/2)\big)$$

The coordinate transformation on V_1 is $y^1 = x^1$, $y^2 = x^2 - \frac{\pi}{2}$ whereas that on V_2 is $y^1 = 2\pi - x^1$, $y^2 = x^2 + \frac{3\pi}{2}$. Finally, let us define a mapping φ_3 by the relations

$$\begin{aligned}
x &= (a + b \sin z^1) \cos z^2, \\
y &= (a + b \sin z^1) \sin z^2, \\
z &= -\, b \cos z^1 \cos (z^2/2), \\
v &= -\, b \cos z^1 \sin (z^2/2),\ 0 < z^1 < 2\pi,\ 0 < z^2 < 2\pi
\end{aligned}$$

where z^1 now denotes the angle translated 90°. The open set U_3 is obtained by deleting from $\mathbb{K}^2$ the circle with radius b centred at the point $x = a$ in xz-plane and the circle with radius a in xy-plane and the circle with radius b in zv-plane both centred at the point O. It is obvious that (U_3, φ_3) is a chart and it contains the set $\{x^1 = 0\}$. We thus obtain $U_1 \cup U_2 \cup U_3 = \mathbb{K}^2$. In the same fashion one can show that coordinate transformations at the overlapping subsets of all these charts are simple analytical functions. Thus, Klein bottle $\mathbb{K}^2$ is an analytical manifold. ■

Example 2.2.7. Real Projective Spaces. Let us consider the space $\mathbb{R}^{n+1}$ whose origin $\mathbf{0} = (0, 0, \dots, 0)$ is deleted. A point of $\mathbb{R}^{n+1}$ is denoted by $\mathbf{x} = (x^1, x^2, \dots, x^{n+1})$. We define a relation R on the set $\mathbb{R}^{n+1} - \{\mathbf{0}\}$ by $\mathbf{x}R\mathbf{y}$ if and only $\mathbf{y} = \lambda\mathbf{x}$, $\lambda \in \mathbb{R} - \{0\}$, or $y^i = \lambda x^i, 1 \leq i \leq n+1$. It is straightforward to see that R is an equivalence relation. The n-dimensional real projective space $\mathbb{RP}^n$ is defined as the quotient space of the topological space $\mathbb{R}^{n+1} - \{\mathbf{0}\}$ with respect to this equivalence relation R: $\mathbb{RP}^n = (\mathbb{R}^{n+1} - \{\mathbf{0}\})/R$. It is clear that the elements of this space that are equivalence classes are straight lines through the origin $\mathbf{0}$ of $\mathbb{R}^{n+1}$. In this case, the canonical projection $\pi : \mathbb{R}^{n+1} - \{\mathbf{0}\} \to \mathbb{RP}^n$ [*see p.* 61] assigns to a non -zero point $\mathbf{x} \in \mathbb{R}^{n+1}$ the line through this point and the origin. Therefore, if we denote a point of the quotient space $\mathbb{RP}^n$ by the equivalence class $[\mathbf{x}] = [x^1, x^2, \dots, x^{n+1}]$, then for each $\lambda \in \mathbb{R}$, $\lambda \neq 0$ the equivalence class $[\lambda\mathbf{x}] = [\lambda x^1, \lambda x^2, \dots, \lambda x^{n+1}]$ specifies the same point, i.e., $[\lambda\mathbf{x}] = [\mathbf{x}]$. The numbers $x^1, x^2, \dots, x^{n+1}$ are called the *homogeneous coordinates* of the point $[\mathbf{x}]$. Employing those coordinates, we can represent the coordinates $\{\xi^1, \dots, \xi^n\}$ of a point in $\mathbb{R}^n$ by the ratios

$$\xi^1 = \frac{x^1}{x^{n+1}},\ \xi^2 = \frac{x^2}{x^{n+1}},\ \dots\ ,\xi^n = \frac{x^n}{x^{n+1}},\ x^{n+1} \neq 0.$$

As corresponding to a point $[\mathbf{x}]$ in the projective space, these coordinates are uniquely determined. We now want to equip the projective space by the quotient topology [*see p.* 62]. Let us choose the sets $U_i, i = 1, 2, \dots, n+1$ in the projective space as follows

$$U_i = \{[\mathbf{x}] \in \mathbb{RP}^n : x^i \neq 0\}.$$

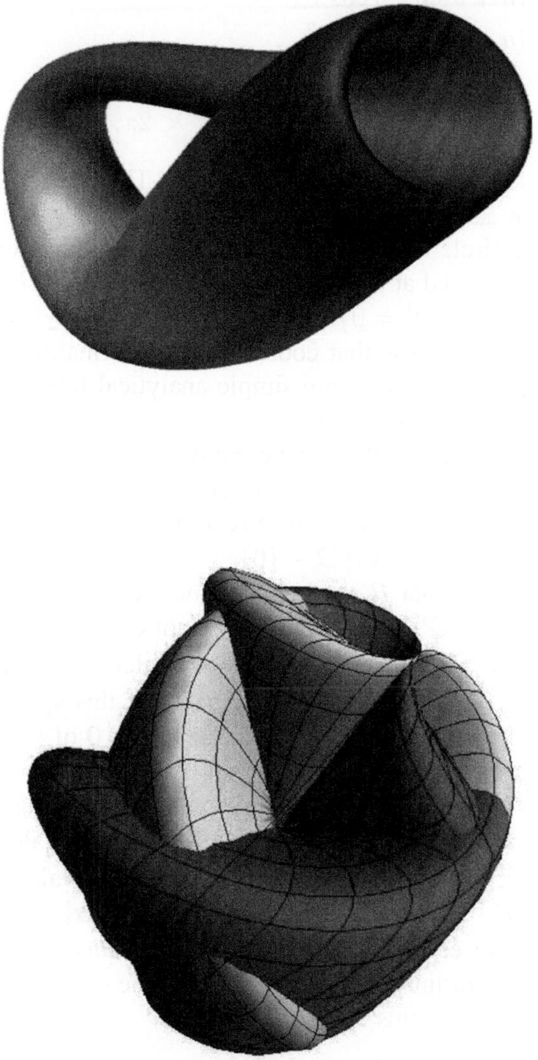

Fig. 2.2.8. Images of Klein bottle in $\mathbb{R}^3$ for two different immersions.

The set U_i consists clearly of the straight lines through the origin of the space $\mathbb{R}^{n+1}$ that do not belong to the n-dimensional subspace determined by the coordinates $(x^1, \dots, x^{i-1}, 0, x^{i+1}, \dots x^{n+1})$ except at the origin. Since the set

$$\pi^{-1}(U_i) = \{\mathbf{x} \in \mathbb{R}^{n+1} - \{\mathbf{0}\} : x^i \neq 0\} \subset \mathbb{R}^{n+1} - \{\mathbf{0}\}$$

is open, the set $U_i \subset \mathbb{RP}^n$ is also open in the quotient topology. Moreover, we see at once that $\bigcup_{i=1}^{n+1} U_i = \mathbb{RP}^n$. We define a mapping $\varphi_i : U_i \to \mathbb{R}^n$ by

$$\varphi_i([\mathbf{x}]) = \Big(\frac{x^1}{x^i}, \dots, \frac{x^{i-1}}{x^i}, \frac{x^{i+1}}{x^i}, \dots, \frac{x^{n+1}}{x^i}\Big), \quad [\mathbf{x}] \in U_i.$$

Evidently this mapping is a homeomorphism. Hence, (U_i, φ_i) is a chart and the collection $\{(U_i, \varphi_i) : i = 1, 2, \dots, n+1\}$ is an atlas for $\mathbb{RP}^n$. On the other hand, in the intersection $U_i \cap U_j$ where charts are overlapping the transition function is easily found to be

$$\begin{aligned}\varphi_j \circ \varphi_i^{-1}\Big(\frac{x^1}{x^i}, \dots, \frac{x^{i-1}}{x^i}, \frac{x^{i+1}}{x^i}, \dots, \frac{x^{n+1}}{x^i}\Big) &= \Big(\frac{x^1}{x^j}, \dots, \frac{x^{j-1}}{x^j}, \frac{x^{j+1}}{x^j}, \dots, \frac{x^{n+1}}{x^j}\Big) \\ &= \frac{x^i}{x^j}\Big(\frac{x^1}{x^i}, \dots, \frac{x^{j-1}}{x^i}, \frac{x^{j+1}}{x^i}, \dots, \frac{x^{n+1}}{x^i}\Big), \; x^i \neq 0, x^j \neq 0.\end{aligned}$$

Since transitions functions are analytic, we conclude that $\mathbb{RP}^n$ is an analytical manifold.

The interest of mathematicians to the real projective plane $\mathbb{RP}^2$ goes rather back in history. It has been observe that this 2-dimensional manifold can be embedded smoothly into $\mathbb{R}^4$. Werner Boy [1879-1914] who was a student of Hilbert had shown in 1901 that this surface can also be immersed in $\mathbb{R}^3$ if it is allowed for the surface to intersect itself. A quite an interesting parametrisation of *Boy's surface* was discovered by American mathematicians Robert B. Kusner and Robert L. Bryant (1953): we define the functions

$$g_1 = -\frac{3}{2}\Im\frac{\zeta(1-\zeta^4)}{\zeta^6 + \sqrt{5}\,\zeta^3 - 1}, \quad g_2 = -\frac{3}{2}\Re\frac{\zeta(1+\zeta^4)}{\zeta^6 + \sqrt{5}\,\zeta^3 - 1}$$

$$g_3 = \Im\frac{1+\zeta^4}{\zeta^6 + \sqrt{5}\,\zeta^3 - 1} - \frac{1}{2}, \qquad g = g_1^2 + g_2^2 + g_3^2$$

where $\zeta = u + iv$ is a complex variable subject to the restriction $|\zeta| \leq 1$ and $\Re$ and $\Im$ denote the real and imaginary parts of a complex number, respectively. Then the Cartesian coordinates of a point on the surface is parametrically given by

$$x(u,v) = \frac{g_1}{g}, \quad y(u,v) = \frac{g_2}{g}, \quad z(u,v) = \frac{g_3}{g}.$$

Boy's surface is depicted in Fig. 2.2.9.

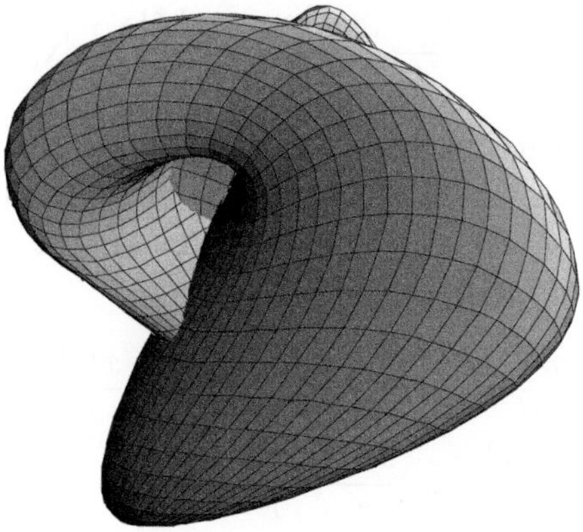

Fig. 2.2.9. Image of $\mathbb{RP}^2$ in 3-dimensional space (Boy's surface). ■

Manifolds with Boundary. In order to define a topological manifold with boundary we need a slightly more generalised concept. Let M_1 be a topological space that is an n-dimensional differentiable manifold. We consider a *closed* subset M of M_1. When M has a boundary ∂M we cannot generate a differentiable structure on the topological subspace M in the usual way because a point $p \in \partial M$ does not have an open neighbourhood remaining entirely inside M that is homeomorphic to an open set of $\mathbb{R}^n$. In order to solve this problem, we propose to consider the following subspace $\mathbb{H}^n$ of $\mathbb{R}^n$:

$$\mathbb{H}^n = \{\mathbf{x} = (x^1, x^2, \dots x^n) \in \mathbb{R}^n : x^n \geq 0\}.$$

The hyperplane $\mathbb{R}^{n-1}$ defined by the relation $x^n = 0$ is the boundary of this *closed half-space*. We know that open sets of the subspace $\mathbb{H}^n$ in the relative topology are intersections of standard open sets in $\mathbb{R}^n$ with $\mathbb{H}^n$. Let

$V \subset \mathbb{H}^n$ be an open set defined this way (Fig. 2.2.10). We denote ***the interior*** of the set V by Int $V = V \cap \{\mathbf{x} \in \mathbb{R}^n : x^n > 0\}$ and its ***boundary*** by $\partial V = V \cap \{\mathbf{x} \in \mathbb{R}^n : x^n = 0\}$. It is clear that $V = \text{Int}\, V \cup \partial V$. We immediately observe that ∂V is not the topological boundary of the set V given on *p.* 55. Actually, ∂V is the intersection of the topological boundary with the boundary $x^n = 0$ of $\mathbb{H}^n$. If this intersection is empty, then V has no boundary according to this definition although the topological boundary may exist in the form $\bar{V} \cap (\mathring{V})'$.

The ***interior*** of M denoted by Int M is the set of points of M that have open neighbourhoods homeomorphic to open subsets of $\mathbb{R}^n$. The ***boundary*** ∂M of M is the complement of Int M with respect to M. The points on ∂M are mapped by homeomorphism to the points on the boundary $x^n = 0$ of $\mathbb{H}^n$. We now define a differentiable structure on M by an atlas $\mathcal{A} = \{(U_\alpha, \varphi_\alpha) : \alpha \in \mathcal{I}\}$ where U_α are open sets in relative topology on M and $\varphi_\alpha : U_\alpha \to V_\alpha$ are homeomorphisms. V_α is an open subset of $\mathbb{H}^n$. Naturally domains of charts will obey the rules (i)-(iii) mentioned on *p.* 53. We can now express the boundary ∂M and the interior Int M of the manifold M by the relations (*see* Fig. 2.2.11)

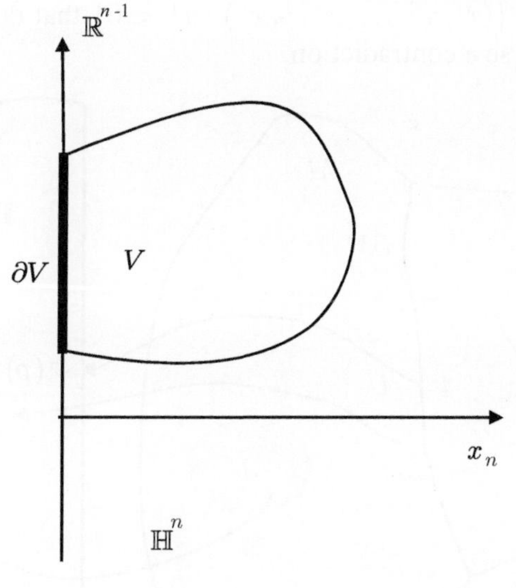

Fig. 2.2.10. An open set in $\mathbb{H}^n$.

$$\partial M = \bigcup_{\alpha \in \mathcal{I}} \varphi_\alpha^{-1}\big(\partial\,(\varphi_\alpha(U_\alpha))\big), \quad \text{Int}\, M = \bigcup_{\alpha \in \mathcal{I}} \varphi_\alpha^{-1}\big(\text{Int}\,(\varphi_\alpha(U_\alpha))\big).$$

If a point $p \in \partial M$ belongs to a chart (U, φ), then its parametrisation is obviously in the form

$$\varphi(p) = (x^1, x^2, \ldots, x^{n-1}, 0).$$

It is clear that Int M *is a n-dimensional manifold without boundary*. We shall show in the sequel that *the boundary ∂M of M is an $(n-1)$-dimensional manifold without boundary*. But we first prove the following lemma.

Lemma 2.2.1. *The position of a point on the boundary of the manifold M is independent of the parametrisation used.*

Let us consider two charts (U_1, φ_1) and (U_2, φ_2) containing a point $p \in \partial M$. We suppose that $\varphi_1(p) = \mathbf{x}_1 = (x^1, x^2, \ldots, x^{n-1}, 0)$ and $\varphi_2(p) = \mathbf{x}_2 = (x^1, x^2, \ldots, x^{n-1}, x^n), x^n > 0$. The transition mapping

$$\varphi_{12}^{-1} = \varphi_1 \circ \varphi_2^{-1} : \varphi_2(U_1 \cap U_2) \to \varphi_1(U_1 \cap U_2)$$

is a homeomorphism on $\mathbb{H}^n$. On the other hand, we assumed that the point $\mathbf{x}_2 \in \mathbb{H}^n$ is an interior point of $\mathbb{R}^n$. Hence, this point has an open neighbourhood $V_{\mathbf{x}_2} \subseteq \varphi_2(U_1 \cap U_2)$ in $\mathbb{R}^n$ that does not intersect the boundary $x^n = 0$. The function φ_{12}^{-1} transforms this open neighbourhood into the open neighbourhood $V_{\mathbf{x}_1} = \varphi_{12}^{-1}(V_{\mathbf{x}_2})$ of $\mathbf{x}_1$ in $\mathbb{R}^n$ (Fig. 2.2.12). But this set contains the points in the form $\{(x^1, x^2, \ldots, x^{n-1}, x^n) : x^n < 0\}$ that does not belong to $\mathbb{H}^n$. This is of course a contradiction. □

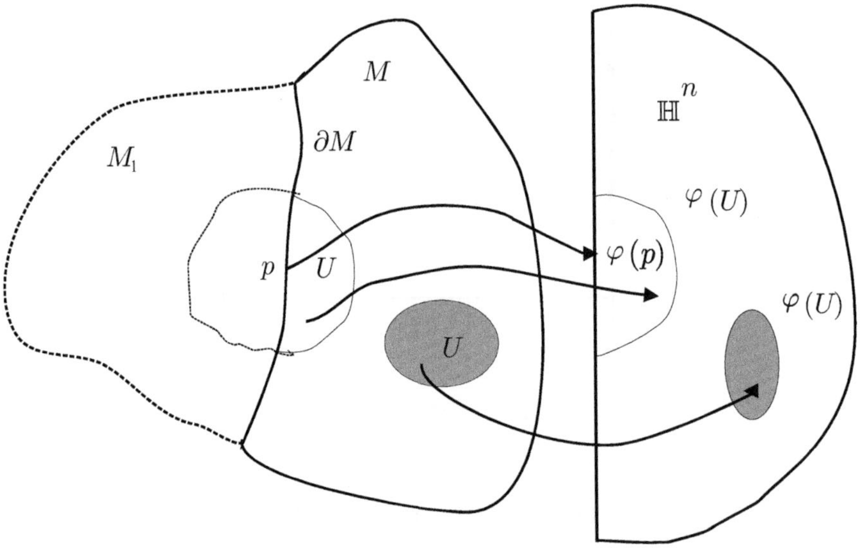

Fig. 2.2.11. A manifold with boundary.

Theorem 2.2.2. *The boundary of an n-dimensional differentiable manifold with boundary is an $(n-1)$-dimensional differentiable manifold.*

Let ∂M be the boundary of the manifold M. If a chart $(U_\alpha, \varphi_\alpha)$ of an atlas $\mathcal{A}$ contains a boundary point $p \in \partial M$, we can then write $\varphi_\alpha(\overline{U}_\alpha) = \varphi_\alpha(U_\alpha) \cap \mathbb{R}^{n-1}$ where we now define $\overline{U}_\alpha = U_\alpha \cap \partial M$ and $\mathbb{R}^{n-1} = \{(x^1, x^2, \ldots, x^{n-1}, x^n) \in \mathbb{R}^n : x^n = 0\}$. The set $\varphi_\alpha(\overline{U}_\alpha)$ is an open set in $\mathbb{R}^{n-1}$ in the relative topology. We denote the restriction of φ_α to the set $\overline{U}_\alpha$ by $\varphi_\alpha|_{\overline{U}_\alpha} = \overline{\varphi}_\alpha : \overline{U}_\alpha \subseteq \partial M \to V_\alpha \subseteq \mathbb{R}^{n-1}$. Evidently, $\overline{\varphi}_\alpha$ is also a homeomorphism. Therefore, the pair $(\overline{U}_\alpha, \overline{\varphi}_\alpha)$ is a chart on ∂M. Since the family $\mathcal{A} = \{(U_\alpha, \varphi_\alpha) : \alpha \in \mathcal{I}\}$ is an atlas on M, it is quite clear that the family $\overline{\mathcal{A}} = \{(\overline{U}_\alpha, \overline{\varphi}_\alpha) : \alpha \in \mathcal{I}\}$ becomes an atlas on ∂M. If this atlas has overlapping charts at a boundary point, these charts will be compatible in view of Lemma 2.2.1. Thus the atlas $\overline{\mathcal{A}}$ gives rise to a differentiable structure on ∂M. Hence the topological space ∂M is an $(n-1)$-dimensional differentiable manifold. □

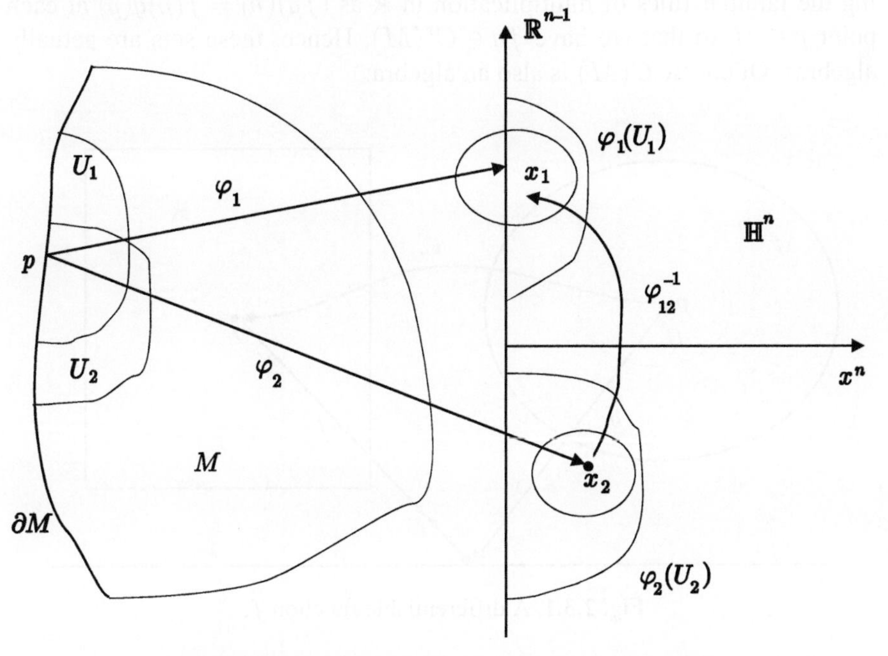

Fig. 2.2.12. A point on the boundary of a manifold.

2.3. DIFFERENTIABLE MAPPINGS

We consider a mapping $f : M \to \mathbb{R}$ on an m-dimensional differentiable manifold $(M, \mathcal{A})$, that is, $f(p) \in \mathbb{R}$ if $p \in M$. Let us assume that the point p is contained in the chart $(U_\alpha, \varphi_\alpha) \in \mathcal{A}$. Then, we can write $f(p) = f\big(\varphi_\alpha^{-1}(\mathbf{x})\big) = (f \circ \varphi_\alpha^{-1})(\mathbf{x})$. If we define a real-valued function of m real variables by $f'_\alpha = f \circ \varphi_\alpha^{-1} : \mathbb{R}^m \to \mathbb{R}$ on the open set $\varphi_\alpha(U_\alpha) \subseteq \mathbb{R}^m$, then the equality $f(p) = f'_\alpha(\mathbf{x})$ becomes valid provided that the condition $\mathbf{x} = \varphi_\alpha(p)$ is satisfied (Fig. 2.3.1). If the function $f'_\alpha(x^1, x^2, \dots, x^m)$ is of class C^r at the point $\mathbf{x} \in \varphi_\alpha(U_\alpha)$, we say that the function f is ***differentiable*** and a ***C^r-function*** at the point $p \in M$ and we usually write $f \in C^r(M, \mathbb{R})$ or just $f \in C^r(M)$. Let us note that $r \leq k$ if the atlas on M is of C^k-class. When we use only the adjectives ***differentiable*** or ***smooth***, we will always mean a function of C^∞-class. If a function f is differentiable at every point of the manifold M, then it is a function differentiable on M. We denote the set of all differentiable functions on M by $C^\infty(M)$ or merely by $C(M)$. We had seen that the set $C^r(M)$ can be equipped with a vector space structure [*see* Example 1.2.2], i.e., we can write $\alpha f + \beta g \in C^r(M)$ where $\alpha, \beta \in \mathbb{R}$. We can also define a product of vectors $f, g \in C^r(M)$ by utilising the familiar rules of multiplication in $\mathbb{R}$ as $(fg)(p) = f(p)g(p)$ at each point $p \in M$ so that we have $fg \in C^r(M)$. Hence, these sets are actually algebras. Of course $C(M)$ is also an algebra.

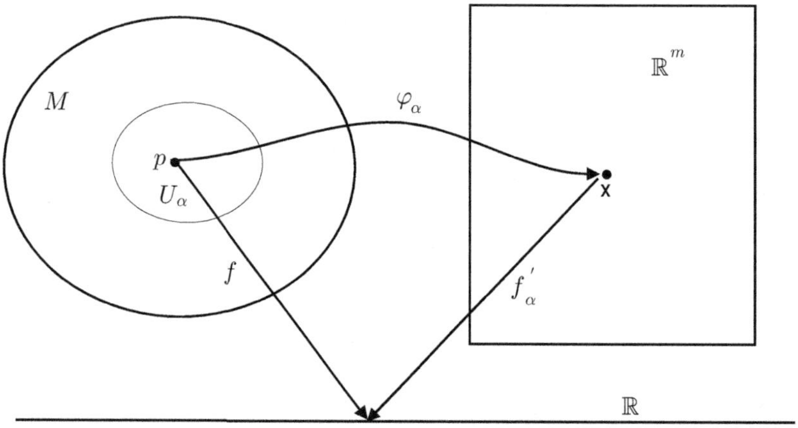

Fig. 2.3.1. A differentiable function f.

We can easily prove that the differentiability of a function $f : M \to \mathbb{R}$ is independent of the chosen atlas among compatible atlases. Let us consider another atlas $\mathcal{B}$ on M and assume that the point $p \in M$ belongs also to the chart $(V_\beta, \psi_\beta) \in \mathcal{B}$. We can thus write

$$f(p) = f'_\alpha(\mathbf{x}) = f'_\beta(\mathbf{y}); \quad p \in U_\alpha \cap V_\beta, \quad \mathbf{x} = \varphi_\alpha(p), \quad \mathbf{y} = \psi_\beta(p)$$

where we have of course defined $f'_\beta = f \circ \psi_\beta^{-1}$. Therefore, we obtain

$$f'_\beta = (f'_\alpha \circ \varphi_\alpha) \circ \psi_\beta^{-1} = f'_\alpha \circ (\varphi_\alpha \circ \psi_\beta^{-1}).$$

Because atlases are compatible, we conclude that if f'_α is differentiable, then the function f'_β must also be differentiable since it is expressed as a composition of differentiable functions. By definition, the partial derivative of a function f at a point $p \in M$ with respect to a coordinate x^i in an open set of $\mathbb{R}^m$ determined by a chart $(U_\alpha, \varphi_\alpha)$ containing the point p will be written at the point $\mathbf{x} = \varphi_\alpha(p)$ as

$$D_i f(p) = \frac{\partial f'_\alpha(\mathbf{x})}{\partial x^i}, \quad i = 1, 2, \dots, m.$$

Higher order derivatives will be represented in the same fashion.

Since a differentiable manifold is actually a topological space, the existence of the partition of unity on this manifold can be discussed. The partition of unity $\{V_i, f_i\}$ on a topological space was discussed on *p.* 62. But, here we further impose the condition that the functions $f_i : M \to [0, 1]$ are to be *smooth*.

It can be shown that if the manifold M is paracompact as a topological space, then for each atlas $\mathcal{A} = \{(U_\lambda, \varphi_\lambda) : \lambda \in \Lambda\}$ there exists a partition of unity subordinate to the open cover $\{U_\lambda : \lambda \in \Lambda\}$.

To prove this proposition in its most generality is beyond the scope of this work. Instead, we shall try to manage it for a paracompact space that is Hausdorff, locally compact and second countable [*see p.* 57]. These properties, however, are enjoyed by many differentiable manifolds encountered in applications. To this end, we start first by demonstrating the existence of a smooth function $\phi : \mathbb{R}^m \to \mathbb{R}$ which is equal to 1 on the closed cube $C[1]$ and is 0 on the complement of the open cube $C(2)$. The *open cube* $C(r)$ with sides of length $2r$ about the origin of $\mathbb{R}^m$ is defined as the subset

$$C(r) = \{\mathbf{x} \in \mathbb{R}^m : |x^i| < r,\, i = 1, \dots, n\} \subset \mathbb{R}^m$$

where $\mathbf{x} = (x^1, x^2, \dots, x^m)$ while the *closed cube* is the subset

$$C[r] = \{x \in \mathbb{R}^m : |x^i| \le r,\, i = 1, \dots, n\} \subset \mathbb{R}^m.$$

Let us consider the function $f : \mathbb{R} \to \mathbb{R}$ defined by

$$f(t) = \begin{cases} e^{-1/t} & t > 0, \\ 0 & t \le 0 \end{cases}$$

which is non-negative, smooth and positive for $t > 0$. Then, we introduce the function

$$g(t) = \frac{f(t)}{f(t) + f(1-t)}$$

depicted in Fig. 2.3.2.

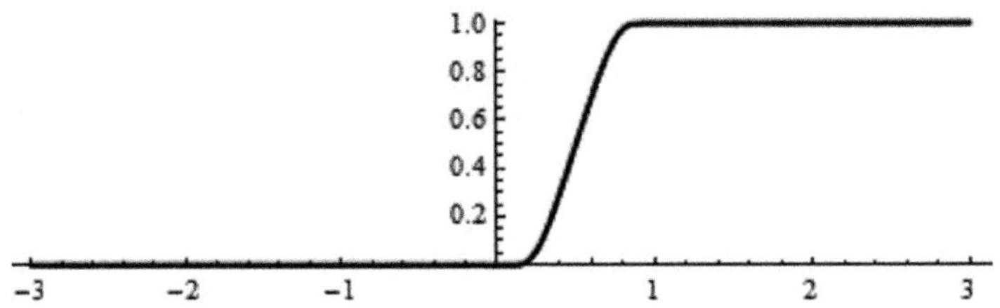

Fig. 2.3.2. The function $g(t)$.

This function is non-negative, smooth, and it is equal to 1 for $t \geq 1$ and to zero for $t \leq 0$.

Next, we construct the function

$$h(t) = g(t+2)g(t-2)$$

shown in Fig. 2.3.3. $h(t)$ is a smooth non-negative function which is equal to 1 on the closed interval $[-1, 1]$ and to zero on the complement of the open interval $(-2, 2)$.

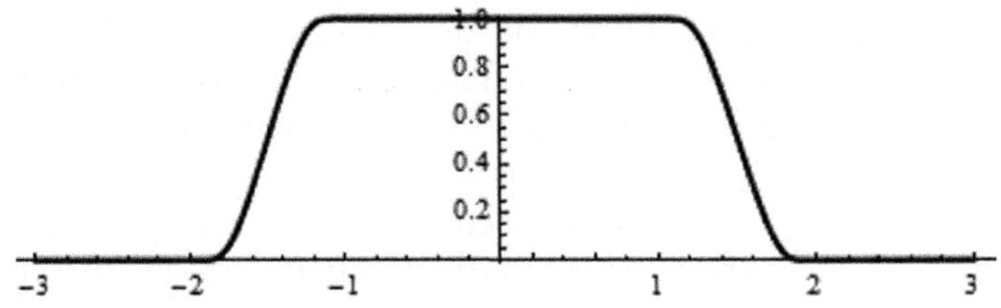

Fig. 2.3.3. The function $h(t)$.

We now define a function $\phi : \mathbb{R}^m \to \mathbb{R}$ by the product

$$\phi(\mathbf{x}) = (h \circ g^1)(\mathbf{x}) \cdots (h \circ g^m)(\mathbf{x}) = h(x^1) \cdots h(x^m)$$

where g^i were defined on *p.* 71. Obviously, this function is equal to 1 on the closed cube $C[1]$ and to zero on the complement of the open cube $C(2)$.

We now consider the relatively compact open cover $\{W_i\}$ of M introduced on *p.* 58. For a point $p \in M$, let i_p be the largest integer such that $p \in M - \bar{W}_{i_p} = (\bar{W}_{i_p})'$. Suppose that for an index $\lambda_p \in \Lambda$ one has $p \in U_{\lambda_p}$.

By definition, we also have $p \in (W_{i_p+1} - \overline{W}_{i_p-1}) = Z_{i_p}$. We consider an open set V in the intersection of the open set of the chart to which the point p belongs with the open set $U_{\lambda_p} \cap Z_{i_p}$. We shall assume that (V, φ) where $V \subseteq U_{\lambda_p} \cap Z_{i_p}$ is a coordinate system centred at the point p chosen in such a way that $\varphi(V) \in \mathbb{R}^m$ contains the closed cube $C[2]$. Next, we define the function $\psi_p : M \to \mathbb{R}$ by

$$\psi_p = \begin{cases} \phi \circ \varphi & \textit{if } p \in V \\ 0 & \textit{otherwise.} \end{cases}$$

Obviously ψ_p is a smooth function on the manifold M. The continuity of ψ_p implies that it is equal to 1 on some open neighbourhood $V_p = \psi_p^{-1}(C(1))$ in V and it has a compact support given by $\psi_p^{-1}(C[2]) \subset V$. We know that $U_\lambda \cap Z_i$ is an open cover for the compact set $K_i = \overline{W}_i - W_{i-1} \subset Z_i$ Thus, for each $i \geq 1$, we can find a finite set of points p_j so that the open sets $U_{\lambda_{p_j}} \cap Z_{i_{p_j}}$ form a finite cover of K_i. Hence, for each i we have a finite family of sets V_{p_j} on which ψ_{p_j} take the value 1 and their supports forms a locally finite family of compact subsets of M. Hence, the set of functions $\{\psi_p\}$ is actually a countable union of finite sets. Therefore they can be enumerated as $\{\psi_i : i \in \mathbb{N}\}$. Thus the function

$$\psi = \sum_{i=1}^{\infty} \psi_i$$

is a well defined smooth function on M and at each point p all but a finitely many functions in this series do vanish. Therefore, we have $\psi(p) > 0$ at each point $p \in M$. Let us now define the functions $f_i : M \to [0, 1]$ as

$$f_i = \frac{\psi_i}{\psi}$$

Hence, the countable family of functions $0 \leq f_i \leq 1$ constitute a partition of unity subordinate to the open cover $\{U_\lambda\}$ with compact supports. □

As we shall see later, this property will prove to be quite significant when we try to define the integration over manifolds.

Example 2.3.1. In the manifold $\mathbb{S}^1$, a partition of unity subordinate to the open cover $\{(0, 2\pi), (-\pi, \pi)\}$ is clearly $\{\sin^2 \frac{\theta}{2}, \cos^2 \frac{\theta}{2}\}$. ■

We shall now give two seemingly different definitions of the *differentiability of a mapping between two differentiable manifolds*. We shall then show that they are actually equivalent.

(i). We consider two differentiable manifolds M and N with dimensions m and n and a continuous mapping $\phi : M \to N$. This mapping will assign a point $q \in N$ to a point $p \in M$ by the relation $q = \phi(p)$. Due to the continuity of ϕ, to each open neighbourhood V of the point q there corresponds an open neighbourhood $U = \phi^{-1}(V)$ of the point p. It is evident that the set inclusion relation $\phi(U) = \phi(\phi^{-1}(V)) \subseteq V$ will be satisfied. Let $g : N \to \mathbb{R}$ be a differentiable function defined on the open set V. We can then define a function on the open set U in the manifold M whose value at the point $p \in M$ is given by the relation $f(p) = g(q) = g(\phi(p))$. Thus each function $g : N \to \mathbb{R}$ defined on V generates a function $f : M \to \mathbb{R}$ defined on U because $\phi(U) \subseteq V$. We can denote the functional relation between them by $f = g \circ \phi = \phi^* g$. The function $\phi^* g$ is called the ***pull-back*** or ***reciprocal image*** of the function g. *If for every differentiable function g defined on N, the function $f = \phi^* g$ is differentiable on M, that is, if for all $g \in C(N)$ one obtains $\phi^* g \in C(M)$, then the mapping $\phi : M \to N$ will be called a **differentiable mapping***. Consequently, a differentiable mapping $\phi : M \to N$ produces a mapping $\phi^* : C(N) \to C(M)$ between algebras $C(N)$ and $C(M)$. The mapping ϕ^* is called the ***dual mapping*** or ***pull-back mapping*** of ϕ. *When ϕ is a homeomorphism and both ϕ and its inverse $\phi^{-1} : N \to M$ are differentiable, then we shall say that the mapping ϕ is a **diffeomorphism***. If we establish a diffeomorphism between two manifolds, they are called ***diffeomorphic manifolds***. Evidently, diffeomorphic manifolds are equivalent as far as their topological and differentiability properties are concerned.

It follows from the definition of pull-back mappings that

$$\phi^*(g_1 + g_2) = \phi^* g_1 + \phi^* g_2, \; \phi^*(g_1 g_2) = (\phi^* g_1)(\phi^* g_2).$$

where $g_1, g_2 \in C(N)$. Hence, we deduce that *the pull-back mapping is an algebra homomorphism*.

If M are N differentiable C^k-manifolds and if there corresponds a $\phi^* g \in C^r(M)$ function for each function $g \in C^r(N)$ for an $r \leq k$, we say that the mapping $\phi : M \to N$ is ***C^r-differentiable***. If ϕ is a homeomorphism and both ϕ and ϕ^{-1} are C^r-differentiable, then we say that ϕ is a ***C^r-diffeomorphism***.

(ii). Let $\phi : M \to N$ be a continuous mapping. This mapping will assign to each point $p \in M$ a point $\phi(p) = q \in N$. These points are located in local charts (U, φ) and (V, ψ), respectively and we can write $\phi(U) \subseteq V$ due to the continuity of ϕ. We denote the local coordinates in those charts by $\mathbf{x} = (x^1, \dots, x^m)$ and $\mathbf{y} = (y^1, \dots, y^n)$, respectively. Hence one writes $\mathbf{x} = \varphi(p) \in \mathbb{R}^m$ and $\mathbf{y} = \psi(q) \in \mathbb{R}^n$. We define by using the transformation $\mathbf{y} = \psi(\phi(\varphi^{-1}(\mathbf{x}))$, a composite mapping

$$\Phi = \psi \circ \phi \circ \varphi^{-1} : \varphi(U) \subseteq \mathbb{R}^m \to \psi(V) \subseteq \mathbb{R}^n$$

so that we can express this relation by $\mathbf{y} = \Phi(\mathbf{x})$ or $y^i = \Phi^i(x^1, \dots, x^m)$, $i = 1, \dots, n$ (Fig. 2.3.4). If the functions Φ^i have continuous derivatives of all orders at the point $\mathbf{x} = \varphi(p)$, namely, if $\Phi \in C^\infty(\mathbb{R}^m, \mathbb{R}^n)$, then we say that ϕ is a *differentiable* or a *smooth* mapping (if Φ is continuously differentiable of order r, then ϕ is a C^r*-differentiable* mapping). If this property is valid for every chart of an atlas, then ϕ *is a differentiable mapping on the manifold* M. If ϕ is a diffeomorphism, then $\phi^{-1} : N \to M$ exists and is differentiable. In this case, it is straightforward to see that ϕ^{-1} is locally represented by a function $\Psi \in C^\infty(\mathbb{R}^n, \mathbb{R}^m)$ given by the inverse relation $\mathbf{x} = \Psi(\mathbf{y})$ such that $\Psi = \varphi \circ \phi^{-1} \circ \psi^{-1}$.

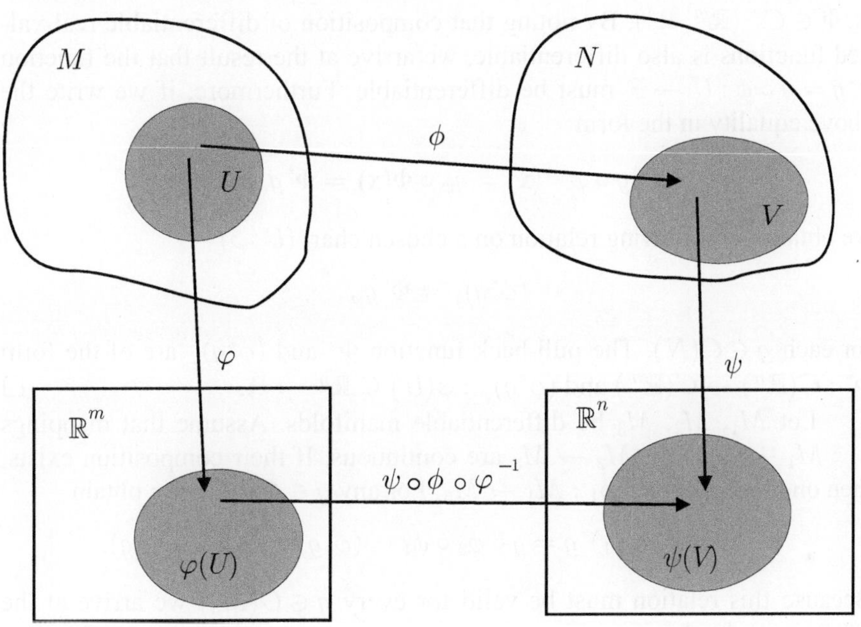

Fig. 2.3.4. A differentiable mapping ϕ.

We shall now try to prove the equivalence of these two definitions:

(i) $\Rightarrow$ ***(ii)*** : We assume that ϕ is differentiable. Let (U, φ) and (V, ψ) be charts enclosing the points $p \in M$ and $\phi(p) = q \in N$, respectively. We define the continuous functions $g_M^i : \mathbb{R}^m \to \mathbb{R}$ and $g_N^i : \mathbb{R}^n \to \mathbb{R}$ by $g_M^i(\mathbf{x}) = x^i$ and $g_N^i(\mathbf{y}) = y^i$. The coordinate functions in those charts are then $\varphi^i = g_M^i \circ \varphi : U \to \mathbb{R}$, $\varphi^i(p) = x^i$, $i = 1, 2, \dots, m$ and $\psi^i = g_N^i \circ \psi : V \to \mathbb{R}$, $\psi^i(q) = \psi^i(\phi(p)) = y^i, i = 1, 2, \dots, n$. Since the function ψ^i is clearly differentiable and the set relation $\phi(U) \subseteq V$ is satisfied, the function $\psi^i \circ \phi : U \to \mathbb{R}$ is also differentiable due to ***(i)***. Since we can write

$\psi^i \circ \phi \circ \varphi^{-1} : \varphi(U) \to \mathbb{R}$ or $g^i_N \circ (\psi \circ \phi \circ \varphi^{-1}) : \varphi(U) \to \mathbb{R}$, we find finally that the function $\Phi = \psi \circ \phi \circ \varphi^{-1} : \varphi(U) \to \psi(V)$ is differentiable at the arbitrary point $p \in M$, i.e., $\Phi \in C^\infty(\mathbb{R}^m, \mathbb{R}^n)$. □

(*ii*) ⇒ (*i*) : We assume again that ϕ is differentiable and we consider an arbitrary function $g : V \to \mathbb{R}$ which is differentiable at a point $q \in N$. Hence the function $g_\psi = g \circ \psi^{-1} : \psi(V) \subseteq \mathbb{R}^n \to \mathbb{R}$ will also be differentiable at the point $\mathbf{y} = \psi(q)$. We can thus write

$$g(q) = g \circ \phi(p) = g_\psi \circ \psi \circ \phi(p) = g_\psi \circ (\psi \circ \phi \circ \varphi^{-1})(\mathbf{x}) = g_\psi \circ \Phi(\mathbf{x}).$$

We have assumed that the function $\Phi = \psi \circ \phi \circ \varphi^{-1}$ is differentiable, that is, $\Phi \in C^\infty(\mathbb{R}^m, \mathbb{R}^n)$. By noting that composition of differentiable real-valued functions is also differentiable, we arrive at the result that the function $\phi^* g = g \circ \phi : U \to \mathbb{R}$ must be differentiable. Furthermore, if we write the above equality in the form

$$\phi^* g \circ \varphi^{-1}(\mathbf{x}) = g_\psi \circ \Phi(\mathbf{x}) = \Phi^* g_\psi(\mathbf{x})$$

we obtain the following relation on a chosen chart (U, φ)

$$(\phi^* g)_\varphi = \Phi^* g_\psi$$

for each $g \in C(N)$. The pull-back function Φ^* and $(\phi^* g)_\varphi$ are of the form $\Phi^* : C(\mathbb{R}^n) \to C(\mathbb{R}^m)$ and $(\phi^* g)_\varphi : \varphi(U) \subseteq \mathbb{R}^m \to \mathbb{R}$. □

Let M_1, M_2, M_3 be differentiable manifolds. Assume that mappings $\phi_1 : M_1 \to M_2$, $\phi_2 : M_2 \to M_3$ are continuous. If their composition exists, then one has $\phi = \phi_2 \circ \phi_1 : M_1 \to M_3$. For any $g \in C(M_3)$, we obtain

$$\phi^* g = (\phi_2 \circ \phi_1)^* g = g \circ \phi_2 \circ \phi_1 = (\phi_2^* g) \circ \phi_1 = \phi_1^*(\phi_2^* g).$$

Because this relation must be valid for every $g \in C(M_3)$ we arrive at the following rule of composition

$$(\phi_2 \circ \phi_1)^* = \phi_1^* \circ \phi_2^*. \tag{2.3.1}$$

This result can of course be extended to an arbitrary number of compositions. Let us now take into account the *identity mapping* $i_M : M \to M$ on the differentiable manifold M. We thus find $i_M(p) = p$ for each $p \in M$. In this case, we obtain $i_M^* g = g \circ i_M = g$ for each $g \in C(M)$. Consequently, we reach to the identity mapping on $C(M)$:

$$i_M^* = i_{C(M)} \tag{2.3.2}$$

Example 2.3.2. Consider the manifold $\mathbb{R}$ with the standard chart $(\mathbb{R}, i_\mathbb{R})$. The function $\phi : \mathbb{R} \to \mathbb{R}$ prescribed by $y = \phi(x) = x^\alpha$, $\alpha > 1$ is a differentiable homeomorphism, but it is not a diffeomorphism. Because the inverse mapping $x = \phi^{-1}(y) = y^{1/\alpha}$ cannot be differentiated at the point $y = 0$. We define now a new differentiable structure on $\mathbb{R}$ by another chart $(\mathbb{R}, \psi = \phi^{-1})$. Let $\mathbb{R}_\phi$ denote the manifold $\mathbb{R}$ equipped by this structure. Hence, for each $y \in \mathbb{R}$ one has $\psi(y) = y^{1/\alpha}$. the local representation of the mapping $\phi : \mathbb{R} \to \mathbb{R}_\phi$ is now given by $\phi^{-1} \circ \phi \circ i_{\mathbb{R}}^{-1} = i_{\mathbb{R}}$, whereas that of the inverse mapping $\phi^{-1} : \mathbb{R}_\phi \to \mathbb{R}$ becomes $i_{\mathbb{R}} \circ \phi^{-1} \circ (\phi^{-1})^{-1} = i_{\mathbb{R}}$. This amounts to say that $\phi : \mathbb{R} \to \mathbb{R}_\phi$ is a diffeomorphism. ■

Example 2.3.3. The manifold $\mathbb{S}^2 = \{\mathbf{x} \in \mathbb{R}^3 : x_1^2 + x_2^2 + x_3^2 = 1\}$ in $\mathbb{R}^3$ will now be considered. We know that this sphere can only be homeomorphic to the plane $\mathbb{R}^2$ by employing two charts of its atlas and two differentiable functions ϕ_1, ϕ_2 given below:

$$\mathbf{y} = \phi_1(\mathbf{x}),\ y_i = \frac{x_i}{1 - x_3};\ \mathbf{y} = \phi_2(\mathbf{x}),\ y_i = \frac{x_i}{1 + x_3};\ \mathbf{y} \in \mathbb{R}^2, \mathbf{x} \in \mathbb{S}^2$$

[*see pp.* 81-82]. Thus, we cannot find a single diffeomorphism $\phi : \mathbb{S}^2 \to \mathbb{R}^2$. Hence, the sphere cannot be diffeomorphic to the plane. On the other hand, when we choose the ellipsoidal surface as another manifold given by $M = \left\{\mathbf{y} \in \mathbb{R}^3 : \frac{y_1^2}{a^2} + \frac{y_2^2}{b^2} + \frac{y_3^2}{c^2} = 1\right\}$, the mapping $\phi : \mathbb{S}^2 \to M$ defined by

$$y_1 = ax_1,\ \ y_2 = bx_2,\ \ y_3 = cx_3$$

is evidently a diffeomorphism. Thus the sphere and the ellipsoid are diffeomorphic manifolds. ■

Example 2.3.4. Let us consider the unit circle $\mathbb{S}^1$ in $\mathbb{R}^2$ and the projective space $\mathbb{RP}^1$. These manifolds will be represented as follows: $\mathbb{S}^1 = \{e^{i\theta}\}$ and $\mathbb{RP}^1 = \left\{\xi = \frac{x_2}{x_1} : \mathbf{x} \in \mathbb{R}^2 - \{\mathbf{0}\}\right\}$. It is easily observed that the single mapping $\phi : S^2 \to \mathbb{RP}^1$ determined by $\xi = \tan\theta$ is a diffeomorphism between two charts (U_1 and U_2) of the projective space $\mathbb{RP}^1$ [*see p.* 87] and two charts of the circle $\mathbb{S}^1$. Hence, these manifolds are diffeomorphic. ■

Example 2.3.5. We define the mapping $\phi : (a, b) \to \mathbb{R}$ by the relation

$$x = \phi(\xi) = \frac{(b - a)(2\xi - a - b)}{4(\xi - a)(b - \xi)},\ \ \xi \in (a, b).$$

The inverse of this function is obtained as

$$\xi = \phi^{-1}(x) = \frac{b^2 - a^2 - 4abx}{\sqrt{(b - a)^2(1 + 4x^2)} - 2(a + b)x + b - a}$$

if we note that ξ must belong to the open interval (a, b). The functions ϕ and ϕ^{-1} are continuous and differentiable former on (a, b) while the latter on $(-\infty, \infty)$. Thus ϕ is a diffeomorphism. This means that every open interval in $\mathbb{R}$ is diffeomorphic to $\mathbb{R}$ itself. ■

Example 2.3.6. A mapping $\phi : \mathbb{T}^2 \to \mathbb{R}^3$ between differentiable manifolds $\mathbb{T}^2$ and $\mathbb{R}^3$ can be defined as follows [*see p.* 82]

$$\phi(\theta, \phi) = \big((a + b\sin\theta)\cos\phi, (a + b\sin\theta)\sin\phi, b\cos\theta\big) = (x, y, z).$$

This mapping is clearly differentiable and smooth. The image of the manifold $\mathbb{T}^2$ in $\mathbb{R}^3$ under the mapping ϕ is the surface

$$x^2 + y^2 + z^2 - 2a\sqrt{x^2 + y^2} + a^2 - b^2 = 0$$

obtained by eliminating parameters θ and ϕ. ■

Let $\phi : M \to N$ be a smooth mapping from the m-dimensional manifold M to the n-dimensional manifold N. We consider points $p \in M$ and $q = \phi(p) \in N$ in the local charts (U, φ) and (V, ψ), respectively. Then the mapping ϕ is represented by the function $\Phi : \varphi(U) \subseteq \mathbb{R}^m \to \psi(V) \subseteq \mathbb{R}^n$ that can be written as $\mathbf{y} = \Phi(\mathbf{x})$ or $y^i = \Phi^i(x^1, \dots, x^m)$, $i = 1, \dots, n$ in terms of local coordinates. We know that Φ^i are smooth functions. *The rank of the mapping ϕ at the point p* is defined as the rank of the following $n \times m$ Jacobian matrix [German mathematician Carl Gustav Jacob Jacobi (1804-1851)]

$$\mathbf{J}(\phi) = \left[\frac{\partial \Phi^i}{\partial x^j}\right] = \begin{bmatrix} \dfrac{\partial \Phi^1}{\partial x^1} & \dfrac{\partial \Phi^1}{\partial x^2} & \cdots & \dfrac{\partial \Phi^1}{\partial x^m} \\ \dfrac{\partial \Phi^2}{\partial x^1} & \dfrac{\partial \Phi^2}{\partial x^2} & \cdots & \dfrac{\partial \Phi^2}{\partial x^m} \\ \vdots & \vdots & & \vdots \\ \dfrac{\partial \Phi^n}{\partial x^1} & \dfrac{\partial \Phi^n}{\partial x^2} & \cdots & \dfrac{\partial \Phi^n}{\partial x^m} \end{bmatrix}.$$

If the rank of ϕ at a point $p \in M$ admits its greatest value, that is, if it is equal to min $\{m, n\}$, then we say that its rank is *maximal* at that point. If the rank of ϕ is maximal every point $p \in S$ of a subset $S \subseteq M$, then its rank is maximal on S.

Theorem 2.3.1. *Let the rank of a mapping $\phi : M \to N$ be maximal at a point $p \in M$. Consider the chart (U, φ) at the point p and the chart (V, ψ) at the point $\phi(p)$ such that $\phi(U) \subseteq V$. Then the local coordinates $(x^1, x^2, \dots, x^m)$ in the neighbourhood of the point $\mathbf{x} = \varphi(p)$ and $(y^1, y^2, \dots, y^m, \dots, y^n)$ in the neighbourhood of the point $\mathbf{y} = \psi(\phi(p)) = \Phi(\mathbf{x})$ can be so chosen that the local representation $\Phi = \psi \circ \phi \circ \varphi^{-1}$ of ϕ*

admits the following forms

$$\mathbf{y} = (x^1, x^2, \dots, x^n) \text{ if } n \le m,$$
$$\mathbf{y} = (x^1, x^2, \dots, x^m, 0, \dots, 0) \text{ if } n > m.$$

In terms of arbitrary coordinates in charts, consider the representation $\eta^i = \Phi^i(\xi^1, \dots, \xi^m), i = 1, \dots, n$. $n \times m$ Jacobian matrix is $[\partial\eta^i/\partial\xi^j]$.

If $n \le m$, the rank of this matrix is n. Let us rearrange the variables in such a way that the determinant of the square matrix $[\partial\eta^i/\partial\xi^j]$, $i, j = 1, \dots, n$ does not vanish. Then according to the well known implicit function theorem, the equations $x^i = \Phi^i(\xi^1, \dots, \xi^n, \xi^{n+1}, \dots, \xi^m)$ have uniquely determined smooth solutions $\xi^i = \Psi^i(x^1, \dots, x^n, \xi^{n+1}, \dots, \xi^m), 1 \le i \le n$ in a sufficiently small neighbourhood. If we now write $\xi^{n+1} = x^{n+1}, \dots, \xi^m = x^m$, the new local coordinates in a neighbourhood of the point p become $(x^1, \dots, x^n, x^{n+1}, \dots, x^m)$. Thus, the local coordinates in the neighbourhood of the image point $\phi(p)$ takes the form $\boldsymbol{\eta} = \mathbf{y} = (x^1, \dots, x^n)$.

If $n > m$, the rank of the Jacobian matrix is m. Let us now rearrange the variables in such a way that a $m \times m$ square submatrix $[\partial\eta^i/\partial\xi^j]$, $i, j = 1, \dots, m$ of the Jacobian matrix has a non-zero determinant. We now choose the new local coordinates in a neighbourhood of the point q as $x^i = \Phi^i(\xi^1, \dots, \xi^m)$, $i = 1, \dots, m$. Then, we can uniquely determine smooth solutions $\xi^i = \Psi^i(x^1, \dots, x^m), i = 1, \dots, m$. Thus, we can define the new local coordinates in a neighbourhood of the point $\phi(p)$ by $\eta^i = y^i = x^i$, $i = 1, \dots, m$ and $y^i = \eta^i - \Phi^i(\xi^1, \dots, \xi^m) = \eta^i - \Omega^i(y^1, \dots, y^m)$, $i = m+1, \dots, n$ where $\Omega = \Phi \circ \Psi$. However, because of the initial relations $\eta^i = \Phi^i(\xi^1, \dots, \xi^m)$, $i = m+1, \dots, n$, we immediately see that we are led to $\mathbf{y} = (x^1, \dots, x^m, 0, \dots, 0)$. □

2.4. SUBMANIFOLDS

Let $\phi : M \to N$ be a smooth mapping between manifolds M^m and N^n. If $m \ge n$ and the rank of ϕ at every point $p \in M$ is n, then the mapping ϕ is called a ***submersion***. In this case, Theorem 2.3.1 indicates that the local representation Φ of ϕ is simply expressible as follows

$$y^1 = x^1,\ y^2 = x^2,\ \dots\ ,\ y^n = x^n$$

with an appropriate choice of coordinates.

Example 2.4.1. The mapping $\phi : \mathbb{R}^3 \to \mathbb{R}^2$ is given by the relations

$$y^1 = x^2 - x^3,\ y^2 = x^1.$$

Jacobian matrix of this mapping is

$$\mathbf{J} = \begin{bmatrix} 0 & 1 & -1 \\ 1 & 0 & 0 \end{bmatrix}$$

and its rank is 2 everywhere. Thus ϕ is a submersion. ■

Example 2.4.2. Let $U \subseteq \mathbb{R}^3$ be an open set. Hence U is a 3-dimensional differentiable manifold. Jacobian matrix of a mapping $\phi : U \to \mathbb{R}$ is of course given by

$$\mathbf{J} = \begin{bmatrix} \dfrac{\partial \phi}{\partial x^1} & \dfrac{\partial \phi}{\partial x^2} & \dfrac{\partial \phi}{\partial x^3} \end{bmatrix}.$$

If ϕ has at least one non-vanishing partial derivative at each point of U, then the rank of this matrix is 1. In this case ϕ is a submersion. As an example let us choose the open set $U = \{\mathbf{x} \in \mathbb{R}^3 : (x^1)^2 + (x^2)^2 + (x^3)^2 > 0\}$ and the mapping given by $\phi(\mathbf{x}) = (x^1)^2 + (x^2)^2 + (x^3)^2$. The Jacobian matrix of this mapping is $\mathbf{J} = 2[\, x^1 \quad x^2 \quad x^3 \,]$ whose entries cannot be all zero in U. Thus ϕ is a submersion. On the other hand, the Jacobian matrix for the mapping $\phi_1(\mathbf{x}) = x^1 x^2 x^3$ is $\mathbf{J} = [\, x^2 x^3 \quad x^1 x^3 \quad x^1 x^2 \,]$. All entries of this matrix may vanish at some points of U (for instance, at $x^1 = x^2 = 0$, $x^3 \neq 0$). At such kind of points the rank of $\mathbf{J}$ is 0. Hence, the mapping $\phi_1 : U \to \mathbb{R}$ is not a submersion. ■

Let $\phi : M \to N$ be a smooth mapping. If $n \geq m$ and the rank of ϕ at every point $p \in M$ is m, then the mapping ϕ is called an ***immersion***. Again, Theorem 2.3.1 implies that the local representation Φ of ϕ is expressible now in the form

$$y^1 = x^1, y^2 = x^2, \ldots, y^m = x^m, y^{m+1} = 0, \ldots y^n = 0$$

with an appropriate choice of coordinates.

Example 2.4.3. The mapping $\phi : \mathbb{R} \to \mathbb{R}^2$ is defined by the relations $y^1 = \cos x^1, y^2 = \sin x^1$. Obviously, this mapping wraps the entire real axis $\mathbb{R}$ on the unit circle $\mathbb{S}^1$. The Jacobian matrix of this mapping becomes $\mathbf{J} = [\, -\sin x^1 \quad \cos x^1 \,]$. The rank of this matrix is 1 everywhere on $\mathbb{R}$. Hence ϕ is an immersion. Since all the points $x^1_n = x^1 + 2n\pi, n \in \mathbb{Z}$, where $\mathbb{Z}$ denotes the set of integers, are mapped on the same point $\mathbf{y} = (y^1, y^2)$ the mapping ϕ is obviously not injective. ■

Example 2.4.4. The mapping $\phi : \mathbb{R}^2 \to \mathbb{R}^3$ revolves the plane curve $x^1 = f(x^2)$ infinitely many times about x^2-axis. $f(x^2) > 0$ is a smooth function. This mapping can be prescribed by the relations

$$y^1 = f(u) \cos v, \;\; y^2 = f(u) \sin v, \;\; y^3 = u; \;\; u, v \in \mathbb{R}$$

where we wrote $x^1 = u, x^2 = v$. The Jacobian matrix is then given by

$$\mathbf{J} = \begin{bmatrix} f'(u)\cos v & -f(u)\sin v \\ f'(u)\sin v & f(u)\cos v \\ 1 & 0 \end{bmatrix}.$$

Since $f(u) > 0$, the rank of this matrix is 2 everywhere. Thus ϕ is an immersion. Clearly, it is not injective. ■

Example 2.4.5. We consider the torus $\mathbb{T}^2 = \mathbb{S}^1 \times \mathbb{S}^1$. The circle $\mathbb{S}^1$ may be represented by complex numbers with constant modulus in the complex plane $\mathbb{C}$. Therefore, we can write

$$\mathbb{T}^2 = \{(z_1, z_2) : z_1, z_2 \in \mathbb{C}, |z_1| = a, |z_2 - a| = b, b < a\} \subset \mathbb{C}^2.$$

We define a mapping $\phi : \mathbb{R} \to \mathbb{T}^2$ by the relations $z_1 = ae^{it}, z_2 - a = be^{irt}$ where r is a *rational number*. We can observe at once that this mapping is an immersion and it produces a closed curve on the torus. In fact, if choose the integer m and n such that $n = mr$ we reach to the same points

$$e^{it+2\pi mi} = z_1,\; be^{irt+2\pi ni} = z_2 - a$$

at all points $t_m = t + 2\pi m$. This means that we reach to the same point on the torus after having revolved m times the point z_1 and n times the point z_2 about O. This immersion is clearly not injective. ■

If $\phi : M \to N$ is an *injective immersion* and if the surjective, consequently, bijective mapping $\phi : M \to \phi(M) \subseteq N$ is a homeomorphism with respect to the relative topology on $\phi(M) \subseteq N$ generated by the topology on the manifold N, then the mapping ϕ is called an ***embedding***.

If the set M^m is a topological subspace of the manifold N^n and the *inclusion mapping* $\mathcal{I} : M \to N$ defined by $\mathcal{I}(p) = p \in N$ for each $p \in M$ is an embedding, then the subpace M is called a ***submanifold*** of dimension $m \le n$ of the manifold N. Indeed, we can readily generate a differentiable structure on M by making use of the differentiable structure on the manifold N. Let us consider a point $p \in M \subseteq N$. This point is located in a chart (U, φ) of the atlas on N. $U' = U \cap M$ is an open set of M in the relative topology. The mapping $\varphi' = \varphi \circ \mathcal{I} : U \cap M \to \mathbb{R}^n$ is a homeomorphism because it is the composition of two homeomorphisms. Let us denote the set of coordinates of the point p in N by $\mathbf{x}$ and the set of coordinates in M by $\mathbf{y}$. As is well known, we write the expression $\mathbf{y} = \mathfrak{I}(\mathbf{x})$ where we define

$$\mathfrak{I} = \varphi' \circ \mathcal{I} \circ \varphi^{-1}.$$

Since the rank of $\mathcal{I}$ is $m < n$ and rank remains invariant under composition of homeomorphisms, the rank of the mapping $\mathfrak{I} : \mathbb{R}^n \to \mathbb{R}^n$ is also $m < n$.

This means that an appropriate choice of coordinates leads to local coordinates $\mathbf{y} = (x^1, \ldots, x^m, 0, \ldots, 0) \in \mathbb{R}^m$ [Theorem 2.3.1]. Hence, one has $\varphi' : U' \to \mathbb{R}^m$. Consequently, the topological subspace M is an m-dimensional differentiable submanifold. Let us now denote $\varphi'(p) = \boldsymbol{\xi} \in \mathbb{R}^m$, $\boldsymbol{\xi} = (\xi^1, \ldots, \xi^m)$ for a point $p \in U' \subseteq M$. Then the equality $p = \varphi^{-1}(\mathbf{x}) = \varphi'^{-1}(\boldsymbol{\xi})$ yields the coordinate transformation $\mathbf{x} = (\varphi \circ \varphi'^{-1})(\boldsymbol{\xi}) = \psi(\boldsymbol{\xi})$ where the mapping $\psi : \mathbb{R}^m \to \mathbb{R}^n$ is expressed by $x^i = \psi^i(\xi^1, \ldots, \xi^m)$, $i = 1, \ldots, n$. These relations describe fully the submanifold M. Evidently, the rank of the matrix $[\partial x^i / \partial \xi^\alpha]$, $\alpha = 1, \ldots, m$ should be m.

If $\phi : M^m \to N^n$ is an embedding, then the subspace $\phi(M) \subseteq N$ is an m-dimensional submanifold of the manifold N.

We take a point $p \in M$ into account and let $q = \phi(p) \in \phi(M) \subseteq N$. Because ϕ is a homeomorphism on its range $\phi(M)$, there exists a chart (U, φ) enclosing the point p of the manifold M and a chart (V, ψ) enclosing the point q of the manifold N such that the open set $\phi(U)$ is contained in the open set V. The rank of the function $\Phi = \psi \circ \phi \circ \varphi^{-1} : \mathbb{R}^m \to \mathbb{R}^n$ which is the local representation of the mapping ϕ is equal to the rank m of the embedding ϕ since φ and ψ are homeomorphisms. Hence, we can rewrite $\Phi : \mathbb{R}^m \to \mathbb{R}^m$ and on the open set $V' = \phi(U) \cap V = \phi(U)$ in the relative topology we have $\psi : V' \to \mathbb{R}^m$. Thus the subspace $\phi(M) \subseteq N$ is an m-dimensional differentiable submanifold of the manifold N. In such a case we sometimes prefer to regard the manifold M as a submanifold of N even if they are actually different manifolds. □

Let the mapping $\phi : M^m \to N^n$ be a submersion. Thus the condition $m \geq n$ will hold and the rank of ϕ will become n. If $Q \subseteq N$ is a submanifold, then the subspace $P = \phi^{-1}(Q) \subseteq M$ is either a submanifold of M or it is empty.

Let us assume that $P = \phi^{-1}(Q)$ is not empty so that $Q \cap \mathcal{R}(\phi) \neq \emptyset$. Since ϕ is a submersion, we can choose the local coordinates $\mathbf{x} = \varphi(p)$ and $\mathbf{y} = \psi(q)$ of the points $p \in P$ and $q = \phi(p) \in Q$ in local charts (U, φ) and (V, ψ) in the form $x^1, x^2, \ldots, x^n, x^{n+1}, \ldots, x^m$ and $y^1 = x^1, y^2 = x^2, \ldots,$ $y^n = x^n$. If the dimension of the submanifold Q is r with $1 \leq r \leq n$, then one can find a coordinate transformation $\mathbf{z} = F(\mathbf{y})$, or $z^1 = F^1(y^1, \ldots, y^n)$, $\ldots, z^n = F^n(y^1, \ldots, y^n)$ such that the local coordinates of the point q can be prescribed by imposing the conditions $z^{r+1} = \cdots = z^n = 0$. We now choose the local coordinates of the point p as follows:

$$\begin{aligned} w^1 &= F^1(x^1, \ldots, x^n), \ldots, w^n = F^n(x^1, \ldots, x^n), \\ w^{n+1} &= x^{n+1}, \ldots, w^m = x^m. \end{aligned}$$

Therefore, the local representation $\mathbf{z} = \Phi(\mathbf{w})$ of the mapping ϕ becomes

$z^1 = w^1, \dots, z^n = w^n$. But the submanifold Q is determined by the conditions $z^{r+1} = \cdots = z^n = 0$. This implies that the subspace $\phi^{-1}(Q)$ in the vicinity of the point p is described by coordinates $(w^1, \dots, w^r, 0, \dots, 0, w^{n+1}, \dots, w^m)$. This is tantamount to say that $\phi^{-1}(Q)$ is an $(m-n+r)$-dimensional submanifold. □

Example 2.4.6. As we have seen before, any open set of a manifold M is an open submanifold [*see p.* 77]. ■

Example 2.4.7. Let us consider a smooth function $\phi : \mathbb{R}^m \to \mathbb{R}$. We further suppose that at a point $\mathbf{x} \in \mathbb{R}^m$, at least one of the partial derivatives $\partial\phi/\partial x^i, i = 1, \dots, n$ does not vanish. Thus the mapping ϕ is a submersion of rank 1. Since we can trivially observe that the singleton $\{0\} \subset \mathbb{R}$ is a 0-dimensional submanifold of the 1-dimensional manifold $\mathbb{R}$, then the subspace $\phi^{-1}(\{0\}) \subset \mathbb{R}^m$, that is, the set $M = \{\mathbf{x} \in \mathbb{R}^m : \phi(\mathbf{x}) = 0\}$ is an $(m-1)$-dimensional submanifold. ■

Example 2.4.8. The function $\phi : (0, \infty) \subset \mathbb{R} \to \mathbb{R}^2$ is given by

$$\phi(t) = \Big(\phi^1(t) = t\cos\frac{1}{t},\ \phi^2(t) = t\sin\frac{1}{t}\Big) \in \mathbb{R}^2.$$

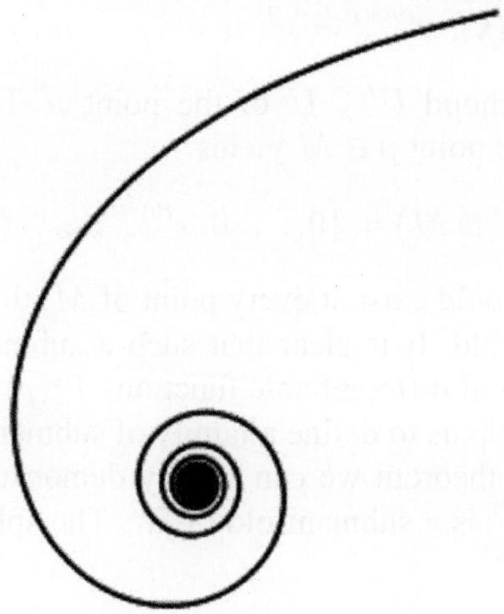

Fig. 2.4.1. Spiral in $\mathbb{R}^2$.

Hence the range $C = \phi\big((0,\infty)\big)$ of the mapping ϕ is a spiral around the point $\mathbf{0}$ in $\mathbb{R}^2$ depicted in Fig. 2.4.1. We obtain $\phi^1(t) \to \infty$ and $\phi^2(t) \to 1$ as $t \to \infty$. We can easily note that this mapping is injective and its rank is 1. Thus it is an injective immersion. The relative topology on C is defined in the usual way by means of open sets $\{C \cap V\}$ where V is an open set in $\mathbb{R}^2$. With respect to these topologies, the mappings ϕ and $\phi^{-1} : C \to (0, \infty)$

are both continuous. Hence ϕ is a homeomorphism, thus it is an embedding. Consequently C is a 1-dimensional submanifold in $\mathbb{R}^2$. ■

Submanifolds can also be determined by means of a set of equations.

Theorem 2.4.1. *We define a subset M of an n-dimensional differentiable manifold N by means of differentiable functions $f^\alpha : N \to \mathbb{R}, \alpha = 1, \dots, m$ where $m \leq n$ as follows*

$$M = \{p \in N : f^\alpha(p) = 0, \alpha = 1, \dots, m\} \subseteq N.$$

We further assume that the rank of the function $f : N \to \mathbb{R}^m$ prescribed by $f(p) = \big(f^1(p), \dots f^m(p)\big)$ is m at each point $p \in M$. In this case M proves to be a submanifold of dimension $n - m$.

Let (U, φ) be a chart containing a point $p \in M$ and let the local coordinates be $\varphi(p) = (x^1, \dots, x^n)$. Since the rank of the mapping f is m on the set M, the matrix $[\partial(f^\alpha \circ \varphi^{-1})/\partial x^i]$ has at least one $m \times m$ square submatrix whose determinant does not vanish. We may rename the variables if necessary so that this square matrix is specified by $[\partial(f^\alpha \circ \varphi^{-1})/\partial x^i], \alpha = 1, \dots, m; i = 1, \dots, m$. Hence, we can perform the following coordinate transformation

$$x'^\alpha = (f^\alpha \circ \varphi^{-1})(\mathbf{x}),\ x'^{m+j} = x^j;\ \alpha = 1, \dots, m, j = 1, \dots, n - m$$

in an open neighbourhood $U' \subseteq U$ of the point p. Thus, the local chart (U', φ') containing the point $p \in M$ yields

$$\varphi'(U' \cap M) = \{0, \dots, 0, x'^{m+1}, \dots, x'^n\}.$$

Since similar charts would exist at every point of M, this set is an $(n - m)$-dimensional submanifold. It is clear that such a submanifold may be also prescribed by a family of differentiable functions $f^\alpha(p) = c^\alpha$ where c^α's are constants. This will help us to define a family of submanifolds. □

By utilising this theorem we can readily demonstrate that $(n-1)$-dimensional sphere $\mathbb{S}^{n-1}$ is a submanifold of $\mathbb{R}^n$. The sphere with a radius R is the subset

$$\mathbb{S}^{n-1} = \{\mathbf{x} \in \mathbb{R}^n : f(\mathbf{x}) = \sum_{i=1}^n (x^i)^2 - R^2 = 0\}.$$

The rank of the function $f : \mathbb{R}^n \to \mathbb{R}$ is 1 at every point $\mathbf{x} \in \mathbb{S}^{n-1}$. Hence, $\mathbb{S}^{n-1}$ is an $(n-1)$-dimensional submanifold. On the other hand, the cone

$$C^{n-1} = \{\mathbf{x} \in \mathbb{R}^n : f(\mathbf{x}) = (x^1)^2 - \sum_{i=2}^n (x^i)^2 = 0\}$$

is not a submanifold of $\mathbb{R}^n$ because the rank of f is 0 at the point $\mathbf{x} = \mathbf{0}$, while it is 1 at all other points. Therefore, if only we delete the point $\mathbf{0}$, then the punctured cone becomes an $(n-1)$-dimensional submanifold of $\mathbb{R}^n$.

If the mapping $\phi : M \to N$ is solely an injective immersion, then the subspace $\phi(M) \subseteq N$ is called an ***immersed manifold***. Unless the mapping ϕ is a homeomorphism on its range, an immersed manifold is obviously not a submanifold.

Example 2.4.9. Let us define the mapping $\phi : \mathbb{R} \to \mathbb{T}^2$ by the relations $z_1 = ae^{it}$ and $z_2 - a = be^{i\alpha t}$ [*see* Example 2.4.5]. Here α is now an irrational number. Hence, we find $t_1 = t_2$ when $\phi(t_1) = \phi(t_2)$. Thus ϕ is injective and its rank is 1. Consequently, it is an injective immersion and $\phi(\mathbb{R})$ becomes an immersed manifold. We can easily show that the set $M = \phi(\mathbb{R})$ is dense in $\mathbb{T}^2$. The mapping ϕ winds the line $\mathbb{R}$ around the torus $\mathbb{T}^2$ without ever traversing the same point on the torus again. In order to prove that the set M is dense in $\mathbb{T}^2$, we have to show that we can find a point in M that is as close as we wish to a given point in $\mathbb{T}^2$. Let us consider an arbitrary point $(ae^{i\omega}, a + be^{i\theta}) \in \mathbb{T}^2$ where $\omega, \theta \in \mathbb{R}$. The distance between the selected point in $\mathbb{T}^2$ and a point in M is given by

$$\begin{aligned}|ae^{i\omega} - ae^{it}| + |a + be^{i\theta} - a - be^{i\alpha t}| &= a|e^{i(\omega-t)} - 1| + b|e^{i(\theta-\alpha t)} - 1|\\ &= a\sqrt{2(1-\cos(\omega-t))} + b\sqrt{2(1-\cos(\theta-\alpha t))}\\ &= 2a\left|\sin\frac{\omega-t}{2}\right| + 2b\left|\sin\frac{\theta-\alpha t}{2}\right|.\end{aligned}$$

Rational numbers are dense in real numbers. Therefore, for each $\epsilon > 0$ and real numbers ω, θ, t, we can find integers p_1, q_1, m and p_2, q_2, n such that the inequalities

$$\left|\frac{\omega-t}{4\pi} - \frac{p_1}{q_1} - m\right| < \epsilon, \quad \left|\frac{\theta-\alpha t}{4\pi} - \frac{p_2}{q_2} - n\right| < \epsilon$$

are satisfied. The integers m and n are so chosen that we ought to have $|p_1/q_1| < 1$ and $|p_2/q_2| < 1$. If we now write $t_1 = t + 4\pi(p_1/q_1)$ and $t_2 = \alpha t + 4\pi(p_2/q_2)$, then the foregoing inequalities take the form

$$|\omega - t_1 - 4\pi m| < 4\pi\epsilon, \quad |\theta - t_2 - 4\pi n| < 4\pi\epsilon.$$

By introducing $t_3 = \max(t_1, t_2)$, these inequalities may be transformed into

$$|\omega - t_3 - 4\pi m| < 4\pi\epsilon, \quad |\theta - t_3 - 4\pi n| < 4\pi\epsilon.$$

Hence, for given real numbers ω, θ we can find a real number t_3 so that one obtains

$$\begin{aligned}
2a\left|\sin\frac{\omega - t}{2}\right| + 2b\left|\sin\frac{\theta - \alpha t}{2}\right| \\
&= 2a\left|\sin\left(\frac{\omega - t_3}{2} - 2\pi m\right)\right| + 2b\left|\sin\left(\frac{\theta - t_3}{2} - 2\pi n\right)\right| \\
&< 2a|\sin 2\pi\epsilon| + 2b|\sin 2\pi\epsilon| < 4\pi\epsilon(a+b)
\end{aligned}$$

It is easy to see that the immersed manifold M is not a submanifold. In fact, under the mapping ϕ the line $\mathbb{R}$ intersects an open set in $\mathbb{T}^2$ infinitely many times. Therefore, an open set in the relative topology on M is the union of infinitely many pieces. Thus it is unbounded. This implies that the image of a bounded open set in $\mathbb{R}$ is unbounded. Hence the mapping ϕ is not continuous with respect to the relative topology, that is, it is not a homeomorphism on its range. ■

2.5. DIFFERENTIABLE CURVES

A *differentiable curve* C on an m-dimensional differentiable manifold M is defined through a differentiable (C^∞) mapping $\gamma : \mathcal{I} \to M$ where $\mathcal{I} = (a, b) \subseteq \mathbb{R}$ is an open interval on the real line. Thus, a point p of the curve $C = \gamma(\mathcal{I}) \subset M$ is given by $p = \gamma(t)$, $t \in \mathcal{I}$. The interval must be open in order to secure differentiability at neighbourhoods of endpoints. If the curve is defined on a closed interval $[a, b]$, then we shall have to assume that the mapping γ admits a C^∞ extension $\overline{\gamma} : (a - \epsilon, b + \epsilon) \to M$ for a number $\epsilon > 0$ so that

$$\overline{\gamma}(t) = \gamma(t), \quad t \in [a, b].$$

To realise the local representation of any point $p = \gamma(t)$ of the curve, it suffices to consider a chart (U, φ) enclosing the point $p \in M$. The locus of the points $\mathbf{x}(t) = \varphi(\gamma(t)) \subset \mathbb{R}^m$ is the local representation of a part of the curve C in the open set $\varphi(U) \subseteq \mathbb{R}^m$. Naturally, when we move on the curve C local representations may change together with charts taken into consideration. By employing the coordinate functions $\varphi^i = g^i \circ \varphi : U \to \mathbb{R}$, $i = 1, \cdots, m$ [*see* p. 71] the parametric representation of the curve C in the open set $\varphi(U)$ is provided by functions $x^i = \varphi^i(\gamma(t)) = \gamma^i(t)$ in local coordinates where we have defined the mappings $\gamma^i = \varphi^i \circ \gamma : \mathcal{I} \subseteq \mathbb{R} \to \mathbb{R}$, $i = 1, \cdots, m$. Since γ is a differentiable mapping, the functions $\gamma^i(t)$ have clearly derivatives of all orders with respect to t. If at every point on the curve, at least one of the first order derivatives does not vanish, then the rank of the mapping γ is 1. In this case, γ becomes an immersion. But the curve may intersect itself, thus we cannot claim that this immersion is injective (Fig. 2.5.1).

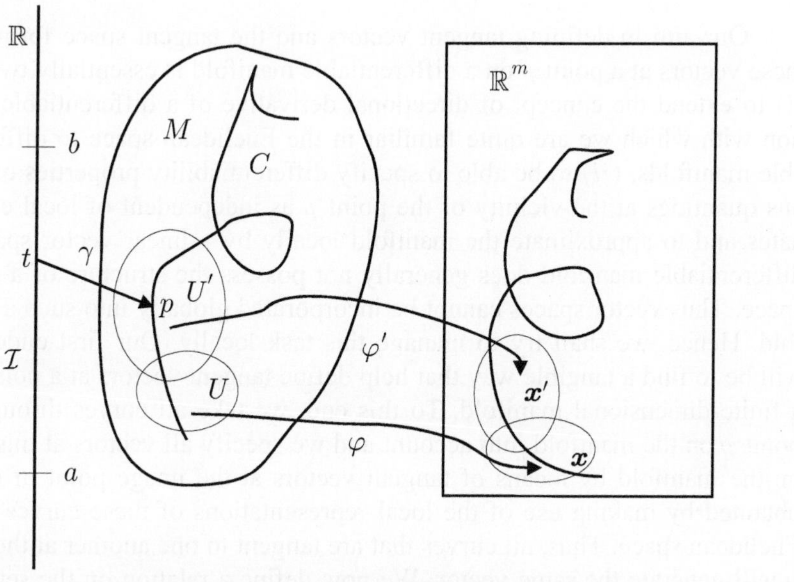

Fig. 2.5.1. A curve on a differentiable manifold.

If the curve C is defined on a closed interval $\mathcal{I} = [a, b]$, we call the points $p_a = \gamma(a)$ and $p_b = \gamma(b)$ the *initial point* and the *end point* of the curve, respectively. We get a *closed curve* if $\gamma(a) = \gamma(b)$. A *simple closed curve* is a closed curve defined on $[a, b]$, however, γ must be an injective mapping on the half-open interval $[a, b)$.

Example 2.5.1. A mapping $\boldsymbol{\gamma} : [0, 2\pi] \to \mathbb{R}^2$ is prescribed by functions $x^1 = \cos t, x^2 = t \sin 2t$. The closed curve in $\mathbb{R}^2$ generated by this mapping is shown in Fig. 2.5.2. We observe that this curve intersects itself. Therefore γ is not an injective mapping. Moreover, it has a corner point.

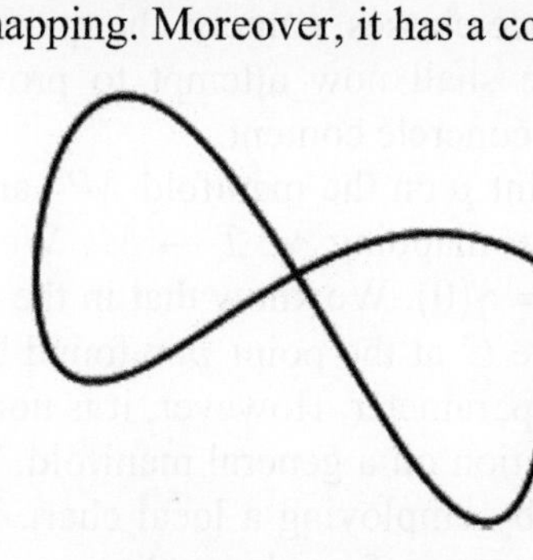

Fig. 2.5.2. A closed curve. ■

2.6. VECTORS. TANGENT SPACES

Our aim in defining tangent vectors and the tangent space formed by these vectors at a point p on a differentiable manifold is essentially twofold: (i) to extend the concept of directional derivative of a differentiable function with which we are quite familiar in the Euclidean space to differentiable manifolds, (ii) to be able to specify differentiability properties of various quantities at the vicinity of the point p as independent of local coordinates and to approximate the manifold locally by a linear vector space. A differentiable manifold does generally not possess the structure of a vector space. Thus vector spaces cannot be incorporated globally into such a manifold. Hence, we shall try to manage this task locally. Our first endeavour will be to find a tangible way that help define tangent vectors at a point p of a finite-dimensional manifold. To this end, we take all curves through the point p on the manifold into account and we specify all vectors at this point on the manifold by means of tangent vectors at the image point of curves obtained by making use of the local representations of these curves in the Euclidean space. Thus, all curves that are tangent to one another at the point p will generate the same vector. We now define a relation on the set of all curves through the point p of the manifold as *being tangent at the point* p. We can readily verify that this is an equivalence relation. Indeed, we see immediately that this relation is *reflexive* (each curve is tangent to itself), *symmetric* (if the curve C_1 is tangent to C_2, then C_2 is tangent to C_1 as well) and *transitive* (if C_1 is tangent to C_2 and C_2 to C_3, then the curve C_1 is obviously tangent to the curve C_3). Hence, all curves through the point p are partitioned into disjoint equivalence classes. All curves in an equivalence class are tangent to one another at the point p, therefore they possess the same tangent vector. We can thus try to identify tangent vectors at a point p of the manifold with equivalence classes of curves through this point. We define the set of equivalence classes, namely, the quotient set as the ***tangent space*** at the point p. We shall now attempt to provide these somewhat abstract ideas with a fully concrete content.

Let us consider a point p on the manifold M^m and a curve C through this point specified by the mapping $\gamma : \mathcal{I} \to M$. We so choose the parameter t of the curve as $p = \gamma(0)$. We know that in the classical analysis, the tangent vector to the curve C at the point p is found by means of differentiation with respect to the parameter. However, it is not possible to apply the usual differentiation operation on a general manifold. Thus we opt to transfer this operation on $\mathbb{R}^m$ by employing a local chart. Let (U, φ) be a chart containing the point p. In terms of local coordinates provided by this chart, local representation C' of the curve C in $\mathbb{R}^m$ is determined parametrically through the differentiable functions $\gamma^i : \mathcal{I} \to \mathbb{R}$ as follows:

$$\overline{x}^i = \gamma^i(t), \; i = 1, \ldots, m. \tag{2.6.1}$$

The local coordinate of the point p is supposed to be $x^i = \gamma^i(0)$. (2.6.1) can now be collectively written as

$$\overline{\mathbf{x}} = \boldsymbol{\gamma}(t) = \gamma^i(t)\mathbf{e}_i$$

where the vectors $\mathbf{e}_i = (0, \ldots, 0, \underset{i}{1}, 0, \ldots 0)$, $i = 1, \ldots, m$ are standard basis vectors for the vector space $\mathbb{R}^m$. As is well known, a tangent vector to the curve C' at a point is specified by its components $\overline{v}^i$ defined by

$$\overline{\mathbf{v}}(t) = \frac{d\overline{\mathbf{x}}}{dt} = \overline{v}^i(t)\mathbf{e}_i, \; \overline{v}^i(t) = \frac{d\overline{x}^i}{dt} = \frac{d\gamma^i}{dt}.$$

Thus, the tangent vector to the curve C' at the point $\mathbf{x} = \varphi(p) \in \mathbb{R}^m$ is given by

$$\mathbf{v} = v^i\mathbf{e}_i, \;\; v^i = \left.\frac{d\gamma^i}{dt}\right|_{t=0}, \; i = 1, \ldots, m. \tag{2.6.2}$$

Since $\gamma^i(t)$ are all smooth functions they can be expanded into a Maclaurin series about the point $t = 0$ [after Scottish mathematician Colin Maclaurin (1698-1746)]. Thus we can write

$$\overline{x}^i = \gamma^i(t) = \gamma^i(0) + \left.\frac{d\gamma^i}{dt}\right|_{t=0} t + \frac{1}{2}\left.\frac{d^2\gamma^i}{dt^2}\right|_{t=0} t^2 + \cdots = x^i + v^i t + o(t).$$

where the *Landau symbol* $o(t)$ [after German mathematician Edmund Georg Hermann Landau (1877-1938)] represents all functions f satisfying the relation $f(t)/t \to 0$ as $t \to 0$. Another curve through the point $\varphi(p)$ can be represented in a similar fashion by expressions

$$\widetilde{\gamma}^{\,i}(t) = x^i + \widetilde{v}^{\,i} t + o(t), \;\;\; \widetilde{v}^{\,i} = \left.\frac{d\widetilde{\gamma}^{\,i}}{dt}\right|_{t=0}.$$

Therefore the *difference* between those two curves is found to be

$$\widetilde{\gamma}^{\,i}(t) - \gamma^i(t) = (\widetilde{v}^{\,i} - v^i)t + o(t).$$

If those two curves are tangent to one another at the point $\varphi(p)$ and have a common tangent vector, then one obtains $\widetilde{v}^{\,i} = v^i$. This, of course, leads to $\widetilde{\gamma}^{\,i}(t) - \gamma^i(t) = o(t)$. Hence, the closeness of two such curves is of second order. It is clear that a relation so defined is an equivalence relation. (2.6.2) implies that tangent vectors at a point $\mathbf{x}$ of $\mathbb{R}^m$ constitute an m-dimensional linear vector space. This vector space is called the *tangent space* at the point $\mathbf{x}$ of $\mathbb{R}^m$ and is denoted by $T_{\mathbf{x}}(\mathbb{R}^m)$. We see at once that the tangent space $T_{\mathbf{x}}(\mathbb{R}^m)$ and $\mathbb{R}^m$ are isomorphic. The isomorphism $\mathbb{R}^m \to T_{\mathbf{x}}(\mathbb{R}^m)$ is provided by the linear mapping that assigns a vector $\mathbf{v} = v^i\mathbf{e}_i \in T_{\mathbf{x}}(\mathbb{R}^m)$ to an ordered m-tuple $(v^1, \dots, v^m) \in \mathbb{R}^m$.

The above approach makes it possible to identify curves tangent to one another at a point p on M as images of curves tangent to one another at the point $\varphi(p)$ in the open set $\varphi(U)$ under the homeomorphism φ^{-1}. We interpret an equivalence class of curves so formed as a tangent vector at a point $p \in M$ to the manifold M. However, since M is generally not endowed with a vector space structure we cannot emplace such vectors into the manifold in the usual sense. In order to achieve this, we have to develop a new but equivalent concept. For this purpose, the classical notion of directional derivative of a function turns out to be very helpful.

We had denoted the set of smooth functions $f : M \to \mathbb{R}$ on a manifold M by $C(M)$. We have seen that this set is an algebra [*see p.* 94]. Henceforth we denote this algebra by $\Lambda^0(M)$.

Let a point $p \in M^m$ be contained in the chart (U, φ). In a neighbourhood of the image point $\mathbf{x} = \varphi(p) \in \varphi(U) \subseteq \mathbb{R}^m$ we define an operator $V'_{\mathbf{x}} : \Lambda^0(\mathbb{R}^m) \to \mathbb{R}$ at that point as follows: this operator will assign a real number to each smooth function $f' \in \Lambda^0(\mathbb{R}^m)$ in association with a given vector $\mathbf{v}(\mathbf{x}) = v^i(\mathbf{x})\mathbf{e}_i$ at that point or, in other words, with a curve C' tangent to this vector at $\mathbf{x}$ by the rule

$$V'_{\mathbf{x}}(f') = \frac{df'(\gamma(t))}{dt}\bigg|_{t=0} = \Big(\frac{d\gamma^i(0)}{dt}\frac{\partial}{\partial x^i}\Big)f' = v^i(\mathbf{x})\frac{\partial f'(\mathbf{x})}{\partial x^i}. \quad (2.6.3)$$

We know that $V'_{\mathbf{x}}(f')$ is the directional derivative of the function f' at the point $\mathbf{x}$ along the curve C', or in the direction of the vector $\mathbf{v}$. Hence the operator $V'_{\mathbf{x}}$ at the point $\mathbf{x}$ can be defined in the following way

$$V'_{\mathbf{x}} = v^i\frac{\partial}{\partial x^i} = \frac{d}{dt}\bigg|_{t=0}. \quad (2.6.4)$$

If there is no ambiguity, we can dispense with the subscript denoting with which point the operator is associated. It is clear from the definition that for every functions $f', g' \in \Lambda^0(\mathbb{R}^n)$ and number $\alpha \in \mathbb{R}$, we can write

$$V'(f' + g') = V'(f') + V'(g'), \quad V'(\alpha f') = \alpha V'(f').$$

Thus V' is a *linear operator* on $\mathbb{R}$. It is also evident that there corresponds a unique operator to each vector $\mathbf{v}$. It is straightforward to see that the set of all these linear operators constitutes a linear vector space. Consider the operators $V_1' = v_1^i \partial/\partial x^i$ and $V_2' = v_2^i \partial/\partial x^i$. We find that

$$\alpha_1 V_1' + \alpha_2 V_2' = (\alpha_1 v_1^i + \alpha_2 v_2^i)\frac{\partial}{\partial x^i} = v^i \frac{\partial}{\partial x^i} = V'$$

for every $\alpha_1, \alpha_2 \in \mathbb{R}$. The mapping $\mathbf{v} \to V'$ between two linear vector spaces is an isomorphism. Indeed, this mapping is linear, because we have $\mathbf{v}_1 + \mathbf{v}_2 \to V_1' + V_2'$, $\alpha\mathbf{v} \to \alpha V'$. This mapping is surjective because each operator V' is generated by a vector $\mathbf{v}$. Let us now suppose that the same operator is associated with two vectors $\mathbf{v}_1$ and $\mathbf{v}_2$. Consequently, for *every* function $f' \in \Lambda^0(\mathbb{R}^n)$ one writes

$$V'(f') = v_1^i \frac{\partial f'}{\partial x^i} = v_2^i \frac{\partial f'}{\partial x^i}.$$

When we choose the function $f' = x^j$, we obtain $V'(x^j) = v_1^i \delta_i^j = v_2^i \delta_i^j$ and $v_1^j = v_2^j$ for $j = 1, \ldots, m$ or $\mathbf{v}_1 = \mathbf{v}_2$. Thus the mapping is injective, hence bijective. In this case the linear vector spaced formed by operators V' is also m-dimensional. Practically, two isomorphic vector spaces can be considered as the same as far as their algebraic properties are concerned. Therefore, instead of the tangent space $T_{\mathbf{x}}(\mathbb{R}^m)$ at a point $\mathbf{x}$ we can take into consideration the isomorphic vector space formed by the operators $V_{\mathbf{x}}'$ at that point.

Let us next consider a curve C on the manifold M through the point $p \in M$ that is determined by a mapping $\gamma : I \to M$, $\gamma(0) = p$. We shall now try to designate similarly an operator V representing the tangent vector of the curve at the point p as a derivative along the curve C. Let us assume that the point p is contained in a chart (U, φ). For each function $f \in \Lambda^0(U)$, we introduce the following operator at the point p

$$V_p(f) = \left.\frac{df\big(\gamma(t)\big)}{dt}\right|_{t=0} = \left.\frac{d(f \circ \gamma)}{dt}\right|_{t=0}. \tag{2.6.5}$$

We determine the function $f' \in \Lambda^0(\mathbb{R}^m)$ such that $f'(\mathbf{x}) = f(p)$ at the point $\mathbf{x} = \varphi(p) \in \mathbb{R}^m$. Hence, this function is given by $f' = f \circ \varphi^{-1}$ and using the relation $f = f' \circ \varphi$, we obtain

$$V_p(f) = \left.\frac{d(f \circ \gamma)}{dt}\right|_{t=0} = \left.\frac{d(f' \circ \varphi \circ \gamma)}{dt}\right|_{t=0} = \left.\frac{df'\big(\gamma(t)\big)}{dt}\right|_{t=0}$$

Therefore, we can write below the defining rule for the operator V_p:

$$V_p(f) = V'_{\mathbf{x}}(f'), \quad \mathbf{x} = \varphi(p). \tag{2.6.6}$$

Thus, the action of the operator V at the point p on a function f is uniquely determined by the components $v^i = dx^i(t)/dt$ of the tangent vector to the curve $C' = \varphi(C)$ at the point $\mathbf{x}$ with local coordinates x^i as follows:

$$V(f) = V_p(f) = \left.\frac{df'}{dt}\right|_{t=0} = v^i\frac{\partial f'}{\partial x^i} = v^i\frac{\partial(f\circ\varphi^{-1})}{\partial x^i}. \tag{2.6.7}$$

(2.6.7) now amply justifies the interpretation that $V(f)$ is the derivative of the function f at a point p along a curve through this point whose tangent vector there is specified by the operator V. We can immediately conclude from the foregoing relations that if the equality $V_1(f) = V_2(f)$ holds for every function $f \in \Lambda^0(U)$, then two curves whose tangent vectors at the point $p \in M$ are given by V_1 and V_2 are tangent to one another at p. Indeed, if we insert coordinate functions $\varphi^j \in \Lambda^0(U), j = 1,\dots,m$ satisfying $\varphi^j(p) = x^j$ into (2.6.7), we find

$$v_1^i\frac{\partial x^j}{\partial x^i} = v_2^i\frac{\partial x^j}{\partial x^i} \text{ and } v_1^i\delta_i^j = v_2^i\delta_i^j$$

leading to $v_1^j = v_2^j, j = 1,\dots,m$. V is a linear operator on $\mathbb{R}$. The relations

$$\begin{aligned}V\big((f+g)(p)\big) &= V\big(f(p)+g(p)\big) = V'\big(f'(\mathbf{x})+g'(\mathbf{x})\big)\\ &= V'\big(f'(\mathbf{x})\big)+V'\big(g'(\mathbf{x})\big) = V\big(f(p)\big)+V\big(g(p)\big)\\ V\big((\alpha f)(p)\big) &= V\big(\alpha f(p)\big) = V'\big(\alpha f'(\mathbf{x})\big) = \alpha V'\big(f'(\mathbf{x})\big) = \alpha V\big(f(p)\big)\end{aligned}$$

imply that $V(f+g) = V(f)+V(g)$ and $V(\alpha f) = \alpha V(f)$. Furthermore, the linear operator V meets the rule given first by German mathematician and philosopher Gottfried Wilhelm von Leibniz (1646-1716):

$$\begin{aligned}V\big((fg)(p)\big) &= V\big(f(p)g(p)\big) = V'\big(f'(\mathbf{x})g'(\mathbf{x})\big) = g'(\mathbf{x})V'\big(f'(\mathbf{x})\big)\\ &+ f'(\mathbf{x})V'\big(g'(\mathbf{x})\big) = g(p)V\big(f(p)\big)+f(p)V\big(g(p)\big)\end{aligned}$$

whence we obtain $V_p(fg) = gV_p(f) + fV_p(g)$ at a point. A linear operator satisfying this *Leibniz rule* on an algebra is called a ***derivation***. When we take notice that the action of the operator V on a function f is specified by (2.6.6), we opt for denoting this operator at the point p by

$$V_p = \left.\frac{d}{dt}\right|_{t=0} = v^i\frac{\partial}{\partial x^i} \tag{2.6.8}$$

with a somewhat slight abuse of notation. As we have mentioned before, the quantity $V_p(f)$ measures the variation in a function $f \in \Lambda^0(M)$ at a point $p \in M$ along a curve C or, in other words, along an equivalence class generated by C, at that point. Let us consider a curve in $\mathbb{R}^m$ defined by

$$\boldsymbol{\gamma}^i(t) = (0, \dots, 0, x^i + t, 0, \dots, 0).$$

This curve is obviously the coordinate line in Cartesian coordinates through the point $(0, \dots, 0, x^i, 0, \dots, 0)$ in $\mathbb{R}^m$. We thus obtain

$$\mathbf{v}_\mathbf{x} = (0, \dots, 0, 1, 0, \dots, 0).$$

We now define a *coordinate line* on M through the point p by the curve $C^i = \varphi^{-1}(\boldsymbol{\gamma}^i(t))$. We then conclude that the operator $\partial/\partial x^i$ helps measure the variation of a function along a coordinate line at the point p.

It is clear that all linear operators V at a point $p \in M$ forms a linear vector space. Due to the relation (2.6.8), this vector space is evidently isomorphic to the tangent space $T_\mathbf{x}(\mathbb{R}^m)$ at the point $\mathbf{x} = \varphi(p)$. Hence, its dimension is m. We call this vector space the ***tangent space*** to the manifold M at the point p and denote it by $T_p(M)$. We also regard the operators V_p as tangent vectors to the manifold M at the point p (Fig.. 2.6.1).

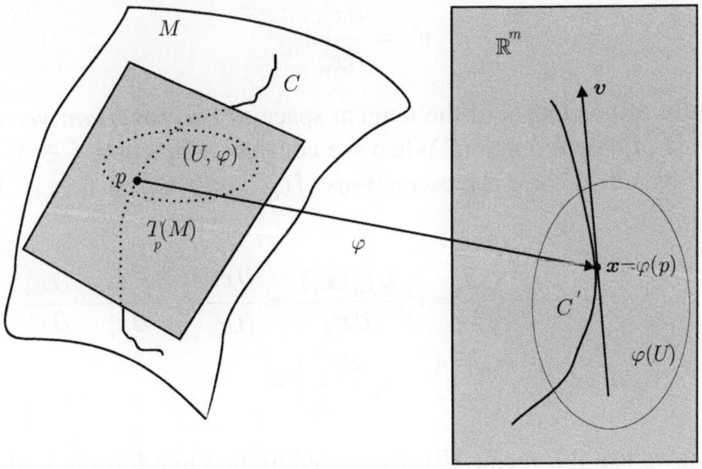

Fig. 2.6.1. Tangent space.

While having defined a vector V at a point $p \in M$ by means of the relation (2.6.8), we utilised the local coordinates provided by a chosen chart at that point. In order that this definition makes sense, we have to prove that the vector, or the operator, V is actually independent of the chosen chart.

Let us take into account two charts $(U_\alpha, \varphi_\alpha)$ and (U_β, φ_β) enclosing the point p. We denote the corresponding local coordinates by $\mathbf{x}_\alpha$ and $\mathbf{x}_\beta$, respectively. The function $\varphi_{\alpha\beta} = \varphi_\beta \circ \varphi_\alpha^{-1} : \varphi_\alpha(U_\alpha) \to \varphi_\beta(U_\beta)$ on the open set $U_\alpha \cap U_\beta$ gives rise to a coordinate transformation $\mathbf{x}_\beta = \varphi_{\alpha\beta}(\mathbf{x}_\alpha)$ (It is obvious that the summation convention will not be valid now on Greek indices). We have then two representations of a curve $C \subset M$ in $\mathbb{R}^m$ through the point p that is determined by the mapping $\gamma : I \to M$:

$$\boldsymbol{\gamma}_\alpha(t) = \varphi_\alpha\big(\gamma(t)\big), \;\; \boldsymbol{\gamma}_\beta(t) = \varphi_\beta\big(\gamma(t)\big).$$

But, in the vicinity of the point p, these two representations are related by

$$\boldsymbol{\gamma}_\beta(t) = \varphi_{\alpha\beta}\big(\boldsymbol{\gamma}_\alpha(t)\big)$$

whence the chain rule leads to

$$\frac{d\gamma_\beta^i}{dt} = \frac{\partial x_\beta^i}{\partial x_\alpha^j}\frac{d\gamma_\alpha^j}{dt}.$$

Thus, at $t = 0$, the components of the tangent vector in two different coordinate systems are connected by the relations

$$v_\beta^i = \frac{\partial x_\beta^i}{\partial x_\alpha^j} v_\alpha^j. \tag{2.6.9}$$

We usually call elements of the tangent space as ***contravariant vectors*** due to this rule of transformation. When we consider a function $f \in \Lambda^0(M)$, it will now have two local representations: $f(p) = f'_\alpha(\mathbf{x}_\alpha) = f'_\beta(\mathbf{x}_\beta)$. We can thus write

$$\begin{aligned} V(f) &= v_\beta^i \frac{\partial f'_\beta(\mathbf{x}_\beta)}{\partial x_\beta^i} = v_\beta^i \frac{\partial f'_\alpha(\mathbf{x}_\alpha)}{\partial x_\beta^i} = \frac{\partial x_\beta^i}{\partial x_\alpha^j} v_\alpha^j \frac{\partial f'_\alpha(\mathbf{x}_\alpha)}{\partial x_\alpha^k}\frac{\partial x_\alpha^k}{\partial x_\beta^i} \\ &= v_\alpha^j \frac{\partial f'_\alpha(\mathbf{x}_\alpha)}{\partial x_\alpha^k}\,\delta_j^k = v_\alpha^j \frac{\partial f'_\alpha(\mathbf{x}_\alpha)}{\partial x_\alpha^j} \end{aligned}$$

which shows that the vector V is expressed in the same form in both charts. Hence, the definition (2.6.8) does not depend on the chosen chart.

Theorem 2.6.1. *m-dimensional tangent space $T_p(M)$ at a point p of an m-dimensional differentiable manifold M has basis vectors, or operators, $\partial/\partial x^i$, $i = 1, \dots, m$ determined by a choice of a local chart.*

Since the vector space $T_p(M)$ is m-dimensional, the set of vectors $\{\dfrac{\partial}{\partial x^i}\}$, where $\{x^i\}$ are local coordinates, must be linearly independent in order to constitute a basis. Let us write

$$V_0 = c^i \frac{\partial}{\partial x^i} = 0$$

where c^i, $i = 1, \dots, m$ are arbitrary constants. Therefore, we ought to get $V_0(f) = 0$ for *every* function $f \in \Lambda^0(M)$. Then, if we introduce the coordinate functions $\varphi^j \in \Lambda^0(M), j = 1, \dots, n$ into that expression, we find that

$$c^i \frac{\partial x^j}{\partial x^i} = c^i \delta_i^j = c^j = 0, \ j = 1, \dots, m.$$

Consequently, the set $\{\partial/\partial x^i\}$ is linearly independent. □

The set $\{\partial/\partial x^i\}$ at the point p is called the ***natural basis*** or ***coordinate basis*** of the tangent space $T_p(M)$. The local coordinates generating this basis will sometimes be called ***natural coordinates***. Let

$$V = v^i \frac{\partial}{\partial x^i}$$

be a tangent vector at the point p. We then obtain for a coordinate function

$$V(\varphi^k) = v^i \frac{\partial x^k}{\partial x^i} = v^i \delta_i^k = v^k. \tag{2.6.10}$$

Thus, we can write

$$V = V(\varphi^i) \frac{\partial}{\partial x^i}. \tag{2.6.11}$$

Evidently, there is an isomorphism between $T_p(M)$ and $\mathbb{R}^m$ provided by the mapping $(v^1, \dots, v^m) \to V_p$.

So far we have defined a tangent space $T_p(M)$ associated with each point of the manifold that contains all "vectors" tangent to the manifold at that point. We can construct a ***vector field*** by a set of vectors formed by choosing a vector $V_p \in T_p(M)$ at each point p of the manifold. We can denote a vector field by $V(p), p \in M$. A vector of the field at a point p can then be enounced as

$$V(p) = v^i(\mathbf{x}) \frac{\partial}{\partial x^i}, \quad \mathbf{x} = \varphi(p) \tag{2.6.12}$$

by employing a chart (U, φ). We have to note that as the point p moves on the manifold, the vector field might be represented by different local coordinates originated from different charts. When we say that the coordinate cover of the manifold M is given by $(x^1, \dots, x^m)$, we actually mean the union of such coordinate systems that might be different in charts covering the manifold. If the functions $v^i(\mathbf{x})$ are all smooth, then we say that V *is a smooth vector field.* When V is a smooth vector field, we deduce that it has the form $V : \Lambda^0(M) \to \Lambda^0(M)$ as a linear operator.

2.7. DIFFERENTIAL OF A MAP BETWEEN MANIFOLDS

Let M^m and N^n be two differentiable manifolds and $\phi : M \to N$ be a differentiable mapping. We know that to each smooth function $g \in \Lambda^0(N)$ there corresponds a smooth function $f = \phi^* g \in \Lambda^0(M)$ [*see* p. 98]. The mapping $\phi^* : \Lambda^0(N) \to \Lambda^0(M)$ is generated by ϕ in the form $\phi^* g = g \circ \phi$ for all $g \in \Lambda^0(N)$. We now try to find a mapping $\phi_* : T_p(M) \to T_{\phi(p)}(N)$ in conjunction with the mapping ϕ that transforms the equivalence class of curves that are tangent at a point $p \in M$ into an equivalence class of curves that are tangent at the point $q = \phi(p) \in N$. Let us now choose a vector $V \in T_p(M)$ and determine a vector $V^* \in T_{\phi(p)}(N)$ such that the equality

$$V(\phi^* g) = V(g \circ \phi) = V^*(g) \tag{2.7.1}$$

is to be satisfied for *all* functions $g \in \Lambda^0(N)$. We can also express this relation for all $g \in \Lambda^0(N)$ as follows:

$$(\phi_* V)(g) = V(\phi^* g), \quad \phi_* : T_p(M) \to T_{\phi(p)}(N) \tag{2.7.2}$$

where $V^* = \phi_* V$. The mapping ϕ_*, which will also be denoted occasionally by $d\phi$, is called the ***differential*** of the mapping ϕ at the point p.

Let us assume that a curve C on a manifold M is specified by a mapping $\gamma : \mathcal{I} \to M$. We also suppose that $0 \in \mathcal{I}$ and $p = \gamma(0)$. The image C^* of the curve C in the manifold N under the mapping ϕ is given by the mapping $\gamma^* = \phi \circ \gamma : \mathcal{I} \to N$. We consider a vector V that is tangent to the curve C at the point p. For any function $g \in \Lambda^0(N)$, we can write

$$\begin{aligned} V(g \circ \phi) &= \left.\frac{d\big((g \circ \phi) \circ \gamma\big)}{dt}\right|_{t=0} = \left.\frac{d\big(g \circ (\phi \circ \gamma)\big)}{dt}\right|_{t=0} \\ &= \left.\frac{d(g \circ \gamma^*)}{dt}\right|_{t=0} = V^*(g). \end{aligned} \tag{2.7.3}$$

Here we make use of the associativity of the composition. We deduce from the relation (2.7.3) that the vector V^* is tangent to the image curve $C^* = \phi(C)$ at the point $\phi(p) \in N$.

ϕ_* is a linear operator on real numbers. In fact, if we consider a real number α and vectors $V_1, V_2 \in T_p(M)$, we see that ϕ_* obeys the rules

$$\begin{aligned}\phi_*(V_1+V_2)(g) &= (V_1+V_2)(\phi^*g) = V_1(\phi^*g)+V_2(\phi^*g)\\ &= \phi_*V_1(g)+\phi_*V_2(g) = (\phi_*V_1+\phi_*V_2)(g)\\ \phi_*(\alpha V)(g) &= \alpha V(\phi^*g) = \alpha\phi_*(V)(g)\end{aligned}$$

for all functions $g \in \Lambda^0(N)$. That proves the linearity of ϕ_* at the point p:

$$\begin{aligned}\phi_*(V_1+V_2) &= \phi_*V_1+\phi_*V_2,\\ \phi_*(\alpha V) &= \alpha\phi_*V.\end{aligned}$$

We now manage to endow the operator ϕ_* so defined in the above with a more concrete structure by utilising local charts in manifolds M and N. Let us assume that the point $p \in M$ belongs to a chart (U, φ), and the point $q = \phi(p) \in N$ belongs to a chart (V, ψ). We denote the local coordinates by $\mathbf{x} = \varphi(p)$, $\mathbf{y} = \psi(q) = (\psi\circ\phi)(p) = (\psi\circ\phi\circ\varphi^{-1})(\mathbf{x}) = \Phi(\mathbf{x})$ from which we can deduce that $\phi = \psi^{-1}\circ\Phi\circ\varphi$. Thus, the local coordinates of corresponding points under the mapping ϕ are functionally related by $y^\alpha = \Phi^\alpha(x^1, \ldots, x^m), \alpha = 1, \ldots, n$. By means of functions

$$\begin{aligned}(g\circ\phi)' &= g\circ\phi\circ\varphi^{-1} \in \Lambda^0(\mathbb{R}^m),\\ g' &= g\circ\psi^{-1} \in \Lambda^0(\mathbb{R}^n)\end{aligned}$$

where $g \in \Lambda^0(N)$, we find that $(g\circ\phi)' = g'\circ\psi\circ\phi\circ\varphi^{-1} = g'\circ\Phi$. Thus, for every function $g' \in \Lambda^0(\mathbb{R}^n)$, the expression (2.7.1) takes the form

$$v^{*\alpha}\frac{\partial g'(\mathbf{y})}{\partial y^\alpha} = v^i\frac{\partial g'(\Phi(\mathbf{x}))}{\partial x^i} = v^i\frac{\partial g'}{\partial y^\alpha}\frac{\partial \Phi^\alpha}{\partial x^i}.$$

which leads to the relation

$$V^* = \phi_*V = v^{*\alpha}\frac{\partial}{\partial y^\alpha} = v^i\frac{\partial\Phi^\alpha}{\partial x^i}\frac{\partial}{\partial y^\alpha} \in T_{\phi(p)}(N) \qquad (2.7.4)$$

where $V = v^i\partial/\partial x^i \in T_p(M)$. Consequently, we deduce that the mapping $\phi_* : T_p(M) \to T_q(N)$ transforms a vector at the point $p \in M$ with components v^i in local coordinates to a vector at the point $q = \phi(p) \in N$ with components

$$v^{*\alpha}(\phi(p)) = \Big(\frac{\partial\Phi^\alpha}{\partial x^i}v^i\Big)(p) \qquad (2.7.5)$$

in local coordinates. This transformation is governed by the Jacobian matrix $\mathbf{J}(\phi) = [\partial\Phi^\alpha/\partial x^i]$ of the mapping ϕ. If only the mapping ϕ has an inverse $\phi^{-1} : N \to M$, then the relation (2.7.5) is expressible as dependent of the point $q \in N$ so that one will then be able to write

$$v^{*\alpha}(q) = \Big[\Big(\frac{\partial \Phi^\alpha}{\partial x^i} v^i\Big) \circ \phi^{-1}\Big](q).$$

If such is the case, one readily observes that the following relation is valid

$$\phi_*(fV_1 + gV_2) = \big((\phi^{-1})^* f\big)(\phi_* V_1) + \big((\phi^{-1})^* g\big)(\phi_* V_2)$$

for any $f, g \in \Lambda^0(M)$ and $V_1, V_2 \in T_p(M)$.

A basis vector

$$\frac{\partial}{\partial x^i} = \delta_i^j \frac{\partial}{\partial x^j}$$

in $T_p(M)$ is transformed in view of (2.7.4) by the operator ϕ_* to a vector

$$\phi_*\Big(\frac{\partial}{\partial x^i}\Big) = \delta_i^j \frac{\partial \Phi^\alpha}{\partial x^j} \frac{\partial}{\partial y^\alpha} = \frac{\partial \Phi^\alpha}{\partial x^i} \frac{\partial}{\partial y^\alpha} \tag{2.7.6}$$

in $T_{\phi(p)}(N)$. Therefore, the matrix representing the linear operator ϕ_* with respect to *natural bases* at the points p and q is the Jacobian matrix $\mathbf{J}(\phi)$. Obviously, the rank of the matrix $\mathbf{J}(\phi)$ at a point $p \in M$ gives the number of linearly independent vectors in the tangent space $T_{\phi(p)}(N)$. If the linear operator $\phi_* = d\phi$ at the point $p \in M$ is surjective, then the rank of $\mathbf{J}(\phi)$ is n. If ϕ_* is injective, the rank of $\mathbf{J}(\phi)$ is m. In that case, ϕ is a submersion if ϕ_* is surjective at every point $p \in M$, whereas ϕ is an immersion if ϕ_* is injective everywhere. When $m = n$ and $\det \mathbf{J}(\phi) \neq 0$, then ϕ_* is an isomorphism and there is an inverse $(\phi_*)^{-1} : T_q(N) \to T_{\phi^{-1}(q)}(M)$ at the point $q \in N$ which is clearly represented with respect to natural bases by the inverse matrix $\mathbf{J}^{-1}$. This means that the equation $\mathbf{y} = \Phi(\mathbf{x})$ has a differentiable inverse $\mathbf{x} = \Phi^{-1}(\mathbf{y})$ in a neighbourhood of the point q in accordance with the celebrated inverse mapping theorem. We can now introduce the mapping $\psi = \varphi^{-1} \circ \Phi^{-1} \circ \varphi' : N \to M$. Then we immediately obtain the composition $\phi \circ \psi = \varphi'^{-1} \circ \Phi \circ \varphi \circ \varphi^{-1} \circ \Phi^{-1} \circ \varphi' = i_N$. Similarly, we come up with $\psi \circ \phi = i_M$ implying that $\psi = \phi^{-1}$ and ψ is differentiable. We thus conclude that *the mapping ϕ becomes a local diffeomorphism at the point $p \in M$ if ϕ_* is an isomorphism at p.*

A point $q \in N$ is called a ***regular value*** of the smooth mapping ϕ if $d\phi : T_p(M) \to T_q(N)$ is surjective at every point p such that $q = \phi(p)$. A point $p \in M$ is then called a ***regular point*** of ϕ if $d\phi : T_p(M) \to T_{\phi(p)}(N)$ is surjective. A point $q \in N$ that is not a regular value is called a ***critical value*** of ϕ. If q is such a point, then the rank of $\mathbf{J}(\phi)$ at points p satisfying $q = \phi(p)$ is less than n. A point $p \in M$ is then called a ***critical point*** of ϕ if

$d\phi$ is not surjective at that point. An important theorem known as the ***Sard theorem*** [after American mathematician Arthur Sard (1909-1980)] states that for second countable manifolds critical values constitute a *null* subset (a set of measure zero) of the manifold N.

Let an m-dimensional smooth manifold M be ***second countable****, and consequently,* ***separable****. It can be demonstrated that such a manifold can be immersed in at most $2m$-dimensional Euclidean space $\mathbb{R}^{2m}$ ($\mathbb{R}^{2m-1}$ if $m > 1$), or it can be embedded in at most $(2m+1)$-dimensional Euclidean space $\mathbb{R}^{2m+1}$ ($\mathbb{R}^{2m}$ if M is not an analytical manifold). These results are known as* ***Whitney's theorems***. We confine ourselves only to say a little bit about the proof. We assume that an m-dimensional manifold M has transversal self-intersections. The main idea of the proof is the possibility of removing self-intersections by embedding the space $\mathbb{R}^m$ into a higher dimensional Euclidean space. Whitney has shown that one can construct an immersion $\phi : M^m \to \mathbb{R}^{2m}$ by removing all self-intersections or double-points and then resorting to the Sard theorem. Since M is locally homeomorphic to $\mathbb{R}^m$, Whitney has introduced a local immersion $\psi_m : \mathbb{R}^m \to \mathbb{R}^{2m}$ that is approximately linear outside of the unit ball containing a single double-point. He has further assumed that the local chart is so parametrised by $(u_1, u_2, \dots, u_m) \in \mathbb{R}^m$ that the double point is given by

$$\mathbf{x}(1, 0, \dots, 0) = \mathbf{x}(-1, 0, \dots, 0).$$

Then we easily verify that the local mapping defined by

$$\psi_m(u_1, u_2, \dots, u_m) = \Big(\frac{1}{u}, u_1 - \frac{2u_1}{u}, \frac{u_1 u_2}{u}, u_2, \cdots, \frac{u_1 u_m}{u}\Big) \in \mathbb{R}^{2m}$$

where $u = (1 + u_1^2)(1 + u_2^2)\cdots(1 + u_m^2)$ is an immersion for all $m \geq 1$ removing the double-point. In fact, we observe that

$$\psi_m(1, 0, \dots, 0) = (1, -1, 0, \cdots, 0),$$
$$\psi_m(-1, 0, \dots, 0) = (1, 1, 0, \cdots, 0).$$

Actually, it can be verified that ψ_m is an embedding except for the double-point. Furthermore, if the norm $\|\mathbf{x}(u_1, u_2, \dots, u_m)\|$ is large, then ψ_m becomes approximately the linear embedding

$$\psi_m(u_1, u_2, \dots, u_m) \approx (0, u_1, 0, u_2, \cdots, 0, u_m).$$

Let M_1, M_2, M_3 be three differentiable manifolds and $\phi_1 : M_1 \to M_2$, $\phi_2 : M_2 \to M_3$ be two differentiable mappings. Let us consider the composition $\phi_2 \circ \phi_1 : M_1 \to M_3$. For every $h \in \Lambda^0(M_3)$ and $V \in T_p(M_1)$, we can write

$$(\phi_2 \circ \phi_1)_* V(h) = V(h \circ \phi_2 \circ \phi_1) = V\big((h \circ \phi_2) \circ \phi_1\big) = (\phi_1)_* V(h \circ \phi_2)$$
$$= \big[(\phi_2)_* \big((\phi_1)_* V\big)\big](h) = \big((\phi_2)_* \circ (\phi_1)_*\big) V(h).$$

We thus conclude that

$$(\phi_2 \circ \phi_1)_* = (\phi_2)_* \circ (\phi_1)_* \ \text{ or } \ d(\phi_2 \circ \phi_1) = d\phi_2 \circ d\phi_1. \qquad (2.7.7)$$

This is known as the ***chain rule***. Let us note that the relation (2.7.7) actually implies that

$$d(\phi_2 \circ \phi_1)(p) = d\phi_2\big(\phi_1(p)\big) \circ d\phi_1(p)$$

at a point p of M.

Let $i_M : M \to M$ be the identity mapping so that we have $i_M(p) = p$ for all $p \in M$. Accordingly one has $di_M : T_p(M) \to T_p(M)$. Since

$$di_M V(f) = V(f \circ i_M) = V(f)$$

for all $f \in \Lambda^0(M)$, we obtain $di_M V = V$ and finally $di_M = i_{T_p(M)}$. $i_{T_p(M)}$ is the identity operator on the vector space $T_p(M)$.

Let the mapping $\phi : M \to N$ be a diffeomorphism so that the mapping $\phi^{-1} : N \to M$ also exists and differentiable. Hence, we get $\phi^{-1} \circ \phi = i_M$, $\phi \circ \phi^{-1} = i_N$ and differentials of these mappings yield in view of (2.7.7)

$$d(\phi^{-1} \circ \phi) = d\phi^{-1} \circ d\phi = i_{T_p(M)},$$
$$d(\phi \circ \phi^{-1}) = d\phi \circ d\phi^{-1} = i_{T_{\phi(p)}(N)}.$$

We thus infer that $d\phi^{-1} = (d\phi)^{-1}$. This result implies that *the linear operator $d\phi$ between tangent spaces $T_p(M)$ and $T_{\phi(p)}(N)$ is an isomorphism since it is a regular operator if ϕ is a diffeomorphism.* If we recall the statement made in p. 122 we can obviously say that *a differentiable mapping $\phi : M \to N$ is a local diffeomorphism at a point $p \in M$ if and only if the linear operator $d\phi(p) : T_p(M) \to T_{\phi(p)}(N)$ that is the differential of ϕ is an isomorphism.* Of course, this statement will make sense if only if tangent spaces $T_p(M)$ and $T_{\phi(p)}(N)$ have the same dimension.

While defining the differential of a mapping between manifolds M and N, we come up with a rather special situation if one of these manifolds is $\mathbb{R}$. Let us first take $M = \mathcal{I}$ where $\mathcal{I} \subseteq \mathbb{R}$ is an open interval and define a curve C on the manifold N by the differentiable mapping $\gamma : \mathcal{I} \to N$. Therefore, the differential of the mapping γ at a point $t \in \mathcal{I}$ is a linear operator $d\gamma = \gamma_* : T_t(\mathcal{I}) \to T_p(N)$ where $p = \gamma(t)$. Since the tangent vector in $\mathbb{R}$ is of the form d/dt, the tangent vector to the curve C at the point $p = \gamma(t) \in N$ is given by

$$V = \gamma_*\Big(\frac{d}{dt}\Big),\ t \in \mathcal{I}$$

In view of (2.7.3), the tangent vector, say, at a point $\gamma(0) \in N$ will satisfy the relation

$$V(g) = \frac{d(g \circ \gamma)}{dt}\bigg|_{t=0},\ \forall g \in \Lambda^0(N).$$

If we make use of the equality (2.7.4) and notice that the chart on M is simply $(\mathcal{I}, i_{\mathbb{R}})$, we obtain the tangent vector V in terms of local coordinates $y^\alpha(t) = \varphi'^\alpha(\gamma(t)), \alpha = 1, \dots, n$ as

$$V = \gamma_*\Big(\frac{d}{dt}\Big) = \frac{dy^\alpha}{dt}\frac{\partial}{\partial y^\alpha}. \tag{2.7.8}$$

Let us now take $N = \mathbb{R}$ and let $\phi : M \to \mathbb{R}$ be a differentiable mapping. The chart on N is now $(\mathbb{R}, i_{\mathbb{R}})$ so it follows that $\Phi = i_{\mathbb{R}} \circ \phi \circ \varphi^{-1} = \phi \circ \varphi^{-1}$. Thus (2.7.4) yields for a vector $V \in T_p(M), p \in M$

$$V^* = \phi_* V = v^i \frac{\partial \Phi}{\partial x^i}\frac{d}{dt} = V(\phi)\frac{d}{dt},\ t \in \mathbb{R}. \tag{2.7.9}$$

Since the tangent space $T_t(\mathbb{R})$ is isomorphic to the linear vector space $\mathbb{R}$, we can take as a basis vector $d/dt \mapsto 1$ and write $\phi_* : T_p(M) \to \mathbb{R}$ so that we obtain $\phi_* V = V(\phi)$. Thus this interpretation allows us to say that the number $d\phi(p)V = \phi_*(p)V$ gives the derivative of the function ϕ at the point $p \in M$ in the direction of V. In this case the operator ϕ_* assigns a real number to every vector in the tangent space $T_p(M)$. Hence, the linear operator $\phi_* = d\phi$ turns out to be actually a linear functional on $T_p(M)$ and, consequently, it can be regarded as an element of the dual space $T_p^*(M)$. Let us now consider the vector $V = \partial/\partial x^i$ whose components are simply $v^i = 1, v^j = 0, j \neq i$. We thus conclude that

$$d\phi\Big(\frac{\partial}{\partial x^i}\Big) = \frac{\partial \Phi}{\partial x^i} = \frac{\partial(\phi \circ \varphi^{-1})}{\partial x^i}.$$

We now insert the coordinate function $\phi = \varphi^j$ into the foregoing general expression. Since $\varphi^j(p) = x^j$, we obtain

$$d\varphi^j\Big(\frac{\partial}{\partial x^i}\Big) = dx^j\Big(\frac{\partial}{\partial x^i}\Big) = \frac{\partial x^j}{\partial x^i} = \delta^j_i.$$

This means that the elements $\{dx^1, \dots, dx^m\}$ constitute a *reciprocal basis* for the dual vector space $T_p^*(M)$. This results in

$$d\phi = \alpha_i dx^i, \ \alpha_i = d\phi\Big(\frac{\partial}{\partial x^i}\Big) = \frac{\partial \Phi}{\partial x^i} \text{ and } d\phi = \frac{\partial \Phi}{\partial x^i} dx^i$$

which coincides with the classical definition of differential of a function. If we consider a vector $V = v^i \partial/\partial x^i$, then we find that $dx^i(V) = v^i$ and $d\phi(V) = v^i \partial\Phi/\partial x^i = V(\phi)$.

Finally, let us consider the mapping $\varphi : U \to \mathbb{R}^m$ of a chart (U, φ) at a point $p \in M$. Since $\varphi(p) = \mathbf{x}$, we get $\varphi_* = d\varphi : T_p(M) \to T_{\mathbf{x}}(\mathbb{R}^m) \simeq \mathbb{R}^m$. Consequently, we can write

$$d\varphi V(f) = V(f \circ \varphi)$$

for any $f \in \Lambda^0(\mathbb{R}^m)$ and $V \in T_p(U)$. On the other hand, due to (2.6.7) and (2.6.6) we obtain

$$d\varphi V(f) = V(f \circ \varphi) = v^i \frac{\partial (f \circ \varphi \circ \varphi^{-1})}{\partial x^i} = v^i \frac{\partial f}{\partial x^i} = V'(f),$$

so we find that $d\varphi V = V'$. Thus, the operator $d\varphi$ assigns an element $\mathbf{v} = (v^1, \dots, v^m) \in \mathbb{R}^m$ to a vector $V \in T_p(U)$. It is straightforward to verify that the operator $d\varphi : T_p(U) \to \mathbb{R}^m$ is an isomorphism.

We now take into account the inverse mapping $\varphi^{-1} : \mathbb{R}^m \to U$. Then we get $d\varphi^{-1} : \mathbb{R}^m \to T_p(U)$ and we obtain $d\varphi^{-1} V'(f) = V'(f \circ \varphi^{-1})$ $= V(f)$ for all $f \in \Lambda^0(M)$ and $V' \in T_p(\mathbb{R}^m)$ yielding the relation $d\varphi^{-1} V'$ $= V$. We thus obtain $d\varphi^{-1} = (d\varphi)^{-1}$. Hence, we conclude that the mapping φ of a chart is a diffeomorphism.

2.8. VECTOR FIELDS. TANGENT BUNDLE

We have seen that we can construct a vector field on a manifold M^m by associating a vector $V(p)$ in the tangent space $T_p(M)$ to each point $p \in M$. If we choose the natural basis in each tangent space $T_p(M)$ a vector field is now expressible as

$$V(p) = v^i(\mathbf{x}) \frac{\partial}{\partial x^i}, \quad \mathbf{x} = \varphi_\alpha(p), \ p \in U_\alpha \tag{2.8.1}$$

where $(U_\alpha, \varphi_\alpha)$ is a chart and $\cup\, U_\alpha = M$. We know that if $v^i : \mathbb{R}^m \to \mathbb{R}$, $i = 1, \dots, m$ are all smooth functions, $V(p)$ is called a smooth vector field. Evidently, a smooth vector field is built by a linear combination of natural basis vectors with functions chosen from the set $C^\infty(\mathbb{R}^m)$. It is known that the set $C^\infty(\mathbb{R}^m)$ is a commutative ring. If we consider a non-zero function its inverse $1/f(\mathbf{x})$ does not exist at points satisfying the equation $f(\mathbf{x}) = 0$. Therefore, although the vector field V designates an element of a vector space at every point $p \in M$, it constitutes actually a ***module*** $\mathfrak{V}(M)$ on the manifold M. In fact, sum of two vector fields and multiplication of a vector field by a smooth function are again vector fields. It goes without saying that $\mathfrak{V}(M)$ reduces to a linear vector space on real numbers.

Let us define a set $T(M)$ as the union of *disjoint* tangent spaces at all points of a manifold M:

$$T(M) = \bigcup_{p\in M} T_p(M) = \{(p, V) : p \in M, V \in T_p(M)\}. \qquad (2.8.2)$$

It is obvious that this set is produced as the union of sets each of which is obtained by attaching to each point $p \in M$ the linear vector space $T_p(M)$ at that point. We shall now try to equip the set $T(M)$ with a differentiable structure of $2m$-dimension. The differentiable manifold $T(M)$ so structured will be called the ***tangent bundle*** of the manifold M. The set M is named as the ***base*** and tangent spaces as the ***fibres*** of the fibre bundle. The ***natural projection***

$$\pi : T(M) \to M, \quad \pi(p, V) = p, \quad V \in T_p(M) \qquad (2.8.3)$$

projects every vector in a tangent space to its base point $p \in M$ to which a particular fibre is attached. It is clear that we can write $T_p(M) = \pi^{-1}(\{p\})$. Moreover, let us consider the set $\mathcal{V} = \pi^{-1}(U) \subset T(M)$ corresponding to an open set $U \in \mathfrak{M}$ where $\mathfrak{M}$ is the topology on M. Because of the properties of the set function π^{-1} we can write obviously

$$\bigcup_{U\in\mathfrak{M}} \mathcal{V} = \bigcup_{U\in\mathfrak{M}} \pi^{-1}(U) = \pi^{-1}\big(\bigcup_{U\in\mathfrak{M}} U\big) = \pi^{-1}(M) = T(M),$$
$$\emptyset = \pi^{-1}(\emptyset).$$

Furthermore, if Λ is an index set, we have the relations

$$\bigcup_{\lambda\in\Lambda} \mathcal{V}_\lambda = \bigcup_{\lambda\in\Lambda} \pi^{-1}(U_\lambda) = \pi^{-1}\big(\bigcup_{\lambda\in\Lambda} U_\lambda\big), U_\lambda \in \mathfrak{M}$$
$$\bigcap_{i=1}^{n} \mathcal{V}_{\lambda_i} = \bigcap_{i=1}^{n} \pi^{-1}(U_{\lambda_i}) = \pi^{-1}\big(\bigcap_{i=1}^{n} U_{\lambda_i}\big), \lambda_i \in \Lambda, U_{\lambda_i} \in \mathfrak{M}.$$

Therefore the class $\mathfrak{T} = \{\mathcal{V} = \pi^{-1}(U) : U \in \mathfrak{M}\}$ is a topology on the set $T(M)$ and $\mathcal{V}$ is an open set in $\mathfrak{T}$. It is clear that the projection π is continuous in this topology. The structure of the tangent bundle is schematically depicted in Fig. 2.8.1.

We suppose that an atlas on the manifold M is given by the family of charts $\mathcal{A} = \{(U_\alpha, \varphi_\alpha) : \alpha \in \mathcal{I}\}$. The set

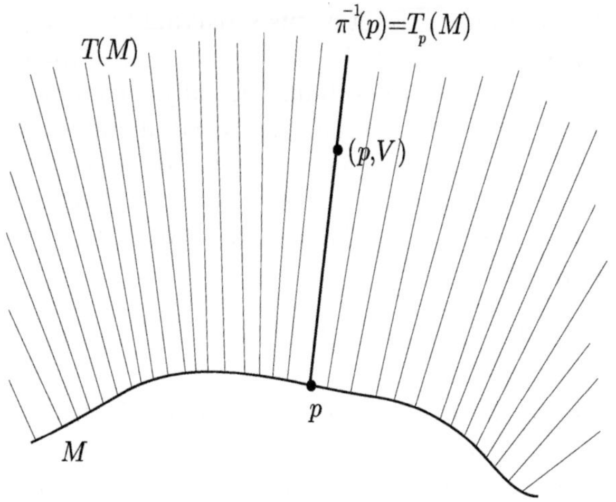

Fig. 2.8.1. Tangent bundle.

$$\mathcal{V}_\alpha = T(U_\alpha) = \pi^{-1}(U_\alpha) \subseteq T(M)$$

will be open in the topology $\mathfrak{T}$. Let us consider a point $(p, V) \in T(M)$. The point $p \in M$ will be located inside a chart $(U_\alpha, \varphi_\alpha)$ of the manifold M and the point (p, V) will be in the open set $\mathcal{V}_\alpha = \pi^{-1}(U_\alpha)$. Hence, in terms of local coordinates $\mathbf{x} = (x^1, \dots, x^m) \in \mathbb{R}^m$ in the open set $\varphi_\alpha(U_\alpha) \subseteq \mathbb{R}^m$, a vector $V \in \mathcal{V}_\alpha$ is expressible as

$$V = v^i \frac{\partial}{\partial x^i}, \ \mathbf{v} = (v^1, \dots, v^m) \in \mathbb{R}^m.$$

We define the mapping $\psi_\alpha : \mathcal{V}_\alpha \to \varphi_\alpha(U_{\underline{\alpha}}) \times \mathbb{R}^m \subseteq \mathbb{R}^{2m}$ in such a way that, for all points $(p, V) \in \mathcal{V}_\alpha$ we get

$$\begin{aligned} \psi_\alpha(p, V) &= \big(\varphi_\alpha(p), d\varphi_\alpha V\big) \\ &= (x^1, \dots, x^m, v^1, \dots, v^m) \in \mathbb{R}^{2m}. \end{aligned} \tag{2.8.4}$$

It is clear that the mapping ψ_α is a homeomorphism. We shall now demonstrate that the family $\{\big(\mathcal{V}_\alpha = \pi^{-1}(U_\alpha), \psi_\alpha\big) : \alpha \in \mathcal{I}\}$ constitutes an atlas on the topological space $T(M)$. We know that $T(M) = \bigcup_{\alpha \in \mathcal{I}} \mathcal{V}_\alpha$. Let us now consider two charts $\big(\mathcal{V}_\alpha, \psi_\alpha\big)$ and $\big(\mathcal{V}_\beta, \psi_\beta\big)$ (the summation convention will of course be suspended on Greek indices). We have to prove that the transition mapping

$$\psi_{\alpha\beta} = \psi_\beta \circ \psi_\alpha^{-1} : \psi_\alpha(\mathcal{V}_\alpha \cap \mathcal{V}_\beta) \subseteq \mathbb{R}^{2m} \to \psi_\beta(\mathcal{V}_\alpha \cap \mathcal{V}_\beta) \subseteq \mathbb{R}^{2m}$$

is smooth. It follows from the relation

$$\begin{aligned}(p, V) = \psi_\alpha^{-1}(\mathbf{x}, \mathbf{v}) &= \big(\varphi_\alpha^{-1}(\mathbf{x}), (d\varphi_\alpha)^{-1}(\mathbf{v})\big) \\ &= \big(\varphi_\alpha^{-1}(\mathbf{x}), d\varphi_\alpha^{-1}(\mathbf{v})\big)\end{aligned}$$

that

$$\begin{aligned}\psi_{\alpha\beta}(\mathbf{x}, \mathbf{v}) &= \big(\varphi_\beta \circ \varphi_\alpha^{-1}(\mathbf{x}), d\varphi_\beta \circ d\varphi_\alpha^{-1}(\mathbf{v})\big) \\ &= \big(\varphi_\beta \circ \varphi_\alpha^{-1}(\mathbf{x}), d(\varphi_\beta \circ \varphi_\alpha^{-1})(\mathbf{v})\big).\end{aligned}$$

Since $\varphi_{\alpha\beta} = \varphi_\beta \circ \varphi_\alpha^{-1} : \mathbb{R}^m \to \mathbb{R}^m$ is smooth, the differential mapping $d\varphi_{\alpha\beta}$ is also smooth. Thus $T(M)$ acquires a structure of a $2m$-dimensional differentiable manifold with the atlas $\{(\pi^{-1}(U_\alpha), \psi_\alpha) : \alpha \in \mathcal{I}\}$. Local coordinates of this manifold is given by $(x^1, \dots, x^m, v^1, \dots, v^m)$.

Due to the relation (2.6.9), the linear operator $d\varphi_{\alpha\beta}$ is represented by the matrix

$$d\varphi_{\alpha\beta} = \mathbf{K}_{\alpha\beta} = \left[\frac{\partial x_\beta^i}{\partial x_\alpha^j}\right]. \tag{2.8.5}$$

Hence $d\varphi_{\alpha\beta} : T_{\mathbf{x}}(\mathbb{R}^m) \to T_{\mathbf{x}}(\mathbb{R}^m)$ is an *automorphism, an isomorphism mapping a vector space onto itself* at a point $p \in M$. We know that we can take $T_{\mathbf{x}}(\mathbb{R}^m) = \mathbb{R}^m$. Therefore, denoting the *general linear group* formed by $m \times m$ regular matrices on fibres $\mathbb{R}^m$ by $GL(m, \mathbb{R})$, we infer that

$$d\varphi_{\alpha\beta} \in GL(m, \mathbb{R}).$$

$GL(m, \mathbb{R})$, or one of its subgroups $G \subseteq GL(m, \mathbb{R})$, is called the ***structural group*** of the tangent bundle. This group ascertains the global character of $T(M)$ and helps us distinguish different bundles defined over the same base space. Then we deduce that in an intersection $\mathcal{V}_\alpha \cap \mathcal{V}_\beta$ on the fibre bundle, the coordinate transformation is determined through the relations

$$\mathbf{x}_\beta = \varphi_{\alpha\beta}(\mathbf{x}_\alpha), \;\; \mathbf{v}_\beta = \mathbf{K}_{\alpha\beta}\mathbf{v}_\alpha, \;\; \mathbf{K}_{\alpha\beta} \in G.$$

If the bundle $T(M)$ is diffeomorphic to the product manifold $M \times \mathbb{R}^m$, it is then called *a globally trivial bundle*. Since every tangent bundle is locally represented as $U \times \mathbb{R}^m$, it is *locally trivial*. Whether this property is also valid globally depends on the structural group. If the tangent bundle is trivial, we can always choose points (p, V_0) in $M \times \mathbb{R}^m$ for all points $p \in M$ where V_0 is a *constant* vector. Hence, the inverse mapping creates a vector field in $T(M)$ that vanishes nowhere on M. This means that *the tangent bundle cannot be trivial if it is not possible to find a vector field that vanishes nowhere on the manifolds.*

A ***local cross section*** of the tangent bundle is the smooth mapping $\sigma : U \to T(U)$ where $U \subseteq M$ is an open set. σ must possess the property $\pi \circ \sigma = i_U$, that is, one has $\pi(\sigma(p)) = p$ for all $p \in U$. If $U = M$, then σ is called a ***global cross section***. The mapping σ will clearly assigns a vector to each point of an open submanifold of M, or M itself. Hence, it prescribes a vector field (Fig. 2.8.2).

Example 2.8.1. As the base manifold, let us choose the circle. It is straightforward to observe that one finds easily a vector field that vanishes nowhere on $\mathbb{S}^1$. Therefore $T(\mathbb{S}^1)$ is a trivial bundle and it can be represented as $\mathbb{S}^1 \times \mathbb{R}$. As a matter of fact if we choose fibres as shown in Fig. 2.8.3(a), then the tangent bundle becomes the Cartesian product of $\mathbb{S}^1$ and $\mathbb{R}$. Since $\mathbb{S}^1$ is designated by a single coordinate, the transformation of coordinates at a point p in overlapping charts are given by $x_\beta = \varphi_{\alpha\beta}(x_\alpha)$, $v_\beta = K_{\alpha\beta}\, v_\alpha$

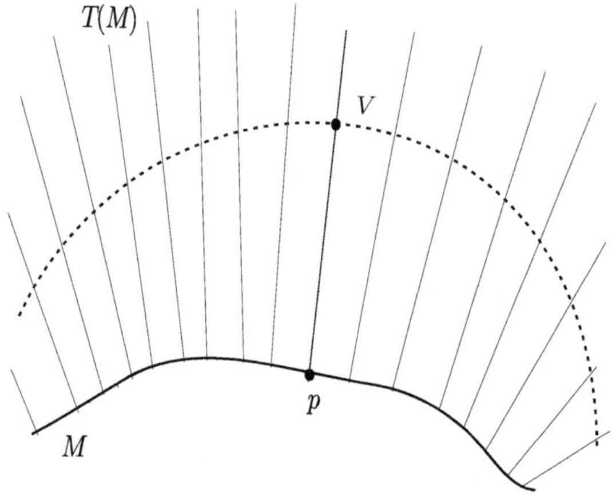

Fig. 2.8.2. Vector field as a cross section.

where the constant $K_{\alpha\beta}$ is the value of $d\varphi_{\alpha\beta}/dx_\alpha$ at p. This number is a member of the multiplication group on $\mathbb{R}$ which is also the structural group of the bundle. In order to find a simple representation let us cut the circle at a point p, unwrap the bundle and make it lie on $\mathbb{R}^2$. To assemble the bundle again all we have to do is to identify p with p', u with u' and v with v'. In this case, the transition mapping in overlapping charts is simply found as the identity mapping $(p, v) \to (p, v)$ and the structural group of the tangent bundle becomes just $\{1\}$. However, we can reassemble the bundle to form the ***Möbius band*** if we identify u with v' and v with u' by twisting the strip. In this case the tangent bundle is no longer trivial. Transition mapping in some overlapping charts is again given by $(p, v) \to (p, v)$ whereas in some others by $(p, v) \to (p, -v)$ and the structural group of the bundle is now $\{1, -1\}$.

Let us consider a Möbius band whose middle circle is situated at the plane $z = 0$, centred at the origin with radius R and its half-width is w. Its parametric equations are given by

$$x = [R + v\cos(u/2)]\cos u, y = [R + v\cos(u/2)]\sin u, z = v\sin(u/2)$$

where $0 \le u < 2\pi$ and $-w \le v \le w$. Indeed for $u = 0$ we get $x = R + v$, $y = z = 0$ while for $u = 2\pi$ we obtain $x = R - v, y = z = 0$. Thus we obtain the description described in Fig.. 2.8.3(b).

Möbius band, or strip, is named after German mathematician August Ferdinand Möbius (1790-1868) who had introduced it on September 1858. Strictly speaking, the band had already been found a little bit earlier by German mathematician Johann Benedict Listing (1808-1882) on July 1858.

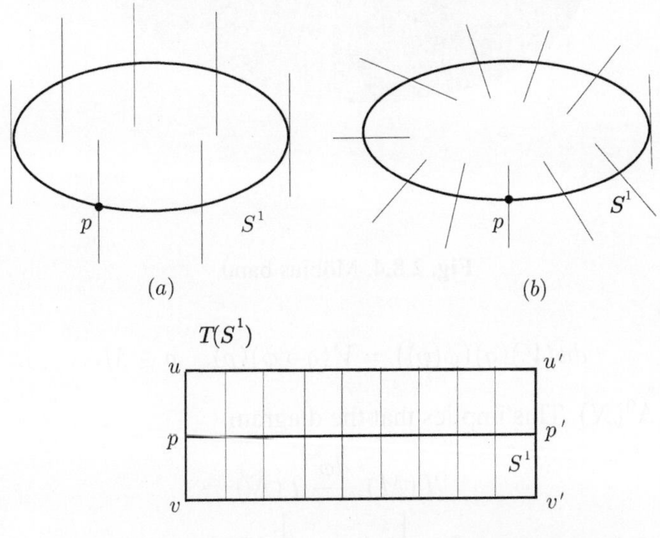

Fig. 2.8.3. Fibre bundles: (a) circle, (b) Möbius band.

Therefore, it would have been more appropriate to call it as *Listing band*. Möbius band is perhaps the most prominent example to *one-sided* and *one-edged surfaces*. In fact, when we start moving on the surface we pass eventually under the surface without crossing the edge. The representation of Möbius band in $\mathbb{R}^3$ is depicted in Fig. 2.8.4.

Let $\phi : M \to N$ be a differentiable mapping between two differentiable manifolds. The differential of ϕ can now be written as an operator between tangent bundles as $\phi_* = d\phi : T(M) \to T(N)$. However, we have to keep in mind that the linear operator $d\phi$ transforms pointwise the vector space $T_p(M)$ into the vector space $T_{\phi(p)}(N)$ and its action can only be embodied through the local charts at points p and $\phi(p)$. Let a smooth vector field in the tangent bundle $T(M)$ be V. Then we define $V^* = d\phi(V)$ by the following relation again

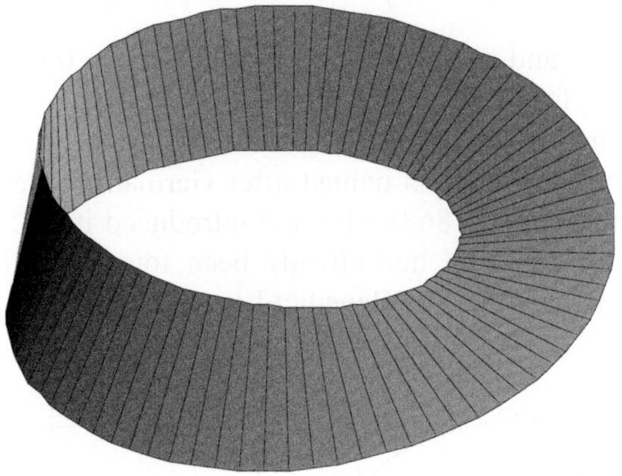

Fig. 2.8.4. Möbius band. ■

$$d\phi(V)(g)\big(\phi(p)\big) = V(g \circ \phi)(p), \quad p \in M$$

for all $g \in \Lambda^0(N)$. This implies that the diagram

$$\begin{array}{ccc} T(M) & \xrightarrow{d\phi} & T(N) \\ \pi_M \downarrow & & \downarrow \pi_N \\ M & \xrightarrow{\phi} & N \end{array}$$

is commutative, that is, $\phi \circ \pi_M = \pi_N \circ d\phi$ where $\pi_M : T(M) \to M$ and $\pi_N : T(N) \to N$ are natural projections,

If only the mapping ϕ has an inverse, then one can write

$$V^*(g)(q) = \big[V(g \circ \phi)\big] \circ \phi^{-1}(q), \quad q \in N. \tag{2.8.6}$$

Thus only for invertible mappings, their differentials are able to assign a vector $V^*(q)$ at every point $q \in N$. In other words, if ϕ^{-1} does not exist, then the image of a vector field on M is generally not a vector field on N. *If ϕ is a diffeomorphism, then the image of a smooth vector field becomes also a smooth vector field . If ϕ^{-1} exists but not smooth, then the image is a vector field but it may not be smooth.*

2.9. FLOWS OVER MANIFOLDS

Let M^m be a smooth manifold and let $V \in T(M)$ be a given vector field. A differentiable curve described by the smooth mapping $\gamma : \mathcal{I} \to M$, $\mathcal{I} = (a, b) \subseteq \mathbb{R}$ will be called an ***integral curve*** of the vector field V, if it is tangent to the field V, i.e., if the relation

$$\gamma_*\Big(\frac{d}{dt}\Big) = V|_{\gamma(\mathcal{I})}$$

is satisfied. In dynamical system, this curve is also called a ***trajectory*** or an ***orbit***. This relation is symbolically expressed as follows:

$$\frac{d\gamma(t)}{dt} = V\big(\gamma(t)\big), \quad t \in (a, b). \tag{2.9.1}$$

We know that the image of this curve in $\mathbb{R}^m$ is determined by expressions

$$x^i = \gamma^i(t) = \varphi^i\big(\gamma(t)\big) \in \mathbb{R}, \; i = 1, \dots, m$$

in local coordinates.

Theorem 2.9.1. *Let a vector field V on a differentiable manifold M^m be given by*

$$V(p) = v^i(\mathbf{x})\frac{\partial}{\partial x^i}, \quad p = \varphi^{-1}(\mathbf{x}).$$

where (U, φ) is the chart to which $p \in M$ belongs. A curve $\gamma : \mathcal{I} \to M$ is an integral curve of the vector field V if and only if the coordinate functions $x^i(t)$ are solutions of the following system of local ordinary differential equations in $\mathbb{R}^m$

$$\frac{dx^i}{dt} = v^i\big(\mathbf{x}(t)\big), \; i = 1, \dots, m. \tag{2.9.2}$$

Indeed, if we take into consideration the relation (2.7.8), we can transform (2.9.1) into the form

$$\frac{dx^i}{dt}\frac{\partial}{\partial x^i} = v^i\big(\mathbf{x}(t)\big)\frac{\partial}{\partial x^i}$$

which can be satisfied if and only if the differential equations

$$\frac{dx^i}{dt} = v^i\big(\mathbf{x}(t)\big), \quad i = 1, \dots, m$$

are held. □

We see that in order to find integral curves of a vector field on a manifold M^m we need to generate curves in $\mathbb{R}^m$ as solutions of differential equations (2.9.2) and then carry them on M by making use of local charts.

Let V be a smooth vector field on M. Hence, all components $v^i(\mathbf{x})$ are smooth functions. If M is also a smooth manifold, then the functions $V^i(p) = v^i(\varphi(p))$ will be smooth, either. Next, we consider a point $p_0 \in M$ and a chart (U, φ) enclosing this point. It is known from the theory of system of ordinary differential equations that [*see* Coddington and Levinson (1955), *p.* 22, Theorem 7.1] *for each point* $\mathbf{x}_0 = \varphi(p_0) \in \mathbb{R}^m$ *there exist an open set* $\mathcal{U}(\mathbf{x}_0) \subseteq \mathbb{R}^m$ *containing this point and an open interval* $\mathcal{I}(\mathbf{x}_0) \subseteq \mathbb{R}$ *so that for all* $\mathbf{x} \in \mathcal{U}(\mathbf{x}_0)$ *and* $t \in \mathcal{I}(\mathbf{x}_0)$ *the following system of ordinary differential equations*

$$\frac{d\phi^i}{dt} = v^i(\boldsymbol{\phi}) \tag{2.9.3}$$

has a unique solution $\boldsymbol{\phi}(t; \mathbf{x})$ *satisfying the initial condition* $\boldsymbol{\phi}|_{t=t_0} = \mathbf{x}_0$ *where* $t_0 \in \mathcal{I}(\mathbf{x}_0)$ *and* $\boldsymbol{\phi}$ *is a vector-valued smooth function of variables* t *and* $\mathbf{x} = (x^1, \dots, x^m)$. If $0 \in \mathcal{I}(\mathbf{x}_0)$, then we usually choose $t_0 = 0$. Thus the function $\boldsymbol{\phi}(t; \mathbf{x})$ designate a curve in $\mathbb{R}^m$ through the point $\overline{\mathbf{x}}(t_0) = \mathbf{x}$ whose equation is parametrically given by

$$\overline{\mathbf{x}}(t) = \boldsymbol{\phi}(t; \mathbf{x}) \in \mathcal{U}(\mathbf{x}_0), \quad \boldsymbol{\phi}(t_0; \mathbf{x}) = \mathbf{x}$$

where $\mathbf{x} \in \mathcal{U}(\mathbf{x}_0)$ and $t \in \mathcal{I}(\mathbf{x}_0)$. If we fix t and write $\boldsymbol{\phi}_t(\mathbf{x}) = \boldsymbol{\phi}(t; \mathbf{x})$, then $\boldsymbol{\phi}_t : \mathcal{U}(\mathbf{x}_0) \to \mathbb{R}^m$ denotes a *family of smooth mappings* depending on the parameter $t \in \mathcal{I}(\mathbf{x}_0)$. For a fixed t, each point $\mathbf{x} \in \mathcal{U}(\mathbf{x}_0)$ is transported along the integral curve of the vector field V to the point $\boldsymbol{\phi}_t(\mathbf{x}) \in \mathbb{R}^m$ determined by this value of the parameter t. *Because of the uniqueness of the solutions such curves cannot intersect.* An open neighbourhood $U_{p_0} = \varphi^{-1}\big(\mathcal{U}(\mathbf{x}_0)\big) \subseteq M$ is associated with each point $p_0 = \varphi^{-1}(\mathbf{x}_0) \in M$ and an integral curve through a point $p \in U_{p_0}$ is characterised by

$$\phi(t; p) = \varphi^{-1} \circ \boldsymbol{\phi}\big(t; \varphi(p)\big), \ t \in \mathcal{I}_{p_0} = \mathcal{I}\big(\varphi(p_0)\big) \subseteq \mathbb{R}. \tag{2.9.4}$$

This function must of course satisfy the relation $\phi(t_0; p) = p$. Points on this curve are found by the transformation

$$\overline{p}(t) = \phi(t;p) = \phi_t(p), \ \ t \in \mathcal{I}_{p_0}.$$

We have to point out that the definition of ϕ by (2.9.4) is valid only for points p and $\overline{p}(t)$ situated in the same chart of the manifold. A new solution is required for a different chart. Therefore, the family of *local* smooth mappings $\{\phi_t : U_{p_0} \to M, t \in \mathcal{I}_{p_0}\}$ transports each point $p \in U_{p_0}$ of the manifold M along an integral curve of the vector field V through this point to the point $\phi_t(p) \in M$. Thus, in essence, the mapping ϕ should be written in the form

$$\phi : U_{p_0} \times \mathcal{I}_{p_0} \to M$$

where the set $U_{p_0} \times \mathcal{I}_{p_0}$ is an open subset of $(m+1)$-dimensional smooth manifold $M \times \mathbb{R}$. Hence, it is an $(m+1)$-dimensional smooth open submanifold. Let us now consider open neighbourhoods U_{p_λ} defined as above covering the manifold M so that $M = \bigcup\limits_{\lambda\in\Lambda} U_{p_\lambda}$. Next, we define the interval $\mathcal{I} = \bigcap\limits_{\lambda\in\Lambda} \mathcal{I}_{p_\lambda} \subseteq \mathbb{R}$. Whenever $\mathcal{I}$ is not empty, ϕ_t becomes a global mapping for all $t \in \mathcal{I}$ so that one is able to write $\phi_t : M \to M$. If M is a compact manifold, then it would be covered by finitely many open sets of the above family. In this case, $\mathcal{I}$ becomes, of course, the intersection of finitely many open intervals. Hence, it cannot be empty. Such a family of mappings generated by a vector field on the manifold is called the ***flow*** of that vector field. If $\mathcal{I} = \mathbb{R}$, then we say that $V \in T(M)$ is a ***complete vector field***. It can be shown that if the vector field is bounded, that is, if there exists a constant $K > 0$ such that $\sum\limits_{i=1}^{m} |v^i(\mathbf{x})| \leq K$ for all $\mathbf{x} \in \mathbb{R}^m$, then the solution of the system of differential equations (2.9.2) will be valid on the entire real axis [*see* Cronin (1980), *p.* 53]. When M is taken as a compact manifold, then all continuous functions defined on M ought to be bounded. Consequently, *smooth vector fields defined on compact manifolds are always complete.*

We now shall try to better understand the structure of the mapping ϕ_t. The functions $\phi^i(t;\mathbf{x})$ are to satisfy

$$\frac{d\phi^i}{dt} = v^i(\boldsymbol{\phi}), \ \ i = 1, \dots, m,$$
$$\phi^i(0;\mathbf{x}) = x^i.$$

We have assumed without loss of generality that $0 \in \mathcal{I}$. Since functions ϕ^i are smooth, they can be expanded into a Maclaurin series in a sufficiently small neighbourhood of the point $t = 0$:

$$\overline{x}^i(t) = \phi^i(t;\mathbf{x}) = \phi^i(0;\mathbf{x}) + \frac{d\phi^i}{dt}\bigg|_{t=0} t + \frac{1}{2!}\frac{d^2\phi^i}{dt^2}\bigg|_{t=0} t^2 + \cdots + \frac{1}{n!}\frac{d^n\phi^i}{dt^n}\bigg|_{t=0} t^n + \cdots.$$

We can evaluate the coefficients of this series at $t = 0$ by using the foregoing ordinary differential equations. As a matter of fact, if we note that we can write

$$\frac{d\phi^i}{dt} = v^i = v^j\delta^i_j = v^j\frac{\partial\phi^i}{\partial\phi^j} = \Big(v^j\frac{\partial}{\partial\phi^j}\Big)\phi^i$$

we easily obtain the following sequence

$$\begin{aligned}
\frac{d^2\phi^i}{dt^2} &= \frac{\partial v^i}{\partial\phi^j}\frac{d\phi^j}{dt} = \frac{\partial v^i}{\partial\phi^j}v^j = \Big(v^j\frac{\partial}{\partial\phi^j}\Big)v^i = \Big(v^j\frac{\partial}{\partial\phi^j}\Big)^2\phi^i\\
\frac{d^3\phi^i}{dt^3} &= \frac{\partial}{\partial\phi^k}\Big(v^j\frac{\partial v^i}{\partial\phi^j}\Big)\frac{d\phi^k}{dt} = \Big(v^j\frac{\partial}{\partial\phi^j}\Big)^2 v^i = \Big(v^j\frac{\partial}{\partial\phi^j}\Big)^3\phi^i\\
&\ \ \vdots\\
\frac{d^n\phi^i}{dt^n} &= \Big(v^j\frac{\partial}{\partial\phi^j}\Big)^{n-1} v^i = \Big(v^j\frac{\partial}{\partial\phi^j}\Big)^n\phi^i\\
&\ \ \vdots
\end{aligned}$$

We know that the vector field $V' \in T(\mathbb{R}^m)$ representing the vector field $V \in T(M)$ in local coordinates is denoted by

$$V'(\mathbf{x}) = v^j(\mathbf{x})\frac{\partial}{\partial x^j}.$$

Thus, after having evaluated the foregoing relations at the point $t = 0$, we arrive at the following result:

$$\begin{aligned}
\frac{d\phi^i}{dt}\bigg|_{t=0} &= v^i(\mathbf{x}) = V'(x^i)\\
\frac{d^2\phi^i}{dt^2}\bigg|_{t=0} &= V'^2(x^i)\\
\frac{d^3\phi^i}{dt^3}\bigg|_{t=0} &= V'^3(x^i)\\
&\ \ \vdots
\end{aligned}$$

$$\left.\frac{d^n\phi^i}{dt^n}\right|_{t=0} = V'^n(x^i)$$
$$\vdots$$

The operator V'^n denotes n times composition $V' \circ V' \circ \cdots \circ V'$ of the operator $V' : \Lambda^0(\mathbb{R}^m) \to \Lambda^0(\mathbb{R}^m)$ by itself. Hence the Taylor series above [English mathematician Brook Taylor (1685-1731)] can be cast into the following series

$$\begin{aligned}\overline{x}^i(t) = \phi^i(t;\mathbf{x}) &= x^i + tV'(x^i) + \frac{t^2}{2!}V'^2(x^i) + \cdots + \frac{t^n}{n!}V'^n(x^i) + \cdots \\ &= \Big(I + tV' + \frac{t^2}{2!}V'^2 + \cdots + \frac{t^n}{n!}V'^n + \cdots\Big)(x^i).\end{aligned}$$

We shall now define the ***exponential operator*** by the absolutely convergent operator series

$$\exp(tV') = e^{tV'} = \sum_{n=0}^{\infty}\frac{t^n}{n!}V'^n \tag{2.9.5}$$

where we have adopted the convention $V'^0 = I$. Thus, we attain at the formula

$$\phi^i(t;\mathbf{x}) = \exp\Big(tv^j(\mathbf{x})\frac{\partial}{\partial x^j}\Big)(x^i) = e^{tV'}(x^i)$$

or

$$\overline{\mathbf{x}}(t) = \boldsymbol{\phi}(t;\mathbf{x}) = e^{tV'}(\mathbf{x}) \tag{2.9.6}$$

where the operator $e^{tV'} : \mathbb{R}^m \to \mathbb{R}^m$ is introduced by

$$e^{tV'}(\mathbf{x}) = (e^{tV'}(x^1), e^{tV'}(x^2), \dots, e^{tV'}(x^m)) \in \mathbb{R}^m.$$

If the operators V_1', V_2' are commutative, namely, if they satisfy the relation $V_1' \circ V_2' = V_2' \circ V_1'$, we find that

$$e^{V_1'+V_2'} = e^{V_1'} \circ e^{V_2'} = e^{V_2'} \circ e^{V_1'}.$$

In effect, if these operators commute the classical binomial expansion yields

$$(V_1' + V_2')^n = \sum_{k=0}^{n}\binom{n}{k}V_1'^kV_2'^{n-k} = \sum_{k=0}^{n}\frac{n!}{k!\,(n-k)!}V_1'^kV_2'^{n-k}.$$

We thus obtain

$$\begin{aligned} e^{V_1'+V_2'} &= \sum_{n=0}^{\infty} \frac{(V_1'+V_2')^n}{n!} = \sum_{n=0}^{\infty}\sum_{k=0}^{n} \frac{V_1'^k V_2'^{n-k}}{k!\,(n-k)!} \\ &= \sum_{n=0}^{\infty} \frac{V_1'^n}{n!} \sum_{m=0}^{\infty} \frac{V_2'^m}{m!} = e^{V_1'} e^{V_2'}. \end{aligned}$$

Since the vector addition is a commutative operation, we infer at once the commutativity of exponential operators. It then follows from (2.9.6) that

$$\boldsymbol{\phi}(t+s;\mathbf{x}) = e^{(t+s)V'}(\mathbf{x}) = e^{tV'} \circ e^{sV'}(\mathbf{x}) = \boldsymbol{\phi}\big(t;\boldsymbol{\phi}(s;\mathbf{x})\big). \quad (2.9.7)$$

Next, we employ the expression (2.9.4) by assuming that $t, s, t+s \in \mathcal{I}_{p_0}$ to reach to the relation

$$\begin{aligned} \phi(t+s;p) &= \varphi^{-1} \circ \boldsymbol{\phi}\big(t+s;\varphi(p)\big) = \varphi^{-1} \circ \boldsymbol{\phi}\big(t;\boldsymbol{\phi}(s;\varphi(p))\big) \\ &= \varphi^{-1} \circ \boldsymbol{\phi}\big(t;\varphi(\phi(s;p))\big) = \phi\big(t;\phi(s;p)\big). \end{aligned}$$

This relation is actually independent of the chart in question. Indeed, according to the definition of the integral curve, both curves $t \mapsto \phi(t+s;p)$ and $t \mapsto \phi\big(t;\phi(s;p)\big)$ satisfy the same differential equations. The initial conditions at $t=0$ are also the same: $\phi(s;p) = \phi\big(0;\phi(s;p)\big) = \phi(s;p)$.

Hence, the uniqueness of solutions leads also to the conclusion

$$\phi(t+s;p) = \phi\big(t;\phi(s;p)\big). \quad (2.9.8)$$

Consequently, we can write

$$\phi_{t+s}(p) = \phi_t\big(\phi_s(p)\big) = \phi_t \circ \phi_s(p)$$

for all $p \in U_{p_0}$ whenever $t, s, t+s \in \mathcal{I}_{p_0}$. If the interval $\mathcal{I}$ is not empty, then (2.9.8) becomes valid for all $p \in M$ and the *global transformation operator* $\phi_t : M \to M$ satisfies the relation

$$\phi_{t+s} = \phi_t \circ \phi_s \quad (2.9.9)$$

if $t, s, t+s \in \mathcal{I}$. This implies that the composition of smooth functions ϕ_t and ϕ_s is again a smooth function provided that the parameters t and s are sufficiently small if $\mathcal{I} \neq \mathbb{R}$. Furthermore, if we take $s = -t$, then we find $\phi_0 = \phi_t \circ \phi_{-t} = i_M$ implying that $(\phi_t)^{-1} = \phi_{-t}$. Hence, the inverse mapping ϕ_t^{-1} is also smooth. This amounts to say that $\{\phi_t : t \in \mathcal{I}\}$ is a family of diffeomorphisms. It is clear that *this set constitutes a group under the operation of composition of mappings*. However, since the group structure prevails only for limited values of the parameter t including 0, this group is called **1-*parameter group of local diffeomorphisms*** of the manifold M. If $\mathcal{I} = \mathbb{R}$, the group is named **1-*parameter group of diffeomorphisms***. The flow ϕ_t is represented by a family of curves that are tangent to a given vector field V at every point of the manifold M. These curves are obtained as images of solutions of the set of differential equations (2.9.2) on M by means of charts. Due to the uniqueness of solutions of equations (2.9.2), the curves of this family cannot intersect except at the *critical points* satisfying the condition $V(p) = 0$ and they fill the manifold. Such a family of curves is called a ***congruence.***

The vector field V that help determine the flow is sometimes called an ***infinitesimal generator*** of the flow.

The flow $\phi(t;p)$ can be endowed with an appearance which makes its group structure more pronounced. Provided that the points $\overline{\mathbf{x}}(t)$ and $\mathbf{x} \in \mathbb{R}^m$ belong to the same chart, we then deduce from the relation $\overline{\mathbf{x}}(t) = e^{tV'}(\mathbf{x})$ that

$$\overline{p}(t) = \phi(t;p) = \varphi^{-1}(\overline{\mathbf{x}}(t)) = (\varphi^{-1} \circ e^{tV'} \circ \varphi)(p).$$

We can now locally define an ***exponential mapping*** $e^{tV} : M \to M$ by

$$e^{tV} = \varphi^{-1} \circ e^{tV'} \circ \varphi. \tag{2.9.10}$$

It is straightforward to demonstrate that this operator possesses the following properties:

$$\begin{aligned} e^{(t+s)V} &= \varphi^{-1} \circ e^{(t+s)V'} \circ \varphi = \varphi^{-1} \circ e^{tV'} \circ e^{sV'} \circ \varphi \\ &= \varphi^{-1} \circ e^{tV'} \circ \varphi \circ \varphi^{-1} \circ e^{sV'} \circ \varphi = e^{tV} \circ e^{sV}, \\ e^{tV} \circ e^{-tV} &= e^{0V} = \varphi^{-1} \circ i_{\mathbb{R}^m} \circ \varphi = i_M. \end{aligned}$$

These properties justify our calling e^{tV} as the exponential mapping and our using the familiar notation. Moreover, for two commutative operators V_1 and V_2 we again obtain

$$\begin{aligned} e^{(V_1+V_2)} &= \varphi^{-1} \circ e^{(V_1'+V_2')} \circ \varphi = \varphi^{-1} \circ e^{V_1'} \circ \varphi \circ \varphi^{-1} \circ e^{V_2'} \circ \varphi \\ &= e^{V_1} \circ e^{V_2} = e^{V_2} \circ e^{V_1}. \end{aligned}$$

We can now express the flow generated by the vector field V on the manifold M also in the form

$$\phi(t;p) = \phi_t(p) = e^{tV}(p). \tag{2.9.11}$$

Naturally, as the parameter t varies, (2.9.11) might tangibly be specified only through different charts.

Let us now take a function $f \in \Lambda^0(M)$ into consideration. We know that the function $f' \in \Lambda^0(\mathbb{R}^m)$ is related to f by the equality $f(p) = f'(\mathbf{x})$, that is, we get $f' = f \circ \varphi^{-1}$. Our task is to evaluate the value of the smooth function f' at the point $\overline{\mathbf{x}}(t)$. To this end, let us consider the expansion

$$f'\big(\overline{\mathbf{x}}(t)\big) = f'\big(e^{tV'}(\mathbf{x})\big) = f'\big|_{t=0} + t\frac{df'}{dt}\bigg|_{t=0} + \cdots + \frac{t^n}{n!}\frac{d^n f'}{dt^n}\bigg|_{t=0} + \cdots.$$

Introducing the relations

$$f'\big|_{t=0} = f'(\mathbf{x}), \quad \frac{df'}{dt}\bigg|_{t=0} = \frac{\partial f'}{\partial \overline{x}^i}\frac{d\overline{x}^i}{dt}\bigg|_{t=0} = v^i(\mathbf{x})\frac{\partial f'}{\partial x^i} = V'\big(f'(\mathbf{x})\big)$$

$$\frac{d^2 f'}{dt^2}\bigg|_{t=0} = \frac{\partial \overline{V}'(f')}{\partial \overline{x}^j}\frac{d\overline{x}^j}{dt}\bigg|_{t=0} = V'^2\big(f'(\mathbf{x})\big),$$

$$\vdots$$

$$\frac{d^n f'}{dt^n}\bigg|_{t=0} = \frac{\partial \overline{V}'^{(n-1)}(f')}{\partial \overline{x}^j}\frac{d\overline{x}^j}{dt}\bigg|_{t=0} = V'^n\big(f'(\mathbf{x})\big), \ldots$$

into that expression we arrive at

$$\begin{aligned} f'\big(e^{tV'}(\mathbf{x})\big) &= f'(\mathbf{x}) + tV'\big(f'(\mathbf{x})\big) + \cdots + \frac{t^n}{n!}V'^n\big(f'(\mathbf{x})\big) + \cdots \\ &= \Big(f' + tV'(f') + \cdots + \frac{t^n}{n!}V'^n(f') + \cdots\Big)(\mathbf{x}) \end{aligned}$$

from which we conclude that

$$\begin{aligned} f'\big(\overline{\mathbf{x}}(t)\big) = f'\big(e^{tV'}(\mathbf{x})\big) &= \sum_{n=0}^{\infty}\frac{t^n}{n!}V'^n(f')(\mathbf{x}) \\ &= e^{tV'(f')}(\mathbf{x}) = e^{tV'}f'(\mathbf{x}) \end{aligned} \tag{2.9.12}$$

On the other hand, if the equalities $f(p) = f'(\mathbf{x}), V^n(f)(p) = V'^n(f')(\mathbf{x})$, $n = 1, 2, \ldots$ are utilised in the expression

$$f\big(\overline{p}(t)\big) = f\big(e^{tV}(p)\big) = f'\big(e^{tV'}(\mathbf{x})\big) = \sum_{n=0}^{\infty}\frac{t^n}{n!}V'^n(f')(\mathbf{x})$$

we find that

$$f\big(\overline{p}(t)\big) = f\big(e^{tV}(p)\big) = \sum_{n=0}^{\infty}\frac{t^n}{n!}V^n(f)(p) \tag{2.9.13}$$

$$= e^{tV(f)}(p) = e^{tV} f(p)$$

Since $V^n(f)(p) \in \mathbb{R}$, the foregoing expression makes sense. If a function f satisfies the equality $f(\overline{p}(t)) = f(p)$ for every point $p \in M$, we say that it is ***invariant*** under the flow. It immediately follows from (2.9.13) that *the necessary and sufficient condition for a function f to remain invariant under the flow generated by a vector field V is*

$$V(f) = 0. \tag{2.9.14}$$

Next, we consider a vector field $V \in T(M)$, its local representation $V' \in T(\mathbb{R}^m)$, and the integral curve through a point $\mathbf{x} \in \mathbb{R}^m$. In view of (2.9.6), we can write $\overline{\mathbf{x}}(t) = e^{tV'}(\mathbf{x})$. Hence, we obtain

$$\begin{aligned}
\overline{v}^i(\overline{\mathbf{x}}) &= \frac{d\overline{x}^i}{dt} = \frac{d}{dt}\Big[x^i + tV'(x^i) + \frac{t^2}{2!}V'^2(x^i) + \cdots + \frac{t^n}{n!}V'^n(x^i) + \cdots\Big] \\
&= V'(x^i) + tV'^2(x^i) + \cdots + \frac{t^{n-1}}{(n-1)!}V'^n(x^i) + \cdots \\
&= \Big[I + tV' + \frac{t^2}{2!}V'^2 + \cdots + \frac{t^{n-1}}{(n-1)!}V'^{n-1} + \cdots\Big]V'(x^i) \\
&= e^{tV'}\big(V'(x^i)\big) = e^{tV'}v^i(\mathbf{x}) = v^i\big(e^{tV'}(\mathbf{x})\big) = v^i(\overline{\mathbf{x}}).
\end{aligned}$$

Here, in the last line we used (2.9.12). We thus conclude that

$$\frac{d}{dt}e^{tV'}(\mathbf{x}) = V'\big(e^{tV'}(\mathbf{x})\big) \text{ and } \frac{d}{dt}e^{tV'}(\mathbf{x})\Big|_{t=0} = V'(\mathbf{x}). \tag{2.9.15}$$

To summarise, one notes that a flow generated by a vector field V on a manifold M is determined as a solution of symbolic differential equation

$$\frac{d\phi_t(p)}{dt} = V\big(\phi_t(p)\big), \quad \phi_0(p) = p, \; p \in M$$

(operation of differentiation can only be realised by means of charts) in the form $\phi_t(p) = e^{tV}(p)$. We can also write

$$\frac{d}{dt}e^{tV}(p) = V\big(e^{tV}(p)\big) \text{ and } \frac{d}{dt}e^{tV}(p)\Big|_{t=0} = V(p) \tag{2.9.16}$$

in accordance with relations (2.9.15).

Let us now consider a function $f \in \Lambda^0(M)$ and try to specify its variation along the flow generated by a vector field V. If the local representation of this function is $f' = f \circ \varphi^{-1} \in \Lambda^0(\mathbb{R}^m)$ subordinate to a chart, we can then write

$$\frac{df'(\overline{\mathbf{x}}(t))}{dt} = \frac{df'(e^{tV'}(\mathbf{x}))}{dt} = \overline{v}^i(\overline{\mathbf{x}})\frac{\partial f'}{\partial \overline{x}^i} = V'(f')|_{\overline{\mathbf{x}}(t)} = V'(f')|_{e^{tV'}(\mathbf{x})}.$$

Since $V(f) = V'(f')$, the above relation leads to

$$\frac{df(e^{tV}(p))}{dt} = V(f)|_{e^{tV}(p)} \quad \text{and} \quad \left.\frac{df(e^{tV}(p))}{dt}\right|_{t=0} = V(f)(p).$$

Let $\psi : M \to N$ be a *diffeomorphism* between two differentiable manifolds. We denote the flow brought out by a vector field V on M by the relation $\overline{p}(t) = e^{tV}(p)$. We get $f = g \circ \psi \in \Lambda^0(M)$ for any $g \in \Lambda^0(N)$. If $q = \psi(p)$ we obtain

$$\begin{aligned} g\big[\psi(e^{tV}(p))\big] &= (g \circ \psi)(e^{tV}(p)) = f(e^{tV}(p)) = \sum_{n=0}^{\infty}\frac{t^n}{n!}V^n(f)(p) \\ &= \sum_{n=0}^{\infty}\frac{t^n}{n!}V^n(g\circ\psi)(p) = \sum_{n=0}^{\infty}\frac{t^n}{n!}\big[V^n(g\circ\psi)\circ\psi^{-1}\big](q) \\ &= \sum_{n=0}^{\infty}\frac{t^n}{n!}V^{*n}(g)(q) = g(e^{tV^*}(q)), \qquad V^* = \psi_*(V) \end{aligned}$$

after having employed (2.9.13) and (2.8.6). Since this relation is in effect for every smooth function g, we infer that

$$\psi(e^{tV}(p)) = e^{t\,d\psi(V)}\psi(p) = e^{t\,\psi_*(V)}\psi(p). \tag{2.9.17}$$

This simply means that *a diffeomorphism between manifolds M and N transforms a flow on M onto a flow on N.*

2.10. LIE DERIVATIVE

Let us assume that we are given two vector fields $U, V \in T(M)$ on a manifold M and the U- and V-congruences generated by those fields are determined. We consider a curve of V-congruence through a point $p \in M$ for the value of the parameter $t = 0$, the point of which corresponding to the value t is the point $q \in M$. U-congruence has now two curves through the points p and q. This situation is depicted schematically in Fig. 2.10.1.

Hence, we can write $q = e^{tV}(p)$. Our aim is to establish a procedure that is able to measure the variation in vectors of the vector field U while one moves along a V-curve. In order to realise this, we have to suggest a

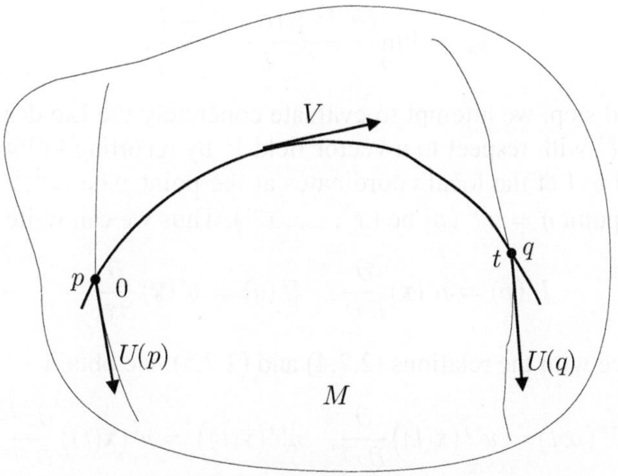

Fig. 2.10.1. Two congruences on a manifold.

scheme that makes it possible to compare vectors $U(p) \in T_p(M)$ and $U(q) \in T_q(M)$ which reside on disjoint vector spaces. In other words, we have to transport the vector $U(q)$ without changing its properties into the tangent space $T_p(M)$. To this end, we introduce a vector $U^* \in T_p(M)$ depending on the parameter t of the V-curve by the following relation

$$U^*(p;t) = (e^{-tV})_* U\big(e^{tV}(p)\big) = (e^{-tV})_*(e^{tV})^* U(p) \qquad (2.10.1)$$

where the linear operator $(\phi_t^{-1})_* = (e^{-tV})_* : T_q(M) \to T_p(M)$ is the differential of the *inverse flow* ϕ_t^{-1} at the point q and places the vector $U(q)$ into the tangent space at the point p. The operator $(e^{tV})^*$ is defined as usual by $(e^{tV})^* U = U \circ e^{tV}$. The vectors $U^*(p;t)$ and $U(p)$ now lie in the same tangent space. Therefore, their difference can now be calculated without any difficulty. We shall next define the ***Lie derivative*** of a vector field U with respect to the vector field V at the point p by the following limiting process

$$\pounds_V U = \lim_{t\to 0} \frac{U^*(p;t) - U(p)}{t} = \lim_{t\to 0} \frac{(e^{-tV})_*(e^{tV})^* - I}{t} U(p). \qquad (2.10.2)$$

[Although it is always referred to the name of Norwegian mathematician Marius Sophus Lie (1842-1899), this concept was first introduced in 1931 by Polish mathematician Władisław Ślebodziński (1884-1972). However, the term *Lie derivative* was coined by Dutch mathematician David van Dantzig (1900-1959) in 1932]. Thus the *Lie derivative operator* can be expressed in the form

$$\pounds_V = \lim_{t \to 0} \frac{(e^{-tV})_*(e^{tV})^* - I}{t}. \tag{2.10.3}$$

In the second step, we attempt to evaluate concretely the Lie derivative of a vector field U with respect to a vector field V by resorting to local charts at points p and q. Let the local coordinates at the point p be $(x^1, \dots, x^m)$ and those at the point $q = e^{tV}(p)$ be $(\overline{x}^1, \dots, \overline{x}^m)$. Thus we can write

$$U(p) = u^i(\mathbf{x}) \frac{\partial}{\partial x^i}, \quad U(q) = u^i(\overline{\mathbf{x}}) \frac{\partial}{\partial \overline{x}^i}.$$

In accordance with the relations (2.7.4) and (2.7.5), we obtain

$$U^*(p; t) = u^{*j}\big(\overline{\mathbf{x}}(t)\big) \frac{\partial}{\partial x^j}, \quad u^{*j}\big(\overline{\mathbf{x}}(t)\big) = u^i\big(\overline{\mathbf{x}}(t)\big) \frac{\partial x^j}{\partial \overline{x}^i}.$$

For very small values of the parameter t, the expression $\overline{\mathbf{x}}(t) = e^{tV'}(\mathbf{x})$ can be approximated by

$$\overline{x}^i(t) = x^i + tv^i(\mathbf{x}) + o(t)$$

where the vector field V is represented as $V = v^i \partial / \partial x^i$. Hence, we are led to a matrix whose elements are given by

$$\frac{\partial \overline{x}^i}{\partial x^j} = \delta^i_j + t \frac{\partial v^i}{\partial x^j} + o(t).$$

It follows from the chain rule of differentiation that the inverse of this matrix is prescribed by $[\partial x^i / \partial \overline{x}^j]$ from which we find approximately

$$\frac{\partial x^i}{\partial \overline{x}^j} = \delta^i_j - t \frac{\partial v^i}{\partial x^j} + o(t).$$

Indeed, it is straightforward to verify that

$$\begin{aligned} \delta^i_j = \frac{\partial \overline{x}^i}{\partial x^k} \frac{\partial x^k}{\partial \overline{x}^j} &= \Big(\delta^i_k + t \frac{\partial v^i}{\partial x^k}\Big)\Big(\delta^k_j - t \frac{\partial v^k}{\partial x^j}\Big) + o(t) \\ &= \delta^i_j + t\Big(\frac{\partial v^i}{\partial x^j} - \frac{\partial v^i}{\partial x^j}\Big) + o(t) = \delta^i_j + o(t). \end{aligned}$$

Furthermore, the Taylor series around the point $\mathbf{x}$ yields

$$u^i\big(\overline{\mathbf{x}}(t)\big) = u^i\big(\mathbf{x} + t\mathbf{v}(\mathbf{x}) + \boldsymbol{o}(t)\big) = u^i(\mathbf{x}) + t \frac{\partial u^i}{\partial x^k} v^k + o(t).$$

We thus find

$$u^{*j} = \Big(u^i + t\frac{\partial u^i}{\partial x^k}v^k + o(t)\Big)\Big(\delta_i^j - t\frac{\partial v^j}{\partial x^i} + o(t)\Big)$$
$$= u^j + t\Big(\frac{\partial u^j}{\partial x^k}v^k - \frac{\partial v^j}{\partial x^i}u^i\Big) + o(t)$$

and obtain finally

$$U^*(p;t) - U(p) = t\Big(v^i\frac{\partial u^j}{\partial x^i} - u^i\frac{\partial v^j}{\partial x^i}\Big)\frac{\partial}{\partial x^j} + o(t).$$

Since $\lim_{t\to 0} o(t)/t = 0$, we conclude that

$$\pounds_V U = w^i\frac{\partial}{\partial x^i} = w^i\partial_i = W, \quad w^i = v^j u^i_{,j} - u^j v^i_{,j} \tag{2.10.4}$$

where we employed the abbreviations $(\,\cdot\,)_{,i} = \partial(\,\cdot\,)/\partial x^i$ and $\partial_i = \partial/\partial x^i$. We observe that *the Lie derivative of a vector field U with respect to a vector field V at every point p is again a vector in the tangent space $T_p(M)$ and the vector field $\pounds_V U$ so created measures the rate of change of the vector U at every point in the manifold along the congruence generated by the vector field V.*

We readily obtain from (2.10.4) the following results

$$\pounds_{\partial_j} U = \frac{\partial u^i}{\partial x^j}\frac{\partial}{\partial x^i}, \quad \pounds_V\Big(\frac{\partial}{\partial x^j}\Big) = -\frac{\partial v^i}{\partial x^j}\frac{\partial}{\partial x^i}. \tag{2.10.5}$$

We can attribute another meaning to the Lie derivative evoking algebraic connotations. We take two vector fields $U, V \in T(M)$ into account on a manifold M. For any function $f \in \Lambda^0(M)$, we get $V(f) \in \Lambda^0(M)$ and also $U\big(V(f)\big) \in \Lambda^0(M)$ so that we can write

$$U\big(V(f)\big) = u^j\frac{\partial}{\partial x^j}\Big(v^i\frac{\partial f}{\partial x^i}\Big) = u^j(v^i f_{,i})_{,j} = u^j v^i_{,j} f_{,i} + u^j v^i f_{,ij}.$$

In a similar way, we arrive at

$$V\big(U(f)\big) = v^j u^i_{,j} f_{,i} + u^i v^j f_{,ij}.$$

Hence, we find that

$$V\big(U(f)\big) - U\big(V(f)\big) = (v^j u^i_{,j} - u^j v^i_{,j}) f_{,i} + (u^i v^j - u^j v^i) f_{,ij}.$$

Second order derivatives $f_{,ij}$ are symmetric with respect to the indices i, j due to the well known relation $f_{,ij} = f_{,ji}$ whereas their coefficients are antisymmetric with respect to the same indices. This means that the second sum in the foregoing relation turns out to be zero. As a result, for every function $f \in \Lambda^0(M)$, the following equality holds

$$(VU - UV)(f) = W(f)$$

where the vector field W is given by (2.10.4). This of course implies that

$$VU - UV = W \in T(M).$$

We now define the ***commutator*** of two vector fields as

$$[V', U'] = [V, U] = VU - UV. \tag{2.10.6}$$

This is tantamount to say that the Lie derivative of a vector field U with respect to the vector field V is expressible as

$$\pounds_V U = [V, U]. \tag{2.10.7}$$

It clearly follows from the definition that the commutation rule $[V, U] = -[U, V]$ is valid. Therefore, Lie derivatives of two vector fields with respect to one another are related by

$$\pounds_V U = -\pounds_U V. \tag{2.10.8}$$

The commutator $[V, U]$ is also called ***Lie bracket*** or ***Lie product***. *Lie product is antisymmetric and one naturally has* $[V, V] = 0$. It might be instructive to evaluate the Lie derivative given by (2.10.4) this time by means of the commutator:

$$\begin{aligned}[V, U] &= [v^i\partial_i, u^j\partial_j] = v^i\partial_i(u^j\partial_j) - u^j\partial_j(v^i\partial_i)\\ &= v^i(\partial_i u^j)\partial_j - u^j(\partial_j v^i)\partial_i + (v^i u^j - u^j v^i)\partial_{ij}\\ &= (v^j\partial_j u^i - u^j\partial_j v^i)\partial_i = \pounds_V U.\end{aligned}$$

Let us now take $V = \partial_i, U = \partial_j$. It is then immediately seen that for all indices i, j, we find

$$[\partial_i, \partial_j] = 0. \tag{2.10.9}$$

Thus Lie derivatives of all natural basis vectors, produced by local charts, with respect to one another vanish.

Another geometrical meaning can easily be attributed to the Lie bracket, namely, the Lie derivative. Suppose that we are given two vector fields and U- and V-congruences generated by them on a manifold M are determined. Let these families of curves are parametrised by t_1 and t_2, respectively. We consider U- and V-curves through a point $p \in M$. Let the points

$q, r \in M$ be determined along respective flows for the values t_1 and t_2 of parameters as

$$q = e^{t_1 U}(p), \quad r = e^{t_2 V}(p).$$

Let us also consider the points $\overline{q} = e^{t_2 V}(q), \overline{r} = e^{t_1 U}(r)$ along flows. In this case, we write

$$\overline{q} = e^{t_2 V} \circ e^{t_1 U}(p), \quad \overline{r} = e^{t_1 U} \circ e^{t_2 V}(p). \tag{2.10.10}$$

We denote images of these points in $\mathbb{R}^m$ by $\mathbf{x}(p), \mathbf{x}(\overline{q}), \mathbf{x}(\overline{r})$. We obtain from relations (2.9.6) the expression

$$\mathbf{x}(\overline{q}) - \mathbf{x}(\overline{r}) = (e^{t_2 V'} e^{t_1 U'} - e^{t_1 U'} e^{t_2 V'})\mathbf{x}(p) = [e^{t_2 V'}, e^{t_1 U'}]\mathbf{x}$$

that can be thought as measuring the "difference" between the points $\overline{q}$ and $\overline{r}$ where we wrote $\mathbf{x}(p) = \mathbf{x}$. On choosing t_1, t_2 sufficiently small, the expansions of exponential mappings yields

$$\begin{aligned}
\mathbf{x}(\overline{q}) - \mathbf{x}(\overline{r}) &= \big\{(I + t_2 V' + \tfrac{1}{2}t_2^2 V'^2 + \cdots)(I + t_1 U' + \tfrac{1}{2}t_1^2 U'^2 + \cdots) \\
&\quad - (I + t_1 U' + \tfrac{1}{2}t_1^2 U'^2 + \cdots)(I + t_2 V' + \tfrac{1}{2}t_2^2 V'^2 + \cdots)\big\}\mathbf{x} \\
&= \big\{I + t_1 U' + t_2 V' + t_1 t_2 V'U' + \tfrac{1}{2}t_2^2 V'^2 + \tfrac{1}{2}t_1^2 U'^2 + \cdots \\
&\quad - I - t_1 U' - t_2 V' - t_1 t_2 U'V' - \tfrac{1}{2}t_1^2 U'^2 - \tfrac{1}{2}t_2^2 V'^2 - \cdots\big\}\mathbf{x} \\
&= \big\{t_1 t_2 (V'U' - U'V') + o(t_1^2, t_2^2, t_1 t_2)\big\}\mathbf{x}.
\end{aligned}$$

Next, we take the parameters t_1, t_2 of the order ϵ where ϵ is a small number. We thus conclude that

$$\begin{aligned}
\mathbf{x}(\overline{q}) - \mathbf{x}(\overline{r}) &= \epsilon^2 [V, U]\big(\mathbf{x}(p)\big) + \boldsymbol{o}(\epsilon^2) \\
&= \epsilon^2 \pounds_V U\big(\mathbf{x}(p)\big) + \boldsymbol{o}(\epsilon^2).
\end{aligned} \tag{2.10.11}$$

In view of (2.10.4), this expression can be cast into the shape

$$x^i(\overline{q}) - x^i(\overline{r}) = t_1 t_2 w^j \frac{\partial x^i}{\partial x^j} + o(\epsilon^2) = t_1 t_2 w^i + o(\epsilon^2) \sim \epsilon^2 w^i + o(\epsilon^2)$$

in terms of components. It is seen that even if we consider rather close points p, q, r, the points $\overline{q}$ and $\overline{r}$ formed as above do not coincide in general. But the difference is of second order and its magnitude is governed by the Lie bracket at the point p (Fig. 2.10.2).

If vector fields U, V commute, then we have $VU = UV$ and $[V, U] = 0$. We know in this case that $e^{t_2 V} \circ e^{t_1 U} = e^{t_1 U} \circ e^{t_2 V} = e^{t_1 U + t_2 V}$. Hence (2.10.10) yields exactly $\overline{q} = \overline{r}$. In other words, the congruence curves

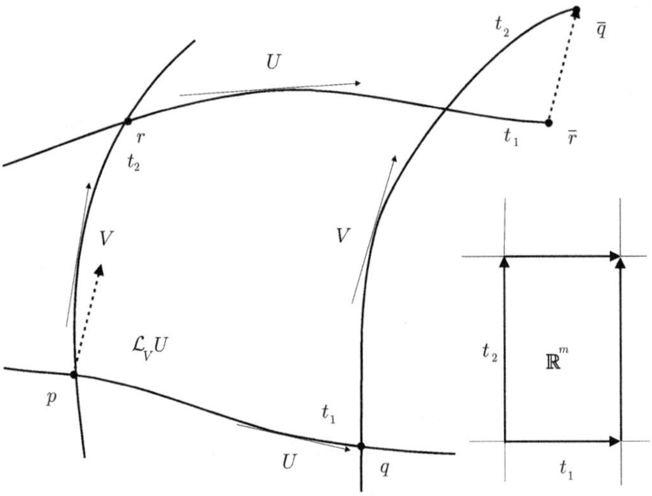

Fig. 2.10.2. The geometrical meaning of the Lie derivative.

through the points q and r intersects at the point $\overline{q}$ for the parameter values t_1 and t_2. This amounts to say that the U- and V- congruences play the part of two families of coordinate lines on M because t_1 and t_2 can now be regarded as two Cartesian coordinates in $\mathbb{R}^m$.

Conversely, we can immediately deduce from (2.10.11) that if we get $[e^{tV}, e^{tU}] = 0$ for all t and $\mathbf{x}$, then we must have $[V, U] = 0$.

It follows directly from the relation (2.10.4) that the Lie product is distributive:

$$[V_1 + V_2, U] = [V_1, U] + [V_2, U], \quad [V, U_1 + U_2] = [V, U_1] + [V, U_2]$$

Therefore, for all vector fields U, V and U_1, U_2, V_1, V_2 we can write

$$£_{V_1+V_2} U = £_{V_1} U + £_{V_2} U, \quad £_V (U_1 + U_2) = £_V U_1 + £_V U_2$$

whence we reach to the operator equality

$$£_{V_1+V_2} = £_{V_1} + £_{V_2}. \tag{2.10.12}$$

Moreover, Lie product satisfies the ***Jacobi identity***. $U, V, W \in T(M)$ are arbitrary three vector fields. Then, the following identity holds

$$J = \big[U, [V, W]\big] + \big[V, [W, U]\big] + \big[W, [U, V]\big] = 0. \tag{2.10.13}$$

To verify this, let us begin with the calculation of the first term:

$$
\begin{aligned}
[U,[V,W]] &= \Big\{u^j(v^k w^i_{,k} - w^k v^i_{,k})_{,j} - (v^k w^j_{,k} - w^k v^j_{,k})u^i_{,j}\Big\}\frac{\partial}{\partial x^i} \\
&= \Big\{u^j v^k_{,j} w^i_{,k} + u^j v^k w^i_{,jk} - u^j v^i_{,k} w^k_{,j} \\
&\qquad - u^j v^i_{,jk} w^k - u^i_{,j} v^k w^j_{,k} + u^i_{,j} v^j_{,k} w^k\Big\}\frac{\partial}{\partial x^i}
\end{aligned}
$$

After having evaluated the other terms in (2.10.13) in a similar fashion, we consider their sum and by eliminating terms cancelling each other we reach to the result

$$
J = \Big\{u^i_{,jk}(v^j w^k - v^k w^j) + v^i_{,jk}(w^j u^k - w^k u^j) + w^i_{,jk}(u^j v^k - u^k v^j)\Big\}\frac{\partial}{\partial x^i}.
$$

However, in the above sums, the terms within parentheses are antisymmetric whereas mixed derivatives are symmetric with respect to relevant indices so that we finally obtain $J = 0$. This result can also be found by resorting to commutators. If we write (2.10.13) explicitly, we obtain

$$
\begin{aligned}
&U[V,W] - [V,W]U + V[W,U] - [W,U]V + W[U,V] - [U,V]W \\
&\quad = UVW - UWV - VWU + WVU + VWU - VUW - WUV \\
&\qquad + UWV + WUV - WVU - UVW + VUW = 0.
\end{aligned}
$$

We can now equip the module $\mathfrak{V}(M)$ that consists of all vector fields on a manifold M with a closed binary operation provided by the Lie bracket that assigns a vector field to every pair of vector fields. This way $\mathfrak{V}(M)$ acquires an algebraic structure. With this structure that is *anticommutative* but *not associative* as is clearly implied by the Jacobi identity (2.10.13), we can now venture to say, with a slight abuse of the term, that the module $\mathfrak{V}(M)$ has become a ***Lie algebra***. In fact, a Lie algebra is usually defined on a vector space. But $\mathfrak{V}(M)$ is a vector space only on the field of real numbers. Thus, strictly speaking, a Lie algebra can be formed on a real vector space by defining the product of two vectors as the Lie bracket. In this case, the Lie product turns out to be a bilinear operations so that for real numbers α_1, α_2, we can write

$$
\begin{aligned}
&[\alpha_1 U_1 + \alpha_2 U_2, V] = \alpha_1[U_1, V] + \alpha_2[U_2, V] \\
&[U, \alpha_1 V_1 + \alpha_2 V_2,] = \alpha_1[U, V_1] + \alpha_2[U, V_2].
\end{aligned}
$$

Obviously, tangent spaces $T_p(M)$ at every point $p \in M$ are Lie algebras in the true sense of the word.

We shall now attempt to measure the change in a vector field U along a V-curve through the point p. We can transport all vectors U in different tangent spaces at every point of the curve into the tangent space $T_p(M)$ utilising the mapping (2.10.1). Since we can add vectors in the same tangent space, the rate of change of the vector field U can be measured directly by the derivative

$$\frac{dU^*(p;t)}{dt} = \lim_{\tau\to 0}\frac{U^*(p;t+\tau) - U^*(p;t)}{\tau}. \tag{2.10.14}$$

We know that the diffeomorphism $e^{tV} : M \to M$ generated by a vector field V on the manifold M will satisfy $e^{(t+\tau)V} = e^{tV} \circ e^{\tau V} = e^{\tau V} \circ e^{tV}$. It then follows from the rule (2.7.7) concerning the composition of differentials that one can write

$$\begin{aligned}
U^*(p;t+\tau) &= (e^{-(t+\tau)V})_*(e^{(t+\tau)V})^*U(p)\\
&= (e^{-\tau V})_* \circ (e^{-tV})_* \circ U \circ e^{\tau V} \circ e^{tV}(p)\\
&= (e^{-\tau V})_* \circ (e^{-tV})_* \circ (e^{tV})^* \circ \big(U \circ e^{\tau V}\big)(p)\\
&= (e^{-\tau V})_* \circ U^*(p;t) \circ e^{\tau V}(p)\\
&= (e^{-\tau V})_* \circ (e^{\tau V})^* U^*(p;t).
\end{aligned}$$

Hence, the derivative (2.10.14) is expressible in the form

$$\frac{dU^*(p;t)}{dt} = \lim_{\tau\to 0}\frac{(e^{-\tau V})_*(e^{\tau V})^* - I}{\tau}U^*(p;t).$$

If we recall the relation (2.10.3), we conclude that

$$\frac{dU^*(p;t)}{dt} = \pounds_V U^*(p;t). \tag{2.10.15}$$

This is a differential equation satisfied by the operator U^* with the initial condition $U^*(p;0) = U(p)$. The solution of this equation is formally expanded into a Maclaurin series around $t = 0$ as follows

$$\begin{aligned}
U^*(p;t) = U^*(p;0) + \left.\frac{dU^*(p;t)}{dt}\right|_{t=0} t + \frac{1}{2}\left.\frac{d^2U^*(p;t)}{dt^2}\right|_{t=0} t^2 + \cdots\\
+ \frac{1}{n!}\left.\frac{d^nU^*(p;t)}{dt^n}\right|_{t=0} t^n + \cdots.
\end{aligned}$$

Since the operator $\pounds_V$ does not depend on the parameter t, we find that

$$\left.\frac{dU^*(p;t)}{dt}\right|_{t=0} = \pounds_V U^*(p;0) = \pounds_V U(p),$$

$$\left.\frac{d^2U^*(p;t)}{dt^2}\right|_{t=0} = \pounds_V \left.\frac{dU^*(p;t)}{dt}\right|_{t=0} = \pounds_V^2 U(p),.$$
$$\vdots$$

where $U^*(p;0) = U(p)$. We thus arrive at the formal operator series

$$U^*(p;t) = U(p) + t\pounds_V U(p) + \frac{t^2}{2}\pounds_V^2 U(p) + \cdots + \frac{t^n}{n!}\pounds_V^n U(p) + \cdots.$$

We now define the exponential operator $e^{t\pounds_V}$ in the usual way as the absolutely convergent series

$$e^{t\pounds_V} = I + t\pounds_V + \frac{t^2}{2!}\pounds_V^2 + \cdots + \frac{t^n}{n!}\pounds_V^n + \cdots$$

whence we are led to the result

$$U^*(p;t) = e^{t\pounds_V} U(p) \in T_p(M), \;\; p \in M. \tag{2.10.16}$$

We deduce from his relation an important property of vector fields. If $\pounds_V U = [V,U] = 0$, then we get $U^*(p;t) = U(p)$ implying that the vector field U does not change on V-congruence. In other words, the vector field U remains ***invariant*** with respect to the vector field V. On the other hand, if $\pounds_V U = 0$, then we have $\pounds_U V = 0$ due to (2.10.8). Therefore, we understand that *if the field U is invariant with respect to the field V, then the field V becomes necessarily invariant with respect to the field U.*

We can now write the Jacobi identity in the form

$$\pounds_U \pounds_V W + \pounds_V \pounds_W U + \pounds_W \pounds_U V = 0.$$

Then properties of Lie derivative allows us to transform this relation into

$$(\pounds_U \pounds_V - \pounds_V \pounds_U)W = \pounds_{\pounds_U V} W$$

or $[\pounds_U, \pounds_V]W = \pounds_{[U,V]}W$. Since this equality must hold for every vector field W, we arrive at the following rather elegant result between two Lie derivative operators

$$[\pounds_U, \pounds_V] = \pounds_{[U,V]}. \tag{2.10.17}$$

In exactly same way, we can define the Lie derivative of a function $f \in \Lambda^0(M)$ as follows

$$\pounds_V f = \lim_{t\to 0} \frac{f(e^{tV}p) - f(p)}{t}.$$

On using charts, we have $f(p) = f'(\mathbf{x})$, $f(e^{tV}p) = f'\big(\overline{\mathbf{x}}(t)\big)$. We can now approximately write $\overline{x}^i(t) = x^i + tv^i(\mathbf{x}) + o(t)$. Hence, Taylor series about the point $\mathbf{x}$ yields

$$f'\big(\overline{\mathbf{x}}(t)\big) - f'(\mathbf{x}) = f'(\mathbf{x}) + tv^i(\mathbf{x})\frac{\partial f}{\partial x^i} + o(t) - f'(\mathbf{x}) = tV'(f') + o(t)$$

and we finally obtain

$$\pounds_V f = V(f) = v^i \frac{\partial f}{\partial x^i} \tag{2.10.18}$$

Thus the Lie derivative of a function f is nothing but the directional derivative of f along the vector V. If $\pounds_V f = 0$, then the function f remains constant on every curve of V-congruence. *Naturally this constant may be different on each curve of the congruence.*

Suppose now that we are given two vector fields $U, V \in T(M)$ and two smooth functions $f, g \in \Lambda^0(M)$. For any function $h \in \Lambda^0(M)$, we can write

$$\begin{aligned}[][fU, gV](h) &= fU\big(gV(h)\big) - gV\big(fU(h)\big) = fU(g)V(h) + fgUV(h) \\ &- gV(f)U(h) - gfVU(h) = \big[fg\,[U,V] + fU(g)V - gV(f)U\big](h).\end{aligned}$$

where we have taken into account that vector fields are actually derivations. We thus obtain

$$[fU, gV] = fg\,[U,V] + fU(g)V - gV(f)U \tag{2.10.19}$$

or equivalently

$$\pounds_{fU}(gV) = fg\,\pounds_U V + f\pounds_U(g)V - g\pounds_V(f)U. \tag{2.10.20}$$

Let $\phi : M \to N$ be a differentiable mapping between manifolds M and N. We know that the differential of ϕ at a point $p \in M$ is the linear operator $\phi_* : T_p(M) \to T_{\phi(p)}(N)$. Consider two vector fields U and V on the manifold M. The Lie bracket of these vector fields at p is given by the vector

$$\begin{aligned}[][U, V] &= w^i \frac{\partial}{\partial x^i} \in T_p(M), \\ w^i &= u^j \frac{\partial v^i}{\partial x^j} - v^j \frac{\partial u^i}{\partial x^j}\end{aligned}$$

in the local coordinates. In view of (2.7.4), the vector $\phi_*[U, V] \in T_{\phi(p)}(M)$ is expressed in the form

$$\phi_*[U,V] = w^{*\alpha}\frac{\partial}{\partial y^\alpha} = w^i\frac{\partial \Phi^\alpha}{\partial x^i}\frac{\partial}{\partial y^\alpha}.$$

Here, $\mathbf{y} = (y^1,\ldots,y^n)$ are the local coordinates at the point $\phi(p) \in N$ and are related to the local coordinates $\mathbf{x} = (x^1,\ldots,x^m)$ at the point $p \in M$ by a functional relation $\mathbf{y} = \Phi(\mathbf{x})$ or functions $y^\alpha = \Phi^\alpha(x^1,\ldots,x^m), \alpha = 1,\ldots,n$ associated with the mapping ϕ. Let us now explicitly evaluate components $w^{*\alpha}$:

$$\begin{aligned}
w^{*\alpha} &= w^i\frac{\partial \Phi^\alpha}{\partial x^i} = \left(u^j\frac{\partial v^i}{\partial x^j} - v^j\frac{\partial u^i}{\partial x^j}\right)\frac{\partial \Phi^\alpha}{\partial x^i}\\
&= u^j\frac{\partial}{\partial x^j}\left(v^i\frac{\partial \Phi^\alpha}{\partial x^i}\right) - v^j\frac{\partial}{\partial x^j}\left(u^i\frac{\partial \Phi^\alpha}{\partial x^i}\right) + (v^j u^i - u^j v^i)\frac{\partial^2 \Phi^\alpha}{\partial x^i \partial x^j}\\
&= u^j\frac{\partial v^{*\alpha}}{\partial x^j} - v^j\frac{\partial u^{*\alpha}}{\partial x^j} = u^j\frac{\partial v^{*\alpha}}{\partial y^\beta}\frac{\partial \Phi^\beta}{\partial x^j} - v^j\frac{\partial u^{*\alpha}}{\partial y^\beta}\frac{\partial \Phi^\beta}{\partial x^j}\\
&= u^{*\beta}\frac{\partial v^{*\alpha}}{\partial y^\beta} - v^{*\beta}\frac{\partial u^{*\alpha}}{\partial y^\beta} = [\phi_* U, \phi_* V]^\alpha.
\end{aligned}$$

We thus conclude that

$$\phi_*[U,V] = [\phi_* U, \phi_* V] \qquad (2.10.21)$$

or $[U,V]^* = [U^*, V^*]$, or $\phi_*(\pounds_U V) = \pounds_{\phi_* U}(\phi_* V)$.

2.11. DISTRIBUTIONS. THE FROBENIUS THEOREM

Let M be an m-dimensional differentiable manifold. Let us consider a subspace $\mathcal{D}_p = \mathcal{T}_p(M) \subset T_p(M)$ of dimension $k < m$ of the tangent space $T_p(M)$ at every point $p \in M$. We may constitute a *tangent subbundle* by union of disjoint subspaces $\mathcal{T}_p(M)$:

$$\mathcal{T}(M) = \bigcup_{p\in M} \mathcal{T}_p(M) = \{(p,V) : p \in M, V \in \mathcal{T}_p(M)\} \subset T(M) \quad (2.11.1)$$

This subbundle is called a ***k-dimensional distribution***. We denote it by $\mathcal{D} = \mathcal{T}(M)$. Thus a k-dimensional distribution really attaches to every point of the manifold a k-dimensional subspace of the tangent space at that point. In order to construct such a distribution, all we have to do is to select k linearly independent vector fields. If vector fields U_α, $\alpha = 1,\ldots,k$ are linearly independent, then the relation

$$a^\alpha(p)U_\alpha(p) = 0$$

with $a^\alpha \in \Lambda^0(M)$ can be satisfied if and only if $a^\alpha(p) = 0$ for $\alpha = 1, \ldots, k$. Such vector fields U_α constitute a *basis* of the distribution.

A distribution $\mathcal{D}$ is called an ***involutive distribution*** *if for every vector fields $U, V \in \mathcal{D}$ one has $[U, V] \in \mathcal{D}$, namely, if $\mathcal{D}$ is closed under the Lie product.* It is clear that all Lie brackets remain in $\mathcal{D}$ if and only if it is possible to find functions $c^\alpha \in \Lambda^0(M)$ such that

$$[U, V] = c^\alpha(p) U_\alpha$$

for all $U, V \in \mathcal{D}$. Since basis vectors U_α are also in $\mathcal{D}$, a necessary condition for the distribution $\mathcal{D}$ to be involutive is that the relations

$$[U_\alpha, U_\beta] = c^\gamma_{\alpha\beta}(p) U_\gamma \tag{2.11.2}$$

should be satisfied for some functions $c^\gamma_{\alpha\beta} \in \Lambda^0(M)$. One can readily shows that this condition is also sufficient. Let us consider vectors $U = \lambda^\alpha U_\alpha$ and $V = \mu^\alpha U_\alpha$. It follows from (2.10.19) that

$$\begin{aligned}[U, V] &= [\lambda^\alpha U_\alpha, \mu^\beta U_\beta] = \lambda^\alpha \mu^\beta [U_\alpha, U_\beta] + \lambda^\alpha U_\alpha(\mu^\beta) U_\beta \\ &\quad - \mu^\beta U_\beta(\lambda^\alpha) U_\alpha = \{c^\gamma_{\alpha\beta} \lambda^\alpha \mu^\beta + \lambda^\alpha U_\alpha(\mu^\gamma) - \mu^\alpha U_\alpha(\lambda^\gamma)\} U_\gamma \\ &\quad = c^\gamma(p) U_\gamma \in \mathcal{D}.\end{aligned}$$

Due to the antisymmetry of Lie brackets, the coefficients $c^\gamma_{\alpha\beta}$ must be antisymmetric with respect to the subscripts:

$$c^\gamma_{\alpha\beta} = -\, c^\gamma_{\beta\alpha}. \tag{2.11.3}$$

Moreover, Lie brackets of vectors in $\mathcal{D}$ ought to satisfy the Jacobi identity. For basis vectors U_α, this identity is reduced to the form

$$\big[U_\alpha, [U_\beta, U_\gamma]\big] + \big[U_\beta, [U_\gamma, U_\alpha]\big] + \big[U_\gamma, [U_\alpha, U_\beta]\big] = 0.$$

On using (2.11.2), this identity yields

$$\begin{aligned}&[U_\alpha, c^\delta_{\beta\gamma} U_\delta] + [U_\beta, c^\delta_{\gamma\alpha} U_\delta] + [U_\gamma, c^\delta_{\alpha\beta} U_\delta] = c^\delta_{\beta\gamma}[U_\alpha, U_\delta] + c^\delta_{\gamma\alpha}[U_\beta, U_\delta] \\ &\quad + c^\delta_{\alpha\beta}[U_\gamma, U_\delta] + U_\alpha(c^\delta_{\beta\gamma}) U_\delta + U_\beta(c^\delta_{\gamma\alpha}) U_\delta + U_\gamma(c^\delta_{\alpha\beta}) U_\delta \\ &= \{c^\delta_{\beta\gamma} c^\lambda_{\alpha\delta} + c^\delta_{\gamma\alpha} c^\lambda_{\beta\delta} + c^\delta_{\alpha\beta} c^\lambda_{\gamma\delta} + U_\alpha(c^\lambda_{\beta\gamma}) + U_\beta(c^\lambda_{\gamma\alpha}) + U_\gamma(c^\lambda_{\alpha\beta})\} U_\lambda = 0.\end{aligned}$$

Since vectors U_λ are linearly independent, we deduce that the coefficients $c^\gamma_{\alpha\beta}$ must satisfy the following relations

$$c^\delta_{\beta\gamma} c^\lambda_{\alpha\delta} + c^\delta_{\gamma\alpha} c^\lambda_{\beta\delta} + c^\delta_{\alpha\beta} c^\lambda_{\gamma\delta} + U_\alpha(c^\lambda_{\beta\gamma}) + U_\beta(c^\lambda_{\gamma\alpha}) + U_\gamma(c^\lambda_{\alpha\beta}) = 0 \tag{2.11.4}$$

for all values of indices $\alpha, \beta, \gamma, \lambda = 1, \dots, k$. Because of the symmetry properties of these coefficients the number of independent relations in (2.11.4) is considerably smaller.

We have discussed in Sec. 2.4 some techniques to specify a submanifold of a given manifold M. We now propose another method to achieve that purpose. Let S be a k-dimensional submanifold of M determined by the relations $x^i = x^i(u^\alpha), i = 1, \dots, m$ and $\alpha = 1, \dots, k$. Then at a point $p \in S$ there will be a k-dimensional tangent space $T_p(S)$. But p is a point of M as well and all vectors at that point belong also to $T_p(M)$. Hence, we can write $T_p(S) \subset T_p(M)$, i.e., $T_p(S)$ is a subspace of $T_p(M)$. Since the inclusion map $\mathcal{I} : S \to M$ is an embedding, its differential $d\mathcal{I} : T(S) \to T(M)$ is an injective linear operator. Because $\mathcal{I} : S \to \mathcal{I}(S)$ is an identity mapping, we can write $d\mathcal{I}(V) = V, V \in T_p(S)$. Thus, if we consider a vector V in the tangent space of S at a point p, its components in tangent spaces $T_p(S)$ and $T_p(M)$ are related by

$$V = v^i \frac{\partial}{\partial x^i} = v^\alpha \frac{\partial}{\partial u^\alpha} = v^\alpha \frac{\partial x^i}{\partial u^\alpha} \frac{\partial}{\partial x^i} \quad \text{or} \quad v^i = \frac{\partial x^i}{\partial u^\alpha} v^\alpha. \qquad (2.11.5)$$

Let $U, V \in T_p(S)$. Due to (2.10.21) we obtain

$$d\mathcal{I}([U,V]) = [d\mathcal{I}(U), d\mathcal{I}(V)] = [U,V]$$

that results in $[U,V] \in T_p(S)$. This means that as long as $S \subseteq M$ is a submanifold, Lie products of vectors in $T_p(S)$ stay in $T_p(S)$. Therefore, such a subspace $T_p(S)$ is a Lie subalgebra of the Lie algebra $T_p(M)$.

Now, conversely, let us suppose that we are given k linearly independent vector fields on the manifold M. In other words, we choose a k-dimensional subspace of the tangent space at every point of the manifold. We then take congruences that are integral curves of those vector fields. Therefore, we can construct a local piece of the manifold which is tangent to a linear vector space formed by the chosen k vectors at every point of the manifold M. Next we have to ask the following question: under which conditions these small pieces of manifolds can be patched together *smoothly* in order to produce a smooth hypersurface forming a submanifold? This question can be quite easily answered qualitatively. When moving on an integral curve of a vector field, the variations of other vector fields are measured by Lie derivatives. In order that these integral curves stay on the hypersurface, Lie derivatives of vector fields must lie in the chosen subspace.

Let us consider a k-dimensional distribution $\mathcal{D}$ on a manifold M. If the tangent space at every point $p \in S$ of a k-dimensional submanifold $S \subseteq M$

is identical with the subspace $\mathcal{D}_p$ of the tangent space $T_p(M)$ of M, that is, if we have

$$d\mathcal{I}(T_p(S)) = \mathcal{D}_{p=\mathcal{I}(p)}, \quad \forall p \in S \tag{2.11.6}$$

where $\mathcal{I} : S \to M$ is the embedding mapping determining the submanifold S, then S is called an ***integral submanifold*** of the distribution $\mathcal{D}$. Sometimes instead of (2.11.6), we may prefer the weaker condition $d\mathcal{I}(T_p(S)) \subseteq \mathcal{D}_{p=i(p)}$ at each point $p \in S$. In this case the dimension of S may be less than k. *If a k-dimensional distribution $\mathcal{D}$ possesses a k-dimensional integral submanifold through every point $p \in M$, then $\mathcal{D}$ is called a **completely integrable distribution***. A fundamental theorem concerning such distributions is provided by German mathematician Ferdinand Georg Frobenius (1849-1917).

Theorem 2.11.1 (The Frobenius Theorem). *A distribution $\mathcal{D}$ on a manifold is completely integrable if and only it is involutive.*

If we assume that the distribution $\mathcal{D}$ is completely integrable, then there exists an integral submanifold S through every point $p \in M$ and at that point the subspace $\mathcal{D}_p \subset T_p(M)$ corresponds to the tangent space of S. Therefore, for each $U, V \in \mathcal{D}_p$ one finds $[U, V] \in \mathcal{D}_p$, namely, $\mathcal{D}$ is involutive.

For the proof of the converse statement, we consider a k-dimensional involutive distribution $\mathcal{D}$ on an m-dimensional manifold M. This distribution is specified by $k \leq m$ linearly independent vector fields $U_\alpha, \alpha = 1, 2, \ldots, k$ in the m-dimensional tangent bundle $T(M)$. Since $\mathcal{D}$ is an involutive distribution, there exist smooth functions $c_{\alpha\beta}^{\gamma} \in \Lambda^0(M)$ satisfying the relations $[U_\alpha, U_\beta] = c_{\alpha\beta}^{\gamma}(p)U_\gamma$ and verifying the conditions (2.11.3) and (2.11.4). Let us choose a new set of linearly independent vector fields by means of the transformation

$$V_\alpha(p) = A_\alpha^\beta(p)U_\beta(p), \quad \alpha, \beta = 1, \ldots, k \tag{2.11.7}$$

where $A_\alpha^\beta(p) \in \Lambda^0(M)$. The only restriction imposed on $k \times k$ matrix $\mathbf{A} = \left[A_\beta^\alpha\right]$ is that $\det \mathbf{A}(p) \neq 0$ at each point $p \in S$. Thus, we can write

$$\begin{aligned}[V_\gamma, V_\delta] = [A_\gamma^\alpha U_\alpha, A_\delta^\beta U_\beta] = A_\gamma^\alpha A_\delta^\beta [U_\alpha, U_\beta] + A_\gamma^\alpha U_\alpha(A_\delta^\beta) U_\beta \\ - A_\delta^\beta U_\beta(A_\gamma^\alpha) U_\alpha = \left[c_{\alpha\beta}^{\mu} A_\gamma^\alpha A_\delta^\beta + A_\gamma^\alpha U_\alpha(A_\delta^\mu) - A_\delta^\alpha U_\alpha(A_\gamma^\mu)\right] U_\mu\end{aligned}$$

Let us denote the inverse matrix by $\mathbf{A}^{-1} = \mathbf{B} = \left[B_\beta^\alpha\right]$, namely, the relations

$$A_\gamma^\alpha B_\beta^\gamma = B_\gamma^\alpha A_\beta^\gamma = \delta_\beta^\alpha$$

will hold. Hence, (2.11.7) gives

$$U_\alpha(p) = B_\alpha^\beta(p) V_\beta(p).$$

The commutators of vectors V_α then become

$$\begin{aligned}[V_\gamma, V_\delta] &= \big[c_{\alpha\beta}^\mu A_\gamma^\alpha A_\delta^\beta + A_\gamma^\alpha U_\alpha(A_\delta^\mu) - A_\delta^\alpha U_\alpha(A_\gamma^\mu)\big] B_\mu^\lambda V_\lambda \qquad (2.11.8)\\ &= C_{\gamma\delta}^\lambda V_\lambda\end{aligned}$$

as it should be expected. We thus find that $[V_\gamma, V_\delta] \in \mathcal{D}$. Here, the functions $C_{\gamma\delta}^\lambda \in \Lambda^0(M)$ are given by

$$\begin{aligned}C_{\gamma\delta}^\lambda &= B_\mu^\lambda \big[c_{\alpha\beta}^\mu A_\gamma^\alpha A_\delta^\beta + A_\gamma^\alpha U_\alpha(A_\delta^\mu) - A_\delta^\alpha U_\alpha(A_\gamma^\mu)\big]\\ &= B_\mu^\lambda \big[c_{\alpha\beta}^\mu A_\gamma^\alpha A_\delta^\beta + V_\gamma(A_\delta^\mu) - V_\delta(A_\gamma^\mu)\big].\end{aligned}$$

The vector fields U_α are prescribed by

$$U_\alpha = u_\alpha^i(\mathbf{x}) \frac{\partial}{\partial x^i}, \quad \mathbf{x} = \varphi(p),$$
$$i = 1, \dots, m; \quad \alpha = 1, \dots, k$$

in a chart (U, φ) containing the point p. Therefore, the vector fields V_α are given by

$$V_\alpha = A_\alpha^\beta u_\beta^i \frac{\partial}{\partial x^i}. \qquad (2.11.9)$$

Since k number of vectors U_α are linearly independent, the rank of the rectangular matrix

$$[u_\alpha^i] = \begin{bmatrix} u_1^1 & u_1^2 & \cdots & u_1^k & \cdots & u_1^m \\ u_2^1 & u_2^2 & \cdots & u_2^k & \cdots & u_2^m \\ \vdots & \vdots & & \vdots & & \vdots \\ u_k^1 & u_k^2 & \cdots & u_k^k & \cdots & u_k^m \end{bmatrix}$$

is k. We rename the coordinates x^i if necessary to arrange this matrix in such a way that $[u_\beta^\alpha]$ can be chosen as the $k \times k$ square matrix with non-vanishing determinant. Then (2.11.9) can be written as follows

$$V_\alpha = A_\alpha^\beta u_\beta^\gamma \frac{\partial}{\partial x^\gamma} + A_\alpha^\beta u_\beta^a \frac{\partial}{\partial x^a}, \; a = k+1, \dots, m. \qquad (2.11.10)$$

So far the matrix $\mathbf{A}$ was arbitrary. We now select it as the inverse of the matrix $[u_\beta^\alpha]$:

$$A^{\beta}_{\alpha}u^{\gamma}_{\beta} = \delta^{\gamma}_{\alpha}, \quad A^{\alpha}_{\beta} = \overset{-1}{u}{}^{\alpha}_{\beta}$$

where the smooth functions $\overset{-1}{u}{}^{\alpha}_{\beta} \in \Lambda^0(M)$ are elements of the inverse matrix $[u^{\alpha}_{\beta}]^{-1}$. With this choice the structure of the expressions (2.11.10) reduces to a much simpler form

$$V_\alpha = \frac{\partial}{\partial x^\alpha} + v^a_\alpha \frac{\partial}{\partial x^a} \tag{2.11.11}$$

where we have introduced the functions v^a_α by

$$v^a_\alpha = u^a_\beta \overset{-1}{u}{}^{\beta}_{\alpha}. \tag{2.11.12}$$

On recalling that $[\partial_i, \partial_j] = 0$, we readily find

$$\begin{aligned}
[V_\alpha, V_\beta] &= [\partial_\alpha + v^a_\alpha \partial_a, \partial_\beta + v^b_\beta \partial_b] = \partial_\alpha\partial_\beta + v^b_{\beta,\alpha}\partial_b + v^b_\beta\, \partial_\alpha\partial_b \\
&\quad + v^a_\alpha \partial_a\partial_\beta + v^a_\alpha v^b_{\beta,a}\partial_a + v^a_\alpha v^b_\beta\, \partial_a\partial_b - \partial_\beta\partial_\alpha - v^a_{\alpha,\beta}\, \partial_a \\
&\quad - v^a_\alpha\, \partial_\beta\partial_a - v^b_\beta\, \partial_b\partial_a - v^b_\beta v^a_{\alpha,b}\partial_a - v^b_\beta v^a_\alpha \partial_b\partial_a \\
&= \{(v^a_{\beta,\alpha} + v^b_\alpha v^a_{\beta,b}) - (v^a_{\alpha,\beta} + v^b_\beta v^a_{\alpha,b})\}\partial_a
\end{aligned}$$

or

$$[V_\alpha, V_\beta] = \{V_\alpha(v^a_\beta) - V_\beta(v^a_\alpha)\}\frac{\partial}{\partial x^a}. \tag{2.11.13}$$

Next we insert (2.11.11) into (2.11.8) and rearrange the terms to obtain

$$[V_\alpha, V_\beta] = C^{\gamma}_{\alpha\beta}V_\gamma = C^{\gamma}_{\alpha\beta}\frac{\partial}{\partial x^\gamma} + C^{\gamma}_{\alpha\beta}v^a_\gamma\frac{\partial}{\partial x^a}.$$

If we compare this expression with (2.11.13) we deduce that all coefficient functions $C^{\gamma}_{\alpha\beta}$ must vanish. Hence, we conclude that

$$[V_\alpha, V_\beta] = 0. \tag{2.11.14}$$

Furthermore, (2.11.13) then implies that the following conditions should also be satisfied

$$V_\alpha(v^a_\beta) = V_\beta(v^a_\alpha). \tag{2.11.15}$$

(2.11.14) means that in an involutive distribution $\mathcal{D}$ one is always able to find k linearly independent vector fields V_α generating this distribution that commute with respect to the Lie product. Consequently, congruences produced by those vector fields constitute a k-dimensional net of coordinate lines at the vicinity of each point of the manifold M. In other words, they form an integral manifold. □

Making use of the information provided by the above theorem, we can determine in a concrete way the integral manifold of an involutive distribution S. Let $(\xi^1, \xi^2, \ldots, \xi^k) \in \mathbb{R}^k$ denote the local coordinates that give rise to *natural basis* $\{V_\alpha\}$ of the tangent bundle $T(S)$. So we can write

$$V_\alpha = \frac{\partial}{\partial \xi^\alpha} = \frac{\partial}{\partial x^\alpha} + v_\alpha^a \frac{\partial}{\partial x^a}, \; \alpha = 1, \ldots, k; a = k+1, \ldots, m.$$

Therefore, one has

$$V_\beta(x^i) = \frac{\partial x^i}{\partial \xi^\beta} = \frac{\partial x^i}{\partial x^\beta} + v_\beta^a \frac{\partial x^i}{\partial x^a} = \delta_\beta^i + v_\beta^a \delta_a^i$$

whence we conclude that

$$\frac{\partial x^\alpha}{\partial \xi^\beta} = \delta_\beta^\alpha, \quad \frac{\partial x^a}{\partial \xi^\alpha} = v_\alpha^a(\mathbf{x}). \tag{2.11.16}$$

Solutions of equations $(2.11.16)_1$ are trivially found as

$$x^\alpha = \xi^\alpha + c^\alpha, \quad \alpha = 1, \ldots, k. \tag{2.11.17}$$

Since equations $(2.11.16)_2$ are generally non-linear, it is usually much more difficult to obtain their solutions. Utilising (2.11.17), we can put these equations into the form

$$\begin{aligned} \frac{\partial x^a}{\partial \xi^\alpha} &= v_\alpha^a(x^1, \ldots, x^k, x^{k+1}, \ldots, x^m) \\ &= v_\alpha^a(\xi^1 + c^1, \ldots, \xi^k + c^k, x^{k+1}, \ldots, x^m), \\ a &= k+1, \ldots, m. \end{aligned}$$

Let us then calculate derivatives of equations $(2.11.16)_2$ with respect to variables ξ^β:

$$\frac{\partial^2 x^a}{\partial \xi^\alpha \partial \xi^\beta} = \frac{\partial v_\alpha^a}{\partial \xi^\beta} = \frac{\partial v_\alpha^a}{\partial x^\gamma}\frac{\partial x^\gamma}{\partial \xi^\beta} + \frac{\partial v_\alpha^a}{\partial x^b}\frac{\partial x^b}{\partial \xi^\beta} = \frac{\partial v_\alpha^a}{\partial x^\beta} + v_\beta^b \frac{\partial v_\alpha^a}{\partial x^b} = V_\beta(v_\alpha^a).$$

This implies that equations $(2.11.16)_2$ can only be solved if the compatibility conditions $V_\alpha(v_\beta^a) = V_\beta(v_\alpha^a)$, that are naturally brought about by the symmetries of second order derivatives, are satisfied. However, these are none other than conditions (2.11.15) that must be elicited by functions v_α^a. Thus, equations $(2.11.16)_2$ can be integrated in principle and the set of coordinates $\{x^a\}$ are expressible in terms of variables ξ^α as below:

$$x^a = f^a(\xi^1, \xi^2, \dots, \xi^k) + c^a$$

where $a = k + 1, \dots, m$ and c^a are $m - k$ arbitrary constants. Let us now define the new local coordinates by means of the relations

$$y^\alpha = \xi^\alpha,\ y^a = x^a - f^a(\xi^1, \xi^2, \dots, \xi^k)$$

where $\alpha = 1, \dots, k$ and $a = k + 1, \dots, m$. In the light of the above developments, we can thus rephrase the Frobenius theorem as follows: *Let $\mathcal{D}$ be a k-dimensional involutive distribution on an m-dimensional manifold. Then there exists local coordinates $y^i, 1 \le i \le m$ such that the vector fields $\partial/\partial y^1 = \partial/\partial \xi^1, \dots, \partial/\partial y^k = \partial/\partial \xi^k$ constitute a local basis of the distribution $\mathcal{D}$ and submanifolds determined by $y^a =$ constant, $k + 1 \le a \le m$ are integral manifolds of $\mathcal{D}$.*

It is now seen that a k-dimensional involutive distribution on an m-dimensional manifold M generates a k-dimensional smooth integral manifold through each point $p \in M$. Therefore, the manifold M can be reconstructed as the union of a family of k-dimensional submanifolds stacked on top of one another. Such a case is called a k-dimensional ***foliation*** of the class C^∞ on the manifold M. Each submanifold is known as a ***leaf*** of the foliation.

Example 2.11.1. Let $M = \mathbb{R}^3$ with a coordinate cover x, y, z. We define a 2-dimensional distribution $\mathcal{D}$ by the vector fields

$$U_1 = -y\frac{\partial}{\partial x} + x\frac{\partial}{\partial y},$$
$$U_2 = -z\frac{\partial}{\partial y} + y\frac{\partial}{\partial z}$$

where we take $x^1 = x, x^2 = y, x^3 = z$. It is easily verified that these vector fields are linearly independent if $y \neq 0$. In fact, we write with $f, g \in \Lambda^0(M)$

$$fU_1 + gU_2 = -yf\partial_x + (xf - zg)\partial_y + yg\partial_z = 0.$$

This relation is satisfied if and only if $f = g = 0$ when $y \neq 0$. On the other hand, the commutator of these vector fields becomes

$$[U_1, U_2] = -z\frac{\partial}{\partial x} + x\frac{\partial}{\partial z} = \frac{z}{y}U_1 + \frac{x}{y}U_2 \in \mathcal{D}.$$

Thus $\mathcal{D}$ is an involutive distribution. Let us first determine the congruences produced by vector fields U_1 and U_2. The solutions of the following simple ordinary differential equations associated with vector fields U_1 and U_2, respectively

$$U_1: \frac{d\overline{x}}{dt} = -\overline{y}, \quad \frac{d\overline{y}}{dt} = \overline{x}, \quad \frac{d\overline{z}}{dt} = 0; \ \overline{\mathbf{x}}(0) = \mathbf{x}$$
$$U_2: \frac{d\overline{x}}{ds} = 0, \quad \frac{d\overline{y}}{ds} = -\overline{z}, \quad \frac{d\overline{z}}{ds} = \overline{y}; \ \overline{\mathbf{x}}(0) = \mathbf{x}$$

yield the U_1-congruence by the following parametric equations

$$\overline{x}(t) = x\cos t - y\sin t, \ \overline{y}(t) = x\sin t + y\cos t, \ \overline{z}(t) = z,$$

and U_2-congruence by equations

$$\overline{x}(s) = x, \ \overline{y}(s) = y\cos s - z\sin s, \ \overline{z}(s) = y\sin s + z\cos s.$$

It is immediately seen that both equations satisfy

$$\overline{x}(t)^2 + \overline{y}(t)^2 + \overline{z}(t)^2 = \overline{x}(s)^2 + \overline{y}(s)^2 + \overline{z}(s)^2 = x^2 + y^2 + z^2.$$

Hence, the 2-dimensional integral manifold through the point $\mathbf{x} = (x, y, z)$ is a sphere whose radius is equal to the distance of this point from the origin 0. But these congruences cannot form a coordinate net on the sphere. Indeed, let us move along U_1 integral curve through the point $\mathbf{x}$ to the point $\mathbf{x}_1$ by the parameter t, then along U_2 integral curve from the point $\mathbf{x}$ to the point $\mathbf{x}_2$ by the parameter s. We find that

$$x_1 = x\cos t - y\sin t, \ y_1 = x\sin t + y\cos t, \ z_1 = z$$
$$x_2 = x, \ y_2 = y\cos s - z\sin s, \ z_2 = y\sin s + z\cos s.$$

Next, we go along U_1 integral curve from $\mathbf{x}_2$ to the point $\mathbf{x}_3$ by t, and along U_2 integral curve from $\mathbf{x}_1$ to the point $\mathbf{x}_4$ by s. We obtain

$$x_3 = x_2\cos t - y_2\sin t, \ y_3 = x_2\sin t + y_2\cos t, \ z_3 = z_2$$
$$x_4 = x_1, \ y_4 = y_1\cos s - z_1\sin s, \ z_4 = y_1\sin s + z_1\cos s$$

or

$$\begin{aligned} x_3 &= x\cos t - y\sin t\cos s + z\sin t\sin s, & x_4 &= x\cos t - y\sin t \\ y_3 &= y\cos t\cos s - z\cos t\sin s, & y_4 &= x\sin t\cos s + y\cos t\cos s - z\sin s \\ z_3 &= z\cos s + y\sin s, & z_4 &= x\sin t\sin s + y\cos t\sin s + z\cos s \end{aligned}$$

It is evident that $\mathbf{x}_3 \neq \mathbf{x}_4$. For instance, for $t = s = \pi/2$ we have $\mathbf{x}_3 = (z, x, y)$, $\mathbf{x}_4 = (-y, -z, x)$. In this case, it would be necessary to produce two commutative vector fields generating the distribution $\mathcal{D}$. We write

$$V_1 = A_1^1 U_1 + A_1^2 U_2 = -A_1^1 y\frac{\partial}{\partial x} + (A_1^1 x - A_1^2 z)\frac{\partial}{\partial y} + A_1^2 y\frac{\partial}{\partial z}$$

$$V_2 = A_2^1 U_1 + A_2^2 U_2 = -A_2^1 y \frac{\partial}{\partial x} + (A_2^1 x - A_2^2 z)\frac{\partial}{\partial y} + A_2^2 y \frac{\partial}{\partial z}$$

and choose

$$A_1^1 = -1/y,\ A_1^2 = 0,\ A_2^1 = 0,\ A_2^2 = 1/y$$

with $\det \mathbf{A} = -1/y^2 \neq 0$. We thus obtain vectors

$$V_1 = \frac{\partial}{\partial x} - \frac{x}{y}\frac{\partial}{\partial y},\quad V_2 = \frac{\partial}{\partial z} - \frac{z}{y}\frac{\partial}{\partial y}.$$

We see at once that $[V_1, V_2] = 0$. The congruences generated by these vector fields are found as solutions of ordinary differential equations

$$\frac{d\overline{x}}{dt} = 1,\ \frac{d\overline{y}}{dt} = -\frac{\overline{x}}{\overline{y}},\ \frac{d\overline{z}}{dt} = 0;\ \overline{\mathbf{x}}(0) = \mathbf{x},$$
$$\frac{d\overline{x}}{ds} = 0,\ \frac{d\overline{y}}{ds} = -\frac{\overline{z}}{\overline{y}},\ \frac{d\overline{z}}{ds} = 1;\ \overline{\mathbf{x}}(0) = \mathbf{x}.$$

These are respectively

$$\overline{x}(t) = x + t,\ \ \overline{y}(t)^2 = y^2 - 2xt - t^2,\ \ \overline{z}(t) = z,$$
$$\overline{x}(s) = x,\ \overline{y}(s)^2 = y^2 - 2zs - s^2,\ \overline{z}(s) = z + s.$$

As above, we now determine again the points $\mathbf{x}_1, \mathbf{x}_2, \mathbf{x}_3$ and $\mathbf{x}_4$ starting from a point $\mathbf{x}$:

$$x_1 = x + t,\quad y_1^2 = y^2 - 2xt - t^2,\quad z_1 = z,$$
$$x_2 = x,\quad y_2^2 = y^2 - 2zs - s^2,\quad z_2 = z + s,$$
$$x_3 = x_2 + t,\ y_3^2 = y_2^2 - 2x_2 t - t^2,\ z_3 = z_2,$$
$$x_4 = x_1,\ y_4^2 = y_1^2 - 2z_1 s - s^2,\ z_4 = z_1 + s.$$

A short calculation then leads to

$$x_3 = x_4 = x + t, y_3^2 = y_4^2 = y^2 - 2xt - 2zs - t^2 - s^2, z_3 = z_4 = z + s.$$

Consequently V_1- and V_2-congruences form a 2-dimensional coordinate net on the sphere.

Let us now parametrise the integral manifolds by variables ξ and η via the general scheme that was given above. We thus write

$$\frac{\partial}{\partial \xi} = \frac{\partial}{\partial x} - \frac{x}{y}\frac{\partial}{\partial y},\ \frac{\partial}{\partial \eta} = \frac{\partial}{\partial z} - \frac{z}{y}\frac{\partial}{\partial y}$$

to obtain

$$\frac{\partial x}{\partial \xi} = 1, \ \frac{\partial x}{\partial \eta} = 0; \quad \frac{\partial y}{\partial \xi} = -\frac{x}{y}, \ \frac{\partial y}{\partial \eta} = -\frac{z}{y}; \quad \frac{\partial z}{\partial \xi} = 0, \ \frac{\partial z}{\partial \eta} = 1$$

the integration of which yields

$$\begin{aligned} x &= \xi + c_1(\eta), \ c_1'(\eta) = 0 \ \text{ and } \ x = \xi + c_1, \\ z &= c_2(\eta), \ c_2'(\eta) = 1 \ \text{ and } \ z = \eta + c_2, \\ \frac{\partial y^2}{\partial \xi} &= -2(\xi + c_1), y^2 = -\xi^2 - 2c_1\xi + c_3(\eta), \frac{\partial c_3}{\partial \eta} = -2(\eta + c_2), \\ c_3(\eta) &= -\eta^2 - 2c_2\eta + c_3 \ \text{ and } \ y^2 = -\xi^2 - \eta^2 - 2c_1\xi - 2c_2\eta + c_3 \end{aligned}$$

where c_1, c_2 and c_3 are arbitrary constants. We define the new coordinates (ξ, η, r) by

$$\xi, \ \eta, \ r^2 = y^2 + \xi^2 + \eta^2 + 2c_1\xi + 2c_2\eta - c_3.$$

If we eliminate variables ξ and η in the expression for r^2 we find

$$r^2 = x^2 + y^2 + z^2 + c_1^2 + c_2^2 - c_3.$$

Hence $r = \textit{constant}$ corresponds to a spherical integral manifold. ■

Let $f : M \to \mathbb{R}$ be a smooth function. The differential of this function is the linear operator $f_* = df : T(M) \to \mathbb{R}$, or a linear functional defined by the relation $f_*(V) = V(f)$ [*see* p. 126]. The vectors in the null space $\mathcal{N}(f_*)$ of the operator f_* that is a subbundle of $T(M)$ satisfy the condition $f_*(V) = V(f) = 0$. If $U, V \in \mathcal{N}(f_*)$, then we have $f_*(U) = f_*(V) = 0$ so that due to (2.10.21), we find $f_*[U, V] = [f_*U, f_*V] = [0, 0] = 0$ and thus $[U, V] \in \mathcal{N}(f_*)$. Hence, the distribution $\mathcal{N}(f_*)$ induced by the function f is involutive.

We next consider a k-dimensional distribution $\mathcal{D}$ of the tangent bundle $T(M)$. We know that this distribution is determined by k linearly independent vector fields U_α. A function $f : M \to \mathbb{R}$ is annihilated by the distribution $\mathcal{D}$ if the relations $U_\alpha(f) = 0$, $\alpha = 1, \ldots, k$ are met. In this case, we obtain $U(f) = 0$ for all vector fields $U \in \mathcal{D}$. This of course implies that such a distribution must satisfy $\mathcal{D} \subseteq \mathcal{N}(f_*)$. Let us then consider the equalities $U_\alpha(f) = 0$, $U_\beta(f) = 0$ with $\alpha \neq \beta$. Utilising these relations, we arrive at the result

$$[U_\alpha, U_\beta](f) = U_\alpha\big(U_\beta(f)\big) - U_\beta\big(U_\alpha(f)\big) = 0. \tag{2.11.18}$$

If $[U_\alpha, U_\beta] \notin \mathcal{D}$ for $\alpha \neq \beta$ where $\alpha, \beta \in \{1, \ldots, k\}$, then relations (2.11.18) provide additional conditions needed for the function f to be annihilated by the distribution $\mathcal{D}$. On the other hand, if the distribution $\mathcal{D}$ is involutive, the conditions (2.11.18) will be satisfied automatically:

$$[U_\alpha, U_\beta](f) = c^\gamma_{\alpha\beta} U_\gamma(f) = 0.$$

In this case, the relations $U_\alpha(f) = 0$ would be sufficient to determine functions annihilated by $\mathcal{D}$. In an involutive distribution, we can always choose *normal basis vectors* satisfying the conditions $[V_\alpha, V_\beta] = 0$ instead of arbitrary basis vectors $U_\alpha = u^i_\alpha(\mathbf{x})\,\partial/\partial x^i$. The vectors $V_\alpha, \alpha = 1, \dots, k$ are given by (2.11.11). Therefore,, we can take the equivalent relations $V_\alpha(f) = 0$ in lieu of $U_\alpha(f) = 0$. Thus, in order to determine functions annihilated by an involutive distribution $\mathcal{D}$, we have to solve the following set of first order partial differential equations

$$v^i_\alpha(\mathbf{x})\,\frac{\partial f}{\partial x^i} = 0,\ \ \alpha = 1, \dots, k. \tag{2.11.19}$$

The components $v^i_\alpha = \delta^i_\alpha + v^a_\alpha \delta^i_a$, where $a = k+1, \dots m$ are given by (2.11.12). These equations can be solved by the usual method of characteristics. We start with the first equation. Its characteristics are obtained as usual by solving the set of autonomous ordinary differential equations below

$$\frac{dx^1}{v^1_1(\mathbf{x})} = \frac{dx^2}{v^2_1(\mathbf{x})} = \cdots = \frac{dx^m}{v^m_1(\mathbf{x})}. \tag{2.11.20}$$

Evidently, characteristics are nothing but the integral curves of the vector field V_1 that are found by integrating the ordinary differential equations

$$\frac{dx^i}{dt} = v^i_1(\mathbf{x}).$$

It is well known that the solution of equations (2.11.20) is expressible in the form

$$g^1(\mathbf{x}) = c^1,\ g^2(\mathbf{x}) = c^2,\ \dots,\ g^{m-1}(\mathbf{x}) = c^{m-1} \tag{2.11.21}$$

where $g^1, g^2, \dots, g^{m-1}$ are given smooth functions and $c^1,\ c^2, \dots, c^{m-1}$ are arbitrary constants. It follows from (2.11.21) that

$$0 = \frac{\partial g^r}{\partial x^i}\frac{dx^i}{dt} = v^i_1 \frac{\partial g^r}{\partial x^i} = V_1(g^r)$$
$$r = 1, 2, \dots, m-1.$$

We can thus see that the following equations

$$
\begin{aligned}
v_1^1\frac{\partial f}{\partial x^1}+v_1^2\frac{\partial f}{\partial x^2}+\cdots+v_1^m\frac{\partial f}{\partial x^m}&=0,\\
v_1^1\frac{\partial g^1}{\partial x^1}+v_1^2\frac{\partial g^1}{\partial x^2}+\cdots+v_1^m\frac{\partial g^1}{\partial x^m}&=0,\\
v_1^1\frac{\partial g^2}{\partial x^1}+v_1^2\frac{\partial g^2}{\partial x^2}+\cdots+v_1^m\frac{\partial g^2}{\partial x^m}&=0,\\
&\vdots\\
v_1^1\frac{\partial g^{m-1}}{\partial x^1}+v_1^2\frac{\partial g^{m-1}}{\partial x^2}+\cdots+v_1^m\frac{\partial g^{m-1}}{\partial x^m}&=0
\end{aligned}
$$

are to be held. Since $V_1 \neq 0$, this homogeneous set of linear equations in terms of m coefficient functions v_1^i can have a nontrivial solution if and only if the determinant of the coefficient functions vanishes:

$$
\begin{vmatrix}
\frac{\partial f}{\partial x^1} & \frac{\partial f}{\partial x^2} & \cdots & \frac{\partial f}{\partial x^m}\\
\frac{\partial g^1}{\partial x^1} & \frac{\partial g^1}{\partial x^2} & \cdots & \frac{\partial g^1}{\partial x^m}\\
\frac{\partial g^2}{\partial x^1} & \frac{\partial g^2}{\partial x^2} & \cdots & \frac{\partial g^2}{\partial x^m}\\
\vdots & \vdots & \cdots & \vdots\\
\frac{\partial g^{m-1}}{\partial x^1} & \frac{\partial g^{m-1}}{\partial x^2} & \cdots & \frac{\partial g^{m-1}}{\partial x^m}
\end{vmatrix}
= \frac{\partial(f, g^1, \dots, g^{m-1})}{\partial(x^1, x^2, \dots, x^m)} = 0.
$$

This means that the function f is not independent of functions $g^1, \dots, g^{m-1}$. We thus conclude that

$$
f = F(g^1, \dots, g^{m-1}). \tag{2.11.22}
$$

Let us now take the equation $V_2(f) = 0$ into account. Inserting (2.11.22) into this equation, we obtain

$$
0 = v_2^i\frac{\partial f}{\partial x^i} = v_2^i\frac{\partial F}{\partial g^r}\frac{\partial g^r}{\partial x^i} = V_2(g^r)\frac{\partial F}{\partial g^r}.
$$

On the other hand, commutativity of vectors V_α results in

$$
V_1\big(V_2(g^r)\big) = V_2\big(V_1(g^r)\big) = V_2(0) = 0.
$$

Hence, functions $V_2(g^r)$ are solutions of the equation

$$
V_1(g) = 0.
$$

Thus, we must write

$$V_2(g^r) = h^r(g^1, \dots, g^{m-1}).$$

Consequently, we find that

$$V_2(f) = h^r(g^1, \dots, g^{m-1}) \frac{\partial F}{\partial g^r} = 0.$$

The solution of this differential equation is similarly expressed as

$$F = \mathcal{F}(m^1, m^2, \dots, m^{m-2})$$

where the functions

$$\begin{aligned} m^s &= m^s(g^1, \dots, g^{m-1}), \\ s &= 1, 2, \dots, m-2 \end{aligned}$$

are determined just as in the previous step. If we continue this way, we observe that every function annihilated by a k-dimensional involutive distribution is represented in the form

$$f = \mathfrak{F}(\mathfrak{g}^1, \mathfrak{g}^2, \dots, \mathfrak{g}^{m-k}). \tag{2.11.23}$$

$m - k$ functions $\mathfrak{g}^1, \mathfrak{g}^2, \dots, \mathfrak{g}^{m-k}$ are definite functions of variables $x^1, x^2, \dots, x^m$ obtained through all the foregoing steps. These functions constitute a set of *maximal solutions* if they are functionally independent, that is, if the following Jacobian with an appropriate ordering of local coordinates does not vanish

$$\frac{\partial(x^1, x^2, \dots, x^k, \mathfrak{g}^1, \mathfrak{g}^2, \dots, \mathfrak{g}^{m-k})}{\partial(x^1, x^2, \dots, x^m)} =$$

$$= \begin{vmatrix} 1 & 0 & \cdots & 0 & \cdots & 0 \\ \vdots & \vdots & \vdots & \vdots & \vdots & \vdots \\ 0 & 0 & \cdots & 1 & \cdots & 0 \\ \dfrac{\partial \mathfrak{g}^1}{\partial x^1} & \dfrac{\partial \mathfrak{g}^1}{\partial x^2} & \cdots & \dfrac{\partial \mathfrak{g}^1}{\partial x^k} & \cdots & \dfrac{\partial \mathfrak{g}^1}{\partial x^n} \\ \vdots & \vdots & \vdots & \vdots & \vdots & \vdots \\ \dfrac{\partial \mathfrak{g}^{n-k}}{\partial x^1} & \dfrac{\partial \mathfrak{g}^{n-k}}{\partial x^2} & \cdots & \dfrac{\partial \mathfrak{g}^{n-k}}{\partial x^k} & \cdots & \dfrac{\partial \mathfrak{g}^{n-k}}{\partial x^n} \end{vmatrix} \neq 0.$$

Such functions $\mathfrak{g}^I$, $I = 1, \dots, m-k$ are named as the *first integrals* or *integral functions* of the distribution $\mathcal{D}$. Since we must have $V_\alpha(f) = 0$ for every function in the form (2.11.23), we find that

$$0 = V_\alpha(f) = v^i_\alpha \frac{\partial f}{\partial \mathfrak{g}^J}\frac{\partial \mathfrak{g}^J}{\partial x^i} = V_\alpha(\mathfrak{g}^J)\frac{\partial f}{\partial \mathfrak{g}^J}.$$

If we select $f = \mathfrak{g}^I$, we then obtain

$$V_\alpha(\mathfrak{g}^J)\,\delta^I_J = V_\alpha(\mathfrak{g}^I) = 0,\ \alpha = 1,\dots,k,\ \ I = 1,\dots,m-k.$$

Hence, for each vector $V \in \mathcal{D}$ one gets

$$d\mathfrak{g}^I(V) = \mathfrak{g}^I_*(V) = V(\mathfrak{g}^I) = 0,\ \ I = 1,\dots,m-k. \quad (2.11.24)$$

Let us now define a subset $\mathcal{M}$ of the differentiable manifold M with the help of local charts as follows

$$\mathcal{M} = \{p \in M : \mathfrak{g}^1(p) = c^1, \mathfrak{g}^2(p) = c^2, \dots, \mathfrak{g}^{m-k}(p) = c^{m-k}\}$$

where $c^1, c^2, \dots, c^{m-k}$ are arbitrary constants. Because of Theorem 2.4.1, we understand that $\mathcal{M}$ is a submanifold. We generate a family of submanifolds, namely, a foliation of the manifold M by giving different values to these constants. If we take into consideration the relations (2.11.24), it becomes clear that the distribution is now specified by

$$\mathcal{D} = \{V \in T(M) : d\mathfrak{g}^1(V) = 0, d\mathfrak{g}^2(V) = 0, \dots, d\mathfrak{g}^{m-k}(V) = 0\}.$$

Hence the family $\mathcal{M}$ are actually integral manifolds of the involutive distribution $\mathcal{D}$. The linear operators $d\mathfrak{g}^I$ are now expressible as

$$dx^\alpha = dx^\alpha,$$

$$d\mathfrak{g}^I = \frac{\partial \mathfrak{g}^I}{\partial x^\alpha}\,dx^\alpha + \frac{\partial \mathfrak{g}^I}{\partial x^a}\,dx^a,\ I = 1,\dots,m-k$$

where $\alpha = 1,\dots,k,\ a = k+1,\dots,m$. Since we have assumed that the Jacobian defined above does not vanish, then the operators $(dx^\alpha, d\mathfrak{g}^I)$ are linearly independent. Let us now reconsider Example 2.11.1. We know that the normalised basis vectors are

$$V_1 = \frac{\partial}{\partial x} - \frac{x}{y}\frac{\partial}{\partial y},\ \ V_2 = \frac{\partial}{\partial z} - \frac{z}{y}\frac{\partial}{\partial y}$$

Hence, for a function $f = f(x, y, z)$, the solution of the equation

$$V_1(f) = \frac{\partial f}{\partial x} - \frac{x}{y}\frac{\partial f}{\partial y} = 0$$

is obtainable through characteristic equations

$$\frac{dx}{1} = -\frac{y\,dy}{x} = \frac{dz}{0}$$

whose integrals are given as

$$g^1 = x^2 + y^2 = c^1, \;\; g^2 = z = c^2.$$

We thus find $f = F(g^1, g^2)$. Since

$$V_2(g^1) = -2z = -2g^2, \;\; V_2(g^2) = 1$$

the function F must satisfy

$$-2g^2 \frac{\partial F}{\partial g^1} + \frac{\partial F}{\partial g^2} = 0.$$

Solution of the ordinary differential equation

$$-dg^1/2g^2 = dg^2/1$$

is $\mathfrak{g}^1 = g^1 + (g^2)^2 = x^2 + y^2 + z^2 = C^1$. Therefore, we arrive at the result $f = F(\mathfrak{g}^1) = F(x^2 + y^2 + z^2)$. Thus, integral manifolds, or leaves, of that 2-dimensional involutive distribution are spheres centred at the origin 0.

II. EXERCISES

2.1. Show that a *discreet topology* can be generated on a set M by choosing every point in M as an open set and this topological space has the structure of a 0-dimensional manifold.

2.2. The standard topology on $\mathbb{R}^2$ is given as unions of open rectangles $(a,b) \times (c,d)$. Discuss whether the mapping $f : [0, 2\pi) \to \mathbb{S}^1$ defined by the rule $f(t) = (\cos t, \sin t)$ is bijective, continuous and it is a homeomorphism with respect to relative topologies on $[0, 2\pi)$ and $\mathbb{S}^1$.

2.3. Two differentiable structures on $\mathbb{R}$ are provided by atlases $\varphi_1(x) = x$ and $\varphi_2(x) = x^3$. We know that these atlases are not compatible [*see* Example 2.2.1]. Yet show that they are diffeomorphic.

2.4. An equivalence relation $\sim$ is defined on the set $S = \{(x,y) \in \mathbb{R}^2 : y = \pm 1\}$ by $x \neq 0, (x,1) \sim (x, -1)$. Show that the quotient space $M = S/\sim$ is a locally Euclidean and second countable space, but not a Hausdorff space (This example is known as *straight line with two centres*).

2.5. $\mathbb{S}^2$ is the sphere given by the equation $x^2 + y^2 + z^2 = 1$. Let us consider its open upper hemisphere $U_z^+ = \{\mathbf{x} \in \mathbb{S}^2 : z > 0\}$, the open set $V_z = \{(x,y) \in \mathbb{R}^2 : x^2 + y^2 < 1\}$ and the mapping $\varphi_z^+ : U_z^+ \to V_z$ determined by $\varphi_z^+(x, y, \sqrt{1 - x^2 - y^2}) = (x, y)$. Similarly, on the open lower hemisphere $U_z^- = \{\mathbf{x} \in \mathbb{S}^2 : z < 0\}$, we define the mapping $\varphi_z^- : U_z^- \to V_z$ by the relation $\varphi_z^-(x,y,-\sqrt{1-x^2-y^2})=(x,y)$. Show that the pairs (U_z^+,φ_z^+) and (U_z^-,φ_z^-) are charts. Prove that we obtain an atlas with six charts when we add to these two those charts (U_x^+,φ_x^+), (U_x^-,φ_x^-) and (U_y^+,φ_y^+), (U_y^-,φ_y^-) involving left, right and front, rear hemispheres constructed in the similar fashion.

2.6. Let $U\subset\mathbb{R}^2$ be an open set and $f:U\to\mathbb{R}$ be a smooth mapping. Show that the *graph* $\{\mathbf{x},f(\mathbf{x})\}$ of this function is a 2-dimensional submanifold of $\mathbb{R}^3$.

2.7. Let $M_i, i=1,2,\ldots,n$ be differentiable manifolds. We take submanifolds $N_i\subset M_i$ into account. Show that the Cartesian product $N_1\times N_2\times\cdots\times N_n$ is a submanifold of the product manifold $M_1\times M_2\times\cdots\times M_n$.

2.8. Discuss whether the following curves defined by mappings $\phi_i:\mathbb{R}\to\mathbb{R}^2$ are immersion or embedding:

$$\phi_1(t)=(t^2-1,t^3-t),\quad 1<t<\infty,$$
$$\phi_2(t)=\Big(\frac{t+1}{2t}\cos t,\frac{t+1}{2t}\sin t\Big),$$
$$\phi_3(t)=(2\cos t,\sin 2t),\quad \phi_4(t)=\big(2\cos(2\arctan t),\,\sin(4\arctan t)\big),$$
$$\phi_5(t)=(at-b\sin t,a-b\cos t),\quad a,b\in\mathbb{R},$$
$$\phi_6(t)=\big(2\sin(at+b),c\cos dt\big),\quad a,b,c,d\in\mathbb{R}.$$

2.9. Discuss whether the following mappings $\phi_1:\mathbb{R}^2\to\mathbb{R}^3$ and $\phi_2:\mathbb{R}^2\to\mathbb{R}^4$ are immersions or submanifolds:

$$\phi_1(u,v)=(R\sin u\cos v,R\sin u\sin v,R\cos u),$$
$$\phi_2(u,v)=\big((a+b\cos u)\cos v,(a+b\cos u)\sin v,b\sin u\cos(v/2),b\sin u\cos(v/2)\big).$$

2.10. Discuss whether the mappings $\phi_1:(0,\infty)^2\to\mathbb{R}^3$, $\phi_2:(0,\infty)^3\to\mathbb{R}^3$ and $\phi_3:(0,\infty)^2\to\mathbb{R}^2$ defined below are immersions or submersions

$$\phi_1(u,v)=\big(u,u^2,v^2/u\big),\qquad \phi_2(u,v,w)=(uvw,uv,w),$$
$$\phi_3(u,v,w)=(vw-u,v-uw).$$

2.11. The mapping $\phi:\mathbb{R}^3\to\mathbb{R}^4$ is given by $\phi(u,v,w)=(u^2-v^2,uv,uw,vw)$. Show that the restriction $\phi|_{\mathbb{S}^2}$ of this mapping satisfies the relation $\phi|_{\mathbb{S}^2}(p)=\phi|_{\mathbb{S}^2}(-p)$ for all $p\in\mathbb{S}^2$. Let us define the mapping $\psi:\mathbb{RP}^2\to\mathbb{R}^4$ by $\psi(\{p,-p\})=\phi|_{\mathbb{S}^2}(p)$. Show that the mapping ψ is an embedding.

2.12. Let us consider the manifold $\mathbb{R}^6$ with the coordinate cover $(x_1,x_2,x_3,x_4,x_5,x_6)$. We define the following subsets:

$$M=\{\mathbf{x}\in\mathbb{R}^6:x_1^2-x_2^2-x_3^2=1,x_4^2-x_5^2-x_6^2=1\}\subset\mathbb{R}^6,$$
$$N=\{\mathbf{x}\in\mathbb{R}^6:x_2^2+x_3^2=1,x_5^2+x_6^2=1\}\subset\mathbb{R}^6,$$
$$P=\{\mathbf{x}\in\mathbb{R}^6:x_1^2-x_2^2+x_3^2-x_4^2\le 0\}\subset\mathbb{R}^6$$

Investigate whether (a) M,N,P are submanifolds of $\mathbb{R}^6$, (b) the set $M\cap N$ is a submanifold of $\mathbb{R}^6,M,N$, (c) P is a submanifold of N with boundary.

2.13. Show that the composition of two immersions is an immersion and the composition of two embeddings is an embedding.

2.14. Show that the subset $GL^+(n,\mathbb{R})$ of matrices with positive determinants and the set $SL(n,\mathbb{R})$ of matrices whose determinants are 1 constitute submanifolds of the manifold $GL(n,\mathbb{R})$.

2.15. Let us denote by $s(n,\mathbb{R})$ the subset of symmetric matrices of the manifold $gl(n,\mathbb{R})$. We define a mapping $\phi : gl(n,\mathbb{R}) \to s(n,\mathbb{R})$ by the rule $\phi(\mathbf{A}) = \mathbf{A}\mathbf{A}^{\mathrm{T}}$. Let $\mathbf{I}_n$ be $n \times n$ identity matrix. Then show that the mapping ϕ is a submersion on the subset $\phi^{-1}(\mathbf{I}_n) \subset gl(n,\mathbb{R})$ and it constitutes a submanifold of the subset $O(n,\mathbb{R})$ of *orthogonal matrices* $\mathbf{A} \in gl(n,\mathbb{R})$ satisfying the condition $\mathbf{A}\mathbf{A}^{\mathrm{T}} = \mathbf{I}_n$.

2.16. Let us take a fixed vector $\mathbf{v}_0 \in \mathbb{R}^n$ into consideration and define a mapping $f : GL^+(n,\mathbb{R}) \to \mathbb{R}^n$ by the relation $f(\mathbf{A}) = \mathbf{A}\mathbf{v}_0$. We naturally assume that $\mathbf{v}_0 \neq \mathbf{0}$. Show that this mapping and its restriction $g = f|_{SO(n,\mathbb{R})}$ to the set of orthogonal matrices $SO(n,\mathbb{R})$ with unit determinants are submersions. Show further that inverse mapping $M = g^{-1}(\{\mathbf{v}_0\}) = \{\mathbf{A} \in SO(n,\mathbb{R}) : \mathbf{A}\mathbf{v}_0 = \mathbf{v}_0\}$ of the set $\{\mathbf{v}_0\}$ under g is a submanifold of the manifold $SO(n,\mathbb{R})$ that is isomorphic to the manifold $SO(n-1,\mathbb{R})$.

2.17. Let $\phi : M \to N$ be an injective immersion between two smooth manifolds. Show that the mapping ϕ is a submersion when M is a compact manifold.

2.18. Let $\phi : M \to N$ be an immersion. If $M_1 \subset M$ is a submanifold, then show that the restriction $\phi|_{M_1}$ is also an immersion.

2.19. We define the mapping $f : GL^+(n,\mathbb{R}) \to GL^+(n+m,\mathbb{R})$ in the form

$$f(\mathbf{A}) = \begin{bmatrix} \mathbf{A} & \mathbf{0} \\ \mathbf{0} & \mathbf{B} \end{bmatrix}, \quad \mathbf{B} \in SO(m,\mathbb{R}).$$

Show that the restriction $f|_{SO(n,\mathbb{R})}$ is an embedding into $SO(n+m,\mathbb{R})$.

2.20. The mapping $\phi : \mathbb{R} \to \mathbb{R}^2$ is defined by $\phi(t) = (t, t^2)$. Determine the image $\phi_* U$ of the vector $U = d/dt$.

2.21. The curve $\gamma : \mathbb{R} \to \mathbb{S}^2$ is given by the relations

$$\gamma(t) = \big(x(t), y(t), z(t)\big) = \Big(\cos t \sin\big(t + \frac{\pi}{3}\big), \sin t \sin\big(t + \frac{\pi}{3}\big), \cos\big(t + \frac{\pi}{3}\big)\Big).$$

Let V denote the vector tangent to this curve at the point $t = 0$. Determine images of the point $\gamma(0)$ and the vector V under the stereographic projection.

2.22. We define a cylinder by $\mathbb{S}^1 \times \mathbb{R} = \{(x,y,z) \in \mathbb{R}^3 : x^2 + y^2 = 1\}$. Its coordinate cover can be taken as (ϕ, z) in polar coordinates. On using spherical coordinates we introduce a mapping $\Phi : \mathbb{S}^2 \to \mathbb{S}^1 \times \mathbb{R}$ by the relation $\Phi(\phi,\theta) = (\phi, \sin\theta)$. Evaluate the differential $d\Phi = \Phi_*$.

2.23. $U, V \in T(M)$ are two vector fields. Their flows are denoted by ϕ_t and ψ_s, respectively. Show that $\phi_t \circ \psi_s = \psi_s \circ \phi_t$ if and only if $[U,V] = 0$.

2.24. The vector field $U \in T(\mathbb{R}^2)$ is given by $U = \partial_x + \partial_y$. Find the flow generated by this vector field and show that this vector field is complete. Does this vector field retain its completeness when it is defined on the manifold $M = \mathbb{R}^2 - \{\mathbf{0}\}$?

2.25. The vector field $U \in T(\mathbb{R}^2 - \{\mathbf{0}\})$ is given by $U = -y\,\partial_x + x\,\partial_y$. Find the flow generated by this vector field and check whether it is a complete vector field.

2.26. Find the integral curves of the vector field $(1 + x^2)\partial_x \in T(\mathbb{R})$ and check whether it is a complete vector field.

2.27. The vector fields $U_1, U_2, U_3 \in T(\mathbb{R}^3)$ are given by

$$U_1 = z\frac{\partial}{\partial y} - y\frac{\partial}{\partial z},$$
$$U_2 = x\frac{\partial}{\partial z} - z\frac{\partial}{\partial x},$$
$$U_3 = y\frac{\partial}{\partial x} - x\frac{\partial}{\partial y}$$

Show that $[U_1, U_2] = U_3$, $[U_2, U_3] = U_1$ and $[U_3, U_1] = U_2$.

2.28. $f \in \mathbb{R}^2 \to \mathbb{R}$ is a smooth function. We define the vector field $U_f \in T(\mathbb{R}^2)$ by

$$U_f = \frac{\partial f}{\partial y}\frac{\partial}{\partial x} - \frac{\partial f}{\partial x}\frac{\partial}{\partial y}.$$

Show that the set formed by such kind of vector fields is closed under the Lie product.

2.29. Let ϕ_t and ψ_t be flows of vector fields $U, V \in T(M)$, respectively. We consider the curve

$$\gamma(t) = \psi_{-\sqrt{t}} \circ \phi_{-\sqrt{t}} \circ \psi_{\sqrt{t}} \circ \phi_{\sqrt{t}}(p)$$

through the point $p \in M$. We assume that $t \in [0, \epsilon]$ for a sufficiently small $\epsilon > 0$. Let $f : M \to \mathbb{R}$ be a smooth function. Show that we can write

$$[U, V](f)\big|_p = \lim_{t\to 0}\frac{f\big(\gamma(t)\big) - f\big(\gamma(0)\big)}{t}$$

and we get $\gamma'(0) = [U, V]$. Verify this property in $T(\mathbb{R}^3)$ for vector fields $U = \partial/\partial y$ and $V = \partial/\partial x + y\,\partial/\partial z$.

2.30. Let $\Phi : M \to N$ be a diffeomorphism. We denote flows generated by vector fields $U \in T(M)$ and $V \in T(N)$ by $\phi_t : M \to M$ and $\psi_t : N \to N$, respectively. We say that vector fields U and V are *Φ-related* if the relation $\Phi_* U = V$, or more explicitly $\Phi_* U(p) = V\big(\Phi(p)\big)$ for all $p \in M$, is satisfied. Show that U and V are Φ-related if and only if $\Phi \circ \phi_t = \psi_t \circ \Phi$. If we take $\Phi = \phi_t$, this relation is satisfied identically so that we find $(\phi_t)_* U = U$. This means that vector fields are conserved under their own flows.

2.31. Let $\Phi : M \to N$ be a diffeomorphism and $U \in T(M)$. Suppose that at every points $p_1, p_2 \in M$ satisfying the condition $\Phi(p_1) = \Phi(p_2)$ we have $\Phi_* U(p_1)$

$= \Phi_* U(p_2) \in T(N)$. Is there a vector field $V \in T(N)$ that is Φ-related with the vector field U?

2.32. Let $\Phi : M \to N$ be a diffeomorphism, $U_1, U_2 \in T(M)$ and $V_1, V_2 \in T(N)$. If vector fields U_1 and V_1, U_2 and V_2 are Φ-related, then show that Lie products $[U_1, U_2]$ and $[V_1, V_2]$ are also Φ-related.

2.33. Let $\Phi : M \to N$ be a diffeomorphism. We assume that vector fields $U \in T(M)$ and $V \in T(N)$ are Φ-related. Show that

$$\pounds_U(\Phi^* g) = \Phi^* \pounds_V(g) = \Phi^* \pounds_{\Phi_* U}(g)$$

for a function g $\in \Lambda^0(N)$.

2.34. Let us consider $f \in \Lambda^0(M)$ and $U \in T(M)$. The flow generated by the vector field U is $\phi_t : M \to M$. Show that the function $\phi_t^* f = f \circ \phi_t$ satisfies the following differential equation

$$\frac{d(\phi_t^* f)}{dt} = \phi_t^* \pounds_U f$$

along the flow.

2.35. The function $f : \mathbb{R}^n \times \mathbb{R} \to \mathbb{R}$ satisfies the following partial differential equation and initial condition

$$\frac{\partial f(\mathbf{x}, t)}{\partial t} = u^i(\mathbf{x}) \frac{\partial f(\mathbf{x}, t)}{\partial x^i}, \quad f(\mathbf{x}, 0) = g(\mathbf{x})$$

where $\mathbf{x} = (x^1, x^2, \ldots, x^n) \in \mathbb{R}^n$. If the vector field $U = u^i(\mathbf{x}) \partial / \partial x^i$ is complete and its flow is $\phi_t : \mathbb{R}^n \to \mathbb{R}^n$, then show that the function $f(\mathbf{x}, t) = g(\phi_t(\mathbf{x}))$ is the solution.

2.36. Find the solution of initial value problem given below:

$$\frac{\partial f}{\partial t} = 2 \frac{\partial f}{\partial x}, \quad f(x, 0) = \sin x.$$

2.37. Find the solution of initial value problem given below:

$$\frac{\partial f}{\partial t} = (x + y) \Big(\frac{\partial f}{\partial x} - \frac{\partial f}{\partial y} \Big), \quad f(x, y, 0) = xy.$$

2.38. Find the solution of initial value problem given below:

$$\frac{\partial f}{\partial t} = -y \frac{\partial f}{\partial x} + x \frac{\partial f}{\partial y}, \quad f(x, y, 0) = x + y.$$

2.39. We consider the vector fields $U, V \in T(M)$. $\phi_t : M \to M$ is the flow of the vector field U. Show that the following relation is valid:

$$\frac{d}{dt} (\phi_t^{-1})_* V = (\phi_t^{-1})_* (\pounds_U V)$$

2.40. Vector fields $U, V \in T(\mathbb{R}^n)$ depending also on a parameter t are given as follows:

$$U = u^i(\mathbf{x}, t)\frac{\partial}{\partial x^i}, \quad V = v^i(\mathbf{x}, t)\frac{\partial}{\partial x^i}.$$

We assume that the functions $v^i(\mathbf{x}, t)$ are satisfying the initial value problem

$$\frac{\partial v^i}{\partial t} = u^j\frac{\partial v^i}{\partial x^j} - \frac{\partial u^i}{\partial x^j}v^j, \; v^i(\mathbf{x}, 0) = g^i(\mathbf{x})$$

for prescribed functions $u^i(\mathbf{x}, t)$. If

$$G = g^i(\mathbf{x})\frac{\partial}{\partial x^i}$$

and ϕ_t is the flow generated by the vector field U, then show that the vector $V = (\phi_t^{-1})_* G$ represents the solution of the initial value problem.

2.41. Find the solution of initial value problem given below:

$$\frac{\partial v^1}{\partial t} = (x+y)\frac{\partial v^1}{\partial x} - (x+y)\frac{\partial v^1}{\partial y} - v^1 - v^2, \; v^1(x, y, 0) = y$$
$$\frac{\partial v^2}{\partial t} = (x+y)\frac{\partial v^2}{\partial x} - (x+y)\frac{\partial v^2}{\partial y} + v^1 + v^2, \; v^2(x, y, 0) = \sin x$$

2.42. Find the solution of initial value problem given below:

$$\frac{\partial v^1}{\partial t} = y\frac{\partial v^1}{\partial x} + x\frac{\partial v^1}{\partial y} - v^2, \; v^1(x, y, 0) = x^2$$
$$\frac{\partial v^2}{\partial t} = y\frac{\partial v^2}{\partial x} + x\frac{\partial v^2}{\partial y} - v^1, \; v^2(x, y, 0) = y$$

2.43. M is an m-dimensional smooth manifold. A k-dimensional involutive distribution $\mathcal{D} \subset T(M)$ is specified by linearly independent vector fields $U_\alpha \in T(M), \alpha = 1, \dots, k$ satisfying the conditions $[U_\alpha, U_\beta] = c_{\alpha\beta}^\gamma(p)U_\gamma$. Smooth functions $F_\alpha : M \times \mathbb{R} \to \mathbb{R}$ are denoted by $F_\alpha(p, t), p \in M, \alpha = 1, \dots, k$. We consider the differential equation

$$U_\alpha(f) = F_\alpha(\mathbf{x}, f), \; \mathbf{x} = (x^1, \dots, x^m), \; \alpha = 1, \dots, k$$

where $f : M \to \mathbb{R}$. Show that the solution $f(\mathbf{x})$ of this system of differential equations may only exists if the functions F_α satisfy the relations

$$\Big(U_\alpha + F_\alpha\frac{\partial}{\partial f}\Big)(F_\beta) - \Big(U_\beta + F_\beta\frac{\partial}{\partial f}\Big)(F_\alpha) = c_{\alpha\beta}^\gamma F_\gamma, \; \alpha, \beta = 1, \dots, k.$$

Show further that the solution is found as the solution of the following differential equations

$$\left(U_\alpha + F_\alpha \frac{\partial}{\partial f}\right)\mathcal{F} = 0, \quad \alpha = 1, \dots, k$$

when the above relations are satisfied.

2.44. We consider the manifold $M = \mathbb{R}^3 - \{\mathbf{0}\}$. Show that the vector fields $V^1 = z\partial_y - y\partial_z$, $V^2 = x\partial_z - z\partial_x$ and $V^3 = y\partial_x - x\partial_y$ in $T(M)$ give rise to a 2-dimensional involutive distribution. Determine its integral manifold.

2.45. Show that the distribution generated by vector fields $V^1 = \partial_y + x\partial_z$ and $V^2 = \partial_x + y\partial_t$ in $T(\mathbb{R}^4)$ does not possess a 2-dimensional integral manifold.

CHAPTER III

LIE GROUPS

3.1. SCOPE OF THE CHAPTER

This chapter is devoted to a concise exposition of Lie groups that help illuminate various structural peculiarities of mappings on manifolds. These groups are so named because it was M. S. Lie who has first studied family of continuous functions forming a group and recognised their effectiveness in revealing some very important and fundamental properties of differential equations. We first define in Sec. 3.2 a Lie group as a smooth manifold endowed with a group operation in which multiplication and inversion operations are supposed to be smooth functions. Some of the salient features of Lie groups are then briefly examined. Next, in Sec. 3.3 we discuss left and right translations generated by an element of the group that are diffeomorphisms mapping the manifold onto itself. Left- and right-invariant vector fields are introduced by means of differentials of these mappings and it is shown that they constitute Lie algebras. After that we briefly investigate in Sec. 3.4 the group homomorphism between Lie groups that preserve group operations. We then consider in Sec. 3.5 one-parameter subgroups of a Lie group that are homomorphisms between the commutative Lie group of real numbers and an abstract Lie group. We then discuss the exponential mapping that may help characterise such one-parameter subgroups. Afterwards in Sec. 3.6 the group of automorphisms mapping the Lie group onto itself and generated by elements of the Lie group itself is defined and it is shown that this group, which is called adjoint representation, is isomorphic to the Lie group. In Sec. 3.5 we examine some notable properties of Lie transformation groups that map a smooth manifold onto itself and form also a Lie group. Finally, Killing vector fields were introduced.

3.2. LIE GROUPS

We assume that a binary operation $* : G \times G \to G$ on a set G, which

Exterior Analysis, DOI: 10.1016/B978-0-12-415902-0.50003-7

will be called briefly as a *product*, satisfy the following conditions:

(i). Operation is closed: $g_1{*}g_2 \in G$ for all $g_1, g_2 \in G$.
(ii). Operation is associative: $g_1{*}(g_2{*}g_3) = (g_1{*}g_2){*}g_3$ for all $g_1, g_2, g_3 \in G$.
(iii). There is an identity element $e \in G$: $e{*}g = g{*}e = g$ for all $g \in G$.
(iv). For each $g \in G$ there is an inverse $g^{-1} \in G$: $g{*}g^{-1} = g^{-1}{*}g = e$.

Then $(G, *)$ is called a ***group***. It is easily observed that the identity element e and the inverse element g^{-1} of an element $g \in G$ are *uniquely specified. A **Lie group** G is also a smooth manifold and the mappings*

$$\sigma : G \times G \to G \quad \textit{and} \quad \iota : G \to G$$

defined by $\sigma(g_1, g_2) = g_1{*}g_2$ *and* $\iota(g) = g^{-1}$ *are smooth mappings*.

These two last conditions can be combined into a single one imposing that *the mapping* $\overline{\sigma} : G \times G \to G$ *defined by the rule* $\overline{\sigma}(g_1, g_2) = g_1{*}g_2^{-1}$ *is smooth*. To prove this proposition, let us first introduce the smooth mapping $\mathcal{I} : G \to G \times G$ by the simple rule $\mathcal{I}(g) = (e, g)$. We see that $\overline{\sigma} \circ \mathcal{I} = \iota$. Indeed, we find at once that $(\overline{\sigma} \circ \mathcal{I})(g) = \overline{\sigma}(e, g) = e{*}g^{-1} = g^{-1} = \iota(g)$ for all $g \in G$. Since ι is now written as the composition of two smooth mappings, it turns out to be a smooth mapping as well. Similarly, let us introduce the smooth mapping $\mathfrak{I} : G \times G \to G \times G$ through the relation

$$\mathfrak{I}(g_1, g_2) = (g_1, g_2^{-1}) = \big(g_1, \iota(g_2)\big)$$

from which it follows that $\overline{\sigma}\big(\mathfrak{I}(g_1, g_2)\big) = \overline{\sigma}(g_1, g_2^{-1}) = g_1{*}g_2 = \sigma(g_1, g_2)$ for all $g_1, g_2 \in G$. Thus the mapping $\sigma = \overline{\sigma} \circ \mathfrak{I}$ is also smooth. If G is a finite m-dimensional manifold, then it is called an *m-parameter Lie group*.

Let $(G, *)$ and $(H, \diamond)$ be two Lie groups. The Cartesian product $G \times H$ of the manifolds G and H can easily be equipped with a group structure by defining the product of elements (g_1, h_1) and (g_2, h_2) of the product manifold $G \times H$ where $g_1, g_2 \in G$ and $h_1, h_2 \in H$ in the following fashion

$$(g_1, h_1) \bullet (g_2, h_2) = (g_1{*}g_2, h_1 \diamond h_2) \in G \times H.$$

One checks readily that the binary operation $\bullet$ is a group operation since it is solely determined by group operations on the Lie groups G and H and smoothness requirements are clearly met. If G and H are m- and n-parameter Lie groups, respectively, then the product manifold $G \times H$ turns out to be an $(m+n)$-parameter Lie group. Such a group is called a ***direct product*** of groups G and H.

Let us now consider some examples to Lie groups.

Example 3.2.1. The smooth manifold $\mathbb{R}^n$ (*see* Example 2.2.1) is a commutative Lie group with respect to the operation of addition in $\mathbb{R}^n$. If $\mathbf{x}, \mathbf{y} \in \mathbb{R}^n$, then we have $\mathbf{y}^{-1} = -\mathbf{y}$ so that we obtain $\mathbf{x} * \mathbf{y}^{-1} = \mathbf{x} - \mathbf{y} = (x^1 - y^1, \ldots, x^n - y^n)$. This is obviously a smooth function.

Example 3.2.2. Let us consider the manifold $GL(n, \mathbb{R})$ which we had introduced in Example 2.2.2 and we had already called the ***general linear group of degree n***. It is immediately seen that this manifold becomes also a non-commutative group with respect to the usual matrix multiplication. Let $\mathbf{A}, \mathbf{B} \in GL(n, \mathbb{R})$. With the coordinates $a^i_j,\ b^i_j \in \mathbb{R}, i, j = 1, \ldots, n$ these matrices are represented by $\mathbf{A} = [a^i_j], \mathbf{B} = [b^i_j]$ and we know that the matrix $\mathbf{AB}^{-1}$ is expressed as follows

$$\mathbf{AB}^{-1} = [a^i_k {b^{-1}}^k_j] = [a^i_k\,(\text{cofactor } b^k_j)^{\mathrm{T}} / \det \mathbf{B}].$$

Nevertheless, this is a smooth function because it is obviously the ratio of two polynomials. Hence $GL(n, \mathbb{R})$ is a Lie group of dimension n^2.

Let us now define a subset of the general linear group given by

$$SL(n, \mathbb{R}) = \{\mathbf{A} \in GL(n, \mathbb{R}) : \det \mathbf{A} = 1\}$$

It is clear that this subset is also a group with respect to matrix multiplication. In view of Theorem 2.4.1, $SL(n, \mathbb{R})$ is a submanifold of dimension $n^2 - 1$ of the general linear group. Hence, it is a Lie group. This group is called the ***special linear group*** or the ***unimodular group***.

We now consider the following subset

$$O(n) = \{\mathbf{A} \in GL(n, \mathbb{R}) : \mathbf{AA}^{\mathrm{T}} = \mathbf{I}\}$$

of the group $GL(n, \mathbb{R})$ which is formed by orthogonal matrices. Since the product of two orthogonal matrices is again an orthogonal matrix, $O(n)$ is a group and Theorem 2.4.1 implies that it is a submanifold of $GL(n, \mathbb{R})$ with the dimension $n^2 - n(n+1)/2 = n(n-1)/2$. Thus, it is a Lie group. $O(n)$ is called the ***orthogonal group***. If $\mathbf{A} \in O(n)$, then $(\det \mathbf{A})^2 = 1$ so that $\det \mathbf{A} = \pm 1$. The Lie group

$$SO(n) = \{\mathbf{A} \in O(n) : \det \mathbf{A} = 1\}$$

whose dimension is also $n(n-1)/2$ is known as the ***special orthogonal group*** because it preserves the length of a vector $\mathbf{x}$ and volumes in $\mathbb{R}^n$. In fact, we obtain for $\mathbf{A} \in O(n)$

$$(\mathbf{Ax})^{\mathrm{T}}\mathbf{Ax} = \mathbf{x}^{\mathrm{T}}\mathbf{A}^{\mathrm{T}}\mathbf{Ax} = \mathbf{x}^{\mathrm{T}}\mathbf{x}.$$

The orthogonal group is in fact a disconnected Lie group that is expressible as the union of two disjoint connected groups as

$$O(n) = SO(n) \cup \mathbf{\Omega} SO(n)$$

where $\mathbf{\Omega}$ is the $n \times n$ matrix

$$\mathbf{\Omega} = \begin{bmatrix} -1 & 0 & \cdots & 0 \\ 0 & 1 & \cdots & 0 \\ \vdots & \vdots & \vdots & \vdots \\ 0 & 0 & \cdots & 1 \end{bmatrix}$$

so that $\det \mathbf{\Omega} = -1$.

Example 3.2.3. The complex plane $\mathbb{C} - \{0\}$ is the 2-dimensional smooth manifold $\mathbb{R}^2 - \{(0,0)\}$. This manifold is also a group with respect to the complex multiplication. On the other hand, if $z_1, z_2 \in \mathbb{C} - \{0\}$, then $z_1 z_2^{-1}$ is a smooth function of real coordinates. Hence, this manifold is a Lie group.

Example 3.2.4. Let us consider the smooth manifold $\mathbb{S}^1$, the unit circle. The points of this manifold can be determined by complex numbers with unit moduli such as $|z| = 1$. If $z_1, z_2 \in \mathbb{S}^1$, then $|z_1 z_2| = |z_1||z_2| = 1$ and this means that $z_1 z_2 \in \mathbb{S}^1$. This is tantamount to say that the manifold $\mathbb{S}^1$ is a Lie group.

Example 3.2.5. The m-torus defined as $\mathbb{T}^m = (\mathbb{S}^1)^m$ is a Lie group because it is the m-fold Cartesian product of a Lie group.

Subgroup. A *submanifold* H of a Lie group G is called a ***subgroup*** if for all elements $h_1, h_2 \in H$ one finds $h_1 {*} h_2 \in H$ and $h_1^{-1} \in H$. Therefore, a subgroup is a submanifold of a Lie group that is closed with respect to operations of group multiplication and inversion.

If a Lie group is *connected*, then the following theorem states that it can be generated by an open neighbourhood of its identity element.

Theorem 3.2.1. *Let G be a connected Lie group and U be an open neighbourhood of the identity element e. We denote the set of all n-fold products of elements of U by $U^n = \{u_1 {*} u_2 {*} \cdots {*} u_n : u_i \in U\}$. Then one can write*

$$G = \bigcup_{n=1}^{\infty} U^n.$$

In other words, each group element $g \in G$ is expressible as a finite product of some elements in the open set U. Hence, we can say that U generates the group G.

Let us choose a fixed $g \in G$, and define a function $\sigma_g : G \to G$ by the rule $\sigma_g(h) = \sigma(g, h) = g {*} h$. σ_g is a diffeomorphism [*see* Sec. 3.3]. Hence, if U is an open set, then the set $\sigma_g(U) = \{g {*} u : u \in U\} \subset G$ will also be open. Consequently, the set U^n is open for all n. Since ι is a diffeomorphism, the set $U^{-1} = \{u^{-1} = \iota(u) : u \in U\}$ is also open. We then conclude that the sets $V = U \cap U^{-1} \subset U$ and V^n are all open. Furthermore, the obvious relationship $V = V^{-1}$ would be valid. Because $e \in U$ and $e = e^{-1}$, we see at once that $e \in V$, i.e., V is not empty. Let us now define the set

$$H = \bigcup_{n=1}^{\infty} V^n \subseteq \bigcup_{n=1}^{\infty} U^n \subseteq G.$$

H is an open set since it is the union of countably many open sets, and it is, consequently, an open submanifold [*see* p. 77]. Due to the property of the set V, H will be a subgroup. We now consider the family of open sets $\sigma_g(H) = \{g*h : h \in H\}$ defined for all $g \in G$. One has evidently the relation $H = \sigma_{g\in H}(H)$. Thus we can obviously write

$$G = H \cup \bigcup_{g\in G, g\notin H} \sigma_g(H).$$

But the open set H is the complement of the open set $\bigcup_{g\in G, g\notin H} \sigma_g(H)$ with respect to G so it must also be a closed set. In a connected topological space only the empty set or the space itself can be both open and closed. H cannot be empty since $e \in H$ so it must be equal to G. We therefore reach to the conclusion that $G = \bigcup_{n=1}^{\infty} U^n$. □

The above theorem indicates that if a Lie group is a connected topological space, then an open neighbourhood of the identity element determines the entire group.

A subgroup H of the group G is called a ***normal*** or ***invariant subgroup*** if for all $h \in H$ we get $h_g = g^{-1}*h*g \in H$ for all $g \in G$ so that H is invariant under ***conjugation***. In other words, if H is a normal subgroup, a ***conjugate*** element $h_g \in H$ corresponds to each element $h \in H$ so that

$$g*h_g = h*g \quad \text{for each } g \in G.$$

This property is symbolically reflected by the notation $g*H = H*g$ for all $g \in G$. Let H be a normal subgroup, the ***quotient group*** is defined as the set $G/H = \{g*H : g \in G\}$. The ***coset*** $g*H$ is the subset of G defined by $\{g*h : \forall h \in H\}$. It is easy to verify that G/H is actually a group. Let us consider the direct product which can be written as follows

$$(g_1*H)*(g_2*H) = (g_1*g_2)*(H*H) = (g_1*g_2)*H \in G/H$$

since one obviously observe the symbolic relation $H*H = H$ because H is a subgroup.

3.3. LIE ALGEBRAS

Let G be a Lie group. We choose a fixed element $g \in G$ to define a mapping $L_g : G \to G$ in such a way that

$$L_g(h) = \sigma(g, h) = g{*}h \tag{3.3.1}$$

for all $h \in G$. L_g is evidently a smooth mapping on the manifold G. The mapping L_g is called the ***left translation*** of the Lie group G by the element $g \in G$. We can obviously define a left translation for each element g of the group G. It can easily be seen that the relation $(L_g)^{-1} = L_{g^{-1}}$ is valid. Indeed, for each $h \in G$ we can write

$$L_g\big(L_{g^{-1}}(h)\big) = g{*}g^{-1}{*}h = e{*}h = h$$

so that we obtain $L_g \circ L_{g^{-1}} = i_G$. Similarly, it is found that $L_{g^{-1}} \circ L_g = i_G$. Hence, the inverse mapping $(L_g)^{-1} = L_{g^{-1}}$ is also smooth. Consequently, the left translation L_g *is a diffeomorphism.* The set of mappings

$$G_1 = \{L_g : g \in G\}$$

constitutes a group with respect to the operation of composition of mappings. In fact, if $L_{g_1}, L_{g_2} \in G_1$, then owing to the relation

$$L_{g_1}\big(L_{g_2}(h)\big) = g_1{*}g_2{*}h = L_{g_1{*}g_2}(h)$$

for all $h \in G$, we obtain $L_{g_1} \circ L_{g_2} = L_{g_1{*}g_2} \in G_1$ since $g_1{*}g_2 \in G$. Because $L_e = i_G$, it then follows that

$$L_e \circ L_g = L_g \circ L_e = L_g\,.$$

Thus, the identity element of G_1 is L_e and the inverse of L_g in G_1 is clearly $L_{g^{-1}}$. Since the composition is an associative binary operation, we finalise the realisation of the group structure of G_1. Therefore, there exists a mapping $\mathcal{L} : G \to G_1$ such that $\mathcal{L}(g) = L_g$. This mapping $\mathcal{L}$ is evidently surjective. Let us further suppose that $\mathcal{L}(g_1) = \mathcal{L}(g_2)$. If $L_{g_1}(h) = L_{g_2}(h)$ for all $h \in G$, the relation $g_1{*}h = g_2{*}h$ then leads to $g_1 = g_2$ if we multiply both sides by h^{-1} from left which means that $\mathcal{L}$ is injective, and consequently is *bijective*. On the other hand, due to the relation $\mathcal{L}(g_1{*}g_2) = L_{g_1} \circ L_{g_2} = \mathcal{L}(g_1) \circ \mathcal{L}(g_2)$, we infer that that the mapping $\mathcal{L}$ preserves group operations. In other words, it is a *group isomorphism*. Hence, *the groups G and G_1 are isomorphic*.

In exactly same fashion, we can define the ***right translation*** of the Lie group G by the element $g \in G$ as the mapping $R_g : G \to G$ such that

$$R_g(h) = h{*}g \tag{3.3.2}$$

for all $h \in G$. We can readily verify that a right translation is also a diffeomorphism and due to the relation $R_{g_1}(R_{g_2}(h)) = h{*}g_2{*}g_1 = R_{g_2*g_1}(h)$ for all $h \in G$, one obtains $R_{g_1} \circ R_{g_2} = R_{g_2*g_1}$. It is then straightforward to observe that the set of mappings $G_2 = \{R_g : g \in G\}$ constitutes a group with respect to the operation of composition. The identity element of this group is $R_e = i_G$ and the inverse of an element is given by $(R_g)^{-1} = R_{g^{-1}}$. It is clear that this group is also isomorphic to G. Therefore, the groups G_1 and G_2 are isomorphic to one another as well. It is now evident that left and right translations are connected through the following relation

$$R_g(h) = g^{-1}{*}g{*}h{*}g = g^{-1}{*}L_g(h){*}g.$$

Therefore, a right translation of an element of the group G is conjugate to its left translation, and vice versa. Moreover, it follows from $(L_{g_1} \circ R_{g_2})(h) = g_1{*}(h{*}g_2) = (g_1{*}h){*}g_2 = (R_{g_2} \circ L_{g_1})(h)$ for all $h \in G$ that these mappings commute, that is,

$$L_{g_1} \circ R_{g_2} = R_{g_2} \circ L_{g_1}. \tag{3.3.3}$$

In case G is a commutative group, we find that $L_g(h) = g{*}h = h{*}g = R_g(h)$ for all $h \in G$. Hence, we deduce that $L_g = R_g$ for all $g \in G$ in such an *Abelian group*.

Inasmuch as the mapping L_g is a diffeomorphism on G, its differential $dL_g|_h : T_h(G) \to T_{g*h}(G)$ is an isomorphism [*see* p. 124] transforming vector fields onto vector fields. A vector field V on the Lie group G is called a ***left-invariant vector field*** if it satisfies the equality

$$dL_g(V(h)) = V(L_g(h)) = V(g{*}h) \tag{3.3.4}$$

for all $g, h \in G$. This means that the image of a vector of such a field at the point h under the linear operator dL_g will be a vector of the same field at the point $g{*}h$. Thus the operator dL_g transforms a left-invariant vector field onto itself. So it is permissible to write symbolically

$$dL_g(V) = V$$

for all $g \in G$. If we take $h = e$ in (3.3.4), we obtain

$$dL_g(V(e)) = V(g) \tag{3.3.5}$$

for all $g \in G$. This relation implies that a left-invariant vector field on G is completely determined by *a vector in the tangent space* $T_e(G)$ *of the* *identity element e of the Lie group G.* So it becomes quite reasonable to interpret left-invariant vector fields as '*constant vector fields*' on the manifold G [Fig. 3.3.1].

Conversely, let us suppose that the relation $dL_g\big(V(e)\big) = V(g)$ is satisfied for all $g \in G$. We then easily deduce that

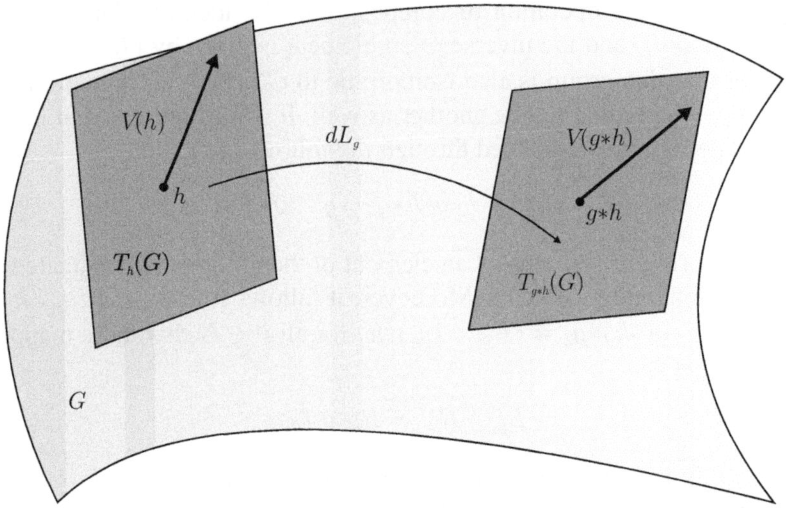

Fig. 3.3.1. A left-invariant vector field.

$$\begin{aligned} V(g{*}h) &= dL_{g*h}\big(V(e)\big) = d(L_g \circ L_h)\big(V(e)\big) \\ &= dL_g\big[dL_h\big(V(e)\big)\big] = dL_g\big(V(h)\big). \end{aligned} \tag{3.3.6}$$

According to (3.3.4), such a vector field V is a left-invariant vector field. We now denote the set of all left-invariant vector fields by $\mathfrak{g}$. It is seen at once that $\mathfrak{g}$ is a linear vector space on real numbers. Indeed, if $V_1, V_2 \in \mathfrak{g}$ and $\alpha_1, \alpha_2 \in \mathbb{R}$, the linearity of the operator dL_g on real numbers leads to the result

$$dL_g(\alpha_1 V_1 + \alpha_2 V_2) = \alpha_1 dL_g(V_1) + \alpha_2 dL_g(V_2) = \alpha_1 V_1 + \alpha_2 V_2$$

from which $\alpha_1 V_1 + \alpha_2 V_2 \in \mathfrak{g}$ follows. If we assume, instead, α_1 are α_2 are smooth functions on G, we realise that the invariance requirement can only be fulfilled if admissible functions are merely constant. The foregoing observations bring to mind the possibility of the existence of a bijective mapping between $\mathfrak{g}$ and $T_e(G)$. To this end, we presently introduce a mapping $\mathcal{G} : \mathfrak{g} \to T_e(G)$ by the rule $\mathcal{G}(V) = V(e)$. Owing to (3.3.5), the operator $\mathcal{G}$

must be linear. Indeed, one can write

$$\begin{aligned}(\alpha_1 V_1 + \alpha_2 V_2)(g) &= dL_g\big(\alpha_1 V_1(e) + \alpha_2 V_2(e)\big)\\ &= \alpha_1 dL_g\big(V_1(e)\big) + \alpha_2 dL_g\big(V_2(e)\big) = \alpha_1 V_1(g) + \alpha_2 V_2(g)\end{aligned}$$

for all $g \in G$. Thus we find that $\mathcal{G}(\alpha_1 V_1 + \alpha_2 V_2) = \alpha_1 \mathcal{G}(V_1) + \alpha_2 \mathcal{G}(V_2)$. The mapping $\mathcal{G}$ is *injective*. Suppose that $\mathcal{G}(V_1) = \mathcal{G}(V_2)$. We then have

$$V_1(g) = dL_g\big(V_1(e)\big) = dL_g\big(V_2(e)\big) = V_2(g)$$

for all $g \in G$ and we conclude that $V_1 = V_2$. $\mathcal{G}$ is *surjective*. Let us consider a vector $V(e) \in T_e(G)$. The vector field defined by $dL_g\big(V(e)\big) = V(g)$ for all $g \in G$ is a left-invariant vector field in view of (3.3.6), hence it is an element of $\mathfrak{g}$. In conclusion, $\mathcal{G}$ is an isomorphism and the vector spaces $\mathfrak{g}$ and $T_e(G)$ are isomorphic. This result dictates that the dimension of $\mathfrak{g}$ will be the same as that of $T_e(G)$. It is, of course, the same as the dimension of the manifold G.

At the identity element e, one writes $dL_g : T_e(G) \to T_g(G)$ so that we have $(dL_g)^{-1} : T_g(G) \to T_e(G)$. Because of the relation $L_{g^{-1}}(g) = e$, we obtain $dL_{g^{-1}} : T_g(G) \to T_e(G)$. On the other hand, the identities $L_g \circ L_{g^{-1}} = L_{g^{-1}} \circ L_g = i_G$ will result in the relations $dL_g \circ dL_{g^{-1}} = I_{T_g(G)}$ and $dL_{g^{-1}} \circ dL_g = I_{T_e(G)}$. It then follow that $(dL_g|_e)^{-1} = dL_{g^{-1}}|_g$.

Since G is a smooth manifold of dimension m, each point of G is contained in an open neighbourhood in G and there is a homeomorphism φ mapping this open set onto an open set of $\mathbb{R}^m$. If local coordinates of a point $h \in G$ are prescribed by $\mathbf{x} = (x^1, \dots, x^m)$ and local coordinates of a point $L_g(h) = g{*}h \in G$ are given by $\mathbf{y} = (y^1, \dots, y^m)$, then we know that there exists a functional relationship in the form $\mathbf{y} = (\varphi \circ L_g \circ \varphi^{-1})(\mathbf{x}) = \mathrm{L}_g(\mathbf{x})$, or $y^i = L^i_g(\mathbf{x})$. Hence the definition (3.3.4) implies that the local components of a left-invariant vector field must satisfy the following expressions

$$v^i(\mathbf{y}) = \frac{\partial L^i_g(\mathbf{x})}{\partial x^j} v^j(\mathbf{x}), \quad \mathbf{y} = \mathrm{L}_g(\mathbf{x}) \tag{3.3.7}$$

for all $\mathbf{x} \in \mathbb{R}^m$ in respective charts.

We now demonstrate that the Lie bracket of vector fields $V_1, V_2 \in \mathfrak{g}$ is also a left-invariant vector field. If we recall (2.10.21) we find that

$$dL_g([V_1, V_2]) = [dL_g(V_1), dL_g(V_2)] = [V_1, V_2],$$

hence, $[V_1, V_2] \in \mathfrak{g}$. As a result of this, we see that *left-invariant vector fields constitute a Lie algebra*. $\mathfrak{g}$ or $T_e(G)$ that is isomorphic to $\mathfrak{g}$ is called the ***Lie algebra*** of the Lie group G. Indeed, since we have $[V_1, V_2] \in \mathfrak{g}$ if

$V_1, V_2 \in \mathfrak{g}$, we understand that the relation

$$\mathcal{G}([V_1, V_2]) = [V_1(e), V_2(e)] = [\mathcal{G}(V_1), \mathcal{G}(V_2)]$$

would also be valid. If the dimension of the manifold G is m, a basis of the vector space $\mathfrak{g}$ are determined by m linearly independent left-invariant vector fields $\{V_i : i = 1, \dots, m\}$. Properties of a Lie algebra will impose the following restriction on these vectors for all $i, j, k = 1, \dots, m$

$$\begin{aligned} [V_i, V_j] + [V_j, V_i] = 0, \ . \\ \big[V_i, [V_j, V_k]\big] + \big[V_j, [V_k, V_i]\big] + \big[V_k, [V_i, V_j]\big] = 0 \end{aligned} \tag{3.3.8}$$

Since $\mathfrak{g}$ is a Lie algebra, there must exist *constants* c_{ij}^k so that one has

$$[V_i, V_j] = c_{ij}^k V_k. \tag{3.3.9}$$

These constants are called ***structure constants*** of the Lie algebra $\mathfrak{g}$ with respect to the basis $\{V_i\}$. Because of the relations (3.3.9) and (3.3.8), the structure constants should meet the conditions

$$\begin{aligned} c_{ij}^k + c_{ji}^k = 0, \\ c_{jk}^n c_{in}^l + c_{ki}^n c_{jn}^l + c_{ij}^n c_{kn}^l = 0 \end{aligned} \tag{3.3.10}$$

for all $i, j, k, l = 1, \dots, m$ [*see* (2.11.4)]. Structure constants holding the conditions (3.3.10) completely determines the Lie algebra. It is clear that the structure constants depend on the selected basis. Let us choose another basis by the transformation $V_j' = a_j^i V_i$ where $\mathbf{A} = [a_j^i]$ is a regular matrix. If we write $[V_i', V_j'] = c_{ij}'^k V_k'$, we easily find that the following expressions must be satisfied

$$c_{ij}'^k a_k^r V_r = [a_i^p V_p, a_j^q V_q] = a_i^p a_j^q [V_p, V_q] = a_i^p a_j^q c_{pq}^r V_r.$$

Since the vectors V_r are linearly independent, we conclude that

$$c_{ij}'^k = a_i^p \, a_j^q \, b_r^k \, c_{pq}^r \tag{3.3.11}$$

where $\mathbf{B} = \mathbf{A}^{-1} = [b_j^i]$. (3.3.11) clearly indicates that structure constants are components of a third order mixed tensor. This tensor is called the ***structure tensor*** of the Lie algebra. We have seen that the Lie algebra of left-invariant vector fields is isomorphic to the tangent space $T_e(G)$ at the identity element e and the integral manifold of that tangent space locally determines the manifold G. This is tantamount to say that the Lie algebra fully determines the Lie group locally in a neighbourhood of e. However, the correspondence between the Lie groups and the Lie algebras is not unique. Although a given

Lie group determines uniquely its Lie algebra, several Lie groups may generate the same Lie algebra. But, it can be shown that among all the Lie groups with the same Lie algebra, there is only one Lie group that is simply connected. Therefore, a given Lie algebra gives rise to a unique simply connected Lie group locally in a neighbourhood of e. Then in view of Theorem 3.2.1 it determines the Lie group globally if the manifold G is connected. Because features of a Lie algebra are entirely elucidated by its structure constants, to investigate the properties of constants satisfying the algebraic relations (3.3.10) provides quite a significant information about the associated Lie group itself.

If structure constants are all zero, we then have $[V_i, V_j] = 0$ so that $\mathfrak{g}$ becomes a commutative Lie algebra. Such algebras are named as *Abelian Lie algebras*.

In exactly the same fashion as we have introduced the left-invariant vectors, we can define the ***right-invariant*** vector fields through the relation $dR_g(V) = V$. We immediately observe that these vector fields constitute a Lie algebra that is isomorphic to the vector space $T_e(G)$. Let us denote Lie algebras of left- and right-invariant vectors by $\mathfrak{g}_L$ and $\mathfrak{g}_R$, respectively. Since both algebras are isomorphic to the tangent space $T_e(G)$, they are of course isomorphic to one another through the isomorphism $\mathcal{G}_R^{-1} \circ \mathcal{G}_L$.

In view of (2.7.7), the relation (3.3.3) yields

$$dL_g \circ dR_g = dR_g \circ dL_g.$$

If V is a left-invariant vector field, we find

$$dL_g\big(dR_g(V)\big) = dR_g\big(dL_g(V)\big) = dR_g(V)$$

for all $g \in G$. This result means that the vector field $dR_g(V)$ turns out also to be a left- invariant vector field. Conversely, if V is a right-invariant vector field, then the same expression implies that the vector field $dL_g(V)$ is a right-invariant vector field.

Example 3.3.1. Consider the affine space $\mathbb{R}^n$ [*see* Example 2.2.1]. This smooth manifold is obviously a commutative Lie group with respect to the following addition operation

$$x + y = (x^1 + y^1, \dots, x^n + y^n)$$

for all $x, y \in \mathbb{R}$. In this case left and right translations are not different and they are given by

$$L_x(y) = R_x(y) = x + y.$$

Let us denote a left-invariant vector field by $V(x) = v^i(x)\,\partial_i$. Then (3.3.7) leads to the relation

$$v^i(x+y) = \frac{\partial(y^i + x^i)}{\partial x^j} v^j(x) = \delta^i_j v^j(x) = v^i(x).$$

Hence the left-invariant vector fields are constant vector fields whose components merely $v^i \in \mathbb{R}$. Of course, they generate a commutative Lie algebra with vanishing structure constant. ■

Example 3.3.2. We wish to compute the Lie algebra of the Lie group $GL(n, \mathbb{R})$. Inasmuch as $GL(n, \mathbb{R})$ is an open submanifold of the manifold $gl(n, \mathbb{R})$, its dimension is n^2. Hence, the tangent space at the identity element $e = \mathbf{I}$ is an n^2-dimensional vector space. We can thus identify the associated Lie algebra with the space $gl(n, \mathbb{R})$ that consists of all $n \times n$ matrices. We can choose as basis vectors the set of following linearly independent $n \times n$ matrices whose only one entry is 1 and all the other entries are 0:

$$V^i_j(e) = \delta^i_k \delta^l_j \frac{\partial}{\partial x^l_k}, \quad i, j, k, l = 1, \dots, n$$

where n^2 matrix entries x^l_k represent the local coordinates of $GL(n, \mathbb{R})$. Left translation is naturally defined as the matrix product $L_g(h) = \mathbf{GH}$ or $\big(L_g(h)\big)^k_l = g^k_m h^m_l$ in terms of components of $\mathbf{G} = [g^i_j]$ and $\mathbf{H} = [h^i_j]$. Hence, according to (3.3.7), the components of a left-invariant vector field must obey the equality

$$\big(V^i_j(g)\big)^k_l = \frac{\partial(g^k_m x^m_l)}{\partial x^p_q} \big(V^i_j(e)\big)^p_q = g^k_m \delta^m_p \delta^q_l \delta^i_q \delta^p_j = g^k_j \delta^i_l .$$

Consequently we can construct left-invariant vector fields by making use of the basis vectors

$$V^i_j(g) = g^k_j \delta^i_l \frac{\partial}{\partial g^k_l} = g^k_j \frac{\partial}{\partial g^k_i}$$

for all $g \in GL(n, \mathbb{R})$. An element of the Lie algebra $\mathfrak{gl}(n)$ will now be expressible as

$$V_{\mathbf{A}} = a^j_i V^i_j(e) = a^j_i \delta^i_k \delta^l_j \frac{\partial}{\partial x^l_k} = a^j_i \frac{\partial}{\partial x^j_i}$$

where the numbers a^j_i are entries of a matrix $\mathbf{A}$. Next, we determine the structure constants of the Lie algebra by evaluating

$$
\begin{aligned}
[V_j^i, V_l^k] &= \Big[g_j^p \frac{\partial}{\partial g_i^p}, g_l^q \frac{\partial}{\partial g_k^q}\Big] = g_j^p \frac{\partial}{\partial g_i^p}\Big(g_l^q \frac{\partial}{\partial g_k^q}\Big) - g_l^q \frac{\partial}{\partial g_k^q}\Big(g_j^p \frac{\partial}{\partial g_i^p}\Big) \\
&= g_j^p \delta_p^q \delta_l^i \frac{\partial}{\partial g_k^q} + g_j^p g_l^q \frac{\partial^2}{\partial g_i^p \partial g_k^q} - g_l^q \delta_q^p \delta_j^k \frac{\partial}{\partial g_i^p} - g_l^q g_j^p \frac{\partial^2}{\partial g_k^q \partial g_i^p} \\
&= \delta_l^i g_j^p \frac{\partial}{\partial g_k^p} - \delta_j^k g_l^p \frac{\partial}{\partial g_i^p} = \delta_l^i V_j^k - \delta_j^k V_l^i.
\end{aligned}
$$

It then follows that

$$[V_j^i, V_l^k] = \big(\delta_l^i \delta_p^k \delta_j^q - \delta_j^k \delta_p^i \delta_l^q\big) V_q^p = c_{jlp}^{ikq} V_q^p.$$

Since $\big(V_{\mathbf{A}}(e)\big)_j^i = a_j^i$, the left-invariant vector field generated by a vector $V_{\mathbf{A}}$ becomes

$$V_{\mathbf{A}}(g) = \big(V_{\mathbf{A}}(g)\big)_j^i \frac{\partial}{\partial g_j^i} = \frac{\partial (g_m^i x_j^m)}{\partial x_l^k} a_l^k \frac{\partial}{\partial g_j^i} = g_k^i a_j^k \frac{\partial}{\partial g_j^i}.$$

Therefore the Lie product (bracket) of left-invariant matrices corresponding to matrices **A** and **B** is found to be

$$
\begin{aligned}
[V_{\mathbf{A}}, V_{\mathbf{B}}](g) &= \Big[g_m^i a_j^m \frac{\partial}{\partial g_j^i}, g_p^k b_l^p \frac{\partial}{\partial g_l^k}\Big] = g_m^i a_j^m b_l^j \frac{\partial}{\partial g_l^i} - g_p^i b_l^p a_j^l \frac{\partial}{\partial g_j^i} \\
&= (a_j^m b_l^j - b_j^m a_l^j) g_m^i \frac{\partial}{\partial g_l^i} = g_m^i [\mathbf{A}, \mathbf{B}]_l^m \frac{\partial}{\partial g_l^i} = V_{[\mathbf{A},\mathbf{B}]}(g).
\end{aligned}
$$

where $[\mathbf{A}, \mathbf{B}] = \mathbf{AB} - \mathbf{BA}$ is the *matrix commutator*. These results clearly indicate that the Lie algebra $\mathfrak{gl}(n)$ is actually generated by the elements of the vector space $gl(n, \mathbb{R})$ on which the Lie product of matrices $\mathbf{A}, \mathbf{B}$ is defined as the matrix commutator $[\mathbf{A}, \mathbf{B}]$. ■

3.4. LIE GROUP HOMOMORPHISMS

Let $(G, *)$ and $(H, \diamond)$ be Lie groups, and $\phi : G \to H$ be a *smooth* function. If, for all $g_1, g_2 \in G$, the relation $\phi(g_1 * g_2) = \phi(g_1) \diamond \phi(g_2)$ is valid, then the function ϕ is called a ***Lie group homomorphism***. Moreover, if the homomorphism ϕ is also a diffeomorphism, ϕ is then a ***Lie group isomorphism***. For the identity element $e \in G$, we simply obtain

$$\phi(g) = \phi(e * g) = \phi(e) \diamond \phi(g).$$

Hence, the unique identity element e' of the Lie group H will necessarily be

$e' = \phi(e)$. Moreover, we can write $e' = \phi(g{*}g^{-1}) = \phi(g) \diamond \phi(g^{-1})$ so that we deduce the relation $\big(\phi(g)\big)^{-1} = \phi(g^{-1})$. Thus, $\phi(G) \subseteq H$ is a subgroup.

If a left translation on G is L_g, then we obtain

$$\phi\big(L_g(g_1)\big) = \phi(g{*}g_1) = \phi(g) \diamond \phi(g_1) = L_{\phi(g)}\big(\phi(g_1)\big)$$

for all $g, g_1 \in G$ from which it follows that $\phi \circ L_g = L_{\phi(g)} \circ \phi : G \to H$ for all $g \in G$. The expression (2.7.7) now leads to the rule

$$d\phi \circ dL_g = dL_{\phi(g)} \circ d\phi. \tag{3.4.1}$$

Let us consider a left-invariant vector field V on G. Since V satisfies the relation $dL_g(V) = V$, (3.4.1) now yields the result

$$d\phi\big(dL_g(V)\big) = d\phi(V) = dL_{\phi(g)}\big(d\phi(V)\big)$$

valid for all $g \in G$. This means that the vector field $d\phi(V)$ is a left-invariant vector field of H on the subgroup $\phi(G) \subseteq H$. Let V_1, V_2 be left-invariant vector fields on G. On taking into account the relation $dL_g([V_1, V_2]) = [V_1, V_2]$, (3.4.1) leads to the conclusion

$$d\phi\big(dL_g([V_1, V_2])\big) = d\phi([V_1, V_2]) = dL_{\phi(g)}\big(d\phi([V_1, V_2])\big)$$

which expresses the fact that $d\phi([V_1, V_2]) = [d\phi(V_1), d\phi(V_2)]$ is a left-invariant vector field on H. In other words, images of left-invariant vector fields under the differential mapping $d\phi$ where ϕ is a homomorphism are elements of a Lie algebra on H. Since the homomorphism ϕ transports the identity element e in G to the identity element $\phi(e)$ in H, we find that $d\phi : T_e(G) \to T_{\phi(e)}(H)$. We denote the Lie algebras on G and H by $\mathfrak{g}$ and $\mathfrak{h}$, respectively. Via isomorphisms $\mathcal{G} : \mathfrak{g} \to T_e(G)$ and $\mathcal{H} : \mathfrak{h} \to T_{\phi(e)}(H)$, which we have discussed on p. 182, we can introduce a *linear operator* $\psi = \mathcal{H}^{-1} \circ d\phi \circ \mathcal{G} : \mathfrak{g} \to \mathfrak{h}$. It is straightforward to see that this operator fulfil the relation

$$\psi([V_1, V_2]) = [\psi(V_1), \psi(V_2)] \tag{3.4.2}$$

for all $V_1, V_2 \in \mathfrak{g}$, that is, ψ preserves the Lie product. We thus conclude that ψ so defined is a Lie algebra homomorphism. The image $\psi(\mathfrak{g})$ of $\mathfrak{g}$ is clearly a *subalgebra* of $\mathfrak{h}$.

When ϕ is an isomorphism, ψ turns out to be likewise an isomorphism and we find that $\mathfrak{h} = \psi(\mathfrak{g})$. In that situation, if the set of vector fields $\{V_i\}$ is a basis for the Lie algebra $\mathfrak{g}$, then the set of vector fields $\{\psi(V_i)\}$ becomes a basis for the Lie algebra $\mathfrak{h}$. Because of relations $[V_i, V_j] = c_{ij}^k V_k$ it follows from (3.4.2) that

$$[\psi(V_i), \psi(V_j)] = \psi([V_i, V_j]) = \psi(c_{ij}^k V_k) = c_{ij}^k \psi(V_k). \quad (3.4.3)$$

Hence, *such an isomorphism preserves structure constants.*

3.5. ONE-PARAMETER SUBGROUPS

We consider a Lie group G. As is well known, the set $\mathbb{R}$ is an Abelian Lie group with respect to the operation of addition. *Let $\phi : \mathbb{R} \to G$ be a Lie group homomorphism. The subset $\{\phi(t) : t \in \mathbb{R}\} = \phi(\mathbb{R}) \subseteq G$ is called a one-parameter subgroup of G.* By definition, the function ϕ must satisfy the condition

$$\phi(t+s) = \phi(t)*\phi(s) = \phi(s)*\phi(t) \quad (3.5.1)$$

for all $s, t \in \mathbb{R}$ because $t + s = s + t$. Therefore, one-parameter subgroups would necessarily be commutative. Inasmuch as ϕ is a homomorphism, we observe that $e = \phi(0)$ and $\big(\phi(t)\big)^{-1} = \phi(-t)$. The smooth function ϕ will evidently describe a smooth curve on the manifold G through the point e.

Theorem 3.5.1. *A curve on a Lie group G is a one-parameter subgroup if and only if it is an integral curve of a left-invariant or a right-invariant vector field through the identity element e.*

Let $\phi : \mathbb{R} \to G$ give rise to a one-parameter subgroup. As in (2.9.1), we represent symbolically a tangent vector at an element $g = \phi(t)$ in the following manner

$$V\big(\phi(t)\big) = \frac{d\phi(t)}{dt}. \quad (3.5.2)$$

Owing to the formula $L_{\phi(t)}\big(\phi(s)\big) = \phi(t)*\phi(s) = \phi(t+s)$, the vector field $V\big(\phi(t)\big)$ under the differential operator $dL_{\phi(t)}$ must satisfy the relation

$$dL_{\phi(t)}\Big(\frac{d\phi(s)}{ds}\Big) = \frac{d\phi(t+s)}{ds} = \frac{d\phi(t+s)}{dt}$$

If we insert $s = 0$ into this expression, we obtain

$$dL_{\phi(t)}\big(V(e)\big) = V\big(\phi(t)\big).$$

which indicates that (3.5.2) is a left-invariant vector field. It is evident that $\phi(t)$ is an integral curve of this vector field through the point $e \in G$. If we associate each point $g \in G$ with a curve defined by

$$\phi_g(t) = g*\phi(t)$$

we produce a congruence on G that is tangent to the left-invariant vector field V. However, it is evident that only the curve of this congruence through the point e corresponds to a one-parameter subgroup.

Conversely, let us now consider a left-invariant vector field $V \in \mathfrak{g}$. This vector field associated with a vector in $T_e(G)$ generates a flow on G whose member through the identity element $e \in G$ will be given just like in (2.9.11) by

$$g_t(e) = e^{tV}(e) \in G. \tag{3.5.3}$$

If we make use of the relation (2.9.17) it follows from (3.5.3) that

$$\begin{aligned} g_t * g_s &= L_{g_t}(g_s) = L_{g_t}\big(e^{sV}(e)\big) = e^{s\,dL_{g_t}(V)}L_{g_t}(e) = e^{sV}(g_t) \\ &= e^{sV}e^{tV}(e) = e^{tV}(e) * e^{sV}(e) = e^{(t+s)V}(e) = g_{t+s}. \end{aligned}$$

This clearly shows that the subset (3.5.3) is a one-parameter subgroup, and we have $e = g_0$ and $(g_t)^{-1} = g_{-t}$.

The case of right-invariant vector fields can be treated in exactly the same manner. □

Let $\phi : \mathbb{R} \to G$ be a one-parameter subgroup. If we write $g(t) = \phi(t)$, this subgroup gives rise to a one-parameter group of transformations of left translations $\{L_{g(t)} : G \to G : t \in \mathbb{R}\}$. At $t = 0$ or equivalently at $g = e$, the tangent vector is determined by $V(e) = dg/dt|_{t=0}$. Hence, the vector field generating this group is found to be

$$\frac{dL_{g(t)}(h)}{dt}\bigg|_{t=0} = \frac{dR_h\big(g(t)\big)}{dt}\bigg|_{t=0} = dR_h\frac{dg}{dt}\bigg|_{t=0} = dR_h\big(V(e)\big) = V^R(h).$$

Thus, *it is a right-invariant vector field.* Similarly, one demonstrates that the generator of a one-parameter group of transformations of right translations $\{R_{g(t)} : G \to G : t \in \mathbb{R}\}$ is a left-invariant vector field:

$$\frac{dR_{g(t)}(h)}{dt}\bigg|_{t=0} = \frac{dL_h(g(t))}{dt}\bigg|_{t=0} = dL_h\big(V(e)\big) = V^L(h). \tag{3.5.4}$$

Exponential mapping $\exp : \mathfrak{g} \to G$ is defined by taking $t = 1$ in the one-parameter group (3.5.3) generated by a vector field $V \in \mathfrak{g}$ as follows:

$$\exp(V) = g_1(V) = e^V(e) \in G.$$

This definition leads automatically to $\exp(\mathbf{0}) = e$. If we regard the vector space $\mathfrak{g}$ as a manifold, its tangent spaces $T_g(\mathfrak{g})$ will be the same everywhere and they will be isomorphic to $\mathfrak{g}$. On the other hand, the tangent space

$T_e(G)$ at the point e is isomorphic to $\mathfrak{g}$. Since the tangent vector field of the curve defined by (3.5.3) is V, the differential of the exponential mapping at the vicinity of the vector $V = 0$ becomes

$$d\exp : \mathfrak{g} \to T_e(G) \simeq \mathfrak{g}$$

yielding $d\exp(V) = V$. The symbol $\simeq$ denotes isomorphism. We thus obtain the identity mapping $d\exp|_{V=0} = i_{\mathfrak{g}}$. We then conclude that at $V = 0$, $d\exp$ is a regular linear operator. This, of course, indicates that the function exp is a *local* diffeomorphism from the Lie algebra $\mathfrak{g}$ to an open neighbourhood of the identity element e of the Lie group G. Therefore, in a neighbourhood U_e of e, a group element g may be expressible in the form

$$g = \exp(V) = \exp(t^i V_i) \in U_e \subseteq G$$

where the set $\{V_1, , \dots, V_n\}$ is a basis for the Lie algebra. The ordered n-tuple of real numbers $(t^1, \dots, t^n) \in \mathbb{R}^n$ are called the *canonical coordinates* of g and they must be sufficiently small in order that $g \in U_e$. Owing to some properties of the exponential mapping illustrated in p. 139, g can also be written in the following way for sufficiently small canonical coordinates t_i, $i = 1, \dots, n$

$$g = \exp(t^1 V_1) {*} \exp(t^2 V_2) {*} \cdots {*} \exp(t^n V_n).$$

because we can always choose commuting basis vectors for the Lie algebra. This amounts to say that the Lie algebra determines locally the Lie group at a neighbourhood of the group's identity element. That is the reason why a basis of a Lie algebra is called as ***infinitesimal generators*** of a Lie group. As we have mentioned before, it cannot be claimed that a given Lie algebra generates a uniquely determined global Lie group. However, if a Lie group is a connected manifold in which e has a simply connected neighbourhood, then the Lie algebra determines globally this group [*see* Theorem 3.2.1].

We now try to get the isomorphism between Lie algebras $\mathfrak{g}_L$ and $\mathfrak{g}_R$ whose existence was established on p. 185 to acquire a more concrete structure and we shall show that this isomorphism is provided by the differential $d\iota : T_g(G) \to T_{g^{-1}}(G)$ of the *inversion diffeomorphism* $\iota : G \to G$ that was defined by $\iota(g) = g^{-1}$. Since we can write

$$(\iota \circ L_g)(h) = (g{*}h)^{-1} = h^{-1}{*}g^{-1} = R_{g^{-1}}(h^{-1})$$

for all $h \in G$, we obtain $(\iota \circ L_g)(e) = R_{g^{-1}}(e)$ for $h = h^{-1} = e$ from which it follows that $d\iota \circ dL_g|_e = dR_{g^{-1}}|_e$ for all $g \in G$. For a vector $V \in T_e(G)$, this equality naturally implies that $d\iota \circ dL_g(V|_e) = dR_{g^{-1}}(V|_e)$ resulting in the relation

$$d\iota\big(V^L(g)\big) = V^R(g^{-1})$$

for all $g \in G$ where V^L and V^R are left and right invariant vectors. Hence, despite an apparent problem in the arguments, we may expect that the operator $d\iota : \mathfrak{g}_L \to \mathfrak{g}_R$ can be a possible candidate for the isomorphism that we are hoping to find. On the other hand, if a vector field V generates the one-parameter subgroup by $g(t) = \exp{(tV)}$, we have $g(t)^{-1} = \exp{(-tV)}$. Thus the tangent vectors to curves $g(t)$ and $g(t)^{-1}$ at the identity element e are prescribed by

$$\left.\frac{dg(t)}{dt}\right|_{t=0} = V\big|_e, \quad \left.\frac{dg(t)^{-1}}{dt}\right|_{t=0} = -V\big|_e.$$

Hence, at the identity element the operator $d\iota\big|_e : T_e(G) \to T_e(G)$ acts in the manner $d\iota(V\big|_e) = -V\big|_e$. It is then straightforward to realise that a right-invariant vector field produced by a vector $V \in T_e(G)$ has to satisfy the relation $V^R(g^{-1}) = -V^R(g)$. Therefore, the isomorphism between $\mathfrak{g}_L$ and $\mathfrak{g}_R$ is now provided by

$$d\iota\big(V^L(g)\big) = -V^R(g). \tag{3.5.5}$$

Whenever $V_1, V_2 \in T_e(G)$, we can define a vector $W = [V_1, V_2] \in T_e(G)$. We know that the left-invariant vector fields associated with these vectors will satisfy the relation $W^L = [V_1^L, V_2^L]$. We thus find

$$-W^R = d\iota(W^L) = d\iota([V_1^L, V_2^L]) = [d\iota(V_1^L), d\iota(V_2^L)] = [V_1^R, V_2^R]$$

that leads easily to the result $[V_1, V_2]_R = -[V_1, V_2]_L$ from which we deduce that if the structure constants of the left Lie algebra are c_{ij}^k, then the structure constants of the right Lie algebra has to be $-c_{ij}^k$.

Let $\mathfrak{g}$ be an n-dimensional Lie algebra of a Lie group G. A subalgebra $\mathfrak{h}$ of this algebra with dimension $m < n$ is again a Lie algebra. In other words, it is an involutive distribution. Therefore, according to the Frobenius theorem it generates an m-dimensional smooth submanifold through the point e. This submanifold is locally an m-parameter Lie group that is a subgroup of G.

Example 3.5.1. We know that the Lie algebra $\mathfrak{gl}(n)$ of the general linear group $GL(n,\mathbb{R})$ consists of $n \times n$ matrices. Hence, we can express a matrix $\mathbf{X} \in GL(n,\mathbb{R})$ in a neighbourhood of the identity element $\mathbf{I}$ by

$$\mathbf{X} = \exp{(\mathbf{A})} = e^{\mathbf{A}}(\mathbf{I})$$

where $\mathbf{A} \in \mathfrak{gl}(n)$. Let us now consider the function $\det : GL(n,\mathbb{R}) \to \mathbb{R}$. In view of the relation (2.9.17), we have

$$\det\left(e^{\mathbf{A}}(\mathbf{I})\right) = e^{d\det(\mathbf{A})}\det(\mathbf{I}) = e^{d\det(\mathbf{A})}.$$

Let us write $\mathbf{A} = a^i_j\,\partial/\partial a^i_j$. Then the relation (2.7.9) yields

$$d\det(\mathbf{A}) = a^i_j \left.\frac{\partial\det(\mathbf{X})}{\partial a^i_j}\right|_{\mathbf{X}=\mathbf{I}}.$$

Due to the equality $\partial\det(\mathbf{X})/\partial x^k_l = \text{cofactor}\,(x^k_l) = X^l_k = \det(\mathbf{X})\,(\mathbf{X}^{-1})^l_k$, we easily arrive at the expression

$$a^i_j\frac{\partial\det(\mathbf{X})}{\partial a^i_j} = \det(\mathbf{X})\,(\mathbf{X}^{-1})^l_k\,\frac{\partial x^k_l}{\partial a^i_j}a^i_j.$$

Inasmuch as we define $\mathbf{X}$ as the following series

$$\mathbf{X} = \exp(\mathbf{A}) = \mathbf{I} + \mathbf{A} + \frac{1}{2!}\mathbf{A}^2 + \cdots + \frac{1}{n!}\mathbf{A}^n + \cdots,$$

then its entries are prescribed by

$$x^k_l = \delta^k_l + a^k_l + \frac{1}{2!}a^k_m a^m_l + \cdots + \frac{1}{n!}a^k_{m_1}a^{m_1}_{m_2}\cdots a^{m_{n-2}}_{m_{n-1}}a^{m_{n-1}}_l + \cdots.$$

Taking into account the relation

$$\begin{aligned}\frac{\partial a^k_{m_1}a^{m_1}_{m_2}\cdots a^{m_{n-1}}_l}{\partial a^i_j}a^i_j &= \delta^k_i\delta^j_{m_1}a^{m_1}_{m_2}\cdots a^{m_{n-1}}_l a^i_j + \delta^{m_1}_i\delta^j_{m_2}a^k_{m_1}\cdots a^{m_{n-1}}_l a^i_j \\ &\cdots + \delta^{m_{n-1}}_i\delta^j_l a^k_{m_1}a^{m_1}_{m_2}\cdots a^i_j = a^k_{m_1}a^{m_1}_{m_2}\cdots a^{m_{n-1}}_l + a^k_{m_1}a^{m_1}_{m_2}\cdots a^{m_{n-1}}_l \\ &+ a^k_{m_1}a^{m_1}_{m_2}\cdots a^{m_{n-1}}_l + \cdots + a^k_{m_1}a^{m_1}_{m_2}\cdots a^{m_{n-1}}_l = n(\mathbf{A}^n)^k_l,\end{aligned}$$

we finally find

$$\frac{\partial x^k_l}{\partial a^i_j}a^i_j = \left[\mathbf{A} + \mathbf{A}^2 + \cdots\frac{1}{(n-1)!}\mathbf{A}^n + \cdots\right]^k_l = (\mathbf{X}\mathbf{A})^k_l = (\mathbf{X})^k_m(\mathbf{A})^m_l$$

and reach to the conclusion

$$d\det(\mathbf{A}) = \det(\mathbf{X})\,(\mathbf{X}^{-1})^l_k(\mathbf{X})^k_m(\mathbf{A})^m_l\big|_{\mathbf{X}=\mathbf{I}} = \delta^l_m\boldsymbol{a}^m_l = a^m_m = \text{tr}\,(\mathbf{A}).$$

We thus obtain the rather elegant result

$$\det(e^{\mathbf{A}}) = e^{\text{tr}\,(\mathbf{A})}.$$

If the matrix $\mathbf{X}$ belongs to the subgroup $SL(n,\mathbb{R})$, then we must have $\det(\mathbf{X}) = 1$. Hence, if the matrix $\mathbf{A}$ is an element of the Lie subalgebra $\mathfrak{sl}(n)$, the condition

$$\det(e^{\mathbf{A}}) = e^{\operatorname{tr}(\mathbf{A})} = 1$$

must hold. This requires that $\operatorname{tr}(\mathbf{A}) = 0$. Consequently, the Lie algebra $\mathfrak{sl}(n)$ consists $n \times n$ traceless matrices.

Next, we consider the orthogonal group $O(n)$. If the matrix $\mathbf{X}$ belongs to that subgroup, the relation $\mathbf{X}\mathbf{X}^{\mathrm{T}} = \mathbf{X}^{\mathrm{T}}\mathbf{X} = \mathbf{I}$ must be satisfied. Let us take again $\mathbf{X} = e^{\mathbf{A}}$. It can easily be verified that $\mathbf{X}^{\mathrm{T}} = (e^{\mathbf{A}})^{\mathrm{T}} = e^{\mathbf{A}^{\mathrm{T}}}$. We thus obtain the condition

$$e^{\mathbf{A}}e^{\mathbf{A}^{\mathrm{T}}} = e^{\mathbf{A}^{\mathrm{T}}}e^{\mathbf{A}} \quad \text{or} \quad [e^{\mathbf{A}}, e^{\mathbf{A}^{\mathrm{T}}}] = \mathbf{0}.$$

But this leads to the conclusion $[\mathbf{A}, \mathbf{A}^{\mathrm{T}}] = \mathbf{0}$ [*see p.* 148]. Hence the relation

$$\mathbf{X}\mathbf{X}^{\mathrm{T}} = e^{\mathbf{A}}e^{\mathbf{A}^{\mathrm{T}}} = e^{\mathbf{A}+\mathbf{A}^{\mathrm{T}}} = \mathbf{I}$$

requires that $\mathbf{A} + \mathbf{A}^{\mathrm{T}} = \mathbf{0}$, or $\mathbf{A}^{\mathrm{T}} = -\mathbf{A}$. Therefore, the Lie algebra $\mathfrak{o}(n)$ of the orthogonal group consists of antisymmetric $n \times n$ matrices. ■

3.6. ADJOINT REPRESENTATION

Let G be a Lie group. We choose an element $g \in G$ and define a mapping $\mathcal{I}_g : G \to G$ by the operation of *conjugation* prescribed by

$$\mathcal{I}_g(h) = g{*}h{*}g^{-1} \in G \tag{3.6.1}$$

for all $h \in G$. It is clear that this mapping is a diffeomorphism, Moreover, because it satisfies the relation

$$\mathcal{I}_g(h_1{*}h_2) = g{*}h_1{*}g^{-1}{*}g{*}h_2{*}g^{-1} = \mathcal{I}_g(h_1){*}\mathcal{I}_g(h_2)$$

for all $h_1, h_2 \in G$, it preserves the group operation. Hence, $\mathcal{I}_g$ is an automorphism on G called the ***inner automorphism***. All other automorphisms of G are named as *outer automorphisms*. The composition of two inner automorphisms yield

$$\mathcal{I}_{g_1} \circ \mathcal{I}_{g_2}(h) = g_1{*}g_2{*}h{*}g_2^{-1}{*}g_1^{-1} = (g_1{*}g_2){*}h{*}(g_1{*}g_2)^{-1} = \mathcal{I}_{g_1{*}g_2}(h)$$

for all $h \in G$ from which we deduce that $\mathcal{I}_{g_1} \circ \mathcal{I}_{g_2} = \mathcal{I}_{g_1{*}g_2}$. We immediately see that $\mathcal{I}_e = i_G$ is the identity mapping. Since $\mathcal{I}_g \circ \mathcal{I}_{g^{-1}} = \mathcal{I}_{g^{-1}} \circ \mathcal{I}_g = \mathcal{I}_e = i_G$ we realise that $(\mathcal{I}_g)^{-1} = \mathcal{I}_{g^{-1}}$. Furthermore, we obtain $\mathcal{I}_g(e) = e$ for all $g \in G$. Therefore all inner automorphisms $\{\mathcal{I}_g : g \in G\}$ transform any curve on the manifold G through the identity element e to another curve passing again through e. The definition (3.6.1) leads to

$$\mathcal{I}_g(h) = L_g\big(R_{g^{-1}}(h)\big) = R_{g^{-1}}\big(L_g(h)\big)$$

for all $h \in G$. We then conclude that

$$\mathcal{I}_g = L_g \circ R_{g^{-1}} = R_{g^{-1}} \circ L_g. \tag{3.6.2}$$

It is now obvious that the set $\mathcal{G} = \{\mathcal{I}_g : g \in G\}$ constitutes a group with respect to the composition of mappings. On taking into account properties of the mappings L_g and R_g, it is easily understood that the expressions (3.6.2) indicate the existence of an isomorphism between this group and the Lie group G.

If G is an Abelian group, then we obtain $\mathcal{I}_g(h) = h$ for each $g \in G$ so that we get $\mathcal{I}_g = i_G$. Hence, in commutative groups the mapping $\mathcal{I}_g$ acquires quite a trivial structure.

Let us now consider the differential $d\mathcal{I}_g$. (3.6.2) yields naturally

$$d\mathcal{I}_g = dL_g \circ dR_{g^{-1}} = dR_{g^{-1}} \circ dL_g. \tag{3.6.3}$$

If $V \in \mathfrak{g}$ is a left-invariant vector field, then it follows from (3.6.3) that

$$d\mathcal{I}_g(V) = dR_{g^{-1}} \circ dL_g(V) = dR_{g^{-1}}(V) = dL_g \circ dR_{g^{-1}}(V)$$

that may be expressed in the way

$$dL_g\big(d\mathcal{I}_g(V)\big) = d\mathcal{I}_g(V).$$

Thus $d\mathcal{I}_g(V)$ becomes also a left-invariant vector field so that we can write $d\mathcal{I}_g(V) \in \mathfrak{g}$ and conclude that $d\mathcal{I}_g : \mathfrak{g} \to \mathfrak{g}$. Since $\mathcal{I}_g$ is a diffeomorphism, its differential $d\mathcal{I}_g$ is a regular linear operator, i.e., an isomorphism. For all vectors $V_1, V_2 \in \mathfrak{g}$, we have $d\mathcal{I}_g([V_1, V_2]) = [d\mathcal{I}_g(V_1), d\mathcal{I}_g(V_2)]$. Therefore, the isomorphism $d\mathcal{I}_g$ preserves the Lie product. In other words, it is an automorphism on the Lie algebra $\mathfrak{g}$. Thus, to each element $g \in G$, there corresponds an automorphism on the Lie algebra $\mathfrak{g}$. Let us denote the linear vector space formed by these automorphism, or to be more concrete, by regular matrices representing these automorphisms, as $Aut(\mathfrak{g})$. We now rename the operator $d\mathcal{I}_g$ as $d\mathcal{I}_g = Ad_g : \mathfrak{g} \to \mathfrak{g}$ for convenience. Let us next introduce the mapping

$$Ad : G \to Aut(\mathfrak{g})$$

in the following manner: $Ad(g) = Ad_g \in Aut(\mathfrak{g})$ for each $g \in G$. On the other hand, one can easily verify that the equality $d\mathcal{I}_{g_1} \circ d\mathcal{I}_{g_2} = d\mathcal{I}_{g_1 * g_2}$ entails the relation

$$Ad(g_1 * g_2) = Ad_{g_1} \circ Ad_{g_2}.$$

Hence Ad is a group homomorphism assigning to each element of the group G a matrix representing an automorphism. That is the reason why it is called the ***adjoint representation*** of the Lie group G over the Lie algebra $\mathfrak{g}$. One the most outstanding successes of the group theory was to predict that every abstract group is homomorphic to a general linear group $GL(n, \mathbb{R})$ which is called a ***representation*** or more precisely an ***unfaithful representation*** of the group. Whenever this homomorphism is an isomorphism, we obtain a ***faithful representation***. *The theory of group representation* deals with the quite difficult, but practically very important problem of determining the number n and the specific form of matrices involved in such a representation.

It is straightforward to verify that what we have discussed above would be equally valid when we replace an element g by its inverse g^{-1} in case Lie algebra is derived from right-invariant vector fields.

Let V be a *left-invariant* vector field. We consider the one-parameter subgroup $\exp(tV)$ produced by V. If we recall (2.9.17), we observe that we can write

$$\begin{aligned} \mathcal{I}_g(\exp(tV)) &= g * \exp(tV) * g^{-1} = \mathcal{I}_g(e^{tV}(e)) \\ &= e^{t\, d\mathcal{I}_g(V)}(\mathcal{I}_g(e)) = \exp(tAd_g(V)) \end{aligned} \quad (3.6.4)$$

for all $g \in G$. This result simply means that under the mapping $\mathcal{I}_g$, the one-parameter subgroup generated by the vector field V is transformed into the one-parameter subgroup generated by the vector field $Ad_g(V)$.

Let us now consider another one-parameter subgroup generated by a left-invariant vector field U whose elements are, of course, given by $g(s) = \exp(sU)$. If we resort to the relation (2.10.16), we arrive at the following expression

$$Ad_{g(s)}(V) = e^{s£_U} V, \; U, V \in \mathfrak{g} \quad (3.6.5)$$

which measures the change in the vector field $Ad_g(V)$ over the subgroup $\exp(sU)$. By employing (3.6.5), we can evaluate the following expression at the point e:

$$\frac{d}{ds} Ad_{g(s)}(V)\bigg|_{s=0} = ad_U(V) = £_U V = [U, V], \; V \in \mathfrak{g}. \quad (3.6.6)$$

We have already seen that $\mathcal{I}_g = i_G$ if G is an Abelian group. Hence, in this case, we obtain $Ad_g = I_{\mathfrak{g}}$ for all $g \in G$ and (3.6.6) leads to $[U, V] = 0$. Thus the Lie algebra of such a Lie group becomes also Abelian. Conversely, it can be shown that if G is a *connected* Lie group whose Lie algebra is Abelian then G, too, will be an Abelian group.

3.7. LIE TRANSFORMATION GROUPS

We assume that we are given a Lie group G of r-parameters and an m-dimensional smooth manifold M. Let us consider a *differentiable* mapping $\Psi : G \times M \to M$ on the product manifold that manifests the *action* of the group G on the manifold M. We thus obtain $\Psi(g, p) = g \blacklozenge p \in M$ for all $g \in G$ and $p \in M$. We can now form a function $\Psi_g : M \to M$ mapping the manifold M onto itself by the relation $\Psi_g(p) = \Psi(g, p)$ where $g \in G$ is a fixed element of the group G. The set $\{\Psi_g : g \in G\}$ will be called a ***Lie transformation group*** if it possesses group properties with respect to composition, that is, when the conditions

(*i*) $\Psi_{g_1} \circ \Psi_{g_2} = \Psi_{g_1 * g_2}$ *or* $g_1 \blacklozenge (g_2 \blacklozenge p) = (g_1 * g_2) \blacklozenge p$ *for* $g_1, g_2 \in G$
(*ii*) *if* $e \in G$ *is the identity element, then* $\Psi_e = i_M$ *is the identity mapping on* M *so that one can write* $\Psi_e(p) = p$ *or* $e \blacklozenge p = p$

are satisfied. Hence, the following properties are valid

$$\Psi_{g_1}(\Psi_{g_2}(p)) = \Psi(g_1, \Psi(g_2, p)) = \Psi(g_1 * g_2, p), \; \Psi_e(p) = \Psi(e, p) = p.$$

It is easy to observe that the foregoing expressions lead to relations

$$\Psi_g^{-1} = \Psi_{g^{-1}}, \;\; \Psi(g^{-1}, \Psi(g, p)) = p \;\text{ or }\; (g^{-1} * g) \blacklozenge p = p.$$

We say that the group G acts ***effectively*** on the manifold M if the relation $\Psi_g(p) = p$ for all $p \in M$ implies $g = e$. If the stronger condition $\Psi_g(p) \neq p$ unless $g = e$ holds, then the group G acts ***freely*** (*without a fixed point*) on the manifold M. If for all $p, q \in M$, there exists an element $g \in G$ such that $\Psi_g(p) = g \blacklozenge p = q$, then the group G acts ***transitively*** on the manifold M.

We now define the set

$$G_p = \{g \in G : \Psi_g(p) = g \blacklozenge p = p\} \subset G$$

for a *fixed point* $p \in M$. It can be demonstrated that G_p is a subgroup of G. If $g \in G_p$, then one has $g \blacklozenge p = p$ from which $g^{-1} \blacklozenge (g \blacklozenge p) = g^{-1} \blacklozenge p$ follows at once. On the other hand, we can write $g^{-1} \blacklozenge (g \blacklozenge p) = (g^{-1} * g) \blacklozenge p = p$ so that we find $g^{-1} \blacklozenge p = p$. Therefore, we find that $g^{-1} \in G_p$. Next, let us consider $g, h \in G_p$, then we obtain $(g{*}h)\blacklozenge p = g\blacklozenge(h\blacklozenge p) = g\blacklozenge p = p$ implying that $g{*}h \in G_p$. Moreover, we observe that $e \in G_p$ since $\Psi_e(p) = p$. Thus G_p is a subgroup of G. The subgroup G_p so defined is known as the ***isotropy group*** of a point $p \in M$. The isotropy groups at the points $p \in M$ and $g\blacklozenge p \in M$ are connected by the conjugation relation

$$G_{g\blacklozenge p} = g{*}G_p{*}g^{-1}, \text{ for any } g \in G.$$

Indeed, for any $h \in G_p$, one deduce $(g{*}h{*}g^{-1})\blacklozenge(g\blacklozenge p) = g\blacklozenge(h\blacklozenge p) = g\blacklozenge p$ so that $g{*}h{*}g^{-1} \in G_{g\blacklozenge p}$. This, of course, means that $g{*}G_p{*}g^{-1} \subseteq G_{g\blacklozenge p}$. Now, consider an element $h \in G_{g\blacklozenge p}$ so that $h\blacklozenge(g\blacklozenge p) = (h{*}g)\blacklozenge p = g\blacklozenge p$ or $(g^{-1}{*}h{*}g)\blacklozenge p = p$ implying that $g^{-1}{*}h{*}g \in G_p$ and $g^{-1}{*}G_{g\blacklozenge p}{*}g \subseteq G_p$ from which we immediately obtain $G_{g\blacklozenge p} \subseteq g{*}G_p{*}g^{-1}$. Thus we arrive at the desired equality given above. However, the statement $g^{-1}{*}h{*}g \in G_p$ for all $h \in G_p$ and for all $g \in G$ implies that G_p is a *normal subgroup*. If G is a freely acting group, then it is clear that $G_p = \{e\}$ at each point $p \in M$.

The ***orbit*** of the group G at a point $p_0 \in M$ is defined as the set

$$\{g\blacklozenge p_0 : g \in G\} = \mathcal{O}_{p_0} \subseteq M.$$

When $p, q \in \mathcal{O}_{p_0}$, then there are $g_1, g_2 \in G$ such that one has $p = g_1\blacklozenge p_0$ and $q = g_2\blacklozenge p_0$. Consequently, we can write $q = (g_2{*}g_1^{-1})\blacklozenge p$. Thus, the group G acts transitively on any orbit $\mathcal{O}_{p_0}$.

Example 3.7.1. Let us consider the smooth manifold $M = \mathbb{R}^n$ and the Lie group $G = GL(n, \mathbb{R})$. If $\mathbf{x} = (x^1, x^2, \ldots, x^n) \in \mathbb{R}^n$ and $\mathbf{A} \in GL(n, \mathbb{R})$, we define the group action on the manifold by $\Psi(\mathbf{A}, \mathbf{x}) = \mathbf{A}\mathbf{x} \in \mathbb{R}^n$. Hence, the isotropy group of a point $\mathbf{x}_0 \in \mathbb{R}^n$ is determined by the following set

$$G_{\mathbf{x}_0} = \{\mathbf{A} \in GL(n, \mathbb{R}) : \mathbf{A}\mathbf{x}_0 = \mathbf{x}_0\}.$$

Thus, elements of the isotropy group can only be $n \times n$ matrices with an eigenvalue 1 and admitting the vector $\mathbf{x}_0$ as an eigenvector associated with that eigenvalue. Therefore, the necessary condition imposed on matrices $\mathbf{A}$ should be $\det(\mathbf{A} - \mathbf{I}) = 0$. For instance, the isotropy group of the point $\mathbf{x}_0 = (1, 0, \ldots, 0)$ consists of matrices of the form

$$\mathbf{A} = \begin{bmatrix} 1 & a_{12} & \cdots & a_{1n} \\ 0 & a_{22} & \cdots & a_{2n} \\ \vdots & \vdots & \vdots & \vdots \\ 0 & a_{n2} & \cdots & a_{nn} \end{bmatrix}, \quad \det\mathbf{A} = \begin{vmatrix} a_{22} & \cdots & a_{2n} \\ \vdots & \vdots & \vdots \\ a_{n2} & \cdots & a_{nn} \end{vmatrix} \neq 0.$$

Obviously, the condition $\det(\mathbf{A} - \mathbf{I}) = 0$ are satisfied. We can easily verify that the orbit of a point $\mathbf{x}_0$ is $\mathcal{O}_{\mathbf{x}_0} = M - \{\mathbf{0}\}$. Indeed, if $\mathbf{x} \neq \mathbf{0} \in M$ is an arbitrary point, we can always construct a matrix $\mathbf{A} \in GL(n, \mathbb{R})$ so as the relation $\mathbf{A}\mathbf{x}_0 = \mathbf{x}$ is satisfied. If we choose $n^2 - n$ entries of that matrix arbitrarily, then the foregoing n equation will help determine the remaining n entries. For example, if $x^1 x^2 \cdots x^n \neq 0$ and $x_0^1 x_0^2 \cdots x_0^n \neq 0$, then we may choose a diagonal matrix such that $a_{11} = x^1/x_0^1, \ldots, a_{nn} = x^n/x_0^n$. ■

By employing the smooth function $\Psi : G \times M \to M$ represented by $\Psi(g, p) = g\blacklozenge p$, we can now introduce two functions for a fixed $g \in G$ and a fixed $p \in M$, respectively, and their differentials in the following manner

$$\begin{aligned} \Psi_g &: M \to M,\ \Psi_g(p) = g\blacklozenge p,\ \ d\Psi_g : T_p(M) \to T_{g\blacklozenge p}(M) \quad (3.7.1)\\ \Psi_p &: G \to M,\ \Psi_p(g) = g\blacklozenge p,\ \ d\Psi_p : T_g(G) \to T_{g\blacklozenge p}(M). \end{aligned}$$

Consider a member $V \in T_e(G)$ of the Lie algebra on G. By means of the linear operator $d\Psi_p : T_e(G) \to T_p(M)$, we can construct a vector field $V^K(p) \in T_p(M)$ on M through the relation

$$V^K(p) = d\Psi_p(V). \tag{3.7.2}$$

Such vector fields V^K are called ***Killing vector fields*** after German mathematician Wilhelm Karl Joseph Killing (1847-1923). To attribute a more concrete meaning to Killing vectors, let us consider a right-invariant vector field $V^R \in \mathfrak{g}_R$ produced by the vector $V \in T_e(G)$ through the usual relation $V^R(g) = dR_g(V)$. For each $h \in G$, we can evidently write

$$\Psi_p \circ R_g(h) = \Psi_p(h{*}g) = (h{*}g)\blacklozenge p = h\blacklozenge(g\blacklozenge p) = \Psi_{g\blacklozenge p}(h).$$

Thus, it follows from (3.7.2) that

$$V^K(g \cdot p) = d\Psi_{g\cdot p}(V) = d\Psi_p \circ dR_g(V) = d\Psi_p\big(V^R(g)\big). \tag{3.7.3}$$

Hence, the Lie product of two Killing vector fields becomes

$$[V_i^K, V_j^K]\big|_{g\blacklozenge p} = [d\Psi_p(V_i^R), d\Psi_p(V_j^R)]\big|_g = d\Psi_p[V_i^R, V_j^R]\big|_g.$$

Inasmuch as $[V_i^R, V_j^R] = -\,c_{ij}^k V_k^R$ [*see p. 192*], then we find that

$$[V_i^K, V_j^K] = -\,c_{ij}^k d\Psi_p(V_k^R) = -\,c_{ij}^k V_k^K,\quad i, j, k = 1, \ldots, r.$$

This relation states that Killing vectors, too, constitute a Lie algebra. If an r-dimensional Lie algebra with structure constants c_{ij}^k is given on a smooth manifold M, then we conclude, conversely, that there exists a Lie group G whose Lie algebra has those structure constants with respect to a basis V_1, $\ldots, V_r$ and the local action of G on M is described by the vector fields $V_i^K = d\Psi_p(V_i), i = 1, \ldots, r$.

We know that a particular vector $V \in T_{\mathrm{e}}(G)$ determines a one-parameter subgroup of the group G via the curve $g(t) = \exp(tV) \in G$ and the relation $V = dg(t)/dt|_{t=0}$ is satisfied. The curve $g(t)$ generates a group of mappings on M through $\{\Psi(g(t), p) = \Psi_{g(t)}(p) = \Psi_p(g(t)) : t \in \mathbb{R}\}$. On the other hand, the integral curve passing through the point $p \in M$ has to satisfy the relation

$$\frac{d\Psi_{g(t)}(p)}{dt} = \frac{d\Psi_p(g(t))}{dt} = W[\Psi_p(g(t))], \quad \Psi_p(e) = p.$$

where W is the vector field tangent to the integral curve. Consequently, the tangent vector to that curve at $t = 0$ should be given by

$$W(p) = \left.\frac{d\Psi_p(g(t))}{dt}\right|_{t=0} = d\Psi_p\Big(\left.\frac{dg(t)}{dt}\right|_{t=0}\Big) = d\Psi_p(V) = V^K(p).$$

Therefore the vector field $W(p)$ is a Killing vector field. The dimension s of the Lie algebra of Killing vectors depends on the rank of the linear operator $d\Psi_p$. If the dimension of the vector space $T_{\mathrm{e}}(G)$ is r, then it is known that one can write $r = n(d\Psi_p) + s$ where $n(d\Psi_p)$ is the dimension of the null space of $d\Psi_p$. Thus this relation implies that $s \leq r$. *If only $d\Psi_p$ is injective, namely, if $n(d\Psi_p) = 0$, we obtain $s = r$.* In this case, we have $V^K = 0$ if and only if $V = 0$. This becomes possible if only the group G acts effectively on the manifold M. To demonstrate this statement, let us consider an effectively acting group G and assume that $V^K = 0$ for a vector $V \neq 0$. The vector V now determines a subgroup $g(t) = \exp(tV)$ in G. This subgroup then generates a curve $p(t) = \Psi_{g(t)}(p_0)$ on the manifold M going through the point $p_0 \in M$. Then we find of course

$$V^K(p(t)) = \frac{d\Psi_{g(t)}(p_0)}{dt} = 0$$

implying that $\Psi_{g(t)}(p_0)$ is *constant*. Hence, we obtain $\Psi_{g(t)}(p_0) = \Psi_e(p_0) = p_0$. However, this contradicts the effectiveness of the group G. As a result of this, we find $s = r$. In other words, *if the group G is acting effectively on the manifold M, then $d\Psi_p$ is an isomorphism if $d\Psi_p$ is injective.*

III. EXERCISES

3.1. The circle $\mathbb{S}^1$ is given by $|z| = 1, z \in \mathbb{C}$. Let us consider the smooth manifold $G = \mathbb{R} \times \mathbb{R} \times \mathbb{S}^1$ and define an operation $* : G \times G \to G$ by

$$(x_1, y_1, z_1) * (x_2, y_2, z_2) = (x_1 + x_2, y_1 + y_2, e^{iy_1x_2} z_1 z_2).$$

Show that $(G, *)$ is a Lie group.

3.2. The mapping $\phi : (\mathbb{R}, +) \to SO(2)$ is defined by

$$\phi(\theta) = \begin{bmatrix} \cos\theta & -\sin\theta \\ \sin\theta & \cos\theta \end{bmatrix}.$$

Show that ϕ is a Lie group homomorphism.

3.3. G is a Lie group. Show that $T(G)$ is also a Lie group.

3.4. We define the set of *symplectic matrices* by the following relation

$$Sp(n, \mathbb{R}) = \{\mathbf{A} \in GL(2n, \mathbb{R}) : \mathbf{A}^{\mathrm{T}} \boldsymbol{J} \mathbf{A} = \boldsymbol{J}\},\ \boldsymbol{J} = \begin{bmatrix} \mathbf{0} & \mathbf{I}_n \\ -\mathbf{I}_n & \mathbf{0} \end{bmatrix}$$

where $\mathbf{I}_n$ is the $n \times n$ unit matrix. Let $\mathbf{A}_1, \mathbf{A}_2, \mathbf{A}_3$ and $\mathbf{A}_4$ be $n \times n$ matrices. We introduce a matrix $\mathbf{A}$ by

$$\mathbf{A} = \begin{bmatrix} \mathbf{A}_1 & \mathbf{A}_2 \\ \mathbf{A}_3 & \mathbf{A}_4 \end{bmatrix}.$$

In order that $\mathbf{A} \in Sp(n, \mathbb{R})$, show that the matrices $\mathbf{A}_1^{\mathrm{T}}\mathbf{A}_3$ and $\mathbf{A}_2^{\mathrm{T}}\mathbf{A}_4$ must be symmetric, and the relation $\mathbf{A}_1^{\mathrm{T}}\mathbf{A}_4 - \mathbf{A}_3^{\mathrm{T}}\mathbf{A}_2 = \mathbf{I}_n$ must be satisfied. Prove that $Sp(n, \mathbb{R})$ is a Lie group. Find the dimension of this group. Show that if $\mathbf{A} \in Sp(n, \mathbb{R})$, then one also finds $\mathbf{A}^{\mathrm{T}} \in Sp(n, \mathbb{R})$.

3.5. Show that the Le algebra of the Lie group $Sp(n, \mathbb{R})$ is determined as follows

$$\mathfrak{sp}(n, \mathbb{R}) = \{\mathbf{B} \in gl(2n, \mathbb{R}) : \mathbf{B}^{\mathrm{T}} \boldsymbol{J} + \boldsymbol{J}\mathbf{B} = \mathbf{0}\}.$$

If we choose a matrix $\mathbf{B} \in \mathfrak{sp}(n, \mathbb{R})$ in the following manner

$$\mathbf{B} = \begin{bmatrix} \mathbf{B}_1 & \mathbf{B}_2 \\ \mathbf{B}_3 & \mathbf{B}_4 \end{bmatrix},$$

show that the relations $\mathbf{B}_1^{\mathrm{T}} = -\mathbf{B}_4$, $\mathbf{B}_2^{\mathrm{T}} = \mathbf{B}_2$ and $\mathbf{B}_3^{\mathrm{T}} = \mathbf{B}_3$ must hold.

3.6. Show that all upper triangular $n \times n$ matrices whose entries on the principal diagonal are all 1 constitute a Lie group $T^u(n, \mathbb{R})$. Evaluate the Lie algebra of $T^u(3, \mathbb{R})$.

3.7. Let $\mathbf{S} \in GL(n, \mathbb{R})$ be a given symmetric matrix and consider the set $R_S = \{\mathbf{A} \in GL(n, \mathbb{R}) : \mathbf{A}^{\mathrm{T}}\mathbf{S}\mathbf{A} = \mathbf{S}\}$. (a) Show that R_S is a Lie group with respect to the matrix product. (b) Show that the mapping $\phi : GL(n, \mathbb{R}) \to GL(n, \mathbb{R})$ prescribed by the relation $\phi(\mathbf{A}) = \mathbf{A}^{\mathrm{T}}\mathbf{S}\mathbf{A} - \mathbf{S}$ is a submersion and the set $R_S = \phi^{-1}(\mathbf{0})$ is a submanifold. Find the dimension of R_S. (c) Show that the Lie algebra of R_S is $\mathfrak{r}_{\mathfrak{s}} = \{\mathbf{B} \in gl(n, \mathbb{R}) : \mathbf{B}^{\mathrm{T}}\mathbf{S} + \mathbf{S}\mathbf{B} = \mathbf{0}\}$. Show further that if $\mathbf{B}_1, \mathbf{B}_2 \in \mathfrak{r}_{\mathfrak{s}}$, then the commutator $[\mathbf{B}_1, \mathbf{B}_2] = \mathbf{B}_1\mathbf{B}_2 - \mathbf{B}_2\mathbf{B}_1$ is also an element of the algebra. (d) Which is the group that will be obtained if one chooses $\mathbf{S} = \mathbf{I}$?

3.8. Find bases and structure constants of Lie algebras $\mathfrak{sl}(n,\mathbb{R})$, $\mathfrak{so}(n)$ and $\mathfrak{o}(n)$ of the Lie groups $SL(n,\mathbb{R})$, $SO(n)$ and $O(n)$ for $n=2$ and $n=3$.

3.9. Show that the vector space $\mathbb{R}^3$ acquires a Lie algebra structure with the usual vectorial product $\times$. We define a mapping $\phi:\mathbb{R}^3\to\mathfrak{o}(3)$ by the relation

$$\phi(\mathbf{v})=\begin{bmatrix}0 & -v_3 & v_2\\ v_3 & 0 & -v_1\\ -v_2 & v_1 & 0\end{bmatrix}$$

where $\mathbf{v}=(v_1,v_2,v_3)\in\mathbb{R}^3$. Show that ϕ is a Lie algebra isomorphism and the equalities

$$\phi(\mathbf{u}\times\mathbf{v})=[\phi(\mathbf{u}),\phi(\mathbf{v})],\quad \phi(\mathbf{u})\,\mathbf{v}=\mathbf{u}\times\mathbf{v},\quad \mathbf{u}\cdot\mathbf{v}=-\frac{1}{2}\mathrm{tr}\,[\phi(\mathbf{u})\,\phi(\mathbf{v})]$$

are satisfied where $\cdot$ denotes the standard scalar product. Show that the *length* of a vector is given by $\|\mathbf{u}\|=\sqrt{\mathbf{u}\cdot\mathbf{u}}=-\,\mathrm{tr}\left(\phi(\mathbf{u})^2\right)/2$.

3.10. Show that the mapping $\phi:\mathbb{R}^3\to\mathfrak{o}(3)$ in Ex. **3.9** satisfies the relation

$$e^{\phi(\mathbf{u})}=\mathbf{I}+\frac{\sin\|\mathbf{u}\|}{\|\mathbf{u}\|}\phi(\mathbf{u})+\frac{1-\cos\|\mathbf{u}\|}{\|\mathbf{u}\|^2}\phi(\mathbf{u})^2$$

which is known as *Rodrigues' equality* [after French mathematician Benjamin Olinde Rodrigues (1795-1851)].

3.11. Let $\mathbf{A},\mathbf{B}\in O(3)$. We define the inner automorphism on $O(3)$ by the usual relation $\mathcal{I}_{\mathbf{A}}(\mathbf{B})=\mathbf{A}\mathbf{B}\mathbf{A}^{-1}=\mathbf{A}\mathbf{B}\mathbf{A}^{\mathrm{T}}$. By employing the foregoing mapping ϕ, show that one can write $Ad_{\mathbf{A}}(\phi(\mathbf{u}))=\phi(\mathbf{A}\mathbf{u})$ and the relation [*see* (3.6.6)] $ad_{\phi(\mathbf{u})}(\phi(\mathbf{v}))=[\phi(\mathbf{u}),\phi(\mathbf{v})]$ leads to $\phi(\mathbf{u}\times\mathbf{v})=[\phi(\mathbf{u}),\phi(\mathbf{v})]$.

3.12. Show that an eigenvalue of a matrix $\mathbf{A}\in SO(3)$ must be 1. Exploiting this fact, prove that every matrix $\mathbf{A}\in SO(3)$ corresponds to a rotation by an amount θ about a vector $\mathbf{u}$ in $\mathbb{R}^3$ and by choosing an appropriate basis in $\mathbb{R}^3$ this matrix can be reduced to the form

$$\mathbf{A}=\begin{bmatrix}1 & 0 & 0\\ 0 & \cos\theta & -\sin\theta\\ 0 & \sin\theta & \cos\theta\end{bmatrix}.$$

3.13. The matrices $\mathbf{A},\mathbf{B}\in gl(n,\mathbb{R})$ are satisfying the condition $\mathbf{A}\mathbf{B}+\mathbf{B}\mathbf{A}^{\mathrm{T}}=\mathbf{0}$. Show that

$$e^{t\mathbf{A}}\,\mathbf{B}\,e^{t\mathbf{A}^{\mathrm{T}}}=\mathbf{B}.$$

Discuss the special cases $\mathbf{B}=\mathbf{I}$ and $\mathbf{B}=\boldsymbol{J}$.

3.14. Show that the set of all ***unitary matrices*** that are defined by the relation $U(n)=\{\mathbf{A}\in GL(n,\mathbb{C}):\mathbf{A}\overline{\mathbf{A}}^{\mathrm{T}}=\mathbf{A}\mathbf{A}^{\dagger}=\mathbf{I}_n\}$ constitutes a Lie group whose Lie algebra is given by $\mathfrak{u}(n,\mathbb{C})=\{\mathbf{A}\in gl(n,\mathbb{C}):\mathbf{A}+\mathbf{A}^{\dagger}=\mathbf{0}\}$. Show also that members of this group preserve the standard inner product $z_i\overline{w}_i$ in $\mathbb{C}^n$.

3.15. Show that the set $SU(n) = U(n) \cap SL(n, \mathbb{C})$ is also a Lie group and its Lie algebra is given by $\mathfrak{su}(n) = \{\mathbf{A} \in gl(n, \mathbb{C}) : \mathbf{A} + \mathbf{A}^\dagger = \mathbf{0}, \operatorname{tr} \mathbf{A} = 0\}$.

3.16. Show that the group $U(n)$ is diffeomorphic to $\mathbb{S}^1 \times SU(n)$.

3.17. Show that the $SU(2)$ is a connected manifold and the mapping $\phi : \mathbb{S}^3 \to SU(2)$ between the sphere $\mathbb{S}^3 \subset \mathbb{R}^4$ and $SU(2)$ defined by

$$\phi(\mathbf{x}) = \begin{bmatrix} x_1 + ix_2 & x_3 + ix_4 \\ -x_3 + ix_4 & x_1 - ix_2 \end{bmatrix}, \quad x_1^2 + x_2^2 + x_3^2 + x_4^4 = 1$$

where $\mathbf{x} = (x_1, x_2, x_3, x_4) \in \mathbb{R}^4$ is a diffeomorphism.

3.18. Show that every matrix $\mathbf{A} \in GL(n, \mathbb{R})$ can be represented in the form $\mathbf{A} = \mathbf{Q}_1\mathbf{S}_1 = \mathbf{S}_2\mathbf{Q}_2$ where $\mathbf{S}_1$ and $\mathbf{S}_2$ are positive definite symmetric matrices, and $\mathbf{Q}_1$ and $\mathbf{Q}_2$ are orthogonal matrices. Prove that this operation called ***polar decomposition*** is uniquely determined and $\mathbf{Q}_1 = \mathbf{Q}_2 = \mathbf{Q}$ so that

$$\mathbf{S}_2 = \mathbf{Q}\mathbf{S}_1\mathbf{Q}^{\mathrm{T}}.$$

3.19. Show that every matrix $\mathbf{A} \in GL(n, \mathbb{C})$ can be represented in the form $\mathbf{A} = \mathbf{U}_1\mathbf{S}_1 = \mathbf{S}_2\mathbf{U}_2$ where $\mathbf{S}_1$ and $\mathbf{S}_2$ are Hermitean matrices [A matrix satisfying the condition $\mathbf{A} = \overline{\mathbf{A}}^{\mathrm{T}} = \mathbf{A}^\dagger$ is called a *Hermitean matrix* after French mathematician Charles Hermite (1822-1901)] and $\mathbf{U}_1$, $\mathbf{U}_2$ are unitary matrices. Prove that this operation is uniquely determined and $\mathbf{U}_1 = \mathbf{U}_2 = \mathbf{U}$ so that one gets

$$\mathbf{S}_2 = \mathbf{U}\mathbf{S}_1\mathbf{U}^\dagger.$$

3.20. Show that if a matrix $\mathbf{A}$ satisfies the equality $\mathbf{A} = \mathbf{A}^\dagger$, then the matrix $\mathbf{B} = i\mathbf{A}$ holds the relation $\mathbf{B} = -\mathbf{B}^\dagger$. Utilising this property show that a basis for the Lie algebra $\mathfrak{su}(2)$ can be chosen as $(i\boldsymbol{\sigma}_1, i\boldsymbol{\sigma}_2, i\boldsymbol{\sigma}_3)$ where *Pauli spin matrices* [Austrian physicist Wolfgang Ernst Pauli (1900-1958)] are given as follows

$$\boldsymbol{\sigma}_1 = \begin{bmatrix} 0 & 1 \\ 1 & 0 \end{bmatrix}, \quad \boldsymbol{\sigma}_2 = \begin{bmatrix} 0 & -i \\ i & 0 \end{bmatrix}, \quad \boldsymbol{\sigma}_3 = \begin{bmatrix} 1 & 0 \\ 0 & -1 \end{bmatrix}.$$

Find the structure constants of the Lie algebra (in Quantum mechanics the conventional basis is taken as $i\boldsymbol{\sigma}_k/2$).

3.21. $(\boldsymbol{\sigma}_1, \boldsymbol{\sigma}_2, \boldsymbol{\sigma}_3)$ are the foregoing Pauli spin matrices. We define a mapping $\phi : \mathbb{R}^3 \to \mathfrak{su}(2)$ in the following manner

$$\phi(\mathbf{u}) = \frac{1}{2i} u^k \boldsymbol{\sigma}_k = \frac{1}{2} \begin{bmatrix} -iu^3 & -iu^1 - u^2 \\ -iu^1 + u^2 & iu^3 \end{bmatrix}, \mathbf{u} = (u^1, u^2, u^3) \in \mathbb{R}^3.$$

Show that (a) the inverse mapping $\phi^{-1} : \mathfrak{su}(2) \to \mathbb{R}^3$ is provided by

$$u^1 = i\operatorname{tr}(\phi(\mathbf{u})\,\boldsymbol{\sigma}_1), \; u^2 = i\operatorname{tr}(\phi(\mathbf{u})\,\boldsymbol{\sigma}_2), \; u^3 = i\operatorname{tr}(\phi(\mathbf{u})\,\boldsymbol{\sigma}_3),$$

(b) the mapping ϕ is a Lie algebra isomorphism so that the relation $\phi(\mathbf{u} \times \mathbf{v})$

$= [\phi(\mathbf{u}), \phi(\mathbf{v})]$ and (c) the equalities $\|\mathbf{u}\|^2 = -\det(u^k \sigma_k)$ and $\mathbf{u}\cdot\mathbf{v} = 2\mathrm{tr}\,\big(\phi(\mathbf{u})\,\phi(\mathbf{v})\big)$ are satisfied.

3.22. We define a mapping $\phi : SU(2) \to GL(3,\mathbb{R})$ in such a way that for each vector $\mathbf{u} \in \mathbb{R}^3$, the relation

$$\left[\phi(\mathbf{A})\mathbf{u}\right]^i \boldsymbol{\sigma}_i = \mathbf{A}(u^i\boldsymbol{\sigma}_i)\mathbf{A}^{-1}, \quad \mathbf{A} \in SU(2)$$

will be satisfied. Show that (a) ϕ has the properties $\phi(\mathbf{A}) = \phi(-\mathbf{A})$, $\phi(\mathbf{I}_2) = \mathbf{I}_3$ and $\phi(\mathbf{A}) \in SO(3)$ and (b) the mapping $\phi : SU(2) \to SO(3)$ is a submersion.

3.23. According to the celebrated *Hamilton-Cayley theorem* [Irish mathematician Sir William Rowan Hamilton (1805-1865) and English mathematician Arthur Cayley (1821-1895)] every 2×2 matrix $\mathbf{A}$ satisfies its characteristic equation

$$\mathbf{A}^2 - \mathrm{tr}\,(\mathbf{A})\,\mathbf{A} + (\det\mathbf{A})\,\mathbf{I} = \mathbf{0}.$$

Utilising this equation, show that if $\mathbf{A} \in \mathfrak{sl}(2,\mathbb{R})$ and $\det\mathbf{A} = \delta$ the relations

$$e^{\mathbf{A}} = \begin{cases} \cos\sqrt{\delta}\,\mathbf{I} + \frac{1}{\sqrt{\delta}}\sin\sqrt{\delta}\,\mathbf{A}, & \delta > 0 \\ \mathbf{I} + \mathbf{A} & \delta = 0 \\ \cosh\sqrt{-\delta}\,\mathbf{I} + \frac{1}{\sqrt{-\delta}}\sinh\sqrt{-\delta}\,\mathbf{A} & \delta < 0 \end{cases}$$

are valid. Show further that $\mathrm{tr}\,e^{\mathbf{A}} \geq -2$. Verify whether the mapping

$$\exp : \mathfrak{sl}(2,\mathbb{R}) \to SL(2,\mathbb{R})$$

is surjective (*Hint*: consider the matrix $\begin{bmatrix} a & 0 \\ 0 & 1/a \end{bmatrix} \in SL(2,\mathbb{R})$).

3.24. $\phi : G \to H$ is a Lie group isomorphism, and $\psi : \mathfrak{g} \to \mathfrak{h}$ is the Lie algebra isomorphism produced by the mapping ϕ [*see p.* 188]. Show that the exponential mappings $\exp_G : \mathfrak{g} \to G$ and $\exp_H : \mathfrak{h} \to H$ satisfy the equality $\phi \circ \exp_G = \exp_H \circ \psi$. [*Hint*: Define two one-parameter groups and evaluate tangents of the curves $\gamma_1 : t \to \phi\big(\exp_G(tU)\big)$ and $\gamma_2 : t \to \exp_H\big(t\psi(U)$ at $t = 0$]. This means that the following diagram

$$\begin{array}{ccc} G & \xrightarrow{\phi} & H \\ \exp_G \uparrow & & \uparrow \exp_H \\ \mathfrak{g} & \xrightarrow{\psi} & \mathfrak{h} \end{array}$$

commutes.

3.25. We define the length of a vector $V \in T(\mathbb{R}^n)$ by $\|V\|^2 = \sum_{i=1}^{n}(v^i)^2$ where v^i are the components of V. A mapping $\phi : \mathbb{R}^n \to \mathbb{R}^n$ is called an ***isometry*** if the condition $\|\phi_* V\| = \|V\|$ is met for all vectors $V \in T(\mathbb{R}^n)$. Show that a vector field $V = v^i(\mathbf{x})\,\partial_i$ generates a one-parameter group of isometries if and only if its components satisfy the following partial differential equations

$$\frac{\partial v^i}{\partial x^j} + \frac{\partial v^j}{\partial x^i} = 0.$$

Show further that this group is the ***Euclidian group*** $E(n)$ of dimension $n(n+1)/2$ which consists of rotations and translations.

3.26. Show that the set of matrices $G = \left\{ \begin{bmatrix} x & y \\ 0 & 1 \end{bmatrix} \in GL(2,\mathbb{R}) : x, y \in \mathbb{R}, x \neq 0 \right\}$ forms a Lie group with respect to the matrix product. Utilise this group to define an appropriate multiplication on $\mathbb{R}^2$ so that a homomorphism between them can be constructed. Consider two curves with tangent vectors $\partial/\partial x$ and $\partial/\partial y$ at $(1,0) \in \mathbb{R}^2$ and verify that these vectors become $x\partial/\partial x$ and $x\partial/\partial y$ under a left translation so that they generate left-invariant vector fields. Hence, they constitute the Lie algebra $\mathfrak{g}$. Employing the fact that the one-parameter subgroup associated with the vector $c_1\partial/\partial x + c_2\partial/\partial y \in \mathfrak{g}$ must be tangent to the vector field $c_1 x\,\partial/\partial x + c_2 x\,\partial/\partial y$, we deduce that it is expressible by the relation

$$x = e^{c_1 t}, \;\; y = \frac{c_2}{c_1}(e^{c_1 t} - 1)$$

which is also obtainable by evaluating the matrix $\exp t \begin{bmatrix} c_1 & c_2 \\ 0 & 0 \end{bmatrix}$.

3.27. We define the action of the Lie group $SO(3)$ on the manifold $\mathbb{R}^3$ with the mapping $\Psi : SO(3) \times \mathbb{R}^3 \to \mathbb{R}^3$ where Ψ is prescribed by

$$\Psi(\mathbf{A}, \mathbf{u}) = \mathbf{A}\mathbf{u}.$$

Here, $\mathbf{A} \in SO(3)$, $\mathbf{u} \in \mathbb{R}^3$. Discuss the properties of this mapping (freely, effectively or transitively acting) and show that orbits are submanifolds $\mathbb{S}^2$. Determine the Killing vector fields.

3.28. We consider the product manifold $Aff(n,\mathbb{R}) = GL(n,\mathbb{R}) \times \mathbb{R}^n$. We define an operation of multiplication $*$ on $Aff(n,\mathbb{R})$ as follows

$$(\mathbf{A}, \mathbf{u}) * (\mathbf{B}, \mathbf{v}) = (\mathbf{AB}, \mathbf{A}\mathbf{v} + \mathbf{u})$$

where $(\mathbf{A}, \mathbf{u}), (\mathbf{B}, \mathbf{v}) \in Aff(n,\mathbb{R})$. Show that $(Aff(n,\mathbb{R}), *)$ is a Lie group called the ***group of affine motions***. Let us further introduce the mapping $\Psi : Aff(n,\mathbb{R}) \times \mathbb{R}^n \to \mathbb{R}^n$ representing the action of this group on the manifold $\mathbb{R}^n$ by the relation $\Psi\big((\mathbf{A}, \mathbf{u}), \mathbf{v}\big) = \mathbf{A}\mathbf{v} + \mathbf{u}$. Ψ is called an ***affine mapping***. Discuss its properties (freely, effectively or transitively acting).

3.29. $\mathbb{Z}$ is the set of integers. Let us consider the group $G = \mathbb{Z}^2$ and the manifold $M = \mathbb{R}^2$, and define the mapping $\Psi : G \times M \to M$ by

$$\Psi\big((a,b),(x,y)\big) = (ax - by, bx).$$

Discuss the properties of this mapping (freely, effectively or transitively acting)

3.30. The action of the Lie group G on the manifold M is given by the mapping $\Psi : G \times M \to M$. Show that each orbit $\mathcal{O}_p$ is a submanifold of M and is diffeomorphic to the quotient manifold G/G_p. G_p is the isotropy group of a point $p \in M$.

3.31. We define the action of the Lie group G on its Lie algebra manifold $M = \mathfrak{g}$ by the mapping $\Psi : G \times \mathfrak{g} \to \mathfrak{g}$ by means of the relation $\Psi(g, V) = Ad_g(V)$. Show that the Killing vector field is then given by $V^K(U) = [V, U]$ where $U \in M$.

CHAPTER IV

TENSOR FIELDS ON MANIFOLDS

4.1. SCOPE OF THE CHAPTER

In this chapter, tensors[1] that were defined previously on linear vector spaces and their duals will be restructured as tensor fields in such a way that they would inhabit in a natural fashion on differentiable manifolds. To this end, we first construct in Sec. 4.2 the *cotangent bundle* by conjoining the dual space of the tangent space at each point of the manifold to this point. That fibre bundle is then equipped with a differentiable structure to make it a smooth manifold. Afterwards it is demonstrated in Sec. 4.3 that multilinear functionals on certain Cartesian products of tangent spaces and their duals at a point of the manifold are represented by elements called contravariant and covariant tensors of a vector space defined as some tensor products of these spaces. The basis of a tensor product space is determined as usual as tensor products of natural bases for a tangent space and its dual. A tensor bundle is built by attaching the associated tensor product vector space to each point of the manifold. Tensor fields are obtained as sections of the tensor bundle. A tensor is now being completely determined through its components on natural bases in tangent spaces and their duals. Transformation rules of these components under the change of local coordinates are then derived quite easily. An exterior form field on a manifold will then reasonably be defined as a completely antisymmetric covariant tensor field and, as it should be, the concept of exterior products is linked to the alternation of tensor products. The contraction is defined as an operation that produces an associated tensor to a given tensor whose order is reduced by two compared to the original tensor. After that, the quotient rule that helps us to recognise whether a given indicial quantity are actually components of a particular tensor is discussed. Finally, the Lie derivative of

[1]The term 'tensor' was first used in the present context by the German physicist Woldemar Voigt (1850-1919) in 1898 while he was studying crystal elasticities.

Exterior Analysis, DOI: 10.1016/B978-0-12-415902-0.50004-9

tensor products of finitely many vector fields on the tangent bundle is calculated.

Tensor analysis is today an indispensable tool in many branches of mathematics and physics. It was mainly developed by Italian mathematicians Gregorio Ricci-Curbastro (1853-1925) and Levi-Civita, and it has turned out to be a great impetus in the development of the theory of general relativity. A sentence from a letter of Einstein to Levi-Civita around 1917 reflects his appraisal of the tensor analysis: "I admire the elegance of your method of computation; it must be nice to ride through these fields upon the horse of true mathematics while the like of us have to make our way laboriously on foot."

4.2. COTANGENT BUNDLE

We consider an m-dimensional smooth manifold M and the tangent space $T_p(M)$ at a point $p \in M$. As is well known, the dual of the tangent space is a linear vector space formed by all linear functionals on the tangent space [*see* p. 11]. We denote this m-dimensional dual space by $T_p^*(M)$ and we also call it the ***cotangent space*** at the point p. When we choose the natural basis of the tangent space at the point p as the vectors $\{\partial/\partial x^i : i = 1,\dots,m\}$ generated by the local coordinates in the chart containing the point p, we have seen on p. 125 that reciprocal basis vectors in the dual space are given by linear functionals as differentials $\{dx^i : i = 1,\dots,m\}$ so that the following relations

$$dx^i\Big(\frac{\partial}{\partial x^j}\Big) = \Big\langle dx^i, \frac{\partial}{\partial x^j}\Big\rangle = \delta^i_j \tag{4.2.1}$$

are satisfied. Hence, at a point $p \in M$, a vector $V \in T_p(M)$ and a linear functional $\omega \in T_p^*(M)$ can be expressed as

$$V = v^i\frac{\partial}{\partial x^i},\quad \omega = \omega_i dx^i,\quad v^i, \omega_i \in \mathbb{R}. \tag{4.2.2}$$

The value of the functional ω on the vector V at p then happens to be

$$\omega(V) = \langle \omega, V\rangle = \Big\langle \omega_i dx^i, v^j\frac{\partial}{\partial x^j}\Big\rangle \tag{4.2.3}$$
$$= \omega_i v^j \delta^i_j = \omega_i v^i \in \mathbb{R}.$$

We shall call elements of the dual space $T_p^*(M)$ as **1-*forms*** at the point p. Next, we define the set

$$T^*(M) = \bigcup_{p\in M} T_p^*(M) = \{(p,\omega) : p \in M, \omega \in T_p^*(M)\}. \quad (4.2.4)$$

By repeating exactly our approach in Sec. 2.8, we see that $T^*(M)$ can be endowed with a differentiable structure making it a $2m$-dimensional smooth manifold which will be called henceforth as the ***cotangent bundle***. The local coordinates of $T^*(M)$ are evidently given by $\{x^1, \ldots, x^m, \omega_1, \ldots, \omega_m\}$. A *section* of the bundle $T^*(M)$ as we have already done in p. 130 characterises this time a 1-form field on the smooth manifold M. In terms of local coordinates in the relevant chart, this field is of course expressible as follows

$$\omega(p) = \omega_i(\mathbf{x})dx^i \in T^*(M), \;\; \mathbf{x} = \varphi(p). \quad (4.2.5)$$

Different charts containing the point p gives rise to a coordinate transformation given by invertible functions $y^i = y^i(x^j)$. When we write the 1-form ω in different local coordinates, the relation

$$\omega(p) = \omega_j dx^j = \omega'_i dy^i = \omega'_i \frac{\partial y^i}{\partial x^j} dx^j$$

leads to the following relations between components of ω in two different coordinate systems

$$\omega_j = \omega'_i \frac{\partial y^i}{\partial x^j} \quad \text{or} \quad \omega'_i = \frac{\partial x^j}{\partial y^i}\omega_j. \quad (4.2.6)$$

Because of this transformation rule, the elements of the cotangent bundle are usually called ***covariant vector*** or ***covector fields***. We have already seen that the transformation rule between components of vectors in two different charts in the tangent bundle are given by [*see* (2.6.9)]

$$v'^i = \frac{\partial y^i}{\partial x^j} v^j. \quad (4.2.7)$$

That is the reason why we call vectors in the tangent bundle as ***contravariant vector fields***.

4.3. TENSOR FIELDS

Let us consider an m-dimensional smooth manifold M and vector spaces $T_p(M)$ and $T_p^*(M)$ at a point $p \in M$. We introduce the following Cartesian product set whose two parts are k-times and l-times cartesian products of $T_p^*(M)$ and $T_p(M)$, respectively

$$\mathrm{T}_p(M)_l^k = \underbrace{T_p^*(M) \times \cdots \times T_p^*(M)}_{k} \times \underbrace{T_p(M) \times \cdots \times T_p(M)}_{l}.$$

We now specify a multilinear functional $\mathcal{T} : \mathrm{T}_p(M)_l^k \to \mathbb{R}$ as a ***k-contravariant*** and ***l-covariant mixed tensor***. We next define a vector space $\mathfrak{T}_p(M)_l^k$ as a tensor product of two vector spaces formed by k-times tensor products of $T_p(M)$ and l-times tensor product of $T_p^*(M)$:

$$\mathfrak{T}_p(M)_l^k = \underbrace{T_p(M) \otimes \cdots \otimes T_p(M)}_{k} \otimes \underbrace{T_p^*(M) \otimes \cdots \otimes T_p^*(M)}_{l}. \quad (4.3.1)$$

The tensor $\mathcal{T}$ can then be expressible as an element of $\mathfrak{T}(M)_l^k$ and we say that it is a $\binom{k}{l}$-tensor. With respect to the basis vectors produced by natural local coordinates in $T_p(M)$ and $T_p^*(M)$ we can write $\mathcal{T}$ in the form

$$\mathcal{T} = t^{i_1 i_2 \cdots i_k}_{j_1 j_2 \cdots j_l} \frac{\partial}{\partial x^{i_1}} \otimes \frac{\partial}{\partial x^{i_2}} \otimes \cdots \otimes \frac{\partial}{\partial x^{i_k}} \otimes dx^{j_1} \otimes dx^{j_2} \otimes \cdots \otimes dx^{j_l}$$

[*see* p. 23]. Here the repeated indices, i.e., dummy indices, $i_1, i_2, \ldots, i_k$ and $j_1, j_2, \ldots, j_l$ are all taking the values from 1 to m. The coefficients $t^{i_1 \cdots i_k}_{j_1 \cdots j_l}$ are called the components of the tensor $\mathcal{T}$. *We frequently identify a tensor $\mathcal{T}$ with its components*. The value of the tensor, or multilinear functional, $\mathcal{T}$ on $\mathrm{T}_p(M)_l^k$, or on *k linear functionals (covariant vectors)* $\omega^{(1)}, \omega^{(2)}, \ldots, \omega^{(k)}$ and *l contravariant vectors* $V_{(1)}, V_{(2)}, \ldots, V_{(l)}$ are prescribed by

$$\mathcal{T}(\,\omega^{(1)}, \ldots, \omega^{(k)}, V_{(1)}, \ldots, V_{(l)}) = t^{i_1 \cdots i_k}_{j_1 \cdots j_l} \omega^{(1)}_{i_1} \cdots \omega^{(k)}_{i_k} v^{j_1}_{(1)} \cdots v^{j_l}_{(l)}.$$

It is straightforward to verify that the m^{k+l} *tensor components* $t^{i_1 \cdots i_k}_{j_1 \cdots j_l}$ are determined by

$$t^{i_1 \cdots i_k}_{j_1 \cdots j_l} = \mathcal{T}(\,dx^{i_1}, \ldots, dx^{i_k}, \partial_{j_1}, \ldots, \partial_{j_l}).$$

It is obvious that we end up with different types of k-contravariant and l-covariant tensors if we change the ordering of spaces $T_p(M)$ and $T_p^*(M)$ in the tensor product keeping the numbers of the component spaces constant. When we take into account a coordinate transformation $y^i = y^i(x^j)$ at the point p, we can write

$$\frac{\partial}{\partial y^i} = \frac{\partial x^j}{\partial y^i} \frac{\partial}{\partial x^j}, \quad dy^i = \frac{\partial y^i}{\partial x^j} dx^j$$

to obtain

$$\begin{aligned}\mathcal{T} &= t'^{i_1\cdots i_k}_{j_1\cdots j_l}\frac{\partial}{\partial y^{i_1}}\otimes\cdots\otimes\frac{\partial}{\partial y^{i_k}}\otimes dy^{j_1}\otimes\cdots\otimes dy^{j_l}\\ &= t'^{i_1\cdots i_k}_{j_1\cdots j_l}\frac{\partial x^{m_1}}{\partial y^{i_1}}\cdots\frac{\partial x^{m_k}}{\partial y^{i_k}}\frac{\partial y^{j_1}}{\partial x^{n_1}}\cdots\frac{\partial y^{j_l}}{\partial x^{n_l}}\frac{\partial}{\partial x^{m_1}}\otimes\cdots\otimes\frac{\partial}{\partial x^{m_k}}\\ &\qquad\otimes dx^{n_1}\otimes\cdots\otimes dx^{n_l}\\ &= t^{m_1\cdots m_k}_{n_1\cdots n_l}\frac{\partial}{\partial x^{m_1}}\otimes\cdots\otimes\frac{\partial}{\partial x^{m_k}}\otimes dx^{n_1}\otimes\cdots\otimes dx^{n_l}\end{aligned}$$

whence we deduce the following relations between components of the same tensor in different coordinate systems

$$t'^{i_1\cdots i_k}_{j_1\cdots j_l}\frac{\partial x^{m_1}}{\partial y^{i_1}}\cdots\frac{\partial x^{m_k}}{\partial y^{i_k}}\frac{\partial y^{j_1}}{\partial x^{n_1}}\cdots\frac{\partial y^{j_l}}{\partial x^{n_l}} = t^{m_1\cdots m_k}_{n_1\cdots n_l}.$$

If we recall the chain rule $(\partial x^{j_r}/\partial y^{i_r})(\partial y^{i_r}/\partial x^{j_s}) = \partial x^{j_r}/\partial x^{j_s} = \delta^{j_r}_{j_s}$, we finally find out that the transformation rule for the components of the tensor $\mathcal{T}$ under the change of coordinates $y^i = y^i(x^j)$ is given by

$$t'^{i_1 i_2\cdots i_k}_{j_1 j_2\cdots j_l} = t^{m_1 m_2\cdots m_k}_{n_1 n_2\cdots n_l}\frac{\partial y^{i_1}}{\partial x^{m_1}}\frac{\partial y^{i_2}}{\partial x^{m_2}}\cdots\frac{\partial y^{i_k}}{\partial x^{m_k}}\frac{\partial x^{n_1}}{\partial y^{j_1}}\frac{\partial x^{n_2}}{\partial y^{j_2}}\cdots\frac{\partial x^{n_l}}{\partial y^{j_l}}$$

We can immediately realise that the set

$$\mathfrak{T}(M)^k_l = \bigcup_{p\in M}\mathfrak{T}_p(M)^k_l = \{(p,\mathcal{T}) : p\in M, \mathcal{T}\in\mathfrak{T}_p(M)^k_l\} \quad (4.3.2)$$

can be endowed with a differentiable structure as was done in Sec. 2.8 so as it becomes an $m + m^{k+l}$-dimensional smooth manifold. This manifold will be called the ***tensor bundle of order k + l*** whose local coordinates are given by $\{x^1,\ldots,x^m,t^{i_1\cdots i_k}_{j_1\cdots j_l} : i_1,\ldots,i_k,j_1,\ldots,j_l = 1,\ldots,m\}$. A *section* of the bundle $\mathfrak{T}(M)^k_l$ characterises a tensor field on the manifold M. In terms of standard local coordinates this tensor field is expressible as

$$\mathcal{T}(p) = t^{i_1\cdots i_k}_{j_1\cdots j_l}(\mathbf{x})\frac{\partial}{\partial x^{i_1}}\otimes\cdots\otimes\frac{\partial}{\partial x^{i_k}}\otimes dx^{j_1}\otimes\cdots\otimes dx^{j_l}. \quad (4.3.3)$$

The sum of two tensor fields $\mathcal{T}_1,\mathcal{T}_2\in\mathfrak{T}(M)^k_l$ of the same type is the tensor field $\mathcal{T} = \mathcal{T}_1 + \mathcal{T}_2 \in \mathfrak{T}(M)^k_l$ whose components are given by

$$t^{i_1\cdots i_k}_{j_1\cdots j_l}(\mathbf{x}) = t^{i_1\cdots i_k}_{(1)j_1\cdots j_l}(\mathbf{x}) + t^{i_1\cdots i_k}_{(2)j_1\cdots j_l}(\mathbf{x}).$$

Similarly, if we choose $f\in\Lambda^0(M)$ and $\mathcal{T}\in\mathfrak{T}(M)^k_l$, then the tensor field $f\mathcal{T}\in\mathfrak{T}(M)^k_l$ is determined by its components $f(\mathbf{x})t^{i_1\cdots i_k}_{j_1\cdots j_l}(\mathbf{x})$. Hence, all tensor fields of the same order and of the same type constitute a module on the commutative ring $\Lambda^0(M)$. It is obvious that one can use the representations $\mathfrak{T}(M)^1_0 = T(M)$ and $\mathfrak{T}(M)^0_1 = T^*(M)$.

The ***operation of contraction*** on a tensor field is defined as in Sec. 1.3. If we remove in (4.3.3) the tensor product between dx^{j_s} and ∂_{i_r}, and notice that $dx^{j_s}(\partial_{i_r}) = \delta^{j_s}_{i_r}$, we obtain the ***contracted tensor***

$$\mathcal{T}_c = {}_c t^{i_1 \cdots i_{k-1}}_{j_1 \cdots j_{l-1}} \frac{\partial}{\partial x^{i_1}} \otimes \cdots \otimes \frac{\partial}{\partial x^{i_{k-1}}} \otimes dx^{j_1} \otimes \cdots \otimes dx^{j_{l-1}}$$

whose order is now $k + l - 2$. The components of a once contracted tensor are given, for instance, by

$${}_c t^{i_1 \cdots i_{k-1}}_{j_1 \cdots j_{l-1}} = t^{i_1 \cdots i_{r-1} i i_{r+1} \cdots i_{k-1}}_{j_1 \cdots j_{s-1} i j_{s+1} \cdots j_{l-1}}.$$

The contraction operation makes it possible for us to propose a rather simple test to recognise whether a given array of coefficients as an indicial quantity in a particular coordinate system are components of a tensor.

Quotient Rule. *Let the coefficients $t^{i_1 \cdots i_k}_{j_1 \cdots j_l}$ be the components of an arbitrary $\binom{k}{l}$-tensor $\mathcal{T}$ of order $k + l$ in a given coordinate system and let $s^{m_1 \cdots m_r}_{n_1 \cdots n_s}$ be an array of numbers considered to be the components of a quantity $\mathcal{S}$ of order $r + s$. We introduce the quantity $\mathcal{R} = \mathcal{T} \times \mathcal{S}$ with the components $r^{i_1 \cdots i_k m_1 \cdots m_r}_{j_1 \cdots j_l n_1 \cdots n_s} = t^{i_1 \cdots i_k}_{j_1 \cdots j_l} s^{m_1 \cdots m_r}_{n_1 \cdots n_s}$. If any contraction of this indicial quantity such as, for instance, $r^{i_1 \cdots i \cdots i_k m_1 \cdots m_r}_{j_1 \cdots j_l n_1 \cdots i \cdots n_s}$ is found to be components of a tensor ${}_c\mathcal{R}$, then the coefficients $s^{m_1 \cdots m_r}_{n_1 \cdots n_s}$ are also components of a $\binom{r}{s}$-tensor $\mathcal{S}$ of order $r + s$.*

Since ${}_c\mathcal{R}$ and $\mathcal{T}$ are tensors, their components transform according to the well known rule under any coordinate transformations. Therefore, we can write

$$\begin{aligned} & t'^{a_1 \cdots j \cdots a_k}_{b_1 \cdots b_l} s'^{p_1 \cdots p_r}_{q_1 \cdots j \cdots q_s}(\mathbf{y}) = \\ & \frac{\partial y^{a_1}}{\partial x^{i_1}} \cdots \frac{\partial y^{j}}{\partial x^{i}} \cdots \frac{\partial y^{a_k}}{\partial x^{i_k}} \frac{\partial x^{j_1}}{\partial y^{b_1}} \cdots \frac{\partial x^{j_l}}{\partial y^{b_l}} t^{i_1 \cdots i \cdots i_k}_{j_1 \cdots j_l} s'^{p_1 \cdots p_r}_{q_1 \cdots j \cdots q_s} = \\ & \frac{\partial y^{a_1}}{\partial x^{i_1}} \cdots \frac{\partial y^{a_k}}{\partial x^{i_k}} \frac{\partial y^{p_1}}{\partial x^{m_1}} \cdots \frac{\partial y^{p_r}}{\partial x^{m_r}} \frac{\partial x^{j_1}}{\partial y^{b_1}} \cdots \frac{\partial x^{j_l}}{\partial y^{b_l}} \frac{\partial x^{n_1}}{\partial y^{q_1}} \cdots \frac{\partial x^{n_s}}{\partial y^{q_s}} t^{i_1 \cdots i \cdots i_k}_{j_1 \cdots j_l} s^{m_1 \cdots m_r}_{n_1 \cdots i \cdots n_s}(\mathbf{x}) \end{aligned}$$

whence we deduce that

$$t^{i_1 \cdots i \cdots i_k}_{j_1 \cdots j_l} \Big(\frac{\partial y^{j}}{\partial x^{i}} s'^{p_1 \cdots p_r}_{q_1 \cdots j \cdots q_s} - \frac{\partial y^{p_1}}{\partial x^{m_1}} \cdots \frac{\partial y^{p_r}}{\partial x^{m_r}} \frac{\partial x^{n_1}}{\partial y^{q_1}} \cdots \frac{\partial x^{n_s}}{\partial y^{q_s}} s^{m_1 \cdots m_r}_{n_1 \cdots i \cdots n_s} \Big) = 0$$

since we can obviously cancel regular matrices $[\partial y^k/\partial x^l]$ in both sides because their determinants do not vanish. Inasmuch as these relations should be satisfied for every tensor $\mathcal{T}$, the expressions within parentheses must be equal to zero from which it follows that

$$s'^{p_1\cdots p_r}_{q_1\cdots j\cdots q_s}(\mathbf{y}) = \frac{\partial y^{p_1}}{\partial x^{m_1}}\cdots\frac{\partial y^{p_r}}{\partial x^{m_r}}\frac{\partial x^{n_1}}{\partial y^{q_1}}\cdots\frac{\partial x^{i}}{\partial y^{j}}\cdots\frac{\partial x^{n_s}}{\partial y^{q_s}}s^{m_1\cdots m_r}_{n_1\cdots i\cdots n_s}(\mathbf{x})$$

This relation shows clearly that $\mathcal{S}$ is a $\binom{r}{s}$-tensor of order $r+s$. □

Let us now consider *a completely antisymmetric* k-covariant tensor field $\omega \in \mathfrak{T}(M)^0_k$. In local natural coordinates, we may represent this tensor in the following form:

$$\omega(p) = \omega_{j_1 j_2\cdots j_k}(\mathbf{x})\, dx^{j_1}\otimes dx^{j_2}\otimes\cdots\otimes dx^{j_k}.$$

Here, the components of this tensor are completely antisymmetric, namely, they must obey the rule $\omega_{i_1\cdots i_p\cdots i_q\cdots i_k}(\mathbf{x}) = -\,\omega_{i_1\cdots i_q\cdots i_p\cdots i_k}(\mathbf{x})$ for every pair of indices. Therefore, by making use of the generalised Kronecker deltas we can write as in (1.4.8)

$$\delta^{i_1 i_2\cdots i_k}_{j_1 j_2\cdots j_k}\omega_{i_1 i_2\cdots i_k}(\mathbf{x}) = k!\,\omega_{[j_1 j_2\cdots j_k]}(\mathbf{x}) = k!\,\omega_{j_1 j_2\cdots j_k}(\mathbf{x})$$

to obtain

$$\omega = \frac{1}{k!}\,\omega_{i_1 i_2\cdots i_k}(\mathbf{x})\,\delta^{i_1 i_2\cdots i_k}_{j_1 j_2\cdots j_k}\,dx^{j_1}\otimes dx^{j_2}\otimes\cdots\otimes dx^{j_k}.$$

Just like we did in Sec. 1.4, we can again define the ***exterior product*** of basis vectors in the tensor bundle $\mathfrak{T}(M)^0_k$ as follows

$$\begin{aligned} dx^{i_1}\wedge dx^{i_2}\wedge\cdots\wedge dx^{i_k} &= \delta^{i_1 i_2\cdots i_k}_{j_1 j_2\cdots j_k}dx^{j_1}\otimes dx^{j_2}\otimes\cdots\otimes dx^{j_k} \\ &= k!\,dx^{[i_1}\otimes dx^{i_2}\otimes\cdots\otimes dx^{i_k]}. \end{aligned}$$

Hence, we reach to the conclusion

$$\omega(p) = \frac{1}{k!}\,\omega_{i_1 i_2\cdots i_k}(\mathbf{x})\,dx^{i_1}\wedge dx^{i_2}\wedge\cdots\wedge dx^{i_k}. \tag{4.3.4}$$

At the point $p \in M$, the tensor $\omega(p)$ is an *alternating k-linear functional* assigning a scalar number to k vectors $V_1, V_2, \dots, V_k$ in the tangent space $T_p(M)$, and consequently, a smooth function to vector fields $V_1(p), V_2(p), \dots, V_k(p)$. If we write

$$V_\alpha(p) = v^i_\alpha(\mathbf{x})\,\partial/\partial x^i,\, \alpha = 1, 2, \dots, k,$$

this function is determined by the expression

$$\begin{aligned}\omega(V_1, V_2, \ldots, V_k)(p) &= \omega_{i_1 i_2 \cdots i_k}(\mathbf{x})\, dx^{[i_1}(V_1)\, dx^{i_2}(V_2) \cdots dx^{i_k]}(V_k) \\ &= \omega_{i_1 i_2 \cdots i_k}(\mathbf{x})\, v_1^{[i_1}(\mathbf{x}) v_2^{i_2}(\mathbf{x}) \cdots v_k^{i_k]}(\mathbf{x}) \\ &= \omega_{i_1 i_2 \cdots i_k}(\mathbf{x})\, v_1^{i_1}(\mathbf{x}) v_2^{i_2}(\mathbf{x}) \cdots v_k^{i_k}(\mathbf{x}) \\ &= \frac{1}{k!}\, \omega_{i_1 i_2 \cdots i_k}(\mathbf{x}) \begin{vmatrix} v_1^{i_1}(\mathbf{x}) & v_2^{i_1}(\mathbf{x}) & \cdots & v_k^{i_1}(\mathbf{x}) \\ v_1^{i_2}(\mathbf{x}) & v_2^{i_2}(\mathbf{x}) & \cdots & v_k^{i_2}(\mathbf{x}) \\ \vdots & \vdots & & \vdots \\ v_1^{i_k}(\mathbf{x}) & v_2^{i_k}(\mathbf{x}) & \cdots & v_k^{i_k}(\mathbf{x}) \end{vmatrix}\end{aligned}$$

Since coefficient functions $\omega_{i_1 i_2 \cdots i_k}$ are completely antisymmetric, the second and the third lines above will of course yield the same numerical value. The completely antisymmetric k-covariant tensor given by (4.3.4) will be called henceforth a ***k-exterior differential form*** or a ***k-exterior form*** or simply a ***k-form*** on the manifold M.

It is evident that the exterior product will not be confined solely on cotangent spaces. A *completely antisymmetric* k-contravariant tensor field $\mathcal{V} \in \mathfrak{T}(M)_0^k$ is expressible as follows in local natural coordinates

$$\mathcal{V}(p) = v^{i_1 i_2 \cdots i_k}(\mathbf{x}) \frac{\partial}{\partial x^{i_1}} \otimes \frac{\partial}{\partial x^{i_2}} \otimes \cdots \otimes \frac{\partial}{\partial x^{i_k}}$$

where the coefficient functions $v^{i_1 i_2 \cdots i_k}(\mathbf{x})$ are completely antisymmetric. Thus, making use of generalised Kronecker deltas, we can again write

$$\delta^{i_1 i_2 \cdots i_k}_{j_1 j_2 \cdots j_k} v^{j_1 j_2 \cdots j_k}(\mathbf{x}) = k!\, v^{[i_1 i_2 \cdots i_k]}(\mathbf{x}) = k!\, v^{i_1 i_2 \cdots i_k}(\mathbf{x}).$$

This of course leads to the representation

$$\mathcal{V}(p) = \frac{1}{k!}\, v^{i_1 i_2 \cdots i_k}(\mathbf{x}) \frac{\partial}{\partial x^{i_1}} \wedge \frac{\partial}{\partial x^{i_2}} \wedge \cdots \wedge \frac{\partial}{\partial x^{i_k}}.$$

Once more, we define the ***exterior product*** of basis vectors in the tensor bundle $\mathfrak{T}(M)_0^k$ in the following way:

$$\frac{\partial}{\partial x^{i_1}} \wedge \frac{\partial}{\partial x^{i_2}} \wedge \cdots \wedge \frac{\partial}{\partial x^{i_k}} = \delta^{i_1 i_2 \cdots i_k}_{j_1 j_2 \cdots j_k} \frac{\partial}{\partial x^{i_1}} \otimes \frac{\partial}{\partial x^{i_2}} \otimes \cdots \otimes \frac{\partial}{\partial x^{i_k}}.$$

At the point $p \in M$, the tensor $\mathcal{V}(p)$ is an *alternating k-linear functional* assigning a scalar number to k covectors, or linear functionals, $\omega_1, \omega_2, \ldots, \omega_k$ in the cotangent space $T_p^*(M)$, and consequently a smooth function to fields $\omega_1(p), \omega_2(p), \ldots, \omega_k(p)$. It is clear that tensor fields of this form constitute a module $\mathfrak{X}^k(M)$. If we take $\mathfrak{X}^0(M) = \Lambda^0(M)$, we immediately observe that the direct sum $\mathfrak{X}(M) = \overset{m}{\underset{k=0}{\oplus}} \mathfrak{X}^k(M)$ is a 2^m-dimensional Grassmann algebra with respect to the exterior product operation $\wedge$. It is evident that one has $\mathfrak{X}^1(M) = T(M)$.

It is obvious that a mixed completely antisymmetric tensor is now expressible in the form

$$\mathcal{T}(p) = t^{i_1\cdots i_k}_{j_1\cdots j_l}(\mathbf{x})\frac{\partial}{\partial x^{i_1}} \wedge \cdots \wedge \frac{\partial}{\partial x^{i_k}} \wedge dx^{j_1} \wedge \cdots \wedge dx^{j_l}.$$

The components $t^{i_1\cdots i_k}_{j_1\cdots j_l}(\mathbf{x})$ must be completely antisymmetric functions with respect to its subscripts and superscripts.

Although we would mainly be interested in exterior forms and their exterior products in this work, it becomes now clear that the use of exterior products are not restricted to such types of entities only.

Finally, we shall try to calculate the Lie derivative of elements in an arbitrary tensor bundle with respect to a vector field V in the tangent bundle of a manifold M. To this end, we illustrate a fundamental property of Lie derivatives. Let us consider the tensor product

$$U_1(p) \otimes U_2(p) \otimes \cdots \otimes U_k(p)$$

of vectors $U_1, U_2, \cdots, U_k$. Its Lie derivative at a point $p \in M$ can now be evaluated as in (2.10.2):

$$\pounds_V(U_1 \otimes \cdots \otimes U_k) = \lim_{t\to 0}\frac{U_1^*(p;t) \otimes \cdots \otimes U_k^*(p;t) - U_1(p) \otimes \cdots \otimes U_k(p)}{t}.$$

In view of (2.10.4), we can write

$$U_i^*(p;t) = U_i(p) + t\pounds_V U_i + o(t), \; i = 1, \ldots, k.$$

Hence, employing the rules of the tensor product, we obtain

$$\begin{aligned} U_1^*(p;t) \otimes U_2^*(p;t) \otimes \cdots \otimes U_k^*(p;t) = \\ \big(U_1(p) + t\pounds_V U_1(p) + o(t)\big) \otimes \\ \big(U_2(p) + t\pounds_V U_2(p) + o(t)\big) \otimes \cdots \otimes \big(U_k(p) + t\pounds_V U_k(p) + o(t)\big) = \\ U_1(p) \otimes U_2(p) \otimes \cdots \otimes U_k(p) \\ + t\big[\pounds_V U_1(p) \otimes U_2(p) \otimes \cdots \otimes U_k(p) \\ + U_1(p) \otimes \pounds_V U_2(p) \otimes \cdots \otimes U_k(p) + \\ \cdots + U_1(p) \otimes U_2(p) \otimes \cdots \otimes \pounds_V U_k(p)\big] + o(t). \end{aligned}$$

We thus conclude that

$$\pounds_V(U_1 \otimes U_2 \otimes \cdots \otimes U_k) = \pounds_V U_1 \otimes U_2 \otimes \cdots \otimes U_k + U_1 \otimes \pounds_V U_2 \otimes \cdots \otimes U_k + \cdots + U_1 \otimes U_2 \otimes \cdots \otimes \pounds_V U_k. \tag{4.3.5}$$

This clearly means that the Lie derivative obeys the classical Leibniz' rule. We utilise this property in Sec. 5.11 to evaluate quite easily the Lie derivative of any tensor.

IV. EXERCISES

4.1. The tensor field $\mathcal{T} \in \mathfrak{T}(\mathbb{R}^2)^0_2$ is given by $\mathcal{T} = dx \otimes dx + dy \otimes dy$. (a) Find the value of this covariant tensor field of order 2 on the vector fields given below

$$U = u_x \frac{\partial}{\partial x} + u_y \frac{\partial}{\partial y}, \quad V = v_x \frac{\partial}{\partial x} + v_y \frac{\partial}{\partial y}.$$

(b) Show that under the coordinate transformation $x = r\cos\theta, y = r\sin\theta$, the same tensor can be written as $\mathcal{T} = dr \otimes dr + r^2\, d\theta \otimes d\theta$.

4.2. Let the tensor $\mathcal{T} \in \mathfrak{T}(\mathbb{R}^3)^0_2$ be given by $\mathcal{T} = dx \otimes dx + dy \otimes dy + dz \otimes dz$. (a) Find the value of $\mathcal{T}$ on vector fields given below

$$U = u_x \frac{\partial}{\partial x} + u_y \frac{\partial}{\partial y} + u_z \frac{\partial}{\partial z}, \quad V = v_x \frac{\partial}{\partial x} + v_y \frac{\partial}{\partial y} + v_z \frac{\partial}{\partial z},$$
$$W = w_x \frac{\partial}{\partial x} + w_y \frac{\partial}{\partial y} + w_z \frac{\partial}{\partial z}.$$

(b) Show that under the coordinate transformation $x = r\cos\theta,\ y = r\sin\theta,\ z = z$ this tensor takes the form $\mathcal{T} = dr \otimes dr + r^2\, d\theta \otimes d\theta + dz \otimes dz$, and (c) under the transformation $x = r\cos\varphi\sin\theta,\ y = r\sin\varphi\sin\theta,\ z = r\cos\theta$ it becomes $\mathcal{T} = dr \otimes dr + r^2 d\theta \otimes d\theta + r^2 \sin\theta\, d\varphi \otimes d\varphi$.

4.3. Components of a tensor $\mathcal{T} \in \mathfrak{T}(M)^3_2$ are given by $t^{ijk}_{lm}(p)$. How many $\binom{2}{1}$-tensors are obtainable through contraction operations? Further contraction operations result in how many $\binom{1}{0}$-tensors?

4.4. Show that the components of the tensor $\mathcal{T} = V \otimes \omega \in \mathfrak{T}(M)^1_1$ are given by $t^i_j(p) = v^i(p)\,\omega_j(p)$.

4.5. Let us consider $\mathfrak{T}(M)^k = \underbrace{T(M) \otimes \cdots \otimes T(M)}_{k}$. What type of a tensor can be regarded as representing a multilinear mapping $\mathfrak{T}(M)^k \to T(M)$?

4.6. If the components of a $\binom{1}{1}$-tensor are the same with respect to every basis, then show that they should be written in the form $t^i_j(p) = t(p)\,\delta^i_j$.

4.7. If the components of a $\binom{k}{l}$-tensor are the same with respect to every basis, then show that either $\mathcal{T} = 0$ or $k = l$.

4.8. If the components of a $\binom{1}{1}$-tensor is symmetric with respect to their indices, that is, if the equalities $t^i_j = t^j_i$ are numerically valid, then show that $t^i_j(p) = t(p)\,\delta^i_j$.

4.9. Show that the generalised Kronecker deltas $\delta^{i_1 i_2 \cdots i_k}_{j_1 j_2 \cdots j_k}$ are components of a $\binom{k}{k}$-tensor and verify that these components remain unchanged in every set of coordinates.

4.10. Let the structure constants of a Lie algebra $\mathfrak{g}$ be c^k_{ij}. Show that $\mathfrak{c}_{ij} = c^k_{il} c^l_{kj}$ are components of a symmetric tensor, whereas $\mathfrak{c}_{ijk} = c^l_{ij}\mathfrak{c}_{lk}$ are components of a completely antisymmetric tensor.

4.11. Assume that $\mathcal{T} \in \mathfrak{T}(M)^0_2$ is a symmetric tensor. We define the components of a tensor $\mathcal{S} \in \mathfrak{T}(M)^0_4$ by the relations

$$s_{ijkl} = t_{ik}t_{jl} - t_{il}t_{jk}.$$

Verify that the following equalities

$$s_{ijkl} = -\, s_{jikl} = -\, s_{ijlk},\ s_{ijkl} + s_{ijkl} + s_{ijkl} = 0$$

are satisfied. Let $U, V \in T(M)$. Show that

$$\mathcal{S}(U,V,U,V) = \mathcal{T}(U,U)\,\mathcal{T}(V,V) - \mathcal{T}(U,V)^2$$

and, if vectors U and V are linearly independent, then one finds for $U \neq \mathbf{0}$ $\mathcal{S}(U,V,U,V) > 0$ whenever $\mathcal{T}(U,U) > 0$.

4.12. A mapping $\phi : \mathbb{R}^3 \to \mathbb{R}^3$ is prescribed by $\phi(x,y,z) = (x+y, 2y-x, z^3)$. Evaluate the action of this mapping on the tensor

$$\mathcal{T} = 3x\frac{\partial}{\partial x} \otimes dy \otimes dz + y\,\frac{\partial}{\partial y} \otimes dx \otimes dz + \sin x\,\frac{\partial}{\partial x} \otimes dx \otimes dy.$$

4.13. A tensor field $\mathcal{T} \in \mathfrak{T}(\mathbb{R}^2)^2_0$ and a vector field $V \in T(\mathbb{R}^2)$ are given, respectively, by

$$\mathcal{T} = x\frac{\partial}{\partial x} \otimes \frac{\partial}{\partial x} \otimes \frac{\partial}{\partial y} - y^2\frac{\partial}{\partial x} \otimes \frac{\partial}{\partial y} \otimes \frac{\partial}{\partial y},\quad V = y\frac{\partial}{\partial x} - x^2\frac{\partial}{\partial y}.$$

Evaluate the Lie derivative $\pounds_V \mathcal{T}$.

4.14. Prove that elements $(v_1, v_2, \ldots, v_k)$ of a vector space are linearly independent if and only if $v_1 \wedge v_2 \wedge \cdots \wedge v_k \neq \mathbf{0}$.

4.15. Prove that the linearly independent sets $(u_1, u_2, \ldots, u_k)$ and $(v_1, v_2, \ldots, v_k)$ are bases of the same k-dimensional subspace of a vector space if and only if

$$u_1 \wedge u_2 \wedge \cdots \wedge u_k = A v_1 \wedge v_2 \wedge \cdots \wedge v_k$$

where $A \neq 0$. Show further that there exist a regular $k \times k$ matrix $\mathbf{A} = [a^i_j]$ such that $u_i = a^j_i v_j$ and $A = \det \mathbf{A}$.

4.16. If $\mathcal{U}, \mathcal{V} \in \mathfrak{X}(M)$ and $V \in \mathfrak{X}^1(M)$, then show that one can write

$$\pounds_V(\mathcal{U}\wedge\mathcal{V})=\pounds_V\mathcal{U}\wedge\mathcal{V}+\mathcal{U}\wedge\pounds_V\mathcal{V}.$$

4.17. Let us consider vector fields $U_i\in\mathfrak{X}^1(M), i=1,\dots,k$, and let us denote the exterior product of these vectors by $\boldsymbol{\mathcal{U}}=U_1\wedge\cdots\wedge U_k\in\mathfrak{X}^k(M)$. We define the ***Schouten-Nijenhuis bracket*** [Dutch mathematician Albert Nijenhuis] $\langle\,,\rangle:\mathfrak{X}(M)\times\mathfrak{X}(M)\to\mathfrak{X}(M)$ through the following expression:

$$\begin{aligned}\langle\boldsymbol{\mathcal{U}},\boldsymbol{\mathcal{V}}\rangle&=\sum_{i=1}^{k}(-1)^{i+1}U_1\wedge\cdots\wedge U_{i-1}\wedge U_{i+1}\wedge\cdots\wedge U_k\wedge\pounds_{U_i}\mathcal{V}\\&=\sum_{i=1}^{k}\sum_{j=1}^{l}(-1)^{i+j}[U_i,V_j]\wedge U_1\wedge\cdots\wedge U_{i-1}\wedge U_{i+1}\wedge\cdots\wedge U_k\\&\qquad\wedge V_1\wedge\cdots\wedge V_{j-1}\wedge V_{j+1}\wedge\cdots\wedge V_l.\end{aligned}$$

where $\boldsymbol{\mathcal{U}}\in\mathfrak{X}^k(M)$ and $\boldsymbol{\mathcal{V}}\in\mathfrak{X}^l(M)$. Assume that $\boldsymbol{\mathcal{U}}\in\mathfrak{X}^k(M),\boldsymbol{\mathcal{V}}\in\mathfrak{X}^l(M)$, $\boldsymbol{\mathcal{W}}\in\mathfrak{X}^m(M),U,V\in\mathfrak{X}^1(M)$ and $f,g\in C^\infty(M)$. Then show that Schouten -Nijenhuis bracket satisfies the following relations:

(a) $\langle\boldsymbol{\mathcal{U}},\boldsymbol{\mathcal{V}}\rangle\in\mathfrak{X}^{k+l-1}(M)$, (b) $\langle f,g\rangle=0$, (c) $\langle U,f\rangle=U(f)$,

(d) $\langle U,V\rangle=[U,V]$,

(e) $\langle\boldsymbol{\mathcal{U}},\boldsymbol{\mathcal{V}}\wedge\boldsymbol{\mathcal{W}}\rangle=\langle\boldsymbol{\mathcal{U}},\boldsymbol{\mathcal{V}}\rangle\wedge\boldsymbol{\mathcal{W}}+(-1)^{(k+1)l}\boldsymbol{\mathcal{V}}\wedge\langle\boldsymbol{\mathcal{U}},\boldsymbol{\mathcal{W}}\rangle$,

(f) $\langle\boldsymbol{\mathcal{U}},\boldsymbol{\mathcal{V}}\rangle=(-1)^{kl}\langle\boldsymbol{\mathcal{V}},\boldsymbol{\mathcal{U}}\rangle$,

(g) *the generalised Jacobi identity*

$$(-1)^{km}\langle\langle\boldsymbol{\mathcal{U}},\boldsymbol{\mathcal{V}}\rangle,\boldsymbol{\mathcal{W}}\rangle+(-1)^{kl}\langle\langle\boldsymbol{\mathcal{V}},\boldsymbol{\mathcal{W}}\rangle,\boldsymbol{\mathcal{U}}\rangle+(-1)^{lm}\langle\langle\boldsymbol{\mathcal{W}},\boldsymbol{\mathcal{U}}\rangle,\boldsymbol{\mathcal{V}}\rangle=0$$

4.18. The fields $\mathcal{U}\in\mathfrak{X}^k(M)$ and $\mathcal{V}\in\mathfrak{X}^l(M)$ are given by

$$\mathcal{U}(p)=\frac{1}{k!}u^{i_1\cdots i_k}(\mathbf{x})\frac{\partial}{\partial x^{i_1}}\wedge\cdots\wedge\frac{\partial}{\partial x^{i_k}},\ \mathcal{V}(p)=\frac{1}{l!}v^{i_1\cdots i_l}(\mathbf{x})\frac{\partial}{\partial x^{i_1}}\wedge\cdots\wedge\frac{\partial}{\partial x^{i_l}}.$$

Show that

$$\langle\mathcal{U},\mathcal{V}\rangle=\frac{1}{(k+l-1)!}\langle\mathcal{U},\mathcal{V}\rangle^{i_1\cdots i_{k+l-1}}(\mathbf{x})\frac{\partial}{\partial x^{i_1}}\wedge\cdots\wedge\frac{\partial}{\partial x^{i_{k+l-1}}}$$

and the coefficient functions $\langle\mathcal{U},\mathcal{V}\rangle^{i_1\cdots i_{k+l-1}}(\mathbf{x})$ are determined by the expressions

$$\begin{aligned}\langle\mathcal{U},\mathcal{V}\rangle^{i_1\cdots i_{k+l-1}}(\mathbf{x})&=\frac{(-1)^k}{k!(l-1)!}\delta^{i_1\cdots\cdots i_{k+l-1}}_{j_1\cdots j_k k_1\cdots k_{l-1}}v^{ik_1\cdots k_{l-1}}\frac{\partial u^{j_1\cdots j_k}}{\partial x^i}\\&+\frac{1}{l!(k-1)!}\delta^{i_1\cdots\cdots i_{k+l-1}}_{j_1\cdots j_{k-1}k_1\cdots k_l}u^{ij_1\cdots j_{k-1}}\frac{\partial v^{k_1\cdots k_l}}{\partial x^i}.\end{aligned}$$

CHAPTER V

EXTERIOR DIFFERENTIAL FORMS

5.1. SCOPE OF THE CHAPTER

Studies of differential forms has started with the works of Grassmann and efforts to extend the integral theorems in classical vector analysis has played a significant part in the development of the theory. Several elemental concepts, for instance the exterior product, has been introduced by French mathematician Jules Henri Poincaré (1854-1912). However, it was French mathematician Élie Joseph Cartan (1869-1951) who enormously contributed in the period from 1899 to 1926 to the establishment of the theoretical framework of exterior forms on differentiable manifolds by identifying exterior differential forms as exterior products of differentials of coordinates (exterior derivatives) and thus equipping them with an algebraic structure.

In Sec. 5.2, the exterior differential forms on differentiable manifolds and exterior algebra formed by them are defined and it is shown that they constitute a module. Sec. 5.3 deals with some useful algebraic properties concerning 1-forms. In Sec. 5.4 the interior product of a vector with an exterior form is defined, various properties of this operation that reduces the degree of the form by one are revealed and criteria for the existence of a divisor of a form are established by making use of the interior product. To replace the natural basis of the exterior algebra, we consider in Sec. 5.5 a top-down generation of a new basis from the volume form, which has the highest degree on a given manifold, by its appropriate interior products with natural basis vectors of the tangent bundle. We examine relations between these bases in detail. In some cases, the use of these bases turns out to be quite advantageous. Sec. 5.6 is concerned with certain subalgebras of the exterior algebra called ideals and characteristic vectors of an exterior form and also of an ideal are introduced. It is shown in Sec. 5.7 that a smooth mapping between two differentiable manifolds gives rise to an additive pull-back operator that transports exterior forms on the range of the mapping to forms on its domain by preserving their degrees. Moreover certain

Exterior Analysis, DOI: 10.1016/B978-0-12-415902-0.50005-0

properties of this operator are emphasised. The exterior derivative which is one of the fundamental operators acting on exterior forms is defined in Sec. 5.8 and its properties are discussed there. Closed and exact forms are introduced as well in this section. Sec. 5.9 deals with Riemannian manifolds endowed with a metric tensor that makes it possible to measure distances between points of a manifold. Metric tensor also helps us to relate covariant components of a tensor with its contravariant components and vice versa. Utilising this opportunity, we define the Hodge dual of a form and the Hodge star operation generating this form. Then, we discuss its properties and scrutinise the co-differential, Laplace-de Rham and Laplace-Beltrami operators. Sec. 5.10 is concerned with closed ideals, the forms belonging to which have exterior derivatives remaining in the ideal and conditions leading to a closed ideal are examined. The Lie derivative of an exterior form that measures the variation in this form along the flow generated by a vector field on a manifold is considered in Sec. 5.11 and the Cartan formula that makes it possible to calculate Lie derivatives of forms relatively easily is derived. We define in Sec. 5.12 isovector fields of an ideal and show that the ideal remains invariant under the flow produced by an isovector field and prove that isovectors constitute a Lie subalgebra of the tangent bundle. Finally, we investigate in Sec. 5.13 the mappings, or submanifolds, annihilating an ideal. The notion of complete integrability is introduced, conditions providing its existence are discussed and the theorems of Cartan and Frobenius, that play a pivotal part in comprehending this concept, are proven. Sec. 5.14 is devoted to an overview of some properties of exterior forms defined on a Lie group which is also a smooth manifold. Left- and right-invariant 1-forms are defined by using certain pull-back mappings on the exterior algebra built on the Lie group. These mappings are generated by left and right translations in the group. It is shown further that left-invariant 1-forms called Maurer-Cartan forms constitute the dual of the Lie algebra of the Lie group and they satisfy a system of exterior differential equations depending on structure constants of the Lie algebra.

5.2. EXTERIOR DIFFERENTIAL FORMS

We have seen in Sec. 4.3 that a k-exterior differential form field on an m-dimensional smooth manifold M is defined as a completely antisymmetric k-covariant tensor field or as an alternating k-linear functional and it can be represented in natural coordinates $\mathbf{x} = \varphi(p)$ in a chosen chart as follows

$$\omega(p) = \frac{1}{k!}\,\omega_{i_1 i_2 \cdots i_k}(\mathbf{x})\,dx^{i_1} \wedge dx^{i_2} \wedge \cdots \wedge dx^{i_k} \tag{5.2.1}$$

where smooth functions $\omega_{i_1 i_2 \cdots i_k} \in \Lambda^0(M)$ are completely antisymmetric in their indices. We call k as the degree of the form. If we identify the sum $\omega = \omega_1 + \omega_2$ of two forms ω_1 and ω_2 of the same degree k by employing the following completely antisymmetric components

$$\omega_{i_1 i_2 \cdots i_k}(\mathbf{x}) = \omega^1_{i_1 i_2 \cdots i_k}(\mathbf{x}) + \omega^2_{i_1 i_2 \cdots i_k}(\mathbf{x}) \in \Lambda^0(M),$$

then we deduce that ω is a k-form as well. Similarly the scalar multiplication $f\omega$ where $f \in \Lambda^0(M)$ is a k-form specified by smooth functions

$$f(\mathbf{x})\omega_{i_1 i_2 \cdots i_k}(\mathbf{x}) \in \Lambda^0(M).$$

Therefore, k-exterior differential forms constitute a module over the commutative ring $\Lambda^0(M)$. Henceforth, we denote this module by $\Lambda^k(M)$. Naturally, $\Lambda^k(M)$ reduces to a vector space over the field of real numbers. When $k > m$, *it is evident that exterior forms vanish identically*. The basis of this module are the following linearly independent k-forms:

$$\{dx^{i_1} \wedge dx^{i_2} \wedge \cdots \wedge dx^{i_k} : i_1, \ldots, i_k = 1, \ldots, m\}$$

whose number is $\binom{m}{k} = \dfrac{m!}{k!\,(m-k)!}$. This basis is expressed more concretely in terms of *essential components* through ordered indices in the form $\{dx^{i_1} \wedge dx^{i_2} \wedge \cdots \wedge dx^{i_k} : 1 \le i_1 < i_2 < \cdots < i_k \le m\}$. In this case (5.2.1) can also be written as

$$\omega(p) = \sum_{1 \le i_1 < i_2 < \cdots < i_k \le m} \omega_{i_1 i_2 \cdots i_k}(\mathbf{x})\, dx^{i_1} \wedge dx^{i_2} \wedge \cdots \wedge dx^{i_k}.$$

Instead of m natural basis dx^j of $T^*(M)$ associated with local coordinates x^j in local charts at every points of the manifold we can of course choose m linearly independent 1-forms prescribed by

$$\theta^i = \theta^i_j(\mathbf{x})\, dx^j \in \Lambda^1(M),\ i, j = 1, \ldots, m;\ \ \det\left[\theta^i_j(\mathbf{x})\right] \neq 0$$

as a basis and represent a k-form in terms of this basis in the following manner

$$\omega(p) = \frac{1}{k!}\,\Omega_{i_1 i_2 \cdots i_k}(\mathbf{x})\, \theta^{i_1} \wedge \theta^{i_2} \wedge \cdots \wedge \theta^{i_k}$$

where

$$\Omega_{i_1 i_2 \cdots i_k}(\mathbf{x}) = \omega_{j_1 j_2 \cdots j_k}(\mathbf{x}) \Theta^{[j_1}_{i_1} \Theta^{j_2}_{i_1} \cdots \Theta^{j_k]}_{i_1}.$$

Here $[\Theta^i_j(\mathbf{x})]$ is the inverse of the matrix $\left[\theta^i_j(\mathbf{x})\right]$.

Just like in Sec. 1.5 we can define the operation of the exterior product of exterior differential forms $\alpha \in \Lambda^k(M), \beta \in \Lambda^l(M)$ by

$$\begin{aligned}\gamma = \alpha \wedge \beta &= \frac{1}{k!\,l!}\alpha_{i_1\cdots i_k}\beta_{j_1\cdots j_l}dx^{i_1}\wedge\cdots\wedge dx^{i_k}\wedge dx^{j_1}\wedge\cdots\wedge dx^{j_l}\\ &= \frac{1}{(k+l)!}\gamma_{i_1\cdots i_k j_1\cdots j_l}dx^{i_1}\wedge\cdots\wedge dx^{i_k}\wedge dx^{j_1}\wedge\cdots\wedge dx^{j_l}\end{aligned}$$

where $\wedge : \Lambda^k(M)\times\Lambda^l(M)\to\Lambda^{k+l}(M)$ assigns now a $(k+l)$-form to k- and l-forms. Here the functions $\gamma_{i_1\cdots i_k j_1\cdots j_l}(\mathbf{x}) \in \Lambda^0(M)$ are given by

$$\gamma_{i_1\cdots i_k j_1\cdots j_l} = \frac{(k+l)!}{k!\,l!}\alpha_{[i_1\cdots i_k}\beta_{j_1\cdots j_l]}$$

[*see* (1.5.1)]. If we regard a function $f \in \Lambda^0(M)$ as a 0-form, we can write

$$f\wedge\omega = f\omega \in \Lambda^k(M)$$

for a k-form ω. It is straightforward to observe that the exterior product possesses the following properties:

$$\begin{aligned}\alpha\wedge(\beta+\gamma) &= \alpha\wedge\beta+\alpha\wedge\gamma,\\ (\alpha+\beta)\wedge\gamma &= \alpha\wedge\gamma+\beta\wedge\gamma,\\ \alpha\wedge(\beta\wedge\gamma) &= (\alpha\wedge\beta)\wedge\gamma = \alpha\wedge\beta\wedge\gamma,\\ \beta\wedge\alpha &= (-1)^{kl}\alpha\wedge\beta,\ \ \alpha\in\Lambda^k(M),\ \beta\in\Lambda^l(M).\end{aligned} \tag{5.2.2}$$

It is thus seen that the exterior product is associative and distributive, but it is generally not commutative. Whenever kl is an even number one has $\beta\wedge\alpha = \alpha\wedge\beta$, whereas $\beta\wedge\alpha = -\alpha\wedge\beta$ when it is an odd number. *If* $\omega\in\Lambda^k(M)$ *and k is an odd number*, then we find that

$$\omega\wedge\omega = (-1)^{k^2}\omega\wedge\omega = -\omega\wedge\omega$$

since k^2 is also an odd number. Thus the *square* of such a form vanishes

$$\omega^2 = \omega\wedge\omega = 0.$$

The set of exterior differential forms of all degrees on a manifold M constitute the ***exterior algebra*** $\Lambda(M)$ with the binary operation of exterior product. The exterior algebra is expressible as the direct sum

$$\begin{aligned}\Lambda(M) &= \Lambda^0(M)\oplus\Lambda^1(M)\oplus\cdots\oplus\Lambda^k(M)\oplus\cdots\oplus\Lambda^m(M)\\ &= \bigoplus_{k=0}^{m}\Lambda^k(M)\end{aligned}$$

of modules $\Lambda^k(M), k = 0, 1, \ldots, m$. Hence $\Lambda(M)$ is a ***graded algebra***. Of course, only the sum of forms of the same degree is really meaningful. Smooth coefficient functions belong to the ring $\Lambda^0(M)$ and the natural basis of the exterior algebra $\Lambda(M)$ is given by

$$\{1\} \cup \{dx^i\} \cup \{dx^i \wedge dx^j, i < j\} \cup \cdots \cup \{dx^{i_1} \wedge \cdots \wedge dx^{i_k}, i_1 < \cdots < i_k\} \\ \cup \cdots \cup \{dx^1 \wedge dx^2 \wedge \cdots \wedge dx^m\}.$$

Thus the dimension of the exterior algebra is

$$\sum_{k=0}^{m} \binom{m}{k} = 2^m.$$

The value of a form $\omega \in \Lambda^k(M)$ on vectors $U_1, U_2, \ldots, U_k \in T(M)$ is computed as we have mentioned in *p.* 26 [*see* (1.4.4)] by the relation

$$\omega(U_1, U_2, \ldots, U_k) = \omega_{i_1 i_2 \cdots i_k} u_1^{i_1} u_2^{i_2} \cdots u_k^{i_k} \tag{5.2.3}$$

where we wrote $U_\alpha = u_\alpha^i(\mathbf{x}) \dfrac{\partial}{\partial x^i}, i = 1, 2, \ldots, m; \alpha = 1, 2, \ldots, k$. It then immediately follows that coefficient functions are determined by

$$\omega_{i_1 i_2 \cdots i_k} = \omega\Big(\frac{\partial}{\partial x^{i_1}}, \frac{\partial}{\partial x^{i_2}}, \ldots, \frac{\partial}{\partial x^{i_k}}\Big). \tag{5.2.4}$$

On an m-dimensional manifold M, the module $\Lambda^m(M)$ is 1-dimensional. Hence, every m-form is represented as

$$\omega = f(\mathbf{x})\, dx^1 \wedge dx^2 \wedge \cdots \wedge dx^m, \;\; f \in \Lambda^0(M),$$

The form

$$\mu = dx^1 \wedge dx^2 \wedge \cdots \wedge dx^m \in \Lambda^m(M) \tag{5.2.5}$$

is called the ***volume form***. Indeed if we consider m linearly independent vector fields $V_1 = \Delta v^1 \dfrac{\partial}{\partial x^1}, \ldots, V_m = \Delta v^m \dfrac{\partial}{\partial x^m}$, we obtain

$$\mu(V_1, V_2, \ldots, V_m) = \begin{vmatrix} \Delta v^1 & 0 & \cdots & 0 \\ 0 & \Delta v^2 & \cdots & 0 \\ \vdots & \vdots & \vdots & \vdots \\ 0 & 0 & \cdots & \Delta v^m \end{vmatrix} = \Delta v^1 \Delta v^2 \cdots \Delta v^m$$

and this is the volume of a rectangular parallelepiped in $\mathbb{R}^m$.

We are not compelled to employ *the natural basis* $\{dx^i\} \subset T^*(M)$

and *its reciprocal basis* $\{\partial/\partial x^i\} \subset T(M)$. Let us introduce *a reciprocal basis* $\{\theta^i\} \subset T^*(M)$ and *a basis* $\{V_i\} \subset T(M)$. Therefore the relations $\theta^i(V_j) = \delta^i_j, i, j = 1, \ldots, m$ are to be satisfied. A form $\omega \in \Lambda^k(M)$ can now be represented by

$$\omega(p) = \frac{1}{k!}\,\omega_{i_1 i_2\cdots i_k}(\mathbf{x})\,\theta^{i_1} \wedge \theta^{i_2} \wedge \cdots \wedge \theta^{i_k}$$

where coefficients $\omega_{i_1 i_2\cdots i_k}$ must of course be completely antisymmetric. Then we obtain

$$\begin{aligned}\omega(V_{i_1}, V_{i_2}, \ldots, V_{i_k}) &= \frac{1}{k!}\,\omega_{j_1 j_2\cdots j_k}\begin{vmatrix} \theta^{j_1}(V_{i_1}) & \theta^{j_1}(V_{i_2}) & \cdots & \theta^{j_1}(V_{i_k}) \\ \theta^{j_2}(V_{i_1}) & \theta^{j_2}(V_{i_2}) & \cdots & \theta^{j_2}(V_{i_k}) \\ \vdots & \vdots & & \vdots \\ \theta^{j_k}(V_{i_1}) & \theta^{j_k}(V_{i_2}) & \cdots & \theta^{j_k}(V_{i_k}) \end{vmatrix} \\ &= \frac{1}{k!}\,\omega_{j_1 j_2\cdots j_k}\begin{vmatrix} \delta^{j_1}_{i_1} & \delta^{j_1}_{i_2} & \cdots & \delta^{j_1}_{i_k} \\ \delta^{j_2}_{i_1} & \delta^{j_2}_{i_2} & \cdots & \delta^{j_2}_{i_k} \\ \vdots & \vdots & & \vdots \\ \delta^{j_k}_{i_1} & \delta^{j_k}_{i_2} & \cdots & \delta^{j_k}_{i_k} \end{vmatrix} = \frac{1}{k!}\,\omega_{j_1 j_2\cdots j_k}\delta^{j_1 j_2\cdots j_k}_{i_1 i_2\cdots i_k}\end{aligned}$$

Therefore, we again conclude that

$$\omega(V_{i_1}, V_{i_2}, \ldots, V_{i_k}) = \omega_{i_1 i_2\cdots i_k}. \tag{5.2.6}$$

5.3. SOME ALGEBRAIC PROPERTIES

We say that a k-form $\Omega \in \Lambda^k(M)$ is a ***simple form*** if it is expressible as an exterior product of k linearly independent 1-forms [*see p.* 36]. Hence, if we can write

$$\Omega = \omega^1 \wedge \omega^2 \wedge \cdots \wedge \omega^k \in \Lambda^k(M)$$

where $\omega^r \in \Lambda^1(M), r = 1, \ldots, k \le m$ are linearly independent, then Ω is a simple k-form.

Theorem 5.3.1. *$\omega^1, \omega^2, \ldots, \omega^k \in \Lambda^1(M)$ are linearly independent 1-forms if and only if $\Omega = \omega^1 \wedge \omega^2 \wedge \cdots \wedge \omega^k \neq 0$.*

Let us suppose first that $\Omega \neq 0$. We consider the linear combination $c_r\omega^r = c_1\omega^1 + c_2\omega^2 + \cdots + c_k\omega^k = 0$ where $c_1, c_2, \ldots, c_k \in \Lambda^0(M)$ are arbitrary coefficient functions. The exterior product of this form by the $(k-1)$-form $\omega^2 \wedge \cdots \wedge \omega^k$ yields $c_1\Omega = 0$ because square of a 1-form vanishes. We thus find $c_1 = 0$. In a similar fashion, we deduce that all coefficients must be zero. Hence, 1-forms $\omega^1, \omega^2, \dots, \omega^k$ are linearly independent. Conversely, let us choose k linearly independent 1-forms $\omega^1, \omega^2, \dots, \omega^k$ that are represented by

$$\omega^r = a_i^r \, dx^i, \; r = 1, \dots, k \leq m; \; i = 1, \dots, m.$$

Hence, the rank of the $k \times m$ matrix $[a_i^r]$ should be k so that this matrix must have at least one $k \times k$ submatrix with non-vanishing determinant. On the other hand, the k-form that is the exterior products of these 1-forms can be written as follows:

$$\begin{aligned}\Omega &= \omega^1 \wedge \omega^2 \wedge \cdots \wedge \omega^k \\ &= a_{i_1}^1 a_{i_2}^2 \cdots a_{i_k}^k dx^{i_1} \wedge dx^{i_2} \wedge \cdots \wedge dx^{i_k} \\ &= a_{[i_1}^1 a_{i_2}^2 \cdots a_{i_k]}^k \, dx^{i_1} \wedge dx^{i_2} \wedge \cdots \wedge dx^{i_k}.\end{aligned}$$

One immediately sees that for a particular choice of indices $i_1, \dots, i_k$, the coefficient of the form $dx^{i_1} \wedge \cdots \wedge dx^{i_k}$ will be the determinant of a $k \times k$ submatrix of the matrix $[a_i^r]$. Therefore, the form Ω is the sum of such k-forms. However, in this sum at least one term is different from zero. Hence, we obtain $\Omega \neq 0$. □

Theorem 5.3.2. *If the forms $\alpha^r, \beta^r \in \Lambda^1(M), r = 1, \dots, k$ are connected by the expression*

$$\beta^r = c_s^r \alpha^s, c_s^r \in \Lambda^0(M), \; r, s = 1, \dots, k,$$

then there exists the relation

$$\beta^1 \wedge \beta^2 \wedge \cdots \wedge \beta^k = (\det [c_s^r]) \alpha^1 \wedge \alpha^2 \wedge \cdots \wedge \alpha^k$$

among them.

In fact, it is readily found that the relation

$$\begin{aligned}\beta^1 \wedge \beta^2 \wedge \cdots \wedge \beta^k &= c_{s_1}^1 c_{s_2}^2 \cdots c_{s_k}^k \alpha^{s_1} \wedge \alpha^{s_2} \wedge \cdots \wedge \alpha^{s_k} \\ &= c_{s_1}^1 c_{s_2}^2 \cdots c_{s_k}^k \delta_{12\cdots k}^{s_1 s_2 \cdots s_k} \alpha^1 \wedge \alpha^2 \wedge \cdots \wedge \alpha^k \\ &= (\det [c_s^r]) \alpha^1 \wedge \alpha^2 \wedge \cdots \wedge \alpha^k\end{aligned}$$

is obtained. □

Theorem 5.3.3. *If 1-forms $\omega^r \in \Lambda^1(M), r = 1, \dots, k$ are linearly independent and if 1-forms $\gamma_r \in \Lambda^1(M)$, $r = 1, \dots, k$ satisfy the relation*

$$\gamma_r \wedge \omega^r = \gamma_1 \wedge \omega^1 + \gamma_2 \wedge \omega^2 + \cdots + \gamma_k \wedge \omega^k = 0,$$

then every form γ_r belongs to the submodule generated by the forms $\omega^1, \omega^2, \dots, \omega^k$. Hence one is able to write

$$\gamma_r = a_{rs}\,\omega^s,\;\; a_{rs} \in \Lambda^0(M),\;\; r,s = 1,\dots,k$$

where the matrix $\mathbf{A} = [a_{rs}]$ *is symmetric, namely,* $a_{rs} = a_{sr}$.

Exterior product of the relation $\gamma_r \wedge \omega^r = 0$ with the $(k-1)$-form $\omega^2 \wedge \cdots \wedge \omega^k$ yields $\gamma_1 \wedge \Omega = 0$. $\Omega = \omega^1 \wedge \omega^2 \wedge \cdots \wedge \omega^k \neq 0$ because 1-forms ω^r are linearly independent. It then follows that the form γ_1 is linearly dependent on the forms $\omega^1, \omega^2, \dots, \omega^k$. In a similar fashion, we find $\gamma_r \wedge \Omega = 0$ for each r. Therefore, the forms γ_r are linear combinations of the forms ω^r. Thus, one can write

$$\gamma_r = a_{rs}\,\omega^s.$$

On the other hand, the relation

$$0 = \gamma_r \wedge \omega^r = a_{rs}\,\omega^s \wedge \omega^r = a_{[rs]}\,\omega^s \wedge \omega^r$$

leads to $a_{[rs]} = 0$, and consequently to the symmetry relation $a_{rs} = a_{sr}$. This theorem is also known as the ***Cartan lemma***. □

5.4. INTERIOR PRODUCT

We have seen that new elements of the exterior algebra $\Lambda(M)$ over an m-dimensional manifold M are generated by exterior products of its elements. But the exterior product is an operation that raises the degrees of forms. Nevertheless, we can obtain at most forms of degree m with an operation raising degrees because we know that forms of degrees higher than m vanish identically. Since it is evident that it is not possible to obtain a form with a lesser degree than a given form by resorting to the exterior product, we need to introduce a new operation to achieve this task. We further wish that this operation possesses appropriate properties. We devise this operation by means of a vector field. We call it the ***interior product*** of a vector field $V \in T(M)$ with an exterior form field $\omega \in \Lambda(M)$. To this end, we introduce the interior product operator $\mathbf{i}$ in the following form

$$\mathbf{i} : T(M) \times \Lambda^k(M) \to \Lambda^{k-1}(M),$$

or

$$\mathbf{i}_V : \Lambda^k(M) \to \Lambda^{k-1}(M)$$

where the vector V is now specified. We further impose the conditions that the operator $\mathbf{i}_V$ has to satisfy the following rules:

$$(i).\, \mathbf{i}_V(f) = 0, V \in T(M), f \in \Lambda^0(M). \tag{5.4.1}$$
$$(ii).\, \mathbf{i}_V(\omega) = \omega(V) = \langle \omega, V \rangle = \omega_i v^i \in \Lambda^0(M), V \in T(M), \omega \in \Lambda^1(M).$$
$$(iii).\, \mathbf{i}_V(\alpha + \beta) = \mathbf{i}_V(\alpha) + \mathbf{i}_V(\beta),\ \ V \in T(M),\ \alpha, \beta \in \Lambda^k(M).$$
$$(iv).\, \mathbf{i}_V(\alpha \wedge \beta) = \mathbf{i}_V(\alpha) \wedge \beta + (-1)^{deg(\alpha)} \alpha \wedge \mathbf{i}_V(\beta),$$
$$V \in T(M),\ \ \alpha, \beta \in \Lambda(M).$$

Here k can only take the values $1, \ldots, m$. Since we can interpret the function $f \in \Lambda^0(M)$ as a 0-degree form so that we can write $f \wedge \omega = f\omega$, the rules (i) and (iv) result in $\mathbf{i}_V(f\omega) = f\mathbf{i}_V(\omega)$. It is readily verified that the above rules would suffice to determine the operator $\mathbf{i}_V$ uniquely. Let us assume that there exists a second operator $\mathbf{i}'_V$ accommodating to these rules. Then, it would be necessary to write $\mathbf{i}_V(f) = \mathbf{i}'_V(f) = 0, \mathbf{i}_V(\omega) = \mathbf{i}'_V(\omega) = \omega(V)$ for each $f \in \Lambda^0(M)$ and $\omega \in \Lambda^1(M)$. We thus find that $\mathbf{i}'_V|_{\Lambda^0(M)} = \mathbf{i}_V|_{\Lambda^0(M)}$, $\mathbf{i}'_V|_{\Lambda^1(M)} = \mathbf{i}_V|_{\Lambda^1(M)}$. But, the rules (iii) and (iv) assure us that actions of these two operators will also be the same on 2-, 3-, $\ldots, m$-forms. Consequently, we write $\mathbf{i}'_V|_{\Lambda(M)} = \mathbf{i}_V|_{\Lambda(M)}$ over the entire exterior algebra so that we get $\mathbf{i}_V = \mathbf{i}'_V$. The rule (iv) indicates clearly that the interior product is an ***antiderivation***. The interior product operator $\mathbf{i}_V$ is sometimes symbolised by the *hook operator* $\rfloor$. In that case, the form $\mathbf{i}_V(\omega)$ will be denoted by $V\rfloor\omega$.

Let $f \in \Lambda^0(M)$. We take $\omega = df \in \Lambda^1(M)$ so $\big(5.4.1\ (ii)\big)$ results in

$$\mathbf{i}_V(df) = df(V) = f_{,i}v^i = V(f).$$

We shall now try to evaluate explicitly the action of the interior product $\mathbf{i}_V : \Lambda(M) \to \Lambda(M)$, which maps the exterior algebra into itself, by the aid of the above rules. Suppose that a form field $\omega \subset \Lambda^k(M)$ and a vector field $V \in T(M)$ are given by

$$\omega = \frac{1}{k!}\, \omega_{i_1 i_2 \cdots i_k}(\mathbf{x})\, dx^{i_1} \wedge dx^{i_2} \wedge \cdots \wedge dx^{i_k},$$
$$V = v^i(\mathbf{x}) \frac{\partial}{\partial x^i}.$$

Because of the relation $\mathbf{i}_V\big(\omega_{i_1 i_2 \cdots i_k}(\mathbf{x})\big) = 0$ we can write

$$\mathbf{i}_V(\omega) = \frac{1}{k!}\, \omega_{i_1 i_2 \cdots i_k} \mathbf{i}_V(dx^{i_1} \wedge dx^{i_2} \wedge \cdots \wedge dx^{i_k}).$$

On the other hand, the rule (ii) dictates that $\mathbf{i}_V(dx^{i_r}) = V(dx^{i_r}) = v^{i_r}$. Hence, according to (iv) we get

$$\begin{aligned}
\mathbf{i}_V(dx^{i_1}\wedge dx^{i_2}\wedge\cdots\wedge dx^{i_k}) &= \mathbf{i}_V(dx^{i_1})\wedge dx^{i_2}\wedge\cdots\wedge dx^{i_k}\\
-\,dx^{i_1}\wedge\mathbf{i}_V(dx^{i_2}\wedge\cdots\wedge dx^{i_k}) &= \mathbf{i}_V(dx^{i_1})\wedge dx^{i_2}\wedge\cdots\wedge dx^{i_k}\\
&\quad -\,dx^{i_1}\wedge\mathbf{i}_V(dx^{i_2})\wedge dx^{i_3}\wedge\cdots\wedge dx^{i_k}\\
&\quad +\,dx^{i_1}\wedge dx^{i_2}\wedge\mathbf{i}_V(dx^{i_3}\wedge\cdots\wedge dx^{i_k})\\
&= \cdots = v^{i_1}dx^{i_2}\wedge\cdots\wedge dx^{i_k} - v^{i_2}dx^{i_1}\wedge dx^{i_3}\wedge\cdots\wedge dx^{i_k}+\cdots\\
&\quad +(-1)^{k-1}v^{i_k}dx^{i_1}\wedge dx^{i_2}\wedge\cdots\wedge dx^{i_{k-1}}\\
&= \sum_{l=1}^{k}(-1)^{l-1}v^{i_l}dx^{i_1}\wedge dx^{i_2}\wedge\cdots\wedge dx^{i_{l-1}}\wedge dx^{i_{l+1}}\wedge\cdots\wedge dx^{i_k}.
\end{aligned}$$

In the last line above, we adopted the convention $dx^{i_0}=1$. So we find that

$$\begin{aligned}
\mathbf{i}_V(\omega) &=\\
&\frac{1}{k!}\sum_{l=1}^{k}(-1)^{l-1}\,\omega_{i_1\cdots i_{l-1}i_l i_{l+1}\cdots i_k}v^{i_l}dx^{i_1}\wedge\cdots\wedge dx^{i_{l-1}}\wedge dx^{i_{l+1}}\wedge\cdots\wedge dx^{i_k}\\
&=\frac{1}{k!}\sum_{l=1}^{k}(-1)^{2(l-1)}\,v^{i_l}\omega_{i_l i_1\cdots i_{l-1}i_{l+1}\cdots i_k}dx^{i_1}\wedge\cdots\wedge dx^{i_{l-1}}\wedge dx^{i_{l+1}}\wedge\cdots\wedge dx^{i_k}\\
&=\frac{1}{k!}\sum_{l=1}^{k}v^{i}\omega_{ii_1i_2\cdots i_{k-1}}dx^{i_1}\wedge dx^{i_2}\wedge\cdots\wedge dx^{i_{k-1}}\\
&=\frac{k}{k!}v^{i}\omega_{ii_1i_2\cdots i_{k-1}}dx^{i_1}\wedge dx^{i_2}\wedge\cdots\wedge dx^{i_{k-1}}.
\end{aligned}$$

In the third line, on making use of the complete antisymmetry of coefficients, we have written $\omega_{i_1\cdots i_{l-1}i_l i_{l+1}\cdots i_k}=(-1)^{l-1}\omega_{i_l i_1\cdots i_{l-1}i_{l+1}\cdots i_k}$. We have gone into the fourth line by appropriately renaming the dummy indices. We finally deduce that, by means of the operator $\mathbf{i}_V$, a k-form $\omega\in\Lambda^k(M)$ is transformed into a $(k-1)$- form $\mathbf{i}_V(\omega)\in\Lambda^{k-1}(M)$ defined by

$$\mathbf{i}_V(\omega)=\frac{1}{(k-1)!}v^{i}\omega_{ii_1i_2\cdots i_{k-1}}dx^{i_1}\wedge dx^{i_2}\wedge\cdots\wedge dx^{i_{k-1}}. \qquad (5.4.2)$$

This expression can also be rewritten in term of essential components as

$$\mathbf{i}_V(\omega)=\sum_{1\le i_1<\cdots<i_{k-1}\le m}v^{i}\omega_{ii_1i_2\cdots i_{k-1}}dx^{i_1}\wedge dx^{i_2}\wedge\cdots\wedge dx^{i_{k-1}}.$$

When we recall, together with the rule (5.4.1.(iii)), that $\mathbf{i}_V(f\omega)=f\mathbf{i}_V(\omega)$ for a function $f\in\Lambda^0(M)$, we immediately see that *the operator* $\mathbf{i}_V$ *is linear over the module* $\mathfrak{V}(M)$. Next, let us consider k arbitrary vector fields V,

$V_1, V_2, \ldots, V_{k-1} \in T(M)$. We know that the value of a form $\omega \in \Lambda^k(M)$ on these vectors is given by

$$\omega(V, V_1, V_2, \ldots, V_{k-1}) = \omega_{ii_1\cdots i_{k-1}} v^i v_1^{i_1} \cdots v_{k-1}^{i_{k-1}}.$$

On the other hand, according to (5.4.2) the value of the form $\mathbf{i}_V(\omega) \in \Lambda^{k-1}(M)$ on vectors $V_1, V_2, \ldots, V_{k-1}$ is found as

$$\mathbf{i}_V(\omega)(V_1, V_2, \ldots, V_{k-1}) = \omega_{ii_1\cdots i_{k-1}} v^i v_1^{i_1} \cdots v_{k-1}^{i_{k-1}}.$$

Therefore, for every vector fields $V, V_1, V_2, \ldots, V_{k-1}$ the equality

$$\mathbf{i}_V(\omega)(V_1, V_2, \ldots, V_{k-1}) = \omega(V, V_1, V_2, \ldots, V_{k-1}) \tag{5.4.3}$$

holds. *Actually, it can be shown that this relation may be employed to define the interior product operator.*

Example 5.4.1. Let the form $\omega \in \Lambda^2(M)$ be given by

$$\omega = \frac{1}{2}\,\omega_{ij}\, dx^i \wedge dx^j, \ \omega_{ji} = -\,\omega_{ij}.$$

Interior product of this form with a vector field V becomes

$$\mathbf{i}_V(\omega) = v^i \omega_{ij}\, dx^j \in \Lambda^1(M) \qquad \blacksquare$$

Let us now calculate the interior product of the form $\omega \in \Lambda^k(M)$ with two vector fields V_1 and V_2 successively. It follows from (5.4.2) by renaming dummy indices that

$$\mathbf{i}_{V_2}\big(\mathbf{i}_{V_1}(\omega)\big) = (\mathbf{i}_{V_2} \circ \mathbf{i}_{V_1})(\omega) = \frac{1}{(k-2)!} v_2^j v_1^i \omega_{iji_1\cdots i_{k-2}}\, dx^{i_1} \wedge \cdots \wedge dx^{i_{k-2}}.$$

It is clear that $(\mathbf{i}_{V_2} \circ \mathbf{i}_{V_1})(\omega) \in \Lambda^{k-2}(M)$. Let us now change the order of the vectors in the interior product. On recalling that the coefficients $\omega_{iji_3\cdots i_k}$ are antisymmetric with respect to indices i and j, we get

$$\begin{aligned}(\mathbf{i}_{V_2} \circ \mathbf{i}_{V_1})(\omega) &= -\frac{1}{(k-2)!} v_1^i v_2^j \omega_{jii_1\cdots i_{k-2}}\, dx^{i_1} \wedge \cdots \wedge dx^{i_{k-2}} \\ &= -\,(\mathbf{i}_{V_1} \circ \mathbf{i}_{V_2})(\omega).\end{aligned}$$

Since this relation must be valid for every form $\omega \in \Lambda(M)$, we arrive at the anticommutativity property of the interior product:

$$\mathbf{i}_{V_1} \circ \mathbf{i}_{V_2} = -\,\mathbf{i}_{V_2} \circ \mathbf{i}_{V_1}. \tag{5.4.4}$$

Thus for every vector V, we get the result

$$\mathbf{i}_V \circ \mathbf{i}_V = \mathbf{i}_V^2 = 0. \tag{5.4.5}$$

The successive interior products of a k-form with l vector fields where $l \leq k$ is the $(k-l)$-form given below:

$$(\mathbf{i}_{V_l} \circ \cdots \circ \mathbf{i}_{V_1})(\omega) = \frac{1}{(k-l)!} v_1^{i_1} \cdots v_l^{i_l} \omega_{i_1 \cdots i_l i_{l+1} \cdots i_k} dx^{i_{l+1}} \wedge \cdots \wedge dx^{i_k}.$$

Evidently the operator $\mathbf{i}_{V_l} \circ \cdots \circ \mathbf{i}_{V_1}$ is completely antisymmetric:

$$\mathbf{i}_{V_l} \circ \cdots \circ \mathbf{i}_{V_p} \circ \cdots \circ \mathbf{i}_{V_q} \circ \cdots \circ \mathbf{i}_{V_1} = -\mathbf{i}_{V_l} \circ \cdots \circ \mathbf{i}_{V_q} \circ \cdots \circ \mathbf{i}_{V_p} \circ \cdots \circ \mathbf{i}_{V_1}.$$

It is readily observed that for k vector fields $V_1, \ldots, V_l, V_{l+1}, \ldots, V_k$, we obtain

$$(\mathbf{i}_{V_l} \circ \cdots \circ \mathbf{i}_{V_1})(\omega)(V_{l+1}, \ldots, V_k) = \omega(V_1, \ldots, V_l, V_{l+1}, \ldots, V_k). \tag{5.4.6}$$

If we take $l = k$, we conclude that

$$(\mathbf{i}_{V_k} \circ \cdots \circ \mathbf{i}_{V_1})(\omega) = v_1^{i_1} v_2^{i_2} \cdots v_k^{i_k} \omega_{i_1 i_2 \cdots i_k} = \omega(V_1, V_2, \ldots, V_k).$$

Thus the successive interior products of a k-form with k ordered vector fields yields the value of this form on these vectors. If $l > k$, then the successive interior products of a k-form with l vectors vanishes identically.

It follows from the definition (5.4.2) that

$$\begin{aligned}
\mathbf{i}_{V_1+V_2}(\omega) &= \frac{1}{(k-1)!}(v_1^i + v_2^i)\, \omega_{i i_1 i_2 \cdots i_{k-1}} dx^{i_1} \wedge dx^{i_2} \wedge \cdots \wedge dx^{i_{k-1}} \\
&= \mathbf{i}_{V_1}(\omega) + \mathbf{i}_{V_2}(\omega), \\
\mathbf{i}_{fV}(\omega) &= \frac{1}{(k-1)!} f v^i \omega_{i i_1 i_2 \cdots i_{k-1}}\, dx^{i_1} \wedge dx^{i_2} \wedge \cdots \wedge dx^{i_{k-1}} = f\mathbf{i}_V(\omega).
\end{aligned}$$

Since these relations are valid for every form $\omega \in \Lambda(M)$, then we reach to the following properties:

$$\mathbf{i}_{V_1+V_2} = \mathbf{i}_{V_1} + \mathbf{i}_{V_2}, \quad \mathbf{i}_{fV} = f\mathbf{i}_V. \tag{5.4.7}$$

Next, let us assume that the forms ω and Ω satisfy the degree condition $deg\,(\Omega) \leq deg\,(\omega)$. If we can find a form ω_1 so that one is able to write $\omega = \omega_1 \wedge \Omega$, the form Ω is called a ***divisor*** of the form ω. It is obvious that $deg\,(\omega_1) = deg\,(\omega) - deg\,(\Omega)$.

Theorem 5.4.1. *A* 1*-form* $\Omega \neq 0$ *is a divisor of a form* $\omega \in \Lambda(M)$ *with non-vanishing degree if and only if* $\omega \wedge \Omega = 0$.

Evidently, this is the necessary condition. If we can write $\omega = \omega_1 \wedge \Omega$, then we obtain $\omega \wedge \Omega = \omega_1 \wedge \Omega \wedge \Omega = 0$ since $\Omega \in \Lambda^1(M)$. We now prove that it is also the sufficient condition. Let us write $\Omega = \Omega_i\, dx^i, i = 1, \ldots, m$. Since $\Omega \neq 0$, at least one of the coefficients should be different from zero. By changing the ordering, if necessary, we take $\Omega_1 \neq 0$. Let us choose a new basis in $T^*(M)$ as follows

$$\theta^1 = \Omega,\, \theta^2 = dx^2,\, \ldots\, ,\, \theta^m = dx^m.$$

The transformation of bases is designated by

$$\begin{bmatrix} \theta^1 \\ \theta^2 \\ \vdots \\ \theta^m \end{bmatrix} = \begin{bmatrix} \Omega_1 & \Omega_2 & \cdots & \Omega_m \\ 0 & 1 & \cdots & 0 \\ \vdots & \vdots & \vdots & \vdots \\ 0 & 0 & \cdots & 1 \end{bmatrix} \begin{bmatrix} dx^1 \\ dx^2 \\ \vdots \\ dx^m \end{bmatrix}.$$

Since $\Omega_1 \neq 0$, the determinant of the matrix of transformation does not vanish. Hence, the inverse transformation becomes

$$dx^1 = \frac{1}{\Omega_1}\,\Omega - \frac{\Omega_2}{\Omega_1}\,\theta^2 - \cdots - \frac{\Omega_m}{\Omega_1}\,\theta^m,\; dx^i = \theta^i, i = 2, \ldots, m$$

On inserting these 1-forms into ω and noting that the square of a 1-form is zero, we arrive at the expression

$$\omega = \omega_1 \wedge \Omega + \omega_2$$

where we must have $deg\,(\omega_1) = deg\,(\omega) - 1$ and $deg\,(\omega_2) = deg\,(\omega)$. The form Ω is not included in forms ω_1 and ω_2. We thus get

$$0 = \omega \wedge \Omega = \omega_1 \wedge \Omega \wedge \Omega + \omega_2 \wedge \Omega = \omega_2 \wedge \Omega$$

whence we deduce that $\omega_2 = 0$. Hence, one writes $\omega = \omega_1 \wedge \Omega$. □

An immediate corollary of this theorem can be expressed in the following manner: *If linearly independent forms $\Omega^1, \Omega^2, \ldots, \Omega^r \in \Lambda^1(M)$ are divisors of a form $\omega \in \Lambda^k(M)$, then the form $\Omega^1 \wedge \Omega^2 \wedge \cdots \wedge \Omega^r$ is also a divisor of ω.*

Indeed if Ω^1 is a divisor, then we write $\omega \wedge \Omega^1 = 0$ and $\omega = \omega_1 \wedge \Omega^1$. Since Ω^2 is also a divisor, the relation $0 = \omega \wedge \Omega^2 = \omega_1 \wedge \Omega^1 \wedge \Omega^2$ should be satisfied. But Ω^1 and Ω^2 are linearly independent so that $\Omega^1 \wedge \Omega^2 \neq 0$. Consequently, we find $\omega_1 \wedge \Omega^2 = 0$. Thus Ω^2 must be a divisor of ω_1. Hence, we have to write $\omega_1 = \omega_2 \wedge \Omega^2$. Continuing this way, we reach to the result

$$\omega = \omega_{\underline{r}} \wedge \Omega^1 \wedge \Omega^2 \wedge \cdots \wedge \Omega^{\underline{r}}.$$ □

If $\omega \in \Lambda^k(M)$, then the condition $\omega \wedge \Omega = 0$ which secures that 1-form Ω is a divisor of ω is cast into the relation

$$\frac{1}{k!}\omega_{i_1 i_2\cdots i_k}\Omega_i\, dx^{i_1}\wedge dx^{i_2}\wedge\cdots\wedge dx^{i_k}\wedge dx^i = 0$$

whence we deduce that the following $\begin{pmatrix} m \\ k+1 \end{pmatrix}$ expressions

$$\Omega_{[i}\omega_{i_1 i_2\cdots i_k]} = 0. \tag{5.4.8}$$

should be satisfied.

We can easily identify through the interior product whether a given k-form is simple.

Theorem 5.4.2. *Let $\omega\in\Lambda^k(M)$ be a non-zero form. We construct a form $\Omega\in\Lambda^1(M)$ as follows*

$$\Omega = (\mathbf{i}_{V_{k-1}}\circ\cdots\circ\mathbf{i}_{V_2}\circ\mathbf{i}_{V_1})(\omega)$$

where $V_1, V_2, \ldots, V_{k-1}\in T(M)$. The form ω is simple if and only if $\omega\wedge\Omega = 0$ for all vector fields $V_1, V_2, \ldots, V_{k-1}\in T(M)$.

To show that this is the necessary condition, let us suppose that ω is a simple form, in other words, it is expressible as $\omega = \omega^1\wedge\omega^2\wedge\cdots\wedge\omega^k$ where $\omega^r\in\Lambda^1(M)$, $r = 1,\ldots,k$. Next, we shall try to determine a basis $\{U_1,\ldots,\ U_k, U_{k+1},\ \ldots, U_m\}$ of the tangent bundle $T(M)$ in such a way that they possess the following properties:

$$\mathbf{i}_{U_\alpha}(\omega^r) = \delta^r_\alpha,\ r = 1,\ldots,k;\alpha = 1,\ldots,k,k+1,\ldots,m.$$

To this end, let us write $U_\alpha = u^i_\alpha\partial_i$ and $\omega^r = \omega^r_i\, dx^i$ in terms of local coordinates. Since $\omega\neq 0$, the forms ω^r are linearly independent. Therefore, the rank of the $k\times m$ matrix $[\omega^r_i]$ is k. We then split the relation $\mathbf{i}_{U_\alpha}(\omega^r) = \omega^r_i u^i_\alpha = \delta^r_\alpha, i = 1,\ldots,m$ into following expressions

$$\begin{aligned} &\omega^r_A u^A_s + \omega^r_\Gamma u^\Gamma_s = \delta^r_s,\ r,s,A = 1,\ldots,k;\ \Gamma = k+1,\ldots,m, \\ &\omega^r_A u^A_\Delta + \omega^r_\Gamma u^\Gamma_\Delta = 0,\ r,A = 1,\ldots,k;\ \Gamma,\Delta = k+1,\ldots,m. \end{aligned} \tag{5.4.9}$$

We may assume without loss of generality that $\det[\omega^r_A]\neq 0$. We thus obtain from (5.4.9) that

$$\begin{aligned} u^A_s &= (\omega^{-1})^A_s - (\omega^{-1})^A_r\,\omega^r_\Gamma u^\Gamma_s, \\ u^A_\Delta &= -\,(\omega^{-1})^A_r\,\omega^r_\Gamma u^\Gamma_\Delta. \end{aligned}$$

On defining

$$\Omega^A_\Gamma = -\,(\omega^{-1})^A_r\,\omega^r_\Gamma,$$

we find that

$$u_r^A = (\omega^{-1})_r^A + \Omega_\Gamma^A u_r^\Gamma, \quad u_\Delta^A = \Omega_\Gamma^A u_\Delta^\Gamma.$$

Hence, the basis vectors U_α meeting the desired conditions can now be expressed as

$$\begin{aligned} U_r &= u_r^i \frac{\partial}{\partial x^i} = u_r^A \frac{\partial}{\partial x^A} + u_r^\Gamma \frac{\partial}{\partial x^\Gamma} \\ &= \big[(\omega^{-1})_r^A + \Omega_\Gamma^A u_r^\Gamma\big] \frac{\partial}{\partial x^A} + u_r^\Gamma \frac{\partial}{\partial x^\Gamma}, \\ U_\Gamma &= u_\Gamma^i \frac{\partial}{\partial x^i} = u_\Gamma^A \frac{\partial}{\partial x^A} + u_\Gamma^\Delta \frac{\partial}{\partial x^\Delta} \\ &= u_\Gamma^\Delta \big[\frac{\partial}{\partial x^\Delta} + \Omega_\Delta^A \frac{\partial}{\partial x^A}\big]. \end{aligned}$$

If we introduce vectors W_A and W_Γ by

$$W_A = \frac{\partial}{\partial x^A}, \quad W_\Gamma = \frac{\partial}{\partial x^\Gamma} + \Omega_\Gamma^A \frac{\partial}{\partial x^A}$$

we obtain

$$U_r = (\omega^{-1})_r^A W_A + u_r^\Gamma W_\Gamma, \quad U_\Gamma = u_\Gamma^\Delta W_\Delta$$

where $[u_r^\Gamma]$ and $[u_\Gamma^\Delta]$ are arbitrary matrices. We observe at once that m vectors $\{W_A, W_\Gamma\}$ are linearly independent. If we restrict the arbitrariness of the square matrix $[u_\Gamma^\Delta]$ such that it has a non-zero determinant, then the vectors $\{U_\alpha\}$ turn out to be linearly independent. Consequently, any vector field V_A with $A = 1, \ldots, k$ can now be expressed as a linear combination

$$V_A = c_A^\alpha U_\alpha = c_A^1 U_1 + \cdots + c_A^m U_m$$

where $c_A^\alpha, \alpha = 1, \ldots, m; A = 1, \ldots, k$ are arbitrary coefficient functions from which we get

$$\begin{aligned} \mathbf{i}_{V_A}(\omega) &= \sum_{r=1}^{k} (-1)^{r-1} \mathbf{i}_{V_A}(\omega^r)\, \omega^1 \wedge \cdots \wedge \omega^{r-1} \wedge \omega^{r+1} \wedge \cdots \wedge \omega^k \\ &= \sum_{r=1}^{k} (-1)^{r-1} c_A^\alpha \delta_\alpha^r\, \omega^1 \wedge \cdots \wedge \omega^{r-1} \wedge \omega^{r+1} \wedge \cdots \wedge \omega^k \\ &= \sum_{r=1}^{k} (-1)^{r-1} c_A^r\, \omega^1 \wedge \cdots \wedge \omega^{r-1} \wedge \omega^{r+1} \wedge \cdots \wedge \omega^k. \end{aligned}$$

Therefore, the $(k-1)$-form $\mathbf{i}_{V_1}(\omega)$ is now a linear combination of k *simple* $(k-1)$-forms. When we apply the operator $\mathbf{i}_{V_2}$ to this form, we see that the $(k-2)$-form $(\mathbf{i}_{V_2}\circ\mathbf{i}_{V_1})(\omega)$ is the linear combination of k *simple* $(k-2)$-forms. On continuing this way by applying the operators $\mathbf{i}_{V_1},\dots,\mathbf{i}_{V_{k-1}}$ successively to the form ω, we reduce the form Ω to the linear combination of k number of 1-forms ω^r:

$$\Omega=(\mathbf{i}_{V_{k-1}}\circ\cdots\circ\mathbf{i}_{V_1})(\omega^1\wedge\cdots\wedge\omega^k)=\lambda_r\omega^r=\lambda_1\omega^1+\cdots+\lambda_k\omega^k.$$

We thus conclude that

$$\omega\wedge\Omega=\omega^1\wedge\cdots\wedge\omega^k\wedge(\lambda_1\omega^1+\cdots+\lambda_k\omega^k)=0.$$

In order to show sufficiency, we consider the k-form

$$\omega=\frac{1}{k!}\,\omega_{i_1\cdots i_k}(\mathbf{x})\,dx^{i_1}\wedge\cdots\wedge dx^{i_k}\in\Lambda^k(M)$$

and the 1-form

$$\Omega=(\mathbf{i}_{V_{k-1}}\circ\cdots\circ\mathbf{i}_{V_1})(\omega)=\omega_{i_1\cdots i_{k-1}i_k}v_1^{i_1}\cdots v_{k-1}^{i_{k-1}}dx^{i_k}\in\Lambda^1(M)$$

which is made up by interior products with arbitrary vector fields $V_1,\dots,V_{k-1}$. Let us then write

$$\omega\wedge\Omega=\frac{1}{k!}\,\omega_{i_1\cdots i_k}\omega_{j_1\cdots j_{k-1}j_k}v_1^{j_1}\cdots v_{k-1}^{j_{k-1}}\,dx^{i_1}\wedge\cdots\wedge dx^{i_k}\wedge dx^{j_k}=0.$$

Since this equality must be satisfied for all vector fields $V_1,\dots,V_{k-1}$, we arrive at the conditions

$$\omega_{i_1\cdots i_k}\omega_{j_1\cdots j_{k-1}j_k}dx^{i_1}\wedge\cdots\wedge dx^{i_k}\wedge dx^{j_k}=0$$

leading to

$$\omega_{j_1\cdots j_{k-1}[j_k}\omega_{i_1\cdots i_k]}=0. \tag{5.4.10}$$

These conditions require that the completely antisymmetric coefficients $\omega_{i_1\cdots i_k}$ have to satisfy certain quadratic equations whose number is clearly $\binom{m}{k-1}\binom{m}{k+1}$. We shall now attempt to recognise the result brought about by these equation in a somewhat indirect way. Since we have presumed that $\omega\neq 0$, we can select $\omega_{12\cdots k}\neq 0$ by renaming, if necessary, reciprocal basis vectors. We then define the following 1-forms

$$\Omega^1 = \omega_{i23\cdots k}\, dx^i,\ \Omega^2 = \omega_{1i3\cdots k}\, dx^i, \ldots, \Omega^k = \omega_{123\cdots k-1i}\, dx^i.$$

Therefore, we can write with $r = 1, \ldots, k$ and $\Gamma = k+1, \ldots, m$

$$\Omega^r = \omega_{123\cdots k}\, dx^r + \omega_{123\cdots \underset{r}{\Gamma} \cdots k}\, dx^\Gamma \tag{5.4.11}$$
$$= \omega_{123\cdots k}\, dx^r + \omega_{123\cdots \underset{r}{k+1} \cdots k}\, dx^{k+1} + \cdots + \omega_{123\cdots \underset{r}{m} \cdots k}\, dx^m$$

These forms are linearly independent. In fact, if we write $c_r \Omega^r = 0$ where $c_r, r = 1, \ldots, k$ are arbitrary coefficient functions, the relation

$$c_r \Omega^r = \omega_{123\cdots k} c_r\, dx^r + \sum_{r=1}^{k} \omega_{123\cdots \underset{r}{\Gamma} \cdots k} c_r\, dx^\Gamma = 0$$

requires that $c_r = 0, r = 1, \ldots, k$. On the other hand, a proper choice of indices $j_1, j_2, \ldots, j_{k-1}$ in (5.4.10) leads to the relations

$$\omega_{23\cdots k[i}\omega_{i_1\cdots i_k]} = 0, \quad \omega_{13\cdots k[i}\omega_{i_1\cdots i_k]} = 0, \quad \ldots \quad ,$$
$$\omega_{123\cdots (k-1)[i}\omega_{i_1\cdots i_k]} = 0.$$

In view of (5.4.8), we infer that the 1-forms $\Omega^1, \Omega^2, \ldots, \Omega^k$ are divisors of the form ω. Since these forms are linearly independent, we conclude that $\omega = \lambda \Omega^1 \wedge \cdots \wedge \Omega^k$. The factor λ can be found by equating coefficients of the form $dx^1 \wedge \cdots \wedge dx^k$ in both sides of this expression. Utilising (5.4.11), we end up with

$$\lambda = \frac{1}{(\omega_{123\cdots k})^{k-1}}.$$

Hence, on defining $\omega^1 = \lambda \Omega^1, \omega^2 = \Omega^2, \ldots, \omega^k = \Omega^k$, we get

$$\omega = \omega^1 \wedge \omega^2 \wedge \cdots \wedge \omega^k \qquad \square$$

Example 5.4.2. We consider the form $\omega = \frac{1}{2}\omega_{ij}\, dx^i \wedge dx^j \in \Lambda^2(M)$. The requirement that this form is to be a simple form can be written from (5.4.10) as follows

$$\omega_{i[j}\,\omega_{kl]} = 0 \ \text{ or } \ \omega_{ij}\,\omega_{kl} + \omega_{ik}\,\omega_{lj} + \omega_{il}\,\omega_{jk} = 0.$$

When this condition is met, we obtain

$$\Omega^1 = \omega_{i2}\, dx^i, \quad \Omega^2 = \omega_{1i}\, dx^i$$

if we take $\omega_{12} \neq 0$. Then we find that

$$\begin{aligned}\Omega^1 \wedge \Omega^2 &= \omega_{i2}\omega_{1j}\, dx^i \wedge dx^j = \omega_{[i2}\omega_{1j]}\, dx^i \wedge dx^j \\ &= \frac{1}{2}(\omega_{i2}\omega_{1j} - \omega_{j2}\omega_{1i})\, dx^i \wedge dx^j.\end{aligned}$$

On the other hand, the coefficients ω_{ij} are satisfying the relations

$$\omega_{i2}\omega_{1j} + \omega_{i1}\omega_{j2} + \omega_{ij}\omega_{21} = \omega_{i2}\omega_{1j} - \omega_{j2}\omega_{1i} - \omega_{ij}\omega_{12} = 0$$

so that we obtain $\omega_{i2}\omega_{1j} - \omega_{j2}\omega_{1i} = \omega_{12}\omega_{ij}$. This yields

$$\omega = \Omega^1 \wedge \Omega^2 / \omega_{12}.$$

Hence, if we choose

$$\omega^1 = \Omega^1/\omega_{12} \quad \text{and} \quad \omega^2 = \Omega^2,$$

we find that

$$\omega = \omega^1 \wedge \omega^2. \qquad \blacksquare$$

5.5. BASES INDUCED BY THE VOLUME FORM

The non-zero m-volume form μ on an m-dimensional manifold M was introduced by (5.2.5). On using Levi-Civita symbols defined in *p.* 31, this form can also be expressed as

$$\begin{aligned}\mu &= dx^1 \wedge dx^2 \wedge \cdots \wedge dx^m \\ &= \frac{1}{m!}\, e_{i_1 i_2 \cdots i_m}\, dx^{i_1} \wedge dx^{i_2} \wedge \cdots \wedge dx^{i_m}.\end{aligned}$$

Our aim is to derive a new set of basis forms for the exterior algebra that may prove to be more advantageous in certain cases than the natural basis. However, to fulfil this task, we have to reveal some novel properties of the generalised Kronecker deltas introduced previously by the expression (1.4.6):

$$\delta^{i_1 i_2 \cdots i_k}_{j_1 j_2 \cdots j_k} = \begin{vmatrix} \delta^{i_1}_{j_1} & \delta^{i_1}_{j_2} & \cdots & \delta^{i_1}_{j_k} \\ \delta^{i_2}_{j_1} & \delta^{i_2}_{j_2} & \cdots & \delta^{i_2}_{j_k} \\ \vdots & \vdots & & \vdots \\ \delta^{i_k}_{j_1} & \delta^{i_k}_{j_2} & \cdots & \delta^{i_k}_{j_k} \end{vmatrix}. \tag{5.5.1}$$

If we expand the $k \times k$ symbolic determinant (5.5.1) with respect to its first row, we obtain the following expression by adopting the convention that $\delta^{i_r}_{j_0}$

does not exist

$$\delta^{i_1 i_2 \cdots i_k}_{j_1 j_2 \cdots j_k} = \sum_{l=1}^{k} (-1)^{l-1} \delta^{i_1}_{j_l} \begin{bmatrix} \delta^{i_2}_{j_1} & \cdots & \delta^{i_2}_{j_{l-1}} & \delta^{i_2}_{j_{l+1}} & \cdots & \delta^{i_2}_{j_k} \\ \vdots & & \vdots & \vdots & & \vdots \\ \delta^{i_k}_{j_1} & \cdots & \delta^{i_k}_{j_{l-1}} & \delta^{i_k}_{j_{l+1}} & \cdots & \delta^{i_k}_{j_k} \end{bmatrix}$$

$$= \sum_{l=1}^{k} (-1)^{l-1} \delta^{i_1}_{j_l} \delta^{i_2 \cdots i_{l-1} i_l i_{l+1} \cdots i_k}_{j_1 \cdots j_{l-1} j_{l+1} \cdots j_k}$$

$$= \delta^{i_1}_{j_1} \delta^{i_2 \cdots i_k}_{j_2 \cdots j_k} + \sum_{l=2}^{k} (-1)^{l-1} \delta^{i_1}_{j_l} \delta^{i_2 \cdots i_{l-1} i_l i_{l+1} \cdots i_k}_{j_1 \cdots j_{l-1} j_{l+1} \cdots j_k}.$$

On the other hand, for $l \geq 2$ we can write

$$\begin{aligned} (-1)^{l-1} \delta^{i_2 \cdots i_{l-1} i_l i_{l+1} \cdots i_k}_{j_1 j_2 \cdots j_{l-1} j_{l+1} \cdots j_k} &= (-1)^{l-1+l-2} \delta^{i_2 \cdots i_{l-1} i_l i_{l+1} \cdots i_k}_{j_2 \cdots j_{l-1} j_1 j_{l+1} \cdots j_k} \\ &= (-1)^{2l-3} \delta^{i_2 \cdots i_{l-1} i_l i_{l+1} \cdots i_k}_{j_2 \cdots j_{l-1} j_1 j_{l+1} \cdots j_k} \\ &= -\, \delta^{i_2 \cdots i_{l-1} i_l i_{l+1} \cdots i_k}_{j_2 \cdots j_{l-1} j_1 j_{l+1} \cdots j_k} \end{aligned}$$

and find

$$\begin{aligned} \delta^{i_1 i_2 \cdots i_k}_{j_1 j_2 \cdots j_k} &= \delta^{i_1}_{j_1} \delta^{i_2 \cdots i_k}_{j_2 \cdots j_k} - \sum_{l=2}^{k} \delta^{i_1}_{j_l} \delta^{i_2 \cdots i_{l-1} i_l i_{l+1} \cdots i_k}_{j_2 \cdots j_{l-1} j_1 j_{l+1} \cdots j_k} \qquad (5.5.2) \\ &= \delta^{i_1}_{j_1} \delta^{i_2 \cdots i_k}_{j_2 \cdots j_k} - \delta^{i_1}_{j_2} \delta^{i_2 i_3 \cdots i_k}_{j_1 j_3 \cdots j_k} - \delta^{i_1}_{j_3} \delta^{i_2 i_3 i_4 \cdots i_k}_{j_2 j_1 j_4 \cdots j_k} - \cdots - \delta^{i_1}_{j_k} \delta^{i_2 i_3 \cdots i_{k-1} i_k}_{j_2 j_3 \cdots j_{k-1} j_1}. \end{aligned}$$

On making a contraction on the indices i_1 and j_1 in (5.5.2) by taking $i_1 = j_1$, we arrive at

$$\begin{aligned} \delta^{i_1 i_2 \cdots i_k}_{i_1 j_2 \cdots j_k} &= m \delta^{i_2 \cdots i_k}_{j_2 \cdots j_k} - (k-1) \delta^{i_2 \cdots i_k}_{j_2 \cdots j_k} \qquad (5.5.3) \\ &= (m-k+1) \delta^{i_2 \cdots i_k}_{j_2 \cdots j_k}. \end{aligned}$$

When we repeat this operation r times, we conclude that

$$\delta^{i_1 \cdots i_r i_{r+1} \cdots i_k}_{i_1 \cdots i_r j_{r+1} \cdots j_k} = \qquad (5.5.4)$$

$$(m-k+1)(m-k+2) \cdots (m-k+r) \delta^{i_{r+1} \cdots i_k}_{j_{r+1} \cdots j_k}.$$

Let us next take $k = m$ in the expression above. We thus conclude that (5.5.4) then yields

$$\delta^{i_1 \cdots i_r i_{r+1} \cdots i_m}_{i_1 \cdots i_r j_{r+1} \cdots j_m} = r!\, \delta^{i_{r+1} \cdots i_m}_{j_{r+1} \cdots j_m} \qquad (5.5.5)$$

so one deduces that

$$\delta^{i_1 i_2 \cdots i_m}_{i_1 i_2 \cdots i_m} = m!. \tag{5.5.6}$$

We know from (1.4.16) that we can write

$$\delta^{i_1 i_2 \cdots i_m}_{j_1 j_2 \cdots j_m} = e^{i_1 i_2 \cdots i_m} e_{j_1 j_2 \cdots j_m}. \tag{5.5.7}$$

Hence, making use of (5.5.5) we can reach to the relation

$$\delta^{i_1 \cdots i_r}_{j_1 \cdots j_r} = \frac{1}{(m-r)!} e^{i_1 \cdots i_r i_{r+1} \cdots i_m} e_{j_1 \cdots j_r i_{r+1} \cdots i_m}.$$

We now define m number of $(m-1)$-forms as follows

$$\mu_i = \mathbf{i}_{\partial_i}(\mu) = \frac{1}{(m-1)!}\, e_{i i_2 \cdots i_m} dx^{i_2} \wedge \cdots \wedge dx^{i_m} \in \Lambda^{m-1}(M). \tag{5.5.8}$$

Let us next evaluate the exterior product of a form μ_i with dx^j to obtain

$$\begin{aligned} dx^j \wedge \mu_i &= \frac{1}{(m-1)!}\, e_{i i_2 \cdots i_m}\, dx^j \wedge dx^{i_2} \wedge \cdots \wedge dx^{i_m} \\ &= \frac{1}{(m-1)!}\, e_{i i_2 \cdots i_m} e^{j i_2 \cdots i_m}\, dx^1 \wedge dx^2 \wedge \cdots \wedge dx^m \\ &= \frac{1}{(m-1)!} \delta^{j i_2 \cdots i_m}_{i i_2 \cdots i_m} \mu = \frac{(m-1)!}{(m-1)!} \delta^j_i \mu. \\ &= \delta^j_i \mu \in \Lambda^m(M) \end{aligned} \tag{5.5.9}$$

We now write $c^i \mu_i = 0$ where c^i are arbitrary functions. The exterior product of this zero form with dx^j is

$$0 = c^i dx^j \wedge \mu_i = c^i \delta^j_i \mu = c^j \mu.$$

Since, μ does not vanish we deduce that $c^j = 0, j = 1, \ldots, m$. Thus m forms $\mu_i \in \Lambda^{m-1}(M)$ are linearly independent and they constitute a basis for the module $\Lambda^{m-1}(M)$.

We shall now try to determine top-down generated bases for the modules $\Lambda^{m-k}(M)$ for $k = 0, 1, \ldots, m$ in an exactly similar fashion. To this end, we introduce the forms

$$\begin{aligned} \mu_{i_k i_{k-1} \cdots i_1} &= (\mathbf{i}_{\partial_{i_k}} \circ \mathbf{i}_{\partial_{i_{k-1}}} \circ \cdots \circ \mathbf{i}_{\partial_{i_1}})(\mu) \\ &= \frac{1}{(m-k)!}\, e_{i_1 \cdots i_k i_{k+1} \cdots i_m}\, dx^{i_{k+1}} \wedge \cdots \wedge dx^{i_m} \in \Lambda^{m-k}(M). \end{aligned} \tag{5.5.10}$$

Because of the properties of the interior product, these forms have to be completely antisymmetric:

$$\mu_{i_k i_{k-1}\cdots i_1} = \mu_{[i_k i_{k-1}\cdots i_1]}.$$

Therefore, the number of their independent components is $\binom{m}{k} = \binom{m}{m-k}$ which is equal to the dimension of the module $\Lambda^{m-k}(M)$. By adopting the convention $\mu_{i_0} = \mu$, the definition (5.5.10) leads to

$$\mu_{i_k i_{k-1}\cdots i_1} = \mathbf{i}_{\partial_{i_k}}(\mu_{i_{k-1}\cdots i_1}), \quad 1 \le k \le m. \tag{5.5.11}$$

On using Levi-Civita symbols, we obtain from (5.5.10) that

$$\begin{aligned} e^{i_1\cdots i_k j_{k+1}\cdots j_m}\mu_{i_k\cdots i_1} &= \frac{1}{(m-k)!}\,\delta^{i_1\cdots i_k j_{k+1}\cdots j_m}_{i_1\cdots i_k i_{k+1}\cdots i_m} dx^{i_{k+1}} \wedge \cdots \wedge dx^{i_m} \\ &= \frac{k!}{(m-k)!}\,\delta^{j_{k+1}\cdots j_m}_{i_{k+1}\cdots i_m} dx^{i_{k+1}} \wedge \cdots \wedge dx^{i_m} \\ &= k!\, dx^{[j_{k+1}} \wedge \cdots \wedge dx^{j_m]} \\ &= k!\, dx^{j_{k+1}} \wedge \cdots \wedge dx^{j_m} \end{aligned}$$

where we have employed (1.4.8). We thus find the inverse relation

$$dx^{i_{k+1}} \wedge \cdots \wedge dx^{i_m} = \frac{1}{k!}\, e^{i_1\cdots i_k i_{k+1}\cdots i_m}\mu_{i_k\cdots i_1}. \tag{5.5.12}$$

Let us now choose $m - (k - l) \le m$, namely, $l \le k$. In this case the form

$$dx^{j_1} \wedge \cdots \wedge dx^{j_l} \wedge \mu_{i_k\cdots i_1}$$

becomes obviously a $(m - k + l)$-form. The explicit evaluation of that form by making use of (5.5.10) and (5.5.12) gives

$$\begin{aligned} &dx^{j_1} \wedge \cdots \wedge dx^{j_l} \wedge \mu_{i_k\cdots i_1} \\ &\quad= \frac{1}{(m-k)!}\, e_{i_1\cdots i_k i_{k+1}\cdots i_m} dx^{j_1} \wedge \cdots \wedge dx^{j_l} \wedge dx^{i_{k+1}} \wedge \cdots \wedge dx^{i_m} \\ &\quad= \frac{1}{(m-k)!}\frac{1}{(k-l)!}\, e_{i_1\cdots i_k i_{k+1}\cdots i_m} e^{s_1\cdots s_{k-l} j_1\cdots j_l i_{k+1}\cdots i_m}\mu_{s_{k-l}\cdots s_1} \\ &\quad= \frac{1}{(k-l)!}\,\delta^{s_1\cdots s_{k-l} j_1\cdots j_l}_{i_1 i_2\cdots i_{k-2} i_{k-1} i_k}\mu_{s_{k-l}\cdots s_1}. \end{aligned} \tag{5.5.13}$$

If we take $l = k$, then (5.5.13) leads to

$$dx^{j_1} \wedge dx^{j_2} \wedge \cdots \wedge dx^{j_k} \wedge \mu_{i_k\cdots i_2 i_1} = \delta^{j_1 j_2\cdots j_k}_{i_1 i_2\cdots i_k}\mu \tag{5.5.14}$$

since we have assumed that $\mu_{s_0} = \mu$. After having this relation on hand, we can easily demonstrate that the forms $\mu_{i_k \cdots i_1}$ constitute a basis for the module $\Lambda^{m-k}(M)$. Let us write

$$c^{i_1 \cdots i_k} \mu_{i_k \cdots i_1} = 0$$

where $c^{i_1 \cdots i_k}$ are arbitrary smooth functions. It is obvious that we can select the coefficient functions $c^{i_1 \cdots i_k}$ as being completely antisymmetric, that is, satisfying relations

$$c^{i_1 \cdots i_k} = c^{[i_1 \cdots i_k]}$$

without loss of generality. The exterior product of the above linear combination with the form $dx^{j_1} \wedge \cdots \wedge dx^{j_k}$ yields due to (5.5.14)

$$\begin{aligned} \delta^{j_1 j_2 \cdots j_k}_{i_1 i_2 \cdots i_k} c^{i_1 \cdots i_k} \mu &= k!\, c^{[i_1 \cdots i_k]} \mu \\ &= k!\, c^{i_1 \cdots i_k} \mu = 0. \end{aligned}$$

Since $\mu \neq 0$, we then deduce that all coefficients vanish, i.e., $c^{i_1 \cdots i_k} = 0$. Therefore, the forms $\mu_{i_k \cdots i_1}$ are linearly independent so they constitute a basis of the module $\Lambda^{m-k}(M)$. Consequently, we obtain the following sequence of ***top down generated bases*** for modules $\Lambda^m(M), \Lambda^{m-1}(M), \ldots, \Lambda^2(M), \Lambda^1(M), \Lambda^0(M)$ from the volume form μ:

$$\begin{aligned}
\Lambda^m(M) &: \mu = \frac{1}{m!} e_{i_1 i_2 \cdots i_m} dx^{i_1} \wedge dx^{i_2} \wedge \cdots \wedge dx^{i_m}, \\
\Lambda^{m-1}(M) &: \mu_i = \mathbf{i}_{\partial_i}(\mu) = \frac{1}{(m-1)!} e_{i i_2 \cdots i_m} dx^{i_2} \wedge \cdots \wedge dx^{i_m}, \\
\Lambda^{m-2}(M) &: \mu_{ji} = \mathbf{i}_{\partial_j}(\mu_i) = \frac{1}{(m-2)!} e_{i j i_3 \cdots i_m} dx^{i_3} \wedge \cdots \wedge dx^{i_m}, \\
&\vdots \\
\Lambda^{m-k}(M) &: \mu_{i_k i_{k-1} \cdots i_1} = \mathbf{i}_{\partial_{i_k}}(\mu_{i_{k-1} \cdots i_1}) \\
&\quad = \frac{1}{(m-k)!} e_{i_1 \cdots i_k i_{k+1} \cdots i_m} dx^{i_{k+1}} \wedge \cdots \wedge dx^{i_m}, \\
&\vdots \\
\Lambda^1(M) &: \mu_{i_{m-1} \cdots i_1} = \mathbf{i}_{\partial_{i_{m-1}}}(\mu_{i_{m-2} \cdots i_1}) = e_{i_1 \cdots i_{m-1} i_m}\, dx^{i_m}, \\
\Lambda^0(M) &: \mu_{i_m \cdots i_1} = e_{i_1 \cdots i_m} = \pm 1.
\end{aligned}$$

If we take $l = 1$ in (5.5.13) and utilise (5.5.2) the following result comes out

$$
\begin{aligned}
dx^i \wedge \mu_{i_k\cdots i_1} &= \frac{1}{(k-1)!}\delta^{j_1\cdots j_{k-1}i}_{i_1\cdots i_{k-1}i_k}\mu_{j_{k-1}\cdots j_1} \\
&= \frac{1}{(k-1)!}\delta^{ij_1\cdots j_{k-1}}_{i_k i_1\cdots i_{k-1}}\mu^{j_1 j_2\cdots j_{k-1}}_{j_{k-1}\cdots j_1} \\
&= \frac{1}{(k-1)!}\Big[\delta^i_{i_k}\delta^{j_1 j_2\cdots j_{k-1}}_{i_1 i_2\cdots i_{k-1}} - \delta^i_{i_1}\delta_{i_k i_2\cdots i_{k-1}} \\
&\quad - \delta^i_{i_2}\delta^{j_1 j_2\cdots j_{k-1}}_{i_1 i_k\cdots i_{k-1}} - \cdots - \delta^i_{i_{k-1}}\delta^{j_1 j_2\cdots j_{k-1}}_{i_1 i_2\cdots i_k}\Big]\mu_{j_{k-1}\cdots j_1} = \\
&\delta^i_{i_k}\mu_{[i_{k-1}\cdots i_2 i_1]} - \delta^i_{i_1}\mu_{[i_{k-1}\cdots i_2 i_k]} - \delta^i_{i_2}\mu_{[i_{k-1}\cdots i_k i_1]} - \cdots - \delta^i_{i_{k-1}}\mu_{[i_k\cdots i_2 i_1]} \\
&= \delta^i_{i_k}\mu_{i_{k-1}\cdots i_2 i_1} - \delta^i_{i_1}\mu_{i_{k-1}\cdots i_2 i_k} \\
&\quad - \delta^i_{i_2}\mu_{i_{k-1}\cdots i_k i_1} - \cdots - \delta^i_{i_{k-1}}\mu_{i_k\cdots i_2 i_1}.
\end{aligned}
$$

Finally, we observe that we can write

$$
dx^i \wedge \mu_{i_k\cdots i_1} = k\,\delta^i_{[i_k}\mu_{i_{k-1}\cdots i_2 i_1]} \tag{5.5.15}
$$

because of the complete antisymmetry of forms $\mu_{i_{k-1}\cdots i_2 i_1}$ with respect to its $k-1$ indices. Indeed, we find that

$$
\begin{aligned}
k\,\delta^i_{[i_k}\mu_{i_{k-1}\cdots i_2 i_1]} &= \frac{k}{k!}\,\delta^{j_1\cdots j_{k-1}j_k}_{i_1\cdots i_{k-1}i_k}\,\delta^i_{j_k}\mu_{j_{k-1}\cdots j_1} \\
&= \frac{1}{(k-1)!}\,\delta^{j_1\cdots j_{k-1}i}_{i_1\cdots i_{k-1}i_k}\mu_{j_{k-1}\cdots j_1}.
\end{aligned}
$$

For instance, we have the relations

$$
\begin{aligned}
dx^i \wedge \mu_{jk} &= 2\delta^i_{[j}\mu_{k]} = \delta^i_j\mu_k - \delta^i_k\mu_j, \\
dx^l \wedge \mu_{kji} &= 3\delta^l_{[k}\mu_{ji]} = \delta^l_k\mu_{ji} + \delta^l_j\mu_{ik} + \delta^l_i\mu_{kj}.
\end{aligned} \tag{5.5.16}
$$

Thus, a form $\omega \in \Lambda^{m-k}(M)$ is also expressible as

$$
\omega = \frac{1}{k!}\,\omega^{i_1 i_2\cdots i_k}(\mathbf{x})\,\mu_{i_k\cdots i_2 i_1} \tag{5.5.17}
$$

where the functions $\omega^{i_1 i_2\cdots i_k} \in \Lambda^0(M)$ are completely antisymmetric, that is, they satisfy the relation $\omega^{i_1 i_2\cdots i_k} = \omega^{[i_1 i_2\cdots i_k]}$.

On utilising this representation, we can readily prove that *every form in* $\Lambda^{m-1}(M)$ *is simple*. A non-zero form $\omega \in \Lambda^{m-1}(M)$ can now be expressed as $\omega = \omega^i\mu_i$. If 1-form $\Omega = \Omega_j\,dx^j$ is a divisor of the form ω, then the relation $\Omega \wedge \omega = 0$ or $\Omega_j\omega^i dx^j \wedge \mu_i = \Omega_j\omega^i\delta^j_i\mu = \Omega_i\omega^i\mu = 0$ must hold. This means that $\Omega_i\omega^i = \Omega_1\omega^1 + \Omega_2\omega^2 + \cdots + \Omega_m\omega^m = 0$. Since we have supposed that $\omega \neq 0$, then at least one coefficient does not vanish. Without loss of generality, we may choose that the coefficient ω^m is different from zero. We thus obtain

$$\Omega_m = -\frac{\omega^1}{\omega^m}\Omega_1 - \frac{\omega^2}{\omega^m}\Omega_2 - \cdots - \frac{\omega^{m-1}}{\omega^m}\Omega_{m-1}$$

and inserting this expression into the form Ω, we get

$$\Omega = \Omega_1(dx^1 - \frac{\omega^1}{\omega^m}\,dx^m) + \cdots + \Omega_{m-1}(dx^{m-1} - \frac{\omega^{m-1}}{\omega^m}\,dx^m).$$

Next, we define $m-1$ linearly independent 1-forms by

$$\Omega^1 = \omega^m\,dx^1 - \omega^1\,dx^m, \Omega^2 = dx^2 - \frac{\omega^2}{\omega^m}\,dx^m, \ldots, \Omega^{m-1} = dx^{m-1} - \frac{\omega^{m-1}}{\omega^m}\,dx^m$$

Each one of these forms divides the form ω. Hence, we can write

$$\omega = \Omega^1 \wedge \Omega^2 \wedge \cdots \wedge \Omega^{m-1}. \qquad \square$$

The interior product of a vector $V = v^i\partial_i$ with a form $\omega \in \Lambda^{m-k}(M)$ can now be expressed as follows

$$\begin{aligned}\mathbf{i}_V(\omega) &= \frac{1}{k!}\,v^i\omega^{i_1 i_2\cdots i_k}\mathbf{i}_{\partial_i}(\mu_{i_k\cdots i_2 i_1}) = \frac{1}{k!}\,v^i\omega^{i_1 i_2\cdots i_k}\mu_{i i_k\cdots i_2 i_1}\\ &= \frac{k+1}{(k+1)!}\,v^{[i}\omega^{i_1 i_2\cdots i_k]}\mu_{i i_k\cdots i_2 i_1} \in \Lambda^{m-(k+1)}(M).\end{aligned}$$

It is clear that a form $\omega \in \Lambda^{m-k}(M)$ can hereby be represented by resorting to two different bases as given below:

$$\omega = \frac{1}{k!}\,\omega^{i_1\cdots i_k}\mu_{i_k\cdots i_1} = \frac{1}{(m-k)!}\,\omega_{i_{k+1}\cdots i_m}dx^{i_{k+1}} \wedge \cdots \wedge dx^{i_m}.$$

When we employ (5.5.10) it follows from this expression that

$$\frac{1}{k!}\frac{1}{(m-k)!}\,\omega^{i_1\cdots i_k}\,e_{i_1\cdots i_k i_{k+1}\cdots i_m}dx^{i_{k+1}} \wedge \cdots \wedge dx^{i_m} = \frac{1}{(m-k)!}\,\omega_{i_{k+1}\cdots i_m}dx^{i_{k+1}} \wedge \cdots \wedge dx^{i_m}$$

so that coefficient functions are interrelated by

$$\omega_{i_{k+1}\cdots i_m} = \frac{1}{k!}\,e_{i_1\cdots i_k i_{k+1}\cdots i_m}\omega^{i_1\cdots i_k} \qquad (5.5.18)$$

After having performed some operations involving Levi-Civita symbols, we readily get

$$\begin{aligned}\omega_{i_{k+1}\cdots i_m} e^{j_1\cdots j_k i_{k+1}\cdots i_m} &= \frac{1}{k!}\,\delta^{j_1\cdots j_k i_{k+1}\cdots i_m}_{i_1\cdots i_k i_{k+1}\cdots i_m}\,\omega^{i_1\cdots i_k}\\ &= \frac{1}{k!}(m-k)!\,\delta^{j_1\cdots j_k}_{i_1\cdots i_k}\,\omega^{i_1\cdots i_k}\\ &= (m-k)!\,\omega^{[j_1\cdots j_k]}\\ &= (m-k)!\,\omega^{j_1\cdots j_k}\end{aligned}$$

and we finally reach to the inverse relation

$$\omega^{j_1\cdots j_k} = \frac{1}{(m-k)!}\,e^{j_1\cdots j_k i_{k+1}\cdots i_m}\omega_{i_{k+1}\cdots i_m}. \tag{5.5.19}$$

Let us consider a form $\omega \in \Lambda^k(M)$ given by

$$\omega = \frac{1}{k!}\,\omega_{i_1\cdots i_k}(\mathbf{x})\,dx^{i_1}\wedge\cdots\wedge dx^{i_k}$$

in the natural basis. On using the same functions $\omega_{i_1\cdots i_k}$, but transferring lower indices to upper indices to comply with the Einstein summation convention in its usual fashion, we may define a form $*\omega \in \Lambda^{m-k}(M)$ associated with the form $\omega \in \Lambda^k(M)$ by the relation

$$*\omega = \frac{1}{k!}\,\omega^{i_1 i_2\cdots i_k}\,\mu_{i_k\cdots i_2 i_1}. \tag{5.5.20}$$

The form $*\omega$ so obtained will called the ***Hodge dual*** of the form ω. This concept was first introduced by English mathematician William Vallance Douglas Hodge (1903-1975). We investigate properties of the Hodge dual a little bit later within the context of the Riemannian manifolds in detail and put the operation of raising the indices of component functions on a more solid foundation. Let us just point out that, according to (5.5.14) one is able to write

$$\begin{aligned}\omega\wedge *\omega &= \Big(\frac{1}{k!}\Big)^2\,\omega_{j_1\cdots j_k}\,\omega^{i_1\cdots i_k} dx^{j_1}\wedge\cdots\wedge dx^{j_k}\wedge\mu_{i_k\cdots i_1}\\ &= \Big(\frac{1}{k!}\Big)^2\,\delta^{j_1\cdots j_k}_{i_1\cdots i_k}\,\omega_{j_1\cdots j_k}\,\omega^{i_1\cdots i_k}\,\mu\\ &= \frac{1}{k!}\omega_{[i_1\cdots i_k]}\,\omega^{i_1\cdots i_k}\,\mu\\ &= \frac{1}{k!}\omega_{i_1\cdots i_k}\omega^{i_1\cdots i_k}\,\mu.\end{aligned}$$

As an example, consider a 1-form $\omega = \omega_i dx^i$. We then obtain

$$*\omega = \omega^i \mu_i = \frac{1}{(m-1)!}\, e_{ii_2\cdots i_m} \omega^i dx^{i_2} \wedge \cdots \wedge dx^{i_m} \in \Lambda^{m-1}(M),$$

and consequently

$$\omega \wedge *\omega = \omega^i \omega_i \mu.$$

5.6. IDEALS OF THE EXTERIOR ALGEBRA $\Lambda(M)$

Since $\Lambda(M)$ is an algebra, it is quite natural that we look for its ideals. A subset, or more precisely a subalgebra, of the exterior algebra $\Lambda(M)$ is called an ***ideal*** $\mathcal{I}$ (*homogeneous ideal*) of $\Lambda(M)$ if it satisfies the conditions below:

(i). *For every forms $\alpha, \beta \in \mathcal{I}$ of the same degree, one has $\alpha + \beta \in \mathcal{I}$.*

(ii). *If $\alpha \in \mathcal{I}$, then one has $\gamma \wedge \alpha = (-1)^{(deg\,\gamma)(deg\,\alpha)} \alpha \wedge \gamma \in \mathcal{I}$ for all $\gamma \in \Lambda(M)$.*

We see that only the sum of forms of the same degree in $\mathcal{I}$ is allowed. That is the reason why we call the ideal $\mathcal{I}$ as a *homogeneous ideal*. It is quite obvious that it is not possible for elements of the ideal to escape outside this subalgebra by means of exterior product.

Let us now consider some r members $\alpha_1, \alpha_2, \dots, \alpha_r$ of the exterior algebra $\Lambda(M)$ that can be of diverse degrees and construct all forms in the following shape

$$\beta = \gamma^1 \wedge \alpha_1 + \cdots + \gamma^r \wedge \alpha_r = \gamma^a \wedge \alpha_a, \;\; \gamma^a \in \Lambda(M), \; a = 1, \dots, r.$$

If the degree of the form β is p, then it is evident that the degree conditions given below must hold

$$deg\,\gamma^a + deg\,\alpha_a = p, \quad deg\,\alpha_a \le p, \quad a = 1, \dots, r.$$

We denote the collection of all members of $\Lambda(M)$ constructed this way by $\mathcal{I}(\alpha_1, \alpha_2, \dots, \alpha_r)$. Let two forms β_1 and β_2 of the same degree belong to $\mathcal{I}$. Hence, we can write

$$\beta_1 = \gamma^a_{(1)} \wedge \alpha_a, \quad \beta_2 = \gamma^a_{(2)} \wedge \alpha_a, \;\; \gamma^a_{(1)}, \gamma^a_{(2)} \in \Lambda(M)$$

so that we obtain

$$\beta_1 + \beta_2 = (\gamma^a_{(1)} + \gamma^a_{(2)}) \wedge \alpha_a.$$

Since $\gamma^a_{(1)} + \gamma^a_{(2)} \in \Lambda(M)$, we see that $\beta_1 + \beta_2 \in \mathcal{I}$. Similarly, if $\beta \in \mathcal{I}$ and $\sigma \in \Lambda(M)$ we have to write

$$\sigma \wedge \beta = \sigma \wedge (\gamma^a \wedge \alpha_a) = (\sigma \wedge \gamma^a) \wedge \alpha_a$$

where $\gamma^a \in \Lambda(M)$. Since $\sigma \wedge \gamma^a \in \Lambda(M)$, we find that $\sigma \wedge \beta \in \mathcal{I}$. These clearly results indicate that the set $\mathcal{I}(\alpha_1, \alpha_2, \dots, \alpha_r)$ so constructed by given forms that may be of various degrees is an ideal of the exterior algebra $\Lambda(M)$. The forms α_1, $\alpha_2, \dots, \alpha_r$ are then naturally called the ***generators*** of the ideal $\mathcal{I}$.

We say that an ideal $\mathcal{I}$ is generated by the forms $\alpha_1, \alpha_2, \dots, \alpha_r$ if each member of which is expressible as the sum of terms admitting at least one member of the set $\{\alpha_1, \alpha_2, \dots, \alpha_r\}$ as an exterior factor.

Example 5.6.1. Let us consider the exterior algebra $\Lambda(\mathbb{R}^4)$ and the coordinate cover $\{x^i\} = \{x, y, z, t\}$ for the manifold $\mathbb{R}^4$. We want to determine the members of the ideal generated by the forms

$$\begin{aligned}
\alpha_1 &= 2\,dx - 3y\,dz,\\
\alpha_2 &= x\,dy - z\,dt,\\
\alpha_3 &= x^2 t\,dx \wedge dt - t\,dy \wedge dz.
\end{aligned}$$

Since the lowest degree of the generating forms is 1, then this ideal cannot contain 0-forms, namely, smooth functions. Forms with degrees higher than 4 are identically zero. We can classify the forms in the ideal according to their degrees as follows:

1-forms: $\beta = f(2\,dx - 3y\,dz) + g(x\,dy - z\,dt), \;\; f, g \in \Lambda^0(\mathbb{R}^4)$

2-forms:

$$\beta = \gamma^1 \wedge \alpha_1 + \gamma^2 \wedge \alpha_2 + f\alpha_3$$

where

$$\gamma^a = f^a dx + g^a dy + h^a dz + k^a dt, \;\; f, f^a, g^a, h^a, k^a \in \Lambda^0(\mathbb{R}^4), a = 1, 2$$

so that we get

$$\begin{aligned}
\beta = &-(2g^1 - xf^2)\,dx \wedge dy - (3yf^1 + 2h^1)\,dx \wedge dz\\
&- (zf^2 + 2k^1 - x^2 tf)\,dx \wedge dt - (3yg^1 + xh^2 + tf)\,dy \wedge dz\\
&- (zg^2 + xk^2)\,dy \wedge dt + (3k^1 y - zh^2)\,dz \wedge dt
\end{aligned}$$

3-forms:

$$\beta = \gamma^1 \wedge \alpha_1 + \gamma^2 \wedge \alpha_2 + \gamma \wedge \alpha_3$$

where

$$\begin{aligned}\gamma^a &= f^a dx \wedge dy + g^a dx \wedge dz + h^a dx \wedge dt + k^a dy \wedge dz \\ &\quad + l^a dy \wedge dt + m^a dz \wedge dt, f^a, g^a, h^a, k^a, l^a, m^a \in \Lambda^0(\mathbb{R}^4), a = 1, 2, \\ \gamma &= f dx + g\, dy + h\, dz + k\, dt, \quad f, g, h, k \in \Lambda^0(\mathbb{R}^4),\end{aligned}$$

so that

$$\begin{aligned}\beta = &-(3yf^1 - 2k^1 + xg^2 + tf)\, dx \wedge dy \wedge dz, \\ &+ (2l^1 - zf^2 - xh^2 - x^2tg)\, dx \wedge dy \wedge dt, \\ &+ (3yh^1 + 2m^1 - zg^2 - x^2th)\, dx \wedge dz \wedge dt, \\ &+ (3yl^1 - zk^2 + xm^2 - tk)\, dy \wedge dz \wedge dt.\end{aligned}$$

4-*forms*:

$$\beta = \gamma^1 \wedge \alpha_1 + \gamma^2 \wedge \alpha_2 + \gamma \wedge \alpha_3$$

where

$$\begin{aligned}\gamma^a &= f^a dx \wedge dy \wedge dz + g^a dx \wedge dy \wedge dt + h^a dx \wedge dz \wedge dt \\ &\quad + k^a dy \wedge dz \wedge dt, \quad f^a, g^a, h^a, k^a \in \Lambda^0(\mathbb{R}^4), a = 1, 2, \\ \gamma &= f dx \wedge dy + g\, dx \wedge dz + h\, dx \wedge dt + k\, dy \wedge dz + l\, dy \wedge dt \\ &\quad + m\, dz \wedge dt + l\, dy \wedge dt + m\, dz \wedge dt, \; f, g, h, k, l, m \in \Lambda^0(\mathbb{R}^4),\end{aligned}$$

so that

$$\beta = (3yg^1 - 2k^1 - zf^2 + xh^2 - th + x^2tk)\, dx \wedge dy \wedge dz \wedge dt. \qquad \blacksquare$$

Let $\mathcal{I}$ be an ideal. If two forms $\alpha, \beta \in \Lambda(M)$ of the same degree are related by $\alpha - \beta \in \mathcal{I}$, we write $\alpha = \beta \bmod \mathcal{I}$ or, amounting to the same thing, $\alpha - \beta = 0 \bmod \mathcal{I}$. When we consider such kind of forms α and β, it becomes clear that we may use the representation $\gamma \wedge (\alpha - \beta) = 0 \bmod \mathcal{I}$ for all forms $\gamma \in \Lambda(M)$.

The ***characteristic vector fields*** of a form $\omega \in \Lambda(M)$ are defined as vector fields satisfying the condition

$$\mathbf{i}_V(\omega) = 0. \tag{5.6.1}$$

These vectors belong to a subbundle of the tangent bundle $T(M)$. Indeed, in view of (5.4.7), if $\mathbf{i}_V(\omega) = 0$ we then obtain $\mathbf{i}_{fV}(\omega) = f\mathbf{i}_V(\omega) = 0$ for all $f \in \Lambda^0(M)$. Likewise, if $\mathbf{i}_{V_1}(\omega) = \mathbf{i}_{V_2}(\omega) = 0$ we get $\mathbf{i}_{V_1+V_2}(\omega) = \mathbf{i}_{V_1}(\omega) + \mathbf{i}_{V_2}(\omega) = 0$. Therefore vectors fV and $V_1 + V_2$ are also characteristic vectors of the form ω. We can easily demonstrate that if the rank of the form defined in Sec. 1.6 is r, the number of linearly independent characteristic vector fields turns out to be $m - r$. Let us take the form $\omega \in \Lambda^k(M)$ into account. Then the relation

$$\mathbf{i}_V(\omega) = \frac{1}{(k-1)!} v^i \omega_{ii_1i_2\cdots i_{k-1}} dx^{i_1} \wedge dx^{i_2} \wedge \cdots \wedge dx^{i_{k-1}} = 0$$

results in $v^i \omega_{ii_1i_2\cdots i_{k-1}} = 0, 1 \leq i_1, i_2, \ldots, i_{k-1} \leq m$. If we note that these relations are identical with equations (1.6.3), we arrive at the fact that if the form possesses $m - r$ linearly independent characteristic vector fields, then its rank must be r. This amounts to say that there are exactly r linearly independent forms $\theta^\alpha \in \Lambda^1(M), \alpha = 1, \ldots, r$ so that ω is represented just as in (1.6.6) by the expression

$$\omega = \frac{1}{k!}\, \omega_{\alpha_1\alpha_2\cdots\alpha_k} \theta^{\alpha_1} \wedge \theta^{\alpha_2} \wedge \cdots \wedge \theta^{\alpha_k}. \tag{5.6.2}$$

When the rank r is equal to m, then the characteristic vector can only be the zero vector.

Let $\mathcal{I}$ be an ideal of the exterior algebra $\Lambda(M)$. If a vector field $V \in T(M)$ satisfies the condition $\mathbf{i}_V(\omega) \in \mathcal{I}$ for all forms $\omega \in \mathcal{I}$, then it is called a ***characteristic vector field of the ideal***[1]. If we recall the definition of an ideal and properties of the interior product, we immediately recognise that characteristic vector fields of an ideal form a submodule $\mathcal{S}(\mathcal{I}) \subseteq \mathfrak{V}(M)$ that is called the ***characteristic subspace of the ideal***. We thus symbolically write $\mathbf{i}_V(\mathcal{I}) \subseteq \mathcal{I}$ whenever $V \in \mathcal{S}(\mathcal{I})$.

Theorem 5.6.1. *Let $\mathcal{I}(\omega^1, \omega^2, \ldots, \omega^r)$ be an ideal of the exterior algebra $\Lambda(M)$ generated by the forms $\omega^1, \omega^2, \ldots, \omega^r \in \Lambda^k(M)$ of the same degree. A vector field $V \in T(M)$ is a characteristic vector field of the ideal $\mathcal{I}$ if and only if $\mathbf{i}_V(\omega^a) = 0, a = 1, 2, \ldots, r$.*

We suppose that $\mathbf{i}_V(\omega^a) = 0, a = 1, \ldots, r$. If $\alpha \in \mathcal{I}$, then we need to write $\alpha = \gamma_a \wedge \omega^a$ where all forms $\gamma_a \in \Lambda(M)$ ought to have the same degree. We thus obtain

$$\mathbf{i}_V(\alpha) = \mathbf{i}_V(\gamma_a) \wedge \omega^a + (-1)^{deg\,\gamma_a} \gamma_a \wedge \mathbf{i}_V(\omega^a) = \mathbf{i}_V(\gamma_a) \wedge \omega^a \in \mathcal{I}.$$

Conversely, let us assume that $\mathbf{i}_V(\alpha) \in \mathcal{I}$ for all $\alpha \in \mathcal{I}$. Consequently, this property is also valid for the forms $\alpha = f_a\, \omega^a \in \Lambda^k(M)$ where the functions $f_a \in \Lambda^0(M)$ are arbitrary. However, it is not possible for $(k-1)$-forms to belong to the ideal. Therefore, we can only write $\mathbf{i}_V(\alpha) = 0$. Hence, we conclude that

[1]Sometimes it is called a ***Cauchy characteristic vector field*** after Cauchy who had introduced the concept of characteristics to partial differential equations.

$$\mathbf{i}_V(\alpha) = f_a\, \mathbf{i}_V(\omega^a) = 0$$

and $\mathbf{i}_V(\omega^a) = 0, a = 1, \dots, r$ because the functions f_a are arbitrary. □

Naturally, Theorem 5.6.1 would also prevail for an ideal generated by the forms $\omega^1, \omega^2, \dots, \omega^r \in \Lambda^1(M)$. The characteristic vectors of such a special ideal will be called *the characteristic vectors of the exterior system* $\{\omega^a \in \Lambda^1(M), a = 1, \dots, r\}$.

Theorem 5.6.2. *The characteristic vectors of an exterior system* $\{\omega^a \in \Lambda^1(M), a = 1, \dots, r\}$ *engender a submodule* $\mathcal{S}$ *of* $\mathfrak{V}(M$. *If the forms* ω^a *are linearly independent, namely, if* $\Omega = \omega^1 \wedge \omega^2 \wedge \dots \wedge \omega^r \neq 0$, *then the dimension of* $\mathcal{S}$ *is* $m - r$.

We know that characteristic vectors of any ideal constitute a *characteristic subspace* $\mathcal{S}$. If $\Omega = \omega^1 \wedge \omega^2 \wedge \dots \wedge \omega^r \neq 0$, then the 1-forms $\omega^1, \dots, \omega^r$ are linearly independent. If we add $m - r$ linearly independent 1-forms $\omega^{r+1}, \dots, \omega^m \in \Lambda^1(M)$ to those forms, then the forms $\omega^1, \dots, \omega^m$ can now be chosen as a basis for $\Lambda^1(M) = T^*(M)$. As is well known, we can select a basis $\{V_i\}$ in $T(M)$ so that $\{\omega^i\}$ becomes reciprocal basis satisfying the relations

$$\mathbf{i}_{V_j}(\omega^i) = \omega^i(V_j) = \delta^i_j, \; i, j = 1, \dots, m.$$

We thus get

$$\mathbf{i}_{V_j}(\omega^i) = 0, \; i = 1, \dots, r; \; j = r + 1, \dots, m.$$

Therefore, $m - r$ linearly independent vectors $V_{r+1}, \dots, V_m$ are actually characteristic vectors of the exterior system. On the other hand, because of the relations

$$\mathbf{i}_{V_1}(\omega^1) = \mathbf{i}_{V_2}(\omega^2) = \cdots = \mathbf{i}_{V_r}(\omega^r) = 1$$

the vectors $V_1, \dots, V_r$ cannot be characteristic vectors of the exterior system. Hence, the dimension of the characteristic subspace $\mathcal{S}$ becomes $m - r$. □

It is seen right away from above that the relations

$$\mathbf{i}_{V_j}(\omega^i) = 0, \; i = r + 1, \dots, m; \; j = 1, \dots, r$$

together with

$$\mathbf{i}_{V_{r+1}}(\omega^{r+1}) = \mathbf{i}_{V_{r+2}}(\omega^{r+2}) = \cdots = \mathbf{i}_{V_m}(\omega^m) = 1$$

are satisfied as well. *This amounts to say that the vector fields* $V_1, \dots, V_r$ *are in turn characteristic vectors of the exterior system* $\{\omega^{r+1}, \dots, \omega^m\}$ *while vectors* $V_{r+1}, \dots, V_m$ *cannot be characteristic vectors of that system.* This means that the dimension of the characteristic subspace of the exterior system $\{\omega^{r+1},\dots,\omega^m\}$ is r. We can summarise the foregoing results by the symbolic relations

$$\Lambda^1(M) = \Lambda^1_{(r)}(M) \oplus \Lambda^1_{(m-r)}(M), \quad T(M) = T_{(m-r)}(M) \oplus T_{(r)}(M).$$

Moreover, if we denote the interior product by the hook operator $\rfloor$, we can also write

$$T_{(m-r)}(M) \rfloor \Lambda^1_{(r)}(M) = 0, \quad T_{(r)}(M) \rfloor \Lambda^1_{(m-r)}(M) = 0$$

whence we readily reach to the following conclusion:

Let $\mathcal{I}(\omega^a)$ be an ideal generated by 1-forms and let V be a characteristic vector field of this ideal. If one has $\mathbf{i}_V(\omega) \neq 0$ for a form $\omega \in \Lambda^1(M)$, then this form cannot belong to the ideal $\mathcal{I}(\omega^a)$ or, conversely, it is not possible to get $\mathbf{i}_V(\omega) = 0$ if $\omega \notin \mathcal{I}(\omega^a)$.

Let the ideal $\mathcal{I}$ be generated by forms $\omega^1,\dots,\omega^r \in \Lambda(M)$ of diverse degrees. Then we can provide the theorem below for a systematic determination of its characteristic vectors.

Theorem 5.6.3. *The necessary and sufficient conditions for a vector $V \in T(M)$ to be a characteristic vector of the ideal $\mathcal{I}(\omega^1,\omega^2,\dots,\omega^r)$ is the existence of forms $\lambda^a_b \in \Lambda(M)$ of suitable degrees such that the relations*

$$\mathbf{i}_V(\omega^a) = \lambda^a_b \wedge \omega^b, \quad a, b = 1, 2, \dots, r$$

are satisfied.

Let us suppose the vector field V holds the foregoing conditions. If ω is a member of the ideal, we can write $\omega = \gamma_a \wedge \omega^a$, $\gamma_a \in \Lambda(M)$. Clearly, one must have $deg\,(\gamma_a) + deg\,(\omega^a) = deg\,(\omega)$. We thus deduce that

$$\begin{aligned}\mathbf{i}_V(\omega) &= \mathbf{i}_V(\gamma_a) \wedge \omega^a + (-1)^{deg\,(\gamma_a)}\gamma_a \wedge \mathbf{i}_V(\omega^a)\\ &= \big(\mathbf{i}_V(\gamma_b) + (-1)^{deg\,(\gamma_a)}\gamma_a \wedge \lambda^a_b\big) \wedge \omega^b \in \mathcal{I}\end{aligned}$$

which means that V is a characteristic vector. Conversely, if V is a characteristic vector, then its interior product with any form in the ideal should lie within the ideal. This rule will of course be valid for the generators ω^a so that one must find forms λ^a_b so much so that the relations $\mathbf{i}_V(\omega^a) = \lambda^a_b \wedge \omega^b$ will hold. □

If $\mathcal{S}(\mathcal{I}) \subseteq T(M)$ is an r-dimensional characteristic subspace of an ideal $\mathcal{I}$, then for all *linearly independent* vectors $V_1,\dots,V_k \in \mathcal{S}, 1 \le k \le r$ and a form $\omega \in \mathcal{I}$ we clearly get

$$(\mathbf{i}_{V_k} \circ \cdots \circ \mathbf{i}_{V_1})(\omega) \in \mathcal{I}, \quad 1 \le k \le r.$$

We consider an ideal $\mathcal{I}(\omega^1, \omega^2, \dots, \omega^s)$ of $\Lambda(M)$ generated by forms of diverse degrees and assume that $\mathcal{S}(\mathcal{I})$ is its characteristic subspace with dimension $m - r$. $\mathcal{S}(\mathcal{I})$ is brought forth by linearly independent vector fields $V_{r+1}, \dots, V_m$. We can supply this set with arbitrary linearly independent vector fields $V_1, \dots, V_r$ to obtain a basis in the tangent bundle $T(M)$. We now pursue the path used in proving Theorem 5.6.2 to determine the reciprocal basis $\theta^1, \dots, \theta^m \in \Lambda^1(M)$ in the cotangent bundle $T^*(M)$ in such a way that we have

$$\mathbf{i}_{V_j}(\theta^i) = \theta^i(V_j) = \delta^i_j, \; i, j = 1, 2, \dots, m.$$

We thus obtain

$$\mathbf{i}_{V_a}(\theta^\alpha) = \delta^\alpha_a = 0, \; a = r + 1, \dots, m, \; \alpha = 1, \dots, r. \tag{5.6.3}$$

This means that the same vectors $V_a, a = r + 1, \dots, m$ span the $(m - r)$-dimensional characteristic subspace of the ideal $\mathcal{J}(\theta^\alpha)$ generated by 1-forms $\theta^\alpha, \alpha = 1, \dots, r$. In other words, we conclude that $\mathcal{S}(\mathcal{I}) = \mathcal{S}(\mathcal{J})$. *The number r is called the rank of the ideal $\mathcal{I}$.* Within this context, we can prove the following theorem.

Theorem 5.6.4. *Let $\mathcal{S}(\mathcal{I})$ be the $(m - r)$-dimensional characteristic subspace of an ideal $\mathcal{I}(\omega^A)$ generated by forms ω^A, $A = 1, \dots, s$ of various degrees. There exist linearly independent 1-forms $\theta^\alpha, \alpha = 1, \dots, r$ and if the ideal generated by these 1-forms is $\mathcal{J}(\theta^\alpha)$, then one finds $\mathcal{I}(\omega^A) \subseteq \mathcal{J}(\theta^\alpha)$.*

If $V_{r+1}, \dots, V_m \in T(M)$ is a basis of the characteristic subspace $\mathcal{S}(\mathcal{I})$, we first complete to a full basis of $T(M)$ as we have mentioned above, then we can construct the reciprocal basis $\theta^1, \dots, \theta^m \in \Lambda^1(M)$ of $T^*(M$. We define $m - r$ degree preserving mappings $h_a : \Lambda(M) \to \Lambda(M)$ where $a = r + 1, \dots, m$ by the rule

$$\sigma_a = h_a(\omega) = \omega - \theta^{\underline{a}} \wedge \mathbf{i}_{V_{\underline{a}}}(\omega) \tag{5.6.4}$$

Let us remember that the summation convention will be disabled on underscored indices. It is clear that $\sigma_a = h_a(\omega) \in \mathcal{I}$ whenever $\omega \in \mathcal{I}$. Next, we consider a generator ω^A of the ideal $\mathcal{I}$. Let us now introduce the forms $\sigma^A_a = h_a(\omega^A) = \omega^A - \theta^{\underline{a}} \wedge \mathbf{i}_{V_{\underline{a}}}(\omega^A) \in \mathcal{I}$ to find

$$\mathbf{i}_{V_{\underline{a}}}(\sigma^A_{\underline{a}}) = \mathbf{i}_{V_a}(\omega^A) - \mathbf{i}_{V_{\underline{a}}}(\theta^{\underline{a}})\, \mathbf{i}_{V_{\underline{a}}}(\omega^A) + \theta^{\underline{a}} \wedge \mathbf{i}^2_{V_{\underline{a}}}(\omega^A) = 0$$

where we have employed the relations $\mathbf{i}_{V_{\underline{a}}}(\theta^{\underline{a}}) = 1$ and $\mathbf{i}^2_{V_{\underline{a}}} = 0$. We see that the definition $\sigma^A_{ba} = h_b \circ h_a(\omega^A) = h_b(\sigma^A_a) = \sigma^A_a - \theta^{\underline{b}} \wedge \mathbf{i}_{V_{\underline{b}}}(\sigma^A_a) \in \mathcal{I}$ leads similarly to $\mathbf{i}_{V_{\underline{b}}}(\sigma^A_{\underline{ba}}) = 0$. Furthermore, since $\mathbf{i}_{V_{\underline{a}}}(\sigma^A_{\underline{a}}) = 0$ we obtain

$$\begin{aligned}\mathbf{i}_{V_{\underline{a}}}(\sigma^A_{b\underline{a}}) &= \mathbf{i}_{V_{\underline{a}}}(\sigma^A_{\underline{a}}) - \delta^b_a\,\mathbf{i}_{V_b}(\sigma^A_a) + \theta^{\underline{b}} \wedge \mathbf{i}_{V_{\underline{a}}} \circ \mathbf{i}_{V_{\underline{b}}}(\sigma^A_a)\\ &= -\,\theta^{\underline{b}} \wedge \mathbf{i}_{V_{\underline{b}}} \circ \mathbf{i}_{V_{\underline{a}}}(\sigma^A_a) = 0\end{aligned}$$

for $b \neq a$. These results clearly indicate that the forms

$$\sigma^A = \sigma^A_{m\cdots r+1} = (h_m \circ \cdots \circ h_{r+1})(\omega^A) \in \mathcal{I} \tag{5.6.5}$$

will satisfy the relations

$$\mathbf{i}_{V_a}(\sigma^A) = 0, \quad a = r+1, \ldots, m, \; A = 1, \ldots, s.$$

Thus, for all vectors $V \in \mathcal{S}(\mathcal{I})$ we find that

$$\mathbf{i}_V(\sigma^A) = 0, \quad A = 1, \ldots, s. \tag{5.6.6}$$

The rule of formation of the forms σ^A, which are of the same degree as the forms ω^A implies that $\mathcal{I}(\omega^A) = \mathcal{I}(\sigma^A)$. We now assume that $\sigma^A \in \Lambda^k(M)$. When we choose the 1-forms $\{\theta^i : i = 1, \ldots, m\}$ as a basis of $T^*(M)$, we can of course write

$$\sigma^A = \frac{1}{k!}\sigma^A_{i_1\cdots i_k}\theta^{i_1} \wedge \cdots \wedge \theta^{i_k}.$$

If we express a vector $V \in T(M)$ as $V = v^i V_i$ and pay attention that the vectors $\{V_i\}$ and the forms $\{\theta^i\}$ are reciprocal bases in $T(M)$ and $T^*(M)$, respectively, then we can describe the interior product of the form σ^A with the vector V as follows

$$\mathbf{i}_V(\sigma^A) = \frac{1}{(k-1)!}v^i\sigma^A_{ii_1i_2\cdots i_{k-1}}\theta^{i_1} \wedge \theta^{i_2} \wedge \cdots \wedge \theta^{i_{k-1}}$$

just as expressed in (5.4.2). On the other hand, when $V \in \mathcal{S}(\mathcal{I})$ we have to write $V = v^a V_a$. We thus get

$$\mathbf{i}_V(\sigma^A) = \frac{1}{(k-1)!}v^a\sigma^A_{ai_1i_2\cdots i_{k-1}}\theta^{i_1} \wedge \theta^{i_2} \wedge \cdots \wedge \theta^{i_{k-1}} = 0$$

since $v^i = 0$ for $i = 1, \ldots, r$. That yields $v^a\sigma^A_{ai_1i_2\cdots i_{k-1}} = 0$. Because this equality must be valid for every choice of functions $v^a \in \Lambda^0(M)$, we find at last that $\sigma^A_{ai_1i_2\cdots i_{k-1}} = 0$. Due to the complete antisymmetry of these functions with respect to its k indices, these relations would be met for all positions of indices. This is tantamount to say that

$$\sigma^A_{i_1i_2\cdots i_k} = 0, \; r+1 \leq i_1, i_2, \ldots, i_k \leq m.$$

Therefore the forms σ^A have to possess the following structure

$$\sigma^A = \frac{1}{k!}\,\sigma^A_{\alpha_1\cdots\alpha_k}\theta^{\alpha_1}\wedge\cdots\wedge\theta^{\alpha_k},\ \ 1 \leq \alpha_1, \alpha_2\ldots, \alpha_k \leq r$$

which implies that $\sigma^A \in \mathcal{J}(\theta^\alpha)$. This result means of course $\mathcal{I}(\sigma^A) \subseteq \mathcal{J}(\theta^\alpha)$ and consequently $\mathcal{I}(\omega^A) \subseteq \mathcal{J}(\theta^\alpha)$. This proves the theorem. □

Example 5.6.1. Let us take the form $\omega = \omega_i(\mathbf{x})\,dx^i \in \Lambda^1(M)$ into account. A vector $V = v^i(\mathbf{x})\,\partial_i \in T(M)$ is a characteristic vector of the form ω if it meets the condition $\mathbf{i}_V(\omega) = v^i\omega_i = 0$. If we take $\omega_1 \neq 0$, we see that there are $m-1$ linearly independent vectors

$$V_k = \omega_1\frac{\partial}{\partial x^k} - \omega_k\frac{\partial}{\partial x^1},\ \ k = 2, 3, \ldots, m$$

satisfying this condition. ■

Example 5.6.2. An exterior system is given by the forms

$$\omega^1 = dx - y\,dz \in \Lambda^1(\mathbb{R}^4),\ \ \omega^2 = dx - x\,dy + t\,dz \in \Lambda^1(\mathbb{R}^4).$$

If $V = v^x\partial_x + v^y\partial_y + v^z\partial_z + v^t\partial_t$ is a characteristic vector of this system, then the following equations should be satisfied:

$$v^x - yv^z = 0,\ \ v^x - xv^y + tv^z = 0.$$

We thus obtain

$$v^x = yv^z,\ \ v^y = \frac{y+t}{x}v^z.$$

Hence, two linearly independent characteristic vectors are found to be

$$V_1 = y\frac{\partial}{\partial x} + \frac{y+t}{x}\frac{\partial}{\partial y} + \frac{\partial}{\partial z},\ \ V_2 = \frac{\partial}{\partial t}.$$ ■

Example 5.6.3. We consider the ideal generated by the forms

$$\omega^1 = dx - y\,dz \in \Lambda^1(\mathbb{R}^4),\ \ \omega^2 = t\,dx\wedge dz - x\,dy\wedge dt\ \in \Lambda^1(\mathbb{R}^4)$$

Its characteristic vector field V must satisfy the relations $\mathbf{i}_V(\omega^1) = 0$ and $\mathbf{i}_V(\omega^2) = \lambda(\mathbf{x})\omega^1$ where $\lambda \in \Lambda^0(\mathbb{R}^4)$ that can be written explicitly as

$$v^x - yv^z = 0,\ tv^x dz - tv^z dx - xv^y dt + xv^t dy = \lambda\,dx - \lambda y\,dz$$

whence we find that

$$\lambda = -\,tv^z,\ v^x = yv^z,\ v^y = v^t = 0.$$

Thus 1-dimensional characteristic subspace of the ideal is spanned by the vector field

$$V = y\frac{\partial}{\partial x} + \frac{\partial}{\partial z}.$$

On the other hand, the characteristic vectors of the forms ω^1 and ω^2 are determined through the relations $\mathbf{i}_V(\omega^1) = 0$ and $\mathbf{i}_U(\omega^2) = 0$ leading to

$$v^x - yv^z = 0,\ u^x = u^z = u^y = u^t = 0.$$

Hence, characteristic vectors are

$$V_1 = y\frac{\partial}{\partial x} + \frac{\partial}{\partial z},\ V_2 = \frac{\partial}{\partial y},\ V_3 = \frac{\partial}{\partial t};\ U = 0. \quad \blacksquare$$

5.7. EXTERIOR FORMS UNDER MAPPINGS

Let M^m and N^n be two differentiable manifolds and $\phi : M \to N$ be a smooth mapping. We know that the mapping $\phi^* : \Lambda^0(N) \to \Lambda^0(M)$ derived from ϕ via the rule $\phi^* g = g \circ \phi$ assigns a smooth function $f = \phi^* g \in \Lambda^0(M)$ to a smooth function $g \in \Lambda^0(N)$ [*see p.* 98]. We shall now show that ϕ gives rise in general to a mapping $\phi^* : \Lambda(N) \to \Lambda(M)$. Let us take a form $\omega \in \Lambda^k(N)$ into consideration. If we denote local coordinates associated with a chart at the point $q \in N$ by $\mathbf{y} = \{y^\alpha\} = \{y^1, y^2, \dots, y^n\}$, we may write

$$\omega(q) = \frac{1}{k!}\,\omega_{\alpha_1\alpha_2\cdots\alpha_k}(\mathbf{y})\, dy^{\alpha_1} \wedge dy^{\alpha_2} \wedge \cdots \wedge dy^{\alpha_k} \in \Lambda^k(N).$$

Here the indices $\alpha_1, \dots, \alpha_k$ take values $1, \dots, n$. On the other hand, if local coordinates in a chart at a point $p \in M$ are $\mathbf{x} = \{x^i\} = \{x^1, x^2, \dots, x^m\}$, we know that the mapping $q = \phi(p)$ elicits a mapping $\Phi : \mathbb{R}^m \to \mathbb{R}^n$ in the functional form $\mathbf{y} = \Phi(\mathbf{x})$ or $y^\alpha = \Phi^\alpha(x^1, \dots, x^m), \alpha = 1, \dots, n$. The differential $d\phi : T_p(M) \to T_{\phi(p)}(N)$ of ϕ at the point p carries a vector at that point p over a vector at the point $q = \phi(p)$. We now define a form $\omega^* = \phi^*\omega$ at the point p corresponding to a form ω at the point $\phi(p)$ in such a way that the numerical equality

$$(\phi^*\omega)(V_1, \dots, V_k) = \omega\big(d\phi(V_1), \dots, d\phi(V_k)\big) \tag{5.7.1}$$

will be satisfied for all vectors $V_1, V_2, \dots, V_k \in T_p(M)$. This relation will actually determine a mapping in the form $\phi^* : \Lambda^k(N) \to \Lambda^k(M)$. In fact, the vector $V^* = d\phi(V)$ is represented in view of (2.7.4) by

$$V^* = v^i \frac{\partial \Phi^\alpha}{\partial x^i}\frac{\partial}{\partial y^\alpha} = v^{*\alpha}\frac{\partial}{\partial y^\alpha}$$

where $V = v^i \dfrac{\partial}{\partial x^i}$. Therefore, we obtain

$$\omega^*(V_1, \dots, V_k) = \omega^*_{i_1\cdots i_k} v_1^{i_1}\cdots v_k^{i_k} =$$
$$\omega(V_1^*, \dots, V_k^*) = \omega_{\alpha_1\cdots\alpha_k} v_1^{*\alpha_1}\cdots v_k^{*\alpha_k} = \omega_{\alpha_1\cdots\alpha_k}\frac{\partial \Phi^{\alpha_1}}{\partial x^{i_1}}\cdots\frac{\partial \Phi^{\alpha_k}}{\partial x^{i_k}} v_1^{i_1}\cdots v_k^{i_k}.$$

Since, this expression would be valid for all vectors $V_1, \dots, V_k$, we reach to the conclusion

$$\begin{aligned}\omega^*_{i_1\cdots i_k}(\mathbf{x}) &= \omega_{\alpha_1\cdots\alpha_k}\big(\Phi(\mathbf{x})\big)\frac{\partial \Phi^{\alpha_1}}{\partial x^{i_1}}\cdots\frac{\partial \Phi^{\alpha_k}}{\partial x^{i_k}} \\ &= \omega_{\alpha_1\cdots\alpha_k}\big(\Phi(\mathbf{x})\big)\frac{\partial \Phi^{[\alpha_1}}{\partial x^{i_1}}\cdots\frac{\partial \Phi^{\alpha_k]}}{\partial x^{i_k}}.\end{aligned} \tag{5.7.2}$$

We have to note that the complete antisymmetry on indices α causes the complete antisymmetry on indices i. Accordingly, the ***pull-back***, or ***reciprocal image*** $\omega^*(p)$ of a form $\omega(q) \in \Lambda^k(N)$, where $q = \phi(p) \in N$ and $p \in M$, is the k-form given by

$$\begin{aligned}\omega^*(p) = \phi^*\omega(q) &= \frac{1}{k!}\,\omega_{\alpha_1\cdots\alpha_k}\big(\Phi(\mathbf{x})\big)\frac{\partial \Phi^{\alpha_1}}{\partial x^{i_1}}\cdots\frac{\partial \Phi^{\alpha_k}}{\partial x^{i_k}}\,dx^{i_1}\wedge\cdots\wedge dx^{i_k} \\ &= \frac{1}{k!}\,\omega^*_{i_1\cdots i_k}(\mathbf{x})\,dx^{i_1}\wedge\cdots\wedge dx^{i_k} \in \Lambda^k(M).\end{aligned}$$

ϕ^* is called the ***pull-back operator*** and it can also be expressed in the usual form $\phi^*\omega = \omega \circ \phi$. However, this operation must be interpreted this time in a broader sense. We simply realise that the form $\phi^*\omega$ is obtainable from the form ω by inserting into ω the differential transformation

$$dy^\alpha = \frac{\partial y^\alpha}{\partial x^i}dx^i = \frac{\partial \Phi^\alpha}{\partial x^i}dx^i$$

in addition to the mapping $\omega_{\alpha_1\cdots\alpha_k} \circ \phi$. It is clear that ϕ^* is a degree preserving mapping. If $n \geq k > m$, then it is evident that $\phi^*\omega = 0$ identically.

Let us consider the forms $\alpha, \beta \in \Lambda^k(N)$. If we notice the relation (5.7.2) we find that

$$\phi^*(\alpha + \beta) = \phi^*\alpha + \phi^*\beta. \tag{5.7.3}$$

Hence the operator ϕ^* is additive. Furthermore, if $\omega \in \Lambda^k(N)$, $\sigma \in \Lambda^l(N)$, then the form $\gamma = \omega \wedge \sigma \in \Lambda^{k+l}(N)$ becomes

$$\gamma = \frac{1}{k!\,l!}\,\omega_{\alpha_1\cdots\alpha_k}\sigma_{\beta_1\cdots\beta_l}\,dy^{\alpha_1}\wedge\cdots\wedge dy^{\alpha_k}\wedge dy^{\beta_1}\wedge\cdots\wedge dy^{\beta_l}$$

and the form $\phi^*\gamma \in \Lambda^{k+l}(M)$ is cast into

$$\phi^*\gamma = \frac{1}{k!\,l!}\,\omega^*_{i_1\cdots i_k}\sigma^*_{j_1\cdots j_l}\,dx^{i_1}\wedge\cdots\wedge dx^{i_k}\wedge dx^{j_1}\wedge\cdots\wedge dx^{j_l}.$$

We thus reach to the conclusion

$$\phi^*(\omega\wedge\sigma) = \phi^*\omega\wedge\phi^*\sigma. \tag{5.7.4}$$

When $g \in \Lambda^0(N)$, we get from (5.7.4)

$$\phi^*(g\,\omega) = (\phi^*g)\phi^*\omega.$$

If $g \in \mathbb{R}$, one finds $\phi^*(g\omega) = g\,\phi^*\omega$. Therefore, ϕ^* *reduces to a linear operator only on the field of real numbers*. On recalling (5.7.4), we recognise that the mapping ϕ^* is a homomorphism on the exterior algebra $\Lambda(N)$. If ϕ is a diffeomorphism, then it becomes clear that the operator ϕ^* will be an algebra isomorphism.

Let M_1, M_2 and M_3 be smooth manifolds, and $\phi : M_1 \to M_2$ and $\psi : M_2 \to M_3$ be smooth mappings. These mappings give rise to pull-back operators $\psi^* : \Lambda(M_3) \to \Lambda(M_2)$ and $\phi^* : \Lambda(M_2) \to \Lambda(M_1)$ so that one has $\psi^*\omega \in \Lambda^k(M_2)$ and $\phi^*(\psi^*\omega) \in \Lambda^k(M_1)$ for a form $\omega \in \Lambda^k(M_3)$. On the other hand, it is straightforward to see that we can write $\psi\circ\phi : M_1 \to M_3$ and $(\psi\circ\phi)^* : \Lambda(M_3) \to \Lambda(M_1)$. In appropriate local coordinates, we have

$$\begin{aligned}
\omega &= \frac{1}{k!}\,\omega_{a_1\cdots a_k}(\mathbf{z})\,dz^{a_1}\wedge\cdots\wedge dz^{a_k},\\
\psi^*\omega &= \frac{1}{k!}\,\omega_{a_1\cdots a_k}\big(\mathbf{z}(\mathbf{y})\big)\frac{\partial z^{a_1}}{\partial y^{\alpha_1}}\cdots\frac{\partial z^{a_k}}{\partial y^{\alpha_k}}\,dy^{\alpha_1}\wedge\cdots\wedge dy^{\alpha_k},\\
\phi^*(\psi^*\omega) &= \frac{1}{k!}\,\omega_{a_1\cdots a_k}\big[\mathbf{z}(\mathbf{y}(\mathbf{x}))\big]\frac{\partial z^{a_1}}{\partial y^{\alpha_1}}\cdots\frac{\partial z^{a_k}}{\partial y^{\alpha_k}}\frac{\partial y^{\alpha_1}}{\partial x^{i_1}}\cdots\frac{\partial y^{\alpha_k}}{\partial x^{i_k}}\,dx^{i_1}\wedge\cdots\wedge dx^{i_k}.
\end{aligned}$$

But, the chain rule of differentiation

$$\frac{\partial z^{a_r}}{\partial y^{\alpha_r}}\frac{\partial y^{\alpha_r}}{\partial x^{i_r}} = \frac{\partial z^{a_r}}{\partial x^{i_r}}$$

implies that

$$\phi^*(\psi^*\omega) = \frac{1}{k!}\,\omega_{a_1\cdots a_k}\big(\mathbf{z}(\mathbf{x})\big)\frac{\partial z^{a_1}}{\partial x^{i_1}}\cdots\frac{\partial z^{a_k}}{\partial x^{i_k}}\,dx^{i_1}\wedge\cdots\wedge dx^{i_k} = (\psi\circ\phi)^*\omega.$$

Since this relation must be valid for all forms $\omega \in \Lambda(M_3)$, we arrive at the composition rule

$$(\psi \circ \phi)^* = \phi^* \circ \psi^*. \tag{5.7.5}$$

If the mapping $\phi : M \to N$ is a *diffeomorphism*, then the mapping $\phi^{-1} : N \to M$ is also a diffeomorphism. Thus the relations $\phi^{-1} \circ \phi = i_M$, $\phi \circ \phi^{-1} = i_N$ and $i_M^* = i_{\Lambda(M)}$, $i_N^* = i_{\Lambda(N)}$ leads, according to (5.7.5), to

$$(\phi^{-1} \circ \phi)^* = i_{\Lambda(M)} = \phi^* \circ (\phi^{-1})^*, \ (\phi \circ \phi^{-1})^* = i_{\Lambda(N)} = (\phi^{-1})^* \circ \phi^*$$

which implies in this case that $(\phi^*)^{-1} = (\phi^{-1})^* : \Lambda(M) \to \Lambda(N)$.

We have so far seen that the mapping $\phi : M \to N$ generates both the differential mapping $d\phi = \phi_* : T(M) \to T(N)$ and the pull-back operator $\phi^* : \Lambda(N) \to \Lambda(M)$. Let us now consider a form $\omega \in \Lambda^k(N)$ at a point $\phi(p) \in N$ corresponding to a point $p \in M$ and a vector $V = v^i \partial_i \in T_p(M)$. We know that the vector $V^* = \phi_*(V) = d\phi(V) \in T_{\phi(p)}(N)$ is given by

$$d\phi(V) = v^\alpha \frac{\partial}{\partial y^\alpha}, \quad v^\alpha = v^i \frac{\partial \Phi^\alpha}{\partial x^i}.$$

The interior product of the form ω with this vector is of course

$$\mathbf{i}_{d\phi(V)}(\omega) = \frac{1}{(k-1)!} \omega_{\alpha_1 \alpha_2 \cdots \alpha_k}(\mathbf{y}) v^{\alpha_1} \, dy^{\alpha_2} \wedge \cdots \wedge dy^{\alpha_k}.$$

The pull-back of that form then becomes

$$\begin{aligned}
\phi^*\big(\mathbf{i}_{d\phi(V)}(\omega)\big) &= \frac{1}{(k-1)!} \omega_{\alpha_1 \alpha_2 \cdots \alpha_k} v^{i_1} \frac{\partial \Phi^{\alpha_1}}{\partial x^{i_1}} \frac{\partial \Phi^{\alpha_2}}{\partial x^{i_2}} \cdots \frac{\partial \Phi^{\alpha_k}}{\partial x^{i_k}} \, dx^{i_2} \wedge \cdots \wedge dx^{i_k} \\
&= \frac{1}{(k-1)!} \omega_{i_1 i_2 \cdots i_k}(\mathbf{x}) v^{i_1} \, dx^{i_2} \wedge \cdots \wedge dx^{i_k} \\
&= \mathbf{i}_V(\phi^* \omega).
\end{aligned}$$

Since this relation would be true for all forms $\omega \in \Lambda(N)$, we conclude that for all vectors $V \in T(M)$ we get the rule

$$\phi^* \circ \mathbf{i}_{\phi_*(V)} = \phi^* \circ \mathbf{i}_{V^*} = \mathbf{i}_V \circ \phi^* : \Lambda^k(N) \to \Lambda^{k-1}(M). \tag{5.7.6}$$

If the operator ϕ_*^{-1} exists, then (5.7.6) means that the relation

$$\phi^* \circ \mathbf{i}_U = \mathbf{i}_{\phi_*^{-1} U} \circ \phi^* \tag{5.7.7}$$

will also be valid for all vectors $U \in T(N)$.

If $\phi : M \to M$, then ϕ maps the manifold M into itself. When ϕ is a diffeomorphism, it produces a coordinate transformation on M that can be represented locally by functions $\mathbf{y} = \Phi(\mathbf{x})$ or $y^\alpha = \Phi^\alpha(x^1, \dots, x^m)$, $\alpha = 1, \dots, m$ where $\Phi = \varphi \circ \phi \circ \varphi^{-1}$ with φ being the local homeomorphism of the associated chart. If we denote the Jacobian determinant by $J = \det \phi = \det [\partial \Phi^\alpha / \partial x^i]$, then we must have, in this case, $J \neq 0$. We would like now to investigate the transformation of bases induced by the volume form under such a mapping ϕ. The transformation of a generic basis form

$$\mu_{\alpha_k \cdots \alpha_1} = \frac{1}{(m-k)!}\, e_{\alpha_1 \cdots \alpha_k \alpha_{k+1} \cdots \alpha_m} dy^{\alpha_{k+1}} \wedge \cdots \wedge dy^{\alpha_m} \in \Lambda^{m-k}(M)$$

yields

$$\phi^* \mu_{\alpha_k \cdots \alpha_1} = \frac{1}{(m-k)!}\, e_{\alpha_1 \cdots \alpha_k \alpha_{k+1} \cdots \alpha_m} \frac{\partial y^{\alpha_{k+1}}}{\partial x^{i_{k+1}}} \cdots \frac{\partial y^{\alpha_m}}{\partial x^{i_m}} dx^{i_{k+1}} \wedge \cdots \wedge dx^{i_m}$$

from which we write

$$\frac{\partial y^{\alpha_1}}{\partial x^{i_1}} \cdots \frac{\partial y^{\alpha_k}}{\partial x^{i_k}} \phi^* \mu_{\alpha_k \cdots \alpha_1} = \frac{1}{(m-k)!}\, e_{\alpha_1 \cdots \alpha_k \alpha_{k+1} \cdots \alpha_m} \frac{\partial y^{\alpha_1}}{\partial x^{i_1}} \cdots \frac{\partial y^{\alpha_k}}{\partial x^{i_k}} \frac{\partial y^{\alpha_{k+1}}}{\partial x^{i_{k+1}}} \cdots \frac{\partial y^{\alpha_m}}{\partial x^{i_m}} dx^{i_{k+1}} \wedge \cdots \wedge dx^{i_m}.$$

According to (1.4.18), we have

$$e_{\alpha_1 \cdots \alpha_m} \frac{\partial y^{\alpha_1}}{\partial x^{i_1}} \cdots \frac{\partial y^{\alpha_m}}{\partial x^{i_m}} = e_{i_1 \cdots i_m} \det \Big[\frac{\partial y^\alpha}{\partial x^i} \Big].$$

Therefore, we find that

$$\frac{\partial y^{\alpha_1}}{\partial x^{i_1}} \cdots \frac{\partial y^{\alpha_k}}{\partial x^{i_k}} \phi^* \mu_{\alpha_k \cdots \alpha_1} = \det \Big[\frac{\partial y^\alpha}{\partial x^i} \Big] \frac{1}{(m-k)!}\, e_{i_1 \cdots i_k i_{k+1} \cdots i_m} dx^{i_{k+1}} \wedge \cdots \wedge dx^{i_m}$$

and finally, owing to (5.5.10)

$$\mu_{i_k \cdots i_1} = \Big(\det \Big[\frac{\partial y^\alpha}{\partial x^i} \Big] \Big)^{-1} \frac{\partial y^{\alpha_1}}{\partial x^{i_1}} \cdots \frac{\partial y^{\alpha_k}}{\partial x^{i_k}} \phi^* \mu_{\alpha_k \cdots \alpha_1}. \tag{5.7.8}$$

Let us now consider a form $\omega \in \Lambda^{m-k}(M)$ described by

$$\omega = \frac{1}{k!} \omega^{\alpha_1 \cdots \alpha_k}(\mathbf{y}) \mu_{\alpha_k \cdots \alpha_1}. \tag{5.7.9}$$

The pull-back of ω thus becomes

$$\phi^*\omega = \frac{1}{k!}\,\omega^{i_1\cdots i_k}(\mathbf{x})\,\mu_{i_k\cdots i_1} = \frac{1}{k!}\,\phi^*\omega^{\alpha_1\cdots\alpha_k}(\mathbf{y})\,\phi^*\mu_{\alpha_k\cdots\alpha_1}$$

and (5.7.8) gives

$$(\phi^*\omega^{\alpha_1\cdots\alpha_k})(\mathbf{x}) = \Big(\det\Big[\frac{\partial y^\alpha}{\partial x^i}\Big]\Big)^{-1}\frac{\partial y^{\alpha_1}}{\partial x^{i_1}}\cdots\frac{\partial y^{\alpha_k}}{\partial x^{i_k}}\,\omega^{i_1\cdots i_k}(\mathbf{x}). \quad (5.7.10)$$

In the module $\Lambda^m(M)$ bases are the volume forms

$$\mu_{\mathbf{x}} = dx^1\wedge\cdots\wedge dx^m,\;\; \mu_{\mathbf{y}} = dy^1\wedge\cdots\wedge dy^m$$

and (5.7.8) leads to

$$\phi^*\mu_{\mathbf{y}} = (\det\phi)\mu_{\mathbf{x}} = \det\Big[\frac{\partial y^\alpha}{\partial x^i}\Big]\mu_{\mathbf{x}} = J\,dx^1\wedge\cdots\wedge dx^m. \quad (5.7.11)$$

Conversely, if the relation (5.7.11) is valid, then we must find $\det\phi \neq 0$. In consequence, the celebrated implicit function theorem states that the mapping ϕ is locally a diffeomorphism. Any form $\omega(\mathbf{y}) \in \Lambda^m(M)$ is now expressible as $\omega(\mathbf{y}) = g(\mathbf{y})\mu_{\mathbf{y}}$. Thus, under coordinates transformation we obtain the form $\phi^*\omega = (g\circ\phi)\det\big[\partial y^\alpha/\partial x^i\big]\mu_{\mathbf{x}}$.

Next, we consider a submanifold S of dimension $r < m$ of the manifold M. We suppose that we describe this submanifold by a smooth mapping $\phi : S \to M$. In local coordinates, this mapping will be prescribed as a coordinate transformation

$$x^i = \Phi^i(u^\alpha),\; i = 1,\ldots,m;\; \alpha = 1,\ldots,r. \quad (5.7.12)$$

The pull-back $\phi^*\omega \in \Lambda^k(S)$ of a form $\omega \in \Lambda^k(M)$ on S is given by

$$\phi^*\omega = \frac{1}{k!}\,\omega_{\alpha_1\cdots\alpha_k}(\mathbf{u})\,du^{\alpha_1}\wedge\cdots\wedge du^{\alpha_k} \quad (5.7.13)$$

where the coefficients $\omega_{\alpha_1\cdots\alpha_k}(\mathbf{u})$ are determined through the relations

$$\omega_{\alpha_1\cdots\alpha_k}(\mathbf{u}) = \omega_{i_1\cdots i_k}\big(\Phi(\mathbf{u})\big)\frac{\partial\Phi^{i_1}}{\partial u^{\alpha_1}}\cdots\frac{\partial\Phi^{i_k}}{\partial u^{\alpha_k}}. \quad (5.7.14)$$

If the form ω does not vanish identically on M, then the submanifold S, consequently the mapping $\phi : S \to M$, satisfying the condition $\phi^*\omega = 0$ is called *a solution of the exterior equation* $\omega = 0$. When $k > r$, then $\phi^*\omega \equiv 0$ identically, that is, any submanifold whose dimension is less than k is automatically a solution of this equation. If $k \leq r$, then the mapping ϕ that gives rise to an r-dimensional solution submanifold is determined, in view of

(5.7.13) and (5.7.14), through the equations $\omega_{\alpha_1\cdots\alpha_k}(\mathbf{u}) = 0$. We then call ϕ as the ***resolvent mapping*** for the exterior equation.

We can introduce another interpretation to a solution of an exterior equation. The differential $d\phi : T(S) \to T(M)$ of the mapping $\phi : S \to M$ push a vector field in the tangent bundle of S up to a vector field in the tangent bundle of M. Let $V \in T(S)$, then we can write

$$V = v^\alpha \frac{\partial}{\partial u^\alpha}, \; d\phi(V) = v^i \frac{\partial}{\partial x^i} \in T(M),$$

where

$$v^i = \frac{\partial \Phi^i}{\partial u^\alpha} v^\alpha.$$

According to (5.7.1), every k linearly independent vector fields selected from $T(S)$ of dimension $r \geq k$ must satisfy the relation

$$\omega\big(d\phi(V_1), \ldots, d\phi(V_k)\big) = (\phi^*\omega)(V_1, \ldots, V_k) = 0$$

since $\phi^*\omega = 0$. Hence, in order to determine *locally* an r-dimensional solution submanifold through a point $p \in M$, all we need to do is to find a subspace $T_p(S)$ of the tangent space $T_p(M)$ annihilating the form ω. We know from the Frobenius theorem that the distribution made up by those local subspaces should be involutive so that the local tangent spaces can be patched together to generate a smooth submanifold.

Example 5.7.1. We take $M = \mathbb{R}^2$ and $\omega = x\,dy - 3y\,dx \in \Lambda^1(\mathbb{R}^2)$. Our aim is to determine a mapping $\phi : \mathbb{R} \to \mathbb{R}^2$ so as $\phi^*\omega = 0$. Let us write

$$x = \alpha(u), \;\; y = \beta(u).$$

Then we get $\phi^*\omega = (\alpha\beta' - 3\beta\alpha')du = 0$ and the condition $\alpha\beta' = 3\beta\alpha'$. This differential equation can be cast into the form

$$\frac{\beta'}{\beta} = 3\frac{\alpha'}{\alpha} \quad \text{or} \quad (\log\beta)' = 3(\log\alpha)'$$

so that we obtain $\beta(u) = C\alpha(u)^3$. Therefore, the curves prescribed by parametric equations $x = \alpha(u)$, $y = C\alpha(u)^3$ where $\alpha(u)$ is an arbitrary function solve the exterior equation $\omega = 0$.

Let us now consider a vector $V = v^u(u)\dfrac{\partial}{\partial u} \in T(\mathbb{R})$. We then have

$$d\phi(V) = \alpha' v^u \frac{\partial}{\partial x} + \beta' v^u \frac{\partial}{\partial y} \in T(\mathbb{R}^2).$$

Hence the equation $\omega\big(d\phi(V)\big) = xv^y - 3yv^x = (\alpha\beta' - 3\beta\alpha')v^u = 0$ leads similarly to the above expression and to

$$v^x = \alpha' f(u), \quad v^y = 3C\alpha^2\alpha' f(u) = C\big(\alpha(u)^3\big)' f(u).$$

where we defined $v^u(u) = f(u)$. ■

Example 5.7.2. We consider $M = \mathbb{R}^3$ and the form

$$\omega = P\,dx^1 \wedge dx^2 + Q\,dx^1 \wedge dx^3 + R\,dx^2 \wedge dx^3 \in \Lambda^2(\mathbb{R}^3)$$

where $P, Q, R \in \Lambda^0(\mathbb{R}^3)$. We define a 2-dimensional solution submanifold by the parametric equations $x^i = \phi^i(u^1, u^2), i = 1, 2, 3$. We denote the functional determinant by

$$\frac{\partial(\phi^i, \phi^j)}{\partial(u^\alpha, u^\beta)} = \frac{\partial\phi^i}{\partial u^\alpha}\frac{\partial\phi^j}{\partial u^\beta} - \frac{\partial\phi^i}{\partial u^\beta}\frac{\partial\phi^j}{\partial u^\alpha}$$

we then attain at the result

$$\phi^*\omega = \left[P\frac{\partial(\phi^1, \phi^2)}{\partial(u^1, u^2)} + Q\frac{\partial(\phi^1, \phi^3)}{\partial(u^1, u^2)} + R\frac{\partial(\phi^2, \phi^3)}{\partial(u^1, u^2)}\right] du^1 \wedge du^2.$$

Therefore, in order to satisfy $\phi^*\omega = 0$ we have to find the solution of the following non-linear partial differential equation

$$P\frac{\partial(\phi^1, \phi^2)}{\partial(u^1, u^2)} + Q\frac{\partial(\phi^1, \phi^3)}{\partial(u^1, u^2)} + R\frac{\partial(\phi^2, \phi^3)}{\partial(u^1, u^2)} = 0$$

where $P = P(\phi^1, \phi^2, \phi^3), Q = Q(\phi^1, \phi^2, \phi^3), R = R(\phi^1, \phi^2, \phi^3)$. ■

5.8. EXTERIOR DERIVATIVE

We define an operator $d : \Lambda(M) \to \Lambda(M)$ on a smooth manifold M mapping the exterior algebra $\Lambda(M)$ into itself in such a way that it holds the following rules:

(i). $d(\omega + \sigma) = d\omega + d\sigma, \;\; d(\lambda\omega) = \lambda\,d\omega; \;\; \omega, \sigma \in \Lambda(M), \lambda \in \mathbb{R}.$
(ii). $d(\omega \wedge \sigma) = d\omega \wedge \sigma + (-1)^{deg\,(\omega)}\omega \wedge d\sigma.$
(iii). $d^2 = d \circ d = 0$, *i.e.*, $d(d\omega) = d^2\omega = 0$ *for all* $\omega \in \Lambda(M)$.
(iv). *If* $f \in \Lambda^0(M)$, *then* $df = f_{,i}\,dx^i \in \Lambda^1(M)$.

The rule (i) means that d is a linear operator on $\mathbb{R}$ whereas the rule (iv) implies that the 1-form df is the classical differential of the smooth function $f \in \Lambda^0(M)$. Here, we have introduced the notation

$$\frac{\partial(\,\cdot\,)}{\partial x^i} \doteq (\,\cdot\,)_{,i} \tag{5.8.1}$$

which we shall employ frequently henceforth. The rule (iii) shows that d is a ***nilpotent operator***. d so defined is called the ***exterior derivative operator*** and the form $d\omega$ is the ***exterior derivative*** of the form ω.

Theorem 5.8.1. *The foregoing rules* (i) - (iv) *determine the exterior derivative operator d uniquely.*

We know that an exterior form $\omega \in \Lambda^k(M)$ on a manifold M is expressible in local coordinates on an open set $U \subseteq M$ as follows

$$\omega = \frac{1}{k!}\,\omega_{i_1 i_2 \cdots i_k} dx^{i_1} \wedge dx^{i_2} \wedge \cdots \wedge dx^{i_k}, \;\; \omega_{i_1 i_2 \cdots i_k} \in \Lambda^0(M).$$

Since $\omega_{i_1 i_2 \cdots i_k}$ is a 0-form, we obtain

$$d\omega = \frac{1}{k!}\big[d\omega_{i_1 \cdots i_k} \wedge dx^{i_1} \wedge \cdots \wedge dx^{i_k} + \omega_{i_1 i_2 \cdots i_k} d(dx^{i_1} \wedge \cdots \wedge dx^{i_k})\big]$$

in view of (ii). We shall now demonstrate by mathematical induction that

$$d(dx^{i_1} \wedge \cdots \wedge dx^{i_k}) = 0.$$

If $k = 1$, because of $(iii$-$iv)$ we find $d(dx^{i_1}) = d^2 x^{i_1} = 0$. Let us assume that the above relation is valid for $k-1$. Hence, we deduce from the rules of exterior differentiation

$$\begin{aligned} d(dx^{i_1} \wedge \cdots \wedge dx^{i_k}) = \\ d^2 x^{i_1} \wedge (dx^{i_2} \wedge \cdots \wedge dx^{i_k}) - dx^{i_1} \wedge d(dx^{i_2} \wedge \cdots \wedge dx^{i_k}) \\ = -\, dx^{i_1} \wedge d(dx^{i_2} \wedge \cdots \wedge dx^{i_k}) = 0. \end{aligned}$$

so that this relation is also valid for k. Therefore, the exterior derivative of the form $\omega \in \Lambda^k(M)$ is designated *uniquely* in local coordinates as follows

$$\begin{aligned} d\omega &= \frac{1}{k!}\, d\omega_{i_1 \cdots i_k} \wedge dx^{i_1} \wedge \cdots \wedge dx^{i_k} \\ &= \frac{1}{k!}\frac{\partial \omega_{i_1 \cdots i_k}}{\partial x^i}\, dx^i \wedge dx^{i_1} \wedge \cdots \wedge dx^{i_k} \\ &= \frac{1}{k!}\omega_{[i_1 \cdots i_k, i]}\, dx^i \wedge dx^{i_1} \wedge \cdots \wedge dx^{i_k} \in \Lambda^{k+1}(M). \end{aligned} \tag{5.8.2}$$

Thus the operator d is of the form $d : \Lambda^k(M) \to \Lambda^{k+1}(M)$ and increases the degree of the form by one. The form $d\omega \in \Lambda^{k+1}(M)$ can be written in the standard form in the following manner

$$d\omega = \frac{1}{(k+1)!}\,\omega_{ii_1\cdots i_k} dx^i \wedge dx^{i_1} \wedge \cdots \wedge dx^{i_k}$$

where we obviously have

$$\omega_{ii_1\cdots i_k} = (k+1)\,\omega_{[i_1\cdots i_k,i]} \in \Lambda^0(M). \qquad (5.8.3)$$

In order that this definition of the exterior derivative to be meaningful it should not depend on the chosen local coordinates, namely, the chosen chart of the atlas. To observe this property, let us consider the coordinate transformation $x^i = x^i(y^j)$ in overlapping charts. We thus write

$$\begin{aligned}\omega &= \frac{1}{k!}\,\omega_{i_1\cdots i_k}(\mathbf{x})\,dx^{i_1} \wedge \cdots \wedge dx^{i_k}\\ &= \frac{1}{k!}\,\omega_{i_1\cdots i_k}(\mathbf{x}(\mathbf{y}))\,dx^{i_1}(\mathbf{y}) \wedge \cdots \wedge dx^{i_k}(\mathbf{y})\end{aligned}$$

so that the exterior derivatives with respect to **y**- and **x**-coordinates are found to be related by

$$\begin{aligned}d_{\mathbf{y}}\omega &= \frac{1}{k!}\frac{\partial\omega_{i_1\cdots i_k}\big(\mathbf{x}(\mathbf{y})\big)}{\partial y^j}\,dy^j \wedge dx^{i_1}(\mathbf{y}) \wedge \cdots \wedge dx^{i_k}(\mathbf{y})\\ &= \frac{1}{k!}\frac{\partial\omega_{i_1\cdots i_k}(\mathbf{x})}{\partial x^i}\frac{\partial x^i}{\partial y^j}\,dy^j \wedge dx^{i_1}(\mathbf{y}) \wedge \cdots \wedge dx^{i_k}(\mathbf{y})\\ &= \frac{1}{k!}\frac{\partial\omega_{i_1\cdots i_k}(\mathbf{x})}{\partial x^i}\,dx^i \wedge dx^{i_1} \wedge \cdots \wedge dx^{i_k} = d_{\mathbf{x}}\omega.\end{aligned}$$

This relation is valid for all $\omega \in \Lambda(M)$. Hence, we obtain $d_{\mathbf{y}} = d_{\mathbf{x}}$ showing that the operator d is intrinsically defined. □

After having defined the exterior derivative by the expression (5.8.2), it is straightforward to see that the rules (i)-(iv) are automatically satisfied. That (i) becomes valid is obvious. To show (ii), let us consider the forms $\omega \in \Lambda^k(M)$ and $\sigma \in \Lambda^l(M)$ given by

$$\begin{aligned}\omega &= \frac{1}{k!}\,\omega_{i_1\cdots i_k} dx^{i_1} \wedge \cdots \wedge dx^{i_k},\\ \sigma &= \frac{1}{l!}\,\sigma_{i_1\cdots i_l}\,dx^{i_1} \wedge \cdots \wedge dx^{i_l}\end{aligned}$$

and evaluate the exterior derivative of $\omega \wedge \sigma$. We obtain

$$d(\omega \wedge \sigma) = d\Big[\frac{1}{k!}\frac{1}{l!}\,\omega_{i_1\cdots i_k}\sigma_{j_1\cdots j_l}\,dx^{i_1} \wedge \cdots \wedge dx^{i_k} \wedge dx^{j_1} \wedge \cdots \wedge dx^{j_l}\Big]$$

$$
\begin{aligned}
&= \frac{1}{k!}\, d\omega_{i_1\cdots i_k} \wedge dx^{i_1} \wedge \cdots \wedge dx^{i_k} \wedge \frac{1}{l!}\,\sigma_{j_1\cdots j_l}\, dx^{j_1} \wedge \cdots \wedge dx^{j_l} \\
&\quad + \frac{1}{k!}\,\omega_{i_1\cdots i_k} \frac{1}{l!}\, d\sigma_{j_1\cdots j_l} \wedge dx^{i_1} \wedge \cdots \wedge dx^{i_k} \wedge dx^{j_1} \wedge \cdots \wedge dx^{j_l} \\
&= \frac{1}{k!}\, d\omega_{i_1\cdots i_k} \wedge dx^{i_1} \wedge \cdots \wedge dx^{i_k} \wedge \frac{1}{l!}\,\sigma_{j_1\cdots j_l}\, dx^{j_1} \wedge \cdots \wedge dx^{j_l} \\
&\qquad + (-1)^k \frac{1}{k!}\,\omega_{i_1\cdots i_k}\, dx^{i_1} \wedge \cdots \wedge dx^{i_k} \wedge \frac{1}{l!}\, d\sigma_{j_1\cdots j_l} \wedge dx^{j_1} \wedge \cdots \wedge dx^{j_l}
\end{aligned}
$$

and we thus get

$$d(\omega \wedge \sigma) = d\omega \wedge \sigma + (-1)^k \omega \wedge d\sigma.$$

Similarly, we find

$$d^2\omega = \frac{1}{k!}\,\omega_{i_1\cdots i_k,ij}\, dx^i \wedge dx^j \wedge dx^{i_1} \wedge \cdots \wedge dx^{i_k} \in \Lambda^{k+2}(M).$$

But the exterior product is antisymmetric with respect to indices i and j, while the second partial derivatives are symmetric. Therefore, summations over these indices from 1 to m become zero and we get $d^2\omega = 0$. The rule (iv) is retrieved immediately form the definition (5.8.2).

We can provide a more explicit expression for coefficient functions $\omega_{ii_1\cdots i_k} \in \Lambda^0(M)$ specifying the form $d\omega \in \Lambda^{k+1}(M)$. If we take notice of the relation (5.5.2) we readily arrive at

$$
\begin{aligned}
\omega_{ii_1\cdots i_k} &= \frac{k+1}{(k+1)!}\,\delta^{j_1 j_2\cdots j_k j}_{i_1 i_2\cdots i_k i}\,\omega_{j_1\cdots j_k,j} = \frac{1}{k!}\Big[\delta^j_i \delta^{j_1 j_2\cdots j_k}_{i_1 i_2\cdots i_k} - \delta^j_{i_1}\delta^{j_1 j_2\cdots j_k}_{i i_2\cdots i_k} \\
&\quad - \delta^j_{i_2}\delta^{j_1 j_2\cdots j_k}_{i_1 i\cdots i_k} - \cdots - \delta^j_{i_k}\delta^{j_1 j_2\cdots j_k}_{i_1 i_2\cdots i_{k-1} i}\Big]\omega_{j_1\cdots j_k,j} \\
&= \frac{1}{k!}\Big[\delta^{j_1 j_2\cdots j_k}_{i_1 i_2\cdots i_k}\,\omega_{j_1\cdots j_k,i} - \delta^{j_1 j_2\cdots j_k}_{i i_2\cdots i_k}\,\omega_{j_1\cdots j_k,i_1} - \delta^{j_1 j_2\cdots j_k}_{i_1 i\cdots i_k}\,\omega_{j_1\cdots j_k,i_2} \\
&\quad - \cdots - \delta^{j_1 j_2\cdots j_k}_{i_1 i_2\cdots i_{k-1} i}\,\omega_{j_1\cdots j_k,i_k}\Big].
\end{aligned}
$$

Since $\omega_{j_1\cdots j_k}$ is completely antisymmetric, this expression may be transformed into the following form:

$$
\begin{aligned}
\omega_{ii_1\cdots i_k} &= \omega_{i_1\cdots i_k,i} - \omega_{ii_2\cdots i_k,i_1} - \omega_{i_1 i\cdots i_k,i_2} - \cdots - \omega_{i_1 i_2\cdots i,i_k} \qquad (5.8.4) \\
&= \omega_{i_1\cdots i_k,i} - \sum_{r=1}^{k} \omega_{i_1\cdots i_{r-1} i i_{r+1}\cdots i_k,i_r}.
\end{aligned}
$$

Example 5.8.1. The exterior derivative of the form $\omega = \omega_j\, dx^j \in \Lambda^1(M)$ will be

$$d\omega = \omega_{j,i}\, dx^i \wedge dx^j = \frac{1}{2}\,\omega_{ij}\, dx^i \wedge dx^j,$$
$$\omega_{ij} = 2\,\omega_{[j,i]} = \omega_{j,i} - \omega_{i,j}.$$

Let us take $M = \mathbb{R}^3$. In this case, the number of the independent components of the coefficients ω_{ij} is three and this matrix can be represented by an axial vector. One can then write $\omega = \mathbf{V} \cdot d\mathbf{r} = X_1\, dx^1 + X_2\, dx^2 + X_3\, dx^3$ where we employed the notation of the classical vector algebra to denote $\mathbf{V} = X_1\mathbf{e}_1 + X_2\mathbf{e}_2 + X_3\mathbf{e}_3$ and $d\mathbf{r} = dx_1\mathbf{e}_1 + dx_2\mathbf{e}_2 + dx_3\mathbf{e}_3$. $(\,\cdot\,)$ is the usual scalar product and $\mathbf{e}_1, \mathbf{e}_2, \mathbf{e}_3$ are orthonormal basis vectors of $\mathbb{R}^3$. The exterior derivative of the form ω becomes

$$d\omega = \Big(\frac{\partial X_3}{\partial x^2} - \frac{\partial X_2}{\partial x^3}\Big)\, dx^2 \wedge dx^3 + \Big(\frac{\partial X_1}{\partial x^3} - \frac{\partial X_3}{\partial x^1}\Big)\, dx^3 \wedge dx^1$$
$$+ \Big(\frac{\partial X_2}{\partial x^1} - \frac{\partial X_1}{\partial x^2}\Big)\, dx^1 \wedge dx^2.$$

Evidently the coefficients of the form $d\omega$ is nothing but the components of ***the curl*** of the vector $\mathbf{V}$, i.e., $\mathbf{W} = \operatorname{curl}\mathbf{V} = \boldsymbol{\nabla} \times \mathbf{V}$. This vector is also expressible as

$$\mathbf{W} = W_i\mathbf{e}_i = e_{ijk}\frac{\partial X_k}{\partial x^j}\mathbf{e}_i = e_{ijk}X_{k,j}\mathbf{e}_i, \quad i, j, k = 1, 2, 3.$$

On the other hand, if we consider the forms

$$\omega_1 = \mathbf{V}_1 \cdot d\mathbf{r} \;\text{ and }\; \omega_2 = \mathbf{V}_2 \cdot d\mathbf{r}$$

we see that their exterior product is

$$\omega_1 \wedge \omega_2 = (X_2Y_3 - X_3Y_2)\, dx^2 \wedge dx^3 + (X_3Y_1 - X_1Y_3)\, dx^3 \wedge dx^1$$
$$+ (X_1Y_2 - X_2Y_1)\, dx^1 \wedge dx^2$$

the coefficients of which are components of the usual vectorial product $\mathbf{W} = \mathbf{V}_1 \times \mathbf{V}_2$. This vector can also be written as follows

$$\mathbf{W} = W_i\mathbf{e}_i = e_{ijk}X_jY_k\mathbf{e}_i, \quad i, j, k = 1, 2, 3.$$

Let us next calculate the exterior derivative of the form $\omega = f\mathbf{V} \cdot d\mathbf{r}$ where $f \in \Lambda^0(\mathbb{R}^3)$, we easily reach to the relation

$$\operatorname{curl} f\mathbf{V} = \operatorname{grad} f \times \mathbf{V} + f\operatorname{curl}\mathbf{V}. \qquad ■$$

Example 5.8.2. We consider the form $\omega = \dfrac{1}{2!}\,\omega_{jk}\, dx^j \wedge dx^k \in \Lambda^2(M)$ whose exterior derivative becomes

$$d\omega = \frac{1}{2!}\,\omega_{jk,i}\,dx^i \wedge dx^j \wedge dx = \frac{1}{3!}\,\omega_{ijk}\,dx^i \wedge dx^j \wedge dx^k \in \Lambda^3(M).$$

According to (5.8.4), the coefficients of this form are given by

$$\omega_{ijk} = \omega_{jk,i} - \omega_{ik,j} - \omega_{ji,k} = \omega_{jk,i} + \omega_{ki,j} + \omega_{ij,k}.$$

Let us choose again $M = \mathbb{R}^3$ and write

$$\omega = X_1\,dx^2 \wedge dx^3 + X_2\,dx^3 \wedge dx^1 + X_3\,dx^1 \wedge dx^2 \in \Lambda^2(\mathbb{R}^3)$$

in terms of essential components. We observe at once that

$$d\omega = \Big(\frac{\partial X_1}{\partial x^1} + \frac{\partial X_2}{\partial x^2} + \frac{\partial X_3}{\partial x^3}\Big)dx^1 \wedge dx^2 \wedge dx^3 \in \Lambda^3(\mathbb{R}^3),$$

namely, the coefficient of this form is just the ***divergence*** $\boldsymbol{\nabla} \cdot \mathbf{V} = \operatorname{div}\mathbf{V}$ of the vector field $\mathbf{V} = X_1\mathbf{e}_1 + X_2\mathbf{e}_2 + X_3\mathbf{e}_3$ which can also be written as follows

$$\boldsymbol{\nabla} \cdot \mathbf{V} = \frac{\partial X_i}{\partial x^i} = X_{i,i}.$$

If we take into account the forms ω_1 and ω_2 defined in Example 5.8.1, then the relation $d(\omega_1 \wedge \omega_2) = d\omega_1 \wedge \omega_2 - \omega_1 \wedge d\omega_2$ yields the equality

$$\operatorname{div}(\mathbf{V}_1 \times \mathbf{V}_2) = \mathbf{V}_2 \cdot \operatorname{curl}\mathbf{V}_1 - \mathbf{V}_1 \cdot \operatorname{curl}\mathbf{V}_2. \qquad \blacksquare$$

We know that a form $\omega \in \Lambda^{m-k}(M)$ is expressible as in (5.5.17) by using a basis induced by the volume form. Since $d\mu_{i_k\cdots i_2 i_1} = 0$, the exterior derivative of this form is given by

$$d\omega = \frac{1}{k!}\,\omega^{i_1 i_2\cdots i_k}{}_{,i}\,dx^i \wedge \mu_{i_k\cdots i_2 i_1} = \frac{k}{k!}\,\omega^{i_1 i_2\cdots i_k}{}_{,i}\,\delta^i_{[i_k}\mu_{i_{k-1}\cdots i_2 i_1]}$$

where we employed the relation (5.5.15). Because of the antisymmetry of $\omega^{i_1 i_2\cdots i_k}$, we conclude that

$$\begin{aligned} d\omega &= \frac{1}{(k-1)!}\,\omega^{i_1 i_2\cdots i_k}{}_{,i}\,\delta^i_{i_k}\mu_{i_{k-1}\cdots i_2 i_1} \\ &= \frac{1}{(k-1)!}\,\omega^{i_1 i_2\cdots i_{k-1} i}{}_{,i}\,\mu_{i_{k-1}\cdots i_2 i_1} \in \Lambda^{m-(k-1)}(M). \end{aligned} \tag{5.8.5}$$

It is clear that one has

$$\omega^{i_1 i_2\cdots i_{k-1} i}{}_{,i} = \frac{\partial \omega^{i_1 i_2\cdots i_{k-1} 1}}{\partial x^1} + \frac{\partial \omega^{i_1 i_2\cdots i_{k-1} 2}}{\partial x^2} + \cdots + \frac{\partial \omega^{i_1 i_2\cdots i_{k-1} m}}{\partial x^m}.$$

We thus see that the coefficients of the form $d\omega$ is evaluated as a kind of *divergence*.

Let $\phi : M \to N$ be a differentiable mapping between the smooth manifolds M and N. We know that this mapping conduces toward the pull-back mapping $\phi^* : \Lambda(N) \to \Lambda(M)$ which assigns a form $\phi^*\omega \in \Lambda^k(M)$ to a form $\omega \in \Lambda^k(N)$.

Theorem 5.8.2. *If $\phi : M \to N$ is a smooth mapping, then we have the relation $d(\phi^*\omega) = \phi^* d\omega$ for all forms $\omega \in \Lambda(N)$. Consequently, one has the following rule of composition*

$$d \circ \phi^* = \phi^* \circ d : \Lambda^k(N) \to \Lambda^{k+1}(M)$$

which means that the operators d and ϕ^* commute.

We prove this theorem by explicitly calculating both sides. Let us consider a form

$$\omega = \frac{1}{k!}\,\omega_{\alpha_1\alpha_2\cdots\alpha_k} dy^{\alpha_1} \wedge dy^{\alpha_2} \wedge \cdots \wedge dy^{\alpha_k} \in \Lambda^k(N).$$

Its exterior derivative is

$$d\omega = \frac{1}{k!}\,\omega_{\alpha_1\alpha_2\cdots\alpha_k,\alpha}\, dy^{\alpha} \wedge dy^{\alpha_1} \wedge dy^{\alpha_2} \wedge \cdots \wedge dy^{\alpha_k}.$$

We thus obtain

$$\phi^* d\omega =$$
$$\frac{1}{k!}\Big(\frac{\partial \omega_{\alpha_1\alpha_2\cdots\alpha_k}}{\partial y^{\alpha}} \circ \phi\Big) \frac{\partial \Phi^{\alpha}}{\partial x^{i}} \frac{\partial \Phi^{\alpha_1}}{\partial x^{i_1}} \cdots \frac{\partial \Phi^{\alpha_k}}{\partial x^{i_k}}\, dx^{i} \wedge dx^{i_1} \wedge \cdots \wedge dx^{i_k}.$$

where the functions $y^{\alpha} = \Phi^{\alpha}(x^i)$ is generated by the mapping ϕ through local charts at the points $p \in M$ and $q = \phi(p) \in N$. On the other hand, due to the symmetry of second derivatives and antisymmetry of exterior products, we get

$$\begin{aligned}
d(\phi^*\omega) &= \frac{1}{k!}\, d\Big(\omega_{\alpha_1\cdots\alpha_k}(\Phi(\mathbf{x}))\frac{\partial \Phi^{\alpha_1}}{\partial x^{i_1}} \cdots \frac{\partial \Phi^{\alpha_k}}{\partial x^{i_k}}\Big)\, dx^{i_1} \wedge \cdots \wedge dx^{i_k} \\
&= \frac{1}{k!}\Big[\frac{\partial \omega_{\alpha_1\cdots\alpha_k}}{\partial y^{\alpha}} \frac{\partial \Phi^{\alpha}}{\partial x^{i}} \frac{\partial \Phi^{\alpha_1}}{\partial x^{i_1}} \cdots \frac{\partial \Phi^{\alpha_k}}{\partial x^{i_k}} + \omega_{\alpha_1\cdots\alpha_k} \frac{\partial^2 \Phi^{\alpha_1}}{\partial x^{i}\partial x^{i_1}} \cdots \frac{\partial \Phi^{\alpha_k}}{\partial x^{i_k}} \\
&\quad + \cdots + \omega_{\alpha_1\cdots\alpha_k} \frac{\partial \Phi^{\alpha_1}}{\partial x^{i_1}} \cdots \frac{\partial^2 \Phi^{\alpha_k}}{\partial x^{i}\partial x^{i_k}}\Big] dx^{i} \wedge dx^{i_1} \wedge \cdots \wedge dx^{i_k} \\
&= \frac{1}{k!} \frac{\partial \omega_{\alpha_1\cdots\alpha_k}(\Phi(\mathbf{x}))}{\partial y^{\alpha}} \frac{\partial \Phi^{\alpha}}{\partial x^{i}} \frac{\partial \Phi^{\alpha_1}}{\partial x^{i_1}} \cdots \frac{\partial \Phi^{\alpha_k}}{\partial x^{i_k}}\, dx^{i} \wedge dx^{i_1} \wedge \cdots \wedge dx^{i_k}
\end{aligned}$$

Therefore, we find that $d(\phi^*\omega) = \phi^* d\omega$ for any $\omega \in \Lambda(N)$. □

If the exterior derivative of a form $\omega \in \Lambda(M)$ vanishes, that is, if $d\omega = 0$, then ω is called a ***closed form***. Thus, closed forms constitute the *null space* or *kernel* of the operator d:

$$\mathcal{N}(d) = \mathrm{Ker}\,(d) = \{\omega \in \Lambda(M) : d\omega = 0\}.$$

If for a form $\omega \in \Lambda^k(M)$, there exists a form $\sigma \in \Lambda^{k-1}(M)$ such that $\omega = d\sigma$, then ω is called an ***exact form***. Obviously, this means that exact forms occupy the range of the operator d:

$$\mathcal{R}(d) = \mathrm{Im}\,(d) = \{\omega \in \Lambda(M) : \omega = d\gamma, \gamma \in \Lambda(M)\}$$

If ω is an exact form, we have $d\omega = d^2\sigma = 0$. Hence, an exact form is naturally a closed form. However, the converse statement is not always true. This subject will be investigated in detail in Chapter VI through the homotopy operator.

If $\omega \in \Lambda^m(M)$ we get $d\omega = 0$ because every $(m+1)$-form is identically zero. Therefore, *every m-form will be closed* on an m-dimensional manifold.

Theorem 5.8.3. *The closed and exact forms in the module $\Lambda^k(M)$ constitute linear vector spaces $\mathcal{C}^k(M)$ and $\mathcal{E}^k(M)$, respectively, over real numbers.*

Let us consider the closed forms $\omega, \sigma \in \Lambda^k(M)$ satisfying $d\omega = d\sigma = 0$. Let $f, g \in \Lambda^0(M$ be arbitrary functions. Then we find that

$$d(f\omega + g\sigma) = df \wedge \omega + dg \wedge \sigma.$$

Hence, this expression vanishes if and only if $df = dg = 0$. Thus if only if f and g are constants, then the form $f\omega + g\sigma$ is closed. In other words, closed forms constitute a linear vector space $\mathcal{C}^k(M)$ only on $\mathbb{R}$.

This time, let us take the exact forms $\omega, \sigma \in \Lambda^k(M)$ into consideration. Hence, there are forms $\alpha, \beta \in \Lambda^{k-1}(M)$ such that $\omega = d\alpha, \sigma = d\beta$. Since we can write

$$\omega + \sigma = d\alpha + d\beta = d(\alpha + \beta)$$

we see that the form $\omega + \sigma$ is exact. Next, let us consider the form

$$f\omega = f d\alpha = d(f\alpha) - df \wedge \alpha$$

for an arbitrary function $f \in \Lambda^0(M)$. This means that the form $f\omega$ can be exact if only $df = 0$, or f is a constant. Thus exact forms constitute a linear vector space $\mathcal{E}^k(M)$ only on $\mathbb{R}$. Since every exact form is closed, it is evident that $\mathcal{E}^k(M) \subseteq \mathcal{C}^k(M)$. □

Next, we define the sets $\mathcal{C}(M) = \overset{m}{\underset{k=0}{\oplus}} \mathcal{C}^k(M)$ and $\mathcal{E}(M) = \overset{m}{\underset{k=1}{\oplus}} \mathcal{E}^k(M)$. We can easily verify that *they form graded subalgebras of the exterior algebra* $\Lambda(M)$ *on* $\mathbb{R}$. In fact, if $\omega, \sigma \in \mathcal{C}(M)$, we have $d\omega = d\sigma = 0$ and consequently, $d(\omega \wedge \sigma) = d\omega \wedge \sigma + (-1)^{deg(\omega)}\omega \wedge d\sigma = 0$ so we find that $\omega \wedge \sigma \in \mathcal{C}(M)$. On the other hand, if $\omega, \sigma \in \mathcal{E}(M)$, then we have to write $\omega = d\alpha,\ \sigma = d\beta$ so that we obtain $\omega \wedge \sigma = d\alpha \wedge d\beta = d(\alpha \wedge d\beta)$ leading to $\omega \wedge \sigma \in \mathcal{E}(M)$.

Example 5.8.3. We consider a form $\omega \in \Lambda^1(M)$. If $\omega \in \mathcal{E}^1(M)$, then there must exist a function $\Omega \in \Lambda^0(M)$ so that we can write $\omega = d\Omega$ or

$$\omega_i\, dx^i = \Omega_{,i}\, dx^i.$$

Hence, the relations $\omega_i = \Omega_{,i}$ must hold. Thus, the coefficients ω_i have to verify the integrability conditions $\omega_{i,j} - \omega_{j,i} = 0$ in order to be able to determine Ω. On the other hand, if the form ω is closed, then we get

$$d\omega = \omega_{i,j}\, dx^j \wedge dx^i = \omega_{[i,j]}\, dx^j \wedge dx^i = 0$$

from which we deduce that $\omega_{[i,j]} = 0$ or $\omega_{i,j} - \omega_{j,i} = 0$. Thus, if the form is exact, then the conditions to be closed is satisfied automatically. However, in order that a closed 1-form is to be exact we have to find the solution of $m(m-1)/2$ first order partial differential equations satisfied by m unknowns ω_i in the form $\omega_i = \Omega_{,i}$. The existence of the solution is, however, strongly dependent on the topology of the manifold. ■

Example 5.8.4. We consider a form $\omega \in \Lambda^2(M)$. This form will be exact if there exists a form $\alpha \in \Lambda^1(M)$ such that $\omega = d\alpha$. Let us then take $\omega = \frac{1}{2}\omega_{ij}\, dx^i \wedge dx^j$ and $\alpha = \alpha_j\, dx^j$. The relation

$$\frac{1}{2}\,\omega_{ij}\, dx^i \wedge dx^j = \alpha_{j,i}\, dx^i \wedge dx^j = \alpha_{[j,i]}\, dx^i \wedge dx^j$$

leads to $\omega_{ij} = 2\alpha_{[j,i]} = \alpha_{j,i} - \alpha_{i,j}$. In order that the functions α_i satisfying these conditions could be determined the 2-form ω must be closed. This becomes possible if the condition

$$d\omega = \frac{1}{2}\,\omega_{ij,k}\, dx^k \wedge\, dx^i \wedge dx^j = \frac{1}{2}\,\omega_{[ij,k]}\, dx^k \wedge\, dx^i \wedge dx^j = 0$$

is met. Therefore, the coefficients ω_{ij} must satisfy the following differential equations

$$\omega_{[ij,k]} = 0 \quad \text{or} \quad \frac{\partial \omega_{ij}}{\partial x^k} + \frac{\partial \omega_{jk}}{\partial x^i} + \frac{\partial \omega_{ki}}{\partial x^j} = 0.$$ ■

Example 5.8.5. Let us consider the form $\omega \in \Lambda^{m-k}(M)$. This form is exact if $\omega = d\alpha$ so that (5.8.5) yields

$$\omega = \frac{1}{k!}\,\omega^{i_1\cdots i_k}\,\mu_{i_k\cdots i_1} = d\Big(\frac{1}{(k+1)!}\alpha^{i_1\cdots i_k i_{k+1}}\,\mu_{i_{k+1}i_k\cdots i_1}\Big)$$
$$= \frac{1}{k!}\alpha^{i_1\cdots i_k i}{}_{,i}\,\mu_{i_k\cdots i_1}.$$

Hence, the coefficients must satisfy a relation like

$$\omega^{i_1\cdots i_k} = \alpha^{i_1\cdots i_k i}{}_{,i}$$

whence we conclude that

$$\omega^{i_1\cdots i_{k-1}j}{}_{,j} = \alpha^{i_1\cdots i_{k-1}ji}{}_{,ij} = 0.$$

If $\omega \in \Lambda^{m-1}(M)$, the above conditions obviously reduce to

$$\omega^i = \alpha^{ij}{}_{,j} \quad \text{and} \quad \omega^i{}_{,i} = \alpha^{ij}{}_{,ij} = 0. \qquad \blacksquare$$

Let us finally consider the sequence of modules

$$\Lambda^0(M) \xrightarrow{d} \cdots \xrightarrow{d} \Lambda^k(M) \xrightarrow{d} \Lambda^{k+1}(M) \xrightarrow{d} \cdots \xrightarrow{d} \Lambda^m(M) \xrightarrow{d} 0 \qquad (5.8.6)$$

where homomorphisms between successive linear vector spaces are provided by the exterior derivative d on real numbers. *Since* $d \circ d = d^2 = 0$, *this sequence is evidently a cochain complex.* As we shall see later in Chapter VIII, this cochain complex will play quite a significant part in revealing some fundamental properties of closed and exact forms that connect some topological and analytical features.

5.9. RIEMANNIAN MANIFOLDS. HODGE DUAL

A 2-covariant tensor field $\mathcal{G} \in \mathfrak{T}(M)^0_2$ on a smooth manifold M will be called a ***metric tensor*** if it obeys the following requirements:

(i). *$\mathcal{G}$ is a symmetric tensor.*
(ii). *The bilinear form $\mathcal{G}_p$ is not degenerate at every point $p \in M$, that is, $\mathcal{G}_p(U, V) = 0$ for all $U \in T_p(M)$ if and only if $V = 0$ at the point p.*

A manifold equipped with such a metric tensor will be called a ***Riemannian manifold***. In local coordinates, the metric tensor is expressible as

$$\mathcal{G} = g_{ij}(\mathbf{x})\,dx^i \otimes dx^j, \;\; g_{ij} = g_{ji}. \qquad (5.9.1)$$

Consequently, the condition $\mathcal{G}(U,V) = g_{ij}u^i v^j = 0$ for all vectors $U = u^i\partial_i$ where $V = v^i\partial_i$ results in $g_{ij}v^j = 0$. Whenever this homogeneous system of linear equations is satisfied if and only if $V = 0$, then the matrix $\mathbf{G} = [g_{ij}]$ must be regular at every point, namely, its inverse must exist. Let us denote the inverse matrix by $\mathbf{G}^{-1} = [(g^{-1})^{ij}] = [g^{ij}]$. Hence, the relations $g^{ik}g_{kj} = g_{jk}g^{ki} = \delta^i_j$ will hold. By means of the metric tensor $\mathcal{G}$, we can assign a field of 1-form in $T^*(M)$ to every vector field $V \in T(M)$ prescribed by $V = v^i\partial/\partial x^i$ where $v^i(\mathbf{x})$ denote the contravariant components of V through the relation

$$\omega_V = \mathcal{G}(V) = g_{ij}v^j\,dx^i = v_i\,dx^i \in T^*(M) = \Lambda^1(M).$$

Thus the metric tensor gives rise to a linear mapping $\mathcal{G} : T(M) \to T^*(M)$. The coefficients of the form ω_V given by

$$v_i = g_{ij}v^j \in \Lambda^0(M) \tag{5.9.2}$$

is called the *covariant components* of the vector V. If we make use of the inverse matrix $\mathbf{G}^{-1}$, (5.9.2) can be transformed into

$$v^i = g^{ij}v_j. \tag{5.9.3}$$

Thus a vector V can also be expressed as

$$V = v^j\frac{\partial}{\partial x^j} = g^{ji}v_i\frac{\partial}{\partial x^j} = v_i e^i.$$

Since the matrix $\mathbf{G}$ is regular, the vectors

$$e^i = g^{ij}\frac{\partial}{\partial x^j},\quad i = 1,\dots,m \tag{5.9.4}$$

constitute a basis for the tangent space as well. It then easily follows from (5.9.1) and (5.9.4) that

$$\mathcal{G}(\partial_i,\partial_j) = g_{ij},\ \mathcal{G}(e^i,e^j) = g_{kl}g^{ik}g^{jl} = g^{ij}. \tag{5.9.5}$$

Let us now consider a form $\omega = \omega_i\,dx^i \in \Lambda^1(M)$ and introduce a vector through the relation

$$V_\omega = g^{ij}\omega_j\frac{\partial}{\partial x^i} = \omega^i\frac{\partial}{\partial x^i} \in T(M),\ \ \omega^i = g^{ij}\omega_j.$$

We can readily verify that $\mathcal{G}(V_\omega) = \omega$. Moreover, we can write

$$\mathcal{G}(V_\omega, V_\sigma) = g_{ij}\omega^i\sigma^j = g_{ij}g^{ik}g^{jl}\omega_k\sigma_l = g^{kl}\omega_k\sigma_l. \tag{5.9.6}$$

These results reveal the fact that the metric tensor furnishes an isomorphism between bundles $T(M)$ and $T^*(M)$. The inverse operator is procured by the inverse matrix g^{ij}. Let us define a new set of basis vectors in $T^*(M)$ by

$$f_i = g_{ij}\,dx^j. \tag{5.9.7}$$

We then obtain

$$f_i(e^j) = g_{ik}\,dx^k(g^{jl}\partial_l) = g_{ik}g^{jl}\delta_l^k = g_{ik}g^{kj} = \delta_i^j$$

which means that $\{e^i\}$ and $\{f_i\}$ are reciprocal bases. On making use of (5.9.7) we can also write $dx^i = g^{ij}f_j$. Utilising (5.9.7), we easily get another representation of the metric tensor

$$\begin{aligned} g^{ij}f_i \otimes f_j &= g^{ij}g_{ik}g_{jl}\,dx^k \otimes dx^l = \delta_k^j g_{jl}\,dx^k \otimes dx^l \\ &= g_{kl}\,dx^k \otimes dx^l = \mathcal{G}. \end{aligned}$$

When we consider a coordinate transformation such as $y^i = y^i(x^j)$ in a neighbourhood of a point $p \in M$ we arrive at the following rule of transformation

$$\begin{aligned} f_i'(\mathbf{y}) = g_{ij}'\,dy^j &= \frac{\partial x^k}{\partial y^i}\frac{\partial x^l}{\partial y^j}g_{kl}\frac{\partial y^j}{\partial x^m}dx^m = \frac{\partial x^k}{\partial y^i}g_{kl}dx^l \\ &= \frac{\partial x^k}{\partial y^i}f_k(\mathbf{x}). \end{aligned}$$

The inverse relation then obviously becomes

$$f_k(\mathbf{x}) = \frac{\partial y^i}{\partial x^k}f_i'(\mathbf{y})$$

Hence, the relation

$$\mathcal{G} = g'^{ij}f_i' \otimes f_j' = g^{kl}f_k \otimes f_l = \frac{\partial y^i}{\partial x^k}\frac{\partial y^j}{\partial x^l}g^{kl}f_i' \otimes f_j'$$

leads to the transformation

$$g'^{ij} = \frac{\partial y^i}{\partial x^k}\frac{\partial y^j}{\partial x^l}g^{kl}$$

meaning that the coefficients g^{ij} are actually contravariant components of the tensor $\mathcal{G}$.

If the tensor $\mathcal{G}$ is positive definite, namely, if for every non-zero vector field $V \in T(M)$ one has

$$\mathcal{G}(V,V) = g_{ij}v^i v^j > 0 \tag{5.9.8}$$

we say that the *Riemannian manifold* is ***complete*** and the *metric* is ***definite.*** If this condition does not hold, then M is a ***pseudo-Riemannian manifold*** or an ***incomplete Riemannian manifold*** and the metric is ***indefinite.*** When the metric on a Riemannian manifold verifies the constraint (5.9.8), then it becomes possible to define an ***inner product*** or, if we put it another way, a ***scalar product of two vectors*** on the tangent bundle $T(M)$ of the manifold through the relation

$$(U,V) = \mathcal{G}(U,V) = g_{ij}u^i v^j, \quad U,V \in T(M). \tag{5.9.9}$$

It is a simple exercise to show that the above definition entirely complies with the rules concerning an inner product on a vector space. Hence, the finite-dimensional vector space $T_p(M)$ then becomes a real Hilbert space. $T(M)$ will then be the union of Hilbert spaces. The relations (5.9.9) and (5.9.8) makes it possible to associate with a vector a positive number that vanishes if and only if the vector is zero. We call this number as the ***length*** or the ***norm*** of the vector V:

$$\|V\| = \sqrt{(V,V)} = \sqrt{g_{ij}v^i v^j} > 0. \tag{5.9.10}$$

In like fashion, we can define an inner product on the dual space $T^*(M)$ by the relation

$$(\omega,\sigma) = g^{ij}\omega_i\sigma_j, \quad \omega,\sigma \in \Lambda^1(M).$$

If $(U,V) = g_{ij}u^i v^j = 0$ for distinct vectors U and V, namely, if their inner product vanishes, we say that these vectors constitute an ***orthogonal*** set. When, in addition, their norms is equal to 1, then they form an ***orthonormal*** set. When we are provided with a set of orthogonal vectors, this set can obviously be cast into a set of orthonormal vectors by dividing each vector by its norm. In a finite-dimensional complete Riemannian manifold, we can always construct an orthonormal basis for $T(M)$ inductively. Let $U_i, i = 1,\dots,m$ be a linearly independent set of vectors. Let us start by taking $W_1 = U_1$ and construct the following sequence of vectors

$$W_i = U_i - \sum_{j=1}^{i-1} \frac{W_j(U_i,W_j)}{\|W_j\|^2}, \quad V_i = \frac{W_i}{\|W_i\|}, \quad i = 1,\dots,m.$$

It is straightforward to verify that the vectors $V_1, V_2, \dots, V_m$ form an orthonormal basis, that is, they possess the property

$$(V_i, V_j) = g_{kl} v_i^k v_j^l = \delta_{ij}.$$

This method that generates generally a set of orthonormal vectors from a given countable set of linearly independent vectors spanning the same subspace is known as the ***Gram-Schmidt orthonormalisation process*** after Danish mathematician Jørgen Pedersen Gram (1850-1916) and German mathematician Erhard Schmidt (1876-1959). They had developed it independently. However, it must be fair to mention that French mathematician Pierre-Simon Laplace (1749-1827) had presented this process much earlier than either Gram or Schmidt albeit in a somewhat limited context. Thus, we can always choose an orthonormal basis in the finite-dimensional $T(M)$ such that the components of the metric tensor become simply

$$\mathcal{G}(V_i, V_j) = (V_i, V_j) = \delta_{ij}.$$

Indeed, if we choose a reciprocal basis $\{\theta^i\}$ in $T^*(M)$ in such way that the relations $\theta^i(V_j) = \delta^i_j$ are satisfied, then the metric tensor will be represented in the following form

$$\mathcal{G} = \delta_{ij}\theta^i \otimes \theta^j = \theta^1 \otimes \theta^1 + \theta^2 \otimes \theta^2 + \cdots + \theta^m \otimes \theta^m.$$

We thus conclude that in a complete Riemannian manifold, there is always a *local basis* in $T(M)$ such that the metric tensor is locally given by an identity matrix. Such a manifold is also called ***locally Euclidean*** as far as the inner product properties are concerned.

If the metric is indefinite, we can still define a kind of inner product by (5.9.9), but, this time, the so-called *norm* of a vector V defined by

$$\|V\| = \sqrt{(V, V)} = \sqrt{g_{ij} v^i v^j} = \sqrt{g^{ij} v_i v_j}$$

may be a real or an imaginary number because the term $(V, V) = g_{ij} v^i v^j$ may be positive, negative or zero. If $g_{ij} v^i v^j = 0$, then $V \neq 0$ is called a ***null vector***. However, metric tensor is still symmetric and non-degenerate. Hence, its real eigenvalues cannot be zero and it has m linearly independent orthogonal eigenvectors $V_1, V_2, \ldots, V_m$ so normalised that $(V_i, V_j) = 0$ if $i \neq j$ and $|(V_{\underline{i}}, V_{\underline{i}})| = 1$, or $(V_{\underline{i}}, V_{\underline{i}}) = \pm 1$. This means that we can write the relation

$$(V_i, V_j) = g_{kl} v_i^k v_j^l = \pm\, \delta_{ij}.$$

According to this definition a null vector will be orthogonal to itself. Hence, the components of the metric tensor with respect to such a basis are prescribed by

$$\mathcal{G}(V_i, V_j) = g_{kl}v_i^k v_j^l = (V_i, V_j) = \pm\,\delta_{ij}.$$

This amount to say that there is always a basis $\{V_i\}$ of $T(M)$ with respect to which the metric tensor is designated by a diagonal matrix whose entries are either $+1$ or -1. We then choose the reciprocal basis $\{\theta^i\}$ in $T^*(M)$ to express the metric tensor in the form

$$\mathcal{G} = \theta^1 \otimes \theta^1 + \cdots + \theta^r \otimes \theta^r - \theta^{r+1} \otimes \theta^{r+1} - \cdots - \theta^m \otimes \theta^m$$

by changing the ordering of basis vectors if necessary. The number $s = m - r$ is called the ***index*** of the metric tensor. We say that the sequence $+\cdots+\ -\cdots-$ that consists of r number of $+$ and s number of $-$ is the ***signature of this tensor***. *The signature is even if s is an even number and is odd if s is an odd number*. A manifold endowed with such a metric is named as a ***locally Minkowskian manifold*** after German mathematician Hermann Minkowski (1864-1909) who had explored such manifolds within the context of the theory of general relativity. If the metric tensor is positive definite, we evidently have $s = 0$ and $r = m$.

The metric tensor provide a means to calculate the arc length of a curve on a manifold. We know that a curve on a manifold M is a differentiable mapping $\gamma : [a, b] \to M$ and the point $p(t) \in M$ on the curve are described by $p(t) = \gamma(t)$, $a \le t \le b$. If the tangent vector of the curve at a point p is $V\big(p(t)\big)$, then the elementary arc length may be defined as

$$ds^2 = \|V(t)\|^2 dt^2 = g_{ij}v^i(t)v^j(t)\,dt^2 = g_{ij}\,dx^i dx^j$$

and the arc length of the curve between the points $p(a)$ and $p(b)$ is consequently given by

$$l = \int_a^b \|V(t)\|dt = \int_a^b \sqrt{g_{ij}v^i(t)v^j(t)}\,dt.$$

If the Riemannian manifold is complete, then l is always a positive number.

The metric tensor also helps convert covariant components of a tensor to its contravariant components and vice versa. Let us consider the covariant tensor

$$\mathcal{T} = t_{j_1 j_2 \cdots j_k} dx^{j_1} \otimes dx^{j_2} \otimes \cdots \otimes dx^{j_k}$$

that can also be written in the form

$$\mathcal{T} = g^{i_1 j_1} \cdots g^{i_k j_k} t_{j_1 \cdots j_k} f_{i_1} \otimes \cdots \otimes f_{i_k} = t^{i_1 \cdots i_k} f_{i_1} \otimes \cdots \otimes f_{i_k}$$

if we use the inverse relation (5.9.7) as $dx^i = g^{ij} f_j$. Here we define

$$t^{i_1\cdots i_k} = g^{i_1 j_1}\cdots g^{i_k j_k} t_{j_1\cdots j_k}. \tag{5.9.11}$$

The coefficients $t^{i_1\cdots i_k}$ are obtained by performing k contractions on a tensor $\mathfrak{T}(M)^{2k}_k$ formed as the product of a $\mathfrak{T}(M)^0_k$ tensor and k times of a $\mathfrak{T}(M)^2_0$ tensor which is the inverse metric tensor. Hence, the quotient rule [*see p.* 212] states that they are nothing but the *contravariant components* of the same tensor $\mathcal{T}$. Thus the components of the inverse metric tensor prove to be useful in *raising* the indices in the tensorial components. Similarly, we can show that the components of the metric tensor can be instrumental in *lowering* indices in the tensorial components. Indeed, if a tensor $\mathcal{T}$ is given in the form

$$\mathcal{T} = t^{j_1\cdots j_k}\frac{\partial}{\partial x^{j_1}}\otimes\cdots\otimes\frac{\partial}{\partial x^{j_k}}$$

then inserting $\partial_i = g_{ij}\, e^j$ that follows from (5.9.4) into the above expression we find that

$$\mathcal{T} = g_{i_1 j_1}\cdots g_{i_k j_k} t^{j_1\cdots j_k}\, e^{i_1}\otimes\cdots\otimes e^{i_k} = t_{i_1\cdots i_k} e^{i_1}\otimes\cdots\otimes e^{i_k}$$

where the covariantly transforming coefficients

$$t_{i_1\cdots i_k} = g_{i_1 j_1}\cdots g_{i_k j_k} t^{j_1\cdots j_k} \tag{5.9.12}$$

are called the *covariant components* of the tensor $\mathcal{T}$. It is seen that the existence of the metric tensor effectively abolishes the distinction between covariant and contravariant tensors and provides a natural transition between components of such kind of tensors. It is clear that this procedure is applicable to any index of mixed components of a tensor.

Suppose that a tensor is defined as a contraction of a product of two tensors. In terms of components we can write for example

$$\begin{aligned} t_{i i_1\cdots i_k}\tau^{i j_1\cdots j_l} &= g_{ij} g^{ik} t^{j}{}_{i_1\cdots i_k}\tau_k{}^{j_1\cdots j_l} = \delta^k_j\, t^j{}_{i_1\cdots i_k}\tau_k{}^{j_1\cdots j_l} \\ &= t^j{}_{i_1\cdots i_k}\tau_j{}^{j_1\cdots j_l}. \end{aligned}$$

We thus reach to the conclusion that such a tensor does not change if we arbitrarily lower one and raise the other of contracted indices.

If we can find a form $\Omega\in\Lambda^m(M)$ on an m-dimensional manifold M such that $\Omega\neq 0$ at every point $p\in M$, then we say that M is an ***orientable manifold*** and Ω is a ***volume form***. In that case, it is clear that one is able to write $\Omega = f(\mathbf{x})\, dx^1\wedge\cdots\wedge dx^m$ where we must have $f\neq 0$ everywhere on M. When M is a *complete Riemannian manifold*, we get $g = \det[g_{ij}] > 0$. Under a coordinate transformation $y^i = y^i(x^j)$, we readily obtain in general

$$\det[g'_{ij}(\mathbf{y})] = \det\Big[\frac{\partial x^k}{\partial y^i}\frac{\partial x^l}{\partial y^j}g_{kl}(\mathbf{x})\Big] = \frac{\det[g_{kl}(\mathbf{x})]}{J^2}$$

where $J = \det[\partial y^i/\partial x^j] \neq 0$. Let us now define $g = |\det[g_{ij}]| > 0$ so that we can write $g'(\mathbf{y}) = g(\mathbf{x})/J^2$. We now introduce a volume form as follows

$$\mu(\mathbf{x}) = \sqrt{g}\,dx^1 \wedge \cdots \wedge dx^m. \tag{5.9.13}$$

If the Riemannian manifold is not complete, then $\det[g_{ij}]$ may be positive or negative although it cannot be zero because we have assumed that the metric tensor is non-degenerate. In that case, we always have $g = |\det[g_{ij}]| > 0$ in (5.9.13). Such a g has obviously the same transformation rule as that of given above. The form $\mu \in \Lambda^m(M)$ will be called the ***Riemannian volume form***. Under a coordinate transformation $y^i = y^i(x^j)$, this form is transformed in the following manner

$$\begin{aligned}
\mu(\mathbf{y}) &= \sqrt{g'}\,dy^1 \wedge \cdots \wedge dy^m \\
&= \frac{\sqrt{g}}{|J|}\frac{\partial y^1}{\partial x^{i_1}}\cdots\frac{\partial y^m}{\partial x^{i_m}}\,dx^{i_1} \wedge \cdots \wedge dx^{i_m} \\
&= \frac{\sqrt{g}}{|J|}e^{i_1\cdots i_m}\frac{\partial y^1}{\partial x^{i_1}}\cdots\frac{\partial y^m}{\partial x^{i_m}}\,dx^1 \wedge \cdots \wedge dx^m \\
&= \sqrt{g}\frac{J}{|J|}\,dx^1 \wedge \cdots \wedge dx^m \\
&= (\operatorname{sgn} J)\sqrt{g}\,dx^1 \wedge \cdots \wedge dx^m \\
&= (\operatorname{sgn} J)\,\mu(\mathbf{x})
\end{aligned}$$

where $\operatorname{sgn} J = J/|J|$ is $+1$ if $J > 0$ and -1 if $J < 0$. Clearly, this volume form remains invariant under coordinate transformations if $J > 0$. The form (5.9.13) can also be written as

$$\begin{aligned}
\mu &= \frac{1}{m!}\sqrt{g}\,e_{i_1\cdots i_m}\,dx^{i_1} \wedge \cdots \wedge dx^{i_m} \\
&= \frac{1}{m!}\,\epsilon_{i_1\cdots i_m}\,dx^{i_1} \wedge \cdots \wedge dx^{i_m}
\end{aligned} \tag{5.9.14}$$

where we defined the ***covariant Levi-Civita permutation tensor*** by the relation

$$\epsilon_{i_1\cdots i_m} = \sqrt{g}\,e_{i_1\cdots i_m}. \tag{5.9.15}$$

On the other hand, the expression

$$
\begin{aligned}
e^{j_1\cdots j_m}\mu &= \frac{1}{m!}\sqrt{g}\,\delta^{j_1\cdots j_m}_{i_1\cdots i_m}\,dx^{i_1}\wedge\cdots\wedge dx^{i_m}\\
&= \sqrt{g}\,dx^{[j_1}\wedge\cdots\wedge dx^{j_m]}\\
&= \sqrt{g}\,dx^{j_1}\wedge\cdots\wedge dx^{j_m}
\end{aligned}
$$

yields

$$dx^{i_1}\wedge\cdots\wedge dx^{i_m} = \frac{e^{i_1\cdots i_m}}{\sqrt{g}}\mu = \epsilon^{i_1\cdots i_m}\mu$$

where the ***contravariant Levi-Civita permutation tensor*** is defined by

$$\epsilon^{i_1\cdots i_m} = \frac{e^{i_1\cdots i_m}}{\sqrt{g}}. \tag{5.9.16}$$

In order to identify the tensorial character of these quantities let us start with the relations

$$
\begin{aligned}
e_{i_1i_2\cdots i_n}(\boldsymbol{y})J^{-1} &= e_{j_1j_2\cdots j_n}(\boldsymbol{x})\frac{\partial x^{j_1}}{\partial y^{i_1}}\frac{\partial x^{j_2}}{\partial y^{i_2}}\cdots\frac{\partial x^{j_n}}{\partial y^{i_n}}\\
e^{i_1i_2\cdots i_n}(\boldsymbol{y})J &= e^{j_1j_2\cdots j_n}(\boldsymbol{x})\frac{\partial y^{i_1}}{\partial x^{j_1}}\frac{\partial y^{i_2}}{\partial x^{j_2}}\cdots\frac{\partial y^{i_n}}{\partial x^{j_n}}
\end{aligned}
$$

from which we deduce the transformation rules of Levi-Civita symbols as

$$
\begin{aligned}
e_{i_1i_2\cdots i_n}(\boldsymbol{y}) &= J\frac{\partial x^{j_1}}{\partial y^{i_1}}\frac{\partial x^{j_2}}{\partial y^{i_2}}\cdots\frac{\partial x^{j_n}}{\partial y^{i_n}}e_{j_1j_2\cdots j_n}(\boldsymbol{x}),\\
e^{i_1i_2\cdots i_n}(\boldsymbol{y}) &= J^{-1}\frac{\partial y^{i_1}}{\partial x^{j_1}}\frac{\partial y^{i_2}}{\partial x^{j_2}}\cdots\frac{\partial y^{i_n}}{\partial x^{j_n}}e^{j_1j_2\cdots j_n}(\boldsymbol{x}).
\end{aligned}
$$

This means that $e_{i_1i_2\cdots i_n}$ and $e^{i_1i_2\cdots i_n}$ are actually ***tensor densities*** because the transformation rule depends on the Jacobian of the coordinate transformation. Since we can write $J = \operatorname{sgn} J|J|$, Levi-Civita tensors will satisfy

$$
\begin{aligned}
\epsilon_{i_1i_2\cdots i_n}(\boldsymbol{y}) &= \operatorname{sgn} J\frac{\partial x^{j_1}}{\partial y^{i_1}}\frac{\partial x^{j_2}}{\partial y^{i_2}}\cdots\frac{\partial x^{j_n}}{\partial y^{i_n}}\epsilon_{j_1j_2\cdots j_n}(\boldsymbol{x}),\\
\epsilon^{i_1i_2\cdots i_n}(\boldsymbol{y}) &= \operatorname{sgn} J\frac{\partial y^{i_1}}{\partial x^{j_1}}\frac{\partial y^{i_2}}{\partial x^{j_2}}\cdots\frac{\partial y^{i_n}}{\partial x^{j_n}}\epsilon^{j_1j_2\cdots j_n}(\boldsymbol{x}).
\end{aligned}
$$

So *Levi-Civita tensors* $\epsilon_{i_1i_2\cdots i_n}$ *and* $\epsilon^{i_1i_2\cdots i_n}$ *are* ***pseudotensors*** because the transformation rule changes sign depending on the Jacobian of the coordinate transformation. They behave like absolute tensors if $J > 0$. In order to understand how they are related, let us consider the relation

$$\begin{aligned} g^{i_1 j_1} \cdots g^{i_n j_n} \epsilon_{j_1 \cdots j_n} &= \sqrt{g}\, e_{j_1 \cdots j_n} g^{i_1 j_1} \cdots g^{i_n j_n} = \sqrt{g} \det [g^{ij}] e^{i_1 \cdots i_n} \\ &= \frac{g}{\det [g_{ij}]} \epsilon^{i_1 i_2 \cdots i_n} = (\operatorname{sgn} \det [g_{ij}])\, \epsilon^{i_1 i_2 \cdots i_n} \end{aligned}$$

Similarly, we find that

$$g_{i_1 j_1} \cdots g_{i_n j_n} \epsilon^{j_1 \cdots j_n} = (\operatorname{sgn} \det [g_{ij}])\, \epsilon_{i_1 \cdots i_n}.$$

Hence, they represent covariant and contravariant components of the same tensor if $\det [g_{ij}] > 0$. We also easily observe that we get the absolute tensor

$$\delta^{i_1 \cdots i_m}_{j_1 \cdots j_m} = e^{i_1 \cdots i_m} e_{j_1 \cdots j_m} = \epsilon^{i_1 \cdots i_m} \epsilon_{j_1 \cdots j_m}.$$

We can now fulfil the task of the top down generation of ordered bases for the exterior algebra $\Lambda(M)$ just like we have done in Sec. 5.5 by using the volume form defined by (5.9.14). Let us introduce similarly the ordered forms

$$\begin{aligned} \mu_{i_k i_{k-1} \cdots i_1} &= (\mathbf{i}_{\partial_{i_k}} \circ \mathbf{i}_{\partial_{i_{k-1}}} \circ \cdots \circ \mathbf{i}_{\partial_{i_1}})(\mu) \\ &= \mathbf{i}_{\partial_{i_k}} (\mu_{i_{k-1} \cdots i_1}) \\ &= \frac{1}{(m-k)!} \epsilon_{i_1 \cdots i_k i_{k+1} \cdots i_m} dx^{i_{k+1}} \wedge \cdots \wedge dx^{i_m} \in \Lambda^{m-k}(M) \end{aligned} \tag{5.9.17}$$

where $1 \le k \le m$. Following the path we have pursued in obtaining the relation (5.5.12), we easily deduce from (5.9.17) that

$$dx^{i_{k+1}} \wedge \cdots \wedge dx^{i_m} = \frac{1}{k!} \epsilon^{i_1 \cdots i_k i_{k+1} \cdots i_m} \mu_{i_k \cdots i_1}. \tag{5.9.18}$$

It is straightforward to see that all expressions appearing between (5.5.13) and (5.5.18) remain without change if we replace μ by (5.9.14) and Levi-Civita symbols by Levi-Civita tensors. In like fashion, we can verify at once that the forms $\mu_{i_k \cdots i_1}$ defined in (5.9.17) constitute a basis of the module $\Lambda^{m-k}(M)$. Thus a form $\omega \in \Lambda^{m-k}(M)$ may be written again as

$$\omega = \frac{1}{k!} \omega^{i_1 \cdots i_k} \mu_{i_k \cdots i_1}.$$

But, the exterior derivative of this form is now rather different from what is given in (5.8.5). This derivative is of course

$$d\omega = \frac{1}{k!} (\omega^{i_1 \cdots i_k}{}_{,i}\, dx^i \wedge \mu_{i_k \cdots i_1} + \omega^{i_1 \cdots i_k} d\mu_{i_k \cdots i_1}).$$

On the other hand, an explicit calculation leads to

$$\begin{aligned}
d\mu_{i_k\cdots i_1} \\
&= \frac{1}{(m-k)!}\, e_{i_1\cdots i_k i_{k+1}\cdots i_m}(\sqrt{g})_{,i}\, dx^i \wedge dx^{i_{k+1}} \wedge \cdots \wedge dx^{i_m} \\
&= \frac{1}{(m-k)!}\frac{(\sqrt{g})_{,i}}{\sqrt{g}}\, \epsilon_{i_1\cdots i_k i_{k+1}\cdots i_m} dx^i \wedge dx^{i_{k+1}} \wedge \cdots \wedge dx^{i_m} \\
&= \frac{1}{(m-k)!}\frac{(\sqrt{g})_{,i}}{\sqrt{g}}\, \epsilon_{i_1\cdots i_k i_{k+1}\cdots i_m}\frac{1}{(k-1)!}\epsilon^{j_1\cdots j_{k-1} i i_{k+1}\cdots i_m}\mu_{j_{k-1}\cdots j_1} \\
&= \frac{1}{(k-1)!}\frac{1}{(m-k)!}\frac{(\sqrt{g})_{,i}}{\sqrt{g}}\delta^{j_1\cdots j_{k-1} i i_{k+1}\cdots i_m}_{i_1\cdots i_{k-1} i_k i_{k+1}\cdots i_m}\mu_{j_{k-1}\cdots j_1} \\
&= \frac{1}{(k-1)!}\frac{(\sqrt{g})_{,i}}{\sqrt{g}}\delta^{j_1\cdots j_{k-1} i}_{i_1\cdots i_{k-1} i_k}\mu_{j_{k-1}\cdots j_1} = k\frac{(\sqrt{g})_{,i}}{\sqrt{g}}\delta^{i}_{[i_k}\mu_{i_{k-1}\cdots i_2 i_1]}.
\end{aligned}$$

Hence, according to (5.5.15) and due to the complete antisymmetry of functions $\omega^{i_1\cdots i_k}$ we obtain

$$\begin{aligned}
d\omega &= \frac{1}{(k-1)!}\Big(\omega^{i_1\cdots i_k}{}_{,i} + \frac{(\sqrt{g})_{,i}}{\sqrt{g}}\,\omega^{i_1\cdots i_k}\Big)\delta^i_{[i_k}\mu_{i_{k-1}\cdots i_1]} \\
&= \frac{1}{(k-1)!}\frac{1}{\sqrt{g}}(\sqrt{g}\,\omega^{i_1\cdots i_k})_{,i}\delta^i_{[i_k}\mu_{i_{k-1}\cdots i_1]} \\
&= \frac{1}{(k-1)!}\frac{1}{\sqrt{g}}(\sqrt{g}\,\omega^{i_1\cdots i_{k-1} i})_{,i}\mu_{i_{k-1}\cdots i_1} = \frac{1}{(k-1)!}\,\omega^{i_1\cdots i_{k-1} i}{}_{;i}\,\mu_{i_{k-1}\cdots i_1}
\end{aligned}$$

where we introduced the definition

$$\omega^{i_1\cdots i_{k-1} i}{}_{;i} = \frac{1}{\sqrt{g}}(\sqrt{g}\,\omega^{i_1\cdots i_{k-1} i})_{,i} \tag{5.9.19}$$

A semicolon in front of an index denotes the *covariant derivative* with respect to a variable depicted by this index. We discuss the concept of covariant derivative in Chapter VII in detail. Here we just confine ourselves to indicate that although the quantities $\omega^{i_1\cdots i_{k-1} i}{}_{,i}$ are not generally components of a tensor, the coefficients $\omega^{i_1\cdots i_{k-1} i}{}_{;i}$ of the form $d\omega$ are components of a $(k-1)$-contravariant tensor. We now suppose that a form

$$\omega = \frac{1}{k!}\,\omega_{i_1\cdots i_k}(\mathbf{x})\, dx^{i_1} \wedge \cdots \wedge dx^{i_k} \in \Lambda^k(M)$$

is given on an orientable Riemannian manifold. The ***Hodge dual*** or just simply the ***dual*** of this form is defined by

$$*\omega = \frac{1}{k!}\,\omega^{i_1\cdots i_k}(\mathbf{x})\,\mu_{i_k\cdots i_1} \in \Lambda^{m-k}(M) \tag{5.9.20}$$

where *contravariant components* are of course now prescribed by

$$\omega^{i_1\cdots i_k} = g^{i_1 j_1}\cdots g^{i_k j_k}\omega_{j_1\cdots j_k}. \tag{5.9.21}$$

The operator $* : \Lambda^k(M) \to \Lambda^{m-k}(M)$ is known as the ***Hodge star operator***. The form (5.9.20) is expressible in the natural basis as

$$\begin{aligned} *\omega &= \frac{1}{k!}\frac{1}{(m-k)!}\,\epsilon_{i_1\cdots i_k i_{k+1}\cdots i_m}\omega^{i_1\cdots i_k}\,dx^{i_{k+1}}\wedge\cdots\wedge dx^{i_m} \\ &= \frac{1}{(m-k)!}\,*\omega_{i_{k+1}\cdots i_m}dx^{i_{k+1}}\wedge\cdots\wedge dx^{i_m} \end{aligned} \tag{5.9.22}$$

where we have defined

$$*\omega_{i_{k+1}\cdots i_m} = \frac{1}{k!}\,\epsilon_{i_1\cdots i_k i_{k+1}\cdots i_m}\omega^{i_1\cdots i_k}. \tag{5.9.23}$$

Hodge star operator is evidently a linear operator on the graded exterior algebra. On applying $*$ operator successively, it follows from (5.9.22) that

$$\begin{aligned} **\omega &= \frac{1}{(m-k)!}\,*\omega^{i_{k+1}\cdots i_m}\mu_{i_m\cdots i_{k+1}} \\ &= \frac{1}{(m-k)!}\frac{1}{k!}\epsilon^{i_1\cdots i_k i_{k+1}\cdots i_m}\omega_{i_1\cdots i_k}\mu_{i_m\cdots i_{k+1}} \\ &= \frac{1}{(m-k)!}\frac{1}{k!}(-1)^{k(m-k)}\epsilon^{i_{k+1}\cdots i_m i_1\cdots i_k}\omega_{i_1\cdots i_k}\mu_{i_m\cdots i_{k+1}} \\ &= (-1)^{k(m-k)}\frac{1}{k!}\,\omega_{i_1\cdots i_k}dx^{i_1}\wedge\cdots\wedge dx^{i_k} = (-1)^{k(m-k)}\omega. \end{aligned}$$

In order to reach to this result, we have raised and lowered the indices appropriately utilising the metric tensor. Consequently, if applied on k-forms, the inverse of the operator $*$ becomes

$$*^{-1} = (-1)^{k(m-k)}* = (-1)^{k(m-1)}* \tag{5.9.24}$$

because $k^2 - k$ is always an even number. It easily verified that the dual of the volume form (5.9.14) is

$$*\mu = \frac{1}{m!}\,\epsilon^{i_1\cdots i_m}\mu_{i_m\cdots i_1} = \frac{1}{m!}\,\epsilon^{i_1\cdots i_m}\epsilon_{i_1\cdots i_m} = 1. \tag{5.9.25}$$

If we take $k = m$, then (5.9.24) yields $*^{-1} = *$ and we obtain

$$*1 = **\mu = \mu. \tag{5.9.26}$$

Let us now consider the forms $\omega, \sigma \in \Lambda^k(M)$ given by

$$\omega = \frac{1}{k!}\,\omega_{i_1\cdots i_k} dx^{i_1} \wedge \cdots \wedge dx^{i_k},$$
$$\sigma = \frac{1}{k!}\,\sigma_{i_1\cdots i_k} dx^{i_1} \wedge \cdots \wedge dx^{i_k}.$$

In this situation, we have $\omega \wedge *\sigma \in \Lambda^m(M)$. If we evaluate this form explicitly, we obtain

$$\begin{aligned}\omega \wedge *\sigma &= \Big(\frac{1}{k!}\Big)^2 \omega_{i_1\cdots i_k}\sigma^{j_1\cdots j_k} dx^{i_1} \wedge \cdots \wedge dx^{i_k} \wedge \mu_{j_k\cdots j_1}\\ &= \Big(\frac{1}{k!}\Big)^2 \omega_{i_1\cdots i_k}\sigma^{j_1\cdots j_k}\delta^{i_1\cdots i_k}_{j_1\cdots j_k}\mu = \frac{1}{k!}\,\omega_{i_1\cdots i_k}\sigma^{[i_1\cdots i_k]}\mu\\ &= \frac{1}{k!}\,\omega_{i_1\cdots i_k}\sigma^{i_1\cdots i_k}\mu.\end{aligned}$$

On the other hand, since the same expression may be directly transformed into the form $\omega \wedge *\sigma = \dfrac{1}{k!}\,\sigma_{i_1\cdots i_k}\omega^{i_1\cdots i_k}\mu$, we arrive at the identity

$$\omega \wedge *\sigma = \sigma \wedge *\omega. \tag{5.9.27}$$

For a form $\omega \in \Lambda^k(M)$, we similarly find

$$\omega \wedge *\omega = \frac{1}{k!}\,\omega_{i_1\cdots i_k}\omega^{i_1\cdots i_k}\mu.$$

Next, we take a form $\omega \in \Lambda^k(M)$ into account and calculate the exterior derivative of its dual. Recalling the definition (5.9.19), we obtain

$$\begin{aligned}d(*\omega) &= \frac{1}{k!}\,d\big(\omega^{i_1\cdots i_k}\mu_{i_k\cdots i_1}\big) = \frac{1}{(k-1)!}\,\omega^{i_1\cdots i_{k-1}i}{}_{;i}\,\mu_{i i_{k-1}\cdots i_1} \qquad (5.9.28)\\ &= \frac{1}{(k-1)!}\frac{1}{(m-k+1)!}\omega^{i_1\cdots i_{k-1}i}{}_{;i}\,\epsilon_{i_1\cdots i_{k-1}i_k\cdots i_m} dx^{i_k} \wedge \cdots \wedge dx^{i_m}.\end{aligned}$$

It is clear that $d(*\omega) \in \Lambda^{m-k+1}(M)$. An operator $\delta : \Lambda^k(M) \to \Lambda^{k-1}(M)$ will now be defined as follows

$$\delta\omega = (-1)^{m(k+1)+1} * d(*\omega) = (-1)^k *^{-1} d(*\omega) \tag{5.9.29}$$

where we adopted the convention $\delta f = 0$ for $f \in \Lambda^0(M)$. Since δ is the composition of linear operators, it is a linear operator on $\mathbb{R}$. According to (5.9.29) we can write $\delta = \pm * d *$. If m is even or if m and k are odd, the sign is $-$, if m is odd and k is even, the sign is $+$. (5.9.29) is then expressed as

$$\delta\omega = \frac{(-1)^{m(k+1)+1}}{(m-k+1)!}\Big[\frac{1}{(k-1)!}\epsilon^{i_1\cdots i_{k-1}i_k\cdots i_m}\omega_{i_1\cdots i_{k-1}i;}{}^{i}\Big]\mu_{i_m\cdots i_k}$$

where we naturally define

$$\omega_{i_1\cdots i_{k-1}i;}{}^{i} = g_{i_1j_1}\cdots g_{i_{k-1}j_{k-1}}\omega^{j_1\cdots j_{k-1}i}{}_{;i}\,.$$

Since we can write

$$\begin{aligned}\frac{1}{(m-k+1)!}\epsilon^{i_1\cdots i_{k-1}i_k\cdots i_m}\mu_{i_m\cdots i_k} &= \frac{(-1)^{(k-1)(m-k+1)}}{(m-k+1)!}\epsilon^{i_k\cdots i_m i_1\cdots i_{k-1}}\mu_{i_m\cdots i_k}\\ &= (-1)^{(k-1)(m-k+1)}dx^{i_1}\wedge\cdots\wedge dx^{i_{k-1}}\end{aligned}$$

on using (5.9.18), we finally reach to the result

$$\delta\omega = \frac{(-1)^k}{(k-1)!}\omega_{i_1\cdots i_{k-1}i;}{}^{i}\,dx^{i_1}\wedge\cdots\wedge dx^{i_{k-1}} \tag{5.9.30}$$

after having omitted even numbers in the exponent $(k-1)(m-k+1)+m(k+1)+1$ of -1 in the above expression. Thus we can regard δ as a sort of *divergence operator*. Hence, the form $(-1)^k\delta\omega\in\Lambda^{k-1}(M)$ will be called the *divergence* of the form $\omega\in\Lambda^k(M)$. We shall call δ as the ***co-differential operator***. Various properties of this operator can easily be identified:

(i). *We have* $\delta\circ\delta\omega=\delta^2\omega=\pm *^{-1}d**^{-1}d*\omega=\pm *^{-1}d^2*\omega=0$ *for all* $\omega\in\Lambda(M)$ *so that we obtain* $\delta^2=0$.

(ii). *If* $\omega\in\Lambda^k(M)$, *we have* $*(\delta\omega)=(-1)^k\,d(*\omega)$.

Indeed (5.9.30) and (5.9.17) yield

$$\begin{aligned}*(\delta\omega) &= \frac{(-1)^k}{(k-1)!}\omega^{i_1\cdots i_{k-1}i}{}_{;i}\mu_{i_{k-1}\cdots i_1}\\ &= \frac{(-1)^k}{(k-1)!(m-k+1)!}\omega^{i_1\cdots i_{k-1}i}{}_{;i}\,\epsilon_{i_1\cdots i_{k-1}i_k\cdots i_m}dx^{i_k}\wedge\cdots\wedge dx^{i_m}.\end{aligned}$$

We then obtain the desired result in view of (5.9.28). We can also arrive at this result directly from the definition of the operator δ. Let us consider a form $\omega\in\Lambda^{k+1}(M)$. We find that

$$*\delta\omega = (-1)^{m(k+2)+1}**d*\omega = (-1)^{mk+1+k(m-1)}d*\omega = (-1)^{k+1}d*\omega.$$

Since the number $1\le k\le m$ is arbitrary, when we apply this operator to the form $\omega \in \Lambda^k(M)$, we get $*\delta = (-1)^k\, d*$.

(iii). *If* $\omega \in \Lambda^k(M)$, *we have* $\delta(*\omega) = (-1)^{k+1}*d(\omega)$.

In fact, discarding even numbers in the exponent of -1 we find

$$\begin{aligned}\delta(*\omega) &= (-1)^{m(m-k+1)+1}*d**\omega = (-1)^{-mk+1+k(m-1)}*d(\omega)\\ &= (-1)^{-k+1}*d(\omega) = (-1)^{k+1}*d(\omega).\end{aligned}$$

Hence, we get $\delta* = (-1)^{k+1}*d$ when applied to the form $\omega \in \Lambda^k(M)$.

(iv). *The relations* $*\delta d = d\delta*$ *and* $*d\delta = \delta d*$ *are valid*:

Let us take $\omega \in \Lambda^k(M)$. By direct calculations, we find

$$\begin{aligned}*\delta d(\omega) &= (-1)^{m(k+2)+1}**d*d\omega = (-1)^{k+1}d*d\omega,\\ d\delta(*\omega) &= (-1)^{m(m-k+1)+1}d*d**\omega = (-1)^{k+1}d*d\omega.\end{aligned}$$

We thus conclude that $*\delta d(\omega) = d\delta*(\omega)$ for all $\omega \in \Lambda(M)$. Similarly, we obtain

$$\begin{aligned}*d\delta(\omega) &= (-1)^{m(k+1)+1}*d*d*\omega,\\ \delta d*(\omega) &= (-1)^{m(m-k+2)+1}*d*d*\omega = (-1)^{m(k+1)+1}*d*d*\omega\end{aligned}$$

where $\omega \in \Lambda^k(M)$. This implies that $*d\delta(\omega) = \delta d*(\omega)$ for all $\omega \in \Lambda(M)$ since it is valid for all degrees.

(v). *The relations* $\delta*d = d*\delta = 0$ *are valid.*

If $\omega \in \Lambda^k(M)$, we get

$$\begin{aligned}\delta*d(\omega) &= (-1)^{m(k+2)+1}*d**d\omega = (-1)^{k+1}*d^2(\omega) = 0,\\ d*\delta(\omega) &= (-1)^{m(k+1)+1}d**d*\omega = (-1)^{m-k+1}d^2(*\omega) = 0\end{aligned}$$

so that $\delta*d(\omega) = d*\delta(\omega) = 0$ for all $\omega \in \Lambda(M)$.

For a form $\omega \in \Lambda^1(M)$ we obtain $*(\delta\omega) = -\,\omega^i{}_{;i}\mu$ and $\delta\omega \in \Lambda^0(M)$ is given by $\delta\omega = -\,\omega^i{}_{;i}$. Let us define the form $\omega = \omega_i dx^i \in \Lambda^1(M)$ associated with a vector field $V = v^i\,\partial_i \in T(M)$ by taking $\omega_i = g_{ij}v^j$. Then, we naturally find $\omega^i = g^{ij}\omega_j = v^i$ so that we are able to write

$$v^i{}_{;i} = \operatorname{div} V = -\,\delta\omega$$

We now define an operator $\Delta : \Lambda^k(M) \to \Lambda^k(M)$ that is linear on $\mathbb{R}$ by the following relation

$$\Delta = \delta d + d\delta. \qquad (5.9.31)$$

Δ is called the ***Laplace-de Rham operator*** after Laplace and Swiss mathematician Georges de Rham (1903-1990). If we take a function $f \in \Lambda^0(M)$ into account, application of this operator yields

$$\Delta f = \delta df + d\delta f = \delta df = \nabla^2 f \tag{5.9.32}$$

where $\nabla^2 = \delta d : \Lambda^0(M) \to \Lambda^0(M)$ is called the ***Laplace-Beltrami operator*** [Italian mathematician Eugenio Beltrami (1835-1900)]. Since we write $df = f_{,i}\, dx^i$, according to (5.9.30) and (5.9.19) we get

$$\nabla^2 f = -\,(f_{,i})_{;}{}^{i} = -\,\frac{1}{\sqrt{g}}(\sqrt{g}\, g^{ij} f_{,j})_{,i}. \tag{5.9.33}$$

In Cartesian coordinates, this expression takes the form

$$\nabla^2 f = -\sum_{i=1}^{m} \frac{\partial^2 f}{\partial x^{i^2}}.$$

We have to note that this operator is differing only in sign from the familiar one encountered in partial differential equations. The Laplace-Beltrami operator Δ possesses the following properties that can easily be verified:

(i). *One has* $\Delta = (d+\delta)^2$.

$$\Delta = (d+\delta)\circ(d+\delta) = d^2 + d\delta + \delta d + \delta^2 = d\delta + \delta d.$$

(ii). *One has* $d\Delta = \Delta d = d\delta d$.

$$d\Delta = d\delta d\, + d^2\delta = d\delta d, \qquad \Delta d = \delta d^2 + d\delta d = d\delta d.$$

(iii). *One has* $\delta\Delta = \Delta\delta = \delta d\delta$.

$$\delta\Delta = \delta^2 d + \delta d\delta = \delta d\delta, \qquad \Delta\delta = \delta d\delta + d\delta^2 = \delta d\delta.$$

(iv). *One has* $*\Delta = \Delta *$.

$$\begin{aligned} *\Delta &= *(\delta d + d\delta) = *\delta d + *d\delta = d\delta * + \delta d * = (d\delta + \delta d)* \\ &= \Delta *. \end{aligned}$$

A form $\omega \in \Lambda^k(M)$ satisfying the equation $\Delta\omega = 0$ will be called a ***harmonic form***. The set

$$\mathrm{H}^k(M) = \{\omega \in \Lambda^k(M) : \Delta\omega = 0\} = \mathcal{N}(\Delta)$$

is a subspace of $\Lambda^k(M)$ on $\mathbb{R}$.

Example 5.9.1. Let us take $M = \mathbb{R}^3$ and we introduce the spherical coordinates (r, θ, ϕ) connected to Cartesian coordinates by the relations

$$x = r\sin\theta\cos\phi,\; y = r\sin\theta\sin\phi,\; z = r\cos\theta,\; 0 \le \theta \le \pi, 0 \le \phi \le 2\pi.$$

Since the arc element is determined by

$$ds^2 = dr^2 + r^2 d\theta^2 + r^2 \sin^2\theta \, d\phi^2,$$

the components of the metric tensor and its inverse are given by

$$g_{rr} = 1, \; g_{\theta\theta} = r^2, \quad g_{\phi\phi} = r^2 \sin^2\theta;$$
$$g^{rr} = 1, \; g^{\theta\theta} = 1/r^2, \; g^{\phi\phi} = 1/r^2 \sin^2\theta.$$

Thus the volume form becomes

$$\mu = r^2 \sin\theta \, dr \wedge d\theta \wedge d\phi$$

whence we produce the basis for $\Lambda^2(\mathbb{R}^3)$

$$\mu_r = \mathbf{i}_{\partial_r}(\mu) = r^2 \sin\theta \, d\theta \wedge d\phi, \;\; \mu_\theta = \mathbf{i}_{\partial_\theta}(\mu) = \; - r^2 \sin\theta \, dr \wedge d\phi,$$
$$\mu_\phi = \mathbf{i}_{\partial_\phi}(\mu) = r^2 \sin\theta \, dr \wedge d\theta$$

We can now represent a form $\omega \in \Lambda^1(\mathbb{R}^3)$ by

$$\omega = \omega_r \, dr + \omega_\theta \, d\theta + \omega_\phi \, d\phi$$

where coefficients are functions of variables r, θ, ϕ. The Hodge dual of the form ω will be given by

$$*\omega = \omega^r \mu_r + \omega^\theta \mu_\theta + \omega^\phi \mu_\phi$$

where the coefficients are calculated as follows

$$\omega^r = \omega_r, \;\; \omega^\theta = \frac{1}{r^2} \omega_\theta, \;\; \omega^\phi = \frac{1}{r^2 \sin^2\theta} \omega_\phi.$$

Therefore, we get

$$*\omega = \frac{\omega_\phi}{\sin\theta} \, dr \wedge d\theta - \omega_\theta \sin\theta \, dr \wedge d\phi + \omega_r r^2 \sin\theta \, d\theta \wedge d\phi.$$

We readily see that we obtain

$$\omega \wedge *\omega = \Big(\omega_r^2 + \frac{1}{r^2} \omega_\theta^2 + \frac{1}{r^2 \sin^2\theta} \omega_\phi^2\Big) \mu.$$

Let us now evaluate the exterior derivatives of the forms ω and $*\omega$. We find

$$d\omega = (\omega_{\theta,r} - \omega_{r,\theta}) dr \wedge d\theta + (\omega_{\phi,r} - \omega_{r,\phi}) dr \wedge d\phi + (\omega_{\phi,\theta} - \omega_{\theta,\phi}) d\theta \wedge d\phi$$

$$d{*\omega} = \big[(\omega_r r^2 \sin\theta)_{,r} + (\omega_\theta \sin\theta)_{,\theta} + \big(\frac{\omega_\phi}{\sin\theta}\big)_{,\phi}\big] dr \wedge d\theta \wedge d\phi$$
$$= \Big(\omega_{r,r} + \frac{2}{r} \omega_r + \frac{1}{r^2} \omega_{\theta,\theta} + \frac{\cos\theta}{r^2 \sin\theta} \omega_\theta + \frac{1}{r^2 \sin^2\theta} \omega_{\phi,\phi}\Big) \mu.$$

Since $*\mu = 1$, $m = 3$, $k = 1$, the co-differential of ω becomes

$$\delta\omega = -*d*\omega = -\Big(\frac{\partial\omega_r}{\partial r} + \frac{2}{r}\,\omega_r + \frac{1}{r^2}\frac{\partial\omega_\theta}{\partial\theta} + \frac{\cos\theta}{r^2\sin\theta}\,\omega_\theta + \frac{1}{r^2\sin^2\theta}\frac{\partial\omega_\phi}{\partial\phi}\Big).$$

Let us now consider the function $f \in \Lambda^0(\mathbb{R}^3)$. Its differential is the 1-form

$$df = f_{,r}\,dr + f_{,\theta}\,d\theta + f_{,\phi}\,d\phi.$$

Hence, if we write $\omega_r = f_{,r}$, $\omega_\theta = f_{,\theta}$, $\omega_\phi = f_{,\phi}$ the above relation leads to

$$\nabla^2 f = \delta df = -\Big(\frac{\partial^2 f}{\partial r^2} + \frac{2}{r}\frac{\partial f}{\partial r} + \frac{1}{r^2}\frac{\partial^2 f}{\partial\theta^2} + \frac{\cos\theta}{r^2\sin\theta}\frac{\partial f}{\partial\theta} + \frac{1}{r^2\sin^2\theta}\frac{\partial^2 f}{\partial\phi^2}\Big)$$

that is the known result apart from a sign difference. ■

5.10. CLOSED IDEALS

Let $\mathcal{I}$ be an ideal of the exterior algebra $\Lambda(M)$. $\mathcal{I}$ is called a ***closed ideal*** if $d\omega \in \mathcal{I}$ for all forms $\omega \in \mathcal{I}$. This situation is symbolically expressed as $d\mathcal{I} \subset \mathcal{I}$. Sometimes, a closed ideal is also named as a ***differential ideal***. Let us consider the ideal $\mathcal{I}(\omega_1, \dots, \omega_r)$ generated by forms $\omega_1, \dots, \omega_r \in \Lambda(M)$. If the ideal $\mathcal{I}$ is not a closed one, then we can construct an *extended ideal* $\overline{\mathcal{I}}(\omega_1, \dots, \omega_r, d\omega_1, \dots, d\omega_r)$, which is called the ***closure*** of $\mathcal{I}$, that will be closed. Indeed, if $\omega \in \overline{\mathcal{I}}$, then there are appropriate forms γ^α and Γ^α, $\alpha = 1, \dots, r$ such that we can write $\omega = \gamma^\alpha \wedge \omega_\alpha + \Gamma^\alpha \wedge d\omega_\alpha$. We then obtain

$$d\omega = d\gamma^\alpha \wedge \omega_\alpha + \big(d\Gamma^\alpha + (-1)^{deg\,\gamma^\alpha}\gamma^\alpha\big) \wedge d\omega_\alpha \in \overline{\mathcal{I}}.$$

Naturally, if the exterior derivatives of some generating forms are already inside the ideal, we have to discard these exterior derivatives as generators in determining the closure. More generally, let us denote the set of forms $d\omega$ corresponding to all forms $\omega \in \mathcal{I}$ by $d\mathcal{I}$. We immediately observe that the set $\overline{\mathcal{I}} = \mathcal{I} \cup d\mathcal{I}$ is a closed ideal in $\Lambda(M)$.

Next, we discuss the necessary and sufficient conditions for an ideal generated by finitely many forms to be closed.

Theorem 5.10.1. *Let $\mathcal{I}(\omega_1, \dots, \omega_r)$ be an ideal of the exterior algebra $\Lambda(M)$. The ideal $\mathcal{I}$ is closed if and only if appropriate forms $\Gamma_\alpha^\beta \in \Lambda(M)$, $\alpha, \beta = 1, \dots, r$ can be so found that the relations $d\omega_\alpha = \Gamma_\alpha^\beta \wedge \omega_\beta$ are satisfied.*

It is clear that the conditions $deg\,\omega_\alpha + 1 = deg\,\Gamma_\alpha^\beta + deg\,\omega_\beta$ or

$$deg\,\Gamma^{\beta}_{\alpha} = deg\,\omega_\alpha - deg\,\omega_\beta + 1 \geq 0$$

should be satisfied if the forms Γ^{β}_{α} exist. Hence, only the generating forms whose degrees are less than or equal to $deg\,\omega_\alpha + 1$ can take place in the sum. Let us first assume that the ideal $\mathcal{I}$ is closed. Then we must get $d\omega_\alpha \in \mathcal{I}$ when $\omega_\alpha \in \mathcal{I}$. This means that there exists forms Γ^{β}_{α} so that the relations $d\omega_\alpha = \Gamma^{\beta}_{\alpha} \wedge \omega_\beta$ are satisfied. For sufficiency, let us assume the existence of the relations $d\omega_\alpha = \Gamma^{\beta}_{\alpha} \wedge \omega_\beta$. If $\omega \in \mathcal{I}$, then we can find forms γ^α so that one is able to write $\omega = \gamma^\alpha \wedge \omega_\alpha$. In this case, the exterior derivative of ω is evaluated as

$$\begin{aligned} d\omega &= d\gamma^\alpha \wedge \omega_\alpha + (-1)^{deg\,\gamma^{\underline{\alpha}}}\gamma^\alpha \wedge d\omega_\alpha \\ &= \big(d\gamma^\beta + (-1)^{deg\,\gamma^{\underline{\alpha}}}\gamma^\alpha \wedge \Gamma^{\beta}_{\alpha}\big) \wedge \omega_\beta \end{aligned}$$

implying that $d\omega \in \mathcal{I}$. However, the forms Γ^{β}_{α} should be restricted because they have to satisfy the following compatibility conditions:

$$\begin{aligned} d^2\omega_\alpha &= d\Gamma^{\beta}_{\alpha} \wedge \omega_\beta + (-1)^{deg\,\Gamma^{\underline{\beta}}_{\underline{\alpha}}}\Gamma^{\beta}_{\underline{\alpha}} \wedge d\omega_\beta \\ &= \big(d\Gamma^{\beta}_{\alpha} + (-1)^{deg\,\Gamma^{\underline{\gamma}}_{\underline{\alpha}}}\Gamma^{\gamma}_{\underline{\alpha}} \wedge \Gamma^{\beta}_{\gamma}\big) \wedge \omega_\beta = 0. \end{aligned}$$

Evidently, in the above sums only forms complying the degree conformity can take place. □

Example 5.10.1. Let us consider the ideal $\mathcal{I}(\omega_1, \omega_2)$ of $\Lambda(\mathbb{R}^4)$ generated by the forms $\omega_1 = dx - y\,dz$, $\omega_2 = t\,dx \wedge dz - x\,dy \wedge dt$. We write

$$d\omega_1 = -\,dy \wedge dz = \Gamma^1_1 \wedge (dx - y\,dz) + \Gamma^2_1(t\,dx \wedge dz - x\,dy \wedge dt)$$

where $\Gamma^1_1 \in \Lambda^1(\mathbb{R}^4), \Gamma^2_1 \in \Lambda^0(\mathbb{R}^4)$. If we choose

$$\Gamma^1_1 = \gamma_1\,dx + \gamma_2\,dy + \gamma_3\,dz + \gamma_4\,dt$$

then we find

$$\begin{aligned} dy \wedge dz = (y\gamma_1 + \gamma_3 - t\Gamma^2_1)\,dx \wedge dz + \gamma_2\,dx \wedge dy + \gamma_4\,dx \wedge dt \\ + y\gamma_2\,dy \wedge dz + x\Gamma^2_1\,dy \wedge dt - y\,\gamma_4 dz \wedge dt. \end{aligned}$$

Comparing both sides, we see that the following equations must hold

$$y\gamma_1 + \gamma_3 - t\Gamma^2_1 = \gamma_2 = \gamma_4 = x\Gamma^2_1 = y\,\gamma_4 = 0,\;\; y\gamma_2 = 1$$

from which we obtain $\Gamma^2_1 = \gamma_4 = 0$, $\gamma_3 = -\,y\gamma_1$. But, to satisfy the relations $\gamma_2 = 0$ and $y\gamma_2 = 1$ simultaneously is not possible. Hence, the form Γ^1_1 does not exist implying that $d\omega_1$ does not belong to $\mathcal{I}$. On the other hand, we have

$$d\omega_2 = dt \wedge dx \wedge dz - dx \wedge dy \wedge dt = \Big(\frac{1}{x}dx + \frac{1}{t}dt\Big) \wedge \omega_2.$$

Thus $d\omega_2$ is inside the ideal. In this case the closure of the ideal $\mathcal{I}$ should be designated by $\overline{\mathcal{I}}(\omega_1, \omega_2, dy \wedge dz)$. ■

When we are dealing with ideals whose generators are 1-forms, the condition of being closed is reduced to a much simpler form.

Theorem 5.10.2. *Let an ideal of the exterior algebra $\Lambda(M)$ generated by linearly independent 1-forms $\omega^1, \dots, \omega^r$ be $\mathcal{I}(\omega^1, \dots, \omega^r)$. If $r < m - 1$, then the ideal $\mathcal{I}$ is closed if and only if the conditions $d\omega^\alpha \wedge \Omega = 0, \alpha = 1, \dots, r$ are satisfied where we defined $\Omega = \omega^1 \wedge \cdots \wedge \omega^r \neq 0$.*

If $\mathcal{I}$ is closed, that is, if there exist forms $\Gamma^\alpha_\beta \in \Lambda^1(M^m)$ so that we can write $d\omega^\alpha = \Gamma^\alpha_\beta \wedge \omega^\beta$, then it is evident that the relations $d\omega^\alpha \wedge \Omega = 0$ are automatically satisfied. Conversely, let us suppose that we get $d\omega^\alpha \wedge \Omega = 0$ for $1 \leq \alpha \leq r$. Next, we add $m - r$ linearly independent 1-forms $\omega^{r+1}, \dots, \omega^m$ to the forms $\omega^1, \dots, \omega^r$ to make a basis $\{\omega^i\} = \{\omega^\alpha, \omega^a\}, a = r + 1, \dots, m, i = 1, \dots, m$ of $\Lambda^1(M)$. In this situation, a basis for the module $\Lambda^2(M)$ becomes $\omega^i \wedge \omega^j, i < j$ and so long as $\lambda^\alpha_{ij} = -\lambda^\alpha_{ji}$, we can write

$$\begin{aligned} d\omega^\alpha &= \lambda^\alpha_{ij}\,\omega^i \wedge \omega^j = \lambda^\alpha_{\beta\gamma}\,\omega^\beta \wedge \omega^\gamma + \lambda^\alpha_{a\beta}\,\omega^a \wedge \omega^\beta + \lambda^\alpha_{\beta a}\,\omega^\beta \wedge \omega^a + \lambda^\alpha_{ab}\,\omega^a \wedge \omega^b \\ &= \lambda^\alpha_{\beta\gamma}\,\omega^\beta \wedge \omega^\gamma + 2\lambda^\alpha_{a\beta}\,\omega^a \wedge \omega^\beta + \lambda^\alpha_{ab}\,\omega^a \wedge \omega^b. \end{aligned}$$

whence we deduce that

$$d\omega^\alpha \wedge \Omega = \lambda^\alpha_{ab}\,\omega^a \wedge \omega^b \wedge \Omega, \;\; \lambda^\alpha_{ab} = -\lambda^\alpha_{ba}.$$

When $r \leq m - 2$, the foregoing expression is a $(r + 2)$-form which is the sum of simple $(r + 2)$-forms. Since the forms ω^α and ω^a are linearly independent none of the forms $\omega^a \wedge \omega^b \wedge \Omega$ vanishes if $a \neq b$. Thus the condition $d\omega^\alpha \wedge \Omega = 0$ can only be realised when $\lambda^\alpha_{ab} = 0$. In this case, we obtain

$$d\omega^\alpha = (\lambda^\alpha_{\gamma\beta}\,\omega^\gamma + 2\lambda^\alpha_{a\beta}\,\omega^a) \wedge \omega^\beta = \Gamma^\alpha_\beta \wedge \omega^\beta \in \mathcal{I}$$

Hence, the ideal $\mathcal{I}$ is closed. □

When $r \geq m - 1$ the forms $d\omega^\alpha \wedge \Omega$ are identically zero because their degrees is higher than m. Therefore, they cannot provide a criterion to identify a closed ideal. However, the theorem below fills this gap.

Theorem 5.10.3. *An ideal of the exterior algebra $\Lambda(M^m)$ is closed if it is generated either by $r = m$ or $r = m - 1$ linearly independent 1-forms.*

When $r = m$, the linearly independent 1-forms $\omega^1, \dots, \omega^m$ generating an ideal $\mathcal{I}$ constitute a basis of $\Lambda^1(M)$. Consequently, we can write

$$d\omega^\alpha = \lambda^\alpha_{\gamma\beta}\,\omega^\gamma \wedge \omega^\beta = \Gamma^\alpha_\beta \wedge \omega^\beta \in \mathcal{I}$$

where $\Gamma^\alpha_\beta = \lambda^\alpha_{\gamma\beta}\,\omega^\gamma$. Hence, the ideal $\mathcal{I}$ is closed. When $r = m-1$, we can choose a 1-form σ that is independent of those $m-1$ forms. Thus $\omega^1, \dots, \omega^{m-1}, \sigma$ become a basis of $\Lambda^1(M)$. If we consider an ideal $\mathcal{I}$ generated by these forms, we get

$$d\omega^\alpha = \lambda^\alpha_{\gamma\beta}\,\omega^\gamma \wedge \omega^\beta + \lambda^\alpha_\beta\,\sigma \wedge \omega^\beta = (\lambda^\alpha_{\gamma\beta}\,\omega^\gamma + \lambda^\alpha_\beta\,\sigma) \wedge \omega^\beta = \Gamma^\alpha_\beta \wedge \omega^\beta \in \mathcal{I}.$$

Hence, the ideal again becomes closed. □

The following theorem is concerned with the closure $\overline{\mathcal{I}}(\omega^1, \dots, \omega^r, d\omega^1, \dots, d\omega^r)$ of an ideal $\mathcal{I}(\omega^1, \dots, \omega^r)$.

Theorem 5.10.4. *The exterior derivative $d\omega$ of a form $\omega \in \Lambda^k(M)$ remains inside the closure $\overline{\mathcal{I}}$ of the ideal $\mathcal{I}$ if and only if we can find forms $\alpha \in \Lambda^k(M)$ and $\beta \in \mathcal{C}^{k+1}(M)$ in the ideal $\mathcal{I}$ such that $d(\omega + \alpha) = \beta$.*

If $\alpha, \beta \in \mathcal{I}$, then we can write $\alpha = \gamma_\alpha \wedge \omega^\alpha$, $\beta = \lambda_\alpha \wedge \omega^\alpha$ for appropriate forms γ_α and λ_α where $\alpha = 1, 2, \dots, r$. Thus, we readily obtain for a k-form ω satisfying the relation $d(\omega + \alpha) = \beta$, the following expression

$$d\omega = -d\alpha + \beta = (-d\gamma_\alpha + \lambda_\alpha) \wedge \omega^\alpha + (-1)^{deg(\gamma_\alpha)}\gamma_\alpha \wedge d\omega^\alpha \in \overline{\mathcal{I}}.$$

The above equality requires that $d\beta = 0$, thus we must have $\beta \in \mathcal{C}^{k+1}(M)$. Conversely, let us assume that $d\omega \in \overline{\mathcal{I}}$. Consequently, we can write

$$d\omega = \lambda_\alpha \wedge \omega^\alpha + \mu_\alpha \wedge d\omega^\alpha$$

where $\lambda_\alpha \in \Lambda^{k+1-deg(\omega^\alpha)}(M)$, $\mu_\alpha \in \Lambda^{k-deg(\omega^\alpha)}(M)$. Because of the relation $d^2\omega = 0$, the forms λ_α and μ_α ought to meet the condition

$$d\lambda_\alpha \wedge \omega^\alpha + (d\mu_\alpha + (-1)^{deg(\lambda_\alpha)}\lambda_\alpha) \wedge d\omega^\alpha = 0.$$

We now define the forms ϕ_α as follows

$$(-1)^{deg(\lambda_\alpha)}\phi_\alpha = d\mu_\alpha + (-1)^{deg(\lambda_\alpha)}\lambda_\alpha, \quad deg(\phi_\alpha) = deg(\lambda_\alpha).$$

If we insert the form $\lambda_\alpha = \phi_\alpha + (-1)^{deg(\lambda_\alpha)-1} d\mu_\alpha$ into above expressions and note that $deg(\mu_\alpha) = deg(\lambda_\alpha) - 1$ by definition, we obtain

$$\begin{aligned} d\omega &= \phi_\alpha \wedge \omega^\alpha + (-1)^{deg(\lambda_\alpha)-1} d\mu_\alpha \wedge \omega^\alpha + \mu_\alpha \wedge d\omega^\alpha \\ &= \phi_\alpha \wedge \omega^\alpha + (-1)^{deg(\lambda_\alpha)-1} d(\mu_\alpha \wedge \omega^\alpha), \\ 0 &= d\phi_\alpha \wedge \omega^\alpha + (-1)^{deg(\phi_\alpha)}\phi_\alpha \wedge d\omega^\alpha = d(\phi_\alpha \wedge \omega^\alpha). \end{aligned}$$

It will suffice now to introduce the forms $\alpha = (-1)^{deg(\lambda_\alpha)}\mu_\alpha \wedge \omega^\alpha \in \mathcal{I}$ and $\beta = \phi_\alpha \wedge \omega^\alpha \in \mathcal{I}$ to reach to the result $d(\omega + \alpha) = \beta$ and $d\beta = 0$. □

5.11. LIE DERIVATIVES OF EXTERIOR FORMS

Let us consider a congruence on a manifold M brought out by a vector field V and the *flow* $\phi_t : M \to M$ induced by this congruence. As is well known, this mapping carries a point $p \in M$ to a point $\overline{p}(t) = \phi_t(p) \in M$. On recalling the relation (2.9.11), we represent this mapping by $\overline{p}(t) = \phi_t(p) = e^{tV}(p)$. We can also write

$$\overline{x}^i(t) = e^{tV}(x^i) = x^i + tV(x^i) + \frac{t^2}{2!}V^2(x^i) + \cdots + \frac{t^n}{n!}V^n(x^i) + \cdots$$

in terms of local coordinates. We employed here only the symbol V for the vector field believing that it will no longer cause any ambiguity.

We suppose that a form field $\omega \in \Lambda^k(M)$ is given. Let us transport the form $\omega(\overline{p}(t))$ at a point $\overline{p}(t)$ to a point p by pulling it back by the mapping ϕ_t^*. We thus obtain

$$\omega^*(p; t) = \omega \circ \phi_t(p) = (\phi_t^*\omega)(p) = (e^{tV})^*\omega.$$

As we have done before, we will now define the ***Lie derivative*** of a form field ω at a point p by the following limiting process:

$$\pounds_V\omega = \lim_{t\to 0}\frac{(e^{tV})^*\omega - \omega}{t} = \lim_{t\to 0}\frac{(e^{tV})^* - i_\Lambda}{t}\,\omega \in \Lambda^k(M) \quad (5.11.1)$$

where $i_\Lambda : \Lambda(M) \to \Lambda(M)$ is the identity operator on the exterior algebra. This definition reveals immediately certain important properties of the Lie derivative.

(i). *We can write*

$$(e^{tV})^*\omega = \omega + t\pounds_V\omega + o(t).$$

(ii). *When* $f \in \Lambda^0(M)$, *we have* [*see* (2.10.18)]

$$\pounds_V f = v^i f_{,i} = V(f) = \mathbf{i}_V(df).$$

In fact, for small values of the parameter t we obtain

$$\pounds_V f = \lim_{t\to 0}\frac{f(\overline{p}(t)) - f(p)}{t} = \lim_{t\to 0}\frac{f(\mathbf{x} + t\mathbf{v} + \boldsymbol{o}(t)) - f(\mathbf{x})}{t} = v^i f_{,i}.$$

(iii). *We have*

$$\pounds_V(\omega + \sigma) = \pounds_V\omega + \pounds_V\sigma.$$

This is observed at once by noting he relation

$$(e^{tV})^*(\omega + \sigma) = (e^{tV})^*\omega + (e^{tV})^*\sigma.$$

(iv). The Leibniz rule

$$£_V(\omega \wedge \sigma) = (£_V\omega) \wedge \sigma + \omega \wedge (£_V\sigma)$$

is in effect.

Recalling the relation $(e^{tV})^*(\omega \wedge \sigma) = (e^{tV})^*\omega \wedge (e^{tV})^*\sigma$, we arrive at the desired result

$$\begin{aligned} £_V(\omega \wedge \sigma) &= \lim_{t\to 0} \frac{(e^{tV})^*\omega \wedge (e^{tV})^*\sigma - \omega \wedge \sigma}{t} \\ &= \lim_{t\to 0} \frac{(\omega + t£_V\omega + o(t)) \wedge (\sigma + t£_V\sigma + o(t)) - \omega \wedge \sigma}{t} \\ &= \lim_{t\to 0} \frac{t(£_V\omega \wedge \sigma + \omega \wedge £_V\sigma) + o(t)}{t} = £_V\omega \wedge \sigma + \omega \wedge £_V\sigma. \end{aligned}$$

This expression can easily be generalised to an arbitrary number of forms so that one is able to write

$$\begin{aligned} £_V(\omega_1 \wedge \omega_2 \wedge \ldots \wedge \omega_r) &= £_V\omega_1 \wedge \omega_2 \wedge \ldots \wedge \omega_r \\ &\quad + \omega_1 \wedge £_V\omega_2 \wedge \ldots \wedge \omega_r + \omega_1 \wedge \omega_2 \wedge \ldots \wedge £_V\omega_r. \end{aligned}$$

This relation offers essentially an approach to calculate the Lie derivative of any form once we determine the Lie derivatives of only 0- and 1-forms. We have already found the Lie derivative of 0-forms. We now try to evaluate the Lie derivative of a 1-form. Let us take

$$\omega = \omega_i\, dx^i \in \Lambda^1(M), \;\; V = v^i \frac{\partial}{\partial x^i} \in T(M).$$

Since we can write. $\overline{x}^i = x^i + tv^i + o(t)$, then the Taylor series about the point $\mathbf{x}$ yields

$$\begin{aligned} (e^{tV})^*\omega &= \omega_i\big(x^j + tv^j + o(t)\big)\big(dx^i + tv^i_{,k}\, dx^k + o(t)\big) \\ &= \Big(\omega_i(\mathbf{x}) + t\frac{\partial \omega_i}{\partial x^j} v^j + o(t)\Big)\Big(dx^i + t\frac{\partial v^i}{\partial x^k}\, dx^k + o(t)\Big) \\ &= \omega_i(\mathbf{x})\, dx^i + t\Big(\frac{\partial \omega_i}{\partial x^j} v^j\, dx^i + \omega_i \frac{\partial v^i}{\partial x^k}\, dx^k\Big) + o(t). \end{aligned}$$

On changing properly the names of dummy indices, we finally get

$$£_V\omega = \Big(\frac{\partial \omega_i}{\partial x^j} v^j + \omega_j \frac{\partial v^j}{\partial x^i}\Big)\, dx^i = (\omega_{i,j} v^j + \omega_j v^j_{,i})\, dx^i. \qquad (5.11.2)$$

The coefficients $(\pounds_V\omega)_i = \omega_{i,j}v^j + \omega_j v^j_{,i} \in \Lambda^0(M)$ totally specifies the 1-form $\pounds_V\omega$. As a special example, let us consider the form $df = f_{,i}\,dx^i$ where $f \in \Lambda^0(M)$. Then, with $\omega_i = f_{,i}$ (5.11.2) leads to

$$\pounds_V\,df = (f_{,ij}v^j + f_{,j}v^j_{,i})\,dx^i = (f_{,j}v^j)_{,i}\,dx^i = \big(V(f)\big)_{,i}\,dx^i = dV(f).$$

If we now select $f = x^k$, we reach to quite a significant conclusion

$$\pounds_V\,dx^k = dV(x^k) = dv^k = v^k_{,i}\,dx^i. \tag{5.11.3}$$

Next, we take a form $\omega \in \Lambda^k(M)$ into account denoted by

$$\omega = \frac{1}{k!}\,\omega_{i_1\cdots i_k}(\mathbf{x})\,dx^{i_1}\wedge\cdots\wedge dx^{i_k}.$$

On utilising the above properties, we can now calculate the Lie derivative of this form as follows:

$$\begin{aligned}
\pounds_V\omega &= \frac{1}{k!}\big[(\pounds_V\omega_{i_1\cdots i_k})\,dx^{i_1}\wedge\cdots\wedge dx^{i_k} + \omega_{i_1i_2\cdots i_k}(\pounds_V\,dx^{i_1})\wedge\cdots\wedge dx^{i_k}\\
&\quad + \cdots + \omega_{i_1\cdots i_k}dx^{i_1}\wedge\cdots\wedge(\pounds_V\,dx^{i_k})\big]\\
&= \frac{1}{k!}\big[\omega_{i_1\cdots i_k,i}v^i\,dx^{i_1}\wedge\cdots\wedge dx^{i_k} + \omega_{i_1i_2\cdots i_k}v^{i_1}_{,i}\,dx^i\wedge dx^{i_2}\wedge\cdots\wedge dx^{i_k}\\
&\quad + \cdots + \omega_{i_1\cdots i_{k-1}i_k}v^{i_k}_{,i}\,dx^{i_1}\wedge\cdots\wedge dx^{i_{k-1}}\wedge dx^i\big]\\
&= \frac{1}{k!}\big[\omega_{i_1\cdots i_k,i}v^i + \omega_{ii_2\cdots i_k}v^i_{,i_1} + \cdots + \omega_{i_1\cdots i_{k-1}i}v^i_{,i_k}\big]dx^{i_1}\wedge\cdots\wedge dx^{i_k}.
\end{aligned}$$

Hence, the Lie derivative of a form $\omega \in \Lambda^k(M)$ is expressible as

$$\pounds_V\omega = \frac{1}{k!}(\pounds_V\omega)_{i_1i_2\cdots i_k}\,dx^{i_1}\wedge dx^{i_2}\wedge\cdots\wedge dx^{i_k} \in \Lambda^k(M)$$

where the *completely antisymmetric* coefficients $(\pounds_V\omega)_{i_1i_2\cdots i_k} \in \Lambda^0(M)$ are determined by

$$\begin{aligned}
(\pounds_V\omega)_{i_1i_2\cdots i_k} &= \omega_{i_1i_2\cdots i_k,i}v^i + \omega_{ii_2\cdots i_k}v^i_{,i_1} + \omega_{i_1ii_3\cdots i_k}v^i_{,i_2} + \cdots + \omega_{i_1i_2\cdots i_{k-1}i}v^i_{,i_k}\\
&= v^i\frac{\partial\omega_{i_1i_2\cdots i_k}}{\partial x^i} + \sum_{r=1}^{k}\omega_{i_1\cdots i_{r-1}ii_{r+1}\cdots i_k}\frac{\partial v^i}{\partial x^{i_r}}
\end{aligned} \tag{5.11.4}$$

It is clear that the complete antisymmetry in the coefficients $\omega_{i_1\cdots i_k}$ renders the coefficients in (5.11.4) completely antisymmetric. It is now clear that the Lie derivative $\pounds_V : \Lambda(M) \to \Lambda(M)$ is a *degree preserving* mapping.

The expression (5.11.2) for Lie derivatives of 1-forms can be transformed into the following identical shape

$$£_V\omega = \big[(\omega_{i,j} - \omega_{j,i})v^j + (v^j\omega_j)_{,i})\big]\,dx^i, \;\; \omega \in \Lambda^1(M).$$

On the other hand, since one has $d\omega = \omega_{i,j}\,dx^j \wedge dx^i$ we obtain

$$\mathbf{i}_V(d\omega) = \omega_{i,j}v^j\,dx^i - \omega_{i,j}v^i\,dx^j = (\omega_{i,j} - \omega_{j,i})v^j\,dx^i,$$
$$d\mathbf{i}_V(\omega) = (v^j\omega_j)_{,i}\,dx^i.$$

We thus arrive at the expression

$$£_V\omega = \mathbf{i}_V(d\omega) + d\mathbf{i}_V(\omega), \;\; \omega \in \Lambda^1(M)$$

known as the ***Cartan magic formula***. We shall now prove that this formula is valid for any form in the exterior algebra.

Theorem 5.11.1. *For any form $\omega \in \Lambda(M)$ and vector field $V \in T(M)$, the Lie derivative of this form is calculated by $£_V\omega = \mathbf{i}_V(d\omega) + d\mathbf{i}_V(\omega)$.*

Let us consider a form $\omega = \dfrac{1}{k!}\,\omega_{i_1\cdots i_k}(\mathbf{x})\,dx^{i_1} \wedge \cdots \wedge dx^{i_k} \in \Lambda^k(M)$ and a vector field $V = v^i(\mathbf{x})\dfrac{\partial}{\partial x^i}$. The exterior derivative of ω is given by

$$d\omega = \frac{1}{k!}\,\omega_{i_1\cdots i_k,i}\,dx^i \wedge dx^{i_1} \wedge \cdots \wedge dx^{i_k}$$
$$= \frac{1}{(k+1)!}(k+1)\,\omega_{[i_1\cdots i_k,i]}\,dx^i \wedge dx^{i_1} \wedge \cdots \wedge dx^{i_k}.$$

Therefore, we obtain

$$\mathbf{i}_V(d\omega) = \frac{1}{k!}(k+1)\,\omega_{[i_1\cdots i_k,i]}v^i\,dx^{i_1} \wedge \cdots \wedge dx^{i_k}.$$

On the other hand, we can write

$$\mathbf{i}_V(\omega) = \frac{1}{(k-1)!}\,\omega_{ii_2\cdots i_k}v^i\,dx^{i_2} \wedge \cdots \wedge dx^{i_k},$$
$$d\mathbf{i}_V(\omega) = \frac{1}{k!}k(\,\omega_{i[i_2\cdots i_k}v^i)_{,i_1]}\,dx^{i_1} \wedge dx^{i_2} \wedge \cdots \wedge dx^{i_k}.$$

Hence, we find that

$$\mathbf{i}_V(d\omega) + d\mathbf{i}_V(\omega) = \frac{1}{k!}\Omega_{i_1i_2\cdots i_k}\,dx^{i_1} \wedge dx^{i_2} \wedge \cdots \wedge dx^{i_k}$$

where the smooth functions $\Omega_{i_1i_2\cdots i_k} \in \Lambda^0(M)$ are defined by

$$\Omega_{i_1i_2\cdots i_k} = (k+1)\,\omega_{[i_1i_2\cdots i_k,i]}v^i + k\,\omega_{i[i_2\cdots i_k,i_1]}v^i + k\,\omega_{i[i_2\cdots i_k}v^i_{,i_1]}.$$

In order to evaluate explicitly the coefficients $\Omega_{i_1 i_2\cdots i_k}$, we resort to the relations (5.8.3) and (5.8.4) to get

$$\begin{aligned}\Omega_{i_1 i_2\cdots i_k} &= \omega_{i_1\cdots i_k,i}v^i - \sum_{r=1}^{k}\omega_{i_1\cdots i_{r-1}ii_{r+1}\cdots i_k,i_r}v^i\\ &+ \omega_{ii_2\cdots i_k,i_1}v^i - \sum_{r=2}^{k}\omega_{ii_2\cdots i_{r-1}i_1i_{r+1}\cdots i_k,i_r}v^i\\ &+ \omega_{ii_2\cdots i_k}v^i_{,i_1} - \sum_{r=2}^{k}\omega_{ii_2\cdots i_{r-1}i_1i_{r+1}\cdots i_k}v^i_{,i_r}\end{aligned}$$

Moreover, since one has $(-1)^{r-2+r-1} = (-1)^{2r-3} = -1$, we can write

$$\omega_{ii_2\cdots i_{r-1}i_1i_{r+1}\cdots i_k} = -\,\omega_{i_1i_2\cdots i_{r-1}ii_{r+1}\cdots i_k}.$$

We thus find

$$\omega_{ii_2\cdots i_k,i_1}v^i - \sum_{r=2}^{k}\omega_{ii_2\cdots i_{r-1}i_1i_{r+1}\cdots i_k,i_r}v^i = \sum_{r=1}^{k}\omega_{i_1\cdots i_{r-1}ii_{r+1}\cdots i_k,i_r}v^i$$

and see, consequently, that the second line above cancels the second term in the first line. If we arrange as well the last line in the similar way, we finally conclude that

$$\begin{aligned}\Omega_{i_1 i_2\cdots i_k} &= \omega_{i_1\cdots i_k,i}v^i + \sum_{r=1}^{k}\omega_{i_1i_2\cdots i_{r-1}ii_{r+1}\cdots i_k}v^i_{,i_r}\\ &= (\pounds_V\omega)_{i_1i_2\cdots i_k}.\end{aligned}$$

Thus for any form $\omega \in \Lambda(M)$, the Cartan magic formula

$$\pounds_V\omega = \mathbf{i}_V(d\omega) + d\mathbf{i}_V(\omega) \tag{5.11.5}$$

becomes valid. In operator form, we can express this relation as follows

$$\pounds_V = \mathbf{i}_V \circ d + d \circ \mathbf{i}_V : \Lambda^k(M) \to \Lambda^k(M),\ 0 \le k \le m. \qquad \square$$

We now consider a form $\omega \in \Lambda^k(M)$ and vector fields $U, V \in T(M)$ and let us calculate the form $\pounds_U\big(\mathbf{i}_V(\omega)\big) \in \Lambda^{k-1}(M)$. Since we have

$$\begin{aligned}\omega &= \frac{1}{k!}\,\omega_{i_1\cdots i_k}dx^{i_1}\wedge\cdots\wedge dx^{i_k},\\ \mathbf{i}_V(\omega) &= \frac{1}{(k-1)!}\,\omega_{ji_2\cdots i_k}v^j\,dx^{i_2}\wedge\cdots\wedge dx^{i_k}\end{aligned}$$

we obtain from (5.11.4) that

$$\pounds_U\big(\mathbf{i}_V(\omega)\big)=\frac{1}{(k-1)!}\Big[\omega_{ji_2\cdots i_k,i}v^j u^i+\omega_{ji_2\cdots i_k}v^j_{,i}u^i + \sum_{r=2}^{k}\omega_{ji_2\cdots i_{r-1}ii_{r+1}\cdots i_k}v^j u^i_{,i_r}\Big]\,dx^{i_2}\wedge\cdots\wedge dx^{i_k}.$$

By adding and subtracting the terms $\omega_{ji_2\cdots i_k}v^i u^j_{,i}$ to the coefficients within brackets above and changing dummy indices appropriately we cast this expression into the equivalent form given below

$$\begin{aligned}\pounds_U\big(\mathbf{i}_V(\omega)\big)&=\frac{1}{(k-1)!}\,\omega_{ji_2\cdots i_k}(u^i v^j_{,i}-v^i u^j_{,i})\,dx^{i_2}\wedge\cdots\wedge dx^{i_k}\\&\quad+\frac{1}{(k-1)!}\Big[\omega_{ji_2\cdots i_k,i}u^i+\omega_{ii_2\cdots i_k}u^i_{,j}\\&\quad+\sum_{r=2}^{k}\omega_{ji_2\cdots i_{r-1}ii_{r+1}\cdots i_k}u^i_{,i_r}\Big]v^j\,dx^{i_2}\wedge\cdots\wedge dx^{i_k}\\&=\frac{1}{(k-1)!}\,\omega_{ji_2\cdots i_k}w^j\,dx^{i_2}\wedge\cdots\wedge dx^{i_k}+\frac{1}{(k-1)!}\Big[\omega_{ji_2\cdots i_k,i}u^i\\&\quad+\sum_{r=1}^{k}\omega_{i_1i_2\cdots i_{r-1}ii_{r+1}\cdots i_k}u^i_{,i_r}\Big]v^{i_1}\,dx^{i_2}\wedge\cdots\wedge dx^{i_k}\end{aligned}$$

where

$$w^j=(u^i v^j_{,i}-v^i u^j_{,i})=[U,V]^j=(\pounds_U V)^j$$

are components of the Lie derivative of the vector field V with respect to the vector field U whereas the expression within brackets are nothing but the coefficients of the Lie derivative of the form ω with respect to the vector field U. Consequently, the above expression is now transformed into

$$\begin{aligned}\pounds_U\big(\mathbf{i}_V(\omega)\big)&=\frac{1}{(k-1)!}\,\omega_{ii_2\cdots i_k}(\pounds_U V)^i\,dx^{i_2}\wedge\cdots\wedge dx^{i_k}\\&\quad+\frac{1}{(k-1)!}\,(\pounds_U\omega)_{ii_2\cdots i_k}v^i\,dx^{i_2}\wedge\cdots\wedge dx^{i_k},\end{aligned}$$

Thus for any form $\omega\in\Lambda(M)$, we obtain

$$\pounds_U\big(\mathbf{i}_V(\omega)\big)=\mathbf{i}_{\pounds_U V}(\omega)+\mathbf{i}_V\big(\pounds_U(\omega)\big).\tag{5.11.6}$$

Hence, we realised that we have managed to establish the following connection between the operators of Lie derivative and interior product

$$\mathbf{i}_{\pounds_U V} = \mathbf{i}_{[U,V]} = \pounds_U \circ \mathbf{i}_V - \mathbf{i}_V \circ \pounds_U = [\pounds_U, \mathbf{i}_V]. \tag{5.11.7}$$

Since the interior product with zero vector vanishes, if $[U, V] = 0$ or $U = V$, namely, if vectors are commutative, then (5.11.7) yields

$$\pounds_U \circ \mathbf{i}_V = \mathbf{i}_V \circ \pounds_U \quad \text{or} \quad \pounds_V \circ \mathbf{i}_V = \mathbf{i}_V \circ \pounds_V. \tag{5.11.8}$$

This means that *the operators $\pounds_U$ and $\mathbf{i}_V$ or $\pounds_V$ and $\mathbf{i}_U$ commute if vector fields U and V are commutative.*

Let us apply (5.11.5) to the form $d\omega$. Since $d^2 = 0$, we get

$$\pounds_V\, d\omega = \mathbf{i}_V(d^2\omega) + d\mathbf{i}_V(d\omega) = d\mathbf{i}_V(d\omega) = d\big(\pounds_V\, \omega - d\mathbf{i}_V(\omega)\big) = d\pounds_V\, \omega.$$

This equality is valid for every form. We thus conclude that

$$\pounds_V \circ d = d \circ \pounds_V. \tag{5.11.9}$$

Hence, *the operators $\pounds_V$ and d commute.*

Let us take $f \in \Lambda^0(M)$ and $V \in T(M)$. If we pay attention to the relations (5.4.7), we deduce that the Lie derivative of a form ω with respect to the vector fV is found to be

$$\begin{aligned} \pounds_{fV}\omega &= \mathbf{i}_{fV}(d\omega) + d\mathbf{i}_{fV}(\omega) = f\mathbf{i}_V(d\omega) + d\big(f\mathbf{i}_V(\omega)\big) \\ &= f\mathbf{i}_V(d\omega) + df \wedge \mathbf{i}_V(\omega) + fd\mathbf{i}_V(\omega) \\ &= f\pounds_V\, \omega + df \wedge \mathbf{i}_V(\omega). \end{aligned} \tag{5.11.10}$$

We immediately see due to (5.4.7) and (5.11.5) that

$$\pounds_{U+V}\, \omega = \pounds_U\, \omega + \pounds_V\, \omega \quad \text{or} \quad \pounds_{U+V} = \pounds_U + \pounds_V. \tag{5.11.11}$$

But, if only $f = c =$ *constant*, then we get $\pounds_{cV}\omega = c\pounds_V\omega$. In this case, it is clear that the addition and scalar multiplication of Lie operators are again Lie operators. Therefore, Lie operators form a linear vector space over $\mathbb{R}$.

Next, we would like to discuss the action of the operator $\pounds_{[U,V]}$, where $U, V \in T(M)$, on a form $\omega \in \Lambda(M)$. In view of (5.11.6), we can write

$$\begin{aligned} \mathbf{i}_{\pounds_U V}(d\omega) &= \pounds_U\big(\mathbf{i}_V(d\omega)\big) - \mathbf{i}_V\big(\pounds_U(d\omega)\big), \\ \mathbf{i}_{\pounds_U V}(\omega) &= \pounds_U\big(\mathbf{i}_V(\omega)\big) - \mathbf{i}_V\big(\pounds_U(\omega)\big). \end{aligned}$$

Let us then introduce these expressions into the Cartan formula

$$\pounds_{[U,V]}\omega = \pounds_{\pounds_U V}\omega = \mathbf{i}_{\pounds_U V}(d\omega) + d\mathbf{i}_{\pounds_U V}(\omega).$$

If we note that the operators $\pounds_U$ and d commute, we reach to the following relation

$$\begin{aligned}
\pounds_{[U,V]}\omega &= \pounds_U\big(\mathbf{i}_V(d\omega)\big) - \mathbf{i}_V\big(\pounds_U(d\omega)\big) + d\pounds_U\big(\mathbf{i}_V(\omega)\big) - d\mathbf{i}_V\big(\pounds_U(\omega)\big)\\
&= \pounds_U\big(\mathbf{i}_V(d\omega)\big) - \mathbf{i}_V\big(d\pounds_U(\omega)\big) + \pounds_U d\big(\mathbf{i}_V(\omega)\big) - d\mathbf{i}_V\big(\pounds_U(\omega)\big)\\
&= \pounds_U\big(\mathbf{i}_V(d\omega) + d\big(\mathbf{i}_V(\omega)\big) - \big[\mathbf{i}_V\big(d\pounds_U(\omega) + d\mathbf{i}_V\big(\pounds_U(\omega)\big)\big]\\
&= \pounds_U\pounds_V\omega - \pounds_V\pounds_U\omega\\
&= (\pounds_U\pounds_V - \pounds_V\pounds_U)\omega.
\end{aligned}$$

Since this relation will be satisfied for all $\omega \in \Lambda(M)$, we get the operator identity given below [*see* (2.10.17)]

$$\pounds_{[U,V]} = \pounds_U\pounds_V - \pounds_V\pounds_U = [\pounds_U, \pounds_V]. \tag{5.11.12}$$

We now assume that an involutive distribution $\mathcal{D} \subseteq T(M)$ is prescribed by linearly independent vector fields $V_\alpha \in T(M), \alpha = 1, \dots, r \leq m$ satisfying the conditions

$$[V_\alpha, V_\beta] = c^\gamma_{\alpha\beta} V_\gamma$$

Let us now associate a Lie operator $\pounds_{V_\alpha}$ to each vector V_α. Then, it follows from (5.11.12) and (5.11.10) that

$$[\pounds_{V_\alpha}, \pounds_{V_\beta}]\,\omega = \pounds_{[V_\alpha,V_\beta]}\,\omega = \pounds_{c^\gamma_{\alpha\beta}V_\gamma}\,\omega = c^\gamma_{\alpha\beta}\,\pounds_{V_\gamma}\omega + dc^\gamma_{\alpha\beta} \wedge \mathbf{i}_{V_\gamma}(\omega)$$

for any $\omega \in \Lambda(M)$ so that we obtain

$$[\pounds_{V_\alpha}, \pounds_{V_\beta}] = c^\gamma_{\alpha\beta}\,\pounds_{V_\gamma} + dc^\gamma_{\alpha\beta} \wedge \mathbf{i}_{V_\gamma}. \tag{5.11.13}$$

Hence, if only the coefficients $c^\gamma_{\alpha\beta}$ are constants, then we are able to write $[\pounds_{V_\alpha}, \pounds_{V_\beta}] = c^\gamma_{\alpha\beta}\,\pounds_{V_\gamma}$. Only in this situation, the operators $\pounds_{V_\alpha}, \alpha = 1, \dots, r$ constitute as well a Lie algebra of operators on the exterior algebra and $c^\gamma_{\alpha\beta}$ becomes structure constants of that algebra. We know that this Lie algebra generates a r-parameter Lie group [*see p.* 191].

We now consider a flow $e^{tV} : M \to M$ on a manifold M generated by a vector field V and the pull-back $\omega^*(t) = (e^{tV})^*\omega$ of a form $\omega \in \Lambda(M)$. The derivative of the form $\omega^*(t)$ with respect to the parameter t can be evaluated as

$$\begin{aligned}
\frac{d\omega^*(t)}{dt} &= \lim_{\tau\to 0}\frac{\omega^*(t+\tau) - \omega^*(t)}{\tau} = \lim_{\tau\to 0}\frac{(e^{(t+\tau)V})^*\omega - (e^{tV})^*\omega}{\tau}\\
&= \lim_{\tau\to 0}\frac{(e^{\tau V}\circ e^{tV})^*\omega - (e^{tV})^*\omega}{\tau}
\end{aligned}$$

$$
\begin{aligned}
&= \lim_{\tau\to 0}\frac{(e^{\tau V})^*\circ(e^{tV})^*\omega - (e^{tV})^*\omega}{\tau}\\
&= \lim_{\tau\to 0}\frac{(e^{\tau V})^* - I}{\tau}(e^{tV})^*\omega\\
&= \pounds_V\omega^*(t).
\end{aligned}
$$

We have seen earlier that the formal solution of this ordinary differential equation under the initial condition $\omega^*(p;0)=\omega(p)$ is given by

$$
\begin{aligned}
\omega^*(p;t) &= e^{t\pounds_V}\omega(p) \qquad (5.11.14)\\
&= \omega + t\pounds_V\omega + \frac{t^2}{2!}\pounds_V^2\omega + \cdots + \frac{t^n}{n!}\pounds_V^n\omega + \cdots
\end{aligned}
$$

The above relation implies that we can write $\omega^* = (e^{tV})^*\omega = e^{t\pounds_V}\omega$ for all forms $\omega\in\Lambda(M)$. Therefore, we formally arrive at the result $(e^{tV})^* = e^{t\pounds_V}$. If $\omega^*(p;t)=\omega(p)$ for all t, we say that the form ω remains ***invariant*** under the flow generated by the vector field V. Evidently, (5.11.14) implies that $\pounds_V\omega=0$ is the necessary and sufficient condition for ω to be invariant.

Let us now suppose that a submodule $\mathcal{L}$ of $\Lambda(M)$ has the following property: $\omega^* = (e^{tV})^*\omega\in\mathcal{L}$ for every form $\omega\in\mathcal{L}$ under the flow e^{tV} generated by a vector field V. We then say that $\mathcal{L}$ is ***stable*** or ***invariant*** submodule under the ***Lie transport*** with respect to the vector field V. It is quite clear that $\mathcal{L}$ is stable if and only if one has $\pounds_V\omega\in\mathcal{L}$ for every form $\omega\in\mathcal{L}$. We symbolically depict this property as $\pounds_V\mathcal{L}\subset\mathcal{L}$. In fact, let us first assume that $\pounds_V\omega\in\mathcal{L}$ for all $\omega\in\mathcal{L}$. We then obtain $\pounds_V(\pounds_V\omega)=\pounds_V^2\omega\in\mathcal{L}$ and similarly $\pounds_V^n\omega\in\mathcal{L}$ for all $n\in\mathbb{N}$. Since $\mathcal{L}$ is a submodule, (5.11.14) implies that $\omega^*\in\mathcal{L}$. Conversely, let us suppose that $\omega^*\in\mathcal{L}$ or all $\omega\in\mathcal{L}$. Since $\omega^*-\omega\in\mathcal{L}$ and t is an arbitrary parameter, we deduce from (5.11.14) that the conditions $\pounds_V\omega\in\mathcal{L}$, $\pounds_V^2\omega\in\mathcal{L}$, $\ldots$, $\pounds_V^n\omega\in\mathcal{L}$, $\ldots$ must be satisfied for all $\omega\in\mathcal{L}$. These conditions are automatically satisfied when $\pounds_V\omega\in\mathcal{L}$. We see that *if a submodule $\mathcal{L}$ of $\Lambda(M)$ is stable under a vector field V, then it is not possible for a form $\omega\in\mathcal{L}$ to escape from that submodule through the action of the Lie derivative.*

Theorem 5.11.2. *The subalgebra $\mathcal{C}(M)$ of closed forms and the subalgebra $\mathcal{E}(M)$ of exact forms of the exterior algebra $\Lambda(M)$ are stable under every vector field $V\in T(M)$.*

If $\omega\in\mathcal{C}(M)$, then $d\omega=0$. Hence, for all vector fields we get $d\pounds_V\omega = \pounds_V d\omega = 0$ and $\pounds_V\omega\in\mathcal{C}(M)$. In like fashion, if $\omega\in\mathcal{E}(M)$, then there is a form $\sigma\in\Lambda(M)$ such that $\omega=d\sigma$. We thus obtain

$$\pounds_V\omega = \pounds_V d\sigma = d\pounds_V\sigma$$

implying that $\pounds_V\omega \in \mathcal{E}(M)$. □

Example 5.11.1. We want to calculate the Lie derivative of the volume form $\mu \in \Lambda^m(M^m)$ given by (5.9.14). Since $d\mu = 0$, we get

$$\pounds_V\mu = \mathbf{i}_V(d\mu) + d\mathbf{i}_V(\mu) = d\mathbf{i}_V(\mu).$$

On recalling (5.5.9) and the exterior derivatives of top down generated bases given on p. 279, it follows from $\mathbf{i}_V(\mu) = v^i\mu_i$ that

$$\begin{aligned}\pounds_V\mu &= v^i_{,j}\,dx^j \wedge \mu_i + v^i d\mu_i = v^i_{,j}\,\delta^j_i\mu + v^i\frac{(\sqrt{g})_{,i}}{\sqrt{g}}\mu \\ &= \Big(v^i_{,i} + v^i\frac{(\sqrt{g})_{,i}}{\sqrt{g}}\Big)\mu = v^i_{;i}\mu.\end{aligned}$$

Thus the volume form μ is invariant under divergenceless, or *solenoidal*, vector fields satisfying the condition $v^i_{;i} = 0$.

As another example, let us calculate the Lie derivatives of the basis forms $\mu_i \in \Lambda^{m-1}(M)$. Since we can write

$$\pounds_V\mu_i = d(v^j\mu_{ji}) + v^j\mathbf{i}_{\partial_j}(d\mu_i) = v^j_{,k}dx^k \wedge \mu_{ji} + v^j d\mu_{ji} + v^j\frac{(\sqrt{g})_{,i}}{\sqrt{g}}\mu_j$$

on taking notice of relations

$$\begin{aligned}dx^k \wedge \mu_{ji} &= \delta^k_j\mu_i - \delta^k_i\mu_j, \\ d\mu_{ji} &= \frac{(\sqrt{g})_{,k}}{\sqrt{g}}\delta^{lk}_{ij}\mu_l = \frac{(\sqrt{g})_{,j}}{\sqrt{g}}\mu_i - \frac{(\sqrt{g})_{,i}}{\sqrt{g}}\mu_j\end{aligned}$$

we finally get the result

$$\begin{aligned}\pounds_V\mu_i &= v^j_{,j}\mu_i - v^j_{,i}\mu_j + v^j\frac{(\sqrt{g})_{,j}}{\sqrt{g}}\mu_i - v^j\frac{(\sqrt{g})_{,i}}{\sqrt{g}}\mu_j + v^j\frac{(\sqrt{g})_{,i}}{\sqrt{g}}\mu_j \\ &= \Big(v^j_{,j} + v^j\frac{(\sqrt{g})_{,j}}{\sqrt{g}}\Big)\mu_i - v^j_{,i}\mu_j = v^j_{;j}\mu_i - v^j_{,i}\mu_j = (v^k_{;k}\,\delta^j_i - v^j_{,i})\mu_j.\end{aligned}$$

Thus the forms μ_i are invariant under vector fields satisfying the relation $v^j_{,i} = v^k_{;k}\,\delta^j_i$. On contracting this expression, we obtain

$$v^k_{,k} = mv^k_{;k} \quad \text{and} \quad mv^j_{,i} = v^k_{,k}\,\delta^j_i. \qquad ■$$

We are now ready to evaluate the Lie derivative of any tensor if we take notice of the relations $(2.10.5)_2$ and (5.11.3) and recall that Lie derivative of tensor products verify the Leibniz rule as emphasised in (4.3.5). Let a tensor field $\mathcal{T} \in \mathfrak{T}(M)^k_l$ be designated by

$$\mathcal{T} = t^{i_1\cdots i_k}_{j_1\cdots j_l}\frac{\partial}{\partial x^{i_1}}\otimes\cdots\otimes\frac{\partial}{\partial x^{i_k}}\otimes dx^{j_1}\otimes\cdots\otimes dx^{j_l}.$$

The Lie derivative of this tensor with respect to a vector field V can then be expressed as

$$\begin{aligned}
\pounds_V\mathcal{T} &= (\pounds_V t^{i_1\cdots i_k}_{j_1\cdots j_l})\frac{\partial}{\partial x^{i_1}}\otimes\cdots\otimes\frac{\partial}{\partial x^{i_k}}\otimes dx^{j_1}\otimes\cdots\otimes dx^{j_l}\\
&+\sum_{r=1}^{k} t^{i_1\cdots i_k}_{j_1\cdots j_l}\frac{\partial}{\partial x^{i_1}}\otimes\cdots\otimes\pounds_V(\frac{\partial}{\partial x^{i_r}})\otimes\cdots\otimes\frac{\partial}{\partial x^{i_k}}\otimes dx^{j_1}\otimes\cdots\otimes dx^{j_l}\\
&+\sum_{r=1}^{l} t^{i_1\cdots i_k}_{j_1\cdots j_l}\frac{\partial}{\partial x^{i_1}}\otimes\cdots\otimes\frac{\partial}{\partial x^{i_k}}\otimes dx^{j_1}\otimes\cdots\otimes\pounds_V(dx^{j_r})\otimes\cdots\otimes dx^{j_l}.
\end{aligned}$$

We thus obtain

$$\begin{aligned}
\pounds_V\mathcal{T} &= t^{i_1\cdots i_k}_{j_1\cdots j_l,i}v^i\frac{\partial}{\partial x^{i_1}}\otimes\cdots\otimes\frac{\partial}{\partial x^{i_k}}\otimes dx^{j_1}\otimes\cdots\otimes dx^{j_l} \qquad (5.11.15)\\
&-\sum_{r=1}^{k} t^{i_1\cdots i_r\cdots i_k}_{j_1\cdots j_l}v^i_{,i_r}\frac{\partial}{\partial x^{i_1}}\otimes\cdots\otimes\frac{\partial}{\partial x^{i}}\otimes\cdots\otimes\frac{\partial}{\partial x^{i_k}}\otimes dx^{j_1}\otimes\cdots\otimes dx^{j_l}\\
&+\sum_{r=1}^{l} t^{i_1\cdots i_k}_{j_1\cdots j_r\cdots j_l}v^{j_r}_{,j}\frac{\partial}{\partial x^{i_1}}\otimes\cdots\otimes\frac{\partial}{\partial x^{i_k}}\otimes dx^{j_1}\otimes\cdots\otimes dx^{j}\otimes\cdots\otimes dx^{j_l}\\
&= (\pounds_V\mathcal{T})^{i_1\cdots i_k}_{j_1\cdots j_l}\frac{\partial}{\partial x^{i_1}}\otimes\cdots\otimes\frac{\partial}{\partial x^{i_k}}\otimes dx^{j_1}\otimes\cdots\otimes dx^{j_l}
\end{aligned}$$

where the components of the tensor $\pounds_V\mathcal{T}$ are given by

$$\begin{aligned}
(\pounds_V\mathcal{T})^{i_1\cdots i_k}_{j_1\cdots j_l} = t^{i_1\cdots i_k}_{j_1\cdots j_l,i}v^i &-\sum_{r=1}^{k} t^{i_1\cdots i_{r-1}ii_{r+1}\cdots i_k}_{j_1\cdots j_l}v^{i_r}_{,i} \qquad (5.11.16)\\
&+\sum_{r=1}^{l} t^{i_1\cdots i_k}_{j_1\cdots j_{r-1}jj_{r+1}\cdots j_l}v^{j}_{,j_r}.
\end{aligned}$$

Let M and N be smooth manifolds and $\phi : M \to N$ be a smooth mapping. Let us consider a form $\omega \in \Lambda(N)$ and a vector field $V \in T(M)$. Let us calculate the Lie derivative of the form ω with respect to the vector field $V^* = \phi_* V \in T(N)$:

$$\pounds_{\phi_* V}\,\omega = \mathbf{i}_{\phi_* V}(d\omega) + d\mathbf{i}_{\phi_* V}(\omega) \in \Lambda(N).$$

We then pull the above form back to $\Lambda(M)$. On making use of (5.7.6) and Theorem 5.8.2, we can write

$$\phi^* \pounds_{\phi_* V}\, \omega = \phi^* \mathbf{i}_{\phi_* V}(d\omega) + \phi^* d\, \mathbf{i}_{\phi_* V}(\omega) = \mathbf{i}_V\big(d(\phi^*\omega)\big) + d\mathbf{i}_V(\phi^*\omega).$$

Therefore, for all forms $\omega \in \Lambda(N)$ we are led to

$$\phi^* \pounds_{\phi_* V}(\omega) = \pounds_V(\phi^*\omega) \in \Lambda(M) \tag{5.11.17}$$

and, consequently, to the relation

$$\phi^* \circ \pounds_{\phi_* V} = \phi^* \circ \pounds_{V^*} = \pounds_V \circ \phi^*. \tag{5.11.18}$$

5.12. ISOVECTOR FIELDS OF IDEALS

Let $\mathcal{I}$ be an ideal of the exterior algebra $\Lambda(M)$. If this ideal is stable under the flow generated by a vector field $V \in T(M)$, namely, if $\pounds_V \omega \in \mathcal{I}$ for every $\omega \in \mathcal{I}$ so that $\pounds_V \mathcal{I} \subset \mathcal{I}$, in other words, if the ideal $\mathcal{I}$ becomes *invariant* under the flow generated by V, then this vector field is called an ***isovector field*** of the ideal $\mathcal{I}$.

Theorem 5.12.1. *Let $\mathcal{I}(\omega^\alpha)$ be the ideal generated by the forms $\omega^\alpha \in \Lambda(M), \alpha = 1, \dots, r$. A vector field $V \in T(M)$ is an isovector field of $\mathcal{I}$ if and only if $\pounds_V \omega^\alpha \in \mathcal{I}$ for every generator ω^α of the ideal.*

If V is an isovector, then one has $\pounds_V \omega \in \mathcal{I}$ for every form $\omega \in \mathcal{I}$ so the generators ω^α must also fulfil the condition $\pounds_V \omega^\alpha \in \mathcal{I}$. This means that there exist appropriate forms $\lambda^\alpha_\beta \in \Lambda(M)$ such that $\pounds_V \omega^\alpha = \lambda^\alpha_\beta \wedge \omega^\beta$. Conversely, let us assume that $\pounds_V \omega^\alpha \in \mathcal{I}$ so that the relations $\pounds_V \omega^\alpha = \lambda^\alpha_\beta \wedge \omega^\beta$ are satisfied. Because of the restriction $deg\,(\lambda^\alpha_\beta) = deg\,(\omega^\alpha) - deg\,(\omega^\beta) \geq 0$, the forms whose degrees higher than that of ω^α cannot take place in the above sum. If $\omega \in \mathcal{I}$, then we can find forms $\gamma_\alpha \in \Lambda(M)$ so that we are able to write $\omega = \gamma_\alpha \wedge \omega^\alpha$. Therefore, we obtain

$$\pounds_V \omega = (\pounds_V \gamma_\alpha) \wedge \omega^\alpha + \gamma_\alpha \wedge \pounds_V \omega^\alpha = (\pounds_V \gamma_\alpha + \gamma_\beta \wedge \lambda^\beta_\alpha) \wedge \omega^\alpha$$

implying that $\pounds_V \omega \in \mathcal{I}$. □

If the ideal $\mathcal{I}$ is generated by forms of the same degree, then the vector field V is an isovector field of that ideal if we can find smooth functions $\lambda^\alpha_\beta \in \Lambda^0(M)$ that enable us to write $\pounds_V \omega^\alpha = \lambda^\alpha_\beta\, \omega^\beta$.

Theorem 5.12.2. *Isovectors of an ideal $\mathcal{I}$ of the exterior algebra $\Lambda(M)$ constitute a Lie algebra that is a subalgebra of the module $\mathfrak{V}(M)$.*

It is easy to show that isovectors form a subspace of the linear vector space $\mathfrak{V}(M)$ over the field of real numbers $\mathbb{R}$. Let V_1 and V_2 be two isovectors. We thus have $\sigma_1 = \pounds_{V_1}\omega \in \mathcal{I}$ and $\sigma_2 = \pounds_{V_2}\omega \in \mathcal{I}$ for every form $\omega \in \mathcal{I}$. On the other hand, (5.11.11) allows us to write $\pounds_{V_1+V_2}\omega = \pounds_{V_1}\omega + \pounds_{V_2}\omega = \sigma_1 + \sigma_2 \in \mathcal{I}$. Hence, $V_1 + V_2$ is an isovector as well. Similarly, Let V be an isovector so that one finds $\sigma = \pounds_V\omega \in \mathcal{I}$ for all $\omega \in \mathcal{I}$. For all functions $f \in \Lambda^0(M)$, the expression (5.11.10) yields

$$\pounds_{fV}\,\omega = f\pounds_V\,\omega + df \wedge \mathbf{i}_V(\omega) = f\sigma + df \wedge \mathbf{i}_V(\omega).$$

Hence, if only $df = 0$, that is, $f = constant$, then one gets $\pounds_{fV}\,\omega = f\sigma \in \mathcal{I}$. Accordingly, isovectors form a vector space only over $\mathbb{R}$. If U and V are isovectors, then (5.11.12) leads to $\pounds_{[U,V]}\omega = \pounds_U\pounds_V\omega - \pounds_V\pounds_U\omega \in \mathcal{I}$ for all $\omega \in \mathcal{I}$ which means that the Lie product $[U, V]$ is also an isovector. Thus, isovectors constitute a Lie algebra over $\mathbb{R}$. □

The, say, r-dimensional Lie algebra formed by isovectors is, of course, determined by linearly independent vectors $V_\alpha, \alpha = 1, \dots, r \leq m$ and there exist *structure constants* $c_{\alpha\beta}^{\gamma}$ so that the conditions $[V_\alpha, V_\beta] = c_{\alpha\beta}^{\gamma} V_\gamma$ hold. Then, on recalling Sec. 3.8, we reach to the conclusion that isovectors generate an r-parameter Lie transformation group on the manifold M and the ideal $\mathcal{I}$ remains invariant under this mapping. In other words, a flow generated by an isovector transforms a form in the ideal to another form also in the ideal.

Theorem 5.12.3. *If the vector field $V \in T(M)$ is an isovector field of an ideal $\mathcal{I}(\omega^\alpha)$ of the exterior algebra $\Lambda(M)$, then it is also an isovector field of its closure $\overline{\mathcal{I}}(\omega^\alpha, d\omega^\alpha)$.*

If V is an isovector field of the ideal $\mathcal{I}$, then there are appropriate forms $\lambda_\beta^\alpha \in \Lambda(M)$ such that one is able to write $\pounds_V\omega^\alpha = \lambda_\beta^\alpha \wedge \omega^\beta$. Employing this relation, we get

$$\pounds_V d\omega^\alpha = d\pounds_V\omega^\alpha = d\lambda_\beta^\alpha \wedge \omega^\beta + (-1)^{deg(\lambda_\beta^\alpha)}\lambda_\beta^\alpha \wedge d\omega^\beta.$$

We consider a form $\sigma \in \overline{\mathcal{I}}$ that can be written as $\sigma = \gamma_\alpha \wedge \omega^\alpha + \Gamma_\alpha \wedge d\omega^\alpha$. Hence, we obtain

$$\begin{aligned}\pounds_V\sigma &= \pounds_V\gamma_\alpha \wedge \omega^\alpha + \gamma_\alpha \wedge \pounds_V\omega^\alpha + \pounds_V\Gamma_\alpha \wedge d\omega^\alpha + \Gamma_\alpha \wedge \pounds_V d\omega^\alpha \\ &= (\pounds_V\gamma_\alpha + \gamma_\beta \wedge \lambda_\alpha^\beta + \Gamma_\beta \wedge d\lambda_\alpha^\beta) \wedge \omega^\alpha \\ &\quad + (\pounds_V\Gamma_\alpha + (-1)^{deg(\lambda_\alpha^\beta)}\Gamma_\beta \wedge \lambda_\alpha^\beta) \wedge d\omega^\alpha \in \overline{\mathcal{I}}\end{aligned}$$

This expression means that V is also an isovector of the closure $\overline{\mathcal{I}}$ of the ideal $\mathcal{I}$. □

Evidently, this theorem does not imply that isovectors of the ideals $\mathcal{I}$ and $\overline{\mathcal{I}}$ are the same. Some isovectors of the closed ideal $\overline{\mathcal{I}}$ may not belong to the set of isovectors of the ideal $\mathcal{I}$. This situation will be remedied to some extent by the following theorem.

Theorem 5.12.4. *If an ideal $\mathcal{I}(\omega^\alpha)$ is generated by forms of the same degree, then isovectors of the ideals $\mathcal{I}$ and $\overline{\mathcal{I}}$ are coincident.*

We have demonstrated in Theorem 5.12.3 that isovectors of $\mathcal{I}$ are also isovectors of $\overline{\mathcal{I}}$. In order prove the present theorem, we have to show that the converse statement is also true. If V is an isovector of $\overline{\mathcal{I}}$, then there are suitable forms λ^α_β and Λ^α_β so that we can write

$$\pounds_V\omega^\alpha = \lambda^\alpha_\beta \wedge \omega^\beta + \Lambda^\alpha_\beta \wedge d\omega^\beta$$

whence we deduce that

$$\pounds_V d\omega^\alpha = d\pounds_V\omega^\alpha = d\lambda^\alpha_\beta \wedge \omega^\beta + \big((-1)^{deg\,(\lambda^\alpha_\beta)}\lambda^\alpha_\beta + d\Lambda^\alpha_\beta\big) \wedge d\omega^\beta.$$

However, if all forms ω^α possess the same degree, say k, then the degree of all forms $d\omega^\alpha$ is $k+1$ implying that we have to take $\Lambda^\alpha_\beta = 0$ and $\lambda^\alpha_\beta \in \Lambda^0(M)$. In this case, the above relations reduce to

$$\pounds_V\omega^\alpha = \lambda^\alpha_\beta\omega^\beta, \;\; \pounds_V d\omega^\alpha = d\lambda^\alpha_\beta \wedge \omega^\beta + \lambda^\alpha_\beta\, d\omega^\beta$$

from which we conclude that an isovector V of the ideal $\overline{\mathcal{I}}$ is also an isovector of the ideal $\mathcal{I}$. □

The following theorem provides a somewhat simplified approach to evaluate isovectors of an ideal.

Theorem 5.12.5. *Let $\mathcal{I}(\omega^\alpha)$ be an ideal of $\Lambda(M)$ generated by forms $\omega^\alpha, \alpha = 1, \ldots, r$ whose degrees satisfy the condition $deg\,\omega^\alpha < k$. We then consider forms $\sigma^a, a = 1, \ldots, s$ such that $deg\,\sigma^a \geq k$. A vector field V is an isovector of the ideal $\mathcal{I}(\omega^\alpha, \sigma^a)$ if and only if*
(i) it is an isovector of the ideal $\mathcal{I}(\omega^\alpha)$,
(ii) $\pounds_V\sigma^a \in \mathcal{I}(\omega^\alpha, \sigma^a)$.

Let us first assume that the vector field V is an isovector of the ideal $\mathcal{I}(\omega^\alpha)$ so that one has $\pounds_V\omega^\alpha = \lambda^\alpha_\beta \wedge \omega^\beta$. We further assume that $\pounds_V\sigma^a = \lambda^a_\alpha \wedge \omega^\alpha + \lambda^a_b \wedge \sigma^b$. If $\omega \in \mathcal{I}(\omega^\alpha, \sigma^a)$, then $\omega = \gamma_\alpha \wedge \omega^\alpha + \gamma_a \wedge \sigma^a$ and its Lie derivative with respect to V is found to be

$$\begin{aligned}\pounds_V\omega &= \pounds_V\gamma_\alpha \wedge \omega^\alpha + \gamma_\alpha \wedge \pounds_V\omega^\alpha + \pounds_V\gamma_a \wedge \sigma^a + \gamma_a \wedge \pounds_V\sigma^a\\ &= (\pounds_V\gamma_\alpha + \gamma_\beta \wedge \lambda^\beta_\alpha + \gamma_a \wedge \lambda^a_\alpha) \wedge \omega^\alpha\\ &\quad + (\pounds_V\gamma_a + \gamma_b \wedge \lambda^b_a) \wedge \sigma^a \in \mathcal{I}(\omega^\alpha, \sigma^a).\end{aligned}$$

Hence V is an isovector of the ideal $\mathcal{I}(\omega^\alpha, \sigma^a)$. Conversely, let us suppose that V is an isovector of the ideal $\mathcal{I}(\omega^\alpha, \sigma^a)$ implying that $\pounds_V\omega \in \mathcal{I}(\omega^\alpha, \sigma^a)$ for all $\omega \in \mathcal{I}(\omega^\alpha, \sigma^a)$. Hence, the above relation requires that the condition

$\gamma_\alpha \wedge \pounds_V\omega^\alpha + \gamma_a \wedge \pounds_V\sigma^a \in \mathcal{I}(\omega^\alpha, \sigma^a)$ must hold. This last expression should be valid of course for all forms ω in the ideal $\mathcal{I}(\omega^\alpha, \sigma^a)$, and consequently, for all forms $\gamma_\alpha, \gamma_a \in \Lambda(M)$ implying that we must have $\pounds_V\omega^\alpha \in \mathcal{I}(\omega^\alpha, \sigma^a)$ and $\pounds_V\sigma^a \in \mathcal{I}(\omega^\alpha, \sigma^a)$. We thus conclude that there must be suitable forms $\lambda^\alpha_\beta, \lambda^\alpha_a, \lambda^a_\alpha, \lambda^a_b$ so that we can write

$$\pounds_V\omega^\alpha = \lambda^\alpha_\beta \wedge \omega^\beta + \lambda^\alpha_a \wedge \sigma^a, \quad \pounds_V\sigma^a = \lambda^a_\alpha \wedge \omega^\alpha + \lambda^a_b \wedge \sigma^b.$$

But, due to the restrictions $deg\,\omega^\alpha < k$ and $deg\,\sigma^a \geq k$, we get $\lambda^\alpha_a = 0$ and we find that $\pounds_V\omega^\alpha = \lambda^\alpha_\beta \wedge \omega^\beta$. Thus V must also be an isovector of the ideal $\mathcal{I}(\omega^\alpha)$. □

Based on the Theorem (5.12.5), we may propose quite an effective method to determine isovector fields of an ideal generated by forms of different degrees. Let us arrange the generators of the ideal according to increasing degrees and collate all forms of the same degree into a set so that let us write $\mathcal{I}(\omega^\alpha, \sigma^a, \gamma^A, \dots)$. The degrees of the forms in each set $\{\omega^\alpha\}$, $\{\sigma^a\}$, $\{\gamma^A\}, \dots$ are the same and they are ordered as follows: $deg\,\omega^\alpha < deg\,\sigma^a < deg\,\gamma^A < \cdots$. In this case, in order to determine the isovector fields, we have to ensure that the conditions

$$\pounds_V\omega^\alpha \in \mathcal{I}(\omega^\alpha), \ \pounds_V\sigma^a \in \mathcal{I}(\omega^\alpha, \sigma^a), \ \pounds_V\gamma^A \in \mathcal{I}(\omega^\alpha, \sigma^a, \gamma^A), \ \dots$$

are satisfied. Since we deal with a lesser number of forms in each set with uniform degrees, calculations turn out to be relatively simpler. Besides, if degrees in two sets differ just 1, and if some generators in one set happen to be exterior derivatives of some forms in the other set, then we can disregard these generators in view of Theorem 5.12.4.

Example 5.12.1. Let us determine the isovector fields of the ideal $\mathcal{I}(\omega^1)$ of the exterior algebra $\Lambda(\mathbb{R}^3)$ generated by $\omega^1 = x\,dy + y\,dz$. We denote a vector field by $V = v^x\partial_x + v^y\partial_y + v^z\partial_z$. We have to show that there exists a function $\lambda \in \Lambda^0(\mathbb{R}^3)$ such that $\pounds_V\omega^1 = \lambda\,\omega^1$. Let us write $d\omega^1 = dx \wedge dy + dy \wedge dz$, $\mathbf{i}_V(d\omega^1) = -\,v^y dx + (v^x - v^z)dy + v^y dz$ and $\mathbf{i}_V(\omega^1) = xv^y + yv^z = F(x, y, z)$. We thus obtain

$$\pounds_V\omega^1 = (F_x - v^y)dx + (F_y + v^x - v^z)dy + (F_z + v^y)dz = \lambda x\,dy + \lambda y\,dz$$

yielding $F_x - v^y = 0$, $F_y + v^x - v^z = \lambda x$ and $F_z + v^y = \lambda y$. Solution of these equations gives $\lambda = (F_x + F_z)/y$ and the isovector field specified by an arbitrary function F becomes

$$V_F = \frac{1}{y}(F + xF_z - yF_y)\frac{\partial}{\partial x} + F_x\frac{\partial}{\partial y} + \frac{1}{y}(F - xF_x)\frac{\partial}{\partial z}.$$

If the isovector fields produced by functions F and G are denoted by V_F and V_G, then their Lie product must be given by $[V_F, V_G] = V_H$. It is rather straightforward to verify that the function $H(x, y, z)$ is obtainable as

$$H = F_x G_y - G_x F_y + \frac{1}{y}(FG_x - GF_x + FG_z - GF_z) + \frac{x}{y}(F_z G_x - G_z F_x).$$

It is plainly seen that isovectors of the ideal $\mathcal{I}(\omega^1)$ constitute an infinite dimensional Lie algebra. ■

We have the following theorem if some of the isovectors of an ideal of $\Lambda(M)$ are also characteristic vectors of the same ideal.

Theorem 5.12.6. *If some of the isovectors of an ideal $\mathcal{I}$ are at the same time characteristic vectors of this ideal, then they form a Lie subalgebra of the Lie algebra of isovectors.*

If U and V are isovectors of an ideal $\mathcal{I}$, then we have $\pounds_U \omega, \pounds_V \omega \in \mathcal{I}$ for all $\omega \in \mathcal{I}$. If these vectors are also characteristic vectors of $\mathcal{I}$, they must satisfy $\mathbf{i}_U(\omega), \mathbf{i}_V(\omega) \in \mathcal{I}$. On making use of (5.11.7), we get

$$\mathbf{i}_{[U,V]}(\omega) = \pounds_U\big(\mathbf{i}_V(\omega)\big) - \mathbf{i}_V\big(\pounds_U(\omega)\big) \in \mathcal{I}.$$

That means that the Lie product $[U, V]$ which is known to be an isovector is also a characteristic vector of the ideal. Therefore, such a subset of isovectors that are also the characteristic vectors of $\mathcal{I}$, is closed under the Lie product, that is, it is a Lie subalgebra. □

We can reach to a more interesting result in closed ideals.

Theorem 5.12.7. *If an ideal $\mathcal{I}$ of $\Lambda(M)$ is closed, then the subspace formed by its isovectors contains the characteristic subspace $\mathcal{S}(\mathcal{I})$.*

Let us assume that the ideal $\mathcal{I}$ is generated by forms $\omega^1, \omega^2, \ldots, \omega^r \in \Lambda(M)$ of various degrees. Since $\mathcal{I}$ is closed, then there are suitable forms $\lambda^\alpha_\beta \in \Lambda(M), \alpha, \beta = 1, \ldots, r$ such that $d\omega^\alpha = \lambda^\alpha_\beta \wedge \omega^\beta$. On the other hand, if $V \in \mathcal{S}(\mathcal{I})$, then there exist appropriate forms $\mu^\alpha_\beta \in \Lambda(M)$ such that $\mathbf{i}_V(\omega^\alpha) = \mu^\alpha_\beta \wedge \omega^\beta$. Hence, according to $(5.4.1)_4$ we find that

$$\begin{aligned}\mathbf{i}_V(d\omega^\alpha) &= \mathbf{i}_V(\lambda^\alpha_\beta) \wedge \omega^\beta + (-1)^{deg(\lambda^\alpha_\beta)} \lambda^\alpha_\beta \wedge \mathbf{i}_V(\omega^\beta)\\ &= \big[\mathbf{i}_V(\lambda^\alpha_\beta) + (-1)^{deg(\lambda^\alpha_\gamma)} \lambda^\alpha_\gamma \wedge \mu^\gamma_\beta)\big] \wedge \omega^\beta \in \mathcal{I}.\end{aligned}$$

But the exterior derivative of the form $\mathbf{i}_V(\omega^\alpha)$ gives

$$\begin{aligned}d\mathbf{i}_V(\omega^\alpha) &= d\mu^\alpha_\beta \wedge \omega^\beta + (-1)^{deg(\mu^\alpha_\beta)} \mu^\alpha_\beta \wedge d\omega^\beta\\ &= \big[d\mu^\alpha_\beta + (-1)^{deg(\mu^\alpha_\gamma)} \mu^\alpha_\gamma \wedge \lambda^\gamma_\beta)\big] \wedge \omega^\beta \in \mathcal{I}\end{aligned}$$

from which we deduce that

$$\pounds_V\omega^\alpha = \mathbf{i}_V(d\omega^\alpha) + d\mathbf{i}_V(\omega^\alpha) \in \mathcal{I}.$$

Then Theorem 5.12.1 states that the characteristic vector V is also an isovector of the closed ideal $\mathcal{I}$, that is, the characteristic subspace of the ideal $\mathcal{I}$ belongs to the subspace generated by isovectors of this closed ideal. □

When we combine this theorem with Theorem 5.12.6 we arrive immediately at the following result: *characteristic vectors of a closed ideal constitute a Lie algebra*. However, we have to stress the fact that the converse of Theorem 5.12.7 is in general not true, i.e., all isovectors of a closed ideal are not necessarily characteristic vectors of this ideal.

5.13. EXTERIOR SYSTEMS AND THEIR SOLUTIONS

We have seen in *p*. 258 how we can engender a nontrivial, $r \geq k$ dimensional solution of an exterior equation $\omega = 0$ where $\omega \in \Lambda^k(M)$. We shall now explore the notion of exterior equations in a more general context.

Let us consider a set $\{\omega^\alpha, \alpha = 1, \dots, N\}$ of forms that might be of different degrees. We specify an r-dimensional submanifold S by the mapping $\phi : S \to M$. *If we get* $\phi^*\omega^\alpha = 0, \alpha = 1, \dots, N$, *namely, if the mapping* $\phi^* : \Lambda(M) \to \Lambda(S)$ *annihilates the forms* $\{\omega^\alpha\}$ *then the mapping* ϕ, *in other words, the submanifold* S *is said to be a solution of the system of exterior equations* $\{\omega^\alpha = 0, \alpha = 1, \dots, N\}$. A submanifold whose dimension is less than the lowest degree of the forms ω^α is of course a trivial solution of the exterior system. Let us now take the ideal $\mathcal{I}(\omega^\alpha)$ into consideration. The mapping ϕ will be the solution of every form $\omega \in \mathcal{I}(\omega^\alpha)$ as well. In fact, if we write $\omega = \lambda_\alpha \wedge \omega^\alpha$, we find from (5.7.4) that $\phi^*\omega = \phi^*\lambda_\alpha \wedge \phi^*\omega^\alpha = 0.$ Conversely, we can easily demonstrate that the forms annihilated on a submanifold S prescribed by a mapping $\phi : S \to M$, or amounting to the same thing, all forms which annihilates the subbundle $T(S) \subset T(M)$ constitute an ideal of the exterior algebra $\Lambda(M)$. Let us consider the pull-back mapping $\phi^* : \Lambda(M) \to \Lambda(S)$ induced by the mapping ϕ. All forms annihilated on the submanifold S satisfy the relation $\phi^*\omega = 0$. We denote the set of all forms ω such that $\phi^*\omega = 0$ by $\mathcal{I} \subset \Lambda(M)$. If $\omega_1, \omega_2 \in \mathcal{I}$ are two forms with the same degree, then we have $\phi^*(\omega_1 + \omega_2) = \phi^*(\omega_1) + \phi^*(\omega_2) = 0$ implying that $\omega_1 + \omega_2 \in \mathcal{I}$. Similarly, if $\omega \in \mathcal{I}$ and $\gamma \in \Lambda(M)$ is an arbitrary form, then $\phi^*(\gamma \wedge \omega) = \phi^*(\gamma) \wedge \phi^*(\omega) = 0$ which means that $\gamma \wedge \omega \in \mathcal{I}$. Hence, $\mathcal{I}$ is an ideal of the exterior algebra.

If all forms of the exterior algebra $\Lambda(M)$ that are annihilated by every solution of exterior equations $\{\omega^\alpha = 0\}$ belong to the ideal $\mathcal{I}(\omega^\alpha)$ generated by forms ω^α, then $\mathcal{I}$ is called a ***complete ideal***.

Theorem 5.13.1. *An ideal of the exterior algebra* $\Lambda(M)$ *generated by* *linearly independent* 1*-forms is complete.*

Let us assume that the ideal is generated by linearly independent forms $\omega^\alpha \in \Lambda^1(M)$, $\alpha = 1, \dots, N \leq m$. As we have mentioned above, the solutions of the exterior equations $\{\omega^\alpha = 0\}$ annihilate every form within the ideal. We now suppose that solutions of the system $\{\omega^\alpha = 0\}$ annihilate a form $\omega \in \Lambda(M)$ as well. By adding suitable linearly independent 1-forms, we can determine a basis of $T^*(M)$ as follows: $\omega^1, \dots, \omega^N, \omega^{N+1}, \dots, \omega^m$. The form ω can now be constructed as a combination of exterior products of these forms. However, we have assumed that $\omega = 0$ whenever $\omega^1 = \dots = \omega^N = 0$. Therefore, at least one of the factors $\omega^1, \dots, \omega^N$ must be present in each term. Hence, we conclude that ω is expressible as

$$\omega = \lambda_1 \wedge \omega^1 + \lambda_2 \wedge \omega^2 + \dots + \lambda_N \wedge \omega^N \in \mathcal{I}(\omega^\alpha) \qquad \square$$

Let us next consider two exterior systems $\{\omega^\alpha, \alpha = 1, \dots, N_1\}$ and $\{\sigma^a, a = 1, \dots, N_2\}$, and the ideals $\mathcal{I}_1 = \mathcal{I}(\omega^\alpha)$ and $\mathcal{I}_2 = \mathcal{I}(\sigma^a)$ generated by them. If these ideals are equal, namely, if they satisfy the relations $\mathcal{I}_1 \subseteq \mathcal{I}_2$ and $\mathcal{I}_2 \subseteq \mathcal{I}_1$, we say that these two exterior systems are *algebraically equivalent*. In this situation, there are appropriate forms λ^α_a and Λ^a_α so that we can write $\omega^\alpha = \lambda^\alpha_a \wedge \sigma^a$ and $\sigma^a = \Lambda^a_\alpha \wedge \omega^\alpha$.

Example 5.13.1. Let us consider a system of exterior equations of the exterior algebra $\Lambda(\mathbb{R}^4)$ specified by the forms $\omega^1 = dx^1 \wedge dx^3$, $\omega^2 = dx^1 \wedge dx^4$, $\omega^3 = dx^1 \wedge dx^2 - dx^3 \wedge dx^4$. A 2-dimensional submanifold of $\mathbb{R}^4$ is determined by the mapping $x^i = \phi^i(u^1, u^2), 1 \leq i \leq 4$. We now impose the condition that this mapping must satisfy

$$\phi^*\omega^1 = \phi^1_{,\alpha}\phi^3_{,\beta}\, du^\alpha \wedge du^\beta = 0, \quad \phi^*\omega^2 = \phi^1_{,\alpha}\phi^4_{,\beta}\, du^\alpha \wedge du^\beta = 0,$$
$$\phi^*\omega^3 = (\phi^1_{,\alpha}\phi^2_{,\beta} - \phi^3_{,\alpha}\phi^4_{,\beta})\, du^\alpha \wedge du^\beta = 0, \ \alpha, \beta = 1, 2.$$

We immediately discover a solution by just inspection as $\phi^1 = \textit{constant}$ and $\phi^3 = \textit{constant}$. We then consider the form $\omega = dx^1 \wedge dx^2$. We find that $\phi^*\omega = \phi^1_{,\alpha}\phi^2_{,\beta}\, du^\alpha \wedge du^\beta = 0$. But, we realise at once that this form does not belong to the ideal $\mathcal{I}(\omega^1, \omega^2, \omega^3)$. Hence, this ideal is not complete. ■

Certain significant properties of ideals of the exterior algebra can be discussed by means of Lie derivatives. An effective tool implementing this approach is provided by the following Cartan theorem.

Theorem 5.13.2 (The Cartan Theorem). *Let $\mathcal{I}$ be an ideal of the exterior algebra $\Lambda(M)$ and let $\mathcal{S}(\mathcal{I}) \subset T(M)$ be the characteristic subspace of constant dimension of this ideal. If $\mathcal{I}$ is a closed ideal, then the subspace $\mathcal{S}(\mathcal{I})$ is an involutive distribution of $T(M)$.*

We know that the characteristic subspace of the ideal $\mathcal{I}$ is defined by $\mathcal{S}(\mathcal{I}) = \{V \in T(M) : \mathbf{i}_V(\mathcal{I}) \subseteq \mathcal{I}\}$. Since we have assumed that $\mathcal{S}$ has the same dimension, say, r at every point of the manifold M, the characteristic subspace is spanned by r linearly independent vector fields $V_\alpha \in T(M), \alpha = 1, \dots, r$. It follows from (5.11.7) that

$$\begin{aligned}\mathbf{i}_{[U,V]}(\omega) &= £_U\big(\mathbf{i}_V(\omega)\big) - \mathbf{i}_V\big(£_U(\omega)\big)\\ &= \mathbf{i}_U\big[d\big(\mathbf{i}_V(\omega)\big)\big] + d\big[\mathbf{i}_U\big(\mathbf{i}_V(\omega)\big)\big] - \mathbf{i}_V\big(\mathbf{i}_U(d\omega)\big) - \mathbf{i}_V\big[d\big(\mathbf{i}_U(\omega)\big)\big]\end{aligned}$$

for all $\omega \in \Lambda(M)$ and $U, V \in T(M)$. Thus we obtain

$$\begin{aligned}&\mathbf{i}_{[V_\alpha,V_\beta]}(\omega) =\\ &\qquad \mathbf{i}_{V_\alpha}\big[d\big(\mathbf{i}_{V_\beta}(\omega)\big)\big] + d\big[\mathbf{i}_{V_\alpha}\big(\mathbf{i}_{V_\beta}(\omega)\big)\big] - \mathbf{i}_{V_\beta}\big(\mathbf{i}_{V_\alpha}(d\omega)\big) - \mathbf{i}_{V_\beta}\big[d\big(\mathbf{i}_{V_\alpha}(\omega)\big)\big]\end{aligned}$$

for all $\omega \in \mathcal{I}$ and $V_\alpha, V_\beta \in \mathcal{S}$. Since V_α and V_β are characteristic vectors of the closed ideal, we can write $\mathbf{i}_{V_\alpha}(\mathcal{I}) \subseteq \mathcal{I}, \alpha = 1, \dots, r$ and $d\mathcal{I} \subseteq \mathcal{I}$. This implies that each term in the right hand side of the above expression is in the ideal. Hence, we get $\mathbf{i}_{[V_\alpha,V_\beta]}(\omega) \in \mathcal{I}$ for all $\omega \in \mathcal{I}$. This amounts to say that $[V_\alpha, V_\beta] \in \mathcal{S}$. In other words, the characteristic subspace is closed under the Lie product. Thus $\mathcal{S}$ is an involutive distribution. Therefore, the characteristic vector fields of a closed ideal engender a smooth r-dimensional submanifold of M. □

Let us now consider the exterior system $D_r = \{\omega^\alpha\}$ comprised of r linearly independent 1-forms. The exterior equations $\omega^\alpha = 0, \alpha = 1, \dots, r$ constitute a ***Pfaff system*** [German mathematician Johann Friedrich Pfaff (1765-1825)]. According to Theorem 5.13.1, the ideal $\mathcal{I}(D_r)$ generated by these forms is complete. The exterior system D_r is ***completely integrable*** if it is annihilated on every one of the $(m - r)$-dimensional submanifolds prescribed by equations of the form

$$g^\alpha(\mathbf{x}) = c^\alpha, \ \ \alpha = 1, \dots, r$$

with r parameter. c^α are arbitrary real constants. Since $\mathcal{I}(D_r)$ is a complete ideal, all forms annihilated by those submanifolds, which are called ***characteristic manifolds***, must belong to this ideal.

Theorem 5.13.3. *An exterior system D_r is completely integrable if and only if it is possible to find a regular $r \times r$ matrix function $\mathbf{A}(\mathbf{x})$ and r independent functions $g^\alpha(\mathbf{x})$ such that the following relations are valid*:

$$\omega^\alpha = A^\alpha_\beta\, dg^\beta,\ \alpha, \beta = 1, \dots, r,\ \mathbf{A}(\mathbf{x}) = [A^\alpha_\beta(\mathbf{x})]. \tag{5.13.1}$$

If the forms $\{\omega^\alpha\}$ are given by the relations (5.13.1), when $g^\beta = c^\beta =$ *constant* we find $dg^\beta = 0$ and consequently $\omega^\alpha = 0$. Thus the exterior system D_r is completely integrable. Conversely, let us assume that the exterior system D_r is completely integrable. Hence, there are r independent functions $g^\alpha(\mathbf{x})$ and the ideal $\mathcal{I}(D_r)$ is annihilated by hypersurfaces $g^\alpha(\mathbf{x}) = c^\alpha$. Next, we form the ideal $\mathcal{I}(dg^\alpha)$ by the forms $dg^\alpha \in \Lambda^1(M)$. Since $dg^\alpha = 0$ on these hypersurfaces, this ideal is also annihilated by them. Because of the fact that both ideals are complete, we arrive at the result $\mathcal{I}(D_r) = \mathcal{I}(dg^\alpha)$. This implies that there are functions $A^\alpha_\beta \in \Lambda^0(M)$ such that $\omega^\alpha = A^\alpha_\beta\, dg^\beta$. The forms ω^α and dg^α are linearly independent. Therefore, we ought to have $\omega^1 \wedge \cdots \wedge \omega^r \neq 0$ and $dg^1 \wedge \cdots \wedge dg^r \neq 0$. Thus, the relation

$$\omega^1 \wedge \cdots \wedge \omega^r = \det(A^\alpha_\beta)\, dg^1 \wedge \cdots \wedge dg^r \neq 0$$

requires that $\det(A^\alpha_\beta) \neq 0$. □

If we calculate the exterior derivative of the expression (5.13.1), we get

$$d\omega^\alpha = dA^\alpha_\beta \wedge dg^\beta = (A^{-1})^\beta_\gamma\, dA^\alpha_\beta \wedge \omega^\gamma \in \mathcal{I}(D_r),$$

Hence, *if the exterior system D_r is completely integrable, then the ideal $\mathcal{I}(D_r)$ must be closed.* That the converse proposition is also true is provided by the following theorem referred to Frobenius.

Theorem 5.13.4 (The Frobenius Theorem). *An exterior system D_r is completely integrable if and only if the ideal $\mathcal{I}(D_r)$ generated by r linearly independent 1-forms $\{\omega^\alpha\}$ is closed, that is, if $d\mathcal{I}(D_r) \subseteq \mathcal{I}(D_r)$ or if there exist r^2 forms $\Gamma^\alpha_\beta \in \Lambda^1(M)$ such that the relations $d\omega^\alpha = \Gamma^\alpha_\beta \wedge \omega^\beta$ are satisfied or if we verify that $d\omega^\alpha \wedge \omega^1 \wedge \cdots \wedge \omega^r = 0$ for $\alpha = 1, \dots, r$.*

We have already seen that the ideal $\mathcal{I}(D_r)$ will be closed if the exterior system D_r is completely integrable. Let us assume, this time, that the ideal $\mathcal{I}(D_r)$ is closed. We know that the dimension of the characteristic subspace $\mathcal{S}(D_r)$ of this ideal is $m - r$ [*see* Theorem 5.6.2]. Let the linearly independent vectors $U_a, a = r+1, \dots, m$ be a basis of that subspace. According to the Cartan theorem 5.13.2, $\mathcal{S}(D_r)$ is an involutive distribution, i.e., there are functions $c^c_{ab} \in \Lambda^0(M)$ such that $[U_a, U_b] = c^c_{ab} U_c$. In this situation, we can choose, as we have done in Theorem 2.11.1, a new basis set as vectors V_a, $a = r+1, \dots, m$ of $\mathcal{S}(D_r)$ such that $[V_a, V_b] = 0$. We shall now show that this property guaranties the existence of independent functions $g^\alpha(\mathbf{x}), \alpha = 1, \dots, r$ satisfying the relations $V_a(g^\alpha) = \mathbf{i}_{V_a}(dg^\alpha) = 0$. To this end, we look for the solutions of the system of differential equations $V_a(f) = 0$. On repeating our approach in Sec. 2.11, we start with $V_{r+1}(f) = 0$. It is known that the *independent solutions* of the first order partial differential equation

$$v^i_{r+1}(\mathbf{x}) \frac{\partial f}{\partial x^i} = 0$$

can be determined through the method of characteristics as follows

$$h^1(\mathbf{x}) = C^1,\ h^2(\mathbf{x}) = C^2,\ \ldots\ , h^{m-1}(\mathbf{x}) = C^{m-1}$$

where $C^1, \ldots, C^{m-1}$ are constants. We then find

$$0 = \frac{\partial h^{\mathfrak{a}}}{\partial x^i}\frac{dx^i}{dt} = v^i_{r+1}(\mathbf{x})\frac{\partial h^{\mathfrak{a}}}{\partial x^i} = V_{r+1}(h^{\mathfrak{a}})$$

where $\mathfrak{a} = 1, \ldots, m-1$. We thus write

$$\begin{aligned} V_{r+1}(f) &= v^1_{r+1}(\mathbf{x})\frac{\partial f}{\partial x^1} + \cdots + v^m_{r+1}(\mathbf{x})\frac{\partial f}{\partial x^m} = 0 \\ V_{r+1}(h^1) &= v^1_{r+1}(\mathbf{x})\frac{\partial h^1}{\partial x^1} + \cdots + v^m_{r+1}(\mathbf{x})\frac{\partial h^1}{\partial x^m} = 0 \\ &\vdots \\ V_{r+1}(h^{m-1}) &= v^1_{r+1}(\mathbf{x})\frac{\partial h^{m-1}}{\partial x^1} + \cdots + v^m_{r+1}(\mathbf{x})\frac{\partial h^{m-1}}{\partial x^m} = 0. \end{aligned}$$

Since $V_{r+1} \neq 0$, it is only possible to find a nontrivial solution to this homogeneous system of equations if the Jacobian, or the functional determinant, of the functions $f, h^1, \ldots, h^{m-1}$ vanishes

$$\frac{\partial(f, h^1, \ldots h^{m-1})}{\partial(x^1, x^2, \ldots, x^m)} = 0.$$

It is known that the general solution of the foregoing equation is

$$f = f(h^1, h^2, \ldots, h^{m-1}). \tag{5.13.2}$$

In the second step, let us apply the operator V_{r+2} on the function (5.13.2) to obtain

$$0 = V_{r+2}(f) = v^i_{r+2}\frac{\partial f}{\partial x^i} = v^i_{r+2}\frac{\partial f}{\partial h^{\mathfrak{a}}}\frac{\partial h^{\mathfrak{a}}}{\partial x^i} = V_{r+2}(h^{\mathfrak{a}})\frac{\partial f}{\partial h^{\mathfrak{a}}}. \tag{5.13.3}$$

On the other hand, because of the relation $V_{r+1}V_{r+2} = V_{r+2}V_{r+1}$, we find

$$V_{r+1}\big(V_{r+2}(h^{\mathfrak{a}})\big) = V_{r+2}\big(V_{r+1}(h^{\mathfrak{a}})\big) = V_{r+2}(0) = 0$$

which means that the functions $V_{r+2}(h^{\mathfrak{a}})$ become solutions of the equation $V_{r+1}(u) = 0$. We can thus write as in (5.13.2)

$$V_{r+2}(h^{\mathfrak{a}}) = H^{\mathfrak{a}}(h^1, h^2, \ldots, h^{m-1})$$

and the equation (5.13.3) takes the form

$$H^{\mathfrak{a}}(h^{\mathfrak{b}})\frac{\partial f}{\partial h^{\mathfrak{a}}}=0,\ \mathfrak{a},\mathfrak{b}=1,\dots,m-1.$$

Hence, the number of independent variables reduces to $m-1$ from m. By repeating the same procedure as above we obtain $f=f(k^1,k^2,\dots,k^{m-2})$ where $k^s=k^s(h^1,h^2,\dots,h^{m-1}), s=1,\dots,m-2$. On applying the operators $V_{r+1},\dots,V_m$, respectively, on the function f, we see that f is dependent on $m-(m-r)=r$ independent functions $g^\alpha\in\Lambda^0(M),\alpha=1,\dots,r$ as follows:

$$f=f(g^1,g^2,\dots,g^r).\tag{5.13.4}$$

The functions g^α are clearly determined by successively solving a sequence of ordinary differential equations with ever decreasing number of dependent variables. We can then write

$$V_a(f)=v_a^i\frac{\partial f}{\partial g^\alpha}\frac{\partial g^\alpha}{\partial x^i}=V_a(g^\alpha)\frac{\partial f}{\partial g^\alpha}=0,\ a=r+1,\dots,m.$$

This relation would of course be valid for all functions in the form (5.13.4). If we choose $f=g^\beta$, we find

$$V_a(g^\alpha)\,\delta_\alpha^\beta=V_a(g^\beta)=0,\ a=r+1,\dots,m,\beta=1,\dots,r$$

implying that

$$V_a(g^\alpha)=\mathbf{i}_{V_a}(dg^\alpha)=0.\tag{5.13.5}$$

Since the functions g^α are independent, the forms $dg^\alpha\in\Lambda^1(M)$ must be linearly independent so that one gets $\Omega=dg^1\wedge\cdots\wedge dg^r\neq 0$. According to Theorem 5.6.1 the relations (5.13.5) express the fact that the vectors $\{V_a\}$ are also characteristic vectors of the ideal $\mathcal{I}(dg^\alpha)$. We can now readily prove that $\mathcal{I}(D_r)=\mathcal{I}(\omega^\alpha)\subseteq\mathcal{I}(dg^\alpha)$. Let us assume that one of the generators of the ideal $\mathcal{I}(\omega^\alpha)$, say ω^α, does not belong to the ideal $\mathcal{I}(dg^\alpha)$. On referring to the statement on p. 249, we are thus compelled to assume that $\mathbf{i}_{V_a}(\omega^\alpha)\neq 0$. However, V_a is a characteristic vector of the ideal $\mathcal{I}(\omega^\alpha)$ as well and the condition $\mathbf{i}_{V_a}(\omega^\alpha)=0$ must be satisfied. In order to remove this contradiction, we have to take $\omega^\alpha\in\mathcal{I}(dg^\alpha)$. Hence, all generators of $\mathcal{I}(\omega^\alpha)$ must belong to $\mathcal{I}(dg^\alpha)$. This means that $\mathcal{I}(\omega^\alpha)\subseteq\mathcal{I}(dg^\alpha)$. Therefore, there exists a regular matrix $[A_\beta^\alpha(\mathbf{x})]$ such that the relations $\omega^\alpha=A_\beta^\alpha\,dg^\beta$ are to be satisfied. Thus, the exterior system is completely integrable. □

We have to pay attention to the fact that the functions g^α and the matrix $[A_\beta^\alpha]$ cannot be determined uniquely. Provided that the functions

$h^\alpha = h^\alpha(g^1, g^2, \ldots, g^r)$ are so chosen that the condition det $(\partial h^\alpha/\partial g^\beta) \neq 0$ is satisfied, the forms dh^α become linearly independent and we find that

$$\mathbf{i}_{V_a}(dh^\alpha) = \mathbf{i}_{V_a}\Big(\frac{\partial h^\alpha}{\partial g^\beta} dg^\beta\Big) = \frac{\partial h^\alpha}{\partial g^\beta}\mathbf{i}_{V_a}(dg^\beta) = 0.$$

Hence, we can write $\omega^\alpha = B^\alpha_\beta\, dh^\beta$. But, it is easily verified that the relation

$$A^\alpha_\beta = B^\alpha_\gamma \frac{\partial h^\gamma}{\partial g^\beta}$$

must be satisfied.

The generalisation of the Frobenius theorem to ideals generated by forms of diverse degrees is given below.

Theorem 5.13.5. *Let $\mathcal{I}$ be a closed ideal of the exterior algebra $\Lambda(M)$ generated by forms of various degrees. If the dimension of the characteristic subspace $\mathcal{S}(\mathcal{I})$ of $\mathcal{I}$ is $m - r$, then there exist r functionally independent functions $g^\alpha(\mathbf{x}) \in \Lambda^0(M), \alpha = 1, \ldots, r$ and the ideal $\mathcal{I}$ is contained in the closed ideal generated by forms $dg^\alpha \in \Lambda^1(M), \alpha = 1, \ldots, r$.*

Since $\mathcal{I}$ is closed, its characteristic subspace is an involutive distribution in view of Theorem 5.13.2. Hence, the characteristic basis vectors $V_a = v^i_a \partial_i, a = r + 1, \ldots, m$ can be so chosen that $[V_a, V_b] = 0$. Thereby, following the path leading to Theorem 5.13.4 we can determine independent functions $g^\alpha(\mathbf{x}), \alpha = 1, \ldots, r$ satisfying the relations $V_a(g^\alpha) = \mathbf{i}_{V_a}(dg^\alpha) = 0$. Let $\mathcal{J}(dg^\alpha)$ denote the completely integrable closed ideal generated by forms $dg^\alpha \in \Lambda^1(M)$. Then Theorem 5.6.4 implies that $\mathcal{I} \subseteq \mathcal{J}(dg^\alpha)$. Since the ideal $\mathcal{J}(dg^\alpha)$ is generated by 1-forms, it is the largest ideal admitting $\mathcal{S}(\mathcal{I})$ as its characteristic subspace. In this case, if $\omega \in \mathcal{I}$ then there are suitable forms $\gamma_\alpha \in \Lambda(M)$ so that one is able to write $\omega = \gamma_\alpha \wedge dg^\alpha$. Consequently, if we introduce $(m - r)$-dimensional *characteristic submanifolds* prescribed by the relations $g^\alpha(\mathbf{x}) = c^\alpha = \mathit{constant}, \alpha = 1, \ldots, r$ obtained through integration of the following sets of ordinary differential equations

$$\frac{dx^1}{v^1_a} = \frac{dx^2}{v^2_a} = \cdots = \frac{dx^m}{v^m_a} \quad \text{or} \quad \frac{dx^i}{dt} = v^i_a;\ a = r + 1, \ldots, m$$

which determine the integral curves of characteristic vector fields, then it is quite clear that those manifolds are also a solution of the ideal $\mathcal{I}$. □

It is now obvious that a solution of a closed ideal $\mathcal{I}$ provided by Theorem 5.13.5 corresponds to a solution determined by maximal number of independent functions g^α. Hence, this approach cannot usually reveal all solutions of the ideal $\mathcal{I}$. It might be quite possible that there exist submanifolds annihilating the ideal $\mathcal{I}$ whose dimensions are larger than $m - r$ so that they can be determined by means of a smaller amount of functions, but not solving the ideal $\mathcal{J}$. However, it is impossible to offer a systematic approach based on the above procedure to access such kinds of solutions corresponding, most probably, to much more realistic situations. Unfortunately, we can frequently produce only rather trivial solutions by applying Theorem 5.13.5.

Example 5.13.2. We build an ideal of the exterior algebra $\Lambda(\mathbb{R}^4)$ by forms $\omega^1 = dx^1 + x^2\,dx^3 + dx^4 \in \Lambda^1(\mathbb{R}^4)$, $\omega^2 = x^2\,dx^2 \wedge dx^3 \in \Lambda^2(\mathbb{R}^4)$. Since $d\omega^1 = dx^2 \wedge dx^3 = \omega^2/x^2$ and $d\omega^2 = 0$, the ideal $\mathcal{I}(\omega^1, \omega^2)$ is closed and its characteristic vectors must satisfy the relations

$$\begin{aligned}\mathbf{i}_V(\omega^1) &= v^1 + x^2\,v^3 + v^4 = 0\\ \mathbf{i}_V(\omega^2) &= x^2\,(v^2 dx^3 - v^3 dx^2) = \lambda(dx^1 + x^2\,dx^3 + dx^4)\end{aligned}$$

from which we obtain

$$\lambda = v^2 = 0,\ v^3 = 0,\ v^4 = -\,v^1$$

and $V = v^1(\partial_1 - \partial_4)$. Thus the basis vector of 1-dimensional characteristic subspace can be chosen as $V_4 = \partial_1 - \partial_4$. Therefore, the solution of the differential equation

$$V_4(f) = \frac{\partial f}{\partial x^1} - \frac{\partial f}{\partial x^4} = 0$$

yields $f = f(x^1 + x^4, x^2, x^3)$ and we have $g^1 = x^1 + x^4$, $g^2 = x^2, g^3 = x^3$. Hence, 1-dimensional solution submanifolds are determined by $x^1 + x^4 = c^1$, $x^2 = c^2, x^3 = c^3$. We immediately observe that if we define the forms $dg^1 = dx^1 + dx^4$, $dg^2 = dx^2$, $dg^3 = dx^3$ we can write $\omega^1 = dg^1 + x^2\,dg^3$, $\omega^2 = x^2\,dg^2 \wedge dg^3$ meaning that $\mathcal{I}(\omega^1, \omega^2) \subset \mathcal{J}(dg^1, dg^2, dg^3)$. However, we can easily check that $\mathcal{I} \neq \mathcal{J}$. For instance, forms like $g(\mathbf{x})\,dg^2$ does not belong to $\mathcal{I}(\omega^1, \omega^2)$.

Let us now search for a larger, say 2-dimensional solution submanifold of the same ideal. We designate the mapping $\phi : \mathbb{R}^2 \to \mathbb{R}^4$ by functions $x^i = f^i(x, y), i = 1, \dots, 4$. The exterior equations

$$\begin{aligned}\phi^*\omega^1 &= \Big(\frac{\partial f^1}{\partial x} + f^2\frac{\partial f^3}{\partial x} + \frac{\partial f^4}{\partial x}\Big)\,dx + \Big(\frac{\partial f^1}{\partial y} + f^2\frac{\partial f^3}{\partial y} + \frac{\partial f^4}{\partial y}\Big)\,dy = 0\\ \phi^*\omega^2 &= f^2\Big(\frac{\partial f^2}{\partial x}\frac{\partial f^3}{\partial y} - \frac{\partial f^2}{\partial y}\frac{\partial f^3}{\partial x}\Big)\,dx \wedge dy = 0\end{aligned}$$

can only be satisfied if we choose the functions f^i as solutions of the first order partial differential equations

$$\frac{\partial f^1}{\partial x}+f^2\frac{\partial f^3}{\partial x}+\frac{\partial f^4}{\partial x}=0,\ \frac{\partial f^1}{\partial y}+f^2\frac{\partial f^3}{\partial y}+\frac{\partial f^4}{\partial y}=0,$$
$$f^2\Big(\frac{\partial f^2}{\partial x}\frac{\partial f^3}{\partial y}-\frac{\partial f^2}{\partial y}\frac{\partial f^3}{\partial x}\Big)=0.$$

For a simple example, we choose to take $f^2 = 0$. Then the solution is easily found to be

$$f^1 = f(x,y),\ f^2 = 0,\ f^3 = g(x,y),\ f^4 = c - f(x,y)$$

where$f$ and g are arbitrary functions and c is an arbitrary constant. ■

We know that if the ideal $\mathcal{I}(\omega^\alpha)$ is not closed, then a closed ideal containing $\mathcal{I}$ is its closure $\overline{\mathcal{I}}(\omega^\alpha, d\omega^\alpha)$.

Theorem 5.13.6. *Let an ideal of the exterior algebra $\Lambda(M)$ be $\mathcal{I}$ and its closure be $\overline{\mathcal{I}} = \mathcal{I} \cup d\mathcal{I}$. If a mapping $\phi : S \to M$ is a solution of the ideal $\mathcal{I}$, then it is likewise a solution of its closure $\overline{\mathcal{I}}$.*

When $\omega \in \mathcal{I}$, we have $\omega, d\omega \in \overline{\mathcal{I}}$. If $\phi^*\omega = 0$, then we get $\phi^*(d\omega) = d(\phi^*\omega) = 0$ according to Theorem 5.8.2. Thus the ideal $\overline{\mathcal{I}}$ is also annihilated under this mapping. In other words, characteristic manifolds of an ideal $\mathcal{I}$ and characteristic manifolds of its closure are the same. □

Theorem 5.13.5 and 5.13.6 help us to specify some solutions of a system of exterior equations generating an ideal that is not closed by means of characteristic manifolds. Let us suppose that the dimension of the characteristic subspace $\mathcal{S}(\overline{\mathcal{I}})$ of the closure $\overline{\mathcal{I}}$ of the ideal $\mathcal{I}$ is $m-r$. Then we can find in the usual way functions $g^\alpha(\mathbf{x}) \in \Lambda^0(M), \alpha = 1,\dots,r$ enabling us to write $\mathcal{I} \subset \overline{\mathcal{I}} \subseteq \mathcal{J}(dg^\alpha)$. Hence the equations $g^\alpha(\mathbf{x}) = c^\alpha$ produce $(m-r)$-dimensional characteristic manifolds annihilating the ideal $\mathcal{I}$. But, since $d\mathcal{I} \not\subset \mathcal{I}$, we are required to enlarge the ideal in order to close it, and consequently, to reduce the dimension of the characteristic subspace. Thus, we are compelled to keep the completely integrable system, in which the ideal $\mathcal{I}$ is embedded, larger than it was necessary.

Even if an ideal $\mathcal{I}$ is not closed, it can be placed into a completely integrable system if its characteristic subspace is 1-dimensional because of the fact that such a subspace constitutes trivially a Lie algebra.

Example 5.13.3. We construct an ideal of the exterior algebra $\Lambda(\mathbb{R}^4)$ by the forms $\omega^1 = dx^1 - x^2\,dx^3$, $\omega^2 = x^4\,dx^1 \wedge dx^3 - x^1 dx^2 \wedge dx^4$. We then have

$$d\omega^1 = -\,dx^2 \wedge dx^3,\ d\omega^2 = \Big(\frac{dx^1}{x^1}+\frac{dx^4}{x^4}\Big)\wedge\omega^2.$$

We obviously get $d\omega^2 \in \mathcal{I}(\omega^1,\omega^2)$. However, we can easily verify that we find $d\omega^1 \notin \mathcal{I}(\omega^1, \omega^2)$. Hence, the closure of $\mathcal{I}$ is $\overline{\mathcal{I}}(\omega^1, \omega^2, d\omega^1)$. Thus, the characteristic subspace of $\overline{\mathcal{I}}$ is prescribed by the equations

$$\begin{aligned} v^1 - x^2 v^3 &= 0 \\ x^4(v^1 dx^3 - v^3 dx^1) - x^1(v^2 dx^4 - v^4 dx^2) &= \lambda(dx^1 - x^2\, dx^3) \\ -\,(v^2 dx^3 - v^3 dx^2) &= \mu(dx^1 - x^2\, dx^3) \end{aligned}$$

whose solution yields $\lambda = \mu = 0$ and $v^1 = v^2 = v^3 = v^4 = 0$. We thus obtain $V = 0$ so that the characteristic subspace is the zero space. The ideal generated by functions $g^i = x^i$ is just $\mathcal{J}(dx^i) = \Lambda(\mathbb{R}^4)$. Hence, we can only get trivial information about the solution. On the other hand, the characteristic subspace of $\mathcal{I}$ is prescribed by the equations

$$\begin{aligned} v^1 - x^2 v^3 &= 0, \\ x^4(v^1 dx^3 - v^3 dx^1) - x^1(v^2 dx^4 - v^4 dx^2) &= \lambda(dx^1 - x^2\, dx^3) \end{aligned}$$

whose solution is $\lambda = -\,x^4 v^3$ and $v^1 = x^2 v^3$, $v^2 = v^4 = 0$. Thus the characteristic subspace is 1-dimensional and is spanned by the vector $V_4 = x^2\partial_1 + \partial_3$. The solution of the partial differential equation $V_4(f) = 0$ is readily obtained as $f = f(g^1, g^2, g^3)$ where we define $g^1 = x^1 - x^2 x^3$, $g^2 = x^2$, $g^3 = x^4$. In this case, we can write $\mathcal{I}(\omega^1, \omega^2) \subset \mathcal{J}(dg^1, dg^2, dg^3)$. Indeed, the relations

$$\begin{aligned} \omega^1 &= dg^1 + x^3 dg^2, \\ \omega^2 &= -\,\frac{x^4}{x^2}\, dx^1 \wedge dg^1 - \frac{x^3 x^4}{x^2}\, dx^1 \wedge dg^2 - x^1 dg^2 \wedge dg^3 \end{aligned}$$

can easily be verified. ■

If we have managed to determine a resolvent mapping for an ideal, new resolvent mappings may be created via an isovector field of that ideal.

Theorem 5.13.7. *Let $\mathcal{I}$ be an ideal of the exterior algebra $\Lambda(M)$ and $\phi : S \to M$ be a resolvent mapping for that ideal. If the vector field V is an isovector field of the ideal, then the flow generated by V transforms ϕ into a 1-parameter family of resolvent mappings.*

If $\phi : S \to M$ is a resolvent mapping, then we have $\omega|_S = \phi^*\omega = 0$ for all $\omega \in \mathcal{I}$. If we further assume that V is an isovector, this implies that $\pounds_V \omega \in \mathcal{I}$ for all $\omega \in \mathcal{I}$. We denote the flow $\psi_V(t) : M \to M$ generated by the isovector field V by $\psi_V(t)(p) = e^{tV}(p)$ and define the 1-parameter family of mappings $\phi_V(t) : S \to M$ as follows

$$\phi_V(t) = \psi_V(t) \circ \phi = e^{tV} \circ \phi.$$

On utilising (5.11.14), we obtain

$$\begin{aligned}\phi_V(t)^*\omega &= (e^{tV}\circ\phi)^*\omega = \left[\phi^*\circ(e^{tV})^*\right]\omega = \phi^*\circ(e^{tV})^*\omega\\ &= \phi^*\omega^* = \phi^*(e^{t\pounds_V}\omega)\end{aligned}$$

for all $\omega\in\mathcal{I}$. However, due to the relation $e^{t\pounds_V}\omega\in\mathcal{I}$ we find $\phi_V(t)^*\omega = 0$. Therefore, each member of the 1-parameter family of mappings $\phi_V(t)$ is also a solution of the ideal $\mathcal{I}$. □

Example 5.13.4. We have already determined the isovector fields of the ideal $\mathcal{I}(x\,dy + y\,dz)$ of the exterior algebra $\Lambda(\mathbb{R}^3)$ in Example 5.12.1. For a tangible example, let us choose $F = xz$. In this case the components of the isovector field become

$$v^x = \frac{x(x+z)}{y},\quad v^y = z,\quad v^z = 0.$$

The flow created by this vector field is found as the solution of the ordinary differential equations

$$\frac{d\overline{x}}{dt} = \frac{\overline{x}(\overline{x}+\overline{z})}{\overline{y}},\quad \frac{d\overline{y}}{dt} = \overline{z},\quad \frac{d\overline{z}}{dt} = 0$$

under the initial conditions $\overline{x}(0) = x, \overline{y}(0) = y, \overline{z}(0) = z$. Hence, the mapping $\psi_V(t)$ is determined by

$$\overline{x}(t) = \frac{x(y+zt)}{y-xt},\quad \overline{y}(t) = zt + y,\quad \overline{z}(t) = z.$$

We shall now look for a 1-dimensional solution of the exterior equation $\omega = x\,dy + y\,dz = 0$ in the form $x = \phi^1(u), y = \phi^2(u), z = \phi^3(u)$. Then $\phi^*\omega = 0$ ends up in the equation

$$\phi^1\frac{d\phi^2}{du} + \phi^2\frac{d\phi^3}{du} = 0.$$

In this situation, the family of resolvent mappings $\phi_V(t) = \psi_V(t)\circ\phi$ is designated by

$$\begin{aligned}&\phi_V^1(u;t) = \phi^1(u)\frac{\phi^2(u)+t\phi^3(u)}{\phi^2(u)-t\phi^1(u)},\\ &\phi_V^2(u;t) = \phi^2(u)+t\phi^3(u),\qquad \phi_V^3(u;t) = \phi^3(u).\end{aligned}$$

The mapping described by $x = \phi_V^1(u;t)$, $y = \phi_V^2(u;t)$, $z = \phi_V^3(u;t)$ is also a solution of the exterior equation $\omega = 0$ for each t. In fact, if we insert these relation into that equation, we obtain

$$\phi_V^1 \frac{d\phi_V^2}{du} + \phi_V^2 \frac{d\phi_V^3}{du} = \frac{\phi^2 + t\phi^3}{\phi^2 - t\phi^1}\Big(\phi^1 \frac{d\phi^2}{du} + \phi^2 \frac{d\phi^3}{du}\Big) = 0.$$

As a simple example, let us take $\phi^1 = -2c_1u^2, \phi^2 = c_2u, \phi^3 = c_1u^2$ where c_1 and c_2 are constants. The new family of solutions is then found to be

$$\phi_V^1 = -\frac{2c_1(c_2 + c_1tu)u^2}{c_2 + 2c_1tu}, \quad \phi_V^2 = (c_2 + c_1tu)u, \quad \phi_V^3 = c_1u^2. \qquad \blacksquare$$

5.14. FORMS DEFINED ON A LIE GROUP

Let G be a finite m-dimensional Lie group. We denote the exterior algebra on this smooth manifold by $\Lambda(G)$. We consider the left and right translations L_g and R_g on G defined by (3.3.1) and (3.3.2), respectively. These diffeomorphisms give rise to the mappings $L_g^* : \Lambda(G) \to \Lambda(G)$ and $R_g^* : \Lambda(G) \to \Lambda(G)$. If a form $\omega \in \Lambda^k(G)$ satisfies the relation

$$L_g^*\omega(g*h) = \omega(h) \text{ or } L_g^*\omega = \omega \tag{5.14.1}$$

for all $g, h \in G$, it is called a ***left-invariant form***. Because of the equality $L_g^{-1} = L_{g^{-1}}$, we infer that $(L_g^{-1})^* = L_{g^{-1}}^*$. Hence, it follows from (5.14.1) that we obtain

$$\omega(g*h) = L_{g^{-1}}^*\omega(h) \tag{5.14.2}$$

for a left-invariant form ω and for all $g, h \in G$. If we take $h = e$, (5.14.2) leads to

$$\omega(g) = L_{g^{-1}}^*\omega(e) \tag{5.14.3}$$

for all $g \in G$. Consequently, all left-invariant k-forms are generated by forms $\omega(e) \in \Lambda^k(G)$ defined on the tensor product $\bigotimes_k T_e^*(G)$ at the identity element $e \in G$. Thus, left-invariant 1-forms are produced by 1-forms in the dual space $T_e^*(G)$. Since the dimension of the vector space $T_e^*(G)$ is m, then there are exactly m linearly independent left-invariant 1-forms and the entire left-invariant 1-forms are expressible as their linear combinations. If we denote a basis of $T_e^*(G)$ by $\omega^1, \omega^2, \ldots, \omega^m$, we can then express a form $\omega(e) \in \Lambda^k(G)$ as follows

$$\omega(e) = \frac{1}{k!}\,\omega_{i_1i_2\cdots i_k}\omega^{i_1} \wedge \omega^{i_2} \wedge \cdots \wedge \omega^{i_k}$$

where $\omega_{i_1 i_2 \cdots i_k} \in \mathbb{R}$ are completely antisymmetric *constant* coefficients. According to the relation (5.14.3), any left-invariant k-form is extracted from the foregoing form with constant coefficients. Similarly, right-invariant forms are defined as

$$R_g^* \omega = \omega \tag{5.14.4}$$

for all $g \in G$ and we write $\omega(g) = R_{g^{-1}}^* \omega(e)$. Hence, right-invariant forms are also generated by m linearly independent 1-forms chosen from the dual space $T_e^*(G)$. The relation $L_g \circ R_g = R_g \circ L_g$ [*see* (3.3.3)] leads of course to $R_g^* \circ L_g^* = L_g^* \circ R_g^*$. Therefore, if ω is a left-invariant form we find that

$$L_g^*(R_g^* \omega) = R_g^*(L_g^* \omega) = R_g^* \omega.$$

Thus $R_g^* \omega$ is a left-invariant form. In the same way, If ω is a right-invariant form, then $L_g^* \omega$ turns out to be a right-invariant form.

Theorem 5.14.1. *If ω is a left (right) invariant form, then $d\omega$ is also a left (right) invariant form.*

According to Theorem 5.8.2, we obtain

$$L_g^* d\omega = d L_g^* \omega = d\omega$$

for all $g \in G$. Similarly, we get $R_g^* d\omega = d\omega$. □

Theorem 5.14.2. *Let G and H be Lie groups and $\phi : G \to H$ be a Lie group homomorphism. Then the pull-back operator $\phi^* : \Lambda(H) \to \Lambda(G)$ transports the left-invariant forms in H to the left-invariant forms in G.*

Let $\omega \in \Lambda(H)$ be a left-invariant form. Since ϕ is a group homomorphism, we readily obtain

$$L_g^*(\phi^* \omega) = (\phi \circ L_g)^* \omega = (L_{\phi(g)} \circ \phi)^* \omega = \phi^*(L_{\phi(g)}^* \omega) = \phi^* \omega$$

[*see* p. 188]. This implies that the form $\phi^* \omega$ is left-invariant. The same property is also valid for right-invariant forms. □

The Lie algebra $\mathfrak{g}$ of the Lie group G that consist of left-invariant vectors is designated by the tangent space $T_e(G)$ and left-invariant 1-forms are elements of the dual space $T_e^*(G)$. Hence, when we choose a basis V_1, $V_2, \dots, V_m$ in $\mathfrak{g} = T_e(G)$, we can find a reciprocal basis $\omega^1, \omega^2, \dots, \omega^m$ in $\mathfrak{g}^* = T_e^*(G)$ such that we get $\omega^i(V_j) = \delta_j^i$.

Theorem 5.14.3. *A form $\omega \in \Lambda^k(G)$ is left-invariant if and only if the function $\omega(V_1, V_2, \dots, V_k)$ is constant for every k left-invariant vector fields $V_1, V_2, \dots, V_k$.*

Let ω be a left-invariant k-form. We can thus write $L_g^* \omega(g{*}h) = \omega(h)$ and $dL_g(V_i|_h) = V_i|_{g*h}$, $i = 1, \dots, k$. (5.7.1) then leads to

$$L_g^*\omega|_{g*h}(V_1|_h,\dots,V_k|_h) = \omega|_{g*h}\big(dL_g(V_1|_h),\dots,dL_g(V_k|_h)\big) \tag{5.14.5}$$

from which we obtain

$$\omega|_h(V_1|_h,\dots,V_k|_h) = \omega|_{g*h}(V_1|_{g*h},\dots,V_k|_{g*h})$$

since $V_1,\dots,V_k$ are left-invariant vectors. If we take $h = e$, then for every $g \in G$ we find that

$$\omega|_g(V_1|_g,\dots,V_k|_g) = \omega|_e(V_1|_e,\dots,V_k|_e) = \mathit{constant}. \tag{5.14.6}$$

Conversely, if the function $\omega(V_1, V_2,\dots,V_k)$ is constant for every k left-invariant vector fields $V_1, V_2,\dots,V_k$, then (5.14.5) yields

$$L_g^*\omega(g)(V_1|_e,\dots,V_k|_e) = \omega(e)(V_1|_e,\dots,V_k|_e)$$

whence we deduce that the relation $L_g^*\omega(g) = \omega(e)$, that is, ω is a left-invariant form. □

The left-invariant 1-forms engendering the dual $\mathfrak{g}^*$ of the Lie algebra $\mathfrak{g}$ of the Lie group G are called ***Maurer-Cartan forms*** [German mathematician Ludwig Maurer (1859-1927)]. So Theorem 5.14.3 implies that the function $\omega(V)$ remains constant for fields $\omega \in \mathfrak{g}^*$ and $V \in \mathfrak{g}$.

Theorem 5.14.4. *Let G be a Lie group and $\theta^i \in \mathfrak{g}^*, i = 1,\dots,m$ be a basis for left-invariant 1-forms. In this case, the following Maurer-Cartan structure equations are satisfied*

$$d\theta^k = -\frac{1}{2}c_{ij}^k\,\theta^i\wedge\theta^j = -\sum_{1\le i<j\le m} c_{ij}^k\,\theta^i\wedge\theta^j. \tag{5.14.7}$$

where $c_{ij}^k = -c_{ji}^k$ are real constants. The constants c_{ij}^k are the same as the structure constants of Lie algebra $\mathfrak{g}$.

According to Theorem 5.14.1, if a basis form θ^k is left-invariant, then its exterior derivative $d\theta^k$ is likewise left-invariant. Therefore, in terms of basis in the dual space $\mathfrak{g}^*$ we can write

$$d\theta^k = -\frac{1}{2}b_{ij}^k\,\theta^i\wedge\theta^j,\quad i,j,k = 1,\dots,m$$

with constant coefficients b_{ij}^k. These numbers ought to satisfy naturally the antisymmetry conditions $b_{ij}^k = -b_{ji}^k$. On the other hand, we get

$$0 = d^2\theta^k = -\frac{1}{2}b_{ij}^k(d\theta^i\wedge\theta^j - \theta^i\wedge d\theta^j)$$

$$
\begin{aligned}
&= \frac{1}{4} b_{ij}^k (b_{lm}^i \theta^l \wedge \theta^m \wedge \theta^j - b_{lm}^j \theta^i \wedge \theta^l \wedge \theta^m) \\
&= \frac{1}{4} b_{ij}^k b_{lm}^i \theta^l \wedge \theta^m \wedge \theta^j - \frac{1}{4} b_{ji}^k b_{lm}^i \theta^l \wedge \theta^m \wedge \theta^j \\
&= \frac{1}{2} b_{ij}^k b_{lm}^i \theta^l \wedge \theta^m \wedge \theta^j = \frac{1}{2} b_{i[j}^k b_{lm]}^i \theta^l \wedge \theta^m \wedge \theta^j.
\end{aligned}
$$

Thus the coefficients b_{ij}^k must satisfy the relations

$$
\frac{3!}{2} b_{i[j}^k b_{lm]}^i = b_{lm}^i b_{ij}^k + b_{mj}^i b_{il}^k + b_{jl}^i b_{im}^k = 0
$$

dictated by the Jacobi identity. Let $V_i \in \mathfrak{g}, i = 1, \ldots, m$ be the reciprocal basis of the Lie algebra with respect to the forms θ^i, that is, the relations $\theta^i(V_j) = \delta_j^i, i, j = 1, \ldots, m$ are to be satisfied. This basis vectors have to verify the relations $[V_i, V_j] = c_{ij}^k V_k$ where c_{ij}^k are structure constants of the Lie algebra $\mathfrak{g}$ with respect to the basis $\{V_i\}$ [*see* (3.3.9)]. In view of the relation (5.2.6), we can write $b_{ij}^k = -\,d\theta^k(V_i, V_j)$. Consider a 1-form $\omega = \omega_i dx^i$. The value of the form $d\omega = \omega_{i,j}\, dx^j \wedge dx^i$ on vector fields $U, V \in T(M)$ is given by

$$
d\omega(U, V) = \omega_{i,j}(u^j v^i - u^i v^j) = (\omega_{i,j} - \omega_{j,i}) u^j v^i.
$$

On the other hand, the relation

$$
U\big(\omega(V)\big) - V\big(\omega(U)\big) = (\omega_{i,j} - \omega_{j,i}) u^j v^i + \omega_i (v^i_{,j} u^j - u^i_{,j} v^j)
$$

leads immediately to

$$
d\omega\,(U, V) = U\big(\omega(V)\big) - V\big(\omega(U)\big) - \omega\,([U, V]). \qquad (5.14.8)
$$

Consequently, because of $\theta^k(V_i) = \delta_i^k, \theta^k(V_j) = \delta_j^k$ we obtain

$$
b_{ij}^k = -\,d\theta^k(V_i, V_j) = \theta^k([V_i, V_j]) = \theta^k(c_{ij}^l V_l) = c_{ij}^l \delta_l^k = c_{ij}^k. \qquad \square
$$

We can now prove the following theorem.

Theorem 5.14.5. *The structure constants of an m-dimensional Lie group vanish if and only if it is locally isomorphic to the group $\mathbb{R}^m$.*

(i). Let the Lie group G be isomorphic to the group $\mathbb{R}^m$. We have seen in Example 3.3.1 that the structure constants of $\mathbb{R}^m$ are zero. The relation (3.4.3) then requires that the structure constants of G are also zero so that G becomes an Abelian group.

(ii). Let the structure constants of the Lie group G be zero. Therefore, (5.14.6) gives $d\theta^k = 0, k = 1, \ldots, m$. *According to the Poincaré lemma,* *there are m smooth functions $\vartheta^k : U \to \mathbb{R}$ on the domain U of a local chart (U, φ) such that $\theta^k = d\vartheta^k$* [*see p.* 334]. We can choose those functions ϑ^k as coordinate functions. Since the forms θ^k are left-invariant, we obtain

$$L_g^*\theta^k(g*h) = \theta^k(h) = d\vartheta^k(h) = dh^k$$

for all $g, h \in G$. $g^k = \vartheta^k(g), h^k = \vartheta^k(h), k = 1, \ldots, m$ are coordinates of g and h. Furthermore, we can readily write $(g*h)^k = \vartheta^k(g*h) = \vartheta^k(L_g h) = \vartheta^k \circ L_g(h) = L_g^k(h)$. Then, on making use of Theorem 5.8.2 we get

$$\begin{aligned} L_g^*\theta^k(g*h) &= L_g^* d\vartheta^k(g*h) = L_g^* dL_g^k(h) = dL_g^* L_g^k(h) \\ &= d\big(L_g^k(h) \circ L_g\big) = dL_g^k(h)\big|_h = \frac{\partial L_g^k(h)}{\partial h^l}\, dh^l. \end{aligned}$$

If we compare the two expressions which we have found for $L_g^*\theta^k(g*h)$, then we deduce that

$$\frac{\partial L_g^k(h)}{\partial h^l}\, dh^l = dh^k \text{ or } \frac{\partial L_g^k(h)}{\partial h^l} = \delta_l^k.$$

It is quite easy to integrate these differential equations to obtain

$$L_g^k(h) = \Theta^k(g) + h^k. \tag{5.14.9}$$

$\Theta^k(g)$ are arbitrary functions. Since the functions ϑ^k are to be determined up to a constant, we can impose the restriction $\vartheta^k(e) = 0, k = 1, \ldots, m$ without loss of generality. Because $L_g(e) = g$, we get $L_g^k(e) = \vartheta^k(g) = g^k$ and when we evaluate the expression (5.14.9) for $h = e$, we end up with the relation $\Theta^k(g) = g^k$. Hence, we find that $\vartheta^k(g*h) = L_g^k(h) = g^k + h^k$. Let us next define the smooth function $\boldsymbol{\vartheta} = (\vartheta^1, \ldots, \vartheta^m) : U \to \mathbb{R}^m$ and the elements $\mathbf{g} = (g^1, \ldots, g^m) \in \mathbb{R}^m$ and $\mathbf{h} = (h^1, \ldots, h^m) \in \mathbb{R}^m$. We thus conclude that

$$\boldsymbol{\vartheta}(g*h) = \mathbf{g} + \mathbf{h} = \boldsymbol{\vartheta}(g) + \boldsymbol{\vartheta}(h). \tag{5.14.10}$$

This implies that the Lie group G is locally isomorphic to the group $\mathbb{R}^m$. $\square$

V. EXERCISES

5.1. We define on the manifold $\mathbb{R}^4$ with the coordinate cover (x, y, z, t) the following exterior forms

$$\omega^1 = y \cos t\, dx + e^x\, dy + t\, dz + (y - z)\, dt \in \Lambda^1(\mathbb{R}^4),$$

$$\omega^2 = \tan x\, dx \wedge dz + (y - z^3)\, dx \wedge dt + \sinh z\, dy \wedge dz \in \Lambda^2(\mathbb{R}^4),$$
$$\omega^3 = e^y\, dy \wedge dz \wedge dt - \cos y\, dx \wedge dy \wedge dz + x\ dx \wedge dz \wedge dt \in \Lambda^3(\mathbb{R}^4),$$
$$\omega^4 = (x^2 + t^3)\, dx \wedge dy \wedge dz \wedge dt \in \Lambda^4(\mathbb{R}^4).$$

Evaluate the exterior forms $\omega^1 \wedge \omega^3, \omega^1 \wedge \omega^2 + \omega^3$, $\omega^3 \wedge \omega^1 - \omega^2 \wedge \omega^2 + \omega^4$, $d\omega^2 - \omega^3 + \omega^2 \wedge \omega^1$, $d\omega^1 \wedge \omega^2 + d\omega^1 \wedge d\omega^1$, $d\omega^3 + d(\omega^1 \wedge \omega^2)$. The vector fields $U, V \in T(\mathbb{R}^4)$ are given by

$$U = y\frac{\partial}{\partial x} - z\frac{\partial}{\partial z} + \frac{\partial}{\partial t}, \quad V = x\frac{\partial}{\partial y} - t\frac{\partial}{\partial z}.$$

Find the forms $\mathbf{i}_U\omega^1$, $\mathbf{i}_V\omega^2$, $\mathbf{i}_U\omega^3$, $\mathbf{i}_V\omega^4$, $\mathbf{i}_U(d\omega^1 + \omega^2)$, $\mathbf{i}_V\mathbf{i}_U\omega^4 + \mathbf{i}_V(d\omega^2)$, $d(\mathbf{i}_U\omega^2) + \mathbf{i}_U(d\omega^2)$, $\pounds_V\omega^1$, $\pounds_U\omega^2$, $\pounds_V\omega^3$, $\pounds_U\omega^4$, $\pounds_U\mathbf{i}_V\omega^2 - \mathbf{i}_U\pounds_V\omega$.

5.2. Consider an exterior form $\omega = z\ dx - x\ dy + x\ dz \in \Lambda^1(\mathbb{R}^3)$ and a vector field $V = y\,\partial_x + z\,\partial_y + x\partial_z \in T(\mathbb{R}^3)$. Evaluate the forms $\pounds_V\omega$, $\pounds_V\pounds_V\omega$, $\pounds_V\pounds_V\pounds_V\omega$, $\pounds_V\pounds_V\pounds_V\pounds_V\omega$ and $\exp(t\pounds_V)\,\omega$.

5.3. Determine vector fields $V \in T(\mathbb{R}^4)$ in such a way that they satisfy the relations (a) $\mathbf{i}_V\omega^1 = 0$, (b) $\mathbf{i}_V\omega^2 = 0$, (c) $\mathbf{i}_V\omega^3 = 0$, (d) $\mathbf{i}_V\omega^4 = 0$. This amounts to say that they will be characteristic vectors of those forms. Forms $\omega^1, \omega^2, \omega^3, \omega^4$ are defined in Exercise **5.1**.

5.4. Express the forms $\omega^1, \omega^2, \omega^3, \omega^4$ in Exercise **5.1** in terms of bases induced by the volume form $\mu = dx \wedge dy \wedge dz \wedge dt$.

5.5. Let $\{\theta^i\} \subset T^*(M)$ and $\{V_i\} \subset T(M), i = 1, \ldots, m$ be reciprocal basis vectors. Verify the equality

$$\mathbf{i}_{V_i}(\theta^{i_1} \wedge \cdots \wedge \theta^{i_k}) = \begin{cases} 0, & \text{if } i \neq i_r, r = 1, \ldots, k \\ (-1)^{r-1}\theta^{i_1} \wedge \cdots \wedge \theta^{i_{r-1}} \wedge \theta^{i_{r+1}} \wedge \cdots \wedge \theta^{i_k}, & \text{if } i = i_r \end{cases}$$

5.6. We define the mapping $\phi : \mathbb{R}^3 \to \mathbb{R}^4$ by the relations

$$x = u\cos v, \quad y = u\sin v, \quad z = w - 2, \quad t = uw.$$

(a) Find the pulled back forms $\phi^*\omega^1$, $\phi^*\omega^2$, $\phi^*\omega^3$, $\phi^*\omega^4$ [*see* Exercise **5.1**], (b) determine the range $\mathcal{R}(\phi) \subset \mathbb{R}^4$, (c) evaluate the inverse mapping $\phi^{-1} : \mathcal{R}(\phi) \to \mathbb{R}^3$, (d) find the vectors $\phi_*\partial_u$, $\phi_*\partial_v$ and $\phi_*\partial_w$, (d) If $\omega = dx \wedge dy \wedge dz$, then evaluate the forms $\phi^*\big(\mathbf{i}_{\phi_*\partial_u}(\omega)\big)$ and $\phi^*\big(\mathbf{i}_{\phi_*\partial_v}(\omega)\big)$.

5.7. A mapping $\phi : \mathbb{R}^2 \to \mathbb{R}^3$ is described by the relations $u = y^2$, $v = xy$, $w = x^3$. The vector fields $U, V \in T(\mathbb{R}^2)$ and the form $\omega \in \Lambda^2(\mathbb{R}^3)$ are given as follows:

$$U = y\frac{\partial}{\partial x} + x\frac{\partial}{\partial y}, \quad V = -\frac{\partial}{\partial x} + y^2\frac{\partial}{\partial y}, \quad \omega = u\,du \wedge dv - vw\,dv \wedge dw.$$

Evaluate the quantities $\phi^*\omega$, ϕ_*U, ϕ_*V, $(\phi^*\omega)(U, V)$, $\omega(\phi_*U, \phi_*V)$, $\mathbf{i}_U(\phi^*\omega), \mathbf{i}_V(\phi^*\omega)$, $\phi^*(\mathbf{i}_{\phi_*U}\omega)$, $\phi^*(\mathbf{i}_{\phi_*V}\omega)$.

5.8. Determine all mappings $\phi : \mathbb{R}^k \to \mathbb{R}^4, 1 \leq k \leq 4$ satisfying the relations (a)

$\phi^*\omega^1 = 0$, (b) $\phi^*\omega^2 = 0$, (c) $\phi^*\omega^3 = 0$, (d) $\phi^*\omega^4 = 0$ where the forms ω^1, $\omega^2, \omega^3, \omega^4$ are those given in Exercise **5.1**

5.9. Show that the form $\omega = z\,dx \wedge dy + yz\,dy \wedge dz + y\,dx \wedge dz \in \Lambda^2(\mathbb{R}^3)$ is closed. Determine a form $\Omega \in \Lambda^1(\mathbb{R}^3)$ such that $\omega = d\Omega$.

5.10. We define the isomorphisms $\phi : \mathbb{R}^3 \to \Lambda^1(\mathbb{R}^3)$, $\psi : \Lambda^0(\mathbb{R}^3) \to \Lambda^3(\mathbb{R}^3)$ by

$$\phi(\mathbf{U}) = \omega_{\mathbf{U}} = u_x\,dx + u_y\,dy + u_z\,dz,$$
$$\psi(f) = \omega_f = f\,dx \wedge dy \wedge dz$$

where $\mathbf{U} = (u_x, u_y, u_z) \in \mathbb{R}^3$ and $f \in \Lambda^0(\mathbb{R}^3)$. Verify that (a) $\omega_{\mathbf{U}\cdot\mathbf{V}} = \omega_{\mathbf{U}} \wedge {*\omega_{\mathbf{V}}} = {*\omega_{\mathbf{U}}} \wedge \omega_{\mathbf{V}}$, (b) $\omega_{\mathbf{U}\times\mathbf{V}} = *(\omega_{\mathbf{U}} \wedge \omega_{\mathbf{V}})$, (c) $\omega_{\operatorname{div}\mathbf{U}} = *\,d * \omega_{\mathbf{U}}$, (d) $\omega_{\operatorname{curl}\mathbf{U}} = *\,d\omega_{\mathbf{U}}$ and show that (e) $\mathbf{U} \cdot \operatorname{curl}\mathbf{U} = 0$ if $d\omega_{\mathbf{U}} \wedge \omega_{\mathbf{U}} = 0$.

5.11. Verify the following relations in $\mathbb{R}^3$:

(a) $\mathbf{U} \times (\mathbf{V} \times \mathbf{W}) = (\mathbf{U} \cdot \mathbf{W})\mathbf{V} - (\mathbf{U} \cdot \mathbf{V})\mathbf{W}$

(b) $\mathbf{U} \cdot (\mathbf{V} \times \mathbf{W}) = \mathbf{V} \cdot (\mathbf{W} \times \mathbf{U}) = \mathbf{W} \cdot (\mathbf{U} \times \mathbf{V})$

(c) $\boldsymbol{\nabla}(fg) = g\boldsymbol{\nabla} f + f\boldsymbol{\nabla} g$

(d) $\boldsymbol{\nabla}(\mathbf{U} \cdot \mathbf{V}) = \mathbf{U} \times \operatorname{curl}\mathbf{V} + \mathbf{V} \times \operatorname{curl}\mathbf{U} + (\mathbf{V} \cdot \boldsymbol{\nabla})\mathbf{U} + (\mathbf{U} \cdot \boldsymbol{\nabla})\mathbf{V}$

(e) $\operatorname{div}(f\mathbf{U}) = f \operatorname{div}\mathbf{U} + \mathbf{U} \cdot \boldsymbol{\nabla} f$

(f) $\operatorname{curl}(\mathbf{U} \times \mathbf{V}) = (\operatorname{div}\mathbf{V})\mathbf{U} - (\operatorname{div}\mathbf{U})\mathbf{V} + (\mathbf{V} \cdot \boldsymbol{\nabla})\mathbf{U} - (\mathbf{U} \cdot \boldsymbol{\nabla})\mathbf{V}$

(g) $\operatorname{div}(\operatorname{curl}\mathbf{U}) = 0$, $\operatorname{curl}(\boldsymbol{\nabla} f) = \mathbf{0}$, $\operatorname{div}(\boldsymbol{\nabla} f \times \boldsymbol{\nabla} g) = 0$

5.12. If M is a Riemannian manifold and $V_1, V_2 \in T(M)$, show that

$$\operatorname{div}[V_1, V_2] = V_1(\operatorname{div} V_2) - V_2(\operatorname{div} V_1).$$

5.13. Let $\omega \in \Lambda^k(M)$ and $V_0, V_1, \dots, V_k \in T(M)$. Verify the relation

$$d\omega(V_0, V_1, \dots, V_k) = \sum_{i=0}^{k} (-1)^i V_i\big(\omega(V_0, V_1, \dots, V_{i-1}, V_{i+1}, \dots, V_k)\big)$$
$$+ \sum_{0 \le i < j \le k} (-1)^{i+j} \omega([V_i, V_j], V_0, V_1, \dots, V_{i-1}, V_{i+1}, \dots, V_{j-1}, V_{j+1}, \dots, V_k).$$

5.14. Let $\omega \in \Lambda^k(M)$ and $V, V_1, \dots, V_k \in T(M)$. Verify the relation

$$\pounds_V\big(\omega(V_1, \dots, V_k)\big) =$$
$$(\pounds_V \omega)(V_1, \dots, V_k) + \sum_{i=1}^{k} \omega(V_1, \dots, V_{i-1}, [V, V_i], V_{i+1}, \dots, V_k).$$

5.15. When $U, V \in T(M)$, verify the validity of the operator identity

$$\pounds_U \circ \mathbf{i}_V - \pounds_V \circ \mathbf{i}_U - \mathbf{i}_{[U,V]} = [d, \mathbf{i}_U \circ \mathbf{i}_V].$$

5.16. Provided that $g \in \Lambda^0(M)$, $dg \neq 0$, show that a function $f \in \Lambda^0(M)$ can be expressed in the form $f(p) = F\big(g(p)\big)$ if it meets the condition $df \wedge dg = 0$. F is a smooth function.

5.17. Let us assume that $g^1, g^2, \dots, g^r \in \Lambda^0(M)$, $dg^1 \wedge dg^2 \wedge \cdots \wedge dg^r \neq 0$. Show that if a function $f \in \Lambda^0(M)$ satisfies the relation $df \wedge dg^1 \wedge \cdots \wedge dg^r = 0$, then it is expressible in the form $f = F(g^1, g^2, \ldots, g^r)$. F is a smooth function of its arguments.

5.18. Let us assume that $g^1, \ldots, g^r \in \Lambda^0(M)$ and $dg^1 \wedge \cdots \wedge dg^r \neq 0$. If we can write for a function $h \in \Lambda^0(M)$ the relation $dh = f_1 dg^1 + \cdots + f_r dg^r$ with functions $f_1, \ldots, f_r \in \Lambda^0(M)$, then show that the relations $h = h(g^1, \ldots, g^r)$ and $f_i = \dfrac{\partial h}{\partial g^i}$, $i = 1, \ldots, r$ must be valid.

5.19. Show that $d{*}df = -{*}\Delta f = -(\Delta f)\mu$ if $f \in \Lambda^0(M)$.

5.20. Show that $d\big(f{*}(dg)\big) = df \wedge {*}(dg) - (f\Delta g)\mu$ if $f, g \in \Lambda^0(M)$.

5.21. Let $\boldsymbol{\mathcal{V}} = V^1 \wedge \cdots \wedge V^k \in \mathfrak{X}^k(M)$ and $\omega \in \Lambda^{k+l}(M)$. We define the *interior product* of the form ω with $\boldsymbol{\mathcal{V}}$ in such a manner that the following relation would be satisfied for all vectors $V^{k+1}, \ldots, V^{k+l} \in \mathfrak{X}(M)$:

$$(\boldsymbol{i}_{\mathcal{V}}\,\omega)(V^{k+1}, \ldots, V^{k+l}) = \omega(V^1, \ldots, V^k, V^{k+1}, \ldots, V^{k+l}).$$

Show that this interior product is well defined and prove the operator equality $\boldsymbol{i}_{\mathcal{U}\wedge\mathcal{V}} = \boldsymbol{i}_{\mathcal{U}} \circ \boldsymbol{i}_{\mathcal{V}}$.

5.22. For $\boldsymbol{\mathcal{U}} \in \mathfrak{X}^k(M)$ and $\boldsymbol{\mathcal{V}} \in \mathfrak{X}^l(M)$ verify the equality

$$\boldsymbol{i}_{\langle\mathcal{U},\mathcal{V}\rangle}\omega = (-1)^{(k+1)\,l}\boldsymbol{i}_{\mathcal{U}} d\boldsymbol{i}_{\mathcal{V}}\omega + (-1)^k \boldsymbol{i}_{\mathcal{V}} d\boldsymbol{i}_{\mathcal{U}}\omega - \boldsymbol{i}_{\mathcal{U}}\boldsymbol{i}_{\mathcal{V}} d\omega.$$

5.23. Assume that $\boldsymbol{\mathcal{V}} = V^1 \wedge V^2 \in \mathfrak{X}^2(M)$ and $f, g, h \in \Lambda^0(M)$. (a) Show that

$$\boldsymbol{i}_{\mathcal{V}}\,(df \wedge dg \wedge dh) = \boldsymbol{i}_{\mathcal{V}}\,(df \wedge dg)\,dh + \boldsymbol{i}_{\mathcal{V}}\,(dg \wedge dh)\,df + \boldsymbol{i}_{\mathcal{V}}\,(dh \wedge df)\,dg.$$

(b) We define the mapping $\{,\} : \Lambda^0(M) \times \Lambda^0(M) \to \Lambda^0(M)$ by the relation

$$\{f, g\} = \boldsymbol{i}_{\mathcal{V}}\,(df \wedge dg).$$

We also name this mapping [*see* p. 707] as the *Poisson bracket* [French mathematician Siméon Denis Poisson (1781-1840)]. Show that this mapping is bilinear and antisymmetric. Prove the identity

$$\{\{f, g\}, h\} + \{\{g, h\}, f\} + \{\{h, f\}, g\} = \boldsymbol{i}_{\langle\mathcal{V},\mathcal{V}\rangle}(df \wedge dg \wedge dh)$$

and then demonstrate that the condition $\langle\boldsymbol{\mathcal{V}}, \boldsymbol{\mathcal{V}}\rangle = \mathbf{0}$ should be satisfied [*see* Exercise **4.17**] in order for this bracket to satisfy the Jacobi identity, and consequently, $\Lambda^0(M)$ endowed with the *product* $\{,\}$ to form a Lie algebra. (c) Show further that the bracket satisfies the equality

$$\{f, gh\} = \{f, g\}h + \{f, h\}g.$$

5.24. Let the vectors U and V be characteristic vectors of an exterior form $\omega \in \Lambda(M)$. Show that $\mathbf{i}_{[U,V]}(\omega) = \mathbf{i}_U \circ \mathbf{i}_V(d\omega)$. Thus, prove that characteristic vector fields of a form ω constitute a Lie subalgebra if and only if the condition $\mathbf{i}_U \circ \mathbf{i}_V(d\omega) = 0$ is satisfied for every pair of characteristic vectors U and V.

5.25. Let $\omega^1, \omega^2, \omega^3, \omega^4$ be the forms given in Exercise **5.1**. Determine the characteristic and isovector fields of the ideals $I(\omega^1,\omega^2)$, $I(\omega^1,\omega^2,\omega^3)$, $I(\omega^1,\omega^2,\omega^4)$. Find maximal solutions annihilating these ideals.

5.26. We define the forms $\omega^1,\omega^2 \in \Lambda^1(\mathbb{R}^4)$ as $\omega^1 = y\,dx + z\,dt, \omega^2 = z\,dy - y\,dz$. Show that the ideal $I(\omega^1,\omega^2)$ is closed. Determine its characteristic and isovector fields. Find the maximal solution annihilating this ideal.

5.27. Determine the characteristic subspaces and isovector fields of ideals

$$\begin{aligned}&I(y\,dx + x\,dy + y\,dz),\\&I\big((1+y^2)\,dx + x\,dy, x^3dz\big),\\&I(y\,dx + xz\,dy, dy\wedge dz)\end{aligned}$$

of $\Lambda(\mathbb{R}^3)$. Find maximal solutions annihilating these ideals.

5.28. M is a Riemannian manifold with a metric tensor $\mathcal{G}$. Show that any submanifold N of M can be made a Riemannian manifold equipped with a metric tensor $\mathcal{G}'$ defined by the relation $\mathcal{G}'(U,V) = \mathcal{G}(U,V)$ for all pair of vectors $U,V \in T(N) \subseteq T(M)$.

5.29. We consider a 4-dimensional manifold M with a coordinate cover $(x^i, f^i : i = 1,2)$ and define the following 1-forms

$$\begin{aligned}&\omega^i = df^i + f^j\alpha^i_j - \beta^i,\\&\alpha^i_j = \alpha^i_{jk}dx^k,\ \ \beta^i = \beta^i_j dx^j\end{aligned}$$

where $\alpha^i_{jk} = \alpha^i_{jk}(x^1,x^2)$ and $\beta^i_j = \beta^i_j(x^1,x^2)$ are given functions.

(a) Let S be a submanifold with the coordinate cover (x^1,x^2). Show that the requirements $\phi^*\omega^i = 0$ that a resolvent mapping $\phi : S \to M$ must satisfy give rise to the first order partial differential equations

$$\frac{\partial f^i}{\partial x^j} + \alpha^i_{kj}f^k = \beta^i_j$$

determining the functions $f^i = f^i(x^1,x^2)$.

(b) Show that the ideal $\mathcal{I}(\omega^1,\omega^2)$ is closed if only the relations

$$\begin{aligned}&d\alpha^i_j - \alpha^k_j\wedge\alpha^i_k = 0,\\&d\,\beta^i - \beta^j\wedge\alpha^i_j = 0\end{aligned}$$

are satisfied and these relations conduce to the integrability conditions

$$\begin{aligned}&\frac{\partial\alpha^i_{jn}}{\partial x^m} - \frac{\partial\alpha^i_{jm}}{\partial x^n} + \alpha^i_{kn}\alpha^k_{jm} - \alpha^i_{km}\alpha^k_{jn} = 0,\\&\frac{\partial\beta^i_j}{\partial x^k} - \frac{\partial\beta^i_k}{\partial x^j} + \beta^l_j\alpha^i_{lk} - \beta^l_k\alpha^i_{lj} = 0.\end{aligned}$$

(c) Show that if the conditions for the ideal $\mathcal{I}(\omega^1,\omega^2)$ to be closed are satisfied, then there exist functions $\Omega^i_j, u^j \in \Lambda^0(M)$ so that one can write

$$\omega^i = \Omega^i_j du^j$$

and solutions of the differential equations are found as

$$u^i(x^1, x^2, f^1, f^2) = constant.$$

5.30. G is a Lie group, $\omega \in \Lambda^1(G)$ is a left-invariant form, U and V are left-invariant vector fields. Show that

$$d\omega(U, V) = -\omega([U, V]).$$

CHAPTER VI

HOMOTOPY OPERATOR

6.1. SCOPE OF THE CHAPTER

In this section, we shall attempt to investigate certain fundamental properties of exterior differential forms in depth. The most powerful tool that we can employ for this purpose is the homotopy operator. However, this operator can only be defined on manifolds possessing a particular structure. This structure is treated in Sec. 6.2. A manifold is called locally contractible if every open set in its atlas can be smoothly shrunk to one of its interior points. This situation is realised if the homeomorphic image of that open set is a star-shaped region in the Euclidean space. In Sec. 6.3, the homotopy operator mapping the exterior algebra into itself is defined, its various properties are unravelled and the Poincaré lemma stating that every closed form is locally exact is demonstrated as a very important application of this operator. Sec. 6.4 is concerned with the proof that every exterior form is locally expressible as the sum of an exact form and an antiexact form occupying the kernel of the homotopy operator. Then the basic properties of antiexact forms are studied in detail. This leads to the conclusion that the entire exterior algebra is actually generated by antiexact forms. In Sec. 6.5, we inquire the effect of the change of the centre of contraction on the homotopy operator. We define in Sec. 6.6 the Darboux classes of 1-forms and introduce their canonical forms. Canonical forms of closed 2-forms are elicited by making use of the Poincaré lemma. We obtain the solution of an exterior differential equation in Sec. 6.7 and a system of exterior differential equations in Sec. 6.8 by resorting to properties of antiexact forms and the homotopy operator.

6.2. STAR-SHAPED REGIONS

Let M be a differentiable manifold. Let us take a point $p_0 \in M$ into account. If we can find a *smooth*, i.e., C^∞ function $h : M \times I \to M$ where $I = [0,1]$ denoted by $h(p;t) = h_t(p) \in M$ on which we shall impose the

Exterior Analysis, DOI: 10.1016/B978-0-12-415902-0.50006-2

restriction $h(p;0) = p_0$ and $h(p;1) = p$ for each point $p \in M$, then we say that the manifold M is ***contractible*** to the point p_0. Contractibility can also be defined locally. Let us consider a local chart (U, φ). We know that $U \subseteq M$ is an open set and $\varphi : U \to \mathbb{R}^m$ is a homeomorphism so that $V = \varphi(U) \subseteq \mathbb{R}^m$ is also an open set. Let us assume that the set U is contractible to a point $p_0 \in U$. If all charts of an atlas have this property, then the manifold M is called a ***locally contractible manifold***. Such a manifold cannot be shrunk smoothly to a point, but each one of the open sets covering this manifold is contractible to a point inside it. If the open set V, which is the homeomorphic image of the open set U, has a suitable structure in the manifold $\mathbb{R}^m$, then we can easily show that U is contractible. To this end, let us assume that we can find a mapping $h' : V \times I \to V$ and a point $\mathbf{x}_0 \in V$ such that we are able to write $h'(\mathbf{x};t) = h'_t(\mathbf{x}) = (1-t)\mathbf{x}_0 + t\mathbf{x} \in V$ for all points $\mathbf{x} \in V$. This expression signifies that a straight line joining any point $\mathbf{x}$ in V to the ***centre point*** $\mathbf{x}_0$ stays entirely in V. Such a region is called a ***star-shaped region*** (Fig. 6.2.1).

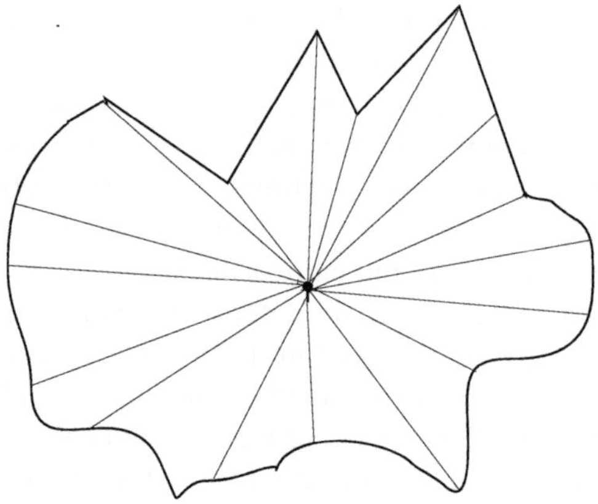

Fig. 6.2.1. Star-shaped region in the Euclidean space.

Evidently, every convex set in $\mathbb{R}^m$ is star-shaped and it is easily shown that open balls in $\mathbb{R}^m$ are convex. Let us consider an open ball in $\mathbb{R}^m$ given by $B_r(\mathbf{x}_0) = \|\mathbf{x} - \mathbf{x}_0\| < r$ where $\mathbf{x}, \mathbf{x}_0 \in \mathbb{R}^m$ and $r > 0$. By using the triangle inequality, we obtain for points $\mathbf{x}, \mathbf{y} \in B_r(\mathbf{x}_0)$ in $\mathbb{R}^m$ and a parameter t satisfying $0 \le t \le 1$

$$
\begin{aligned}
\|(1-t)\mathbf{x} + t\mathbf{y} - \mathbf{x}_0\| &= \|(1-t)(\mathbf{x} - \mathbf{x}_0) + t(\mathbf{y} - \mathbf{x}_0)\| \\
&\le (1-t)\|(\mathbf{x} - \mathbf{x}_0)\| + t\|(\mathbf{y} - \mathbf{x}_0)\| \\
&< (1-t)r + tr = r
\end{aligned}
$$

This result shows that $(1-t)\mathbf{x}+t\mathbf{y}\in B_r(\mathbf{x}_0)$. Therefore, any open ball in $\mathbb{R}^m$ is a convex set.

The open set U is homeomorphic to an open set of $\mathbb{R}^m$ that is expressible as some union of open balls. Hence, U itself is the union of inverse images of some open balls implying that a component open subset of U is homeomorphic to a *convex open ball* with centre at a point $\mathbf{x}$. We thus conclude that every manifold is locally contractible and is locally homeomorphic to a star-shaped region. Conversely, when $V\in\mathbb{R}^m$ is a star-shaped open set, if we define on an open set $U=\varphi^{-1}(V)$ of the manifold M a mapping $h_t=\varphi^{-1}\circ h'_t\circ\varphi$ such that $h_t(p)\in U$ for all points $p\in U$ and $t\in[0,1]$, then we immediately observe that the set U can be contracted to the point $p_0=\varphi^{-1}(\mathbf{x}_0)$ by the mapping h_t.

The entire manifold $\mathbb{R}^m$ is star-shaped with respect to the origin $\mathbf{0}$, in fact to every point of $\mathbb{R}^m$. Hence, a manifold M is contractible if it is homeomorphic to the manifold $\mathbb{R}^m$. That the converse statement is not generally true can be demonstrated by constructing a counter example. Three dimensional Whitehead manifold is obtained by embedding a *solid torus* T_1 (a solid torus is a filled-in torus $\mathbb{T}^2$) inside *three dimensional sphere* $\mathbb{S}^3$, then a solid torus T_2 inside T_1 and continuing this way *ad infinitum* [discovered by English mathematician John Henry Constantine Whitehead (1904-1960)]. Hence, we can formally represent the Whitehead manifold by $\bigcap_{i=1}^{\infty}T_i$. A rather small part of the Whitehead manifold is depicted in Fig. 6.2.2. This manifold is contractible but it is not homeomorphic to $\mathbb{R}^3$.

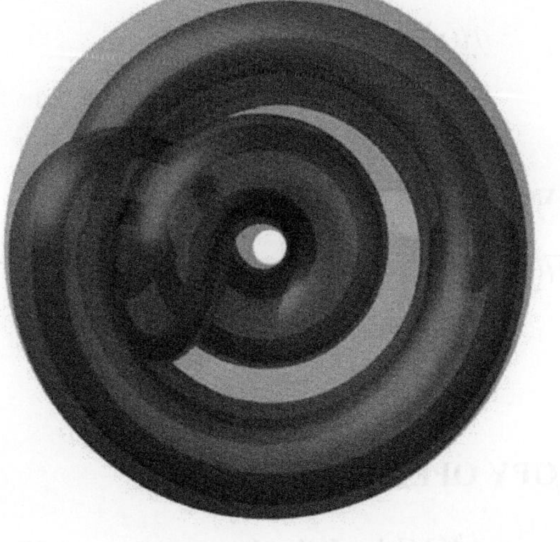

Fig. 6.2.2. Whitehead manifold.

Let a form field $\omega \in \Lambda^k(M)$ be represented in a local chart by

$$\omega(\mathbf{x}) = \frac{1}{k!}\,\omega_{i_1\cdots i_k}(\mathbf{x})dx^{i_1} \wedge \cdots \wedge dx^{i_k}$$

This form will of course be defined on an open set U of the manifold M. We can define a new k-form $\overline{\omega}$ depending on a parameter $t \in [0,1]$ in the following manner

$$\overline{\omega}(\mathbf{x};t) = \frac{1}{k!}\,\omega_{i_1\cdots i_k}\big[\mathbf{x}_0 + t(\mathbf{x}-\mathbf{x}_0)\big]dx^{i_1} \wedge \cdots \wedge dx^{i_k}. \qquad (6.2.1)$$

If U is contractible, then $\overline{\omega}$ is specified everywhere in U. It is clear that $\overline{\omega}(\mathbf{x};0) = \omega(\mathbf{x}_0)$ and $\overline{\omega}(\mathbf{x};1) = \omega(\mathbf{x})$. Let us now define the new independent variables by $u^i = x_0^i + t(x^i - x_0^i),\ t \in [0,1],\ i = 1,\dots,m$. If we write $\overline{\omega}_{i_1\cdots i_k}(\mathbf{u}) = \omega_{i_1\cdots i_k}\big[\mathbf{x}_0 + t(\mathbf{x}-\mathbf{x}_0)\big]$, it then follows from (6.2.1) that

$$\begin{aligned} d\overline{\omega} &= \frac{1}{k!}\,d\overline{\omega}_{i_1\cdots i_k}(\mathbf{u}) \wedge dx^{i_1} \wedge \cdots \wedge dx^{i_k} \qquad (6.2.2)\\ &= \frac{1}{k!}\frac{\partial \overline{\omega}_{i_1\cdots i_k}}{\partial u^i}\,du^i \wedge dx^{i_1} \wedge \cdots \wedge dx^{i_k}\\ &= t\,\frac{1}{k!}\frac{\partial \overline{\omega}_{i_1\cdots i_k}}{\partial u^i}\,dx^i \wedge dx^{i_1} \wedge \cdots \wedge dx^{i_k}\\ &= t\,\overline{d\omega} \end{aligned}$$

We denote the radius vector in the region V which is the homeomorphic image of U with respect to the point $\mathbf{x}_0$ by the relation

$$\begin{aligned} \mathcal{H}(\mathbf{x}) &= (x^i - x_0^i)\frac{\partial}{\partial x^i} \qquad (6.2.3)\\ &= \frac{d}{dt}\big[x_0^i + t(x^i - x_0^i)\big]\frac{\partial}{\partial x^i} \end{aligned}$$

We thus get $\mathcal{H}(\mathbf{x}_0) = 0$. It is clear that one finds

$$\begin{aligned} \overline{\mathcal{H}}(\mathbf{x};t) &= \mathcal{H}\big[\mathbf{x}_0 + t(\mathbf{x}-\mathbf{x}_0)\big] = t(x^i - x_0^i)\frac{\partial}{\partial x^i} \qquad (6.2.4)\\ &= t\mathcal{H}(\mathbf{x}) \end{aligned}$$

for $t \in [0,1]$.

6.3. HOMOTOPY OPERATOR

Let a form $\omega \in \Lambda^k(M)$ be defined on an open set $U \subseteq M$ that is contractible to a point $p_0 \in M$. We will assume that the homeomorphic image

$V \subseteq \mathbb{R}^m$ of the set U is a star-shaped region. We define *the linear operator* $H : \Lambda^k(U) \to \Lambda^{k-1}(U)$ by the following expression in local coordinates

$$H\omega = \int_0^1 \mathbf{i}_{\mathcal{H}}\big(\overline{\omega}(\mathbf{x};t)\big)t^{k-1}dt \tag{6.3.1}$$
$$= \frac{1}{(k-1)!}\int_0^1 t^{k-1}(x^{i_1}-x_0^{i_1})\,\omega_{i_1i_2\cdots i_k}\big[\mathbf{x}_0 + t(\mathbf{x}-\mathbf{x}_0)\big]\,dt\,dx^{i_2}\wedge\cdots\wedge dx^{i_k}.$$

Since V is star-shaped, the form ω is prescribed at every point of the open set U. Therefore, the operator H introduced by (6.3.1) is well defined on the exterior algebra $\Lambda(U)$. H is called ***the homotopy operator***. This definition will automatically lead to the result $Hf = 0$ for $f \in \Lambda^0(M)$.

If we choose $\mathbf{x}_0$ at the origin $\mathbf{0}$ of the local coordinate system without loss of generality, then the homotopy operator takes the form

$$H\omega(\mathbf{x}) = \int_0^1 \mathbf{i}_{\mathcal{H}}\omega(t\mathbf{x})t^{k-1}dt$$
$$= \frac{1}{(k-1)!}\int_0^1 t^{k-1}x^{i_1}\,\omega_{i_1i_2\cdots i_k}(t\mathbf{x})\,dt\,dx^{i_2}\wedge\cdots\wedge dx^{i_k}$$

Let us now consider vector fields $V_1, V_2, \ldots, V_{k-1} \in T(M)$. The above expression implies that

$$H\omega(\mathbf{x})(V_1,\ldots,V_{k-1}) = \int_0^1 \omega(t\mathbf{x})(\mathbf{x},V_1,\ldots,V_{k-1})t^{k-1}dt$$

The main properties of the homotopy operator are embodied in the following theorem.

Theorem 6.3.1. *The homotopy operator H has the properties listed below*:

(i). $dH + Hd = i_\Lambda$ *if* $k \geq 1$.
$Hdf(\mathbf{x}) = f(\mathbf{x}) - f(\mathbf{x}_0)$ *if* $k = 0$.
(ii). $H \circ H = H^2 = 0$ *and* $H\omega(\mathbf{x}_0) = 0$.
(iii). $HdH = H$ *and* $dHd = d$.
(iv). $HdHd = (Hd)^2 = Hd$ *and* $dHdH = (dH)^2 = dH$,
$(dH)(Hd) = dH^2d = 0$ *and* $(Hd)(dH) = Hd^2H = 0$.
(v). $\mathbf{i}_{\mathcal{H}} \circ H = 0$ *and* $H \circ \mathbf{i}_{\mathcal{H}} = 0$.

(i). We consider a form $\omega \in \Lambda^k(M), k \geq 1$. We shall try to evaluate explicitly the action of the operator $d \circ H + H \circ d$ on this form. At the first step, we obtain

$$\begin{aligned}(dH+Hd)\omega &= d(H\omega)+H(d\omega)\\ &= \int_0^1 d\big(\mathbf{i}_{\mathcal{H}}(\overline{\omega})\big)t^{k-1}dt+\int_0^1 \mathbf{i}_{\mathcal{H}}(\overline{d\omega})t^k dt\\ &= \int_0^1 t^{k-1}\big[d\big(\mathbf{i}_{\mathcal{H}}(\overline{\omega})\big)+\mathbf{i}_{\mathcal{H}}(d\overline{\omega})\big]dt\\ &= \int_0^1 t^{k-1}\pounds_{\mathcal{H}}\overline{\omega}\,dt\end{aligned}$$

where we have employed the Cartan magic formula. On the other hand, Lie derivative with respect to the vector $\mathcal{H}$ yields

$$\pounds_{\mathcal{H}}\overline{\omega}=\frac{1}{k!}\,(\pounds_{\mathcal{H}}\overline{\omega})_{i_1\cdots i_k}\,dx^{i_1}\wedge\cdots\wedge dx^{i_k}$$

where the coefficients follow from (5.11.4) as

$$\begin{aligned}(\pounds_{\mathcal{H}}\overline{\omega})_{i_1\cdots i_k} &= (x^i-x_0^i)\frac{\partial\overline{\omega}_{i_1i_2\cdots i_k}}{\partial x^i}+\sum_{r=1}^{k}\overline{\omega}_{i_1\cdots i_{r-1}ii_{r+1}\cdots i_k}\frac{\partial(x^i-x_0^i)}{\partial x^{i_r}}\\ &= (x^i-x_0^i)\frac{\partial\overline{\omega}_{i_1\cdots i_k}}{\partial x^i}+\sum_{r=1}^{k}\overline{\omega}_{i_1\cdots i_{r-1}ii_{r+1}\cdots i_k}\delta^i_{i_r}\\ &= t(x^i-x_0^i)\frac{\partial\overline{\omega}_{i_1\cdots i_k}}{\partial u^i}+k\overline{\omega}_{i_1\cdots i_k}=t\frac{d\overline{\omega}_{i_1\cdots i_k}}{dt}+k\overline{\omega}_{i_1\cdots i_k}.\end{aligned}$$

Hence, we get

$$\begin{aligned}(dH+Hd)\omega &= \frac{1}{k!}\int_0^1\Big[t^k\frac{d\overline{\omega}_{i_1\cdots i_k}}{dt}+kt^{k-1}\overline{\omega}_{i_1\cdots i_k}\Big]dt\,dx^{i_1}\wedge\cdots\wedge dx^{i_k}\\ &= \frac{1}{k!}\int_0^1\frac{d}{dt}(t^k\overline{\omega}_{i_1\cdots i_k})\,dt\,dx^{i_1}\wedge\cdots\wedge dx^{i_k}\\ &= \frac{1}{k!}\big(t^k\omega_{i_1\cdots i_k}\big[\mathbf{x}_0+t(\mathbf{x}-\mathbf{x}_0)\big]\big)\Big|_{t=0}^{t=1}dx^{i_1}\wedge\cdots\wedge dx^{i_k}\\ &= \frac{1}{k!}\omega_{i_1\cdots i_k}(\mathbf{x})\,dx^{i_1}\wedge\cdots\wedge dx^{i_k}.\end{aligned}$$

This means that for every form $\omega\in\Lambda(M)$ with non-zero degree, we find

$$(dH+Hd)\omega=\omega\quad\text{or}\quad dH+Hd=i_\Lambda \tag{6.3.2}$$

where i_Λ denotes the identity mapping on the exterior algebra $\Lambda(M)$. When $k=0$, on resorting to (6.2.2) for every function $f\in\Lambda^0(M)$ we arrive at the result

$$\begin{aligned}
dHf + Hdf = Hdf \qquad\qquad\qquad\qquad\qquad\qquad\qquad\qquad\qquad (6.3.3)\\
&= \int_0^1 \mathbf{i}_{\mathcal{H}}(\overline{df})\,dt = \int_0^1 \frac{1}{t}\,\mathbf{i}_{\mathcal{H}}(d\overline{f})\,dt\\
&= \int_0^1 \frac{1}{t}\frac{\partial \overline{f}}{\partial x^i}\,\mathbf{i}_{\mathcal{H}}(dx^i)\,dt = \int_0^1 \frac{x^i - x_0^i}{t}\,t\,\frac{\partial \overline{f}}{\partial u^i}\,dt\\
&= \int_0^1 (x^i - x_0^i)\,\frac{\partial \overline{f}}{\partial u^i}\,dt = \int_0^1 \frac{d\overline{f}}{dt}\,dt = \overline{f}\big|_0^1 = f(\mathbf{x}) - f(\mathbf{x}_0).
\end{aligned}$$

(ii). Since $H\omega(\mathbf{x}_0) = \int_0^1 \mathbf{i}_{\mathcal{H}(\mathbf{x}_0)}\big(\omega(\mathbf{x}_0)\big)t^{k-1}dt$ and $\mathcal{H}(\mathbf{x}_0) = 0$, we obtain $H\omega(\mathbf{x}_0) = 0$. On the other hand, for a form $\omega \in \Lambda^k(M)$ we find

$$\begin{aligned}
H^2\omega &= \int_0^1 s^{k-2}\,\mathbf{i}_{\mathcal{H}}\Big[\int_0^1 t^{k-1}\,\mathbf{i}_{\mathcal{H}}(\overline{\omega}(t))\,dt\Big](s)\,ds\\
&= \int_0^1\!\!\int_0^1 t^{k-1}s^{k-2}\,\mathbf{i}_{\mathcal{H}}\big[\mathbf{i}_{\overline{\mathcal{H}(s)}}\big(\overline{\overline{\omega}(t)}\big)(s)\big]\,dt\,ds\\
&= \int_0^1\!\!\int_0^1 t^{k-1}s^{k-2}\,\mathbf{i}_{\mathcal{H}}\big[\mathbf{i}_{s\mathcal{H}}\big(\overline{\omega}(t)\big)(s)\big]\,dt\,ds\\
&= \int_0^1\!\!\int_0^1 t^{k-1}s^{k-1}\,\mathbf{i}_{\mathcal{H}}^2\big(\overline{\omega}(t)\big)(s)\,dt\,ds = 0
\end{aligned}$$

where we have employed (6.2.4) and the relation $\mathbf{i}_{\mathcal{H}}^2 = 0$.

(iii). Since $d^2 = 0$ and $H^2 = 0$, the relation (6.3.2) leads right away to $dHd = d$ and $HdH = H$.

(iv). If we make use of the property (iii) in expressions $(Hd)^2 = HdHd$ and $(dH)^2 = dHdH$, we find that $(Hd)^2 = Hd$ and $(dH)^2 = dH$.

(v). This property can also be demonstrated quite easily. If we consider a form $\omega \in \Lambda^k(M)$, we obtain

$$\mathbf{i}_{\mathcal{H}}(H\omega) = \mathbf{i}_{\mathcal{H}}\Big[\int_0^1 \mathbf{i}_{\mathcal{H}}(\overline{\omega})\,t^{k-1}dt\Big] = \int_0^1 \mathbf{i}_{\mathcal{H}}\big(\mathbf{i}_{\mathcal{H}}(\overline{\omega})\big)t^{k-1}dt = 0$$

$$\begin{aligned}
H\mathbf{i}_{\mathcal{H}}(\omega) &= \int_0^1 t^{k-2}\,\mathbf{i}_{\mathcal{H}}\big(\overline{\mathbf{i}_{\mathcal{H}}(\omega)}\big)\,dt = \int_0^1 t^{k-2}\,\mathbf{i}_{\mathcal{H}}\big(\mathbf{i}_{\overline{\mathcal{H}}}(\overline{\omega})\big)\,dt\\
&= \int_0^1 t^{k-1}\,\mathbf{i}_{\mathcal{H}}\big(\mathbf{i}_{\mathcal{H}}(\overline{\omega})\big)\,dt = 0
\end{aligned}$$

because of the relation $\mathbf{i}_{\mathcal{H}} \circ \mathbf{i}_{\mathcal{H}} = 0$. □

In case we can define the homotopy operator, the celebrated Poincaré lemma can readily be proven.

Theorem 6.3.2 (The Poincaré Lemma). *An exterior form defined on an open set $U \subseteq M$ contractible to one of its interior points is closed if and only if it is exact on U.*

If a form ω is exact, that is, if one is able to write $\omega = d\Omega$, this form is closed because $d\omega = 0$. Conversely, let us now assume that $\omega \in \Lambda^k(M)$ is a closed form. When the homeomorphic image of U in $\mathbb{R}^m$ is a star-shaped region, we will be free to employ the homotopy operator. Since $d\omega = 0$, we then obtain

$$\omega = dH\omega + H d\omega = d(H\omega) = d\Omega$$

where we have defined the form $\Omega = H\omega \in \Lambda^{k-1}(M)$. Thus, the closed form ω is likewise an exact form on U. Since every chart of an m-dimensional differentiable manifold M is homeomorphic to an open set of $\mathbb{R}^m$, the Poincaré Lemma is locally valid. Therefore, *every closed form on M is locally, in other words, in an open neighbourhood of every point $p \in M$, is an exact form*. However, this statement is generally not true globally. This means that we cannot be sure in general the existence of a form Ω defined over the entire manifold M so that a closed form is expressed as $\omega = d\Omega$. For instance, if we have prescribed a closed form on the punctured differentiable manifold $\mathbb{R}^m - \{\mathbf{0}\}$, we cannot validate the Poincaré Lemma on any open set containing the point $\{\mathbf{0}\}$. □

If we take the manifold $\mathbb{R}^m, m > 0$ into consideration, we know that the whole manifold can be contracted, say, to the point $\mathbf{0}$. Hence, according to the Poincaré lemma *every closed form defined on the entire $\mathbb{R}^m$ is globally exact*. Similarly, we can say that *every closed form on a contractible manifold M is globally exact.*

Example 6.3.1. A form $\omega \in \Lambda^2(\mathbb{R}^3)$ is given by

$$\omega = -2(x+y)z\,dx \wedge dy + x^2 dy \wedge dz + y^2 dz \wedge dx.$$

We observe at once that $d\omega = 0$. $\mathbb{R}^3$ is star-shaped with respect to the centre $\mathbf{0}$. Thus, the radius vector can be taken as $\mathcal{H} = x\partial_x + y\partial_y + z\partial_z$. We can then evaluate the form $H\omega$ easily as

$$\begin{aligned}\Omega = H\omega = \int_0^1 t\big[-2t^2(x+y)z(x\,dy - y\,dx) + t^2x^2(y\,dz - z\,dy) \\ + t^2y^2(z\,dx - x\,dz)\big]dt \\ = \frac{1}{4}\big[yz(3y+2x)\,dx - xz(3x+2y)\,dy + xy(x-y)\,dz\big] \in \Lambda^1(\mathbb{R}^3).\end{aligned}$$

We can readily verify that the relation $\omega = d\Omega$ holds.

Let us now consider a more general 2-form defined by

$$\omega = R\,dx \wedge dy + P\,dy \wedge dz + Q\,dz \wedge dx \in \Lambda^2(\mathbb{R}^3).$$

If ω is a closed form, that is, if $d\omega = 0$, then we have to impose the following restriction on the functions $P(x,y,z), Q(x,y,z), R(x,y,z)$:

$$\frac{\partial P}{\partial x} + \frac{\partial Q}{\partial y} + \frac{\partial R}{\partial z} = 0.$$

In this situation, on resorting to the homotopy operator, we can determine the form $\Omega = H\omega$ as follows

$$\begin{aligned}\Omega = &\Big(\int_0^1 \big[tzQ(tx,ty,tz) - tyR(tx,ty,tz)\big]dt\Big)dx \\ &+ \Big(\int_0^1 \big[txR(tx,ty,tz) - tzP(tx,ty,tz)\big]dt\Big)dy \\ &+ \Big(\int_0^1 \big[tyP(tx,ty,tz) - txQ(tx,ty,tz)\big]dt\Big)dz.\end{aligned}$$

If we recall the restriction imposed of the functions P, Q, R, we can verify at once that we get the relation $\omega = d\Omega$. This is of course valid on the entire manifold $\mathbb{R}^3$. ■

It is clear that the form Ω introduced in the foregoing theorem cannot be determined uniquely. Evidently, for an arbitrary form $\sigma \in \Lambda^{k-2}(M)$, the form $\Omega' = \Omega + d\sigma$ will also satisfy the relation $\omega = d\Omega'$.

We had denoted the graded algebra $\mathcal{E}(U)$ of exact forms on an open subset $U \subseteq M$. For a form $\omega \in \Lambda^k(U)$, we get $dH\omega \in \mathcal{E}^k(U)$ implying that $dH : \Lambda(U) \to \mathcal{E}(U)$. But, the restriction $dH|_{\mathcal{E}^k(U)} : \mathcal{E}^k(U) \to \mathcal{E}^k(U)$ satisfies the relation $\omega = dH|_{\mathcal{E}^k(U)}\omega$. Hence, we may regard the operator d as the inverse of the operator H on $\mathcal{E}^k(U)$.

Let M and N be, respectively, m- and n-dimensional differentiable manifolds with $n \geq m$. $\phi : M \to N$ is a smooth mapping. We consider an open subset $U \subseteq M$. Let us assume that the mapping ϕ is a diffeomorphism on U. Thus $\phi^{-1} : \phi(U) \to U$ is a smooth mapping. If U is contractible to a point $p_0 \in U$, then the region $\phi(U) \subset N$ can also be contracted to the point $\phi(p_0) \in \phi(U)$ and since ϕ^{-1} is continuous on $\phi(U) \subseteq N$, we see that $\phi(U)$ is also an open subset. The mappings ϕ and ϕ^{-1} give obviously rise to pull-back mappings $\phi^* : \Lambda(\phi(U)) \to \Lambda(U)$ and $(\phi^*)^{-1} : \Lambda(U) \to \Lambda(\phi(U))$. Let H be the homotopy operator defined on the region U. If $\omega \in \Lambda(\phi(U))$, we have $\phi^*\omega \in \Lambda(U)$ and we can write

$$dH\phi^*\omega + Hd\phi^*\omega = \phi^*\omega.$$

According to Theorem 5.8.2, it is possible to write $Hd\phi^*\omega = H\phi^* d\omega$. Let us now define an operator $H^* : \Lambda^k\big(\phi(U)\big) \to \Lambda^{k-1}\big(\phi(U)\big)$ through the relation

$$\phi^* H^* = H\phi^* \quad \text{or} \quad H^* = (\phi^*)^{-1} H\phi^*. \tag{6.3.4}$$

We thus obtain

$$d\phi^* H^*\omega + H\phi^* d\omega = \phi^* dH^*\omega + \phi^* H^* d\omega = \phi^*(dH^*\omega + H^* d\omega) = \phi^*\omega.$$

By applying the operator $(\phi^*)^{-1}$ on this expression, we find that

$$dH^*\omega + H^* d\omega = \omega \ \text{ or } \ dH^* + H^* d = i_{\Lambda(\phi(U))}.$$

H^* is then called the homotopy operator generated by the mapping ϕ.

6.4. EXACT AND ANTIEXACT FORMS

Let $U \subseteq M$ be a contractible open set on which the homotopy operator can be defined where M is an m-dimensional smooth manifold. Thus, on taking heed of the relation (6.3.2) it becomes possible to express a form $\omega \in \Lambda(U)$ in the following manner

$$\omega = dH\omega + Hd\omega = \omega_e + \omega_a \tag{6.4.1}$$

where we introduce the following forms with degree preserving operations

$$\omega_e = dH\omega, \quad \omega_a = Hd\omega = \omega - \omega_e. \tag{6.4.2}$$

They will be called as the ***exact*** and ***antiexact parts*** of the form ω, respectively. (6.4.2) then leads to the result $H\omega_a = H^2 d\omega = 0$. Hence, antiexact forms are located in the null space or the kernel of the linear operator H. Let us denote the set of all antiexact forms of the module $\Lambda^k(U)$ by $\mathcal{A}^k(U)$. $\mathcal{E}^0(U)$ is of course empty. On the other hand, we can write $f(\mathbf{x}) - f(\mathbf{x}_0) = Hdf = f_a$ for all $f \in \Lambda^0(U)$. So there will be no harm in assuming that $\mathcal{A}^0(U) = \Lambda^0(U)$. We can now easily demonstrate the following lemmas.

Lemma 6.4.1. *The operator dH maps $\mathcal{E}^k(U)$ onto $\mathcal{E}^k(U)$ and $\Lambda(U)$ onto $\mathcal{E}(U)$. Furthermore, the operator d is the inverse of the operator H when the domain of H is restricted to $\mathcal{E}^k(U)$.*

In view of (6.4.2), $dH\omega$ is exact for every $\omega \in \Lambda^k(U)$ thus dH maps $\Lambda^k(U)$ into $\mathcal{E}^k(U)$. If $\omega \in \mathcal{E}^k(U)$, then $\omega = d\alpha$ where $\alpha \in \Lambda^{k-1}(U)$ so we get $dH\omega = dHd\alpha = d\alpha = \omega$. Hence dH restricted to $\mathcal{E}^k(U)$ is the identity operator. This also shows that dH is a surjective mapping.

Lemma 6.4.2. *The necessary and sufficient conditions to completely determine the set $\mathcal{A}^k(U), k \geq 1$ of antiexact forms are given as follows*

$$\mathcal{A}^k(U) = \{\alpha \in \Lambda^k(U) : \mathbf{i}_{\mathcal{H}}(\alpha) = 0, \alpha(\mathbf{x}_0) = 0, k > 0\}.$$

For all $\omega \in \Lambda^k(U)$, according to Theorem 6.3.1 (v) and (ii) we find that antiexact parts satisfy $\mathbf{i}_{\mathcal{H}}(\omega_a) = 0$ and $\omega_a(\mathbf{x}_0) = 0$. Conversely, let us assume that a form $\alpha \in \Lambda^k(U)$ satisfies the relations $\mathbf{i}_{\mathcal{H}}(\alpha) = 0, \alpha(\mathbf{x}_0) = 0$. For an arbitrary form $\beta \in \Lambda^{k-1}(U)$, let us write $\omega = d\beta + \alpha$. However, we have $H\alpha = \int_0^1 t^{k-1}\mathbf{i}_{\mathcal{H}}(\overline{\alpha})\,dt = \int_0^1 t^{k-2}\mathbf{i}_{\overline{\mathcal{H}}}(\overline{\alpha})\,dt = \int_0^1 t^{k-2}\overline{\mathbf{i}_{\mathcal{H}}(\alpha)}\,dt = 0$ so that we get $H\omega = Hd\beta + H\alpha = Hd\beta$ and $\omega_e = dH\omega = dHd\beta = d\beta$. Hence, we obtain $\alpha = \omega - \omega_e = \omega_a = Hd\omega$. This equality does not lead to a contradiction if only $\alpha(\mathbf{x}_0) = 0$. Thus we find $\alpha \in \mathcal{A}^k(U)$. □

Lemma 6.4.3. *The operator Hd maps $\Lambda^k(U)$ onto $\mathcal{A}^k(U)$ and $\mathcal{A}^m(U) = 0$ on the m-dimensional open set U. Furthermore, the operator H is the inverse of the operator d when the domain of H is restricted to $\mathcal{A}^k(U)$.*

We obviously have $Hd : \Lambda^k(U) \to \mathcal{A}^k(U)$. Let us consider the form $\omega_a = Hd\omega \in \mathcal{A}^k(U)$ where $\omega \in \Lambda^k(U)$. We then obtain $Hd\omega_a = (Hd)^2\omega = Hd\omega = \omega_a$. This also shows that Hd restricted to $\mathcal{A}^k(U)$ is the identity operator for $k \geq 1$. If $k = 0$, then the same situation is also realised up to a constant: $f(\mathbf{x}) = Hdf(\mathbf{x}) + f(\mathbf{x}_0)$. If we pay attention to the sequence $\Lambda^k(U) \xrightarrow{d} \Lambda^{k+1}(U) \xrightarrow{H} \Lambda^k(U)$, we observe at once that $\mathcal{A}^m(U) = 0$ on the m-dimensional open set U of the manifold M^m. □

Various properties of antiexact forms are collected in the theorem below.

Theorem 6.4.1. *Antiexact forms possess the following properties*:

(i). $\mathcal{A}^k(U) \subseteq \mathcal{N}(H) = \mathrm{Ker}\,(H), k \geq 0$.

(ii). *If $\alpha \in \mathcal{A}^k(U)$ and $\beta \in \mathcal{A}^l(U)$, then $\alpha \wedge \beta \in \mathcal{A}^{k+l}(U)$.*

(iii). *For $k \geq 1$, $\mathcal{A}^k(U)$ is a module on $\Lambda^0(U)$.*

(i). We have seen above that $H\alpha = 0$ because of $\mathbf{i}_{\mathcal{H}}(\alpha) = 0$. Hence, we find that $\mathcal{A}^k(U) \subseteq \mathcal{N}(H)$.

(ii). For $k = 0$, this statement becomes true automatically. Therefore, we take the case $\min\{k, l\} \geq 1$ into account. Since, the antiexact form factors vanish at the point $\mathbf{x}_0$, we naturally obtain $(\alpha \wedge \beta)(\mathbf{x}_0) = 0$. On the other hand, we get

$$\mathbf{i}_{\mathcal{H}}(\alpha \wedge \beta) = (\mathbf{i}_{\mathcal{H}}\alpha) \wedge \beta + (-1)^k \alpha \wedge (\mathbf{i}_{\mathcal{H}}\beta) = 0 + 0 = 0.$$

We thus conclude that $\alpha \wedge \beta \in \mathcal{A}^{k+l}(U)$.

(iii). The set $\mathcal{A}^k(U)$ is a submodule of the module $\Lambda^k(U)$. If $\alpha, \beta \in \mathcal{A}^k(U)$, we get $\mathbf{i}_{\mathcal{H}}(\alpha + \beta) = \mathbf{i}_{\mathcal{H}}(\alpha) + \mathbf{i}_{\mathcal{H}}(\beta) = 0, (\alpha + \beta)(\mathbf{x}_0) = \alpha(\mathbf{x}_0) + \beta(\mathbf{x}_0) = 0$ and $\mathbf{i}_{\mathcal{H}}(f\alpha) = f\mathbf{i}_{\mathcal{H}}(\alpha) = 0$ and $f(\mathbf{x}_0)\alpha(\mathbf{x}_0) = 0$ for all

$f \in \Lambda^0(U)$. We thus have $\alpha + \beta \in \mathcal{A}^k(U)$ and $f\alpha \in \mathcal{A}^k(U)$. □

When $\omega \in \mathcal{A}^k(U)$, if we write $d\omega = \alpha \in \Lambda^{k+1}(U)$ then we are led to $\omega = H\alpha$. Likewise, when $f \in \Lambda^0(U)$, if we write $df = \alpha \in \Lambda^1(U)$ we get $f(\mathbf{x}) = H\alpha + f(\mathbf{x}_0)$. We can immediately observe that $\mathcal{A}(U) = \overset{m}{\underset{k=0}{\oplus}} \mathcal{A}^k(U)$ is a graded algebra that is a subalgebra of $\Lambda(U)$. Furthermore, for any form $\omega \in \Lambda(U)$ we obtain $Hd\omega \in \mathcal{A}(U)$ so that we can symbolically write the relation $\mathcal{A}(U) = Hd\big(\Lambda(U)\big)$. Because of the identity $Hd = (Hd)^2$, we are led to the conclusion that Hd is a ***projection operator***. Hence, we can say that *the algebra $\mathcal{A}(U)$ is a Hd-projection of the algebra $\Lambda(U)$*.

With the information we have acquired so far, we can now manage to better identify the characteristics of the operator H. For $k \geq 0$, it is possible to express $H : \Lambda^{k+1}(U) \to \mathcal{A}^k(U)$ implying that $\mathcal{A}^k(U) = H\big(\Lambda^{k+1}(U)\big)$. Indeed, If $\omega \in \Lambda^{k+1}(U)$, then we find that $H\omega \in \Lambda^k(U)$ and $\mathbf{i}_{\mathcal{H}}(H\omega) = 0$, $H\omega(\mathbf{x}_0) = 0$ because of Theorem (6.3.1) (v) and (ii) and consequently $H\omega \in \mathcal{A}^k(U)$. Conversely, let us suppose that $\alpha \in \mathcal{A}^k(U)$. This means that $\alpha = Hd\alpha$. Next, we introduce the form $\beta = d\alpha \in \Lambda^{k+1}(U)$ so we get $\alpha = H\beta$.

Theorem 6.4.2. *If $\alpha \in \mathcal{A}^k(U)$, there exists a form $\widehat{\alpha} \in \Lambda^{k+1}(U)$ such that α is expressible as $\alpha = \mathbf{i}_{\mathcal{H}}\widehat{\alpha}$.*

When $\alpha \in \mathcal{A}^k(U)$, there is a form $\beta \in \Lambda^{k+1}(U)$ such that one is able to write $\alpha = H\beta$ and thus it has the following expression

$$\alpha = \int_0^1 t^k\, \mathbf{i}_{\mathcal{H}}\overline{\beta}\, dt =$$
$$= \mathbf{i}_{\mathcal{H}}\Big(\frac{1}{(k+1)!}\int_0^1 t^k \beta_{i_1\cdots i_{k+1}}\big[\mathbf{x}_0 + t(\mathbf{x}-\mathbf{x}_0)\big]\,dt\, dx^{i_1}\wedge\cdots\wedge dx^{i_{k+1}}\Big) = \mathbf{i}_{\mathcal{H}}\widehat{\alpha}.$$

Conversely, if $\alpha = \mathbf{i}_{\mathcal{H}}\widehat{\alpha}$, then we find $\mathbf{i}_{\mathcal{H}}\alpha = \mathbf{i}_{\mathcal{H}}^2\widehat{\alpha} = 0$ and $\alpha(\mathbf{x}_0) = 0$ since $\mathcal{H}(\mathbf{x}_0) = 0$. □

Next, as an application of Theorem 6.4.2, let us show once more that *the exterior product of two antiexact forms is again an antiexact form.* If α, β are antiexact forms, then they are expressible as $\alpha = \mathbf{i}_{\mathcal{H}}\widehat{\alpha}$ and $\beta = \mathbf{i}_{\mathcal{H}}\widehat{\beta}$. We thus get

$$\alpha \wedge \beta = \mathbf{i}_{\mathcal{H}}\widehat{\alpha} \wedge \mathbf{i}_{\mathcal{H}}\widehat{\beta} = \mathbf{i}_{\mathcal{H}}(\widehat{\alpha} \wedge \mathbf{i}_{\mathcal{H}}\widehat{\beta}).$$

Recalling that $H \circ \mathbf{i}_{\mathcal{H}} = 0$, we obtain

$$Hd\mathbf{i}_{\mathcal{H}}(\widehat{\alpha} \wedge \mathbf{i}_{\mathcal{H}}\widehat{\beta}) = \mathbf{i}_{\mathcal{H}}(\widehat{\alpha} \wedge \mathbf{i}_{\mathcal{H}}\widehat{\beta}) - dH\mathbf{i}_{\mathcal{H}}(\widehat{\alpha} \wedge \mathbf{i}_{\mathcal{H}}\widehat{\beta}) = \mathbf{i}_{\mathcal{H}}(\widehat{\alpha} \wedge \mathbf{i}_{\mathcal{H}}\widehat{\beta})$$

Introducing the form $\gamma = \mathbf{i}_{\mathcal{H}}(\widehat{\alpha} \wedge \mathbf{i}_{\mathcal{H}}\widehat{\beta})$, we finally find

$$\alpha \wedge \beta = Hd\mathbf{i}_{\mathcal{H}}(\widehat{\alpha} \wedge \mathbf{i}_{\mathcal{H}}\widehat{\beta}) = Hd\gamma \in \mathcal{A}(U).$$

Let us consider a form $\omega \in \Lambda^k(U)$. This form may be expressed as

$$\omega = \alpha + \beta, \quad \alpha = dH\omega \in \mathcal{E}^k(U), \quad \beta = Hd\omega \in \mathcal{A}^k(U).$$

This implies that one is allowed to write $\Lambda^k(U) = \mathcal{E}^k(U) + \mathcal{A}^k(U)$. But we can readily show that $\mathcal{E}^k(U) \cap \mathcal{A}^k(U) = \{0\}$. Let $\omega \in \mathcal{E}^k(U) \cap \mathcal{A}^k(U)$ so that this form has to satisfy both $\omega = d\sigma$ and $\omega = Hd\omega$. This leads to the result $\omega = Hd^2\sigma = 0$ which amounts to say that we have a direct sum at hand: $\Lambda^k(U) = \mathcal{E}^k(U) \oplus \mathcal{A}^k(U)$. We then conclude that the exterior algebra on U may be represented as the direct sum $\Lambda(U) = \mathcal{E}(U) \oplus \mathcal{A}(U)$ of graded algebras of exact and antiexact forms.

Actually, we can show that the algebra of antiexact forms generates almost the entire exterior algebra on U.

Theorem 6.4.3. *A form $\omega \in \Lambda^k(U), k \geq 1$ has a unique representation $\omega = d\alpha + \beta$ where $\alpha \in \mathcal{A}^{k-1}(U)$ and $\beta \in \mathcal{A}^k(U)$.*

Since we have assumed that U is contractible, any form ω can be expressed as $\omega = dH\omega + Hd\omega$. We then introduce the antiexact forms $\alpha = H\omega \in \mathcal{A}^{k-1}(U)$ and $\beta = Hd\omega \in \mathcal{A}^k(U)$ to represent ω as $\omega = d\alpha + \beta$. However, it remains now to demonstrate that this representation is unique. To this end, let us suppose that there exists another representation in the shape $\omega = d\alpha_1 + \beta_1$ where $\alpha_1 \in \mathcal{A}^{k-1}(U)$ and $\beta_1 \in \mathcal{A}^k(U)$. We then find $d(\alpha - \alpha_1) + (\beta - \beta_1) = 0$ and the exterior derivative of this form gives $d(\beta - \beta_1) = 0$. Because $\beta - \beta_1 \in \mathcal{A}^k(U)$ and Hd is the identity operator on $\mathcal{A}^k(U)$, we obtain at once $0 = Hd(\beta - \beta_1) = \beta - \beta_1$, or $\beta_1 = \beta$. Therefore, we get $d(\alpha - \alpha_1) = 0$ and the Poincaré lemma leads to $\alpha - \alpha_1 = d\gamma$ where $\gamma \in \Lambda^{k-2}(U)$ whenever $k > 1$. Since $\alpha - \alpha_1 \in \mathcal{A}^{k-1}(U)$, we find that $H(\alpha - \alpha_1) = Hd\gamma = 0$. Hence, the relation $\gamma = dH\gamma + Hd\gamma = dH\gamma$ gives rise to $\alpha - \alpha_1 = d^2H\gamma = 0$, or $\alpha_1 = \alpha$. Thus, this representation is unique.

But, if $k = 1$, then we have $\alpha - \alpha_1 \in \mathcal{A}^0(U) = \Lambda^0(U)$ and the condition $d(\alpha - \alpha_1) = 0$ results in $\alpha_1 = \alpha + \mathit{constant}$. Namely, in this case the form α can only be determined uniquely up to a constant.

This theorem can be symbolically expressed in the form

$$\Lambda^k(U) = d\big(\mathcal{A}^{k-1}(U)\big) \oplus \mathcal{A}^k(U), \ k \geq 1 \qquad \square$$

Example 6.4.1. $\omega \in \Lambda^1(\mathbb{R}^3)$ is given by $\omega = 2x\,dx + z\,dy - y^2\,dz$ so that we get $d\omega = -(1 + 2y)\,dy \wedge dz$. If we choose the point $\mathbf{x}_0 = (0, 0, 0)$ as the centre, the radius vector becomes $\mathcal{H} = x\,\partial_x + y\,\partial_y + z\,\partial_z$. Then, by applying the homotopy operator, we obtain

$$\begin{aligned}H\omega &= \int_0^1 \mathbf{i}_{\mathcal{H}}(2tx\,dx + tz\,dy - t^2y^2\,dz)\,dt \\ &= \int_0^1 (2tx^2 + tyz - t^2y^2z)\,dt = x^2 + \tfrac{1}{2}yz - \tfrac{1}{3}y^2z \\ Hd\omega &= -\int_0^1 t(1+2ty)\mathbf{i}_{\mathcal{H}}(dy\wedge dz)dt \\ &= -\int_0^1 t(1+2ty)(ydz - zdy)dt = \left(\tfrac{1}{2} + \tfrac{2}{3}y\right)(z\,dy - y\,dz)\end{aligned}$$

Hence, the form ω is expressible as

$$\omega = d\left(x^2 + \tfrac{1}{2}yz - \tfrac{1}{3}y^2z\right) + \left(\tfrac{1}{2} + \tfrac{2}{3}y\right)(z\,dy - y\,dz) \qquad \blacksquare$$

Let us now consider two antiexact forms $\alpha \in \mathcal{A}^k(U)$ and $\beta \in \mathcal{A}^l(U)$. Since we know that $\alpha\wedge\beta \in \mathcal{A}^{k+l}(U)$, we can write $Hd(\alpha\wedge\beta) = \alpha\wedge\beta$ whence we deduce that $H(d\alpha\wedge\beta) + (-1)^k H(\alpha\wedge d\beta) = \alpha\wedge\beta$. Hence, we obtain

$$H(d\alpha\wedge\beta) = \alpha\wedge\beta + (-1)^{k+1}H(\alpha\wedge d\beta). \qquad (6.4.3)$$

This relation can be interpreted as a sort of integration by parts.

6.5. CHANGE OF CENTRE

The open set $U \subseteq M$ may be contractible with respect to several points. Therefore, its homeomorphic image in $\mathbb{R}^m$ may appear to be star-shaped with respect to various centres. Since the homotopy operator is explicitly dependent on the location of the centre, we shall then try to establish the connection between homotopy operators associated with different centres.

Theorem 6.5.1. *According to a local chart, let $V = \varphi(U)$ be a star-shaped region of the Euclidean space where $U \subseteq M$ is an open set. If H_1 and H_2 are two homotopy operators associated with centres $\mathbf{x}_1$ and $\mathbf{x}_2$, respectively, then they are related by*

$$\begin{aligned}H_1\omega &= H_2\omega + \gamma + d\lambda \;\; \textit{if } deg\,\omega > 1, \\ H_1\omega &= H_2\omega + \gamma + c \quad \textit{if } deg\,\omega = 1\end{aligned}$$

where we define $\gamma = -H_2H_1d\omega \in \mathcal{A}_2(U)$ and $\lambda = H_2H_1\omega \in \mathcal{A}_2(U)$. c is a constant.

For a form $\omega \in \Lambda^k(U)$, we can of course write

$$\omega = dH_1\omega + H_1 d\omega = dH_2\omega + H_2 d\omega$$

from which we find that $(H_2 - H_1)d\omega = d(H_1 - H_2)\omega$. If we replace ω by $d\omega$ in this expression, we get $d(H_2 - H_1)d\omega = 0$. In view of the Poincaré lemma, there exists a form $\alpha \in \Lambda^{k-1}(U)$ so that one is able to write $(H_2 - H_1)d\omega = d\alpha$. According to Theorem 6.4.3 the form α is given by $\alpha = d\beta + \gamma$ where $\gamma \in \mathcal{A}_2^{k-1}(U)$. The relation $(H_2 - H_1)\,d\omega = d\gamma$ then leads to $(H_2^2 - H_2H_1)\,d\omega = H_2 d\gamma$ and $-\,H_2H_1 d\omega = \gamma$. On the other hand, the equality $d(H_1 - H_2)\,\omega = d\gamma$ or $d\big[(H_1 - H_2)\,\omega - \gamma\big] = 0$ gives rise to

$$(H_1 - H_2)\,\omega = \gamma + d\sigma,\ \ \sigma \in \Lambda^{k-2}(U)$$

if $k > 1$. If we write $\sigma = d\nu + \lambda$ where $\lambda \in \mathcal{A}_2^{k-2}(U)$, we obtain

$$H_1\omega = H_2\omega + \gamma + d\lambda.$$

On applying the operator H_2 on the above equality, we eventually find that $\lambda = H_2H_1\omega$. However, if $k = 1$, then it follows from the foregoing relation that $(H_1 - H_2)\omega - \gamma = c = \mathit{constant}$ and, consequently,

$$H_1\omega = H_2\omega + \gamma + c. \qquad \square$$

Example 6.5.1. $\omega \in \Lambda^1(\mathbb{R}^3)$ is given by $\omega = 2x\,dx + z\,dy - y^2\,dz$. Let us consider two centres $\mathbf{x}_1 = (1,0,1)$ and $\mathbf{x}_2 = (0,0,0)$. With the radius vector $\mathcal{H}_2$ we have already found in Example 6.4.1 that

$$H_2\omega = x^2 + \tfrac{1}{2}yz - \tfrac{1}{3}y^2 z.$$

With the radius vector $\mathcal{H}_1 = (x-1)\,\partial_x + y\,\partial_y + (z-1)\,\partial_z$, we are led to the relation

$$\begin{aligned} H_1\omega &= \int_0^1 \mathbf{i}_{\mathcal{H}_1}\big[2\big(1 + t(x-1)\big)dx + \big(1 + t(z-1)\big)dy - t^2y^2\,dz\big]dt \\ &= \int_0^1 \big[2(x-1)\big(1 + t(x-1)\big) + y\big(1 + t(z-1)\big) - t^2y^2(z-1)\big]dt \\ &= 2(x-1) + (x-1)^2 + y + \tfrac{1}{2}y(z-1) - \tfrac{1}{3}y^2(z-1) \\ &= H_2\omega + \tfrac{1}{2}y + \tfrac{1}{3}y^2 - 1. \end{aligned}$$

We see that $\gamma = \frac{1}{2}y + \frac{1}{3}y^2$ and $c = -\,1$. ■

6.6. CANONICAL FORMS OF 1-FORMS, CLOSED 2- FORMS

We consider a form $\omega = \omega_i(\mathbf{x})\,dx^i \in \Lambda^1(M)$ on an m-dimensional smooth manifold M. Starting with this form, let us construct the following sequence of forms of increasing degrees:

$$\begin{aligned}
&I_1 = \omega \in \Lambda^1(M),\ I_2 = d\omega \in \Lambda^2(M),\ I_3 = \omega \wedge d\omega \in \Lambda^3(M),\\
&I_4 = d\omega \wedge d\omega \in \Lambda^4(M), \dots, I_{2n} = (d\omega)^n = \underbrace{d\omega \wedge \cdots \wedge d\omega}_{n} \in \Lambda^{2n}(M),\\
&I_{2n+1} = \omega \wedge I_{2n} \in \Lambda^{2n+1}(M), \dots,\ n = 0, 1, 2, \dots .
\end{aligned}$$

Here, we adopt the convention that $I_0 = 1$. It is clearly observable from this sequence that we can write the recurrence relations

$$\begin{aligned}
I_{2n} &= d\omega \wedge I_{2n-2} = I_2 \wedge I_{2n-2},\\
I_{2n+1} &= I_1 \wedge I_{2n}, \qquad n = 0, 1, \dots .
\end{aligned} \tag{6.6.1}$$

By definition, we evidently have

$$dI_{2n} = 0,\quad dI_1 = I_2,\quad dI_{2n+1} = dI_1 \wedge I_{2n} = I_2 \wedge I_{2n} = I_{2n+2}.$$

This sequence must be finite because all forms whose degrees is greater than m are identically zero. We can thus write $I_{m+k} \equiv 0$ when $k = 1, 2, \dots$. However, this sequence may vanish beginning from a number $K \leq m$. Since $d\omega$ is a 2-form, its rank is $r = 2k \leq m$ [*see* p. 39]. This means that we can express $d\omega$ as follows

$$d\omega = f^1 \wedge g^1 + f^2 \wedge g^2 + \cdots + f^k \wedge g^k,\ f^i, g^i \in \Lambda^1(M), i = 1, \dots, k$$

and we get $(d\omega)^k \neq 0, (d\omega)^{k+n} = 0, n = 1, 2, \dots$ [*see* p. 44].

Let us now suppose that we have succeeded in determining an integer $K(\mathbf{x}, \omega) > 0$ such that

$$I_{K(\mathbf{x},\omega)} \neq 0,\quad I_{K(\mathbf{x},\omega)+n} = 0,\ n = 1, 2, \dots . \tag{6.6.2}$$

This number may generally be dependent on the points of the manifold. We assume that forms are defined on an open set $U \subseteq M$. We denote the homeomorphic image of U through appropriate charts by $V \subseteq \mathbb{R}^m$. The positive integer $K(\omega) = \sup_{\mathbf{x} \in V} K(\mathbf{x}, \omega)$ is called the ***Darboux class*** of the 1-form ω relative to the set U [French mathematician Jean Gaston Darboux (1842 -1917)]. The points at which $K(\mathbf{x}) < K(\omega)$ is said to be *critical points* of the form ω relative to the set U while the points at which $K(\mathbf{x}) = K(\omega)$ are called *regular points* relative to U. *We shall here assume that all points in the region U are regular.* The ***rank*** $r(\omega)$ of the form ω relative to U is defined as the greatest *even integer* less than or equal to $K(\omega)$ whereas the number $\epsilon(\omega) = K(\omega) - r(\omega)$ is called the ***index*** of ω relative to U. According to this definition $\epsilon(\omega)$ is either 0 or 1.

Let us first prove the following lemma.

Lemma 6.6.1. *Let $\Omega \in \Lambda^k(M)$ be a closed simple k-form. Then the form Ω is expressible as exterior products of so called gradients, that is, as $\Omega = dg^1 \wedge dg^2 \wedge \cdots \wedge dg^k$ where $g^i \in \Lambda^0(M)$, $i = 1, \ldots, k$.*

Let us take the closed form $\Omega = \omega^1 \wedge \omega^2 \wedge \cdots \wedge \omega^k \neq 0$ where the forms $\omega^i \in \Lambda^1(M)$, $i = 1, \ldots, k$ are linearly independent. We now consider the ideal $\mathcal{I}(\omega^1, \ldots, \omega^k)$. We will show that this ideal is also closed. Since $d\Omega = 0$, we readily obtain

$$\begin{aligned} d\Omega &= \sum_{i=1}^{k} (-1)^{i-1} \omega^1 \wedge \cdots \wedge \omega^{i-1} \wedge d\omega^i \wedge \omega^{i+1} \wedge \cdots \wedge \omega^k \\ &= \sum_{i=1}^{k} d\omega^i \wedge \omega^1 \wedge \cdots \wedge \omega^{i-1} \wedge \omega^{i+1} \wedge \cdots \wedge \omega^k = 0 \end{aligned}$$

where we have adopted the convention $\omega^0 = 1$. It follows from the above expression that we arrive at the relation

$$\begin{aligned} 0 = \omega^j \wedge d\Omega &= \sum_{i=1}^{k} \omega^j \wedge d\omega^i \wedge \omega^1 \wedge \cdots \wedge \omega^{i-1} \wedge \omega^{i+1} \wedge \cdots \wedge \omega^k \\ &= \omega^j \wedge d\omega^j \wedge \omega^1 \wedge \cdots \wedge \omega^{j-1} \wedge \omega^{j+1} \wedge \cdots \wedge \omega^k \\ &= (-1)^{j+1} d\omega^j \wedge \omega^1 \wedge \cdots \wedge \omega^{j-1} \wedge \omega^j \wedge \omega^{j+1} \wedge \cdots \wedge \omega^k \end{aligned}$$

for every form $\omega^j \in \Lambda^1(M)$. This implies that

$$d\omega^i \wedge \Omega = 0, \; i = 1, \ldots, k.$$

These relations show according to Theorem 5.10.2 that the ideal $\mathcal{I}(\omega^1, \ldots, \omega^k)$ is closed. Hence, as we have demonstrated in Theorem 5.13.5 we can find *independent* functions $f^i \in \Lambda^0(M)$, $i = 1, \ldots, k$ so that this ideal is equivalent to the ideal $\mathcal{I}(df^1, \ldots, df^k)$. Furthermore we have to write

$$\omega^i = A^i_j \, df^j, \;\; A^i_j \in \Lambda^0(M)$$

whence we deduce that

$$\Omega = \omega^1 \wedge \cdots \wedge \omega^k = \alpha df^1 \wedge \cdots \wedge df^k, \; \alpha = \det(A^i_j) \in \Lambda^0(M).$$

On the other hand, we must have $d\Omega = d\alpha \wedge df^1 \wedge \cdots \wedge df^k = 0$. Let us now choose new local coordinates for the manifold M as f^i, x^a where $i = 1, \ldots, k$ and $a = k+1, \ldots, m$. Because the function α may presently be written as

$$\alpha = \alpha(f^1, \ldots, f^k, x^{k+1}, \ldots, x^m),$$

the expression $d\Omega = 0$ can be cast into the form

$$\Big(\frac{\partial \alpha}{\partial f^i}\, df^i + \frac{\partial \alpha}{\partial x^a}\, dx^a\Big) \wedge df^1 \wedge \cdots \wedge df^k$$
$$= \frac{\partial \alpha}{\partial x^a}\, dx^a \wedge df^1 \wedge \cdots \wedge df^k = 0.$$

These equations yield

$$\frac{\partial \alpha}{\partial x^a} = 0,$$

where $a = k+1, \ldots, m$ so that we find $\alpha = \alpha(f^1, \ldots, f^k)$. Let us now define functions

$$g^1 = g^1(f^1, \ldots, f^k),\ g^2 = f^2,\ \ldots,\ g^k = f^k$$

such that we can express α without loss of generality as

$$\alpha = \frac{\partial g^1}{\partial f^1} \neq 0.$$

Since we now have

$$\det\Big(\frac{\partial g^i}{\partial f^j}\Big) = \begin{vmatrix} \dfrac{\partial g^1}{\partial f^1} & \dfrac{\partial g^1}{\partial f^2} & \cdots & \dfrac{\partial g^1}{\partial f^k} \\ 0 & 1 & \cdots & 0 \\ \vdots & \vdots & \vdots & \vdots \\ 0 & 0 & \cdots & 1 \end{vmatrix} = \frac{\partial g^1}{\partial f^1} = \alpha \neq 0,$$

the functions $g^1, \ldots, g^k$ are independent and we get

$$dg^1 \wedge \cdots \wedge dg^k = \det\,(\partial g^i/\partial f^j)\, df^1 \wedge \cdots \wedge df^k.$$

Hence, we are led to the conclusion

$$\Omega = dg^1 \wedge \cdots \wedge dg^k,\ \ g^i \in \Lambda^0(M),\ i = 1, \ldots, k. \qquad \square$$

We can now prove the following theorem.

Theorem 6.6.1 (The Darboux Theorem). *If the Darboux class of the form $\omega \in \Lambda^1(M)$ is K, then the canonical form of ω is given by*

$$\omega = u^\alpha dv_\alpha + \epsilon(\omega)\, dv_{k+1},\ \ u^\alpha, v_\alpha, v_{k+1} \in \Lambda^0(M),\ \alpha = 1, \ldots, k$$

where $k = [K/2]$ *denotes the greatest integer that is less than or equal to* $K/2$. *If* K *is even, then* $\epsilon(\omega) = 0$ *whereas* $\epsilon(\omega) = 1$ *if* K *is odd.*

We shall prove this theorem in two steps.

(i). *Darboux class is an even integer.* Thus, we can write $K = 2k$ so that we have to take $k = [K/2] = K/2$. According to the definition of the Darboux class, we obviously find

$$I_{2k} \neq 0,\ I_{2k+1} = 0,\ I_{2k+2} = 0, \ldots .$$

Hence, the rank of the form $d\omega \in \Lambda^2(M)$ is $2k$ implying that $I_{2k} \in \Lambda^{2k}(M)$ is a simple form. Since $dI_{2k} = 0$, in view of Lemma 6.6.1 the form I_{2k} is a gradient product, i.e., by means of independent functions $u^1, v_1, u^2, v_2, \ldots,$ u^k, v_k the form I_{2k} can be depicted as follows

$$I_{2k} = k!\, du^1 \wedge dv_1 \wedge du^2 \wedge dv_2 \wedge \cdots \wedge du^k \wedge dv_k.$$

Accordingly the form $d\omega$ has the following structure [*see* Sec. 1.6]

$$d\omega = du^\alpha \wedge dv_\alpha,\ \ u^\alpha, v_\alpha \in \Lambda^0(M),\ \alpha = 1, \ldots, k.$$

On the other hand, the satisfaction of the condition

$$I_{2k+1} = \omega \wedge I_{2k} = k!\, \omega \wedge du^1 \wedge dv_1 \wedge du^2 \wedge dv_2 \wedge \cdots \wedge du^k \wedge dv_k = 0$$

suggests due to Theorem 5.3.1 that the 1-forms $\omega, du^1, dv_1, \ldots, du^k, dv_k$ are linearly dependent. Therefore, there exist functions $f_\alpha, g^\alpha \in \Lambda^0(M)$ that enable us to write

$$\omega = f_\alpha\, du^\alpha + g^\alpha\, dv_\alpha.$$

Hence, the form ω belongs to the ideal $\mathcal{I}(du^\alpha, dv_\alpha)$. In this situation, we naturally get

$$d\omega = du^\alpha \wedge dv_\alpha = df_\alpha \wedge du^\alpha + dg^\alpha \wedge dv_\alpha,$$

Since $2k \leq m$, we are free to choose new local coordinates as follows: u^α, $v_\alpha, \alpha = 1, \ldots, k; x^a, a = 2k+1, \ldots, m$. In the equation just above, let us evaluate differentials of functions f_α, g^α with respect to new coordinates. On comparing both sides, we find that

$$\frac{\partial f_\alpha}{\partial x^a} = 0, \quad \frac{\partial g^\alpha}{\partial x^a} = 0 \tag{6.6.3}$$

since there are no terms like $dx^a \wedge du^\alpha$ and $dx^a \wedge dv_\alpha$ at the left hand side of that expression. We thus conclude that these function must have the forms $f_\alpha = f_\alpha(u^\beta, v_\beta)$ and $g^\alpha = g^\alpha(u^\beta, v_\beta)$. Remaining terms then conduce to the relations

$$\frac{\partial f_\alpha}{\partial u^\beta}du^\beta\wedge du^\alpha+\frac{\partial f_\alpha}{\partial v_\beta}dv_\beta\wedge du^\alpha+\frac{\partial g^\alpha}{\partial u^\beta}du^\beta\wedge dv_\alpha$$
$$+\frac{\partial g^\alpha}{\partial v_\beta}dv_\beta\wedge dv_\alpha=du^\alpha\wedge dv_\alpha.$$

However, on utilising the antisymmetry of the exterior product we can transform this expression into the form

$$\frac{1}{2}\Big(\frac{\partial f_\alpha}{\partial u^\beta}-\frac{\partial f_\beta}{\partial u^\alpha}\Big)du^\beta\wedge du^\alpha+\frac{1}{2}\Big(\frac{\partial g^\alpha}{\partial v_\beta}-\frac{\partial g^\beta}{\partial v_\alpha}\Big)dv_\beta\wedge dv_\alpha$$
$$+\Big(\frac{\partial g^\beta}{\partial u^\alpha}-\frac{\partial f_\alpha}{\partial v_\beta}\Big)du^\alpha\wedge dv_\beta=\delta^\beta_\alpha\,du^\alpha\wedge dv_\beta$$

which result in the equations

$$\frac{\partial f_\alpha}{\partial u^\beta}=\frac{\partial f_\beta}{\partial u^\alpha},\quad \frac{\partial g^\alpha}{\partial v_\beta}=\frac{\partial g^\beta}{\partial v_\alpha},\quad \frac{\partial g^\beta}{\partial u^\alpha}-\frac{\partial f_\alpha}{\partial v_\beta}=\delta^\beta_\alpha. \tag{6.6.4}$$

The equations $(6.6.4)_{1-2}$ ensure the existence of functions ϕ_1 and ψ_1 that make it possible for us to write

$$f_\alpha=\frac{\partial\phi_1}{\partial u^\alpha},\ g^\alpha=\frac{\partial\psi_1}{\partial v_\alpha},\quad \phi_1,\psi_1=\phi_1,\psi_1(u^\alpha,v_\alpha,x^a). \tag{6.6.5}$$

But, because of (6.6.3) we get

$$\frac{\partial^2\phi_1}{\partial u^\alpha\partial x^a}=\frac{\partial^2\psi_1}{\partial v_\alpha\partial x^a}=0$$

whence we obtain

$$\begin{aligned}\phi_1(\mathbf{u},\mathbf{v})&=\phi(\mathbf{u},\mathbf{v})+\overline{\phi}(\mathbf{v},x^a),\\ \psi_1(\mathbf{u},\mathbf{v})&=\psi(\mathbf{u},\mathbf{v})+\overline{\psi}(\mathbf{u},x^a).\end{aligned} \tag{6.6.6}$$

If we insert the expressions (6.6.5) and (6.6.6) into the equations $(6.6.4)_3$, we find that

$$\frac{\partial^2\psi_1}{\partial v_\beta\partial u^\alpha}-\frac{\partial^2\phi_1}{\partial u^\alpha\partial v_\beta}=\frac{\partial^2}{\partial v_\beta\partial u^\alpha}(\psi_1-\phi_1)=\frac{\partial^2}{\partial v_\beta\partial u^\alpha}(\psi-\phi)=\delta^\beta_\alpha.$$

The integration of these equations yields readily

$$\psi-\phi=u^\alpha v_\alpha+\Phi(\mathbf{u})+\Psi(\mathbf{v}) \tag{6.6.7}$$

and it follows from (6.6.5) that

$$f_\alpha = \frac{\partial \phi_1}{\partial u^\alpha} = \frac{\partial \phi}{\partial u^\alpha}, \quad g^\alpha = \frac{\partial \psi_1}{\partial v_\alpha} = \frac{\partial \psi}{\partial v_\alpha}$$

On the other hand, we obtain from the expression (6.6.7) that

$$\frac{\partial \psi}{\partial v_\alpha} = u^\alpha + \frac{\partial \phi}{\partial v_\alpha} + \frac{\partial \Psi}{\partial v_\alpha},$$

Hence, the form ω is expressible as

$$\begin{aligned}\omega &= \frac{\partial \phi}{\partial u^\alpha} du^\alpha + \frac{\partial \psi}{\partial v_\alpha} dv_\alpha = u^\alpha dv_\alpha + \frac{\partial \phi}{\partial u^\alpha} du^\alpha + \frac{\partial \phi}{\partial v_\alpha} dv_\alpha + \frac{\partial \Psi}{\partial v_\alpha} dv_\alpha \\ &= u^\alpha dv_\alpha + d\widetilde{\phi}\,,\ \widetilde{\phi} = \phi + \Psi\end{aligned}$$

where the function $\widetilde{\phi}\,(\mathbf{u}, \mathbf{v})$ can be selected arbitrarily. Therefore, if we take $\widetilde{\phi} = constant$, the canonical form of ω is found to be

$$\omega = u^\alpha dv_\alpha = u^1 dv_1 + u^2 dv_2 + \cdots + u^k dv_k. \tag{6.6.8}$$

It is clear that this representation is not unique. For instance, because of the identity

$$u^\alpha dv_\alpha = d(u^\alpha v_\alpha) - v_\alpha du^\alpha$$

we get $\omega = \ -v_\alpha du^\alpha$ if we choose $\widetilde{\phi} \ = \ -u^\alpha v_\alpha$.

(ii). *Darboux class is an odd integer.* Thus we can write $K = 2k + 1$. Hence, we can take $k = [K/2] = (K-1)/2$ which requires that

$$I_{2k} \neq 0,\ I_{2k+1} \neq 0,\ I_{2k+2} = 0,\ \ldots.$$

Since $I_{2k} \neq 0$ and $I_{2k+2} = 0$, the rank of the form $d\omega \in \Lambda^2(M)$, that must be an even number, is again $2k$. This implies that I_{2k} is still a simple form and a gradient product. On the other hand, we can write $I_{2k+1} = \omega \wedge I_{2k}$. Since I_{2k} is a simple form and ω is a 1-form, the form $I_{2k+1} \in \Lambda^{2k+1}(M)$ is also a simple form. I_{2k+1} is a closed form because $dI_{2k+1} = I_{2k+2} = 0$. Consequently, I_{2k+1} is likewise a gradient product. Since $I_{2k+1} = \omega \wedge I_{2k}$ $\neq 0$, the form $\omega \in \Lambda^1(M)$ cannot be expressed as a linear combination of factor forms of the simple form I_{2k}. Therefore the form I_{2k} is a divisor of the form I_{2k+1}. Since I_{2k+1} is a gradient product, this form is expressible as

$$I_{2k+1} = d\eta \wedge I_{2k}, \quad \eta \in \Lambda^0(M).$$

Thus, just like in the part (i), we can write

$$I_{2k} = k!\,du^1 \wedge dv_1 \wedge \cdots \wedge du^k \wedge dv_k,$$
$$d\omega = du^\alpha \wedge dv_\alpha,\ u^\alpha, v_\alpha \in \Lambda^0(M),\ \alpha = 1, \dots, k.$$

Let us now introduce a form $\omega' \in \Lambda^1(M)$ through a gradient transformation given as

$$\omega' = \omega - d\lambda$$

where $\lambda \in \Lambda^0(M)$. We obviously get $d\omega' = d\omega$ so that we obtain $I'_{2k} = I_{2k}$ while I'_{2k+1} is found to be

$$I'_{2k+1} = \omega' \wedge I'_{2k} = \omega \wedge I_{2k} - d\lambda \wedge I_{2k} = I_{2k+1} - d\lambda \wedge I_{2k}.$$

We thus arrive at the relation

$$I'_{2k+1} = d\eta \wedge I_{2k} - d\lambda \wedge I_{2k} = d(\eta - \lambda) \wedge I_{2k}.$$

On choosing the arbitrary function λ as $\lambda = \eta$, we conclude that

$$I'_{2k} \neq 0 \ \text{ and } \ I'_{2k+1} = 0.$$

This amounts to say that the Darboux class of the form ω' is $K = 2k$ and its canonical form turns out to be $\omega' = u^\alpha dv_\alpha$ as in (i). Hence, we obtain $\omega = u^\alpha dv_\alpha + d\eta$. Since the functions u^α, v_α are independent, we can write $\eta = \eta(\mathbf{u}, \mathbf{v}, x^a)$ by a choice of local coordinates as above. Thus the form $d\lambda$ is not expressible as a linear combination of the forms du^α, dv_α so it does not belong to the ideal generated by these forms. Hence, the function $\eta = v_{k+1}$ is independent of the functions u^α, v_α. Ultimately, in terms of $2k+1$ independent functions the canonical form of ω now becomes

$$\omega = u^\alpha dv_\alpha + dv_{k+1} = u^1 dv_1 + \cdots + u^k dv_k + dv_{k+1}. \tag{6.6.9}$$

This finishes the proof of the theorem showing that we are now able to write

$$\omega = u^\alpha dv_\alpha + \epsilon(\omega) dv_{k+1} = u^1 dv_1 + \cdots + u^k dv_k + \epsilon(\omega) dv_{k+1}. \quad \square$$

Example 6.6.1. We take the form $\omega = 2x\,dx + z\,dy - y^2\,dz \in \Lambda^1(\mathbb{R}^3)$ into consideration. Let us construct the sequence

$$I_1 = \omega, I_2 = -(1+2y)dy \wedge dz, I_3 = -2x(1+2y)dx \wedge dy \wedge dz, I_4 = 0.$$

We thus find $K = 3$ and $k = 1$. Let us choose $v_2 = x^2$ and assume that the functions u^1 and v_1 depend only on y and z. Then the relations

$$u^1 \frac{\partial v_1}{\partial y} = z, \quad u^1 \frac{\partial v_1}{\partial z} = -\,y^2$$

yield the partial differential equation $y^2(\partial v_1/\partial y) + z(\partial v_1/\partial z) = 0$ whose solution is $v_1 = f(y^{-1} + \log z)$. For simplicity, let us choose the function as $v_1 = y^{-1} + \log z$. Then we find that $u^1 = -\,y^2 z$. Hence ω can now be expressed in the following canonical form

$$\omega = -\,y^2 z\, d(y^{-1} + \log z) + d(x^2) \qquad \blacksquare$$

Example 6.6.2. Let us consider the form

$$\omega = (x - y^2)dx + (y^3 - z^2)dy + t^2 dt \in \Lambda^1(\mathbb{R}^4).$$

Since we obtain

$$I_1 = \omega, \quad I_2 = -\,2ydy \wedge dx - 2zdz \wedge dy,$$
$$I_3 = -\,2z(x - y^2)\, dx \wedge dz \wedge dy, \qquad I_4 = 0$$

we find that $K = 3$ and $k = 1$. Hence one can write $\omega = u^1 dv_1 + dv_2$. We then readily show that

$$u^1 = 2xy - z^2, \; v_1 = y, \; v_2 = \frac{x^2}{2} + \frac{y^4}{4} - xy^2 + \frac{t^3}{3}. \qquad \blacksquare$$

The number K can be equal at most to the dimension m of the manifold. Therefore, a form in $\Lambda^1(M)$ is expressible at most $k = [m/2] + \epsilon$ number of terms. For instance, we can write

$$m = 1 : \omega = dv_1, \qquad v_1 \in \Lambda^0(\mathbb{R}),$$
$$m = 2 : \omega = u^1 dv_1, \qquad u^1, v_1 \in \Lambda^0(\mathbb{R}^2),$$
$$m = 3 : \omega = u^1 dv_1 + dv_2, \quad u^1, v_1, v_2 \in \Lambda^0(\mathbb{R}^3).$$

We can now discuss the second Darboux theorem concerning closed 2-forms which is in fact an almost trivial corollary of the Darboux theorem.

Theorem 6.6.2. *Let U be an open set of m-dimensional manifold M contractible to one of its points. The homeomorphic image of this set in $\mathbb{R}^m$ through an appropriate chart is a star-shaped region. The canonical form of a closed form $\omega \in \Lambda^2(U)$ is given by*

$$\omega = du^\alpha \wedge dv_\alpha \qquad (6.6.10)$$

where the functions $u^\alpha, v_\alpha \in \Lambda^0(U), \alpha = 1, \dots, k$ are independent and $k = [K/2]$.

Poincaré lemma states that there exists a form $\Omega \in \Lambda^1(U)$ such that the relation $\omega = d\Omega$ is satisfied. If the Darboux class of Ω is K, then we can write $\Omega = u^\alpha dv_\alpha + \epsilon dv_{k+1}$ where $\alpha = 1, \dots, k, k = [K/2]$. We thus find $\omega = du^\alpha \wedge dv_\alpha$. Since every differentiable manifold is locally contractible, then every form $\omega \in \Lambda^2(M)$ becomes locally expressible in the *canonical form* (6.6.10). □

We know that a form $\omega \in \Lambda^1(M)$ is completely integrable if only it can be written as $\omega = \xi\, d\eta$ where $\xi, \eta \in \Lambda^0(M)$. This is only possible if $k = 0$ and $k = 1$ or Darboux classes $K = 1$ and $K = 2$. In those cases, we get $\omega = dv_1$ and $\omega = u^1 dv_1$, respectively. We thus conclude that *1-forms whose Darboux classes are $K \geq 3$ are not completely integrable.* This result coincides with the concept of accessibility propounded by Greek-German mathematician Constantin Carathéodory (1873-1950). Let us consider a form $\omega \in \Lambda^1(M)$. We say that a form ω has the ***inaccessibility*** property if and only if a sufficiently small neighbourhood of any point $p \in M$ contains a point $q \in M$ that cannot be reached by a path (a curve) through p entirely on M satisfying the exterior equation $\omega = 0$. If a 1-form ω does not have the inaccessibility property, namely, if in a neighbourhood of any point p there is no point that cannot be reached by such paths, then the form ω possesses the ***accessibility*** property.

Theorem 6.6.3 (The Carathéodory Theorem). *A form $\omega \in \Lambda^1(M)$ that is not identically zero has the inaccessibility property if and only if its Darboux class is less than three. If its Darboux class is greater than or equal to three, then ω has the accessibility property.*

If $K = 1$, then $\omega = dv_1$ and the exterior equation $\omega = 0$ holds only on $(m-1)$-dimensional submanifolds $\mathcal{S}_c$ described by $v_1(\mathbf{x}) = c$ where c's are arbitrary constants. Let us assume that a point $p \in M$ is located on $\mathcal{S}_{c_1}$. For a sufficiently small number δ, we immediately see that there is a point q in any neighbourhood of the point p that belongs to the submanifold $\mathcal{S}_{c_1+\delta}$. Since these two hypersurfaces cannot intersect, no curve through the point p lying on $\mathcal{S}_{c_1}$ thereby satisfying the equation $\omega = 0$ can reach to the point q. If $K = 2$, then $\omega = u^1 dv_1$ and the exterior equation $\omega = 0$ holds on $(m-1)$-dimensional submanifolds $\mathcal{S}_c$ described by $v_1(\mathbf{x}) = c$. Thus, we naturally arrive at the same conclusion about inaccessibility.

Let us now consider the case $K \geq 3$. In this situation we have the representations

$$\omega = u^\alpha dv_\alpha \text{ if } K \text{ is even}; \quad \omega = u^\alpha dv_\alpha + dv_{k+1} \text{ if } K \text{ is odd}.$$

where $\alpha = 1, \dots, k, k = [K/2]$. Since the form ω is not identically zero, at least one of the functions u^α does not vanish in a neighbourhood of a point $p \in M$. On dividing the form ω by this function and renaming the indices if necessary, we can cast the exterior equation $\omega = 0$ into the form

$$u^1 dv_1 + \cdots + u^{r-1} dv_{r-1} + dv_r = 0 \tag{6.6.11}$$

where $r = k$ if K is even and $r = k + 1$ if K is odd. A solution of the equation (6.6.11) can evidently be taken as

$$v_\alpha(\mathbf{x}) = c_\alpha,\ \alpha = 1, \ldots, r-1;\ v_r(\mathbf{x}) = c_r. \tag{6.6.12}$$

The equations (6.6.12) prescribe obviously a family of $(m-r)$-dimensional submanifolds of M.

Let us now consider the points $p = \varphi^{-1}(\mathbf{x}_1)$ and $q = \varphi^{-1}(\mathbf{x}_2)$ on the manifold M and we introduce the constants

$$v_\alpha^1 = v_\alpha(\mathbf{x}_1),\ v_\alpha^2 = v_\alpha(\mathbf{x}_2),\ v_r^1 = v_r(\mathbf{x}_1),\ v_r^2 = v_r(\mathbf{x}_2);\ u_1^\alpha = u^\alpha(\mathbf{x}_1).$$

Then let us assume that we can choose a path $\mathbf{x} = \boldsymbol{\xi}(t)$ on M from the point p to the point q such that the following relations are satisfied

$$v_\alpha\big(\boldsymbol{\xi}(t)\big) = v_\alpha(t) = v_\alpha^1 + (v_\alpha^2 - v_\alpha^1)t,$$
$$u^\alpha\big(\boldsymbol{\xi}(t)\big) = u^\alpha(t) = u_1^\alpha + h^\alpha t$$

where $\boldsymbol{\xi}(0) = \mathbf{x}_1$ and $\boldsymbol{\xi}(1) = \mathbf{x}_2$. $h^\alpha, \alpha = 1, \ldots, r-1$ are presently arbitrary constants. Such a path can be chosen, for instance, by first introducing the functions

$$\xi^{\mathfrak{a}}(t) = x_1^{\mathfrak{a}} + t(x_2^{\mathfrak{a}} - x_1^{\mathfrak{a}}),\ \mathfrak{a} = 2r-1, \ldots, m$$

and then employing the $2r-2$ number of equations

$$v_\alpha\big(\xi^a(t), \xi^{\mathfrak{a}}(t)\big) = v_\alpha^1 + (v_\alpha^2 - v_\alpha^1)t,\ \ u^\alpha\big(\xi^a(t), \xi^{\mathfrak{a}}(t)\big) = u_1^\alpha + h^\alpha t$$

to determine the remaining functions $\xi^a(t), a = 1, \ldots, 2r-2$ in a neighbourhood of the point p. On the path from p to q, the equation (6.6.11) takes the form

$$\frac{dv_r}{dt} + u^\alpha \frac{dv_\alpha}{dt} = \frac{dv_r}{dt} + (u_1^\alpha + h^\alpha t)(v_\alpha^2 - v_\alpha^1) = 0.$$

The integration of this simple differential equation for v_r with the initial condition $v_r(0) = v_r^1$ gives

$$v_r(t) = v_r^1 + (v_\alpha^1 - v_\alpha^2)(u_1^\alpha + \frac{1}{2}h^\alpha t)t.$$

We now need to select the constants h^α in such a way that the relation

$$v_r^2 = v_r^1 + (v_\alpha^1 - v_\alpha^2)(u_1^\alpha + \frac{1}{2}h^\alpha)$$

will hold at $t = 1$. If $v_{\alpha_0}^1 \neq v_{\alpha_0}^2$ for an index α_0, then we reach to our objective by taking $h^\alpha = 0$ for all $\alpha \neq \alpha_0$ and

$$h^{\alpha_0} = 2\frac{v_r^2 - v_r^1 - (v_\alpha^1 - v_\alpha^2)u_1^\alpha}{v_{\alpha_0}^1 - v_{\alpha_0}^2}.$$

If $v_\alpha^1 = v_\alpha^2$ for all $\alpha = 1, \dots, r-1$, we first determine a path such that

$$v_r(t) = v_r^1 + at, \quad v_\alpha(t) = v_\alpha^1 + b_\alpha t, \quad u^\alpha(t) = u_1^\alpha.$$

Since the form ω does not vanish, $u_1^{\alpha_0} \neq 0$ at least for an index α_0. Along the path, the exterior equation $\omega = 0$ can only be satisfied if $a + b_\alpha u_1^\alpha = 0$. Let us choose $b_{\alpha_0} = -a/u_1^{\alpha_0}$ and $b_\alpha = 0$ for all $\alpha \neq \alpha_0$. In this case, we obtain $v_\alpha^1 = v_\alpha^2$ for all $\alpha \neq \alpha_0, v_{\alpha_0}(1) = v_{\alpha_0}^1 + b_{\alpha_0} \neq v_{\alpha_0}^1, v_r(1) = v_r^1 + a$. If we now choose $\big(v_r(1), v_\alpha(1) = v_\alpha^1\ (\alpha \neq \alpha_0), v_{\alpha_0}(1), u^\alpha(1) = u_1\big)$ as the new initial point, we can find a path on which $\omega = 0$ from this point to the point (v_r^2, v_α^2) because $v_{\alpha_0}(1) \neq v_{\alpha_0}^1$. Therefore, when $K \geq 3$ we can always find a path from the point $p \in M$ to reach to a point in a neighbourhood of p such that the exterior equation $\omega = 0$ is satisfied along this path. □

6.7. AN EXTERIOR DIFFERENTIAL EQUATION

Exterior equations involving exterior derivatives of exterior forms will be called *exterior differential equations*. This section is devoted to finding the solution of the exterior differential equation

$$d\Omega = \Gamma \wedge \Omega + \Sigma \tag{6.7.1}$$

defined on a *contractible open set* U of a manifold M by making use of the homotopy operator. $\Gamma \in \Lambda^1(U)$ and $\Sigma \in \Lambda^{k+1}(U)$ are given exterior forms. We look for all forms $\Omega \in \Lambda^k(U)$ satisfying the equation (6.7.1). For the existence of a solution, it is clear that the forms Γ and Σ cannot be assigned arbitrarily. The closure condition $d^2\Omega = 0$ requires clearly that the equality $d\Gamma \wedge \Omega - \Gamma \wedge d\Omega + d\Sigma = 0$ or

$$d\Sigma = \Gamma \wedge \Gamma \wedge \Omega + \Gamma \wedge \Sigma - d\Gamma \wedge \Omega = \Gamma \wedge \Sigma - d\Gamma \wedge \Omega$$

must be satisfied since $\Gamma \in \Lambda^1(U)$. By introducing the form $\Theta = d\Gamma$ where we obviously have $d\Theta = 0$, we can transform the system to be solved into

$$d\Omega = \Gamma \wedge \Omega + \Sigma, \;\; d\Sigma = \Gamma \wedge \Sigma - \Theta \wedge \Omega, \;\; \Theta = d\Gamma, \;\; d\Theta = 0.$$

Our aim is to determine all forms Ω, Σ, Γ and Θ satisfying the above relations. According to the Poincaré lemma, we can locally write $\Theta = d\theta$ where $\theta \in \Lambda^1(U)$. Since the form θ is incorporated in the above equations through only its exterior derivative, we can choose $\theta \in \mathcal{A}^1(U)$ without loss of generality according to Theorem 6.4.3. Let H be the homotopy operator on the contractible open set U. Then we get $H\theta = 0$ and find that

$$\theta = dH\theta + Hd\theta = H\Theta = Hd\Gamma.$$

On the other hand, the form $\Gamma \in \Lambda^1(U)$ can be expressed as

$$\Gamma = dH\Gamma + Hd\Gamma = dH\Gamma + \theta = d\gamma + \theta, \ \gamma = H\Gamma \in \mathcal{A}^0(U).$$

Let us now consider the transformations $\Omega = e^\rho\omega$ and $\Sigma = e^\rho\sigma$ where $\rho \in \Lambda^0(U)$. The relations

$$\begin{aligned} d\Omega &= e^\rho d\rho \wedge \omega + e^\rho d\omega = e^\rho(d\gamma + \theta) \wedge \omega + e^\rho\sigma \\ d\Sigma &= e^\rho d\rho \wedge \sigma + e^\rho d\sigma = e^\rho(d\gamma + \theta) \wedge \sigma - e^\rho d\theta \wedge \omega \end{aligned}$$

lead to

$$d\omega = (d\gamma - d\rho + \theta) \wedge \omega + \sigma, \ d\sigma = (d\gamma - d\rho + \theta) \wedge \sigma - d\theta \wedge \omega.$$

We now choose the arbitrary function ρ as $\rho = \gamma = H\Gamma \in \mathcal{A}^0(U)$. We then reach to the equations

$$d\omega = \theta \wedge \omega + \sigma, \quad d\sigma = \theta \wedge \sigma - d\theta \wedge \omega.$$

With the definition $\beta = \theta \wedge \omega \in \Lambda^{k+1}(U)$, we obtain

$$d\beta = d\theta \wedge \omega - \theta \wedge d\omega = d\theta \wedge \omega - \theta \wedge (\theta \wedge \omega + \sigma) = d\theta \wedge \omega - \theta \wedge \sigma.$$

Hence, our system is reduced to a much simpler system

$$d\omega = \beta + \sigma, \quad d\sigma = \ -d\beta.$$

The identities $\omega = dH\omega + Hd\omega$, $\sigma = dH\sigma + Hd\sigma$ then yield

$$\omega = dH\omega + H(\beta + \sigma), \ \sigma = dH\sigma - Hd\beta = dH\sigma - \beta + dH\beta.$$

If we define the forms $\phi = H\omega \in \mathcal{A}^{k-1}(U)$, $\eta = H\sigma \in \mathcal{A}^k(U)$, we get

$$\omega = d\phi + \eta + H\beta, \quad \sigma = d\eta - \beta + dH\beta.$$

On using the above relations, we can easily determine the form β. If we write

$$\beta = \theta \wedge \omega = \theta \wedge d\phi + \theta \wedge \eta + \theta \wedge H\beta$$

and note that $\theta \wedge \eta + \theta \wedge H\beta \in \mathcal{A}^{k+1}(U)$, we then find

$$H\beta = H(\theta \wedge d\phi) + H(\theta \wedge \eta + \theta \wedge H\beta) = H(\theta \wedge d\phi).$$

We thus obtain

$$\beta = \theta \wedge d\phi + \theta \wedge \eta + \theta \wedge H(\theta \wedge d\phi)$$

and consequently

$$\begin{aligned}\omega &= d\phi + \eta + H(\theta \wedge d\phi),\\ \sigma &= d\big(\eta + H(\theta \wedge d\phi)\big) - \theta \wedge \big(d\phi + \eta + H(\theta \wedge d\phi)\big).\end{aligned}$$

Hence, the solution of the exterior differential equation is found to be

$$\begin{aligned}\Omega &= e^{\gamma}\big[d\phi + \eta + H(\theta \wedge d\phi)\big],\\ \Sigma &= e^{\gamma}\big[d\big(\eta + H(\theta \wedge d\phi)\big) - \theta \wedge \big(d\phi + \eta + H(\theta \wedge d\phi)\big)\big]\\ \Gamma &= d\gamma + \theta, \quad \Theta = d\Gamma = d\theta\end{aligned} \tag{6.7.2}$$

where $\gamma \in \mathcal{A}^0(U)$, $\theta \in \mathcal{A}^1(U)$, $\phi \in \mathcal{A}^{k-1}(U)$ and $\eta \in \mathcal{A}^k(U)$ are arbitrary forms. Now, introducing the arbitrary form $\chi = \eta + H(\theta \wedge d\phi) \in \mathcal{A}^k(U)$, we can express the above solution in a much simpler fashion as

$$\Omega = e^{\gamma}(d\phi + \chi), \;\; \Sigma = e^{\gamma} d\chi - \theta \wedge \big(d\phi + \chi\big). \tag{6.7.3}$$

If we take $\Sigma = 0$ in (6.7.1), we arrive at the following exterior differential equation

$$d\Omega = \Gamma \wedge \Omega \tag{6.7.4}$$

together with compatibility conditions $\Theta \wedge \Omega = 0, \Theta = d\Gamma$. Since we now have $\sigma = 0$, we find of course $\eta = H\sigma = 0$ and it follows from $(6.7.2)_2$ that $dH(\theta \wedge d\phi) - \theta \wedge \big(d\phi + H(\theta \wedge d\phi)\big) = 0$. We first calculate the exterior derivative of this expression, then consider its exterior product with the form $\theta \in \mathcal{A}^1(U)$ to get

$$d\big[\theta \wedge \big(d\phi + H(\theta \wedge d\phi)\big)\big] = d^2 H(\theta \wedge d\phi) = 0, \theta \wedge dH(\theta \wedge d\phi) = 0,$$

respectively. But, because we can write

$$d\big[\theta \wedge \big(d\phi + H(\theta \wedge d\phi)\big)\big] = d\theta \wedge \big(d\phi + H(\theta \wedge d\phi)\big) - \theta \wedge dH(\theta \wedge d\phi)$$

we must conclude that

$$d\theta \wedge \big(d\phi + H(\theta \wedge d\phi)\big) = 0. \tag{6.7.5}$$

Therefore, the solution of (6.7.4) is represented by

$$\Omega = e^{\gamma}\big[d\phi + H(\theta \wedge d\phi)\big], \; \gamma = H\Gamma, \; \theta = Hd\Gamma \tag{6.7.6}$$

subject to the condition (6.7.5). The forms $\Omega \in \Lambda^k(U)$ satisfying the equation (6.7.4) are called *recursive forms* with coefficient forms Γ.

A recursive form Ω is called a *gradient recursive form* if the coefficient form Γ is exact. In this case, we have $\theta = 0$ and the solution reduces to

$$\Omega = e^{\gamma}\, d\phi,\ \gamma = H\Gamma,\ \ d\Omega = d\gamma \wedge \Omega. \tag{6.7.7}$$

6.8. A SYSTEM OF EXTERIOR DIFFERENTIAL EQUATIONS

We shall now try to deal with a significantly more difficult problem. Let us consider a system of exterior differential equations

$$d\Omega^i = \ -\Gamma^i_j \wedge \Omega^j + \Sigma^i,\ i,j = 1,2,\dots,r \tag{6.8.1}$$

prescribed on a contractible open set U of a manifold M. Here $\Omega^i \in \Lambda^k(U)$ are forms to be determined, and $\Gamma^i_j \in \Lambda^1(U)$ and $\Sigma^i \in \Lambda^{k+1}(U)$ are assumed to be given forms. Minus sign in (6.8.1) is chosen for convenience. It would be rather advantageous to employ a matrix notation in order to discuss this problem more efficiently. Let us denote a *matrix form* whose all entries consist of forms Φ^p_q of the same degree by $\mathbf{\Phi} = [\Phi^p_q]$. If another matrix form with different degree is $\mathbf{\Psi} = [\Psi^p_q]$, we can define the exterior product of these two matrix forms by applying the usual rule of matrix multiplication, but replacing the ordinary multiplications by exterior products as follows

$$\mathbf{\Phi} \wedge \mathbf{\Psi} = [\Phi^p_k \wedge \Psi^k_q].$$

Obviously, we have $deg\,(\mathbf{\Phi} \wedge \mathbf{\Psi}) = deg\,(\mathbf{\Phi}) + deg\,(\mathbf{\Psi})$. We can easily verify that the transpose relation $(\mathbf{\Phi} \wedge \mathbf{\Psi})^{\mathrm{T}} = (-1)^{deg\,(\mathbf{\Phi})\,deg\,(\mathbf{\Psi})}\mathbf{\Psi}^{\mathrm{T}} \wedge \mathbf{\Phi}^{\mathrm{T}}$ will be satisfied. Therefore, we are now able to write the equations (6.8.1) in the form

$$d\mathbf{\Omega} = \ -\mathbf{\Gamma} \wedge \mathbf{\Omega} + \mathbf{\Sigma}. \tag{6.8.2}$$

As matrix forms, we shall use the notations $\mathbf{\Omega} \in \mathbf{\Lambda}^k(U)$, $\mathbf{\Gamma} \in \mathbf{\Lambda}^1(U)$ and $\mathbf{\Sigma} \in \mathbf{\Lambda}^{k+1}(U)$. The compatibility equations are naturally found by taking the exterior derivative of (6.8.2). We get $\mathbf{0} = \ -d\mathbf{\Gamma} \wedge \mathbf{\Omega} + \mathbf{\Gamma} \wedge d\mathbf{\Omega} + d\mathbf{\Sigma}$ that induces the relation

$$d\mathbf{\Sigma} = (d\mathbf{\Gamma} + \mathbf{\Gamma} \wedge \mathbf{\Gamma}) \wedge \mathbf{\Omega} - \mathbf{\Gamma} \wedge \mathbf{\Sigma}. \tag{6.8.3}$$

Let us now define the following forms

$$\mathbf{\Theta} = d\mathbf{\Gamma} + \mathbf{\Gamma} \wedge \mathbf{\Gamma},\ \ \Theta^i_j = d\Gamma^i_j + \Gamma^i_k \wedge \Gamma^k_j \tag{6.8.4}$$

where $\mathbf{\Theta} = [\Theta^i_j] \in \mathbf{\Lambda}^2(U)$. Thus, the relation

$$d\boldsymbol{\Sigma} = \boldsymbol{\Theta} \wedge \boldsymbol{\Omega} - \boldsymbol{\Gamma} \wedge \boldsymbol{\Sigma} \tag{6.8.5}$$

must be satisfied. On the other hand, the exterior derivative of (6.8.4) yields

$$\begin{aligned} d\boldsymbol{\Theta} &= d\boldsymbol{\Gamma} \wedge \boldsymbol{\Gamma} - \boldsymbol{\Gamma} \wedge d\boldsymbol{\Gamma} = -\boldsymbol{\Gamma} \wedge \boldsymbol{\Gamma} \wedge \boldsymbol{\Gamma} + \boldsymbol{\Theta} \wedge \boldsymbol{\Gamma} + \boldsymbol{\Gamma} \wedge \boldsymbol{\Gamma} \wedge \boldsymbol{\Gamma} - \boldsymbol{\Gamma} \wedge \boldsymbol{\Theta} \\ &= \boldsymbol{\Theta} \wedge \boldsymbol{\Gamma} - \boldsymbol{\Gamma} \wedge \boldsymbol{\Theta}. \end{aligned}$$

It is readily checked that the exterior derivative of the above expression vanishes identically. Consequently, the exterior differential equations to be treated take finally the shapes

$$\begin{aligned} d\boldsymbol{\Omega} &= -\boldsymbol{\Gamma} \wedge \boldsymbol{\Omega} + \boldsymbol{\Sigma}, \quad d\boldsymbol{\Sigma} = \boldsymbol{\Theta} \wedge \boldsymbol{\Omega} - \boldsymbol{\Gamma} \wedge \boldsymbol{\Sigma}, \\ d\boldsymbol{\Gamma} &= -\boldsymbol{\Gamma} \wedge \boldsymbol{\Gamma} + \boldsymbol{\Theta}, \quad d\boldsymbol{\Theta} = \boldsymbol{\Theta} \wedge \boldsymbol{\Gamma} - \boldsymbol{\Gamma} \wedge \boldsymbol{\Theta}. \end{aligned} \tag{6.8.6}$$

Our task is to find the admissible forms of $\boldsymbol{\Gamma}$ and $\boldsymbol{\Sigma}$, and to determine $\boldsymbol{\Omega}$. According to Theorem 6.4.3, we can take

$$\boldsymbol{\Gamma} = d\boldsymbol{\gamma} + \boldsymbol{\Gamma}_a, \ \ \boldsymbol{\gamma} \in \boldsymbol{\mathcal{A}}^0(U), \ \ \boldsymbol{\Gamma}_a \in \boldsymbol{\mathcal{A}}^1(U) \tag{6.8.7}$$

We can of course represent (6.8.7) explicitly as

$$\Gamma^i_j = d\gamma^i_j + (\Gamma_a)^i_j$$

where $\gamma^i_j \in \mathcal{A}^0(U)$ and $(\Gamma_a)^i_j \in \mathcal{A}^1(U)$. If the centre of the star-shaped homeomorphic image of the region U in $\mathbb{R}^m$ is $\mathbf{x}_0$, then we can take without loss of generality $\boldsymbol{\gamma}(\mathbf{x}_0) = \mathbf{0}$ since $\boldsymbol{\gamma}$ enters the equations through its differential. If we insert the expression (6.8.7) into the equation $(6.8.6)_3$, we find

$$d\boldsymbol{\Gamma}_a = -d\boldsymbol{\gamma} \wedge d\boldsymbol{\gamma} - \boldsymbol{\Gamma}_a \wedge d\boldsymbol{\gamma} - d\boldsymbol{\gamma} \wedge \boldsymbol{\Gamma}_a - \boldsymbol{\Gamma}_a \wedge \boldsymbol{\Gamma}_a + \boldsymbol{\Theta}. \tag{6.8.8}$$

Let us assume that $\mathbf{B} \in \boldsymbol{\mathcal{A}}^0(U)$ is an arbitrary regular $r \times r$ matrix, that is, $\det \mathbf{B} \neq 0$. We define the forms $\overline{\boldsymbol{\Gamma}}$ and $\widetilde{\boldsymbol{\Theta}}$ by

$$\overline{\boldsymbol{\Gamma}} = \mathbf{B}^{-1}\boldsymbol{\Gamma}_a\mathbf{B}, \qquad \widetilde{\boldsymbol{\Theta}} = \mathbf{B}^{-1}\boldsymbol{\Theta}\mathbf{B}. \tag{6.8.9}$$

On inverting these expressions, we get $\boldsymbol{\Gamma}_a = \mathbf{B}\overline{\boldsymbol{\Gamma}}\mathbf{B}^{-1}$ and $\boldsymbol{\Theta} = \mathbf{B}\widetilde{\boldsymbol{\Theta}}\mathbf{B}^{-1}$. It is clear that $\overline{\boldsymbol{\Gamma}} \in \boldsymbol{\mathcal{A}}^1(U)$. The exterior derivative of $(6.8.9)_1$ gives

$$d\overline{\boldsymbol{\Gamma}} = d\mathbf{B}^{-1} \wedge \boldsymbol{\Gamma}_a\mathbf{B} + \mathbf{B}^{-1}(d\boldsymbol{\Gamma}_a)\mathbf{B} - \mathbf{B}^{-1}\boldsymbol{\Gamma}_a \wedge d\mathbf{B}.$$

Differentiating $\mathbf{B}^{-1}\mathbf{B} = \mathbf{I}$, we find that $d\mathbf{B}^{-1}\mathbf{B} + \mathbf{B}^{-1}d\mathbf{B} = \mathbf{0}$ from which we deduce that $d\mathbf{B}^{-1} = -\mathbf{B}^{-1}(d\mathbf{B})\mathbf{B}^{-1}$. On using $(6.8.9)_1$, we finally get

$$d\overline{\boldsymbol{\Gamma}} = \mathbf{B}^{-1}(d\boldsymbol{\Gamma}_a)\mathbf{B} - \mathbf{B}^{-1}d\mathbf{B} \wedge \overline{\boldsymbol{\Gamma}} - \overline{\boldsymbol{\Gamma}}\mathbf{B}^{-1} \wedge d\mathbf{B}.$$

On the other hand, it follows from (6.8.8) that

$$\mathbf{B}^{-1}(d\boldsymbol{\Gamma}_a)\mathbf{B} = -\mathbf{B}^{-1}(d\boldsymbol{\gamma}\wedge d\boldsymbol{\gamma})\mathbf{B} - \mathbf{B}^{-1}\mathbf{B}\overline{\boldsymbol{\Gamma}}\mathbf{B}^{-1}\wedge(d\boldsymbol{\gamma})\mathbf{B}$$
$$- \mathbf{B}^{-1}d\boldsymbol{\gamma}\wedge\mathbf{B}\overline{\boldsymbol{\Gamma}}\mathbf{B}^{-1}\mathbf{B} - \mathbf{B}^{-1}\mathbf{B}\overline{\boldsymbol{\Gamma}}\mathbf{B}^{-1}\wedge\mathbf{B}\overline{\boldsymbol{\Gamma}}\mathbf{B}^{-1}\mathbf{B} + \widetilde{\boldsymbol{\Theta}}.$$

Consequently, we obtain

$$d\overline{\boldsymbol{\Gamma}} = -\mathbf{B}^{-1}(d\boldsymbol{\gamma}\wedge d\boldsymbol{\gamma})\mathbf{B} - \overline{\boldsymbol{\Gamma}}\wedge\mathbf{B}^{-1}(d\boldsymbol{\gamma})\mathbf{B} - \mathbf{B}^{-1}(d\boldsymbol{\gamma})\mathbf{B}\wedge\overline{\boldsymbol{\Gamma}} - \overline{\boldsymbol{\Gamma}}\wedge\overline{\boldsymbol{\Gamma}}$$
$$+\boldsymbol{\Theta} - \mathbf{B}^{-1}d\mathbf{B}\wedge\overline{\boldsymbol{\Gamma}} - \overline{\boldsymbol{\Gamma}}\wedge\mathbf{B}^{-1}d\mathbf{B} = -\overline{\boldsymbol{\Gamma}}\wedge\mathbf{B}^{-1}(d\boldsymbol{\gamma}\mathbf{B} + d\mathbf{B})$$
$$-\mathbf{B}^{-1}(d\boldsymbol{\gamma}\mathbf{B} + d\mathbf{B})\wedge\overline{\boldsymbol{\Gamma}} - \mathbf{B}^{-1}(d\boldsymbol{\gamma}\wedge d\boldsymbol{\gamma})\mathbf{B} - \overline{\boldsymbol{\Gamma}}\wedge\overline{\boldsymbol{\Gamma}} + \widetilde{\boldsymbol{\Theta}}.$$

We shall now try to remove the arbitrariness of the matrix $\mathbf{B}$ in such a way that the relation

$$d\boldsymbol{\gamma}\,\mathbf{B} + d\mathbf{B} = -\mathbf{B}\boldsymbol{\mu} \quad \text{or} \quad d\gamma^i_k\, B^k_j + dB^i_j = B^i_k\mu^k_j \qquad (6.8.10)$$

is satisfied and the matrix form $\boldsymbol{\mu}\in\boldsymbol{\Lambda}^1(U)$ belongs to the set $\boldsymbol{\mathcal{A}}^1(U)$. The exterior derivative of (6.8.10) yields

$$-d\boldsymbol{\gamma}\wedge d\mathbf{B} + d\mathbf{B}\wedge\boldsymbol{\mu} + \mathbf{B}\,d\boldsymbol{\mu} = \mathbf{0} \text{ and } d\boldsymbol{\mu} = \mathbf{B}^{-1}d\boldsymbol{\gamma}\wedge d\mathbf{B} - \mathbf{B}^{-1}d\mathbf{B}\wedge\boldsymbol{\mu}.$$

Therefore, on employing (6.8.10) we conclude that

$$d\boldsymbol{\mu} = -\mathbf{B}^{-1}(d\boldsymbol{\gamma}\wedge d\boldsymbol{\gamma})\mathbf{B} - \mathbf{B}^{-1}d\boldsymbol{\gamma}\mathbf{B}\wedge\boldsymbol{\mu} + \mathbf{B}^{-1}d\boldsymbol{\gamma}\mathbf{B}\wedge\boldsymbol{\mu} + \boldsymbol{\mu}\wedge\boldsymbol{\mu}$$
$$= -\mathbf{B}^{-1}(d\boldsymbol{\gamma}\wedge d\boldsymbol{\gamma})\mathbf{B} + \boldsymbol{\mu}\wedge\boldsymbol{\mu}.$$

If $\boldsymbol{\mu}\in\boldsymbol{\mathcal{A}}^1(U)$, then one has $\boldsymbol{\mu}\wedge\boldsymbol{\mu}\in\boldsymbol{\mathcal{A}}^2(U)$ so the relations $\boldsymbol{\mu} = Hd\boldsymbol{\mu}$ and $H(\boldsymbol{\mu}\wedge\boldsymbol{\mu}) = \mathbf{0}$ hold. Thus, we can choose

$$\boldsymbol{\mu} = Hd\boldsymbol{\mu}$$
$$= -H\big(\mathbf{B}^{-1}(d\boldsymbol{\gamma}\wedge d\boldsymbol{\gamma})\mathbf{B}\big)$$

so that we obtain

$$d\mathbf{B} = -d\boldsymbol{\gamma}\mathbf{B} + \mathbf{B}H\big(\mathbf{B}^{-1}(d\boldsymbol{\gamma}\wedge d\boldsymbol{\gamma})\mathbf{B}\big).$$

Because $\mathbf{B}\in\boldsymbol{\mathcal{A}}^0(U)$, we know that we can write $\mathbf{B} - \mathbf{B}_0 = Hd\mathbf{B}$ where $\mathbf{B}_0 = \mathbf{B}(\mathbf{x}_0)$. $\mathbf{B}H\big(\mathbf{B}^{-1}(d\boldsymbol{\gamma}\wedge d\boldsymbol{\gamma})\mathbf{B}\big)\in\boldsymbol{\mathcal{A}}^1(U)$ implies that this form is in the null space of the operator H. This means that on applying the operator H on $d\mathbf{B}$, the matrix $\mathbf{B}$ will have to satisfy the equation

$$\mathbf{B} = \mathbf{B}_0 - H(d\boldsymbol{\gamma}\mathbf{B}) = \mathbf{B}_0 - H(\boldsymbol{\Gamma}\mathbf{B}) + H(\boldsymbol{\Gamma}_a\mathbf{B}).$$

But since $\boldsymbol{\Gamma}_a\in\boldsymbol{\mathcal{A}}^1(U)$, we get $H(\boldsymbol{\Gamma}_a\mathbf{B}) = \mathbf{0}$. If we write $\mathbf{B} = \mathbf{A}\mathbf{B}_0$, we see at once that $\mathbf{A}$ is a regular matrix and $\mathbf{A}(\mathbf{x}_0) = \mathbf{I}$. We are thus led to the conclusion that the matrix $\mathbf{A}$ has to satisfy the integral equation

$$\mathbf{A} + H(\boldsymbol{\Gamma}\mathbf{A}) = \mathbf{A} + H(d\boldsymbol{\gamma}\mathbf{A}) = \mathbf{I}, \ \mathbf{A}(\mathbf{x}_0) = \mathbf{I}. \tag{6.8.11}$$

Let us now denote the forms $\boldsymbol{\Gamma}$ by $\Gamma^i_k = \Gamma^i_{kl}dx^l$. Then (6.8.11) is explicitly expressed in indicial notation as

$$a^i_j(\mathbf{x}) + \int_0^1 (x^l - x^l_0)\Gamma^i_{kl}[\mathbf{x}_0 + t(\mathbf{x} - \mathbf{x}_0)]a^k_j[\mathbf{x}_0 + t(\mathbf{x} - \mathbf{x}_0)]dt = \delta^i_j.$$

$\mathbf{A} = [a^i_j]$ is called the ***attitude matrix***. With the present choice of the matrix $\mathbf{B}$, we immediately see that one is able to write

$$d\overline{\boldsymbol{\Gamma}} = \overline{\boldsymbol{\Gamma}} \wedge \boldsymbol{\mu} + \boldsymbol{\mu} \wedge \overline{\boldsymbol{\Gamma}} + d\boldsymbol{\mu} - \boldsymbol{\mu} \wedge \boldsymbol{\mu} - \overline{\boldsymbol{\Gamma}} \wedge \overline{\boldsymbol{\Gamma}} + \widetilde{\boldsymbol{\Theta}}.$$

Since $\overline{\boldsymbol{\Gamma}}, \boldsymbol{\mu} \in \boldsymbol{\mathcal{A}}^1(U)$, when we apply the homotopy operator H to the foregoing equation we find $\overline{\boldsymbol{\Gamma}} = \boldsymbol{\mu} + H\widetilde{\boldsymbol{\Theta}}$ and introducing all these results into (6.8.7) we arrive at the expression

$$\begin{aligned}
\boldsymbol{\Gamma} &= d\boldsymbol{\gamma} + \mathbf{B}\boldsymbol{\mu}\mathbf{B}^{-1} + \mathbf{B}H(\widetilde{\boldsymbol{\Theta}})\mathbf{B}^{-1} \\
&= d\boldsymbol{\gamma} - d\boldsymbol{\gamma} - d\mathbf{B}\mathbf{B}^{-1} + \mathbf{B}H(\widetilde{\boldsymbol{\Theta}})\mathbf{B}^{-1} \\
&= (\mathbf{B}H\widetilde{\boldsymbol{\Theta}} - d\mathbf{B})\mathbf{B}^{-1} = \big(\mathbf{B}H(\mathbf{B}^{-1}\boldsymbol{\Theta}\mathbf{B}) - d\mathbf{B}\big)\mathbf{B}^{-1}.
\end{aligned}$$

On inserting now $\mathbf{B} = \mathbf{A}\mathbf{B}_0$ and $\mathbf{B}^{-1} = \mathbf{B}_0^{-1}\mathbf{A}^{-1}$ into the above relation, we finally conclude that

$$\boldsymbol{\Gamma} = \big[\mathbf{A}H(\mathbf{A}^{-1}\boldsymbol{\Theta}\mathbf{A}) - d\mathbf{A}\big]\mathbf{A}^{-1}. \tag{6.8.12}$$

Let us now take the equation $(6.8.6)_4$ into account. We then introduce a matrix form $\overline{\boldsymbol{\Theta}} = \mathbf{A}^{-1}\boldsymbol{\Theta}\mathbf{A}$ so that one can write $\boldsymbol{\Theta} = \mathbf{A}\overline{\boldsymbol{\Theta}}\mathbf{A}^{-1}$ whose exterior derivative has to satisfy

$$d\mathbf{A} \wedge \overline{\boldsymbol{\Theta}}\mathbf{A}^{-1} + \mathbf{A}\,d\overline{\boldsymbol{\Theta}}\,\mathbf{A}^{-1} + \mathbf{A}\overline{\boldsymbol{\Theta}} \wedge d\mathbf{A}^{-1} = \mathbf{A}\overline{\boldsymbol{\Theta}}\mathbf{A}^{-1} \wedge \boldsymbol{\Gamma} - \boldsymbol{\Gamma} \wedge \mathbf{A}\overline{\boldsymbol{\Theta}}\mathbf{A}^{-1}.$$

On recalling the equality $d\mathbf{A}^{-1}\mathbf{A} = -\mathbf{A}^{-1}d\mathbf{A}$, this equation leads to

$$\begin{aligned}
d\overline{\boldsymbol{\Theta}} &= -\mathbf{A}^{-1}\boldsymbol{\Gamma}\mathbf{A} \wedge \overline{\boldsymbol{\Theta}} + \overline{\boldsymbol{\Theta}} \wedge \mathbf{A}^{-1}\boldsymbol{\Gamma}\mathbf{A} - \mathbf{A}^{-1}d\mathbf{A} \wedge \overline{\boldsymbol{\Theta}} - \overline{\boldsymbol{\Theta}} \wedge d\mathbf{A}^{-1}\mathbf{A} \\
&= -(\mathbf{A}^{-1}\boldsymbol{\Gamma}\mathbf{A} + \mathbf{A}^{-1}d\mathbf{A}) \wedge \overline{\boldsymbol{\Theta}} + \overline{\boldsymbol{\Theta}} \wedge (\mathbf{A}^{-1}\boldsymbol{\Gamma}\mathbf{A} + \mathbf{A}^{-1}d\mathbf{A}).
\end{aligned}$$

On the other hand, it follows from (6.8.12) that

$$\mathbf{A}^{-1}\boldsymbol{\Gamma}\mathbf{A} + \mathbf{A}^{-1}d\mathbf{A} = H(\overline{\boldsymbol{\Theta}}).$$

Therefore, we obtain

$$d\overline{\boldsymbol{\Theta}} = -H(\overline{\boldsymbol{\Theta}}) \wedge \overline{\boldsymbol{\Theta}} + \overline{\boldsymbol{\Theta}} \wedge H(\overline{\boldsymbol{\Theta}}).$$

In view of Theorem 6.4.3, we can use the representation $\bar{\boldsymbol{\Theta}} = d\boldsymbol{\theta} + \bar{\boldsymbol{\Theta}}_a$ in which $\boldsymbol{\theta} = H\bar{\boldsymbol{\Theta}} \in \boldsymbol{\mathcal{A}}^1(U)$ and $\bar{\boldsymbol{\Theta}}_a = Hd\bar{\boldsymbol{\Theta}} \in \boldsymbol{\mathcal{A}}^2(U)$ are, respectively, 1- and 2-antiexact forms. We thus find $H(\bar{\boldsymbol{\Theta}}) = Hd\boldsymbol{\theta} + H\bar{\boldsymbol{\Theta}}_a = \boldsymbol{\theta}$. Therefore, we obtain

$$\begin{aligned} d\bar{\boldsymbol{\Theta}}_a = dHd\bar{\boldsymbol{\Theta}} = d\bar{\boldsymbol{\Theta}} &= -\boldsymbol{\theta} \wedge d\boldsymbol{\theta} + d\boldsymbol{\theta} \wedge \boldsymbol{\theta} - \boldsymbol{\theta} \wedge \bar{\boldsymbol{\Theta}}_a + \bar{\boldsymbol{\Theta}}_a \wedge \boldsymbol{\theta} \\ &= d(\boldsymbol{\theta} \wedge \boldsymbol{\theta}) - \boldsymbol{\theta} \wedge \bar{\boldsymbol{\Theta}}_a + \bar{\boldsymbol{\Theta}}_a \wedge \boldsymbol{\theta}. \end{aligned}$$

When we apply the operator H to that expression, we get $Hd\bar{\boldsymbol{\Theta}}_a = \bar{\boldsymbol{\Theta}}_a = \boldsymbol{\theta} \wedge \boldsymbol{\theta}$ and we obtain the following representation

$$\bar{\boldsymbol{\Theta}} = d\boldsymbol{\theta} + \boldsymbol{\theta} \wedge \boldsymbol{\theta}, \quad \boldsymbol{\Theta} = \mathbf{A}(d\boldsymbol{\theta} + \boldsymbol{\theta} \wedge \boldsymbol{\theta})\mathbf{A}^{-1} \tag{6.8.13}$$

whereas (6.8.12) takes the shape

$$\boldsymbol{\Gamma} = (\mathbf{A}\boldsymbol{\theta} - d\mathbf{A})\mathbf{A}^{-1} \text{ and } \boldsymbol{\theta} = \mathbf{A}^{-1}(\boldsymbol{\Gamma}\mathbf{A} + d\mathbf{A}). \tag{6.8.14}$$

Let us now define the matrix forms $\boldsymbol{\omega} \in \boldsymbol{\Lambda}^k(U)$ and $\boldsymbol{\sigma} \in \boldsymbol{\Lambda}^{k+1}(U)$ through the relations

$$\boldsymbol{\Omega} = \mathbf{A}\boldsymbol{\omega}, \quad \boldsymbol{\Sigma} = \mathbf{A}\boldsymbol{\sigma}.$$

So the equation $(6.8.6)_1$ is transformed into

$$d\boldsymbol{\Omega} = d\mathbf{A} \wedge \boldsymbol{\omega} + \mathbf{A}d\boldsymbol{\omega} = -\boldsymbol{\Gamma}\mathbf{A} \wedge \boldsymbol{\omega} + \mathbf{A}\boldsymbol{\sigma}$$

from which we extract the expression

$$d\boldsymbol{\omega} = -\mathbf{A}^{-1}(\boldsymbol{\Gamma}\mathbf{A} + d\mathbf{A}) \wedge \boldsymbol{\omega} + \boldsymbol{\sigma} = -\boldsymbol{\theta} \wedge \boldsymbol{\omega} + \boldsymbol{\sigma}. \tag{6.8.15}$$

Similarly, the equation $(6.8.6)_2$ becomes

$$d\boldsymbol{\Sigma} = d\mathbf{A} \wedge \boldsymbol{\sigma} + \mathbf{A}d\boldsymbol{\sigma} = \mathbf{A}(d\boldsymbol{\theta} + \boldsymbol{\theta} \wedge \boldsymbol{\theta})\mathbf{A}^{-1} \wedge \mathbf{A}\boldsymbol{\omega} - \boldsymbol{\Gamma}\mathbf{A} \wedge \boldsymbol{\sigma}$$

and one obtains

$$d\boldsymbol{\sigma} = -\boldsymbol{\theta} \wedge \boldsymbol{\sigma} + (d\boldsymbol{\theta} + \boldsymbol{\theta} \wedge \boldsymbol{\theta}) \wedge \boldsymbol{\omega}. \tag{6.8.16}$$

After having resorted to Theorem 6.4.3, we can write

$$\boldsymbol{\omega} = d\boldsymbol{\phi} + \boldsymbol{\omega}_a, \quad \boldsymbol{\sigma} = d\boldsymbol{\eta} + \boldsymbol{\sigma}_a, \quad \boldsymbol{\beta} = \boldsymbol{\theta} \wedge \boldsymbol{\omega} \tag{6.8.17}$$

where we introduced the matrix forms $\boldsymbol{\phi} = H\boldsymbol{\omega} = H(\mathbf{A}^{-1}\boldsymbol{\Omega}) \in \boldsymbol{\mathcal{A}}^{k-1}(U)$, $\boldsymbol{\omega}_a \in \boldsymbol{\mathcal{A}}^k(U)$ and $\boldsymbol{\eta} = H\boldsymbol{\sigma} = H(\mathbf{A}^{-1}\boldsymbol{\Sigma}) \in \boldsymbol{\mathcal{A}}^k(U)$, $\boldsymbol{\sigma}_a \in \boldsymbol{\mathcal{A}}^{k+1}(U)$. It then follows from (6.8.17), (6.8.15) and (6.8.16) that

$$\begin{aligned} d\boldsymbol{\omega} &= d\boldsymbol{\omega}_a = \boldsymbol{\sigma} - \boldsymbol{\beta} = d\boldsymbol{\eta} + \boldsymbol{\sigma}_a - \boldsymbol{\beta}, \\ d\boldsymbol{\sigma} &= d\boldsymbol{\sigma}_a = d\boldsymbol{\theta} \wedge \boldsymbol{\omega} - \boldsymbol{\theta} \wedge (\boldsymbol{\sigma} - \boldsymbol{\beta}) = d\boldsymbol{\theta} \wedge \boldsymbol{\omega} - \boldsymbol{\theta} \wedge d\boldsymbol{\omega} = d\boldsymbol{\beta}. \end{aligned}$$

On applying the operator H to these equations, we get

$$\omega_a = -H\boldsymbol{\beta} + \boldsymbol{\eta}, \quad \sigma_a = Hd\boldsymbol{\beta} = \boldsymbol{\beta} - dH\boldsymbol{\beta}$$

and, consequently, we arrive at the result

$$\omega = d\boldsymbol{\phi} - H\boldsymbol{\beta} + \boldsymbol{\eta}, \quad \sigma = d\boldsymbol{\eta} - dH\boldsymbol{\beta} + \boldsymbol{\beta}. \tag{6.8.18}$$

In the relation

$$\boldsymbol{\beta} = \boldsymbol{\theta} \wedge d\boldsymbol{\phi} - \boldsymbol{\theta} \wedge H\boldsymbol{\beta} + \boldsymbol{\theta} \wedge \boldsymbol{\eta}$$

$\boldsymbol{\theta} \wedge H\boldsymbol{\beta}$ and $\boldsymbol{\theta} \wedge \boldsymbol{\eta}$ are antiexact forms. Therefore, we can write

$$H\boldsymbol{\beta} = H(\boldsymbol{\theta} \wedge d\boldsymbol{\phi}) \quad \text{and} \quad \boldsymbol{\beta} = \boldsymbol{\theta} \wedge d\boldsymbol{\phi} - \boldsymbol{\theta} \wedge H(\boldsymbol{\theta} \wedge d\boldsymbol{\phi}) + \boldsymbol{\theta} \wedge \boldsymbol{\eta}.$$

On the other hand, if we take notice of the relation $dH\boldsymbol{\beta} = dH(\boldsymbol{\theta} \wedge d\boldsymbol{\phi}) = \boldsymbol{\theta} \wedge d\boldsymbol{\phi} - Hd(\boldsymbol{\theta} \wedge d\boldsymbol{\phi}) = \boldsymbol{\theta} \wedge d\boldsymbol{\phi} - H(d\boldsymbol{\theta} \wedge d\boldsymbol{\phi})$, then (6.8.18) leads to

$$\begin{aligned} \omega &= d\boldsymbol{\phi} + \boldsymbol{\eta} - H(\boldsymbol{\theta} \wedge d\boldsymbol{\phi}), \\ \sigma &= d\boldsymbol{\eta} + \boldsymbol{\theta} \wedge \boldsymbol{\eta} + H(d\boldsymbol{\theta} \wedge d\boldsymbol{\phi}) - \boldsymbol{\theta} \wedge H(\boldsymbol{\theta} \wedge d\boldsymbol{\phi}). \end{aligned}$$

Thus, the solution of the system of exterior differential equations (6.8.6) is provided by

$$\begin{aligned} \boldsymbol{\Omega} &= \mathbf{A}\big[d\boldsymbol{\phi} + \boldsymbol{\eta} - H(\boldsymbol{\theta} \wedge d\boldsymbol{\phi})\big], \\ \boldsymbol{\Sigma} &= \mathbf{A}\big[d\boldsymbol{\eta} + \boldsymbol{\theta} \wedge \boldsymbol{\eta} + H(d\boldsymbol{\theta} \wedge d\boldsymbol{\phi}) - \boldsymbol{\theta} \wedge H(\boldsymbol{\theta} \wedge d\boldsymbol{\phi})\big], \\ \boldsymbol{\Gamma} &= (\mathbf{A}\boldsymbol{\theta} - d\mathbf{A})\mathbf{A}^{-1}, \\ \boldsymbol{\theta} &= \mathbf{A}^{-1}(\boldsymbol{\Gamma}\mathbf{A} + d\mathbf{A}), \\ \boldsymbol{\Theta} &= \mathbf{A}(d\boldsymbol{\theta} + \boldsymbol{\theta} \wedge \boldsymbol{\theta})\mathbf{A}^{-1} \end{aligned} \tag{6.8.19}$$

where $\boldsymbol{\phi} \in \boldsymbol{\mathcal{A}}^{k-1}(U)$ and $\boldsymbol{\eta} \in \boldsymbol{\mathcal{A}}^{k}(U)$ are arbitrary matrix forms and the matrix $\mathbf{A}$ is determined by solving the integral equation (6.8.11) once the matrix form $\boldsymbol{\Gamma}$ is prescribed. The matrix form $\boldsymbol{\theta} \in \boldsymbol{\mathcal{A}}^{1}(U)$ is then found from $(6.8.19)_4$. On the contrary, if we choose a matrix form $\boldsymbol{\theta}$, then the admissible matrix form $\boldsymbol{\Gamma}$ is deduced from $(6.8.19)_3$. The matrix $\mathbf{A}$ has to be the solution of the integrodifferential equation

$$\mathbf{A} + H(\mathbf{A}\boldsymbol{\theta} - d\mathbf{A}) = \mathbf{I}$$

obtained from (6.8.11) by replacing $\boldsymbol{\Gamma}$ by $(6.8.19)_3$. Let us now define a matrix form $\boldsymbol{\psi} \in \boldsymbol{\mathcal{A}}^{k}(U)$ by $\boldsymbol{\psi} = \boldsymbol{\eta} - H(\boldsymbol{\theta} \wedge d\boldsymbol{\phi})$ whose exterior derivative is expressible as $d\boldsymbol{\psi} = d\boldsymbol{\eta} - \boldsymbol{\theta} \wedge d\boldsymbol{\phi} + H(d\boldsymbol{\theta} \wedge d\boldsymbol{\phi})$. Then we easily verify that the relations $(6.8.19)_{1-2}$ are reduced to simpler forms given below

$$\boldsymbol{\Omega} = \mathbf{A}(d\boldsymbol{\phi} + \boldsymbol{\psi}), \quad \boldsymbol{\Sigma} = \mathbf{A}\big[d\boldsymbol{\psi} + \boldsymbol{\theta} \wedge (d\boldsymbol{\phi} + \boldsymbol{\psi})\big]. \tag{6.8.20}$$

But, in order that these equations are to be legitimate, we have to demonstrate that the matrix $\mathbf{A}$ determined through (6.8.11) is regular, that is, $A = \det \mathbf{A} \neq 0$. We rewrite (6.8.10) as $d\boldsymbol{\gamma}\mathbf{A} + d\mathbf{A} = -\mathbf{A}\mathbf{B}_0\boldsymbol{\mu}\mathbf{B}_0^{-1}$. We can then easily find that dA is expressible as

$$\begin{aligned} dA &= (\partial A/\partial a^i_j)da^i_j = (\mathit{Cofactor}\, a^i_j)\, da^i_j \\ &= A\overset{-1j}{a}{}_i\, da^i_j = A\,\mathrm{tr}\,(d\mathbf{A}\mathbf{A}^{-1}). \end{aligned}$$

Hence, we obtain $dA = -A\,\mathrm{tr}\,(d\boldsymbol{\gamma} + \mathbf{A}\mathbf{B}_0\boldsymbol{\mu}\mathbf{B}_0^{-1}\mathbf{A}^{-1}) = -A\,\mathrm{tr}\,(d\boldsymbol{\gamma} + \boldsymbol{\mu})$ and consequently

$$d\mathrm{log}A = -d\mathrm{tr}\,\boldsymbol{\gamma} - \mathrm{tr}\,\boldsymbol{\mu}, \quad Hd\mathrm{log}A = -\mathrm{tr}\,Hd\boldsymbol{\gamma} - \mathrm{tr}\,H\boldsymbol{\mu}.$$

Since $H\boldsymbol{\mu} = \mathbf{0}$ and $Hd\boldsymbol{\gamma} = \boldsymbol{\gamma}$, we get $\log A(\mathbf{x}) - \log 1 = -\mathrm{tr}\,\boldsymbol{\gamma}(\mathbf{x})$. We thus conclude that

$$A(\mathbf{x}) = e^{-\mathrm{tr}\,\boldsymbol{\gamma}} = e^{-\mathrm{tr}\,H\boldsymbol{\Gamma}} \neq 0$$

proving that $\mathbf{A}$ is a regular matrix.

Next, we define two systems of exterior differential equations on an open set $U \subseteq M$:

$$\begin{aligned} d\boldsymbol{\Omega} &= -\boldsymbol{\Gamma} \wedge \boldsymbol{\Omega} + \boldsymbol{\Sigma}, & d\boldsymbol{\Omega}' &= -\boldsymbol{\Gamma}' \wedge \boldsymbol{\Omega}' + \boldsymbol{\Sigma}', \\ d\boldsymbol{\Sigma} &= \boldsymbol{\Theta} \wedge \boldsymbol{\Omega} - \boldsymbol{\Gamma} \wedge \boldsymbol{\Sigma}, & d\boldsymbol{\Sigma}' &= \boldsymbol{\Theta}' \wedge \boldsymbol{\Omega}' - \boldsymbol{\Gamma}' \wedge \boldsymbol{\Sigma}', \\ d\boldsymbol{\Gamma} &= -\boldsymbol{\Gamma} \wedge \boldsymbol{\Gamma} + \boldsymbol{\Theta}, & d\boldsymbol{\Gamma}' &= -\boldsymbol{\Gamma}' \wedge \boldsymbol{\Gamma}' + \boldsymbol{\Theta}', \\ d\boldsymbol{\Theta} &= \boldsymbol{\Theta} \wedge \boldsymbol{\Gamma} - \boldsymbol{\Gamma} \wedge \boldsymbol{\Theta}, & d\boldsymbol{\Theta}' &= \boldsymbol{\Theta}' \wedge \boldsymbol{\Gamma}' - \boldsymbol{\Gamma}' \wedge \boldsymbol{\Theta}'. \end{aligned}$$

We say that these two systems are *equivalent* if we have $\boldsymbol{\Omega} = \boldsymbol{\Omega}'$ on U, namely, if they lead to the same solution. In such a case we first observe that the relation

$$\boldsymbol{\Sigma}' = \boldsymbol{\Sigma} + (\boldsymbol{\Gamma}' - \boldsymbol{\Gamma}) \wedge \boldsymbol{\Omega}$$

must be satisfied. Actually both systems involve same kind of solutions as (6.8.19). But these solutions should be interrelated in order to obtain the same $\boldsymbol{\Omega}$ from those two systems. These relations turns out to be quite complicated. That is the reason why they are not included here. Notwithstanding, a specific situation bears a particular importance. In the second system, let us take

$$\boldsymbol{\Gamma}' = -\,d\mathbf{A}\,\mathbf{A}^{-1}.$$

Then $(6.8.19)_3$ gives rise to $\mathbf{\Gamma} - \mathbf{\Gamma}' = \mathbf{A}\boldsymbol{\theta}\mathbf{A}^{-1}$ and $\mathbf{\Sigma}' = \mathbf{\Sigma} - \mathbf{A}\boldsymbol{\theta}\mathbf{A}^{-1} \wedge \mathbf{\Omega}$. We know that $\boldsymbol{\theta} \in \boldsymbol{\mathcal{A}}^1(U)$. We thus get $\mathbf{A}\boldsymbol{\theta}\mathbf{A}^{-1} \in \boldsymbol{\mathcal{A}}^1(U)$ so we find that

$$\boldsymbol{\gamma}' = H(\mathbf{\Gamma}') = -H(d\mathbf{A}\,\mathbf{A}^{-1}) = H(\mathbf{A}\boldsymbol{\theta}\mathbf{A}^{-1} - d\mathbf{A}\,\mathbf{A}^{-1}) = H(\mathbf{\Gamma}) = \boldsymbol{\gamma}.$$

This implies that $\mathbf{A}$ and $\mathbf{A}'$ satisfy the same matrix integral equation (6.8.11) so that we can take $\mathbf{A}' = \mathbf{A}$. On the other hand, it follows from the relation $d\mathbf{A}^{-1} = -\mathbf{A}^{-1}d\mathbf{A}\mathbf{A}^{-1}$ that

$$d\mathbf{\Gamma}' = d\mathbf{A} \wedge d\mathbf{A}^{-1} = -d\mathbf{A}\mathbf{A}^{-1} \wedge d\mathbf{A}\mathbf{A}^{-1} = -\mathbf{\Gamma}' \wedge \mathbf{\Gamma}'.$$

We thus obtain $\mathbf{\Theta}' = \mathbf{0}$ and $\boldsymbol{\theta}' = H(\mathbf{A}^{-1}\mathbf{\Theta}'\mathbf{A}) = \mathbf{0}$. Furthermore, we find

$$\begin{aligned}
\boldsymbol{\phi}' &= H(\mathbf{A}^{-1}\mathbf{\Omega}) = \boldsymbol{\phi}, \\
\boldsymbol{\eta}' &= H(\mathbf{A}^{-1}\mathbf{\Sigma}') = H(\mathbf{A}^{-1}\mathbf{\Sigma} - \boldsymbol{\theta}\mathbf{A}^{-1} \wedge \mathbf{\Omega}) = \boldsymbol{\eta} - H(\boldsymbol{\theta} \wedge \boldsymbol{\omega}) \\
&= \boldsymbol{\eta} - H(\boldsymbol{\theta} \wedge d\boldsymbol{\phi}) - H(\boldsymbol{\theta} \wedge \boldsymbol{\omega}_a) = \boldsymbol{\eta} - H(\boldsymbol{\theta} \wedge d\boldsymbol{\phi}).
\end{aligned}$$

We thus understand that any system of exterior differential equations is equivalent to the system

$$d\mathbf{\Omega}' = -\mathbf{\Gamma}' \wedge \mathbf{\Omega}' + \mathbf{\Sigma}',\ d\mathbf{\Sigma}' = -\mathbf{\Gamma}' \wedge \mathbf{\Sigma}',\ d\mathbf{\Gamma}' = -\mathbf{\Gamma}' \wedge \mathbf{\Gamma}',\ \mathbf{\Theta}' = \mathbf{0}.$$

As to the solution of the equivalent system, it is easily found that

$$\mathbf{\Omega} = \mathbf{A}(d\boldsymbol{\phi} + \boldsymbol{\eta}'), \quad \mathbf{\Sigma}' = \mathbf{A}d\boldsymbol{\eta}', \quad \mathbf{\Gamma}' = -d\mathbf{A}\,\mathbf{A}^{-1}$$

from which we reach to a sort of generalisation of the Frobenius theorem: a representation $\mathbf{\Omega} = \mathbf{A}d\boldsymbol{\phi}$ for a matrix k-form $\mathbf{\Omega}$ becomes possible if and only if $\boldsymbol{\eta}' = \mathbf{0}$. This entails of course the condition $\mathbf{\Sigma}' = \mathbf{0}$.

Example 6.8.1. The matrix forms $\mathbf{\Omega} \in \mathbf{\Lambda}^k(U)$ and $\mathbf{\Gamma} \in \mathbf{\Lambda}^1(U)$ are given by

$$\mathbf{\Omega} = \begin{bmatrix} \Omega_1 \\ \Omega_2 \end{bmatrix},\ \mathbf{\Gamma} = -\mathbf{G}\,df,\ \mathbf{G} = \begin{bmatrix} a & a \\ 2a & 0 \end{bmatrix},\ f \in \Lambda^0(U),\ f(\mathbf{x}_0) = 0$$

where a is a constant. We search for the solution of the system of exterior differential equations $d\mathbf{\Omega} = -\mathbf{\Gamma} \wedge \mathbf{\Omega}$. In terms of the component forms, this system is expressed as follows

$$d\Omega_1 = a\,df \wedge (\Omega_1 + \Omega_2),\ d\Omega_2 = 2a\,df \wedge \Omega_1.$$

We immediately observe that

$$d\mathbf{\Gamma} = \mathbf{0},\ \mathbf{\Gamma} \wedge \mathbf{\Gamma} = \mathbf{G}^2 df \wedge df = \mathbf{0}.$$

In this case (6.8.6) yields $\mathbf{\Theta} = \mathbf{0}$ and, consequently, $\boldsymbol{\theta} = \mathbf{0}$. It then follows from (6.8.19) that $\mathbf{\Gamma} = -d\mathbf{A}\,\mathbf{A}^{-1}$. The integral equation (6.8.11) now takes the form $\mathbf{A} - H(d\mathbf{A}) = \mathbf{I}$. This in turn gives $\mathbf{A}(\mathbf{x}) - \mathbf{A}(\mathbf{x}) + \mathbf{A}(\mathbf{x}_0) = \mathbf{I}$, that is, the integral equation is satisfied identically. Hence, the matrix $\mathbf{A}$ is determined from the solution of the differential equation $d\mathbf{A} = -\boldsymbol{\Gamma}\mathbf{A} = \mathbf{G}\mathbf{A}\,df$. We know that the solution is expressible as

$$\mathbf{A}(\mathbf{x}) = e^{f(\mathbf{x})\mathbf{G}},\ \mathbf{A}(\mathbf{x}_0) = e^{f(\mathbf{x}_0)\mathbf{G}} = \mathbf{I}.$$

Since $f\mathbf{G}$ is a 2×2 matrix, we can write $e^{f(\mathbf{x})\mathbf{G}} = \alpha_0(\mathbf{x})\mathbf{I} + \alpha_1(\mathbf{x})f(\mathbf{x})\mathbf{G}$ according to the celebrated Hamilton-Cayley theorem which states that every square matrix satisfies its characteristic equation. The eigenvalues of the matrix $\mathbf{G}$ are $2a$ and $-a$ so that the coefficient functions $\alpha_0(\mathbf{x})$ and $\alpha_1(\mathbf{x})$ are found from the equations

$$e^{2af} = \alpha_0 + 2af\alpha_1,\ e^{-af} = \alpha_0 - af\alpha_1$$

as the following expressions

$$\alpha_0(\mathbf{x}) = \frac{e^{2af(\mathbf{x})} + 2e^{-af(\mathbf{x})}}{3},\ \alpha_1(\mathbf{x}) = \frac{e^{2af(\mathbf{x})} - e^{-af(\mathbf{x})}}{3af(\mathbf{x})}.$$

Therefore, the matrix $\mathbf{A}$ is given by

$$\mathbf{A} = \begin{bmatrix} \dfrac{2e^{2af(\mathbf{x})} + e^{-af(\mathbf{x})}}{3} & \dfrac{e^{2af(\mathbf{x})} - e^{-af(\mathbf{x})}}{3} \\ \dfrac{2(e^{2af(\mathbf{x})} - e^{-af(\mathbf{x})})}{3} & \dfrac{e^{2af(\mathbf{x})} + 2e^{-af(\mathbf{x})}}{3} \end{bmatrix}.$$

Since, in the present example we have $\boldsymbol{\Sigma} = \mathbf{0}$, the solution will be in the form $\boldsymbol{\Omega} = \mathbf{A}d\boldsymbol{\phi}$. We thus obtain the solution

$$\Omega_1 = \frac{1}{3}(2e^{2af(\mathbf{x})} + e^{-af(\mathbf{x})})\,d\phi_1 + \frac{1}{3}(e^{2af(\mathbf{x})} - e^{-af(\mathbf{x})})\,d\phi_2$$
$$\Omega_2 = \frac{2}{3}(e^{2af(\mathbf{x})} - e^{-af(\mathbf{x})})\,d\phi_1 + \frac{1}{3}(e^{2af(\mathbf{x})} + 2e^{-af(\mathbf{x})})\,d\phi_2$$

where $\boldsymbol{\phi}(\mathbf{x}) = [\phi_1(\mathbf{x})\ \phi_2(\mathbf{x})]^{\mathrm{T}}$ is an arbitrary vector function.

VI. EXERCISES

6.1. For forms $\omega \in \Lambda(\mathbb{R}^4)$ given below evaluate the forms $H\omega$ and their ω_e exact and ω_a antiexact parts. H is the homotopy operator with the centre $(x, y, z, t) = (0, 0, 0, 0)$:

$(a)\ \omega = (1 + t^2)\,dx + z\,dy + x^3 dz + xyz\,dt.$

$(b)\ \omega = t^2 dx \wedge dy + y\ dx \wedge dz + z^3 dx \wedge dt + x^2\ dy \wedge dz + xy\ dz \wedge dt.$

$(c)\ \omega = x^2 t\,dx \wedge dy \wedge dz + (x^2 + z^2)\,dx \wedge dy \wedge dt + yt\,dy \wedge dt \wedge dz.$

$(d)\ \omega = (x^2 + y^2 + z^2 + t^2)\, dx \wedge dy \wedge dz \wedge dt.$

6.2. Repeat the above operations by shifting the centre of the homotopy operator to the point $(1, 0, 1, 0)$.

6.3. For the form $\omega = \cos(3t)\, dx + \sin(2z)\, dy \in \Lambda^1(\mathbb{R}^4)$ evaluate the forms $H\omega, \omega_e, \omega_a$. The centre of the homotopy operator H is the point $(0, 0, 0, 0)$. Determine the same forms when the centre is changed to the point $(1, 1, 1, 1)$.

6.4. Determine the Darboux classes, ranks and indices of the forms $\omega \in \Lambda^1(\mathbb{R}^4)$ given below:

$$\begin{aligned}
&(a)\ \omega = y^2 dx + x^2 dy\\
&(b)\ \omega = yz\, dx + x^2 yt\, dt\\
&(c)\ \omega = (1 - t^2)\, dx + (x^2 + y^2 - z^3)\, dy + xyzt\, dt\\
&(d)\ \omega = (t + z^3)\, dx + (x^2 - y^2 + 1)\, dz + (y^2 - z)\, dt\\
&(e)\ \omega = yt\, dx + (x^2 + t^2 - z)\, dy + (1 + y)\, dz + (z^2 - x)\, dt
\end{aligned}$$

6.5. Let $\omega \in \Lambda(M)$. Show that the following relations

$$(e^{tdH})\omega = Hd\omega + e^t dH\omega, \quad (e^{tHd})\omega = dH\omega + e^t Hd\omega$$

can locally be validated.

6.6. The function $u : M \to \mathbb{R}$ vanishes at the centre of the homotopy operator, that is, it satisfies the condition $u(\mathbf{x}_0) = 0$. Consider the following integral equation for the function $f : M \to \mathbb{R}$:

$$f = 1 + H(f\, du).$$

Show that the solution of this integral equation is given by $f = e^u$.

6.7. Investigate the same problem for the integral equation

$$f = k + H(f\, du)$$

where $k \neq 0$ is a given constant. Discuss the case $u(\mathbf{x}_0) \neq 0$.

6.8. Assume that $\Omega \in \Lambda^1(\mathbb{R}^3), \Sigma \in \Lambda^2(\mathbb{R}^3)$ and

$$\Gamma = (2x + z)\, dx + (2y + z)\, dy - (x + y)\, dz \in \Lambda^1(\mathbb{R}^3).$$

Find the solution of the exterior differential equation $d\Omega = \Gamma \wedge \Omega + \Sigma$. The centre of the homotopy operator will be taken as the point $\mathbf{0} \in \mathbb{R}^3$.

CHAPTER VII

LINEAR CONNECTIONS

7.1. SCOPE OF THE CHAPTER

Derivatives of components of vector and tensor fields on differentiable manifolds with respect to local coordinates are not generally components of a tensor. However, it becomes possible to define a sort of derivative of a tensor field which proves to be also a tensor through a geometrical structure imposed on a manifold called a ***linear*** or ***affine connection***. In this way, we can properly accomplish the task of extending the classical differential geometry to higher dimensions. From another standpoint, the affine connection can be interpreted as a suitable structure interconnecting neighbouring tangent spaces of a smooth manifold and thus enabling us to differentiate tensor fields. Although the inception of the concept of the linear connection goes back to the developments in 19. century in the geometry and tensor calculus, its formal structure is merely established in early 1920s by Élie Cartan and Herman Weyl. The term *connection* was first used by Cartan. In Sec. 7.2, we define a third order linear connection that is not a tensor but whose coefficients transform obeying certain rules under change of coordinates. Except for this restriction this geometrical object on a manifold can be chosen arbitrarily. We then discuss the covariant derivative of a tensor preserving the tensorial properties by means of a linear connection and its characteristics. We further scrutinise the torsion and curvature tensors of a manifold introduced through the linear connection. Sec. 7.3 is concerned with the Cartan connection engendered by choosing an arbitrary basis and its dual in the tangent and cotangent bundles, respectively, instead of the natural basis and its dual. The torsion and curvature tensors are then defined via that connection. Cartan connection enables us to study the differential geometry of a manifold by employing a ***moving frame*** (***repère mobile***). We define the Levi-Civita connection on a Riemannian manifold in Sec. 7.4 as a connection that causes the covariant derivative of the metric tensor and the torsion tensor to vanish. It is shown that such a connection is determined uniquely. Finally, Sec. 7.5 is devoted to study the special structures of the

Exterior Analysis, DOI: 10.1016/B978-0-12-415902-0.50007-4

operators d, δ, Δ introduced in Secs. 5.8 and 5.9 acquired by means of covariant derivatives within the context of the Levi-Civita connection.

7.2. CONNECTIONS ON MANIFOLDS

We know that vector and tensor fields on smooth manifolds are specified by their components that are differentiable functions depending on local coordinates in charts of an atlas. But derivatives of those components do not usually constitute components of a new tensor. Notwithstanding, we can manage to create new tensor fields by some kind of differentiation of vector and tensor fields on tangent and cotangent bundles by endowing the manifold with a new structure. Let us start by considering a rather simple example. A vector field $V \in T(M)$ on the manifold M is designated by

$$V = v^i(\mathbf{x})\frac{\partial}{\partial x^i}$$

in natural coordinates. We know that a coordinate transformation in the local chart like $y^i = y^i(x^j)$ gives rise to a transformation between *contravariant components* of the vector at the point $\mathbf{x}$ as follows

$$v'^i(\mathbf{y}) = \frac{\partial y^i}{\partial x^k}v^k(\mathbf{x}).$$

If we calculate the gradient of the components v'^i by using the chain rule of differentiation, we obtain

$$\frac{\partial v'^i}{\partial y^j} = \frac{\partial y^i}{\partial x^k}\frac{\partial x^l}{\partial y^j}\frac{\partial v^k}{\partial x^l} + \frac{\partial^2 y^i}{\partial x^k \partial x^l}\frac{\partial x^l}{\partial y^j}v^k. \tag{7.2.1}$$

This transformation rule under a general change of coordinates shows clearly that the quantities $\partial v^k/\partial x^l$ cannot be covariant and contravariant components of a second order tensor due to the existence of the second part in (7.2.1). If only the coordinate transformation satisfy the relations

$$\frac{\partial^2 y^i}{\partial x^k \partial x^l} = 0,$$

that is, if it is an *affine transformation* given by $y^i = a^i_j x^j + b^i$ with constant coefficients, only then the gradient behaves like a tensor. We shall now try to modify m^2 quantities $\partial v^k/\partial x^l$ in such a way that it will acquire the properties of the components of a $\binom{1}{1}$-tensor. To this end, we introduce a new operation of differentiation that will be called the ***covariant derivative***. This derivative of the component v^i with respect to the variable x^j is identified by means of presently arbitrarily chosen m^3 functions $\Gamma^i_{jk}(\mathbf{x})$ as follows

$$\nabla_j v^i = v^i{}_{;j} = v^i{}_{,j} + \Gamma^i_{jk} v^k \tag{7.2.2}$$

where the functions $\Gamma^i_{jk}(\mathbf{x})$ will be called the ***coefficients of linear*** or ***affine connection***. Next, we shall attempt to determine these coefficients in such a way that $v^i{}_{;j} = \nabla_j v^i$ become components of a second order $\binom{1}{1}$-tensor. We thus wish that the following relation must be satisfied in a coordinate transformation $y^i = y^i(x^j)$:

$$\nabla'_j v'^i = \frac{\partial y^i}{\partial x^k}\frac{\partial x^l}{\partial y^j}\nabla_l v^k.$$

On utilising (7.2.1), we easily obtain the transformation rule

$$\frac{\partial y^i}{\partial x^k}\frac{\partial x^l}{\partial y^j}\frac{\partial v^k}{\partial x^l} + \frac{\partial^2 y^i}{\partial x^k \partial x^l}\frac{\partial x^l}{\partial y^j}v^k + \Gamma'^i_{jk}\frac{\partial y^k}{\partial x^l}v^l = \frac{\partial y^i}{\partial x^k}\frac{\partial x^l}{\partial y^j}\Big(\frac{\partial v^k}{\partial x^l} + \Gamma^k_{ln}v^n\Big)$$

where $\Gamma'^i_{jk}(\mathbf{y})$ denote coefficients of the linear connection in the new coordinate system. After having cancelled similar terms in both sides and modified the dummy indices appropriately, we are led to the conclusion

$$v^n\Big(\Gamma'^i_{jk}\frac{\partial y^k}{\partial x^n} - \Gamma^k_{ln}\frac{\partial y^i}{\partial x^k}\frac{\partial x^l}{\partial y^j} + \frac{\partial^2 y^i}{\partial x^n \partial x^l}\frac{\partial x^l}{\partial y^j}\Big) = 0.$$

In order that this expression holds for every component function v^n we have to choose the coefficients Γ^i_{jk} and Γ'^i_{jk} in such a way that they must obey the transformation rule

$$\Gamma'^i_{jk}(\mathbf{y}) = \frac{\partial y^i}{\partial x^m}\frac{\partial x^l}{\partial y^j}\frac{\partial x^n}{\partial y^k}\Gamma^m_{ln}(\mathbf{x}) - \frac{\partial^2 y^i}{\partial x^n \partial x^l}\frac{\partial x^l}{\partial y^j}\frac{\partial x^n}{\partial y^k}.$$

On the other hand, differentiating the expression

$$\frac{\partial y^i}{\partial x^l}\frac{\partial x^l}{\partial y^j} = \delta^i_j$$

resulting from the chain rule, with respect to the variable x^n we find that

$$\frac{\partial^2 y^i}{\partial x^n \partial x^l}\frac{\partial x^l}{\partial y^j} = -\frac{\partial y^i}{\partial x^l}\frac{\partial^2 x^l}{\partial y^j \partial x^n} = -\frac{\partial y^i}{\partial x^l}\frac{\partial^2 x^l}{\partial y^j \partial y^k}\frac{\partial y^k}{\partial x^n}.$$

Hence, we can write

$$-\frac{\partial^2 y^i}{\partial x^n \partial x^l}\frac{\partial x^l}{\partial y^j}\frac{\partial x^n}{\partial y^k} = \frac{\partial y^i}{\partial x^l}\frac{\partial^2 x^l}{\partial y^j \partial y^l}\frac{\partial y^l}{\partial x^n}\frac{\partial x^n}{\partial y^k} = \frac{\partial^2 x^l}{\partial y^j \partial y^k}\frac{\partial y^i}{\partial x^l}$$

and, consequently, we obtain

$$\Gamma'^i_{jk}(\mathbf{y}) = \frac{\partial y^i}{\partial x^m}\frac{\partial x^l}{\partial y^j}\frac{\partial x^n}{\partial y^k}\Gamma^m_{ln}(\mathbf{x}) + \frac{\partial^2 x^l}{\partial y^j \partial y^k}\frac{\partial y^i}{\partial x^l}. \tag{7.2.3}$$

Every choice of coefficients Γ^i_{jk} verifying the transformation rule (7.2.3) gives rise to a covariant derivative and conduces to a *linear connection.* Thus, it appears that it is possible to have many, probably infinitely many, choices for linear connections. The coefficients Γ^i_{jk} are usually named as the ***Christoffel symbols of the second kind***, because they were employed for the first time by German mathematician Elwin Bruno Christoffel (1829-1900) in tensor analysis within the context of the Riemannian geometry. The rule (7.2.3) indicates clearly that *the Christoffel symbols cannot be the components of a third order tensor*. Nevertheless, the symmetry of mixed partial derivatives leads us to the result

$$\begin{aligned}\Gamma'^i_{jk} - \Gamma'^i_{kj} &= \Big(\frac{\partial y^i}{\partial x^m}\frac{\partial x^l}{\partial y^j}\frac{\partial x^n}{\partial y^k} - \frac{\partial y^i}{\partial x^m}\frac{\partial x^l}{\partial y^k}\frac{\partial x^n}{\partial y^j}\Big)\Gamma^m_{ln} \\ &= \frac{\partial y^i}{\partial x^m}\frac{\partial x^l}{\partial y^j}\frac{\partial x^n}{\partial y^k}(\Gamma^m_{ln} - \Gamma^m_{nl}).\end{aligned}$$

This clearly shows that $\Gamma^i_{[jk]} = \frac{1}{2}(\Gamma^i_{jk} - \Gamma^i_{kj})$ which are the antisymmetric parts of Christoffel symbols with respect to their subscripts, behave like the components of a third order $\binom{1}{2}$-tensor antisymmetric with respect to covariant indices.

When $f \in \Lambda^0(M)$, then we already know that $f_{,i}$ are components of a covariant vector since $df \in \Lambda^1(M)$. Hence, we can take $f_{,i} = f_{;i} = \nabla_i f$. Let us now represent the vectors $\nabla_j V, j = 1, \dots, m$ denoting covariant derivatives of a vector field V with respect to variables x^j by the vector field

$$\nabla_j V = (\nabla_j v^i)\frac{\partial}{\partial x^i} = v^i{}_{;j}\frac{\partial}{\partial x^i}$$

We can then introduce a second order $\binom{1}{1}$- tensor by

$$\boldsymbol{\nabla} V = dx^j \otimes \nabla_j V = (\nabla_j v^i)\, dx^j \otimes \frac{\partial}{\partial x^i} = v^i{}_{;j}\, dx^j \otimes \frac{\partial}{\partial x^i}. \tag{7.2.4}$$

The second order tensor ∇V is characterised as the ***gradient*** of the vector field V. We also call it the ***covariant derivative of the vector field*** V. The operator $\nabla = dx^j \otimes \nabla_j$ is in the form $\nabla : T(M) = \mathfrak{T}(M)^1_0 \to \mathfrak{T}(M)^1_1$ and it assigns a second order tensor field to a vector field. It follows from the definition (7.2.4) that if $V_1, V_2 \in T(M)$, we get

$$\nabla(V_1 + V_2) = \nabla V_1 + \nabla V_2.$$

On the other hand, if $f \in \Lambda^0(M)$, then we find that

$$\begin{aligned}\nabla(fV) &= \nabla_j(fv^i)\, dx^j \otimes \frac{\partial}{\partial x^i} = (f_{,j}v^i + f\nabla_j v^i)\, dx^j \otimes \frac{\partial}{\partial x^i} \\ &= df \otimes V + f\nabla V.\end{aligned}$$

Therefore, the operator ∇ *is linear only on real numbers*. Let us now especially choose $V = \partial/\partial x^k$ whose components are $v^i = \delta^i_k$. Then (7.2.2) gives $\nabla_j v^i = \Gamma^i_{jl}\delta^l_k = \Gamma^i_{jk}$ and it follows from (7.2.4) that

$$\nabla_j \frac{\partial}{\partial x^k} = \Gamma^i_{jk}\frac{\partial}{\partial x^i}, \quad \nabla\frac{\partial}{\partial x^k} = \Gamma^i_{jk}\, dx^j \otimes \frac{\partial}{\partial x^i}. \tag{7.2.5}$$

These relations geometrically expresses the fact that the coefficients of a linear connection measure the change between basis vectors of a tangent space and those of an adjacent tangent space on a manifold. It seems that the operator ∇ interconnects neighbouring tangent spaces by means of the coefficients of connection providing thus a tool for transporting vectors in one tangent space into another tangent space. In this way, it becomes possible to differentiate vectors and endow these derivatives with tensorial properties independent of coordinate transformations, and to find a counterpart of the concept of parallel transport on differentiable manifolds that is almost trivial in the Euclidean space. That are the reasons why the operator ∇ is sometimes called a ***linear*** or ***affine connection*** on a manifold.

We postulate that the covariant derivative specified by the operator ∇_j will still obey the classical Leibniz rule. With the help of this postulate we can easily evaluate covariant derivative of a covariant vector, or in other words of a 1-form. Let us consider $\omega \in \Lambda^1(M)$ and $V \in T(M)$ so that we write $\omega = \omega_i dx^i$ and $V = v^i\partial_i$. We thus obtain $\omega(V) = \omega_i v^i \in \Lambda^0(M)$. Therefore, the relation

$$\begin{aligned}(\omega_i v^i)_{,j} &= \omega_{i,j}v^i + \omega_i v^i_{,j} = (\omega_i v^i)_{;j} = \omega_{i;j}v^i + \omega_i v^i_{;j} \\ &= \omega_{i;j}v^i + \omega_i\big(v^i_{,j} + \Gamma^i_{jk}v^k\big) = \omega_{i;j}v^i + \omega_i v^i_{,j} + \omega_k\Gamma^k_{ji}v^i\end{aligned}$$

yields $\big[\omega_{i;j} - \big(\omega_{i,j} - \Gamma^k_{ji}\omega_k\big)\big]v^i = 0$. Because the vector V is arbitrary, we finally obtain

$$\nabla_j \omega_i = \omega_{i;j} = \omega_{i,j} - \Gamma^k_{ji}\omega_k. \tag{7.2.6}$$

We can easily verify that $\omega_{i;j}$ are components of a second order covariant $\binom{0}{2}$-tensor. If we recall the transformation

$$\omega'_i = \frac{\partial x^k}{\partial y^i}\omega_k$$

and employ the relation (7.2.3) and the chain rule, we get

$$\begin{aligned}\frac{\partial \omega'_i}{\partial y^j} - \Gamma'^k_{ji}\omega'_k &= \frac{\partial^2 x^k}{\partial y^i \partial y^j}\omega_k + \frac{\partial x^k}{\partial y^i}\frac{\partial \omega_k}{\partial x^l}\frac{\partial x^l}{\partial y^j} \\ &\quad - \frac{\partial y^k}{\partial x^m}\frac{\partial x^l}{\partial y^j}\frac{\partial x^n}{\partial y^i}\Gamma^m_{ln}\frac{\partial x^p}{\partial y^k}\omega_p - \frac{\partial^2 x^l}{\partial y^j \partial y^i}\frac{\partial y^k}{\partial x^l}\frac{\partial x^p}{\partial y^k}\omega_p \\ &= \frac{\partial^2 x^k}{\partial y^i \partial y^j}\omega_k - \frac{\partial^2 x^l}{\partial y^i \partial y^j}\omega_l + \frac{\partial x^k}{\partial y^i}\frac{\partial x^l}{\partial y^j}\frac{\partial \omega_k}{\partial x^l} \\ &\quad - \frac{\partial x^l}{\partial y^j}\frac{\partial x^n}{\partial y^i}\Gamma^m_{ln}\omega_m = \frac{\partial x^k}{\partial y^i}\frac{\partial x^l}{\partial y^j}\Big(\frac{\partial \omega_k}{\partial x^l} - \Gamma^m_{lk}\omega_m\Big)\end{aligned}$$

Hence, we obtain

$$\nabla'_j \omega'_i = \frac{\partial x^l}{\partial y^j}\frac{\partial x^k}{\partial y^i}\nabla_l \omega_k$$

as it should be. Let us now define 1-forms $\nabla_j \omega, j = 1, \ldots, m$ by

$$\nabla_j \omega = \nabla_j \omega_i \, dx^i = \omega_{i;j} \, dx^i.$$

Then the covariant derivative of a 1-form ω can be written as

$$\boldsymbol{\nabla}\omega = dx^j \otimes \nabla_j \omega = \nabla_j \omega_i \, dx^j \otimes dx^i = \omega_{i;j} \, dx^j \otimes dx^i. \tag{7.2.7}$$

If we choose a form $\omega = dx^k$, we have $\omega_i = \delta^k_i$ and (7.2.6) yields $\nabla_j \omega_i = -\Gamma^l_{ji}\delta^k_l = -\Gamma^k_{ji}$ so that we find

$$\nabla_j \, dx^k = -\Gamma^k_{ji} \, dx^i, \;\; \boldsymbol{\nabla} dx^k = -\Gamma^k_{ji} \, dx^j \otimes dx^i. \tag{7.2.8}$$

Hence, the same coefficients of a connection measure also the changes in basis forms in the cotangent bundle between adjacent dual spaces.

We can now proceed to calculate the covariant derivative of a tensor field $\mathcal{T} \in \mathfrak{T}(M)^k_l$ by taking into account the equalities $(7.2.5)_1$ and $(7.2.8)_1$ associated with basis vectors and the Leibniz rule. Let the tensor field be given by

$$\mathcal{T} = t^{i_1 \cdots i_k}_{j_1 \cdots j_l} \frac{\partial}{\partial x^{i_1}} \otimes \cdots \otimes \frac{\partial}{\partial x^{i_k}} \otimes dx^{j_1} \otimes \cdots \otimes dx^{j_l}.$$

The covariant derivative of this tensor is found to be a tensor specified by the expression

$$\begin{aligned}\nabla_j \mathcal{T} &= t^{i_1 \cdots i_k}_{j_1 \cdots j_l, j} \frac{\partial}{\partial x^{i_1}} \otimes \cdots \otimes \frac{\partial}{\partial x^{i_k}} \otimes dx^{j_1} \otimes \cdots \otimes dx^{j_l} \\ &+ \sum_{r=1}^{k} t^{i_1 \cdots i_k}_{j_1 \cdots j_l} \frac{\partial}{\partial x^{i_1}} \otimes \cdots \otimes \nabla_j \frac{\partial}{\partial x^{i_r}} \otimes \cdots \otimes \frac{\partial}{\partial x^{i_k}} \otimes dx^{j_1} \otimes \cdots \otimes dx^{j_l} \\ &+ \sum_{r=1}^{l} t^{i_1 \cdots i_k}_{j_1 \cdots j_l} \frac{\partial}{\partial x^{i_1}} \otimes \cdots \otimes \frac{\partial}{\partial x^{i_k}} \otimes dx^{j_1} \otimes \cdots \otimes \nabla_j dx^{j_r} \otimes \cdots \otimes dx^{j_l} \\ &= \nabla_j t^{i_1 \cdots i_k}_{j_1 \cdots j_l} \frac{\partial}{\partial x^{i_1}} \otimes \cdots \otimes \frac{\partial}{\partial x^{i_k}} \otimes dx^{j_1} \otimes \cdots \otimes dx^{j_l}.\end{aligned}$$

It is easily verified that the covariant derivatives of the components of that tensor with respect to the variable x^j are expressed by the relation

$$\begin{aligned}\nabla_j t^{i_1 \cdots i_k}_{j_1 \cdots j_l} = t^{i_1 \cdots i_k}_{j_1 \cdots j_l ; j} = \frac{\partial t^{i_1 \cdots i_k}_{j_1 \cdots j_l}}{\partial x^j} &+ \sum_{r=1}^{k} \Gamma^{i_r}_{jn} t^{i_1 \cdots i_{r-1} n i_{r+1} \cdots i_k}_{j_1 \cdots j_l} \\ &- \sum_{r=1}^{l} \Gamma^{n}_{j j_r} t^{i_1 \cdots i_k}_{j_1 \cdots j_{r-1} n j_{r+1} \cdots j_l}.\end{aligned}$$

Thus the covariant derivative $\boldsymbol{\nabla} \mathcal{T} \in \mathfrak{T}(M)^k_{l+1}$ of a tensor field $\mathcal{T} \in \mathfrak{T}(M)^k_l$ is defined as

$$\begin{aligned}\boldsymbol{\nabla} \mathcal{T} &= dx^j \otimes \nabla_j \mathcal{T} \qquad (7.2.9) \\ &= t^{i_1 \cdots i_k}_{j_1 \cdots j_l ; j}\, dx^j \otimes \frac{\partial}{\partial x^{i_1}} \otimes \cdots \otimes \frac{\partial}{\partial x^{i_k}} \otimes dx^{j_1} \otimes \cdots \otimes dx^{j_l}\end{aligned}$$

This time, the operator $\boldsymbol{\nabla}$ is in the form $\boldsymbol{\nabla} : \mathfrak{T}(M)^k_l \to \mathfrak{T}(M)^k_{l+1}$. Let $\mathcal{T}$ and $\mathcal{S}$ be two tensor fields. It is straightforward to check that

$$\boldsymbol{\nabla}(\mathcal{T} \otimes \mathcal{S}) \neq \boldsymbol{\nabla} \mathcal{T} \otimes \mathcal{S} + \mathcal{T} \otimes \boldsymbol{\nabla} \mathcal{S}.$$

Indeed, the two tensor products in the right hand side are actually different tensors because the forms dx^j appearing in covariant derivatives do not occupy the same place in them. So, it is not possible to add those tensors. Hence, covariant derivative of tensor products cannot satisfy the Leibniz rule. But, we can readily verify that the Leibniz rule holds for covariant derivative of tensor components. Let us now consider $\mathcal{T} \in \mathfrak{T}(M)^k_l$ and $\mathcal{S} \in \mathfrak{T}(M)^p_q$. It follows from the definition of the covariant derivative ∇_j that we find the relation below for the components of the tensor product $\mathcal{T} \otimes \mathcal{S}$:

$$\begin{aligned}(t^{i_1\cdots i_k}_{j_1\cdots j_l} s^{k_1\cdots k_p}_{l_1\cdots l_q})_{;j} &= \frac{\partial t^{i_1\cdots i_k}_{j_1\cdots j_l}}{\partial x^j} s^{k_1\cdots k_p}_{l_1\cdots l_q} + t^{i_1\cdots i_k}_{j_1\cdots j_l} \frac{\partial s^{k_1\cdots k_p}_{l_1\cdots l_q}}{\partial x^j} \\ &\quad + \sum_{r=1}^{k} \Gamma^{i_r}_{jn} t^{i_1\cdots i_{r-1} n i_{r+1}\cdots i_k}_{j_1\cdots j_l} s^{k_1\cdots k_p}_{l_1\cdots l_q} \\ &\quad + \sum_{r=1}^{p} t^{i_1\cdots i_k}_{j_1\cdots j_l} \Gamma^{k_r}_{jn} s^{k_1\cdots k_{r-1} n k_{r+1}\cdots k_p}_{l_1\cdots l_q} \\ &\quad - \sum_{r=1}^{l} \Gamma^{n}_{j j_r} t^{i_1\cdots i_k}_{j_1\cdots j_{r-1} n j_{r+1}\cdots j_l} s^{k_1\cdots k_p}_{l_1\cdots l_q} \\ &\quad - \sum_{r=1}^{l} t^{i_1\cdots i_k}_{j_1\cdots j_l} \Gamma^{n}_{j l_r} s^{k_1\cdots k_p}_{l_1\cdots l_{r-1} n l_{r+1}\cdots l_q} \\ &= t^{i_1\cdots i_k}_{j_1\cdots j_l;j} s^{k_1\cdots k_p}_{l_1\cdots l_q} + t^{i_1\cdots i_k}_{j_1\cdots j_l} s^{k_1\cdots k_p}_{l_1\cdots l_q;j}.\end{aligned}$$

Hence, we are now allowed to write the following relation

$$\nabla_j(\mathcal{T} \otimes \mathcal{S}) = \nabla_j\mathcal{T} \otimes \mathcal{S} + \mathcal{T} \otimes \nabla_j\mathcal{S}. \tag{7.2.10}$$

We can now define on the tangent bundle of a manifold the ***covariant derivative of a vector field*** V ***in the direction of a vector field*** U by means of the affine connection $\boldsymbol{\nabla}$ as follows

$$\begin{aligned}\nabla_U V &= \mathbf{i}_U(dx^j)\nabla_j V = u^j \nabla_j V = (v^i{}_{;j}\, u^j)\frac{\partial}{\partial x^i} \\ &= \left[U(v^i) + \Gamma^i_{jk} u^j v^k\right]\frac{\partial}{\partial x^i} \in T(M).\end{aligned} \tag{7.2.11}$$

where we have obviously defined the operator $\nabla_U = u^i \nabla_i$. We immediately observe that one obtains $\nabla_{\partial_j} V = \nabla_j V$ and

$$\nabla_{\partial_j}\partial_k = \Gamma^i_{jk}\partial_i. \tag{7.2.12}$$

Thus the connection coefficient Γ^i_{jk} clearly denotes the ith component of the covariant derivative of the kth natural basis vector in $T(M)$ in the direction of the jth natural basis vector. Therefore, the affine connection can also be interpreted as an operator $\boldsymbol{\nabla} : T(M) \times T(M) \to T(M)$. For $f \in \Lambda^0(M)$, we simply get

$$\nabla_U f = u^i f_{,i} = U(f). \tag{7.2.13}$$

On resorting to the definition (7.2.11), we can easily demonstrate that the following relations are satisfied:

$$\nabla_{U_1+U_2} V = \nabla_{U_1} V + \nabla_{U_2} V, \quad \nabla_U (V_1 + V_2) = \nabla_U V_1 + \nabla_U V_2.$$

Moreover, for all functions $f \in \Lambda^0(M)$ we obtain for all $U, V \in T(M)$

$$\nabla_{fU} V = f \nabla_U V, \quad \nabla_U (fV) = f \nabla_U V + U(f) V. \tag{7.2.14}$$

Thus, the operator ∇_U proves to be linear with respect to the vector U on the module $\Lambda^0(M)$. On the other hand, ∇_U becomes linear with respect to the vector V if only $U(f) = 0$, consequently, only on $\mathbb{R}$.

Let us consider a curve $C = \gamma(t)$ on the manifold M described by the mapping $\gamma : \mathcal{I} \to M$. If $U(t)$ is the tangent vector to this curve, we know that we can write

$$U = u^i \frac{\partial}{\partial x^i}, \quad u^i(t) = \frac{dx^i}{dt} = \frac{d\gamma^i}{dt}$$

where $\gamma^i = \varphi^i \circ \gamma$. We say that a vector field V is ***parallel*** along the curve C if $\nabla_U V = \mathbf{0}$. In view of (7.2.11), the components of such a vector field V must satisfy

$$\frac{\partial v^i}{\partial x^j}\frac{dx^j}{dt} + \Gamma^i_{jk}\frac{dx^j}{dt} v^k = \frac{dv^i}{dt} + \Gamma^i_{jk}\big(\gamma(t)\big)\frac{dx^j}{dt} v^k = 0. \tag{7.2.15}$$

(7.2.15) comprises a system of m first order linear ordinary differential equations to determine m dependent variables v^i. With a prescribed initial condition $V(t_0) = V_0$, the solution $V(t)$ of (7.2.15) is called the ***parallel translation*** of V_0 along the curve C. If $\Gamma^i_{jk} = 0$ on the manifold, then equations (7.2.15) yields $v^i = constant$ on C. A curve C is called a ***geodesic*** of the manifold if its tangent vectors are parallel along C. Therefore, the condition $\nabla_U U = \mathbf{0}$ must be satisfied on a geodesic.. Hence, the family of geodesics on a manifold are integral curves of the following system of second order, generally non-linear ordinary differential equations

$$\frac{d^2x^i}{dt^2} + \Gamma^i_{jk}(\mathbf{x})\frac{dx^j}{dt}\frac{dx^k}{dt} = 0. \tag{7.2.16}$$

Obviously, they are heavily dependent on connection coefficients. We can transform these equations to a system of first order differential equations by introducing auxiliary variables as follows

$$\frac{dx^i}{dt} = u^i, \quad \frac{du^i}{dt} = -\Gamma^i_{jk}(\mathbf{x})u^j u^k$$

from which we conclude that there is a unique geodesic on a smooth manifold M through a point $p \in M$ and tangent to a vector U at that point. If $\Gamma^i_{jk} = 0$ on a manifold, the solution of (7.2.16) reduces merely to the family of *straight lines* $x^i = a^i t + b^i$.

If a vector field V satisfies the condition $\nabla_U V = \mathbf{0}$ for every vector field U, we say that it is a ***parallel vector field*** *on the manifold.* In this case, (7.2.11) leads to $(v^i{}_{,j} + \Gamma^i_{jk}v^k)u^j = 0$ for all $u^j \in \Lambda^0(M)$ or

$$v^i{}_{,j} + \Gamma^i_{jk}v^k = 0; \quad i,j,k = 1,\dots,m. \tag{7.2.17}$$

(7.2.17) is a system of m^2 first order, linear partial differential equations involving only m variables v^i. Thus, it is usually no avail to expect to find a parallel vector field on a manifold unless its linear connection has a particular structure. It is quite easy to establish the integrability conditions of these differential equations. It follows from (7.2.17) that

$$v^i{}_{,jl} = -\Gamma^i_{jk,l}v^k - \Gamma^i_{jk}v^k_{,l} = -(\Gamma^i_{jn,l} + \Gamma^i_{jk}\Gamma^k_{ln})v^n.$$

Therefore, the compatibility relation $v^i{}_{,jl} = v^i{}_{,lj}$ can only be satisfied if

$$\Gamma^i_{jn,l} - \Gamma^i_{ln,j} + \Gamma^i_{jk}\Gamma^k_{ln} - \Gamma^i_{lk}\Gamma^k_{jn} = 0.$$

We shall show a little later that a connection whose coefficients are satisfying the above relations is curvature-free.

For a tensor $\mathcal{T} \in \mathfrak{T}(M)^k_l$, we define in a similar way

$$\nabla_U \mathcal{T} = \mathbf{i}_U(dx^j)\nabla_j \mathcal{T} = u^j \nabla_j \mathcal{T} \in \mathfrak{T}(M)^k_l. \tag{7.2.18}$$

The tensor $\nabla_U \mathcal{T}$ is called the ***covariant derivative of a tensor field*** $\mathcal{T}$ ***in the direction of the vector field*** U. Due to (7.2.10), we observe at once that the following rules are obeyed

$$\begin{aligned}\nabla_U(\mathcal{T}_1 + \mathcal{T}_2) &= \nabla_U \mathcal{T}_1 + \nabla_U \mathcal{T}_2,\\ \nabla_U(\mathcal{T}_1 \otimes \mathcal{T}_2) &= \nabla_U \mathcal{T}_1 \otimes \mathcal{T}_2 + \mathcal{T}_1 \otimes \nabla_U \mathcal{T}_2.\end{aligned}$$

It is also clear that *the operator ∇_U commutes with any operation of contraction on a tensor $\mathcal{T}$.*

Torsion and Curvature Tensors. We know that for any function $f \in \Lambda^0(M)$ partial derivatives are order-independent so that the symmetry relation $f_{,ij} = f_{,ji}$ is met. Since $f_{,i}$ are components of a covariant vector, we may ask whether this property is also preserved for covariant derivatives, that is, we may question the validity of the equality $\nabla_j f_{,i} = \nabla_i f_{,j}$. The relation (7.2.6) results then in

$$\nabla_j f_{,i} - \nabla_i f_{,j} = f_{,ij} - \Gamma^k_{ji} f_{,k} - f_{,ji} + \Gamma^k_{ij} f_{,k} = (\Gamma^k_{ij} - \Gamma^k_{ji}) f_{,k} = T^k_{ij} f_{,k}.$$

We have seen on *p.* 368 that

$$T^k_{ij} = -T^k_{ji} = \Gamma^k_{ij} - \Gamma^k_{ji} \tag{7.2.19}$$

is a third order $\binom{1}{2}$-tensor which is antisymmetric with respect to covariant indices. It is called the ***torsion tensor*** of the manifold. If the connection is symmetric, that is, if $\Gamma^k_{ij} = \Gamma^k_{ji}$, then the torsion tensor vanishes. For a nonzero T^k_{ij}, we necessarily get $\nabla_j f_{,i} \neq \nabla_i f_{,j}$. Let us now try to repeat the operation above associated with a scalar function for a vector V this time. On utilising (7.2.2) we obtain

$$\begin{aligned}\nabla_k \nabla_j v^i &= (v^i{}_{,j} + \Gamma^i_{jl} v^l)_{,k} + \Gamma^i_{kn}(v^n{}_{,j} + \Gamma^n_{jl} v^l) - \Gamma^n_{kj}(v^i{}_{,n} + \Gamma^i_{nl} v^l)\\ &= v^i{}_{,jk} + (\Gamma^i_{jl,k} + \Gamma^i_{kn}\Gamma^n_{jl}) v^l - \Gamma^l_{kj} \nabla_l v^i + \Gamma^i_{jl} v^l_{,k} + \Gamma^i_{kl} v^l_{,j}.\end{aligned}$$

Hence, we are easily led to the conclusion

$$\nabla_k \nabla_j v^i - \nabla_j \nabla_k v^i = R^i_{lkj} v^l - T^l_{kj} \nabla_l v^i. \tag{7.2.20}$$

where we have defined

$$R^i_{jkl} = \Gamma^i_{lj,k} - \Gamma^i_{kj,l} + \Gamma^i_{kn}\Gamma^n_{lj} - \Gamma^i_{ln}\Gamma^n_{kj}. \tag{7.2.21}$$

Since the left hand side of (7.2.20) involves the components of a third order tensor, (7.2.21) are components of a fourth order tensor according to the quotient rule. This tensor is called the ***curvature tensor*** of the manifold. Hence, the second covariant derivatives of a vector commute if only the torsion and curvature tensors of a manifold vanish. It is evident that the curvature tensor is antisymmetric with respect to its last two covariant indices:

$$R^i_{jkl} = -R^i_{jlk} \tag{7.2.22}$$

We consider two vector fields $U, V \in T(M)$. We then obtain

$$\begin{aligned}\nabla_U V - \nabla_V U &= \left[(v^i{}_{,j}\, u^j + \Gamma^i_{jk} u^j v^k)\right] \frac{\partial}{\partial x^i} - \left[(u^i{}_{,j}\, v^j + \Gamma^i_{jk} v^j u^k)\right] \frac{\partial}{\partial x^i}\\ &= (\Gamma^i_{jk} - \Gamma^i_{kj}) u^j v^k \frac{\partial}{\partial x^i} + (v^i{}_{,j}\, u^j - u^i{}_{,j}\, v^j) \frac{\partial}{\partial x^i}\end{aligned}$$

$$= \left[T^i_{jk}u^j v^k + [U,V]^i\right]\frac{\partial}{\partial x^i}.$$

Thus the ***torsion operator***

$$\boldsymbol{\tau}(U,V) = \nabla_U V - \nabla_V U - [U,V] = T^i_{jk}u^j v^k \frac{\partial}{\partial x^i} \tag{7.2.23}$$

assigns obviously a vector field to two vector fields through (7.2.23). As such it is of the form $\boldsymbol{\tau} : T(M) \times T(M) \to T(M)$. Let us now consider three vector fields $U, V, W \in T(M)$. We just obtain

$$\nabla_U \nabla_V W = \Big[u^l v^j_{,l} w^i_{;j} + u^l v^j w^i_{,jl} + (\Gamma^i_{jk,l} + \Gamma^i_{ln}\Gamma^n_{jk})u^l v^j w^k + \Gamma^i_{jk}(u^l v^j + u^j v^l) w^k_{;l}\Big]\frac{\partial}{\partial x^i}.$$

If we recall the relation (7.2.20) and pay attention to symmetric terms with respect to vectors U and V, we then arrive at the result

$$\nabla_U \nabla_V W - \nabla_V \nabla_U W = \big([U,V]^j w^i_{;j} + R^i_{klj}u^l v^j w^k\big)\frac{\partial}{\partial x^i}.$$

Thus the ***curvature operator***

$$\boldsymbol{\rho}(U,V) = \nabla_U \nabla_V - \nabla_V \nabla_U - \nabla_{[U,V]} = [\nabla_U, \nabla_V] - \nabla_{[U,V]} \tag{7.2.24}$$

is instrumental in assigning a vector field to three vector fields U, V, W by the relation

$$\boldsymbol{\rho}(U,V)W = (R^i_{klj}u^l v^j)w^k \frac{\partial}{\partial x^i} = \rho^i_k(U,V)\, w^k \frac{\partial}{\partial x^i}. \tag{7.2.25}$$

It is of the form $\boldsymbol{\rho} : T(M) \times T(M) \to \mathfrak{T}(M)^1_1$. It follows from (7.2.23) and (7.2.24) that

$$\boldsymbol{\tau}(U,V) = -\boldsymbol{\tau}(V,U), \quad \boldsymbol{\rho}(U,V) = -\boldsymbol{\rho}(V,U).$$

Making use of (2.10.19) and (7.2.14), we readily observe that we can write

$$\boldsymbol{\tau}(fU,gV) = fg\boldsymbol{\tau}(U,V), \quad \boldsymbol{\rho}(fU,gV)hW = fgh\,\boldsymbol{\rho}(U,V)W$$

for all $f, g, h \in \Lambda^0(M)$.

Some results obtained in this section by employing Christoffel symbols can be reversed in order to define the connection in the sense of Koszul [French mathematician Jean-Louis Koszul (1921)]. Koszul has abstractly defined the affine connection ∇ as a mapping that assigns a vector $\nabla_U V$ to every pair of vectors (U, V). This mapping is desired to satisfy the rules mentioned on *p.* 373. With this definition Christoffel symbols are prescribed by the relation (7.2.12) and the covariant derivative is introduced in a similar way. The expressions (7.2.23) and (7.2.24) are employed to define the torsion and curvature tensors, respectively.

7.3. CARTAN CONNECTION

In the previous section, we have employed natural basis vectors in $T(M)$ and $T^*(M)$ determined by local charts. We now prefer a more general approach in introducing the linear connection. Let us construct a basis for the tangent bundle $T(M)$ by collecting arbitrarily chosen m linearly independent vectors $e_1, e_2, \dots, e_m \in T_p(M)$ associated with every point $p \in M$. Since $T(M)$ is closed under the Lie product, there exist some functions $c_{ij}^k \in \Lambda^0(M)$ satisfying the relations

$$[e_i, e_j] = c_{ij}^k e_k, \quad i, j, k = 1, \dots, m. \tag{7.3.1}$$

We know that the conditions (2.11.3) and (2.11.4) must be imposed on these functions. They may now be written as

$$\begin{aligned} c_{ij}^k &= -\, c_{ji}^k, \\ c_{ij}^q c_{pq}^k &+ c_{jp}^q c_{iq}^k + c_{pi}^q c_{jq}^k + e_p(c_{ij}^k) + e_i(c_{jp}^k) + e_j(c_{pi}^k) = 0. \end{aligned}$$

Cartan has called the basis vectors $\{e_i\}$ attached to every point of the manifold as the ***moving frame***. Let us now denote the reciprocal basis vectors, namely, 1-forms in $T^*(M)$ by $\theta^1, \dots, \theta^m$. They of course satisfy the relations $\theta^i(e_j) = \mathbf{i}_{e_j}(\theta^i) = \delta_j^i$. Resorting to the path followed in Sec. 5.14, we immediately discern that the reciprocal basis vector must obey the rules

$$d\theta^k = -\,\frac{1}{2}\, c_{ij}^k\, \theta^i \wedge \theta^j. \tag{7.3.2}$$

We shall now introduce an affine connection on a manifold M through a mapping $\boldsymbol{\nabla} : T(M) = \mathfrak{T}(M)_0^1 \to \mathfrak{T}(M)_1^1$ satisfying the following rules:

$$\begin{aligned} \boldsymbol{\nabla}(U + V) &= \boldsymbol{\nabla} U + \boldsymbol{\nabla} V, \\ \boldsymbol{\nabla}(fV) &= df \otimes V + f\boldsymbol{\nabla} V, \ \boldsymbol{\nabla} f = df, \forall f \in \Lambda^0(M), \end{aligned} \tag{7.3.3}$$

The last rule specifies the action of the mapping $\boldsymbol{\nabla}$ on a scalar function f. We call $\boldsymbol{\nabla}$ the ***Cartan connection*** and the tensor $\boldsymbol{\nabla} V$ the *covariant derivative* of a vector V with respect to that connection. If we express 1-form df as $\boldsymbol{\nabla} f = df = \alpha_i \theta^i$, then the familiar properties imply that

$\mathbf{i}_{e_j}(df) = e_j(f) = \alpha_j$ so that $\boldsymbol{\nabla} f$ is expressible as

$$\boldsymbol{\nabla} f = df = e_i(f)\theta^i \in \Lambda^1(M).$$

Next, let us employ the description $\boldsymbol{\nabla} f = (\nabla_i f)\theta^i$ that is tantamount to say that the covariant derivative of a function $f \in \Lambda^0(M)$ in the direction of the vector e_i with respect to Cartan connection is given by

$$\nabla_i f = e_i(f). \tag{7.3.4}$$

Covariant derivatives of basis vector can be expressed in the form

$$\boldsymbol{\nabla} e_i = \gamma^j_{ki}\theta^k \otimes e_j, \quad \gamma^j_{ki} \in \Lambda^0(M) \tag{7.3.5}$$

where the functions γ^j_{ki} play now the part of *connection coefficients*. Then, the covariant derivative of a vector V is evaluated from (7.3.3)$_2$, (7.3.4) and (7.3.5) as the following expression

$$\boldsymbol{\nabla} V = \boldsymbol{\nabla}(v^i e_i) = dv^i \otimes e_i + v^i \boldsymbol{\nabla} e_i = \big[e_j(v^i) + \gamma^i_{jk}v^k\big]\theta^j \otimes e_i.$$

If we denote the components of this tensor by $\nabla_j v^i$, we can write

$$\boldsymbol{\nabla} V = \nabla_j v^i\, \theta^j \otimes e_i, \quad \nabla_j v^i = e_j(v^i) + \gamma^i_{jk}v^k. \tag{7.3.6}$$

Let us further define the vectors $\nabla_j V = (\nabla_j v^i)e_i$ that allows us to write $\boldsymbol{\nabla} V = \theta^j \otimes \nabla_j V$. Then, we obviously draw the conclusion

$$\nabla_j e_i = \gamma^k_{ji} e_k. \tag{7.3.7}$$

Let us now choose a new basis in $T(M)$ via a regular matrix $\mathbf{B}(\mathbf{x})$ as follows: $e'_i = b^j_i e_j$. If we suppose that reciprocal basis vectors transform in the form $\theta'^i = a^i_j\,\theta^j$, then we find that

$$\delta^i_j = \theta'^i(e'_j) = a^i_k b^l_j\,\theta^k(e_l) = a^i_k b^l_j \delta^k_l = a^i_k b^k_j$$

implying that $\mathbf{A} = \mathbf{B}^{-1}$. Components of a vector with respect to the new basis will become $v'^i = a^i_j v^j$. In this case, we can write

$$dv^i = d(b^i_j v'^j) = v'^j db^i_j + b^i_j\, dv'^j = b^i_j\, dv'^j + v'^j e_k(b^i_j)\theta'^k.$$

Hence, we obtain

$$\begin{aligned}
\boldsymbol{\nabla} V &= dv^i \otimes e_i + \gamma^i_{jk}v^k\,\theta^j \otimes e_i \\
&= a^l_i\big[b^i_j\, dv'^j + v'^j e_k(b^i_j)\theta'^k\big] \otimes e'_l + \gamma^i_{jk}b^k_l b^j_m a^n_i v'^l \theta'^m \otimes e'_n
\end{aligned}$$

$$= \big[dv'^j + v'^k a^j_i e'_l(b^i_k)\theta'^l\big] \otimes e'_j + \gamma^i_{jk} b^k_l b^j_m a^n_i v'^l \theta'^m \otimes e'_n$$
$$= dv'^i \otimes e'_i + \gamma'^i_{jk} v'^k \theta'^j \otimes e'_i$$

from which we deduce after properly changing the dummy indices that the connection coefficients in the new basis must satisfy the relation

$$\gamma'^i_{jk} = a^i_l b^m_j b^n_k \gamma^l_{mn} + a^i_l e'_j(b^l_k) = a^i_l b^m_j b^n_k \gamma^l_{mn} + a^i_l b^m_j e_m(b^l_k).$$

Because of the last terms, we understand that the coefficients γ^i_{jk} cannot be components of a third order tensor.

Let us now consider a vector $V = v^i e_i$ and a 1-form $\omega = \omega_i \theta^i$. Since we have assumed that the operator ∇_j satisfies the Leibniz rule, the expression $\omega(V) = \omega_i v^i \in \Lambda^0(M)$ yields

$$\nabla_j(\omega_i v^i) = e_j(\omega_i v^i) = e_j(\omega_i)v^i + \omega_i e_j(v^i) = v^i \nabla_j \omega_i + \omega_i \nabla_j v^i$$
$$= v^i \nabla_j \omega_i + \omega_i \big[e_j(v^i) + \gamma^i_{jk} v^k\big]$$

whence we deduce that $v^i\big[\nabla_j \omega_i - \big(e_j(\omega_i) - \gamma^k_{ji}\omega_k\big)\big] = 0$ and since V is an arbitrary vector, we are led to the conclusion

$$\nabla_j \omega_i = e_j(\omega_i) - \gamma^k_{ji}\omega_k. \tag{7.3.8}$$

We can thus write $\nabla_j \omega = (\nabla_j \omega_i)\,\theta^i$ and $\boldsymbol{\nabla}\omega = \theta^j \otimes \nabla_j \omega = \nabla_j \omega_i\, \theta^j \otimes \theta^i$. Hence, the operator $\boldsymbol{\nabla}$ is now a mapping $\boldsymbol{\nabla} : T^*(M) = \mathfrak{T}(M)^0_1 \to \mathfrak{T}(M)^0_2$. On the other hand, we can readily reach to the following relations

$$\nabla_j\, \theta^k = -\,\gamma^k_{ji}\,\theta^i, \quad \boldsymbol{\nabla}\theta^k = -\,\gamma^k_{ji}\,\theta^j \otimes \theta^i. \tag{7.3.9}$$

Covariant derivative with respect to Cartan connection that we have dealt with so far can easily be extended to any tensor $\mathcal{T} \in \mathfrak{T}(M)^k_l$. As an example, let us take the tensor

$$\mathcal{T} = t^{i_1\cdots i_k}_{j_1\cdots j_l} e_{i_1} \otimes \cdots \otimes e_{i_k} \otimes \theta^{j_1} \otimes \cdots \otimes \theta^{j_l}$$

into consideration. We then find $\boldsymbol{\nabla}\mathcal{T} = \theta^j \otimes \nabla_j \mathcal{T} \in \mathfrak{T}(M)^k_{l+1}$ and

$$\nabla_j \mathcal{T} = \nabla_j t^{i_1\cdots i_k}_{j_1\cdots j_l} e_{i_1} \otimes \cdots \otimes e_{i_k} \otimes \theta^{j_1} \otimes \cdots \otimes \theta^{j_l}$$

where the components $\nabla_j t^{i_1\cdots i_k}_{j_1\cdots j_l}$ are given by

$$\nabla_j t^{i_1\cdots i_k}_{j_1\cdots j_l} = e_j\big(t^{i_1\cdots i_k}_{j_1\cdots j_l}\big) + \sum_{r=1}^{k} \gamma^{i_r}_{jn} t^{i_1\cdots i_{r-1} n i_{r+1}\cdots i_k} - \sum_{r=1}^{l} \gamma^n_{jj_r} t^{i_1\cdots i_k}_{j_1\cdots j_{r-1} n j_{r+1}\cdots j_l}$$

We observe at once that the relation (7.2.10) is also valid in Cartan connection as well.

Covariant derivative of a vector field $V = v^i e_i \in T(M)$ in the direction of a vector field $U = u^i e_i \in T(M)$ can be defined as in Sec. 7.2 by the vector

$$\nabla_U V = \mathbf{i}_U(\theta^j)\nabla_j V = u^j \nabla_j V = \big[U(v^i) + \gamma^i_{jk} u^j v^k\big] e_i. \quad (7.3.10)$$

Similarly, the covariant derivative of a tensor field $\mathcal{T} \in \mathfrak{T}(M)^k_l$ in the direction of a vector field U is designated by

$$\nabla_U \mathcal{T} = u^j \nabla_j \mathcal{T} \in \mathfrak{T}(M)^k_l.$$

It is also evident that one is able to write

$$\nabla_{e_k} e_j = \nabla_k e_j = \gamma^i_{kj} e_i, \quad \nabla_{e_k}\theta^j = \nabla_k\, \theta^j = -\gamma^j_{ki}\theta^i. \quad (7.3.11)$$

In order to define the torsion tensor $\boldsymbol{\mathcal{T}}$ and the curvature tensor $\boldsymbol{\mathcal{R}}$ in Cartan connection, we can make use of the relations (7.2.23) and (7.2.24). Consider a form field $\omega \in \Lambda^1(M)$ and vector fields $U, V, W \in T(M)$ and write

$$\begin{aligned} \boldsymbol{\mathcal{T}}(\omega, U, V) &= \omega\big(\boldsymbol{\tau}(U, V)\big) = T^i_{jk}\omega_i u^j v^k, \\ \boldsymbol{\mathcal{R}}(\omega, U, V, W) &= \omega\big(\boldsymbol{\rho}(U, V)W\big) = R^i_{klj}\omega_i u^l v^j w^k. \end{aligned}$$

It follows from (7.2.23) that

$$\begin{aligned} \nabla_{u^j e_j} v^k e_k - \nabla_{v^k e_k} u^j e_j - [u^j e_j, v^k e_k] &= \big(\nabla_{e_j} e_k - \nabla_{e_k} e_j - [e_j, e_k]\big) u^j v^k \\ &= (\gamma^i_{jk} - \gamma^i_{kj} - c^i_{jk}) u^j v^k e_i. \end{aligned}$$

Therefore, the third order torsion tensor is found as

$$\boldsymbol{\mathcal{T}} = T^i_{jk} e_i \otimes \theta^j \otimes \theta^k, \quad T^i_{jk} = -T^i_{kj} = \gamma^i_{jk} - \gamma^i_{kj} - c^i_{jk}. \quad (7.3.12)$$

In the like fashion, the relation

$$\boldsymbol{\rho}(U, V)W = u^j v^k w^l \boldsymbol{\rho}(e_j, e_k) e_l$$

leads us to

$$\begin{aligned} \boldsymbol{\rho}(e_j, e_k) e_l &= \big([\nabla_{e_j}, \nabla_{e_k}] - \nabla_{[e_j, e_k]}\big) e_l = \big([\nabla_{e_j}, \nabla_{e_k}] - c^m_{jk}\nabla_{e_m}\big) e_l \\ &= \nabla_{e_j}\nabla_{e_k} e_l - \nabla_{e_k}\nabla_{e_j} e_l - c^m_{jk}\nabla_{e_m} e_l \\ &= \nabla_{e_j}\gamma^i_{kl} e_i - \nabla_{e_k}\gamma^i_{jl} e_i - c^m_{jk}\gamma^i_{ml} e_i \end{aligned}$$

$$= \gamma^i_{kl}\nabla_{e_j}e_i + e_j(\gamma^i_{kl})\,e_i - \gamma^i_{jl}\nabla_{e_k}e_i - e_k(\gamma^i_{jl})\,e_i - c^m_{jk}\gamma^i_{ml}e_i$$
$$= \big[e_j(\gamma^i_{kl}) - e_k(\gamma^i_{jl}) + \gamma^i_{jm}\gamma^m_{kl} - \gamma^i_{km}\gamma^m_{jl} - c^m_{jk}\gamma^i_{ml}\big]e_i = R^i_{ljk}e_i$$

from which we can manage to extract the components of the fourth order curvature tensor $\mathcal{R} = R^i_{ljk}e_i \otimes \theta^l \otimes \theta^j \otimes \theta^k$ as follows

$$R^i_{ljk} = e_j(\gamma^i_{kl}) - e_k(\gamma^i_{jl}) + \gamma^i_{jm}\gamma^m_{kl} - \gamma^i_{km}\gamma^m_{jl} - c^m_{jk}\gamma^i_{ml}. \qquad (7.3.13)$$

It is easily seen from the definition that this tensor possesses the antisymmetry property $R^i_{ljk} = -R^i_{lkj}$.

When we choose a connection determined by the coefficients γ^i_{jk}, this makes it possible to generate a new connection without a torsion. Let us define the new connection coefficients by

$$\gamma'^i_{jk} = \gamma^i_{jk} - \frac{1}{2}T^i_{jk},$$

we then find that

$$T'^i_{jk} = \gamma'^i_{jk} - \gamma'^i_{kj} - c^i_{jk} = \gamma^i_{jk} - \gamma^i_{kj} - c^i_{jk} - T^i_{jk} = 0.$$

We shall now try to discuss the action of the commutator $[\nabla_j, \nabla_k]$ on diverse tensor fields. Let us first consider the scalar function $f \in \Lambda^0(M)$. We obtain from (7.3.4) and (7.3.8) that

$$\begin{aligned}\nabla_k\nabla_j f &= \nabla_k e_j(f)\\ &= e_k\big(e_j(f)\big) - \gamma^m_{kj}e_m(f).\end{aligned}$$

Because of the commutation relation $e_ke_j - e_je_k = c^m_{kj}\,e_m$, we get

$$\begin{aligned}(\nabla_k\nabla_j - \nabla_j\nabla_k)f &= -\,(\gamma^m_{kj} - \gamma^m_{jk} - c^m_{kj})\,e_m(f) \qquad (7.3.14)\\ &= -\,T^m_{kj}\,e_m(f).\end{aligned}$$

Next, we consider a vector field $V \in T(M)$. Since we have

$$\nabla_j v^i = e_j(v^i) + \gamma^i_{jm}v^m$$

we find that

$$\begin{aligned}\nabla_k\nabla_j v^i = {}& e_k\big(e_j(v^i)\big) + \big[e_k(\gamma^i_{jm}) + \gamma^i_{kn}\gamma^n_{jm} - \gamma^n_{kj}\gamma^i_{nm}\big]v^m\\ &+ \gamma^i_{jm}e_k(v^m) + \gamma^i_{km}e_j(v^m) - \gamma^m_{kj}e_m(v^i).\end{aligned}$$

Hence, we arrive at

$$(\nabla_k\nabla_j-\nabla_j\nabla_k)\,v^i=\big[e_k(\gamma^i_{jm})-e_j(\gamma^i_{km})+\gamma^i_{kn}\gamma^n_{jm}-\gamma^i_{jn}\gamma^n_{km}-c^n_{kj}\gamma^i_{nm}\big]v^m \\ -T^m_{kj}\big(e_m(v^i)+\gamma^i_{mn}v^n\big).$$

Consequently, we obtain

$$[\nabla_k,\nabla_j]v^i=(\nabla_k\nabla_j-\nabla_j\nabla_k)v^i=R^i_{mkj}v^m-T^m_{kj}\nabla_m v^i. \quad (7.3.15)$$

Similarly, for a form $\omega\in\Lambda^1(M)$, we have $\nabla_j\omega_i=e_j(\omega_i)-\gamma^m_{ji}\omega_m$ and we thus find

$$\nabla_k\nabla_j\omega_i=e_k\big(e_j(\omega_i)\big)+\big[-e_k(\gamma^m_{ji})+\gamma^n_{kj}\gamma^m_{ni}+\gamma^n_{ki}\gamma^m_{jn}\big]\omega_m \\ -\gamma^m_{ji}e_k(\omega_m)-\gamma^m_{ki}e_j(\omega_m)-\gamma^m_{kj}e_m(\omega_i).$$

This relation gives rise to

$$(\nabla_k\nabla_j-\nabla_j\nabla_k)\,\omega_i=\big[e_j(\gamma^m_{ki})-e_k(\gamma^m_{ji})+\gamma^n_{ki}\gamma^m_{jn}-\gamma^n_{ji}\gamma^m_{kn}-c^n_{jk}\gamma^m_{ni}\big]\omega_m \\ +T^m_{jk}\big(e_m(\omega_i)-\gamma^n_{mi}\omega_n\big)$$

from which we obtain at once

$$[\nabla_k,\nabla_j]\omega_i=(\nabla_k\nabla_j-\nabla_j\nabla_k)\,\omega_i=R^m_{ijk}\omega_m+T^m_{jk}\nabla_m\omega_i \quad (7.3.16) \\ =-R^m_{ikj}\omega_m-T^m_{kj}\nabla_m\omega_i.$$

With these information at hand, we can easily evaluate the action of the operator $[\nabla_k,\nabla_j]$ on any tensor. But, let us first verify that the operator $[\nabla_k,\nabla_j]$ obeys the Leibniz rule on tensor products. If $\mathcal{T}$ and $\mathcal{S}$ are two tensor fields, we can obviously write

$$\nabla_j(\mathcal{T}\otimes\mathcal{S})=\nabla_j\mathcal{T}\otimes\mathcal{S}+\mathcal{T}\otimes\nabla_j\mathcal{S}$$
$$\nabla_k\nabla_j(\mathcal{T}\otimes\mathcal{S})=\nabla_k\nabla_j\mathcal{T}\otimes\mathcal{S}+\nabla_j\mathcal{T}\otimes\nabla_k\mathcal{S}+\nabla_k\mathcal{T}\otimes\nabla_j\mathcal{S}+\mathcal{T}\otimes\nabla_k\nabla_j\mathcal{S}$$

whence we draw at once the conclusion

$$[\nabla_k,\nabla_j](\mathcal{T}\otimes\mathcal{S})=[\nabla_k,\nabla_j](\mathcal{T})\otimes\mathcal{S}+\mathcal{T}\otimes[\nabla_k,\nabla_j](\mathcal{S}). \quad (7.3.17)$$

Let us consider a tensor $\mathcal{T}$ of order $k+l$. Then we can produce a scalar function $f=t^{i_1\cdots i_k}_{j_1\cdots j_l}v^{j_1}_{(1)}\cdots v^{j_l}_{(l)}\omega^{(1)}_{i_1}\cdots\omega^{(k)}_{i_k}$ by considering its action on l vectors $V_{(1)},\dots V_{(l)}$ and k 1-forms $\omega^{(1)},\dots,\omega^{(k)}$. On utilising (7.3.14), we have

$$[\nabla_k,\nabla_j](f)=-T^m_{kj}\,e_m(f)=-T^m_{kj}\,e_m\big(t^{i_1\cdots i_k}_{j_1\cdots j_l}\big)v^{j_1}_{(1)}\cdots v^{j_l}_{(l)}\omega^{(1)}_{i_1}\cdots\omega^{(k)}_{i_k} \\ -T^m_{kj}\,t^{i_1\cdots i_k}_{j_1\cdots j_l}e_m\big(v^{j_1}_{(1)}\cdots v^{j_l}_{(l)}\omega^{(1)}_{i_1}\cdots\omega^{(k)}_{i_k}\big).$$

On the other hand, because of the Leibniz rule (7.3.17), on applying the commutator $[\nabla_k, \nabla_j]$ on the function f defined above, we get

$$
\begin{aligned}
[\nabla_k, \nabla_j](f) &= [\nabla_k, \nabla_j]\big(t^{i_1\cdots i_k}_{j_1\cdots j_l} v^{j_1}_{(1)}\cdots v^{j_l}_{(l)}\omega^{(1)}_{i_1}\cdots\omega^{(k)}_{i_k}\big)\\
&= [\nabla_k, \nabla_j]\big(t^{i_1\cdots i_k}_{j_1\cdots j_l}\big) v^{j_1}_{(1)}\cdots v^{j_l}_{(l)}\omega^{(1)}_{i_1}\cdots\omega^{(k)}_{i_k}\\
&\quad + t^{i_1\cdots i_k}_{j_1\cdots j_l}\Big[\sum_{r=1}^{l} v^{j_1}_{(1)}\cdots[\nabla_k, \nabla_j]\, v^{j_r}_{(r)}\cdots v^{j_l}_{(l)}\omega^{(1)}_{i_1}\cdots\omega^{(k)}_{i_k}\\
&\quad + \sum_{r=1}^{k} v^{j_1}_{(1)}\cdots v^{j_l}_{(l)}\omega^{(1)}_{i_1}\cdots[\nabla_k, \nabla_j]\,\omega^{(r)}_{i_r}\cdots\omega^{(k)}_{i_k}\Big].
\end{aligned}
$$

Let us now employ (7.3.15) and (7.3.16) to transform the terms within brackets in the third and fourth lines of the above expression into the following form

$$
\begin{aligned}
&\sum_{r=1}^{l} v^{j_1}_{(1)}\cdots\big(R^{j_r}_{mkj}v^{m}_{(r)} - T^{m}_{kj}\nabla_m v^{j_r}_{(r)}\big)\cdots v^{j_l}_{(l)}\omega^{(1)}_{i_1}\cdots\omega^{(k)}_{i_k}\\
&\quad + \sum_{r=1}^{k} v^{j_1}_{(1)}\cdots v^{j_l}_{(l)}\omega^{(1)}_{i_1}\cdots\big(-R^{m}_{i_r kj}\omega^{(r)}_{m} - T^{m}_{kj}\nabla_m\omega^{(r)}_{i_r}\big)\cdots\omega^{(k)}_{i_k}.
\end{aligned}
$$

Next, we write

$$
\begin{aligned}
\nabla_m v^{j_r}_{(r)} &= e_m(v^{j_r}_{(r)}) + \gamma^{j_r}_{mn}v^{n}_{(r)},\\
\nabla_m \omega^{(r)}_{i_r} &= e_m(\omega^{(r)}_{i_r}) - \gamma^{n}_{m i_r}\omega^{(r)}_{n}
\end{aligned}
$$

and change the dummy indices properly to obtain

$$
\begin{aligned}
[\nabla_k, \nabla_j](f) &= \Big\{[\nabla_k, \nabla_j]\big(t^{i_1\cdots i_k}_{j_1\cdots j_l}\big) + \sum_{r=1}^{l} R^{m}_{j_r kj}t^{i_1\cdots i_k}_{j_1\cdots j_{r-1} m j_{r+1}\cdots j_l} -\\
&\quad \sum_{r=1}^{k} R^{i_r}_{mkj}t^{i_1\cdots i_{r-1} m i_{r+1}\cdots i_k}_{j_1\cdots j_l} - T^{m}_{kj}\Big[\sum_{r=1}^{l}\gamma^{n}_{m j_r}t^{i_1\cdots i_k}_{j_1\cdots j_{r-1} n j_{r+1}\cdots j_l}\\
&\quad - \sum_{r=1}^{l}\gamma^{i_r}_{mn}t^{i_1\cdots i_{r-1} n i_{r+1}\cdots i_k}_{j_1\cdots j_l}\Big]\Big\} v^{j_1}_{(1)}\cdots v^{j_l}_{(l)}\omega^{(1)}_{i_1}\cdots\omega^{(k)}_{i_k}\\
&= -T^{m}_{kj}\, e_m\big(t^{i_1\cdots i_k}_{j_1\cdots j_l}\big) v^{j_1}_{(1)}\cdots v^{j_l}_{(l)}\omega^{(1)}_{i_1}\cdots\omega^{(k)}_{i_k}\\
&\quad - T^{m}_{kj}\, t^{i_1\cdots i_k}_{j_1\cdots j_l} e_m\big(v^{j_1}_{(1)}\cdots v^{j_l}_{(l)}\omega^{(1)}_{i_1}\cdots\omega^{(k)}_{i_k}\big).
\end{aligned}
$$

Consequently, on collecting terms suitably, we end up with the relations

$$\Big\{[\nabla_k,\nabla_j](t^{i_1\cdots i_k}_{j_1\cdots j_l})+\sum_{r=1}^{l}R^{m}_{j_rkj}t^{i_1\cdots i_k}_{j_1\cdots j_{r-1}mj_{r+1}\cdots j_l}-\sum_{r=1}^{k}R^{i_r}_{mkj}t^{i_1\cdots i_{r-1}mi_{r+1}\cdots i_k}_{j_1\cdots j_l}$$
$$+T^m_{kj}\nabla_m t^{i_1\cdots i_k}_{j_1\cdots j_l}\Big\}v^{j_1}_{(1)}\cdots v^{j_l}_{(l)}\omega^{(1)}_{i_1}\cdots\omega^{(k)}_{i_k}=0.$$

Since this expression must be satisfied for arbitrary vectors and 1-forms, we finally reach to the desired relation

$$(\nabla_k\nabla_j-\nabla_j\nabla_k)\big(t^{i_1\cdots i_k}_{j_1\cdots j_l}\big)=\sum_{r=1}^{k}R^{i_r}_{mkj}\,t^{i_1\cdots i_{r-1}mi_{r+1}\cdots i_k}_{j_1\cdots j_l} \tag{7.3.18}$$
$$-\sum_{r=1}^{l}R^{m}_{j_rkj}\,t^{i_1\cdots i_k}_{j_1\cdots j_{r-1}mj_{r+1}\cdots j_l}-T^m_{kj}\nabla_m\big(t^{i_1\cdots i_k}_{j_1\cdots j_l}\big).$$

Next, we define m^2 ***connection* 1*-forms*** by

$$\Gamma^i_j=\gamma^i_{kj}\,\theta^k, \tag{7.3.19}$$

m ***torsion* 2*-forms*** by

$$\Sigma^i=\frac{1}{2}T^i_{jk}\theta^j\wedge\theta^k, \tag{7.3.20}$$

and m^2 ***curvature* 2*-forms*** by

$$\Theta^i_j=\frac{1}{2}R^i_{jkl}\theta^k\wedge\theta^l. \tag{7.3.21}$$

Theorem 7.13.1. *Torsion and curvature forms satisfy the Cartan structural equations*

$$d\theta^i=-\Gamma^i_j\wedge\theta^j+\Sigma^i,\ \ \Theta^i_j=d\Gamma^i_j+\Gamma^i_k\wedge\Gamma^k_j \tag{7.3.22}$$

in the moving frame.

Indeed, it follows from (7.3.2) and (7.3.12) that

$$\begin{aligned}d\theta^i&=-\frac{1}{2}c^i_{jk}\,\theta^j\wedge\theta^k\\&=\frac{1}{2}T^i_{jk}\,\theta^j\wedge\theta^k-\frac{1}{2}(\gamma^i_{jk}-\gamma^i_{kj})\theta^j\wedge\theta^k\\&=\Sigma^i-\gamma^i_{kj}\theta^k\wedge\theta^j\\&=-\Gamma^i_j\wedge\theta^j+\Sigma^i.\end{aligned}$$

On the other hand, the relation (7.13.14) leads to

$$\begin{aligned} d\Gamma^i_j &= d\gamma^i_{lj} \wedge \theta^l + \gamma^i_{mj}\, d\theta^m \\ &= \big[e_k(\gamma^i_{lj}) - \frac{1}{2}\gamma^i_{mj}c^m_{kl}\big]\theta^k \wedge \theta^l \\ \Gamma^i_m \wedge \Gamma^m_j &= \gamma^i_{km}\gamma^m_{lj}\,\theta^k \wedge \theta^l. \end{aligned}$$

We thus conclude that

$$\begin{aligned} d\Gamma^i_j + \Gamma^i_m \wedge \Gamma^m_j &= \big[e_k(\gamma^i_{lj}) + \gamma^i_{km}\gamma^m_{lj} - \frac{1}{2}\gamma^i_{mj}c^m_{kl}\big]\theta^k \wedge \theta^l \\ &= \frac{1}{2}\big[e_k(\gamma^i_{lj}) - e_l(\gamma^i_{kj}) + \gamma^i_{km}\gamma^m_{lj} - \gamma^i_{lm}\gamma^m_{kj} - c^m_{kl}\gamma^i_{mj}\big]\theta^k \wedge \theta^l \\ &= \frac{1}{2}R^i_{jkl}\theta^k \wedge \theta^l = \Theta^i_j. \qquad \square \end{aligned}$$

The expressions (7.3.22) provide us with a quite an effective tool to discover relatively easily some interesting relations between the curvature and torsion tensors. From $(7.3.22)_1$, we can write

$$\begin{aligned} d\Sigma^i &= d\Gamma^i_j \wedge \theta^j - \Gamma^i_j \wedge d\theta^j \\ &= (d\Gamma^i_j + \Gamma^i_k \wedge \Gamma^k_j) \wedge \theta^j - \Gamma^i_j \wedge \Sigma^j \\ &= \Theta^i_j \wedge \theta^j - \Gamma^i_j \wedge \Sigma^j. \end{aligned}$$

Employing the definitions (7.3.20) and (7.3.21), we get

$$\Theta^i_j \wedge \theta^j - \Gamma^i_j \wedge \Sigma^j = \frac{1}{2}(R^i_{jkl} - \gamma^i_{jm}T^m_{kl})\,\theta^j \wedge \theta^k \wedge \theta^l.$$

By appropriately renaming the dummy indices, we obtain from (7.3.20) and (7.3.2) that

$$\begin{aligned} d\Sigma^i &= \frac{1}{2}(dT^i_{jk} \wedge \theta^j \wedge \theta^k + T^i_{jk}d\theta^j \wedge \theta^k - T^i_{jk}\theta^j \wedge d\theta^k) \\ &= \frac{1}{2}\big[e_j(T^i_{kl}) - \frac{1}{2}\,c^m_{jk}T^i_{ml} + \frac{1}{2}\,c^m_{kl}T^i_{jm}\big]\theta^j \wedge \theta^k \wedge \theta^l. \end{aligned}$$

Hence, we get

$$\Big[R^i_{jkl} - e_j(T^i_{kl}) - \gamma^i_{jm}T^m_{kl} + \frac{1}{2}\,c^m_{jk}T^i_{ml} - \frac{1}{2}\,c^m_{kl}T^i_{jm}\Big]\theta^j \wedge \theta^k \wedge \theta^l = 0.$$

Since the covariant derivative of the torsion tensor is

$$\nabla_j T^i_{kl} = e_j(T^i_{kl}) + \gamma^i_{jm}T^m_{kl} - \gamma^m_{jk}T^i_{ml} - \gamma^m_{jl}T^i_{km},$$

the foregoing equality can be transformed into

$$\big[R^i_{jkl} - \nabla_j T^i_{kl} + (\gamma^m_{jk} - \tfrac{1}{2}c^m_{jk})T^i_{lm} - \gamma^m_{jl}T^i_{km} - \tfrac{1}{2}\,c^m_{kl}T^i_{jm}\big]\theta^j \wedge \theta^k \wedge \theta^l = 0.$$

If we add and subtract the terms $\frac{1}{2}c^m_{jl}T^i_{km}$ into the brackets above and note that $\frac{1}{2}(c^m_{jl}T^i_{km} + c^m_{kl}T^i_{jm})\,\theta^j \wedge \theta^k = 0$, then we get

$$\big[R^i_{jkl} - \nabla_j T^i_{kl} + (\gamma^m_{jk} - \tfrac{1}{2}c^m_{jk})T^i_{lm} - (\gamma^m_{jl} - \tfrac{1}{2}\,c^m_{jl})T^i_{km}\big]\theta^j \wedge \theta^k \wedge \theta^l = 0.$$

On utilising the antisymmetry with respect to indices k and l, we obtain

$$\big[R^i_{jkl} - \nabla_j T^i_{kl} + (2\gamma^m_{[jk]} - c^m_{jk})T^i_{lm}\big]\theta^j \wedge \theta^k \wedge \theta^l = 0.$$

Let us now insert (7.3.12) into the above expression to cast it into the form

$$(R^i_{jkl} - \nabla_j T^i_{kl} + T^m_{jk}T^i_{lm})]\theta^j \wedge \theta^k \wedge \theta^l = 0.$$

Then we finally reach to the following identity

$$R^i_{[jkl]} = \nabla_{[j}T^i_{kl]} - T^m_{[jk}T^i_{l]m}. \tag{7.3.23}$$

The explicit form of the expression (7.3.23) becomes

$$\begin{aligned} &R^i_{jkl} + R^i_{klj} + R^i_{ljk} = \\ &\qquad \nabla_j T^i_{kl} + \nabla_k T^i_{lj} + \nabla_l T^i_{jk} - T^m_{jk}T^i_{lm} - T^m_{kl}T^i_{jm} - T^m_{lj}T^i_{km}. \end{aligned}$$

Next, we evaluate the exterior derivative of $(7.3.22)_2$ to obtain

$$\begin{aligned} d\Theta^i_j &= d\Gamma^i_k \wedge \Gamma^k_j - \Gamma^i_k \wedge d\Gamma^k_j = \Theta^i_k \wedge \Gamma^k_j - \Gamma^i_m \wedge \Gamma^m_k \wedge \Gamma^k_j \\ &- \Gamma^i_k \wedge \Theta^k_j + \Gamma^i_k \wedge \Gamma^k_m \wedge \Gamma^m_j = \Theta^i_k \wedge \Gamma^k_j - \Gamma^i_k \wedge \Theta^k_j. \end{aligned}$$

As is easily seen, we can write

$$\begin{aligned} \Theta^i_k \wedge \Gamma^k_j - \Gamma^i_k \wedge \Theta^k_j &= \frac{1}{2}(R^i_{nlm}\gamma^n_{kj} - R^n_{jlm}\gamma^i_{kn})\,\theta^k \wedge \theta^l \wedge \theta^m, \\ d\Theta^i_j &= \frac{1}{2}\big[e_k(R^i_{jlm}) - \frac{1}{2}\,c^n_{kl}R^i_{jnm} + \frac{1}{2}\,c^n_{mk}R^i_{jln}\big]\theta^k \wedge \theta^l \wedge \theta^m. \end{aligned}$$

We thus obtain

$$\big[e_k(R^i_{jlm}) + \gamma^i_{kn}R^n_{jlm} - \gamma^n_{kj}R^i_{nlm} - \frac{1}{2}\,c^n_{kl}R^i_{jnm} + \frac{1}{2}\,c^n_{mk}R^i_{jln}\big]\theta^k \wedge \theta^l \wedge \theta^m = 0.$$

On account of the expression

$$\nabla_k R^i_{jlm} = e_k(R^i_{jlm}) + \gamma^i_{kn}R^n_{jlm} - \gamma^n_{kj}R^i_{nlm} - \gamma^n_{kl}R^i_{jnm} - \gamma^n_{km}R^i_{jln}$$

the above equality can be transformed into

$$\big[\nabla_k R^i_{jlm} - (\gamma^n_{kl} - \tfrac{1}{2}c^n_{kl})R^i_{jmn} + (\gamma^n_{km} - \tfrac{1}{2}c^n_{km})R^i_{jln}\big]\theta^k \wedge \theta^l \wedge \theta^m = 0$$

If we take notice the antisymmetry of the exterior products of 1-forms in this expression and properly rename the dummy indices, we get

$$\big[\nabla_k R^i_{jlm} - (2\gamma^n_{[kl]} - c^n_{kl})R^i_{jmn}\big]\theta^k \wedge \theta^l \wedge \theta^m = 0$$

and

$$(\nabla_k R^i_{jlm} - T^n_{kl}R^i_{jmn})\,\theta^k \wedge \theta^l \wedge \theta^m = 0$$

after having inserted (7.3.12). We then finally obtain the following identity

$$\nabla_{[k}R^i_{|j|lm]} = T^n_{[kl}R^i_{|j|m]n} \tag{7.3.24}$$

where the operation of alternation will be suspended on the index j occupying the space inside two vertical bars. The expressions (7.3.23) and (7.3.24) are called the **1*st and* 2*nd Bianchi identities***, respectively, because they were first discovered by Italian mathematician Luigi Bianchi (1856-1928) albeit in a different framework. The explicit form of the relations (7.3.24) becomes

$$\nabla_k R^i_{jlm} + \nabla_l R^i_{jmk} + \nabla_m R^i_{jkl} = T^n_{kl}R^i_{jmn} + T^n_{lm}R^i_{jkn} + T^n_{mk}R^i_{jln}.$$

When the torsion tensor vanishes, then we would necessarily get $R^i_{[jkl]} = 0$ and $\nabla_{[k}R^i_{|j|lm]} = 0$.

Let us now introduce the matrix forms $\mathbf{\Omega} = [\theta^i] \in \mathbf{\Lambda}^1(M)$ where all 1-forms θ^i are linearly independent, $\mathbf{\Gamma} = [\Gamma^i_j] \in \mathbf{\Lambda}^1(M)$, $\mathbf{\Sigma} = [\Sigma^i] \in \mathbf{\Lambda}^2(M)$ and $\mathbf{\Theta} = [\Theta^i_j] \in \mathbf{\Lambda}^2(M)$. Hence, the Cartan structural equations (7.3.22) can now be expressed as follows in the matrix form

$$d\mathbf{\Omega} = -\mathbf{\Gamma} \wedge \mathbf{\Omega} + \mathbf{\Sigma}, \quad \mathbf{\Theta} = d\mathbf{\Gamma} + \mathbf{\Gamma} \wedge \mathbf{\Gamma}.$$

The exterior derivatives of these forms satisfy the relations

$$d\mathbf{\Sigma} = \mathbf{\Theta} \wedge \mathbf{\Omega} - \mathbf{\Gamma} \wedge \mathbf{\Sigma}, \quad d\mathbf{\Theta} = \mathbf{\Theta} \wedge \mathbf{\Gamma} - \mathbf{\Gamma} \wedge \mathbf{\Theta}.$$

These equations coincide with the system of exterior differential equations given by (6.8.6). Thus the local solutions of these differential equations on an open set $U \subseteq M$ contractible to one of its interior points are provided by the expressions (6.8.19). By employing these expressions for various purposes we can determine the basis forms and connection coefficients generating certain torsion and curvature tensors with desired properties.

7.4. LEVI-CIVITA CONNECTION

As we have cited in Sec. 5.9, a Riemannian manifold is equipped with a symmetric, second order covariant metric tensor

$$\mathcal{G} = g_{ij}(\mathbf{x})\,\theta^i \otimes \theta^j, \quad g_{ij} = g_{ji}.$$

We assume that the basis in the cotangent bundle $T^*(M)$ can be so chosen that the matrix $\mathbf{G} = [g_{ij}]$ is regular at every point of the manifold, that is, the inverse matrix denoted by $\mathbf{G}^{-1} = [(g^{-1})^{ij}] = [g^{ij}]$ exists. The elementary arc length on this manifold will be measured by the relation

$$ds^2 = g_{ij}\,\theta^i\theta^j$$

similar to that given on *p.* 274.

Theorem 7.4.1. *There is a unique torsion-free affine connection on a Riemannian manifold with respect to which the covariant derivative of the metric tensor vanishes.*

The conditions $\nabla\mathcal{G} = \mathbf{0}$ and $\mathcal{T} = \mathbf{0}$ necessitate, respectively

$$\begin{aligned}\nabla_k g_{ij} = e_k(g_{ij}) - \gamma^l_{ki}g_{lj} - \gamma^l_{kj}g_{il} = 0,\\ T^i_{jk} = \gamma^i_{jk} - \gamma^i_{kj} - c^i_{jk} = 0.\end{aligned}$$

Let us now write $\gamma^i_{jk} = \gamma^i_{[jk]} + \gamma^i_{(jk)}$ where the symmetric and antisymmetric parts with respect to the subscripts of the connection coefficients are denoted, respectively, by

$$\gamma^i_{(jk)} = \frac{1}{2}(\gamma^i_{jk} + \gamma^i_{kj}), \quad \gamma^i_{[jk]} = \frac{1}{2}(\gamma^i_{jk} - \gamma^i_{kj}).$$

Then the condition $\nabla_k g_{ij} = 0$ yields

$$\gamma^i_{[jk]} = \frac{1}{2}\,c^i_{jk} \quad \text{and} \quad \gamma^l_{(ki)}g_{lj} + \gamma^l_{(kj)}g_{il} = e_k(g_{ij}) - \frac{1}{2}(c^l_{ki}g_{lj} + c^l_{kj}g_{il}).$$

On introducing the definitions

$$\gamma_{ijk} = g_{il}\gamma^l_{jk}, \quad c_{ijk} = g_{il}c^l_{jk} = -\,c_{ikj} \tag{7.4.1}$$

we can cast the above expressions into the forms

$$\gamma_{i[jk]} = \frac{1}{2}\,c_{ijk}, \quad \gamma_{j(ki)} + \gamma_{i(kj)} = e_k(g_{ij}) - \frac{1}{2}(c_{jki} + c_{ikj}). \tag{7.4.2}$$

Since the relations $(7.4.2)_2$ must be valid for all values of indices, we can employ cyclic permutations to write

$$\begin{aligned}\gamma_{j(ki)} + \gamma_{i(kj)} &= e_k(g_{ij}) - \tfrac{1}{2}(c_{jki} + c_{ikj}),\\ \gamma_{k(ij)} + \gamma_{j(ik)} &= e_i(g_{jk}) - \tfrac{1}{2}(c_{kij} + c_{jik}),\\ \gamma_{i(jk)} + \gamma_{k(ji)} &= e_j(g_{ki}) - \tfrac{1}{2}(c_{ijk} + c_{kji}).\end{aligned}$$

If we add the first and the third lines and subtract from the resulting expression the second line, consider the symmetries in indices and recall that we must write $\gamma_{ijk} = \gamma_{i[jk]} + \gamma_{i(jk)}$, we obtain

$$\gamma_{ijk} = \frac{1}{2}\big[e_j(g_{ki}) + e_k(g_{ij}) - e_i(g_{jk})\big] + \frac{1}{2}(c_{ijk} + c_{jik} + c_{kij}). \quad (7.4.3)$$

As to the connection coefficients, they are found from (7.4.1) as

$$\gamma^i_{jk} = g^{il}\gamma_{ljk}. \quad (7.4.4)$$

It is clear that when the metric tensor is specified the unique connection satisfying the conditions $\boldsymbol{\mathcal{T}} = \mathbf{0}$ and $\boldsymbol{\nabla\mathcal{G}} = \mathbf{0}$ is given by (7.4.4). □

Although this connection is known as the ***Levi-Civita connection***, some authors prefer to use term the ***Riemannian connection***. (7.4.4) is then explicitly written as follows

$$\gamma^i_{jk} = \frac{1}{2}g^{il}\big[e_j(g_{kl}) + e_k(g_{lj}) - e_l(g_{jk})\big] + \frac{1}{2}(c^i_{jk} + g_{jm}g^{il}c^m_{lk} + g_{km}g^{il}c^m_{lj}).$$

In natural coordinates we should take $e_i = \partial_i = \partial/\partial x^i$. Let us denote the connection coefficients corresponding to this case by $\Gamma^i_{jk} = \gamma^i_{jk}$. Since, in this case we have $c^i_{jk} = 0$, we find that $\Gamma^i_{jk} = \Gamma^i_{kj}$ and the *symmetric connection* is determined by the coefficients

$$\begin{aligned}\Gamma^i_{jk} &= g^{il}\Gamma_{ljk}, \quad (7.4.5)\\ \Gamma_{ljk} &= \frac{1}{2}\Big(\frac{\partial g_{kl}}{\partial x^j} + \frac{\partial g_{lj}}{\partial x^k} - \frac{\partial g_{jk}}{\partial x^l}\Big).\end{aligned}$$

the quantities Γ_{ijk} are called the ***Christoffel symbols of the first kind*** while we know that Γ^i_{jk} are the *Christoffel symbols of the second kind*. We thus conclude that the natural coordinates and the metric tensor on a Riemannian manifold create in a concrete way the linear connection whose existence was anticipated in Sec. 7.2.

If we choose vectors $\{e_i\}$ in the tangent spaces describing the moving frame as an orthonormal basis and forms $\{\theta^i\}$ as their reciprocal basis in the dual spaces, we know that the metric tensor reduces to $g_{ij} = \mp\,\delta_{ij}$. Then (7.4.3) gives in view of $(7.4.1)_2$

$$\gamma_{ijk} = \frac{1}{2}(c_{ijk} + c_{jik} + c_{kij}) = -\gamma_{kji}.$$

Because of the fact that we have selected the Levi-Civita connection in such a way that $\nabla_k g_{ij} = 0$, the relation $g^{il}g_{lj} = \delta^i_j$ leads immediately to $\nabla_k g^{ij} = 0$. Since the torsion tensor is zero, the curvature tensor, which will be called henceforth in this section as the ***Riemann curvature tensor*** or the ***Riemann-Christoffel tensor***, will satisfy the relations

$$\begin{aligned} R^i_{[jkl]} &= 0 \text{ or } R^i_{jkl} + R^i_{klj} + R^i_{ljk} = 0 \qquad (7.4.6) \\ \nabla_{[k} R^i_{|j|lm]} &= 0 \text{ or } \nabla_k R^i_{jlm} + \nabla_l R^i_{jmk} + \nabla_m R^i_{jkl} = 0 \end{aligned}$$

in accordance with (7.3.23) and (7.3.24). We can now define the *covariant curvature tensor* by

$$R_{ijkl} = g_{im} R^m_{jkl.}$$

Some properties of this tensor can be revealed most easily in natural coordinates. In these coordinates, it follows from the relation (7.2.21) that

$$\begin{aligned} R_{ijkl} &= g_{im}(g^{mn}\Gamma_{nlj})_{,k} - g_{im}(g^{mn}\Gamma_{nkj})_{,l} + \Gamma_{ikn}\Gamma^n_{lj} - \Gamma_{iln}\Gamma^n_{kj} \\ &= g_{im}g^{mn}{}_{,k}\Gamma_{nlj} + \Gamma_{ilj,k} - g_{im}g^{mn}{}_{,l}\Gamma_{nkj} - \Gamma_{ikj,l} + \Gamma_{ikn}\Gamma^n_{lj} - \Gamma_{iln}\Gamma^n_{kj}. \end{aligned}$$

On the other hand, if we insert the relation

$$g^{mn}{}_{,k} = -\Gamma^m_{kr}g^{rn} - \Gamma^n_{kr}g^{mr}$$

obtained from the condition $\nabla_k g^{mn} = 0$ into the above expression, we find

$$\begin{aligned} R_{ijkl} &= -\Gamma_{ikn}\Gamma^n_{lj} - \Gamma^n_{ki}\Gamma_{nlj} + \Gamma_{iln}\Gamma^n_{kj} + \Gamma^n_{li}\Gamma_{nkj} + \Gamma_{ilj,k} - \Gamma_{ikj,l} \\ &\qquad + \Gamma_{ikn}\Gamma^n_{lj} - \Gamma_{iln}\Gamma^n_{kj} \\ &= \Gamma_{ilj,k} - \Gamma_{ikj,l} + \Gamma^n_{li}\Gamma_{nkj} - \Gamma^n_{ki}\Gamma_{nlj} \end{aligned}$$

and on making use of (7.4.5) we finally arrive at the relation

$$R_{ijkl} = \frac{1}{2}(g_{il,jk} - g_{ik,jl} + g_{jk,il} - g_{jl,ik}) + \Gamma^n_{li}\Gamma_{nkj} - \Gamma^n_{ki}\Gamma_{nlj}. \qquad (7.4.7)$$

Because of the symmetry properties $g_{ij} = g_{ji}$ and $\Gamma_{ijk} = \Gamma_{ikj}$, we readily observe the existence of the block symmetries

$$R_{ijkl} = R_{klij} = -R_{lkij}. \qquad (7.4.8)$$

Since R_{ijkl} is a tensor, this symmetry properties will be valid in every coordinate systems.

The ***Ricci tensor*** R_{ij} [after Italian mathematician Gregorio Ricci-Curbastro who is rightly considered one of the principal founders of the absolute differential calculus connected with covariant differentiation, or tensor analysis as we call it today] is defined as a contraction of the curvature tensor R^i_{jkl}:

$$R_{ij} = R^k_{ikj} = \Gamma^k_{ji,k} - \Gamma^k_{ki,j} + \Gamma^k_{km}\Gamma^m_{ji} - \Gamma^k_{jm}\Gamma^m_{ki}. \tag{7.4.9}$$

If we note that in view of (7.4.8) we can write $R_{kilj} = R_{ljki}$, and consequently $R^k_{ilj} = R_{lj}{}^k{}_i$ we find $R^k_{ikj} = R_{kj}{}^k{}_i = R^k_{jki}$ and finally we arrive at the relation

$$R_{ij} = R_{ji}.$$

Hence the *Ricci tensor is symmetric*. Moreover, we can easily deduce from (7.4.5) that a contraction on the Christoffel symbols of the second kind is found to be

$$\Gamma^j_{ji} = \frac{1}{2} g^{kl} \frac{\partial g_{kl}}{\partial x^i}.$$

Let $g = \det [g_{kl}]$. We then obtain

$$\frac{\partial g}{\partial x^i} = \frac{\partial g}{\partial g_{kl}} \frac{\partial g_{kl}}{\partial x^i} = g g^{kl} \frac{\partial g_{kl}}{\partial x^i}.$$

Therefore, we get

$$\Gamma^j_{ji} = \frac{1}{2g}\frac{\partial g}{\partial x^i} = \frac{1}{\sqrt{|g|}}\frac{\partial \sqrt{|g|}}{\partial x^i} = \frac{\partial \log|g|^{1/2}}{\partial x^i}. \tag{7.4.10}$$

Because of the symmetry of the Ricci tensor, $(7.4.6)_1$ is satisfied identically. As to $(7.4.6)_2$, the contraction on indices i and l yields

$$\nabla_k R^i_{jim} + \nabla_i R^i_{jmk} + \nabla_m R^i_{jki} = \nabla_k R_{jm} - \nabla_m R_{jk} + \nabla_i R^i_{jmk} = 0.$$

We now raise the index j by recalling that the covariant derivative of the tensor g^{ij} vanishes to obtain

$$\nabla_k R^j_m - \nabla_m R^j_k + \nabla_i R^{ij}{}_{mk} = 0.$$

By taking notice of $R^{ij}{}_{mk} = R^{ji}{}_{km}$ and contracting indices j and k, we get

$$\nabla_j R^j_m - \nabla_m R^j_j + \nabla_i R^i_m = 0 \quad \text{or} \quad 2\nabla_i R^i_m - \nabla_m R = 0$$

where $R = R^i_i$ is a scalar quantity. Accordingly, the relation

$$\nabla_i(R^i_j - \tfrac{1}{2}\delta^i_j R) = 0 \tag{7.4.11}$$

must be satisfied. (7.4.11) is called the ***contracted Bianchi identity***. The second order tensor

$$G^i_j = R^i_j - \tfrac{1}{2}\delta^i_j R \tag{7.4.12}$$

is sometimes named as the ***Einstein tensor*** because of its association with the theory of general relativity. If we recall (7.4.10), we see that this tensor satisfies the following relation

$$\nabla_i G^i_j = G^i_{j,i} + \Gamma^k_{ki} G^i_j - \Gamma^k_{ij} G^i_k = \frac{1}{\sqrt{|g|}}\frac{\partial}{\partial x^i}(\sqrt{|g|}\,G^i_j) - \Gamma^k_{ij} G^i_k = 0.$$

Since a Riemannian manifold equipped with the Levi-Civita connection is torsion-free, then $(7.3.22)_1$ takes the shape $d\theta^i = -\Gamma^i_j \wedge \theta^j$. On the other hand, we know that we can write

$$e_k(g_{ij})\,\theta^k = \gamma_{ikj}\theta^k + \gamma_{jki}\theta^k$$

since $\nabla_k g_{ij} = 0$. If we define the forms $\Gamma_{ij} = g_{il}\Gamma^l_j \in \Lambda^1(M)$, the above relation may be cast into the form

$$e_k(g_{ij})\,\theta^k = \Gamma_{ij} + \Gamma_{ji}.$$

If the metric tensor is *constant*, we find that $e_k(g_{ij}) = 0$ and $\Gamma_{ji} = -\Gamma_{ij}$, namely, the matrix 1-form $\mathbf{\Gamma} = [\Gamma_{ij}]$ is antisymmetric. This property will always exist in an orthonormal frame since one then has $g_{ij} = \delta_{ij}$. Moreover the inverse matrix g^{ij} is also the identity matrix. Hence, we can write $\Gamma_{ij} = \Gamma^i_j = -\Gamma_{ji} = -\Gamma^j_i$.

Example 7.4.1. We define a spherically symmetric *indefinite* metric in the 4-dimensional space-time manifold by the relation

$$ds^2 = e^{2\lambda}dt^2 - e^{2\mu}dr^2 - r^2 d\theta^2 - r^2\sin^2\theta\, d\phi^2$$

where (r,θ,ϕ) are spherical space coordinates and t is the time coordinate. This metric was proposed by German astronomer and mathematician Karl Schwarzschild (1873-1916) in order to obtain the first exact analytical solution of Einstein's field equations of the general relativity in the vacuum. The exponents functions in this expression are taken as $\lambda = \lambda(r,t)$ and $\mu = \mu(r,t)$. The field equations of the general relativity connect the Einstein's tensor to the energy-momentum tensor reflecting physical properties of the medium involved through the relation

$$G^i_j = -\kappa T^i_j, \quad i,j = 0,1,2,3$$

where κ is a constant. Due to (7.4.11), the energy-momentum tensor must satisfy $\nabla_i T^i_j = 0$, that is, it must be divergence-free. Since $T^i_j = 0$ in the vacuum, we ought to find $G^i_j = 0$. The relation $G^i_i = -R = 0$ then implies that the following equations must also hold:

$$R_{ij} = 0.$$

Let us now define four linearly independent 1-forms by the expressions given below

$$\theta^0 = e^\lambda dt, \ \theta^1 = e^\mu dr, \ \theta^2 = r\,d\theta, \ \theta^3 = r\sin\theta\,d\phi.$$

With this choice of basis, we can write

$$\mathcal{G} = g_{ij}\theta^i \otimes \theta^j = \theta^0 \otimes \theta^0 - \theta^1 \otimes \theta^1 - \theta^2 \otimes \theta^2 - \theta^3 \otimes \theta^3$$

where $0 \le i,j \le 3$. Hence, all components g_{ij} of the metric tensor become constant being equal to 0, 1 or -1. Thus, the connection forms Γ_{ij} have to be antisymmetric. If we denote the partial derivatives with respect to the variables r and t by subscripts r and t, we then obtain

$$d\theta^0 = -\lambda_r\,e^{-\mu}\,\theta^0 \wedge \theta^1, \ d\theta^1 = \mu_t\,e^{-\lambda}\,\theta^0 \wedge \theta^1, \ d\theta^2 = \frac{e^{-\mu}}{r}\,\theta^1 \wedge \theta^2,$$

$$d\theta^3 = \frac{e^{-\mu}}{r}\,\theta^1 \wedge \theta^3 + \frac{\cot\theta}{r}\,\theta^2 \wedge \theta^3.$$

In this case, the coefficients $c^i_{jk} = -c^i_{kj}$ are found to be

$$c^0_{01} = \lambda_r\,e^{-\mu}, \ c^1_{01} = -\mu_t\,e^{-\lambda}, \ c^2_{12} = -\frac{e^{-\mu}}{r},$$
$$c^3_{13} = -\frac{e^{-\mu}}{r}, \ c^3_{23} = -\frac{\cot\theta}{r}.$$

All other coefficients are zero. When we carefully scrutinise the Cartan structural equations $d\theta^i = -\Gamma^i_j \wedge \theta^j$, we realise that connection forms must be designated as follows

$$\Gamma^0_1 = \lambda_r\,e^{-\mu}\,\theta^0 + \mu_t\,e^{-\lambda}\,\theta^1 = -\Gamma^1_0, \ \Gamma^2_1 = \frac{e^{-\mu}}{r}\,\theta^2 = -\Gamma^1_2,$$

$$\Gamma^3_1 = \frac{e^{-\mu}}{r}\,\theta^3 = -\Gamma^1_3, \ \Gamma^3_2 = \frac{\cot\theta}{r}\,\theta^3 = -\Gamma^2_3, \ \Gamma^2_0 = -\Gamma^0_2 = 0,$$

$$\Gamma^3_0 = -\Gamma^0_3 = 0.$$

The curvature forms $\Theta^i_j = d\Gamma^i_j + \Gamma^i_k \wedge \Gamma^k_j$ will then possess the property $\Theta^i_j = -\Theta^j_i$. These relations between the forms $\mathbf{\Theta}$ and $\mathbf{\Gamma}$ result readily in the curvature forms below

$$\Theta^0_1 = d\Gamma^0_1 = \big[e^{-2\lambda}(\mu_{tt} - \mu_t\lambda_t + \mu_t^2) - e^{-2\mu}(\lambda_{rr} - \mu_r\lambda_r + \lambda_r^2)\big]\theta^0 \wedge \theta^1,$$

$$\Theta^0_2 = \Gamma^0_1 \wedge \Gamma^1_2 = -\frac{e^{-2\mu}\lambda_r}{r}\,\theta^0 \wedge \theta^2 - \frac{e^{-(\lambda+\mu)}\mu_t}{r}\,\theta^1 \wedge \theta^2,$$

$$\Theta^0_3 = \Gamma^0_1 \wedge \Gamma^1_3 = -\frac{e^{-2\mu}\lambda_r}{r}\,\theta^0 \wedge \theta^3 - \frac{e^{-(\lambda+\mu)}\mu_t}{r}\,\theta^1 \wedge \theta^3,$$

$$\Theta^1_2 = d\Gamma^1_2 + \Gamma^1_3 \wedge \Gamma^3_2 = \frac{e^{-(\lambda+\mu)}\mu_t}{r}\,\theta^0 \wedge \theta^2 + \frac{e^{-2\mu}\mu_r}{r}\,\theta^1 \wedge \theta^2,$$

$$\Theta^1_3 = d\Gamma^1_3 + \Gamma^1_2 \wedge \Gamma^2_3 = \frac{e^{-(\lambda+\mu)}\mu_t}{r}\,\theta^0 \wedge \theta^3 + \frac{e^{-2\mu}\mu_r}{r}\,\theta^1 \wedge \theta^3,$$

$$\Theta^2_3 = d\Gamma^2_3 + \Gamma^2_1 \wedge \Gamma^1_3 = \frac{1 - e^{-2\mu}}{r^2}\,\theta^2 \wedge \theta^3,$$

$$\Theta^1_0 = -\Theta^0_1,\ \Theta^2_0 = -\Theta^0_2,\ \Theta^3_0 = -\Theta^0_3,$$

$$\Theta^2_1 = -\Theta^1_2,\ \Theta^3_1 = -\Theta^1_3,\ \Theta^3_2 = -\Theta^2_3.$$

By making use of these forms, the components of the curvature tensor are found as

$$\begin{aligned}
R^0_{101} &= -R^1_{001} = -R^0_{110} = R^1_{010} \\
&= e^{-2\lambda}(\mu_{tt} - \mu_t\lambda_t + \mu_t^2) - e^{-2\mu}(\lambda_{rr} - \mu_r\lambda_r + \lambda_r^2) \\
R^0_{202} &= -R^2_{002} = -R^0_{220} = R^2_{020} = R^0_{303} = -R^3_{003} = -R^0_{330} \\
&= R^3_{030} = -\frac{e^{-2\mu}\lambda_r}{r} \\
R^0_{212} &= -R^2_{012} = -R^0_{221} = R^2_{021} = R^0_{313} = -R^3_{013} = -R^0_{331} \\
&= R^3_{031} = R^1_{202} = -R^2_{102} = -R^2_{220} = R^2_{120} = -R^1_{303} \\
&= R^3_{103} = -R^3_{130} = R^1_{330} = -\frac{e^{-(\lambda+\mu)}\mu_t}{r} \\
R^1_{212} &= -R^2_{112} = R^2_{121} = -R^1_{221} = R^1_{313} = -R^3_{113} = R^3_{131} \\
&= -R^1_{331} = \frac{e^{-2\mu}\mu_r}{r} \\
R^2_{323} &= -R^3_{223} = R^3_{232} = -R^2_{332} = \frac{1 - e^{-2\mu}}{r^2}
\end{aligned}$$

The components of the Ricci tensor are easily obtained as

$$R_{02} = R^1_{012} + R^3_{032} = 0, \qquad R_{03} = R^1_{013} + R^2_{023} = 0,$$

$$
\begin{aligned}
R_{13} &= R^0_{103} + R^2_{123} = 0, \qquad R_{12} = R^0_{102} + R^3_{132} = 0,\\
R_{23} &= R^0_{203} + R^1_{213} = 0,\\
R_{00} &= R^1_{010} + R^2_{020} + R^3_{030} = e^{-2\lambda}(\mu_{tt} - \mu_t\lambda_t + \mu_t^2)\\
&\quad - e^{-2\mu}(\lambda_{rr} - \mu_r\lambda_r + \lambda_r^2) - 2\frac{e^{-2\mu}\lambda_r}{r},\\
R_{11} &= R^0_{101} + R^2_{121} + R^3_{131} = e^{-2\lambda}(\mu_{tt} - \mu_t\lambda_t + \mu_t^2)\\
&\quad - e^{-2\mu}(\lambda_{rr} - \mu_r\lambda_r + \lambda_r^2) + 2\frac{e^{-2\mu}\mu_r}{r},\\
R_{22} &= R^0_{202} + R^1_{212} + R^3_{232} = \frac{e^{-2\mu}(\mu_r - \lambda_r)}{r} + \frac{1 - e^{-2\mu}}{r^2},\\
R_{33} &= R^0_{303} + R^1_{313} + R^2_{323} = \frac{e^{-2\mu}(\mu_r - \lambda_r)}{r} + \frac{1 - e^{-2\mu}}{r^2},\\
R_{01} &= R^1_{011} + R^2_{021} + R^3_{031} = -\frac{e^{-(\lambda+\mu)}\mu_t}{r}.
\end{aligned}
$$

In this situation, the Einstein equations $R_{ij} = 0$ yield $\mu_t = 0$ and $\mu = \mu(r)$ for the component $R_{01} = 0$. When we employ this property in the other equations, we conclude that

$$
\begin{aligned}
\lambda_{rr} - \mu_r\lambda_r + \lambda_r^2 - \frac{2\lambda_r}{r} &= 0,\\
\lambda_{rr} - \mu_r\lambda_r + \lambda_r^2 + \frac{2\mu_r}{r} &= 0,\\
\frac{e^{-2\mu}(\mu_r - \lambda_r)}{r} + \frac{1 - e^{-2\mu}}{r^2} &= 0.
\end{aligned}
$$

Let us subtract the first equation in the above list from the second one to find $\mu_r + \lambda_r = 0$, and consequently $\lambda(r,t) = -\mu(r) + f(t)$. As to the last equation, it yields

$$
2e^{-2\mu}\mu_r = -\frac{de^{-2\mu}}{dr} = \frac{e^{-2\mu} - 1}{r}.
$$

By integrating this equation, we obtain

$$
e^{-2\mu} = 1 - \frac{K}{r}.
$$

Hence, we can deduce the expression

$$
e^{2\lambda} = e^{2f(t)}\Big(1 - \frac{K}{r}\Big) = F(t)\Big(1 - \frac{K}{r}\Big), \quad F(t) > 0.
$$

We can immediately observe that the first and the second equations will be satisfied as well with these expressions for λ and μ. Let us introduce a new variable τ by the relation $d\tau/dt = \sqrt{F(t)}$. We then see that we can take $F(t) = 1$ without loss of generality. Therefore, the metric satisfying the Einstein equations takes the form

$$ds^2 = \Big(1 - \frac{K}{r}\Big)dt^2 - \Big(1 - \frac{K}{r}\Big)^{-1}dr^2 - r^2(d\theta^2 + \sin^2\theta\, d\phi^2).$$

For physical reasons, we choose $K > 0$. The characteristic components of the curvature tensor become

$$R^0{}_{101} = R^2{}_{323} = \frac{K}{r^3},\ R^0{}_{202} = R^1{}_{212} = -\frac{K}{2r^3},\ R^0{}_{212} = 0.$$

This metric that was obtained by Schwarzschild in 1915 constitutes the first and simplest exact solution of the Einstein equations. It determines the curvature of space-time, in other words the gravitational field, created by spherical symmetric static body. Let us choose

$$\begin{aligned}\sqrt{F(t)} &= c = \text{speed of light},\\ K &= r_S = 2Gm/c^2 = \text{Schwarzschild radius}\end{aligned}$$

where m is the mass of the body, G is the universal gravitation constant. With these physical parameters, the Schwarzschild metric takes the form

$$ds^2 = c^2\Big(1 - \frac{2Gm}{c^2r}\Big)dt^2 - \Big(1 - \frac{2Gm}{c^2r}\Big)^{-1}dr^2 - r^2(d\theta^2 + \sin^2\theta\, d\phi^2).$$

This metric involving singularities at $r = 0$ is the first ever solution that predicts the existence of black holes. ■

Since the torsion tensor vanishes in Levi-Civita connection, (7.3.18) takes the form

$$[\nabla_k, \nabla_j]\, t^{i_1\cdots i_k}_{j_1\cdots j_l} = \sum_{r=1}^{k} R^{i_r}{}_{mkj} t^{i_1\cdots i_{r-1}mi_{r+1}\cdots i_k}_{j_1\cdots j_l} - \sum_{r=1}^{l} R^m{}_{j_rkj} t^{i_1\cdots i_k}_{j_1\cdots j_{r-1}mj_{r+1}\cdots j_l}.$$

In addition, because of the symmetries (7.4.8), the effect of the curvature tensor on a vector $\boldsymbol{a}$ can be written as

$$R^m{}_{ikj}\, a_m = R_{mikj}\, a^m = -R_{imkj}\, a^m = -R_i{}^m{}_{kj}\, a_m.$$

Hence, the effect of the commutator $[\nabla_k, \nabla_j]$ is expressible by the relation

$$[\nabla_k, \nabla_j]\, t^{i_1\cdots i_k}_{j_1\cdots j_l} = (\nabla_k\nabla_j - \nabla_j\nabla_k)\, t^{i_1\cdots i_k}_{j_1\cdots j_l} = \tag{7.4.13}$$

$$\sum_{r=1}^{k} R^{i_r}{}_{mkj}\, t^{i_1\cdots i_{r-1} m i_{r+1}\cdots i_k}_{j_1\cdots j_l} + \sum_{r=1}^{l} R_{j_r}{}^{m}{}_{kj}\, t^{i_1\cdots i_k}_{j_1\cdots j_{r-1} m j_{r+1}\cdots j_l}$$

In natural coordinates, the covariant derivative of a tensor $\mathcal{T}$ in the direction of a vector V is given by

$$v^i \nabla_i t^{i_1\cdots i_k}_{j_1\cdots j_l} = \\ t^{i_1\cdots i_k}_{j_1\cdots j_l,i} v^i + \sum_{r=1}^{k} v^i \Gamma^{i_r}_{in} t^{i_1\cdots i_{r-1} n i_{r+1}\cdots i_k}_{j_1\cdots j_l} - \sum_{r=1}^{l} v^i \Gamma^{n}_{ij_r} t^{i_1\cdots i_k}_{j_1\cdots j_{r-1} n j_{r+1}\cdots j_l}.$$

Since the connection is symmetric, we can obviously write

$$v^i \Gamma^{i_r}_{in} = \Gamma^{i_r}_{ni} v^i = v^{i_r}_{;n} - v^{i_r}_{,n}.$$

We are thus led to the relation

$$v^i \nabla_i t^{i_1\cdots i_k}_{j_1\cdots j_l} = \\ t^{i_1\cdots i_k}_{j_1\cdots j_l,i} v^i - \sum_{r=1}^{k} t^{i_1\cdots i_{r-1} n i_{r+1}\cdots i_k}_{j_1\cdots j_l} v^{i_r}_{,n} + \sum_{r=1}^{l} t^{i_1\cdots i_k}_{j_1\cdots j_{r-1} n j_{r+1}\cdots j_l} v^{n_r}_{,j_r} \\ + \sum_{r=1}^{k} t^{i_1\cdots i_{r-1} n i_{r+1}\cdots i_k}_{j_1\cdots j_l} v^{i_r}_{;n} - \sum_{r=1}^{l} t^{i_1\cdots i_k}_{j_1\cdots j_{r-1} n j_{r+1}\cdots j_l} v^{n}_{;j_r}$$

where the first line is none other than the Lie derivative of the tensor $\mathcal{T}$ with respect to the vector field V [*see* (5.11.16)]. We thereby obtain the relation

$$(\nabla_V \mathcal{T} - £_V \mathcal{T})^{i_1\cdots i_k}_{j_1\cdots j_l} = \sum_{r=1}^{k} t^{i_1\cdots i_{r-1} n i_{r+1}\cdots i_k}_{j_1\cdots j_l} v^{i_r}_{;n} - \sum_{r=1}^{l} t^{i_1\cdots i_k}_{j_1\cdots j_{r-1} n j_{r+1}\cdots j_l} v^{n}_{;j_r}.$$

If we apply this relation to the metric tensor $\mathcal{G}$, we then find

$$(£_V \mathcal{G})_{ij} = g_{kj} v^k_{;i} + g_{ik} v^k_{;j} = v_{i;j} + v_{j;i}$$

because $\nabla \mathcal{G} = \mathbf{0}$. This means that a vector field V that leaves the metric tensor invariant under the flow created by this vector field, namely, satisfying the relation $£_V \mathcal{G} = \mathbf{0}$ has to verify the partial differential equations

$$v_{i;j} + v_{j;i} = v_{i,j} + v_{j,i} - 2\Gamma^k_{ij} v_k = 0. \qquad (7.4.14)$$

(7.4.14) are known as the ***Killing equations*** and a vector field satisfying these equations is called a ***Killing vector field***. If V is a Killing vector field, we then obtain by cyclic permutation of indices

$$\begin{aligned} v_{i;jk} - v_{i;kj} &= v_{i;jk} + v_{k;ij} = R_{ilkj}v^l, \\ v_{j;ki} + v_{i;jk} &= R_{jlik}v^l, \\ v_{k;ij} + v_{j;ki} &= R_{klji}v^l. \end{aligned}$$

If we subtract the third equation from the sum of the first two equations in the above list, we get

$$2v_{i;jk} = (R_{ilkj} + R_{jlik} - R_{klji})v^l.$$

However, the symmetries of the curvature tensor yield $R_{ilkj} + R_{jlik} - R_{klji} = 2R_{ijkl}$ [*see* (7.4.6)$_1$ with the lowered index i]. Thus a Killing vector field must satisfy the relation

$$v_{i;jk} = R_{ijkl}v^l$$

from which we easily deduct the expressions

$$v^{j}_{;ij} = R_{ij}v^j, \quad v^{i\ j}_{;j} = -R^{i}_{j}v^j.$$

7.5. DIFFERENTIAL OPERATORS

Let us assume that the exterior differential form $\omega \in \Lambda^k(M)$ is defined on a Riemannian manifold M. We have introduced the exterior derivative operator d in Sec. 5.8 while the operators of co-differential δ and Laplace-de Rham $\varDelta$ in Sec. 5.9. In this section, we shall try to present a more detailed discussion of the structure of these operators when such a manifold is endowed with the Levi-Civita connection. We first consider the exterior derivative operator $d : \Lambda^k(M) \to \Lambda^{k+1}(M)$. We know that the exterior derivative of a form $\omega \in \Lambda^k(M)$ is

$$d\omega = \frac{1}{k!}\,\omega_{[i_1\cdots i_k,i]}\,dx^i \wedge dx^{i_1} \wedge \cdots \wedge dx^{i_k}.$$

If we now insert the expression

$$\omega_{i_1\cdots i_k,i} = \nabla_i\omega_{i_1\cdots i_k} + \sum_{r=1}^{k}\Gamma^j_{ii_r}\omega_{i_1\cdots i_{r-1}ji_{r+1}\cdots i_k}$$

into the above relation, pay attention to the complete antisymmetry of the exterior product $dx^i \wedge dx^{i_1} \wedge \cdots \wedge dx^{i_k}$ and the symmetry $\Gamma^j_{ii_r} = \Gamma^j_{i_ri}$ of the connection coefficients, we reach to the conclusion that can be written as

$$d\omega = \frac{k+1}{(k+1)!}\nabla_{[i}\omega_{i_1\cdots i_k]}\,dx^i\wedge dx^{i_1}\wedge\cdots\wedge dx^{i_k}. \tag{7.5.1}$$

The co-differential operator $\delta:\Lambda^k(M)\to\Lambda^{k-1}(M)$ was defined by the expression (5.9.30):

$$\delta\omega = \frac{(-1)^k}{(k-1)!}\,\omega_{[i_1\cdots i_{k-1}]i;}{}^{i}\,dx^{i_1}\wedge\cdots\wedge dx^{i_{k-1}}.$$

The components of this form is determined by

$$\omega^{i_1\cdots i}{}_{;i} = \frac{(\sqrt{g}\,\omega^{i_1\cdots i})_{,i}}{\sqrt{g}}$$

as given in (5.9.19) and through $\omega_{i_1\cdots i_{k-1}i;}{}^{i} = g_{i_1j_1}\cdots g_{i_{k-1}j_{k-1}}\omega^{j_1\cdots j_{k-1}i}{}_{;i}$. We shall now demonstrate that these relations are actually associated with the covariant derivative generated by the Levi-Civita connection. The relation

$$\begin{aligned}\omega^{i_1\cdots i_k}{}_{;i} &= \nabla_i\omega^{i_1\cdots i_k}\\ &= \omega^{i_1\cdots i_k}{}_{,i} + \sum_{r=1}^{k}\Gamma^{i_r}_{ij}\omega^{i_1\cdots i_{r-1}ji_{r+1}\cdots i_k}\end{aligned}$$

leads to

$$\begin{aligned}\nabla_i\omega^{i_1\cdots i_{k-1}i} &= \omega^{i_1\cdots i_{k-1}i}{}_{,i} + \sum_{r=1}^{k-1}\Gamma^{i_r}_{ij}\omega^{i_1\cdots i_{r-1}ji_{r+1}\cdots i_{k-1}i} + \Gamma^{i}_{ij}\omega^{i_1\cdots i_{k-1}j}\\ &= \omega^{i_1\cdots i_{k-1}i}{}_{,i} + \Gamma^{i}_{ij}\omega^{i_1\cdots i_{k-1}j} = \frac{1}{\sqrt{g}}(\sqrt{g}\,\omega^{i_1\cdots i_{k-1}i})_{,i}\end{aligned}$$

where we have employed the complete antisymmetry of the components $\omega^{i_1\cdots i_k}$, the symmetry of the connection coefficients Γ^k_{ij} and the relation (7.4.10). Hence, by raising and lowering indices by means of the metric tensor, the co-differential operator becomes expressible as follows

$$\begin{aligned}\delta\omega &= \frac{(-1)^k}{(k-1)!}\nabla^i\omega_{i_1\cdots i_{k-1}i}\,dx^{i_1}\wedge\cdots\wedge dx^{i_{k-1}} \\ &= -\frac{1}{(k-1)!}\nabla^i\omega_{ii_1\cdots i_{k-1}}\,dx^{i_1}\wedge\cdots\wedge dx^{i_{k-1}}\end{aligned} \tag{7.5.2}$$

where we have evidently defined $\nabla^i = g^{ij}\nabla_j$.

The Laplace-de Rham operator

$$\Delta = \delta d + d\delta : \Lambda^k(M) \to \Lambda^k(M)$$

can now be evaluated by using (7.5.1) and (7.5.2). We can thus write

$$d\delta\omega = -\frac{k}{k!}\nabla_{[i_1}(\nabla^i\omega_{|i|i_2\cdots i_k]})\,dx^{i_1}\wedge dx^{i_2}\wedge\cdots\wedge dx^{i_k},$$

and

$$\delta d\omega = -\frac{k+1}{k!}\nabla^i(\nabla_{[i}\omega_{i_1i_2\cdots i_k]})\,dx^{i_1}\wedge dx^{i_2}\wedge\cdots\wedge dx^{i_k}$$

where the antisymmetries in the coefficients are explicitly described. If we express the form $\Delta\omega$ as

$$\Delta\omega = \frac{1}{k!}(\Delta\omega)_{i_1\cdots i_k}dx^{i_1}\wedge\cdots\wedge dx^{i_k}\in\Lambda^k(M)$$

then its components are determined by the following expression

$$(\Delta\omega)_{i_1i_2\cdots i_k} = -(k+1)\,\nabla^i\nabla_{[i}\omega_{i_1i_2\cdots i_k]} - k\,\nabla_{[i_1}\nabla^i\omega_{|i|i_2\cdots i_k]}.$$

If we utilise the relation (5.5.2), we obtain at once

$$(k+1)\nabla^i\nabla_{[i}\omega_{i_1i_2\cdots i_k]} = \nabla^i\nabla_i\,\omega_{i_1i_2\cdots i_k} - \sum_{r=1}^{k}\nabla^i\nabla_{i_r}\,\omega_{i_1\cdots i_{r-1}ii_{r+1}\cdots i_k}$$

and

$$k\nabla_{[i_1}\nabla^i\omega_{|i|i_2\cdots i_k]} = \nabla_{i_1}\nabla^i\,\omega_{ii_2\cdots i_k} - \sum_{r=2}^{k}\nabla_{i_r}\nabla^i\,\omega_{ii_2\cdots i_{r-1}i_1i_{r+1}\cdots i_k}$$
$$= \sum_{r=1}^{k}\nabla_{i_r}\nabla^i\,\omega_{i_1\cdots i_{r-1}ii_{r+1}\cdots i_k}.$$

Hence, we draw the conclusion

$$(\Delta\omega)_{i_1\cdots i_k} = -\nabla^i\nabla_i\omega_{i_1i_2\cdots i_k} - \sum_{r=1}^{k}(\nabla_{i_r}\nabla^i - \nabla^i\nabla_{i_r})\,\omega_{i_1\cdots i_{r-1}ii_{r+1}\cdots i_k}.$$

On the other hand, let us note that one is able to write

$$[\nabla_{i_r},\nabla^i]\,\omega_{i_1\cdots i_{r-1}ii_{r+1}\cdots i_k} = (-1)^{r-1}[\nabla_{i_r},\nabla_i]\,\omega^i_{i_1\cdots i_{r-1}i_{r+1}\cdots i_k}$$

Employing then the relation (7.4.13), we find that

$$[\nabla_{i_r},\nabla_i]\,\omega^i_{i_1\cdots i_{r-1}i_{r+1}\cdots i_k} = R^i_{mi_ri}\,\omega^m_{i_1\cdots i_{r-1}i_{r+1}\cdots i_k} + \sum_{s=1,s\neq r}^{k} R_{i_s}{}^m{}_{i_ri}\,\omega^i_{i_1\cdots i_{s-1}mi_{s+1}\cdots i_{r-1}i_{r+1}\cdots i_k}$$

$$= -\,R_{mi_r}\,\omega^m_{i_1\cdots i_{r-1}i_{r+1}\cdots i_k} + \sum_{s=1,s\neq r}^{k}(-1)^{s-1}R_{i_s}{}^m{}_{i_ri}\,\omega^i_{mi_1\cdots i_{s-1}i_{s+1}\cdots i_{r-1}i_{r+1}\cdots i_k}$$

$$= -\,R^m_{i_r}\,\omega_{mi_1\cdots i_{r-1}i_{r+1}\cdots i_k} + \sum_{s=1,s\neq r}^{k}(-1)^{s-1}R_{i_s}{}^m{}_{i_r}{}^i\,\omega_{imi_1\cdots i_{s-1}i_{s+1}\cdots i_{r-1}i_{r+1}\cdots i_k}$$

Therefore, we see that the Laplace-de Rham operator is completely determined by the components

$$(\Delta\omega)_{i_1\cdots i_k} = -\,g^{ij}\nabla_j\nabla_i\,\omega_{i_1i_2\cdots i_k} - \sum_{r=1}^{k}(-1)^r R^m_{i_r}\,\omega_{mi_1\cdots i_{r-1}i_{r+1}\cdots i_k} - \sum_{r=1}^{k}\sum_{s=1,s\neq r}^{k}(-1)^{r+s}R_{i_s}{}^m{}_{i_r}{}^i\,\omega_{imi_1\cdots i_{s-1}i_{s+1}\cdots i_{r-1}i_{r+1}\cdots i_k}. \tag{7.5.3}$$

VII. EXERCISES

7.1. A manifold M is equipped with two connections defined by the Christoffel symbols Γ^i_{jk} and Γ'^i_{jk}. Show that the quantities $\Upsilon^i_{jk} = \Gamma^i_{jk} - \Gamma'^i_{jk}$ are components of a $\binom{1}{2}$-tensor.

7.2. Let $\boldsymbol{\nabla}$ be a connection on a manifold M. Show that the operator $\boldsymbol{\nabla}^*$ defined by the relation $\nabla^*_U V = \nabla_U V + \boldsymbol{\tau}(U,V)$ is also a connection on M whose torsion tensor is determined by $-\,\boldsymbol{\tau}$. $\boldsymbol{\nabla}^*$ is called the *conjugate connection.*

7.3. $\boldsymbol{\nabla}$ is a connection on a manifold M. Show that the connection defined by the relation $\boldsymbol{\nabla}^s = \frac{1}{2}(\boldsymbol{\nabla} + \boldsymbol{\nabla}^*)$ is symmetric. Find the connection coefficients.

7.4. Show that the connections $\boldsymbol{\nabla}$, $\boldsymbol{\nabla}^*$ and $\boldsymbol{\nabla}^s$ have the same geodesics on the manifold M.

7.5. A connection on the manifold $\mathbb{R}^2$ whose coordinate cover is (x^1,x^2) is prescribed by Christoffel symbols $\Gamma^1_{12} = \Gamma^1_{21} = 1$ and all other coefficients $\Gamma^i_{jk} = 0$. Determine the geodesics.

7.6. A connection $\boldsymbol{\nabla}$ and a tensor S^i_{jk} that is antisymmetric with respect to its covariant indices are given on a manifold M. Show that there is a unique connection on M with the same geodesics as those of $\boldsymbol{\nabla}$ and its torsion tensor being equal to S^i_{jk}.

7.7. If the parameter of a geodesic curve determined by the equation (7.2.16) is t, then show that a change of the parameter in the form $\tau = \alpha t + \beta$ where α and β are constants still satisfies that equation. Thus, a parameter of a geodesic curve may be named as an *affine parameter*.

7.8. The *indefinite Lorentz metric* on the manifold $\mathbb{R}^{n+1}$ [Dutch physicist Hendrik Antoon Lorentz (1853-1928)] is introduced by the relation

$$\mathcal{G}' = -\,dx_0 \otimes dx_0 + \sum_{i=1}^{n} dx_i \otimes dx_i.$$

Hence, the Lorentz inner product and the *length* of a vector is determined, respectively, by

$$\mathcal{G}'(U,V) = -\,u_0 v_0 + \sum_{i=1}^{n} u_i v_i,$$

$$\mathcal{G}'(U,U) = -\,u_0^2 + \sum_{i=1}^{n} u_i^2.$$

In this case, $(\mathbb{R}^{n+1}, \mathcal{G}')$ becomes obviously a pseudo-Riemannian manifold. Let us now define an n-dimensional submanifold of the manifold $\mathbb{R}^{n+1}$ as follows

$$H^n = \{\mathbf{x} \in \mathbb{R}^{n+1} : x_0^2 - \sum_{i=1}^{n} x_i^2 = 1, x_0 > 0\} \subset \mathbb{R}^{n+1}.$$

H^n is called a *hyperbolic space*. Show that the metric $\mathcal{G}'$ generates a definite metric $\mathcal{G}$ on H^n whose components are given by

$$\mathcal{G} = g_{ij}\, dx_i \otimes dx_j, \quad g_{ij} = \delta_{ij} - \frac{x_i x_j}{x_0^2}$$

and $(H^n, \mathcal{G})$ becomes a complete Riemannian manifold.

7.9. *Hyperbolic plane* H^2 is defined as the submanifold

$$H^2 = \{\mathbf{x} \in \mathbb{R}^3 : x_0^2 - x_1^2 - x_2^2 = 1, x_0 > 0\} \subset \mathbb{R}^3$$

where $\mathbb{R}^3$ is equipped with the Lorentz metric. By using a coordinate transformation

$$x_1 = r\cos\theta, \quad x_2 = r\sin\theta$$

show that the metric tensor of this Riemannian manifold is given by the relation

$$\mathcal{G} = \frac{dr \otimes dr}{1 + r^2} + r^2 d\theta \otimes d\theta$$

Find further the form of this metric tensor under the coordinate transformation $r = \sinh s$.

7.10. Compute the metric tensor of the 3-dimensional hyperbolic space H^3. Use first the spherical coordinates (r, θ, ϕ) given by

$$x = r \sin\theta \cos\phi,$$
$$y = r \sin\theta \sin\phi,$$
$$z = r \cos\theta$$

then the transformation $r = \sinh s$.

7.11. Show that the number of the independent components of the curvature tensor R_{ijkl} satisfying the symmetry relations (7.4.8) in an n-dimensional manifold is given by

$$\frac{1}{12} n^2 (n^2 - 1).$$

7.12. The Weyl tensor is defined by the relation

$$W^{ij}{}_{kl} = R^{ij}{}_{kl} - 2\delta^{[i}{}_{[k} R^{j]}{}_{l]} + \frac{1}{3}\delta^{[i}{}_{[k}\,\delta^{j]}{}_{l]} R.$$

Show that all contractions of this tensor yield zero tensors.

7.13. The metric tensor in a 2-dimensional Riemannian manifold is given in the following form

$$\mathcal{G} = dr \otimes dr + [f(r,\theta)]^2 d\theta \otimes d\theta.$$

Let us choose the basis forms in $T^*(M)$ as

$$\theta^1 = dr, \quad \theta^2 = f(r,\theta)\, d\theta.$$

Find the reciprocal basis vectors e_1, e_2 in $T(M)$ and the coefficients c^k_{ij}. Determine the Christoffel symbols and the curvature tensor.

7.14. Show that the operator of covariant differentiation satisfies the Jacobi identity

$$\big[\nabla_i, [\nabla_j, \nabla_k]\big] + \big[\nabla_j, [\nabla_k, \nabla_i]\big] + \big[\nabla_k, [\nabla_i, \nabla_j]\big] = 0.$$

7.15. Let M be a Riemannian manifold and $\mathcal{R}$ be its curvature tensor. Show that the relation $\mathcal{R}(V_1, V_2, V_3, V_4) = \mathcal{G}\big(\mathcal{R}(V_2, V_3, V_4), V_1\big)$ is satisfied for all vector fields $V_i \in T(M), i = 1, 2, 3, 4$. Verify further the following identities:
(a) $\mathcal{R}(V_1, V_2, V_3, V_4) = -\mathcal{R}(V_2, V_1, V_3, V_4)$,
(b) $\mathcal{R}(V_1, V_2, V_3, V_4) = -\mathcal{R}(V_1, V_2, V_4, V_3)$,
(c) $\mathcal{R}(V_1, V_2, V_3, V_4) = \mathcal{R}(V_3, V_4, V_1, V_2)$,
(d) $\mathcal{R}(V_1, V_2, V_3, V_4) + \mathcal{R}(V_1, V_3, V_4, V_2) + \mathcal{R}(V_1, V_4, V_2, V_3) = 0$.

7.16. Calculate the function Δf in cylindrical coordinates where $f \in \Lambda^0(\mathbb{R}^3)$.

7.17. Calculate the form $\Delta\omega$ in cylindrical and spherical coordinates where $\omega \in \Lambda^1(\mathbb{R}^3)$.

7.18. Let $(M, \mathcal{G})$ and $(N, \boldsymbol{\Gamma})$ be two Riemannian manifolds. If a diffeomorphism $\phi : M \to N$ fulfil the condition $\phi^* \boldsymbol{\Gamma} = \mathcal{G}$, then it is called an ***isometry***. If such an isometry is established, then we say that the manifolds M and N are *isometric*. Show that a diffeomorphism $\phi : M \to M$ is an isometry if and only if the condition

$$\phi^*(\Delta f) = \Delta(\phi^* f)$$

holds for all functions $f \in \Lambda^0(M)$.

CHAPTER VIII

INTEGRATION OF EXTERIOR FORMS

8.1. SCOPE OF THE CHAPTER

In this chapter, the integral of an exterior differential form over a submanifold of a given manifold, whose dimension is equal to the degree of the exterior form, is treated as a linear operator assigning a real number to that form. As is well known, the form reduces to a simple form on such a submanifold and the integral is roughly defined as a multiple Riemann integral of the single scalar function characterising that form. However, in order that this definition acquires a formal content, we have to exert quite a great effort and to equip the manifold with adequate structures such as simplices and chains. We also deal in this chapter with the cohomology and homology groups that are inspired by these structure and prove to be very helpful in revealing some hidden properties of closed forms. Sec. 8.2 introduces the concept of orientability of a manifold by means of a volume form on a manifold. In Sec. 8.3, the integration of forms is discussed on a very simple manifold, the Euclidean space. We treat the simplices in the Euclidean space that can be used as building blocks to generalise this approach to any smooth manifold in Sec. 8.4. We then discuss chains and their boundaries, and cycles. We further define differentiable singular simplices and chains that are images of a standard simplex at the origin of the Euclidean space on a differentiable manifold by means of smooth functions. In Sec. 8.5, we propose two different courses to follow in order to evaluate the integrals of forms on smooth manifolds. If we can manage to cover the manifold with a differentiable singular chain, the form can be pulled piecewise back to the standard simplex on which the integrations can be performed relatively easily, then these integral is summed up to obtain the integral on the manifold. In another approach, we can utilise the partition of unity on the manifold if it exists of course. The Stokes theorem that is one of the corner stones of the theory of integration of exterior forms is proven in Sec. 8.6 on the chains and also on manifolds with boundaries. This theorem matches the integral of the exterior derivative of a form on a

Exterior Analysis, DOI: 10.1016/B978-0-12-415902-0.50008-6

manifold with the integral of this form on the boundary of this manifold. Sec. 8.7 is concerned with the determination of conservation laws corresponding to exact forms in an ideal that are annihilated by a solution submanifolds. Sec. 8.8 deals with the cohomology groups that are the quotient spaces of closed forms with respect to exact forms and homology groups that are quotient spaces of linear spaces of cycles with respect to linear spaces of cycles which are boundaries of chains. It is then tried to reveal important relationships between these two groups. These relationships connect the structure of closed forms on a manifold to the topological structure of that manifold. In Sec. 8.9, we define the inner product of forms on a Riemannian manifold by using the Hodge dual so that the exterior algebra is transformed into an inner product space. On making use of the structure so established, the properties of the Laplace-de Rham operator and the harmonic forms occupying the null space of this operator are investigated, and then the Hodge-de Rham decomposition theorem is explored. Finally, Sec. 8.10 is devoted to the Poincaré duality unravelling quite an interesting relation between cohomology groups in some kind of manifolds.

8.2. ORIENTABLE MANIFOLDS

We have already defined an orientable manifold on p. 275. Let us hence recall that an m-dimensional manifold M is called an ***orientable manifold*** if we can find a form $\mu \in \Lambda^m(M)$ such that $\mu(p) \neq 0$ at every point $p \in M$. Such a form μ will be called a ***volume form***. Since the module $\Lambda^m(M)$ is 1-dimensional, every non-zero m-form Ω, consequently every new volume form is expressible as a multiple of the chosen volume form μ, namely, as $\Omega = f(p)\mu$ where $f \in \Lambda^0(M)$ and $f(p) \neq 0$ at every point of the manifold .

Let us assume that two volume forms μ_1 and μ_2 are related by an expression $\mu_1(p) = f(p)\mu_2(p)$ where $f \in \Lambda^0(M)$ and $f(p) > 0$ for all $p \in M$. This constitutes an equivalence relation on the set of volume forms because it is readily verified that it is reflexive, symmetric and transitive. Thus the set of volume forms is partitioned into equivalence classes $[\mu]$. An ***orientation*** of the manifold M is defined as an equivalence class $[\mu]$. We call the pair $(M, [\mu])$ as an ***oriented manifold***.

An oriented connected differentiable manifold M can possess only two orientations.

Let Ω and μ be volume forms. Hence, we can write $\Omega = f\mu$ for a non-zero function f. However, because M is connected a function $f \neq 0$ will be either $f(p) > 0$ or $f(p) < 0$ at every point $p \in M$. Thus Ω can only be a member of either the orientation $[\mu]$ or $[-\mu]$.

*The **positive orientation** of a connected manifold M is given by the equivalence class $[\mu]$ while its **negative orientation** by the equivalence class $[-\mu]$.* □

Let $e_1(p), e_2(p), \ldots, e_m(p)$ be a basis of the tangent space $T_p(M)$ and μ be a volume form. If $\mu(e_1, e_2, \ldots, e_m) > 0$, it is so at every point p of a connected manifold and for all equivalent volume forms. Such kind of basis vectors constitutes a ***right frame***. Similarly, if $\mu(e_1, e_2, \ldots, e_m) < 0$, then the basis vectors forms a ***left frame***. Since the form μ vanishes at no points of the manifold, it is evident that *the function $\mu(e_1, e_2, \ldots, e_m)$ cannot change its sign in an oriented manifold.* Hence, when moving on an oriented manifold the chosen basis vectors cannot change their *orientation*, in other words, their right or left characters. We can change the basis $e_1, e_2, \ldots, e_m$ to a basis $e'_1, e'_2, \ldots, e'_m$ through a linear transformation $e'_j = a^i_j e_i$, $i, j = 1, \ldots, m$ where $a^i_j \in \Lambda^0(M)$ and the matrix $\mathbf{A}(p) = \left[a^i_j(p)\right]$ must hold the condition $\det \mathbf{A} \neq 0$. On the other hand, because of the relation

$$\mu(e'_1, e'_2, \ldots, e'_m) = (\det \mathbf{A})\, \mu(e_1, e_2, \ldots, e_m)$$

the change of basis does not affect the right or the left character of bases if $\det \mathbf{A} > 0$. If only $\det \mathbf{A} < 0$, then a change of basis alters the orientation by shifting a right frame to the left one and vice versa.

We can immediately deduct from above the following result: *if the left-right character of a frame of basis vectors of the tangent bundle of a manifold changes when this frame is translated along a closed curve of the manifold as to bring it back to the initial point again, then this manifold is non-orientable.*

A non-connected manifold is still called orientable if its connected components are orientable. However, in each component its orientation can be chosen arbitrarily.

Theorem 8.2.1. *An m-dimensional connected paracompact differentiable manifold M is orientable if and only if there exists an atlas $\mathcal{A} = \{(U_\alpha, \varphi_\alpha) : \alpha \in \Gamma\}$ on M such that the differentiable transition mapping $\varphi_{\alpha\beta} = \varphi_\beta \circ \varphi_\alpha^{-1} : \varphi_\alpha(U_\alpha \cap U_\beta) \to \varphi_\beta(U_\alpha \cap U_\beta)$, induced by the overlapping charts $(U_\alpha, \varphi_\alpha)$ and (U_β, φ_β) having local coordinates $\mathbf{x}$ and $\mathbf{y}$, respectively, on the set $U_\alpha \cap U_\beta \neq \emptyset$, has a local representation $y^i = \Phi^i(x^1, x^2, \ldots, x^m)$, $i = 1, 2, \ldots m$ where $\mathbf{x} \in \varphi_\alpha(U_\alpha \cap U_\beta)$ and $\mathbf{y} \in \varphi_\beta(U_\alpha \cap U_\beta)$ possessing a positive Jacobian $J = \det(\partial y^i / \partial x^j)$.*

Let M be an oriented manifold. Hence, there is a volume form μ on M. By taking simple changes in local coordinates that might involve reflections if need be into consideration we may suppose in a chart $(U_\alpha, \varphi_\alpha)$ that $\mu\Big(\dfrac{\partial}{\partial x^1}, \dfrac{\partial}{\partial x^2}, \cdots \dfrac{\partial}{\partial x^m}\Big) > 0$. Such a coordinate system is said to be *positive local coordinates*. If the same kind of changes are performed, if necessary, in the chart (U_β, φ_β) to choose positive coordinates there, then the familiar relation

$$0 < \mu\Big(\frac{\partial}{\partial y^1}, \frac{\partial}{\partial y^2}, \cdots \frac{\partial}{\partial y^m}\Big) = \det\Big(\frac{\partial y^i}{\partial x^j}\Big)\mu\Big(\frac{\partial}{\partial x^1}, \frac{\partial}{\partial x^2}, \cdots \frac{\partial}{\partial x^m}\Big)$$

requires that $J = \det(\partial y^i/\partial x^j) > 0$.

We now conversely assume that the manifold M has an atlas with the above mentioned properties. Let us assume that $\{(V_a, f_a) : a \in A\}$ be a partition of unity subordinate to that atlas where $\{V_a\}$ is a locally finite open cover of M. *The paracompactness of the manifold M assures solely that such a partition of unity can always be found. In fact, if such a partition of unity on M is contrived, then the theorem turns out to be still valid even if M is not paracompact.* Since every open set V_a belongs to an open set U_{α_a} of a chart, the atlas $\{(V_a, \varphi_a) : a \in A\}$ formed by defining the mapping $\varphi_a = \varphi_{\alpha_a}|_{V_a}$ will satisfy the condition of positive Jacobian. Let us denote the positive local coordinates in the chart (V_a, φ_a) by $x_a^1, x_a^2, \dots, x_a^m$ and introduce a form $\omega \in \Lambda^m(M)$ in the following manner

$$\omega = \sum_a f_a \, dx_a^1 \wedge dx_a^2 \wedge \cdots \wedge dx_a^m$$

where each term can be extended to the entire manifold M if we recall that each f_a vanishes outside its support. Any point $p \in M$ is now located in a chart (V, φ) with local coordinates $x^1, x^2, \dots, x^m$ and for all charts such that $V_a \cap V \neq \emptyset$ we will get $\det(\partial x_a^i/\partial x^j) > 0$. We can thus write

$$\begin{aligned}\omega(p) &= \sum_a f_a(p) \, dx_a^1 \wedge \cdots \wedge dx_a^m \\ &= \sum_a f_a(p) \det(\partial x_a^i/\partial x^j) \, dx^1 \wedge \cdots \wedge dx^m.\end{aligned}$$

On the other hand, we know that all functions in the partition must satisfy $f_a(p) \geq 0$ and at each point $p \in M$ at least one function among them should be positive. Since the factor $\det(\partial x_a^i/\partial x^j)$ is positive by assumption, we conclude that $\omega(p) \neq 0$ at each point $p \in M$. Hence, ω is a volume form and the manifold M is orientable. □

Example 8.2.1. A non-zero n-form, namely, a standard volume form on the manifold $\mathbb{R}^n$ can be defined as $\mu = dx^1 \wedge dx^2 \wedge \cdots \wedge dx^n$. Different arrangements of the forms dx^i yield either $+\mu$ or $-\mu$. Therefore, $\mathbb{R}^n$ is an oriented manifold. ■

Example 8.2.2. Let us consider the sphere $\mathbb{S}^2 \subset \mathbb{R}^3$. This submanifold of $\mathbb{R}^3$ is prescribed by the equation $x^2+y^2+z^2=R^2$. We now define a form $\mu \in \Lambda^2(\mathbb{R}^3)$ as follows

$$\mu = \frac{z\,dx\wedge dy + x\,dy\wedge dz + y\,dz\wedge dx}{(x^2+y^2+z^2)^{1/2}}.$$

It is clear that the form μ vanishes nowhere on $\mathbb{S}^2$. Thus μ is a volume, or in the true sense of the term, an area form. The structure of this form is best illustrated in spherical coordinates. The change of coordinates

$$x = R\sin\theta\cos\phi,\ y = R\sin\theta\sin\phi,\ z = R\cos\theta$$

reduces the volume form μ to

$$\mu = R^2\sin\theta\,d\theta\wedge d\phi.$$

If delete the poles and choose $\theta \in (0,\pi)$, that is, if we consider two charts as it should be, we observe $\mu \neq 0$ in both charts. We also easily notice that the orientation of basis vectors in $T(\mathbb{S}^2)$ does not change along a closed curve on $\mathbb{S}^2$. Hence, $\mathbb{S}^2$ is an orientable manifold. ■

Example 8.2.3. As an example to non-orientable manifolds, we take the Möbius band introduced in Example 2.8.1 into consideration. We know that the Möbius band is a 2-dimensional submanifold of $\mathbb{R}^3$ prescribed by the parametric equations

$$x = \big(R+v\cos(u/2)\big)\cos u,\, y = \big(R+v\cos(u/2)\big)\sin u,\, z = v\sin(u/2)$$

where $u\in[0,2\pi)$ and $v\in[-w,w]$. A basis of the tangent bundle of this manifold can be chosen as the following linearly independent vectors

$$\begin{aligned}
V_u(u,v) &= \frac{\partial}{\partial u} = \frac{\partial x}{\partial u}\frac{\partial}{\partial x} + \frac{\partial y}{\partial u}\frac{\partial}{\partial y} + \frac{\partial z}{\partial u}\frac{\partial}{\partial z}\\
&= -\frac{1}{2}\Big[4R\cos\frac{u}{2} + v(2+3\cos u)\Big]\sin\frac{u}{2}\frac{\partial}{\partial x}\\
&\quad + \frac{1}{4}\Big[4R\cos u + v\Big(\cos\frac{u}{2} + 3\cos\frac{3u}{2}\Big)\Big]\frac{\partial}{\partial y} + \frac{1}{2}v\cos\frac{u}{2}\frac{\partial}{\partial z}\\
V_v(u,v) &= \frac{\partial}{\partial v} = \frac{\partial x}{\partial v}\frac{\partial}{\partial x} + \frac{\partial y}{\partial v}\frac{\partial}{\partial y} + \frac{\partial z}{\partial v}\frac{\partial}{\partial z}\\
&= \cos\frac{u}{2}\cos u\frac{\partial}{\partial x} + \cos\frac{u}{2}\sin u\frac{\partial}{\partial y} + \sin\frac{u}{2}\frac{\partial}{\partial z} = V_v(u)
\end{aligned}$$

that are tangent to the curves $v = \textit{constant}$ and $u = \textit{constant}$, respectively. Since the scalar product can be defined on $\mathbb{R}^3$, we can readily verify that the relation $V_u \cdot V_v = 0$ is satisfied. Thus the basis vectors so chosen are orthogonal. In the particular case $v = 0$, the vector field V_u takes the form

$$V_u(u,0) = R\Big(-\sin u\frac{\partial}{\partial x} + \cos u\frac{\partial}{\partial y}\Big).$$

We thus obtain

$$\begin{aligned} V_u(0,0) &= R\frac{\partial}{\partial y}, \quad & V_v(0,0) &= \frac{\partial}{\partial x}; \\ V_u(2\pi,0) &= R\frac{\partial}{\partial y}, \quad & V_v(2\pi,0) &= -\frac{\partial}{\partial x}. \end{aligned}$$

2π in the arguments of the vectors V_u and V_v must be interpreted as the limiting value as $u \to 2\pi$. The above relation clearly show that when we translate the basis vectors at the origin $(0,0)$ along the circle $v = 0$, they change their orientation as we approach again to the origin. Therefore, the Möbius band is not oriented. ■

Example 8.2.4. Let us consider the projective space $\mathbb{RP}^n$. We define open sets U_α and U_β of two charts of the manifold by the rules $x^\alpha \neq 0$, $x^\beta \neq 0$, $\alpha, \beta \in \{1, \dots, n+1\}$ as on *p.* 87. The local coordinates in those charts are, respectively, given by

$$\xi^i_\alpha = \frac{x^i}{x^\alpha}, \quad \xi^i_\beta = \frac{x^i}{x^\beta}, \quad i = 1, 2, \dots, n$$

We of course take $\xi^\alpha_\alpha = \xi^\beta_\beta = 1$ and $\xi^\alpha_\beta = x^\alpha/x^\beta$. The coordinate transformation in the open set $U_\alpha \cap U_\beta$ is depicted by the relations $\xi^i_\alpha = \xi^i_\beta/\xi^\alpha_\beta$, $\xi^\beta_\alpha = 1/\xi^\alpha_\beta$. Thus, the entries of the Jacobian matrix become

$$\frac{\partial \xi^i_\alpha}{\partial \xi^j_\beta} = \frac{\delta^i_j}{\xi^\alpha_\beta}, \ i \neq \alpha, \beta, \ j \neq \alpha, \beta; \quad \frac{\partial \xi^\beta_\alpha}{\partial \xi^\alpha_\beta} = -\frac{1}{(\xi^\alpha_\beta)^2}.$$

Hence, the determinant is found to be

$$\det\Big(\frac{\partial \xi^i_\alpha}{\partial \xi^j_\beta}\Big) = -\frac{1}{(\xi^\alpha_\beta)^{n+1}}.$$

If n is odd, and consequently, $n+1$ is an even number, then the sign of the determinant remains the same regardless of the sign of ξ^α_β and it can be rendered positive by a suitable change of coordinates. But if n is even, hence $n+1$ is an odd number the determinant changes its sign depending on the sign of ξ^α_β. In conclusion, according to Theorem 8.2.1 we understand that *the projective space $\mathbb{RP}^n$ is orientable if n is odd and non-orientable if n is even.* ■

Let us now consider a k-dimensional submanifold S of an m-dimensional oriented manifold M. According to our assumption there is a volume form $\mu \in \Lambda^m(M)$ on M and an equivalence class $[\mu]$. A basis for the tangent bundle $T(S)$ is given by k locally linearly independent vector fields $V_1, V_2, \dots, V_k$. We can then construct a basis $V_1, \dots, V_m$ for the tangent bundle $T(M)$ by supplying $m-k$ locally linearly independent vector fields $V_{k+1}, V_{k+2}, \dots, V_m$ that do not belong to $T(S)$. That is the reason why we may call these supplementary vectors as the *normal vectors* to the tangent bundle $T(S)$. Since μ is a volume form, we get $\mu(V_1, \dots, V_k, V_{k+1}, \dots, V_m) \neq 0$. Next, we define a form $\mu' \in \Lambda^k(S)$ as

$$\mu' = \mathbf{i}_{V_m} \circ \cdots \circ \mathbf{i}_{V_{k+1}}(\mu).$$

It is clear that this form can only be different from zero on the vectors in $T(S)$. Hence, the restriction $\mu' \in \Lambda^k(S)$ becomes meaningful. On the other hand, we have

$$\mu'(V_1, \dots, V_k) = \mu(V_1, \dots, V_k, V_{k+1}, \dots, V_m) \neq 0$$

so that μ' plays the part of a volume form on the submanifold S induced by the volume form μ. We then say that the submanifold S is ***externally oriented*** by the manifold M. However, this generally may not mean that the manifold S is *oriented internally* in the usual way.

Example 8.2.5. Let us consider the sphere $\mathbb{S}^{n-1}$ as a submanifold of the oriented manifold $\mathbb{R}^n$ determined by the equation $\sum\limits_{i=1}^{n}(x^i)^2 = R^2$. A basis of the tangent bundle $T(\mathbb{S}^{n-1})$ can be chosen as the set of the following vector fields

$$V_i = \frac{\partial}{\partial x^i} - \frac{x^i}{x^n}\frac{\partial}{\partial x^n}, \quad i = 1, \dots, n-1$$

on noting that we can write $\partial x^n/\partial x^i = -\,x^i/x^n, i = 1, \dots, n-1$ on $\mathbb{S}^{n-1}$. Let us define a vector $V \in T(\mathbb{R}^n)$ by the relation

$$V = \frac{1}{R}\sum_{k=1}^{n} x^k \frac{\partial}{\partial x^k}.$$

Because of the scalar product $V \cdot V_i = (x^i - x^i)/R = 0$, we realise that the non-zero vector V is orthogonal to all vectors V_i. Therefore, it does not belong to $T(\mathbb{S}^{n-1})$. Consequently, the manifold $\mathbb{R}^n$ induces a volume form

$$\begin{aligned}\mu' &= \mathbf{i}_V(\mu_n) = \mathbf{i}_V(dx^1 \wedge \cdots \wedge dx^n)\\ &= \frac{1}{R}\sum_{k=1}^{n}(-1)^{k-1}x^k dx^1 \wedge \cdots \wedge dx^{k-1} \wedge dx^{k+1} \cdots \wedge dx^n\end{aligned}$$

on the submanifold $\mathbb{S}^{n-1}$ by externally orienting it. The structure of this form can be understood better if we employ hyperspherical coordinates. The ***hyperspherical coordinates*** are defined by the relations

$$\begin{aligned}x^1 &= R\cos\phi_1\\ x^2 &= R\sin\phi_1\cos\phi_2\\ x^3 &= R\sin\phi_1\sin\phi_2\cos\phi_3\\ &\vdots\\ x^k &= R\prod_{i=1}^{k-1}\sin\phi_i\cos\phi_k, \quad 1 \le k \le n-1\\ &\vdots\\ x^{n-1} &= R\sin\phi_1\cdots\sin\phi_{n-2}\cos\phi_{n-1}\\ x^n &= R\sin\phi_1\cdots\sin\phi_{n-2}\sin\phi_{n-1}\end{aligned}$$

where the conditions $0 \le \phi_1, \ldots, \phi_{n-2} \le \pi$ and $0 \le \phi_{n-1} \le 2\pi$ are to be satisfied. It is then immediately observed that the induced volume form on $\mathbb{S}^{n-1}$ can be written as follows

$$\mu' = R^{n-1}\sin^{n-2}\phi_1\sin^{n-3}\phi_2\cdots\sin\phi_{n-2}\, d\phi_1 \wedge d\phi_2 \wedge \cdots \wedge d\phi_{n-1}.$$

We can further realise that the submanifold $\mathbb{S}^{n-1}$ is also internally oriented by the form μ' if we restrict the coordinates $\phi_1, \ldots, \phi_{n-2}$ into the open interval $(0, \pi)$.

The volume form of the circle $\mathbb{S}^1$ $(n = 2)$ is

$$\mu' = \frac{1}{R}(x^1dx^2 - x^2dx^1) = Rd\phi$$

whereas the volume form (area form) of the sphere $\mathbb{S}^2$ $(n = 3)$ becomes

$$\begin{aligned}\mu' &= \frac{1}{R^2}(x^1\, dx^2 \wedge dx^3 + x^2\, dx^3 \wedge dx^1 + x^3\, dx^1 \wedge dx^2)\\ &= R^2\sin\phi_1 d\phi_1 \wedge d\phi_2.\end{aligned}$$

After having obtained the induced volume form, *the area* of the hypersurface $\mathbb{S}^n$ can be found easily by integrating this form. By using the definition of the hyperspherical coordinates, we obtain

$$\begin{aligned}\int_{\mathbb{S}^n}\mu' &= R^n\int_0^{\pi}\cdots\int_0^{\pi}\int_0^{2\pi}\sin^{n-1}\phi_1\sin^{n-2}\phi_2\cdots\sin\phi_{n-1}\,d\phi_1\cdots d\phi_{n-1}d\phi_n\\ &= 2\pi R^n\int_0^{\pi}\cdots\int_0^{\pi}\sin^{n-1}\phi_1\sin^{n-2}\phi_2\cdots\sin\phi_{n-1}\,d\phi_1\cdots d\phi_{n-1}\neq 0.\end{aligned}$$

In terms of the Gamma function $\Gamma(z)$, the relation

$$\int_0^{\pi}\sin^k\phi\,d\phi = \frac{\pi^{1/2}\,\Gamma\big(\frac{k+1}{2}\big)}{\Gamma(1+\frac{k}{2})},\quad 1\le k\le n-1$$

leads to the result

$$\begin{aligned}S(\mathbb{S}^n) &= \int_{\mathbb{S}^n}\mu'\\ &= 2\pi^{\frac{n+1}{2}}R^n\frac{\Gamma(1)\,\Gamma(\frac{3}{2})}{\Gamma(\frac{3}{2})\,\Gamma(2)}\cdots\frac{\Gamma(\frac{k+1}{2})}{\Gamma(1+\frac{k}{2})}\frac{\Gamma(1+\frac{k}{2})}{\Gamma(1+\frac{k+1}{2})}\cdots\frac{\Gamma(\frac{n}{2})}{\Gamma(1+\frac{n-1}{2})}\\ &= \frac{2\pi^{\frac{n+1}{2}}R^n}{\Gamma\big(\frac{n+1}{2}\big)}.\end{aligned}$$

In fact, since $\Gamma(1)=1,\Gamma(3/2)=\sqrt{\pi}/2,\Gamma(2)=1,\Gamma(5/2)=3\sqrt{\pi}/4,\cdots$ we find that $S(\mathbb{S}^1)=2\pi R,\ S(\mathbb{S}^2)=4\pi R^2,\ S(\mathbb{S}^3)=2\pi^2R^3,\ S(\mathbb{S}^4)=8\pi^2R^4/3$. ■

Example 8.2.6. We have seen in Example 8.2.3 that the Möbius band is non-orientable. We shall now demonstrate that the Möbius band can be externally oriented by the manifold $\mathbb{R}^3$ and an induced volume form may be defined on it. We can now introduce a vector field $W\in T(\mathbb{R}^3)$ that is not situated in the tangent bundle of the Möbius band on resorting to the vectorial product in $\mathbb{R}^3$ as follows

$$\begin{aligned}W = V_u\times V_v &= \frac{1}{2}\Big[2R\cos u + v\Big(\cos\frac{3u}{2}-\cos\frac{u}{2}\Big)\Big]\sin\frac{u}{2}\frac{\partial}{\partial x}\\ &\quad+\frac{1}{2}\Big[2R\cos u\sin\frac{u}{2}+v(\cos u+\sin^2 u)\Big]\frac{\partial}{\partial y}\\ &\quad-\cos\frac{u}{2}\Big(R+v\cos\frac{u}{2}\Big)\frac{\partial}{\partial z}.\end{aligned}$$

The length of this vector is

$$\|W\|^2 = W\cdot W = \Big[\frac{1}{4}(3+2\cos u)\,v^2 + 2Rv\cos\frac{u}{2}+R^2\Big].$$

Then by employing the unit vector $N = W/\|W\|$, we obtain the 2-dimensional volume form induced by the volume form in $\mathbb{R}^3$ as

$$\begin{aligned}\mu' &= \mathbf{i}_N(dx \wedge dy \wedge dz) \\ &= \sqrt{\frac{1}{4}(3 + 2\cos u)\, v^2 + 2R\, v \cos\frac{u}{2} + R^2}\, du \wedge dv\end{aligned}$$

This form will enable us to calculate the area of the Möbius band. ■

8.3. INTEGRATION OF FORMS IN THE EUCLIDEAN SPACE

We want to begin the study of the theory of integration of exterior forms with some rather simple examples that do not differ much from the classical integration. We first consider a differentiable curve C in $\mathbb{R}^n$ and a form $\omega \in \Lambda^1(\mathbb{R}^n)$. We know that the curve C is described by a smooth mapping $\gamma : [a, b] \to \mathbb{R}^n$. The curve C is a 1-dimensional manifold which is a submanifold of $\mathbb{R}^n$ if certain conditions are met and it is prescribed by the equations $x^i = x^i(t), 0 \le t \le 1$. On the curve C, the form $\omega = \omega_i(\mathbf{x})\, dx^i \in \Lambda^1(\mathbb{R}^n)$ is given by the expression

$$\omega(t) = \omega_i\big(\mathbf{x}(t)\big)\frac{dx^i}{dt}\, dt.$$

The integral of the 1-form ω on the curve C is a linear operator in the form of $\int_C : \Lambda^1(\mathbb{R}^n) \to \mathbb{R}$ that assigns a real number to this 1-form defined as follows

$$\int_C \omega = \int_0^1 \omega_i\big(\mathbf{x}(t)\big)\frac{dx^i}{dt}\, dt.$$

The integral in the right hand side is the well known Riemann integral. Sometimes it is not possible to describe the curve by just one parameter. In such a case, the interval $[a, b]$ is partitioned into subintervals such as $a = t_0 < t_1 < \cdots < t_{m-1} < t_m = b$ making it possible to use a different parametrisation in each interval. If we denote the part of the curve corresponding to the interval $[t_j, t_{j+1}]$ by C_j, the integral may be expressed in the form

$$\int_C \omega = \sum_{j=0}^{m-1}\int_{C_j} \omega = \sum_{j=0}^{m-1}\int_{t_j}^{t_{j+1}} \omega_i\big(\mathbf{x}(t)\big)\frac{dx^i}{dt}\, dt.$$

By generalising this approach, we can define the integral of a form $\omega \in \Lambda^n(\mathbb{R}^n)$ given by $\omega(\mathbf{x}) = \omega_{12\cdots n}(\mathbf{x})\, dx^1 \wedge dx^2 \wedge \cdots \wedge dx^n$ as the linear operator $\Lambda^n(\mathbb{R}^n) \to \mathbb{R}$

$$\int_{\mathbb{R}^n} \omega = \int_{\mathbb{R}^n} \omega_{12\cdots n}(\mathbf{x})\, dx^1 dx^2 \cdots dx^n$$

assigning a real number to the n-form ω. The right hand side of the above expression is the multiple Riemann integral of the function $\omega_{12\cdots n}(\mathbf{x})$ with n variables. Naturally, in order that the form ω can be integrated, this integral must exist. When the support of the form ω is compact, that is, when the smooth function $\omega_{12\cdots n}(\mathbf{x})$ vanishes outside a closed and bounded subset of $\mathbb{R}^n$, then it is bounded on this set and the integral will definitely exist. *It is obvious that the integral changes sign if we change the orientation of the manifold.*

Let us now consider a k-dimensional submanifold S_k of $\mathbb{R}^n$ prescribed by the parametric equations $x^i = x^i(u^1, \dots, u^k), i = 1, \dots, n$. We further assume that the parameters u^α,$\alpha = 1, \dots, k$ vary in the region $\prod_{\alpha=1}^{k} [a^\alpha, b^\alpha]$ of $\mathbb{R}^k$ where $[a^\alpha, b^\alpha] \subset \mathbb{R}$ is a closed interval and the symbol $\prod$ represent the Cartesian product of intervals. This set will be called a ***closed k-rectangle***. We know that the value of a k-form

$$\omega(\mathbf{x}) = \frac{1}{k!}\, \omega_{i_1\cdots i_k}(\mathbf{x})\, dx^{i_1} \wedge \cdots \wedge dx^{i_k} \in \Lambda^k(\mathbb{R}^n)$$

on the submanifold S_k is given by the expression

$$\begin{aligned}\omega(\mathbf{u}) &= \frac{1}{k!}\, \omega_{i_1\cdots i_k}\big(\mathbf{x}(\mathbf{u})\big) \frac{\partial x^{i_1}}{\partial u^{\alpha_1}} \cdots \frac{\partial x^{i_k}}{\partial u^{\alpha_k}}\, du^{\alpha_1} \wedge \cdots \wedge du^{\alpha_k} \\ &= \frac{1}{k!}\, \widetilde{\omega}_{\alpha_1\cdots\alpha_k}(\mathbf{u})\, du^{\alpha_1} \wedge \cdots \wedge du^{\alpha_k} = \widetilde{\omega}_{1\cdots k}(\mathbf{u})\, du^1 \wedge \cdots \wedge du^k\end{aligned}$$

where $a^\alpha \le u^\alpha \le b^\alpha$ with an appropriate ordering of parameters. Consequently, the integral of the k-form ω on the submanifold S_k will be defined as the following multiple Riemann integral

$$\int_{S_k} \omega = \int_{a^1}^{b^1} \int_{a^2}^{b^2} \cdots \int_{a^k}^{b^k} \widetilde{\omega}_{12\cdots k}(\mathbf{u})\, du^1 du^2 \cdots du^k.$$

Generally, the submanifold S_k may not be described by a single parametrisation. In such a case, the domain of integration may be the union of some k-rectangles and the integral is expressed as the sum of integrals over those sets. Naturally, these integrals must be convergent. However, in order to define the integral of a form on a differentiable manifold we shall need to equip the manifold with a much more different formal structure from those introduced sketchily in this section.

Example 8.3.1. The integral of the area form associated with the Möbius band given in Example 8.2.5 can be written as

$$A = \alpha R^2 \int_{-1}^{1}\int_{0}^{2\pi}\sqrt{1 + 2\alpha\nu\cos\frac{u}{2} + \frac{\alpha^2}{4}(3 + 2\cos u)\,\nu^2}\,dud\nu$$

where we defined the variable $\nu = v/w$ and the coefficient $\alpha = w/R$. w is the half width of the band. It is not possible to find the exact value of this integral. So we have to resort to numerical integration. For instance, we find $A = 3.1499R^2$ for $\alpha = 1/4$, $A = 1.2572R^2$ for $\alpha = 1/10$. It is readily verified that $A \to 4\pi\alpha R^2$ when $\alpha \to 0$. ■

8.4. SIMPLICES AND CHAINS

The one of the main building blocks in integrating forms over differentiable manifolds are made up by simplices and chains generated by them in the Euclidean space. Let us consider $k+1$ points $P_0, P_1, \dots P_k$ in the Euclidean space $\mathbb{R}^k$. We suppose that two ordered points (P, Q) in $\mathbb{R}^k$ designate the vector $Q - P$ connecting the first point P to the second point Q. Next, we assume that k vectors $P_1 - P_0, \dots, P_k - P_0$ are linearly independent. Hence, for any point $P \in \mathbb{R}^k$, the vector $P - P_0$ can be represented by

$$P - P_0 = \sum_{i=1}^{k} \xi^i (P_i - P_0),\ \ \xi^i \in \mathbb{R}.$$

If we choose $0 \le \xi^i \le 1$ and $\sum_{i=1}^{k} \xi^i \le 1$ for all $i = 1, \dots, k$, then we observe that the vector $P - P_0$ stays within the *k-dimensional closed and convex polyhedral region* formed by vectors $P_i - P_0, 1 \le i \le k$ as edges. Thus for a point P in this region, we can formally write

$$P = \Big[1 - \sum_{i=1}^{k} \xi^i\Big] P_0 + \sum_{i=1}^{k} \xi^i P_i = \sum_{i=0}^{k} \xi^i P_i \tag{8.4.1}$$

where we define $\xi^0 = 1 - \sum_{i=1}^{k} \xi^i \ge 0$. Therefore, the conditions $\sum_{i=0}^{k} \xi^i = 1$ and $\xi^i \ge 0$ for all $0 \le i \le k$ will be satisfied. We shall now symbolise the *closed and convex set* produced by the *ordered points* $P_0, P_1, \dots P_k$ as

$$s_k = [P_0, P_1, \dots, P_k] \subset \mathbb{R}^k. \tag{8.4.2}$$

s_k will be called a ***k-simplex***. Since it is a closed and bounded subset of $\mathbb{R}^k$, s_k becomes clearly a compact subset. If $P \in s_k$, then this point may now be represented by the formal expression (8.4.1). The non-negative real numbers $(\xi^0, \xi^1, \ldots, \xi^k)$ are called the ***barycentric coordinates*** of a point P inside the simplex s_k. The orientation of s_k is specified by the definite order of the successive generating points. We choose the order in (8.4.2) as the positive orientation of the simplex. A different ordering of these points specifies actually the same set. However, the orientation of the simplex may then change. We immediately recognise that if the new ordering is obtained from (8.4.2) by an even permutation of the order of the points in (8.4.2), then the sense in which the points follow each other, thus the orientation of the simplex, remains unchanged whereas if it is an odd permutation the orientation of the simplex is reversed. Let us denote a permutation of the numbers $0, \ldots, k$ by π. Hence, we can obviously write

$$[P_{\pi(0)}, P_{\pi(1)}, \ldots, P_{\pi(k)}] = \operatorname{sgn}(\pi)\,[P_0, P_1, \ldots, P_k]$$

where $\operatorname{sgn}(\pi) = 1$ if π is an even permutation while $\operatorname{sgn}(\pi) = -1$ if it is an odd permutation. If we make use of the coordinates in $\mathbb{R}^k$ and write $P = \{x^\alpha\}, P_i = \{x_i^\alpha\}, \alpha = 1, \ldots k; i = 0, 1, \ldots, k$, then (8.4.1) can be expressed more concretely as

$$x^\alpha = \sum_{i=0}^{k} \xi^i x_i^\alpha,\ \alpha = 1, \ldots k;\ \xi^i \geq 0,\ \sum_{i=0}^{k} \xi^i = 1.$$

The ***face*** opposite to the point P_i in a simplex s_k is defined as the $(k-1)$-simplex obtained by deleting this point from the k-simplex s_k. But in order to render its orientation compatible with the principal simplex, we first put this point into the first position in the ordering so that we obtain

$$[P_i, P_0, \ldots, P_{i-1}, P_{i+1}, \ldots, P_k] = (-1)^i [P_0, \ldots, P_{i-1}, P_i, P_{i+1}, \ldots, P_k]$$

from which we deduce that the faces of a k-simplex are found to be

$$s_{k-1}^i = (-1)^i [P_0, P_1, \ldots, P_{i-1}, P_{i+1}, \ldots, P_k] \qquad (8.4.3)$$

where $i = 0, 1, \ldots, k$. We now define the ***oriented boundary*** of a simplex s_k as a formal sum of its faces:

$$\partial s_k = \sum_{i=0}^{k} s_{k-1}^i. \qquad (8.4.4)$$

Let a family of k-simplices $\{s_k^a : a \in A\}$ where A is an index set be given. The *formal* linear combination

$$c_k = \sum_{a \in A} \lambda_a s_k^a \tag{8.4.5}$$

where $\lambda_a \in \mathbb{R}$ is called a ***k-chain*** in the space $\mathbb{R}^k$. Thus appending simplices in a repetitive way if necessary and playing with their orientations, it becomes possible to produce rather complicated geometrical structures. According to this definition, *the boundary of a k-simplex becomes a $(k-1)$-chain*. In view of the definition (8.4.5), we may say that *all k-chains on $\mathbb{R}^k$ constitutes a linear vector space denoted by $C_k(\mathbb{R}^k)$*.

Let us now take without loss of generality $0 \le j < i \le k$ and consider the faces s_{k-1}^i and s_{k-1}^j of a simplex s_k:

$$s_{k-1}^i = (-1)^i [P_0, P_1, \dots, P_j, \dots, P_{i-1}, P_{i+1}, \dots, P_k],$$
$$s_{k-1}^j = (-1)^j [P_0, P_1, \dots, P_{j-1}, P_{j+1}, \dots, P_i, \dots, P_k].$$

It then follows from above that the jth face of s_{k-1}^i and the ith face of s_{k-1}^j are expressible as

$$s_{k-2}^{ij} = (-1)^i (-1)^j [P_0, P_1, \dots, P_{j-1}, P_{j+1}, \dots, P_{i-1}, P_{i+1}, \dots, P_k],$$
$$s_{k-2}^{ji} = (-1)^j (-1)^{i-1} [P_0, P_1, \dots, P_{j-1}, P_{j+1}, \dots, P_{i-1}, P_{i+1}, \dots, P_k].$$

These two sets are identical except for their orientations so that we get

$$s_{k-2}^{ij} = -\, s_{k-2}^{ji}.$$

Consequently, we conclude that

$$\partial(\partial s_k) = \partial^2 s_k = \sum_{i=0}^{k} \sum_{j=0}^{k} s_{k-2}^{ij} = 0. \tag{8.4.6}$$

This means that *the boundary of the boundary of a simplex is zero*.

Some low dimensional simplices can easily be visualised. $s_0 = [P_0]$ is just a point whereas $s_1 = [P_0, P_1]$ corresponds to a vector, $s_2 = [P_0, P_1, P_2]$ to an oriented triangle and $s_3 = [P_0, P_1, P_2, P_3]$ to an oriented tetrahedron. These simplices are displayed in Fig. 8.4.1.

The boundaries of simplices s_1, s_2, s_3 shown in Fig. 8.4.1 are then given by

$$\partial s_1 = [P_1] - [P_0]$$
$$\partial s_2 = [P_1, P_2] - [P_0, P_2] + [P_0, P_1]$$
$$\partial s_3 = [P_1, P_2, P_3] - [P_0, P_2, P_3] + [P_0, P_1, P_3] - [P_0, P_1, P_2].$$

whereas $\partial s_0 = 0$.

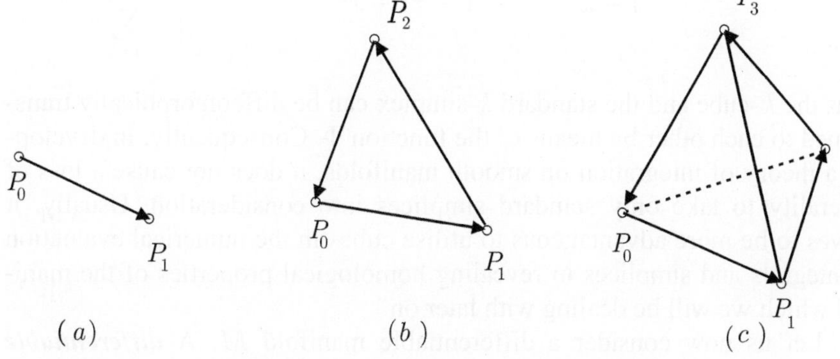

Fig. 8.4.1. Some simplices: (a) s_1-simplex, (b) s_2-simplex, (c) s_3-simplex.

The ***standard k-simplex*** in $\mathbb{R}^k$ is the k-simplex formed by the points $Q_0 = (0, 0, \ldots, 0, 0)$, $Q_1 = (1, 0, \ldots, 0, 0), Q_2 = (0, 1, 0, \ldots, 0), \ldots,$ $Q_k = (0, 0, \ldots, 0, 1)$. Hence, the ***standard k-simplex*** is the set

$$\mathfrak{s}_k = \{(x^1, x^2, \ldots, x^k) \in \mathbb{R}^k : 0 \leq x^i \leq 1, i = 1, \ldots, k; \sum_{i=1}^{k} x^i \leq 1\}.$$

It is straightforward to see that any k-simplex can be generated from the standard k-simplex via an *affine transformation*.

When we are treating in Sec. 8.3 the integration of exterior forms in the Euclidean space we encountered certain subset of $\mathbb{R}^k$ called k-rectangles. We observe at once that these subsets can be reduced to the Cartesian product $[0, 1]^k$ called the ***k-cube*** by a very simple scaling transformation of coordinates. We can further show that a k-cube, or a ***box***, is diffeomorphic to the standard k-simplex. We define a mapping $\Phi : [0, 1]^k \to \mathfrak{s}_k$ on the set

$$[0, 1]^k = \{(y^1, y^2, \ldots, y^k) \in \mathbb{R}^k : 0 \leq y^i \leq 1, i = 1, \ldots, k\}$$

by the following relations

$$\begin{aligned}
x^1 &= y^1, \\
x^2 &= y^2(1 - y^1), \\
x^3 &= y^3(1 - y^1)(1 - y^2), \\
&\vdots \\
x^k &= y^k(1 - y^1)(1 - y^2)\cdots(1 - y^{k-1}).
\end{aligned}$$

We can easily verify that the inverse mapping $\Phi^{-1} : \mathfrak{s}_k \to [0, 1]^k$ is given by

$$y^1 = x^1,\ y^2 = \frac{x^2}{1-x^1},\ y^3 = \frac{x^3}{1-x^1-x^2},\ \dots,\ y^k = \frac{x^k}{1-\sum\limits_{i=1}^{k-1} x^i}.$$

Thus the k-cube and the standard k-simplex can be diffeomorphically transformed to each other by means of the function Φ. Consequently, in developing a theory of integration on smooth manifolds, it does not cause a loss of generality to take only standard simplices into consideration. Usually, it proves to be more advantageous to utilise cubes in the numerical evaluation of integrals and simplices in revealing homological properties of the manifold which we will be dealing with later on.

Let us now consider a differentiable manifold M. A ***differentiable singular k-simplex*** σ_k on M is specified by a *smooth* function $f : V \to M$ mapping the standard k-simplex $\mathfrak{s}_k$ in $\mathbb{R}^k$ into the manifold M. In order to secure the differentiability of this function, its domain V is taken as an open neighbourhood of $\mathfrak{s}_k$. Since $\mathfrak{s}_k$ is compact, $\sigma_k = f(\mathfrak{s}_k)$ will necessarily be a compact subset of M. Thus a singular k-simplex on M is designated by the triple $\sigma_k = (\mathfrak{s}_k, V, f)$. The image points $\mathfrak{Q}_i = f(Q_i) \in M, i = 0, \dots, k$ correspond to the vertices of the singular k-simplex. A family of various singular k-simplices on the manifold M is naturally specified by the set $\{\sigma_k^a = (\mathfrak{s}_k, V_a, f_a : \mathfrak{s}_k \subset V_a), a \in A\}$ where A is an index set (Fig. 8.4.2).

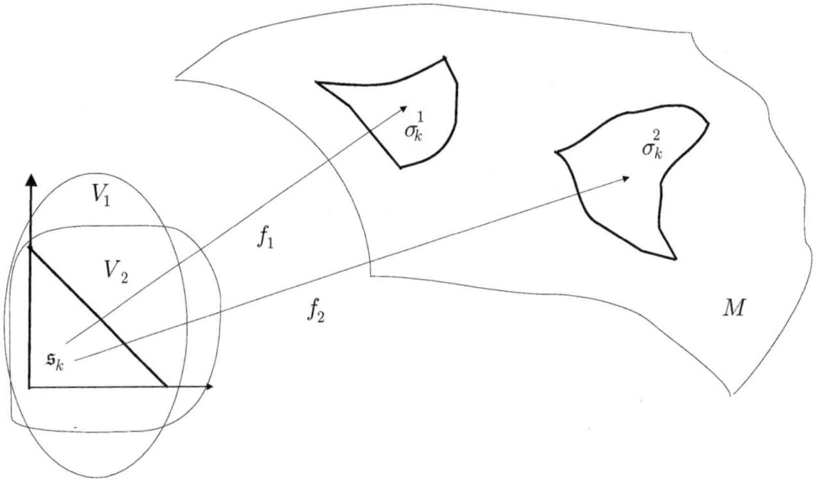

Fig. 8.4.2. Two singular simplices on a manifold M.

With $\lambda_a \in \mathbb{R}$ and $a \in A$, the *formal* linear combination

$$c_k = \sum_{a \in A} \lambda_a \, \sigma_k^a \tag{8.4.7}$$

is called a ***differentiable singular k-chain on M***. It is clear that a singular chain is the union of some singular simplices. If λ is a positive integer, this will imply that we pass over that simplex λ times. If λ is negative the orientation will be reversed. A single simplex σ_k can be regarded as a chain in the form $1 \cdot \sigma_k$. In accordance with this definition, we may say that the sum and multiplication with real numbers of chains is again a chain. Hence, we may think that all k-chains on a manifold M constitute a linear vector space $C_k(M)$.

Let us consider a singular k-simplex σ_k. Let the faces of the standard k-simplex $\mathfrak{s}_k$ be $\mathfrak{s}_{k-1}^i, i = 0, \ldots, k$. The restriction of the function f to the set $\mathfrak{s}_{k-1}^i$ is expressible as $f|_{\mathfrak{s}_{k-1}^i} : V_i \to M$ where the subset $V_i \subset \mathbb{R}^{k-1}$ of V is an open neighbourhood of $\mathfrak{s}_{k-1}^i$. We characterise the following sets

$$\sigma_{k-1}^i = f(\mathfrak{s}_{k-1}^i) \text{ or } \sigma_{k-1}^i = (\mathfrak{s}_{k-1}^i, V_i, f), i = 0, \ldots, k \tag{8.4.8}$$

as the ***faces*** of the singular k-simplex σ_k. The ***boundary*** of σ_k is the image of the boundary of $\mathfrak{s}_k$ under the function f. We thus get $\partial \sigma_k = f(\partial \mathfrak{s}_k)$ showing the validity of the commutation relation $\partial f(\mathfrak{s}_k) = f(\partial \mathfrak{s}_k)$. On the other hand, the boundary of σ_k may also be defined as

$$\partial \sigma_k = \sum_{i=0}^{k} \sigma_{k-1}^i. \tag{8.4.9}$$

Hence, it is a singular $(k-1)$-chain. Therefore, the function f must formally satisfy the relation

$$f\Big(\sum_{i=0}^{k} \mathfrak{s}_{k-1}^i\Big) = \sum_{i=0}^{k} f(\mathfrak{s}_{k-1}^i). \tag{8.4.10}$$

The boundary operator $\partial : C_k(M) \to C_{k-1}(M)$ introduced in (8.4.9) can be extended to an arbitrary chain by the following definition

$$\partial c_k = \partial\Big(\sum_{a \in A} \lambda_a \, \sigma_k^a\Big) = \sum_{a \in A} \lambda_a \, \partial \sigma_k^a. \tag{8.4.11}$$

This definition indicates clearly that ∂ is a linear operator. This operator can be applied for $k \geq 1$ without any problem. Since the boundary of a 0-simplex cannot be defined, we adopt the convention $\partial = 0$ on $C_0(M)$. We can state the theorem below concerning the boundary operator.

Theorem 8.4.1. *The boundary operator ∂ is linear and we have $\partial \circ \partial = \partial^2 = 0$ on $C_k(M)$.*

The linearity of the operator ∂ originates directly from the definition. On the other hand, the image of a zero simplex under f is obviously zero. Thus, we find that

$$\partial^2 \sigma_k = \partial f(\partial \mathfrak{s}_k) = f(\partial^2 \mathfrak{s}_k) = f(0) = 0.$$

Because of the linearity of the operator ∂, we immediately reach to the conclusion that

$$\partial(\partial c_k) = \partial^2 c_k = 0$$

for any chain. □

If the boundary of a chain c_k is zero, i.e., if we can write $\partial c_k = 0$, this chain is called a ***cycle***. Hence, the boundary of every chain is a cycle.

Let M and N be smooth manifolds and $\phi : M \to N$ be a smooth function. We consider a singular k-simplex $\sigma_k = (\mathfrak{s}_k, V, f)$ on the manifold M. The image of σ_k on the manifold N under the mapping ϕ is the set $\phi(\sigma_k) = \phi\big(f(\mathfrak{s}_k)\big) = (\phi \circ f)(\mathfrak{s}_k)$. But the set $\phi(\sigma_k)$ is a singular k-simplex on N because $f' = \phi \circ f : V \to N$ is a smooth function. In this case, we are led to the result

$$\partial\big(\phi(\sigma_k)\big) = f'(\partial \mathfrak{s}_k) = (\phi \circ f)(\partial \mathfrak{s}_k) = \phi\big(f(\partial \mathfrak{s}_k)\big) = \phi(\partial \sigma_k)$$

implying that the operators ∂ and ϕ commute. So we get the relation

$$\partial \circ \phi = \phi \circ \partial. \tag{8.4.12}$$

Because of the linearity of ∂ this result will equally be valid for a chain c_k:

$$\partial\big(\phi(c_k)\big) = \phi(\partial c_k).$$

Let S be an k-dimensional submanifold of an m-dimensional smooth manifold M. The usual coordinates in the standard k-simplex $\mathfrak{s}_k$ in $\mathbb{R}^k$ will be denoted by $u^\alpha, \alpha = 1, \dots, k$. *Let us assume that a singular k-chain $c_k = \{\sigma_k^a = (\mathfrak{s}_k, V_a, f_a) : a \in A\}$ can be found as satisfying the following conditions*:

(a). *Each singular k-simplex σ_k^a parametrizes a region $S_a = \sigma_k^a = f_a(\mathfrak{s}_k)$ of S by $\mathbf{x} = (\varphi \circ f_a)(\mathbf{u})$ where φ is the homeomorphism in a local chart.*

(b). *We have $S = \bigcup\limits_{a \in A} S_a$.*

(c). *Each f_a is injective and the rank of the differential df_a is k. Furthermore, for every $a \neq b$ we have $f_a(\mathring{\mathfrak{s}}_k) \cap f_b(\mathring{\mathfrak{s}}_k) = \emptyset$. Hence,* *the singular k-simplices can touch each other solely along their boundaries.*

Then we say that the chain c_k parametrizes the submanifold S by $u^1, \dots, u^k$.

All singular chains under the operator $\partial : C_{k+1}(M) \to C_k(M)$ constitute a ***chain complex*** specified by the following *decreasing* sequence

$$\cdots \longrightarrow C_{k+1}(M) \xrightarrow{\partial} C_k(M) \xrightarrow{\partial} C_{k-1}(M) \longrightarrow \cdots \quad (8.4.13)$$

because of the fact that $\partial \circ \partial = \partial^2 = 0$. This implies that $\mathcal{R}_k(\partial) \subseteq \mathcal{N}_k(\partial)$. $\mathcal{N}_k(\partial) = \mathrm{Ker}\,(\partial) \subseteq C_k(M)$ is called the space of ***k-cycles***, and $\mathcal{R}_k(\partial) = \mathrm{Im}\,(\partial) \subseteq C_k(M)$ is called the space of ***k-boundaries***.

Let us now consider the dual space $C_k^*(M)$ of the vector space $C_k(M)$. If $f_k \in C_k^*(M)$, then it is a *linear functional* $f_k : C_k(M) \to \mathbb{R}$. Such an f_k will be instumental in creating a ***singular k-cochain on M***. In order to justify this terminology, we shall introduce the ***coboundary operator*** $\mathfrak{d}$ acting on the dual space $C_k^*(M)$ by the relation

$$f_{k+1}(c_{k+1}) = (\mathfrak{d} f_k)(c_{k+1}) = f_k(\partial c_{k+1}) \quad (8.4.14)$$

for all $c_{k+1} \in C_{k+1}(M)$. Obviously $\mathfrak{d} : C_k^*(M) \to C_{k+1}^*(M)$ is a homomorphism and one writes $f_{k+1} = \mathfrak{d} f_k$. Moreover, it is straightforward to see that for any $f_k \in C_k^*(M)$ we obtain

$$(\mathfrak{d} \circ \mathfrak{d}) f_k(c_{k+2}) = \mathfrak{d}^2 f_k(c_{k+2}) = \mathfrak{d} f_k(\partial c_{k+2}) = f_k(\partial^2 c_{k+2}) = 0$$

implying that $\mathfrak{d} \circ \mathfrak{d} = \mathfrak{d}^2 = 0$. Hence, we find that $\mathcal{R}_k(\mathfrak{d}) \subseteq \mathcal{N}_{k+1}(\mathfrak{d})$. This means that all singular cochains under the coboundary operator $\mathfrak{d}$ constitute a ***cochain complex*** given by

$$\cdots \longrightarrow C_{k-1}^*(M) \xrightarrow{\mathfrak{d}} C_k^*(M) \xrightarrow{\mathfrak{d}} C_{k+1}^*(M) \longrightarrow \cdots. \quad (8.4.15)$$

$\mathcal{N}_k(\mathfrak{d}) = \mathrm{Ker}\,(\mathfrak{d}) \subseteq C_k^*(M)$ is called the space of ***k-cocycles***, and $\mathcal{R}_k(\mathfrak{d}) = \mathrm{Im}\,(\mathfrak{d}) \subseteq C_k^*(M)$ is called the space of ***k-coboundaries***.

8.5. INTEGRATION OF FORMS ON MANIFOLDS

We assume that M is an m-dimensional differentiable manifold. Let us consider a form $\omega \in \Lambda^k(M)$. It is known that this form is expressed in local coordinates as

$$\omega(\mathbf{x}) = \frac{1}{k!}\, \omega_{i_1 \cdots i_k}(\mathbf{x})\, dx^{i_1} \wedge \cdots \wedge dx^{i_k}.$$

We shall now try to define the integral of this form on a $k \leq m$ dimensional submanifold S of M. To this end, we first assume that there exists a singular k-chain c_k that parametrizes S just as we have depicted at the end of Sec. 8.4. If we recall that a chain is a linear combination of singular simplices, we realise at once that it would be entirely sufficient to define the integral on a single singular k-simplex $\sigma_k = (\mathfrak{s}_k, V, f)$. The smooth function $f : V \to M$ enables us to establish a smooth relationship between the natural coordinates $\mathbf{x} \in \mathbb{R}^m$ and the parameters $\mathbf{u} \in \mathfrak{s}_k \subset \mathbb{R}^k$ in the form $\mathbf{x} = \mathbf{x}(\mathbf{u})$. Let $f^* : \Lambda(M) \to \Lambda(V)$ be the pull-back operator induced by the mapping f. This operator pulls the k-form ω defined on the simplex σ_k back to the form $\omega^* = f^*\omega$ on the standard simplex $\mathfrak{s}_k$ as follows

$$\begin{aligned}\omega^*(\mathbf{u}) &= \frac{1}{k!}\,\omega_{i_1\cdots i_k}\big(\mathbf{x}(\mathbf{u})\big)\frac{\partial x^{i_1}}{\partial u^{\alpha_1}}\cdots\frac{\partial x^{i_k}}{\partial u^{\alpha_k}}\,du^{\alpha_1}\wedge\cdots\wedge du^{\alpha_k}\\ &= \frac{1}{k!}\,\widetilde{\omega}_{\alpha_1\cdots\alpha_k}(\mathbf{u})\,du^{\alpha_1}\wedge\cdots\wedge du^{\alpha_k} = \widetilde{\omega}_{1\cdots k}(\mathbf{u})\,du^1\wedge\cdots\wedge du^k.\end{aligned}$$

The integral on σ_k is now defined by the relation

$$\int_{\sigma_k}\omega = \int_{\mathfrak{s}_k} f^*\omega = \int_{\mathfrak{s}_k}\widetilde{\omega}_{1\cdots k}(\mathbf{u})\,du^1\cdots du^k \in \mathbb{R} \tag{8.5.1}$$

reducing this integral to a multiple Riemann integral on the standard k-simplex $\mathfrak{s}_k$ in the Euclidean space. Since we have assumed that the k-chain $c_k = \sum\limits_{a\in A}\lambda_a\,\sigma_k^a$ parametrizes the submanifold S, the integral of the k-form ω on S is eventually given by the sum

$$\int_S\omega = \int_{c_k}\omega = \sum_{a\in A}\lambda_a\int_{\sigma_k^a}\omega = \sum_{a\in A}\lambda_a\int_{\mathfrak{s}_k^a} f_a^*\omega. \tag{8.5.2}$$

In order this definition to be consistent, we have to show that this integral is independent of the choice of parametrisation of S. Without loss of generality, we may suppose that S is subject to two different parametrisations by two chains $c_k = \{(\mathfrak{s}_k, V_a, f_a) : a \in A\}$ and $c_k' = \{(\mathfrak{s}_k, U_\mathfrak{a}, g_\mathfrak{a}) : \mathfrak{a} \in \mathfrak{A}\}$ with all real coefficients are $\lambda_a = \lambda_\mathfrak{a} = 1$. Because we can write

$$S = \bigcup_{a\in A} f_a(V_a) = \bigcup_{\mathfrak{a}\in\mathfrak{A}} g_\mathfrak{a}(U_\mathfrak{a})$$

we evidently obtain

$$S = \big(\bigcup_{a\in A} f_a(V_a)\big)\cap\big(\bigcup_{\mathfrak{a}\in\mathfrak{A}} g_\mathfrak{a}(U_\mathfrak{a})\big) = \bigcup_{a\in A,\mathfrak{a}\in\mathfrak{A}} f_a(V_a)\cap g_\mathfrak{a}(U_\mathfrak{a}).$$

Since the mappings f_a and $g_\mathfrak{a}$ are injective, they are bijective mappings over their ranges. Consequently, their inverses exist so that one is able to write

$$f_a^{-1}\circ g_\mathfrak{a} : (g_\mathfrak{a}^{-1}\circ f_a)(V_a)\cap U_\mathfrak{a} \to V_a\cap (f_a^{-1}\circ g_\mathfrak{a})(U_\mathfrak{a})$$

We thus reach to the desired result as follows

$$\begin{aligned}\int_{c_k}\omega &= \sum_{a\in A}\int_{V_a} f_a^*\omega = \sum_{a\in A,\mathfrak{a}\in\mathfrak{A}}\int_{V_a\cap(f_a^{-1}\circ g_\mathfrak{a})(U_\mathfrak{a})} f_a^*\omega \\ &= \sum_{a\in A,\mathfrak{a}\in\mathfrak{A}}\int_{(g_\mathfrak{a}^{-1}\circ f_a)(V_a)\cap U_\mathfrak{a}} (f_a^{-1}\circ g_\mathfrak{a})^* f_a^*\omega \\ &= \sum_{a\in A,\mathfrak{a}\in\mathfrak{A}}\int_{(g_\mathfrak{a}^{-1}\circ f_a)(V_a)\cap U_\mathfrak{a}} g_\mathfrak{a}^*\circ (f_a^{-1})^* f_a^*\omega = \sum_{\mathfrak{a}\in\mathfrak{A}}\int_{U_\mathfrak{a}} g_\mathfrak{a}^*\omega = \int_{c_k'}\omega.\end{aligned}$$

We thus realise that the integral of a k-form on a k-dimensional submanifold can be evaluated as the sum of some multiple integrals over a simple standard k-simplex once we manage to parametrize this submanifold by a suitable chain. If the chain is finite, then this procedure does not cause undue difficulties. But if the chain is infinite, we may then have to face up with a serious problem of convergence. In such a case, if the support of the form ω, i.e., the set *supp* $(\omega) = \overline{\{p\in M : \omega(p)\neq 0\}}$ is compact so that it can be covered with finitely many open sets, then surely no problems occur.

Let M, N be smooth manifolds and $\phi : M\to N$ be a smooth mapping. If c_k is a k-chain on M, we know that $c_k' = \phi(c_k)$ is a k-chain on N. Hence, if $\omega\in\Lambda^k(N)$ we immediately observe that

$$\int_{c_k'}\omega = \int_{\phi(c_k)}\omega = \int_{c_k}\phi^*\omega. \tag{8.5.3}$$

Example 8.5.1. We want to calculate the integral of the form $\omega = xyz^2dx\wedge dy\wedge dz\in\Lambda^3(\mathbb{R}^3)$ on the standard 3-simplex $\mathfrak{s}_3$. On using the familiar method of calculation of multiple integrals, we obtain

$$\begin{aligned}\int_{\mathfrak{s}_3}\omega &= \int_{\mathfrak{s}_3} xyz^2dxdydz = \int_0^1 dx\int_0^{1-x}dy\int_0^{1-x-y}xyz^2dz \\ &= \int_0^1 dx\int_0^{1-x}\frac{1}{3}xy(1-x-y)^3dy \\ &= \frac{1}{60}\int_0^1 x(1-x)^5dx = \frac{1}{2520}\end{aligned}$$

■

Example 8.5.2. We shall calculate the integral of the 2-form

$$\omega = (x^2 + z^2)dx \wedge dy + (x^2 + y^2)dy \wedge dz + (y^2 + z^2)dz \wedge dx \in \Lambda^2(\mathbb{R}^3)$$

on the 2-chain $c_2 = \sum_{i=0}^{3} \sigma_2^i$ made up of the *faces* of the tetrahedron formed by the points $Q_0 = (0,0,0)$, $Q_1 = (a,0,0)$, $Q_2 = (0,b,0)$, $Q_3 = (0,0,c)$ in $\mathbb{R}$. The simplices of the chain are given by

$$\begin{aligned}\sigma_2^0 &= [Q_1, Q_2, Q_3], \quad \sigma_2^1 = -[Q_0, Q_2, Q_3],\\ \sigma_2^2 &= [Q_0, Q_1, Q_3], \quad \sigma_2^3 = -[Q_0, Q_1, Q_2].\end{aligned}$$

We define the standard 2-simplex by

$$\mathfrak{s}_2 = \{(u,v) \in \mathbb{R}^2 : 0 \le u, v \le 1,\ u + v \le 1\}.$$

Then the functions $f_i(u,v) = (x,y,z), i = 0,1,2,3$ identifying singular 2-simplices σ_2^i become

$$\begin{aligned}f_1(u,v) &= (0, bu, cv),\\ f_2(u,v) &= (au, 0, cv),\\ f_3(u,v) &= (au, bv, 0)\\ f_0(u,v) &= (au, bv, c(1-u-v)).\end{aligned}$$

Indeed, we can readily verify that these functions provide the following mappings

$$\begin{aligned}f_1 &: \mathfrak{s}_2 \to [Q_0, Q_2, Q_3],\\ f_2 &: \mathfrak{s}_2 \to [Q_0, Q_1, Q_3],\\ f_3 &: \mathfrak{s}_2 \to [Q_0, Q_1, Q_2],\\ f_0 &: \mathfrak{s}_2 \to [Q_1, Q_2, Q_3].\end{aligned}$$

When we pull the form ω from those faces back to $\mathfrak{s}_2$, we obtain the forms

$$\begin{aligned}f_1^*\omega &= b^3c\, u^2 du \wedge dv,\\ f_2^*\omega &= -ac^3v^2 du \wedge dv,\\ f_3^*\omega &= a^3b\, u^2 du \wedge dv,\\ f_0^*\omega &= \left[a^2b(a+c)u^2 + b^2c(a+b)v^2 + ac^2(c+b)(1-u-v)^2\right] du \wedge dv.\end{aligned}$$

We thus find

$$\int_{\sigma_2^1} \omega = -b^3c \int_{u=0}^{1} \int_{v=0}^{1-u} u^2 du dv = -\frac{b^3c}{12},$$

$$\int_{\sigma_2^2}\omega = -\,ac^3\int_{u=0}^{1}\int_{v=0}^{1-u}v^2dudv = -\,\frac{ac^3}{12},$$
$$\int_{\sigma_2^3}\omega = -\,a^3b\int_{u=0}^{1}\int_{v=0}^{1-u}u^2dudv = -\,\frac{a^3b}{12},$$
$$\int_{\sigma_2^0}\omega =$$
$$\int_{u=0}^{1}\int_{v=0}^{1-u}\big[a^2b(a+c)u^2+b^2c(a+b)v^2+c^2a(c+b)(1-u-v)^2\big]dudv$$
$$=\frac{1}{12}\big[a^3b+a^2bc+b^3c+ac(b^2+bc+c^2)\big]$$

whence we arrive at the result

$$\int_{c_2}\omega = \frac{abc}{12}(a+b+c). \qquad \blacksquare$$

The approach we have followed above to evaluate the integral of a k-form on a k-dimensional manifold consists of decomposing a complicated region to much simpler regions by means of k-chains and summing all integrals calculated relatively easily on those regions. We shall now discuss a second approach that may prove to be more effective in certain cases. In that approach, we decompose the form into some forms that vanish outside of some simple regions covering the manifold and we add the integrals of these forms together to obtain the final result.

We consider a k-dimensional smooth submanifold S of an m-dimensional smooth manifold M. Let $\mathcal{A}_M = \{(U'_\lambda, \varphi'_\lambda) : \lambda \in \Lambda\}$ be an atlas of M. We know that this atlas induces an atlas $\mathcal{A}_S = \{(U_\lambda, \varphi_\lambda) : \lambda \in \Lambda\}$ on S [*see p.* 105] where $U_\lambda = U'_\lambda \cap S, \varphi_\lambda = \varphi'_\lambda \circ i : U_\lambda \to \mathbb{R}^k$ and $\mathcal{I} : S \to M$ is the inclusion mapping $\mathcal{I}(p) = p$ for all $p \in S$. Let us now assume that there exists a partition of unity $\{V_a, f_a : a \in A\}$ on the submanifold S subordinate to the atlas $\mathcal{A}_S$ [*see p.* 62]. Each set V_a belongs to an open set U_{λ_a} of a chart of this atlas. We now consider a form $\omega \in \Lambda^k(M)$ and try to evaluate its integral over S. Since the partition of unity implies that $\sum_{a\in A} f_a(p) = 1$ for all $p \in S$, we can write

$$\omega|_S = \omega(p) = \sum_{a\in A}\omega_a(p),\quad \omega_a(p) = f_a(p)\,\omega(p) \in \Lambda^k(U_{\lambda_a}),$$
$$supp\,(\omega_a) \subseteq supp\,(f_a) \subset V_a \subseteq U_{\lambda_a}.$$

We thus obtain

$$\int_S \omega = \sum_{a\in A} \int_{supp\,(f_a)} f_a\,\omega = \sum_{a\in A} \int f_a\,\omega. \tag{8.5.4}$$

If the sum at the right hand side is convergent, the integral of the form ω on S is expressed as the sum integrals of forms that vanish outside of certain regions. When S is a paracompact manifold, we had mentioned before [*see p.* 95] that a partition of unity can be found subordinate to every atlas. We know that there exist merely finitely many functions f_a in a neighbourhood of each point $p \in S$. However, if $\mathcal{A}_S$ does not contain a finite number of open sets, infinitely many terms may nevertheless be involved in the sum and we naturally have to face up with a problem of convergence. When the support of the form ω on the submanifold S is compact, it can always be covered by a finitely many open sets, so the expression (8.5.4) becomes a finite sum in this case. Therefore, the problem of convergence disappears naturally. If the submanifold S itself is compact, this situation will always occur.

In order that the integral of a form given by (8.5.4) has a meaning, it should not be dependent on the chosen atlas and the partition of unity. To show this, let us consider two atlases and their two charts $\{U_\lambda, \varphi_\lambda : \lambda \in \Lambda\}$ and $\{W_\gamma, \psi_\gamma : \gamma \in \Gamma\}$ on S and two partitions of unity $\{V_a, f_a : a \in A\}$ and $\{Z_b, g_b : b \in B\}$ on S subordinate to those atlases, respectively. Since $S = \bigcup_{\lambda\in\Lambda} U_\lambda = \bigcup_{\gamma\in\Gamma} W_\gamma$, we can obviously write $S = \bigcup_{\lambda\in\Lambda,\gamma\in\Gamma} (U_\lambda \cap W_\gamma)$. Thus, the family $\{U_\lambda \cap W_\gamma : \lambda \in \Lambda, \gamma \in \Gamma\}$ is likewise an open cover of S. We then realise that $\{V_a \cap Z_b, f_a g_b : a \in A, b \in B\}$ is the partition of unity subordinate to the open cover $\{U_\lambda \cap W_\gamma\}$. Accordingly, the integral of the form ω can be written in two different ways as follows

$$\begin{aligned}\int_S \omega &= \sum_{a\in A} \int f_a \omega = \sum_{a\in A} \sum_{b\in B} \int f_a g_b\,\omega \\ &= \sum_{b\in B} \int g_b \omega = \sum_{b\in B} \sum_{a\in A} \int g_b f_a \omega.\end{aligned}$$

since we can write $\omega(p) = \sum_{a\in A} f_a(p)\omega(p) = \sum_{b\in B} g_b(p)\omega(p)$.

As a matter of fact, if the above sums converge absolutely, then we are allowed to interchange freely the order of summations in the above expressions. Furthermore, if the support of the form ω is compact, then this will happen naturally. Hence, if we consider two partitions of unity we obtain

$$\int_S \omega = \sum_{a \in A} \int f_a \, \omega = \sum_{b \in B} \int g_b \, \omega.$$

Hence, the integral is independent of the chosen charts and partitions of unity subordinate to them.

8.6. THE STOKES THEOREM

We had defined a manifold with boundary in *pp.* 90-93. We had seen there that the boundary ∂S of such a k-dimensional differentiable manifold S is a $(k-1)$-dimensional differentiable manifold and the local coordinates $(x^1, x^2, \dots, x^{k-1}, x^k) \in \mathbb{R}^k$ can be so chosen that the boundary in $\mathbb{R}^{k-1}$ is represented by $(x^1, x^2, \dots, x^{k-1}, x^k = 0)$. The Stokes theorem that is rather simple looking at a first glance but having a great potential in provoking very important developments [it is commemorated by the name of English mathematician Sir George Gabriel Stokes (1819-1903) who utilised a similar theorem in the context of classical vector analysis[1]] states that the following relation

$$\int_S d\omega = \int_{\partial S} \omega \tag{8.6.1}$$

is valid for every form $\omega \in \Lambda^{k-1}(S)$. This theorem is very important because it helps derive the classical theorems of Green-Gauss and Kelvin-Stokes as well as the fundamental theorem of calculus. It also links topology and analysis because the boundary operator ∂ on the right hand side is purely geometric whereas the integral and the exterior derivative on the left hand side are purely analytic. We shall first prove this theorem for a manifold with boundary prescribed by a singular chain c_k whose boundary is given by ∂c_k.

Theorem 8.6.1 (The Stokes Theorem on Chains). *Let M be a differentiable manifold. We assume that there exists a k-chain $c_k \in C_k(M)$ and consider an exterior differential form $\omega \in \Lambda^{k-1}(M)$. We then have the equality*

[1]It was actually Sir William Thomson (Lord Kelvin) (1824-1907) who discovered this relation within the context of classical vector analysis and communicated it to Stokes in July 1850. However, Stokes is identified with this theorem because he asked its proof on 1854 Smith's Prize examination in Cambridge University. It is not known whether the students were able to answer that question. That is the reason why some authors call this theorem as the Kelvin-Stokes theorem.

$$\int_{c_k} d\omega = \int_{\partial c_k} \omega$$

provided that the integrals converge.

For a chain given by $c_k = \sum_a \lambda_a \sigma_k^a$, its boundary is expressed as $\partial c_k = \sum_a \lambda_a \, \partial\sigma_k^a$. Hence, it would suffice to show that the above relation is valid for a single singular k-simplex σ_k. Since $\sigma_k = (\mathfrak{s}_k, V, f)$ or, in short, $\sigma_k = f(\mathfrak{s}_k)$, we can write

$$\int_{\sigma_k} d\omega = \int_{\mathfrak{s}_k} f^* d\omega = \int_{\mathfrak{s}_k} d(f^*\omega)$$

on resorting to the pull-back operation where we make use of the property $f^* \circ d = d \circ f^*$ in accordance with Theorem 5.8.2. The form $\theta = f^*\omega \in \Lambda^{k-1}(\mathbb{R}^k)$ will be defined on an open neighbourhood V of the standard k-simplex $\mathfrak{s}_k$ in $\mathbb{R}^k$. Let us now denote the local coordinates in $\mathbb{R}^k$ by $u^1, \dots, u^k$. Hence, the form θ becomes expressible *in terms of its essential components* as

$$\begin{aligned}\theta &= \sum_{i=1}^{k} (-1)^{i-1} \theta_i(\mathbf{u})\, du^1 \wedge \cdots \wedge du^{i-1} \wedge du^{i+1} \wedge \cdots \wedge du^k \\ &= \sum_{i=1}^{k} \vartheta_i\end{aligned}$$

where we have obviously introduced the forms $\vartheta_i \in \Lambda^{k-1}(\mathbb{R}^k), i = 1, 2, \dots, k$ as follows

$$\vartheta_i = (-1)^{i-1} \theta_i(\mathbf{u})\, du^1 \wedge \cdots \wedge du^{i-1} \wedge du^{i+1} \wedge \cdots \wedge du^k.$$

The factor $(-1)^{i-1}$ is inserted for convenience. Thus the exterior derivative $d\theta$ may be expressed in the following manner

$$\begin{aligned}d\theta &= \sum_{i=1}^{k} (-1)^{i-1} \frac{\partial \theta_i}{\partial u^j} du^j \wedge du^1 \wedge \cdots \wedge du^{i-1} \wedge du^{i+1} \wedge \cdots \wedge du^k \\ &= \frac{\partial \theta_i}{\partial u^i} du^1 \wedge \cdots \wedge du^k = \sum_{r=1}^{k} \frac{\partial \theta_{\underline{r}}}{\partial u^{\underline{r}}} du^1 \wedge \cdots \wedge du^r \wedge \cdots \wedge du^k\end{aligned}$$

where the summation convention is suspended as usual on underscored indices. Therefore, we can write

$$\int_{\mathfrak{s}_k} d\theta = \sum_{r=1}^{k} \int_{\mathfrak{s}_k} \frac{\partial \theta_{\underline{r}}}{\partial u^{\underline{r}}}\, du^1 \cdots du^r \cdots du^k. \tag{8.6.2}$$

These integrals can now easily be evaluated by consulting to the standard simplex represented in Fig. 8.6.1.

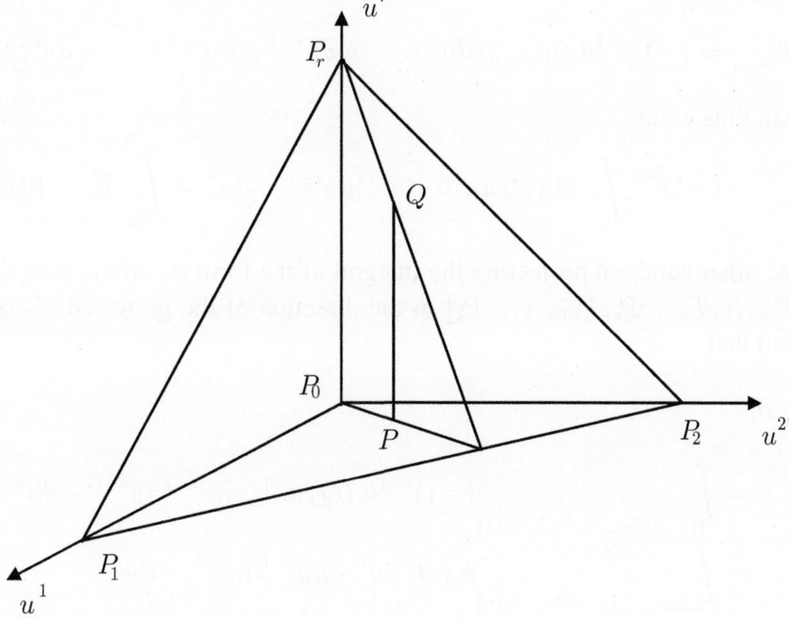

Fig. 8.6.1. Standard simplex $\mathfrak{s}_k$ positioned with respect to the preferred variable u^r.

In fact, on recalling the relation

$$[P_0, P_1, \ldots, P_{r-1}, P_{r+1}, \ldots, P_k] = (-1)^r \mathfrak{s}^r_{k-1}$$

[*see* (8.4.3)], we obtain

$$\begin{aligned} &\int_{\mathfrak{s}_k} \frac{\partial \theta_{\underline{r}}}{\partial u^{\underline{r}}}\, du^1 \cdots du^{r-1} du^r du^{r+1} \cdots du^k \\ &\quad = (-1)^r \int_{\mathfrak{s}^r_{k-1}} \big[\theta_r(Q) - \theta_r(P)\big]\, du^1 \cdots du^{r-1} du^{r+1} \cdots du^k \end{aligned} \tag{8.6.3}$$

where we define on relevant faces

$$\theta_r(P) = \theta_r(u^1, \cdots, u^{r-1}, 0, u^{r+1}, \cdots, u^k),$$

$$\theta_r(Q) = \theta_r\Big(u^1, \cdots, u^{r-1}, 1 - \sum_{i=1, i\neq r}^{k} u^i, u^{r+1}, \cdots, u^k\Big).$$

Since $u^r = 0$ on the face $\mathfrak{s}^r_{k-1}$, we get $du^r = 0$ there and it follows from the definition of the form θ that only one term in the expression for θ survives on $\mathfrak{s}^r_{k-1}$:

$$\theta|_{\mathfrak{s}^r_{k-1}} = (-1)^{r-1}\theta_r(\mathbf{u}|_{u^r=0})\, du^1 \wedge \cdots \wedge du^{r-1} \wedge du^{r+1} \wedge \cdots \wedge du^k.$$

We can thus write

$$(-1)^{r-1}\int_{\mathfrak{s}^r_{k-1}} \theta_r(P)\, du^1 \cdots du^{r-1} du^{r+1} \cdots du^k = \int_{\mathfrak{s}^r_{k-1}} \theta. \tag{8.6.4}$$

On the other hand, on projecting the integral of the form ϑ_r on the face $\mathfrak{s}^0_{k-1} = [P_1, \ldots, P_{r-1}, P_r, P_{r+1}, \ldots, P_k]$ in the direction of the preferred u^r-axis, we find that

$$\begin{aligned}
&\int_{\mathfrak{s}^0_{k-1}} \vartheta_r \\
&\quad = \int_{[P_1,\ldots,P_{r-1},P_0,P_{r+1},\ldots,P_k]} (-1)^{r-1}\theta_r(Q)\, du^1 \cdots du^{r-1} du^{r+1} \cdots du^k \\
&\quad = \int_{[P_0,P_1,\ldots,P_{r-1},P_{r+1},\ldots,P_k]} \theta_r(Q)\, du^1 \cdots du^{r-1} du^{r+1} \cdots du^k \\
&\quad = (-1)^r \int_{\mathfrak{s}^r_{k-1}} \theta_r(Q)\, du^1 \cdots du^{r-1} du^{r+1} \cdots du^k.
\end{aligned} \tag{8.6.5}$$

In order to facilitate the computation of the integral in the fourth line above, let us introduce the change of variables $(u^1, u^2, \ldots, u^{r-1}, u^{r+1}, \ldots, u^k) \to (v^1, \ldots, v^{k-1})$ through the relations

$$\begin{aligned}
&v^1 = u^2, \ldots, v^{r-2} = u^{r-1}, \;\; v^{r-1} = 1 - \sum_{i=1, i\neq r}^{k} u^i, \\
&v^r = u^{r+1}, \ldots, v^{k-1} = u^k.
\end{aligned}$$

We readily observe that the inverse relations are then given by

$$u^1 = 1 - \sum_{i=1}^{k-1} v^i, u^2 = v^1, \ldots u^{r-1} = v^{r-2}, u^{r+1} = v^r, \ldots, u^k = v^{k-1}.$$

The relation

$$du^1 \wedge \cdots \wedge du^{r-1} \wedge du^{r+1} \wedge \cdots \wedge du^k = (-1)^{r-1} dv^1 \wedge dv^2 \wedge \cdots \wedge dv^{k-1}$$

implies that the Jacobian of the transformation is $J = (-1)^{r-1}$. We thus get $|J| = 1$ and

$$du^1 \cdots du^{r-1} du^{r+1} \cdots du^k = dv^1 dv^2 \cdots dv^{k-1}.$$

Since $0 \le u^i \le 1$, $\sum_{i=1}^{k} u^i \le 1$ in the standard k-simplex $\mathfrak{s}_k$, the new variables $v^1, \ldots, v^{k-1}$ will evidently satisfy the conditions $\sum_{i=1}^{k-1} v^i \le 1$ and $0 \le v^i \le 1$, $i = 1, \ldots, k-1$. Therefore, they depict the standard $(k-1)$-simplex $\mathfrak{s}_{k-1}$ in $\mathbb{R}^{k-1}$. Hence, we can write

$$\begin{aligned}
\int_{\mathfrak{s}^r_{k-1}} \theta_r(Q)\, du^1 \cdots du^{r-1} du^{r+1} \cdots du^k
&= \int_{\mathfrak{s}_{k-1}} \overline{\theta}_r(v^1, \ldots, v^{k-1})\, dv^1 \cdots dv^{k-1} \\
&= \int_{\mathfrak{s}_{k-1}} \theta_r\Big(1 - \sum_{i=1}^{k-1} v^i, v^1, \ldots, v^{k-1}\Big)\, dv^1 \cdots dv^{k-1}.
\end{aligned}$$

On inserting the expressions (8.6.4) and (8.6.5) into the relations (8.6.3) and (8.6.2), we reach to the conclusion

$$\int_{\mathfrak{s}_k} d\theta = \sum_{r=1}^{k} \int_{\mathfrak{s}^r_{k-1}} \theta + \sum_{r=1}^{k} \int_{\mathfrak{s}^0_{k-1}} \vartheta_r = \sum_{r=0}^{k} \int_{\mathfrak{s}^r_{k-1}} \theta = \int_{\partial \mathfrak{s}_k} \theta.$$

If we take the relations (8.4.12) and (8.5.3) into consideration, the above equality leads to the following expression

$$\int_{\sigma_k} d\omega = \int_{\partial \sigma_k} \omega$$

whence we deduce the Stokes theorem on k-chains in the following form

$$\int_{c_k} d\omega = \int_{\partial c_k} \omega \tag{8.6.6}$$

where $\omega \in \Lambda^{k-1}(M)$. □

If the chain c_k is the boundary of a chain b_{k+1}, i.e., if ∂b_{k+1} then $\partial c_k = \partial^2 b_{k+1} = 0$ and consequently, we obtain

$$\int_{c_k} d\omega = \int_{\partial b_{k+1}} d\omega = \int_{\partial^2 b_{k+1}} \omega = 0.$$

Similarly, if a chain c_k is a cycle, we then have $\partial c_k = 0$ and we clearly find this time

$$\int_{c_k} d\omega = 0.$$

On the other hand, if ω is a *closed form*, namely, if $d\omega = 0$, then on the boundary of every chain, we find

$$\int_{\partial c_k} \omega = 0.$$

But, we have to warn that satisfaction of this condition on the boundary of every k-chain does not generally mean that the $(k-1)$-form ω is closed.

If the difference of two k-chains is the boundary of a $(k+1)$-chain, that is, if $c_k - c'_k = \partial b_{k+1}$, we then get $\partial c_k - \partial c'_k = \partial(c_k - c'_k) = \partial^2 b_{k+1} = 0$. Hence, the difference of such kind of chains is a cycle and the relation (8.6.6) yields

$$\int_{c_k} d\omega = \int_{c'_k} d\omega.$$

On the other hand, if $\omega \in \Lambda^k(M)$, we then find

$$\int_{c_k} \omega - \int_{c'_k} \omega = \int_{c_k - c'_k} \omega = \int_{\partial b_{k+1}} \omega = \int_{b_{k+1}} d\omega$$

Thus, if ω is a closed form, i.e., if $d\omega = 0$, we also observe the following equality for this sort of chains

$$\int_{c_k} \omega = \int_{c'_k} \omega.$$

If $\omega \in \Lambda^{k-1}(M)$ is an *exact form*, we have to write $\omega = d\theta$ and (8.6.6) leads to the identity

$$0 = \int_{c_k} d^2\theta = \int_{\partial c_k} d\theta = \int_{\partial^2 c_k} \theta = 0.$$

Example 8.6.1. We now want to evaluate the integral in Example 8.5.2 by means of the relation

$$\sum_{i=0}^{3}\int_{\sigma_2^i}\omega=\int_{\partial\mathfrak{s}_3}\omega=\int_{\mathfrak{s}_3}d\omega.$$

The integral of the form

$$d\omega=2(x+y+z)\,dx\wedge dy\wedge dz$$

on $\mathfrak{s}_3$ can be calculated as

$$\begin{aligned}\int_{\mathfrak{s}_3}d\omega&=2\int_{x=0}^{a}\int_{y=0}^{-(b/a)x+b}\int_{z=0}^{c[1-(x/a)-(y/b)]}(x+y+z)\,dxdydz\\&=\frac{abc}{12}(a+b+c).\end{aligned}$$

We thus arrive at the same result. ■

Let us now consider a k-dimensional differentiable manifold S with a boundary ∂S. We know that we are able to choose the local coordinates $(x^1,x^2,\dots,x^k)$ in a chart in such a way that $x^k=0$ defines the boundary ∂S. Thus local coordinates of any point at the boundary are then given by $(x^1,x^2,\dots,x^{k-1})$. Let the vectors $e_1,e_2,\dots,e_{k-1}$ be a local basis for the tangent bundle $T(\partial S)$. Two vectors that do not belong to $T(\partial S)$ are $\partial/\partial x^k$ and $-\,\partial/\partial x^k$. The former vector is called as the ***interior normal*** of the boundary ∂S while the latter as the ***exterior normal***. We assume that S is positively oriented by the volume form $\mu_k=dx^1\wedge\cdots\wedge dx^k$ so that we have $\mu_k(\partial/\partial x^1,\dots,\partial/\partial x^k)>0$. We shall now adopt the convention that the $(k-1)$-dimensional boundary manifold ∂S will be ***positively oriented*** with respect to its exterior normal if $\mu_k(-\,\partial/\partial x^k,e_1,e_2,\dots,e_{k-1})>0$. We now propose the following form of the Stokes theorem for smooth manifolds with boundary.

Theorem 8.6.2 (Stokes' Theorem on Manifolds with Boundary). *Let S be a k-dimensional smooth manifold with boundary and $\omega\in\Lambda^{k-1}(S)$ be an exterior form with a compact support. If $\mathcal{I}:\partial S\to S$ is the inclusion mapping identifying boundary points as points of the manifold, then the form $\mathcal{I}^*\omega\in\Lambda^{k-1}(\partial S)$ will satisfy the relation*

$$\int_{\partial S}\mathcal{I}^*\omega=\int_S d\omega\quad\textit{or in short}\quad\int_{\partial S}\omega=\int_S d\omega. \tag{8.6.7}$$

The manifold ∂S is supposed to be positively oriented.

Let us first assume that the support D_ω of the form ω lies within a chart (U,φ) whose local coordinates are $(x^1,x^2,\dots,x^k)$. Thus the form ω can be represented as

$$\omega = \sum_{i=1}^{k} (-1)^{i-1} \omega_i(\mathbf{x})\, dx^1 \wedge \cdots \wedge dx^{i-1} \wedge dx^{i+1} \wedge \cdots \wedge dx^k$$

if $\mathbf{x} \in \varphi(D_\omega) \subset \mathbb{R}^k$ and $\omega = 0$ if $\mathbf{x} \notin \varphi(D_\omega)$. We shall further assume that $x^k = 0$ on the boundary ∂S. In this case, the exterior derivative of the form ω can be written as

$$d\omega = \sum_{r=1}^{k} \frac{\partial \omega_{\underline{r}}}{\partial x^{\underline{r}}} dx^1 \wedge \cdots \wedge dx^r \wedge \cdots \wedge dx^k \in \Lambda^k(S)$$

as we have attested previously. Next, we have to distinguish two different situations.

(i). Let $\partial S \cap U = \emptyset$ so that we obviously obtain

$$\int_{\partial S} \omega = 0.$$

On the other hand, the set $\varphi(D_\omega) \subset \varphi(U) = V$ in $\mathbb{R}^k$ is closed and bounded because it is a compact set being the image of a compact set under a homeomorphism. Consequently, we can assume that $\varphi(D_\omega) \subset K_k$ where the k-dimensional ***box*** K_k is actually a k-rectangle defined by $K_k = \prod_{r=1}^{k} [a^r, b^r]$, $0 < a^r < b^r < \infty$. At all of the end points of the box the following conditions will evidently be satisfied since the support of ω is supposed to be compact:

$$\omega_r(x^1, \ldots, a^r, \ldots, x^k) = \omega_r(x^1, \ldots, b^r, \ldots, x^k) = 0, \quad 1 \le r \le k.$$

We thus find that

$$\int_S d\omega = \int_{\varphi(D_\omega)} d\omega = \sum_{r=1}^{k} \int_{K_{k-1}} \Big[\int_{a^r}^{b^r} \frac{\partial \omega_{\underline{r}}}{\partial x^{\underline{r}}}\, dx^r dx^1 \cdots dx^{r-1} dx^{r+1} \cdots dx^k$$

$$= \sum_{r=1}^{k} \int_{K_{k-1}} \omega_r(x^1, \ldots, x^r, \ldots, x^k)\Big|_{x^r = a^r}^{x^r = b^r} dx^1 \cdots dx^{r-1} dx^{r+1} \cdots dx^k = 0$$

which proves that the relation (8.6.7) will hold in this case.

(ii). Let $\partial S \cap U \neq \emptyset$. In this case, with $x^k = 0$ on ∂S we obtain

$$\mathcal{I}^* \omega = (-1)^{k-1} \omega_k(x^1, x^2, \ldots, x^{k-1}, 0)\, dx^1 \wedge dx^2 \wedge \cdots \wedge dx^{k-1}.$$

But, we have to take now $0 = a^k < b^k < \infty$ in the box K_k containing the image $\varphi(D_\omega)$ of the support of the ω. Hence, this time we get

$$\begin{aligned}\int_S d\omega &= \int_{\varphi(D_\omega)} d\omega = \sum_{r=1}^k \int_{K_{k-1}} \Big[\int_{a^r}^{b^r} \frac{\partial \omega_r}{\partial x^r} dx^r dx^1 \cdots dx^{r-1} dx^{r+1} \cdots dx^k \\ &= \sum_{r=1}^k \int_{K_{k-1}} \omega_r(x^1, \ldots, x^r, \ldots, x^k)\Big|_{x^r=a^r}^{x^r=b^r} dx^1 \cdots dx^{r-1} dx^{r+1} \cdots dx^k \\ &= -\int_{K_{k-1}} \omega_k(x^1, x^2, \ldots, x^{k-1}, 0)\, dx^1 dx^2 \cdots dx^{k-1}.\end{aligned}$$

On the other hand, since $\mu_k(-\partial/\partial x^k, \partial/\partial x^1, \ldots, \partial/\partial x^{k-1}) = (-1)^k$, we cannot say that the basis $(\partial/\partial x^1, \ldots, \partial/\partial x^{k-1})$ is positively oriented in $T(\partial S)$. Accordingly, we find

$$\begin{aligned}\int_{\partial S} \mathcal{I}^* \omega &= (-1)^{2k-1} \int_{K_{k-1}} \omega_k(x^1, x^2, \ldots, x^{k-1}, 0)\, dx^1 dx^2 \cdots dx^{k-1} \\ &= -\int_{K_{k-1}} \omega_k(x^1, x^2, \ldots, x^{k-1}, 0)\, dx^1 dx^2 \cdots dx^{k-1}\end{aligned}$$

that results in

$$\int_{\partial S} \mathcal{I}^* \omega = \int_S d\omega.$$

We now wish to relax the condition that the support of the form ω is compact as it appears in the statement of the theorem. However, we shall instead suppose that there is an atlas on S subordinate to which there exists a partition of unity $\{V_\alpha, f_\alpha\}$ where $V_\alpha \subseteq U$ and (U, φ) is a chart of the atlas. We now impose the restriction that $\varphi(\mathit{supp}\, f_\alpha) \subseteq \mathbb{R}^k$ is bounded for each member of the family. Since $\varphi(\mathit{supp}\, f_\alpha)$ is also closed, the image of $\mathit{supp}\, f_\alpha$ is a compact set in $\mathbb{R}^k$. Hence, due to the homeomorphism $\mathit{supp}\, f_\alpha$ becomes a compact subset in S. Let us now define the forms $\omega_\alpha = f_\alpha \omega \in \Lambda^{k-1}(S)$ associated with the form $\omega \in \Lambda^{k-1}(S)$. The support of ω_α is the same as that of f_α, i.e., it is a compact subset. Thus, Stokes' theorem can be applied to such forms. Because of the relation $\sum_\alpha f_\alpha = 1$ we get $\sum_\alpha df_\alpha = 0$. Therefore, we can write

$$\omega = \sum_\alpha f_\alpha \omega = \sum_\alpha \omega_\alpha, \quad d\omega = \sum_\alpha (df_\alpha\, \omega + f_\alpha d\omega) = \sum_\alpha d\omega_\alpha$$

and, consequently, obtain

$$\begin{aligned}\int_S d\omega &= \sum_\alpha \int_S d\omega_\alpha = \sum_\alpha \int_{\partial S} \mathcal{I}^*\omega_\alpha = \int_{\partial S} \sum_\alpha \mathcal{I}^*\omega_\alpha \\ &= \int_{\partial S} \mathcal{I}^* \sum_\alpha \omega_\alpha = \int_{\partial S} \mathcal{I}^*\omega\end{aligned}$$

provided that the above sum is convergent and we are allowed to interchange summation and integration operations. When S is a paracompact manifold and the support of the form ω is compact, the number of the functions f_α involved is finite so these operations can always be performed. □

If ω is a closed form, i.e., if $d\omega = 0$, then we get $\int_{\partial S} \omega = 0$ on a manifold S with boundary ∂S. However, this condition does not imply in general that the form ω is exact, namely, there exists a form σ such that $\omega = d\sigma$ and $\int_{\partial S} d\sigma = 0$.

Example 8.6.2. We consider the form

$$\omega = \frac{-\,y\,dx + x\,dy}{x^2 + y^2} \in \Lambda^1(M)$$

defined on the manifold $M = \mathbb{R}^2 - (0,0)$. Let $D \subset M$ be a region bounded by a closed curve C containing the point $(0,0)$. We can immediately see that $d\omega = 0$. Furthermore, we can easily verify that one is able to write

$$\omega = d\theta, \quad \theta = \arctan\frac{y}{x}.$$

Hence, we find that

$$\int_C \omega = \int_C d\theta = 2\pi \neq 0.$$

It is clear that this result is not in essence in contradiction with Stokes' theorem because it is originated from the fact that the real boundary of the region is described by $C \cup \{(0,0)\}$. ■

We shall now attempt to obtain Stokes' theorem by following a completely different path. This approach will also prove to be rather advantageous from the standpoint of giving rise to new interpretations. We consider a k-dimensional submanifold S of an m-dimensional differentiable manifold M where $k \leq m$. Let us assume that this submanifold is specified by a smooth mapping $\phi : S \to M$. So in local coordinates, the submanifold S is defined by a parametrisation $x^i = \phi^i(u^\alpha), i = 1,\dots,m; \alpha = 1,\dots,k$. Let $U \subseteq S$ be a region with boundary. Its $(k-1)$-dimensional boundary ∂U

may be determined by a mapping $\psi : \partial U \to S$ or through functions $u^\alpha = \psi^\alpha(v^a)$, $a = 2, \ldots, k$. Since the dimension of the boundary ∂U is $k-1$, the rank of the matrix $[\partial u^\alpha / \partial v^a]$ must be $k-1$. This amounts to say that we can take $\det[\partial u^b / \partial v^a] \neq 0$ by changing the ordering of coordinates if necessary. We can thus write $v^a = \xi^a(u^b)$ and the manifold ∂U may be described by the equation $u^1 = \xi^1(u^2, \ldots, u^k)$. Next, we select the new local coordinates (w^1, w^a) for the manifold S by the expressions $w^a = u^a$, $w^1 = u^1 - \xi^1(u^2, \ldots, u^k)$. Hence, the boundary ∂U of the region U is determined by the condition $w^1 = 0$. In this case, the parameters $w^a, a = 2, \ldots, k$ constitute the local coordinates of ∂U. All the vectors $\partial / \partial w^a$ belong to $T(\partial U)$. Only the vector $\partial / \partial w^1$ is not in the tangent bundle of ∂U and lies in $T(S)$. We now introduce a vector field $V = \partial / \partial w^1$. This vector field creates a flow, that is, a one-parameter mapping $e^{tV} : S \to S$ on the submanifold S dragging the region U onto a region $U(t) \subset S$ (Fig. 8.6.2).

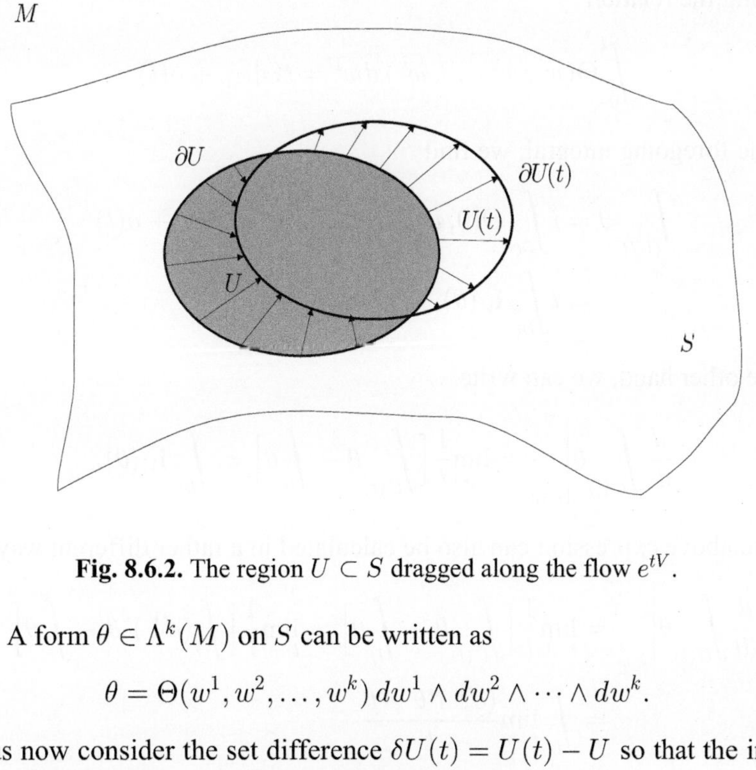

Fig. 8.6.2. The region $U \subset S$ dragged along the flow e^{tV}.

A form $\theta \in \Lambda^k(M)$ on S can be written as

$$\theta = \Theta(w^1, w^2, \ldots, w^k)\, dw^1 \wedge dw^2 \wedge \cdots \wedge dw^k.$$

Let us now consider the set difference $\delta U(t) = U(t) - U$ so that the integration of the form θ over which can be expressed as

$$\int_{\delta U(t)} \theta = \int_{U(t)} \theta - \int_U \theta$$

For a small parameter t, we can choose local coordinates in the vicinity of ∂U as $(t, w^2, \dots, w^k)$. We can then expand the function Θ into a Maclaurin series about $t = 0$ and write

$$\Theta(t, w^2, \dots, w^k) = \Theta|_{t=0} + \frac{\partial\Theta}{\partial t}\bigg|_{t=0} t + \cdots.$$

We can obviously write the following expression for small values of the parameter t

$$\begin{aligned}\int_{\delta U(t)} \theta &= \int_{\delta U(t)} \Theta(w^1, w^2, \dots, w^k)\, dw^1 dw^2 \cdots dw^k \\ &= \int_{\partial U} \Big(\int_0^t \Theta(w^1, w^2, \dots, w^k)\, dw^1\Big) dw^2 \cdots dw^k.\end{aligned}$$

Inserting the relation

$$\int_0^t \Theta(w^1, w^2, \dots, w^k)\, dw^1 = t\Theta|_{t=0} + o(t)$$

into the foregoing integral, we find

$$\begin{aligned}\int_{\delta U(t)} \theta &= t\int_{\partial U} \Theta(0, w^2, \dots, w^k)\, dw^2 \cdots dw^k + o(t) \\ &= t\int_{\partial U} \mathbf{i}_V(\theta) + o(t).\end{aligned}$$

On the other hand, we can write

$$\frac{d}{dt}\int_{U(t)} \theta\bigg|_{t=0} = \lim_{t\to 0}\frac{1}{t}\Big[\int_{U(t)} \theta - \int_U \theta\Big] = \int_{\partial U} \mathbf{i}_V(\theta).$$

But the above expression can also be calculated in a rather different way:

$$\begin{aligned}\frac{d}{dt}\int_{U(t)} \theta\bigg|_{t=0} &= \lim_{t\to 0}\frac{1}{t}\Big[\int_{U(t)} \theta - \int_U \theta\Big] = \lim_{t\to 0}\frac{1}{t}\Big[\int_U (e^{tV})^*\theta - \int_U \theta\Big] \\ &= \int_U \lim_{t\to 0}\frac{(e^{tV})^*\theta - \theta}{t} \\ &= \int_U £_V\theta.\end{aligned}$$

We are thus led to quite an interesting result given below

$$\int_U \pounds_V \theta = \int_{\partial U} \mathbf{i}_V(\theta).$$

However, we know that we can write $\pounds_V \theta = \mathbf{i}_V(d\theta) + d\mathbf{i}_V(\theta)$ in view of (5.11.5). $d\theta \in \Lambda^{k+1}(M)$ vanishes identically on the k-dimensional manifold S so that we get $\pounds_V \theta = d\mathbf{i}_V(\theta)$ on S and arrive at the result

$$\int_U d\mathbf{i}_V(\theta) = \int_{\partial U} \mathbf{i}_V(\theta). \tag{8.6.8}$$

Let us now write $\omega = \mathbf{i}_V(\theta) \in \Lambda^{k-1}(M)$. Since θ and to some extent V are arbitrary, we can take ω as an arbitrary $(k-1)$-form. Therefore, we derive again the Stokes theorem in its familiar form:

$$\int_U d\omega = \int_{\partial U} \omega.$$

Let us now take an m-dimensional *complete* Riemannian manifold M into account. If we denote the local coordinates by $(x^1, \dots, x^m)$, the elementary arc length on the manifold is given by

$$ds^2 = g_{ij}(\mathbf{x})\, dx^i dx^j$$

where we know that g_{ij} is a positive definite, covariant symmetric tensor. The volume form is prescribed by

$$\mu_m = \sqrt{g}\, dx^1 \wedge \cdots \wedge dx^m$$

where $g = \det[g_{ij}] > 0$. Let S be a k-dimensional submanifold of M. This submanifold is parametrically determined by relations $x^i = x^i(u^\alpha), \alpha = 1, \dots, k$. In this circumstance, the elementary arc length on the submanifold S can be introduced as

$$ds^2 = g_{ij}(\mathbf{x}(\mathbf{u})) \frac{\partial x^i}{\partial u^\alpha} \frac{\partial x^j}{\partial u^\beta} du^\alpha du^\beta = a_{\alpha\beta}(\mathbf{u})\, du^\alpha du^\beta > 0 \tag{8.6.9}$$

where the second order, covariant symmetric tensor

$$a_{\alpha\beta} = g_{ij} \frac{\partial x^i}{\partial u^\alpha} \frac{\partial x^j}{\partial u^\beta} \tag{8.6.10}$$

denotes the metric tensor $\boldsymbol{\mathcal{A}}$ on S induced by the metric tensor $\boldsymbol{\mathcal{G}}$ on M. It follows at once from (8.6.9) that $\boldsymbol{\mathcal{A}}$ is also positive definite. The volume form on S can now be defined as

$$\mu_k = \sqrt{a}\, du^1 \wedge \cdots \wedge du^k \tag{8.6.11}$$

where $a = \det[a_{\alpha\beta}] > 0$ due to the positive definiteness of $\mathcal{A}$.

Let us now write a form $\omega \in \Lambda^{m-k}(M)$ as

$$\omega = \frac{1}{k!}\,\omega^{i_1\cdots i_k}\mu_{i_k\cdots i_1}$$

where top down generated basis form $\mu_{i_k\cdots i_1}$ are given [*see* (5.9.17)] by

$$\mu_{i_k\cdots i_1} = \frac{1}{(m-k)!}\,\epsilon_{i_1\cdots i_k i_{k+1}\cdots i_m} dx^{i_{k+1}} \wedge \cdots \wedge dx^{i_m}$$

Let us now evaluate these forms on an $(m-k)$-dimensional submanifold S of M. On supposing that the submanifold S is specified by the parameters $(u^1, \ldots, u^{m-k})$, we get

$$\begin{aligned}
&\mu_{i_k\cdots i_1}\\
&= \frac{1}{(m-k)!}\,\epsilon_{i_1\cdots i_k i_{k+1}\cdots i_m}\frac{\partial x^{i_{k+1}}}{\partial u^{\alpha_1}}\cdots\frac{\partial x^{i_m}}{\partial u^{\alpha_{m-k}}} du^{\alpha_1} \wedge \cdots \wedge du^{\alpha_{m-k}}\\
&= \frac{1}{(m-k)!}\,\epsilon_{i_1\cdots i_k i_{k+1}\cdots i_m} e^{\alpha_1\cdots\alpha_{m-k}}\frac{\partial x^{i_{k+1}}}{\partial u^{\alpha_1}}\cdots\frac{\partial x^{i_m}}{\partial u^{\alpha_{m-k}}} du^1 \wedge \cdots \wedge du^{m-k}
\end{aligned}$$

where Greek indices take the values $1, 2, \ldots, m-k$. On employing the relation (8.6.11) for a volume form μ_k in the form μ_{m-k}, we can write $\mu_{m-k} = \sqrt{a}\, du^1 \wedge \cdots \wedge du^{m-k}$. If we introduce the Levi-Civita tensor

$$\epsilon^{\alpha_1\cdots\alpha_{m-k}} = \frac{e^{\alpha_1\cdots\alpha_{m-k}}}{\sqrt{a}},$$

we end up with the result

$$\mu_{i_k\cdots i_1} = \frac{1}{(m-k)!}\,\epsilon_{i_1\cdots i_k i_{k+1}\cdots i_m}\epsilon^{\alpha_1\cdots\alpha_{m-k}}\frac{\partial x^{i_{k+1}}}{\partial u^{\alpha_1}}\cdots\frac{\partial x^{i_m}}{\partial u^{\alpha_{m-k}}}\mu_{m-k}.$$

We now define a completely antisymmetric covariant tensor on S through the following components

$$n_{i_1\cdots i_k} = \frac{1}{(m-k)!}\,\epsilon_{i_1\cdots i_k i_{k+1}\cdots i_m}\epsilon^{\alpha_1\cdots\alpha_{m-k}}\frac{\partial x^{i_{k+1}}}{\partial u^{\alpha_1}}\cdots\frac{\partial x^{i_m}}{\partial u^{\alpha_{m-k}}}. \tag{8.6.12}$$

We then see that we can write

$$\mu_{i_k\cdots i_1} = n_{i_1\cdots i_k}\mu_{m-k}. \tag{8.6.13}$$

It follows from (8.6.12) that

$$n_{i_1\cdots i_k}\frac{\partial x^{i_1}}{\partial u^{\alpha_{m-k+1}}}\cdots\frac{\partial x^{i_k}}{\partial u^{\alpha_m}}=0.$$

This is true because the expression

$$\epsilon_{i_1\cdots i_k i_{k+1}\cdots i_m}\frac{\partial x^{i_1}}{\partial u^{\alpha_{m-k+1}}}\cdots\frac{\partial x^{i_k}}{\partial u^{\alpha_m}}\frac{\partial x^{i_{k+1}}}{\partial u^{\alpha_1}}\cdots\frac{\partial x^{i_m}}{\partial u^{\alpha_{m-k}}}$$

is completely antisymmetric with respect to indices $i_1,\dots,i_m$ and this entails that it becomes also completely antisymmetric with respect to indices $\alpha_1,\ \dots,\alpha_{m-k}$. However, this latter indices take on only $m-k$ different values. Therefore, it is not possible to avoid getting repeated indices in the set $i_1,\dots,i_m$ taking m different values.

The exterior derivative of the form ω is [*see* (5.9.19)]

$$\begin{aligned} d\omega &= \frac{1}{(k-1)!}\,\omega^{i_1\cdots i_{k-1}i}{}_{;i}\,\mu_{i_{k-1}\cdots i_1} \\ &= \frac{1}{(k-1)!}\,\omega^{i_1\cdots i_{k-1}i}{}_{;i}\,n_{i_1\cdots i_{k-1}}\mu_{m-k+1}. \end{aligned}$$

Similarly, we can find

$$\omega=\frac{1}{k!}\,\omega^{i_1\cdots i_k}n_{i_1\cdots i_k}\mu_{m-k}.$$

In view of (8.6.11), we can introduce the volume element on S as

$$dV_k=\sqrt{a}\,du^1\cdots du^k$$

Let us now consider a form $\omega\in\Lambda^{m-k}(M)$ defined on a region U_{m-k+1} on an $(m-k+1)$-dimensional submanifold S whose boundary is given by the manifold ∂U_{m-k}. Application of the Stokes theorem by using parameters peculiar to those submanifolds yields

$$\begin{aligned}\int_{U_{m-k+1}}\omega^{i_1\cdots i_{k-1}i}{}_{;i}\,n_{i_1\cdots i_{k-1}}dV_{m-k+1} &= \\ \frac{1}{k}\int_{\partial U_{m-k}}&\omega^{i_1\cdots i_k}n_{i_1\cdots i_k}dV_{m-k}.\end{aligned} \tag{8.6.14}$$

An important special case that may be deduced from the above relation corresponds to $k=1$. In this case, one has

$$\omega=\omega^i\mu_i\ \text{ and }\ d\omega=\omega^i{}_{;i}\,\mu$$

so we obtain

$$\int_{U_m} \omega^i{}_{;i}\, dV_m = \int_{\partial U_{m-1}} \omega^i n_i\, dV_{m-1} \tag{8.6.15}$$

where the components $n_i, i = 1, \dots, m$ are defined by

$$n_i = \frac{1}{(m-1)!}\, \epsilon_{ii_1\cdots i_{m-1}} \epsilon^{\alpha_1\cdots\alpha_{m-1}} \frac{\partial x^{i_1}}{\partial u^{\alpha_1}} \cdots \frac{\partial x^{i_{m-1}}}{\partial u^{\alpha_{m-1}}}. \tag{8.6.16}$$

It is clearly seen that the relations

$$n_i \frac{\partial x^i}{\partial u^\alpha} = 0, \ \ \alpha = 1, \dots, m-1$$

will be satisfied. The quantities

$$\frac{\partial x^i}{\partial u^\alpha}, \ \ \alpha = 1, \dots, m-1$$

are contravariant components of $m-1$ vectors in $T(\partial U_{m-1})$. The vector whose covariant components are n_i in $T(M)$ is orthogonal to all those vectors. Hence, it is called the ***exterior normal*** **n** to the boundary ∂U_{m-1} (if U_m is positively oriented). We shall now show that **n** is a unit vector. We first evaluate the expression

$$\begin{aligned}
n_i n^i &= \frac{1}{\big[(m-1)!\big]^2}\, \epsilon_{ii_1\cdots i_{m-1}} \epsilon^i_{j_1\cdots j_{m-1}} \epsilon^{\alpha_1\cdots\alpha_{m-1}} \epsilon^{\beta_1\cdots\beta_{m-1}} \\
&\quad \times \frac{\partial x^{i_1}}{\partial u^{\alpha_1}} \cdots \frac{\partial x^{i_{m-1}}}{\partial u^{\alpha_{m-1}}} \frac{\partial x^{j_1}}{\partial u^{\beta_1}} \cdots \frac{\partial x^{j_{m-1}}}{\partial u^{\beta_{m-1}}} \\
&= \frac{1}{\big[(m-1)!\big]^2}\, g_{j_1k_1} \cdots g_{j_{m-1}k_{m-1}}\, \epsilon_{ii_1\cdots i_{m-1}} \epsilon^{ik_1\cdots k_{m-1}} \\
&\quad \times \epsilon^{\alpha_1\cdots\alpha_{m-1}} \epsilon^{\beta_1\cdots\beta_{m-1}} \frac{\partial x^{i_1}}{\partial u^{\alpha_1}} \cdots \frac{\partial x^{i_{m-1}}}{\partial u^{\alpha_{m-1}}} \frac{\partial x^{j_1}}{\partial u^{\beta_1}} \cdots \frac{\partial x^{j_{m-1}}}{\partial u^{\beta_{m-1}}} \\
&= \frac{1}{\big[(m-1)!\big]^2}\, g_{j_1k_1} \cdots g_{j_{m-1}k_{m-1}}\, \delta^{k_1\cdots k_{m-1}}_{i_1\cdots i_{m-1}} \epsilon^{\alpha_1\cdots\alpha_{m-1}} \epsilon^{\beta_1\cdots\beta_{m-1}} \\
&\quad \times \frac{\partial x^{i_1}}{\partial u^{\alpha_1}} \cdots \frac{\partial x^{i_{m-1}}}{\partial u^{\alpha_{m-1}}} \frac{\partial x^{j_1}}{\partial u^{\beta_1}} \cdots \frac{\partial x^{j_{m-1}}}{\partial u^{\beta_{m-1}}}
\end{aligned}$$

where we have utilised the relations (5.5.7) and (5.5.5). The last line above is completely antisymmetric with respect to indices $\alpha_1, \cdots, \alpha_{m-1}$. Thus it also becomes completely antisymmetric with respect to indices $i_1, \cdots, i_{m-1}$. Hence, according to (1.4.8) we conclude that

$$n_i n^i = \frac{1}{(m-1)!}\,\epsilon^{\alpha_1\cdots\alpha_{m-1}}\epsilon^{\beta_1\cdots\beta_{m-1}} g_{j_1k_1}\cdots g_{j_{m-1}k_{m-1}}\frac{\partial x^{j_1}}{\partial u^{\beta_1}}\frac{\partial x^{k_1}}{\partial u^{\alpha_1}}$$
$$\cdots\frac{\partial x^{j_{m-1}}}{\partial u^{\beta_{m-1}}}\frac{\partial x^{k_{m-1}}}{\partial u^{\alpha_{m-1}}}$$
$$= \frac{1}{(m-1)!}\,\epsilon^{\alpha_1\cdots\alpha_{m-1}}\epsilon^{\beta_1\cdots\beta_{m-1}} a_{\beta_1\alpha_1}\cdots a_{\beta_{m-1}\alpha_{m-1}}.$$

On the other hand, the relation

$$a = \det[a_{\alpha\beta}] = \frac{1}{(m-1)!}\,e^{\alpha_1\cdots\alpha_{m-1}}e^{\beta_1\cdots\beta_{m-1}}a_{\beta_1\alpha_1}\cdots a_{\beta_{m-1}\alpha_{m-1}}$$

yields $n_i n^i = 1$. In a similar fashion, it is a simple exercise to demonstrate the validity of the relation

$$n_{i_1\cdots i_k}n^{i_1\cdots i_k} = k!.$$

The relation (8.6.15) is called the ***Green-Gauss-Ostrogradski*** or ***divergence formula*** generalised to an m-dimensional manifold [after English, German and Russian mathematicians, respectively, George Green (1793-1841), Johann Carl Friedrich Gauss (1777-1855) and Mikhail Vasilevich Ostrogradski (1801-1862)].

Example 8.6.3. We consider a bounded region $U \subset \mathbb{R}^3$ and the 2-form $\omega = X dy \wedge dz + Y dz \wedge dx + Z dx \wedge dy$. In view of (8.6.13), we can write

$$\mu_x = dy \wedge dz = n_x\mu_2,\ \ \mu_y = dz \wedge dx = n_y\mu_2,\ \ \mu_z = dx \wedge dy = n_z\mu_2$$

on ∂U. The components of the unit exterior normal vector ∂U to the closed surface ∂U are given by

$$n_i = e_{ijk}e^{\alpha\beta}\frac{\partial x^j}{\partial u^\alpha}\frac{\partial x^k}{\partial u^\beta},$$

where we denote $x^1 = x, x^2 = y, x^3 = z$. On the region U, we get

$$d\omega = \Big(\frac{\partial X}{\partial x} + \frac{\partial Y}{\partial y} + \frac{\partial Z}{\partial z}\Big)dx \wedge dy \wedge dz.$$

Hence, the Stokes theorem takes the form

$$\int_U \Big(\frac{\partial X}{\partial x} + \frac{\partial Y}{\partial y} + \frac{\partial Z}{\partial z}\Big)dv = \int_{\partial U}(Xn_x + Yn_y + Zn_z)da.$$

Let us introduce vectors $\mathbf{F} = (X, Y, Z)$ and $\mathbf{n} = (n_x, n_y, n_z)$. Then the Stokes theorem leads to the quite familiar formula

$$\int_U \operatorname{div} \mathbf{F}\, dv = \int_U \boldsymbol{\nabla} \cdot \mathbf{F}\, dv = \int_{\partial U} \mathbf{F} \cdot \mathbf{n}\, da. \qquad \blacksquare$$

Example 8.6.4. Let U be a 2-dimensional submanifold, in other words a surface, in $\mathbb{R}^3$ and $\partial U = C$ be the closed curve that supports this surface. We consider a 1-form $\omega = X dx + Y dy + Z dz$. We denote by s the arc length of the curve C. Then the form ω is expressible on C as

$$\omega = \Big(X\frac{dx}{ds} + Y\frac{dy}{ds} + Z\frac{dz}{ds}\Big)\, ds = \mathbf{F} \cdot \mathbf{t}\, ds$$

on C where $\mathbf{t}$ is the unit tangent vector of C. On the other hand, one has

$$d\omega = \Big(\frac{\partial Z}{\partial y} - \frac{\partial Y}{\partial z}\Big)\, dy \wedge dz + \Big(\frac{\partial X}{\partial z} - \frac{\partial Z}{\partial x}\Big)\, dz \wedge dx + \Big(\frac{\partial Y}{\partial x} - \frac{\partial X}{\partial y}\Big)\, dx \wedge dy.$$

Hence, the exterior derivative of the form ω can be written as follows

$$d\omega = \mathbf{n} \cdot \operatorname{curl} \mathbf{F}\, \mu_2.$$

Therefore, the Stokes theorem associated with exterior forms leads to the familiar expression

$$\int_U \mathbf{n} \cdot \operatorname{curl} \mathbf{F}\, da = \int_U \mathbf{n} \cdot (\boldsymbol{\nabla} \times \mathbf{F})\, da = \int_C \mathbf{F} \cdot \mathbf{t}\, ds$$

known as the ***Kelvin-Stokes formula*** in the classical vector analysis. ■

Through the Stokes theorem, we can generalise a relation known as the integration by parts in the classical analysis. We take two forms $\omega \in \Lambda^k(M)$ and $\sigma \in \Lambda^l(M)$ into consideration. Let U be a region on a submanifold of M with dimension $k + l + 1 \le m$. It follows from the exterior derivative of the form $\omega \wedge \sigma$ that

$$\int_U d\omega \wedge \sigma = \int_{\partial U} \omega \wedge \sigma - (-1)^k \int_U \omega \wedge d\sigma. \tag{8.6.17}$$

8.7. CONSERVATION LAWS

Let $\mathcal{I}(\omega^a), a = 1, \ldots, A$ be an ideal of the exterior algebra $\Lambda(M)$. We know that if the mapping $\phi : S \to M$ satisfy the condition $\phi^*\omega = 0$ for all $\omega \in \mathcal{I}$, then it is a solution of this ideal. Here the solution hypersurface S is a submanifold with dimension, say, $k \le m$. We shall now try to determine *non-zero exact k-forms* in the ideal $\mathcal{I}(\omega^a)$ annihilated by the solution submanifold. To this end, we consider a form $\omega \in \Lambda^k(M)$ in the ideal and look for a form $\Omega \in \Lambda^{k-1}(M)$ such that $\omega = d\Omega$. Since $\omega \in \mathcal{I}$, we can write

$$0 = \phi^*\omega = \phi^* d\Omega = d\phi^*\Omega.$$

Let $U_k \subseteq S$ be a smooth k-dimensional region and ∂U_k be its boundary. It follows from the Stokes theorem that

$$\int_{\partial U_k} \phi^*\Omega = \int_{U_k} d\phi^*\Omega = 0.$$

Consequently, the form Ω must satisfy the relation

$$\int_{\partial U_k} \phi^*\Omega = 0 \tag{8.7.1}$$

on every $U_k \subseteq S$ with boundary. (8.7.1) is called a ***conservation law*** in the integral form. Let us now suppose that the mapping ϕ is parametrically prescribed by the relations $x^i = \phi^i(u^1, \ldots, u^k), i = 1, \ldots, m$. We take the volume form on S as $\mu = du^1 \wedge \cdots \wedge du^k$ and define basis $(k-1)$-forms $\mu_\alpha = \mathbf{i}_{\partial_\alpha}\mu, \alpha = 1, \ldots, k$ in $\Lambda^{k-1}(S)$. Since the form Ω will eventually be pulled back on the submanifold S we can choose

$$\Omega = \Omega^\alpha(\mathbf{x})\mu_\alpha(\mathbf{u}) \text{ and } \phi^*\Omega = \Omega^\alpha(\mathbf{x}(\mathbf{u}))\mu_\alpha(\mathbf{u})$$

without loss of generality. Accordingly, in order that a form $\omega \in \mathcal{I}$ is to be exact, we have to find suitable forms $\gamma_a \in \Lambda(M)$ so that Ω satisfies

$$\omega = \gamma_a \wedge \omega^a = d\Omega = d\Omega^\alpha(\mathbf{x}) \wedge \mu_\alpha(\mathbf{u}). \tag{8.7.2}$$

On the other hand, the relation

$$d\phi^*\Omega = d\Omega^\alpha(\mathbf{x}(\mathbf{u})) \wedge \mu_\alpha = \frac{\partial\Omega^\alpha}{\partial x^i}\frac{\partial x^i}{\partial u^\beta}\, du^\beta \wedge \mu_\alpha = \frac{\partial\Omega^\alpha}{\partial x^i}\frac{\partial x^i}{\partial u^\alpha}\mu = 0$$

implies that the functions Ω^α ought to satisfy the divergence equation

$$\frac{\partial\Omega^\alpha}{\partial x^i}\frac{\partial x^i}{\partial u^\alpha} = \frac{\partial\phi^*\Omega^\alpha(\mathbf{u})}{\partial u^\alpha} = 0. \tag{8.7.3}$$

Example 8.7.1. The coordinate cover in the manifold $M = \mathbb{R}^3$ is given by $\mathbf{x} = (x, v, t)$. We consider the ideal $\mathcal{I}$ generated by the following forms:

$$\omega^1 = dv + n^2 x\, dt,$$
$$\omega^2 = dx - v\, dt.$$

On a 1-dimensional solution submanifold prescribed by the mapping $x = x(t), v = v(t)$, these forms will have to satisfy

$$\phi^*\omega^1 = \Big(\frac{dv}{dt} + n^2 x\Big)dt = 0, \ \phi^*\omega^2 = \Big(\frac{dx}{dt} - v\Big)dt = 0.$$

Hence, the solution submanifold is determined through the differential equation below associated with 1-dimensional oscillating systems

$$\frac{d^2x}{dt^2} + n^2 x = 0.$$

Since the solution submanifold is 1-dimensional, we have to look naturally for exact 1-forms. Let $\Omega \in \Lambda^0(M)$. Due to (8.7.3), we find that

$$\frac{d\phi^*\Omega}{dt} = \frac{\partial\Omega}{\partial x}\frac{dx}{dt} + \frac{\partial\Omega}{\partial v}\frac{dv}{dt} + \frac{\partial\Omega}{\partial t} = 0.$$

Thus, we must have $\phi^*\Omega = constant.$

On the other hand, the condition (8.7.2) takes the form

$$\gamma_1(dv + n^2 x\, dt) + \gamma_2(dx - v\, dt) = d\Omega = \Omega_x\, dx + \Omega_v\, dv + \Omega_t\, dt$$

where $\gamma_1, \gamma_2 \in \Lambda^0(M)$. The subscripts indicate the variables with respect to which partial derivatives will be evaluated. We thereby obtain

$$\gamma_1 = \Omega_v, \ \gamma_2 = \Omega_x, \ n^2 x\gamma_1 - v\gamma_2 = \Omega_t$$

or

$$-\, n^2 x\Omega_v + v\Omega_x + \Omega_t = 0.$$

It is obvious that this equation corresponds to the relation $d\phi^*\Omega/dt = 0$ on the solution submanifold. In order to solve the foregoing partial differential equation, we can employ the method of characteristics. To this end, we have to solve the following system of ordinary differential equations

$$\frac{dv}{n^2 x} = -\,\frac{dx}{v} = -\,dt.$$

It is a simple exercise to see that $\Omega = \Omega(\xi, \eta)$ where

$$\xi = \frac{1}{2}(v^2 + n^2 x^2), \quad \eta = t - \int\frac{dx}{v}.$$

Thus, independent conservation laws become $\xi = constant,\ \eta = constant.$ In this case, every function $\Omega(\xi, \eta)$ remains constant on the solution submanifold. If we take $U_1 = [t_1, t_2]$, then we find on ∂U_1 the known result

$$\frac{1}{2}(v^2 + n^2x^2)\Big|_{t_1}^{t_2} = 0, \quad t_2 - t_1 - \int_{x_1}^{x_2} \frac{dx}{v} = 0. \qquad \blacksquare$$

Example 8.7.2. Let $\mathbf{x} = (\theta, u, v, x, t)$ be a coordinate cover on the manifold $M = \mathbb{R}^5$. We consider the ideal $\mathcal{I}$ generated by the following forms

$$\omega^1 = du \wedge dt - v\, dx \wedge dt, \; \omega^2 = d\theta - u\, dx - v\, dt.$$

On a 2-dimensional solution submanifold prescribed by the mapping $\theta = \theta(x,t), u = u(x,t)$ and $v = v(x,t)$, we have to satisfy the conditions below

$$\begin{aligned}\phi^*\omega^1 &= (u_x - v)\, dx \wedge dt = 0,\\ \phi^*\omega^2 &= (\theta_x - u)\, dx + (\theta_t - v)\, dt = 0\end{aligned}$$

whence we deduce that

$$v = u_x, \; u = \theta_x, \; v = \theta_t \text{ or } \theta_t = \theta_{xx}.$$

The last equation describes a one-dimensional heat conduction,. More generally it models a diffusion process. Let us take the volume form in a 2-dimensional solution submanifold as $\mu = dx \wedge dt$. We then get $\mu_1 = dt$, $\mu_2 = -dx$. Hence, we have to look for a form in the following shape

$$\Omega = \Phi\, dt - \Psi\, dx \in \Lambda^1(M)$$

where we have defined $\Omega^1 = \Phi \in \Lambda^0(M)$, $\Omega^2 = \Psi \in \Lambda^0(M)$. The condition (8.7.2) then yields

$$\gamma_1\omega^1 + \gamma_2 \wedge \omega^2 = d\Omega = d\Phi \wedge dt - d\Psi \wedge dx$$

where $\gamma_1 \in \Lambda^0(M), \gamma_2 \in \Lambda^1(M)$. If we express γ_2 as

$$\gamma_2 = a\, d\theta + b\, du + c\, dv + e\, dx + f dt$$

where $a, b, c, e, f \in \Lambda^0(M)$, then the above relation is transformed into

$$\begin{aligned}&(\gamma_1 - bv - \Phi_u)\, du \wedge dt - (\gamma_1 v - fu + ev + \Phi_x + \Psi_t)\, dx \wedge dt + b\, du \wedge d\theta\\ &+ c\, dv \wedge d\theta - (e + au - \Psi_\theta)\, d\theta \wedge dx - (f + av + \Phi_\theta)\, d\theta \wedge dt\\ &- (bu - \Psi_u)\, du \wedge dx - (cu - \Psi_v)\, dv \wedge dx - (cv + \Phi_v)\, dv \wedge dt = 0\end{aligned}$$

whence we extract the relations

$$\begin{aligned}&\gamma_1 - bv = \Phi_u, \quad -\gamma_1 v + fu - ev = \Phi_x + \Psi_t, \quad b = 0, \quad c = 0,\\ &e + au = \Psi_\theta, \; f + av = -\Phi_\theta, \; bu = \Psi_u, \; cu = \Psi_v, \; cv = -\Phi_v.\end{aligned}$$

by equating the coefficients of the linearly independent 2-forms to zero. Hence, we end up with the relations

$$\gamma_1 = \Phi_u,\ e = \Psi_\theta - au,\ f = -\Phi_\theta - av,\ \Psi_u = 0,\ \Psi_v = 0,\ \Phi_v = 0$$
$$\Phi_x + \Psi_t + v(\Phi_u + \Psi_\theta) + u\Phi_\theta = 0$$

implying first that we must have $\Psi = \Psi(\theta, x, t)$, $\Phi = \Phi(\theta, u, x, t)$. Since Φ and Ψ are independent of v, it is required that the coefficient of v in the last equation above must vanish yielding $\Phi_u = -\Psi_\theta$. On noting that Ψ does not depend on u, this expression is easily integrated to give

$$\Phi = -\Psi_\theta u + \phi(\theta, x, t)$$

where ϕ is an arbitrary function. Thus, the expression $\Phi_x + \Psi_t + u\Phi_\theta = 0$ yields the equation

$$\phi_x + \Psi_t + u(\phi_\theta - \Psi_{\theta x}) - u^2\Psi_{\theta\theta} = 0.$$

However, this equation is satisfied if only

$$\Psi_{\theta\theta} = 0,\quad \phi_\theta = \Psi_{\theta x},\quad \Psi_t + \phi_x = 0.$$

The first two equations give

$$\Psi = \alpha(x,t)\,\theta + \beta(x,t), \phi = \Psi_x + \varphi(x,t) \text{ and } \phi = \alpha_x\theta + \beta_x + \varphi.$$

As to the last equation, it yields

$$(\alpha_{xx} + \alpha_t)\theta + \beta_{xx} + \varphi_x + \beta_t = 0.$$

Therefore, the functions Φ and Ψ are finally given by

$$\Phi = -\alpha u + \alpha_x\theta + \beta_x + \varphi,\quad \Psi = \alpha\theta + \beta$$

provided that the functions α, β and φ are to satisfy the equations

$$\alpha_{xx} + \alpha_t = 0,\quad \beta_{xx} + \beta_t + \varphi_x = 0.$$

Thus the conservation law takes the form

$$\frac{\partial}{\partial x}(-\alpha\theta_x + \alpha_x\theta + \beta_x + \varphi) + \frac{\partial}{\partial t}(\alpha\theta + \beta) = 0$$

and we arrive at the integral relation

$$\int_C (\Phi\,dt - \Psi\,dx) = 0$$

on every closed curve C in the (x,t)-plane. ■

Example 8.7.3. The coordinate cover of the manifold $M = \mathbb{R}^4$ is given by $\mathbf{x} = (x, t, u, c)$. Let us consider the ideal $\mathcal{I}$ generated by the following 2-forms

$$\begin{aligned}\omega^1 &= -\,du \wedge dx + u\,du \wedge dt + \alpha c\,dc \wedge dt,\\ \omega^2 &= -\,dc \wedge dx + u\,dc \wedge dt + \frac{1}{\alpha}\,c\,du \wedge dt.\end{aligned}$$

On a 2-dimensional solution submanifold prescribed by the mapping $u = u(x, t)$, $c = c(x, t)$, the following relations must hold

$$\begin{aligned}\phi^*\omega^1 &= (u_t + uu_x + \alpha cc_x)\,dx \wedge dt = 0,\\ \phi^*\omega^2 &= (c_t + uc_x + \frac{1}{\alpha}\,cu_x)\,dx \wedge dt = 0.\end{aligned}$$

Subscripts denote partial derivatives with respect to relevant variables. They of course give rise to partial differential equations

$$\begin{aligned}u_t + uu_x + \alpha cc_x &= 0,\\ c_t + uc_x + \frac{1}{\alpha}\,cu_x &= 0\end{aligned}$$

to determine the functions $u(x, t)$ and $c(x, t)$ prescribing the solution manifold. These equations are modelling the *one-dimensional isentropic gas flow* for the choice

$$\alpha = \frac{2}{\gamma - 1}$$

and the *shallow water theory* in hydrodynamics describing the propagation gravity waves on the free surface of an incompressible fluid of infinite extent in x-direction on a horizontal flat bottom for the choice $\alpha = 2$. In isentropic gas flow, γ denotes the ratio of specific heats of the gas under constant pressure and constant volume. u is the velocity of the gas while c denotes the local sound speed. In the shallow water theory, u is the velocity of the fluid and $c = \sqrt{gh}$ where h is the elevation of the water surface during the propagation of the gravity wave from the horizontal bottom. g denotes the well known gravitational acceleration.

We shall now attempt to find conservation laws by taking into account a form $\Omega = \Phi(\mathbf{x})dt - \Psi(\mathbf{x})dx \in \Lambda^1(M)$. In order that $d\Omega$ is to be in the ideal $\mathcal{I}(\omega^1, \omega^2)$, we have to write

$$\gamma_1\omega^1 + \gamma_2\omega^2 = d\Omega = d\Phi \wedge dt - d\Psi \wedge dx$$

where $\gamma_1, \gamma_2 \in \Lambda^0(M)$. Therefore, the relation

$$
\begin{aligned}
&-\gamma_1 du\wedge dx-\gamma_2\,dc\wedge dx+\Big(u\gamma_1+\frac{c}{\alpha}\gamma_2\Big)du\wedge dt+(\alpha c\gamma_1+u\gamma_2)dc\wedge dt\\
&=(\Phi_x+\Psi_t)dx\wedge dt+\Phi_u du\wedge dt+\Phi_c dc\wedge dt-\Psi_u du\wedge dx-\Psi_c dc\wedge dx
\end{aligned}
$$

leads to the partial differential equations

$$
\begin{aligned}
&\Phi_x+\Psi_t=0, \qquad \gamma_1=\Psi_u, \qquad \gamma_2=\Psi_c\\
&u\Psi_u+\frac{c}{\alpha}\Psi_c=\Phi_u, \quad \alpha c\Psi_u+u\Psi_c=\Phi_c.
\end{aligned}
$$

to determine the functions Φ and Ψ. On the other hand, the symmetry relation $\Phi_{uc}=\Phi_{cu}$ leads from the last two equations above to the second order linear partial differential equation for the function Ψ

$$
\alpha^2\Psi_{uu}-\Psi_{cc}+\frac{\alpha-1}{c}\Psi_c=0.
$$

On the other hand, we can find from the above relations

$$
\Psi_u=\frac{\alpha u\Phi_u-c\Phi_c}{\alpha(u^2-c^2)}, \quad \Psi_c=\frac{u\Phi_c-\alpha c\Phi_u}{u^2-c^2}
$$

from which we obtain

$$
\alpha^2\Phi_{uu}-\Phi_{cc}+\frac{(\alpha-1)(u^2+c^2)}{c(u^2-c^2)}\Phi_c-\frac{2\alpha(\alpha-1)}{c(u^2-c^2)}\Phi_u=0.
$$

Of course the solution functions $\Phi(x,t,u,c)$ and $\Psi(x,t,u,c)$ are interrelated through the relations above. We anticipate that our field equations may possess infinitely many conservation laws since they are originated from solutions of partial differential equations. Indeed, certain particular solutions of those equations justify this expectation[1]. It can be shown that a polynomial type of conservation laws that are independent of x,t can be found as

$$
\Psi_n=c^nC_n^{(\frac{\alpha}{2}-n)}\Big(\frac{u}{\alpha c}\Big),\ \Phi_n=c^{n+1}\Big[\frac{u}{c}C_n^{(\frac{\alpha}{2}-n)}\Big(\frac{u}{\alpha c}\Big)-C_{n+1}^{(\frac{\alpha}{2}-n)}\Big(\frac{u}{\alpha c}\Big)\Big]
$$

where $n=1,2,\dots$ and $n\neq\alpha/2$. $C_n^{(\lambda)}$ denotes a Gegenbauer polynomial [after German mathematician Leopold Bernhard Gegenbauer (1849-1903)]. These sequence of orthogonal polynomials is found from a generating function through the expansion

[1]For a detailed analysis one may consult to Şuhubi, E. S., Conservation laws for one-dimensional isentropic gas flows, *International Journal of Engineering Science*, **22**, 119-126, 1984.

$$\frac{1}{(1-2xt+t^2)^\lambda}=\sum_{n=0}^{\infty}C_n^{(\lambda)}(x)t^n.$$

They can also be obtained by the following recurrence relation

$$C_0^{(\lambda)}(x)=1,\ C_1^{(\lambda)}(x)=2\lambda x$$
$$C_n^{(\lambda)}(x)=\frac{1}{n}\big[2x(n+\lambda-1)C_{n-1}^{(\lambda)}(x)-(n+2\lambda-2)C_{n-2}^{(\lambda)}(x)\big].$$

We confine ourselves here in giving only a few samples of this infinite set:

$$\Psi_1=\frac{\alpha-2}{\alpha}u,\Phi_1=\frac{\alpha-2}{2}\Big(\frac{1}{\alpha}u^2+c^2\Big),$$
$$\Psi_2=-\frac{\alpha-4}{2}\Big(c^2-\frac{\alpha-2}{\alpha^2}u^2\Big),\Phi_2=-\frac{\alpha-4}{\alpha}u\Big(c^2-\frac{\alpha-2}{3\alpha}u^2\Big),$$
$$\Psi_3=\frac{(\alpha-4)(\alpha-6)}{6}u\Big(\frac{\alpha-2}{\alpha^3}u^2-\frac{3}{\alpha}c^2\Big),$$
$$\Phi_3=-\frac{(\alpha-4)(\alpha-6)}{4}\Big(\frac{1}{2}c^4+\frac{\alpha+2}{\alpha^2}u^2c^2-\frac{\alpha-2}{2\alpha^3}u^4\Big),$$
$$\Psi_4=\frac{(\alpha-6)(\alpha-8)}{24}u\Big(\frac{(\alpha-2)(\alpha-4)}{\alpha^4}u^4-6\frac{\alpha-4}{\alpha^2}u^2c^2+3c^4\Big),$$
$$\Phi_4=\frac{(\alpha-6)(\alpha-8)}{2\alpha}u\Big(c^4-\frac{(\alpha+1)(\alpha-4)}{3\alpha^2}u^2c^2+\frac{(\alpha-2)(\alpha-4)}{15\alpha^3}u^4\Big),$$
$$\vdots$$

■

Example 8.7.4. As a final example, we shall try to establish conservation laws associated with the field equations of a hyperelastic body in motion occupying an open region $\Omega\subseteq\mathbb{R}^3$ initially[2]. To facilitate our investigation we employ Cartesian coordinates. The position of a material particle before deformation will be determined by *material coordinates* $X_K, K=1,2,3$ whereas the place of the same particle in $\mathbb{R}^3$ at time t will be denoted by the *spatial coordinates* $x_k, k=1,2,3$. The motion of this continuous medium is determined by the diffeomorphism $x_k=x_k(\mathbf{X},t)$ with parameter t. A homogeneous hyperelastic material is characterised by the *stress potential* $\Sigma(\mathbf{C})$ in which $\mathbf{C}=\mathbf{F}^{\mathrm{T}}\mathbf{F}$ is the deformation tensor where $\mathbf{F}=[x_{k,K}]$ is the deformation gradient tensor, or matrix, whose components are denoted

[2] Şuhubi, E. S., Conservation laws in nonlinear elastodynamics, *International Journal of Engineering Science*, **27**, 441-453, 1989.

by $F_{kK} = \dfrac{\partial x_k}{\partial X_K} = x_{k,K}$. The equations of motion of the body are

$$\frac{\partial T_{Kk}}{\partial X_K} + \rho_0 f_k = \rho_0 \frac{\partial v_k}{\partial t} \tag{8.7.4}$$

where T_{Kk} is the Piola-Kirchhoff stress tensor of the first kind [after Italian mathematician and physicist Gabrio Piola (1794-1850) and German mathematician and physicist Gustav Robert Kirchhoff (1824-1887)] and $v_k = \dfrac{\partial x_k}{\partial t}$ are the components of the velocity vector of a particle. ρ_0 is the constant density of the undeformed medium and $f_k = f_k(\mathbf{X}, t)$ represents the components of the given body force density. Constitutive equations characterising the elastic behaviour of the medium are of the form

$$\begin{aligned} T_{Kk} &= \frac{\partial \Sigma}{\partial F_{kK}}, \\ T_{Kk} F_{lK} &= T_{Kl} F_{kK}. \end{aligned} \tag{8.7.5}$$

The relations $(8.7.5)_2$ arise from the symmetry of the Cauchy stress tensor. Therefore, equations of motion may be reduced to the following system of first order partial differential equations

$$\begin{aligned} C_{kKlL} \frac{\partial F_{lL}}{\partial X_K} - \rho_0 \frac{\partial v_k}{\partial t} + \rho_0 f_k &= 0, \\ \frac{\partial F_{kK}}{\partial X_L} - \frac{\partial F_{kL}}{\partial X_K} &= 0, \\ \frac{\partial F_{kK}}{\partial t} - \frac{\partial v_k}{\partial X_K} &= 0 \end{aligned}$$

where the coefficients

$$C_{kKlL}(\mathbf{F}) = \frac{\partial T_{Kk}}{\partial F_{lL}} = \frac{\partial^2 \Sigma}{\partial F_{kK} \partial F_{lL}} = C_{lLkK}(\mathbf{F}) \tag{8.7.6}$$

are called the *elasticities* of the medium. Let us now consider the 19-dimensional manifold K with a coordinate cover $(X_K, t, x_k, v_k, F_{kK})$. We first introduce the 3- and 2-forms below

$$\begin{aligned} \mu &= dX_1 \wedge dX_2 \wedge dX_3 = \frac{1}{3!} e_{KLM}\, dX_K \wedge dX_L \wedge dX_M, \\ \mu_K &= \mathbf{i}_{\partial_K} \mu = \frac{1}{2} e_{KLM}\, dX_L \wedge dX_M. \end{aligned}$$

We then define the following 4-forms:

$$\begin{aligned}
\omega_k &= \rho_0\, dv_k \wedge \mu + C_{kKlL} dF_{lL} \wedge \mu_K \wedge dt + \rho_0 f_k\, \mu \wedge dt,\\
\omega_{kK} &= dF_{kK} \wedge \mu + dv_k \wedge \mu_K \wedge dt,\\
\pi_{kK} &= e_{KLM}\, dF_{kL} \wedge \mu_M \wedge dt,\\
\sigma_k &= dx_k \wedge \mu + v_k\, \mu \wedge dt,\\
\sigma_{kK} &= dx_k \wedge \mu_K \wedge dt - F_{kK}\, \mu \wedge dt.
\end{aligned}$$

We can readily verify that

$$\begin{aligned}
&d\omega_k = 0, \quad d\omega_{kK} = 0, \quad d\pi_{kK} = 0,\\
&d\sigma_k = \rho_0^{-1} dt \wedge \omega_k, \quad d\sigma_{kK} = -\, dt \wedge \omega_{kK}
\end{aligned}$$

These relations mean that the ideal $\mathcal{I}$ generated by these 4-forms is closed. Let the submanifold with the coordinate cover (X_K, t) be M. We can easily check that the mapping $\phi : M \to K$ annihilating these forms, and consequently the ideal $\mathcal{I}$, provides the solution of the differential field equations. In fact, we find that

$$\begin{aligned}
\phi^*\omega_k &= \rho_0 \frac{\partial v_k}{\partial t} dt \wedge \mu + C_{kKlL} F_{lL,M} dX_M \wedge \mu_K \wedge dt + \rho_0 f_k\, \mu \wedge dt\\
&= \Big(-\rho_0 \frac{\partial v_k}{\partial t} + C_{kKlL} F_{lL,K} + \rho_0 f_k\Big) \mu \wedge dt = 0,\\
\phi^*\omega_{kK} &= \Big(-\frac{\partial F_{kK}}{\partial t} + \frac{\partial v_k}{\partial X_K}\Big) \mu \wedge dt = 0,\\
\phi^*\pi_{kK} &= e_{KLM} \frac{\partial F_{kL}}{\partial X_M} \mu \wedge dt = 0,\\
\phi^*\sigma_k &= \Big(-\frac{\partial x_k}{\partial t} + v_k\Big) \mu \wedge dt = 0,\\
\phi^*\sigma_{kK} &= \Big(\frac{\partial x_k}{\partial X_K} - F_{kK}\Big) \mu \wedge dt = 0.
\end{aligned}$$

We shall now look for the *exact* 4-forms in the ideal $\mathcal{I}$. If $\omega \in \mathcal{I}$, then we can write

$$\omega = \phi_k \omega_k + \phi_{kK} \omega_{kK} + \nu_{kK} \pi_{kK} + \psi_k \sigma_k + \psi_{kK} \sigma_{kK}$$

where $\phi_k, \phi_{kK}, \nu_{kK}, \psi_k, \psi_{kK} \in \Lambda^0(K)$. Let us now introduce a 3-form

$$\Omega = \Phi \mu - \Phi_K\, \mu_K \wedge dt \in \Lambda^3(K)$$

where $\Phi, \Phi_K \in \Lambda^0(K)$. Next, we try to determine the functions $\phi_k, \phi_{kK}, \nu_{kK}, \psi_k, \psi_{kK}$ as to satisfy the relation

$$\omega = d\Omega = d\Phi \wedge \mu - d\Phi_K \wedge \mu_K \wedge dt.$$

Under the solution mapping, we have

$$0 = \phi^*\omega = \phi^* d\Omega = d\phi^*\Omega$$

so that we obtain the conservation equations

$$\frac{\partial(\phi^*\Phi)}{\partial t} + \frac{\partial(\phi^*\Phi_K)}{\partial X_K} = 0.$$

The relation $\omega = d\Omega$ now yields

$$\begin{aligned}
&\Big(\rho_0\phi_k - \frac{\partial\Phi}{\partial v_k}\Big)dv_k \wedge \mu + \Big(C_{kKlL}\phi_k - e_{KLM}\nu_{lM} + \frac{\partial\Phi_K}{\partial F_{lL}}\Big)dF_{lL} \wedge \mu_K \wedge dt \\
&\quad + \Big(\rho_0 f_k\phi_k + \psi_k v_k - \psi_{kK}F_{kK} + \frac{\partial\Phi}{\partial t} + \frac{\partial\Phi_K}{\partial X_K}\Big)\mu \wedge dt \\
&\quad + \Big(\phi_{kK} - \frac{\partial\Phi}{\partial F_{kK}}\Big)dF_{kK} \wedge \mu + \Big(\phi_{kK} + \frac{\partial\Phi_K}{\partial v_k}\Big)dv_k \wedge \mu_K \wedge dt \\
&\quad + \Big(\psi_k - \frac{\partial\Phi}{\partial x_k}\Big)dx_k \wedge \mu + \Big(\psi_{kK} + \frac{\partial\Phi_K}{\partial x_k}\Big)dx_k \wedge \mu_K \wedge dt = 0
\end{aligned}$$

from which we extract the following expressions

$$\begin{aligned}
\rho_0\phi_k &= \frac{\partial\Phi}{\partial v_k}, \\
\phi_{kK} &= \frac{\partial\Phi}{\partial F_{kK}} = -\frac{\partial\Phi_K}{\partial v_k}, \\
\psi_k &= \frac{\partial\Phi}{\partial x_k}, \quad \psi_{kK} = -\frac{\partial\Phi_K}{\partial x_k} \\
C_{kKlL}\phi_k - e_{KLM}\nu_{lM} &+ \frac{\partial\Phi_K}{\partial F_{lL}} = 0, \\
\rho_0 f_k\phi_k + \psi_k v_k - \psi_{kK}F_{kK} &+ \frac{\partial\Phi}{\partial t} + \frac{\partial\Phi_K}{\partial X_K} = 0.
\end{aligned}$$

It follows from the fifth expression above by employing the first one, recalling the relation $e_{KLM}e_{KLN} = 2\delta_{MN}$ and evaluating its symmetric part with respect to indices K and L that

$$\begin{aligned}
&\nu_{lM} = \frac{1}{2}e_{KLM}\Big(\frac{1}{\rho_0}C_{kKlL}\frac{\partial\Phi}{\partial v_k} + \frac{\partial\Phi_K}{\partial F_{lL}}\Big) \\
&\frac{1}{\rho_0}(C_{kKlL} + C_{kLlK})\frac{\partial\Phi}{\partial v_k} + \frac{\partial\Phi_K}{\partial F_{lL}} + \frac{\partial\Phi_L}{\partial F_{lK}} = 0.
\end{aligned}$$

Hence, the equations to be satisfied by the functions Φ and Φ_K depending on 19 variables are reduced to

$$\frac{\partial \Phi}{\partial F_{kK}} + \frac{\partial \Phi_K}{\partial v_k} = 0, \tag{8.7.7}$$

$$\frac{\partial \Phi_K}{\partial F_{lL}} + \frac{\partial \Phi_L}{\partial F_{lK}} + B_{KLkl}\frac{\partial \Phi}{\partial v_k} = 0,$$

$$\frac{\partial \Phi}{\partial t} + \frac{\partial \Phi_K}{\partial X_K} + \frac{\partial \Phi}{\partial x_k}v_k + \frac{\partial \Phi_K}{\partial x_k}F_{kK} + \frac{\partial \Phi}{\partial v_k}f_k = 0.$$

where the functions $B_{KLkl}(\mathbf{F})$ are defined by

$$B_{KLkl} = B_{LKkl} = B_{KLlk} = \frac{1}{\rho_0}(C_{kKlL} + C_{kLlK}). \tag{8.7.8}$$

enjoy several symmetries in subscripts depicted above that can be verified just by inspection. The system (8.7.7) contains 28 equations to determine only four functions Φ and Φ_K for an arbitrary stress potential Σ. In order to find the solution of this system, let us start by differentiating $(8.7.7)_2$ and employing $(8.7.7)_1$ to obtain

$$\frac{\partial^2 \Phi}{\partial F_{lL}\partial F_{mK}} + \frac{\partial^2 \Phi}{\partial F_{lK}\partial F_{mL}} = B_{KLkl}\frac{\partial^2 \Phi}{\partial v_k \partial v_m}$$

The left hand side of this expression is symmetric in indices l and m imposing the following restriction on the right hand side:

$$B_{KLkl}\frac{\partial^2 \Phi}{\partial v_k \partial v_m} = B_{KLkm}\frac{\partial^2 \Phi}{\partial v_k \partial v_l}.$$

For fixed K and L, this implies that the symmetric matrix $\partial^2\Phi/\partial v_k\partial v_l$ commutes with arbitrary symmetric matrices B_{KLkl}. According to the well known *Schur lemma* of the group theory [Russian born German mathematician Issai Schur (1875-1941)] this matrix can only be a multiple of the unit matrix. Therefore, we ought to write that

$$\frac{\partial^2 \Phi}{\partial v_k \partial v_l} = \rho_0 \phi(\mathbf{X}, t, \mathbf{x}, \mathbf{v}, \mathbf{F})\delta_{kl}$$

When $k \neq l$, we are evidently led to $\partial^2\Phi/\partial v_k\partial v_l = 0$.Hence we readily observe that

$$\frac{\partial^3 \Phi}{\partial v_k \partial v_l \partial v_l} = \rho_0 \frac{\partial \phi}{\partial v_k} = 0$$

so that the function ϕ becomes independent of the variables $\mathbf{v} = \{v_k\}$. We thus obtain

$$\Phi = \frac{1}{2}\rho_0\phi(\mathbf{X},t,\mathbf{x},\mathbf{F})v_k v_k + \lambda_k(\mathbf{X},t,\mathbf{x},\mathbf{F})v_k + \mu(\mathbf{X},t,\mathbf{x},\mathbf{F})$$

Let us now insert this expression into $(8.7.7)_1$ to obtain

$$\frac{\partial \Phi_K}{\partial v_k} = -\frac{1}{2}\rho_0\frac{\partial \phi}{\partial F_{kK}}v_m v_m - \frac{\partial \lambda_m}{\partial F_{kK}}v_m - \frac{\partial \mu}{\partial F_{kK}}$$

and

$$\frac{\partial^2 \Phi_K}{\partial v_k \partial v_l} = -\rho_0\frac{\partial \phi}{\partial F_{kK}}v_l - \frac{\partial \lambda_l}{\partial F_{kK}}.$$

The symmetry on the left hand side with respect to indices k and l now requires that

$$\rho_0\Big(\frac{\partial \phi}{\partial F_{kK}}v_l - \frac{\partial \phi}{\partial F_{lK}}v_k\Big) + \frac{\partial \lambda_l}{\partial F_{kK}} - \frac{\partial \lambda_k}{\partial F_{lK}} = 0.$$

Since ϕ and $\boldsymbol{\lambda}$ do not depend on $\mathbf{v}$, we immediately obtain

$$\frac{\partial \phi}{\partial F_{kK}} = 0,$$
$$\frac{\partial \lambda_l}{\partial F_{kK}} = \frac{\partial \lambda_k}{\partial F_{lK}}.$$

Hence, we see that $\phi = \phi(\mathbf{X},t,\mathbf{x})$ We thus conclude that

$$\Phi_K = -\frac{1}{2}\frac{\partial \lambda_l}{\partial F_{kK}}v_k v_l - \frac{\partial \mu}{\partial F_{kK}}v_k + \Psi_K(\mathbf{X},t,\mathbf{x},\mathbf{F})$$

In order to determine the arbitrary functions appearing in Φ and Φ_K, we have to introduce these expressions into $(8.7.7)_2$ and $(8.7.7)_3$. After tedious, but not overly complicated manipulations, which we abstain from repeating them here, we arrive at the following result when $f_k = 0$

$$\begin{aligned}\Phi &= a\Big(\Sigma + \frac{1}{2}\rho_0 v_k v_k\Big) + b_k\rho_0 v_k + c_k e_{klm}\rho_0 x_l v_m + d_L\rho_0 x_{k,L}v_k\\ \Phi_K &= -aT_{Kk}v_k - b_k T_{Kk} - c_k e_{klm}x_l T_{Km}\\ &\quad + d_L\Big[\Big(\Sigma - \frac{1}{2}\rho_0 v_k v_k\Big)\delta_{KL} - T_{Kk}x_{k,L}\Big]\end{aligned}$$

where a, b_k, c_k and d_L are arbitrary constants. A reader interested with details may be referred to the work mentioned above. Therefore, independent conservation laws will be, respectively

$$\frac{\partial}{\partial t}\Big(\Sigma+\frac{1}{2}\rho_0 v_k v_k\Big)=\frac{\partial}{\partial X_K}(T_{Kk}v_k), \qquad \textit{balance of energy} \tag{8.7.9}$$

$$\rho_0\frac{\partial v_k}{\partial t}=\frac{\partial T_{Kk}}{\partial X_K}, \qquad \textit{balance of linear momentum}$$

$$\rho_0\frac{\partial}{\partial t}(e_{klm}x_l v_m)=\frac{\partial}{\partial X_K}(e_{klm}x_l T_{Km}), \ \textit{balance of angular momentum}$$

$$\rho_0\frac{\partial}{\partial t}(x_{k,L}v_k)+\frac{\partial}{\partial X_K}\Big[\Big(\Sigma-\frac{1}{2}\rho_0 v_k v_k\Big)\delta_{KL}-T_{Kk}x_{k,L}\Big]=0$$

The first three expressions corresponds in the framework of the classical mechanics to conservation laws to which every correctly formulated conservative system must obey. However, the last conservation law is of different character and it is peculiar only to the field equations of elasticity. If we integrate the conservation laws in the differential form on the region Ω and employ the divergence theorem we obtain

$$\frac{\partial}{\partial t}\int_\Omega \Phi\,dV+\int_{\partial\Omega}\Phi_K N_K\,dA=0$$

where the vector $\mathbf{N}$ is the unit exterior normal to the boundary $\partial\Omega$ of the region Ω. Hence, conservation laws in integral form are given by

$$\frac{\partial}{\partial t}\int_\Omega\Big(\Sigma+\frac{1}{2}\rho_0|\mathbf{v}|^2\Big)\,dV-\int_{\partial\Omega}T_{Kk}v_k N_K\,dA=0, \tag{8.7.10}$$

$$\frac{\partial}{\partial t}\int_\Omega\rho_0 v_k\,dV-\int_{\partial\Omega}T_{Kk}N_K\,dA=0,$$

$$\frac{\partial}{\partial t}\int_\Omega e_{klm}\rho_0 x_l v_m\,dV-\int_{\partial\Omega}e_{klm}x_l T_{Km}N_K\,dA=0,$$

$$\frac{\partial}{\partial t}\int_\Omega\rho_0 x_{k,L}v_k\,dV+\int_{\partial\Omega}\Big[\Big(\Sigma-\frac{1}{2}\rho_0|\mathbf{v}|^2\Big)N_L-T_{Kk}x_{k,L}N_K\Big]dA=0$$

The last integral is the non-linear dynamical counterpart of the J-integral that is frequently utilised in fracture mechanics ■

8.8. THE COHOMOLOGY OF DE RHAM

In Chapter VI, we had shown through the homotopy operator that all closed forms on a differentiable manifold M are *locally* exact. This property may not valid, however, *globally*, in other words, over the entire manifold. That the character of the connection between closed and exact forms depends only on the topology of the manifold, particularly on the *holes* within, but not on its differentiable structure has been demonstrated by de Rham through the investigation of cohomology groups on the module of exterior forms and homology groups on the topology of the manifold.

We had already seen that all closed and exact forms defined on a differentiable manifold M^m constitute subalgebras $\mathcal{C}(M)$ and $\mathcal{E}(M)$ of the exterior algebra $\Lambda(M)$, respectively, on $\mathbb{R}$ whereas $\mathcal{C}^k(M)$ and $\mathcal{E}^k(M)$ are vector subspaces of the module $\Lambda^k(M)$ on $\mathbb{R}$ [*see* Theorem 5.8.3]. We obviously have $\mathcal{E}^k(M) \subseteq \mathcal{C}^k(M)$, namely, $\mathcal{E}^k(M)$ is a subspace of $\mathcal{C}^k(M)$. *We shall now define a relation* $\sim$ *on the vector space* $\mathcal{C}^k(M)$ *as follows: two closed forms are related if their difference is an exact form.* Hence, for two forms $\omega_1, \omega_2 \in \mathcal{C}^k(M)$ the relation $\omega_1 \sim \omega_2$ implies that $\omega_1 - \omega_2 = d\theta$ where $\theta \in \Lambda^{k-1}(M)$. $\sim$ is an equivalence relation. Indeed, $\omega \sim \omega$ since $0 = \omega - \omega = d0$ *so the relation is reflexive.* If $\omega_1 \sim \omega_2$, then one has $\omega_2 - \omega_1 = d(-\theta)$ and $\omega_2 \sim \omega_1$ *so the relation is symmetric.* If $\omega_1 \sim \omega_2$ and $\omega_2 \sim \omega_3$, then we get $\omega_1 - \omega_2 = d\theta_1$, $\omega_2 - \omega_3 = d\theta_2$ and, consequently, $\omega_1 - \omega_3 = d\theta_1 + d\theta_2 = d(\theta_1 + \theta_2)$ and $\omega_1 \sim \omega_3$ *so the relation is transitive.* Therefore, the vector space $\mathcal{C}^k(M)$ is partitioned into disjoint equivalence classes. An equivalence class associated with a form $\omega \in \mathcal{C}^k(M)$ will be the set

$$[\omega] = \{\omega + \sigma : \sigma \in \mathcal{E}^k(M)\} = \{\omega + d\theta : \theta \in \Lambda^{k-1}(M)\}. \quad (8.8.1)$$

This set is called a ***cohomology class***. All forms belong to the cohomology class of the form ω are called as ***cohomologous forms*** to ω. We have seen on *p.* 5 that the quotient set of these equivalence classes may be equipped with a structure of a linear vector space on $\mathbb{R}$. We shall denote the *quotient space* of $\mathcal{C}^k(M)$ with respect to its subspace $\mathcal{E}^k(M)$ by the *vector space*

$$H^k(M) = \mathcal{C}^k(M)/\mathcal{E}^k(M). \quad (8.8.2)$$

If we consider the cochain complex (5.8.6) given by

$$\Lambda^0(M) \xrightarrow{d} \cdots \xrightarrow{d} \Lambda^k(M) \xrightarrow{d} \Lambda^{k+1}(M) \xrightarrow{d} \cdots \xrightarrow{d} \Lambda^m(M) \to 0$$

we see that this quotient space is also expressible in the equivalent form

$$H^k(M) = \mathcal{N}_k(d)/\mathcal{R}_k(d).$$

The zero element of this vector space is given by $[\mathbf{0}] = \mathcal{E}^k(M)$. Since the linear vector space $H^k(M)$ is known to be an Abelian group, it will thus be named as the ***kth de Rham cohomology group*** of the manifold M. The dimension $b_k(M)$ of the linear vector space $H^k(M)$ which is the number of the linearly independent equivalence classes is called by Poincaré as the ***kth***

Betti number [after Italian mathematician Enrico Betti (1823-1892)] of the manifold M. Evidently b_k is a positive integer that might be infinite. As we shall observe later that Betti numbers are dependent on the topology of the manifold M, particularly on its connectedness and number of holes within M. The sum

$$\chi(M)=\sum_{k=0}^{m}(-1)^k b_k(M) \tag{8.8.3}$$

formed by Betti numbers is called the ***Euler-Poincaré characteristic*** of the manifold M [Swiss mathematician Leonhard Euler (1707-1783)]. In order that all closed k-forms on a manifold are to be exact k-forms, we must clearly have $\mathcal{C}^k(M)=\mathcal{E}^k(M)$. This is of course tantamount to say that $H^k(M)=\mathbf{0}$. Hence, $b_k(M)=0$ in such a case.

Since $\mathcal{E}^0(M)=\{0\}$, we naturally get $H^0(M)=\mathcal{C}^0(M)$. On the other hand, if a function $f\in\Lambda^0(M)$ is closed, that is, if $df=0$, we find that f is constant. When M is a connected manifold, the function f takes of course a unique constant value on M. Hence, we get $H^0(M)=\mathbb{R}$ and consequently $b_0(M)=1$. But, if the manifold M is a disconnected union of r connected components, the function will be allowed to take a different constant value on each component. So we find $H^0(M)=\mathbb{R}^r$ and $b_0(M)=r$. If $k>m$, then all k-forms on M vanish leading to the result $H^k(M)=\mathbf{0}$. Since all closed forms on $\mathbb{R}^m$ with $m>0$ are exact [*see* p. 334], we deduce that $\mathcal{C}^k(\mathbb{R}^m)=\mathcal{E}^k(\mathbb{R}^m)$. Accordingly, we obtain $H^k(\mathbb{R}^m)=\mathbf{0}$ for $1\le k\le m$. Thus Betti numbers become $b_0(\mathbb{R}^m)=1$, $b_k(\mathbb{R}^m)=0$, $1\le k\le m$. When M is a contractible manifold, we similarly have $H^k(M)=\mathbf{0}, 1\le k\le m$ and Betti numbers are found to be $b_0(M)=1$, $b_k(M)=0$, $1\le k\le m$.

The direct sum $H(M)=\overset{m}{\underset{k=0}{\oplus}} H^k(M)$ is a linear vector space on $\mathbb{R}$. Let us take the cohomology classes $[\omega]\in H^k(M)$, $[\sigma]\in H^l(M)$ into consideration where the representatives of classes are $\omega\in\mathcal{C}^k(M)$ and $\sigma\in\mathcal{C}^l(M)$. An operation of multiplication $\sqcup$ on $H(M)$ will now be defined by

$$[\omega]\sqcup[\sigma]=[\omega\wedge\sigma]. \tag{8.8.4}$$

Endowed with this operation, $H(M)$ is named as the ***de Rham algebra***.

Let us now consider smooth manifolds M and N and a smooth mapping $\phi: M\to N$. We know that the mapping ϕ generates the pull-back operator $\phi^*:\Lambda(N)\to\Lambda(M)$. Since the operator ϕ^*, that is linear on $\mathbb{R}$, and d are commutative, ϕ^* transforms closed forms into closed forms and also exact forms into exact forms. In fact, owing to Theorem 5.8.2 for a form $\omega\in\mathcal{C}^k(N)$ we immediately obtain $\phi^*\omega\in\mathcal{C}^k(M)$ because of the relation

$0 = \phi^*(d\omega) = d(\phi^*\omega)$. In the same manner, for a form $\omega \in \mathcal{E}^k(N)$ we have $\omega = d\sigma$ and we obtain $\phi^*\omega = \phi^*(d\sigma) = d(\phi^*\sigma)$, or $\phi^*\omega \in \mathcal{E}^k(M)$. For a closed form $\omega \in \mathcal{C}^k(N)$ the vector $[\omega] \in H^k(N)$ is the set of forms $\omega + d\theta$ for all forms $\theta \in \Lambda^{k-1}(N)$. In this case, we get

$$\phi^*(\omega + d\theta) = \phi^*\omega + d(\phi^*\theta) \in [\phi^*\omega]$$

so that we obtain $\phi^*[\omega] = [\phi^*\omega] \in H^k(M)$ for every $[\omega] \in H^k(N)$. This means that a *linear transformation* $\phi^* : H(N) \to H(M)$ between de Rham algebras arises from the mapping ϕ. ϕ^* *is actually a homomorphism*. Indeed if $[\omega], [\sigma] \in H(N)$, we can easily obtain

$$\begin{aligned}\phi^*([\omega] \sqcup [\sigma]) &= \phi^*[\omega \wedge \sigma] = [\phi^*(\omega \wedge \sigma)] = [\phi^*\omega \wedge \phi^*\sigma]\\ &= [\phi^*\omega] \sqcup [\phi^*\sigma].\end{aligned}$$

If ϕ is a diffeomorphism, then $\phi^* : H(N) \to H(M)$ becomes naturally an isomorphism.

Example 8.8.1. We consider a submanifold in $\mathbb{R}^{n+1}$ given by the unit sphere $\mathbb{S}^n$. We suppose that that the poles are the points defined by $x^{n+1} = \pm 1$. If we employ the hyperspherical coordinates introduced on *p.* 412 satisfying the conditions $0 \leq \phi_1, \dots, \phi_{n-1} \leq \pi$ and $0 \leq \phi_n \leq 2\pi$, we know that the volume form on $\mathbb{S}^n$ can be chosen as

$$\mu = \sin^{n-1}\phi_1 \sin^{n-2}\phi_2 \cdots \sin\phi_{n-1}\, d\phi_1 \wedge d\phi_2 \wedge \cdots \wedge d\phi_n \in \Lambda^n(\mathbb{S}^n)$$

Since $d\mu = 0$, μ is a closed form. But it is not an exact form. Indeed, because one has $\partial\mathbb{S}^n = 0$, an exact form $\omega = d\sigma \in \Lambda^n(\mathbb{S}^n)$ must satisfy the condition

$$\int_{\mathbb{S}^n} \omega = \int_{\mathbb{S}^n} d\sigma = \int_{\partial\mathbb{S}^n} \sigma = 0$$

in accordance with the Stokes theorem. However, we had already seen that [*see p.* 413]

$$\int_{\mathbb{S}^n} \mu = \frac{2\pi^{\frac{n+1}{2}}}{\Gamma\left(\frac{n+1}{2}\right)} \neq 0.$$

We shall now try to demonstrate the following proposition: *A closed form $\omega \in \Lambda^n(\mathbb{S}^n) = \mathcal{C}^n(\mathbb{S}^n)$ is an exact form if and only if* $\int_{\mathbb{S}^n} \omega = 0$. If the form ω is exact, this condition is satisfied straightforwardly as is seen above. We shall use the method of mathematical induction to show that it is also the necessary condition. To this end, we shall first prove this proposition for

$n = 1$. The embedding $\phi : \mathbb{R} \to \mathbb{S}^1 \subset \mathbb{R}^2$ prescribed by $\phi(\theta) = e^{i\theta}$ determines the 1-dimensional manifold $\mathbb{S}^1$. If $\omega \in \Lambda^1(\mathbb{S}^1)$, then we have $\phi^*\omega = f(\theta)\,d\theta$. In order that this form is to be uniquely defined, the function must be 2π-periodic. We then introduce a function $F(\theta) = \int_0^\theta f(\tau)\,d\tau$. If $\int_{\mathbb{S}^1}\omega = 0$, then we get

$$0 = \int_{\mathbb{S}^1}\omega = \int_\theta^{\theta+2\pi} f(\tau)\,d\tau = F(\theta + 2\pi) - F(\theta),\ \ \forall\theta \in \mathbb{R}.$$

This means that $F(\theta)$ ought to be a 2π-periodic function. Thus, a *unique* function $G \in \Lambda^0(\mathbb{S}^1)$ may be defined through the relation $\phi^*G = F$. Hence, we can write $\phi^*\omega = f(\theta)\,d\theta = dF = d(\phi^*G) = \phi^*dG$ from which it follows that $\omega = dG$, i.e., $\omega \in \mathcal{E}^1(\mathbb{S}^1)$. In order to apply the mathematical induction, we shall now suppose that *the proposition in question is true in the manifold* $\mathbb{S}^{n-1}$ and then try to prove that it will also be true in the manifold $\mathbb{S}^n$. We know that the manifold $\mathbb{S}^n$ can be prescribed by an atlas with two charts. The following open sets of these charts

$$U_1 = \{\mathbf{x} \in \mathbb{S}^n : x^{n+1} < 1\} \ \text{ and } \ U_2 = \{\mathbf{x} \in \mathbb{S}^n : x^{n+1} > -1\}$$

yield $U_1 \cup U_2 = \mathbb{S}^n$ and by *a* stereographic projection [*see* p. 81] these sets become homeomorphic to $\mathbb{R}^n$. We define the *north* and *south hemispheres* of $\mathbb{S}^n$ as the closed sets

$$N = \{\mathbf{x} \in \mathbb{S}^n : x^{n+1} \geq 0\} \subset U_2 \ \text{ and } \ S = \{\mathbf{x} \in \mathbb{S}^n : x^{n+1} \leq 0\} \subset U_1,$$

respectively. We see at once $N \cup S = \mathbb{S}^n$ and we observe that $N \cap S = \{\mathbf{x} \in \mathbb{S}^n : x^{n+1} = 0\} = \mathbb{S}^{n-1}$. The latter set can of course be taken as the common boundary of N and S and it should be oriented in reverse directions whether it is considered as the boundary ∂N or ∂S. Since the sets U_1 and U_2 are homeomorphic to $\mathbb{R}^n$, they are contractible sets. Let $\omega \in \mathcal{C}^n(\mathbb{S}^n)$ be a closed form satisfying the condition $\int_{\mathbb{S}^n}\omega = 0$. According to the Poincaré lemma, restrictions of the form ω to regions U_1 and U_2 are exact forms, namely, there exist forms $\sigma_1 \in \Lambda^{n-1}(U_1)$ and $\sigma_2 \in \Lambda^{n-1}(U_2)$ such that the relations $\omega|_{U_1} = d\sigma_1$ and $\omega|_{U_2} = d\sigma_2$ are held. Therefore, if we choose that ∂S is positively oriented, then the Stokes theorem leads to

$$\begin{aligned} 0 &= \int_{\mathbb{S}^n}\omega = \int_S\omega + \int_N\omega = \int_S d\sigma_1 + \int_N d\sigma_2 = \int_{\partial S}\sigma_1 + \int_{\partial N}\sigma_2 \\ &= \int_{\mathbb{S}^{n-1}}\sigma_1 - \int_{\mathbb{S}^{n-1}}\sigma_2 = \int_{\mathbb{S}^{n-1}}(\sigma_1 - \sigma_2). \end{aligned}$$

Hence, the integral of the form $(\sigma_1 - \sigma_2)|_{\mathbb{S}^{n-1}} \in \Lambda^{n-1}(\mathbb{S}^{n-1})$ on $\mathbb{S}^{n-1}$ vanishes. According to our assumption the form $(\sigma_1 - \sigma_2)|_{\mathbb{S}^{n-1}}$ is *exact*. Let us now define a smooth mapping $\psi : U \to \mathbb{S}^{n-1}$ on the open set $U = U_1 \cap U_2$ by assigning to each point $\mathbf{x} \in U$ the point of intersection of the meridian through the point $\mathbf{x}$ with the equator $\mathbb{S}^{n-1}$. In this case, the form $(\sigma_1 - \sigma_2)|_U = \psi^*(\sigma_1 - \sigma_2)|_{\mathbb{S}^{n-1}}$ on U will also be exact. Thus, there exists a form $\alpha \in \Lambda^{n-2}(U)$ such that $(\sigma_1 - \sigma_2)|_U = d\alpha$. Let us choose a form $\beta \in \Lambda^{n-2}(\mathbb{S}^n)$ so that one gets $d\beta|_U = d\alpha$ on U and introduce a form $\sigma \in \Lambda^{n-1}(\mathbb{S}^n)$ as follows

$$\sigma = \begin{cases} \sigma_1, & \text{on } U_1, \\ \sigma_2 + d\beta, & \text{on } U_2. \end{cases}$$

On $U = U_1 \cap U_2$, we find $\sigma_1 = \sigma_2 + d\beta$ and $\sigma_1 - \sigma_2 = d\beta = d\alpha$ whence we conclude that $\omega = d\sigma$ and $\omega \in \mathcal{E}^n(\mathbb{S}^n)$. We have thus shown that if we assume that the proposition is true for $n-1$, then it becomes also true for n. Since we have already seen that the proposition is true for $n = 1$, we are led to the conclusion that it is true for every n.

We shall now demonstrate that if $\omega \in \Lambda^n(\mathbb{S}^n)$ is a closed form, we can always find a number $c \in \mathbb{R}$ such that $\omega - c\mu$ is rendered as an exact form. On employing hyperspherical coordinates, we can generally express this form as follows

$$\omega = f(\phi_1, \ldots, \phi_{n-1}, \phi_n)\, d\phi_1 \wedge \cdots \wedge d\phi_{n-1} \wedge d\phi_n$$

where the function f is π-periodic in variables $\phi_1, \ldots, \phi_{n-1}$ and 2π-periodic in the variable ϕ_n. We thus get

$$\int_{\mathbb{S}^n} \omega = \int_0^{\pi} \cdots \int_0^{\pi} \int_0^{2\pi} f(\phi_1, \ldots, \phi_{n-1}, \phi_n)\, d\phi_1 \cdots d\phi_{n-1} d\phi_n.$$

Let us choose a real number c as follows

$$c = \frac{\displaystyle\int_{\mathbb{S}^n} \omega}{\displaystyle\int_{\mathbb{S}^n} \mu} = \frac{\Gamma\left(\frac{n+1}{2}\right)}{2\pi^{\frac{n+1}{2}}} \int_{\mathbb{S}^n} \omega.$$

With this choice the closed form $\alpha = \omega - c\mu \in \Lambda^n(\mathbb{S}^n)$ will clearly satisfy the condition

$$\int_{\mathbb{S}^n} \alpha = 0.$$

Hence α is an exact form, i.e., $\alpha \in \mathcal{E}^n(\mathbb{S}^n)$. If $c = 0$, then the closed form $\omega \in \Lambda^n(\mathbb{S}^n)$ will obviously be an exact form. According to this result, all closed n-forms on $\mathbb{S}^n$ are to be cohomologous to a real constant multiple of the volume form μ. Hence, we can write $H^n(\mathbb{S}^n) = \mathbb{R}$. On the other hand, because $\mathbb{S}^n$ is a connected manifold we know that $H^0(\mathbb{S}^n) = \mathbb{R}$. Therefore, the corresponding Betti numbers are $b_0(\mathbb{S}^n) = b_n(\mathbb{S}^n) = 1$.

We shall now try to determine the cohomology groups $H^k(\mathbb{S}^n)$ for $1 \le k \le n-1$ of the n-sphere. To this end, we shall resort once more to mathematical induction. Let us first take a closed form $\omega \in \mathcal{C}^1(\mathbb{S}^n)$ into account. Since $d\omega = 0$ on U_1 and U_2, the Poincaré lemma implies that there are functions $f \in \Lambda^0(U_1)$ and $g \in \Lambda^0(U_2)$ such that one writes $\omega|_{U_1} = df$ and $\omega|_{U_2} = dg$. On the open set $U = U_1 \cap U_2$, we thus get $d(f-g)|_U = (df - dg)|_U = (\omega - \omega)|_U = 0$ so that we find $f - g = c = constant$. Let us now define a function $\varphi \in \Lambda^0(\mathbb{S}^n)$ as follows

$$\varphi = \begin{cases} f & \text{on } U_1, \\ g + c & \text{on } U_2. \end{cases}$$

Then we obtain $\omega = d\varphi$, i.e., $\omega \in \mathcal{E}^1(\mathbb{S}^n)$. This means that every closed 1-form on $\mathbb{S}^n$ is exact. In consequence, we find $H^1(\mathbb{S}^n) = 0$. *Let us now assume that every closed* $(k-1)$*-form on* $\mathbb{S}^n$ *is exact.* We consider a closed form $\omega \in \mathcal{C}^k(\mathbb{S}^n)$. Since $d\omega = 0$ again on open sets U_1 and U_2, the Poincaré lemma indicates that there are the forms $\sigma_1 \in \Lambda^{k-1}(U_1)$ and $\sigma_2 \in \Lambda^{k-1}(U_2)$ so that one has $\omega|_{U_1} = d\sigma_1$ and $\omega|_{U_2} = d\sigma_2$. Thereby, we obtain

$$d(\sigma_1 - \sigma_2)|_U = (d\sigma_1 - d\sigma_2)|_U = (\omega - \omega)|_U = 0$$

on $U = U_1 \cap U_2$. We thus conclude that $\sigma_1 - \sigma_2 \in \mathcal{C}^{k-1}(U)$ and our assumption assures us that there exists a form $\alpha \in \Lambda^{k-2}(U)$ such that we have $(\sigma_1 - \sigma_2)|_U = d\alpha$. Let us now choose a form $\beta \in \Lambda^{k-2}(\mathbb{S}^n)$ as to satisfy the relation $d\beta|_U = d\alpha$ on U and define a form $\sigma \in \Lambda^{k-1}(\mathbb{S}^n)$ in the following manner

$$\sigma = \begin{cases} \sigma_1 & \text{on } U_1, \\ \sigma_2 + d\beta & \text{on } U_2. \end{cases}$$

We thus conclude that $\omega = d\sigma$, that is, $\omega \in \mathcal{E}^k(\mathbb{S}^n)$. Hence, the mathematical induction prove the proposition that *every closed k-form on* $\mathbb{S}^n$ *satisfying the condition* $1 \le k \le n-1$ *is exact* so that one obtains $H^k(\mathbb{S}^n) = 0$ for $1 \le k \le n-1$. Betti numbers thus become $b_k(\mathbb{S}^n) = 0, 1 \le k \le n-1$. The relation (8.8.3) then yields the Euler-Poincaré characteristic of the n-sphere as $\chi(\mathbb{S}^n) = b_0 + (-1)^n b_n = 1 + (-1)^n$. Hence $\chi(\mathbb{S}^n) = 0$ if n is an odd number and $\chi(\mathbb{S}^n) = 2$ if n is an even number. ■

The salient property of the sphere $\mathbb{S}^n$ is that it is a connected, oriented and compact manifold.

We had seen that all singular k-chains on a manifold M constitute a linear vector space $C_k(M)$ [*see p.* 421]. We know that the set $\{C_k(M)\}$ constitutes a chain complex under the boundary operator ∂. Let us then consider the subset of $C_k(M)$ formed by k-cycles $\overset{\circ}{C}_k(M) = \{c_k \in C_k(M) : \partial c_k = 0\}$. Because the sum of two k-cycles and a real multiple of k-cycle is also a k-cycle, $\overset{\circ}{C}_k(M)$ is a subspace of the linear vector space $C_k(M,$ hence it is a linear vector space by itself. Let us denote the set of k-cycles that are boundaries of $(k+1)$-chains by

$$B_k(M) = \{c_k = \partial b_{k+1} : \; b_{k+1} \in C_{k+1}(M)\}.$$

Evidently, the set $B_k(M)$ is also a linear vector space and it is clear that $B_k(M) \subseteq \overset{\circ}{C}_k(M)$ since $\partial c_k = 0$ if $c_k \in B_k(M)$. We now define a relation $\sim$ on $\overset{\circ}{C}_k(M)$ as follows: $c_k', c_k'' \in \overset{\circ}{C}_k(M)$ are related if their difference is a boundary of a $(k+1)$-chain, namely $c_k' \sim c_k''$ if only $c_k' - c_k'' = \partial b_{k+1} \in B_k(M)$. Two cycles whose difference is a boundary will be called ***homologous cycles***. It can readily be verified that this relation defining the ***homology*** on the manifold M is an equivalence relation. Equivalence classes, in other words ***homology classes***, are defined by

$$[c_k] = \{c_k + \partial b_{k+1} : c_k \in \overset{\circ}{C}_k(M), b_{k+1} \in C_{k+1}(M)\}$$

Let us denote the quotient space generated by those classes by

$$H_k(M) = \overset{\circ}{C}_k(M)/B_k(M). \tag{8.8.5}$$

$H_k(M)$ is a linear vector space on real numbers $\mathbb{R}$. As such it is an Abelian group and is named as the ***kth differentiable singular homology group*** of the manifold M. If we consider the chain complex (8.4.13), this quotient space can also be expressed equivalently as

$$H_k(M) = \mathcal{N}_k(\partial)/\mathcal{R}_k(\partial).$$

We can roughly say that homology groups illustrate the existence and the distribution of holes in topological spaces. The zero element of the vector space $H_k(M)$ is naturally given as $[\mathbf{0}] = B_k(M)$.

Sometimes, it does not prove to be very convenient to work with a chain complex with decreasing indices. Especially, quite a difficult problem arises if we wish to establish a relationship between the de Rham cohomology groups and the homology groups. To circumvent this obstacle we may employ the cochain complex (8.4.15) with the coboundary operator $\mathfrak{d}$. Then, the ***kth singular cohomology group*** M is defined as the following quotient space of the vector space $\mathcal{N}_k(\mathfrak{d})$ with respect to its subspace $\mathcal{R}_k(\mathfrak{d})$

$$\mathcal{H}_k(M) = \mathcal{N}_k(\mathfrak{d})/\mathcal{R}_k(\mathfrak{d}).$$

Hence, *the vector space $\mathcal{H}_k(M)$ is the quotient space of k-cocycles with respect to k-boundaries*. The equivalence class $[f_k] \in \mathcal{H}_k(M)$ related to an element $f_k \in C_k^*(M)$ is the set of all linear functionals on $C_k(M)$ given by $[f_k] = f_k + \mathfrak{d}g_{k-1}$ for all $g_{k-1} \in C_{k-1}^*(M)$ and $\mathfrak{d}f_k = 0$. $[f_k]$ is known to be a ***singular cohomology class***. We shall now try to demonstrate the following proposition:

The kth singular cohomology group is isomorphic to the dual space $H_k^*(M)$ of the kth singular homology group $H_k(M)$, namely, there is a natural isomorphism $\mathfrak{C}_k{:}\mathcal{H}_k(M) \to H_k^*(M)$ such that $\mathfrak{C}_k([f_k]) \in H_k^*(M)$.

Let us consider an arbitrary equivalence class $[f_k] \in \mathcal{H}_k(M)$ where the linear functional $f_k \in C_k^*(M)$ satisfying $\mathfrak{d}f_k = 0$ is a representative of this class. Consequently, f_k is also a linear functional on the subspace $\overset{\circ}{C}_k(M)$ of k-cycles. On the other hand, the relation $0 = \mathfrak{d}f_k(c_{k+1}) = f_k(\partial c_{k+1})$ implies that f_k vanishes on all k-boundaries in the form $c_k = \partial b_{k+1}$ in $C_k(M)$ forming the subspace $B_k(M)$. Hence, f_k becomes a linear functional defined on the homology group $H_k(M)$ because if $[c_k] \in H_k(M)$ where $c_k \in C_k(M)$ is arbitrary, we get $f_k(c_k + \partial b_{k+1}) = f_k(c_k)$. Thus the value of the functional f_k on an equivalence class $[c_k]$ is independent of the representative of this class. However, in order to say that the linear functional $f_k \in H_k^*(M)$ is well defined, we have to show that its value is also independent of the representative of the equivalence class $[f_k]$. This is, however, easily deduced from

$$\begin{aligned} f_k + \mathfrak{d}g_{k-1})(c_k + \partial b_{k+1}) &= f_k(c_k) + \mathfrak{d}g_{k-1}(c_k) + \mathfrak{d}g_{k-1}(\partial b_{k+1}) \\ &= f_k(c_k) \end{aligned}$$

since $\mathfrak{d}g_{k-1}(c_k) = g_{k-1}(\partial c_k) = 0$ because $\partial c_k = 0$ and $\mathfrak{d}g_{k-1}(\partial b_{k+1}) = g_{k-1}(\partial^2 b_{k+1}) = 0$ because $\partial^2 = 0$. We have thus found that to each element of $\mathcal{H}_k(M)$ we can assign a unique element of $H_k^*(M)$. Obviously, this mapping is linear and in order to prove that it is an isomorphism, we must show that it is both injective and surjective. Since equivalence classes are disjoint, injectivity of the mapping is evident. To show surjectivity let us consider a functional $f_k \in H_k^*(M)$. Since $\mathfrak{d}f_k \in C_{k+1}^*(M)$, we get $0 = \mathfrak{d}f_k(c_{k+1}) = f_k(\partial c_{k+1})$ for all $c_{k+1} \in \overset{\circ}{C}_{k+1}(M)$. We thus find $\mathfrak{d}f_k = 0$. We can then define the set of functionals $[f] = f_k + \mathfrak{d}g_{k-1}$ with $g_{k-1} \in C_{k-1}^*(M)$. Clearly, $[f] \in \mathcal{H}_k(M)$. On the other hand, for $c_k \in \overset{\circ}{C}_k(M)$ we get

$$(f_k + \eth g_{k-1})(c_k) = f_k(c_k) + g_{k-1}(\partial c_k) = f_k(c_k).$$

Therefore, we have shown that a functional f_k in $H_k^*(M)$ is the image of an equivalence class $[f]$ in $\mathcal{H}_k(M)$, namely, the mapping is surjective. Thus, *the mapping* $\mathfrak{C}_k{:}\mathcal{H}_k(M) \to H_k^*(M)$ *is an isomorphism.* □

Let us now define a mapping $\mathcal{B}_k : \Lambda^k(M) \times C_k(M) \to \mathbb{R}$ as the integral of a k-form over a k-chain as follows

$$\mathcal{B}_k(\omega, c_k) = \int_{c_k} \omega \in \mathbb{R}. \tag{8.8.6}$$

Naturally, in order that this definition is justifiable, the integral (8.8.6) must exist. This mapping is obviously linear with respect to both the k-form ω and the k-chain c_k. In other words, $\mathcal{B}_k(\omega, c_k)$ is a bilinear, real valued functional. Whenever we consider a *fixed* form $\omega_0 \in \Lambda^k(M)$, then the real valued function

$$\mathcal{F}_{\omega_0}^{(k)}(c_k) = \mathcal{B}_k(\omega_0, c_k) = \int_{c_k} \omega_0 \tag{8.8.7}$$

turns out to be intrinsically a linear functional on the vector space $C_k(M)$. Thus, (8.8.6) is actually generating a mapping $\mathcal{F}_k : \Lambda^k(M) \to C_k(M)^*$ over real numbers from the *vector space* $\Lambda^k(M)$ into the dual space $C_k(M)^*$ designated by $\mathcal{F}_k(\omega) = \mathcal{F}_\omega^{(k)}$. The definition (8.8.7) signify at once that the mapping $\mathcal{F}_k$ is linear, in other words, it is a homomorphism.

Next, we introduce in similar fashion a real valued and bilinear functional $\overset{\circ}{\mathcal{B}}_k : H^k(M) \times H_k(M) \to \mathbb{R}$ through the relation

$$\overset{\circ}{\mathcal{B}}_k([\omega], [c_k]) = \int_{c_k} \omega \in \mathbb{R} \tag{8.8.8}$$

where the closed form $\omega \in C^k(M)$ and cycle $c_k \in \overset{\circ}{C}_k(M)$ are arbitrarily selected representatives of the equivalence classes $[\omega] \in H^k(M)$ and $[c_k] \in H_k(M)$. On the other hand, in order that the definition (8.8.8) bears a meaning the value of the functional must be independent of the chosen representatives of the equivalence classes. This can be proven quite easily, however, if we recall that

$$[\omega] = \{\omega + d\theta : \omega \in C^k(M), \theta \in \Lambda^{k-1}(M)\},$$
$$[c_k] = \{c_k + \partial b_{k+1} : c_k \in \overset{\circ}{C}_k(M), b_{k+1} \in C_{k+1}(M)\}$$

and then utilise the Stokes theorem to obtain

$$\begin{aligned}\int_{c_k+\partial b_{k+1}}(\omega+d\theta)&=\int_{c_k}\omega+\int_{c_k}d\theta+\int_{\partial b_{k+1}}\omega+\int_{\partial b_{k+1}}d\theta\\&=\int_{c_k}\omega+\int_{\partial c_k}\theta+\int_{b_{k+1}}d\omega+\int_{\partial^2 b_{k+1}}\theta=\int_{c_k}\omega\end{aligned}$$

for all forms $\theta\in\Lambda^{k-1}(M)$ and all boundaries $b_{k+1}\in C_{k+1}(M)$ where we employed the relations $d\omega=0,\partial c_k=0,\partial^2 b_{k+1}=0$. It now clear that the functional (8.8.8) determines a homomorphism $\mathring{\mathcal{F}}_k:H^k(M)\to H_k(M)^*$ defined by

$$\mathring{\mathcal{F}}_k([\omega])([c_k])=\mathring{\mathcal{B}}_k([\omega],[c_k])=\int_{c_k}\omega\in\mathbb{R}$$

from the cohomology group $H^k(M)$ into the vector space $H_k(M)^*$ that is the dual of the homology group $H_k(M)$. We have seen above that the vector spaces $H_k(M)^*$ and $\mathcal{H}_k(M)$ are isomorphic. Therefore, $\mathring{\mathcal{F}}_k$ may as well be regarded as a homomorphism between $H^k(M)$ and $\mathcal{H}_k(M)$. We can now show the simple lemma given below:

Lemma 8.8.1. *M and N are smooth manifolds and $\phi:M\to N$ is a smooth mapping. Then the following diagram commutes:*

$$\begin{array}{ccc}H^k(N)&\xrightarrow{\phi^*}&H^k(M)\\\downarrow\mathring{\mathcal{F}}_k&&\downarrow\mathring{\mathcal{F}}_k\\\mathcal{H}_k(N)&\xrightarrow{\phi^*}&\mathcal{H}_k(M).\end{array}$$

We know that if $\omega\in\Lambda^k(N)$, then $\phi^*\omega=\omega\circ\phi\in\Lambda^k(M)$. Similarly, if $f\in C_k^*(N)$, we then obtain $f(c_k^*)=f\big(\phi(c_k)\big)=(f\circ\phi)(c_k)=\phi^*f(c_k)$ where $c_k\in C_k(M)$. Thus, $f\circ\phi=\phi^*f\in C_k^*(M)$. The relation (8.5.3) then requires that

$$\mathring{\mathcal{F}}_k(\phi^*[\omega])([c_k])=\mathring{\mathcal{F}}_k([\omega])(\phi[c_k])$$

for all $[\omega]\in H^k(N)$ and $[c_k]\in H_k(M)$. □

De Rham's theorem proven in 1931 states that *if M is a Hausdorff, locally compact, second countable and oriented smooth manifold, then this homomorphism $\mathring{\mathcal{F}}_k$, called* ***de Rham homomorphism****, is actually an isomorphism.* In order to prove this theorem, we need first to investigate certain properties of Mayer-Vietoris sequences [after Austrian mathematicians

Walther Mayer (1887-1948) and supercentenarian Leopold Vietoris (1891-2002)].

Theorem 8.8.1 (Mayer-Vietoris). *Let M be an m-dimensional smooth manifold supporting a partition of unity and $U, V \subset M$ open subsets such that $U \cup V = M$. We consider the cochain complex*

$$\Lambda^0(M) \xrightarrow{d} \cdots \xrightarrow{d} \Lambda^k(M) \xrightarrow{d} \Lambda^{k+1}(M) \xrightarrow{d} \cdots \xrightarrow{d} \Lambda^m(M) \to 0$$

and the cohomology groups $H^k(M) = \mathcal{N}_k(d)/\mathcal{R}_k(d)$. Then for all $0 \le k \le m$, there exists a homomorphism $\Gamma : H^k(U \cap V) \to H^{k+1}(M)$ such that the following ***Mayer-Vietoris sequence*** *is exact:*

$$\cdots \xrightarrow{\Gamma} H^k(M) \xrightarrow{\varphi} H^k(U) \oplus H^k(V) \xrightarrow{\psi} H^k(U \cap V) \xrightarrow{\Gamma} H^{k+1}(M) \xrightarrow{\varphi} \cdots$$

The homomorphisms φ and ψ are defined by $\varphi = \mathcal{I}_3^* \oplus \mathcal{I}_4^*$ and $\psi = \mathcal{I}_1^* - \mathcal{I}_2^*$ where $\mathcal{I}_1 : U \cap V \to U, \mathcal{I}_2 : U \cap V \to V, \mathcal{I}_3 : U \to M$ and $\mathcal{I}_4 : V \to M$ are inclusion mappings and $\mathcal{I}_1^* : \Lambda^k(U) \to \Lambda^k(U \cap V)$, $\mathcal{I}_2^* : \Lambda^k(V) \to \Lambda^k(U \cap V)$, $\mathcal{I}_3^* : \Lambda^k(M) \to \Lambda^k(U)$, $\mathcal{I}_4^* : \Lambda^k(M) \to \Lambda^k(V)$ are corresponding pull-back operators..

Although we have proven only for Hausdorff, locally compact and second countable manifolds, we had mentioned that if the manifold M is paracompact, then for each open cover $\{U_\lambda : \lambda \in \Lambda\}$ of M there exists a partition of unity subordinate to this cover. We shall see that only the existence of the partition of unity will be crucial for our proof of this theorem.

In view of Theorem 1.2.3, we merely need to show that the short sequence

$$0 \to \Lambda^k(M) \xrightarrow{\varphi} \Lambda^k(U) \oplus \Lambda^k(V) \xrightarrow{\psi} \Lambda^k(U \cap V) \to 0$$

is exact. For a form $\omega \in \Lambda^k(M)$, we have

$$\varphi(\omega) = \big(\mathcal{I}_3^*(\omega), \mathcal{I}_4^*(\omega)\big) = (\omega|_U, \omega|_V).$$

If $\alpha \in \Lambda^k(U)$, $\beta \in \Lambda^k(V)$, then we get

$$\psi(\alpha, \beta) = \mathcal{I}_1^*(\alpha) - \mathcal{I}_2^*(\beta) = \alpha|_{U \cap V} - \beta|_{U \cap V}$$

We first demonstrate that the sequence is exact at $\Lambda^k(M)$. To this end, we only need to show that φ is injective. Let us take $\varphi(\omega) = (\omega|_U, \omega|_V)$ $= (0, 0)$ that leads to $\omega|_U = 0$ and $\omega|_V = 0$. Since $U \cup V = M$, this implies that $\omega = 0$ which proves the injectivity.

In order to prove the exactness at $\Lambda^k(U) \oplus \Lambda^k(V)$ let us apply the operator $\psi \circ \varphi$ on a form $\omega \in \Lambda^k(M)$ to obtain

$$(\psi \circ \varphi)(\omega) = \psi(\omega|_U, \omega|_V) = \omega|_{U\cap V} - \omega|_{U\cap V} = 0.$$

Hence, $\psi \circ \varphi = 0$ on the module $\Lambda^k(M)$ so that we get $\mathcal{R}(\varphi) \subseteq \mathcal{N}(\psi)$. Next, let us consider $(\alpha, \beta) \in \mathcal{N}(\psi)$ which means that $\alpha|_{U\cap V} = \beta|_{U\cap V}$. Thus, there exists a form $\omega \in \Lambda^k(M)$ such that $\omega|_U = \alpha$ and $\omega|_V = \beta$. We then clearly write $(\alpha, \beta) = \varphi(\omega)$ implying that $\mathcal{N}(\psi) \subseteq \mathcal{R}(\varphi)$. Hence, we get $\mathcal{R}(\varphi) = \mathcal{N}(\psi)$ which proves the exactness.

To prove the exactness at $\Lambda^k(U \cap V)$ we just have to show that ψ is surjective. Since (U, V) is an open cover of M, there exists a partition of unity (f_1, f_2) subordinate to (U,V) such that $supp\,(f_1) \subset U, supp\,(f_2) \subset V$. Let $\sigma \in \Lambda^k(U \cap V)$. We define the forms $\lambda \in \Lambda^k(U)$ and $\mu \in \Lambda^k(V)$ as follows

$$\lambda = \begin{cases} f_1\sigma & \text{on } U \cap V \\ 0 & \text{on } U - supp\,(f_1) \end{cases}, \quad \mu = \begin{cases} -f_2\sigma & \text{on } U \cap V \\ 0 & \text{on } V - supp\,(f_1) \end{cases}.$$

We then obtain $\psi(\lambda, \mu) = \lambda|_{U\cap V} - \mu|_{U\cap V} = f_1\sigma - (-f_2\sigma) = (f_1 + f_2)\sigma = \sigma$ that amounts to say that $\mathcal{R}(\psi) = \Lambda^k(U \cap V)$. We thus conclude that Mayer-Vietoris sequence is exact. □

In exactly the same fashion we can show that Mayer-Vietoris sequence

$$\cdots \xrightarrow{\Gamma'} \mathcal{H}_k(M) \xrightarrow{\varphi} \mathcal{H}_k(U) \oplus \mathcal{H}_k(V) \xrightarrow{\psi} \mathcal{H}_k(U \cap V) \xrightarrow{\Gamma'} \mathcal{H}_{k+1}(M) \xrightarrow{\varphi^*} \cdots$$

based on the cochain complex [*see* (8.4.15)]

$$\cdots \longrightarrow C^*_{k-1}(M) \xrightarrow{\eth} C^*_k(M) \xrightarrow{\eth} C^*_{k+1}(M) \longrightarrow \cdots$$

is exact. $\eth$ is the coboundary operator defined in (8.4.14). To prove the existence, we only have to show that the short sequence

$$0 \to C^*_k(M) \xrightarrow{\varphi} C^*_k(U) \oplus C^*_k(V) \xrightarrow{\psi} C^*_k(U \cap V) \to 0$$

is exact. The inclusion operators $\mathcal{I}_1, \mathcal{I}_2, \mathcal{I}_3$ and $\mathcal{I}_4$ are the same as those given above in Theorem 8.8.1. Pull-back operators

$$\begin{aligned} &\mathcal{I}^*_1 : C^*_k(U) \to C^*_k(U \cap V),\ \mathcal{I}^*_2 : C^*_k(V) \to C^*_k(U \cap V), \\ &\mathcal{I}^*_3 : C^*_k(M) \to C^*_k(U), \quad \mathcal{I}^*_2 : C^*_k(V) \to C^*_k(U \cap V) \end{aligned}$$

simply produce restrictions of functionals. For instance, $\mathcal{I}^*_1(f) = f|_{U\cap V}$ for a functional $f \in C^*_k(U)$. For a functional $f \in C^*_k(M)$, we get

$$\varphi(f) = (\mathcal{I}^*_3(f), \mathcal{I}^*_4(f)) = (f|_U, f|_V)$$

and for $g \in C_k^*(U)$, $h \in C_k^*(V)$

$$\psi(g,h) = \mathcal{I}_1^*(g) - \mathcal{I}_2^*(h) = g|_{U\cap V} - h|_{U\cap V}.$$

A smooth manifold M is called a ***de Rham manifold*** if the homomorphism on M is an isomorphism.

Lemma 8.8.2. *An open convex subset $U \subseteq \mathbb{R}^n$ is a de Rham manifold.*

Since a convex open set in $\mathbb{R}^n$ is star-shaped, the Poincaré lemma is applicable. Hence we find that $H^k(U) = 0$ for $k > 0$ and $H^0(U) = \mathbb{R}$. This automatically implies that $\mathcal{H}_k(U) = 0$ for $k > 0$ and $\mathcal{H}_0(U) = \mathbb{R}$ since the dual space of $\mathbb{R}$ is also $\mathbb{R}$. Therefore we only have to demonstrate that $\mathring{\mathcal{F}}_0 : H^0(U) \to \mathcal{H}_0(U)$ is an isomorphism. But elements of $H^0(U)$ are constant functions and a σ_0 singular simplex is just a single point. Thus, $\mathring{\mathcal{F}}_0$ assigns the same real number to a real number. □

Lemma 8.8.3. *Let $\{U_\lambda : \lambda \in \Lambda\}$ be a class of open, pairwise disjoint, de Rham subsets of a smooth manifold M. Then $U = \bigcup_{\lambda\in\Lambda} U_\lambda$ is also a de Rham manifold.*

In order to prove this lemma, we must show that the following diagram commutes isomorphically:

$$\begin{array}{ccc}
H^k(U) & \xrightarrow{\mathcal{A}} & \bigcup_{\lambda\in\Lambda} H^k(U_\lambda) \\
\downarrow \mathfrak{I}_k & & \downarrow \bigoplus_{\lambda\in\Lambda} \mathring{\mathcal{F}}_{k,\lambda} \\
\mathcal{H}_k(U) & \xrightarrow{\mathcal{B}} & \bigcup_{\lambda\in\Lambda} \mathcal{H}_k(U_\lambda).
\end{array}$$

In view of Lemma 8.8.1 the diagram commutes. To show that the homomorphism $\mathfrak{I}_k$ is an isomorphism, we only need to prove that $\mathcal{A}$ and $\mathcal{B}$ are isomorphisms because $\mathring{\mathcal{F}}_{k,\lambda} : H^k(U_\lambda) \to \mathcal{H}_k(U_\lambda), \lambda \in \Lambda$ are isomorphisms by definition on pairwise disjoint de Rham subsets U_λ. In order to determine the homomorphism $\mathcal{A} : H^k(U) \to \bigcup_{\lambda\in\Lambda} H^k(U_\lambda)$, we first define $\mathcal{I}_\lambda^*\omega$ on the set $\bigcup_{\lambda\in\Lambda} H^k(U_\lambda)$ by

$$\mathcal{I}_\lambda^*\omega = \begin{cases} \mathcal{I}_\lambda^*\omega & \text{on } H^k(U_\lambda) \\ 0 & \text{on } \bigcup_{\mu\in\Lambda,\mu\neq\lambda} H^k(U_\mu) \end{cases}$$

where $\mathcal{I}_\lambda = U_\lambda \to U$ are the inclusion mappings and, thus for $\omega \in \Lambda^k(U)$ we get $\mathcal{I}_\lambda^*\omega \in \Lambda^k(U_\lambda)$ that is none other than the restriction of ω on U_λ. We then take for $[\omega] \in H^k(U)$

$$\mathcal{A}[\omega] = \bigoplus_{\lambda\in\Lambda} [\mathcal{I}_\lambda^* \omega] \in \bigcup_{\lambda\in\Lambda} H^k(U_\lambda)$$

which clearly indicates that $\mathcal{A}$ is an isomorphism. $\mathcal{A}$ is injective because equivalence classes are disjoint. Next, let us choose $\omega_\lambda \in \Lambda^k(U_\lambda)$. Since the sets U_λ are pairwise disjoint, then $\omega = \bigoplus_{\lambda\in\Lambda} \omega_\lambda \in \Lambda^k(U)$ with $\mathcal{I}_\lambda^*\omega = \omega_\lambda$. We thus obtain $\mathcal{A}[\omega] = \bigoplus_{\lambda\in\Lambda} [\mathcal{I}_\lambda^*\omega]$ so that $\mathcal{A}$ is surjective. In exactly similar way, we can show that $\mathcal{B}$ is likewise an isomorphism. Since the diagram commutes, we deduce that $\mathfrak{I}_k : H^k(U) \to \mathcal{H}_k(U)$ is also an isomorphism. This means that $U = \bigcup_{\lambda\in\Lambda} U_\lambda$ is a de Rham manifold.

Lemma 8.8.4. *Let U and V be open subsets of a smooth manifold M. We assume that U, V and $U \cap V$ are de Rham manifolds. Then $U \cup V$ is also a de Rham manifold.*

Let us consider the following Mayer-Vietoris sequences associated with de Rham cohomology and singular cohomology that are exact in view of Theorem 8.8.1:

$$\begin{array}{ccc}
\vdots & & \vdots \\
\downarrow & & \downarrow \\
H^{k-1}(U) \oplus H^{k-1}(V) & \xrightarrow{\cong} & \mathcal{H}^{k-1}(U) \oplus \mathcal{H}^{k-1}(V) \\
\downarrow & & \downarrow \\
H^{k-1}(U \cap V) & \xrightarrow{\cong} & \mathcal{H}^{k-1}(U \cap V) \\
\downarrow & & \downarrow \\
H^k(U \cup V) & \xrightarrow{\mathfrak{I}_k} & \mathcal{H}^k(U \cup V) \\
\downarrow & & \downarrow \\
H^k(U) \oplus H^k(V) & \xrightarrow{\cong} & \mathcal{H}^k(U) \oplus \mathcal{H}^k(V) \\
\downarrow & & \downarrow \\
H^k(U \cap V) & \xrightarrow{\cong} & \mathcal{H}^k(U \cap V) \\
\downarrow & & \downarrow \\
H^{k+1}(U \cup V) & \xrightarrow{\mathfrak{I}_{k+1}} & \mathcal{H}^{k+1}(U \cup V) \\
\downarrow & & \downarrow \\
\vdots & & \vdots
\end{array}$$

Our assumption dictates that the two homomorphisms before the homomorphism $\mathfrak{I}_k : H^k(U \cup V) \to \mathcal{H}^k(U \cup V)$ and the other two after that are isomorphisms denoted by the symbol $\simeq$. Then Theorem 1.2.2 (the five lemma) states that $\mathfrak{I}_k$ must be an isomorphism. Hence $U \cup V$ is a de Rham manifold. □

Finally, we have to prove the following lemma:

Lemma 8.8.5. *Let M be a smooth m-dimensional second countable manifold. Assume that $P(U)$ denotes a property associated with an open subset U of M satisfying the four conditions given below:*

(i). *$P(\emptyset)$ is true.*

(ii). *$P(U)$ is true for any U diffeomorphic to a convex open subset of $\mathbb{R}^m$.*

(iii). *If $P(U), P(V)$ and $P(U \cap V)$ are true, then $P(U \cup V)$ is also true.*

(vi). *If $\{U_i : i \in \mathbb{N}\}$ is a sequence of pairwise disjoint open subsets and $P(U_i)$ is true for each $i \in \mathbb{N}$, then $P\big(\bigcup_{i=1}^{n} U_i\big)$ is also true.*

In that case $P(M)$ will also be true. This property satisfies also the above conditions for all convex open subsets of $\mathbb{R}^m$.

Since M is second countable, then it is expressed as a countable union of open sets. Every open set is covered by open sets of some charts. Hence, every open set $U \subseteq M$ is diffeomorphic to an open set of $\mathbb{R}^m$. Therefore, to prove the lemma it suffices to show that the property is true for an open set in $\mathbb{R}^m$. Since $\mathbb{R}^m$ is second countable [*see p.* 70] an open set is expressible as a countable union of open balls B. We know that open balls in $\mathbb{R}^m$ are convex sets [*see p.* 328]. Thus, $P(\mathrm{B})$ is true on open balls. Moreover, it is straightforward to see that intersection of two open balls is also convex. Hence, (iii) implies that countable unions of open balls are convex. Thus, open sets in $\mathbb{R}^m$ are convex and in view of (ii) $P(U)$ is true. Let us now suppose that $P(U)$ and $P(V)$ are true. Since $U \cap V$ is an open set, it is diffeomorphic to a convex open set in $\mathbb{R}^m$ so $P(U \cap V)$ is true. By (iii), we conclude that $P(U \cup V)$ is also true. Therefore the property must be true for a countable union of open sets. Consequently $P(M)$ is true. □

We can now easily prove the de Rham theorem.

Theorem 8.8.2. *Let M be a locally compact, second countable and oriented smooth manifold. The homomorphism $\mathring{\mathcal{F}}_k : H^k(M) \to \mathcal{H}_k(M)$ is an isomorphism.*

In order to prove this theorem, we have to show that such a manifold is a de Rham manifold. Let us define a property P associated with an open subset of M as being a de Rham manifold. The condition (i) in Lemma 8.8.5 is met due to the fact that $H^k(\emptyset) = 0$ because there are no k-forms on the empty set and $\mathcal{H}_k(\emptyset) = 0$ since there are no homomorphisms from $\emptyset$ to $\mathbb{R}$. Lemmas 8.8.2 and 8.8.4 indicate that P satisfies the conditions (ii) and

(iii) while Lemma 8.8.3 implies that the condition (iv) is also satisfied. Thus M becomes a de Rham manifold. □

Since $\mathcal{H}_k(M)$ and $H_k(M)^*$ are isomorphic, then the homomorphism $H^k(M) \to H_k(M)^*$ is also an isomorphism for such manifolds. Of course, the foregoing results will naturally be valid for compact, second countable manifolds.

An interested reader is suggested to consult to Hodge (1952) and de Rham (1955) for a more detailed proof of the de Rham theorem. For a sheaf-theoretic treatment that is probably the most direct and elegant way to show this theorem we refer to Singer and Thorpe (1967) or Warner (1971). However, to investigate the theory sheaves transcends the intended level of this work.

When $\overset{\circ}{\mathcal{F}}_k$ is an isomorphism and $H^k(M)$ *is finite-dimensional*, then the dual space $H_k(M)^*$ and, consequently, the vector space $H_k(M)$ are of finite and the same dimension. Hence, if $b_k(M) < \infty$, we find

$$b_k(M) = b\big(H^k(M)\big) = b\big(H_k(M)^*\big) = b\big(H_k(M)\big).$$

Isomorphism implies that $\overset{\circ}{\mathcal{F}}_k$ is a bijective, namely, injective and surjective mapping. *Injectiveness requires that if* $\overset{\circ}{\mathcal{F}}_k([\omega]) = \mathbf{0} \in H_k(M)^*$, *then we get* $[\omega] = [\mathbf{0}]$. The ***period*** of a closed form ω, consequently, of the equivalence class produced by this form over a cycle c_k is defined by

$$\pi(c_k) = \int_{c_k} \omega.$$

Therefore, vanishing of the functional $\overset{\circ}{\mathcal{F}}_k([\omega])$ means that all periods of the equivalence class $[\omega]$ are zero. The equality $[\omega] = [\mathbf{0}]$ implies that the form ω is exact. On the other hand, the Stokes theorem indicates that if a form ω is exact, then all of its periods vanish on every cycle c_k. Hence, it follows from de Rham's theorem that *a closed k-form* ω *is exact if and only if all of its periods are zero. That the mapping* $\overset{\circ}{\mathcal{F}}_k$ *is surjective amounts to say that every linear functional on the vector space* $H_k(M)$ *is generated through a closed k-form.* Since such a linear functional is prescribed by its value on every cycle, de Rham's theorem leads to the following conclusion: *when we assign a number* $\pi(c_k) \in \mathbb{R}$ *to every cycle* $c_k \in \overset{\circ}{C}_k(M)$, *there exists a closed k-form* ω_0 *admitting these numbers as its periods, namely, verifying the relation* $\pi(c_k) = \int_{c_k} \omega_0$ *for every k-cycle* c_k, *if only these numbers satisfy the conditions* $\pi\Big(\sum_i a_i c_k^{(i)}\Big) = \sum_i a_i \pi\big(c_k^{(i)}\big)$, $a_i \in \mathbb{R}$ *and if* $c_k \in B_k(M)$, *then one must have* $\pi(c_k) = 0$.

8.9. HARMONIC FORMS. THEORY OF HODGE-DE RHAM

Let $(M,\mathcal{G})$ be an m-dimensional complete Riemannian manifold. A form $\omega\in\Lambda^k(M)$ is given by $\omega=\frac{1}{k!}\,\omega_{i_1\cdots i_k}dx^{i_1}\wedge\cdots\wedge dx^{i_k}$. We know that the *Hodge dual* of this form is the form $*\omega=\frac{1}{k!}\,\omega^{i_1\cdots i_k}\mu_{i_k\cdots i_1}\in\Lambda^{m-k}(M)$ [*see* (5.9.20)]. Contravariant components in this expression are related to the covariant components by $\omega^{i_1\cdots i_k}=g^{i_1j_1}\cdots g^{i_kj_k}\omega_{j_1\cdots j_k}$. If we take two forms $\omega,\sigma\in\Lambda^k(M)$ into consideration, we have already known that the identity (5.9.27) enables us to write

$$\begin{aligned}\omega\wedge*\sigma=\sigma\wedge*\omega&=\frac{1}{k!}\,\omega_{i_1\cdots i_k}\sigma^{i_1\cdots i_k}\mu\\&=\frac{1}{k!}\,\omega^{i_1\cdots i_k}\sigma_{i_1\cdots i_k}\mu\in\Lambda^m(M)\end{aligned}\tag{8.9.1}$$

where μ is the volume form given by (5.9.13) or (5.9.14). Thus, for every form $\omega\in\Lambda^k(M)$ we can write

$$\omega\wedge*\omega=\frac{1}{k!}\,\omega_{i_1\cdots i_k}\omega^{i_1\cdots i_k}\mu.$$

Because the Riemannian manifold is complete, we may assume that the metric tensor is positive definite so that we must have

$$\omega_{i_1\cdots i_k}\omega^{i_1\cdots i_k}=g_{i_1j_1}\cdots g_{i_kj_k}\omega^{i_1\cdots i_k}\omega^{j_1\cdots j_k}>0\tag{8.9.2}$$

if $\omega\neq 0$. Hence, an ***inner product*** on $\Lambda^k(M)$, that is a vector space on real numbers $\mathbb{R}$, may be defined as follows

$$(\omega,\sigma)_k=(\sigma,\omega)_k=\int_M\omega\wedge*\sigma=\frac{1}{k!}\int_M\omega_{i_1\cdots i_k}\sigma^{i_1\cdots i_k}\mu\in\mathbb{R}\tag{8.9.3}$$

due to the property (8.9.1). It is easily verified that (8.9.3) obeys all rules imposed on an inner product [*see p.* 68]. *When M is a compact manifold, the integral* (8.9.3) *will always exist.* We immediately recognise that the mapping $(\,\cdot\,,\,\cdot\,)_k:\Lambda^k(M)\times\Lambda^k(M)\to\mathbb{R}$ so defined is a *symmetric bilinear functional* on real numbers. Because of (8.9.2) we get $(\omega,\omega)_k\geq 0$ which becomes zero if and only if $\omega=0$. The non-negative number

$$\|\omega\|_k=\sqrt{(\omega,\omega)_k}\geq 0\tag{8.9.4}$$

may now be called the ***norm*** of a form $\omega\in\Lambda^k(M)$. Since $\Lambda^k(M)$ equipped with (8.9.3) becomes an ***inner product space***, the well known ***Schwarz***

inequality must be satisfied:

$$(\omega, \sigma)_k \leq \|\omega\|_k \|\sigma\|_k. \tag{8.9.5}$$

This inner product on $\Lambda^k(M)$ can easily be extended to an inner product on the graded algebra $\Lambda(M) = \bigoplus_{k=0}^{m} \Lambda^k(M)$. We take the forms $\omega \in \Lambda^k(M)$ and $\sigma \in \Lambda^l(M)$ into account so that we get $\omega \wedge *\sigma \in \Lambda^{m+k-l}(M)$. This form vanishes identically if $k > l$. Its degree is less than m if $k < l$, hence its integral over M cannot be defined. If we adopt the convention that such an integral also vanishes, we can then define the inner product of two arbitrary k-form ω and l-form σ in $\Lambda(M)$ in the following manner

$$(\omega, \sigma) = \begin{cases} (\omega, \sigma)_k & \textit{if } k = l, \\ = 0 & \textit{if } k > l, \\ = 0 & \textit{if } k < l. \end{cases}$$

This definition amounts to admit that the vector spaces $\Lambda^k(M)$ and $\Lambda^l(M)$ are *orthogonal* with respect to this inner product whenever $k \neq l$. Henceforth, by adopting this definition we shall not designate the inner product as dependent on the index k. In view of the definition (8.9.3), we obtain the following relation for forms $\omega, \sigma \in \Lambda^k(M)$

$$\begin{aligned}(*\omega, *\sigma) &= \int_M *\omega \wedge **\sigma = (-1)^{k(m-k)} \int_M *\omega \wedge \sigma \\ &= (-1)^{2k(m-k)} \int_M \sigma \wedge *\omega = \int_M \sigma \wedge *\omega \\ &= \int_M \omega \wedge *\sigma = (\omega, \sigma).\end{aligned}$$

This means that the Hodge star operator $* : \Lambda^k(M) \to \Lambda^{m-k}(M)$, which is obviously a linear operator, preserves the inner product. It is well known that a linear operator between two inner product spaces that preserves the inner product is called a ***unitary*** or ***conformal operator***. Consequently, *the Hodge star operator $*$ is a unitary or conformal operator on the exterior algebra with respect to the inner product so defined.*

Let us now consider the forms $\omega \in \Lambda^{k-1}(M)$ and $\sigma \in \Lambda^k(M)$ and evaluate the exterior derivative of the form $\omega \wedge *\sigma \in \Lambda^{m-1}(M)$ to obtain:

$$d(\omega \wedge *\sigma) = d\omega \wedge *\sigma + (-1)^{k-1} \omega \wedge d(*\sigma) \in \Lambda^m(M).$$

On making use of the relation $*\delta = (-1)^k d*$ between the operators of co-differential δ and the exterior derivative d [*see* p. 283] we arrive at the expression

$$d(\omega \wedge *\sigma) = d\omega \wedge *\sigma - \omega \wedge *\delta\sigma.$$

Then the Stokes theorem yields

$$\int_M (d\omega \wedge *\sigma - \omega \wedge *\delta\sigma) = \int_M d(\omega \wedge *\sigma) = \int_{\partial M} \omega \wedge *\sigma$$

so that we get

$$\int_M d\omega \wedge *\sigma = \int_M \omega \wedge *\delta\sigma + \int_{\partial M} \omega \wedge *\sigma.$$

On recalling the definition of the inner product, we thus conclude that

$$(d\omega, \sigma) = (\omega, \delta\sigma) + \int_{\partial M} \omega \wedge *\sigma.$$

If M is a manifold without boundary ($\partial M = \emptyset$), we necessarily have to write $\int_{\partial M} \omega \wedge *\sigma = 0$ to obtain

$$(d\omega, \sigma) = (\omega, \delta\sigma). \tag{8.9.6}$$

According to the foregoing relation, we are led to the following conclusion: *let $\Lambda(M)$ be the graded exterior algebra on a compact manifold without boundary. The operators on the exterior algebra $d : \Lambda^k(M) \to \Lambda^{k+1}(M)$ and $\delta : \Lambda^k(M) \to \Lambda^{k-1}(M)$ are* ***adjoint operators*** *on $\Lambda(M)$.* This result will also be valid for all forms with compact support on a manifold with boundary. Because such forms will necessarily vanish on the boundary of the manifold.

The Laplace-de Rham operator $\delta d + d\delta = \Delta : \Lambda^k(M) \to \Lambda^k(M)$ was defined by (5.9.31). When $\omega, \sigma \in \Lambda^k(M)$, (8.9.6) together with the symmetry of the inner product leads to the result

$$\begin{aligned}(\Delta\omega, \sigma) &= (\delta d\omega, \sigma) + (d\delta\omega, \sigma) = (d\omega, d\sigma) + (\delta\omega, \delta\sigma) \\ &= (\omega, \delta d\sigma) + (\omega, d\delta\sigma) = \big(\omega, (\delta d + d\delta)\sigma\big) \\ &= (\omega, \Delta\sigma).\end{aligned}$$

Hence, with respect to this inner product the operator Δ on $\Lambda(M)$ becomes a ***self-adjoint operator*** if M is a manifold without boundary. It follows from the above relation that we obtain

$$(\Delta\omega, \omega) = (d\omega, d\omega) + (\delta\omega, \delta\omega) = \|d\omega\|^2 + \|\delta\omega\|^2 \geq 0 \tag{8.9.7}$$

for all $\omega \in \Lambda(M)$. Since $(\Delta\omega, \omega) \geq 0$ for all $\omega \in \Lambda(M)$ we may describe Δ as a ***positive definite*** or ***elliptic operator***.

A form $\omega \in \Lambda(M)$ is called a ***harmonic form*** if $\Delta\omega = 0$.

Theorem 8.9.1. *Let M be a compact and oriented Riemannian manifold without boundary. We consider a form $\omega \in \Lambda^k(M)$. The form ω is harmonic if and only if $d\omega = 0$ and $\delta\omega = 0$.*

If $d\omega = 0$ and $\delta\omega = 0$, then the definition leads to $\Delta\omega = 0$. Conversely let us assume that $\Delta\omega = 0$. Then it follows from (8.9.7) that $0 = (0, \omega) = \|d\omega\|^2 + \|\delta\omega\|^2$. Hence, we get $\|d\omega\| = \|\delta\omega\| = 0$ from which we deduce that $d\omega = 0$ and $\delta\omega = 0$. □

Consequently, *all harmonic forms are also closed on manifolds to which the above theorem might be applied.* We had previously mentioned that all harmonic forms $\omega \in \Lambda^k(M)$ holding the condition $\Delta\omega = 0$ constitute the following subspace

$$\mathrm{H}^k(M) = \{\omega \in \Lambda^k(M) : \Delta\omega = 0\} = \mathcal{N}(\Delta)$$

on real numbers.

We had seen that $\Delta f = \delta d f = \nabla^2 f$ when $f \in \Lambda^0(M)$. The operator ∇^2 was defined by (5.9.33). Therefore, the solution of the Laplace equation $\nabla^2 f = 0$ on a compact manifold without boundary must satisfy the condition $df = 0$. Hence, *the harmonic function f can only be a constant number*. This result is a sort of generalisation of the well known ***Liouville theorem*** [French mathematician Joseph Liouville (1809-1882)].

Probably the most important theorem concerning harmonic forms has been demonstrated by Hodge. Because the proof of this theorem is quite difficult and requires a rather good knowledge of functional analysis and properties of elliptic operators, we shall not be able to present its proof here in its full generality.

Theorem 8.9.2. (The Hodge Decomposition Theorem). *Let M be a compact and oriented Riemannian manifold without boundary. For each form $\omega \in \Lambda^k(M)$, there exist forms $\alpha \in \Lambda^{k-1}(M)$, $\beta \in \Lambda^{k+1}(M)$ and $\gamma \in \mathrm{H}^k(M)$ so that one can express ω as*

$$\omega = d\alpha + \delta\beta + \gamma \tag{8.9.8}$$

and this representation is unique. We can thus write symbolically

$$\Lambda^k(M) = d\Lambda^{k-1}(M) \oplus \delta\Lambda^{k+1}(M) \oplus \mathrm{H}^k(M).$$

All these subspaces are mutually orthogonal.

We can easily show the orthogonality of subspaces. Since $\delta^2 = 0$, we find that

$$(d\alpha, \delta\beta) = (\alpha, \delta^2\beta) = 0.$$

On the other hand, if $\Delta\gamma = 0$, then we have $d\gamma = 0, \delta\gamma = 0$ so that we are led to the results

$$(d\alpha, \gamma) = (\alpha, \delta\gamma) = 0, \quad (\delta\beta, \gamma) = (\beta, d\gamma) = 0.$$

The difficult part of the theorem is to show the existence of the forms α, β, γ satisfying the relation (8.9.8). As to this part, the interested readers may be referred to Warner (1971, Ch. 6). In order to prove the uniqueness, let us suppose that there are two representations of this form:

$$\omega = d\alpha_1 + \delta\beta_1 + \gamma_1 = d\alpha_2 + \delta\beta_2 + \gamma_2.$$

Hence, if we denote $\alpha = \alpha_1 - \alpha_2, \beta = \beta_1 - \beta_2, \gamma = \gamma_1 - \gamma_2$, we realise that the condition below should be satisfied:

$$d\alpha + \delta\beta + \gamma = 0.$$

On evaluating the exterior derivative of that expression, we get $d\delta\beta = 0$ from which we deduce that $0 = (d\delta\beta, \beta) = (\delta\beta, \delta\beta) = \|\delta\beta\|^2$ and $\delta\beta = 0$. So the foregoing equality is reduced to $d\alpha + \gamma = 0$. The co-differential of this last expression yields $\delta d\alpha = 0$. Thus, we can obtain at once $0 = (\delta d\alpha, \alpha) = (d\alpha, d\alpha) = \|d\alpha\|^2$ and $d\alpha = 0$ implying that $\gamma = 0$. Hence, we find that $d\alpha_1 = d\alpha_2, \delta\beta_1 = \delta\beta_2, \gamma_1 = \gamma_2$. □

Theorem 8.9.3. *Let M be a compact and oriented Riemannian manifold without boundary. The solution of the equation $\Delta\omega = \sigma$ where the form $\sigma \in \Lambda^k(M)$ is prescribed does exist if and only if the form σ is orthogonal to the vector space $\mathrm{H}^k(M)$, in other words, if $(\sigma, \lambda) = 0$ for all $\lambda \in \mathrm{H}^k(M)$.*

Let us first assume that $\Delta\omega = \sigma$ and $\lambda \in \mathrm{H}^k(M)$. We then obtain

$$(\sigma, \lambda) = (\Delta\omega, \lambda) = (\omega, \Delta\lambda) = (\omega, 0) = 0.$$

Conversely, we suppose that the form σ satisfies the condition $(\sigma, \lambda) = 0$ for all $\lambda \in \mathrm{H}^k(M)$. On utilising the Hodge decomposition, we may write $\sigma = d\alpha + \delta\beta + \gamma$ so that the above condition yields

$$\begin{aligned} 0 &= (\sigma, \gamma) = (d\alpha, \gamma) + (\delta\beta, \gamma) + (\gamma, \gamma) \\ &= (\alpha, \delta\gamma) + (\beta, d\gamma) + (\gamma, \gamma) \\ &= (\gamma, \gamma) = \|\gamma\|^2 \end{aligned}$$

and we find that $\gamma = 0$. Hence, the form σ can only take the shape $\sigma = d\alpha + \delta\beta$. Let us now write $\omega = \omega_1 + \omega_2$ and try to determine the solutions of the equations $\Delta\omega_1 = d\alpha$ and $\Delta\omega_2 = \delta\beta$ separately. If we employ the representation $\alpha = d\alpha_1 + \delta\beta_1 + \gamma_1$, we get $d\alpha = d\delta\beta_1$. If we in turn write $\beta_1 = d\alpha_2 + \delta\beta_2 + \gamma_2$ and note that $d^2 = 0$, we obtain

$$d\alpha = d\delta\beta_1 = d\delta\, d\alpha_2 = (d\delta + \delta d)d\alpha_2 = \Delta(d\alpha_2).$$

Therefore, the equation $\Delta\omega_1 = d\alpha$ admits a solution in the form $\omega_1 = d\alpha_2$. Similarly, we take $\beta = d\alpha_3 + \delta\beta_3 + \gamma_3$ to obtain $\delta\beta = \delta d\alpha_3$ by noting that $\delta^2 = 0$. By using the representation $\alpha_3 = d\alpha_4 + \delta\beta_4 + \gamma_4$, we find that

$$\delta\beta = \delta d\alpha_3 = \delta d\delta\beta_4 = (\delta d + d\delta)\delta\beta_4 = \Delta(\delta\beta_4).$$

Therefore, we conclude that the equation $\Delta\omega_2 = \delta\beta$ admits a solution in the form $\omega_2 = \delta\beta_4$. Ultimately, we find that the equation $\Delta\omega = \sigma$ possesses a solution in the form $\omega = d\alpha_2 + \delta\beta_4$. □

We have above touched upon the fact that harmonic forms are closed. Hence, there exists a linear operator $\mathcal{I} : \mathrm{H}^k(M) \to \mathcal{C}^k(M)$ embedding $\mathrm{H}^k(M)$ into $\mathcal{C}^k(M)$. Let $\pi : \mathcal{C}^k(M) \to H^k(M)$ be the linear canonical mapping. We are thus led to the conclusion that there exists a linear transformation $\psi = \pi \circ \mathcal{I} : \mathrm{H}^k(M) \to H^k(M)$ between the vector space of harmonic k. forms and the relevant cohomology group.

Theorem 8.9.4. *Let M be a compact and oriented Riemannian manifold without boundary. The vector spaces $\mathrm{H}^k(M)$ and $H^k(M)$ are isomorphic.*

In order to prove this theorem, we have to show that the linear operator ψ introduced above is bijective. Let us first assume that $\omega \in \mathrm{H}^k(M)$ and $[\omega] = \psi(\omega) = [\mathbf{0}] \in H^k(M)$. This means that ω is an exact form and one writes $\omega = d\sigma$. But, we get $(\omega, d\sigma) = (\delta\omega, \sigma) = 0$ since $\delta\omega = 0$. We thus arrive at the relation $(\omega, \omega) = \|\omega\|^2 = 0$ implying that $\omega = 0$. This amounts to say that ψ is injective. We now consider an arbitrary cohomology class $[\omega] \in H^k(M)$. ω is a representative of this equivalence class. Due to the Hodge decomposition theorem, we can write $\omega = d\alpha + \delta\beta + \gamma$. Since $d\omega = 0$, we find that $d\delta\beta = 0$. It then follows just as above that $\delta\beta = 0$ which implies that *a closed form ω is represented as* $\omega = d\alpha + \gamma$. This of course gives $[\omega] = [\gamma]$ so that one is able to write $[\omega] = \psi(\gamma)$ where $\gamma \in \mathrm{H}^k(M)$. Consequently, we see that ψ is surjective. As a result, ψ is identified as an isomorphism. Hence, the vector spaces $\mathrm{H}^k(M)$ and $H^k(M)$ are isomorphic. Accordingly, we can say that *every cohomology class has a harmonic representative* in manifolds complying with the assumptions of the theorem. □

According to a property that we shall again not be able prove here, the null space of the operator Δ, or more generally of a linear elliptic operator, is finite-dimensional if M is a compact Riemannian manifold [interested readers may be referred to Warner (1971)]. Therefore, on such a manifold the vector space $\mathrm{H}^k(M)$ is finite-dimensional. Since isomorphic spaces must have the same dimension we can now state that *dimensions of cohomology groups, that is, Betti numbers on a compact and oriented Riemannian manifold without boundary are all finite.*

We have seen while proving the above theorem that any closed form $\omega \in \mathcal{C}^k(M)$ on a compact and oriented Riemannian manifold without boundary is expressible as $\omega = d\alpha + \gamma$ where $\alpha \in \Lambda^{k-1}(M), \gamma \in \mathrm{H}^k(M)$. If c_k is a k-cycle, then we can write

$$\begin{aligned}\pi(c_k) &= \int_{c_k} \omega = \int_{c_k} d\alpha + \int_{c_k} \gamma \\ &= \int_{\partial c_k} \alpha + \int_{c_k} \gamma = \int_{c_k} \gamma.\end{aligned}$$

This is tantamount to say that *there exists a unique harmonic k-form γ possessing the same periods as a closed k-form ω on such kind of manifolds.*

8.10. POINCARE DUALITY

Let M be an m-dimensional compact, oriented, smooth Riemannian manifold without boundary. We shall now introduce a bilinear functional $P : H^k(M) \times H^{m-k}(M) \to \mathbb{R}$ through the following relation

$$P([\omega], [\sigma]) = \int_M \omega \wedge \sigma, \tag{8.10.1}$$

where $[\omega] \in H^k(M)$ and $[\sigma] \in H^{m-k}(M)$, and the forms $\omega \in \mathcal{C}^k(M)$ and $\sigma \in \mathcal{C}^{m-k}(M)$ are arbitrary representatives of these cohomology classes. In order that the functional (8.10.1) known as the ***Poincaré form*** proves to be meaningful, it must be independent of the selection of the representatives of equivalence classes. This property can be shown quite easily. We consider the forms $\alpha \in \Lambda^{k-1}(M)$, $\beta \in \Lambda^{m-k-1}(M)$. If we note that $d\omega = 0$ and $d\sigma = 0$, the Stokes theorem results in the expression

$$\begin{aligned}\int_M (\omega + d\alpha) \wedge (\sigma + d\beta) &= \int_M (\omega \wedge \sigma + d\alpha \wedge \sigma + \omega \wedge d\beta + d\alpha \wedge d\beta) \\ = \int_M \omega \wedge \sigma &+ \int_M d(\alpha \wedge \sigma) + (-1)^k \int_M d(\omega \wedge \beta) + \int_M d(\alpha \wedge d\beta) \\ = \int_M \omega \wedge \sigma &+ \int_{\partial M} \alpha \wedge \sigma + (-1)^k \int_{\partial M} \omega \wedge \beta + \int_{\partial M} \alpha \wedge d\beta = \int_M \omega \wedge \sigma.\end{aligned}$$

Boundary integrals vanish because we have assumed that $\partial M = \emptyset$. Hence, we can write $P([\omega],[\sigma]) = P(\omega,\sigma)$ for arbitrary representatives. We shall now demonstrate that this bilinear form is non-degenerate. To this end, it would suffice to determine an equivalence class $[\sigma] \neq [\mathbf{0}]$ such that $P([\omega],[\sigma]) \neq 0$ whenever $[\omega] \neq [\mathbf{0}]$. Let $[\omega] \in H^k(M)$ be a non-zero cohomology class. We choose the form $\omega \in \mathrm{H}^k(M)$ as the harmonic representative of that cohomology class. We thus write $\Delta\omega = 0$. The form ω cannot be identically zero since $[\omega] \neq [\mathbf{0}]$. Let us next consider the Hodge dual $*\omega \in \Lambda^{m-k}(M)$ of the form ω. We had obtained the relation $*\Delta = \Delta *$ on p. 284. We thus find $\Delta *\omega = *\Delta\omega = 0$ implying that $*\omega \in \mathrm{H}^{m-k}(M)$. This form can be chosen as the harmonic representative of the cohomology class $[*\omega] \in H^{m-k}(M)$. Next, we take $[\sigma] = [*\omega]$. Hence, we conclude that

$$P(\omega, *\omega) = \int_M \omega \wedge *\omega = (\omega,\omega) = \|\omega\|^2 \neq 0$$

On the other hand, this relation signifies that $P(\omega, *\omega) = 0$ if and only if $\omega = 0$. Hence, the bilinear form P is non-degenerate. Let us fix a class $[\omega]$ in (8.10.1). Then a linear functional $L([\omega])$ on the vector space $H^{m-k}(M)$ can be introduced by the relation

$$L([\omega])([\sigma]) = P([\omega],[\sigma]).$$

Thus the bilinear form $P([\omega],[\sigma])$ induces a linear transformation

$$L : H^k(M) \to \big[H^{m-k}(M)\big]^*. \tag{8.10.2}$$

We can realise right away that the non-degeneracy of the bilinear form P secures that the linear operator L is injective. On the other hand, we know that the dimensions of $H^k(M)$ and $H^{m-k}(M)$, consequently, that of the dual $\big[H^{m-k}(M)\big]^*$ is finite. In this case, L becomes an isomorphism so that the spaces $H^k(M)$ and $\big[H^{m-k}(M)\big]^*$ are isomorphic. Because a finite-dimensional vector space and its dual are isomorphic, we thus infer that the spaces $H^k(M)$ and $H^{m-k}(M)$ are isomorphic. This property is called the ***Poincaré duality***. Therefore, we can regard these two spaces as the same as far as their algebraic properties are concerned. Hence, the Betti numbers of compact, oriented Riemannian manifolds without boundary must satisfy the relation

$$b_k(M) = b\big(H^k(M)\big) = b\big(H^{m-k}(M)\big) = b_{m-k}(M). \tag{8.10.3}$$

If the dimension of the manifold is an odd number, then its Euler-Poincaré characteristic becomes

$$\chi(M) = \sum_{k=0}^{m} (-1)^k b_k(M) = 0.$$

In fact, we find in this case $(-1)^{m-k} b_{m-k}(M) = (-1)^{k+1} b_k(M)$ and the corresponding terms cancel each other in the above sum.

According to the Poincaré duality, the vector spaces $H^m(M)$, $H^0(M)$ are isomorphic in a compact, oriented Riemannian manifold without boundary. We know that $H^0(M) = \mathbb{R}$ when M is *connected*. Thus, in this sort of manifolds the cohomology group $H^m(M)$ is isomorphic to $\mathbb{R}$. Furthermore, it is possible to show that $H^{m-1}(M) = 0$ if the manifold M is *simply connected*, that is, if every closed curve on M can be contracted smoothly to a point inside the curve. Indeed, due to the Poincaré duality, the vector spaces $H^{m-1}(M)$ and $H^1(M)$ are isomorphic. In local coordinates, let us write a form $\omega \in \Lambda^1(M)$ as $\omega = \omega_i dx^i$ where $i = 1, \ldots, m$. If ω is closed, then the condition

$$d\omega = \omega_{i,j} dx^j \wedge dx^i = 0 \quad \text{or} \quad \omega_{[i,j]} = 0$$

must be satisfied. If M is simply connected, then it is well known that the general solution of the following system of partial differential equations

$$\frac{\partial \omega_i}{\partial x^j} - \frac{\partial \omega_j}{\partial x^i} = 0$$

is provided as follows

$$\omega_i = \frac{\partial f}{\partial x^i}$$

where $f \in \Lambda^0(M)$. Thereby, we get $\omega = f_{,i}\, dx^i = df$. Thus, on such kind of manifolds every closed 1-form is exact. Therefore, we find $H^1(M) = 0$, and consequently, $H^{m-1}(M) = 0$.

VIII. EXERCISES

8.1. Show that a k-dimensional submanifold of a manifold M is orientable if one can find a k-form that vanishes nowhere on this submanifold.

8.2. Show that the Cartesian product of orientable manifolds is also an orientable manifold.

8.3. Show that the Cartesian product of non-orientable manifolds is also a non-orientable manifold.

8.4. Show that the Klein bottle is non-orientable.

8.5. Show that the Lie groups $GL(n, \mathbb{R})$ and $GL(n, \mathbb{C})$ are orientable.

8.6. Show that if a form $\omega \in \Lambda^k(M)$ satisfies the condition $\int_{\sigma_k} \omega = 0$ on *every* singular k-simplex, then one has $\omega = 0$.

8.7. Show that if forms $\omega_1, \omega_2 \in \Lambda^k(M)$ satisfy the condition $\int_{\sigma_k} \omega_1 = \int_{\sigma_k} \omega_2$ on *every* singular k-simplex σ_k, then one has $\omega_1 = \omega_2$.

8.8. Show that a form $\omega \in \Lambda^k(M)$ turns out to be closed if it satisfies the condition $\int_{\partial\sigma_{k+1}} \omega = 0$ on *every* singular $(k+1)$-simplex.

8.9. Show that the volume form on the hyperbolic plane H^2 (*see* Exercise **7.9**) is given by

$$\mu = \frac{r}{\sqrt{1+r^2}}\, dr \wedge d\theta = \sinh s\, ds \wedge d\theta.$$

Find the volume of the subregion of H^2 satisfying the condition $1 \leq x_0 \leq 2$.

8.10. U is an m-dimensional compact submanifold with boundary of an m-dimensional Riemannian manifold. Show that for a vector field $V \in T(M)$ one is able to write

$$\int_U \operatorname{div} V \mu = \int_{\partial U} \mathbf{i}_V \mu.$$

8.11. We consider the simplex $s_2 = (P_0, P_1, P_2)$ in $\mathbb{R}^2$ where $P_0 = (0,0)$, $P_1 = (1,0)$, $P_2 = (0,2)$. Evaluate the integral of the 1-form

$$\omega = (x^2 + 7y)\, dx + (y \sin y^2 - x)\, dy$$

on the cycle ∂s_2.

8.12. Show that the form $\omega = (2x + y\cos xy)\, dx + x\cos xy\, dy \in \Lambda^1(\mathbb{R}^2)$ is exact. Find the integral of this form on the cycle defined in Exercise. **8.11**.

8.13. Show that a form $\omega \in \Lambda^2(\mathbb{S}^2)$ is exact if only the following condition is met

$$\int_{\mathbb{S}^2} \omega = 0.$$

8.14. We consider the form $\omega = x^1 dx^2 \wedge dx^3 \in \Lambda^2(\mathbb{R}^3)$ where $(x^1, x^2, x^3) \in \mathbb{R}^3$. Show that

$$\int_{\mathbb{S}^2} \omega|_{\mathbb{S}^2} = \frac{4\pi}{3} R^3$$

where R is the radius of the sphere $\mathbb{S}^2$. Since the form $\omega|_{\mathbb{S}^2} \in \Lambda^2(\mathbb{S}^2)$ is clearly closed, this result indicates the fact that that every closed 2-form on $\mathbb{S}^2$ is not necessarily exact.

8.15. Show that every closed 1-form on $\mathbb{S}^2$ is exact..

8.16. Show that the restriction $\omega|_{\mathbb{S}^{n-1}} \in \Lambda^{n-1}(\mathbb{S}^{n-1})$ to the sphere $\mathbb{S}^{n-1}$ of the form $\omega = \epsilon_{ii_1\cdots i_{n-1}} x^i dx^{i_1} \wedge \cdots \wedge dx^{i_{n-1}} \in \Lambda^{n-1}(\mathbb{R}^n)$ is a closed form that is not exact and it does vanish nowhere on $\mathbb{S}^{n-1}$.

8.17. Let us consider the manifold $M = \mathbb{R}^n - \{\mathbf{0}\}$ and the form

$$\omega = \frac{x^1 dx^1 + x^2 dx^2 + \cdots + x^n dx^n}{\left((x^1)^2 + (x^2)^2 + \cdots + (x^n)^2\right)^{n/2}} \in \Lambda^1(M).$$

Determine the form $*\omega$ and show that it is closed. Evaluate the integral

$$\int_{\mathbb{S}^{n-1}} *\omega.$$

Is the form $*\omega$ exact?

8.18. Let us consider a form $\omega \in \Lambda^{m-1}(M)$ on an m-dimensional compact and orientable manifold M without boundary ($\partial M = \emptyset$). Show that there exists a point $p \in M$ such that $d\omega(p) = 0$.

8.19. Let G be a compact and oriented Lie group. We define the mapping $\iota(g) = g^{-1}$ for every $g \in G$. Show that we can write the following relation for every continuous function f on G

$$\int_G f = \int_G f \circ \iota.$$

8.20. Let us consider the functions $f, g \in \Lambda^0(\mathbb{R}^n)$ and a finite region $D \subset \mathbb{R}^n$. By employing the Stokes theorem, derive the Green formula given below

$$\int_{\partial D} \big(f*(dg) - g*(df)\big) = -\int_D (f\Delta g - g\Delta f)\mu.$$

8.21. Let us consider the manifold $M = \mathbb{R}^5$ with a coordinate cover (x, t, θ, u, v) and the forms

$$\omega^1 = du \wedge dt + dv \wedge dx \in \Lambda^2(\mathbb{R}^5), \quad \omega^2 = d\theta - u\,dx - v\,dt \in \Lambda^1(\mathbb{R}^5)$$

Let $\mathcal{I}$ be the ideal generated by these forms. Assume that the 2-dimensional solution submanifold of the ideal $\mathcal{I}$ is prescribed by the mapping $\theta = \theta(x, t)$, $u = u(x, t)$ and $v = v(x, t)$. Show that the *wave equation* $\theta_{xx} - \theta_{tt} = 0$ is satisfied on the solution submanifold. Determine the conservation laws of this equation.

8.22. Show that a form $\omega \in \Lambda^1(M)$ is exact if it satisfies the condition $\displaystyle\int_{\mathcal{C}} \omega = 0$ for *every* closed curve $\mathcal{C} \subset M$.

8.23. Show that a connected manifold M is simply connected if and only if one gets $H^1(M) = 0$.

8.24. Determine the de Rham cohomology of the annular region depicted by the condition $1 < \sqrt{x_1^2 + x_2^2} < 2$ in $\mathbb{R}^2$.

CHAPTER IX

PARTIAL DIFFERENTIAL EQUATIONS

9.1. SCOPE OF THE CHAPTER

We can say with a little bit of hyperbolism that to study partial differential equations on smooth manifolds via exterior forms is actually reduced to dealing with a kind of algebraic theory of these equations. The formal treatment of this subject must be based on the theory of jet bundles. However, we prefer here to follow a more direct and concrete path and we attempt to characterise partial differential equations by contact manifolds obtained by extending the main manifold. In Sec. 9.2, we first extend a set of partial differential equations of finite order to a system of first order equations by introducing auxiliary variables. We then show that solutions of this system coincide with solutions of a closed ideal of an exterior algebra defined on an extended manifold. The coordinate cover of this manifold consists of independent and dependent variables, and auxiliary variables corresponding to various order partial derivatives of dependent variables with respect to independent variables. The higher is the order of original system, the huger will be the dimension of the extended manifold. We call 1-forms connecting partial derivatives and auxiliary variables as contact forms and the closure of the ideal generated by them as the contact ideal. The structure of this ideal plays a significant part in the so-called algebraic theory of partial differential equations. The fundamental ideal is constructed through exterior forms describing differential equations together with the contact forms. The first approach that comes to mind to find solutions of the fundamental ideal seems to determine its characteristic vectors in order to be able to apply the Cartan theorem. But, this method proves to be quite unfruitful except for a first order non-linear partial differential equation with one dependent variable. That is the reason why we have chosen to concentrate our efforts to discuss in detail the symmetry transformations that enable us to generate a new family of solutions from a known solution. Since we know that symmetry transformations are generated by isovectors of an ideal, we are first concerned with unravelling the structure of the

Exterior Analysis, DOI: 10.1016/B978-0-12-415902-0.50009-8

isovector fields of the contact ideal in Sec. 9.3. Sec. 9.4 is devoted to the derivation of determining equations of isovector components of the fundamental ideal, especially the balance ideal associated with balance equations. Sec. 9.5 deals with the similarity solution that remains invariant under a symmetry transformation. In order to benefit substantially from a symmetry transformation, we need first to find a solution, albeit simple, of the system. This of course creates a serious problem. To overcome this obstacle to some extent, we present a method of generalised characteristics in Sec. 9.6 by making use of the isovector fields that helps us to generate a solution from given initial data satisfying certain conditions on an initial manifold. We propose another method in Sec. 9.7 by generalising the contact forms as to include undetermined coefficient functions so that one may be able to explore various possibilities to generate a solution. Some closed horizontal ideals of the exterior algebra introduced that way may prove to be instrumental in obtaining certain solutions. Finally, we investigate in Sec. 9.8 the groups of equivalence transformations that are much more general than the symmetry transformations. When we are given a family of partial differential equations, by means of such a transformation we can transform a member of the family to another member of the same family. The general solutions of the determining equations of isovector fields inducing these kind of transformations are also provided.

9.2. IDEALS FORMED BY DIFFERENTIAL EQUATIONS

We consider an n-dimensional smooth manifold M^n. A set of partial differential equations with A number of members of order m involving the dependent variables $u^\alpha,\ \alpha = 1, \ldots, N$ might be locally represented by

$$F^a(x^i, u^\alpha, u^\alpha_{,i}, u^\alpha_{,ij}, \ldots, u^\alpha_{,i_1 i_2 \cdots i_m}) = 0,\ a = 1, \ldots, A \qquad (9.2.1)$$

where the local coordinates $x^i, i = 1, \ldots, n$ in the n-dimensional open set of a chart of the atlas in M denote the independent variables. We assume that all functions F^a are differentiable with respect to their arguments. We define all partial derivatives of order r of u^α with respect to the independent variables x^i as follows

$$\frac{\partial^r u^\alpha}{\partial x^{i_1} \partial x^{i_2} \cdots \partial x^{i_r}} = u^\alpha_{,i_1 i_2 \cdots i_r},\ 1 \le i_1, i_2, \cdots, i_r \le n$$

where $i_1 + i_2 + \cdots + i_r = r$ and $0 \le r \le m$. We adopt the convention that the index i_0 does not exist, hence $u^\alpha_{,i_0} = u^\alpha$ for $r = 0$. In order to identify the global properties of solutions of the system of partial differential equations, we have to solve a rather difficult problem of joining smoothly the results found in local charts, To avoid this problem we shall usually select our manifold as the Euclidean space $M^n = \mathbb{R}^n$ and we shall suppose that the system of differential equations are defined on an open set $\mathcal{D}_n \subseteq \mathbb{R}^n$. In other words, this will mean that all future developments in this chapter will actually be of local character.

In order to study a system of partial differential equations via exterior forms we have to enlarge this system to that of first order partial differential equations by introducing auxiliary variables because of the fact that only the first order exterior derivatives are not identically nil. Introduction of auxiliary variables requires necessarily to enlarge the dimension of the relevant smooth manifold extensively.

The $(n+N)$-dimensional product manifold $G = \mathbb{R}^n \times \mathbb{R}^N = \mathbb{R}^{n+N}$ whose local coordinates are $\{x^i, u^\alpha : 1 \le i \le n, 1 \le \alpha \le N\}$ will be called the ***graph space***. A smooth mapping $\phi : \mathcal{D}_n \to G$ will be propounded as a *regular mapping* if it carries the property

$$\phi^* \mu \neq 0 \tag{9.2.2}$$

where $\mu = dx^1 \wedge \cdots \wedge dx^n$ is the volume form in $\mathbb{R}^n$. This mapping ϕ may be designated by smooth functions $x^i = \Phi^i(\xi^j), u^\alpha = \Phi^\alpha(\xi^j), 1 \le j \le n$ where $(\xi^1, \xi^2, \ldots, \xi^n) \in \mathcal{D}_n$. However, if ϕ is a regular mapping we ought to have

$$\phi^* \mu = \det\Big(\frac{\partial \Phi^i}{\partial \xi^j}\Big) d\xi^1 \wedge d\xi^2 \wedge \cdots \wedge d\xi^n \neq 0$$

due to the condition (9.2.2) which leads to $\det(\partial \Phi^i / \partial \xi^j) \neq 0$. Hence, at least locally the variables ξ^j are expressible in terms of the variables x^i so that the mapping ϕ may be equally represented by

$$u^\alpha = \Phi^\alpha\big(\xi^j(x^i)\big) = \phi^\alpha(x^i) \tag{9.2.3}$$

without loss of generality. *A function in the form* (9.2.3) *constitutes a solution of the system of differential equations* (9.2.1) *when inserted in those expressions the equality*

$$F^a(x^i, \phi^\alpha, \phi^\alpha_{,i}, \phi^\alpha_{,ij}, \ldots, \phi^\alpha_{,i_1 i_2 \cdots i_m}) \equiv 0$$

is satisfied identically.

We shall now try to represent a system of partial differential equations via exterior differential forms. In order to achieve this, we have to transform a system of higher order partial differential equations into a much larger system of first order partial differential equations by introducing auxiliary variables as we had mentioned above. To this end, let us define

$$u^\alpha_{,i_1\cdots i_r} = v^\alpha_{i_1\cdots i_r} = v^\alpha_{i_1\cdots i_{r-1},i_r} \tag{9.2.4}$$

where $0 \leq r \leq m, 1 \leq i_1, \ldots, i_r \leq n$. We take of course $v^\alpha_{i_0} = u^\alpha$. Due to their definition, the auxiliary variables $v^\alpha_{i_1\cdots i_r}$ of order r are completely symmetric in indices $i_1, \ldots, i_r$. Thus their number reduces to $N\binom{n+r-1}{r}$ from Nn^r. Hence, when we incorporate the variables u^α $(r = 0)$ into auxiliary variables, their total number reaches to

$$D = N\sum_{r=0}^{m}\binom{n+r-1}{r} = N\binom{n+m}{m} = N\frac{(n+m)!}{n!\,m!}$$

which may be quite a huge number if m is large. The $(n + D)$-dimensional manifold whose coordinate cover is given by $\{x^i, v^\alpha_{i_1i_2\cdots i_r} : 0 \leq r \leq m\}$ is called the ***jet bundle*** on the base manifold M. The theory of jet bundles that makes it possible to define various order partial derivatives on smooth manifolds has been brought forward first by French mathematician Charles Ehresmann (1905-1979). Since we will be interested in a local approach here, we shall not treat partial differential equations within the formalism of jet bundles. That is the reason why we call this manifold by a more familiar term as the ***mth order contact manifold*** and we denote by $\mathcal{C}_m$. We shall now introduce the following 1-forms on $\mathcal{C}_m$

$$\sigma^\alpha_{i_1i_2\cdots i_r} = dv^\alpha_{i_1i_2\cdots i_r} - v^\alpha_{i_1i_2\cdots i_r i}dx^i \in \Lambda^1(\mathcal{C}_m) \tag{9.2.5}$$

where $0 \leq r \leq m-1$. Their number is obviously given by $N\binom{n+m-1}{m-1}$. We name these forms as ***contact* 1*-forms***. In accordance with our convention, we evidently get $\sigma^\alpha_{i_0} = \sigma^\alpha = du^\alpha - v^\alpha_i dx^i$ for $r = 0$. Since the exterior product of all contact 1-forms may be written as

$$\bigwedge_{1\leq\alpha\leq N;1\leq i_r\leq n,0\leq r\leq m-1} \sigma^\alpha_{i_1i_2\cdots i_r} = \bigwedge_{1\leq\alpha\leq N;1\leq i_r\leq n,0\leq r\leq m-1} dv^\alpha_{i_1i_2\cdots i_r} + \cdots \neq 0$$

we see that they are linearly independent on the manifold $\mathcal{C}_m$.

The system of mth order partial differential equations (9.2.1) is now reduced to a system of first order partial differential equations described by the relations (9.2.4) and the algebraic equations

$$F^a(x^i, u^\alpha, v^\alpha_i, v^\alpha_{ij}, \ldots, v^\alpha_{i_1i_2\cdots i_m}) = 0, \; a = 1, \ldots, A. \tag{9.2.6}$$

(9.2.6) merely represents a functional relation among the coordinates of the contact manifold. Therefore, they only help to define a submanifold of $\mathcal{C}_m$.

We can now lift a regular mapping $\phi : \mathcal{D}_n \to G$ *depicted by* $u^\alpha = \phi^\alpha(x^i)$ *to the regular mapping* $\phi : \mathcal{D}_n \to \mathcal{C}_m$ *if we choose this mapping in such a way that the pull-back relations*

$$\phi^*\sigma^\alpha_{i_1 i_2 \cdots i_r} = (v^\alpha_{i_1 i_2 \cdots i_r, i} - v^\alpha_{i_1 i_2 \cdots i_r i})\, dx^i = 0, \quad \phi^* F^a = 0$$

are satisfied, in other words, we get

$$v^\alpha_{i_1 i_2 \cdots i_r i} = \frac{\partial v^\alpha_{i_1 i_2 \cdots i_r}}{\partial x^i}, \qquad 0 \le r \le m-1.$$

On applying successively the above equality, we immediately observe that the independent coordinates in the manifold $\mathcal{C}_m$ are reduced to the form $v^\alpha_{i_1 i_2 \cdots i_r} = \phi^\alpha_{,i_1 i_2 \cdots i_r}, 0 \le r \le m$ and the mapping ϕ constitutes a solution of the system of partial differential equations. According to Theorem 5.8.2 we find that $\phi^* dF^a = d(\phi^* F^a) = 0$. Thus, this solution is also a solution of the ideal

$$\mathfrak{I}(\sigma^\alpha_{i_1 i_2 \cdots i_r}, 0 \le r \le m-1; dF^a)$$

generated by 1-forms. Since an ideal generated solely by 1-forms is complete (*see* Theorem 5.13.1), then the ideal $\mathfrak{I}$ contains all forms annihilated by the solution of the system (9.2.1). Furthermore, because of the commutation relation $\phi^* d\sigma^\alpha_{i_1 i_2 \cdots i_r} = d(\phi^* \sigma^\alpha_{i_1 i_2 \cdots i_r}) = 0$ the mapping ϕ annihilates also the closure

$$\overline{\mathfrak{I}}(\sigma^\alpha_{i_1 i_2 \cdots i_r}; d\sigma^\alpha_{i_1 i_2 \cdots i_r}; dF^a; 1 \le \alpha \le N, 1 \le i_r \le n, 0 \le r \le m-1, 1 \le a \le A)$$

of $\mathfrak{I}$. Thus, with the purpose of applying the Cartan theorem we can take the closed ideal $\overline{\mathfrak{I}}$ into account instead of the ideal $\mathfrak{I}$. However, due to the symmetries of $v^\alpha_{i_1 i_2 \cdots i_r}$ with respect to their subscripts, we can write

$$\begin{aligned} d\sigma^\alpha_{i_1 i_2 \cdots i_r} &= -\, dv^\alpha_{i_1 i_2 \cdots i_r i} \wedge dx^i = -\, \sigma^\alpha_{i_1 i_2 \cdots i_r i} \wedge dx^i - v^\alpha_{i_1 i_2 \cdots i_r i j} dx^j \wedge dx^i \\ &= -\, \sigma^\alpha_{i_1 i_2 \cdots i_r i} \wedge dx^i \end{aligned}$$

for $0 \le r \le m-2$. This means that the forms $d\sigma^\alpha_{i_1 i_2 \cdots i_r}, 0 \le r \le m-2$ are already in the ideal $\mathfrak{I}$. Therefore, it becomes sufficient to add only the forms $d\sigma^\alpha_{i_1 i_2 \cdots i_{m-1}}$ that cannot be expressed in this way to the ideal to obtain its closure. The closed ideal

$$\mathcal{I}_m = \overline{\mathcal{I}}(\sigma^\alpha, \sigma^\alpha_{i_1}, \sigma^\alpha_{i_1 i_2}, \dots \sigma^\alpha_{i_1 i_2 \cdots i_{m-1}}; d\sigma^\alpha_{i_1 i_2 \cdots i_{m-1}})$$

will henceforth be called the ***m*th order contact ideal**. On the other hand, we have to consider in essence the closed ideal

$$\mathfrak{I}_m = \overline{\mathfrak{I}}(\sigma^\alpha, \sigma^\alpha_{i_1}, \sigma^\alpha_{i_1 i_2}, \ldots, \sigma^\alpha_{i_1 i_2 \cdots i_{m-2}}, \sigma^\alpha_{i_1 i_2 \cdots i_{m-1}}; d\sigma^\alpha_{i_1 i_2 \cdots i_{m-1}}; dF^a)$$

called the ***fundamental ideal***. The most systematic method that we may have recourse to find a solution of this ideal is to determine its characteristic vector fields to utilise Theorem 5.13.5. We first wish to implement this procedure on a rather simple example. Let a first order partial differential equation with a single variable be given by

$$F(x^i, u, u_{,i}) = 0, \;\; 1 \le i \le n \tag{9.2.7}$$

Since $m = 1$, we write $v_i = u_{,i}$. $(2n+1)$-dimensional contact manifold $\mathcal{C}_1$ has the coordinate cover $\{x^i, u, v_i\}$. On this manifold, we define the forms

$$\sigma = du - v_i\, dx^i, d\sigma = -\, dv_i \wedge dx^i, dF = \frac{\partial F}{\partial x^i}\, dx^i + \frac{\partial F}{\partial u}\, du + \frac{\partial F}{\partial v_i}\, dv_i.$$

A vector field

$$V = X^i \frac{\partial}{\partial x^i} + U \frac{\partial}{\partial u} + V_i \frac{\partial}{\partial v_i} \in T(\mathcal{C}_1)$$

is a characteristic vector field of the closed ideal $\mathfrak{I}_1 = \overline{\mathfrak{I}}(\sigma, d\sigma, dF)$ if one is able to find functions $\lambda, \mu \in \Lambda^0(\mathcal{C}_1)$ so that the relations below are satisfied

$$\mathbf{i}_V(\sigma) = 0, \;\; \mathbf{i}_V(dF) = V(F) = 0, \;\; \mathbf{i}_V(d\sigma) = \lambda\sigma + \mu dF$$

from which one obtains the following equations that must be satisfied by the components of the characteristic vector field:

$$U - v_i X^i = 0, \qquad \frac{\partial F}{\partial x^i} X^i + \frac{\partial F}{\partial u} U + \frac{\partial F}{\partial v_i} V_i = 0, \tag{9.2.8}$$

$$-\, V_i\, dx^i + X^i\, dv_i = \Big(\lambda + \mu \frac{\partial F}{\partial u}\Big) du - \Big(\lambda v_i - \mu \frac{\partial F}{\partial x^i}\Big) dx^i + \mu \frac{\partial F}{\partial v_i}\, dv_i.$$

$(9.2.8)_3$ and $(9.2.8)_1$ lead to the result

$$\lambda = -\,\mu \frac{\partial F}{\partial u}, \;\; X^i = \mu \frac{\partial F}{\partial v_i}, \;\; U = \mu \frac{\partial F}{\partial v_i} v_i, \;\; V_i = -\,\mu \Big(\frac{\partial F}{\partial u} v_i + \frac{\partial F}{\partial x^i}\Big).$$

Hence, the characteristic vector field is determined as follows:

$$V = \mu \Big[\frac{\partial F}{\partial v_i} \frac{\partial}{\partial x^i} + v_i \frac{\partial F}{\partial v_i} \frac{\partial}{\partial u} - \Big(\frac{\partial F}{\partial x^i} + \frac{\partial F}{\partial u} v_i\Big) \frac{\partial}{\partial v_i}\Big]. \tag{9.2.9}$$

This vector field is 1-dimensional. We verify at once that $(9.2.8)_2$ becomes identically zero when we insert into it the vector field (9.2.9). As is well known, the solution manifold is produced by the integral curves of the characteristic vector field (9.2.9). If we denote the parameter of the curve by t, then $2n+1$ autonomous ordinary differential equations determining this family of ***characteristic curves*** on the manifold $\mathcal{C}_1$ are given by

$$\frac{dx^i}{dt} = \frac{\partial F}{\partial v_i}, \quad \frac{du}{dt} = v_i \frac{\partial F}{\partial v_i}, \tag{9.2.10}$$
$$\frac{dv_i}{dt} = -\Big(\frac{\partial F}{\partial x^i} + \frac{\partial F}{\partial u} v_i\Big)$$

where we have chosen $\mu = 1$ in (9.2.9) without loss of generality. The variation of the function F along a characteristic curve is found to be

$$\frac{dF}{dt} = \frac{\partial F}{\partial x^i}\frac{dx^i}{dt} + \frac{\partial F}{\partial u}\frac{du}{dt} + \frac{\partial F}{\partial v_i}\frac{dv_i}{dt} = 0$$

when we take (9.2.10) into consideration. Thus, F remains constant along a characteristic curve. This means that if the differential equation is satisfied at a point of the manifold $\mathcal{C}_1$, it is then satisfied along the characteristic curve through that point. A solution in the form $u(x^1, \ldots, x^n) - u = 0$ of the equation (9.2.7) represents an n-dimensional submanifold, or a hypersurface, in the graph space. The normal vector to this hypersurface is determined by its components $(v_i = u_{,i}, 1 \leq i \leq n; \, -1)$. Since we have

$$\frac{du}{dt} = \frac{\partial u}{\partial x^i}\frac{dx^i}{dt} = v_i \frac{dx^i}{dt} = v_i \frac{\partial F}{\partial v_i}$$

on this hypersurface, characteristic curves are also on it. However, to each point of the curve we attach a surface element perpendicular to the normal at that point. Hence, we form a ***characteristic strip*** as was reflected in the classical terminology. In order to find the solution we need to consider characteristic strips emanating from an $(n-1)$*-dimensional initial manifold* S that is not tangent to the characteristic vector field and prescribed by the initial conditions. Let us assume that initial submanifold S is depicted through parameters $\mathbf{s} = (s^1, s^2, \ldots, s^{n-1})$ as follows:

$$x^i = x_0^i(\mathbf{s}), \quad i = 1, \ldots, n.$$

We suppose that the initial data on this manifold are given by the relations

$$u = u_0(\mathbf{s}), \quad v_i = v_i^0(\mathbf{s}), \, i = 1, \ldots, n$$

in terms of parameters $\mathbf{s} = (s^1, s^2, \ldots, s^{n-1})$. But, these data cannot be chosen arbitrarily. They have to satisfy the conditions

$$F(x_0^i, u_0, v_i^0) = 0; \quad \frac{\partial u_0}{\partial s^\alpha} = v_i^0 \frac{\partial x_0^i}{\partial s^\alpha}, \quad \alpha = 1, \ldots, n-1.$$

We thus obtain n equations to determine n initial conditions v_i^0. We shall assume that these equations have at least one solution. Let us now denote the solution of ordinary differential equations (9.2.10) under the initial conditions $x^i(0) = x_0^i(\mathbf{s}), u(0) = u_0(\mathbf{s}), v_i(0) = v_i^0(\mathbf{s})$ by the relations

$$x^i = \mathcal{X}^i(t; \mathbf{s}), \quad u = \mathcal{U}(t; \mathbf{s}), \quad v_i = \mathcal{V}_i(t; \mathbf{s}).$$

Since we have assumed that the characteristic vector field does not belong to the tangent bundle of the initial manifold, we can write

$$\frac{\partial(\mathcal{X}^1, \mathcal{X}^2, \ldots, \mathcal{X}^n)}{\partial(s^1, \ldots, s^{n-1}, t)} = \begin{vmatrix} \frac{\partial \mathcal{X}^1}{\partial s^1} & \cdots & \frac{\partial \mathcal{X}^1}{\partial s^{n-1}} & \frac{\partial \mathcal{X}^1}{\partial t} \\ \frac{\partial \mathcal{X}^2}{\partial s^1} & \cdots & \frac{\partial \mathcal{X}^2}{\partial s^{n-1}} & \frac{\partial \mathcal{X}^2}{\partial t} \\ \vdots & \vdots & \vdots & \vdots \\ \frac{\partial \mathcal{X}^n}{\partial s^1} & \cdots & \frac{\partial \mathcal{X}^n}{\partial s^{n-1}} & \frac{\partial \mathcal{X}^n}{\partial t} \end{vmatrix} = \begin{vmatrix} \frac{\partial \mathcal{X}^1}{\partial s^1} & \cdots & \frac{\partial \mathcal{X}^1}{\partial s^{n-1}} & \frac{\partial F}{\partial v_1} \\ \frac{\partial \mathcal{X}^2}{\partial s^1} & \cdots & \frac{\partial \mathcal{X}^2}{\partial s^{n-1}} & \frac{\partial F}{\partial v_2} \\ \vdots & \vdots & \vdots & \vdots \\ \frac{\partial \mathcal{X}^n}{\partial s^1} & \cdots & \frac{\partial \mathcal{X}^n}{\partial s^{n-1}} & \frac{\partial F}{\partial v_n} \end{vmatrix} \neq 0.$$

Hence, in the neighbourhood of the initial manifold, n variables t, s^α can be expressed in terms of variables x^i by resorting to the inverse mapping theorem whence we arrive at the solution of the partial differential equation (9.2.7) in the following form

$$u = \mathcal{U}(t(\mathbf{x}), \mathbf{s}(\mathbf{x})) = u(x^1, x^2, \ldots, x^n).$$

We shall now deal with some applications of the general solution discussed above.

Example 9.2.1. We consider the equation

$$F = \sum_{i=1}^{n} \Big(\frac{\partial u}{\partial x^i}\Big)^2 - 1 = \sum_{i=1}^{n} v_i^2 - 1 = 0$$

known as the *eiconal equation* in the geometrical optics. Since

$$\frac{\partial F}{\partial x^i} = 0, \ \frac{\partial F}{\partial u} = 0, \ \frac{\partial F}{\partial v_i} = 2v_i$$

the characteristic equations (9.2.10) take the form

$$\frac{dx^i}{dt} = 2v_i, \ \frac{du}{dt} = 2\sum_{i=1}^{n} v_i^2 = 2, \ \frac{dv_i}{dt} = 0$$

from which we reach to the conclusion

$$x^i = 2v_i^0(\mathbf{s})t + x_0^i(\mathbf{s}); \ \ u = 2t + u_0(\mathbf{s}); \ \ v_i = v_i^0(\mathbf{s}), \ \sum_{i=1}^{n}(v_i^0)^2 = 1.$$

Thus, we can express the solution implicitly as

$$x^i = v_i^0(\mathbf{s})\,[u - u_0(\mathbf{s})] + x_0^i(\mathbf{s}), \ \sum_{i=1}^{n}(v_i^0)^2 = 1, \ v_i^0\frac{\partial x_0^i}{\partial s^\alpha} = 0$$

by eliminating the parameter t. Consequently, the solution manifold $x^i(\mathbf{s})$ corresponding to a chosen value for u is obtained by translating the initial manifold by an amount $u - u_0(\mathbf{s})$ along a unit vector field $\mathbf{v}^0(\mathbf{s})$ which is *orthogonal* to that manifold and the solution $u = u(\mathbf{x})$ is determined by this family of $(n-1)$-dimensional *level manifolds* in $\mathbb{R}^n$. ■

Example 9.2.2. *Quasilinear Equations.* Let us consider the equation

$$a^i(\mathbf{x}, u)\frac{\partial u}{\partial x^i} - b(\mathbf{x}, u) = 0.$$

Since $F = a^i(\mathbf{x}, u)v_i - b(\mathbf{x}, u) = 0$, we find

$$\frac{\partial F}{\partial v_i} = a^i, \ \frac{\partial F}{\partial u} = \frac{\partial a^i}{\partial u}v_i - \frac{\partial b}{\partial u}, \ \frac{\partial F}{\partial x^i} = \frac{\partial a^j}{\partial x^i}v_j - \frac{\partial b}{\partial x^i}.$$

Hence, the equations $(9.2.10)_{1-2}$ take the form

$$\frac{dx^i}{dt} = a^i(\mathbf{x}, u), \quad \frac{du}{dt} = a^i(\mathbf{x}, u)v_i = b(\mathbf{x}, u).$$

The solution of a first order quasilinear equation then follows from the solution of the above ordinary differential equations. ■

Example 9.2.3. *Hamilton-Jacobi equation.*

The Hamilton-Jacobi partial differential equations governing the motion of a dynamical system of n degrees of freedom [*see* (11.5.18)] [after mathematicians Hamilton and Jacobi] are given by

$$\frac{\partial S}{\partial t}+H\Big(q^1,\dots,q^n,t,\frac{\partial S}{\partial q^1},\dots,\frac{\partial S}{\partial q^n}\Big)=0$$

where $S=S(\mathbf{q},t)$. We denote the generalised coordinates by $(q^1,\dots,q^n)$, time by t and the action function by S. Generalised momenta are defined by $p_i=\partial S/\partial q^i, i=1,\dots,n$. H is the Hamiltonian function. If we introduce $p=\partial S/\partial t$, we obtain

$$F=p+H(q^1,\dots,q^n,t,p_1,\dots,p_n)=0.$$

If we denote the parameter of a characteristic curve by s, then it follows from $(9.2.10)_1$ that

$$\frac{dt}{ds}=\frac{\partial F}{\partial p}=1.$$

Thus, we can choose $s=t$ without loss of generality. Since $\partial F/\partial S=0$, then equations associated with characteristic strips are found to be

$$\begin{aligned}
&\frac{dq^i}{dt}=\frac{\partial F}{\partial p_i}=\frac{\partial H}{\partial p_i},\\
&\frac{dS}{dt}=p_i\frac{\partial F}{\partial p_i}+p\frac{\partial F}{\partial p}=p_i\frac{\partial H}{\partial p_i}+p=p_i\frac{\partial H}{\partial p_i}-H\\
&\frac{dp}{dt}=-\frac{\partial F}{\partial t}=-\frac{\partial H}{\partial t},\ \frac{dp_i}{dt}=-\frac{\partial F}{\partial q^i}=-\frac{\partial H}{\partial q^i}.
\end{aligned}$$

As a result, we obtain the well known Hamilton equations of analytical mechanics:

$$\frac{dq^i}{dt}=\frac{\partial H}{\partial p_i},\ \frac{dp_i}{dt}=-\frac{\partial H}{\partial q^i},\ \frac{dS}{dt}=p_i\frac{\partial H}{\partial p_i}-H,\ \frac{dp}{dt}=-\frac{\partial H}{\partial t}.\quad\blacksquare$$

The method of characteristics that works quite well for the partial differential equation involving a single dependent variable turns out to be rather ineffective when looking for the solution of the general system (9.2.6). Let us denote the *characteristic vector field* V of the ideal $\mathfrak{I}_m$ generated by that system as follows

$$\begin{aligned}
V&=X^i\frac{\partial}{\partial x^i}+U^\alpha\frac{\partial}{\partial u^\alpha}+V^\alpha_{i_1}\frac{\partial}{\partial v^\alpha_{i_1}}+\dots+V^\alpha_{i_1\cdots i_m}\frac{\partial}{\partial v^\alpha_{i_1\cdots i_m}}\\
&=X^i\frac{\partial}{\partial x^i}+\sum_{r=0}^{m}V^\alpha_{i_1\cdots i_r}\frac{\partial}{\partial v^\alpha_{i_1\cdots i_r}}.
\end{aligned}$$

If we note that

$$dF^a = \frac{\partial F^a}{\partial x^i}\, dx^i + \sum_{r=0}^{m} \frac{\partial F^a}{\partial v^\alpha_{i_1\cdots i_r}}\, dv^\alpha_{i_1\cdots i_r},$$

then the vector field V must satisfy the relations

$$\mathbf{i}_V(\sigma^\alpha_{i_1 i_2\cdots i_r}) = V^\alpha_{i_1\cdots i_r} - v^\alpha_{i_1\cdots i_r i}X^i = 0, \tag{9.2.11}$$

$$\mathbf{i}_V(dF^a) = V(F^a) = \frac{\partial F^a}{\partial x^i}X^i + \sum_{r=0}^{m} \frac{\partial F^a}{\partial v^\alpha_{i_1\cdots i_r}} V^\alpha_{i_1\cdots i_r} = 0$$

$$\begin{aligned}\mathbf{i}_V(d\sigma^\alpha_{i_1 i_2\cdots i_{m-1}}) &= -\,V^\alpha_{i_1\cdots i_{m-1}i}\, dx^i + X^i\, dv^\alpha_{i_1\cdots i_{m-1}i}\\ &= \sum_{s=0}^{m-1} \lambda^{\alpha j_1\cdots j_s}_{\beta i_1\cdots i_{m-1}}\sigma^\beta_{j_1\cdots j_s} + \Lambda^\alpha_{i_1\cdots i_{m-1}a} dF^a\end{aligned}$$

where $\lambda^{\alpha j_1\cdots j_s}_{\beta i_1\cdots i_{m-1}}, \Lambda^\alpha_{i_1\cdots i_{m-1}a} \in \Lambda^0(\mathcal{C}_m),\ \ 0 \le s \le m-1$. $(9.2.11)_{1-2}$ then yields

$$V^\alpha_{i_1\cdots i_r} = v^\alpha_{i_1\cdots i_r i}X^i, \quad 0 \le r \le m-1, \tag{9.2.12}$$

$$\Big(\frac{\partial F^a}{\partial x^i} + \sum_{r=0}^{m-1} \frac{\partial F^a}{\partial v^\alpha_{i_1\cdots i_r}} v^\alpha_{i_1\cdots i_r i}\Big)X^i + \frac{\partial F^a}{\partial v^\alpha_{i_1\cdots i_m}} V^\alpha_{i_1\cdots i_m} = 0$$

whereas $(9.2.11)_3$ results in

$$\begin{aligned}&-\,V^\alpha_{i_1\cdots i_{m-1}i}\, dx^i + X^i\, dv^\alpha_{i_1\cdots i_{m-1}i}\\ &\quad= \sum_{s=0}^{m-1}\Big[\lambda^{\alpha j_1\cdots j_s}_{\beta i_1\cdots i_{m-1}} + \Lambda^\alpha_{i_1\cdots i_{m-1}a}\frac{\partial F^a}{\partial v^\beta_{j_1\cdots j_s}}\Big] dv^\beta_{j_1\cdots j_s}\\ &\qquad+ \Big[\Lambda^\alpha_{i_1\cdots i_{m-1}a}\frac{\partial F^a}{\partial x^i} - \sum_{s=0}^{m-1}\lambda^{\alpha j_1\cdots j_s}_{\beta i_1\cdots i_{m-1}} v^\beta_{j_1\cdots j_s i}\Big] dx^i\\ &\qquad+ \Lambda^\alpha_{i_1\cdots i_{m-1}a}\frac{\partial F^a}{\partial v^\beta_{j_1\cdots j_m}}\, dv^\beta_{j_1\cdots j_m}.\end{aligned}$$

We thus see that the following relations must be satisfied

$$\lambda^{\alpha j_1\cdots j_s}_{\beta i_1\cdots i_{m-1}} = -\,\Lambda^\alpha_{i_1\cdots i_{m-1}a}\frac{\partial F^a}{\partial v^\beta_{j_1\cdots j_s}}, \quad 0 \le s \le m-1 \tag{9.2.13}$$

$$V^\alpha_{i_1\cdots i_{m-1}i} = -\,\Lambda^\alpha_{i_1\cdots i_{m-1}a}\Big[\frac{\partial F^a}{\partial x^i} + \sum_{s=0}^{m-1}\frac{\partial F^a}{\partial v^\beta_{j_1\cdots j_s}} v^\beta_{j_1\cdots j_s i}\Big],$$

$$X^i\,dv^\alpha_{i_1\cdots i_{m-1}i} = \Lambda^\alpha_{i_1\cdots i_{m-1}a}\frac{\partial F^a}{\partial v^\beta_{j_1\cdots j_m}}\,dv^\beta_{j_1\cdots j_m}.$$

It then follows from $(9.2.13)_3$ and $(9.2.12)_2$ that

$$X^i\delta^\alpha_\beta\delta^{j_1}_{i_1}\cdots\delta^{j_{m-1}}_{i_{m-1}} = \Lambda^\alpha_{i_1\cdots i_{m-1}a}\frac{\partial F^a}{\partial v^\beta_{j_1\cdots j_{m-1}i}} \tag{9.2.14}$$

$$\Big(X^i\delta^a_b - \Lambda^\alpha_{i_1\cdots i_{m-1}b}\frac{\partial F^a}{\partial v^\alpha_{i_1\cdots i_{m-1}i}}\Big)\Big(\frac{\partial F^b}{\partial x^i} + \sum_{r=0}^{m-1}\frac{\partial F^b}{\partial v^\beta_{j_1\cdots j_r}}v^\beta_{j_1\cdots j_r i}\Big) = 0$$

After having performed contraction operations on indices $(\alpha,\beta), (j_1, i_1), \ldots (j_{m-1}, i_{m-1})$ of Kronecker deltas on the left hand side of the expression $(9.2.14)_1$ we find that

$$n^{m-1}NX^i = \Lambda^\alpha_{i_1\cdots i_{m-1}a}\frac{\partial F^a}{\partial v^\alpha_{i_1\cdots i_{m-1}i}}. \tag{9.2.15}$$

If we insert the above expression in $(9.2.14)_{1-2}$, we deduce that the functions $\Lambda^\alpha_{i_1\cdots i_{m-1}a}$ must satisfy the equations

$$\Lambda^\gamma_{k_1\cdots k_{m-1}a}\frac{\partial F^a}{\partial v^\gamma_{k_1\cdots k_{m-1}i}}\delta^\alpha_\beta\delta^{j_1}_{i_1}\cdots\delta^{j_{m-1}}_{i_{m-1}} - n^{m-1}N\Lambda^\alpha_{i_1\cdots i_{m-1}a}\frac{\partial F^a}{\partial v^\beta_{j_1\cdots j_{m-1}i}} = 0,$$

$$\Big(\Lambda^\alpha_{i_1\cdots i_{m-1}c}\frac{\partial F^c}{\partial v^\alpha_{i_1\cdots i_{m-1}i}}\,\delta^a_b - n^{m-1}N\Lambda^\alpha_{i_1\cdots i_{m-1}b}\frac{\partial F^a}{\partial v^\alpha_{i_1\cdots i_{m-1}i}}\Big)\times$$
$$\Big(\frac{\partial F^b}{\partial x^i} + \sum_{r=0}^{m-1}\frac{\partial F^b}{\partial v^\beta_{j_1\cdots j_r}}v^\beta_{j_1\cdots j_r i}\Big) = 0.$$

When $N > 1$, we can always pick out the indices α and β as to be $\alpha \neq \beta$. In this case, if all partial differential equations are of order m, then none of the derivatives $\dfrac{\partial F^a}{\partial v^\beta_{j_1\cdots j_{m-1}i}}$ vanish implying that $\Lambda^\alpha_{i_1\cdots i_{m-1}a} = 0$ and $X^i = 0$. Consequently, we find $V^\alpha_{i_1\cdots i_r} = 0$ for $0 \leq r \leq m$. Hence, the dimension of the characteristic vector space is zero and we end up only with the trivial solution that consists of constants satisfying the equations (9.2.1). If $N = 1$ and $m > 1$, then we immediately see that we obtain the same result. If some equations in the system have lesser orders than m, some coefficients $\Lambda^\alpha_{i_1\cdots i_{m-1}a}$ may not be necessarily zero. In the case $N = 1,\ m = 1,\ A > 1$, the first set of equations above are satisfied identically. On arranging the second equations, we obtain

$$F^{ab}\Lambda_b = 0$$

where the antisymmetric $A \times A$ matrix $\mathbf{F}$ is given by

$$F^{ab} = -F^{ba} = \frac{\partial F^a}{\partial x^i}\frac{\partial F^b}{\partial v_i} - \frac{\partial F^b}{\partial x^i}\frac{\partial F^a}{\partial v_i} + v_i\Big(\frac{\partial F^a}{\partial u}\frac{\partial F^b}{\partial v_i} - \frac{\partial F^b}{\partial u}\frac{\partial F^a}{\partial v_i}\Big).$$

If only $\det \mathbf{F} = 0$ (when A is an odd number this determinant will always be zero) then all coefficients Λ_a do not have to vanish and we may have the opportunity to write

$$X^i = \Lambda_a \frac{\partial F^a}{\partial v_i},\; U^\alpha = V_0 = \lambda_a \frac{\partial F^a}{\partial v_i} v_i,\; V_i = -\Lambda_a\Big(\frac{\partial F^a}{\partial x^i} + \frac{\partial F^a}{\partial u} v_i\Big),$$
$$V = \Lambda_a\Big[\frac{\partial F^a}{\partial v_i}\Big(\frac{\partial}{\partial x^i} + v_i\frac{\partial}{\partial u}\Big) - \Big(\frac{\partial F^a}{\partial x^i} + \frac{\partial F^a}{\partial u} v_i\Big)\frac{\partial}{\partial v_i}\Big].$$

The dimension of the characteristic subspace is equal to the number of independent functions Λ_a. On the other hand, if $N = 1$, $m = 1, A = 1$ then we arrive at the previously found solution

$$X^i = \Lambda\frac{\partial F}{\partial v_i},\; U = V_0 = \Lambda\frac{\partial F}{\partial v_i} v_i,\; V_i = -\Lambda\Big(\frac{\partial F}{\partial x^i} + \frac{\partial F}{\partial u} v_i\Big),$$
$$V = \Lambda\Big[\frac{\partial F}{\partial v_i}\Big(\frac{\partial}{\partial x^i} + v_i\frac{\partial}{\partial u}\Big) - \Big(\frac{\partial F}{\partial x^i} + \frac{\partial F}{\partial u} v_i\Big)\frac{\partial}{\partial v_i}\Big].$$

A nontrivial solution is likewise obtained for a system of quasilinear first order partial differential equations with same principal parts

$$a^i(\mathbf{x}, \mathbf{u})\frac{\partial u^\alpha}{\partial x^i} - b^\alpha(\mathbf{x}, \mathbf{u}) = 0.$$

In this case, the characteristic vector field can be written as follows

$$V = X^i\frac{\partial}{\partial x^i} + U^\alpha\frac{\partial}{\partial u^\alpha} + V_i^\alpha\frac{\partial}{\partial v_i^\alpha}.$$

On the other hand, since we have to take $F^\alpha = a^i(\mathbf{x}, \mathbf{u})v_i^\alpha - b^\alpha(\mathbf{x}, \mathbf{u}) = 0$, then the equations (9.2.12-13-14) lead to the relations

$$U^\alpha = v_i^\alpha X^i,\; \lambda_\beta^\alpha = -\Lambda_\gamma^\alpha\frac{\partial F^\gamma}{\partial u^\beta},\; X^i\delta_\beta^\alpha = \Lambda_\gamma^\alpha\frac{\partial F^\gamma}{\partial v_i^\beta} = \Lambda_\gamma^\alpha\delta_\beta^\gamma a^i = \Lambda_\beta^\alpha a^i$$

whence we deduce that $\Lambda_\beta^\alpha = \delta_\beta^\alpha$ and $\lambda_\beta^\alpha = -\partial F^\alpha/\partial u^\beta$. Therefore, the components of the characteristic vector field are found as

$$X^i = a^i, \;\; U^\alpha = v_i^\alpha a^i, \;\; V_i^\alpha = -\Big(\frac{\partial F^\alpha}{\partial x^i} + \frac{\partial F^\alpha}{\partial u^\beta} v_i^\beta\Big).$$

These components satisfy identically the relation $(9.2.12)_2$. Since one must write $a^i v_i^\alpha = b^\alpha$, the solution of the system of partial differential equations is constructed by means of the solutions of the following system of ordinary differential equations

$$\frac{dx^i}{dt} = a^i(\mathbf{x}, \mathbf{u}),$$
$$\frac{du^\alpha}{dt} = b^\alpha(\mathbf{x}, \mathbf{u}).$$

To study general solutions of differential equations we usually make use of Lie transformation groups. In such kind of methods, the isovectors of closed ideals generated by differential equations play quite a significant part. Although symmetry groups have emerged at the beginning of 20th Century, their investigation through exterior differential forms started by a seminal paper published on 1971[1].

9.3. ISOVECTOR FIELDS OF THE CONTACT IDEAL

Let $\mathcal{C}_m$ be the contact manifold defined in Sec. 9.2. We first consider the closed ideal

$$\mathcal{I}_m = \overline{\mathcal{I}}(\sigma^\alpha, \sigma^\alpha_{i_1}, \sigma^\alpha_{i_1 i_2}, \ldots, \sigma^\alpha_{i_1 i_2 \cdots i_{m-1}}; d\sigma^\alpha_{i_1 i_2 \cdots i_{m-1}}) \qquad (9.3.1)$$

which we have called the *mth order contact ideal. The properties of this ideal will remain the same for all system of mth order partial differential equations*. We know that a vector field $V \in T(\mathcal{C}_m)$ is an isovector field of the ideal $\mathcal{I}_m$ if it satisfies the relation $\pounds_V \mathcal{I}_m \subset \mathcal{I}_m$. On the other hand, since the ideal

$$\mathcal{I}(\sigma^\alpha, \sigma^\alpha_{i_1}, \sigma^\alpha_{i_1 i_2}, \ldots, \sigma^\alpha_{i_1 i_2 \cdots i_{m-1}}) \qquad (9.3.2)$$

is generated by forms of the same degree (1-forms in the present case), isovector fields of this ideal will be the same as those of its closure $\mathcal{I}_m$ in accordance with Theorem 5.12.4. We may represent a vector field V by the expression

[1]Harrison, B. K., Estabrook, F. B., Geometric Approach to invariance groups and solution of partial differential systems, *Journal of Mathematical Physics*, **12**, 653-666, 1971.

$$V = X^i \frac{\partial}{\partial x^i} + \sum_{r=0}^{m} V^\alpha_{i_1 \cdots i_r} \frac{\partial}{\partial v^\alpha_{i_1 \cdots i_r}} \in T(\mathcal{C}_m) \tag{9.3.3}$$

where $X^i, V^\alpha_{i_1 \cdots i_r} \in \Lambda^0(\mathcal{C}_m)$ with $0 \leq r \leq m$. Here, we adopt the conventions

$$V^\alpha_{i_0} = V^\alpha = U^\alpha$$

and

$$\sum_{r=0}^{m} V^\alpha_{i_1 \cdots i_r} \frac{\partial}{\partial v^\alpha_{i_1 \cdots i_r}} =$$

$$U^\alpha \frac{\partial}{\partial u^\alpha} + V^\alpha_{i_1} \frac{\partial}{\partial v^\alpha_{i_1}} + V^\alpha_{i_1 i_2} \frac{\partial}{\partial v^\alpha_{i_1 i_2}} + \cdots + V^\alpha_{i_1 \cdots i_m} \frac{\partial}{\partial v^\alpha_{i_1 \cdots i_m}}.$$

There are of course summations on all repeated dummy indices. Since the variables $v^\alpha_{i_1 \cdots i_r}$ are completely symmetric with respect to their subscripts for $r \geq 2$, only the completely symmetric parts of corresponding components $V^\alpha_{i_1 \cdots i_r}$ will survive in summations above. Therefore, without loss of generality we may assume that $V_{i_1 \cdots i_r}$ are completely symmetric with respect to their subscripts for $r \geq 2$. As is well known, the necessary and sufficient conditions for a vector field V to be an isovector of the ideal (9.3.2) are the satisfaction of the following relations

$$\pounds_V \sigma^\alpha_{i_1 \cdots i_k} = \sum_{r=0}^{m-1} \lambda^{\alpha j_1 \cdots j_r}_{\beta i_1 \cdots i_k} \sigma^\beta_{j_1 \cdots j_r}, \quad k = 0, 1, \ldots, m-1 \tag{9.3.4}$$

for certain functions $\lambda^{\alpha j_1 \cdots j_r}_{\beta i_1 \cdots i_k} \in \Lambda^0(\mathcal{C}_m), 0 \leq k, r \leq m-1$. The discussions presented in this section and the subsequent one are borrowed from the work cited below[2]. On employing the formula (5.11.5) to calculate the Lie derivative, we obtain

$$\begin{aligned}
\pounds_V \sigma^\alpha_{i_1 \cdots i_k} &= \mathbf{i}_V (d\sigma^\alpha_{i_1 \cdots i_k}) + d\mathbf{i}_V (\sigma^\alpha_{i_1 \cdots i_k}) \\
&= -V^\alpha_{i_1 \cdots i_k i} dx^i + X^i dv^\alpha_{i_1 \cdots i_k i} + d(V^\alpha_{i_1 \cdots i_k} - X^i v^\alpha_{i_1 \cdots i_k i}) \\
&= -V^\alpha_{i_1 \cdots i_k i} dx^i + dV^\alpha_{i_1 \cdots i_k} - v^\alpha_{i_1 \cdots i_k i} dX^i
\end{aligned}$$

by recalling the relation $d\sigma^\alpha_{i_1 \cdots i_k} = -dv^\alpha_{i_1 \cdots i_k i} \wedge dx^i$. Therefore, (9.3.4) yields

---

[2]Şuhubi, E. S., Isovector fields and similarity solutions for general balance equations, *International Journal of Engineering Science*, **29**, 133-150, 1991.

$$\left[\frac{\partial V^\alpha_{i_1\cdots i_k}}{\partial x^i} - v^\alpha_{i_1\cdots i_k j}\frac{\partial X^j}{\partial x^i} - V^\alpha_{i_1\cdots i_k i}\right]dx^i + \sum_{r=0}^{m}\left[\frac{\partial V^\alpha_{i_1\cdots i_k}}{\partial v^\beta_{j_1\cdots j_r}} - v^\alpha_{i_1\cdots i_k i}\frac{\partial X^i}{\partial v^\beta_{j_1\cdots j_r}}\right]dv^\beta_{j_1\cdots j_r}$$
$$= \sum_{r=0}^{m-1}\lambda^{\alpha j_1\cdots j_r}_{\beta i_1\cdots i_k}(dv^\beta_{j_1\cdots j_r} - v^\beta_{j_1\cdots j_r i}dx^i)$$

On equating the coefficients of linearly independent like 1-forms at both sides of the foregoing expression, we arrive at the following relations

$$-\sum_{r=0}^{m-1}\lambda^{\alpha j_1\cdots j_r}_{\beta i_1\cdots i_k}v^\beta_{j_1\cdots j_r i} = \frac{\partial V^\alpha_{i_1\cdots i_k}}{\partial x^i} - v^\alpha_{i_1\cdots i_k j}\frac{\partial X^j}{\partial x^i} - V^\alpha_{i_1\cdots i_k i}, 0 \le k \le m-1$$
$$\lambda^{\alpha j_1\cdots j_r}_{\beta i_1\cdots i_k} = \frac{\partial V^\alpha_{i_1\cdots i_k}}{\partial v^\beta_{j_1\cdots j_r}} - v^\alpha_{i_1\cdots i_k i}\frac{\partial X^i}{\partial v^\beta_{j_1\cdots j_r}}, \quad 0 \le k, r \le m-1 \quad (9.3.5)$$
$$\frac{\partial V^\alpha_{i_1\cdots i_k}}{\partial v^\beta_{j_1\cdots j_m}} - v^\alpha_{i_1\cdots i_k i}\frac{\partial X^i}{\partial v^\beta_{j_1\cdots j_m}} = 0, \qquad 0 \le k \le m-1$$

Equations $(9.3.5)_2$ determine the functions $\lambda^{\alpha j_1\cdots j_r}_{\beta i_1\cdots i_k}$. Inserting these functions into equations $(9.3.5)_1$, we reach to the recurrence relations given below that relate the components $V^\alpha_{i_1\cdots i_k i}$ to the components $V^\alpha_{i_1\cdots i_k}$ and X^i

$$V^\alpha_{i_1\cdots i_k i} = \frac{\partial V^\alpha_{i_1\cdots i_k}}{\partial x^i} - v^\alpha_{i_1\cdots i_k j}\frac{\partial X^j}{\partial x^i} + \sum_{r=0}^{m-1}\left[\frac{\partial V^\alpha_{i_1\cdots i_k}}{\partial v^\beta_{j_1\cdots j_r}} - v^\alpha_{i_1\cdots i_k j}\frac{\partial X^j}{\partial v^\beta_{j_1\cdots j_r}}\right]v^\beta_{j_1\cdots j_r i}, 0 \le k \le m-1. \quad (9.3.6)$$

Let us now consider equations $(9.3.5)_3$ and start with equations corresponding to $k = m-1$. If we differentiate these equations with respect to the variables $v^\gamma_{k_1\cdots k_m}$, we then find that

$$\frac{\partial^2 V^\alpha_{i_1\cdots i_{m-1}}}{\partial v^\beta_{j_1\cdots j_m}\partial v^\gamma_{k_1\cdots k_m}} - v^\alpha_{i_1\cdots i_{m-1}i}\frac{\partial^2 X^i}{\partial v^\beta_{j_1\cdots j_m}\partial v^\gamma_{k_1\cdots k_m}} - \delta^\alpha_\gamma\delta^{k_1}_{i_1}\cdots\delta^{k_{m-1}}_{i_{m-1}}\delta^{k_m}_i\frac{\partial X^i}{\partial v^\beta_{j_1\cdots j_m}} = 0.$$

When we take into account the symmetry of the second order derivatives with respect to the variables $v^\alpha_{i_1\cdots i_m}$, the above equations give rise to

$$\delta^\alpha_\gamma \delta^{k_1}_{i_1} \cdots \delta^{k_{m-1}}_{i_{m-1}} \frac{\partial X^{k_m}}{\partial v^\beta_{j_1 \cdots j_m}} = \delta^\alpha_\beta \delta^{j_1}_{i_1} \cdots \delta^{j_{m-1}}_{i_{m-1}} \frac{\partial X^{j_m}}{\partial v^\gamma_{k_1 \cdots k_m}} \tag{9.3.7}$$

Contraction operations on indices $(\alpha, \gamma), (k_1, i_1), \ldots, (k_{m-1}, i_{m-1})$ lead to the result

$$N n^{m-1} \frac{\partial X^{k_m}}{\partial v^\beta_{j_1 \cdots j_{m-1} j_m}} = \frac{\partial X^{j_m}}{\partial v^\beta_{j_1 \cdots j_{m-1} k_m}}. \tag{9.3.8}$$

Introducing (9.3.8) into the right hand side of (9.3.7), we get

$$\delta^\alpha_\gamma \delta^{k_1}_{i_1} \cdots \delta^{k_{m-1}}_{i_{m-1}} \frac{\partial X^{k_m}}{\partial v^\beta_{j_1 \cdots j_m}} = \delta^\alpha_\beta \delta^{j_1}_{i_1} \cdots \delta^{j_{m-1}}_{i_{m-1}} N n^{m-1} \frac{\partial X^{k_m}}{\partial v^\gamma_{k_1 \cdots k_{m-1} j_m}}.$$

Contracting this time the indices $(\alpha, \beta), (j_1, i_1), \ldots, (j_{m-1}, i_{m-1})$ above, we finally reach to the conclusion

$$(N^2 n^{2(m-1)} - 1) \frac{\partial X^{k_m}}{\partial v^\gamma_{k_1 \cdots k_{m-1} j_m}} = 0.$$

When we take partial differential equations into consideration, we clearly have $n > 1$. Furthermore, if we assume that $m > 1$, then for $N \geq 1$ we get $N n^{m-1} \neq 1$. *When* $m = 1$ *we will have to distinguish the case* $N = 1$ *from the case* $N > 1$. We thus find

$$\frac{\partial X^i}{\partial v^\beta_{j_1 \cdots j_m}} = 0 \tag{9.3.9}$$

and it follows from equations $(9.3.5)_3$ that

$$\frac{\partial V^\alpha_{i_1 \cdots i_k}}{\partial v^\beta_{j_1 \cdots j_m}} = 0, \quad k = 0, 1, \ldots, m-1. \tag{9.3.10}$$

This means that the functions X^i and $V^\alpha_{i_1 \cdots i_k}$ cannot depend on the variables $v^\alpha_{i_1 \cdots i_m}$. Let us now write the relation (9.3.6) for $k = 0, 1, \ldots, m-2$:

$$\begin{aligned} V^\alpha_{i_1 \cdots i_k i} &= \frac{\partial V^\alpha_{i_1 \cdots i_k}}{\partial x^i} - v^\alpha_{i_1 \cdots i_k j} \frac{\partial X^j}{\partial x^i} + \sum_{r=0}^{m-2} \left[\frac{\partial V^\alpha_{i_1 \cdots i_k}}{\partial v^\beta_{j_1 \cdots j_r}} - v^\alpha_{i_1 \cdots i_k j} \frac{\partial X^j}{\partial v^\beta_{j_1 \cdots j_r}} \right] v^\beta_{j_1 \cdots j_r i} \\ &+ \left[\frac{\partial V^\alpha_{i_1 \cdots i_k}}{\partial v^\beta_{j_1 \cdots j_{m-1}}} - v^\alpha_{i_1 \cdots i_k j} \frac{\partial X^j}{\partial v^\beta_{j_1 \cdots j_{m-1}}} \right] v^\beta_{j_1 \cdots j_{m-1} i}, \quad 0 \leq k \leq m-2 \end{aligned}$$

Because of (9.3.10), the functions $V^{\alpha}_{i_1\cdots i_k i}$ must be independent of the variables $v^{\beta}_{j_1\cdots j_{m-1}i}$, hence their coefficients have to vanish:

$$\frac{\partial V^{\alpha}_{i_1\cdots i_k}}{\partial v^{\beta}_{j_1\cdots j_{m-1}}} - v^{\alpha}_{i_1\cdots i_k j}\frac{\partial X^j}{\partial v^{\beta}_{j_1\cdots j_{m-1}}} = 0,\ 0 \le k \le m-2. \qquad (9.3.11)$$

Equations (9.3.11) carry the same structural properties as equations $(9.3.5)_3$. Therefore, they lead similarly to

$$\frac{\partial X^i}{\partial v^{\beta}_{j_1\cdots j_{m-1}}} = 0,\ \frac{\partial V^{\alpha}_{i_1\cdots i_k}}{\partial v^{\beta}_{j_1\cdots j_{m-1}}} = 0,\ k = 0,1,\ldots,m-2$$

if $Nn^{m-2} \neq 1$. Starting from this result we can verify by mathematical induction that the following relations are to be satisfied if $Nn^{m-s-1} \neq 1$

$$\frac{\partial X^i}{\partial v^{\beta}_{j_1\cdots j_{m-s}}} = 0,\quad \frac{\partial V^{\alpha}_{i_1\cdots i_k}}{\partial v^{\beta}_{j_1\cdots j_{m-s}}} = 0, \qquad (9.3.12)$$

where $k = 0,1,\ldots,m-s-1, s = 0,1,\ldots,m-2$. We have shown above that these relations are true for $s = 0,1$. We shall now assume that they are true for s and try to prove that they are also true for $s+1$. On writing the relation (9.3.6) for $k = 0,1,\ldots,\ m-s-2$, we obtain

$$\begin{aligned}V^{\alpha}_{i_1\cdots i_k i} = {} & \frac{\partial V^{\alpha}_{i_1\cdots i_k}}{\partial x^i} - v^{\alpha}_{i_1\cdots i_k j}\frac{\partial X^j}{\partial x^i} \\ & + \sum_{r=0}^{m-2-s}\left[\frac{\partial V^{\alpha}_{i_1\cdots i_k}}{\partial v^{\beta}_{j_1\cdots j_r}} - v^{\alpha}_{i_1\cdots i_k j}\frac{\partial X^j}{\partial v^{\beta}_{j_1\cdots j_r}}\right]v^{\beta}_{j_1\cdots j_r i} \\ & + \left[\frac{\partial V^{\alpha}_{i_1\cdots i_k}}{\partial v^{\beta}_{j_1\cdots j_{m-s-1}}} - v^{\alpha}_{i_1\cdots i_k j}\frac{\partial X^j}{\partial v^{\beta}_{j_1\cdots j_{m-s-1}}}\right]v^{\beta}_{j_1\cdots j_{m-s-1}i}, 0 \le k \le m-s-2.\end{aligned}$$

But because of (9.3.12), the functions $V^{\alpha}_{i_1\cdots i_k i}$ cannot depend on variables $v^{\beta}_{j_1\cdots j_{m-s-1}i}$ so that their coefficients must vanish:

$$\frac{\partial V^{\alpha}_{i_1\cdots i_k}}{\partial v^{\beta}_{j_1\cdots j_{m-s-1}}} - v^{\alpha}_{i_1\cdots i_k j}\frac{\partial X^j}{\partial v^{\beta}_{j_1\cdots j_{m-s-1}}} = 0,\ k = 0,1,\ldots,m-s-2.$$

We thus obtain in the similar fashion

$$\frac{\partial X^i}{\partial v^{\beta}_{j_1\cdots j_{m-s-1}}} = 0,\quad \frac{\partial V^{\alpha}_{i_1\cdots i_k}}{\partial v^{\beta}_{j_1\cdots j_{m-s-1}}} = 0,\ k = 0,1,\ldots,m-s-2$$

if $Nn^{m-s-2} \neq 1$. This justifies the proposition (9.3.12). However, we have to be a little bit more careful for the case $s = m - 1$. If we write the relation (9.3.6) for $k = 0$, we then find

$$V_i^\alpha = \frac{\partial U^\alpha}{\partial x^i} - v_j^\alpha \frac{\partial X^j}{\partial x^i} + \Big(\frac{\partial U^\alpha}{\partial u^\beta} - v_j^\alpha \frac{\partial X^j}{\partial u^\beta}\Big) v_i^\beta + \Big(\frac{\partial U^\alpha}{\partial v_j^\beta} - v_k^\alpha \frac{\partial X^k}{\partial v_j^\beta}\Big) v_{ji}^\beta.$$

On the other hand, the functions V_i^α must be independent of the variables v_{ji}^β so that one gets

$$\frac{\partial U^\alpha}{\partial v_i^\beta} - v_j^\alpha \frac{\partial X^j}{\partial v_i^\beta} = 0. \tag{9.3.13}$$

Next, we differentiate (9.3.13) with respect to the variables v_k^γ. The symmetry of the second order derivatives leads eventually to the result

$$\delta_\gamma^\alpha \frac{\partial X^k}{\partial v_i^\beta} = \delta_\beta^\alpha \frac{\partial X^i}{\partial v_k^\gamma}.$$

A contraction on indices (α, γ) gives

$$N \frac{\partial X^k}{\partial v_i^\beta} = \frac{\partial X^i}{\partial v_k^\beta}. \tag{9.3.14}$$

On inserting this result into the above expression and contracting this time on indices (α, β), we finally obtain

$$(N^2 - 1) \frac{\partial X^k}{\partial v_i^\gamma} = 0. \tag{9.3.15}$$

In evaluating this inequality, we have to distinguish two cases:

(i). We assume that $N > 1$. Hence the number of dependent variables is more than one. In this case (9.3.15) and (9.3.13) yield

$$\frac{\partial X^i}{\partial v_j^\alpha} = 0 \ \text{ and } \ \frac{\partial U^\alpha}{\partial v_i^\beta} = 0.$$

We thus obtain

$$X^i = X^i(\mathbf{x}, \mathbf{u}), \quad V_{i_0}^\alpha = U^\alpha = U^\alpha(\mathbf{x}, \mathbf{u}). \tag{9.3.16}$$

Thus X^i and U^α components of the isovector field have to depend solely on coordinates x^i and u^α of the graph space. If the components (9.3.16) are inserted into (9.3.6) on taking notice of (9.3.12), we realise that other components of isovector fields of the contact ideal are determined by the following recurrence relations

$$\begin{aligned} V^\alpha_{i_1\cdots i_k i} = & \frac{\partial V^\alpha_{i_1\cdots i_k}}{\partial x^i} - v^\alpha_{i_1\cdots i_k j}\frac{\partial X^j}{\partial x^i} \\ & + \left[\frac{\partial V^\alpha_{i_1\cdots i_k}}{\partial u^\beta} - v^\alpha_{i_1\cdots i_k j}\frac{\partial X^j}{\partial u^\beta}\right] v^\beta_i \\ & + \sum_{r=1}^{k} v^\beta_{j_1\cdots j_r i}\frac{\partial V^\alpha_{i_1\cdots i_k}}{\partial v^\beta_{j_1\cdots j_r}}, \quad k = 0, 1, \ldots, m-1. \end{aligned} \tag{9.3.17}$$

Let us now define a set of vector fields, or differential operators, $D_i^{(k)}$ where $i = 1, \ldots, n$, $k = 0, 1, \ldots, m-1$ as follows

$$\begin{aligned} D_i^{(k)} & = \frac{\partial}{\partial x^i} + v^\alpha_i\frac{\partial}{\partial u^\alpha} + \sum_{r=1}^{k} v^\alpha_{i_1\cdots i_r i}\frac{\partial}{\partial v^\alpha_{i_1\cdots i_r}} \\ & = \frac{\partial}{\partial x^i} + v^\alpha_i\frac{\partial}{\partial u^\alpha} + v^\alpha_{ji}\frac{\partial}{\partial v^\alpha_j} + \cdots + v^\alpha_{i_1\cdots i_k i}\frac{\partial}{\partial v^\alpha_{i_1\cdots i_k}}. \end{aligned} \tag{9.3.18}$$

This operator may also be defined by the recurrence relations

$$D_i = D_i^{(0)} = \frac{\partial}{\partial x^i} + v^\alpha_i\frac{\partial}{\partial u^\alpha}, \qquad D_i^{(k+1)} = D_i^{(k)} + v^\alpha_{i_1\cdots i_{k+1} i}\frac{\partial}{\partial v^\alpha_{i_1\cdots i_{k+1}}}$$

By employing the operator defined in (9.3.18), we can express the recurrence relations (9.3.17) connecting isovector components in the form

$$V^\alpha_{i_1\cdots i_k i} = D_i^{(k)}(V^\alpha_{i_1\cdots i_k}) - v^\alpha_{i_1\cdots i_k j}D_i^{(k)}(X^j) = D_i^{(k)}(V^\alpha_{i_1\cdots i_k} - v^\alpha_{i_1\cdots i_k j}X^j)$$

where $0 \le k \le m-1$. By introducing the functions

$$F^\alpha_{i_1\cdots i_k} = V^\alpha_{i_1\cdots i_k} - v^\alpha_{i_1\cdots i_k j}X^j = \mathbf{i}_V(\sigma^\alpha_{i_1\cdots i_k}) \in \Lambda^0(\mathcal{C}_m)$$

we can also write

$$V^\alpha_{i_1\cdots i_k i} = D_i^{(k)}(F^\alpha_{i_1\cdots i_k}) = D_i^{(k)}\big(\mathbf{i}_V(\sigma^\alpha_{i_1\cdots i_k})\big). \tag{9.3.19}$$

Next, we define vector fields $V_G \in T(G)$ by

$$V_G = X^i(\mathbf{x}, \mathbf{u})\frac{\partial}{\partial x^i} + U^\alpha(\mathbf{x}, \mathbf{u})\frac{\partial}{\partial u^\alpha}. \tag{9.3.20}$$

Since $T(G)$ is a Lie algebra, these vectors generate a Lie group of diffeomorphisms mapping the manifold G onto itself. On the other hand, we know that isovectors of an ideal of the exterior algebra constitute a Lie subalgebra of the tangent bundle, or the module, of the manifold involved. We denote the Lie algebra of the isovectors of the contact ideal $\mathcal{I}_m$ by $\mathfrak{L}_{\mathcal{I}_m} \subset T(\mathcal{C}_m)$. A mere choice of $n+N$ smooth functions $X^i(\mathbf{x},\mathbf{u})$ and $U^\alpha(\mathbf{x},\mathbf{u})$ determines a *uniquely* defined member

$$V = V_G + \sum_{r=1}^{m} D_{i_r}^{(r-1)}(F^\alpha_{i_1\cdots i_{r-1}})\frac{\partial}{\partial v^\alpha_{i_1\cdots i_r}}$$

of the Lie algebra $\mathfrak{L}_{\mathcal{I}_m}$. Therefore, this expression can be regarded as the *lift* of a vector $V_G \in T(G)$ to a vector $V \in \mathfrak{L}_{\mathcal{I}_m} \subset T(\mathcal{C}_m)$. Since $\mathfrak{L}_{\mathcal{I}_m}$ is a Lie algebra, it generates a Lie group of diffeomorphisms on $\mathcal{C}_m$. It is evident that this group is a subgroup of the Lie group of diffeomorphisms on $\mathcal{C}_m$ generated by the Lie algebra $T(\mathcal{C}_m)$. But it is the only group preserving contact 1-forms. If we regard the manifold $\mathcal{C}_m$ as a fibre bundle on the base G, then the isovector V is called the ***mth order prolongation*** of the vector V_G. We adopt the notation $\mathfrak{L}_{\mathcal{I}_m} = \mathrm{pr}^{(m)}\big(T(G)\big)$. The rather complicated structures of prolongations are clearly illustrated in the two examples below:

$$V_i^\alpha = \frac{\partial U^\alpha}{\partial x^i} - v^\alpha_j\frac{\partial X^j}{\partial x^i} + v^\beta_i\frac{\partial U^\alpha}{\partial u^\beta} - v^\alpha_j v^\beta_i\frac{\partial X^j}{\partial u^\beta} \tag{9.3.21}$$

$$\begin{aligned}
V_{ij}^\alpha &= \frac{\partial V_i^\alpha}{\partial x^j} - v^\alpha_{ik}\frac{\partial X^k}{\partial x^j} + \Big(\frac{\partial V_i^\alpha}{\partial u^\beta} - v^\alpha_{ik}\frac{\partial X^k}{\partial u^\beta}\Big)v^\beta_j + v^\beta_{kj}\frac{\partial V_i^\alpha}{\partial v^\beta_k}\\
&= \frac{\partial^2 U^\alpha}{\partial x^i\partial x^j} - v^\alpha_k\frac{\partial^2 X^k}{\partial x^i\partial x^j} + v^\beta_i\frac{\partial^2 U^\alpha}{\partial x^j\partial u^\beta} + v^\beta_j\frac{\partial^2 U^\alpha}{\partial x^i\partial u^\beta} + v^\beta_{ij}\frac{\partial U^\alpha}{\partial u^\beta} -\\
&\quad v^\alpha_{ik}\frac{\partial X^k}{\partial x^j} - v^\alpha_{jk}\frac{\partial X^k}{\partial x^i} - v^\beta_i v^\alpha_k\frac{\partial^2 X^k}{\partial x^j\partial u^\beta} - v^\beta_j v^\alpha_k\frac{\partial^2 X^k}{\partial x^i\partial u^\beta} +\\
&\quad v^\gamma_i v^\beta_j\frac{\partial^2 U^\alpha}{\partial u^\gamma\partial u^\beta} - (v^\beta_j v^\alpha_{ik} + v^\alpha_k v^\beta_{ij} + v^\beta_i v^\alpha_{jk})\frac{\partial X^k}{\partial u^\beta} - v^\beta_i v^\gamma_j v^\alpha_k\frac{\partial^2 X^k}{\partial u^\beta\partial u^\gamma}.
\end{aligned}$$

If we recall that the variables v^α_{ij} are symmetric with respect to the indices i,j , we observe in the above relation the components V^α_{ij} become symmetric with respect to the same indices as it should be.

(ii). We now assume that $N=1$, that is, there is only one dependent variable. If we write $v^1_{i_1\cdots i_r} = v_{i_1\cdots i_r}, r = 0,1,\dots,m,$ then the isovector field may be represented by

$$V = X^i\frac{\partial}{\partial x^i} + U\frac{\partial}{\partial u} + \sum_{r=1}^{m} V_{i_1\cdots i_r}\frac{\partial}{\partial v_{i_1\cdots i_r}}$$

where we denote $V^1_{i_1\cdots i_r} = V_{i_1\cdots i_r}$. In this case, the relation (9.3.14) becomes

$$\frac{\partial X^k}{\partial v_i} = \frac{\partial X^i}{\partial v_k}.$$

The solution of this set of equations is found to be

$$X^i = X^i(\mathbf{x}, u, \mathbf{v}) = -\frac{\partial F}{\partial v_i} \tag{9.3.22}$$

where $F = F(x^i, u, v_j)$ is an arbitrary smooth function of $2n+1$ variables. Due to this structure of functions X^k, equations (9.3.7) are then satisfied automatically. With $U^1 = U$, equations (9.3.13) lead to

$$\begin{aligned}\frac{\partial U}{\partial v_i} = v_j\frac{\partial X^j}{\partial v_i} &= -v_j\frac{\partial^2 F}{\partial v_j \partial v_i}\\ &= -\frac{\partial}{\partial v_i}\Big(v_j\frac{\partial F}{\partial v_j} - F\Big)\end{aligned}$$

The integration of the above differential equations involves an arbitrary function of variables x^i and u. Absorbing this function into the arbitrary function F, we obtain

$$U = U(\mathbf{x}, u, \mathbf{v}) = F - v_j\frac{\partial F}{\partial v_j}. \tag{9.3.23}$$

Other components of the isovector field are clearly given by the relations

$$V_{i_1\cdots i_k i} = D_i^{(k)}(V_{i_1\cdots i_k}) - v_{i_1\cdots i_k j}D_i^{(k)}(X^j) \tag{9.3.24}$$

where the operator $D_i^{(k)}$ of (9.3.18) should now be expressed as

$$D_i^{(k)} = \frac{\partial}{\partial x^i} + v_i\frac{\partial}{\partial u} + \sum_{r=1}^{k} v_{i_1\cdots i_r i}\frac{\partial}{\partial v_{i_1\cdots i_r}}$$

The recurrence relations (9.3.24) make it possible to determine all components of the isovector field uniquely once one chooses a smooth function $F(\mathbf{x}, u, \mathbf{v})$. The components X^i and U are then determined uniquely through the relations (9.3.22) and (9.3.23). The relation (9.3.24) may be explicitly expressed as

$$V_{i_1\cdots i_k i} = \frac{\partial V_{i_1\cdots i_k}}{\partial x^i} + v_{i_1\cdots i_k j}\frac{\partial^2 F}{\partial x^i \partial v_j} +$$

$$+\Big(\frac{\partial V_{i_1\cdots i_k}}{\partial u}+v_{i_1\cdots i_k j}\frac{\partial^2 F}{\partial u\partial v_j}\Big)v_i+v_{ik}v_{i_1\cdots i_k j}\frac{\partial^2 F}{\partial v_k\partial v_j} \tag{9.3.25}$$

$$+\sum_{r=1}^{k}v_{j_1\cdots j_r i}\frac{\partial V_{i_1\cdots i_k}}{\partial v_{j_1\cdots j_r}}, 0\le k\le m-1.$$

In general, the vector

$$V_G = X^i(\mathbf{x},u,\mathbf{v})\frac{\partial}{\partial x^i}+U(\mathbf{x},u,\mathbf{v})\frac{\partial}{\partial u}$$

is no longer dependent only the coordinates of the graph space. Hence, we cannot interpret the isovector field as a prolongation of a vector field V_G in $T(G)$. In order that an isovector is a prolongation of a member of $T(G)$, the functions X^i and U must be independent of variables $\mathbf{v}$. On the other hand, we easily see that in order to be able to obtain $X^i = \mathcal{X}^i(\mathbf{x},u)$, the equation $\partial F/\partial v_i = -\mathcal{X}^i(\mathbf{x},u)$ requires that the function F must have the form

$$F = -\mathcal{X}^i(\mathbf{x},u)v_i + G(\mathbf{x},u)$$

In such a case, (9.3.23) yields $U = G(\mathbf{x},u)$. Hence, *isovectors are found to be mth order prolongations of vectors $V_G\in T(G)$ if only F is an affine function of variables v_i*. Otherwise, isovectors may be interpreted as prolongations of the tangent bundle $T(\mathcal{C}_1)$ and one may then write $\mathfrak{I}_{\mathcal{I}_m} = \mathrm{pr}^{(m-1)}(T(\mathcal{C}_1))$.

The structure of isovectors corresponding to the case $N=1$ might be illustrated to some extent by the following examples:

$$X^i = -\frac{\partial F}{\partial v_i},\quad U = F - v_i\frac{\partial F}{\partial v_i},\quad F = F(\mathbf{x},u,\mathbf{v}) \tag{9.3.26}$$

$$V_i = \frac{\partial F}{\partial x^i}+v_i\frac{\partial F}{\partial u}$$

$$V_{ij} = \frac{\partial^2 F}{\partial x^i\partial x^j}+v_i\frac{\partial^2 F}{\partial u\partial x^j}+v_j\frac{\partial^2 F}{\partial u\partial x^i}+v_{ik}\frac{\partial^2 F}{\partial v_k\partial x^j}+v_{jk}\frac{\partial^2 F}{\partial v_k\partial x^i}$$
$$+v_{ij}\frac{\partial F}{\partial u}+v_iv_j\frac{\partial^2 F}{\partial u^2}+(v_iv_{jk}+v_jv_{ik})\frac{\partial^2 F}{\partial v_k\partial u}+v_{ik}v_{jl}\frac{\partial^2 F}{\partial v_k\partial v_l}$$

We can collect the cases **(*i*)** and **(*ii*)** discussed above in the theorem below:

Theorem 9.3.1. *A vector field $V = X^i\partial/\partial x^i + \sum_{r=0}^{m} V^\alpha_{i_1\cdots i_r}\partial/\partial v^\alpha_{i_1\cdots i_r}$ of $T(\mathcal{C}_m)$ is an isovector field of the contact ideal $\mathcal{I}_m$ if and only if the* *relations* $V^\alpha_{i_1\cdots i_k i} = D^{(k)}_i (V^\alpha_{i_1\cdots i_k} - v^\alpha_{i_1\cdots i_k j} X^j)$ *for* $0 \le k \le m-1$ *are satisfied. The operators* $D^{(k)}_i$ *are given by* (9.3.18). *To determine isovector components completely one has to prescribe* $n+N$ *smooth functions* $X^i = X^i(\mathbf{x},\mathbf{u})$ *and* $U^\alpha = U^\alpha(\mathbf{x},\mathbf{u})$ *when* $N>1$, *whereas a single function* $F = F(\mathbf{x},u,\mathbf{v})$ *would be sufficient when* $N=1$ *through which the components* X^i *and* U *are found as* $X^i = -\partial F/\partial v_i, U = F - v_i(\partial F/\partial v_i)$. □

Since isovectors forming a Lie algebra produce groups of diffeomorphisms, we can state that this theorem is a somewhat generalised version of the celebrated Bäcklund theorem for $N>1$ [Swedish mathematician Albert Victor Bäcklund (1845-1929]: *The most general diffeomorphisms on* $\mathcal{C}_m$ *preserving the contact structure are prolongations of diffeomorphisms of the graph space.* Since this result restricts substantially admissible diffeomorphisms on $\mathcal{C}_m$, it creates a rather significant obstacle one has to surmount in determining solutions of partial differential equations by resorting to transformations preserving contact structures. We shall be able to overcome this obstacle later by choosing a more convenient ideal of $\Lambda(\mathcal{C}_m)$ instead of the contact ideal $\mathcal{I}_m$ [*see* Sec. 9.7].

The next step after having found isovector fields of the contact ideal would be to determine linearly independent isovector fields of the closed ideal generated by the given system of partial differential equations. Thus, it will become possible to obtain *Lie groups of symmetry transformations* that leave the system of partial differential equations invariant through which one can obtain families of new solutions from a given solution. However, this approach proves to be quite fruitful as far as the analytical procedures are concerned in balance equations derived from conservation laws. Since natural laws are generally of this form, many field equations encountered in physics and engineering fall naturally into this category. Thus, we can say that balance equations are come across most frequently in practical applications. This subject will be discussed in detail in the subsequent section. However, we shall try here to elucidate the approach that we use to employ in determining isovectors associated with a given system of first order partial differential equations through a somewhat complicated example.

Example 9.3.1. We consider the partial differential equations introduced in Example 8.7.3. The functions $u(x,t)$ and $c(x,t)$ satisfy the following first order partial differential equations

$$u_t + uu_x + \alpha c c_x = 0,\; c_t + uc_x + \frac{1}{\alpha} c u_x = 0,\;\; \alpha \neq 1$$

where $(x,t) \in \mathbb{R}^2$. The physical origins of these equations was also explained in that example. In order to simplify a little these equations, let us make the transformations

$$r = u + \alpha c, \quad s = u - \alpha c$$

to readily arrive at

$$r_t + \mathfrak{f}(r,s)r_x = 0, \quad s_t + \mathfrak{g}(r,s)s_x = 0$$

where the functions $\mathfrak{f}$ and $\mathfrak{g}$ are defined by

$$\mathfrak{f}(r,s) = \frac{\alpha+1}{2\alpha}r + \frac{\alpha-1}{2\alpha}s, \ \mathfrak{g}(r,s) = \frac{\alpha-1}{2\alpha}r + \frac{\alpha+1}{2\alpha}s.$$

We now introduce the forms $\omega_1, \omega_2 \in \Lambda^2(\mathbb{R}^4)$ as follows

$$\begin{aligned}\omega_1 &= -\, dr \wedge dx + \mathfrak{f}(r,s)\, dr \wedge dt,\\ \omega_2 &= -\, ds \wedge dx + \mathfrak{g}(r,s)\, ds \wedge dt.\end{aligned}$$

The coordinate cover of the manifold $G = \mathbb{R}^4$ is given by (x,t,r,s). If we define a solution mapping $\phi : \mathbb{R}^2 \to \mathbb{R}^4$ by relations $\big(x, t, r(x,t), s(x,t)\big)$, we then obtain

$$\begin{aligned}\phi^*\omega_1 &= \big[r_t + \mathfrak{f}(r,s)r_x\big]dx \wedge dt = 0,\\ \phi^*\omega_2 &= \big[s_t + \mathfrak{g}(r,s)s_x\big]dx \wedge dt = 0.\end{aligned}$$

Thus, the solution mapping ϕ annihilates the ideal generated by the forms ω_1 and ω_2. We can easily check that the exterior derivatives of the forms ω_1 and ω_2 are found to be as

$$\begin{aligned}d\omega_1 = -\, d\omega_2 &= -\,\frac{\alpha-1}{2\alpha}dr \wedge ds \wedge dt\\ &= \frac{\alpha-1}{2(r-s)}(ds \wedge \omega_1 + dr \wedge \omega_2).\end{aligned}$$

Hence, the ideal generated by the forms ω_1 and ω_2 is closed. Since the differential equations are of first order, we can just take the isovector field in the form below

$$V = X\frac{\partial}{\partial x} + T\frac{\partial}{\partial t} + R\frac{\partial}{\partial r} + S\frac{\partial}{\partial s}.$$

The components X, T, R, S are smooth functions of the variables x, t, r, s. In order that V becomes an isovector field we have to find smooth functions $\lambda_{11}, \lambda_{12}, \lambda_{21}, \lambda_{22}(x,t,r,s)$ so that the relations

$$\pounds_V\omega_1 = \lambda_{11}\omega_1 + \lambda_{12}\omega_2, \ \ \pounds_V\omega_2 = \lambda_{21}\omega_1 + \lambda_{22}\omega_2$$

are satisfied. If we employ the expressions

$$\begin{aligned}
\mathbf{i}_V(\omega_1) &= -Rdx + \mathfrak{f}(r,s)\,Rdt + \big[X - \mathfrak{f}(r,s)\,T\big]dr \\
\mathbf{i}_V(\omega_2) &= -Sdx + \mathfrak{g}(r,s)\,Sdt + \big[X - \mathfrak{g}(r,s)\,T\big]ds \\
\mathbf{i}_V(d\omega_1) &= -\mathbf{i}_V(d\omega_2) = -\frac{\alpha-1}{2\alpha}(Rds\wedge dt - Sdr\wedge dt + Tdr\wedge ds)
\end{aligned}$$

in the Cartan magic formula, we obtain

$$\begin{aligned}
&\pounds_V\omega_1 = \big[R_t + \mathfrak{f}(r,s)R_x\big]dx\wedge dt - \big[R_r + X_x - \mathfrak{f}(r,s)T_x\big]dr\wedge dx \\
&- R_s\,ds\wedge dx + \Big[\frac{\alpha+1}{2\alpha}R + \frac{\alpha-1}{2\alpha}S + \mathfrak{f}(r,s)(R_r+T_t) - X_t\Big]dr\wedge dt \\
&\qquad + \mathfrak{f}(r,s)R_s ds\wedge dt + \big[X_s - \mathfrak{f}(r,s)T_s\big]ds\wedge dr = \\
&-\lambda_{11}dr\wedge dx + \lambda_{11}\mathfrak{f}(r,s)\,dr\wedge dt - \lambda_{12}ds\wedge dx + \lambda_{12}\mathfrak{g}(r,s)ds\wedge dt \\
&\pounds_V\omega_2 = \big[S_t + \mathfrak{g}(r,s)S_x\big]dx\wedge dt - \big[S_s + X_x - \mathfrak{g}(r,s)T_x\big]ds\wedge dx \\
&- S_r\,dr\wedge dx + \Big[\frac{\alpha-1}{2\alpha}R + \frac{\alpha+1}{2\alpha}S + \mathfrak{g}(r,s)(S_s+T_t) - X_t\Big]ds\wedge dt \\
&\qquad + \big[X_r - \mathfrak{g}(r,s)T_r\big]dr\wedge ds + \mathfrak{g}(r,s)S_r dr\wedge dt = \\
&-\lambda_{21}dr\wedge dx + \lambda_{21}\mathfrak{f}(r,s)\,dr\wedge dt - \lambda_{22}ds\wedge dx + \lambda_{22}\mathfrak{g}(r,s)ds\wedge dt
\end{aligned}$$

from which we extract the following system of equations

$$\begin{aligned}
&\lambda_{11} = R_r + X_x - \mathfrak{f}(r,s)T_x, \qquad \lambda_{12} = R_s, \\
&\lambda_{21} = S_r\,, \qquad \lambda_{22} = S_s + X_x - \mathfrak{g}(r,s)T_x \\
&R_t + \mathfrak{f}(r,s)R_x = 0,\quad \frac{1}{\alpha}(r-s)R_s = 0,\quad X_s - \mathfrak{f}(r,s)T_s = 0 \\
&\frac{\alpha+1}{2\alpha}R + \frac{\alpha-1}{2\alpha}S - X_t + \mathfrak{f}(r,s)\big[-X_x + T_t + \mathfrak{f}(r,s)T_x\big] = 0 \\
&S_t + \mathfrak{g}(r,s)S_x = 0,\quad -\frac{1}{\alpha}(r-s)S_r = 0,\quad X_r - \mathfrak{g}(r,s)T_r = 0 \\
&\frac{\alpha-1}{2\alpha}R + \frac{\alpha+1}{2\alpha}S - X_t + \mathfrak{g}(r,s)\big[-X_x + T_t + \mathfrak{g}(r,s)T_x\big] = 0
\end{aligned} \tag{9.3.27}$$

where we have noted that

$$\mathfrak{f}(r,s) - \mathfrak{g}(r,s) = \frac{r-s}{\alpha}.$$

Equations $(9.3.27)_6$ and $(9.3.27)_{10}$ yield obviously

$$R = R(x,t,r), \quad S = S(x,t,s).$$

Next, let us differentiate $(9.3.27)_5$ and $(9.3.27)_9$ with respect to s and r to find, respectively

$$\frac{\alpha-1}{2\alpha}R_x = 0, \quad \frac{\alpha-1}{2\alpha}S_x = 0$$

and, consequently, $R_t = 0$ and $S_t = 0$. We thus get

$$R = R(r), \quad S = S(s).$$

On writing the equations $(9.3.27)_7$ and $(9.3.27)_{11}$ in the form

$$\big[X - \mathfrak{f}(r,s)T\big]_s + \frac{\alpha-1}{2\alpha}T = 0, \ \big[X - \mathfrak{g}(r,s)T\big]_r + \frac{\alpha-1}{2\alpha}T = 0,$$

we obtain

$$\big[X - \mathfrak{f}(r,s)T\big]_s = \big[X - \mathfrak{g}(r,s)T\big]_r.$$

This expression implies that the following relations are obtainable

$$X - \mathfrak{f}(r,s)T = \Phi_r, \quad X - \mathfrak{g}(r,s)T = \Phi_s, \tag{9.3.28}$$
$$T = -\frac{2\alpha}{\alpha-1}\Phi_{rs} = \frac{\alpha(\Phi_s - \Phi_r)}{r-s}$$

where $\Phi = \Phi(x,t,r,s)$. Hence, the function Φ must satisfy the partial differential equation

$$2(r-s)\Phi_{rs} + (\alpha-1)(\Phi_s - \Phi_r) = 0. \tag{9.3.29}$$

It follows from $(9.3.27)_8$ and $(9.3.27)_{12}$ that

$$R(r) = X_t + rX_x - rT_t - \frac{1}{4}\Big[(3r-s)(r+s) + \frac{(r-s)^2}{\alpha^2}\Big]T_x,$$
$$S(s) = X_t + sX_x - sT_t + \frac{1}{4}\Big[(r-3s)(r+s) - \frac{(r-s)^2}{\alpha^2}\Big]T_x.$$

By adding the first two expressions in (9.3.28) and using the third one we obtain

$$X = \frac{\big[(\alpha+1)r + (\alpha-1)s\big]\Phi_s - \big[(\alpha-1)r + (\alpha+1)s\big]\Phi_r}{2(r-s)}.$$

Inserting this expression for X together with $(9.3.28)_3$ into $R(r)$ and $S(s)$ given above, we find that

$$2R(r) = -(\alpha-1)\big(\Phi_{ts} + \mathfrak{g}(r,s)\Phi_{xs}\big) + (\alpha+1)\big(\Phi_{tr} + \mathfrak{f}(r,s)\Phi_{xr}\big),$$
$$2S(s) = (\alpha+1)\big(\Phi_{ts} + \mathfrak{g}(r,s)\Phi_{xs}\big) - (\alpha-1)\big(\Phi_{tr} + \mathfrak{f}(r,s)\Phi_{xr}\big).$$

This result means that the derivatives of the right hand side of the first equation with respect to variables x, t, s, and the derivatives of the right hand side of the second equation with respect to variables x, t, r must vanish. The derivatives with respect to t give

$$-(\alpha-1)\big(\Phi_{tts}+\mathfrak{g}(r,s)\Phi_{xts}\big)+(\alpha+1)\big(\Phi_{ttr}+\mathfrak{f}(r,s)\Phi_{xtr}\big)=0,$$
$$(\alpha+1)\big(\Phi_{tts}+\mathfrak{g}(r,s)\Phi_{xts}\big)-(\alpha-1)\big(\Phi_{ttr}+\mathfrak{f}(r,s)\Phi_{xtr}\big)=0$$

whence we deduce that

$$\Phi_{trt}+\mathfrak{f}(r,s)\Phi_{xrt}=0,\quad \Phi_{tst}+\mathfrak{g}(r,s)\Phi_{xst}=0$$

since $\alpha\neq 0$. So we can write

$$\Phi_{tr}+\mathfrak{f}(r,s)\Phi_{xr}=\mathfrak{f}A_x(x,r,s),\ \ \Phi_{ts}+\mathfrak{g}(r,s)\Phi_{xs}=\mathfrak{g}B_x(x,r,s)$$

where A and B are arbitrary functions. We then easily obtain

$$\Phi_r=A(x,r,s)+\phi(\xi,r,s),\ \ \Phi_s=B(x,r,s)+\psi(\eta,r,s)$$

where the characteristic variables are

$$\xi=x-\mathfrak{f}(r,s)t,\quad \eta=x-\mathfrak{g}(r,s)t.$$

Similarly, the following equations must hold

$$\Phi_{trx}+\mathfrak{f}(r,s)\Phi_{xrx}=0,\quad \Phi_{tsx}+\mathfrak{g}(r,s)\Phi_{xsx}=0$$

from which we get

$$A_{xx}=0,\ \ B_{xx}=0.$$

We thus conclude that

$$A(x,r,s)=a(r,s)x+b(r,s),\ \ B(x,r,s)=c(r,s)x+d(r,s).$$

Functions A, B, ϕ, ψ must satisfy the compatibility condition $\Phi_{rs}=\Phi_{sr}$, that is, the following equation must hold

$$(a_s-c_r)x-\frac{\alpha-1}{2\alpha}(\phi_\xi-\psi_\eta)t+b_s-d_r+\phi_s-\psi_r=0.$$

If we calculate the variables x and t in terms of ξ and η, and insert them into the above equation, we obtain

$$\frac{\alpha(\eta\mathfrak{f}-\xi\mathfrak{g})}{r-s}(a_s-c_r)-\frac{(\alpha-1)(\eta-\xi)}{2(r-s)}(\phi_\xi-\psi_\eta)+b_s-d_r+\phi_s-\psi_r=0. \tag{9.3.30}$$

On differentiating this expression successively with respect to variables ξ and η, we find that

$$-\frac{\alpha-1}{2(r-s)}(\phi_{\xi\xi}+\psi_{\eta\eta})=0.$$

We thus have to take

$$\phi_{\xi\xi}(\xi,r,s)=-\psi_{\eta\eta}(\eta,r,s)=2k(r,s)$$

whence we deduce that

$$\begin{aligned}\phi(\xi,r,s)&=k(r,s)\xi^2+m(r,s)\xi+n(r,s),\\ \psi(\eta,r,s)&=-k(r,s)\eta^2+p(r,s)\eta+q(r,s).\end{aligned}$$

If we introduce these functions into (9.3.30) and arrange the resulting terms, we then get the following polynomial in ξ and η

$$\begin{aligned}&\Big[k_s+\frac{\alpha-1}{r-s}k\Big]\xi^2+\Big[k_r-\frac{\alpha-1}{r-s}k\Big]\eta^2+\frac{1}{2(r-s)}\Big[(\alpha-1)(m-p)\\ &-[(\alpha-1)r+(\alpha+1)s](a_s-c_r)+2(r-s)m_s\Big]\xi\\ &+\frac{1}{2(r-s)}\Big[-(\alpha-1)(m-p)+[(\alpha+1)r+(\alpha-1)s](a_s-c_r)\\ &-2(r-s)p_r\Big]\eta+(b+n)_s-(d+q)_r=0.\end{aligned}$$

The coefficients above must be zero so that we obtain

$$k(r,s)=\overline{c}_1(r-s)^{\alpha-1},\ (a+m)_s=(c+p)_r,\ (b+n)_s=(d+q)_r.$$

Therefore, we can write

$$m=\omega_r-a,\ p=\omega_s-c,\ n=\Omega_r-b,\ q=\Omega_s-d$$

where $\omega=\omega(r,s),\Omega=\Omega(r,s)$. Thus the only equation to be satisfied is

$$\begin{aligned}(\alpha-1)(a-c)+2\alpha\mathfrak{f}a_s-2\alpha\mathfrak{g}c_r+(\alpha-1)(\omega_s-\omega_r)&\\ -2(r-s)\omega_{rs}=0&\end{aligned}\tag{9.3.31}$$

so that we arrive at the result

$$\begin{aligned}\Phi_r&=\overline{c}_1(r-s)^{\alpha-1}(x-\mathfrak{f}t)^2+\omega_r x-\mathfrak{f}(\omega_r-a)t+\Omega_r,\\ \Phi_s&=-\overline{c}_1(r-s)^{\alpha-1}(x-\mathfrak{g}t)^2+\omega_s x-\mathfrak{g}(\omega_s-c)t+\Omega_s.\end{aligned}$$

When we insert these relations into (9.3.29), we see first that we have to take $\overline{c}_1=0$ and the remaining terms give rise to the following equations

$$
\begin{aligned}
&2(r-s)\omega_{rs}+(\alpha-1)(\omega_s-\omega_r)=0, &(9.3.32)\\
&2(r-s)\Omega_{rs}+(\alpha-1)(\Omega_s-\Omega_r)=0,\\
&(\alpha-1)\mathfrak{g}(a-c)+(\alpha-1)\mathfrak{g}(\omega_s-\omega_r)-2(r-s)\mathfrak{f}(a_s-\omega_{rs})=0.
\end{aligned}
$$

On the other hand, we can now write

$$
\begin{aligned}
2R(r)&=-(\alpha-1)\mathfrak{g}c+(\alpha+1)\mathfrak{f}a,\\
2S(s)&=(\alpha+1)\mathfrak{g}c-(\alpha-1)\mathfrak{f}a.
\end{aligned}
$$

Because of the relations $R_s=S_r=0$, the equations below must be held

$$-(\alpha-1)(\mathfrak{g}c)_s+(\alpha+1)(\mathfrak{f}a)_s=0,\ (\alpha+1)(\mathfrak{g}c)_r-(\alpha-1)(\mathfrak{f}a)_r=0.$$

If we differentiate the first equation with respect to r and the second one with respect to s, we find that

$$(\mathfrak{f}a)_{rs}=0,\ \ (\mathfrak{g}c)_{rs}=0$$

whence we obtain

$$\mathfrak{f}a=\lambda(r)+\mu(s),\mathfrak{g}c=\frac{\alpha-1}{\alpha+1}\lambda(r)+\frac{\alpha+1}{\alpha-1}\mu(s)+2c_1 \quad (9.3.33)$$

and

$$
\begin{aligned}
R(r)&=\frac{2\alpha}{\alpha+1}\lambda(r)-(\alpha-1)c_1,\\
S(s)&=\frac{2\alpha}{\alpha+1}\mu(s)+(\alpha+1)c_1.
\end{aligned}
$$

If we insert the expressions (9.3.33) into equations (9.3.31) and $(9.3.32)_3$, solve the resulting expressions for $(\alpha-1)(\omega_s-\omega_r)$ and $2(r-s)\omega_{rs}$ and put them into the equation $(9.3.32)_1$ we reach to the equation

$$
\begin{aligned}
&-2(\alpha-1)\lambda+2(\alpha+1)\mu+(r-s)\big[(\alpha-1)\lambda'+(\alpha+1)\mu'\big] &(9.3.34)\\
&\qquad\qquad+2(\alpha^2-1)c_1=0.
\end{aligned}
$$

Differentiating (9.3.34) successively with respect to r and s, we are led to

$$-(\alpha-1)\lambda''(r)+(\alpha+1)\mu''(s)=0$$

from which we find

$$\lambda(r)=c_2r^2+c_3r+c_4,\quad \mu(s)=\frac{\alpha-1}{\alpha+1}c_2s^2+c_5s+c_6$$

On inserting these expressions into (9.3.34) we obtain

$$\lambda(r) = c_2 r^2 + c_3 r + c_4,$$
$$\mu(s) = \frac{\alpha-1}{\alpha+1}c_2 s^2 + \frac{\alpha-1}{\alpha+1}c_3 s + \frac{\alpha-1}{\alpha+1}c_4 - (\alpha-1)c_1.$$

If we employ these relations in (9.3.33) and, (9.3.31) and $(9.3.32)_3$, we come up with the relations

$$\omega_{rs} = -\frac{\alpha(\alpha-1)}{\alpha+1}c_2, \quad \omega_s - \omega_r = \frac{2\alpha}{\alpha+1}c_2(r-s).$$

Integration of the first equation yields

$$\omega = -\frac{\alpha(\alpha-1)}{\alpha+1}c_2 rs + m(r) + n(s)$$

while the second equation then results in

$$\alpha c_2(r-s) + m'(r) - n'(s) = 0.$$

The solution of this equation is easily found as

$$m(r) = -\frac{1}{2}\alpha c_2 r^2 + c_5 r + c_6,$$
$$n(s) = -\frac{1}{2}\alpha c_2 s^2 + c_5 s + c_7.$$

Hence, by replacing the arbitrary constant $c_6 + c_7$ by c_6, we get

$$\omega(r,s) = -\frac{\alpha(\alpha-1)}{\alpha+1}c_2 rs - \frac{1}{2}\alpha c_2(r^2+s^2) + c_5(r+s) + c_6.$$

On making use of these expressions where they are pertinent and defining new arbitrary constants as appropriate combinations of old constant, we ultimately obtain isovector components depending on constants a_1, a_2, a_3, a_4 and a function $\Omega(r,s)$, being a solution of the partial differential equation $(9.3.32)_2$, as follows

$$X = a_4 x + \{a_3 - a_1(\alpha+1)[\alpha(r+s)^2 - (r-s)^2]\}t + \frac{\alpha[\mathfrak{f}(r,s)\Omega_s - \mathfrak{g}(r,s)\Omega_r]}{r-s}, \tag{9.3.35}$$

$$T = 4\alpha^2 a_1 x - [a_2 - a_4 + 4\alpha(\alpha+1)a_1(r+s)]t + \frac{\alpha(\Omega_s - \Omega_r)}{r-s},$$

$$R = 4\alpha a_1 r^2 + a_2 r + a_3, \qquad S = 4\alpha a_1 s^2 + a_2 s + a_3.$$

Therefore, the linearly independent isovectors are given by

$$\begin{aligned}
V_1 = & -(\alpha+1)\big[\alpha(r+s)^2-(r-s)^2\big]t\frac{\partial}{\partial x}\\
& +4\alpha\big[\alpha x-(\alpha+1)(r+s)t\big]\frac{\partial}{\partial t}+4\alpha r^2\frac{\partial}{\partial r}+4\alpha s^2\frac{\partial}{\partial s},\\
V_2 = & -t\frac{\partial}{\partial t}+r\frac{\partial}{\partial r}+s\frac{\partial}{\partial s},\ V_3=t\frac{\partial}{\partial x}+\frac{\partial}{\partial r}+\frac{\partial}{\partial s},\ V_4=x\frac{\partial}{\partial x}+t\frac{\partial}{\partial t}\\
V_\Omega = & \frac{\alpha}{r-s}\Big[\big[\mathfrak{f}(r,s)\Omega_s-\mathfrak{g}(r,s)\Omega_r\big]\frac{\partial}{\partial x}+(\Omega_s-\Omega_r)\frac{\partial}{\partial t}\Big]
\end{aligned}$$

To determine the symmetry groups we have to solve the following autonomous ordinary differential equations

$$\frac{d\overline{x}}{d\epsilon}=X(\overline{x},\overline{t},\overline{r},\overline{s}),\ \frac{d\overline{t}}{d\epsilon}=T(\overline{x},\overline{t},\overline{r},\overline{s}),\ \frac{d\overline{r}}{d\epsilon}=R(\overline{r}),\ \frac{d\overline{s}}{d\epsilon}=S(\overline{s})$$

under the initial conditions $\overline{x}(0)=x$, $\overline{t}(0)=t$, $\overline{r}(0)=r$ and $\overline{s}(0)=s$ where ϵ is taken as the flow parameter. Hence, the one-parameter Lie group generated by the isovector field V_1 becomes

$$\begin{aligned}
\overline{x}(\epsilon) &= \big[(4\alpha r\epsilon-1)(4\alpha s\epsilon-1)\big]^{\frac{\alpha-1}{2}}\big[(2\alpha(\alpha-1)(r+s)\epsilon-16\alpha^3rs\epsilon^2\\
&\quad +1)x+\big((r-s)^2-\alpha^2(r+s)^2-4\alpha rs+8\alpha^2(\alpha+1)rs(r+s)\epsilon\big)\epsilon t\big]\\
\overline{t}(\epsilon) &= \big[(4\alpha r\epsilon-1)(4\alpha s\epsilon-1)\big]^{\frac{\alpha-1}{2}}\big[4\alpha^2x\epsilon+\big(1-2\alpha(\alpha+1)(r+s)\epsilon\big)t\big]\\
\overline{r}(\epsilon) &= -r/(4\alpha r\epsilon-1),\qquad \overline{s}(\epsilon)=-s/(4\alpha s\epsilon-1)
\end{aligned}$$

Similarly, the isovector field V_2 leads up to the Lie group

$$\overline{x}(\epsilon)=x,\ \overline{t}(\epsilon)=t\,e^{-\epsilon},\ \overline{r}(\epsilon)=r\,e^{\epsilon},\ \overline{s}(\epsilon)=s\,e^{\epsilon},$$

the isovector field V_3 to the Lie group

$$\overline{x}(\epsilon)=x+\epsilon t,\ \overline{t}(\epsilon)=t\,,\ \overline{r}(\epsilon)=r+\epsilon,\ \overline{s}(\epsilon)=s+\epsilon,$$

and the isovector field V_4 to the Lie group

$$\overline{x}(\epsilon)=x\,e^{\epsilon},\ \overline{t}(\epsilon)=t\,e^{\epsilon},\ \overline{r}(\epsilon)=r,\ \overline{s}(\epsilon)=s.$$

On the other hand, the function $\Omega(r,s)$ satisfying $(9.3.32)_2$ generates the Lie group

$$\begin{aligned}
\overline{x}(\epsilon) &= x+\frac{\alpha}{r-s}\big[\mathfrak{f}(r,s)\Omega_s-\mathfrak{g}(r,s)\Omega_r\big]\epsilon,\\
\overline{t}(\epsilon) &= t+\frac{\alpha}{r-s}(\Omega_s-\Omega_r)\epsilon,\ \overline{r}(\epsilon)=r,\ \overline{s}(\epsilon)=s.
\end{aligned}$$

If we wish to pass to the physically meaningful dependent variables (u, c), then the isovector field should be depicted by

$$V = X\frac{\partial}{\partial x} + T\frac{\partial}{\partial t} + U\frac{\partial}{\partial u} + C\frac{\partial}{\partial c}.$$

If we take into account the relations

$$\frac{\partial}{\partial r} = \frac{1}{2}\Big(\frac{\partial}{\partial u} + \frac{1}{\alpha}\frac{\partial}{\partial c}\Big), \quad \frac{\partial}{\partial s} = \frac{1}{2}\Big(\frac{\partial}{\partial u} - \frac{1}{\alpha}\frac{\partial}{\partial c}\Big)$$

we readily obtain

$$\begin{aligned} U(u,c) &= \frac{R(u+\alpha c) + S(u-\alpha c)}{2}, \\ C(u,c) &= \frac{R(u+\alpha c) - S(u-\alpha c)}{2\alpha}. \end{aligned}$$

Thus, it follows from (9.3.35) that

$$\begin{aligned} X &= a_4 x + \big[a_3 - 4\alpha(\alpha+1)a_1(u^2 - \alpha c^2)\big]t + 2\Omega_u - \frac{2u}{\alpha c}\Omega_c, \\ T &= 4\alpha^2 a_1 x - \big[a_2 - a_4 + 8\alpha(\alpha+1)a_1 u\big]t - \frac{1}{2\alpha c}\Omega_c, \\ U &= 4\alpha a_1(u^2 + \alpha^2 c^2) + a_2 u + a_3, \\ C &= 8\alpha a_1 uc + a_2 c \end{aligned}$$

where the function $\Omega(u, c)$ has now to be taken as a solution of the partial differential equation

$$\alpha^2\Omega_{uu} - \Omega_{cc} - \frac{\alpha - 1}{c}\Omega_c = 0.$$

Hence, the linearly independent isovectors become

$$\begin{aligned} V_1 = &-4\alpha(\alpha+1)(u^2 - \alpha c^2)t\frac{\partial}{\partial x} + 4\alpha\big[\alpha x - 2(\alpha+1)ut\big]\frac{\partial}{\partial t} \\ &+ 4\alpha(u^2 + \alpha^2 c^2)\frac{\partial}{\partial u} + 8\alpha uc\frac{\partial}{\partial c}, \\ V_2 = &-t\frac{\partial}{\partial t} + u\frac{\partial}{\partial u} + c\frac{\partial}{\partial c}, \; V_3 = t\frac{\partial}{\partial x} + \frac{\partial}{\partial u}, \; V_4 = x\frac{\partial}{\partial x} + t\frac{\partial}{\partial t}, \\ V_\Omega = &\, 2\Big(\Omega_u - \frac{u}{\alpha c}\Omega_c\Big)\frac{\partial}{\partial x} - \frac{1}{2\alpha c}\Omega_c\frac{\partial}{\partial t}\Big]. \end{aligned}$$

It is easily seen that the Lie transformation group generated by the isovector V_1 is now given by

$$\begin{aligned}
\overline{x}(\epsilon) &= \big[1 - 8\alpha u\epsilon + 16\alpha^2(u^2 - \alpha^2c^2)\epsilon^2\big]^{\frac{\alpha-1}{2}} \big[\{1 + 4\alpha[(\alpha - 1)u - \\
&\alpha^2(u^2 - \alpha^2c^2)\epsilon 1)x - 4\alpha(\alpha + 1)\{(u^2 - \alpha c^2) - 4\alpha u(u + \alpha c)^2\epsilon\}\epsilon t\big], \\
\overline{t}(\epsilon) &= \big[1 - 8\alpha u\epsilon + 16\alpha^2(u^2 - \alpha^2c^2)\epsilon^2\big]^{\frac{\alpha-1}{2}} \big[4\alpha^2x\epsilon \\
&\qquad + \{1 - 4\alpha(\alpha - 1)u\epsilon\}t\big], \\
\overline{u}(\epsilon) &= \frac{u - 4\alpha(u^2 - \alpha^2c^2)\epsilon}{1 - 8\alpha u\epsilon + 16\alpha^2(u^2 - \alpha^2c^2)\epsilon^2}, \\
\overline{c}(\epsilon) &= \frac{c}{1 - 8\alpha u\epsilon + 16\alpha^2(u^2 - \alpha^2c^2)\epsilon^2}.
\end{aligned}$$

Similarly, the isovector V_2 gives rise to the Lie group

$$\overline{x}(\epsilon) = x,\ \overline{t}(\epsilon) = t\,e^{-\epsilon},\ \overline{u}(\epsilon) = u\,e^{\epsilon},\ \overline{c}(\epsilon) = c\,e^{\epsilon},$$

the isovector V_3 to the Lie group

$$\overline{x}(\epsilon) = x + \epsilon t,\ \overline{t}(\epsilon) = t\,,\ \overline{u}(\epsilon) = u + \epsilon,\ \overline{c}(\epsilon) = c\,,$$

and the isovector V_4 to the Lie group

$$\overline{x}(\epsilon) = x\,e^{\epsilon},\ \overline{t}(\epsilon) = t\,e^{\epsilon},\ \overline{u}(\epsilon) = u,\ \overline{c}(\epsilon) = c.$$

The function $\Omega(u, c)$ generates the Lie group

$$\begin{aligned}
\overline{x}(\epsilon) &= x + 2\Big(\Omega_u - \frac{u}{\alpha c}\Omega_c\Big)\epsilon, \\
\overline{t}(\epsilon) &= t - \frac{1}{2\alpha c}\Omega_c\epsilon,\quad \overline{u}(\epsilon) = u,\quad \overline{c}(\epsilon) = c.
\end{aligned}$$ ■

9.4. ISOVECTOR FIELDS OF BALANCE IDEALS

Before dealing with partial differential equations in the form of general balance equations, we would like first to consider the system of non-linear partial differential equations given by (9.2.1). This time we shall represent this system via n-forms

$$\omega^a = F^a\mu \in \Lambda^n(\mathcal{C}_m),\ a = 1, \dots, A$$

defined on the mth order contact manifold $\mathcal{C}_m$. The volume form of the manifold M is the n-form $\mu = dx^1 \wedge \cdots \wedge dx^n$. We shall also need the forms $\mu_i = \mathbf{i}_{\partial_i}(\mu) \in \Lambda^{n-1}(M)$. The reason why we use n-forms in association with the field equations instead of 0-forms as before is that they happen to be more beneficial in determining isovector fields. The regular solution mapping $\phi : \mathcal{D}_n \to \mathcal{C}_m$ introduced on p. 489 gives rise to the relation

$$\phi^*\omega^a = (\phi^* F^a)(\phi^*\mu) = 0$$

since $\phi^*\mu \neq 0$ and $\phi^* F^a = 0$. Thus, it annihilates the forms ω^a. Let us now consider the *fundamental ideal*

$$\mathfrak{I}_m = \mathcal{I}(\sigma^\alpha, \sigma^\alpha_{i_1}, \sigma^\alpha_{i_1 i_2}, \ldots, \sigma^\alpha_{i_1 i_2 \cdots i_{m-1}}; d\sigma^\alpha_{i_1 i_2 \cdots i_{m-1}}; \omega^a)$$

This ideal is closed. Indeed, if we make use of the definitions (9.2.5), we obtain

$$\begin{aligned} d\omega^a = dF^a \wedge \mu &= \Big(\frac{\partial F^a}{\partial x^i}\, dx^i + \sum_{r=0}^{m} \frac{\partial F^a}{\partial v^\alpha_{i_1\cdots i_r}}\, dv^\alpha_{i_1\cdots i_r}\Big) \wedge \mu \\ &= \Big[\Big(\frac{\partial F^a}{\partial x^i} + \sum_{r=0}^{m-1} \frac{\partial F^a}{\partial v^\alpha_{i_1\cdots i_r}} v^\alpha_{i_1\cdots i_r i}\Big)\, dx^i + \sum_{r=0}^{m-1} \frac{\partial F^a}{\partial v^\alpha_{i_1\cdots i_r}}\, \sigma^\alpha_{i_1\cdots i_r}\Big] \wedge \mu \\ &\qquad - \frac{\partial F^a}{\partial v^\alpha_{i_1\cdots i_m}}\, d\sigma^\alpha_{i_1\cdots i_{m-1}} \wedge \mu_{i_m} \\ &= (-1)^n \sum_{r=0}^{m-1} \frac{\partial F^a}{\partial v^\alpha_{i_1\cdots i_r}}\, \mu \wedge \sigma^\alpha_{i_1\cdots i_r} - \frac{\partial F^a}{\partial v^\alpha_{i_1\cdots i_m}}\, \mu_{i_m} \wedge d\sigma^\alpha_{i_1\cdots i_{m-1}} \in \mathfrak{I}_m \end{aligned}$$

where we have utilised the known relations $dx^i \wedge \mu = 0$, $dx^j \wedge \mu_i = \delta^j_i \mu$ and $dv^\alpha_{i_1\cdots i_{m-1} i_m} \wedge \mu = dv^\alpha_{i_1\cdots i_{m-1} i} \wedge dx^i \wedge \mu_{i_m} = -\, d\sigma^\alpha_{i_1\cdots i_{m-1}} \wedge \mu_{i_m}$. In order to determine the isovector fields of this ideal, we first consider the isovector field V of the contact ideal whose general structure has been fully revealed in Sec. 9.3. According to Theorem 5.12.5, we have to impose further the condition $\pounds_V \omega^a \in \mathfrak{I}_m$ on this vector. If we evaluate the Lie derivative of ω^a by noting the relations

$$\mathbf{i}_V(\omega^a) = F^a X^i \mu_i, \;\; \mathbf{i}_V(d\omega^a) = V(F^a)\mu - X^i dF^a \wedge \mu_i$$

we thus conclude that the following conditions should be satisfied

$$\begin{aligned} \pounds_V \omega^a &= V(F^a)\mu + F^a dX^i \wedge \mu_i \\ &= \lambda^a_b F^b \mu + \sum_{r=0}^{m-1} \lambda^{a i_1\cdots i_r}_\alpha \wedge \sigma^\alpha_{i_1\cdots i_r} + \Lambda^{a i_1\cdots i_{m-1}}_\alpha \wedge d\sigma^\alpha_{i_1\cdots i_{m-1}}. \end{aligned}$$

But, we must of course show that it is possible to find forms $\lambda^a_b \in \Lambda^0(\mathcal{C}_m)$, $\lambda^{a i_1\cdots i_r}_\alpha \in \Lambda^{n-1}(\mathcal{C}_m), r = 0, 1, \ldots, m, \Lambda^{a i_1\cdots i_{m-1}}_\alpha \in \Lambda^{n-2}(\mathcal{C}_m)$ satisfying the above relations. If we recall that $X^i = X^i(\mathbf{x}, \mathbf{u})$ if $N > 1$, the foregoing expressions require that we have to take

$$\lambda^a_\alpha = (-1)^{n-1}F^a\frac{\partial X^i}{\partial u^\alpha}\mu_i;\ \lambda^{ai_1\cdots i_r}_\alpha = 0, 1 \leq r \leq m-1;\ \Lambda^{ai_1\cdots i_{m\text{-}1}}_\alpha = 0$$

and the coefficient of μ there yields

$$V(F^a) + \Big[\Big(\frac{\partial X^i}{\partial x^i} + \frac{\partial X^i}{\partial u^\alpha}v^\alpha_i\Big)\delta^a_b - \lambda^a_b\Big]F^b = 0.$$

However, we know that the system of differential equations has to comply with the conditions $\phi^* F^a = 0$. Hence, the isovector components must satisfy the relations

$$\phi^* V(F^a) = \phi^*\Big[\frac{\partial F^a}{\partial x^i}X^i + \sum_{r=0}^{m}\frac{\partial F^a}{\partial v^\alpha_{i_1\cdots i_r}}V^\alpha_{i_1\cdots i_r}\Big] = 0,\ \phi^* F^a = 0$$

for $1 \leq a \leq A$. The functions $X^i(\mathbf{x}, \mathbf{u})$ and $U^\alpha(\mathbf{x}, \mathbf{u})$ determining completely the isovector components can be found in principle from the above equations. These equations are exactly the same as the determining equations for infinitesimal generators of Lie symmetry groups obtained by the classical approach [*see* Olver (1986), Ch. 2]. Consequently, it is not possible to get useful information about isovector components without knowing explicitly the structure of functions F^a. The case $N = 1$ can likewise be discussed in a similar manner.

On the other hand, when partial differential equations are of balance type we can attain to much more feasible results than those obtained above. An $(m+1)$th order balance equations with n independent variables x^i and N dependent variables u^α are specified by

$$\frac{\partial \Sigma^{\alpha i}}{\partial x^i} + \Sigma^\alpha = 0,\ i = 1, 2, \ldots, n;\ \alpha = 1, 2, \ldots, N \tag{9.4.1}$$

where $\Sigma^{\alpha i}$ and Σ^α are smooth functions of variables x^i, u^α and partial derivatives $u^\alpha_{,i}, u^\alpha_{,ij}, \ldots, u^\alpha_{,i_1i_2\cdots i_m}$ of functions $u^\alpha = u^\alpha(x^i)$ up to and including mth order. Because of the physical significance, we shall assume that the number of equations are equal to the number of unknowns. However, methods that we shall explore fully in this section and some of subsequent sections will be equally applicable to a case in which the number of equations differs from the number of unknowns, that is, to balance equations in the form

$$\frac{\partial \Sigma^{ai}}{\partial x^i} + \Sigma^a = 0,\ a = 1, 2, \ldots, A.$$

As we have mentioned earlier, we suppose that the differential equations are defined on an open set $\mathcal{D}_n \subseteq \mathbb{R}^n$. If we integrate equations (9.4.1) on the region $\mathcal{D}_n$ whose exterior unit normal is $\mathbf{n}$ and make use of the divergence theorem, we obtain the following integral relation

$$\int_{\partial\mathcal{D}_n} \Sigma^{ai} n_i dS = -\int_{\mathcal{D}_n} \Sigma^{a} dV.$$

We call $\Sigma^{ai} n_i$ as the ***flux*** along the boundary of the region and $-\Sigma^a$ as the ***source*** inside the region. Thus the total flux is balanced by the total source. In order to say that the set (9.4.1) is of $(m+1)$th order, at least one of the functions $\Sigma^{\alpha i}$ must contain an mth order derivative $u^\alpha_{,i_1 i_2\cdots i_m}$. The explicit form of equations (9.4.1) is found by resorting to the chain rule as follows

$$\sum_{r=0}^{m} \frac{\partial \Sigma^{\alpha i}}{\partial u^\beta_{,i_1 i_2\cdots i_r}} u^\beta_{,i_1 i_2\cdots i_r i} + \frac{\partial \Sigma^{\alpha i}}{\partial x^i} + \Sigma^\alpha = 0 \tag{9.4.2}$$

where we have again adopted the convention

$$\sum_{r=0}^{m} \frac{\partial \Sigma^{\alpha i}}{\partial u^\beta_{,i_1 i_2\cdots i_r}} u^\beta_{,i_1 i_2\cdots i_r i} = \frac{\partial \Sigma^{\alpha i}}{\partial u^\beta} u^\beta_{,i} + \frac{\partial \Sigma^{\alpha i}}{\partial u^\beta_{,i_1}} u^\beta_{,i_1 i} + \cdots + \frac{\partial \Sigma^{\alpha i}}{\partial u^\beta_{,i_1\cdots i_m}} u^\beta_{,i_1\cdots i_m i}.$$

In understanding the real extent of above expressions we should recall that all repeated dummy indices indicate summations over their ranges. As we said if the order of this set of partial differential equations is $m+1$, then at least one of the coefficients $\partial\Sigma^{\alpha i}/\partial u^\beta_{,i_1\cdots i_m}$ must be different from zero. Since equations (9.4.2) are linear with respect to $(m+1)$th order derivatives, they constitute a set of *quasilinear* partial differential equations. In order to utilise exterior forms the set (9.4.2) has to be transformed to a system of first order partial differential equations by introducing again auxiliary variables. Through the auxiliary variables $v^\alpha_{i_1 i_2\cdots i_r} = u^\alpha_{,i_1 i_2\cdots i_r}$, $0 \le r \le m$ that are *completely symmetric in its subscripts* defined as in (9.2.4), we can readily enlarge our system to the following first order system

$$v^\alpha_{i_1 i_2\cdots i_r} = v^\alpha_{i_1 i_2\cdots i_{r-1}, i_r}, \quad 0 \le r \le m; \quad v^\alpha_{i_0} = u^\alpha$$

$$\sum_{r=0}^{m} \frac{\partial \Sigma^{\alpha i}}{\partial v^\beta_{i_1 i_2\cdots i_r}} v^\beta_{i_1 i_2\cdots i_r, i} + \frac{\partial \Sigma^{\alpha i}}{\partial x^i} + \Sigma^\alpha = 0$$

Let us now consider the contact 1-forms (9.2.5)

$$\sigma^\alpha_{i_1 i_2\cdots i_r} = dv^\alpha_{i_1 i_2\cdots i_r} - v^\alpha_{i_1 i_2\cdots i_r i} dx^i \in \Lambda^1(\mathcal{C}_m), \quad 0 \le r \le m-1$$

and N ***balance n-forms***

$$\omega^\alpha = d\Sigma^{\alpha i} \wedge \mu_i + \Sigma^\alpha \mu \in \Lambda^n(\mathcal{C}_m). \tag{9.4.3}$$

In this section, we shall frequently find the opportunity of using the relations (5.5.10-13-14). (9.4.3) balance forms may be explicitly written as

$$\omega^\alpha = \Big(\frac{\partial \Sigma^{\alpha i}}{\partial x^i} + \Sigma^\alpha\Big)\mu + \sum_{r=0}^{m} \frac{\partial \Sigma^{\alpha i}}{\partial v^\beta_{i_1\cdots i_r}}\, dv^\beta_{i_1\cdots i_r} \wedge \mu_i. \tag{9.4.4}$$

A regular mapping $\phi : \mathcal{D}_n \to \mathcal{C}_m$ becomes a solution of balance equations if it satisfies the relations

$$\phi^*\sigma^\alpha_{i_1 i_2\cdots i_r} = 0,\ \ 0 \le r \le m-1;\ \ \phi^*\omega^\alpha = 0.$$

In fact, the equations

$$\phi^*\sigma^\alpha_{i_1 i_2\cdots i_r} = (v^\alpha_{i_1 i_2\cdots i_r, i} - v^\alpha_{i_1 i_2\cdots i_r i})\, dx^i = 0,\ \ 0 \le r \le m-1$$
$$\phi^*\omega^\alpha = \Big[\frac{\partial \Sigma^{\alpha i}}{\partial x^i} + \Sigma^\alpha + \sum_{r=0}^{m} \frac{\partial \Sigma^{\alpha i}}{\partial v^\beta_{i_1\cdots i_r}}\, v^\beta_{i_1\cdots i_r, i}\Big]\mu = 0$$

yield $v^\alpha_{i_1 i_2\cdots i_r i} = v^\alpha_{i_1 i_2\cdots i_r, i}$ and $v^\alpha_{i0} = u^\alpha$, $v^\alpha_{i_1 i_2\cdots i_r} = u^\alpha_{,i_1 i_2\cdots i_r}$ from which we recover the differential equations (9.4.2). We shall now consider the ideal below of the exterior algebra $\Lambda(\mathcal{C}_m)$

$$\mathfrak{I}_m = \mathcal{I}(\sigma^\alpha_{i_1 i_2\cdots i_r}, 0 \le r \le m-1; d\sigma^\alpha_{i_1 i_2\cdots i_{m-1}}; \omega^\alpha)$$

where $1 \le \alpha \le N$, $1 \le i_r \le n$, $0 \le r \le m-1$. $\mathfrak{I}_m$ will be called henceforth as the ***balance ideal*** or the ***fundamental ideal***. We can immediately verify that the balance ideal $\mathfrak{I}_m$ is closed. On using the definition of contact 1-forms, we obtain

$$d\omega^\alpha = d\Sigma^\alpha \wedge \mu = \frac{\partial \Sigma^\alpha}{\partial x^i}\, dx^i \wedge \mu + \sum_{r=0}^{m} \frac{\partial \Sigma^\alpha}{\partial v^\beta_{i_1\cdots i_r}}\, dv^\beta_{i_1\cdots i_r} \wedge \mu$$
$$= \Big[(-1)^n \sum_{r=0}^{m-1} \frac{\partial \Sigma^\alpha}{\partial v^\beta_{i_1\cdots i_r}}\, \mu \wedge \sigma^\beta_{i_1\cdots i_r} - \frac{\partial \Sigma^\alpha}{\partial v^\beta_{i_1\cdots i_{m-1} i}}\, \mu_i \wedge d\sigma^\beta_{i_1\cdots i_{m-1}}\Big]$$

that amounts to say that $d\omega^\alpha \in \mathfrak{I}_m$. Solutions of balance equations in question annihilate the ideal $\mathfrak{I}_m$. We shall now attempt to determine isovector fields of the closed balance ideal $\mathfrak{I}_m$. To this end, we resort to Theorem 5.12.5. Let $V \in T(\mathcal{C}_m)$ be an isovector field of the contact ideal obtained in the previous section. We shall now try to specify the particular structure of this vector that permits us to determine appropriate forms $\lambda^\alpha_\beta \in \Lambda^0(\mathcal{C}_m)$;

$\gamma_\beta^{\alpha i_1\cdots i_r} \in \Lambda^{n-1}(\mathcal{C}_m), r = 0, 1, \ldots, m-1; \Gamma_\beta^{\alpha i_1\cdots i_{m-1}} \in \Lambda^{n-2}(\mathcal{C}_m)$ such that the following relations are satisfied

$$\pounds_V \omega^\alpha = \lambda_\beta^\alpha \omega^\beta + \sum_{r=0}^{m-1} \gamma_\beta^{\alpha i_1\cdots i_r} \wedge \sigma_{i_1\cdots i_r}^\beta + \Gamma_\beta^{\alpha i_1\cdots i_{m-1}} \wedge d\sigma_{i_1\cdots i_{m-1}}^\beta. \tag{9.4.5}$$

Since we can write

$$\begin{aligned}\mathbf{i}_V(d\omega^\alpha) &= \mathbf{i}_V(d\Sigma^\alpha)\mu - d\Sigma^\alpha \wedge \mathbf{i}_V(\mu) = V(\Sigma^\alpha)\mu - X^i d\Sigma^\alpha \wedge \mu_i \\ \mathbf{i}_V(\omega^\alpha) &= \mathbf{i}_V(d\Sigma^{\alpha i})\mu_i - d\Sigma^{\alpha i} \wedge \mathbf{i}_V(\mu_i) + \Sigma^\alpha \mathbf{i}_V(\mu) \\ &= V(\Sigma^{\alpha i})\mu_i - X^j d\Sigma^{\alpha i} \wedge \mu_{ji} + X^i \Sigma^\alpha \mu_i,\end{aligned}$$

the Cartan formula $\pounds_V \omega^\alpha = \mathbf{i}_V(d\omega^\alpha) + d\mathbf{i}_V(\omega^\alpha)$ then leads to

$$\begin{aligned}\pounds_V \omega^\alpha = V(\Sigma^\alpha)\mu + \left[dV(\Sigma^{\alpha i}) + \Sigma^\alpha dX^i\right] \wedge \mu_i \\ - dX^j \wedge d\Sigma^{\alpha i} \wedge \mu_{ji}.\end{aligned} \tag{9.4.6}$$

Let us recall that

$$V(f) = X^i \frac{\partial f}{\partial x^i} + \sum_{r=0}^{m} V_{i_1\cdots i_r}^\alpha \frac{\partial f}{\partial v_{i_1\cdots i_r}^\alpha} \in \Lambda^0(\mathcal{C}_m)$$

for a smooth function $f \in \Lambda^0(\mathcal{C}_m)$. We now have to take into consideration two different cases concerning isovectors of the contact ideal.

(i). Let $N > 1$. Therefore we have to choose $X^i = X^i(\mathbf{x}, \mathbf{u})$, $U^\alpha = U^\alpha(\mathbf{x}, \mathbf{u})$ and the components $V_{i_1\cdots i_r}^\alpha, 1 \le r \le m$ are found from (9.3.19). If we evaluate (9.4.6) under this constraint we obtain

$$\begin{aligned}\pounds_V \omega^\alpha = V(\Sigma^\alpha)\mu + \bigg[\frac{\partial V(\Sigma^{\alpha i})}{\partial x^j} dx^j + \sum_{r=0}^{m} \frac{\partial V(\Sigma^{\alpha i})}{\partial v_{i_1\cdots i_r}^\beta} dv_{i_1\cdots i_r}^\beta \\ + \Sigma^\alpha \Big(\frac{\partial X^i}{\partial x^j} dx^j + \frac{\partial X^i}{\partial u^\beta} du^\beta\Big)\bigg] \wedge \mu_i \\ - \Big(\frac{\partial X^j}{\partial x^k} dx^k + \frac{\partial X^j}{\partial u^\gamma} du^\gamma\Big) \wedge \bigg[\frac{\partial \Sigma^{\alpha i}}{\partial x^l} dx^l + \sum_{r=0}^{m} \frac{\partial \Sigma^{\alpha i}}{\partial v_{i_1\cdots i_r}^\beta} dv_{i_1\cdots i_r}^\beta\bigg] \wedge \mu_{ji}.\end{aligned} \tag{9.4.7}$$

By making use of the relation $dx^j \wedge \mu_i = \delta_i^j \mu$ we cast (9.4.7) into

$$\begin{aligned}\pounds_V \omega^\alpha &= A^\alpha \mu + A_\beta^{\alpha i} du^\beta \wedge \mu_i + A_{\beta\gamma}^{\alpha ij} du^\beta \wedge du^\gamma \wedge \mu_{ji} \\ &\quad + \sum_{r=1}^{m} A_\beta^{\alpha i i_1\cdots i_r} dv_{i_1\cdots i_r}^\beta \wedge \mu_i + \sum_{r=1}^{m} A_{\beta\gamma}^{\alpha i j i_1\cdots i_r} dv_{i_1\cdots i_r}^\beta \wedge du^\gamma \wedge \mu_{ji}\end{aligned}$$

where the smooth functions

$$
\begin{aligned}
A^\alpha &= V(\Sigma^\alpha) + \frac{\partial V(\Sigma^{\alpha i})}{\partial x^i} + \Sigma^\alpha \frac{\partial X^i}{\partial x^i} + \frac{\partial \Sigma^{\alpha i}}{\partial x^i}\frac{\partial X^j}{\partial x^j} - \frac{\partial \Sigma^{\alpha i}}{\partial x^j}\frac{\partial X^j}{\partial x^i} \\
A^{\alpha i}_\beta &= \frac{\partial V(\Sigma^{\alpha i})}{\partial u^\beta} + \Sigma^\alpha \frac{\partial X^i}{\partial u^\beta} + \frac{\partial \Sigma^{\alpha j}}{\partial x^j}\frac{\partial X^i}{\partial u^\beta} - \frac{\partial \Sigma^{\alpha i}}{\partial x^j}\frac{\partial X^j}{\partial u^\beta} \\
&\qquad + \frac{\partial \Sigma^{\alpha i}}{\partial u^\beta}\frac{\partial X^j}{\partial x^j} - \frac{\partial \Sigma^{\alpha j}}{\partial u^\beta}\frac{\partial X^i}{\partial x^j} \\
A^{\alpha ij}_{\beta\gamma} &= -A^{\alpha ji}_{\beta\gamma} = -A^{\alpha ij}_{\gamma\beta} = \frac{\partial \Sigma^{\alpha[i}}{\partial u^{[\beta}}\frac{\partial X^{j]}}{\partial u^{\gamma]}} \\
&= \frac{1}{4}\Big[\frac{\partial \Sigma^{\alpha i}}{\partial u^\beta}\frac{\partial X^j}{\partial u^\gamma} - \frac{\partial \Sigma^{\alpha j}}{\partial u^\beta}\frac{\partial X^i}{\partial u^\gamma} + \frac{\partial \Sigma^{\alpha j}}{\partial u^\gamma}\frac{\partial X^i}{\partial u^\beta} - \frac{\partial \Sigma^{\alpha i}}{\partial u^\gamma}\frac{\partial X^j}{\partial u^\beta}\Big] \\
A^{\alpha i i_1\cdots i_r}_\beta &= \frac{\partial V(\Sigma^{\alpha i})}{\partial v^\beta_{i_1\cdots i_r}} + \frac{\partial \Sigma^{\alpha i}}{\partial v^\beta_{i_1\cdots i_r}}\frac{\partial X^j}{\partial x^j} - \frac{\partial \Sigma^{\alpha j}}{\partial v^\beta_{i_1\cdots i_r}}\frac{\partial X^i}{\partial x^j}, \qquad 1 \le r \le m \\
A^{\alpha ij i_1\cdots i_r}_{\beta\gamma} &= -A^{\alpha ji i_1\cdots i_r}_{\beta\gamma} = \frac{\partial \Sigma^{\alpha[i}}{\partial v^\beta_{i_1\cdots i_r}}\frac{\partial X^{j]}}{\partial u^\gamma} \\
&= \frac{1}{2}\Big[\frac{\partial \Sigma^{\alpha i}}{\partial v^\beta_{i_1\cdots i_r}}\frac{\partial X^j}{\partial u^\gamma} - \frac{\partial \Sigma^{\alpha j}}{\partial v^\beta_{i_1\cdots i_r}}\frac{\partial X^i}{\partial u^\gamma}\Big], \qquad 1 \le r \le m
\end{aligned} \tag{9.4.8}
$$

are all elements of $\Lambda^0(\mathcal{C}_m)$. It is obvious that the functions $A^{\alpha i i_1\cdots i_r}_\beta$ and $A^{\alpha ij i_1\cdots i_r}_{\beta\gamma}$ are completely symmetric in indices $i_1, \ldots, i_r$. The antisymmetry in indices i, j arise from the antisymmetry of forms $\mu_{ji} \in \Lambda^{n-2}(M)$ and antisymmetry with respect to indices β, γ in $A^{\alpha ij}_{\beta\gamma}$ from the exterior product $du^\beta \wedge du^\gamma$. If we make the transformations

$$
dv^\alpha_{i_1 i_2\cdots i_r} = \sigma^\alpha_{i_1 i_2\cdots i_r} + v^\alpha_{i_1 i_2\cdots i_r j}\, dx^j, \qquad 0 \le r \le m-1
$$

in the expression above for $\pounds_V \omega^\alpha$ and use the relations

$$
\begin{aligned}
dx^k \wedge \mu_{ji} &= \delta^k_j \mu_i - \delta^k_i \mu_j, \\
dx^k \wedge dx^l \wedge \mu_{ji} &= (\delta^k_i \delta^l_j - \delta^k_j \delta^l_i)\mu
\end{aligned}
$$

we arrive at

$$
\begin{aligned}
\pounds_V \omega^\alpha &= \Big[A^\alpha + A^{\alpha i}_\beta v^\beta_i + 2A^{\alpha ij}_{\beta\gamma} v^\beta_i v^\gamma_j + \sum_{r=1}^{m-1} (A^{\alpha i i_1\cdots i_r}_\beta \\
&\quad + 2A^{\alpha ij i_1\cdots i_r}_{\beta\gamma} v^\gamma_j) v^\beta_{i_1\cdots i_r i}\Big]\mu + (A^{\alpha i i_1\cdots i_m}_\beta + 2A^{\alpha ij i_1\cdots i_m}_{\beta\gamma} v^\gamma_j)\, dv^\beta_{i_1\cdots i_m} \wedge \mu_i
\end{aligned} \tag{9.4.9}
$$

$$+\Big\{(-1)^{n-1}\Big[A^{\alpha i}_{\beta}+4A^{\alpha ij}_{\beta\gamma}v_j-2\sum_{r=1}^{m-1}A^{\alpha iji_1\cdots i_r}_{\gamma\beta}v^{\gamma}_{i_1\cdots i_rj}\Big]\mu_i$$
$$-A^{\alpha ij}_{\beta\gamma}\mu_{ji}\wedge\sigma^{\gamma}+A^{\alpha iji_1\cdots i_m}_{\gamma\beta}\mu_{ji}\wedge dv^{\gamma}_{i_1\cdots i_m}\Big\}\wedge\sigma^{\beta}$$
$$+\sum_{r=1}^{m-1}\Big[(-1)^{n-1}(A^{\alpha ii_1\cdots i_r}_{\beta}+2A^{\alpha iji_1\cdots i_r}_{\beta\gamma}v^{\gamma}_j)\,\mu_i-A^{\alpha iji_1\cdots i_r}_{\beta\gamma}\mu_{ji}\wedge\sigma^{\gamma}\Big]\wedge\sigma^{\beta}_{i_1\cdots i_r}.$$

The functions in (9.4.8) comprise now solely presently arbitrary functions $X^i(\mathbf{x},\mathbf{u})$ and $U^{\alpha}(\mathbf{x},\mathbf{u})$ as unknowns. On the other hand (9.4.4) can thereby be written as

$$\omega^{\alpha}=\Big(\Sigma^{\alpha}+\frac{\partial\Sigma^{\alpha i}}{\partial x^i}+\sum_{r=0}^{m-1}\frac{\partial\Sigma^{\alpha i}}{\partial v^{\beta}_{i_1\cdots i_r}}v^{\beta}_{i_1\cdots i_ri}\Big)\mu+\frac{\partial\Sigma^{\alpha i}}{\partial u^{\beta}}\,\sigma^{\beta}\wedge\mu_i$$
$$+\sum_{r=1}^{m-1}\frac{\partial\Sigma^{\alpha i}}{\partial v^{\beta}_{i_1\cdots i_r}}\,\sigma^{\beta}_{i_1\cdots i_r}\wedge\mu_i+\frac{\partial\Sigma^{\alpha i}}{\partial v^{\beta}_{i_1\cdots i_m}}\,dv^{\beta}_{i_1\cdots i_m}\wedge\mu_i.$$

On inserting this expression together with (9.4.9) into the relation (9.4.5) and equating the coefficients of linearly independent like forms in both sides we end up with the following result

$$\lambda^{\alpha}_{\beta}\Big(\Sigma^{\beta}+\frac{\partial\Sigma^{\beta i}}{\partial x^i}+\sum_{r-0}^{m-1}\frac{\partial\Sigma^{\beta i}}{\partial v^{\gamma}_{i_1\cdots i_r}}v^{\gamma}_{i_1\cdots i_ri}\Big)= \tag{9.4.10}$$
$$A^{\alpha}+A^{\alpha i}_{\beta}v^{\beta}_i+2A^{\alpha ij}_{\beta\gamma}v^{\beta}_iv^{\gamma}_j+\sum_{r=1}^{m-1}\big(A^{\alpha ii_1\cdots i_r}_{\beta}+2A^{\alpha iji_1\cdots i_r}_{\beta\gamma}v^{\gamma}_j\big)v^{\beta}_{i_1\cdots i_ri}$$
$$\gamma^{\alpha}_{\beta}=(-1)^n\lambda^{\alpha}_{\gamma}\frac{\partial\Sigma^{\gamma i}}{\partial u^{\beta}}\mu_i$$
$$+(-1)^{n-1}\Big[A^{\alpha i}_{\beta}+A^{\alpha ij}_{\beta\gamma}v^{\gamma}_j-\sum_{r=1}^{m-1}A^{\alpha iji_1\cdots i_r}_{\gamma\beta}v^{\gamma}_{i_1\cdots i_rj}\Big]\mu_i$$
$$-A^{\alpha ij}_{\beta\gamma}\mu_{ji}\wedge\sigma^{\gamma}+A^{\alpha iji_1\cdots i_m}_{\gamma\beta}\mu_{ji}\wedge dv^{\gamma}_{i_1\cdots i_m}$$
$$\gamma^{\alpha i_1\cdots i_r}_{\beta}=(-1)^n\lambda^{\alpha}_{\gamma}\frac{\partial\Sigma^{\gamma i}}{\partial v^{\beta}_{i_1\cdots i_r}}\,\mu_i$$
$$+(-1)^{n-1}(A^{\alpha ii_1\cdots i_r}_{\beta}+2A^{\alpha iji_1\cdots i_r}_{\beta\gamma}v^{\gamma}_j)\,\mu_i$$
$$-A^{\alpha iji_1\cdots i_r}_{\beta\gamma}\mu_{ji}\wedge\sigma^{\gamma},\quad 1\le r\le m-1$$

whereas the last N equations to be satisfied take the form

$$\Big(A^{\alpha i i_1\cdots i_m}_{\beta}+2A^{\alpha i j i_1\cdots i_m}_{\beta\gamma}v^\gamma_j-\lambda^\alpha_\gamma\frac{\partial\Sigma^{\gamma i}}{\partial v^\beta_{i_1\cdots i_m}}\Big)\,dv^\beta_{i_1\cdots i_m}\wedge\mu_i$$
$$=-\,\Gamma^{\alpha i_1\cdots i_{m-1}}_{\beta}\wedge dv^\beta_{i_1\cdots i_{m-1}i_m}\wedge dx^{i_m}$$

when we write $d\sigma^\beta_{i_1\cdots i_{m-1}}=-\,dv^\beta_{i_1\cdots i_m i}\wedge dx^i$. The composition of this expressions suggests that it would be rather adequate to choose the forms $\Gamma^{\alpha i_1\cdots i_{m-1}}_{\beta}\in\Lambda^{n-2}(\mathcal{C}_m)$ as follows

$$\Gamma^{\alpha i_1\cdots i_{m-1}}_{\beta}=\gamma^{\alpha i_1\cdots i_{m-1}ij}_{\beta}\mu_{ji},\ \ \gamma^{\alpha i_1\cdots i_{m-1}ij}_{\beta}\in\Lambda^0(\mathcal{C}_m).$$

Since the forms μ_{ji} are antisymmetric in indices i,j, we can take without loss of generality $\gamma^{\alpha i_1\cdots i_{m-1}ij}_{\beta}=-\,\gamma^{\alpha i_1\cdots i_{m-1}ji}_{\beta}$. We thus obtain

$$\begin{aligned}
&-\Gamma^{\alpha i_1\cdots i_{m-1}}_{\beta}\wedge dv^\beta_{i_1\cdots i_{m-1}i_m}\wedge dx^{i_m}\\
&\qquad=-\,\gamma^{\alpha i_1\cdots i_{m-1}ij}_{\beta}\mu_{ji}\wedge dv^\beta_{i_1\cdots i_{m-1}i_m}\wedge dx^{i_m}\\
&\qquad=-\,\gamma^{\alpha i_1\cdots i_{m-1}ij}_{\beta}dv^\beta_{i_1\cdots i_{m-1}i_m}\wedge dx^{i_m}\wedge\mu_{ji}\\
&\qquad=-\,\gamma^{\alpha i_1\cdots i_{m-1}ij}_{\beta}dv^\beta_{i_1\cdots i_{m-1}i_m}\wedge(\delta^{i_m}_j\mu_i-\delta^{i_m}_i\mu_j)\\
&=-\,\gamma^{\alpha i_1\cdots i_{m-1}ii_m}_{\beta}dv^\beta_{i_1\cdots i_{m-1}i_m}\wedge\mu_i+\gamma^{\alpha i_1\cdots i_{m-1}i_m j}_{\beta}dv_{i_1\cdots i_{m-1}i_m}\wedge\mu_j\\
&\qquad=2\gamma^{\alpha i_1\cdots i_{m-1}i_m i}_{\beta}dv_{i_1\cdots i_{m-1}i_m}\wedge\mu_i
\end{aligned}$$

whence we deduce that

$$2\gamma^{\alpha i_1\cdots i_{m-1}i_m i}_{\beta}=A^{\alpha i i_1\cdots i_m}_{\beta}+2A^{\alpha i j i_1\cdots i_m}_{\beta\gamma}v^\gamma_j-\lambda^\alpha_\gamma\frac{\partial\Sigma^{\gamma i}}{\partial v^\beta_{i_1\cdots i_m}}.$$

But, because of the antisymmetry of functions $\gamma^{\alpha i_1\cdots i_{m-1}i_m i}_{\beta}$ with respect to indices i,i_m, we see that the following relations must be satisfied

$$\begin{aligned}
4\gamma^{\alpha i_1\cdots i_{m-1}i_m i}_{\beta}&=4\gamma^{\alpha i_1\cdots i_{m-1}[i_m i]}_{\beta}=A^{\alpha i i_1\cdots i_{m-1}i_m}_{\beta}+2A^{\alpha i j i_1\cdots i_{m-1}i_m}_{\beta\gamma}v^\gamma_j\\
&-A^{\alpha i_m i_1\cdots i_{m-1}i}_{\beta}-2A^{\alpha i_m j i_1\cdots i_{m-1}i}_{\beta\gamma}v^\gamma_j-\lambda^\alpha_\gamma\Big(\frac{\partial\Sigma^{\gamma i}}{\partial v^\beta_{i_1\cdots i_{m-1}i_m}}-\frac{\partial\Sigma^{\gamma i_m}}{\partial v^\beta_{i_1\cdots i_{m-1}i}}\Big)\\
0&=A^{\alpha i i_1\cdots i_{m-1}i_m}_{\beta}+2A^{\alpha i j i_1\cdots i_{m-1}i_m}_{\beta\gamma}v^\gamma_j+A^{\alpha i_m i_1\cdots i_{m-1}i}_{\beta}+2A^{\alpha i_m j i_1\cdots i_{m-1}i}_{\beta\gamma}v^\gamma_j\\
&-\lambda^\alpha_\gamma\Big(\frac{\partial\Sigma^{\gamma i}}{\partial v^\beta_{i_1\cdots i_{m-1}i_m}}+\frac{\partial\Sigma^{\gamma i_m}}{\partial v^\beta_{i_1\cdots i_{m-1}i}}\Big)
\end{aligned}$$

Consequently, we can state the theorem below:

Theorem 9.4.1. *In order that an isovector field V of the contact ideal* *becomes also an isovector of the balance ideal,* $n+N+N^2$ *number of functions* $X^i(\boldsymbol{x},\boldsymbol{u}), U^\alpha(\boldsymbol{x},\boldsymbol{u}),\ \lambda^\alpha_\beta \in \Lambda^0(\mathcal{C}_m)$ *ought to satisfy the determining equations*

$$\lambda^\alpha_\beta\Big(\Sigma^\beta + \frac{\partial \Sigma^{\beta i}}{\partial x^i} + \sum_{r=0}^{m-1} \frac{\partial \Sigma^{\beta i}}{\partial v^\gamma_{i_1\cdots i_r}} v^\gamma_{i_1\cdots i_r i}\Big) = \tag{9.4.11}$$

$$A^\alpha + A^{\alpha i}_\beta v^\beta_i + 2A^{\alpha ij}_{\beta\gamma} v^\beta_i v^\gamma_j + \sum_{r=1}^{m-1}\big(A^{\alpha i i_1\cdots i_r}_\beta + 2A^{\alpha i j i_1\cdots i_r}_{\beta\gamma} v^\gamma_j\big) v^\beta_{i_1\cdots i_r i}$$

$$\lambda^\alpha_\gamma\Big(\frac{\partial \Sigma^{\gamma i}}{\partial v^\beta_{i_1\cdots i_{m-1} i_m}} + \frac{\partial \Sigma^{\gamma i_m}}{\partial v^\beta_{i_1\cdots i_{m-1} i}}\Big) =$$

$$A^{\alpha i i_1\cdots i_{m-1} i_m}_\beta + A^{\alpha i_m i_1\cdots i_{m-1} i}_\beta + 2(A^{\alpha i j i_1\cdots i_{m-1} i_m}_{\beta\gamma} + A^{\alpha i_m j i_1\cdots i_{m-1} i}_{\beta\gamma}) v^\gamma_j$$

whenever $N > 1$. □

The number of the equations (9.4.11) that help determine isovector components, or *infinitesimal generators* in the nomenclature of the classical theory of Lie symmetry groups, are considerably less than those in the classical theory because exterior products are quite effective in eliminating some of the redundant equations. However, it is still a large number. It can easily be checked that there can be at most $N + \frac{1}{2}N^2\binom{n+m-2}{m-1}n(n+1)$ number of determining equations. Therefore, we must expect that the number of the determining equations would be much larger than that of unknowns. This property amounts to say that the shape of the solutions would perhaps be restricted to a great extent even if they exist.

If $m = 0$, that is, if functions entering the balance equations are in the form $\Sigma^{\alpha i}(\mathbf{x},\mathbf{u})$ and $\Sigma^\alpha(\mathbf{x},\mathbf{u})$, we get a system of first order equations. In that case we do not need the contact ideal and only the components X^i and U^α of the isovector field survive. One can readily verify that the determining equations are then reduced to

$$\lambda^\alpha_\beta\Big(\Sigma^\beta + \frac{\partial \Sigma^{\beta i}}{\partial x^i}\Big) = A^\alpha,\ \lambda^\alpha_\gamma \frac{\partial \Sigma^{\gamma i}}{\partial u^\beta} = A^{\alpha i}_\beta,\ A^{\alpha ij}_{\beta\gamma} = 0.$$

(ii). Let $N = 1$. Hence, the components of an isovector field V of the contact ideal become $X^i = X^i(\mathbf{x}, u, \mathbf{v})$, $U = U(\mathbf{x}, u, \mathbf{v})$ generated from a function $F = F(\mathbf{x}, u, \mathbf{v})$ via the relations (9.3.22-23) and $V_{i_1\cdots i_r} \in \Lambda^0(\mathcal{C}_m)$, $1 \le r \le m$ determined by (9.3.25). In this case, the single balance equation takes the form

$$\frac{\partial \Sigma^i}{\partial x^i} + \Sigma = 0$$

and the *balance n-form* producing this equation is given by

$$\omega = d\Sigma^i \wedge \mu_i + \Sigma\mu = \Big(\frac{\partial \Sigma^i}{\partial x^i} + \Sigma\Big)\mu + \sum_{r=0}^{m} \frac{\partial \Sigma^i}{\partial v_{i_1\cdots i_r}}\, dv_{i_1\cdots i_r} \wedge \mu_i.$$

The vector V will be an isovector field of the balance ideal if one is able to find suitable forms $\lambda \in \Lambda^0(\mathcal{C}_m); \gamma^{i_1\cdots i_r} \in \Lambda^{n-1}(\mathcal{C}_m), r = 0, 1, \ldots, m-1$; $\Gamma^{i_1\cdots i_{m-1}} \in \Lambda^{n-2}(\mathcal{C}_m)$ so that the relation

$$\pounds_V\omega = \lambda\,\omega + \sum_{r=0}^{m-1} \gamma^{i_1\cdots i_r} \wedge \sigma_{i_1\cdots i_r} + \Gamma^{i_1\cdots i_{m-1}} \wedge d\sigma_{i_1\cdots i_{m-1}} \tag{9.4.12}$$

is satisfied. The Lie derivative of the balance form ω with respect to the vector V follows from (9.4.6) as

$$\pounds_V\omega = V(\Sigma)\mu + \big[dV(\Sigma^i) + \Sigma\, dX^i\big] \wedge \mu_i - dX^j \wedge d\Sigma^i \wedge \mu_{ji}.$$

However, because of the possible dependence of the components X^i on the variables v_i the above expression may now lead to a different result from (9.4.7):

$$\begin{aligned}\pounds_V\omega = V(\Sigma)\mu &+ \Big[\frac{\partial V(\Sigma^i)}{\partial x^j}\, dx^j + \sum_{r=0}^{m} \frac{\partial V(\Sigma^i)}{\partial v_{i_1\cdots i_r}}\, dv_{i_1\cdots i_r} \\ &+ \Sigma\Big(\frac{\partial X^i}{\partial x^j}\, dx^j + \frac{\partial X^i}{\partial u}\, du + \frac{\partial X^i}{\partial v_j}\, dv_j\Big)\Big] \wedge \mu_i - \Big(\frac{\partial X^j}{\partial x^k}\, dx^k \\ &+ \frac{\partial X^j}{\partial u}\, du + \frac{\partial X^j}{\partial v_k}\, dv_k\Big) \wedge \Big[\frac{\partial \Sigma^i}{\partial x^l}\, dx^l + \sum_{r=0}^{m} \frac{\partial \Sigma^i}{\partial v_{i_1\cdots i_r}}\, dv_{i_1\cdots i_r}\Big] \wedge \mu_{ji}.\end{aligned}$$

We can arrange this expression into the following form

$$\begin{aligned}\pounds_V\omega = A\mu &+ A^i\, du \wedge \mu_i + A^{ij}\, dv_j \wedge \mu_i + B^{ijk}\, dv_k \wedge du \wedge \mu_{ji} \\ &+ C^{ijkl} dv_l \wedge dv_k \wedge \mu_{ji} + \sum_{r=2}^{m} A^{ii_1\cdots i_r} dv_{i_1\cdots i_r} \wedge \mu_i \\ &+ \sum_{r=2}^{m} B^{iji_1\cdots i_r} dv_{i_1\cdots i_r} \wedge du \wedge \mu_{ji} + + \sum_{r=2}^{m} C^{ijki_1\cdots i_r} dv_{i_1\cdots i_r} \wedge dv_k \wedge \mu_{ji}\end{aligned} \tag{9.4.13}$$

where the smooth functions

$$A = V(\Sigma) + \frac{\partial V(\Sigma^i)}{\partial x^i} + \Sigma\frac{\partial X^i}{\partial x^i} + \frac{\partial \Sigma^i}{\partial x^i}\frac{\partial X^j}{\partial x^j} - \frac{\partial \Sigma^i}{\partial x^j}\frac{\partial X^j}{\partial x^i} \tag{9.4.14}$$

$$
\begin{aligned}
A^i &= \frac{\partial V(\Sigma^i)}{\partial u} + \Sigma\frac{\partial X^i}{\partial u} + \frac{\partial \Sigma^j}{\partial x^j}\frac{\partial X^i}{\partial u} - \frac{\partial \Sigma^i}{\partial x^j}\frac{\partial X^j}{\partial u} \\
&\qquad + \frac{\partial \Sigma^i}{\partial u}\frac{\partial X^j}{\partial x^j} - \frac{\partial \Sigma^j}{\partial u}\frac{\partial X^i}{\partial x^j} \\
A^{ij} &= \frac{\partial V(\Sigma^i)}{\partial v_j} + \Sigma\frac{\partial X^i}{\partial v_j} + \frac{\partial \Sigma^k}{\partial x^k}\frac{\partial X^i}{\partial v_j} - \frac{\partial \Sigma^i}{\partial x^k}\frac{\partial X^k}{\partial v_j} \\
&\qquad + \frac{\partial \Sigma^i}{\partial v_j}\frac{\partial X^k}{\partial x^k} - \frac{\partial \Sigma^k}{\partial v_j}\frac{\partial X^i}{\partial x^k} \\
B^{ijk} &= -B^{jik} = \frac{\partial \Sigma^{[i}}{\partial v_k}\frac{\partial X^{j]}}{\partial u} - \frac{\partial \Sigma^{[i}}{\partial u}\frac{\partial X^{j]}}{\partial v_k} \\
&= \frac{1}{2}\Big[\frac{\partial \Sigma^i}{\partial v_k}\frac{\partial X^j}{\partial u} - \frac{\partial \Sigma^j}{\partial v_k}\frac{\partial X^i}{\partial u} + \frac{\partial \Sigma^j}{\partial u}\frac{\partial X^i}{\partial v_k} - \frac{\partial \Sigma^i}{\partial u}\frac{\partial X^j}{\partial v_k}\Big] \\
C^{ijkl} &= -C^{jikl} = -C^{ijlk} = \frac{\partial \Sigma^{[i}}{\partial v_{[l}}\frac{\partial X^{j]}}{\partial v_{k]}} \\
&= \frac{1}{4}\Big[\frac{\partial \Sigma^i}{\partial v_l}\frac{\partial X^j}{\partial v_k} - \frac{\partial \Sigma^j}{\partial v_l}\frac{\partial X^i}{\partial v_k} + \frac{\partial \Sigma^j}{\partial v_k}\frac{\partial X^i}{\partial v_l} - \frac{\partial \Sigma^i}{\partial v_k}\frac{\partial X^j}{\partial v_l}\Big] \\
A^{ii_1\cdots i_r} &= \frac{\partial V(\Sigma^i)}{\partial v_{i_1\cdots i_r}} + \frac{\partial \Sigma^i}{\partial v_{i_1\cdots i_r}}\frac{\partial X^j}{\partial x^j} - \frac{\partial \Sigma^j}{\partial v_{i_1\cdots i_r}}\frac{\partial X^i}{\partial x^j}, \\
B^{iji_1\cdots i_r} &= -B^{jii_1\cdots i_r} = \frac{\partial \Sigma^{[i}}{\partial v_{i_1\cdots i_r}}\frac{\partial X^{j]}}{\partial u} \\
&= \frac{1}{2}\Big[\frac{\partial \Sigma^i}{\partial v_{i_1\cdots i_r}}\frac{\partial X^j}{\partial u} - \frac{\partial \Sigma^j}{\partial v_{i_1\cdots i_r}}\frac{\partial X^i}{\partial u}\Big], \\
C^{ijki_1\cdots i_r} &= -C^{jiki_1\cdots i_r} = \frac{\partial \Sigma^{[i}}{\partial v_{i_1\cdots i_r}}\frac{\partial X^{j]}}{\partial v_k} \\
&= \frac{1}{2}\Big[\frac{\partial \Sigma^i}{\partial v_{i_1\cdots i_r}}\frac{\partial X^j}{\partial v_k} - \frac{\partial \Sigma^j}{\partial v_{i_1\cdots i_r}}\frac{\partial X^i}{\partial v_k}\Big]
\end{aligned}
$$

are elements of $\Lambda^0(\mathcal{C}_m)$. By writing again $dv_{i_1\cdots i_r} = \sigma_{i_1\cdots i_r} + v_{i_1\cdots i_r j}\,dx^j$, $0 \leq r \leq m-1$ and arranging suitably the resulting expression, we can express the Lie derivative $£_V\omega$ in the following manner:

$$
\begin{aligned}
£_V\omega = \Big[&A + A^i v_i + A^{ij}v_{ji} + 2B^{ijk}v_j v_{ki} + 2C^{ijkl}v_{kj}v_{li} \qquad (9.4.15) \\
&+ \sum_{r=2}^{m-1}(A^{ii_1\cdots i_r} + 2B^{iji_1\cdots i_r}v_j + 2C^{ijki_1\cdots i_r}v_{kj})v_{i_1\cdots i_r i}\Big]\mu \\
&+ (A^{ii_1\cdots i_m} + 2B^{iji_1\cdots i_m}v_j + 2C^{ijki_1\cdots i_m}v_{kj})\,dv_{i_1\cdots i_m}\wedge\mu_i
\end{aligned}
$$

$$+\Big[A^i - 2B^{ijk}v_{kj} - 2\sum_{r=2}^{m-1} B^{iji_1\cdots i_r}v_{i_1\cdots i_rj}\Big]\sigma\wedge\mu_i$$
$$+ B^{iji_1\cdots i_m}dv_{i_1\cdots i_m}\wedge\sigma\wedge\mu_{ji}$$
$$+\Big[A^{ij} + 2B^{ikj}v_k + 4C^{iklj}v_{lk} - 2\sum_{r=2}^{m-1} C^{ikji_1\cdots i_r}v_{i_1\cdots i_rk}\Big]\sigma_j\wedge\mu_i$$
$$- B^{ikj}\sigma\wedge\sigma_j\wedge\mu_{ki} + C^{ikji_1\cdots i_m}dv_{i_1\cdots i_m}\wedge\sigma_j\wedge\mu_{ki} + C^{ikjl}\sigma_l\wedge\sigma_j\wedge\mu_{ki}$$
$$+\sum_{r=2}^{m-1}(A^{ii_1\cdots i_r} + 2B^{iji_1\cdots i_r}v_j + 2C^{ijki_1\cdots i_r}v_{kj})\,\sigma_{i_1\cdots i_r}\wedge\mu_i$$
$$-\sum_{r=2}^{m-1}B^{iji_1\cdots i_r}\sigma\wedge\sigma_{i_1\cdots i_r}\wedge\mu_{ji} - \sum_{r=2}^{m-1}C^{ijki_1\cdots i_r}\sigma_k\wedge\sigma_{i_1\cdots i_r}\wedge\mu_{ji}$$

Next, we transform (9.4.12) into

$$\pounds_V\omega = \lambda\Big[\Big(\Sigma + \frac{\partial\Sigma^i}{\partial x^i} + \sum_{r=0}^{m-1}\frac{\partial\Sigma^i}{\partial v_{i_1\cdots i_r}}v_{i_1\cdots i_ri}\Big)\mu + \frac{\partial\Sigma^i}{\partial u}\sigma\wedge\mu_i + \frac{\partial\Sigma^i}{\partial v_j}\sigma_j\wedge\mu_i$$
$$+\sum_{r=2}^{m-1}\frac{\partial\Sigma^i}{\partial v_{i_1\cdots i_r}}\sigma_{i_1\cdots i_r}\wedge\mu_i + \frac{\partial\Sigma^i}{\partial v_{i_1\cdots i_m}}dv_{i_1\cdots i_m}\wedge\mu_i\Big] + \gamma\wedge\sigma + \gamma^j\wedge\sigma_j$$
$$+\sum_{r=2}^{m-1}\gamma^{i_1\cdots i_r}\wedge\sigma_{i_1\cdots i_r} - \Gamma^{i_1\cdots i_{m-1}}\wedge dv_{i_1\cdots i_{m-1}i_m}\wedge dx^{i_m}$$

and compare it with (9.4.14) to obtain

$$\lambda\Big(\Sigma + \frac{\partial\Sigma^i}{\partial x^i} + \sum_{r=0}^{m-1}\frac{\partial\Sigma^i}{\partial v_{i_1\cdots i_r}}v_{i_1\cdots i_ri}\Big) \tag{9.4.16}$$
$$= A + A^iv_i + A^{ij}v_{ji} + 2B^{ijk}v_jv_{ki}$$
$$+ 2C^{ijkl}v_{kj}v_{li} + \sum_{r=2}^{m-1}(A^{ii_1\cdots i_r} + 2B^{iji_1\cdots i_r}v_j + 2C^{ijki_1\cdots i_r}v_{kj})v_{i_1\cdots i_ri}$$
$$\gamma = (-1)^{n-1}\Big[A^i - 2B^{ijk}v_{kj} - 2\sum_{r=2}^{m-1}B^{iji_1\cdots i_r}v_{i_1\cdots i_rj} - \lambda\frac{\partial\Sigma^i}{\partial u}\Big]\mu_i$$
$$+ B^{iji_1\cdots i_m}\mu_{ji}\wedge dv_{i_1\cdots i_m}$$
$$\gamma^j = -B^{ikj}\mu_{ki}\wedge\sigma + C^{ikji_1\cdots i_m}\mu_{ki}\wedge dv_{i_1\cdots i_m} + C^{ikjl}\mu_{ki}\wedge\sigma_l$$
$$+\Big[(-1)^{n-1}(A^{ij} + 2B^{ikj}v_k + 4C^{iklj}v_{lk}$$

$$
- 2\sum_{r=2}^{m-1} C^{ikji_1\cdots i_r} v_{i_1\cdots i_r k} - \lambda\frac{\partial \Sigma^i}{\partial v_j}\Big] \mu_i
$$

$$
\begin{aligned}
\gamma^{i_1\cdots i_r} = (-1)^{n-1}\Big(& A^{ii_1\cdots i_r} + 2B^{iji_1\cdots i_r} v_j \\
& + 2C^{ijki_1\cdots i_r} v_{kj} - \lambda\frac{\partial \Sigma^i}{\partial v_{i_1\cdots i_r}}\Big)\mu_i \\
& - B^{iji_1\cdots i_r}\mu_{ji}\wedge\sigma - C^{ijki_1\cdots i_r}\mu_{ji}\wedge\sigma_k, \quad 2\le r\le m-1
\end{aligned}
$$

while the remaining expression is given by

$$
\begin{aligned}
&(A^{ii_1\cdots i_m} + 2B^{iji_1\cdots i_m} v_j + 2C^{ijki_1\cdots i_m} v_{kj} \\
&\quad - \lambda\frac{\partial \Sigma^i}{\partial v_{i_1\cdots i_m}}) dv_{i_1\cdots i_m}\wedge\mu_i \; = \; -\Gamma^{i_1\cdots i_{m-1}}\wedge dv_{i_1\cdots i_{m-1}i_m}\wedge dx^{i_m}
\end{aligned}
$$

whose structure suggests that it would really be appropriate to choose the forms $\Gamma^{i_1\cdots i_{m-1}} \in \Lambda^{n-2}(\mathcal{C}_m)$ as follows

$$
\Gamma^{i_1\cdots i_{m-1}} = \gamma^{i_1\cdots i_{m-1}ij}\mu_{ji}
$$

where $\gamma^{i_1\cdots i_{m-1}ij} \in \Lambda^0(\mathcal{C}_m)$. Due to the antisymmetry of the forms μ_{ji} with respect to the indices i, j, we can take without loss of generality $\gamma_\beta^{\alpha i_1\cdots i_{m-1}ij} = -\gamma_\beta^{\alpha i_1\cdots i_{m-1}ji}$. We thus get

$$
\begin{aligned}
-\Gamma^{i_1\cdots i_{m-1}}\wedge dv_{i_1\cdots i_{m-1}i_m}\wedge dx^{i_m}& \\
&= -\gamma^{i_1\cdots i_{m-1}ij}\mu_{ji}\wedge dv_{i_1\cdots i_{m-1}i_m}\wedge dx^{i_m} \\
&= 2\gamma^{i_1\cdots i_{m-1}i_m i} dv_{i_1\cdots i_{m-1}i_m}\wedge\mu_i
\end{aligned}
$$

whence we conclude that

$$
\begin{aligned}
4\gamma^{i_1\cdots i_{m-1}i_m i} &= 4\gamma^{i_1\cdots i_{m-1}[i_m i]} = A^{ii_1\cdots i_{m-1}i_m} + 2B^{iji_1\cdots i_{m-1}i_m} v_j \\
&+ 2C^{ijki_1\cdots i_{m-1}i_m} v_{kj} - A^{i_m i_1\cdots i_{m-1}i} - 2B^{i_m ji_1\cdots i_{m-1}i} v_j - 2C^{i_m jki_1\cdots i_{m-1}i} v_{kj} \\
&- \lambda\Big(\frac{\partial \Sigma^i}{\partial v_{i_1\cdots i_{m-1}i_m}} - \frac{\partial \Sigma^{i_m}}{\partial v_{i_1\cdots i_{m-1}i}}\Big), \\
0 &= A^{ii_1\cdots i_{m-1}i_m} + 2B^{iji_1\cdots i_{m-1}i_m} v_j + 2C^{ijki_1\cdots i_{m-1}i_m} v_{kj} + A^{i_m i_1\cdots i_{m-1}i} \\
&+ 2B^{i_m ji_1\cdots i_{m-1}i} v_j + 2C^{i_m jki_1\cdots i_{m-1}i} v_{kj} - \lambda\Big(\frac{\partial \Sigma^i}{\partial v_{i_1\cdots i_{m-1}i_m}} + \frac{\partial \Sigma^{i_m}}{\partial v_{i_1\cdots i_{m-1}i}}\Big).
\end{aligned}
$$

Therefore, we can state the theorem below:

Theorem 9.4.2. *An isovector field V of a contact ideal generated by a single dependent variable u can also be an isovector field of the balance* *ideal created by a single balance equation* $(N = 1)$ *if and only if the smooth functions* $F = F(\mathbf{x}, u, \mathbf{v})$ *producing the isovector components and* $\lambda \in \Lambda^0(\mathcal{C}_m)$ *must satisfy the determining equations*

$$\begin{aligned}\lambda\Big(\Sigma + \frac{\partial \Sigma^i}{\partial x^i} + \sum_{r=0}^{m-1} \frac{\partial \Sigma^i}{\partial v_{i_1\cdots i_r}} v_{i_1\cdots i_r i}\Big)\\ &= A + A^i v_i + A^{ij} v_{ji} + 2B^{ijk} v_j v_{ki} + 2C^{ijkl} v_{li} v_{kj}\\ &\quad + \sum_{r=2}^{m-1} (A^{ii_1\cdots i_r} + 2B^{iji_1\cdots i_r} v_j + 2C^{ijki_1\cdots i_r} v_{kj}) v_{i_1\cdots i_r i},\end{aligned}$$

$$\begin{aligned}\lambda\Big(\frac{\partial \Sigma^i}{\partial v_{i_1\cdots i_{m-1} i_m}} + \frac{\partial \Sigma^{i_m}}{\partial v_{i_1\cdots i_{m-1} i}}\Big) &= A^{ii_1\cdots i_{m-1} i_m} + A^{i_m i_1\cdots i_{m-1} i} \\ &\quad + 2(B^{iji_1\cdots i_{m-1} i_m} + B^{i_m j i_1\cdots i_{m-1} i}) v_j\\ &\quad + 2(C^{ijki_1\cdots i_{m-1} i_m} + C^{i_m jki_1\cdots i_{m-1} i}) v_{kj} \quad \square\end{aligned} \tag{9.4.17}$$

The above theorem looses its validity for $m = 1$. In this case, the isovector field is represented by

$$V = X^i \frac{\partial}{\partial x^i} + U \frac{\partial}{\partial u} + V_i \frac{\partial}{\partial v_i}$$

and its components are given by $(9.3.26)_{1-3}$. The expression (9.4.6) takes of course now the form

$$\begin{aligned}\pounds_V \omega = V(\Sigma)\mu &+ \Big[\frac{\partial V(\Sigma^i)}{\partial x^j} dx^j + \frac{\partial V(\Sigma^i)}{\partial u} du + \frac{\partial V(\Sigma^i)}{\partial v_j} dv_j\\ &+ \Sigma\Big(\frac{\partial X^i}{\partial x^j} dx^j + \frac{\partial X^i}{\partial u} du + \frac{\partial X^i}{\partial v_j} dv_j\Big)\Big] \wedge \mu_i - \Big(\frac{\partial X^j}{\partial x^k} dx^k\\ &+ \frac{\partial X^j}{\partial u} du + \frac{\partial X^j}{\partial v_k} dv_k\Big) \wedge \Big[\frac{\partial \Sigma^i}{\partial x^l} dx^l + \frac{\partial \Sigma^i}{\partial u} du + \frac{\partial \Sigma^i}{\partial v_l} dv_l\Big] \wedge \mu_{ji}.\end{aligned}$$

By employing the transformation $du = \sigma + v_i\, dx^i$ on the above relation, we can finally obtain

$$\begin{aligned}\pounds_V \omega = (A + A^i v_i)\mu + A^i \sigma \wedge \mu_i + (A^{ij} + 2B^{ikj} v_k)\, dv_j \wedge \mu_i\\ + B^{ijk} dv_k \wedge \sigma \wedge \mu_{ji} + C^{ijkl} dv_l \wedge dv_k \wedge \mu_{ji}.\end{aligned}$$

On the other hand, (9.4.12) can now be cast into

$$\pounds_V \omega = \lambda\, \omega + \gamma \wedge \sigma + \Gamma \wedge d\sigma = \lambda\Big(\Sigma + \frac{\partial \Sigma^i}{\partial x^i} + \frac{\partial \Sigma^i}{\partial u} v_i\Big)\mu +$$

$$+\lambda\frac{\partial\Sigma^i}{\partial v_j}dv_j\wedge\mu_i+\gamma\wedge\sigma-\Gamma\wedge dv_m\wedge dx^m$$

where $\lambda\in\Lambda^0(\mathcal{C}_m)$, $\gamma\in\Lambda^{n-1}(\mathcal{C}_m)$, $\Gamma\in\Lambda^{n-2}(\mathcal{C}_m)$. In order to be able to compare the two expressions above for $\pounds_V\omega$, let us choose this time

$$\Gamma=\gamma^{ij}\mu_{ij}+\gamma^{ijkl}dv_l\wedge\mu_{ijk},\qquad\gamma^{ij},\gamma^{ijkl}\in\Lambda^0(\mathcal{C}_m).$$

Because of the antisymmetry of the forms $\mu_{ijk}\in\Lambda^3(M)$, we can take the functions γ^{ijkl} as completely antisymmetric with respect to indices i,j,k without loss of generality. Since $(5.5.16)_2$ allows us to write

$$dx^m\wedge\mu_{ijk}=\delta_i^m\,\mu_{jk}+\delta_j^m\,\mu_{ki}+\delta_k^m\,\mu_{ij}$$

we easily obtain

$$\begin{aligned}-\Gamma\wedge dv_m\wedge dx^m&=-\,dv_m\wedge dx^m\wedge\Gamma=-\,\gamma^{ij}dv_m\wedge dx^m\wedge\mu_{ij}\\&\quad+\gamma^{ijkl}dv_m\wedge dv_l\wedge dx^m\wedge\mu_{ijk}=\gamma^{ij}\left(-\,dv_i\wedge\mu_j+dv_j\wedge\mu_i\right)\\&\quad+\gamma^{ijkl}(dv_i\wedge dv_l\wedge\mu_{jk}+dv_j\wedge dv_l\wedge\mu_{ki}+dv_k\wedge dv_l\wedge\mu_{ij})\\&=2\gamma^{ij}dv_j\wedge\mu_i+3\gamma^{ijkl}dv_l\wedge dv_k\wedge\mu_{ji}\end{aligned}$$

But, exterior products appearing in the second form on the right hand side in the last line above are antisymmetric in indices k and l. This entails that the functions γ^{ijkl} should be completely antisymmetric with respect to all superscripts. We thus arrive at the following relations

$$\begin{aligned}\lambda\Big(\Sigma+\frac{\partial\Sigma^i}{\partial x^i}+\frac{\partial\Sigma^i}{\partial u}v_i\Big)&=A+A^iv_i,\\\lambda\frac{\partial\Sigma^i}{\partial v_j}+2\gamma^{ij}&=A^{ij}+2B^{ikj}v_k,\\\gamma&=(-\,1)^{n-1}A^i\mu_i+B^{ijk}\mu_{ji}\wedge dv_k,\\3\gamma^{ijkl}&=C^{ijkl}.\end{aligned}$$

Since the functions $\gamma^{ij},\gamma^{ijkl}$ are completely antisymmetric, we then obtain

$$\begin{aligned}4\gamma^{ij}&=A^{ij}-A^{ji}-(B^{kij}-B^{kji})v_k-\lambda\Big(\frac{\partial\Sigma^i}{\partial v_j}-\frac{\partial\Sigma^j}{\partial v_i}\Big),\\3\gamma^{ijkl}&=C^{[ijkl]}=C^{i[jk]l}\end{aligned}$$

where in the last line, we have made use of antisymmetries of the functions C^{ijkl} with respect to pairs of indices (i,j) and (k,l). Hence, *the determining equations for the isovector components corresponding to the case* $m=1$

take the following special forms

$$\lambda\Big(\Sigma+\frac{\partial\Sigma^i}{\partial x^i}+\frac{\partial\Sigma^i}{\partial u}v_i\Big)=A+A^iv_i, \tag{9.4.18}$$
$$\lambda\Big(\frac{\partial\Sigma^i}{\partial v_j}+\frac{\partial\Sigma^j}{\partial v_i}\Big)=A^{ij}+A^{ji}-2(B^{kij}+B^{kji})v_k,$$
$$C^{ijkl}+C^{ikjl}=0.$$

We know that isovector fields of the balance ideal constitute a Lie algebra, and this algebra in turn induces a Lie group of transformations. This group is called the ***symmetry group*** of the system of differential equations. If we obtain r linearly independent isovectors V_a from the determining equations, then any isovector field may be represented by $V=c^aV_a$ where $c^a\in\mathbb{R}, a=1,\dots,r$ are arbitrary constants. In this case, the symmetry group becomes an r-dimensional Lie group. If arbitrary functions are involved in isovector components, then the Lie group turns out to be infinite dimensional. Let a regular mapping $\phi:\mathcal{D}_n\to\mathcal{C}_m$ be a solution of the balance ideal $\mathfrak{I}_m$ satisfying the conditions $\phi^*\sigma^\alpha_{i_1i_2\cdots i_r}=0, r=0,1,\dots,m-1$; $\phi^*\omega^\alpha=0$. If V is an isovector field of this ideal, we had already shown in Theorem 5.13.7 that the mappings

$$\phi_V(t)=e^{tV}\circ\phi:\mathcal{D}_n\to\mathcal{C}_m,$$

where t is a real parameter, constitute a 1-parameter family of solutions of the ideal $\mathfrak{I}_m$. Let us recall that the mapping $\phi:\mathcal{D}_n\to\mathcal{C}_m$ is obtained by *lifting* the solution mapping $\phi:\mathcal{D}_n\to G$ specified by $u^\alpha=\phi^\alpha(x^i)$. Hence, to determine the family of solutions $\phi_V(t):\mathcal{D}_n\to\mathcal{C}_m$ when $N>1$, we have to solve the following set of autonomous ordinary differential equations

$$\frac{d\overline{x}^i}{dt}=X^i(\overline{\mathbf{x}},\overline{\mathbf{u}}),\qquad \frac{d\overline{u}^\alpha}{dt}=U^\alpha(\overline{\mathbf{x}},\overline{\mathbf{u}}) \tag{9.4.19}$$
$$\frac{d\overline{v}^\alpha_{i_1\cdots i_r}}{dt}=V_{i_1\cdots i_r}(\overline{\mathbf{x}},\overline{\mathbf{u}},\overline{v}^\alpha_{i_1},\dots,\overline{v}^\alpha_{i_1\cdots i_r}), r=1,\dots,m$$

under the initial conditions $\overline{\mathbf{x}}(0)=\mathbf{x}$, $\overline{\mathbf{u}}(0)=\mathbf{u}$, $\overline{v}^\alpha_{i_1\cdots i_r}(0)=v^\alpha_{i_1\cdots i_r}$ where $r=1,\dots,m$. Let us now consider the vector field

$$V_G=X^i(\mathbf{x},\mathbf{u})\frac{\partial}{\partial x^i}+U^\alpha(\mathbf{x},\mathbf{u})\frac{\partial}{\partial u^\alpha}$$

which is the projection of the isovector onto the tangent bundle of the graph space G. We know that this vector induces a Lie group of diffeomorphisms mapping the manifold G onto itself. Let us suppose that a solution $\phi : \mathcal{D}_n \to G$ of the system of differential equations is depicted by the given expressions $u^\alpha = \phi^\alpha(x^i)$. If we represent the solution of $n + N$ ordinary differential equations $(9.4.19)_{1-2}$ under the initial conditions $\overline{\mathbf{x}}(0) = \mathbf{x}$ and $\overline{\mathbf{u}}(0) = \mathbf{u}$ by

$$\overline{x}^i = \psi^i(t; \mathbf{x}, \mathbf{u}), \quad \overline{u}^\alpha = \psi^\alpha(t; \mathbf{x}, \mathbf{u})$$

we find

$$\overline{x}^i = \psi^i(t; \mathbf{x}, \boldsymbol{\phi}(\mathbf{x})) = \Psi^i(t; \mathbf{x}), \quad \overline{u}^\alpha = \psi^\alpha(t; \mathbf{x}, \boldsymbol{\phi}(\mathbf{x})) = \Psi^i(t; \mathbf{x})$$

when we insert the original solution $\mathbf{u} = \boldsymbol{\phi}(\mathbf{x})$ into these relations. On solving the variables x^i in terms of $\overline{x}^i$ from the first set of equations and introduce the result into the second set, we ultimately obtain the family of solutions $\phi_V(t) : \mathcal{D}_n \to G$ in the form $\overline{u}^\alpha = \Phi_t^\alpha(\overline{x}^i)$. Hence, this procedure based on isovectors of the graph space enables us to produce a family of new, probably more complicated, solutions if we have at hand a solution, however simple, of the set of partial differential equations. But, it is clear that if we do not know a particular solution, this approach cannot help us at all to generate any new solution.

In the case of $N = 1$, the components X^i and U of the isovector fields will depend on the variables x^i, u, v_i. Hence, we can project isovectors only on the tangent bundle of the manifold $\mathcal{C}_1$. Consequently, to determine the group of transformations, we have to solve the following set of ordinary differential equations

$$\frac{d\overline{x}^i}{dt} = -\frac{\partial F}{\partial \overline{v}_i}, \quad \frac{d\overline{u}}{dt} = F - \overline{v}_i \frac{\partial F}{\partial \overline{v}_i}, \quad \frac{d\overline{v}_i}{dt} = \frac{\partial F}{\partial \overline{x}^i} + \overline{v}_i \frac{\partial F}{\partial \overline{u}},$$

under the initial conditions $\overline{x}^i(0) = x^i$, $\overline{u}(0) = u$, $\overline{v}_i(0) = v_i$. Here, the function $F = F(\mathbf{x}, u, \mathbf{v})$ determines the isovector components. When we accomplish to integrate these differential equations, we arrive at the result

$$\overline{x}^i = \psi^i(t; \mathbf{x}, u, \mathbf{v}), \quad \overline{u} = \psi(t; \mathbf{x}, u, \mathbf{v}), \quad \overline{v}_i = \psi_i(t; \mathbf{x}, u, \mathbf{v}).$$

Since, the transformations between (x^i, u) and $(\overline{x}^i, \overline{u})$ now involve derivatives $u_{,i}$, they become now ***Bäcklund transformations*** forming a group. A Bäcklund transformation reduces to a Lie transformation if and only if the function F is an affine function of variables v_i.

The approach we have developed so far to determine isovector fields of balance equations may also be used to find isovector fields associated with mth order non-linear partial differential equations given by (9.2.1) and taken into account at the beginning of this section. In this case, we have to take $\Sigma^{ai} \equiv 0$ and we write $\omega^a = \Sigma^a(x^i, u^\alpha, v_i^\alpha, \ldots, v_{i_1 i_2 \cdots i_m}^\alpha)\mu$. Evidently, the *fundamental ideal* induced by n-forms ω^a is closed. Thus, if V is an isovector field of the contact ideal, then the functions X^i and U^α should be found from equations (9.4.11) such that $V(\Sigma^a) = 0$ when $\Sigma^a = 0$ or the function F from equations (9.4.17) or (9.4.18) such that $V(\Sigma) = 0$ when $\Sigma = 0$. This is quite a difficult procedure to accomplish. Nonetheless, if we consider a first order equation in the form $\Sigma(x^i, u, u_{,i}) = 0$, the isovector field can be found easily. In this case, we may choose $F(x^i, u, v_i) = -\Sigma$ to find the isovector components as follows

$$X^i = \frac{\partial \Sigma}{\partial v_i},\ U = -\Sigma + v_i \frac{\partial \Sigma}{\partial v_i},\ V_i = -\Big(\frac{\partial \Sigma}{\partial x^i} + v_i \frac{\partial \Sigma}{\partial u}\Big). \quad (9.4.20)$$

We thus obtain

$$V(\Sigma) = \frac{\partial \Sigma}{\partial v_i}\frac{\partial \Sigma}{\partial x^i} + \Big(v_i \frac{\partial \Sigma}{\partial v_i} - \Sigma\Big)\frac{\partial \Sigma}{\partial u} - \Big(\frac{\partial \Sigma}{\partial x^i} + v_i \frac{\partial \Sigma}{\partial u}\Big)\frac{\partial \Sigma}{\partial v_i} = -\frac{\partial \Sigma}{\partial u}\Sigma$$

implying that $V(\Sigma) = 0$ whenever $\Sigma = 0$. In this situation, we can obviously write

$$\begin{aligned} \mathbf{i}_V(\sigma) &= U - v_i X^i = \Sigma = 0,\ \mathbf{i}_V(\omega) = \Sigma X^i \mu_i = 0, \\ \mathbf{i}_V(d\sigma) &= -V_i dx^i + X^i dv_i = d\Sigma + \frac{\partial \Sigma}{\partial u}\sigma = \frac{\partial \Sigma}{\partial u}\sigma. \end{aligned}$$

Hence, this isovector field is likewise a characteristic vector field of the ideal. Consequently, we again find the previously given solution (9.2.10) by employing this isovector field.

Example 9.4.1. The time-dependent, one-dimensional heat equation in a homogeneous medium, or more generally the one-dimensional diffusion equation modelling various physical phenomena is given by

$$\frac{\partial u}{\partial t} = \frac{\partial}{\partial x}\Big(\kappa(u)\frac{\partial u}{\partial x}\Big) + h(x,t,u) \quad (9.4.21)$$

where u is the temperature, t is the time, x is the spatial variable and $\kappa(u)$ is a constitutive quantity called the *coefficient of thermal diffusivity* that may be dependent on temperature and h is the heat source. Let us denote $x = x^1,\ t = x^2,\ v_1 = u_x,\ v_2 = u_t,\ \mu = dx \wedge dt,\ \mu_1 = dt,\ \mu_2 = -dx$. Then we arrive at the balance equation

$$\frac{\partial \Sigma^1}{\partial x^1} + \frac{\partial \Sigma^2}{\partial x^2} + \Sigma = 0$$

where

$$\Sigma^1 = \kappa(u)\,v_1,\ \Sigma^2 = -\,u,\ \Sigma = h(x^1, x^2, u).$$

In this case, the isovector field must be prescribed by

$$V = X^1\frac{\partial}{\partial x^1} + X^2\frac{\partial}{\partial x^2} + U\frac{\partial}{\partial u} + V_1\frac{\partial}{\partial v_1} + V_2\frac{\partial}{\partial v_2}$$

and its components are specified by relations $(9.3.26)_{1-2}$ through an arbitrary function $F = F(x^1, x^2, u, v_1, v_2) = F(x, t, u, v_1, v_2)$. With these data, non-zero coefficient functions in (9.4.13) become

$$\begin{aligned}
&V(\Sigma^1) = \kappa' v_1 U + \kappa V_1,\\
&V(\Sigma^2) = -\,U,\\
&V(\Sigma) = X^1\frac{\partial h}{\partial x} + X^2\frac{\partial h}{\partial t} + U\frac{\partial h}{\partial u}\\
&\quad A = V(\Sigma) + \kappa' v_1\frac{\partial U}{\partial x} + \kappa\frac{\partial V_1}{\partial x} - \frac{\partial U}{\partial t} + h\Big(\frac{\partial X^1}{\partial x} + \frac{\partial X^2}{\partial t}\Big)\\
&\quad A^1 = \kappa'' v_1 U + \kappa' v_1\frac{\partial U}{\partial u} + \kappa' V_1 + \kappa\frac{\partial V_1}{\partial u} + h\frac{\partial X^1}{\partial u} + \kappa' v_1\frac{\partial X^2}{\partial t} + \frac{\partial X^1}{\partial t}\\
&\quad A^2 = -\frac{\partial U}{\partial u} + h\frac{\partial X^2}{\partial u} - \frac{\partial X^1}{\partial x} - \kappa' v_1\frac{\partial X^2}{\partial x}\\
&\quad A^{11} = \kappa' U + \kappa' v_1\frac{\partial U}{\partial v_1} + \kappa\frac{\partial V_1}{\partial v_1} + h\frac{\partial X^1}{\partial v_1} + \kappa\frac{\partial X^2}{\partial t}\\
&\quad A^{22} = -\frac{\partial U}{\partial v_2} + h\frac{\partial X^2}{\partial v_2},\\
&\quad A^{12} = \kappa' v_1\frac{\partial U}{\partial v_2} + \kappa\frac{\partial V_1}{\partial v_2} + h\frac{\partial X^1}{\partial v_2},\\
&\quad A^{21} = -\frac{\partial U}{\partial v_1} + h\frac{\partial X^2}{\partial v_1} - \kappa\frac{\partial X^2}{\partial x},\\
&\quad 2B^{121} = \kappa\frac{\partial X^2}{\partial u} - \frac{\partial X^1}{\partial v_1} - \kappa' v_1\frac{\partial X^2}{\partial v_1},\\
&\quad 2B^{122} = -\frac{\partial X^1}{\partial v_2} - \kappa' v_1\frac{\partial X^2}{\partial v_2}.
\end{aligned}$$

Hence, the determining equations (9.4.18) are found to be

$$\begin{aligned}
\lambda(\kappa' v_1^2 - v_2 + h) &= A + A^1 v_1 + A^2 v_2,\\
\lambda\kappa &= A^{11} + 2B^{121} v_2,\\
A^{22} - 2B^{122} v_1 &= 0,
\end{aligned}$$

$$A^{12} + A^{21} - 2B^{121} v_1 + 2B^{122} v_2 = 0$$

whence we extract, respectively, the following four equations to be satisfied by the single function F

$$\begin{aligned}
&\lambda(v_2 - h - \kappa' v_1^2) + \Big(\kappa'' v_1^2 + \frac{\partial h}{\partial u}\Big) F + 2\kappa' v_1 \frac{\partial F}{\partial x} + (2\kappa' v_1^2 - v_2) \frac{\partial F}{\partial u} \\
&\quad - \frac{\partial F}{\partial t} - \Big(\kappa'' v_1^3 + \frac{\partial h}{\partial x} + \frac{\partial h}{\partial u} v_1\Big) \frac{\partial F}{\partial v_1} - (\kappa'' v_1^2 v_2 + \frac{\partial h}{\partial t} + \frac{\partial h}{\partial u} v_2) \frac{\partial F}{\partial v_2} \\
&\quad + (v_2 - h - \kappa' v_1^2) \Big(\frac{\partial^2 F}{\partial x \partial v_1} + \frac{\partial^2 F}{\partial t \partial v_2} + v_1 \frac{\partial^2 F}{\partial u \partial v_1} + v_2 \frac{\partial^2 F}{\partial u \partial v_2}\Big) \\
&\quad + \kappa \frac{\partial^2 F}{\partial x^2} + 2\kappa v_1 \frac{\partial^2 F}{\partial x \partial u} + \kappa v_1^2 \frac{\partial^2 F}{\partial u^2} = 0, \\
&\kappa' \Big(F - v_1 \frac{\partial F}{\partial v_1} - v_2 \frac{\partial F}{\partial v_2}\Big) + (v_2 - h - \kappa' v_1^2) \frac{\partial^2 F}{\partial v_1^2} + \kappa \Big(- \lambda + \frac{\partial F}{\partial u} \\
&\quad + \frac{\partial^2 F}{\partial x \partial v_1} - \frac{\partial^2 F}{\partial t \partial v_2} + v_1 \frac{\partial^2 F}{\partial u \partial v_1} - v_2 \frac{\partial^2 F}{\partial u \partial v_2}\Big) = 0, \qquad (9.4.22) \\
&(v_2 - h - \kappa' v_1^2) \frac{\partial^2 F}{\partial v_2^2} = 0, \\
&(v_2 - h - \kappa' v_1^2) \frac{\partial^2 F}{\partial v_1 \partial v_2} + \kappa \Big(\frac{\partial^2 F}{\partial x \partial v_2} + v_1 \frac{\partial^2 F}{\partial u \partial v_2}\Big) = 0.
\end{aligned}$$

$(9.4.22)_3$ yields

$$F = f(x, t, u, v_1)\, v_2 + g(x, t, u, v_1).$$

On inserting this expression into the equation $(9.4.22)_4$, we find that

$$\kappa \Big(\frac{\partial f}{\partial x} + v_1 \frac{\partial f}{\partial u}\Big) - (\kappa' v_1^2 + h) \frac{\partial f}{\partial v_1} + v_2 \frac{\partial f}{\partial v_1} = 0$$

from which we obviously obtain

$$\frac{\partial f}{\partial v_1} = 0, \quad \frac{\partial f}{\partial u} = 0, \quad \frac{\partial f}{\partial x} = 0.$$

Hence, we see that F must take the form

$$F = f(t)\, v_2 + g(x, t, u, v_1).$$

Let us now introduce this function F into equations $(9.4.22)_{1-2}$ and eliminate the function λ between these two equations so obtained. If we then equate the coefficient of the variable v_2^2 to zero in this expression, we are led to the simple partial differential equation $\partial^2 g/\partial v_1^2 = 0$ whose solution is immediately obtained as

$$g = \alpha(x,t,u)v_1 + \beta(x,t,u)$$

The coefficient of v_2 in the same expression gives

$$-\kappa f' + \kappa'\beta + 2\kappa\frac{\partial\alpha}{\partial x} + 2\kappa\frac{\partial\alpha}{\partial u}v_1 = 0$$

whence we deduce $\partial\alpha/\partial u = 0$ and $g = \alpha(x,t)\,v_1 + \beta(x,t,u)$. Let us insert this function g into that expression. Then the vanishing of the coefficients of variables v_1, v_1^2 and v_2 together with the remaining expression lead, respectively, to the equations

$$\begin{aligned}
&-\kappa\frac{\partial\alpha}{\partial t} + \kappa^2\frac{\partial^2\alpha}{\partial x^2} + 2\kappa\kappa'\frac{\partial\beta}{\partial x} + 2\kappa^2\frac{\partial^2\beta}{\partial x\partial u} = 0, \qquad (9.4.23)\\
&\big[\kappa\kappa'' - (\kappa')^2\big]\beta + \kappa\kappa'\frac{\partial\beta}{\partial u} + \kappa^2\frac{\partial^2\beta}{\partial u^2} = 0,\\
&-\kappa f' + \kappa'\beta + 2\kappa\frac{\partial\alpha}{\partial x} = 0,\\
&\Big(h\frac{\kappa'}{\kappa} - \frac{\partial h}{\partial u}\Big)\beta + h\Big(2\frac{\partial\alpha}{\partial x} + \frac{\partial\beta}{\partial u}\Big) + \frac{\partial\beta}{\partial t} + f\frac{\partial h}{\partial t} + \alpha\frac{\partial h}{\partial x} - \kappa\frac{\partial^2\beta}{\partial x^2} = 0,
\end{aligned}$$

The second order differential equation in $(9.4.23)_2$ can be written as

$$\frac{\partial^2\beta}{\partial u^2} + \frac{\kappa'}{\kappa}\frac{\partial\beta}{\partial u} + \Big(\frac{\kappa'}{\kappa}\Big)'\beta = \frac{\partial^2\beta}{\partial u^2} + \frac{\partial}{\partial u}\Big(\frac{\kappa'}{\kappa}\beta\Big) = 0$$

from which we get

$$\frac{\partial\beta}{\partial u} + \frac{\kappa'}{\kappa}\beta = n(x,t)$$

yielding

$$\frac{\partial}{\partial u}(\kappa\beta) = n(x,t)\kappa(u).$$

We thus obtain

$$\beta(x,t,u) = \frac{1}{\kappa(u)}\big[m(x,t) + n(x,t)K(u)\big]$$

where m and n are arbitrary functions of their arguments and we introduce the indefinite integral $K(u) = \int\kappa(u)\,du$ so that we have $\kappa(u) = K'(u)$.

When we insert this expression into the first equation in (9.4.23), we get

$$-\frac{\partial \alpha}{\partial t}+\kappa(u)\Big(2\frac{\partial n}{\partial x}+\frac{\partial^2 \alpha}{\partial x^2}\Big)=0.$$

Whenever the function $\kappa(u)$ is not a constant, the foregoing equation can only be satisfied if

$$\alpha=\alpha(x),\qquad n=-\tfrac{1}{2}\alpha'(x)+n_0(t)$$

Therefore, the function F must be in the form

$$F=f(t)\,v_2+\alpha(x)\,v_1+\frac{1}{K'(u)}\big[m(x,t)+[-\tfrac{1}{2}\alpha'(x)+n_0(t)]K(u)\big]$$

Furthermore, the third and the fourth equations in (9.4.23) should also be satisfied:

$$K'(u)^2\big(2\alpha'(x)+f'(t)\big)+ \\ K''(u)\Big[m(x,t)+\big(n_0(t)-\tfrac{1}{2}\alpha'(x)\big)K(u)\Big]=0$$

$$K'(u)h(x,t,u)\big(2n_0(t)+3\alpha'(x)\big)+K(u)\Big[2n_0'(t)+K'(u)\frac{\partial^3\alpha}{\partial x^3}+ \\ \big(\alpha'(x)-2n_0(t)\big)\frac{\partial h}{\partial u}\Big]+2\Big(\frac{\partial m}{\partial t}-m(x,t)\frac{\partial h}{\partial u}\Big)+2K'(u)\Big(f(t)\frac{\partial h}{\partial t} \\ +\alpha(x)\frac{\partial h}{\partial x}-\frac{\partial^2 m}{\partial x^2}\Big)=0.$$

These equations restrict the admissible forms of functions $f(t), \alpha(x), n_0(t)$, $m(x,t)$ and structures of physical data $\kappa(u)$ and $h(x,t,u)$ so that isovector fields are realisable. Interested readers can determine admissible choices without experiencing too much difficulties by scrutinising these equations. It is clearly seen that the function F is an affine function of the variables v_1 and v_2. Therefore, in this case isovectors will be prolongations of the vectors V_G.

As a special case, we take $\kappa=1, h=0$. Hence, the field equation becomes

$$\frac{\partial u}{\partial t}=\frac{\partial^2 u}{\partial x^2}.$$

This equation is obtained by non-dimensionalising the heat conduction equation in the absence of the heat source and by assuming that the coefficient of thermal diffusivity is constant. Equations (9.4.23) are reduced in this case to the form

$$
\begin{gathered}
-\frac{\partial \alpha}{\partial t}+\frac{\partial^2 \alpha}{\partial x^2}+2\frac{\partial^2 \beta}{\partial x \partial u}=0, \\
\frac{\partial^2 \beta}{\partial u^2}=0, \quad \frac{\partial \beta}{\partial t}-\frac{\partial^2 \beta}{\partial x^2}=0, \\
-f'+2\frac{\partial \alpha}{\partial x}=0.
\end{gathered}
$$

The second and the fourth equations yield

$$\beta=\lambda(x,t)u+\mu(x,t), \quad \alpha=\frac{1}{2}f'(t)x+\gamma(t).$$

If we insert these expressions into the first and the third equations, we obtain

$$-\frac{1}{2}f''(t)x-\gamma'(t)+2\frac{\partial \lambda}{\partial x}=0, \quad \frac{\partial \lambda}{\partial t}-\frac{\partial^2 \lambda}{\partial x^2}=0, \quad \frac{\partial \mu}{\partial t}-\frac{\partial^2 \mu}{\partial x^2}=0$$

We then introduce the function

$$\lambda=\frac{1}{8}f''(t)x^2+\frac{1}{2}\gamma'(t)x+\delta(t),$$

found from integrating the first equation above, into the second equation, we arrive at

$$\frac{1}{8}f'''(t)x^2+\frac{1}{2}\gamma''(t)x+\delta'(t)-\frac{1}{4}f''(t)=0$$

whence we obviously deduce the relations

$$f'''(t)=0, \ \gamma''(t)=0, \ \delta'(t)=\frac{1}{4}f''(t).$$

We therefore find

$$f(t)=4c_1t^2+2c_2t+c_3, \quad \gamma(t)=2c_4t+c_5, \quad \delta(t)=2c_1t+c_6$$

where $c_1, \ldots, c_6$ are arbitrary constants. Hence, the function F is expressible as follows

$$
\begin{aligned}
F &= \big[(4c_1t+c_2)x+2c_4t+c_5\big]v_1+(4c_1t^2+2c_2t+c_3)v_2 \\
&\qquad +\big[c_1(x^2+2t)+c_4x+c_6\big]u+\mu(x,t) \\
&= c_1\big[4txv_1+4t^2v_2+(x^2+2t)u\big] \\
&\qquad +c_2(xv_1+2tv_2)+c_3v_2+c_4(2tv_1+xu)+c_5v_1+c_6u+\mu(x,t).
\end{aligned}
$$

The function $\mu(x,t)$ is any solution of the linear equation $\dfrac{\partial\mu}{\partial t}-\dfrac{\partial^2\mu}{\partial x^2}=0$. Thus, the linearly independent isovectors are found to be

$$\begin{aligned}
V^1 =& -4xt\frac{\partial}{\partial x}-4t^2\frac{\partial}{\partial t}+(x^2+2t)u\frac{\partial}{\partial u}+\big[(x^2+6t)v_1+2xu\big]\frac{\partial}{\partial v_1}\\
&+\big[4xv_1+(x^2+10t)v_2+2u\big]\frac{\partial}{\partial v_2},\\
V^2 =& -x\frac{\partial}{\partial x}-2t\frac{\partial}{\partial t}+v_1\frac{\partial}{\partial v_1}+2v_2\frac{\partial}{\partial v_2},\ V^3=-\frac{\partial}{\partial t},\ V^5=-\frac{\partial}{\partial x},\\
V^4 =& -2t\frac{\partial}{\partial x}+xu\frac{\partial}{\partial u}+(u+xv_1)\frac{\partial}{\partial v_1}+(2v_1+xv_2)\frac{\partial}{\partial v_2},\\
V^6 =& \,u\frac{\partial}{\partial u}+v_1\frac{\partial}{\partial v_1}+v_2\frac{\partial}{\partial v_2},\qquad V_\mu=\mu\frac{\partial}{\partial u}+\frac{\partial\mu}{\partial x}\frac{\partial}{\partial v_1}+\frac{\partial\mu}{\partial t}\frac{\partial}{\partial v_2}.
\end{aligned}$$

That these vectors constitute a Lie algebra as it should be can be observed at once from the relations

$$\begin{aligned}
&[V^1,V^2]=2V^1, [V^1,V^3]=4V^2+2V^6, [V^1,V^4]=0, [V^1,V^5]=2V^4,\\
&[V^1,V^6]=0,\ \ [V^2,V^3]=-2V^3,\ \ [V^2,V^4]=-V^4,\ \ [V^2,V^5]=V^5,\\
&[V^2,V^6]=0,\quad [V^3,V^4]=-2V^5,\quad [V^3,V^5]=0,\quad [V^3,V^6]=0,\\
&[V^4,V^5]=V^6,\quad [V^4,V^6]=0,\quad [V^5,V^6]=0,\quad [V^3,V_\mu]=-V_{\mu_t},\\
&[V^1,V_\mu]=-2V_{x^2\mu+2\mu t+4xt\mu_x+4t^2\mu_t},\qquad [V^2,V_\mu]=-V_{x\mu_x+2t\mu_t},\\
&[V^4,V_\mu]=-V_{x\mu+2t\mu_x}, [V^5,V_\mu]=-V_{\mu_x}, [V^6,V_\mu]=-V_\mu, [V_\mu,V_\nu]=0.
\end{aligned}$$

Here, we have defined $\mu_x=\partial\mu/\partial x$ and $\mu_t=\partial\mu/\partial t$. Since isovectors are prolongations of vectors of the form V_G, it would suffice to integrate the differential equations $(9.4.19)_{1,2}$ in order to determine the associated symmetry groups. To simplify the operations, let us take isovectors into consideration one by one:

The isovector V^1 gives rise to the ordinary differential equations

$$\frac{d\overline{x}}{ds}=-4\overline{x}\,\overline{t},\ \frac{d\overline{t}}{ds}=-4\overline{t}^2,\ \frac{d\overline{u}}{ds}=(\overline{x}^2+2\,\overline{t})\overline{u}$$

whose solutions under the initial conditions $\overline{x}(0)=x,\overline{t}(0)=t,\overline{u}(0)=u$ are readily found to be

$$\overline{x}(s)=\frac{x}{1+4st},\ \overline{t}(s)=\frac{t}{1+4st},\ \overline{u}(s)=u(x,t)\sqrt{1+4st}\,e^{\frac{sx^2}{1+4st}}.$$

It then follows from these relations that

$$x = \frac{\overline{x}}{1 - 4s\overline{t}}, \qquad t = \frac{\overline{t}}{1 - 4s\overline{t}}.$$

We thus manufacture a 1-parameter family of new solutions of the partial differential equation

$$\frac{\partial \overline{u}}{\partial \overline{t}} = \frac{\partial^2 \overline{u}}{\partial \overline{x}^2}$$

from a given solution $u(x, t)$ by the following manner

$$\overline{u}(\overline{x}, \overline{t}; s) = u\Big(\frac{\overline{x}}{1 - 4s\overline{t}}, \frac{\overline{t}}{1 - 4s\overline{t}}\Big)\frac{1}{\sqrt{1 - 4s\overline{t}}}\, e^{\frac{s\overline{x}^2}{1-4s\overline{t}}}.$$

In other words, if a function $u(x, t)$ is a solution of the heat conduction equation under consideration, then the family of functions

$$u\Big(\frac{x}{1 - 4st}, \frac{t}{1 - 4st}\Big)\frac{1}{\sqrt{1 - 4st}}\, e^{\frac{sx^2}{1-4st}}$$

become also solutions of the same equation. For instance, the trivially obtained simple solution $u(x, t) = 1$ gives rise to the family of new solutions $u(x, t) = e^{\frac{sx^2}{1-4st}} / \sqrt{1 - 4st}$.

If we consider the equations

$$\frac{d\overline{x}}{ds} = -\,\overline{x}\,, \quad \frac{d\overline{t}}{ds} = -\,2\overline{t}, \quad \frac{d\overline{u}}{ds} = 0$$

corresponding to the isovector V^2, we obtain

$$\overline{x}(s) = x\, e^{-s}, \;\; \overline{t}(s) = t\, e^{-2s},$$
$$\overline{u}(s) = u.$$

This result implies that a solution is invariant under a *scaling transformation*: $u(x, t) = u(\lambda x, \lambda^2 t)$ where λ is a constant.

The isovector V^4 generates the differential equations

$$\frac{d\overline{x}}{ds} = -\,2\overline{t}, \quad \frac{d\overline{t}}{ds} = 0, \quad \frac{d\overline{u}}{ds} = \overline{x}\,\overline{u}$$

whose solution is

$$\overline{x}(s) = x\, - 2st, \;\; \overline{t}(s) = t, \;\; \overline{u}(s) = u\, e^{sx - s^2 t}.$$

Hence, if $u(x, t)$ is a solution, then the function

$$u(x + 2st, t)\, e^{sx+s^2t}$$

provides a family of solutions. For example, the trivial solution $u = x + a$ leads to

$$u = (x + 2st + a)\, e^{sx+s^2t}.$$

We can easily check that the isovectors V^3, V^5 and V^6, respectively, give rise to transformations

$$\begin{aligned} &\overline{x} = x, \quad \overline{t} = t - s, \quad \overline{u} = u; \\ &\overline{x} = x - s, \quad \overline{t} = t, \quad \overline{u} = u; \\ &\overline{x} = x, \quad \overline{t} = t, \quad \overline{u} = u\, e^s. \end{aligned}$$

These transformations mean that solutions are invariant under translations in the temporal and spatial variables, and by multiplications with constants.

The isovector V_μ gives

$$\overline{x} = x, \quad \overline{t} = t, \quad \overline{u} = u + s\mu(x, t).$$

This is an expected result associated with linear equations reflecting the fact that solutions may be superimposed. ■

Example 9.4.2. As a more complicated example, let us consider the non-homogeneous ***Korteweg-de Vries equation*** [after Dutch mathematicians Diederik Johannes Korteweg (1848-1941) and Gustav de Vries (1866-1934)]

$$\frac{\partial u}{\partial t} + u\frac{\partial u}{\partial x} + \frac{\partial^3 u}{\partial x^3} = f(x, t, u). \tag{9.4.24}$$

where t and x denote the time and the space variables. This equation models the propagation of solitons in a medium. We denote the independent variables by $x^1 = x$, $x^2 = t$. We introduce the auxiliary variables $v_1 = u_{,1}$, $v_2 = u_{,2}$, $v_{11} = u_{,11}$, $v_{22} = u_{,22}$ and $v_{12} = v_{21} = u_{,12}$. Hence, (9.4.24) is transformed into the first order equation

$$\frac{\partial v_{11}}{\partial x} + uv_1 + v_2 - f = 0.$$

We thus have

$$\begin{aligned} \Sigma^1 = v_{11}, \quad \Sigma^2 = 0, \\ \Sigma = uv_1 + v_2 - f. \end{aligned}$$

The isovector field is prescribed by

$$V = X^1\frac{\partial}{\partial x^1} + X^2\frac{\partial}{\partial x^2} + U\frac{\partial}{\partial u} + V_1\frac{\partial}{\partial v_1} + V_2\frac{\partial}{\partial v_2} + V_{11}\frac{\partial}{\partial v_{11}} + V_{12}\frac{\partial}{\partial v_{12}} + V_{22}\frac{\partial}{\partial v_{22}}.$$

The components of this vector field are given by the expressions (9.3.26). Because of the relations

$$\begin{aligned} &V(\Sigma^1) = V_{11}, \quad V(\Sigma^2) = 0,\\ &V(\Sigma) = -X^1\frac{\partial f}{\partial x} - X^2\frac{\partial f}{\partial t} + U\Big(v_1 - \frac{\partial f}{\partial u}\Big) + V_1 u + V_2 \end{aligned}$$

we realise that we need only to know explicit forms of the following relevant components

$$\begin{aligned} X^1 &= -\frac{\partial F}{\partial v_1},\ X^2 = -\frac{\partial F}{\partial v_2},\\ U &= F - v_1\frac{\partial F}{\partial v_1} - v_2\frac{\partial F}{\partial v_2},\\ V_1 &= \frac{\partial F}{\partial x} + v_1\frac{\partial F}{\partial u},\\ V_2 &= \frac{\partial F}{\partial t} + v_2\frac{\partial F}{\partial u},\\ V_{11} &= \frac{\partial^2 F}{\partial x^2} + 2v_1\frac{\partial^2 F}{\partial x\partial u} + v_1^2\frac{\partial^2 F}{\partial u^2}\\ &+ v_{11}\Big(2\frac{\partial^2 F}{\partial x\partial v_1} + 2v_1\frac{\partial^2 F}{\partial u\partial v_1} + \frac{\partial F}{\partial u}\Big) + 2v_{12}\Big(\frac{\partial^2 F}{\partial x\partial v_2} + v_1\frac{\partial^2 F}{\partial u\partial v_2}\Big)\\ &+ v_{11}^2\frac{\partial^2 F}{\partial v_1^2} + v_{12}^2\frac{\partial^2 F}{\partial v_2^2} + 2v_{11}v_{12}\frac{\partial^2 F}{\partial v_1\partial v_2} \end{aligned}$$

where $F = F(x, t, u, v_1, v_2)$ is presently an arbitrary function. The coefficients given in (9.4.13) that are not identically zero can now be evaluated as follows

$$\begin{aligned} A = &-X^1\frac{\partial f}{\partial x} - X^2\frac{\partial f}{\partial t} + U\Big(v_1 - \frac{\partial f}{\partial u}\Big) + V_1 u + V_2 + \frac{\partial V_{11}}{\partial x}\\ &- (uv_1 + v_2 - f)\Big(\frac{\partial^2 F}{\partial x\partial v_1} + \frac{\partial^2 F}{\partial t\partial v_2}\Big);\\ A^1 = &\frac{\partial V_{11}}{\partial u} - (uv_1 + v_2 - f)\frac{\partial^2 F}{\partial u\partial v_1},\ A^2 = -(uv_1 + v_2 - f)\frac{\partial^2 F}{\partial u\partial v_2}; \end{aligned}$$

$$
\begin{aligned}
A^{11} &= \frac{\partial V_{11}}{\partial v_1} - (uv_1 + v_2 - f)\frac{\partial^2 F}{\partial v_1^2}, \qquad A^{22} = -(uv_1 + v_2 - f)\frac{\partial^2 F}{\partial v_2^2},\\
A^{12} &= \frac{\partial V_{11}}{\partial v_2} - (uv_1 + v_2 - f)\frac{\partial^2 F}{\partial v_1 \partial v_2}, A^{21} = -(uv_1 + v_2 - f)\frac{\partial^2 F}{\partial v_1 \partial v_2};\\
A^{111} &= \frac{\partial V_{11}}{\partial v_{11}} - \frac{\partial^2 F}{\partial t \partial v_2}, \qquad A^{112} = A^{121} = \frac{\partial V_{11}}{\partial v_{12}}, \qquad A^{122} = \frac{\partial V_{11}}{\partial v_{22}},\\
A^{211} &= \frac{\partial^2 F}{\partial x \partial v_2}, \qquad A^{212} = A^{221} = A^{222} = 0;\\
B^{ijk} &= 0; \quad C^{ijkl} = 0;\\
B^{1211} &= -B^{2111} = -\frac{\partial^2 F}{\partial u \partial v_2},\\
C^{12111} &= -C^{21111} = -\frac{1}{2}\frac{\partial^2 F}{\partial v_1 \partial v_2}, \; C^{12211} = -C^{21211} = -\frac{1}{2}\frac{\partial^2 F}{\partial v_2^2}.
\end{aligned}
$$

Thus $(9.4.17)_2$ takes the form

$$
\lambda\Big(\frac{\partial \Sigma^i}{\partial v_{mn}} + \frac{\partial \Sigma^n}{\partial v_{mi}}\Big) = A^{imn} + A^{nmi} + 2(B^{ijmn} + B^{njmi})v_j + 2(C^{ijkmn} + C^{njkmi})v_{kj}.
$$

If we introduce the coefficient calculated above into these expressions, we end up with the independent equations given below

$$
\begin{aligned}
\frac{\partial V_{11}}{\partial v_{11}} - \frac{\partial^2 F}{\partial t \partial v_2} - \frac{\partial^2 F}{\partial u \partial v_2}v_2 - \frac{\partial^2 F}{\partial v_1 \partial v_2}v_{12} - \frac{\partial^2 F}{\partial v_2^2}v_{22} &= \lambda \quad (9.4.25)\\
\frac{\partial^2 F}{\partial x \partial v_2} + \frac{\partial^2 F}{\partial u \partial v_2}v_1 + \frac{\partial^2 F}{\partial v_2^2}v_{12} + \frac{\partial^2 F}{\partial v_1 \partial v_2}v_{11} &= 0.
\end{aligned}
$$

$(9.4.25)_2$ yields first

$$
\frac{\partial^2 F}{\partial v_2^2} = \frac{\partial^2 F}{\partial v_1 \partial v_2} = 0
$$

and consequently $F = \alpha(x, t, u)v_2 + \beta(x, t, u, v_1)$. We then get

$$
\frac{\partial \alpha}{\partial x} = \frac{\partial \alpha}{\partial u} = 0
$$

so we find that

$$
F = \alpha(t)v_2 + \beta(x, t, u, v_1).
$$

Hence, it follows from (9.4.25)$_1$ that

$$\lambda = 2\frac{\partial^2\beta}{\partial x\partial v_1} + 2\frac{\partial^2\beta}{\partial u\partial v_1}v_1 + \frac{\partial\beta}{\partial u} + 2\frac{\partial^2\beta}{\partial v_1^2}v_{11} - \alpha'(t).$$

Therefore, the expression

$$\lambda(uv_1 + v_2 - f) = \\ A + A^1 v_1 + A^2 v_2 + A^{11}v_{11} + A^{22}v_{22} + (A^{12} + A^{21})v_{12}$$

given by (9.4.17)$_1$ can be written as

$$\frac{\partial^3\beta}{\partial v_1^3}v_{11}^3 + 3\Big(\frac{\partial^3\beta}{\partial u\partial v_1^2}v_1 + \frac{\partial^2\beta}{\partial u\partial v_1} + \frac{\partial^3\beta}{\partial x\partial v_1^2}\Big)v_{11}^2 + 3\Big(-\frac{\partial^2\beta}{\partial v_1^2}(uv_1 - f) \\ + \frac{\partial^3\beta}{\partial u^2\partial v_1}v_1^2 + \frac{\partial^2\beta}{\partial u^2}v_1 + 2\frac{\partial^3\beta}{\partial x\partial u\partial v_1}v_1 + \frac{\partial^2\beta}{\partial x\partial u} + \frac{\partial^3\beta}{\partial x^2\partial v_1}\Big)v_{11} + \frac{\partial^3\beta}{\partial u^3}v_1^3 \\ + \Big(3\frac{\partial^3\beta}{\partial x\partial u^2} - 3\frac{\partial^2\beta}{\partial u\partial v_1}u - \frac{\partial\beta}{\partial v_1}\Big)v_1^2 + \Big(\beta + \frac{\partial f}{\partial u}\frac{\partial\beta}{\partial v_1} + 3\frac{\partial^2\beta}{\partial u\partial v_1}f - \\ 3\frac{\partial^2\beta}{\partial x\partial v_1}u + 3\frac{\partial^3\beta}{\partial x^2\partial u}\Big)v_1 - \Big(3\frac{\partial^2\beta}{\partial v_1^2}v_{11} + 3\frac{\partial^2\beta}{\partial u\partial v_1}v_1 + 3\frac{\partial^2\beta}{\partial x\partial v_1} - \alpha'\Big)v_2 \\ - \frac{\partial f}{\partial u}\beta + \frac{\partial f}{\partial t}\alpha + \frac{\partial f}{\partial x}\frac{\partial\beta}{\partial v_1} + f\frac{\partial\beta}{\partial u} + \frac{\partial\beta}{\partial t} + \frac{\partial\beta}{\partial x}u + 3f\frac{\partial^2\beta}{\partial x\partial v_1} + \frac{\partial^3\beta}{\partial x^3} = 0.$$

From the coefficient of v_2, we first obtain

$$\frac{\partial^2\beta}{\partial v_1^2} = 0 \quad \text{and} \quad \beta = \phi(x,t,u)v_1 + \psi(x,t,u),$$

then the relation

$$3\frac{\partial\phi}{\partial u}v_1 + \Big(3\frac{\partial\phi}{\partial x} - \alpha'(t)\Big) = 0$$

leads to

$$\frac{\partial\phi}{\partial u} = 0, \quad 3\frac{\partial\phi}{\partial x} - \alpha'(t) = 0$$

from which we obtain $\phi = \phi(x,t)$, and

$$\phi(x,t) = \frac{1}{3}\alpha'(t)x + \gamma(t).$$

If we insert these relations into the above equation, we see that coefficients of v_{11}^3 and v_{11}^2 vanish automatically while the coefficient of v_{11} gives

$$\frac{\partial^2\psi}{\partial u^2}v_1 + \frac{\partial^2\psi}{\partial x\partial u} = 0$$

whence we get

$$\psi(x,t,u) = \mu(t)u + \nu(x,t).$$

If we introduce this function into the remaining expression above and set the coefficient of v_1 to zero, we get

$$\Big(\mu(t) - \frac{2}{3}\alpha'(t)\Big)u + \nu(x,t) + \gamma'(t) + \frac{1}{3}\alpha''(t)x = 0$$

from which we obtain

$$\mu(t) = \frac{2}{3}\alpha'(t),\ \ \nu(x,t) = -\frac{1}{3}\alpha''(t)x - \gamma'(t).$$

The remaining term imposes the following restriction on the admissible forms of the functions f, α and γ.

$$\Big(\frac{1}{3}\alpha'(t)x + \gamma(t)\Big)\frac{\partial f}{\partial x} + \alpha(t)\frac{\partial f}{\partial t} + \Big(\frac{1}{3}\alpha''(t)x + \gamma'(t) - \frac{2}{3}\alpha'(t)u\Big)\frac{\partial f}{\partial u}$$
$$+ \frac{5}{3}\alpha'(t)f + \frac{1}{3}\alpha''(t)u - \frac{1}{3}\alpha'''(t)x - \gamma''(t) = 0$$

in order that a nontrivial symmetry group exists. Together with this side condition, the function F is expressible as

$$F = \Big(\frac{1}{3}\alpha'(t)x + \gamma(t)\Big)v_1 + \alpha(t)v_2 + \frac{2}{3}\alpha'(t)u - \frac{1}{3}\alpha''(t)x - \gamma'(t)$$

depending on somewhat arbitrary functions $\alpha(t)$ and $\gamma(t)$. Therefore, isovectors are prolongation's of vectors V_G in tangent bundle of the graph space. Their components are given by

$$X^1 = -\frac{1}{3}\alpha'(t)x - \gamma(t), \quad X^2 = -\alpha(t),$$
$$U = \frac{2}{3}\alpha'(t)u - \frac{1}{3}\alpha''(t)x - \gamma'(t).$$

In homogeneous Korteweg-de Vries equation we have $f = 0$ so that the functions $\alpha(t)$ and $\gamma(t)$ ought to satisfy the additional constraint

$$\frac{1}{3}\alpha''(t)u - \frac{1}{3}\alpha'''(t)x - \gamma''(t) = 0$$

whence we immediately obtain

$$\alpha(t) = 3c_1 t + c_2, \quad \gamma(t) = c_3 t + c_4.$$

Hence, the relevant isovector components become

$$X^1 = -c_1 x - c_3 t - c_4, \; X^2 = -3c_1 t - c_2, \; U = 2c_1 u - c_3$$

Consequently, parts of linearly independent isovectors in the tangent bundle of the graph space are designated as follows

$$V^1 = -x\frac{\partial}{\partial x} - 3t\frac{\partial}{\partial t} + 2u\frac{\partial}{\partial u}, \quad V^2 = -\frac{\partial}{\partial t},$$
$$V^3 = -t\frac{\partial}{\partial x} - \frac{\partial}{\partial u}, \qquad V^4 = -\frac{\partial}{\partial x}.$$

As an example, let us determine the admissible form of the function f leading to these isovectors. Assuming $c_1 \neq 0$, we define new constants by $c_2/c_1 = a_2$, $c_3/c_1 = a_3$, $c_4/c_1 = a_4$. Then the general solution of the differential equation

$$(x + a_3 t + a_4)\frac{\partial f}{\partial x} + (3t + a_2)\frac{\partial f}{\partial t} + (a_3 - 2u)\frac{\partial f}{\partial u} + 5f = 0$$

is found by resorting the method of characteristics as

$$f(x,t,u) = (a_3 - 2u)^{5/2} g(\xi, \eta)$$

where the characteristic variables are defined by

$$\xi = \frac{x - \dfrac{1}{2}(a_2 a_3 - 2a_4 + a_3 t)}{(3t + a_2)^{1/3}},$$
$$\eta = (3t + a_2)^{2/3}(u - \frac{1}{2}a_3)$$

It is immediately observed that the isovector V^1 generates the *scaling transformation*

$$\overline{x} = \lambda x, \; \overline{t} = \lambda^3 t, \; \overline{u} = u/\lambda^2$$

with $\lambda = e^{-s}$ whereas the isovector V^3 produces the group

$$\overline{x} = x - st, \; \overline{t} = t, \; \overline{u} = u - s.$$

The isovectors V^2 and V^4 induce, respectively, translations in the temporal and spatial variables.

We have to determine the integral curves of the isovector field in order to derive a family of solutions from a known solution $u(x,t)$. The differential equations to be integrated are

$$\frac{d\overline{x}}{ds} = -\,c_2(a\overline{x} + b\overline{t} + c),$$
$$\frac{d\overline{t}}{ds} = -\,c_2(3a\overline{t} + 1),$$
$$\frac{d\overline{u}}{ds} = c_2(2a\overline{u} - b),\ c_2 \neq 0$$

where we have defined $c_1/c_2 = a, c_3/c_2 = b, c_4/c_2 = c$. These equations are to be solved under the initial conditions $\overline{x}(0) = x, \overline{t}(0) = t, \overline{u}(0) = u$. This solution is easily found as

$$\overline{x}(s) = \frac{2(b-3ac) + 3\big[2a^2x + 2ac - b(1+at)\big]e^{-c_2s} + b(1+3at)e^{-3c_2s}}{6a^2}$$
$$\overline{t}(s) = \frac{(1+3at)e^{-3c_2s} - 1}{3a},$$
$$\overline{u}(s) = \frac{\big[2au(x,t) - b\big]e^{2c_2s} + b}{2a}$$

■

Example 9.4.3. As an example to the case $N > 1$, we shall treat the ***boundary layer*** equations associated with a semi-infinite flat plate along x-axis placed in a unidirectional flow of an incompressible viscous fluid. The field equations governing this flow are given by

$$\nu\frac{\partial^2 u}{\partial y^2} - u\frac{\partial u}{\partial x} - v\frac{\partial u}{\partial y} + \mathcal{U}(x)\mathcal{U}'(x) = 0,\quad \frac{\partial u}{\partial x} + \frac{\partial v}{\partial y} = 0$$

where u and v are velocity components along x- and y-axes, respectively, and the constant ν is the kinematic viscosity. $\mathcal{U}(x)$ is the velocity field in the direction of x-axis before the plate is installed into the flow. The second equation above represents the incompressibility condition. Boundary layer equations are highly useful approximations to exact equations of viscous flow known as Navier-Stokes equations [*see* Exercise **9.15**] Let us denote $x^1 = x$, $x^2 = y$, $u^1 = u$, $u^2 = v$, $v_1^1 = u_{,1}$, $v_2^1 = u_{,2}$, $v_1^2 = v_{,1}$ and $v_2^2 = v_{,2}$. Field equations then become

$$\nu\frac{\partial v_2^1}{\partial x^2} - u^1v_1^1 - u^2v_2^1 + f(x^1) = 0,\quad v_1^1 + v_2^2 = 0 \qquad (9.4.26)$$

where $f = \mathcal{U}\mathcal{U}'$. Since the variable v_2^2 is eliminated by $v_2^2 = -\,v_1^1$, the isovector field can be taken as

$$V = X^1\frac{\partial}{\partial x^1} + X^2\frac{\partial}{\partial x^2} + U^1\frac{\partial}{\partial u^1} + U^2\frac{\partial}{\partial u^2} + V_1^1\frac{\partial}{\partial v_1^1} + V_2^1\frac{\partial}{\partial v_1^1} + V_1^2\frac{\partial}{\partial v_1^2}.$$

Indeed, the relation $dv_1^1 + dv_2^2 = 0$ yields $V_2^2 = -V_1^1$. The functions X^1, X^2, U^1 and U^2 depend only on the variables x^1, x^2, u^1, u^2. The other components of the isovector field of the contact ideal follow from $(9.3.21)_1$:

$$\begin{aligned}
V_1^1 &= \frac{\partial U^1}{\partial x^1} - \Big(\frac{\partial X^1}{\partial x^1} - \frac{\partial U^1}{\partial u^1}\Big)v_1^1 + \frac{\partial U^1}{\partial u^2}v_1^2 - \frac{\partial X^2}{\partial x^1}v_2^1 - \frac{\partial X^1}{\partial u^1}(v_1^1)^2 \\
&\quad - \frac{\partial X^2}{\partial u^1}v_1^1 v_2^1 - \frac{\partial X^1}{\partial u^2}v_1^1 v_1^2 - \frac{\partial X^2}{\partial u^2}v_2^1 v_1^2 \\
V_2^2 &= \frac{\partial U^2}{\partial x^2} - \Big(\frac{\partial X^2}{\partial x^2} - \frac{\partial U^2}{\partial u^2}\Big)v_2^2 + \frac{\partial U^2}{\partial u^1}v_2^1 - \frac{\partial X^1}{\partial x^2}v_1^2 - \frac{\partial X^2}{\partial u^2}(v_2^2)^2 \\
&\quad - \frac{\partial X^2}{\partial u^1}v_2^2 v_2^1 - \frac{\partial X^1}{\partial u^2}v_2^2 v_1^2 - \frac{\partial X^1}{\partial u^1}v_2^1 v_1^2 \\
V_2^1 &= \frac{\partial U^1}{\partial x^2} - \frac{\partial X^1}{\partial x^2}v_1^1 - \Big(\frac{\partial X^2}{\partial x^2} - \frac{\partial U^1}{\partial u^1}\Big)v_2^1 + \frac{\partial U^1}{\partial u^2}v_2^2 - \frac{\partial X^1}{\partial u^2}v_1^1 v_2^2 \\
&\quad - \frac{\partial X^2}{\partial u^1}(v_2^1)^2 - \frac{\partial X^1}{\partial u^1}v_1^1 v_2^1 - \frac{\partial X^2}{\partial u^2}v_2^2 v_2^1 \\
V_1^2 &= \frac{\partial U^2}{\partial x^1} - \frac{\partial X^2}{\partial x^1}v_2^2 + \frac{\partial U^2}{\partial u^1}v_1^1 - \Big(\frac{\partial X^1}{\partial x^1} - \frac{\partial U^2}{\partial u^2}\Big)v_1^2 - \frac{\partial X^2}{\partial u^1}v_1^1 v_2^2 \\
&\quad - \frac{\partial X^1}{\partial u^2}(v_1^2)^2 - \frac{\partial X^1}{\partial u^1}v_1^1 v_1^2 - \frac{\partial X^2}{\partial u^2}v_2^2 v_1^2
\end{aligned} \tag{9.4.27}$$

If we replace v_2^2 in these expressions by $-v_1^1$, we see that the condition the $V_2^2 + V_1^1 = 0$ is fulfilled provided that equations below are satisfied

$$\begin{aligned}
\frac{\partial U^1}{\partial x^1} + \frac{\partial U^2}{\partial x^2} &= 0, \\
\frac{\partial X^1}{\partial x^1} - \frac{\partial X^2}{\partial x^2} - \frac{\partial U^1}{\partial u^1} + \frac{\partial U^2}{\partial u^2} &= 0, \\
\frac{\partial X^1}{\partial x^2} - \frac{\partial U^1}{\partial u^2} &= 0, \\
\frac{\partial X^2}{\partial x^1} - \frac{\partial U^2}{\partial u^1} &= 0, \\
\frac{\partial X^1}{\partial u^1} + \frac{\partial X^2}{\partial u^2} &= 0.
\end{aligned} \tag{9.4.28}$$

In view of the balance equation $(9.4.26)_1$, we have to take

$$\Sigma^{1i} = \nu v_2^1 \delta_2^i, \quad \Sigma^1 = -u^1 v_1^1 - u^2 v_2^1 + f.$$

We thus find $V(\Sigma^{11}) = 0$ and

$$\begin{aligned} V(\Sigma^{12}) &= \nu V_2^1, \\ V(\Sigma^1) &= -v_1^1 U^1 - v_2^1 U^2 - u^1 V_1^1 - u^2 V_2^1 + X^1 f'. \end{aligned}$$

We then obtain from (9.4.8) that

$$\begin{aligned} A^1 &= X^1 f' - v_1^1 U^1 - v_2^1 U^2 - u^1 V_1^1 - u^2 V_2^1 + \nu \frac{\partial V_2^1}{\partial x^2} \\ &\quad + (f - u^1 v_1^1 - u^2 v_2^1)\Big(\frac{\partial X^1}{\partial x^1} + \frac{\partial X^2}{\partial x^2}\Big), \\ A_\beta^{1i} &= \nu \frac{\partial V_2^1}{\partial u^\beta}\, \delta_2^i + (f - u^1 v_1^1 - u^2 v_2^1)\frac{\partial X^i}{\partial u^\beta}, \\ A_\beta^{1ij} &= \nu \Big[\frac{\partial V_2^1}{\partial v_j^\beta}\, \delta_2^i + \Big(\frac{\partial X^1}{\partial x^1} + \frac{\partial X^2}{\partial x^2}\Big)\, \delta_2^i \delta_2^j \delta_\beta^1 - \frac{\partial X^i}{\partial x^2}\, \delta_2^j \delta_\beta^1\Big],\ A_{\beta\gamma}^{1ij} = 0, \\ 2A_{\beta\gamma}^{1ijk} &= -2A_{\beta\gamma}^{1jik} = \nu \Big(\frac{\partial X^j}{\partial u^\gamma}\, \delta_2^i - \frac{\partial X^i}{\partial u^\gamma}\, \delta_2^j\Big)\, \delta_2^k \delta_\beta^1. \end{aligned}$$

Hence, the equations (9.4.11) can now be written as

$$\begin{aligned} \lambda \Big(\frac{\partial \Sigma^{1i}}{\partial x^i} + \frac{\partial \Sigma^{1i}}{\partial u^\alpha} v_i^\alpha + \Sigma^1\Big) &= A^1 + A_\beta^{1i} v_i^\beta, \\ \lambda \Big(\frac{\partial \Sigma^{1i}}{\partial v_j^\beta} + \frac{\partial \Sigma^{1j}}{\partial v_i^\beta}\Big) &= A_\beta^{1ij} + A_\beta^{1ji} + 2(A_{\beta\gamma}^{1ikj} + A_{\beta\gamma}^{1jki}) v_k^\gamma \end{aligned}$$

whose explicit forms become

$$\begin{aligned} \lambda(f - u^1 v_1^1 - u^2 v_2^1) &= X^1 f' - v_1^1 U^1 - v_2^1 U^2 - u^1 V_1^1 - u^2 V_2^1 + \nu \frac{\partial V_2^1}{\partial x^2} \\ &\quad + \nu \frac{\partial V_2^1}{\partial u^\beta} v_2^\beta + (f - u^1 v_1^1 - u^2 v_2^1)\Big(\frac{\partial X^1}{\partial x^1} + \frac{\partial X^2}{\partial x^2} + \frac{\partial X^i}{\partial u^\beta} v_i^\beta\Big) \quad (9.4.29) \\ 2\lambda\, \delta_2^i \delta_2^j \delta_\beta^1 &= \frac{\partial V_2^1}{\partial v_j^\beta}\, \delta_2^i + \frac{\partial V_2^1}{\partial v_i^\beta}\, \delta_2^j \\ &\quad + 2\Big(\frac{\partial X^1}{\partial x^1} + \frac{\partial X^2}{\partial x^2} + \frac{\partial X^k}{\partial u^\gamma} v_k^\gamma\Big)\delta_2^i \delta_2^j \delta_\beta^1 \\ &\quad - \Big(\frac{\partial X^i}{\partial x^2} + \frac{\partial X^i}{\partial u^\gamma} v_2^\gamma\Big)\delta_2^j \delta_\beta^1 - \Big(\frac{\partial X^j}{\partial x^2} + \frac{\partial X^j}{\partial u^\gamma} v_2^\gamma\Big)\delta_2^i \delta_\beta^1 \end{aligned}$$

The second set of equations above leads immediately to

$$\begin{aligned}
\lambda &= \frac{\partial V_2^1}{\partial v_2^1} + \frac{\partial X^1}{\partial x^1} + \frac{\partial X^1}{\partial u^1}v_1^1 + \frac{\partial X^1}{\partial u^2}v_1^2 \qquad (9.4.30)\\
&= \frac{\partial U^1}{\partial u^1} + \frac{\partial X^1}{\partial x^1} - \frac{\partial X^2}{\partial x^2} + \frac{\partial X^2}{\partial u^2}v_1^1 - 2\frac{\partial X^2}{\partial u^1}v_2^1 + \frac{\partial X^1}{\partial u^2}v_1^2\\
0 &= \frac{\partial V_2^1}{\partial v_1^1} - \frac{\partial X^1}{\partial x^2} - \frac{\partial X^1}{\partial u^1}v_2^1 + \frac{\partial X^1}{\partial u^2}v_1^1,\\
0 &= \frac{\partial V_2^1}{\partial v_2^2} = \frac{\partial V_2^1}{\partial v_1^2}.
\end{aligned}$$

Because we had replaced v_2^2 by $-v_1^1$ the equations $(9.4.30)_3$ are satisfied identically whereas $(9.4.30)_2$ gives

$$2\frac{\partial X^1}{\partial x^2} + \frac{\partial U^1}{\partial u^2} - 3\frac{\partial X^1}{\partial u^2}v_1^1 + \Big(2\frac{\partial X^1}{\partial u^1} - \frac{\partial X^2}{\partial u^2}\Big)v_2^1 = 0$$

whence we extract the relations

$$\frac{\partial X^1}{\partial u^2} = 0,\ \frac{\partial U^1}{\partial u^2} = -2\frac{\partial X^1}{\partial x^2},\ \frac{\partial X^2}{\partial u^2} = 2\frac{\partial X^1}{\partial u^1}.$$

If we insert these results into equations $(9.4.28)_{3,5}$, we find that

$$\frac{\partial X^1}{\partial x^2} = \frac{\partial X^1}{\partial u^1} = \frac{\partial X^2}{\partial u^2} = \frac{\partial U^1}{\partial u^2} = 0.$$

Consequently, at this stage we reach to the following components

$$\begin{aligned}
&X^1 = \xi(x^1),\ X^2 = X^2(x^1, x^2, u^1),\\
&U^1 = U^1(x^1, x^2, u^1).
\end{aligned}$$

Introducing these together with the relation $(9.4.30)_1$ into $(9.4.29)_1$, taking $v_2^2 = -v_1^1$ and arranging the resulting expression, we conclude that

$$\begin{aligned}
&\nu\frac{\partial^2 X^2}{\partial (u^1)^2}(v_2^1)^3 - \Big(\nu\frac{\partial^2 U^1}{\partial (u^1)^2} - 2u^2\frac{\partial X^2}{\partial u^1} - 2\nu\frac{\partial^2 X^2}{\partial x^2 \partial u^1}\Big)(v_2^1)^2 + 2u^1\frac{\partial X^2}{\partial u^1}v_1^1 v_2^1\\
&-\Big(\nu\frac{\partial^2 X^2}{\partial (x^2)^2} + u^1\frac{\partial X^2}{\partial x^1} - u^2\frac{\partial X^2}{\partial x^2} + 3f\frac{\partial X^2}{\partial u^1} + 2\nu\frac{\partial^2 U^1}{\partial x^2 \partial u^1} - U^2\Big)v_2^1 +\\
&\Big[u^1\Big(2\frac{\partial X^2}{\partial x^2} - \frac{d\xi}{dx^1}\Big) + U^1\Big]v_1^1 - \nu\frac{\partial^2 U^1}{\partial (x^2)^2} + u^1\frac{\partial U^1}{\partial x^1} + u^2\frac{\partial U^1}{\partial x^2}\\
&\qquad + f\frac{\partial U^1}{\partial u^1} - 2f\frac{\partial X^2}{\partial x^2} - \xi f' = 0.
\end{aligned}$$

Equating the coefficients of powers of v^1_1 and v^1_2 to zero, we readily see that the following relations are to be satisfied

$$\begin{aligned}
&\frac{\partial X^2}{\partial u^1} = 0, \qquad \frac{\partial^2 U^1}{\partial (u^1)^2} = 0, \qquad\qquad (9.4.31)\\
&U^1 = \Big(\frac{dX^1}{dx^1} - 2\frac{\partial X^2}{\partial x^2}\Big)u^1,\ U^2 = u^1\frac{\partial X^2}{\partial x^1} - u^2\frac{\partial X^2}{\partial x^2},\\
&\frac{\partial U^1}{\partial x^2} = 0, \qquad \frac{\partial^2 X^2}{\partial (x^2)^2} = 0,\\
&0 = u^1\frac{\partial U^1}{\partial x^1} + u^2\frac{\partial U^1}{\partial x^2} + f\frac{\partial U^1}{\partial u^1} - 2f\frac{\partial X^2}{\partial x^2} - X^1 f'.
\end{aligned}$$

These relations imply that

$$\begin{aligned}
X^2 &= \alpha(x^1)x^2 + \beta(x^1),\\
U^1 &= \big[\xi'(x^1) - 2\alpha(x^1)\big]u^1,\\
U^2 &= \big[\alpha'(x^1)x^2 + \beta'(x^1)\big]u^1 - \alpha(x^1)u^2.
\end{aligned}$$

If we insert these expressions into equations (9.4.28), then the first equation yields

$$\xi''(x^1) - \alpha'(x^1) = 0 \ \text{ and } \ \alpha(x^1) = \xi'(x^1) + c_0.$$

The other equations are satisfied identically. The last equation (9.4.31) takes the form

$$\xi''(u^1)^2 + \xi f' + (3\alpha + c_0)f = 0$$

so that we obtain $\xi'' = 0$ and

$$\xi(x^1) = c_1x^1 + c_2,\ \alpha(x^1) = c_1 + c_0 = c_3.$$

Therefore, the relevant components of the isovector field are found as

$$\begin{aligned}
X^1 &= c_1x^1 + c_2, & X^2 &= c_3x^2 + \beta(x^1),\\
U^1 &= (c_1 - 2c_3)u^1, & U^2 &= \beta'(x^1)u^1 - c_3u^2.
\end{aligned} \qquad (9.4.32)$$

We see that the function f must satisfy $(c_1x^1 + c_2)f' + (4c_3 - c_1)f = 0$. On assuming $c_1 \neq 0$ and writing $c_2/c_1 = a, c_3/c_1 = b$ we realise that f has to be chosen in the form

$$\frac{1}{2}(\mathcal{U}^2)' = f(x^1) = A(x^1 + a)^{1-4b}$$

to be admitted by the symmetry group. Thus, the admissible velocity field is

$$\mathcal{U}(x^1) = \sqrt{B + \frac{A(x^1 + a)^{2(1-2b)}}{1 - 2b}}.$$

If we take $c_1 = 0$, then we get $f(x^1) = Ae^{-cx^1}$ where $c = 4c_3/c_2$. Linearly independent isovectors are then given by

$$V^1 = x^1 \frac{\partial}{\partial x^1} + u^1 \frac{\partial}{\partial u^1},$$
$$V^2 = \frac{\partial}{\partial x^1},$$
$$V^3 = x^2 \frac{\partial}{\partial x^2} - 2u^1 \frac{\partial}{\partial u^1} - u^2 \frac{\partial}{\partial u^2},$$
$$V_\beta = \beta(x^1) \frac{\partial}{\partial x^2} + \beta'(x^1) u^1 \frac{\partial}{\partial u^2}. \blacksquare$$

Example 9.4.4. As an example to a problem to determine symmetries of which proves to be quite difficult by using classical methods, we consider the equations governing the motion of a hyperelastic body[1]. In order to simplify the discussion, we shall employ Cartesian coordinates. We had denoted the location of a particle in the undeformed body by *material coordinates* X_K with $K = 1, 2, 3$ and the location of the same point at time t by *spatial coordinates* x_k with $k = 1, 2, 3$ [*see* p. 453]. The motion of the body was specified by a diffeomorphism represented by relations $x_k = x_k(\mathbf{X}, t)$. We know that a homogeneous hyperelastic medium is characterised by a given *stress potential* $\Sigma(\mathbf{C})$ where $\mathbf{C} = \mathbf{F}^{\mathrm{T}}\mathbf{F}$ is the deformation tensor. $\mathbf{F} = [x_{k,K}]$ is the tensor of deformation gradients. Here, we use the notation

$$F_{kK} = x_{k,K} = \frac{\partial x_k}{\partial X_K}$$

The equations of motion of an hyperelastic material are designated by

$$\frac{\partial}{\partial X_K}\Big(\frac{\partial \Sigma}{\partial x_{k,K}}\Big) - \rho_0 \frac{\partial v_k}{\partial t} = 0, \; v_k = \frac{\partial x_k}{\partial t} \tag{9.4.33}$$

in the absence of body forces [*see* (8.7.4-5)]. These equations will turn out to be an example to the case $m = 1, n = 4, N = 3$. In order to utilise directly the determining equations (9.4.11), let us introduce the notations

---

[1]Adetailed discussion of this problem involving heterogenous materials can be found in the following work: Şuhubi, E. S. and A. Bakkaloğlu, Symmetry groups for arbitrary motions of hyperelastic solids, *International Journal of Engineering Science*, **35**, 637-657, 1997.

$$X_4 = t,\ \ \Sigma_{kK}(\mathbf{F}) = \frac{\partial \Sigma}{\partial F_{kK}},\ \ \Sigma_{k4}(\mathbf{v}) = -\rho_0 v_k.$$

Hence, the equations of motion are reduced to the form

$$\frac{\partial \Sigma_{kK}}{\partial X_K} + \frac{\partial \Sigma_{k4}}{\partial X_4} = 0.$$

We take the isovector field as follows

$$V = -\Phi_K \frac{\partial}{\partial X_K} - \Psi \frac{\partial}{\partial t} + \Omega_k \frac{\partial}{\partial x_k} + V_{kK} \frac{\partial}{\partial F_{kK}} + V_k \frac{\partial}{\partial v_k}$$

where the functions Φ_K, Ψ, Ω_k depend only on the variables X_K, t, x_k and for the components V_{kK} and V_k, we have the expressions below

$$V_{kK} = \frac{\partial \Omega_k}{\partial X_K} + F_{kL}\frac{\partial \Phi_L}{\partial X_K} + v_k \frac{\partial \Psi}{\partial X_K} + F_{lK}\frac{\partial \Omega_k}{\partial x_l} + F_{kL}F_{lK}\frac{\partial \Phi_L}{\partial x_l} + F_{lK}v_k \frac{\partial \Psi}{\partial x_l}$$

$$V_k = \frac{\partial \Omega_k}{\partial t} + F_{kL}\frac{\partial \Phi_L}{\partial t} + v_k \frac{\partial \Psi}{\partial t} + v_l \frac{\partial \Omega_k}{\partial x_l} + F_{kL}v_l \frac{\partial \Phi_L}{\partial x_l} + v_k v_l \frac{\partial \Psi}{\partial x_l}.$$

We thus obtain

$$B_{kK} = V(\Sigma_{kK}) = C_{kKlL}V_{lL},\ V(\Sigma_{k4}) = -\rho_0 V_k$$

where we had defined

$$C_{kKlL} = \frac{\partial \Sigma_{kK}}{\partial F_{lL}} = \frac{\partial^2 \Sigma}{\partial F_{kK}\partial F_{lL}} = C_{lLkK}.$$

We had already called the tensor $C_{kKlL}(\mathbf{F})$ enjoying the block symmetry shown above as the elasticities of the material in Example 8.7.4. Hence, the coefficients appearing in the determining equations (9.4.11) become

$$A_k = B_{kK,K} - \rho_0 \dot{V}_k,\ A_{Kkl} = B_{kK,l},\ A_{4kl} = -\rho_0 V_{k,l},$$

$$A_{KLkl} = \frac{\partial B_{kK}}{\partial F_{lL}} - C_{kKlL}(\Phi_{M,M} + \dot{\Psi}) + C_{kMlL}\Phi_{K,M},$$

$$A_{4Lkl} = -\rho_0 \frac{\partial V_k}{\partial F_{lL}} + C_{kKlL}\Psi_{,K},\ A_{K4kl} = \frac{\partial B_{kK}}{\partial v_l} - \rho_0 \dot{\Phi}_K\,\delta_{kl},$$

$$A_{44kl} = -\rho_0 \frac{\partial V_k}{\partial v_l} + \rho_0 \Phi_{K,K}\,\delta_{kl}, A_{4LMklm} = -A_{L4Mklm} = \frac{1}{2}C_{kLlM}\Psi_{,m},$$

$$A_{KLMklm} = -A_{LKMklm} = \frac{1}{2}(C_{kLlM}\Phi_{K,m} - C_{kKlM}\Phi_{L,m}),$$

$$A_{4L4klm} = -A_{L44klm} = \frac{1}{2}\rho_0 \Phi_{L,m}\,\delta_{kl}$$

where an overdot $(\dot{\cdot})$ denotes the derivative with respect to the time variable t. All other coefficients turn out be zero. Therefore, the equations (9.4.11) take the form

$$\begin{aligned}
\lambda_{kl}\frac{\partial \Sigma_{lK}}{\partial X_k} &= A_k + A_{Kkl}F_{lK} + A_{4kl}v_l \\
\lambda_{km}\Big(\frac{\partial \Sigma_{mK}}{\partial F_{lL}} + \frac{\partial \Sigma_{mL}}{\partial F_{lK}}\Big) &= A_{KLkl} + A_{LKkl} + 2(A_{KMLklm} + A_{LMKklm})F_{mM} \\
&\quad + 2(A_{K4Lklm} + A_{L4Kklm})v_m \\
-\rho_0\lambda_{kl} &= A_{44kl} + 2A_{4M4klm}F_{mM} \\
0 &= A_{K4kl} + A_{4Kkl} + 2A_{4LKklm}F_{mL} + 2A_{K44klm}v_m
\end{aligned}$$

where λ_{km} are arbitrary functions of the coordinates $(X_K, t, x_k, F_{kK}, v_k)$ of the contact manifold. The above equations thus yield the following result

$$\begin{aligned}
&B_{kK,K} - \rho_0\dot{V}_k + B_{kK,l}F_{lK} - \rho_0 V_{k,l}v_l = 0, \qquad (9.4.34)\\
&\lambda_{km}(C_{mKlL} + C_{mLlK}) = \frac{\partial B_{kK}}{\partial F_{lL}} + \frac{\partial B_{kL}}{\partial F_{lK}} - (C_{kKlL} + C_{kLlK})(\Phi_{M,M} + \dot{\Psi} \\
&\qquad + \Phi_{M,m}F_{mM} + \Psi_{,m}v_m) + C_{kMlL}\Phi_{K,M} + C_{kMlK}\Phi_{L,M} \\
&\qquad + (C_{kMlL}\Phi_{K,m} + C_{kMlK}\Phi_{L,m}]F_{mM},\\
&\lambda_{kl} = \frac{\partial V_k}{\partial v_l} - (\Phi_{K,K} + \Phi_{M,m}F_{mM})\,\delta_{kl}\\
&\frac{\partial B_{kK}}{\partial v_l} - \rho_0\dot{\Phi}_K\,\delta_{kl} - \rho_0\frac{\partial V_k}{\partial F_{lK}} + C_{kLlK}(\Psi_{,L} + \Psi_{,m}F_{mL}) - \rho_0\Phi_{K,m}v_m\delta_{kl} = 0
\end{aligned}$$

Because of the relations

$$\frac{\partial B_{kK}}{\partial v_l} = C_{kLlK}(\Psi_{,L} + \Psi_{,m}F_{mL}), \quad \frac{\partial V_k}{\partial F_{lK}} = (\dot{\Phi}_K + \Phi_{K,m}v_m)\delta_{kl}$$

the equation $(9.4.34)_4$ takes the form

$$(C_{kKlL} + C_{kLlK})(\Psi_{,L} + \Psi_{,m}F_{mL}) - 2\rho_0(\dot{\Phi}_K + \Phi_{K,m}v_m)\delta_{kl} = 0$$

Since the components $C_{kKlL} + C_{kLlK}$ are coefficients of the terms like $x_{k,KL}$ in the field equations (9.4.33), they cannot be all zero. It then follows from the above equation that

$$\Phi_{K,m} = 0, \;\; \dot{\Phi}_K = 0, \;\; \Psi_{,L} = 0, \;\; \Psi_{,m} = 0.$$

Thus, we must have $\Phi_K = \Phi_K(\mathbf{X})$, $\Psi = \Psi(t)$. In this case, we get

$$\begin{aligned}B_{kK} &= C_{kKlL}(\Omega_{l,L} + F_{lM}\Phi_{M,L} + F_{mL}\Omega_{l,m}),\\ \lambda_{kl} &= \Omega_{k,l} + (\dot{\Psi} - \Phi_{K,K})\delta_{kl}\end{aligned}$$

so that $(9.4.34)_1$ can be written as

$$\begin{aligned}&C_{kKlL}\big[\Omega_{l,LK} + F_{lM}\Phi_{M,LK} + F_{mL}\Omega_{l,mK} + (\Omega_{l,Lm} + F_{nL}\Omega_{l,mn})F_{mK}\big]\\ &\quad -\rho_0\big[\Omega_{k,lm}v_l v_m + (\ddot{\Psi}\delta_{kl} + 2\dot{\Omega}_{k,l})v_l + \ddot{\Omega}_k\big] = 0. \qquad (9.4.35)\end{aligned}$$

This requires that we have to take

$$\frac{\partial^2\Omega_k}{\partial x_l \partial x_m} = 0, \quad \frac{\partial^2\Omega_k}{\partial x_l \partial t} = -\frac{1}{2}\ddot{\Psi}(t)\,\delta_{kl}.$$

These equations yield easily

$$\Omega_k(\mathbf{X},t,\mathbf{x}) = -\frac{1}{2}\dot{\Psi}(t)x_k + \gamma_{kl}(\mathbf{X})x_l + \Gamma_k(\mathbf{X},t) = \Lambda_{kl}(\mathbf{X},t)x_l + \Gamma_k(\mathbf{X},t)$$

where we have defined

$$\Lambda_{kl}(\mathbf{X},t) = -\frac{1}{2}\dot{\Psi}(t)\delta_{kl} + \gamma_{kl}(\mathbf{X}).$$

Then the equations $(9.4.34)_2$ reduce to the form

$$\begin{aligned}&(C_{mKlL} + C_{mLlK})\Omega_{k,m} - (C_{kKmL} + C_{kLmK})\Omega_{m,l} + \{2(C_{kKlL} + C_{kLlK})\dot{\Psi}\\ &\quad - (C_{kKlM} + C_{kMlK})\Phi_{L,M} - (C_{kMlL} + C_{kLlM})\Phi_{K,M} \qquad (9.4.36)\\ &\quad - (C_{kKlLmM} + C_{kLlKmM})(\Omega_{m,M} + F_{mN}\Phi_{N,M} + F_{nM}\Omega_{m,n})\} = 0\end{aligned}$$

where we have introduced the tensor

$$C_{kKlLmM} = \frac{\partial^3\Sigma}{\partial F_{kK}\partial F_{lL}\partial F_{mM}}.$$

The block symmetries manifested by this tensor are obvious. It is plainly observed that the term within braces in the expression (9.4.36) is symmetric in indices k and l due to the block symmetry of the components C_{kKlL}. Hence, the antisymmetric part of that expression must satisfy the relations

$$(C_{mKlL} + C_{mLlK})(\Omega_{k,m} + \Omega_{m,k}) = (C_{mKkL} + C_{mLkK})(\Omega_{l,m} + \Omega_{m,l}).$$

Let us define matrices $\mathbf{A}^{(KL)}$ via $A_{kl}^{(KL)} = C_{kKlL} + C_{kLlK}$. These matrices are symmetric. Consequently, we obtain

$$\Omega_{(k,m)}A^{(KL)}_{ml} = A^{(KL)}_{km}\Omega_{(m,l)} \quad \text{or} \quad \Lambda_{(km)}A^{(KL)}_{ml} = A^{(KL)}_{km}\Lambda_{(ml)}$$

Thus, the symmetric part of the matrix $\mathbf{\Lambda}$ commutes with every symmetric matrix $\mathbf{A}^{(KL)}$. According to the well known *Schur lemma* of the group theory $\Lambda_{(kl)}$ can only be a multiple of the unit matrix. Therefore, on noting that $\Lambda_{[kl]}$ may be represented by an axial vector a_m, we can write

$$\Lambda_{kl} = \Lambda_{(kl)} + \Lambda_{[kl]} = \Lambda_0(\mathbf{X},t)\delta_{kl} + e_{klm}a_m(\mathbf{X},t)$$

whence we get

$$\gamma_{(kl)}(\mathbf{X}) = \Big[\Lambda_0(\mathbf{X},t) + \frac{1}{2}\dot{\Psi}(t)\Big]\delta_{kl} = \lambda_0(\mathbf{X})\delta_{kl},$$
$$\gamma_{[kl]}(\mathbf{X}) = e_{klm}a_m(\mathbf{X}).$$

We thus conclude that

$$\Omega_k(\mathbf{X},t,\mathbf{x}) = \Big[\lambda_0(\mathbf{X}) - \frac{1}{2}\dot{\Psi}(t)\Big]x_k + e_{klm}a_m(\mathbf{X})x_l + \Gamma_k(\mathbf{X},t).$$

If we insert this expression into (9.4.35-36) and equate the coefficients of x_n to zero, we get $\dddot{\Psi} = 0$ together with

$$\lambda_{0,M}\,\delta_{mn} + e_{mnr}a_{r,M} = 0$$

and consequently $\lambda_{0,M} = 0,\, a_{r,M} = 0$ leading to

$$\lambda_0(\mathbf{X}) = a_0, \quad a_k(\mathbf{X}) = a_k,$$
$$\Psi(t) = b_1t^2 + 2b_2t + b_3.$$

The equation (9.4.35) now takes the form

$$C_{kKlL}(\Gamma_{l,LK} + F_{lN}\Phi_{N,LK}) - \rho_0\ddot{\Gamma}_k = 0.$$

We differentiate this expression with respect to F_{mM} to obtain

$$C_{kKlLmM}(\Gamma_{l,LK} + F_{lN}\Phi_{N,LK}) + C_{kKmL}\Phi_{M,LK} = 0$$

that can be satisfied for any non-linear elastic material if only $\Phi_{M,LK} = 0$ and $\Gamma_{l,LK} = 0$. This of course implies that $\ddot{\Gamma}_k = 0$. We thus easily obtain

$$\Phi_K(\mathbf{X}) = A_{KL}X_L + B_K,$$
$$\Gamma_k(\mathbf{X},t) = (\alpha_{kK}t + \beta_{kK})X_K + \mu_kt + \nu_k.$$

Replacing the functions in (9.4.36) by the above expressions, we see at once that the coefficient of the variable t vanishes if only we take $b_1 = 0$ and

$\alpha_{kK} = 0$. If we take into consideration the identity

$$F_{nN}\, C_{kKlLmM} = F_{nN}\frac{\partial^3\Sigma}{\partial F_{kK}\partial F_{lL}\partial F_{mM}} = \frac{\partial^2}{\partial F_{kK}\partial F_{lL}}\Big(F_{nN}\frac{\partial\Sigma}{\partial F_{mM}}\Big) \\ - C_{kKmM}\,\delta_{nl}\delta_{NL} - C_{lLmM}\,\delta_{nk}\delta_{NK}$$

the remaining terms in the expressions (9.4.36) can be arranged in the following manner

$$\frac{\partial^2\mathcal{F}}{\partial F_{kK}\partial F_{lL}} + \frac{\partial^2\mathcal{F}}{\partial F_{kL}\partial F_{lK}} = 0 \tag{9.4.37}$$

where the function $\mathcal{F}$ is defined as

$$\mathcal{F} = \big[A_{NM}F_{mN} + (a_0 - b_2)F_{mM} + e_{klm}a_l F_{kM} + \beta_{mM}\big]\frac{\partial\Sigma}{\partial F_{mM}} \\ - 2(a_0 + b_2)\Sigma.$$

A rather straightforward but somewhat tedious calculation for details of which we may refer to the work cited above shows that the solution of the equations (9.4.37) is expressible as

$$\mathcal{F} = \frac{1}{3!}\gamma e_{klm}e_{KLM}F_{kK}F_{lL}F_{mM} + \frac{1}{2}\gamma_{mM}e_{klm}e_{KLM}F_{kK}F_{lL} \\ + \delta_{kK}F_{kK} + \delta.$$

If we recall identities

$$J = \det\mathbf{F} = \frac{1}{3!}e_{klm}e_{KLM}F_{kK}F_{lL}F_{mM}$$
$$\frac{\partial J}{\partial F_{kK}} = JF^{-1}_{Kk} = \frac{1}{2}e_{klm}e_{KLM}F_{kK}F_{lL}$$

where F^{-1}_{Kk} are entries of the inverse matrix $\mathbf{F}^{-1}$, $\mathcal{F}$ can also be written in the form

$$\mathcal{F} = \gamma J + \gamma_{kK}JF^{-1}_{Kk} + \delta_{kK}F_{kK} + \delta.$$

We had emphasised the fact that the stress potential Σ is actually dependent on the components $C_{KL} = F_{kK}F_{kL}$ of the deformation tensor. Once this transformation is fulfilled, we see that we have to take $\beta_{mM} = \gamma_{kK} = \delta_{kK} = 0$ in order to remove the dependence on $\mathbf{F}$. The relevant components of the isovector field are then given by

$$\Phi_K = A_{KL}X_L + B_K,$$

$$\Psi = 2b_2 t + b_3,$$
$$\Omega_k = (a_0 - b_2)x_k + e_{klm}a_m x_l + \mu_k t + \nu_k$$

and the admissible functions Σ must be solutions of the equation

$$2\big[A_{ML} + (a_0 - b_2)\delta_{ML}\big]C_{KM}\frac{\partial \Sigma}{\partial C_{KL}} - 2(a_0 + b_2)\Sigma = \gamma\sqrt{\det \mathbf{C}} + \delta.$$

In the components Ω_k, the terms ν_k indicate that the space is homogeneous whereas the terms $e_{klm}a_m x_l = -(\mathbf{a} \times \mathbf{x})_k$ imply that the space is isotropic. The terms $(a_0 - b_2)x_k + \mu_k t$ mean that the field equations are invariant under a Galilean transformation [Italian physicist and astronomer Galileo Galilei (1564-1642)]. Of course, these symmetry groups must be present in all classical mechanical system modelled correctly. ■

9.5. SIMILARITY SOLUTIONS

As we have mentioned several times we can produce a new family of solutions from a known solution of a system of partial differential equations if we possess an isovector field of the fundamental ideal generating a symmetry group of transformations. In this section, however, we shall try to determine structural properties of certain solutions that remain *invariant* under a particular symmetry group. If a mapping $\phi : \mathcal{D}_n \to \mathcal{C}_m$ corresponds to a solution to a system of partial differential equations which remains invariant with respect to an isovector field V, then it has to satisfy the requirement $\phi_V(t) \circ \phi = e^{tV} \circ \phi = \phi$. Let us suppose that such a solution is given in the form $f^\alpha = \phi^\alpha(\mathbf{x}) - u^\alpha = 0$. *If these functions are to be invariant under the flow generated by an isovector V, then it must satisfy the condition*

$$\pounds_V f^\alpha = V(f^\alpha) = 0$$

[*see* (2.9.14)]. This means that a ***group-invariant solution***, in other words, a ***similarity solution*** must satisfy the system of quasilinear partial differential equations

$$X^i\big(x^j, \phi^\beta(\mathbf{x})\big)\frac{\partial \phi^\alpha}{\partial x^i} - U^\alpha\big(x^j, \phi^\beta(\mathbf{x})\big) = 0 \qquad (9.5.1)$$

when $N > 1$, or the non-linear partial differential equation

$$X^i\big(x^j, \phi(\mathbf{x}), \phi(\mathbf{x})_{,j}\big)\frac{\partial \phi}{\partial x^i} - U\big(x^j, \phi(\mathbf{x}), \phi(\mathbf{x})_{,j}\big) = 0 \qquad (9.5.2)$$

when $N = 1$. These partial differential equations usually specify the structure of a similarity solution in the following manner

$$u^\alpha = \phi^\alpha(\xi^a), \quad a = 1, 2, \ldots, p < n$$

where ξ^a are known functions of independent variables x^i. After having installed these functions in the original system of differential equations we are led to a new set of partial differential equations with a smaller number of novel independent variables since x^i are replaced by ξ^a. That is why we expect that to solve them may be somewhat easier compared to original equations. If we manage to find a solution of these equations we then reach to the functional form of a similarity solution. Inserting this form into original field equation we can find an explicit solution. If we denote a solution of a given system of partial differential equations by a regular mapping $\phi : \mathcal{D}_n \subseteq \mathbb{R}^n \to \mathcal{C}_m$, then the relations (9.5.1) or (9.5.2) require that the mapping ϕ must satisfy the condition

$$\phi^*\big(\mathbf{i}_V(\sigma^\alpha)\big) = 0 \tag{9.5.3}$$

to be a similarity solution associated with an isovector V. In fact, this result follows immediately from $\mathbf{i}_V(\sigma^\alpha) = U^\alpha - v_i^\alpha X^i$ and $\phi^* v_i^\alpha = u_{,i}^\alpha$. Actually, we readily observe that a similarity solution satisfies as well the condition $\phi^*\mathbf{i}_V(\mathcal{I}_m) = 0$ where $\mathcal{I}_m$ is the contact ideal defined in (9.3.1). If we take into account the relations $V^\alpha_{i_1\cdots i_k i} = D_i^{(k)}(V^\alpha_{i_1\cdots i_k}) - v^\alpha_{i_1\cdots i_k j} D_i^{(k)}(X^j)$ for isovector components of the contact ideal given by (9.3.19) in the expression $\mathbf{i}_V(\sigma^\alpha_{i_1 i_2\cdots i_r}) = V^\alpha_{i_1\cdots i_r} - v^\alpha_{i_1\cdots i_r i} X^i$, we get

$$\begin{aligned}\phi^*\big(\mathbf{i}_V(\sigma^\alpha_{i_1\cdots i_r})\big) &= (\phi^* V^\alpha_{i_1\cdots i_{r-1}})_{,i_r} - u^\alpha_{,i_1\cdots i_{r-1} i}(\phi^* X^i)_{,i_r} - u^\alpha_{,i_1\cdots i_r i}\phi^* X^i \\ &= (\phi^* V^\alpha_{i_1\cdots i_{r-1}} - u^\alpha_{,i_1\cdots i_{r-1} i}\phi^* X^i)_{,i_r} = \big[\phi^*\big(\mathbf{i}_V(\sigma^\alpha_{i_1\cdots i_{r-1}})\big)\big]_{,i_r}.\end{aligned}$$

If we continue to utilise this recurrence relation successively, we finally obtain

$$\phi^*\big(\mathbf{i}_V(\sigma^\alpha_{i_1\cdots i_r})\big) = \big[\phi^*\big(\mathbf{i}_V(\sigma^\alpha)\big)\big]_{,i_1\cdots i_r} = 0$$

where $r = 0, 1, \ldots, m-1$. We thus conclude that

$$\phi^* V^\alpha_{i_1\cdots i_r} = (\phi^* X^i)\, u^\alpha_{,i_1\cdots i_r i}, \; r = 0, 1, \ldots, m-1.$$

On the other hand, we have

$$\begin{aligned}\phi^*\big(\mathbf{i}_V(d\sigma^\alpha_{i_1\cdots i_{m-1}})\big) &= -\,\phi^*(V^\alpha_{i_1\cdots i_{m-1} i}\, dx^i - X^i dv^\alpha_{i_1\cdots i_{m-1} i}) \\ &= -\,\big(\phi^* V^\alpha_{i_1\cdots i_{m-1} i} - (\phi^* X^j) u^\alpha_{,i_1\cdots i_{m-1} i j}\big) dx^i.\end{aligned}$$

But inserting the relation

$$\phi^* V^\alpha_{i_1\cdots i_{m-1}i} = (\phi^* V_{i_1\cdots i_{m-1}})_{,i} - (\phi^* X^j)_{,i} u^\alpha_{,i_1\cdots i_{m-1}j}$$

into the foregoing expression, we find

$$\phi^*\big(\mathbf{i}_V(d\sigma^\alpha_{i_1\cdots i_{m-1}})\big) = \big(\phi^* V^\alpha_{i_1\cdots i_{m-1}} - (\phi^* X^j)\, u^\alpha_{,i_1\cdots i_{m-1}j}\big)_{,i} dx^i = 0$$

whence we draw the conclusion $\phi^*\mathbf{i}_V(\mathcal{I}_m) = 0$. Next, we consider the balance ideal $\mathfrak{I}_m$ and write

$$\begin{aligned}\phi^*\big(\mathbf{i}_V(\omega^\alpha)\big) &= \phi^*\big(\mathbf{i}_V(d\Sigma^{\alpha i} \wedge \mu_i + \Sigma^\alpha \mu)\big)\\ &= \phi^*\big(V(\Sigma^{\alpha i})\mu_i - X^j d\Sigma^{\alpha i} \wedge \mu_{ji} + X^j \Sigma^\alpha \mu_j)\big).\end{aligned}$$

However, because of the relations

$$\begin{aligned}\phi^*\big(V(\Sigma^{\alpha i})\big) &= \phi^*\Big[X^j \frac{\partial \Sigma^{\alpha i}}{\partial x^j} + \sum_{r=0}^m V^\beta_{i_1\cdots i_r} \frac{\partial \Sigma^{\alpha i}}{\partial v^\beta_{i_1\cdots i_r}}\Big]\\ &= (\phi^* X^j)\Big[\frac{\partial \Sigma^{\alpha i}}{\partial x^j} + \sum_{r=0}^m u^\beta_{,i_1\cdots i_r j} \frac{\partial \Sigma^{\alpha i}}{\partial u^\beta_{,i_1\cdots i_r}}\Big] = (\phi^* X^j) \frac{\partial(\phi^*\Sigma^{\alpha i})}{\partial x^j}\end{aligned}$$

$$\phi^* d\Sigma^{\alpha i} = \frac{\partial(\phi^*\Sigma^{\alpha i})}{\partial x^j}\, dx^j$$

we obtain

$$\begin{aligned}\phi^*\big(\mathbf{i}_V(\omega^\alpha)\big) &= (\phi^* X^j)\Big[\frac{\partial(\phi^*\Sigma^{\alpha i})}{\partial x^j}\mu_i - \frac{\partial(\phi^*\Sigma^{\alpha i})}{\partial x^j}\mu_i + \frac{\partial(\phi^*\Sigma^{\alpha i})}{\partial x^i}\mu_j\\ &\quad + \phi^*(\Sigma^\alpha)\mu_j\Big] = (\phi^* X^j)\Big[\frac{\partial(\phi^*\Sigma^{\alpha i})}{\partial x^i} + \phi^*(\Sigma^\alpha)\Big]\mu_j = 0\end{aligned}$$

so we arrive at the result $\phi^*\mathbf{i}_V(\mathfrak{I}_m) = 0$. It is clear that this property will be equally valid for a fundamental ideal generated by forms $\omega^a = \Sigma^a \mu$.

Example 9.5.1. Let us consider the isovector field obtained previously in Example 9.4.1 for the heat conduction equation

$$-V^1_G = 4xt\frac{\partial}{\partial x} + 4t^2\frac{\partial}{\partial t} - (x^2 + 2t)u\frac{\partial}{\partial u}$$

except for a sign difference. The similarity solution associated with this vector field must satisfy the partial differential equation

$$4xt\frac{\partial u}{\partial x} + 4t^2\frac{\partial u}{\partial t} + (x^2 + 2t)u = 0$$

whose characteristics are described by the ordinary differential equations

$$\frac{dx}{4xt} = \frac{dt}{4t^2} = -\frac{du}{(x^2+2t)u}.$$

Hence the solution becomes

$$u(x,t) = \frac{e^{-\frac{x^2}{4t}}}{\sqrt{t}} T(\xi), \quad \xi = \frac{x}{t}.$$

where $T(\xi)$ is an arbitrary function. On inserting this expression into the field equation $u_t = u_{xx}$, we simply obtain $T'' = 0$. Thus, this similarity solution takes the form

$$u(x,t) = \frac{e^{-\frac{x^2}{4t}}}{\sqrt{t}} \Big(c_1 \frac{x}{t} + c_2\Big)$$ ■

Example 9.5.2. We now consider an isovector field associated with the Korteweg-de Vries equation given by

$$-V^1 - cV^3 = (x+ct)\frac{\partial}{\partial x} + 3t\frac{\partial}{\partial t} - (2u-c)\frac{\partial}{\partial u}$$

where c is a constant. The similarity solution associated with this isovector must satisfy the partial differential equation

$$(x+ct)\frac{\partial u}{\partial x} + 3t\frac{\partial u}{\partial t} + 2u - c = 0$$

whose characteristics are determined via the ordinary differential equations

$$\frac{dx}{x+ct} = \frac{dt}{3t} = -\frac{du}{2u-c}.$$

Hence, the solution is found as

$$u(x,t) = \frac{c}{2} + \phi(\xi)t^{-2/3}, \quad \xi = t^{-1/3}\Big(x - \frac{1}{2}ct\Big).$$

where $\phi(\xi)$ is an arbitrary function. If we insert this expression into the equation $u_t + uu_x + u_{xxx} = 0$, we deduce the following non-linear ordinary differential equation

$$3\phi''' + (3\phi - \xi)\phi' - 2\phi = 0. \tag{9.5.4}$$

In order to get an idea about the structure of solutions of this equation, a numerically obtained solution under the initial conditions $\phi(0) = 0$,

$\phi'(0) = 1,\ \phi''(0) = 1$ is depicted in Fig. 9.5.1.

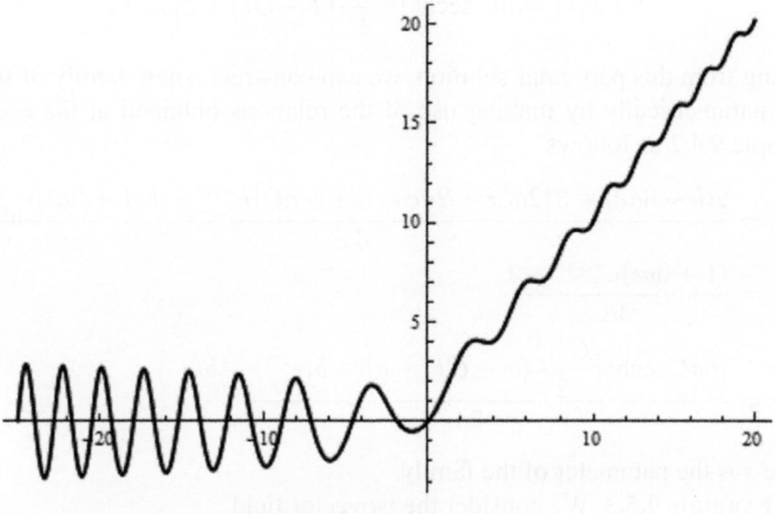

Fig. 9.5.1. A numerical solution of the equation (9.5.4).

As another isovector, we choose

$$-CV^4 - V^2 = C\frac{\partial}{\partial x} + \frac{\partial}{\partial t}.$$

where C is a constant. The similarity solution now satisfies the simple partial differential equation

$$C\frac{\partial u}{\partial x} + \frac{\partial u}{\partial t} = 0$$

whose solution is in the form $u = u(\xi)$ where $\xi = x - Ct$. Hence, we have to solve the non-linear equation

$$u''' + uu' - Cu' = 0$$

or its first integral

$$u'' + \frac{1}{2}u^2 - Cu = c_1.$$

The general solution of this equation can be found in terms of elliptic functions. However, if we impose the condition $u \to 0$ for $\xi \to \pm\infty$, we have to take $c_1 = 0$. In this case, the solution is expressible in elementary functions and the advancing wave type of a solution of Korteweg-de Vries equation yields the well known ***soliton*** solution

$$u(x,t) = 3C\,\mathrm{sech}^2\big[\frac{C^{1/2}}{2}(x - Ct) + \delta\big].$$

Starting from this particular solution, we can construct a new family of solutions parametrically by making use of the relations obtained at the end of Example 9.4.2 as follows

$$\overline{x}(s) = \frac{2(b - 3ac) + 3\big[2a^2x + 2ac - b(1 + at)\big]e^{-c_1 s} + b(1 + 3at)e^{-3c_1 s}}{6a^2}$$

$$\overline{t}(s) = \frac{(1 + 3at)e^{-3c_1 s} - 1}{3a}$$

$$\overline{u}(s) = \frac{\big[6aC\mathrm{sech}^2\big[\frac{C^{1/2}}{2}(x - Ct) + \delta\big] - b\big]e^{2c_1 s} + b}{2a}$$

where s is the parameter of the family. ■

Example 9.5.3. We consider the isovector field

$$V_G = (x + a)\frac{\partial}{\partial x} + by\frac{\partial}{\partial x} + (1 - 2b)u\frac{\partial}{\partial u} - bv\frac{\partial}{\partial v}$$

associated with partial differential equations governing the boundary layer flow past a flat plate discussed in Example 9.4.3. Equations (9.5.1) now take the form

$$(x + a)\frac{\partial u}{\partial x} + by\frac{\partial u}{\partial x} - (1 - 2b)u = 0,\ (x + a)\frac{\partial v}{\partial x} + by\frac{\partial v}{\partial x} + bv = 0$$

the solution of which is easily obtained as

$$u(x,y) = (x + a)^{1-2b}\phi(\xi),\ v(x,y) = (x + a)^{-b}\psi(\xi);\ \xi = y(x + a)^{-b}$$

where $\phi(\xi)$ and $\psi(\xi)$ are arbitrary functions. Introduction of these expressions together with the admissible function $f(x^1) = A(x^1 + a)^{1-4b}$ into the field equations (9.4.26) gives rise to the following set of ordinary differential equations

$$\begin{aligned}&\nu\phi''' + b\xi\phi\phi' - (1 - 2b)\phi^2 - \phi'\psi + A = 0, \\ &b\xi\phi' - \psi' - (1 - 2b)\phi = 0\end{aligned} \tag{9.5.5}$$

A numerical solution of the above equations corresponding to $b = 1/2$, $\nu = 1, A = 1$ under the initial conditions $\phi(0) = 0, \phi'(0) = 0, \phi''(0) = 1.5$ and $\psi(0) = 0$ that may not reflect an actual physical situation is depicted in Fig. 9.5.2.

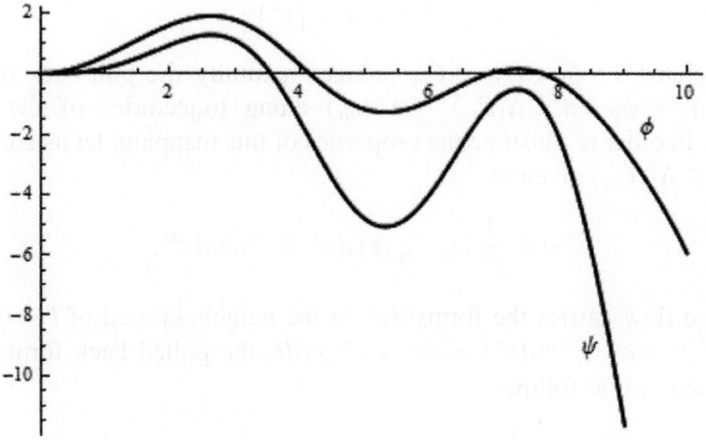

Fig. 9.5.2. Numerical solutions of the equations (9.5.5). ∎

9.6. THE METHOD OF GENERALISED CHARACTERISTICS

We have seen in Sec. 9.2 that the solution of a first order non-linear partial differential equation can be constructed by means of characteristics starting from a given initial submanifold. We shall now try to generalise this method by employing isovectors of the ideal $\mathfrak{I}_m$ of $\Lambda(\mathcal{C}_m)$ associated with a system of partial differential equations. Let us denote the $\mathfrak{n} = n + D$ number of local coordinates $\{x^i, v^\alpha_{i_1 i_2 \cdots i_r} : 0 \le r \le m\}$ of the contact manifold $\mathcal{C}_m$ by $z^\mathfrak{a}$, $\mathfrak{a} = 1, \dots, \mathfrak{n}$. Consider a vector field $V = v^\mathfrak{a}(\mathbf{z})\, \partial/\partial z^\mathfrak{a} \in T(\mathcal{C}_m)$. We know that its integral curves are obtained as solutions of the following ordinary differential equations and initial conditions

$$\frac{d\zeta^\mathfrak{a}}{dt} = v^\mathfrak{a}(\boldsymbol{\zeta}), \quad \zeta^\mathfrak{a}(0) = z^\mathfrak{a}$$

in the form $\zeta^\mathfrak{a} = \phi^\mathfrak{a}(t; \mathbf{z}) = \phi_V(t) z^\mathfrak{a} = e^{tV}(z^\mathfrak{a})$. We wish to get the parameter t acquired a status of a coordinate to appreciate its independent variations. Therefore, we embed the integral curves into the *graph manifold* $L_m = \mathcal{C}_m \times \mathbb{R}$ whose coordinates are prescribed by $\{z^\mathfrak{a}, t\}$. Thus the contact manifold $\mathcal{C}_m$ might be specified as a submanifold of the manifold L_m obtained by $t = 0$. It appears to be advantageous now to extend the mapping ϕ_V describing the flow as $\phi_V : L_m \to L_m$ such that

$$\phi_V(\{z^\mathfrak{a}, t\}) = \{\zeta^\mathfrak{a} = e^{tV}(z^\mathfrak{a}), t\}. \tag{9.6.1}$$

We can naturally define a canonical projection $\pi : L_m \to \mathcal{C}_m$ as follows

$$\pi(\{z^{\mathfrak{a}}, t\}) = \{z^{\mathfrak{a}}\}.$$

The operator $\pi \circ \phi_V : L_m \to \mathcal{C}_m$ induces naturally the pull-back operator $(\pi \circ \phi_V)^* = \phi_V^* \circ \pi^* : \Lambda(\mathcal{C}_m) \to \Lambda(L_m)$ along trajectories of the vector field V. In order to illustrate the properties of this mapping, let us consider a form $\omega \in \Lambda^k(\mathcal{C}_m)$ given by

$$\omega = \frac{1}{k!}\, \omega_{\mathfrak{a}_1 \cdots \mathfrak{a}_k}(\mathbf{z})\, dz^{\mathfrak{a}_1} \wedge \cdots \wedge dz^{\mathfrak{a}_k}.$$

Since the flow carries the forms $dz^{\mathfrak{a}}$ in the neighbourhood of $t = 0$ to the forms $d\zeta^{\mathfrak{a}} = \phi_V^* \circ \pi^*(dz^{\mathfrak{a}}) = dz^{\mathfrak{a}} + v^{\mathfrak{a}}(\mathbf{z})\, dt$, the pulled back form can be written at $t = 0$ as follows

$$\begin{aligned} \omega^*\big|_{t=0} = \pi^*\omega\big|_{t=0} &= \frac{1}{k!}\, \omega_{\mathfrak{a}_1 \cdots \mathfrak{a}_k}(\mathbf{z})\, (dz^{\mathfrak{a}_1} + v^{\mathfrak{a}_1} dt) \wedge \cdots \wedge (dz^{\mathfrak{a}_k} + v^{\mathfrak{a}_k} dt) \\ &= \omega + \frac{1}{(k-1)!}\, \omega_{\mathfrak{a}_1 \cdots \mathfrak{a}_k}(\mathbf{z}) v^{\mathfrak{a}_1} dt \wedge dz^{\mathfrak{a}_2} \wedge \cdots \wedge dz^{\mathfrak{a}_k} \\ &= \omega + dt \wedge \mathbf{i}_V(\omega). \end{aligned}$$

where we have employed the complete antisymmetries of both the coefficients $\omega_{\mathfrak{a}_1 \cdots \mathfrak{a}_k}$ and exterior products. Hence, in view of the relation (5.11.14) the form ω^* can be expressed as

$$\omega^*(t; \mathbf{z}) = \phi_V^* \circ \pi^* \omega(\mathbf{z}) = e^{t\pounds_V}\big(\omega + dt \wedge \mathbf{i}_V(\omega)\big). \tag{9.6.2}$$

Next, we introduce an operator $E_V : \Lambda(\mathcal{C}_m) \to \Lambda(L_m)$ that maps an exterior algebra into a larger exterior algebra by the rule

$$E_V \omega = \omega + dt \wedge \mathbf{i}_V(\omega) \in \Lambda(L_m), \;\; \omega \in \Lambda(\mathcal{C}_m). \tag{9.6.3}$$

The operator E_V has the following properties:

$$\begin{aligned} &(i).\; E_V(\omega_1 + \omega_2) = E_V\omega_1 + E_V\omega_2, \\ &\qquad\quad E_V(f\omega) = f E_V\omega,\; f \in \Lambda^0(\mathcal{C}_m), \\ &(ii).\; E_V(\omega_1 \wedge \omega_2) = E_V\omega_1 \wedge E_V\omega_2, \\ &(iii).\qquad d(E_V\omega) = E_V(d\omega) - dt \wedge \pounds_V\omega, \\ &(iv).\qquad \pounds_V(E_V\omega) = E_V(\pounds_V\omega), \\ &(v).\quad (\phi_V^* \circ \pi^*)\,\omega = e^{t\pounds_V} E_V\omega. \end{aligned} \tag{9.6.4}$$

The relations in (9.6.4) can easily be verified:
(i). This is evident because of the properties of the operator $\mathbf{i}_V$. Hence, E_V is a linear operator on the exterior algebra $\Lambda(\mathcal{C}_m)$.
(ii). To see this, it suffices to note that

$$E_V(\omega_1 \wedge \omega_2) = \omega_1 \wedge \omega_2 + dt \wedge \mathbf{i}_V(\omega_1) \wedge \omega_2 + (-1)^{deg\,\omega_1} dt \wedge \omega_1 \wedge \mathbf{i}_V(\omega_2)$$
$$E_V\omega_1 \wedge E_V\omega_2 = \omega_1 \wedge \omega_2 + (-1)^{deg\,\omega_1} dt \wedge \omega_1 \wedge \mathbf{i}_V(\omega_2) + dt \wedge \mathbf{i}_V(\omega_1) \wedge \omega_2.$$

(iii). This follows from the relation

$$d(E_V\omega) = d\omega - dt \wedge d\mathbf{i}_V(\omega) = d\omega + dt \wedge \mathbf{i}_V(d\omega) - dt \wedge \pounds_V\omega$$

where we have employed the Cartan magic formula.
(iv). This is immediately seen if we take notice of the relations $\pounds_V dt = dV(t) = 0$ and $\pounds_V(\mathbf{i}_V(\omega)) = \mathbf{i}_V(\pounds_V(\omega))$ [*see* (5.11.8)$_2$].
(v). This is in fact just the relation (9.6.2).

According to the properties (9.6.4)$_{1-2}$, we see that the set

$$\Lambda_V(L_m) = \{E_V\omega : \omega \in \Lambda(\mathcal{C}_m)\} \subseteq \Lambda(L_m)$$

becomes an exterior algebra. We can now prove the theorem below.

Theorem 9.6.1. *If a vector field $V \in T(\mathcal{C}_m)$ is an isovector field of a closed ideal $\mathfrak{I}$ of the exterior algebra $\Lambda(\mathcal{C}_m)$, then it is also an isovector field of the ideal $E_V\mathfrak{I}$ of the exterior algebra $\Lambda_V(L_m)$.*

If $\omega_1, \omega_2 \in \mathfrak{I}$, we have of course $\omega_1 + \omega_2 \in \mathfrak{I}$ and then (9.6.4)$_1$ leads to $E_V\omega_1 + E_V\omega_2 = E_V(\omega_1 + \omega_2) \in E_V\mathfrak{I}$. Similarly, if $\omega \in \mathfrak{I}$, we find that $\gamma \wedge \omega \in \mathfrak{I}$ with $\gamma \in \Lambda(\mathcal{C}_m)$. Since a form $\gamma' \in \Lambda_V(L_m)$ must now be written as $\gamma' = E_V\gamma$ where $\gamma \in \Lambda(\mathcal{C}_m)$, we thus get $\gamma' \wedge E_V\omega = E_V\gamma \wedge E_V\omega = E_V(\gamma \wedge \omega) \in E_V\mathfrak{I}$ due to (9.6.4)$_2$. Therefore, the set $E_V\mathfrak{I}$ is an ideal of the exterior algebra $\Lambda_V(L_m)$. However, this ideal is no longer closed since we have the relation $d \circ E_V \neq E_V \circ d$ because of the property (9.6.4)$_3$. If the vector field V is an isovector field of the closed ideal $\mathfrak{I}$, then we get $d\omega \in \mathfrak{I}$ and $\pounds_V\omega \in \mathfrak{I}$ for all forms $\omega \in \mathfrak{I}$. On the other hand, because of (9.6.4)$_4$ and (5.11.9) the relations

$$\pounds_V(E_V\omega) = E_V(\pounds_V\omega) \in E_V\mathfrak{I}$$
$$\pounds_V(E_V d\omega) = E_V(\pounds_V d\omega) = E_V(d\pounds_V\omega) \in E_V\mathfrak{I}$$

are satisfied. Hence V is also an isovector field of the ideal $E_V\mathfrak{I}$. □

Let $D_{n-1} \subseteq \mathbb{R}^{n-1}$ denote a connected, open subset whose local coordinates are provided by $\{s^1, \dots, s^{n-1}\}$. We next suppose that the mapping $\psi : D_{n-1} \to \mathcal{C}_m$ prescribed by smooth functions $z^{\mathfrak{a}} = \psi^{\mathfrak{a}}(s^1, \dots, s^{n-1})$ specifies an *initial data submanifold* in $\mathcal{C}_m$. Let us then determine the integral curves of an isovector field V of a closed ideal $\mathfrak{I}$ of $\Lambda(\mathcal{C}_m)$ as solutions of the ordinary differential equations

$$\frac{d\zeta^{\mathfrak{a}}}{d\tau} = v^{\mathfrak{a}}(\boldsymbol{\zeta}),\ \zeta^{\mathfrak{a}}(0) = \psi^{\mathfrak{a}}(\mathbf{s}),\ \mathfrak{a} = 1, \dots, \mathfrak{n} \qquad (9.6.5)$$

where τ is a real parameter. It becomes now possible to introduce a mapping $\Psi : D_n = D_{n-1} \times \mathbb{R} \to \mathcal{C}_m$ through the relations

$$z^{\mathfrak{a}} = \zeta^{\mathfrak{a}}\big(\psi^{\mathfrak{b}}(\mathbf{s});\tau\big) = \Psi^{\mathfrak{a}}(s^1,\dots,s^{n-1};\tau). \qquad (9.6.6)$$

We have already mentioned that $\mathcal{C}_m$ is a submanifold of L_m specified by $t=0$. We now define a simple *extension* $\widehat{\psi} : D_{n-1} \times \mathbb{R} \longrightarrow L_m$ of the mapping ψ as follows

$$\widehat{\psi}\,(s^1,\dots,s^{n-1};\tau) = \{z^{\mathfrak{a}} = \psi^{\mathfrak{a}}(s^1,\dots,s^{n-1}), t=\tau\}. \qquad (9.6.7)$$

In this case, the mapping $\Psi : D_n \longrightarrow \mathcal{C}_m$ can be expressed as

$$\Psi = \pi \circ \phi_V \circ \widehat{\psi} = \pi \circ e^{\tau V} \circ \widehat{\psi} \qquad (9.6.8)$$

when we recall (9.6.1). In the light of the information acquired so far, the following theorem can be proposed.

Theorem 9.6.2. *Let $V \in T(\mathcal{C}_m)$ be an isovector field of a closed ideal $\mathfrak{I}$ of $\Lambda(\mathcal{C}_m)$ and $D_{n-1} \subseteq \mathbb{R}^{n-1}$ be a connected open set whose local coordinates are given by $\{s^1,\dots,s^{n-1}\}$. The mapping $\psi : D_{n-1} \longrightarrow \mathcal{C}_m$ determines an initial data submanifold in $\mathcal{C}_m$ through the smooth functions $z^{\mathfrak{a}} = \psi^{\mathfrak{a}}(\mathbf{s}), \mathfrak{a} = 1,\dots,\mathfrak{n}$. If the extension $\widehat{\psi}$ of ψ holds the condition*

$$(\widehat{\psi})^*(E_V\mathfrak{I}) = 0, \qquad (9.6.9)$$

then the mapping $\Psi = \pi \circ \phi_V \circ \widehat{\psi}$ satisfies the relation $\Psi^*\mathfrak{I} = 0$. Hence, the mapping $\Psi : D_{n-1} \times \mathbb{R} \to \mathcal{C}_m$ becomes a solution of the ideal $\mathfrak{I}$.

The proof of this theorem is rather straightforward at the first glance. If we keep in mind the relation $(9.6.4)_5$, the pull-back operator Ψ^* may be expressible as

$$\Psi^*\mathfrak{I} = \big((\widehat{\psi})^* \circ \phi_V^* \circ \pi^*\big)\mathfrak{I} = (\widehat{\psi})^*\big(e^{t\pounds_V}(E_V\mathfrak{I})\big). \qquad (9.6.10)$$

But, according to Theorem 9.5.1, the isovector field V is also an isovector field of the ideal $E_V\mathfrak{I}$. We thus find $e^{t\pounds_V}(E_V\mathfrak{I}) \subset E_V\mathfrak{I}$. Hence, it follows from (9.6.9) that $(\widehat{\psi})^*\big(e^{t\pounds_V}(E_V\mathfrak{I})\big) = 0$, and as a consequence $\Psi^*\mathfrak{I} = 0$. It is obvious that the condition (9.6.9) imposes a restriction on admissible forms of initial data. If we pay attention to the definition (9.6.7), we observe that $(\widehat{\psi})^*dz^{\mathfrak{a}} = \psi^*dz^{\mathfrak{a}}$ and $(\widehat{\psi})^*dt = d\tau$. In this case, we can also express the condition (9.6.9) as

$$(\widehat{\psi})^*(E_V\omega) = \psi^*\omega + d\tau \wedge \psi^*\big(\mathbf{i}_V(\omega)\big) = 0 \qquad (9.6.11)$$

for all $\omega \in \mathfrak{I}$. On the other hand, one can write

$$\psi^* dz^a = \frac{\partial \psi^a}{\partial s^1} ds^1 + \frac{\partial \psi^a}{\partial s^2} ds^2 + \cdots + \frac{\partial \psi^a}{\partial s^{n-1}} ds^{n-1}$$

and since 1-forms $ds^1, ds^2, \dots, ds^{n-1}, d\tau$ are all linearly independent, we then conclude that the condition (9.6.11) is satisfied if and only if one has

$$\psi^* \omega = 0, \quad \psi^*\big(\mathbf{i}_V(\omega)\big) = 0 \tag{9.6.12}$$

for all $\omega \in \mathfrak{I}$. Let us now assume that the closed ideal $\mathfrak{I}$ is generated by $2r$ forms $\{\omega^\alpha, d\omega^\alpha : \alpha = 1, 2, \dots, r\}$. Then it becomes clear that the conditions (9.6.12) are satisfied if and only if we have

$$\psi^* \omega^\alpha = 0, \ \psi^*\big(\mathbf{i}_V(\omega^\alpha)\big) = 0; \quad \psi^* d\omega^\alpha = 0, \ \psi^*\big(\mathbf{i}_V(d\omega^\alpha)\big) = 0$$

for $\alpha = \ 1, 2, \dots, r$. Actually, we immediately see that the second set of equations involving exterior derivative are automatically satisfied in case the first set of conditions are met. Indeed, we obtain $\psi^* d\omega^\alpha = d\psi^* \omega^\alpha = 0$ in accordance with Theorem 5.8.2. On taking into account the Cartan formula, we have $\psi^* \mathbf{i}_V(d\omega^\alpha) = \psi^* \pounds_V \omega^\alpha - \psi^* d\mathbf{i}_V(\omega^\alpha)$. Since V is an isovector field of the ideal, we can write $\pounds_V \omega^\alpha = \lambda^\alpha_\beta \wedge \omega^\beta + \Lambda^\alpha_\beta \wedge d\omega^\beta \in \mathfrak{I}$. We thus obtain $\psi^* \pounds_V \omega^\alpha = 0$ and $\psi^* \mathbf{i}_V(d\omega^\alpha) = -\, d\psi^* \mathbf{i}_V(\omega^\alpha) = 0$. In this situation, the conditions (9.6.12) concerning the initial data are recovered if and only if we are assured that the relations

$$\psi^* \omega^\alpha = 0, \ \psi^*\big(\mathbf{i}_V(\omega^\alpha)\big) = 0, \ \alpha = 1, \dots, r \tag{9.6.13}$$

are satisfied. □

We now consider a closed fundamental ideal associated with a given system of partial differential equations and an isovector field V of this ideal. Let the mapping $\psi : D_{n-1} \longrightarrow \mathcal{C}_m$ specify again an initial data submanifold. We assume that the mapping ψ is satisfying the ***transversality condition***

$$\psi^*\big(\mathbf{i}_V(\mu)\big) \neq 0$$

for the volume form μ in D_n. The transversality condition implies that the rank of ψ is $n-1$ since its domain is the $(n-1)$-dimensional region D_{n-1}. Moreover, we find $(\widehat{\psi})^* \mu = \psi^* \mu = 0$ on D_{n-1} since μ is an n-form. Therefore, we can write

$$\Psi^* \mu = (\widehat{\psi})^* \big(e^{t\pounds_V}(E_V \mu)\big) = (\widehat{\psi})^* \big[e^{t\pounds_V}(dt \wedge \mathbf{i}_V(\mu))\big].$$

At $t = 0$ we get $\Psi^* \mu\big|_{t=0} = (\widehat{\psi})^*(dt \wedge \mathbf{i}_V(\mu)) = d\tau \wedge \psi^*\big(\mathbf{i}_V(\mu)\big) \neq 0$. Thus, the condition $\Psi^* \mu \neq 0$ is satisfied about $\tau = 0$ on the set $D_n = D_{n-1} \times \mathbb{R}$. Consequently, the mapping Ψ defined by (9.6.5) or (9.6.6) is regular. In this case, we can enounce the following theorem that turns out to be actually a direct descendant of Theorem 9.6.2.

Theorem 9.6.3. *Let V be an isovector field of a closed fundamental ideal $\mathfrak{I}_m$ associated with a given system of partial differential equations. If the mapping $\psi : D_{n-1} \to \mathcal{C}_m$ specifying an initial data submanifold holds the conditions*

$$\psi^*\big(\mathbf{i}_V(\mu)\big) \neq 0,\ \psi^*\mathfrak{I}_m = 0,\ \psi^*\big(\mathbf{i}_V(\mathfrak{I}_m)\big) = 0, \qquad (9.6.14)$$

then the mapping $\Psi = \pi \circ \phi_V \circ \widehat{\psi} : D_n \to \mathcal{C}_m$ prescribed by the equations (9.6.5) and (9.6.6) on $D_n = D_{n-1} \times \mathbb{R}$ satisfies the condition $\Psi^*\mathfrak{I}_m = 0$, that is, it becomes a solution of the ideal $\mathfrak{I}_m$. □

We had called $\psi(D_{n-1}) \subset \mathcal{C}_m$ the *initial data submanifold.* We regard (9.6.5) as the equations determining the *characteristics* corresponding to the pair (V, ψ) satisfying the transversality condition. $(9.6.14)_{2-3}$ represent *restrictions imposed on initial data* on the relevant submanifold. Ψ is then called as a ***generalised characteristic solution*** associated with a chosen V.

Generally, characteristic solutions of a system of partial differential equations have to satisfy some additional conditions.

Theorem 9.6.4. *If Ψ is the generalised characteristic solution generated by an isovector field V of the closed fundamental ideal $\mathfrak{I}_m$ associated with a system of partial differential equations, then the condition*

$$\Psi^*\big(\mathbf{i}_V(\mathfrak{I}_m)\big) = 0 \qquad (9.6.15)$$

should be satisfied on the domain of Ψ. Therefore, the generalised characteristic solutions have to fulfil a specific set of additional constraints. However, if $\mathbf{i}_V(\mathfrak{I}_m) \subset \mathfrak{I}_m$, namely, if the isovector field is also a characteristic vector field of the ideal, then these conditions are automatically satisfied.

We can realise at once that, we can write

$$E_V\mathbf{i}_V(\omega) = \mathbf{i}_V(\omega) + dt \wedge \mathbf{i}_V \circ \mathbf{i}_V(\omega) = \mathbf{i}_V(\omega)$$

for a form $\omega \in \mathfrak{I}_m$ due to (5.4.5). Therefore (9.6.10) yields

$$\Psi^*\big(\mathbf{i}_V(\omega)\big) = (\widehat{\psi})^*\big[e^{t\pounds_V}\big(\mathbf{i}_V(\omega)\big)\big].$$

But the relation $\pounds_V\big(\mathbf{i}_V(\omega)\big) = \mathbf{i}_V\big(\pounds_V(\omega)$ implies that for every natural number k, we can write $\pounds_V^k\big(\mathbf{i}_V(\omega)\big) = \mathbf{i}_V\big(\pounds_V^k(\omega)\big)$. Hence, we reach to the result

$$e^{t\pounds_V}\big(\mathbf{i}_V(\omega)\big) = \mathbf{i}_V\big(e^{t\pounds_V}(\omega)\big).$$

We have $e^{t\pounds_V}(\omega) \in \mathfrak{I}_m$ since V is an isovector field. Because of $(9.6.12)_2$, we get $\Psi^*\big(\mathbf{i}_V(\omega)\big) = 0$ as well. We have seen in Sec. 9.5 that it suffices to satisfy the relation

$$\Psi^*\big(\mathbf{i}_V(\sigma^\alpha)\big) = 0$$

in order that the constraint (9.6.15) is fulfilled. In the light of this constraint we can say that, the generalised characteristic solutions are nothing but certain group-invariant solutions. However, if the isovector field V is at the same time a characteristic vector of the fundamental ideal, that is, if one has $\mathbf{i}_V(\mathfrak{I}_m) \subset \mathfrak{I}_m$, then $\Psi^*(\mathfrak{I}_m) = 0$ implies that $\Psi^*\big(\mathbf{i}_V(\mathfrak{I}_m)\big) = 0$. In this case the additional constraint is of course redundant. □

The determination of solution of a given system of partial differential equations satisfying prescribed initial conditions by the method of generalised characteristics seems at the first glance the same as the construction of similarity solutions investigated in Sec. 9.5. But, in order to obtain a tangible benefit from a similarity solution we need to solve first analytically partial differential equations (9.5.1) or (9.5.2). Furthermore, boundary and/or initial conditions have to comply totally with the structure of the similarity solution whereas in the method of generalised characteristics the mapping Ψ is determined by solving a system of ordinary differential equations if the initial data manifold is suitably chosen as to comply with the imposed restrictions. It is of course much easier to find numerical solutions of ordinary differential equations to construct a solution of partial differential equations at least approximately.

Example 9.6.1. We consider the non-linear partial differential equation

$$\Sigma = \frac{\partial u}{\partial x^1}\frac{\partial u}{\partial x^2} - ku = 0, \;\; k = constant$$

where $n = 2, N = 1$. Introducing $v_1 = u_{,1}, v_2 = u_{,2}$, we characterise this equation by the following 2-form

$$\omega = \Sigma\mu = (v_1 v_2 - ku)\, dx^1 \wedge dx^2.$$

As we already mentioned on p. 538 in Sec. 9.4, we can determine an isovector field by taking $F(x^1, x^2, u, v_1, v_2) = -\Sigma = ku - v_1 v_2$. Hence, we get

$$X^1 = v_2, \;\; X^2 = v_1, \;\; U = ku + v_1 v_2, \;\; V_1 = kv_1, \;\; V_2 = kv_2.$$

We define the mapping $\psi : \mathbb{R} \to C_1$ specifying the initial data submanifold by the relations below depending on a single parameter s

$$x^1 = s, \, x^2 = 0, \, u = \psi_0(s), \, v_1 = \psi_1(s), \, v_2 = \psi_2(s).$$

Since we have $\sigma = du - v_1 dx^1 - v_2 dx^2$, $\mathbf{i}_V(\sigma) = U - v_1 X^1 - v_2 X^2$, the constraints

$$\begin{aligned}\psi^*\big(\mathbf{i}_V(\mu)\big) &= \psi^*(X^1\,dx^2 - X^2\,dx^1) = \,-\,\psi_1(s)\,ds \neq 0\\ \psi^*\sigma &= (\psi_0' - \psi_1)\,ds = 0,\\ \psi^*\,\mathbf{i}_V(\sigma) &= \psi^*(ku - v_1v_2) = k\psi_0 - \psi_1\psi_2 = 0\end{aligned}$$

on the initial data yield

$$\psi_1 \neq 0, \quad \psi_1 = \psi_0', \qquad \psi_2 = \frac{k\psi_0}{\psi_0'}.$$

Equations (9.6.5) can now be written as

$$\begin{aligned}&\frac{dx^1}{d\tau} = v_2, \;\; \frac{dx^2}{d\tau} = v_1, \;\; \frac{du}{d\tau} = ku + v_1v_2, \;\; \frac{dv_1}{d\tau} = kv_1, \;\; \frac{dv_2}{d\tau} = kv_2,\\ &x^1(0) = s,\; x^2(0) = 0, \qquad u(0) = \psi_0(s),\\ &v_1(0) = \psi_0'(s), \;\; v_2(0) = \frac{k\psi_0(s)}{\psi_0'(s)}\end{aligned}$$

whose solution is easily found to be

$$\begin{aligned}&x^1 = s + \frac{\psi_0(s)}{\psi_0'(s)}(e^{k\tau} - 1), \;\; x^2 = \frac{\psi_0'(s)}{k}(e^{k\tau} - 1),\\ &u = \psi_0(s\,)\,e^{2k\tau}\\ &v_1 = \psi_0'(s)\,e^{k\tau}, \quad v_2 = \frac{k\psi_0(s)}{\psi_0'(s)}\,e^{k\tau}.\end{aligned}$$

These relations create a solution in the form $u = u(x^1, x^2)$ after having expressed the parameters (s, τ) in terms of independent variables x^1 and x^2 by inverting, at least in principle the relations for x^1 and x^2.

As a very simple example, let us suppose that $\psi_0(s) = cs$ where c is a constant. We then obtain

$$x^1 = s + s(e^{k\tau} - 1), \;\; x^2 = \frac{c}{k}(e^{k\tau} - 1)$$

whence we easily deduce that

$$s = \frac{x^1}{1 + \frac{k}{c}x^2}, \quad e^{k\tau} = 1 + \frac{k}{c}x^2.$$

Hence the solution becomes simply

$$u(x^1, x^2) = cx^1\Big(1 + \frac{k}{c}x^2\Big).$$

On the other hand, if we take $\psi_0(s) = cs^2$, we have

$$x^1 = s + \frac{1}{2}s(e^{k\tau} - 1), \ x^2 = \frac{2c}{k}s(e^{k\tau} - 1)$$

and we find

$$s = x^1 - \frac{k}{4c}x^2, \ e^{k\tau} = 1 + \frac{kx^2}{2cx^1 - \frac{k}{2}x^2}$$

so that the corresponding solution becomes

$$u(x^1, x^2) = c\Big(x^1 - \frac{k}{4c}x^2\Big)^2\Big[\frac{2cx^1 + \frac{k}{2}x^2}{2cx^1 - \frac{k}{2}x^2}\Big]^2. \qquad \blacksquare$$

Example 9.6.2. As a more difficult example, let us consider the partial differential equation characterised by the 2-form

$$\omega = df_1 \wedge df_2 \in \Lambda^2(K_1), \ f_1 = (x^1)^2 + v_2^2, \ f_2 = (x^2)^2 + v_1^2$$

where we have again $n = 2, N = 1$ and $v_1 = u_{,1}, v_2 = u_{,2}$. The form ω may explicitly be written as

$$\omega = 4(x^1x^2dx^1 \wedge dx^2 + x^1v_1dx^1 \wedge dv_1 + x^2v_2dv_2 \wedge dx^2 + v_1v_2dv_2 \wedge dv_1)$$

If we choose a mapping $u = \phi(x^1, x^2)$ annihilating the form ω, then $\phi^*\omega = 0$ yields quite a complicated non-linear second order partial differential equation

$$\phi_{,1}\phi_{,2}\phi_{,11}\phi_{,22} - \phi_{,1}\phi_{,2}(\phi_{,12})^2 - (x^1\phi_{,1} + x^2\phi_{,2})\phi_{,12} - x^1x^2 = 0.$$

This partial differential equation is known as the non-homogeneous ***Monge-Ampère equation*** [French mathematicians Gaspard Monge (1746-1818) and André Marie Ampère (1775-1836)]. Let $V \in T(\mathcal{C}_1)$ be an isovector field of the contact ideal. By definition, we get $d\omega = 0$. We thus obtain

$$£_V\omega = d\mathbf{i}_V(\omega) = dV(f_1) \wedge df_2 - dV(f_2) \wedge df_1$$

where

$$V(f_1) = 2x^1X^1 + 2v_2V_2, \ \ V(f_2) = 2x^2X^2 + 2v_1V_1.$$

We now wish so specify a simple isovector field as to be $V(f_1) = V(f_2) = 0$ implying $£_V\omega = 0$. If we take the function $F = F(x^1, x^2, u, v_1, v_2)$

into account, then it follows from (9.3.26) that this special function F must satisfy the equations

$$v_2\frac{\partial F}{\partial x^2}+v_2^2\frac{\partial F}{\partial u}-x^1\frac{\partial F}{\partial v_1}=0,\ v_1\frac{\partial F}{\partial x^1}+v_1^2\frac{\partial F}{\partial u}-x^2\frac{\partial F}{\partial v_2}=0.$$

The solution of the first equation is easily found to be as $F=F(\xi,\eta)$ where characteristic variables are $\xi=x^1x^2+v_1v_2,\ \eta=u-x^2v_2$. Inserting this result into the second equation we obtain $\big((x^2)^2+v_1^2\big)\partial F/\partial\eta=0$ implying that F is independent of η. Thus, we see that each smooth function of the form $F=F(x^1x^2+v_1v_2)$ generates an isovector field. As a simple example, let us just take $F=x^1x^2+v_1v_2$. Hence, we find that

$$V=-v_2\frac{\partial}{\partial x^1}-v_1\frac{\partial}{\partial x^2}+(x^1x^2-v_1v_2)\frac{\partial}{\partial u}+x^2\frac{\partial}{\partial v_1}+x^1\frac{\partial}{\partial v_2}$$

We choose the mapping $\psi:\mathbb{R}\to C_1$ as follows

$$x^1=s,\ x^2=1,\ u=\psi_0(s),\ v_1=\psi_1(s),\ v_2=\psi_2(s).$$

The expression $\psi^*\sigma=0$ yields again $\psi_1=\psi_0'$ while the constraint $\psi^*\mathbf{i}_V(\sigma)$ $=\psi^*(x^1x^2+v_1v_2)=0$ requires that $\psi_2=-s/\psi_0'$. Because of the relation $\mathbf{i}_V(\mu)=-v_2dx^2+v_1dx^1$, the transversality condition $\psi^*\big(\mathbf{i}_V(\mu)\big)\neq 0$ is met if $\psi_1=\psi_0'\neq 0$. To deduce the solution mapping Ψ, we have to solve the ordinary differential equations

$$\begin{aligned}
&\frac{dx^1}{d\tau}=-v_2,\ \frac{dx^2}{d\tau}=-v_1,\ \frac{du}{d\tau}=x^1x^2+v_1v_2,\\
&\frac{dv_1}{d\tau}=x^2,\ \frac{dv_2}{d\tau}=x^1\\
&x^1(0)=s,\ \ x^2(0)=1,\ \ u(0)=\psi_0(s),\\
&v_1(0)=\psi_0'(s),\ \ v_2(0)=-\frac{s}{\psi_0'(s)}
\end{aligned}$$

from which we readily obtain the parametric solution

$$\begin{aligned}
x^1&=s\cos\tau+\frac{s}{\psi_0'(s)}\sin\tau,\ x^2=\cos\tau-\psi_0'(s)\sin\tau,\\
v_1&=\sin\tau+\psi_0'(s)\cos\tau,\ v_2=s\sin\tau-\frac{s}{\psi_0'(s)}\cos\tau\\
u&=\psi_0(s)+s\sin 2\tau-\frac{s}{2}\Big(\psi_0'(s)-\frac{1}{\psi_0'(s)}\Big)(1-\cos 2\tau)
\end{aligned}$$

If we would be able to eliminate the parameters s and τ, we might obtain the corresponding solution in the form $u = u(x^1, x^2)$. ■

Example 9.6.3. Let us take into account Korteweg-de Vries equation studied in Example 9.4.2. The most general isovector field associated with this equation would be

$$V = -(c_1 x + c_3 t + c_4)\frac{\partial}{\partial x} - (3c_1 t + c_2)\frac{\partial}{\partial t} + (2c_1 u - c_3)\frac{\partial}{\partial u} + 3c_1 v_1 \frac{\partial}{\partial v_1} + (c_3 v_1 + 5c_1 v_2)\frac{\partial}{\partial v_2}.$$

We define the mapping $\psi : \mathbb{R} \to C_2$ specifying the initial data submanifold by the relations

$$x = s,\ t = 0,\ u = \psi_0(s),\ v_1 = \psi_1(s),\ v_2 = \psi_2(s),$$
$$v_{11} = \psi_{11}(s),\ \ v_{12} = \psi_{12}(s),\ \ v_{12} = \psi_{11}(s)$$

where $v_1 = u_{,1},\ \ v_2 = u_{,2},\ \ v_{11} = u_{,11},\ \ v_{12} = u_{,12},\ \ v_{22} = u_{,22}$. Because $\psi^*(\mathbf{i}_V(\mu)) = c_2\, ds$, the tranversality condition is satisfied if we take $c_2 \neq 0$. 1-forms generating the contact ideal are

$$\sigma = du - v_1 dx - v_2\, dt,$$
$$\sigma_1 = dv_1 - v_{11}\, dx - v_{12}\, dt,$$
$$\sigma_2 = dv_2 - v_{12}\, dx - v_{22}\, dt.$$

Hence, the expressions

$$\psi^*\sigma - (\psi_0' - \psi_1)\, ds = 0,$$
$$\psi^*(\sigma_1) = (\psi_1' - \psi_{11}) ds = 0,$$
$$\psi^*(\sigma_2) = (\psi_2' - \psi_{12}) ds = 0$$

yield $\psi_1(s) = \psi_0'(s),\ \ \psi_{11}(s) = \psi_1'(s) = \psi_0''(s),\ \ \psi_{12}(s) = \psi_2'(s)$. On the other hand, we find $\psi^*\, \mathbf{i}_V(\sigma) = 2c_1\psi_0 - c_3 + (c_1 s + c_4)\psi_1 + c_2\psi_2 = 0$ and, consequently

$$\psi_2(s) = \frac{1}{c_2}\big[c_3 - 2c_1\psi_0(s) - (c_1 s + c_4)\psi_0'(s)\big].$$

The balance form is given by $\omega = dv_{11} \wedge dt + (uv_1 + v_2 - f)\, dx \wedge dt$. Therefore, the relation $\psi^* dt = 0$ leads to $\psi^*\omega \equiv 0$ and it follows from

$$\psi^*\, \mathbf{i}_V(\omega) = c_2\psi^*\big[dv_{11} + (uv_1 + v_2 - f)\, dx\big] = 0$$

that $\psi_{11}' + \psi_0\psi_1 + \psi_2 - f(s, 0, \psi_0) = 0$. Hence, the admissible initial data $\psi_0(s)$ has to satisfy the following non-linear, third order ordinary differential equation

$$\psi_0'''(s) + \big[\psi_0(s) - as - c\big]\psi_0'(s) - 2a\psi_0(s) + b - f\big(s, 0, \psi_0(s)\big) = 0$$

where new constants are defined as $c_1/c_2 = a, c_3/c_2 = b, c_4/c_2 = c$. In order to obtain the mapping Ψ we must solve the linear ordinary differential equations

$$\frac{dx}{d\tau} = -c_2(ax + bt + c), \; \frac{dt}{d\tau} = -c_2(3at + 1), \; \frac{du}{d\tau} = c_2(2au - b)$$

under the initial conditions $x(0) = s, t(0) = 0, u(0) = \psi_0(s)$. The solution is easily found to be

$$x(s,\tau) = \frac{2(b - 3ac) + 3(2a^2 s + 2ac - b)e^{-c_2 a\tau} + be^{-3c_2 a\tau}}{6a^2}$$

$$t(s,\tau) = \frac{e^{-3c_2 a\tau} - 1}{3a}, \quad u(s,\tau) = \frac{\big(2a\psi_0(s) - b\big)e^{2c_2 a\tau} + b}{2a}.$$

As a simple application, we take $a = b = 0$ and $f = 0$. In this case, ψ_0 must satisfy the non-linear differential equation

$$\psi_0'''(s) + \big(\psi_0(s) - c\big)\psi_0'(s) = 0$$

whose solution is known to be

$$\psi_0(s) = 3c\,\mathrm{sech}^2\Big(\frac{c^{1/2}}{2}s + d\Big)$$

Since, in the limit $a \to 0, b \to 0$ we get

$$x = s - c_2 c\tau, \; t = -c_2\tau, \; u = \psi_0(s)$$

we have $s = x - ct$ and the soliton solution

$$u = \psi_0(x - ct) = 3c\,\mathrm{sech}^2\Big(\frac{c^{1/2}}{2}(x - ct) + d\Big)$$

is obtained as a generalised characteristic solution. ■

Example 9.6.4. This time we choose $n = 2, N = 2$ and consider the partial differential equations

$$\frac{\partial u}{\partial x} + \frac{\partial v}{\partial t} = 1, \quad \frac{\partial u}{\partial t}\frac{\partial v}{\partial x} = c^2$$

where c is a real constant. If we eliminate u or v between these equations, we see that u and v dependent variables have to satisfy separate non-linear wave equations

$$\Big(\frac{\partial u}{\partial t}\Big)^2 \frac{\partial^2 u}{\partial x^2} = c^2 \frac{\partial^2 u}{\partial t^2}, \quad \Big(\frac{\partial v}{\partial x}\Big)^2 \frac{\partial^2 v}{\partial t^2} = c^2 \frac{\partial^2 v}{\partial x^2}.$$

Let us write $x^1 = x$, $x^2 = t$, $u^1 = u$ and $u^2 = v$ so that $v_1^1 = u_{,1}^1, v_2^1 = u_{,2}^1$, $v_1^2 = u_{,1}^2, v_2^2 = u_{,2}^2$. The contact forms in $\mathcal{C}_1$ are given by

$$\sigma^1 = du^1 - v_1^1 dx^1 - v_2^1 dx^2, \ \ \sigma^2 = du^2 - v_1^2 dx^1 - v_2^2 dx^2.$$

The relevant components of the isovector field V of the contact ideal may be extracted from (9.4.27). We know that $\mathbf{X} = \mathbf{X}(\mathbf{x}, \mathbf{u})$, $\mathbf{U} = \mathbf{U}(\mathbf{x}, \mathbf{u})$ appearing in those expressions are arbitrary functions. 0-forms inducing the differential equations become

$$F^1 = v_1^1 + v_2^2 - 1 = 0, \ \ F^2 = v_2^1 v_1^2 - c^2 = 0.$$

We shall be looking for a simpler kind of an isovector field. Hence, we want to satisfy the conditions $\pounds_V dF^1 = 0$ and $\pounds_V dF^2 = 0$. Since $\mathbf{i}_V(F^1) = \mathbf{i}_V(F^2) = 0$, they are reduced to $\mathbf{i}_V(dF^1) = 0$ and $\mathbf{i}_V(dF^2) = 0$. These relations lead to

$$V_1^1 + V_2^2 = 0, \qquad v_1^2 V_2^1 + v_2^1 V_1^2 = 0.$$

If we insert the expressions $v_2^2 = 1 - v_1^1$ and $v_1^2 = c^2/v_2^1$ into (9.4.27) we get the following polynomial identity in terms of the variables v_1^1 and v_2^1

$$\begin{aligned}
&-\Big(\frac{\partial X^1}{\partial u^1} + \frac{\partial X^2}{\partial u^2}\Big)(v_1^1)^2 v_2^1 + \Big(\frac{\partial U^2}{\partial u^1} - \frac{\partial X^2}{\partial u^1} - \frac{\partial X^2}{\partial x^1}\Big)(v_2^1)^2 + \\
&\Big(\frac{\partial U^1}{\partial u^1} - \frac{\partial U^2}{\partial u^2} + 2\frac{\partial X^2}{\partial u^2} - \frac{\partial X^1}{\partial x^1} + \frac{\partial X^2}{\partial x^2}\Big) v_1^1 v_2^1 + \Big(\frac{\partial U^2}{\partial u^2} + \frac{\partial U^1}{\partial x^1} + \frac{\partial U^2}{\partial x^2} \\
&- (1+c^2)\frac{\partial X^2}{\partial u^2} - c^2\frac{\partial X^1}{\partial u^1} - \frac{\partial X^2}{\partial x^2}\Big) v_2^1 + c^2\Big(\frac{\partial U^1}{\partial u^2} - \frac{\partial X^1}{\partial u^2} - \frac{\partial X^1}{\partial x^2}\Big) = 0 \\
&\frac{\partial X^2}{\partial u^1}(v_1^1)^2 (v_2^1)^2 + \Big(\frac{\partial U^2}{\partial u^1} - \frac{\partial X^2}{\partial u^1} + \frac{\partial X^2}{\partial x^1}\Big) v_1^1 (v_2^1)^2 + c^2 \frac{\partial X^1}{\partial u^2}(v_1^1)^2 \\
&\quad + \Big(\frac{\partial U^2}{\partial x^1} - c^2 \frac{\partial X^2}{\partial u^1} - \frac{\partial X^2}{\partial x^1}\Big)(v_2^1)^2 - 2c^2\Big(\frac{\partial X^1}{\partial u^1} - \frac{\partial X^2}{\partial u^2}\Big) v_1^1 v_2^1 \\
&\qquad - c^2\Big(\frac{\partial U^1}{\partial u^2} + \frac{\partial X^1}{\partial u^2} + \frac{\partial X^1}{\partial x^2}\Big) v_1^1 + c^2\Big(\frac{\partial U^1}{\partial u^1} + \frac{\partial U^2}{\partial u^2} - 2\frac{\partial X^2}{\partial u^2} \\
&\qquad\quad - \frac{\partial X^1}{\partial x^1} - \frac{\partial X^2}{\partial x^2}\Big) v_2^1 + c^2\Big(\frac{\partial U^1}{\partial u^2} + \frac{\partial U^1}{\partial x^2} - c^2 \frac{\partial X^1}{\partial u^2}\Big) = 0.
\end{aligned}$$

Hence, we deduce 13 equations below obtained by setting the coefficients of v_1^1 and v_2^1 to zero:

$$\frac{\partial X^1}{\partial u^1}+\frac{\partial X^2}{\partial u^2}=0^1,\ \frac{\partial U^2}{\partial u^1}-\frac{\partial X^2}{\partial u^1}-\frac{\partial X^2}{\partial x^1}=0^2,\ \frac{\partial U^1}{\partial u^2}-\frac{\partial X^1}{\partial u^2}-\frac{\partial X^1}{\partial x^2}=0^3,$$
$$\frac{\partial U^1}{\partial u^1}-\frac{\partial U^2}{\partial u^2}+2\frac{\partial X^2}{\partial u^2}-\frac{\partial X^1}{\partial x^1}+\frac{\partial X^2}{\partial x^2}=0^4,\ \frac{\partial U^2}{\partial u^1}-\frac{\partial X^2}{\partial u^1}+\frac{\partial X^2}{\partial x^1}=0^5,$$
$$\frac{\partial U^2}{\partial u^2}+\frac{\partial U^1}{\partial x^1}+\frac{\partial U^2}{\partial x^2}-(1+c^2)\frac{\partial X^2}{\partial u^2}-c^2\frac{\partial X^1}{\partial u^1}-\frac{\partial X^2}{\partial x^2}=0^6,\ \frac{\partial X^2}{\partial u^1}=0^7,$$
$$\frac{\partial X^1}{\partial u^2}=0^8,\ \frac{\partial U^2}{\partial x^1}-c^2\frac{\partial X^2}{\partial u^1}-\frac{\partial X^2}{\partial x^1}=0^9,\ \frac{\partial X^1}{\partial u^1}-\frac{\partial X^2}{\partial u^2}=0^{10},$$
$$\frac{\partial U^1}{\partial u^2}+\frac{\partial X^1}{\partial u^2}+\frac{\partial X^1}{\partial x^2}=0^{11},\qquad \frac{\partial U^1}{\partial u^2}+\frac{\partial U^1}{\partial x^2}-c^2\frac{\partial X^1}{\partial u^2}=0^{12},$$
$$\frac{\partial U^1}{\partial u^1}+\frac{\partial U^2}{\partial u^2}-2\frac{\partial X^2}{\partial u^2}-\frac{\partial X^1}{\partial x^1}-\frac{\partial X^2}{\partial x^2}=0^{13}.$$

Equations 1 and 10 together with equations 7 and 8 give rise to

$$X^1=X^1(x^1,x^2),\quad X^2=X^2(x^1,x^2)$$

On employing these relations, it follows from equations 3 and 11 and equations 2 and 5 that

$$X^1=g(x^1),\ X^2=h(x^2),$$
$$U^1=U^1(x^1,x^2,u^1),\ U^2=U^2(x^1,x^2,u^2).$$

Thus equations 12 and 9 yield

$$U^1=U^1(x^1,u^1),\ U^2=U^2(x^2,u^2).$$

If we add and subtract equations 4 and 13, we obtain

$$\frac{\partial U^1}{\partial u^1}=g'(x^1),\ \frac{\partial U^2}{\partial u^2}=h'(x^2)$$

whose integrations result in

$$U^1=g'(x^1)u^1+\gamma(x^1),\ U^2=h'(x^2)u^2+\delta(x^2).$$

If we introduce these expressions into the equation 6, we find that

$$g''(x^1)u^1+h''(x^2)u^2+\gamma'(x^1)+\delta'(x^2)=0$$

whence we deduce that

$$g''(x^1)=0,\ h''(x^2)=0,\ \gamma'(x^1)=\ -\delta'(x^2)=\mathit{constant}$$

Hence, we are led to the conclusion

$$g = c_1x^1 + c_2,\ h = c_3x^2 + c_4,\ \gamma = c_5x^1 + c_6,\ \delta = -c_5x^2 + c_7.$$

Thus, the functions determining the isovector components become

$$\begin{aligned} X^1 &= c_1x^1 + c_2, & X^2 &= c_3x^2 + c_4, \\ U^1 &= c_1u^1 + c_5x^1 + c_6, & U^2 &= c_3u^2 - c_5x^2 + c_7. \end{aligned}$$

This means that the isovector field in question is the prolongation of the vector field

$$\begin{aligned} V_G = {} & (c_1x^1 + c_2)\frac{\partial}{\partial x^1} + (c_3x^2 + c_4)\frac{\partial}{\partial x^2} + (c_5x^1 + c_1u^1 + c_6)\frac{\partial}{\partial u^1} \\ & + (-c_5x^2 + c_3u^2 + c_7)\frac{\partial}{\partial u^2}. \end{aligned}$$

In order to easily produce a characteristic solution, we select a particular form of the vector field V_G by taking $c_1 = a, c_3 = 1, c_5 = b, c_2 = c_4 = c_6 = c_7 = 0$:

$$V_G = ax^1\frac{\partial}{\partial x^1} + x^2\frac{\partial}{\partial x^2} + (bx^1 + au^1)\frac{\partial}{\partial u^1} + (-bx^2 + u^2)\frac{\partial}{\partial u^2}$$

We define the mapping $\psi : \mathbb{R} \to C_1$ through the given smooth functions

$$\begin{aligned} & x^1 = s, \quad x^2 = 1, \quad u^1 = \psi_0^1(s), \quad u^2 = \psi_0^2(s) \\ & v_1^1 = \psi_1^1(s),\ v_2^1 = \psi_2^1(s),\ v_1^2 = \psi_1^2(s),\ v_2^2 = \psi_2^2(s). \end{aligned}$$

The transversality condition is met if $\psi^*(\mathbf{i}_V(\mu)) = ax^1dx^2 - x^2dx^1 \neq 0$. The constraints on initial data must satisfy the relations

$$\begin{aligned} & \psi_1^1 = (\psi_0^1)',\ \psi_1^2 = (\psi_0^2)', \\ & \psi_2^1 = bs + a\psi_0^1 - as(\psi_0^1)', \\ & \psi_2^2 = -b + \psi_0^2 - as(\psi_0^2)', \\ & \psi^*F^1 = (\psi_0^1)' - b + \psi_0^2 - as(\psi_0^2)' - 1 = 0, \\ & \psi^*F^2 = \left[bs + a\psi_0^1 - as(\psi_0^1)'\right](\psi_0^2)' = c^2. \end{aligned}$$

This amounts to say that to generate the characteristic solution corresponding to our present choice, the initial data ψ_0^1 and ψ_0^2 must satisfy the ordinary non-linear differential equations

$$(\psi_0^1)' + \psi_0^2 - as(\psi_0^2)' = b + 1,\ \left[bs + a\psi_0^1 - as(\psi_0^1)'\right](\psi_0^2)' = c^2.$$

The solution of this non-linear system is obviously not easy to find. But, we can try out to obtain a particular solution. Let us choose

$$\psi_0^1(s) = \alpha s + \beta s \log s, \;\; \psi_0^2\,(s) = \gamma + \delta \log s.$$

Introducing these functions into the differential equations, we see that the coefficients must satisfy the following relations

$$\alpha + \beta + \gamma - a\delta + (\beta + \delta) \log s = b + 1, \;\; (b - a\beta)\delta = c^2.$$

If we take $\delta = \; - \beta$, then the second equation implies that β ought to be chosen as a root of the quadratic equation

$$a\beta^2 - b\beta - c^2 = 0.$$

We therefore reach to the conclusion

$$\psi_0^1(s) = \alpha s + \beta s \log s, \;\; \psi_0^2\,(s) = b + 1 - \beta(1 + a) - \alpha - \beta \log s$$

where α is also an arbitrary constant. To determine the characteristic solution associated with the isovector field taken into consideration, we have to solve the ordinary linear differential equations

$$\frac{dx^1}{d\tau} = ax^1, \;\; \frac{dx^2}{d\tau} = x^2,$$
$$\frac{du^1}{d\tau} = bx^1 + au^1, \;\; \frac{du^2}{d\tau} = \; - bx^2 + u^2$$

under the initial conditions $x^1(0) = s$, $x^2(0) = 1$, $u^1(0) = \psi_0^1(s)$, $u^2(0) = \psi_0^2(s)$. We thus obtain

$$x^1 = se^{a\tau}, \;\; x^2 = e^{a\tau}, \;\; u^1 = s(\alpha + \beta \log s + b\tau)e^{a\tau},$$
$$u^2 = \big[b + 1 - \beta(1 + a) - \alpha - \beta \log s - b\tau\big]e^{a\tau}$$

describing the solution parametrically. ■

9.7. HORIZONTAL IDEALS AND THEIR SOLUTIONS

The most general transformation preserving the structure of a contact ideal $\mathcal{I}_m = \overline{\mathcal{I}}(\sigma^\alpha, \sigma^\alpha_{i_1}, \sigma^\alpha_{i_1 i_2}, \ldots \sigma^\alpha_{i_1 i_2 \cdots i_{m-1}}, d\sigma^\alpha_{i_1 i_2 \cdots i_{m-1}})$ has been determined in Sec. 9.3. Especially for $N > 1$, we know that this transformation is found as a prolongation of a point transformation in the graph space. This limitation creates, however, a major obstacle in obtaining solutions of a system of partial differential equations by using transformation methods This obstacle can be overcome to some extent by enlarging the contact ideal in an appropriate way. To this end, we would like first to determine linearly independent vector fields $V_i \in T(\mathcal{C}_m), i = 1, \ldots, n$ as to satisfy the relations

$$\mathbf{i}_{V_i}(dx^j) = \delta_i^j, \tag{9.7.1}$$
$$\mathbf{i}_{V_{i_1}} \circ \mathbf{i}_{V_{i_2}} \circ \cdots \circ \mathbf{i}_{V_{i_k}}(\omega) = 0,\ 1 \leq k \leq n,\ \forall\, \omega \in \mathcal{I}_m \cap \Lambda^k(\mathcal{C}_m).$$

We call such a set of vector fields as a ***canonical system***. A canonical system generates an n-dimensional submodule of the tangent bundle $T(\mathcal{C}_m)$, so they constitute a basis for the n-dimensional module of ***Cartan annihilators*** of the contact ideal $\mathcal{I}_m$. It is easily seen that the conditions $(9.7.1)_2$ are fulfilled if and only if the following relations are satisfied:

$$\mathbf{i}_{V_i}(\sigma^\alpha_{i_1\cdots i_r}) = 0,\ 0 \leq r \leq m-1; \tag{9.7.2}$$
$$\mathbf{i}_{V_j} \circ \mathbf{i}_{V_i}(d\sigma^\alpha_{i_1\cdots i_{m-1}}) = 0.$$

Indeed, if the conditions $(9.7.1)_2$ hold, then the conditions (9.7.2) are automatically satisfied since the generators $\sigma^\alpha_{i_1\cdots i_r}$ and $d\sigma^\alpha_{i_1 i_2\cdots i_{m-1}}$ of the ideal $\mathcal{I}_m$ are, respectively 1- and 2- forms. Conversely, let us assume that the conditions (9.7.2) are met. Let $\omega \in \mathcal{I}_m$ be a k-form in the ideal. Therefore, we have to write

$$\omega = \sum_{r=0}^{m-1} \lambda_\alpha^{i_1\cdots i_r} \wedge \sigma^\alpha_{i_1\cdots i_r} + \Lambda_\alpha^{i_1\cdots i_{m-1}} \wedge d\sigma^\alpha_{i_1\cdots i_{m-1}}$$

where $\lambda_\alpha^{i_1\cdots i_r} \in \Lambda^{k-1}(\mathcal{C}_m)$ and $\Lambda_\alpha^{i_1\cdots i_{m-1}} \in \Lambda^{k-2}(\mathcal{C}_m)$. But, we have

$$\mathbf{i}_{V_{i_1}} \circ \cdots \circ \mathbf{i}_{V_{i_k}}(\lambda_\alpha^{i_1\cdots i_r}) = 0,\ \ \mathbf{i}_{V_{i_1}} \circ \cdots \circ \mathbf{i}_{V_{i_{k-1}}}(\Lambda_\alpha^{i_1\cdots i_{m-1}}) = 0$$

because of the degrees of those forms. Then we immediately observe that we get

$$\begin{aligned}\mathbf{i}_{V_{i_1}} \circ \cdots \circ \mathbf{i}_{V_{i_k}}(\omega) = &\sum_{r=0}^{m-1} \mathbf{i}_{V_{i_1}} \circ \cdots \circ \mathbf{i}_{V_{i_k}}(\lambda_\alpha^{i_1\cdots i_r}) \wedge \sigma^\alpha_{i_1\cdots i_r}\\ &+ \mathbf{i}_{V_{i_1}} \circ \cdots \circ \mathbf{i}_{V_{i_{k-1}}}(\Lambda_\alpha^{i_1\cdots i_{m-1}}) \wedge d\sigma^\alpha_{i_1\cdots i_{m-1}} = 0\end{aligned}$$

provided the relations (9.7.2) are satisfied. Let us represent a vector field $V_i \in T(\mathcal{C}_m)$ by

$$V_i = X_i^j \frac{\partial}{\partial x^j} + \sum_{r=0}^{m} V^\alpha_{i i_1\cdots i_r} \frac{\partial}{\partial v^\alpha_{i_1\cdots i_r}}$$

where $X_i^j, V^\alpha_{i i_1\cdots i_r} \in \Lambda^0(\mathcal{C}_m)$. *It is clear that the smooth functions $V^\alpha_{i i_1\cdots i_r}$ are to be taken as completely symmetric in subscripts $i_1, \ldots, i_r$ without loss of generality*. The condition $(9.7.1)_1$ yields simply

$$X_i^j = \delta_i^j$$

whereas we find from $(9.7.2)_1$ that

$$\mathbf{i}_{V_i}(dv^\alpha_{i_1\cdots i_r} - v^\alpha_{i_1\cdots i_r j}\,dx^j) = V^\alpha_{ii_1\cdots i_r} - v^\alpha_{i_1\cdots i_r i} = 0$$

and $V^\alpha_{ii_1\cdots i_r} = v^\alpha_{i_1\cdots i_r i}$ for $0 \le r \le m-1$. On the other hand, we can write $d\sigma^\alpha_{i_1\cdots i_{m-1}} = -\,dv^\alpha_{i_1\cdots i_{m-1}k} \wedge dx^k$ so that we arrive at the following interior products

$$\begin{aligned}\mathbf{i}_{V_i}(d\sigma^\alpha_{i_1\cdots i_{m-1}}) &= -\,V^\alpha_{ii_1\cdots i_{m-1}k}\,dx^k + dv^\alpha_{i_1\cdots i_{m-1}i},\\ \mathbf{i}_{V_j} \circ \mathbf{i}_{V_i}(d\sigma^\alpha_{i_1\cdots i_{m-1}}) &= -\,V^\alpha_{ii_1\cdots i_{m-1}j} + V^\alpha_{ji_1\cdots i_{m-1}i} = 0.\end{aligned}$$

This implies the symmetry property $V^\alpha_{ii_1\cdots i_{m-1}j} = V^\alpha_{ji_1\cdots i_{m-1}i}$ amounting to say that *the coefficients* $V^\alpha_{ii_1\cdots i_m}$ *must be completely symmetric with respect to all their subscripts*. Therefore the general form of a canonical system involving n linearly independent vector fields and satisfying the conditions (9.7.1) is given by

$$V_i = \frac{\partial}{\partial x^i} + \sum_{r=0}^{m-1} v^\alpha_{i_1\cdots i_r i}\frac{\partial}{\partial v^\alpha_{i_1\cdots i_r}} + V^\alpha_{ii_1\cdots i_m}\frac{\partial}{\partial v^\alpha_{i_1\cdots i_m}}. \qquad (9.7.3)$$

Thus a contact manifold of order m admits infinitely many canonical systems associated with $N\binom{n+m}{m+1} = \frac{N(n+m)!}{(n-1)!\,(m+1)!}$ number of arbitrary smooth functions $V^\alpha_{ii_1\cdots i_m} \in \Lambda^0(\mathcal{C}_m)$. We next consider another vector V_j of the canonical system by

$$V_j = \frac{\partial}{\partial x^j} + \sum_{s=0}^{m-1} v^\beta_{j_1\cdots j_s j}\frac{\partial}{\partial v^\beta_{j_1\cdots j_s}} + V^\beta_{jj_1\cdots j_m}\frac{\partial}{\partial v^\beta_{j_1\cdots j_m}}$$

Successive application of the operators V_i and V_j results in the following expression after some manipulations

$$\begin{aligned}V_iV_j = {} & \frac{\partial^2}{\partial x^i\partial x^j} + V^\alpha_{ii_1\cdots i_m}\frac{\partial^2}{\partial v^\alpha_{i_1\cdots i_m}\partial x^j} + V^\alpha_{ji_1\cdots i_m}\frac{\partial^2}{\partial v^\alpha_{i_1\cdots i_m}\partial x^i}\\ & + \sum_{r=0}^{m-1}\Big[v^\alpha_{i_1\cdots i_r i}\frac{\partial^2}{\partial v^\alpha_{i_1\cdots i_r}\partial x^j} + v^\alpha_{i_1\cdots i_r j}\frac{\partial^2}{\partial v^\alpha_{i_1\cdots i_r}\partial x^i}\Big]\\ & + \sum_{r=0}^{m-1}\sum_{s=0}^{m-1} v^\alpha_{i_1\cdots i_r i}v^\beta_{j_1\cdots j_s j}\frac{\partial^2}{\partial v^\alpha_{i_1\cdots i_r}\partial v^\beta_{j_1\cdots j_s}}\end{aligned}$$

$$+\sum_{r=0}^{m-1}\big[V^\alpha_{ii_1\cdots i_m}v^\beta_{j_1\cdots j_rj}+V^\alpha_{ji_1\cdots i_m}v^\beta_{j_1\cdots j_ri}\big]\frac{\partial^2}{\partial v^\alpha_{i_1\cdots i_m}\partial v^\beta_{j_1\cdots j_r}}$$
$$+\sum_{r=1}^{m-1}v^\alpha_{i_1\cdots i_{r-1}ij}\frac{\partial}{\partial v^\alpha_{i_1\cdots i_{r-1}}}+V^\alpha_{ii_1\cdots i_{m-1}j}\frac{\partial}{\partial v^\alpha_{i_1\cdots i_{m-1}}}$$
$$+V_i\big(V^\alpha_{ji_1\cdots i_m}\big)\frac{\partial}{\partial v^\alpha_{i_1\cdots i_m}}$$

where we have renamed the dummy indices whenever necessary. Hence, the Lie product of these two vector fields is easily found to be

$$\begin{aligned}[V_i,V_j]&=V_iV_j-V_jV_i \\ &=\big[V_i\big(V^\alpha_{ji_1\cdots i_m}\big)-V_j\big(V^\alpha_{ii_1\cdots i_m}\big)\big]\frac{\partial}{\partial v^\alpha_{i_1\cdots i_m}}.\end{aligned}\tag{9.7.4}$$

The differential of a function $f\in\Lambda^0(\mathcal{C}_m)$ can now be expressed as

$$\begin{aligned}df&=\frac{\partial f}{\partial x^i}\,dx^i+\sum_{r=0}^{m}\frac{\partial f}{\partial v^\alpha_{i_1\cdots i_r}}\,dv^\alpha_{i_1\cdots i_r} \\ &=\Big[\frac{\partial f}{\partial x^i}+\sum_{r=0}^{m-1}v^\alpha_{i_1\cdots i_ri}\frac{\partial f}{\partial v^\alpha_{i_1\cdots i_r}}+V^\alpha_{ii_1\cdots i_m}\frac{\partial f}{\partial v^\alpha_{i_1\cdots i_m}}\Big]dx^i \\ &\quad+\sum_{r-0}^{m-1}\frac{\partial f}{\partial v^\alpha_{i_1\cdots i_,}}\,\sigma^\alpha_{i_1\cdots i_r}+\frac{\partial f}{\partial v^\alpha_{i_1\cdots i_m}}\,\Sigma^\alpha_{i_1\cdots i_m} \\ &=V_i(f)\,dx^i+\sum_{r=0}^{m-1}\frac{\partial f}{\partial v^\alpha_{i_1\cdots i_r}}\,\sigma^\alpha_{i_1\cdots i_r}+\frac{\partial f}{\partial v^\alpha_{i_1\cdots i_m}}\,\Sigma^\alpha_{i_1\cdots i_m}\in\Lambda^1(\mathcal{C}_m).\end{aligned}\tag{9.7.5}$$

In order to obtain this relation we have first replaced 1-forms $dv_{i_1\cdots i_r}$ with 1-forms $\sigma_{i_1\cdots i_r}+v^\alpha_{i_1\cdots i_ri}\,dx^i$ for $0\le r\le m-1$ and then we have further introduced 1-forms $\Sigma^\alpha_{i_1\cdots i_m}$ appearing in the above expression. However, their definition will be given a little bit later in (9.7.7).

The ***vertical ideal*** of the exterior algebra $\Lambda(\mathcal{C}_m)$ is identified as the closed ideal prescribed as follows

$$\mathcal{V}_m=\mathcal{I}(dx^1,dx^2,\dots,dx^n).\tag{9.7.6}$$

On the D-dimensional submanifold $\{x^i=c^i;i=1,\dots,n\}$ where c^i's are constants, this ideal is obviously annihilated. In other words, the ideal $\mathcal{V}_m$ vanishes when restricted to the fibres of the ideal $\mathcal{C}_m$ over $\mathcal{D}_n$. This, of course, justifies our use of the term '*vertical*'. Let us now take into account the forms

$$\Sigma^\alpha_{i_1\cdots i_m} = dv^\alpha_{i_1\cdots i_m} - V^\alpha_{ii_1\cdots i_m} dx^i \in \Lambda^1(\mathcal{C}_m). \tag{9.7.7}$$

We shall call them as ***horizontal* 1*-forms***. A ***horizontal ideal*** of the exterior algebra $\Lambda(\mathcal{C}_m)$ will now be defined as

$$\mathcal{H}_m = \mathcal{I}(\sigma^\alpha_{i_1\cdots i_r}, 0 \le r \le m-1;\ \Sigma^\alpha_{i_1\cdots i_m}). \tag{9.7.8}$$

We know that $d\sigma^\alpha_{i_1\cdots i_r} \in \mathcal{H}_m$ for $0 \le r \le m-2$. For $r = m-1$, due to the relation $d\sigma^\alpha_{i_1\cdots i_{m-1}} = -\, dv^\alpha_{i_1\cdots i_{m-1}i} \wedge dx^i$ we find immediately that

$$\begin{aligned} d\sigma^\alpha_{i_1\cdots i_{m-1}} &= -\,\Sigma^\alpha_{i_1\cdots i_{m-1}i} \wedge dx^i - V^\alpha_{ii_1\cdots i_{m-1}j}\, dx^i \wedge dx^j \\ &= -\,\Sigma^\alpha_{i_1\cdots i_{m-1}i} \wedge dx^i \in \mathcal{H}_m. \end{aligned} \tag{9.7.9}$$

Moreover on using (9.7.5), the relation

$$d\Sigma^\alpha_{i_1\cdots i_m} = -\, dV^\alpha_{ji_1\cdots i_m} \wedge dx^j$$

leads to

$$\begin{aligned} d\Sigma^\alpha_{i_1\cdots i_m} = -\, V_i(V^\alpha_{ji_1\cdots i_m})\, dx^i \wedge dx^j - \sum_{r=0}^{m-1} \frac{\partial V^\alpha_{ji_1\cdots i_m}}{\partial v^\alpha_{i_1\cdots i_r}} \sigma_{i_1\cdots i_r} \wedge dx^j \\ - \frac{\partial V^\alpha_{ji_1\cdots i_m}}{\partial v^\alpha_{i_1\cdots i_m}} \Sigma^\alpha_{i_1\cdots i_m} \wedge dx^j \end{aligned}$$

from which we deduce that

$$d\Sigma^\alpha_{i_1\cdots i_m} + \frac{1}{2}\big[V_i(V^\alpha_{ji_1\cdots i_m}) - V_j(V^\alpha_{ii_1\cdots i_m})\big] dx^i \wedge dx^j \in \mathcal{H}_m. \tag{9.7.10}$$

Therefore, we get $d\Sigma^\alpha_{i_1\cdots i_m} \in \mathcal{H}_m$ *if and only if the conditions*

$$V_i(V^\alpha_{ji_1\cdots i_m}) = V_j(V^\alpha_{ii_1\cdots i_m}) \tag{9.7.11}$$

are satisfied. In this case, the horizontal ideal $\mathcal{H}_m$ turns out to be a closed ideal. In view of (9.7.4), we see at once that the relation (9.7.11) becomes possible if and only if

$$[V_i, V_j] = 0,\ \ 1 \le i, j \le n \tag{9.7.12}$$

that is, *if the canonical system consists of commuting vector fields.*

We denote the characteristic subspace of the ideal $\mathcal{H}_m$ which will be called henceforth as the ***horizontal module*** by $\mathcal{S}_{\mathcal{H}_m}$. Thus, if $U \in T(\mathcal{C}_m)$ is a characteristic vector of $\mathcal{H}_m$, then the relation $\mathbf{i}_U(\mathcal{H}_m) \subset \mathcal{H}_m$ must be satisfied. Since all generators of the horizontal ideal are 1-forms, it becomes possible to comply with this condition if and only if we have

$$\mathbf{i}_U(\sigma^\alpha_{i_1\cdots i_r}) = 0,\ 0 \le r \le m-1;\ \ \mathbf{i}_U(\Sigma^\alpha_{i_1\cdots i_m}) = 0. \tag{9.7.13}$$

Let us take a vector field

$$U = X^i\frac{\partial}{\partial x^i} + \sum_{r=0}^{m} U^\alpha_{i_1\cdots i_r}\frac{\partial}{\partial v^\alpha_{i_1\cdots i_r}}$$

into consideration. Then (9.7.13) requires that

$$U^\alpha_{i_1\cdots i_r} = v^\alpha_{i_1\cdots i_r i}X^i,\ 0 \le r \le m-1;\ \ U^\alpha_{i_1\cdots i_m} = V^\alpha_{ii_1\cdots i_m}X^i.$$

Hence, any vector $U \in \mathcal{S}_{\mathcal{H}_m}$ can be written as

$$U = X^i\Big(\frac{\partial}{\partial x^i} + \sum_{r=0}^{m-1} v^\alpha_{i_1\cdots i_r i}\frac{\partial}{\partial v^\alpha_{i_1\cdots i_r}} + V^\alpha_{ii_1\cdots i_m}\frac{\partial}{\partial v^\alpha_{i_1\cdots i_m}}\Big) = X^i V_i.$$

Thus, *canonical system constitutes a basis of the horizontal module as well.* Consequently, *to each choice of completely symmetric smooth functions $V^\alpha_{ii_1\cdots i_r} \in \Lambda^0(\mathcal{C}_m)$ there corresponds a horizontal ideal of $\Lambda(\mathcal{C}_m)$. The horizontal module of this ideal coincides with both the modules of characteristic vectors of $\mathcal{H}_m$ and Cartan annihilators of the contact ideal $\mathcal{I}_m$.* It is clear that the vectors V_i satisfy naturally the characteristic conditions (9.7.13). In this situation the canonical system produces the distribution $\mathcal{S}_{\mathcal{H}_m}$. In case the conditions (9.7.11) are also met, this distribution proves to be involutive. We can now show the following theorem.

Theorem 9.7.1. *The horizontal module $\mathcal{S}_{\mathcal{H}_m}$ is the module of isovectors of the horizontal ideal $\mathcal{H}_m$ if and only if the ideal $\mathcal{H}_m$ is closed.*

In order that a vector field $V \in \mathcal{S}_{\mathcal{H}_m}$ is to be an isovector field of the horizontal ideal $\mathcal{H}_m$, the conditions $\pounds_V\sigma^\alpha_{i_1\cdots i_r} \in \mathcal{H}_m$ where $0 \le r \le m-1$ and $\pounds_V\Sigma^\alpha_{i_1\cdots i_m} \in \mathcal{H}_m$ should be satisfied. If we note (9.7.13), we get

$$\pounds_V\sigma^\alpha_{i_1\cdots i_r} = \mathbf{i}_V(d\sigma^\alpha_{i_1\cdots i_r}),\ \ \pounds_V\Sigma^\alpha_{i_1\cdots i_m} = \mathbf{i}_V(d\Sigma^\alpha_{i_1\cdots i_m}).$$

The relations (9.7.9) and (9.7.10) lead to $\pounds_V\sigma^\alpha_{i_1\cdots i_r} \in \mathcal{H}_m$ for $0 \le r \le m-1$ together with

$$\pounds_V\Sigma^\alpha_{i_1\cdots i_m} + \frac{1}{2}\big[V_i(V^\alpha_{ji_1\cdots i_m}) - V_j(V^\alpha_{ii_1\cdots i_m})\big]\mathbf{i}_V(dx^i \wedge dx^j) \in \mathcal{H}_m.$$

If we write $V = X^iV_i$, then we find $\mathbf{i}_V(dx^i \wedge dx^j) = X^i\,dx^j - X^j\,dx^i$. Therefore, the last terms belong obviously to the vertical ideal. Thus, we find $\pounds_V \Sigma^\alpha_{i_1\cdots i_m} \in \mathcal{H}_m$ *if and only if* the conditions

$$V_i(V^\alpha_{j i_1\cdots i_m}) = V_j(V^\alpha_{i i_1\cdots i_m})$$

are again satisfied. (9.7.11) constitute the necessary and sufficient conditions for a horizontal ideal $\mathcal{H}_m$ to be closed. We had seen that they were equivalent to the conditions $[V_i, V_j] = 0$. □

According to Theorem 5.13.4, the closed horizontal ideals are completely integrable. Let $\mathfrak{H}_m$ denote the set of all closed horizontal ideals. We can readily demonstrate that this set is not empty. For instance, we may consider the smooth functions $f^\alpha \in \Lambda^0(M)$ and define the functions

$$V^\alpha_{i i_1\cdots i_m} = \frac{\partial^{m+1} f^\alpha(\mathbf{x})}{\partial x^i \partial x^{i_1}\cdots \partial x^{i_m}}.$$

These function plainly verify both the condition of complete symmetry and the relations (9.7.11). When we consider a member of $\mathfrak{H}_m$, the vectors V_i, $1 \leq i \leq n$ generate n-dimensional integral manifolds in $\mathcal{C}_m$ annihilating the closed ideal $\mathcal{H}_m$. Since the dimension of $\mathcal{C}_m$ is $n + D$, we know that these manifolds are obtainable from the independent solutions $g^a \in \Lambda^0(\mathcal{C}_m)$, $a = 1, \ldots, D$ of the linear partial differential equations

$$V_i(g) = 0,\ i = 1, \ldots, n$$

by setting $g^a = c^a$ where c^a are real constants. The general solution of the above equations may be written as $g = G(g^1, \ldots, g^D) = G(g^a)$. Hence, the closed ideal $\mathcal{H}_m$ provides an n-dimensional *foliation* on the manifold $\mathcal{C}_m$ [*see* Sec. 2.11]. Each choice of constants characterises a *leaf*.

Next, we shall try to calculate all isovector fields of a horizontal ideal $\mathcal{H}_m \in \mathfrak{H}_m$. If U is an isovector, then the relations $\pounds_U \sigma^\alpha_{i_1\cdots i_r} \in \mathcal{H}_m$ for $0 \leq r \leq m-1$ and $\pounds_U \Sigma^\alpha_{i_1\cdots i_m} \in \mathcal{H}_m$ must be satisfied. If we make use of the Cartan formula $\pounds_U \omega = d\big(\mathbf{i}_U(\omega)\big) + \mathbf{i}_U(d\omega)$, we get

$$\begin{aligned}
\pounds_U \sigma^\alpha_{i_1\cdots i_r} &= d\big(\mathbf{i}_U(\sigma^\alpha_{i_1\cdots i_r})\big) - \mathbf{i}_U(\sigma^\alpha_{i_1\cdots i_r i})\, dx^i + \mathbf{i}_U(dx^i)\, \sigma^\alpha_{i_1\cdots i_r i},\\
\pounds_U \sigma^\alpha_{i_1\cdots i_{m-1}} &= d\big(\mathbf{i}_U(\sigma^\alpha_{i_1\cdots i_{m-1}})\big) - \mathbf{i}_U(\Sigma^\alpha_{i_1\cdots i_{m-1} i})\, dx^i + \mathbf{i}_U(dx^i)\, \Sigma^\alpha_{i_1\cdots i_{m-1} i},\\
\pounds_U \Sigma^\alpha_{i_1\cdots i_m} &= d\big(\mathbf{i}_U(\Sigma^\alpha_{i_1\cdots i_m})\big) - \mathbf{i}_U(dV^\alpha_{i i_1\cdots i_{m-1}})\, dx^i + \mathbf{i}_U(dx^i)\, dV^\alpha_{i_1\cdots i_m i}
\end{aligned}$$

where $0 \leq r \leq m-2$. Let us represent the isovector U as

$$U = X^i V_i + \sum_{r=0}^{m} U^\alpha_{i_1\cdots i_r} \frac{\partial}{\partial v^\alpha_{i_1\cdots i_r}}$$

without loss of generality where X^i, $U^\alpha_{i_1\cdots i_r} \in \Lambda^0(\mathcal{C}_m)$ for $0 \leq r \leq m$ are smooth functions. Since members of the canonical system are characteristic vectors of $\mathcal{H}_m$, we eventually obtain

$$\begin{aligned}\pounds_U\sigma^\alpha_{i_1\cdots i_r} &= \big[V_i(U^\alpha_{i_1\cdots i_r}) - U^\alpha_{i_1\cdots i_r i}\big]dx^i + \sum_{s=0}^{m-1}\frac{\partial U^\alpha_{i_1\cdots i_r}}{\partial v^\beta_{i_1\cdots i_s}}\sigma^\beta_{i_1\cdots i_s}\\ &\quad + \frac{\partial U^\alpha_{i_1\cdots i_r}}{\partial v^\beta_{i_1\cdots i_m}}\Sigma^\beta_{i_1\cdots i_m} + X^i\sigma^\alpha_{i_1\cdots i_r i}, \qquad 0\le r\le m-2\\ \pounds_U\sigma^\alpha_{i_1\cdots i_{m-1}} &= \big[V_i(U^\alpha_{i_1\cdots i_{m-1}}) - U^\alpha_{i_1\cdots i_{m-1}i}\big]dx^i + \sum_{r=0}^{m-1}\frac{\partial U^\alpha_{i_1\cdots i_{m-1}}}{\partial v^\beta_{i_1\cdots i_r}}\sigma^\beta_{i_1\cdots i_r}\\ &\quad + \frac{\partial U^\alpha_{i_1\cdots i_{m-1}}}{\partial v^\beta_{i_1\cdots i_m}}\Sigma^\beta_{i_1\cdots i_m} + X^i\Sigma^\alpha_{i_1\cdots i_{m-1}i}.\end{aligned}$$

Thus, in order to get $\pounds_U\sigma^\alpha_{i_1\cdots i_r}\in\mathcal{H}_m, 0\le r\le m-1$ we have to set

$$U^\alpha_{i_1\cdots i_r i} = V_i(U^\alpha_{i_1\cdots i_r}),\ \ 0\le r\le m-1.$$

The solution of this recurrence relation is clearly given by

$$U^\alpha_{i_1 i_2\cdots i_r} = V_{i_1}V_{i_2}\cdots V_{i_r}(U^\alpha),\ \ 0\le r\le m \tag{9.7.14}$$

in terms of N functions $U^\alpha\in\Lambda^0(\mathcal{C}_m)$ where we have adopted the convention that $U^\alpha_{i_0}=U^\alpha$. On the other hand, the relation

$$\begin{aligned}\pounds_U\Sigma^\alpha_{i_1\cdots i_m} &= \big[V_i(U^\alpha_{i_1\cdots i_m}) - U(V^\alpha_{ii_1\cdots i_m}) + X^jV_i(V^\alpha_{ji_1\cdots i_m})\big]dx^i\\ &\quad + \sum_{r=0}^{m-1}\frac{\partial U^\alpha_{i_1\cdots i_m}}{\partial v^\beta_{i_1\cdots i_r}}\sigma^\beta_{i_1\cdots i_r} + \frac{\partial U^\alpha_{i_1\cdots i_m}}{\partial v^\beta_{i_1\cdots i_m}}\Sigma^\beta_{i_1\cdots i_m}\\ &\quad + X^j\Big[\sum_{r=0}^{m-1}\frac{\partial V^\alpha_{ji_1\cdots i_m}}{\partial v^\beta_{i_1\cdots i_r}}\sigma^\beta_{i_1\cdots i_r} + \frac{\partial V^\alpha_{ji_1\cdots i_m}}{\partial v^\beta_{i_1\cdots i_m}}\Sigma^\beta_{i_1\cdots i_m}\Big]\end{aligned}$$

requires that we have to satisfy the following equations

$$V_i(U^\alpha_{i_1\cdots i_m}) - U(V^\alpha_{ii_1\cdots i_m}) + X^jV_i(V^\alpha_{ji_1\cdots i_m}) = 0$$

in order to get $\pounds_U\Sigma^\alpha_{i_1\cdots i_m}\in\mathcal{H}_m$. Thereby we obtain the expressions

$$V_i(U^\alpha_{i_1\cdots i_m}) = \sum_{r=0}^{m}U^\beta_{i_1\cdots i_r}\frac{\partial V^\alpha_{ii_1\cdots i_m}}{\partial v^\beta_{i_1\cdots i_r}} + X^j\big[V_j(V^\alpha_{ii_1\cdots i_m}) - V_i(V^\alpha_{ji_1\cdots i_m})\big].$$

Because $\mathcal{H}_m$ is closed, the relations (9.7.11) are to be satisfied. Hence, the smooth functions U^α ought to verify the restrictions

$$\begin{aligned}V_iV_{i_1}\cdots V_{i_m}(U^\alpha) &= \sum_{r=0}^{m} V_{i_1}\cdots V_{i_r}(U^\beta)\frac{\partial V^\alpha_{ii_1\cdots i_m}}{\partial v^\beta_{i_1\cdots i_r}} \\ &= U^\beta \frac{\partial V^\alpha_{ii_1\cdots i_m}}{\partial u^\beta} + V_{i_1}(U^\beta)\frac{\partial V^\alpha_{ii_1\cdots i_m}}{\partial v^\beta_{i_1}} + \cdots + V_{i_1}\cdots V_{i_m}(U^\beta)\frac{\partial V^\alpha_{ii_1\cdots i_m}}{\partial v^\beta_{i_1\cdots i_m}}.\end{aligned} \tag{9.7.15}$$

For $U^\alpha = 0$, (9.7.15) is trivially fulfilled. Therefore we arrive at the following theorem.

Theorem 9.7.2. *In terms of n functions $X^i \in \Lambda^0(\mathcal{C}_m)$ and N functions $U^\alpha \in \Lambda^0(\mathcal{C}_m)$ satisfying* (9.7.15), *all isovector fields of a closed horizontal ideal $\mathcal{H}_m$ are expressible in the form*

$$U = X^i V_i + \sum_{r=0}^{m} V_{i_1}\cdots V_{i_r}(U^\alpha)\frac{\partial}{\partial v^\alpha_{i_1\cdots i_r}}. \tag{9.7.16}$$

Canonical system constitutes also the module of isovectors of closed $\mathcal{H}_m$ if only one chooses $U^\alpha = 0$. □

When the ideal $\mathcal{H}_m$ is closed, the canonical system is a Lie algebra and n-dimensional submanifold $\mathfrak{S}$ it produces annihilates this ideal. The mapping $\phi : \mathfrak{S} \to \mathcal{C}_m$ prescribing the manifold $\mathfrak{S}$ is a solution mapping of the ideal, that is, one has $\phi^*\mathcal{H}_m = 0$. Since the Lie products of vectors V_i vanish, they generate a coordinate mesh on $\mathfrak{S}$. We can determine the mapping ϕ by means of congruences that are integral curves of vector fields V_i. Let us denote $\mathfrak{n} = n + D$ number of coordinates of the manifold $\mathcal{C}_m$ by

$$\{x^i, v^\alpha_{i_1\cdots i_r} : 0 \le r \le m\} = \{z^{\mathfrak{a}} : 1 \le \mathfrak{a} \le \mathfrak{n}\}$$

as in Sec. 9.6. Let us take into consideration characteristic vector fields, or Cartan annihilators $V_i = v^{\mathfrak{a}}_i(\mathbf{z})\partial/\partial z^{\mathfrak{a}} \in T(\mathcal{C}_m)$ of the horizontal ideal $\mathcal{H}_m$. We know that Lie products of these vector fields vanish. Their trajectories are found as usual by integrating the ordinary differential equations

$$\frac{d\zeta^{\mathfrak{a}}}{dt^i} = v^{\mathfrak{a}}_i(\boldsymbol{\zeta}), \quad \zeta^{\mathfrak{a}}(0) = z^{\mathfrak{a}}$$

where t^i is a real parameter. In order to determine the mapping ϕ, we start with the vector field V_1. We can formally express the solution of the ordinary differential equations

$$\frac{d\zeta^{\mathfrak{a}}_1}{dt^1} = v^{\mathfrak{a}}_1(\boldsymbol{\zeta}_1), \quad \zeta^{\mathfrak{a}}_1(0) = z^{\mathfrak{a}}$$

as $\zeta^{\mathfrak{a}}_1(t^1;\mathbf{z}) = e^{t^1V_1}(z^{\mathfrak{a}})$. In the second step, the solution of the equations

$$\frac{d\zeta_2^{\mathfrak{a}}}{dt^2} = v_2^{\mathfrak{a}}(\boldsymbol{\zeta}_2), \quad \zeta_2^{\mathfrak{a}}(0) = \zeta_1^{\mathfrak{a}}(t^1; \mathbf{z})$$

can be written as $\zeta_2^{\mathfrak{a}}(t^2; \zeta_1) = e^{t^2 V_2}(\zeta_1) = e^{t^2 V_2} e^{t^1 V_1}(z^{\mathfrak{a}})$. Since the vectors V_1 and V_2 commute, we then find that

$$\zeta_2^{\mathfrak{a}}(t^2; \zeta_1(t^1; \mathbf{z})) = e^{t^1 V_1 + t^2 V_2}(z^{\mathfrak{a}}).$$

If we continue in this fashion, the mapping $\boldsymbol{\zeta} = \phi(\mathbf{t}; \mathbf{z})$ is specified by the relation

$$\zeta^{\mathfrak{a}} = e^{t^i V_i}(z^{\mathfrak{a}}). \tag{9.7.17}$$

This expression determines n-dimensional solution manifold of a closed ideal $\mathcal{H}_m$ through any point $\mathbf{z}$ of $\mathcal{C}_m$ or, in other words, a leaf of the foliation annihilating the ideal $\mathcal{H}_m$ passing through a point $z^{\mathfrak{a}}$. The integration parameters $t^1, t^2, \dots, t^n$ form the *natural coordinates* of the solution manifold.

The action of the mapping ϕ on the coordinates x^i of the manifold M can easily be evaluated. Since $v_j^i = \delta_j^i$, the differential equations

$$\frac{dx^i}{dt^j} = \delta_j^i, \quad x^i(0) = x_0^i$$

yield immediately the simple solution

$$x^i = x_0^i + t^i. \tag{9.7.18}$$

Thus, coordinates of the open set $\mathcal{D}_n$ of $\mathbb{R}^n$ over which differential equations are defined and local coordinates of the solution manifold are connected by a simple translation. In this case, we have $\phi^* dx^i = dt^i$ and as a result of this we obtain

$$\phi^* \mu = \phi^*(dx^1 \wedge dx^2 \wedge \cdots \wedge dx^n) = dt^1 \wedge dt^2 \wedge \cdots \wedge dt^n.$$

On writing $\phi^* u^\alpha = \phi^\alpha(\mathbf{t})$ and noting that $\phi^* \sigma_{i_1 \cdots i_r}^\alpha = 0$ and $\phi^* \Sigma_{i_1 \cdots i_m}^\alpha = 0$, we draw the conclusion

$$\phi^* v_{i_1 \cdots i_r}^\alpha = \frac{\partial^r \phi^\alpha(\mathbf{t})}{\partial t^{i_1} \cdots \partial t^{i_r}}, 0 \le r \le m-1; \; \phi^* V_{i i_1 \cdots i_m}^\alpha = \frac{\partial^{m+1} \phi^\alpha(\mathbf{t})}{\partial t^i \partial t^{i_1} \cdots \partial t^{i_m}}.$$

Hence, selected functions $V_{i i_1 \cdots i_m}^\alpha$ provide information about $(m+1)$th order partial derivatives of functions u^α on the solution manifold.

If we take notice of the relation (9.7.9), we immediately realise that ϕ is a solution mapping of the contact ideal. Instead of the closed fundamental ideal $\mathfrak{I}_m$ [*see p.* 521], let us now introduce a *new balance ideal* by

$$\mathfrak{B}_m = \mathcal{I}(\sigma^\alpha_{i_1\cdots i_r}, 0 \le r \le m-1; \Sigma^\alpha_{i_1\cdots i_m}; \omega^\alpha)$$

where n-forms ω^α are given in (9.4.3). *If* $\mathcal{H}_m \in \mathfrak{H}_m$, *then* $\mathfrak{B}_m$ *is closed.* In fact, since $dx^i \wedge \mu = 0$ then (9.7.5) for the function Σ^α implies that

$$d\omega^\alpha = d\Sigma^\alpha \wedge \mu \in \mathcal{H}_m \subset \mathfrak{B}_m.$$

On the other hand, we can similarly write

$$\omega^\alpha = d\Sigma^{\alpha i} \wedge \mu_i + \Sigma^\alpha \mu = \big[V_i(\Sigma^{\alpha i}) + \Sigma^\alpha\big]\mu$$
$$+ \Big[\sum_{r=0}^{m-1} \frac{\partial \Sigma^{\alpha i}}{\partial v^\beta_{i_1\cdots i_r}} \sigma^\beta_{i_1\cdots i_r} + \frac{\partial \Sigma^{\alpha i}}{\partial v^\beta_{i_1\cdots i_m}} \Sigma^\beta_{i_1\cdots i_m}\Big] \wedge \mu_i \in \Lambda^n(\mathcal{C}_m).$$

Introducing the smooth functions

$$\mathcal{F}^\alpha = V_i(\Sigma^{\alpha i}) + \Sigma^\alpha \in \Lambda^0(\mathcal{C}_m), \tag{9.7.19}$$

we readily observe that

$$\omega^\alpha - \mathcal{F}^\alpha \mu \in \mathcal{H}_m. \tag{9.7.20}$$

Therefore, we obtain

$$\phi^*\omega^\alpha = \phi^*\mathcal{F}^\alpha \phi^*\mu = \phi^*\mathcal{F}^\alpha dt^1 \wedge dt^2 \wedge \cdots \wedge dt^n$$

for the solution mapping of the horizontal ideal. Consequently, we conclude that $\phi^*\omega^\alpha = 0$ if and only if $\phi^*\mathcal{F}^\alpha = 0$. Hence, only in this case the solution mapping ϕ of $\mathcal{H}_m$ corresponds to a solution of the new balance ideal as well. Since $\mathcal{F}^\alpha(\mathbf{z}) \in \Lambda^0(\mathcal{C}_m)$ are 0-forms, we get $\phi^*\mathcal{F}^\alpha = 0$ if only if $\mathcal{F}^\alpha = 0$. Let us define the submanifold $\mathcal{P}_m \subseteq \mathcal{C}_m$ by

$$\mathcal{P}_m = \{\mathbf{z} \in \mathcal{C}_m : \mathcal{F}^\alpha(\mathbf{z}) = 0,\ \alpha = 1, \ldots, N\}. \tag{9.7.21}$$

Hence, we see that the relations $\phi^*\mathcal{F}^\alpha = 0$ can only be realised on the region $\mathcal{R}_m \subseteq \mathcal{C}_m$ that is determined by non-empty intersections of submanifold $\mathcal{P}_m$ with the leaves of the foliation generated by the mapping ϕ. Moreover, because the set $\mathcal{D}_n \subseteq \mathbb{R}^n$ over which differential equations are defined is open, the set $\phi^*\mathcal{R}_m$ must also be open. Therefore, it is clear that we can obtain such a solution under rather restricting conditions. However, the dependence of the ideal $\mathcal{H}_m$ on functions $V^\alpha_{ii_1\cdots i_m}$ that offer some freedom of choice despite they have to obey certain rules might offer various alternatives. That makes it possible to find some useful solutions by clever choices.

We now attempt to determine isovector fields of the balance ideal $\mathfrak{B}_m$.

Let U be an isovector field of the closed horizontal ideal $\mathcal{H}_m$ given by (9.7.16). It then follows from (9.7.20) that

$$\pounds_U \omega^\alpha - \pounds_U(\mathcal{F}^\alpha \mu) \in \mathcal{H}_m.$$

Furthermore, we can write

$$\pounds_U(\mathcal{F}^\alpha \mu) = \pounds_U(\mathcal{F}^\alpha)\mu + \mathcal{F}^\alpha \pounds_U(\mu) = U(\mathcal{F}^\alpha)\mu + \mathcal{F}^\alpha dX^i \wedge \mu_i$$

whence we deduce

$$\pounds_U(\mathcal{F}^\alpha \mu) = \big[U(\mathcal{F}^\alpha) + \mathcal{F}^\alpha V_i(X^i)\big]\mu + \mathcal{G}^\alpha, \ \mathcal{G}^\alpha \in \mathcal{H}_m$$

on utilising (9.7.5). Since $\omega^\alpha \notin \mathcal{H}_m$, in order to obtain $\pounds_U \omega^\alpha \in \mathfrak{B}_m$ we may write $\Lambda^\alpha_\beta \omega^\beta = \Lambda^\alpha_\beta \mathcal{F}^\beta \mu \bmod \mathcal{H}_m$ in view of $\omega^\alpha = \mathcal{F}^\alpha \mu \bmod \mathcal{H}_m$ for functions $\Lambda^\alpha_\beta \in \Lambda^0(\mathcal{C}_m)$ so that the conditions

$$\big[U(\mathcal{F}^\alpha) + \mathcal{F}^\alpha V_i(X^i)\big]\mu = \Lambda^\alpha_\beta \mathcal{F}^\beta \mu$$

lead to the result

$$\pounds_U \omega^\alpha = \Lambda^\alpha_\beta \omega^\beta \bmod \mathcal{H}_m \in \mathfrak{B}_m.$$

In other words, we have to find some functions $\lambda^\alpha_\beta \in \Lambda^0(\mathcal{C}_m)$ so that the relations

$$U(\mathcal{F}^\alpha) = X^i V_i(\mathcal{F}^\alpha) + \sum_{r=0}^{m} V_{i_1} \cdots V_{i_r}(U^\beta) \frac{\partial \mathcal{F}^\alpha}{\partial v^\beta_{i_1 \cdots i_r}} = \lambda^\alpha_\beta \mathcal{F}^\beta \quad (9.7.22)$$

must be satisfies. Here, we have defined $\lambda^\alpha_\beta = \Lambda^\alpha_\beta - V_i(X^i)\,\delta^\alpha_\beta$. Equations (9.7.22) help us to determine the admissible functions X^i and U^α complying with the conditions (9.7.15) for isovectors of the closed ideal $\mathcal{H}_m$ to be isovectors of the balance ideal $\mathfrak{B}_m$ as well. Knowing isovectors of the balance ideal makes it possible for us according to Theorem 5.13.7 to elicit new families of solutions if we have a solution at hand. If we take $U^\alpha = 0$, then we have $U = X^i V_i$ and if we write $\lambda^\alpha_\beta = \lambda^\alpha_{\beta i} X^i$ without loss of generality, we must be able to find functions $\lambda^\alpha_{\beta i} \in \Lambda^0(\mathcal{C}_m)$ such that

$$V_i(\mathcal{F}^\alpha) = \lambda^\alpha_{\beta i} \mathcal{F}^\beta$$

in order that canonical system coincides with set of isovectors of the balance ideal. These relations pave the way to produce some solutions of the balance ideal by suitably choosing somewhat arbitrary functions $V^\alpha_{i i_1 \cdots i_m}$ characterising the horizontal ideal $\mathcal{H}_m$.

(i). Let us suppose that the completely symmetric functions $V^{\alpha}_{ii_1\cdots i_m}$ may be so chosen that the conditions (9.7.11) and the relations

$$\begin{aligned}\mathcal{F}^{\alpha} &= V_i(\Sigma^{\alpha i}) + \Sigma^{\alpha} \\ &= \frac{\partial \Sigma^{\alpha i}}{\partial x^i} + \sum_{r=0}^{m-1} v^{\alpha}_{i_1\cdots i_r i}\frac{\partial \Sigma^{\alpha i}}{\partial v^{\alpha}_{i_1\cdots i_r}} + V^{\alpha}_{ii_1\cdots i_m}\frac{\partial \Sigma^{\alpha i}}{\partial v^{\alpha}_{i_1\cdots i_m}} + \Sigma^{\alpha} = 0\end{aligned}$$

are satisfied. In this case every leaf of $\mathcal{H}_m$ proves to be a solution manifold of the balance ideal, and, consequently, of the system of partial differential equations.

(ii). Let us suppose that the completely symmetric functions $V^{\alpha}_{ii_1\cdots i_m}$ in subscripts may be so chosen that the conditions (9.7.11) and the relations

$$V_i(\mathcal{F}^{\alpha}) = V_i V_j(\Sigma^{\alpha j}) + V_i(\Sigma^{\alpha}) = 0$$

are satisfied by taking $\lambda^{\alpha}_{\beta i} = 0$. In this case, we know that we can write

$$\mathcal{F}^{\alpha} = F^{\alpha}(g^a),\;\; V_i(g^a) = 0,\;\; g^a \in \Lambda^0(\mathcal{C}_m),\;\; a = 1, \dots, D.$$

Leaves of the ideal $\mathcal{H}_m$ are obtained by setting $g^a = c^a$ where c^a are real constants. Out of these leaves, those corresponding to solution manifolds can be found by determining the constants satisfying the algebraic equations $F^{\alpha}(c^1, c^2, \dots, c^D) = 0$ where $1 \le \alpha \le N$.

(iii). Let us suppose that the completely symmetric functions $V^{\alpha}_{ii_1\cdots i_m}$ may be so chosen that the conditions (9.7.11) and the relations

$$V_i(\mathcal{F}^{\alpha}) = \lambda^{\alpha}_{\beta i}\mathcal{F}^{\beta}$$

are satisfied for functions $\lambda^{\alpha}_{\beta i} \in \Lambda^0(\mathcal{C}_m)$ that are not all equal to zero. In this case each leaf of the foliation of $\mathcal{H}_m$ intersecting the set $\mathcal{P}_m$ given by (9.7.21) becomes the graph of a solution mapping of the balance ideal.

(iv). Finally, let us assume that the distribution $\mathcal{S}_{\mathcal{H}_m}$ is not involutive, but the restriction $\mathcal{S}_{\mathcal{H}_m}|_{\mathcal{P}_m}$ belongs to the tangent bundle $T(\mathcal{P}_m)$ and is involutive. In this case, although the horizontal ideal $\mathcal{H}_m$ is not completely integrable over the manifold $\mathcal{C}_m$, its restriction on the submanifold $\mathcal{P}_m$ is completely integrable. In order to implement this, we have to choose the completely symmetric functions $V^{\alpha}_{ii_1\cdots i_m}$ in such a way that we might be able to find functions $\Lambda^{\alpha}_{\beta iji_1\cdots i_m}, \lambda^{\alpha}_{\beta i} \in \Lambda^0(\mathcal{C}_m)$ such that the relations

$$[V_i, V_j] = \Lambda^{\alpha}_{\beta iji_1\cdots i_m}\mathcal{F}^{\beta}\frac{\partial}{\partial v^{\alpha}_{i_1\cdots i_m}},\;\; V_i(\mathcal{F}^{\alpha}) = \lambda^{\alpha}_{\beta i}\mathcal{F}^{\beta}$$

or on noting (9.7.4)

$$V_i(V^\alpha_{ji_1\cdots i_m}) - V_j(V^\alpha_{ii_1\cdots i_m}) = \Lambda^\alpha_{\beta iji_1\cdots i_m}\mathcal{F}^\beta,\ V_i(\mathcal{F}^\alpha) = \lambda^\alpha_{\beta i}\mathcal{F}^\beta,$$

are to be satisfied. The functions $\Lambda^\alpha_{\beta iji_1\cdots i_m}$ are antisymmetric in indices i, j and completely symmetric in indices $i_1, \dots, i_m$. In this situation, some solutions of the balance ideal can be found by determining the integral curves of the canonical system passing through $\mathcal{P}_m$.

Example 9.7.1. As an example to the case $N = 1, n = 2$, let us consider the ***Gordon equation*** [German physicist Walter Gordon (1893-1939)]

$$\frac{\partial^2 u}{\partial x \partial t} = \Phi'(u)$$

where Φ is a smooth function of its argument. The reason why this function is introduced into the equation as a derivative is to facilitate the calculations. Let us take

$$x^1 = x,\ x^2 = t,\ u_x = v_1,\ u_t = v_2,$$
$$\mu = dx \wedge dt,\ \mu_1 = dt,\ \mu_2 = -\,dx.$$

Then the ideal $\mathcal{H}_1$ is generated by the following 1-forms

$$\begin{aligned}\sigma &= du - v_1 dx - v_2\, dt,\\ \Sigma_1 &= dv_1 - V_{11} dx - V_{21} dt,\\ \Sigma_2 &= dv_2 - V_{12}\, dx - V_{22}\, dt.\end{aligned}$$

Symmetry condition is met by taking $V_{21} = V_{12}$. We denote thc canonical system by

$$\begin{aligned}V_1 &= \frac{\partial}{\partial x} + v_1\frac{\partial}{\partial u} + V_{11}\frac{\partial}{\partial v_1} + V_{12}\frac{\partial}{\partial v_2},\\ V_2 &= \frac{\partial}{\partial t} + v_2\frac{\partial}{\partial u} + V_{12}\frac{\partial}{\partial v_1} + V_{22}\frac{\partial}{\partial v_2}.\end{aligned}$$

The balance form will now be written as

$$\omega = -\,dv_1 \wedge dx - \Phi'(u)\, dx \wedge dt = d\Sigma^2 \wedge \mu_2 + \Sigma\mu$$

so that we have $\Sigma^1 = 0, \Sigma^2 = v_1, \Sigma = -\,\Phi'(u)$. Consequently, (9.7.19) takes the form

$$\mathcal{F} = V_2(\Sigma^2) + \Sigma = V_{12} - \Phi'(u).$$

Hence, if we choose $V_{12} = \Phi'(u)$, then we get $\mathcal{F} = 0$. Thus, each leaf of $\mathcal{H}_1$ will constitute a solution manifold. Furthermore, the conditions (9.7.11) are reduced to

$$V_1(V_{12}) = V_2(V_{11}),\ V_1(V_{22}) = V_2(V_{12})$$

so that we get the partial differential equations below to determine the functions V_{11} and V_{22}

$$\begin{aligned}\frac{\partial V_{11}}{\partial t} + v_2\frac{\partial V_{11}}{\partial u} + \Phi'(u)\frac{\partial V_{11}}{\partial v_1} + V_{22}\frac{\partial V_{11}}{\partial v_2} &= v_1\Phi''(u),\\ \frac{\partial V_{22}}{\partial x} + v_1\frac{\partial V_{22}}{\partial u} + V_{11}\frac{\partial V_{22}}{\partial v_1} + \Phi'(u)\frac{\partial V_{22}}{\partial v_2} &= v_2\Phi''(u).\end{aligned}$$

Evidently, we will not be able to find the general solution of these non-linear equations for an arbitrary function Φ. However, we may try a particular solution in the form

$$V_{11} = \frac{v_1}{v_2}\Phi'(u),\ V_{22} = \frac{v_2}{v_1}\Phi'(u).$$

It is a very simple exercise to show that this choice satisfies the above equations identically. Therefore, the canonical system corresponding to this case are given by

$$\begin{aligned}V_1 &= \frac{\partial}{\partial x} + v_1\frac{\partial}{\partial u} + \Phi'(u)\Big(\frac{v_1}{v_2}\frac{\partial}{\partial v_1} + \frac{\partial}{\partial v_2}\Big),\\ V_2 &= \frac{\partial}{\partial t} + v_2\frac{\partial}{\partial u} + \Phi'(u)\Big(\frac{\partial}{\partial v_1} + \frac{v_2}{v_1}\frac{\partial}{\partial v_2}\Big)\end{aligned}$$

In order to determine the foliation of the closed ideal $\mathcal{H}_1$ we have to solve the following linear partial differential equations

$$\begin{aligned}V_1(f) &= \frac{\partial f}{\partial x} + v_1\frac{\partial f}{\partial u} + \Phi'(u)\frac{v_1}{v_2}\frac{\partial f}{\partial v_1} + \Phi'(u)\frac{\partial f}{\partial v_2} = 0,\\ V_2(f) &= \frac{\partial f}{\partial t} + v_2\frac{\partial f}{\partial u} + \Phi'(u)\frac{\partial f}{\partial v_1} + \Phi'(u)\frac{v_2}{v_1}\frac{\partial f}{\partial v_2} = 0,\end{aligned}$$

To this end, we apply the method of characteristics. From the first equation, we get the ordinary differential equations

$$\begin{aligned}&\frac{dx}{ds} = 1,\quad \frac{dt}{ds} = 0,\quad \frac{du}{ds} = v_1,\\ &\frac{dv_1}{ds} = \Phi'(u)\frac{v_1}{v_2},\quad \frac{dv_2}{ds} = \Phi'(u).\end{aligned}$$

The trivial characteristic variable is $t = c_0$. From the fourth and fifth equations we obtain the characteristic variable

$$\xi = \frac{v_1}{v_2} = c_1$$

while the third and fourth equations yield

$$\frac{dv_1}{du} = \frac{c_1 \Phi'(u)}{v_1}$$

whose integral provides another characteristic variable

$$\eta = \frac{1}{2} v_1^2 - c_1 \Phi(u) = c_2.$$

Finally, the first and third equations lead to

$$\frac{du}{dx} = v_1 = \sqrt{2\big[c_1 \Phi(u) + c_2\big]}$$

whose solution gives the last characteristic variable

$$\zeta = x - \int \frac{du}{\sqrt{2}\sqrt{c_1 \Phi(u) + c_2}} = \overline{c}_3.$$

Consequently, the general solution of the first partial differential equation becomes $f = F(\xi, \eta, \zeta, t)$. On introducing this function into the second partial differential equation, we find

$$\frac{\partial F}{\partial t} - \frac{1}{c_1} \frac{\partial F}{\partial \zeta} = 0$$

whose solution is obviously $f = F(\xi, \eta, \psi)$ where $\psi = t + c_1 \zeta = c_3$. Therefore, the leaves of the horizontal ideal $\mathcal{H}_1$ are characterised by the functions below

$$g^1 = t + c_1 \left[x - \int \frac{du}{\sqrt{2}\sqrt{c_1 \Phi(u) + c_2}}\right] = c_3,$$

$$g^2 = \frac{v_1}{v_2} = c_1, \qquad g^3 = \frac{1}{2} v_1^2 - c_1 \Phi(u) = c_2.$$

Hence, a solution is given implicitly by

$$\frac{c_1}{\sqrt{2}} \int \frac{du}{\sqrt{c_1 \Phi(u) + c_2}} = t + c_1 x - c_3 \qquad (9.7.23)$$

depending on three arbitrary constants. As a simple example, let us take

$$\frac{\partial^2 u}{\partial x \partial t} = u.$$

We thus have $\Phi(u) = \frac{u^2}{2}$. If we define new constants by

$$a_1 = \sqrt{c_1}, \quad a_2 = \frac{2c_2}{c_1}, \quad a_3 = \frac{c_3}{\sqrt{c_1}},$$

then the expression (9.7.23) assumes the form

$$\int \frac{du}{\sqrt{u^2 + a_2}} = \log\left[u + \sqrt{u^2 + a_2}\right] = \frac{t}{a_1} + a_1 x - a_3$$

whence we arrive at the solution

$$u(x,t) = \frac{1}{2}\left[e^{\frac{t}{a_1} + a_1 x - a_3} - a_2\, e^{-\left(\frac{t}{a_1} + a_1 x - a_3\right)}\right].$$

Finally, let us consider the *sine-Gordon equation*

$$\frac{\partial^2 u}{\partial x \partial t} + \sin u = 0.$$

In this case, we have $\Phi(u) = \cos u$. If we introduce the new constants by $a_1 = \sqrt{c_1}, a_2 = c_2/c_1, a_3 = c_3/\sqrt{c_1}$, then it follows from (9.7.23) that

$$\int \frac{du}{\sqrt{\cos u + a_2}} = \frac{2}{\sqrt{a_2 + 1}} F\Big(\frac{u}{2}, \sqrt{\frac{2}{a_2 + 1}}\Big) = \frac{t}{a_1} + a_1 x - a_3$$

where F is the *Legendre elliptic integral of the first kind* [French mathematician Adrien-Marie Legendre (1752-1833)] defined by

$$F(\phi, k) = \int_0^{\phi} \frac{d\theta}{\sqrt{1 - k^2 \sin^2\theta}}.$$

The solution of the equation $F(\phi, k) = \psi$ for ϕ in terms of ψ is expressible as $\sin\phi = \operatorname{sn}\psi$ where sn denotes the *Jacobi elliptic function* and the function $\eta = \operatorname{sn}\xi$ is found as the solution of the non-linear ordinary differential equation

$$\frac{d\eta}{d\xi} = \sqrt{(1 - \eta^2)(1 - k^2\eta^2)}.$$

Hence, a particular solution of the sine-Gordon equation may be written as

$$u(x,t) = 2\arcsin \mathrm{sn}\left[\frac{\sqrt{a_2+1}}{2}\left(\frac{t}{a_1} + a_1 x - a_3\right)\right]$$

depending on three parameters. ■

Example 9.7.2. We consider n partial differential equations

$$\frac{\partial u}{\partial x^i} = \phi_i(\mathbf{x}, u)$$

involving a single dependent variable where $\phi_i(\mathbf{x}, u)$ are given functions. We look for the solution $u = u(\mathbf{x})$. But, because of the symmetry relations $u_{,ij} = u_{,ji}$, the functions ϕ_i must satisfy the compatibility conditions

$$\frac{\partial \phi_i}{\partial x^j} + \frac{\partial \phi_i}{\partial u}\phi_j = \frac{\partial \phi_j}{\partial x^i} + \frac{\partial \phi_j}{\partial u}\phi_i \tag{9.7.24}$$

for the existence of a solution. The horizontal ideal $\mathcal{H}_1$ is now be generated solely by 1-forms

$$\sigma = du - v_i\,dx^i,\ \ \Sigma_i = dv_i - V_{ji}\,dx^j.$$

The functions $V_{ji} \in \Lambda^0(\mathcal{C}_1)$ are presently arbitrary except for satisfying the symmetry condition $V_{ij} = V_{ji}$. Despite there exists just one dependent variable, namely, $N = 1$, there are n balance equations ($A = n$). Therefore, we choose balance n-forms as

$$\omega_i = \Sigma_i\mu = \big(v_i - \phi_i(\mathbf{x}, u)\big)\mu,\ \ \mu = dx^1 \wedge \cdots \wedge dx^n.$$

The canonical system will now have the form

$$V_i = \frac{\partial}{\partial x^i} + v_i\frac{\partial}{\partial u} + V_{ij}\frac{\partial}{\partial v_j}.$$

From (9.7.19), we obtain

$$\mathcal{F}_i = v_i - \phi_i(\mathbf{x}, u). \tag{9.7.25}$$

Hence, the submanifold $\mathcal{P}_1$ of $\mathcal{C}_1$ is specified by the relations $\mathcal{F}_i = 0$, or

$$v_i = \phi_i(\mathbf{x}, u),\ \ 1 \le i \le n.$$

Next, let us choose the functions V_{ij} as

$$V_{ij} = \frac{\partial \phi_i}{\partial x^j} + \frac{\partial \phi_i}{\partial u}\phi_j. \tag{9.7.26}$$

Because of the compatibility conditions (9.7.24), the symmetry relations are automatically satisfied. However, if we consider the case (iv), the equality (9.7.11) must be satisfied at least on the submanifold $\mathcal{P}_1$. For this purpose, let us evaluate the expressions $V_i(\mathcal{F}_j)$ and $V_i(V_{kj}) - V_j(V_{ki})$ and employ the relations (9.7.24) to obtain

$$V_i(\mathcal{F}_j) = -\frac{\partial \phi_j}{\partial u}\mathcal{F}_i$$

and

$$\begin{aligned} V_i(V_{kj}) - V_j(V_{ki}) &= V_i\Big(\frac{\partial \phi_k}{\partial x^j} + \frac{\partial \phi_k}{\partial u}\phi_j\Big) - V_j\Big(\frac{\partial \phi_k}{\partial x^i} + \frac{\partial \phi_k}{\partial u}\phi_i\Big) \\ &= \Big(\frac{\partial^2 \phi_k}{\partial x^j \partial u} + \frac{\partial^2 \phi_k}{\partial u^2}\phi_j + \frac{\partial \phi_k}{\partial u}\frac{\partial \phi_j}{\partial u}\Big)\mathcal{F}_i \\ &\quad - \Big(\frac{\partial^2 \phi_k}{\partial x^i \partial u} + \frac{\partial^2 \phi_k}{\partial u^2}\phi_i + \frac{\partial \phi_k}{\partial u}\frac{\partial \phi_i}{\partial u}\Big)\mathcal{F}_j. \end{aligned}$$

Since $\mathcal{F}_i = 0$ on $\mathcal{P}_1$, the relations $V_i(\mathcal{F}_j) = 0$ and $V_i(V_{kj}) - V_j(V_{ki}) = 0$ are also satisfied on the same submanifold. Therefore, the solutions to our system of differential equations are obtained via the integral curves of the vector fields

$$V_i = \frac{\partial}{\partial x^i} + v_i\frac{\partial}{\partial u} + \Big(\frac{\partial \phi_i}{\partial x^j} + \frac{\partial \phi_i}{\partial u}\phi_j\Big)\frac{\partial}{\partial v_j}$$

passing through the submanifold $\mathcal{P}_1$. As a special case, let us take $n = 2$, $x^1 = x, x^2 = t$ and choose the functions ϕ_1 and ϕ_2 as follows

$$\phi_1 = \frac{t-u}{x+e^u}, \quad \phi_2 = \frac{x+t}{x+e^u}.$$

We can easily verify that these functions satisfy the compatibility conditions (9.7.24) [*see* Edelen and Wang (1992), *p.* 144]. Hence, the relations (9.7.26) yield at once

$$\begin{aligned} V_{11} &= \frac{(u-t)\big[e^u(t-u+2)+2x\big]}{(x+e^u)^3} \\ V_{22} &= \frac{(x+e^u)^2 - (x+t)^2 e^u}{(x+e^u)^3} \\ V_{12} = V_{21} &= \frac{(x+e^u)^2 - (x+e^u)(x+t) + e^u(u-t)(x+t)}{(x+e^u)^3} \end{aligned}$$

In this case, the solution mapping must be found by solving the differential equations $V_1(f) = 0$ and $V_2(f) = 0$ by using the method of characteristics. However, we might reach to a particular solution by a simple observation. Let us define a mapping $\phi : G \to C_1$ by the relations $x = x,\ t = t,\ u = u,$ $v_1 = \phi_1, v_2 = \phi_2$. We then immediately see that $\phi^*\Sigma_i = 0, i = 1, 2$ whereas the expression $\phi^*\sigma = 0$ gives

$$\begin{aligned}\phi^*\sigma &= du - \frac{t-u}{x+e^u}\,dx - \frac{x+t}{x+e^u}\,dt \\ &= \frac{1}{x+e^u}\big[(x+e^u)\,du + (u-t)\,dx - (x+t)\,dt\big] \\ &= \frac{1}{x+e^u}\,d\Big(e^u + xu - \frac{1}{2}t^2 - xt\Big) = 0.\end{aligned}$$

Therefore, some implicit solutions of the partial differential equations

$$\frac{\partial u}{\partial x} = \frac{t-u}{x+e^u}, \quad \frac{\partial u}{\partial t} = \frac{x+t}{x+e^u}$$

are provided by

$$e^u + xu - \frac{1}{2}t^2 - xt = c$$

where c is an arbitrary constant.

An exemplary plot of this function is depicted in Fig. 9.7.1. ■

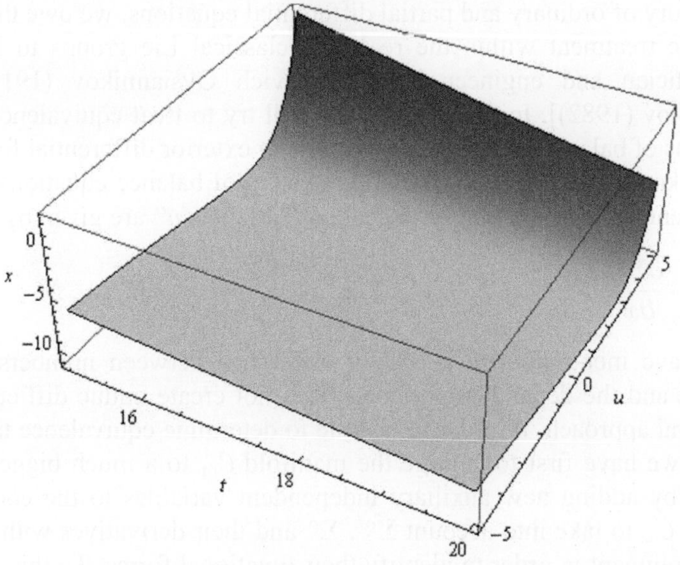

Fig. 9.7.1. A plot of $e^u + xu - \frac{1}{2}t^2 - xt = 10$.

9.8. EQUIVALENCE TRANSFORMATIONS

Several system of partial differential equations, particularly modelling natural laws, contain some arbitrary functions or parameters reflecting physical constitution of materials involved. Almost all field equations of classical continuum physics fall into this category. Thus, such systems are actually family of equations whose fundamental structures remain unchanged but show some differences in their physical constitutions from one material to another through some constitutive functions or parameters. For instance, the field equations of hyperelastic solids are of the same type and only the particular form of the stress potential distinguishes one material from the other. The ***equivalence groups*** are defined as groups of continuous transformations that leave a given family of equations invariant. In contrast to a symmetry transformation that transforms one set of equations into themselves, an equivalence transformation maps an arbitrary member of the family onto another member of the same family which may possess somewhat different physical properties. Meanwhile, it transforms a solution of the one member onto a solution of another member of that family.

In other words, if we manage to determine an equivalence transformation, we can employ a solution corresponding to a certain material to obtain a solution associated with another material of the same sort whose physical properties obey the rules dictated by the appropriate equivalence transformation. Although the concept of equivalence transformations is well-known in the theory of ordinary and partial differential equations, we owe their first systematic treatment within the realm of classical Lie groups to Russian mathematician and engineer Lev Vasil'evich Ovsiannikov (1919) [*see* Ovsiannikov (1982)]. In this section, we will try to treat equivalence transformations of balance equations by employing exterior differential forms.

We know that an $(m+1)$th order system of balance equations with n independent variables x^i and N dependent variables u^α are given by (9.4.1)

$$\frac{\partial \Sigma^{\alpha i}}{\partial x^i} + \Sigma^\alpha = 0, \quad i = 1, 2, \ldots, n; \quad \alpha = 1, 2, \ldots, N.$$

As we have mentioned on p. 522, a difference between numbers of the equations and the dependent variables does not create undue difficulties in our general approach. In order to be able to determine equivalence transformations, we have first to enlarge the manifold $\mathcal{C}_m$ to a much bigger manifold $\mathcal{K}_m$ by adding new auxiliary independent variables to the coordinate cover of $\mathcal{C}_m$ to take into account $\Sigma^{\alpha i}$, Σ^α and their derivatives with respect to their argument in order to identify their functional forms. To this end, we introduce the new auxiliary variables

$$s_j^{\alpha i} = \frac{\partial \Sigma^{\alpha i}}{\partial x^j}, \quad s_\beta^{\alpha i i_1 \cdots i_r} = \frac{\partial \Sigma^{\alpha i}}{\partial v_{i_1 \cdots i_r}^\beta}, \tag{9.8.1}$$
$$t_i^\alpha = \frac{\partial \Sigma^\alpha}{\partial x^i}, \quad t_\beta^{\alpha i_1 \cdots i_r} = \frac{\partial \Sigma^\alpha}{\partial v_{i_1 \cdots i_r}^\beta}, \quad 0 \le r \le m.$$

It is clear that the variables $s_\beta^{\alpha i i_1 \cdots i_r}$ and $t_\beta^{\alpha i_1 \cdots i_r}$ are completely symmetric in the indices $i_1, \cdots, i_r$. Hence, the coordinate cover of the manifold $\mathcal{K}_m$ that is enlarged significantly compared to that of the manifold $\mathcal{C}_m$ are given by

$$\{x^i, \Sigma^{\alpha i}, \Sigma^\alpha, s_j^{\alpha i}, t_i^\alpha, \{v_{i_1 \cdots i_r}^\alpha, s_\beta^{\alpha i i_1 \cdots i_r}, t_\beta^{\alpha i_1 \cdots i_r} : 0 \le r \le m\}\}.$$

We can easily verify that the dimension of the manifold $\mathcal{K}_m$ is at most

$$n + (1+n)^2 N + [1 + (n+1)N] N \frac{(n+m)!}{n! m!}.$$

Let $\Lambda(\mathcal{K}_m)$ be the exterior algebra on the manifold $\mathcal{K}_m$. Contact 1-forms in the manifold $\mathcal{K}_m$ are now defined by

$$\sigma_{i_1 \cdots i_r}^\alpha = dv_{i_1 \cdots i_r}^\alpha - v_{i_1 \cdots i_r i}^\alpha \, dx^i \in \Lambda^1(\mathcal{K}_m), \; 0 \le r \le m-1, \tag{9.8.2}$$
$$\Omega^{\alpha i} = d\Sigma^{\alpha i} - s_j^{\alpha i} dx^j - \sum_{r=0}^m s_\beta^{\alpha i i_1 \cdots i_r} dv_{i_1 \cdots i_r}^\beta \in \Lambda^1(\mathcal{K}_m),$$
$$\Omega^\alpha = d\Sigma^\alpha - t_i^\alpha dx^i - \sum_{r=0}^m t_\beta^{\alpha i_1 \cdots i_r} dv_{i_1 \cdots i_r}^\beta \in \Lambda^1(\mathcal{K}_m).$$

Balance n-forms are again given by

$$\omega^\alpha = d\Sigma^{\alpha i} \wedge \mu_i + \Sigma^\alpha \mu \in \Lambda^n(\mathcal{K}_m).$$

Let $\mathfrak{D}_m$ denote the balance ideal of $\Lambda(\mathcal{K}_m)$ generated by forms ω^α, $\Omega^{\alpha i}$, Ω^α, $d\Omega^{\alpha i}$, $d\Omega^\alpha$, $\{\sigma_{i_1 \cdots i_r}^\alpha : 0 \le r \le m-1\}$ and $d\sigma_{i_1 \cdots i_{m-1}}^\alpha$. Exterior derivatives of 1-forms are

$$d\sigma_{i_1 \cdots i_r}^\alpha = -\, dv_{i_1 \cdots i_r i}^\alpha \wedge dx^i, \quad 0 \le r \le m-1,$$
$$d\Omega^{\alpha i} = -\, ds_j^{\alpha i} \wedge dx^j - \sum_{r=0}^m ds_\beta^{\alpha i i_1 \cdots i_r} \wedge dv_{i_1 \cdots i_r}^\beta,$$
$$d\Omega^\alpha = -\, dt_i^\alpha \wedge dx^i - \sum_{r=0}^m dt_\beta^{\alpha i_1 \cdots i_r} \wedge dv_{i_1 \cdots i_r}^\beta.$$

The ideal $\mathfrak{D}_m$ is closed. Indeed, we can easily verify that

$$d\omega^\alpha = d\Sigma^\alpha \wedge \mu = \Omega^\alpha \wedge \mu + \sum_{r=0}^{m-1} t_\beta^{\alpha i_1\cdots i_r} \sigma^\beta_{i_1\cdots i_r} \wedge \mu$$
$$- t_\beta^{\alpha i_1\cdots i_{m-1} i_m} d\sigma^\beta_{i_1\cdots i_{m-1}} \wedge \mu_{i_m} \in \mathfrak{D}_m.$$

If a mapping $\phi : M \to \mathcal{K}_m$ annihilates the ideal $\mathfrak{D}_m$, that is, if the pull-back operator $\phi^* : \Lambda(\mathcal{K}_m) \to \Lambda(M)$ satisfies the relations

$$\phi^* \sigma^\alpha_{i_1\cdots i_r} = 0, 0 \le r \le m-1, \ \ \phi^* \Omega^{\alpha i} = 0, \ \ \phi^* \Omega^\alpha = 0, \ \ \phi^* \omega^\alpha = 0,$$

then we easily observe that ϕ is a solution mapping of the given system of partial differential equations. In order to determine the equivalence transformations, we have to find isovector fields that leave the closed balance ideal $\mathfrak{D}_m$ invariant. A vector field $V \in T(\mathcal{K}_m)$ can now be represented by the expression

$$V = X^i \frac{\partial}{\partial x^i} + S^{\alpha i} \frac{\partial}{\partial \Sigma^{\alpha i}} + T^\alpha \frac{\partial}{\partial \Sigma^\alpha} + S^{\alpha i}_j \frac{\partial}{\partial s^{\alpha i}_j} + T^\alpha_i \frac{\partial}{\partial t^\alpha_i} \tag{9.8.3}$$
$$+ \sum_{r=0}^{m} \Big(V^\alpha_{i_1\cdots i_r} \frac{\partial}{\partial v^\alpha_{i_1\cdots i_r}} + S_\beta^{\alpha i i_1\cdots i_r} \frac{\partial}{\partial s_\beta^{\alpha i i_1\cdots i_r}} + T_\beta^{\alpha i_1\cdots i_r} \frac{\partial}{\partial t_\beta^{\alpha i_1\cdots i_r}} \Big).$$

All coefficients in the vector field (9.8.3) are smooth functions of the coordinates of the manifold $\mathcal{K}_m$. Let us first take into consideration the ideal $\mathcal{I}(\{\sigma^\alpha_{i_1\cdots i_r} : 0 \le r \le m-1\}, \Omega^{\alpha i}, \Omega^\alpha)$. Since this ideal is produced by 1-forms only, Theorem 5.12.4 secures us that the isovectors of this ideal and its closure

$$\mathfrak{C}_m = \mathcal{I}(\{\sigma^\alpha_{i_1\cdots i_r} : 0 \le r \le m-1\}, d\sigma_{i_1\cdots i_{m-1}}, \Omega^{\alpha i}, \Omega^\alpha, d\Omega^{\alpha i}, d\Omega^\alpha)$$

are the same. Therefore, to determine the isovector fields of the closed contact ideal $\mathfrak{C}_m$, we have to show the existence of smooth functions ,

$$\lambda^{\alpha j_1\cdots j_r}_{\beta i_1\cdots i_k}, \ K^\alpha_{\beta i i_1\cdots i_k}, \ L^\alpha_{\beta i_1\cdots i_k}, \ \mu_\beta^{\alpha i i_1\cdots i_r}, \ M^{\alpha i}_{\beta j}, \ M^{\alpha i}_\beta, \ \nu_\beta^{\alpha i_1\cdots i_r}, \ N^\alpha_{\beta i}, \ N^\alpha_\beta$$

belonging to $\Lambda^0(\mathcal{K}_m)$ such that the following relations are satisfied

$$£_V\, \sigma^\alpha_{i_1\cdots i_k} = \sum_{r=0}^{m-1} \lambda^{\alpha j_1\cdots j_r}_{\beta i_1\cdots i_k} \sigma^\beta_{j_1\cdots j_r} + K^\alpha_{\beta i i_1\cdots i_k} \Omega^{\beta i} + L^\alpha_{\beta i_1\cdots i_k} \Omega^\beta, \tag{9.8.4}$$
$$£_V\, \Omega^{\alpha i} = \sum_{r=0}^{m-1} \mu_\beta^{\alpha i i_1\cdots i_r} \sigma^\beta_{i_1\cdots i_r} + M^{\alpha i}_{\beta j} \Omega^{\beta j} + M^{\alpha i}_\beta \Omega^\beta,$$

$$£_V\,\Omega^\alpha = \sum_{r=0}^{m-1} \nu_\beta^{\alpha i_1\cdots i_r}\sigma^\beta_{i_1\cdots i_r} + N^\alpha_{\beta i}\,\Omega^{\beta i} + N^\alpha_\beta\,\Omega^\beta$$

where $0 \le k \le m-1$. To this end, we start with the equations $(9.8.4)_1$. For $0 \le k \le m-1$, the relations

$$\begin{aligned} F^\alpha_{i_1\cdots i_k} &= \mathbf{i}_V\,\sigma^\alpha_{i_1\cdots i_k} = V^\alpha_{i_1\cdots i_k} - v^\alpha_{i_1\cdots i_k i}X^i, \\ \mathbf{i}_V\,d\sigma^\alpha_{i_1\cdots i_k} &= -\,V^\alpha_{i_1\cdots i_k i}dx^i + X^i dv^\alpha_{i_1\cdots i_k i} \end{aligned} \tag{9.8.5}$$

lead to

$$\begin{aligned} £_V\,\sigma^\alpha_{i_1\cdots i_k} &= dF_{i_1\cdots i_k} - V^\alpha_{i_1\cdots i_k i}dx^i + X^i dv^\alpha_{i_1\cdots i_k i} \\ &= dV^\alpha_{i_1\cdots i_k} - V^\alpha_{i_1\cdots i_k i}dx^i - v^\alpha_{i_1\cdots i_k i}dX^i \end{aligned}$$

If we introduce the above expressions into $(9.8.4)_1$ after having calculated the differentials $dF_{i_1\cdots i_k}$ and $dV_{i_1\cdots i_k}$, and arrange the resulting expression, we then find that

$$\begin{aligned} &\Big[-V^\alpha_{i_1\cdots i_k i} + \frac{\partial F^\alpha_{i_1\cdots i_k}}{\partial x^i} + \sum_{r=0}^{m-1}\lambda^{\alpha j_1\cdots j_r}_{\beta i_1\cdots i_k}v^\beta_{j_1\cdots j_r i} + K^\alpha_{\beta j i_1\cdots i_k}s^{\beta j}_i \\ &\quad + L^\alpha_{\beta i_1\cdots i_k}t^\beta_i\Big]dx^i + \Big[\frac{\partial F^\alpha_{i_1\cdots i_k}}{\partial \Sigma^{\beta i}} - K^\alpha_{\beta i i_1\cdots i_k}\Big]d\Sigma^{\beta i} \\ &\quad + \Big[\frac{\partial F^\alpha_{i_1\cdots i_k}}{\partial \Sigma^\beta} - L^\alpha_{\beta i_1\cdots i_k}\Big]d\Sigma^\beta \\ &\quad + \frac{\partial F^\alpha_{i_1\cdots i_k}}{\partial s^{\beta i}_j}ds^{\beta i}_j + \frac{\partial F^\alpha_{i_1\cdots i_k}}{\partial t^\beta_i}dt^\beta_i \\ &\quad + \sum_{r=0}^{m-1}\Big[\frac{\partial V^\alpha_{i_1\cdots i_k}}{\partial v^\beta_{j_1\cdots j_r}} - v^\alpha_{i_1\cdots i_k i}\frac{\partial X^i}{\partial v^\beta_{j_1\cdots j_r}} - \lambda^{\alpha j_1\cdots j_r}_{\beta i_1\cdots i_k} \\ &\quad + K^\alpha_{\gamma i i_1\cdots i_k}s^{\gamma i j_1\cdots j_r}_\beta + L^\alpha_{\gamma i_1\cdots i_k}t^{\gamma j_1\cdots j_r}_\beta\Big]dv^\beta_{j_1\cdots j_r} \\ &\quad + \sum_{r=0}^{m}\frac{\partial F^\alpha_{i_1\cdots i_k}}{\partial s^{\beta i j_1\cdots j_r}_\gamma}ds^{\beta i j_1\cdots j_r}_\gamma \\ &\quad + \sum_{r=0}^{m}\frac{\partial F^\alpha_{i_1\cdots i_k}}{\partial t^{\beta j_1\cdots j_r}_\gamma}dt^{\beta j_1\cdots j_r}_\gamma + \Big[\frac{\partial V^\alpha_{i_1\cdots i_k}}{\partial v^\beta_{j_1\cdots j_m}} - v^\alpha_{i_1\cdots i_k i}\frac{\partial X^i}{\partial v^\beta_{j_1\cdots j_m}} \\ &\quad + K^\alpha_{\gamma i i_1\cdots i_k}s^{\gamma i j_1\cdots j_m}_\beta + L^\alpha_{\gamma i_1\cdots i_k}t^{\gamma j_1\cdots j_m}_\beta\Big]dv^\beta_{j_1\cdots j_m} = 0 \end{aligned}$$

from which we get the following relations for $k = 0, 1, \ldots, m-1$ and

$r = 0, 1, \ldots, m-1$

$$K^{\alpha}_{\beta i i_1 \cdots i_k} = \frac{\partial F^{\alpha}_{i_1 \cdots i_k}}{\partial \Sigma^{\beta i}}, \quad L^{\alpha}_{\beta i_1 \cdots i_k} = \frac{\partial F^{\alpha}_{i_1 \cdots i_k}}{\partial \Sigma^{\beta}}, \tag{9.8.6}$$

$$\frac{\partial F^{\alpha}_{i_1 \cdots i_k}}{\partial s^{\beta i}_{j}} = 0, \frac{\partial F^{\alpha}_{i_1 \cdots i_k}}{\partial t^{\beta}_{i}} = 0, \frac{\partial F^{\alpha}_{i_1 \cdots i_k}}{\partial s^{\beta i j_1 \cdots j_r}_{\gamma}} = 0, \frac{\partial F^{\alpha}_{i_1 \cdots i_k}}{\partial t^{\beta j_1 \cdots j_r}_{\gamma}} = 0, 0 \le r \le m,$$

$$V^{\alpha}_{i_1 \cdots i_k i} = \frac{\partial F^{\alpha}_{i_1 \cdots i_k}}{\partial x^i} + \sum_{r=0}^{m-1} \lambda^{\alpha j_1 \cdots j_r}_{\beta i_1 \cdots i_k} v^{\beta}_{j_1 \cdots j_r i} + \frac{\partial F^{\alpha}_{i_1 \cdots i_k}}{\partial \Sigma^{\beta j}} s^{\beta j}_{i} + \frac{\partial F^{\alpha}_{i_1 \cdots i_k}}{\partial \Sigma^{\beta}} t^{\beta}_{i},$$

$$\lambda^{\alpha j_1 \cdots j_r}_{\beta i_1 \cdots i_k} = \frac{\partial V^{\alpha}_{i_1 \cdots i_k}}{\partial v^{\beta}_{j_1 \cdots j_r}} - v^{\alpha}_{i_1 \cdots i_k i} \frac{\partial X^i}{\partial v^{\beta}_{j_1 \cdots j_r}} + \frac{\partial F^{\alpha}_{i_1 \cdots i_k}}{\partial \Sigma^{\gamma i}} s^{\gamma i j_1 \cdots j_r}_{\beta}$$
$$+ \frac{\partial F^{\alpha}_{i_1 \cdots i_k}}{\partial \Sigma^{\gamma}} t^{\gamma j_1 \cdots j_r}_{\beta}, \; r = 0, 1, \ldots, m-1,$$

$$\frac{\partial V^{\alpha}_{i_1 \cdots i_k}}{\partial v^{\beta}_{j_1 \cdots j_m}} - v^{\alpha}_{i_1 \cdots i_k i} \frac{\partial X^i}{\partial v^{\beta}_{j_1 \cdots j_m}} + \frac{\partial F^{\alpha}_{i_1 \cdots i_k}}{\partial \Sigma^{\gamma i}} s^{\gamma i j_1 \cdots j_m}_{\beta} + \frac{\partial F^{\alpha}_{i_1 \cdots i_k}}{\partial \Sigma^{\gamma}} t^{\gamma j_1 \cdots j_m}_{\beta} = 0.$$

The equations $(9.8.6)_{3\text{-}6}$ indicate that the functions $F^{\alpha}_{i_1 \cdots i_k}$, $0 \le k \le m-1$ cannot depend on the variables $s^{\beta i}_{j}, t^{\beta}_{i}, s^{\beta i j_1 \cdots j_r}_{\gamma}, t^{\beta j_1 \cdots j_r}_{\gamma}, 0 \le r \le m$. If $k < m-1$, then it follows from $(9.8.5)_1$ that

$$\frac{\partial V^{\alpha}_{i_1 \cdots i_k}}{\partial v^{\beta}_{j_1 \cdots j_m}} - v^{\alpha}_{i_1 \cdots i_k i} \frac{\partial X^i}{\partial v^{\beta}_{j_1 \cdots j_m}} = \frac{\partial F^{\alpha}_{i_1 \cdots i_k}}{\partial v^{\beta}_{j_1 \cdots j_m}}.$$

Hence $(9.8.6.)_5$ takes the form

$$\frac{\partial F^{\alpha}_{i_1 \cdots i_k}}{\partial v^{\beta}_{j_1 \cdots j_m}} + \frac{\partial F^{\alpha}_{i_1 \cdots i_k}}{\partial \Sigma^{\gamma i}} s^{\gamma i j_1 \cdots j_m}_{\beta} + \frac{\partial F^{\alpha}_{i_1 \cdots i_k}}{\partial \Sigma^{\gamma}} t^{\gamma j_1 \cdots j_m}_{\beta} = 0, 0 \le k < m-2$$

whence we deduce, respectively,

$$\frac{\partial F^{\alpha}_{i_1 \cdots i_k}}{\partial \Sigma^{\gamma i}} = 0, \frac{\partial F^{\alpha}_{i_1 \cdots i_k}}{\partial \Sigma^{\gamma}} = 0, \frac{\partial F^{\alpha}_{i_1 \cdots i_k}}{\partial v^{\beta}_{j_1 \cdots j_m}} = 0, \; k = 0, 1, \ldots, m-2.$$

Let us now take $k = m-1$. This time $(9.8.5)_1$ gives rise to the relation

$$\frac{\partial F^{\alpha}_{i_1 \cdots i_{m-1}}}{\partial v^{\beta}_{j_1 \cdots j_m}} =$$
$$\frac{\partial V^{\alpha}_{i_1 \cdots i_{m-1}}}{\partial v^{\beta}_{j_1 \cdots j_m}} - v^{\alpha}_{i_1 \cdots i_{m-1} i} \frac{\partial X^i}{\partial v^{\beta}_{j_1 \cdots j_m}} - \delta^{\alpha}_{\beta} \delta^{j_1}_{i_1} \delta^{j_2}_{i_2} \cdots \delta^{j_{m-1}}_{i_{m-1}} \delta^{j_m}_{i} X^i.$$

We then insert this expression into $(9.8.6)_5$ for $k = m-1$ to obtain

$$\delta^\alpha_\beta \delta^{j_1}_{i_1}\delta^{j_2}_{i_2}\cdots\delta^{j_{m-1}}_{i_{m-1}}\delta^{j_m}_{i} X^i + \frac{\partial F^\alpha_{i_1\cdots i_{m-1}}}{\partial v^\beta_{j_1\cdots j_m}} + \frac{\partial F^\alpha_{i_1\cdots i_{m-1}}}{\partial \Sigma^{\gamma i}} s_\beta^{\gamma i j_1\cdots j_m} + \frac{\partial F^\alpha_{i_1\cdots i_{m-1}}}{\partial \Sigma^{\gamma}} t_\beta^{\gamma j_1\cdots j_m} = 0.$$

When $N > 1$, we can always choose the indices α and β in such a way that $\alpha \neq \beta$. In that case, we find that

$$\frac{\partial F^\alpha_{i_1\cdots i_{m-1}}}{\partial \Sigma^{\gamma i}} = 0,\ \frac{\partial F^\alpha_{i_1\cdots i_{m-1}}}{\partial \Sigma^{\gamma}} = 0;\ \frac{\partial F^\alpha_{i_1\cdots i_{m-1}}}{\partial v^\beta_{j_1\cdots j_m}} = 0,\ \alpha \neq \beta.$$

Thus if $0 \le k \le m-1$, then the functions $F^\alpha_{i_1\cdots i_k}$ and consequently the functions X^i and $V^\alpha_{i_1\cdots i_k} = F^\alpha_{i_1\cdots i_k} + v^\alpha_{i_1\cdots i_k i} X^i$ cannot depend on variables $\Sigma^{\gamma i}$, Σ^γ and $v^\beta_{j_1\cdots j_m}$ whenever $\alpha \neq \beta$. Therefore, $(9.8.6)_{3,5}$ reduce to the expressions

$$V^\alpha_{i_1\cdots i_k i} = \frac{\partial V^\alpha_{i_1\cdots i_k}}{\partial x^i} - v^\alpha_{i_1\cdots i_k j}\frac{\partial X^j}{\partial x^i} + \sum_{r=0}^{m-1}\Big[\frac{\partial V^\alpha_{i_1\cdots i_k}}{\partial v^\beta_{j_1\cdots j_r}} - v^\alpha_{i_1\cdots i_k j}\frac{\partial X^j}{\partial v^\beta_{j_1\cdots j_r}}\Big] v^\beta_{j_1\cdots j_r i},$$

$$\frac{\partial V^\alpha_{i_1\cdots i_k}}{\partial v^\beta_{j_1\cdots j_m}} - v^\alpha_{i_1\cdots i_k i}\frac{\partial X^i}{\partial v^\beta_{j_1\cdots j_m}} = 0,\ k = 0,1,\ldots,m-1.$$

The above equations are exactly the same as the equations $(9.3.5)_3$ and (9.3.6). Hence, their solutions are again given by the recurrence relations

$$X^i = X^i(\mathbf{x},\mathbf{u}),\ U^\alpha = U^\alpha(\mathbf{x},\mathbf{u}), \tag{9.8.7}$$
$$V^\alpha_{i_1\cdots i_k i} = \frac{\partial V^\alpha_{i_1\cdots i_k}}{\partial x^i} - v^\alpha_{i_1\cdots i_k j}\frac{\partial X^j}{\partial x^i} + \Big[\frac{\partial V^\alpha_{i_1\cdots i_k}}{\partial u^\beta} - v^\alpha_{i_1\cdots i_k j}\frac{\partial X^j}{\partial u^\beta}\Big] v^\beta_i + \sum_{r=1}^{k} v^\beta_{j_1\cdots j_r i}\frac{\partial V^\alpha_{i_1\cdots i_k}}{\partial v^\beta_{j_1\cdots j_r}},\quad k = 0,1,\ldots,m-1$$

[*see* (9.3.16) and (9.3.17)]. Let us now take the relations $(9.8.4)_2$ into account. By employing the relations

$$\mathbf{i}_V\Omega^{\alpha i} = S^{\alpha i} - s^{\alpha i}_j X^j - \sum_{r=0}^{m} s_\beta^{\alpha i i_1\cdots i_r} V^\beta_{i_1\cdots i_r} = F^{\alpha i} \tag{9.8.8}$$

$$
\begin{aligned}
\mathbf{i}_V d\Omega^{\alpha i} = &- S_j^{\alpha i} dx^j + X^j ds_j^{\alpha i} \\
&- \sum_{r=0}^{m} \left(S_\beta^{\alpha i i_1 \cdots i_r} dv^\beta_{i_1 \cdots i_r} - V^\beta_{i_1 \cdots i_r} ds_\beta^{\alpha i i_1 \cdots i_r} \right)
\end{aligned}
$$

we can evaluate Lie derivatives through the Cartan formula. Let us then make the transformation $dv^\beta_{i_1 \cdots i_r} = \sigma^\beta_{i_1 \cdots i_r} + v^\beta_{i_1 \cdots i_r i} dx^i$ for $0 \le r \le m-1$, arrange the resulting expression and equate the coefficients of independent 1-forms to zero to obtain

$$
M^{\alpha i}_{\beta j} = \frac{\partial F^{\alpha i}}{\partial \Sigma^{\beta j}}, \quad M^{\alpha i}_\beta = \frac{\partial F^{\alpha i}}{\partial \Sigma^\beta}, \quad \frac{\partial F^{\alpha i}}{\partial t^\beta_j} = 0, \quad \delta^\alpha_\beta \delta^i_k X^j + \frac{\partial F^{\alpha i}}{\partial s^{\beta k}_j} = 0.
$$

$$
\begin{aligned}
&\frac{\partial F^{\alpha i}}{\partial t_\beta^{\gamma i_1 \cdots i_r}} = 0, \quad \delta^\alpha_\gamma \delta^i_j V^\beta_{i_1 \cdots i_r} + \frac{\partial F^{\alpha i}}{\partial s_\beta^{\gamma j i_1 \cdots i_r}} = 0, \quad 0 \le r \le m, \\
&\mu_\beta^{\alpha i i_1 \cdots i_r} = \overline{\mu}_\beta^{\alpha i i_1 \cdots i_r} - S_\beta^{\alpha i i_1 \cdots i_r}, \\
&\overline{\mu}_\beta^{\alpha i i_1 \cdots i_r} = \frac{\partial F^{\alpha i}}{\partial v^\beta_{i_1 \cdots i_r}} + \frac{\partial F^{\alpha i}}{\partial \Sigma^{\gamma j}} s_\beta^{\gamma j i_1 \cdots i_r} + \frac{\partial F^{\alpha i}}{\partial \Sigma^\gamma} t_\beta^{\gamma i_1 \cdots i_r}, \quad 0 \le r \le m-1, \\
&S^{\alpha i}_j = \overline{S}^{\alpha i}_j - \sum_{r=0}^{m-1} S_\beta^{\alpha i i_1 \cdots i_r} v^\beta_{i_1 \cdots i_r j}, \\
&\overline{S}^{\alpha i}_j = \frac{\partial F^{\alpha i}}{\partial x^j} + \sum_{r=0}^{m-1} \overline{\mu}_\beta^{\alpha i i_1 \cdots i_r} v^\beta_{i_1 \cdots i_r j} + \frac{\partial F^{\alpha i}}{\partial \Sigma^{\beta k}} s^{\beta k}_j + \frac{\partial F^{\alpha i}}{\partial \Sigma^\beta} t^\beta_j, \\
&S_\beta^{\alpha i i_1 \cdots i_m} = \frac{\partial F^{\alpha i}}{\partial v^\beta_{i_1 \cdots i_m}} + \frac{\partial F^{\alpha i}}{\partial \Sigma^{\gamma j}} s_\beta^{\gamma j i_1 \cdots i_m} + \frac{\partial F^{\alpha i}}{\partial \Sigma^\gamma} t_\beta^{\gamma i_1 \cdots i_m} = \overline{\mu}_\beta^{\alpha i i_1 \cdots i_m}.
\end{aligned}
$$

As is clearly seen from the above relations, there are no restrictions on functions $S_\beta^{\alpha i i_1 \cdots i_r}, r = 0, 1, \ldots, m-1$, hence they can be selected totally arbitrarily. Finally, let us take the equations $(9.8.4)_3$ into account. The Lie derivatives evaluated by using the relations

$$
\mathbf{i}_V \Omega^\alpha = T^\alpha - t^\alpha_i X^i - \sum_{r=0}^{m} t_\beta^{\alpha i_1 \cdots i_r} V^\beta_{i_1 \cdots i_r} = G^\alpha \tag{9.8.9}
$$

and

$$
\mathbf{i}_V d\Omega^\alpha = - T^\alpha_i dx^i + X^i dt^\alpha_i - \sum_{r=0}^{m} T_\beta^{\alpha i_1 \cdots i_r} dv^\beta_{i_1 \cdots i_r} + \sum_{r=0}^{m} V^\beta_{i_1 \cdots i_r} dt_\beta^{\alpha i_1 \cdots i_r}
$$

lead eventually to the results

$$N^{\alpha}_{\beta i} = \frac{\partial G^{\alpha}}{\partial \Sigma^{\beta i}}, \quad N^{\alpha}_{\beta} = \frac{\partial G^{\alpha}}{\partial \Sigma^{\beta}}, \quad \frac{\partial G^{\alpha}}{\partial s^{\beta i}_{j}} = 0, \quad \delta^{\alpha}_{\beta} X^{i} + \frac{\partial G^{\alpha}}{\partial t^{\beta}_{i}} = 0;$$

$$\frac{\partial G^{\alpha}}{\partial s^{\gamma i i_1\cdots i_r}_{\beta}} = 0, 0 \le r \le m, \; \delta^{\alpha}_{\gamma} V^{\beta}_{i_1\cdots i_r} + \frac{\partial G^{\alpha}}{\partial t^{\gamma i_1\cdots i_r}_{\beta}} = 0, \; 0 \le r \le m,$$

$$\nu^{\alpha i_1\cdots i_r}_{\beta} = \overline{\nu}^{\alpha i_1\cdots i_r}_{\beta} - T^{\alpha i_1\cdots i_r}_{\beta},$$

$$\overline{\nu}^{\alpha i_1\cdots i_r}_{\beta} = \frac{\partial G^{\alpha}}{\partial v^{\beta}_{i_1\cdots i_r}} + \frac{\partial G^{\alpha}}{\partial \Sigma^{\gamma i}} s^{\gamma i i_1\cdots i_r}_{\beta} + \frac{\partial G^{\alpha}}{\partial \Sigma^{\gamma}} t^{\gamma i_1\cdots i_r}_{\beta}, 0 \le r \le m-1,$$

$$T^{\alpha}_{i} = \overline{T}^{\alpha}_{i} - \sum_{r=0}^{m-1} T^{\alpha i_1\cdots i_r}_{\beta} v^{\beta}_{i_1\cdots i_r i},$$

$$\overline{T}^{\alpha}_{i} = \frac{\partial G^{\alpha}}{\partial x^{i}} + \sum_{r=0}^{m-1} \overline{\nu}^{\alpha i_1\cdots i_r}_{\beta} v^{\beta}_{i_1\cdots i_r i} + \frac{\partial G^{\alpha}}{\partial \Sigma^{\beta j}} s^{\beta j}_{i} + \frac{\partial G^{\alpha}}{\partial \Sigma^{\beta}} t^{\beta}_{i},$$

$$T^{\alpha i_1\cdots i_m}_{\beta} = \frac{\partial G^{\alpha}}{\partial v^{\beta}_{i_1\cdots i_m}} + \frac{\partial G^{\alpha}}{\partial \Sigma^{\gamma i}} s^{\gamma i i_1\cdots i_m}_{\beta} + \frac{\partial G^{\alpha}}{\partial \Sigma^{\gamma}} t^{\gamma i_1\cdots i_m}_{\beta} = \overline{\nu}^{\alpha i_1\cdots i_m}_{\beta}.$$

where the functions $T^{\alpha i_1\cdots i_r}_{\beta}, r = 0, 1, \ldots, m-1$ may be chosen totally arbitrarily. Thus, isovector fields of the contact ideal are characterised in the following manner

$$\begin{aligned} V = {} & X^{i}\frac{\partial}{\partial x^{i}} + \sum_{r=0}^{m} V^{\alpha}_{i_1\cdots i_r}\frac{\partial}{\partial v^{\alpha}_{i_1\cdots i_r}} + S^{\alpha i}\frac{\partial}{\partial \Sigma^{\alpha i}} + T^{\alpha}\frac{\partial}{\partial \Sigma^{\alpha}} + \overline{S}^{\alpha i}_{j}\frac{\partial}{\partial s^{\alpha i}_{j}} \\ & + \overline{T}^{\alpha}_{i}\frac{\partial}{\partial t^{\alpha}_{i}} + S^{\alpha i i_1\cdots i_m}_{\beta}\frac{\partial}{\partial s^{\alpha i i_1\cdots i_m}_{\beta}} + T^{\alpha i_1\cdots i_m}_{\beta}\frac{\partial}{\partial t^{\alpha i_1\cdots i_m}_{\beta}} \\ & + V_1 + V_2 \end{aligned}$$

where the vectors fields V_1 and V_2 are defined by

$$V_1 = \sum_{r=0}^{m-1} S^{\alpha i i_1\cdots i_r}_{\beta}\Big(\frac{\partial}{\partial s^{\alpha i i_1\cdots i_r}_{\beta}} - v^{\beta}_{i_1\cdots i_r j}\frac{\partial}{\partial s^{\alpha i}_{j}}\Big),$$

$$V_2 = \sum_{r=0}^{m-1} T^{\alpha i_1\cdots i_r}_{\beta}\Big(\frac{\partial}{\partial t^{\alpha i_1\cdots i_r}_{\beta}} - v^{\beta}_{i_1\cdots i_r i}\frac{\partial}{\partial t^{\alpha}_{i}}\Big)$$

depending on indeterminate coefficients $S^{\alpha i i_1\cdots i_r}_{\beta}$ and $T^{\alpha i_1\cdots i_r}_{\beta}, r = 0, 1, \ldots, m-1$. On the other hand, we can easily demonstrate that the Lie derivatives of the generators of the contact ideal with respect to the vector fields V_1 and V_2 satisfy the relations

$$
\begin{aligned}
\pounds_{V_1}\sigma^{\alpha}_{i_1\cdots i_r} &= 0,\ \pounds_{V_1}\Omega^{\alpha} = 0,\ \pounds_{V_1}\omega^{\alpha} = 0,\\
\pounds_{V_1}\Omega^{\alpha i} &= \sum_{r=0}^{m-1} S^{\alpha i i_1\cdots i_r}_{\beta}(v^{\beta}_{i_1\cdots i_r j}dx^j - dv^{\beta}_{i_1\cdots i_r})\\
&= -\sum_{r=0}^{m-1} S^{\alpha i i_1\cdots i_r}_{\beta}\sigma^{\beta}_{i_1\cdots i_r},\\
\pounds_{V_2}\sigma^{\alpha}_{i_1\cdots i_r} &= 0,\quad \pounds_{V_2}\Omega^{\alpha i} = 0,\ \pounds_{V_2}\omega^{\alpha} = 0,\\
\pounds_{V_2}\Omega^{\alpha} &= \sum_{r=0}^{m-1} T^{\alpha i_1\cdots i_r}_{\beta}(v^{\beta}_{i_1\cdots i_r i}dx^i - dv^{\beta}_{i_1\cdots i_r})\\
&= -\sum_{r=0}^{m-1} T^{\alpha i_1\cdots i_r}_{\beta}\sigma^{\beta}_{i_1\cdots i_r}
\end{aligned}
$$

so that they are automatically isovectors of the contact ideal without imposing any restriction on the coefficient functions. Hence, these trivial isovector fields can be discarded without loss of generality because they will not be operative in determining the equivalence groups. The rather simple differential equations for the functions $F^{\alpha i}$ and G^{α} appearing in the first and second lines in the foregoing sets of relations concerned with the coefficients in equations (9.8.4) can readily be integrated to yield

$$
\begin{aligned}
F^{\alpha i} &= -s^{\alpha i}_{j}X^j - \sum_{r=0}^{m} s^{\alpha i i_1\cdots i_r}_{\beta}V^{\beta}_{i_1\cdots i_r} + \mathcal{F}^{\alpha i},\\
G^{\alpha} &= -t^{\alpha}_{i}X^i - \sum_{r=0}^{m} t^{\alpha i_1\cdots i_r}_{\beta}V^{\beta}_{i_1\cdots i_r} + \mathcal{G}^{\alpha}
\end{aligned}
$$

where the functions $\mathcal{F}^{\alpha i}$ and $\mathcal{G}^{\alpha}$ depend only on the variables

$$
x^j, \Sigma^{\beta j}, \Sigma^{\beta},\ \{v^{\beta}_{i_1\cdots i_r} : r = 0, 1, \dots, m\}.
$$

When we insert the above expressions into the equations $(9.8.8)_1$ and $(9.8.9)_1$, we see at once that we can write

$$
S^{\alpha i} = \mathcal{F}^{\alpha i}, \quad T^{\alpha} = \mathcal{G}^{\alpha}.
$$

Therefore, the isovector fields of the contact ideal are entirely specified by $n + 2N + nN$ functions X^i, U^{α} depending only on x^i, u^{α} and $S^{\alpha i}, T^{\alpha}$ only on $x^j, \Sigma^{\beta j}, \Sigma^{\beta}, v^{\beta}_{i_1\cdots i_r}, 0 \le r \le m$ They are presently chosen arbitrarily.

In order to determine the isovector fields of the balance ideal, we next have to consider the following relations

$$\pounds_V\,\omega^\alpha = \nu^\alpha_\beta\omega^\beta + \sum_{r=0}^{m-1}\sigma^\beta_{i_1\cdots i_r}\wedge A^{\alpha i_1\cdots i_r}_\beta + \sum_{r=0}^{m-1}d\sigma^\beta_{i_1\cdots i_r}\wedge B^{\alpha i_1\cdots i_r}_\beta$$
$$+\,\Omega^{\beta i}\wedge C^\alpha_{\beta i} + d\Omega^{\beta i}\wedge D^\alpha_{\beta i} + \Omega^\beta\wedge C^\alpha_\beta + d\Omega^\beta\wedge D^\alpha_\beta$$

and show that the forms $\nu^\alpha_\beta \in \Lambda^0(\mathcal{K}_m)$; $A^{\alpha i_1\cdots i_r}_\beta, C^\alpha_{\beta i}, C_\beta \in \Lambda^{n-1}(\mathcal{K}_m)$; and $B^{\alpha i_1\cdots i_r}_\beta, D^\alpha_{\beta i}, D^\alpha_\beta \in \Lambda^{n-2}(\mathcal{K}_m)$ can be so found as to satisfy the above expressions. On inserting the expressions

$$\begin{aligned}
\pounds_V\,\omega^\alpha &= \mathbf{i}_V(d\,\omega^\alpha) + d(\mathbf{i}_V\omega^\alpha)\\
&= T^\alpha\mu + (dS^{\alpha i} + \Sigma^\alpha dX^i)\wedge\mu_i + d\Sigma^{\alpha i}\wedge dX^j\wedge\mu_{ji}\\
&= \Big(T^\alpha + \frac{\partial S^{\alpha i}}{\partial x^i} + \Sigma^\alpha\frac{\partial X^i}{\partial x^i}\Big)\mu + \Big(\frac{\partial S^{\alpha i}}{\partial u^\beta} + \Sigma^\alpha\frac{\partial X^i}{\partial u^\beta}\Big)du^\beta\wedge\mu_i\\
&\quad + \sum_{r=1}^{m}\frac{\partial S^{\alpha i}}{\partial v^\beta_{i_1\cdots i_r}}dv^\beta_{i_1\cdots i_r}\wedge\mu_i + \frac{\partial S^{\alpha i}}{\partial\Sigma^{\beta j}}d\Sigma^{\beta j}\wedge\mu_i\\
&\quad + \frac{\partial S^{\alpha i}}{\partial\Sigma^\beta}d\Sigma^\beta\wedge\mu_i - \frac{\partial X^j}{\partial x^i}d\Sigma^{\alpha i}\wedge\mu_j + \frac{\partial X^j}{\partial x^j}d\Sigma^{\alpha i}\wedge\mu_i\\
&\quad - \frac{\partial X^j}{\partial u^\beta}du^\beta\wedge d\Sigma^{\alpha i}\wedge\mu_{ji}
\end{aligned}$$

into $\pounds_V\,\omega^\alpha$, we immediately see that we have to take $D^\alpha_{\beta i} = 0$ and $D^\alpha_\beta = 0$. Let us note that we can write

$$\begin{aligned}
dv^\alpha_{i_1\cdots i_r} &= \sigma^\alpha_{i_1\cdots i_r} + v^\alpha_{i_1\cdots i_r i}\,dx^i,\ 0\le r\le m-1,\\
d\sigma^\alpha_{i_1\cdots i_r} &= -\,dv^\alpha_{i_1\cdots i_r i}\wedge dx^i\\
&= -\,\sigma^\alpha_{i_1\cdots i_r i}\wedge dx^i,\ 0\le r\le m-2,\\
d\sigma^\alpha_{i_1\cdots i_{m-1}} &= -\,dv^\alpha_{i_1\cdots i_{m-1}i}\wedge dx^i
\end{aligned}$$

$$\begin{aligned}
d\Sigma^{\alpha i} &= \Omega^{\alpha i} + \Big(s^{\alpha i}_j + \sum_{r=0}^{m-1}s^{\alpha i i_1\cdots i_r}_\beta v^\beta_{i_1\cdots i_r j}\Big)dx^j\\
&\quad + \sum_{r=0}^{m-1}s^{\alpha i i_1\cdots i_r}_\beta\sigma^\beta_{i_1\cdots i_r} + s^{\alpha i i_1\cdots i_m}_\beta dv^\beta_{i_1\cdots i_m}\\
d\Sigma^\alpha &= \Omega^\alpha + \Big(t^\alpha_i + \sum_{r=0}^{m-1}t^{\alpha i_1\cdots i_r}_\beta v^\beta_{i_1\cdots i_r i}\Big)dx^i\\
&\quad + \sum_{r=0}^{m-1}t^{\alpha i_1\cdots i_r}_\beta\sigma^\beta_{i_1\cdots i_r} + t^{\alpha i_1\cdots i_m}_\beta dv^\beta_{i_1\cdots i_m}
\end{aligned}$$

and then introduce these transformations into proper places in the invariance relations connected with Lie derivatives of forms ω^α with respect to an isovector field V and arrange the resulting expressions suitably to arrive at the equations below

$$
\begin{aligned}
&\Big[\Gamma^\alpha - \nu^\alpha_\beta\Big(\Sigma^\beta + s^{\beta i}_i + \sum_{r=0}^{m-1} s^{\beta i i_1\cdots i_r}_\gamma v^\gamma_{i_1\cdots i_r i}\Big)\Big]\mu \\
&\qquad + \sigma^\beta \wedge \Big[\Gamma^{\alpha i}_\beta \mu_i - \frac{\partial X^j}{\partial u^\beta} s^{\alpha i}_\gamma \sigma^\gamma \wedge \mu_{ji} \\
&\qquad\qquad - \nu^\alpha_\gamma s^{\gamma i}_\beta \mu_i - \frac{\partial X^j}{\partial u^\beta} s^{\alpha i i_1\cdots i_m}_\gamma dv^\gamma_{i_1\cdots i_m} \wedge \mu_{ji} - A^\alpha_\beta\Big] \\
&+ \sum_{r=1}^{m-1} \sigma^\beta_{i_1\cdots i_r} \wedge \Big[\Gamma^{\alpha i i_1\cdots i_r}_\beta \mu_i - \nu^\alpha_\gamma s^{\gamma i i_1\cdots i_r}_\beta \mu_i - A^{\alpha i_1\cdots i_r}_\beta + dx^{i_r} \wedge B^{\alpha i_1\cdots i_{r-1}}_\beta\Big] \\
&\quad + dv^\beta_{i_1\cdots i_{m-1} i_m} \wedge \big(\Gamma^{\alpha i i_1\cdots i_m}_\beta \mu_i - \nu^\alpha_\gamma s^{\gamma i i_1\cdots i_m}_\beta \mu_i + dx^{i_m} \wedge B^{\alpha i_1\cdots i_{m-1}}_\beta\big) \\
&\qquad + \Omega^{\beta j} \wedge \Big(\Gamma^{\alpha i}_{\beta j}\mu_i - \frac{\partial X^i}{\partial u^\gamma}\delta^\alpha_\beta \sigma^\gamma \wedge \mu_{ji} - \nu^\alpha_\beta \delta^i_j \mu_i - C^\alpha_{\beta j}\Big) \\
&\qquad\qquad + \Omega^\beta \wedge \Big(\frac{\partial S^{\alpha i}}{\partial \Sigma^\beta}\mu_i - C^\alpha_\beta\Big) = 0.
\end{aligned}
$$

The coefficient functions Γ^α, $\Gamma^{\alpha i}_\beta$, $\Gamma^{\alpha i}_{\beta j}$ and $\Gamma^{\alpha i i_1\cdots i_r}_\beta, r = 1, \ldots, m$ appearing in these equations and derived from essentially unknown functions X^i, $S^{\alpha i}$, T^α are listed below

$$
\begin{aligned}
\Gamma^\alpha &= T^\alpha + \frac{\partial S^{\alpha i}}{\partial x^i} + \Sigma^\alpha \frac{\partial X^i}{\partial x^i} + \Big(\frac{\partial S^{\alpha i}}{\partial u^\beta} + \Sigma^\alpha \frac{\partial X^i}{\partial u^\beta}\Big) v^\beta_i + \sum_{r=1}^{m-1} \frac{\partial S^{\alpha i}}{\partial v^\beta_{i_1\cdots i_r}} v^\beta_{i_1\cdots i_r i} \\
&\quad + \Big\{\frac{\partial S^{\alpha i}}{\partial \Sigma^{\beta j}} + \Big[\Big(\frac{\partial X^k}{\partial x^k} + \frac{\partial X^k}{\partial u^\gamma} v^\gamma_k\Big)\delta^i_j - \Big(\frac{\partial X^i}{\partial x^j} + \frac{\partial X^i}{\partial u^\gamma} v^\gamma_j\Big)\Big]\delta^\alpha_\beta\Big\} \times \\
&\quad \Big(s^{\beta j}_i + \sum_{r=0}^{m-1} s^{\beta j i_1\cdots i_r}_\gamma v^\gamma_{i_1\cdots i_r i}\Big) + \frac{\partial S^{\alpha i}}{\partial \Sigma^\beta}\Big(t^\beta_i + \sum_{r=0}^{m-1} t^{\beta i_1\cdots i_r}_\gamma v^\gamma_{i_1\cdots i_r i}\Big) \\
\Gamma^{\alpha i}_\beta &= \frac{\partial S^{\alpha i}}{\partial u^\beta} + \Sigma^\alpha \frac{\partial X^i}{\partial u^\beta} + \Big\{\frac{\partial S^{\alpha i}}{\partial \Sigma^{\gamma j}} + \Big[\Big(\frac{\partial X^k}{\partial x^k} + \frac{\partial X^k}{\partial u^\delta} v^\delta_k\Big)\delta_j \\
&\quad - \Big(\frac{\partial X^i}{\partial x^j} + \frac{\partial X^i}{\partial u^\delta} v^\delta_j\Big)\Big]\delta^\alpha_\gamma\Big\} s^{\gamma j}_\beta + \frac{\partial S^{\alpha i}}{\partial \Sigma^\gamma} t^\gamma_\beta + \frac{\partial X^i}{\partial u^\beta}\Big(s^{\alpha j}_j + \sum_{r=0}^{m-1} s^{\alpha j i_1\cdots i_r}_\gamma v^\gamma_{i_1\cdots i_r j}\Big) \\
&\quad - \frac{\partial X^j}{\partial u^\beta}\Big(s^{\alpha i}_j + \sum_{r=0}^{m-1} s^{\alpha i i_1\cdots i_r}_\gamma v^\gamma_{i_1\cdots i_r j}\Big)
\end{aligned}
$$

(9.8.10)

$$\Gamma_{\beta}^{\alpha i i_1\cdots i_r} = \frac{\partial S^{\alpha i}}{\partial v_{i_1\cdots i_r}^{\beta}} + \Big\{\frac{\partial S^{\alpha i}}{\partial \Sigma^{\gamma j}} + \Big[\Big(\frac{\partial X^k}{\partial x^k} + \frac{\partial X^k}{\partial u^\delta}v_k^\delta\Big)\delta_j^i - \Big(\frac{\partial X^i}{\partial x^j} + \frac{\partial X^i}{\partial u^\delta}v_j^\delta\Big)\Big]\delta_\gamma^\alpha\Big\} s_\beta^{\gamma j i_1\cdots i_r} + \frac{\partial S^{\alpha i}}{\partial \Sigma^\gamma} t_\beta^{\gamma i_1\cdots i_r}, \ 1 \le r \le m$$

$$\Gamma_{\beta j}^{\alpha i} = \frac{\partial S^{\alpha i}}{\partial \Sigma^{\beta j}} + \Big[\Big(\frac{\partial X^k}{\partial x^k} + \frac{\partial X^k}{\partial u^\gamma}v_k^\gamma\Big)\delta_j^i - \Big(\frac{\partial X^i}{\partial x^j} + \frac{\partial X^i}{\partial u^\gamma}v_j^\gamma\Big)\Big]\delta_\beta^\alpha$$

To satisfy the foregoing equations, the coefficient functions must verify the conditions

$$\Gamma^\alpha - \nu_\beta^\alpha\Big(\Sigma^\beta + s_i^{\beta i} + \sum_{r=0}^{m-1} s_\gamma^{\beta i i_1\cdots i_r} v_{i_1\cdots i_r i}^\gamma\Big) = 0,$$

$$A_\beta^\alpha = (\Gamma_\beta^{\alpha i} - \nu_\gamma^\alpha s_\beta^{\gamma i})\mu_i - \frac{\partial X^j}{\partial u^\beta}(s_\gamma^{\alpha i}\sigma^\gamma + s_\gamma^{\alpha i i_1\cdots i_m} dv_{i_1\cdots i_m}^\gamma) \wedge \mu_{ji},$$

$$A_\beta^{\alpha i_1\cdots i_r} = (\Gamma_\beta^{\alpha i i_1\cdots i_r} - \nu_\gamma^\alpha s_\beta^{\gamma i i_1\cdots i_r})\mu_i + dx^{i_r} \wedge B_\beta^{\alpha i_1\cdots i_{r-1}},$$

$$C_{\beta j}^\alpha = (\Gamma_{\beta j}^{\alpha i} - \nu_\beta^\alpha \delta_j^i)\mu_i - \frac{\partial X^i}{\partial u^\gamma}\delta_\beta^\alpha \sigma^\gamma \wedge \mu_{ji}, \ C_\beta^\alpha = \frac{\partial S^{\alpha i}}{\partial \Sigma^\beta}\mu_i,$$

$$(\Gamma_\beta^{\alpha i i_1\cdots i_m} - \nu_\gamma^\alpha s_\beta^{\gamma i i_1\cdots i_m})\mu_i + dx^{i_m} \wedge B_\beta^{\alpha i_1\cdots i_{m-1}} = 0.$$

To fully exploit the last relation above, let us take

$$B_\beta^{\alpha i_1\cdots i_{m-1}} = B_\beta^{\alpha i_1\cdots i_{m-1} i j}\mu_{ij}$$

where $B_\beta^{\alpha i_1\cdots i_{m-1} i j} = -B_\beta^{\alpha i_1\cdots i_{m-1} j i} \in \Lambda^0(\mathcal{K}_m)$ are arbitrary functions. On employing the relation

$$\begin{aligned} dx^{i_m} \wedge B_\beta^{\alpha i_1\cdots i_{m-1}} &= B_\beta^{\alpha i_1\cdots i_{m-1} i j} dx^{i_m} \wedge \mu_{ij} \\ &= B_\beta^{\alpha i_1\cdots i_{m-1} i j}(\delta_i^{i_m}\mu_j - \delta_j^{i_m}\mu_i) \\ &= B_\beta^{\alpha i_1\cdots i_{m-1} i_m j}\mu_j - B_\beta^{\alpha i_1\cdots i_{m-1} i i_m}\mu_i \\ &= 2B_\beta^{\alpha i_1\cdots i_{m-1} i_m i}\mu_i \end{aligned}$$

we get

$$2B_\beta^{\alpha i_1\cdots i_{m-1} i_m i} = -(\Gamma_\beta^{\alpha i i_1\cdots i_m} - \nu_\gamma^\alpha s_\beta^{\gamma i i_1\cdots i_m}).$$

However, the left hand side above is antisymmetric with respect to its last two superscripts. Hence, the symmetric part of the right hand side with respect to indices i and i_m must vanish. Therefore, the determining equations for the isovector components X^i, $S^{\alpha i}$, T^α and the functions ν_β^α are found eventually to be

$$\Gamma^\alpha = \nu^\alpha_\beta\Big(\Sigma^\beta + s_i^{\beta i} + \sum_{r=0}^{m-1} s_\gamma^{\beta i i_1\cdots i_r} v^\gamma_{i_1\cdots i_r i}\Big), \tag{9.8.11}$$
$$\Gamma_\beta^{\alpha i i_1\cdots i_m} + \Gamma_\beta^{\alpha i_m i_1\cdots i} = \nu^\alpha_\gamma(s_\beta^{\gamma i i_1\cdots i_m} + s_\beta^{\gamma i_m i_1\cdots i}).$$

These equations do not impose any restriction on the isovector components U^α. *Nonetheless, if some variables do not appear in the coordinate cover of the manifold $\mathcal{K}_m$ due to a particular structure of the balance equations, that might entail some new restrictions on the isovector components because the corresponding isovector components must then be set to zero.* Equivalence transformations are now obtained by solving the following autonomous ordinary differential equations

$$\begin{aligned}
\frac{d\overline{x}^i}{d\epsilon} &= X^i(\overline{x}^j,\overline{u}^\beta),\ \frac{d\overline{u}^\alpha}{d\epsilon} = U^\alpha(\overline{x}^j,\overline{u}^\beta),\\
\frac{d\overline{v}^\alpha_{i_1\cdots i_r}}{d\epsilon} &= V^\alpha_{i_1\cdots i_r}(\overline{x}^j,\overline{v}^\beta_{j_1\cdots j_s}),\ 0\le s\le r,\ 1\le r\le m,\\
\frac{d\overline{\Sigma}^{\alpha i}}{d\epsilon} &= \mathcal{F}^{\alpha i}(\overline{x}^j,\overline{\Sigma}^{\beta j},\overline{\Sigma}^\beta,\overline{v}^\beta_{j_1\cdots j_r}),\ 0\le r\le m,\\
\frac{d\overline{\Sigma}^\alpha}{d\epsilon} &= \mathcal{G}^\alpha(\overline{x}^j,\overline{\Sigma}^{\beta j},\overline{\Sigma}^\beta,\overline{v}^\beta_{j_1\cdots j_r}),\ 0\le r\le m
\end{aligned}$$

under the initial conditions

$$\begin{aligned}
&\overline{x}^i(0) = x^i,\ \overline{v}^\alpha_{i_1\cdots i_r}(0) = v^\alpha_{i_1\cdots i_r},\ 0\le r\le m;\\
&\overline{\Sigma}^{\alpha i}(0) = \Sigma^{\alpha i},\ \ \overline{\Sigma}^\alpha(0) = \Sigma^\alpha.
\end{aligned}$$

For $m = 0$, that is, for the first order balance equations we have to take $\Sigma^{\alpha i}(\mathbf{x},\mathbf{u})$ and $\Sigma^\alpha(\mathbf{x},\mathbf{u})$, and we have to modify our analysis substantially. In this case, the coordinates of $\mathcal{K}_m$ are merely $\{x^i, u^\alpha, \Sigma^{\alpha i}, \Sigma^\alpha, s_j^{\alpha i}, s_\beta^{\alpha i}, t_i^\alpha, t_\beta^\alpha\}$ [*see* (9.8.1)]. The contact ideal is now constructed by the following 1-forms

$$\Omega^{\alpha i} = d\Sigma^{\alpha i} - s_j^{\alpha i}dx^j - s_\beta^{\alpha i}du^\beta,\ \ \Omega^\alpha = d\Sigma^\alpha - t_i^\alpha dx^i - t_\beta^\alpha du^\beta.$$

An isovector field of the balance ideal must then be taken as

$$\begin{aligned}
V = X^i\frac{\partial}{\partial x^i} + U^\alpha\frac{\partial}{\partial u^\alpha} + S^{\alpha i}\frac{\partial}{\partial\Sigma^{\alpha i}} + T^\alpha\frac{\partial}{\partial\Sigma^\alpha} + S_j^{\alpha i}\frac{\partial}{\partial s_j^{\alpha i}} + S_\beta^{\alpha i}\frac{\partial}{\partial s_\beta^{\alpha i}}\\
+ T_i^\alpha\frac{\partial}{\partial t_i^\alpha} + T_\beta^\alpha\frac{\partial}{\partial t_\beta^\alpha}.
\end{aligned}$$

In order that this vector turns out to be an isovector of the contact ideal, one has to satisfy the relations

$$\mathbf{i}_V\Omega^{\alpha i} = L^{\alpha i}_{\beta j}\Omega^{\beta j} + L^{\alpha i}_{\beta}\Omega^{\beta}, \quad \mathbf{i}_V\Omega^{\alpha} = M^{\alpha}_{\beta i}\Omega^{\beta i} + M^{\alpha}_{\beta}\Omega^{\beta}$$

from which we easily deduce that

$$L^{\alpha i}_{\beta j} = \frac{\partial F^{\alpha i}}{\partial \Sigma^{\beta j}},\ L^{\alpha i}_{\beta} = \frac{\partial F^{\alpha i}}{\partial \Sigma^{\beta}},\ M^{\alpha}_{\beta i} = \frac{\partial G^{\alpha}}{\partial \Sigma^{\beta i}},\ M^{\alpha}_{\beta} = \frac{\partial G^{\alpha}}{\partial \Sigma^{\beta}},$$
$$\frac{\partial F^{\alpha i}}{\partial t^{\beta}_{j}} = 0,\ \frac{\partial F^{\alpha i}}{\partial t^{\beta}_{\gamma}} = 0,\ \frac{\partial G^{\alpha}}{\partial s^{\beta i}_{j}} = 0,\ \frac{\partial G^{\alpha}}{\partial s^{\beta i}_{\gamma}} = 0,$$
$$S^{\alpha i}_{j} = \frac{\partial F^{\alpha i}}{\partial x^{j}} + \frac{\partial F^{\alpha i}}{\partial \Sigma^{\beta k}} s^{\beta k}_{j} + \frac{\partial F^{\alpha i}}{\partial \Sigma^{\beta}} t^{\beta}_{j},\ S^{\alpha i}_{\beta} = \frac{\partial F^{\alpha i}}{\partial u^{\beta}} + \frac{\partial F^{\alpha i}}{\partial \Sigma^{\gamma j}} s^{\gamma j}_{\beta} + \frac{\partial F^{\alpha i}}{\partial \Sigma^{\gamma}} t^{\gamma}_{\beta},$$
$$T^{\alpha}_{i} = \frac{\partial G^{\alpha}}{\partial x^{i}} + \frac{\partial G^{\alpha}}{\partial \Sigma^{\beta j}} s^{\beta j}_{i} + \frac{\partial G^{\alpha}}{\partial \Sigma^{\beta}} t^{\beta}_{i},\quad T^{\alpha}_{\beta} = \frac{\partial G^{\alpha}}{\partial u^{\beta}} + \frac{\partial G^{\alpha}}{\partial \Sigma^{\gamma i}} s^{\gamma i}_{\beta} + \frac{\partial G^{\alpha}}{\partial \Sigma^{\gamma}} t^{\gamma}_{\beta},$$
$$X^{j}\delta^{\alpha}_{\beta}\delta^{i}_{k} + \frac{\partial F^{\alpha i}}{\partial s^{\beta k}_{j}} = X^{i}\delta^{\alpha}_{\beta} + \frac{\partial G^{\alpha}}{\partial t^{\beta}_{i}} = 0, U^{\beta}\delta^{\alpha}_{\gamma}\delta^{i}_{j} + \frac{\partial F^{\alpha i}}{\partial s^{\gamma j}_{\beta}} = U^{\beta}\delta^{\alpha}_{\gamma} + \frac{\partial G^{\alpha}}{\partial t^{\gamma}_{\beta}} = 0.$$

Here, we have defined

$$\mathbf{i}_V\Omega^{\alpha i} = F^{\alpha i} = S^{\alpha i} - X^{j}s^{\alpha i}_{j} - U^{\beta}s^{\alpha i}_{\beta},\ \mathbf{i}_V\Omega^{\alpha} = G^{\alpha} = T^{\alpha} - X^{i}t^{\alpha}_{i} - U^{\beta}t^{\alpha}_{\beta}$$

If we scrutinise carefully the above equations, we immediately reach to the conclusion

$$X^{i} = X^{i}(x^{j}, u^{\beta}, \Sigma^{\beta j}, \Sigma^{\beta}),\ \ U^{\alpha} = U^{\alpha}(x^{j}, u^{\beta}, \Sigma^{\beta j}, \Sigma^{\beta})$$

and

$$F^{\alpha i} = -X^{j}s^{\alpha i}_{j} - U^{\beta}s^{\alpha i}_{\beta} + \mathcal{F}^{\alpha i},\ G^{\alpha} = -X^{i}t^{\alpha}_{i} - U^{\beta}t^{\alpha}_{\beta} + \mathcal{G}^{\alpha}$$

where $\mathcal{F}^{\alpha i}$ and $\mathcal{G}^{\alpha}$ depend only on the variables $x^{j}, u^{\beta}, \Sigma^{\beta j}, \Sigma^{\beta}$. Ultimately, let us consider the relations

$$\pounds_V\,\omega^{\alpha} = \nu^{\alpha}_{\beta}\omega^{\beta} + \Omega^{\beta i}\wedge C^{\alpha}_{\beta i} + d\Omega^{\beta i}\wedge D^{\alpha}_{\beta i} + \Omega^{\beta}\wedge C^{\alpha}_{\beta} + d\Omega^{\beta}\wedge D^{\alpha}_{\beta}.$$

We readily observe that $D^{\alpha}_{\beta i} = 0, D^{\alpha}_{\beta} = 0$. We thus obtain

$$\begin{aligned}&\big[\Gamma^{\alpha} - \nu^{\alpha}_{\beta}(\Sigma^{\beta} + s^{\beta i}_{i})\big]\mu + (G^{\alpha i}_{\beta} + \Gamma^{\alpha i}_{\gamma}t^{\gamma}_{\beta} - \nu^{\alpha}_{\gamma}s^{\gamma i}_{\beta})\,du^{\beta}\wedge\mu^{\alpha}_{i}\\ &\quad + \Omega^{\beta j}\wedge\Big[-C_{\beta j} + (\Gamma^{\alpha i}_{\beta j} - \nu^{\alpha}_{\beta}\delta^{i}_{j})\mu_{i} - \Big(\Gamma^{\alpha ik}_{\gamma\beta j}\,du^{\gamma} + \frac{\partial X^{k}}{\partial \Sigma^{\beta j}}\Omega^{\alpha i}\Big)\wedge\mu_{ki}\Big]\\ &\quad + \Gamma^{\alpha ij}_{\beta\gamma}du^{\beta}\wedge du^{\gamma}\wedge\mu_{ji}\end{aligned}$$

$$+\,\Omega^\beta\wedge\Big[-C^\alpha_\beta+\Gamma^{\alpha i}_\beta\mu_i-\frac{\partial X^j}{\partial\Sigma^\beta}(\Omega^{\alpha i}+s^{\alpha i}_\gamma\,du^\gamma)\wedge\mu_{ji}\Big]=0.$$

Smooth functions Γ^α, $\Gamma^{\alpha i}_\beta$, $G^{\alpha i}_\beta$, $\Gamma^{\alpha i}_{\beta j}$, $\Gamma^{\alpha ij}_{\beta\gamma k}$ and $\Gamma^{\alpha ij}_{\beta\gamma}$ in the module $\Lambda^0(\mathcal{K}_0)$ appearing in the above equations are given as follows

$$\begin{aligned}
\Gamma^\alpha&=T^\alpha+\frac{\partial S^{\alpha i}}{\partial x^i}+\Sigma^\alpha\frac{\partial X^i}{\partial x^i}+\Big(\frac{\partial S^{\alpha i}}{\partial\Sigma^{\beta j}}+\Sigma^\alpha\frac{\partial X^i}{\partial\Sigma^{\beta j}}+\frac{\partial X^k}{\partial x^k}\delta^\alpha_\beta\delta^i_j\\
&\quad-\frac{\partial X^i}{\partial x^j}\delta^\alpha_\beta\Big)s^{\beta j}_i+\Big(\frac{\partial S^{\alpha i}}{\partial\Sigma^\beta}+\Sigma^\alpha\frac{\partial X^i}{\partial\Sigma^\beta}\Big)t^\beta_i\\
&\quad-\frac{\partial X^j}{\partial\Sigma^{\beta k}}(s^{\beta k}_i s^{\alpha i}_j-s^{\beta k}_j s^{\alpha i}_i)-\frac{\partial X^j}{\partial\Sigma^\beta}(t^\beta_i s^{\alpha i}_j-t^\beta_j s^{\alpha i}_i),\\
\Gamma^{\alpha i}_\beta&=\frac{\partial S^{\alpha i}}{\partial\Sigma^\beta}+\Sigma^\alpha\frac{\partial X^i}{\partial\Sigma^\beta}-\frac{\partial X^j}{\partial\Sigma^\beta}(s^{\alpha i}_j-s^{\alpha k}_k\delta^i_j),\\
G^{\alpha i}_\beta&=\frac{\partial S^{\alpha i}}{\partial u^\beta}+\Sigma^\alpha\frac{\partial X^i}{\partial u^\beta}\\
&\quad+\Big(\frac{\partial S^{\alpha i}}{\partial\Sigma^{\gamma j}}+\Sigma^\alpha\frac{\partial X^i}{\partial\Sigma^{\gamma j}}+\frac{\partial X^k}{\partial x^k}\delta^\alpha_\gamma\delta^i_j-\frac{\partial X^i}{\partial x^j}\delta^\alpha_\gamma\Big)s^{\gamma j}_\beta\\
&\quad-\frac{\partial X^j}{\partial u^\beta}(s^{\alpha i}_j-s^{\alpha k}_k\delta^i_j)-\frac{\partial X^i}{\partial\Sigma^{\gamma k}}(s^{\alpha j}_\beta s^{\gamma k}_j-s^{\gamma k}_\beta s^{\alpha j}_j)\\
&\quad+\frac{\partial X^j}{\partial\Sigma^{\gamma k}}(s^{\alpha i}_\beta s^{\gamma k}_j-s^{\gamma k}_\beta s^{\alpha i}_j)+\frac{\partial X^j}{\partial\Sigma^\gamma}(t^\gamma_j s^{\alpha i}_\beta-t^\gamma_k s^{\alpha k}_\beta\delta^i_j),\\
\Gamma^{\alpha i}_{\beta j}&=\frac{\partial S^{\alpha i}}{\partial\Sigma^{\beta j}}+\Sigma^\alpha\frac{\partial X^i}{\partial\Sigma^{\beta j}}-\frac{\partial X^k}{\partial x^l}(\delta^i_k\delta^l_j-\delta^l_k\delta^i_j)\delta^\alpha_\beta\\
&\quad-\frac{\partial X^k}{\partial\Sigma^{\beta j}}(s^{\alpha i}_k-s^{\alpha l}_l\delta^i_k)\\
&\quad+\Big(\frac{\partial X^l}{\partial\Sigma^{\gamma k}}s^{\gamma k}_l\delta^i_j-\frac{\partial X^i}{\partial\Sigma^{\gamma k}}s^{\gamma k}_j+\frac{\partial X^k}{\partial\Sigma^\gamma}t^\gamma_k\delta^i_j-\frac{\partial X^i}{\partial\Sigma^\gamma}t^\gamma_j\Big)\delta^\alpha_\beta,\\
\Gamma^{\alpha ij}_{\beta\gamma k}&=\frac{\partial X^j}{\partial\Sigma^{\gamma k}}s^{\alpha i}_\beta-\Big(\frac{\partial X^j}{\partial\Sigma^{\delta l}}s^{\delta l}_\beta+\frac{\partial X^j}{\partial\Sigma^\delta}t^\delta_\beta+\frac{\partial X^j}{\partial u^\beta}\Big)\delta^\alpha_\gamma\delta^i_k,\\
\Gamma^{\alpha ij}_{\beta\gamma}&=-\frac{\partial X^j}{\partial u^\beta}s^{\alpha i}_\gamma+\Big(\frac{\partial X^j}{\partial\Sigma^{\delta k}}s^{\delta k}_\gamma+\frac{\partial X^j}{\partial\Sigma^\delta}t^\delta_\gamma\Big)s^{\alpha i}_\beta.
\end{aligned}$$

Hence, we have to take

$$\begin{aligned}
C^\alpha_{\beta j}&=(\Gamma^{\alpha i}_{\beta j}-\nu^\alpha_\beta\delta^i_j)\mu_i-\Big(\Gamma^{\alpha ik}_{\gamma\beta j}\,du^\gamma+\frac{\partial X^k}{\partial\Sigma^{\beta j}}\Omega^{\alpha i}\Big)\wedge\mu_{ki},\\
C^\alpha_\beta&=\Gamma^{\alpha i}_\beta\mu_i-\frac{\partial X^j}{\partial\Sigma^\beta}(\Omega^{\alpha i}+s^{\alpha i}_\gamma\,du^\gamma)\wedge\mu_{ji}
\end{aligned}$$

in order to satisfy the above equations. Thus, we reach to the following determining equations

$$\Gamma^{\alpha} - \nu^{\alpha}_{\beta}(\Sigma^{\beta} + s^{\beta i}_{i}) = 0, \tag{9.8.12}$$
$$G^{\alpha i}_{\beta} + \Gamma^{\alpha i}_{\gamma} t^{\gamma}_{\beta} - \nu^{\alpha}_{\gamma} s^{\gamma i}_{\beta} = 0,$$
$$\Gamma^{\alpha[ij]}_{[\beta\gamma]} = 0$$

The foregoing results will loose their validity in the case $N = 1$, that is, if there is only a single dependent variable. Evidently, we do not need to employ the Greek indices anymore since they all take only the value 1. In that case, we have to consider obviously just one balance equation which is represented by

$$\frac{\partial \Sigma^i}{\partial x^i} + \Sigma = 0.$$

The coordinate cover of the contact ideal $\mathcal{K}_m$ then becomes

$$\{x^i,\ \Sigma^i,\ \Sigma,\ s^i_j,\ t_i,\ \{v_{i_1\cdots i_r},\ s^{ii_1\cdots i_r},\ t^{i_1\cdots i_r} : 0 \le r \le m\}\}$$

where we have naturally defined the auxiliary variables by

$$s^i_j = \frac{\partial \Sigma^i}{\partial x^j},$$
$$s^{ii_1\cdots i_r} = \frac{\partial \Sigma^i}{\partial v_{i_1\cdots i_r}},$$
$$t_i = \frac{\partial \Sigma}{\partial x^i},$$
$$t^{i_1\cdots i_r} = \frac{\partial \Sigma}{\partial v_{i_1\cdots i_r}},\ 0 \le r \le m.$$

Consequently, an isovector field of $\mathcal{K}_m$ that is a member of the tangent bundle $T(\mathcal{K}_m)$ must be represented by

$$V = X^i \frac{\partial}{\partial x^i} + S^i \frac{\partial}{\partial \Sigma^i} + T \frac{\partial}{\partial \Sigma} + S^i_j \frac{\partial}{\partial s^i_j} + T_i \frac{\partial}{\partial t_i}$$
$$+ \sum_{r=0}^{m} \Big(V_{i_1\cdots i_r} \frac{\partial}{\partial v_{i_1\cdots i_r}} + S^{ii_1\cdots i_r} \frac{\partial}{\partial s^{ii_1\cdots i_r}} + T^{i_1\cdots i_r} \frac{\partial}{\partial t^{i_1\cdots i_r}} \Big).$$

On making use of exactly the same sort of operations as we had done in the case of $N > 1$, we obtain in this situation the following expressions for $0 \le k \le m-1$

$$
\begin{aligned}
V_{i_1\cdots i_k i} &= \frac{\partial F_{i_1\cdots i_k}}{\partial x^i} + \sum_{r=0}^{m-1} \lambda_{i_1\cdots i_k}^{j_1\cdots j_r} v_{j_1\cdots j_r i} + \frac{\partial F_{i_1\cdots i_k}}{\partial \Sigma^j} s_i^j \\
&\quad + \frac{\partial F_{i_1\cdots i_k}}{\partial \Sigma} t_i, \\
\lambda_{i_1\cdots i_k}^{j_1\cdots j_r} &= \frac{\partial V_{i_1\cdots i_k}}{\partial v_{j_1\cdots j_r}} - v_{i_1\cdots i_k i} \frac{\partial X^i}{\partial v_{j_1\cdots j_r}} + \frac{\partial F_{i_1\cdots i_k}}{\partial \Sigma^i} s^{i j_1\cdots j_r} \\
&\quad + \frac{\partial F_{i_1\cdots i_k}}{\partial \Sigma} t^{j_1\cdots j_r}, \; 0 \le r \le m-1, \qquad (9.8.13) \\
\frac{\partial V_{i_1\cdots i_k}}{\partial v_{j_1\cdots j_m}} &- v_{i_1\cdots i_k i} \frac{\partial X^i}{\partial v_{j_1\cdots j_m}} + \frac{\partial F_{i_1\cdots i_k}}{\partial \Sigma^i} s^{i j_1\cdots j_m} \\
&\quad + \frac{\partial F_{i_1\cdots i_k}}{\partial \Sigma} t^{j_1\cdots j_m} = 0, \\
\delta_j^i V_{i_1\cdots i_r} &+ \frac{\partial F^i}{\partial s^{j i_1\cdots i_r}} = V_{i_1\cdots i_r} + \frac{\partial G}{\partial t^{i_1\cdots i_r}} = 0, \qquad 0 \le r \le m, \\
\delta_k^i X^j &+ \frac{\partial F^i}{\partial s_j^k} = X^i + \frac{\partial G}{\partial t_i} = 0
\end{aligned}
$$

where the functions $F_{i_1\cdots i_k}, F^i, G \in \Lambda^0(\mathcal{K}_m)$ are defined by the relations

$$
\begin{aligned}
F_{i_1\cdots i_k} &= V_{i_1\cdots i_k} - v_{i_1\cdots i_k i} X^i, \qquad (9.8.14) \\
F^i &= S^i - s_j^i X^j - \sum_{r=0}^{m} s^{i i_1\cdots i_r} V_{i_1\cdots i_r}, \\
G &= T - t_i X^i - \sum_{r=0}^{m} t^{i_1\cdots i_r} V_{i_1\cdots i_r}.
\end{aligned}
$$

Furthermore, we immediately observe that $F_{i_1\cdots i_k}$ cannot depend on the variables s_j^i, t_i and $\{s^{i i_1\cdots i_r}, t^{i_1\cdots i_r} : 0 \le r \le m\}$ while F^i is independent of the variables t_i and $\{t^{i_1\cdots i_r} : 0 \le r \le m\}$, and G is independent of the variables s_j^i and $\{s^{i i_1\cdots i_r} : 0 \le r \le m\}$. For $k < m-1$, it follows from $(9.8.13)_3$ and $(9.8.14)_1$ that

$$
\frac{\partial F_{i_1\cdots i_k}}{\partial v_{j_1\cdots j_m}} + \frac{\partial F_{i_1\cdots i_k}}{\partial \Sigma^i} s^{i j_1\cdots j_m} + \frac{\partial F_{i_1\cdots i_k}}{\partial \Sigma} t^{j_1\cdots j_m} = 0
$$

which gives rise to

$$
\frac{\partial F_{i_1\cdots i_k}}{\partial \Sigma^i} = 0, \quad \frac{\partial F_{i_1\cdots i_k}}{\partial \Sigma} = 0, \quad \frac{\partial F_{i_1\cdots i_k}}{\partial v_{j_1\cdots j_m}} = 0, \;\; 0 \le k \le m-2.
$$

By taking $k = m - 1$ in the relation $(9.8.13)_3$, we get

$$\delta^{j_1}_{i_1}\delta^{j_2}_{i_2}\cdots\delta^{j_{m-1}}_{i_{m-1}}\delta^{j_m}_{i}X^i + \frac{\partial F_{i_1\cdots i_{m-1}}}{\partial v_{j_1\cdots j_m}} + \frac{\partial F_{i_1\cdots i_{m-1}}}{\partial \Sigma^i}s^{ij_1\cdots j_m} + \frac{\partial F_{i_1\cdots i_{m-1}}}{\partial \Sigma}t^{j_1\cdots j_m} = 0.$$

On the other hand, according to the relations $(9.8.13)_{4,5}$, the functions X^i and $V_{i_1\cdots i_r}$ cannot depend on the variables s^i_j, t_i and $\{s^{ii_1\cdots i_r}, t^{i_1\cdots i_r}\}$ where $0 \le r \le m$. But, this leads us to the conclusion

$$\frac{\partial F_{i_1\cdots i_k}}{\partial \Sigma^i} = 0, \quad \frac{\partial F_{i_1\cdots i_k}}{\partial \Sigma} = 0, \quad 0 \le k \le m-1.$$

Therefrom, we arrive at the previously derived solutions (9.3.22), (9.3.23) and (9.3.25) for the components X^i, U and $V_{i_1\cdots i_k}$. In a similar fashion, we deduce the following expressions

$$F^i = -\, s^i_j X^j - \sum_{r=0}^{m} s^{ii_1\cdots i_r}V_{i_1\cdots i_r} + \mathcal{F}^i,$$
$$G = -\, t_i X^i - \sum_{r=0}^{m} t^{i_1\cdots i_r}V_{i_1\cdots i_r} + \mathcal{G}$$

from $(9.8.13)_{4,5}$ where the functions $\mathcal{F}^i$ and $\mathcal{G}$ depend only on the variables x^j, Σ^j, Σ and $\{v_{i_1\cdots i_r} : 0 < r < m\}$. In conclusion, isovector fields of the contact ideal are to be expressed as

$$V = X^i\frac{\partial}{\partial x^i} + \sum_{r=0}^{m} V_{i_1\cdots i_r}\frac{\partial}{\partial v_{i_1\cdots i_r}} + S^i\frac{\partial}{\partial \Sigma^i} + T\frac{\partial}{\partial \Sigma} + \overline{S}^i_j\frac{\partial}{\partial s^i_j} + \overline{T}_i\frac{\partial}{\partial t_i} + S^{ii_1\cdots i_m}\frac{\partial}{\partial s^{ii_1\cdots i_m}} + T^{i_1\cdots i_m}\frac{\partial}{\partial t^{i_1\cdots i_m}} + V_1 + V_2.$$

Trivial isovector fields that can be discarded without loss of generality are given by

$$V_1 = \sum_{r=0}^{m-1} S^{ii_1\cdots i_r}\Big(\frac{\partial}{\partial s^{ii_1\cdots i_r}} - v_{i_1\cdots i_r j}\frac{\partial}{\partial s^i_j}\Big),$$
$$V_2 = \sum_{r=0}^{m-1} T^{i_1\cdots i_r}\Big(\frac{\partial}{\partial t^{i_1\cdots i_r}} - v_{i_1\cdots i_r i}\frac{\partial}{\partial t_i}\Big)$$

as dependent upon arbitrarily selected functions $S^{ii_1\cdots i_r}$ and $T^{i_1\cdots i_r}$ with

$0 \leq r \leq m-1$. The remaining isovector components are listed below

$$S^i = \mathcal{F}^i, \quad T = \mathcal{G}, \tag{9.8.15}$$
$$\begin{aligned}
\overline{S}^i_j &= \frac{\partial F^i}{\partial x^j} + \sum_{r=0}^{m-1}\Big(\frac{\partial F^i}{\partial v_{i_1\cdots i_r}} + \frac{\partial F^i}{\partial \Sigma^j} s^{ji_1\cdots i_r} + \frac{\partial F^i}{\partial \Sigma} t^{i_1\cdots i_r}\Big) v_{i_1\cdots i_r j} \\
&\quad + \frac{\partial F^i}{\partial \Sigma^k} s^k_j + \frac{\partial F^i}{\partial \Sigma} t_j, \\
\overline{T}_i &= \frac{\partial G}{\partial x^i} + \sum_{r=0}^{m-1}\Big(\frac{\partial G}{\partial v_{i_1\cdots i_r}} + \frac{\partial G}{\partial \Sigma^i} s^{ji_1\cdots i_r} + \frac{\partial G}{\partial \Sigma} t^{i_1\cdots i_r}\Big) v_{i_1\cdots i_r i} \\
&\quad + \frac{\partial G}{\partial \Sigma^j} s^j_i + \frac{\partial G}{\partial \Sigma} t_i, \\
S^{ii_1\cdots i_m} &= \frac{\partial F^i}{\partial v_{i_1\cdots i_m}} + \frac{\partial F^i}{\partial \Sigma^j} s^{ji_1\cdots i_m} + \frac{\partial F^i}{\partial \Sigma} t^{i_1\cdots i_m}, \\
T^{i_1\cdots i_m} &= \frac{\partial G}{\partial v_{i_1\cdots i_m}} + \frac{\partial G}{\partial \Sigma^i} s^{ii_1\cdots i_m} + \frac{\partial G}{\partial \Sigma} t^{i_1\cdots i_m}.
\end{aligned}$$

They are entirely determined in terms of $n+2$ functions

$$F(x^i, u, v_i),\ S^i = \mathcal{F}^i(x^j, \Sigma^j, \Sigma, v_{i_1\cdots i_r}),\ T = \mathcal{G}(x^j, \Sigma^j, \Sigma, v_{i_1\cdots i_r})$$

with $0 \leq r \leq m$. In order that this vector field becomes also an isovector field of the balance ideal, we have to show the existence of the forms $\nu \in \Lambda^0(\mathcal{K}_m)$; $A^{i_1\cdots i_r},\ B_i,\ B \in \Lambda^{n-1}(\mathcal{K}_m)$; $C^{i_1\cdots i_r},\ D_i,\ D \in \Lambda^{n-2}(\mathcal{K}_m)$, $0 \leq r \leq m-1$ such that the following relation is satisfied

$$\begin{aligned}
\pounds_V\omega &= T\mu - X^i d\Sigma \wedge \mu_i + d(S^i\mu_i - X^j d\Sigma^i \wedge \mu_{ji} + \Sigma X^i \mu_i) \\
&= T\mu + (dS^i + \Sigma dX^i) \wedge \mu_i - dX^j \wedge d\Sigma^i \wedge \mu_{ji} = T\mu \\
&\quad + \Big(\frac{\partial S^i}{\partial x^j} + \Sigma\frac{\partial X^i}{\partial x^j}\Big) dx^j \wedge \mu_i + \Big(\frac{\partial S^i}{\partial u} + \Sigma\frac{\partial X^i}{\partial u}\Big) du \wedge \mu_i \\
&\quad + \Big(\frac{\partial S^i}{\partial v_j} + \Sigma\frac{\partial X^i}{\partial v_j}\Big) dv_j \wedge \mu_i \\
&\quad + \sum_{r=2}^{m} \frac{\partial S^i}{\partial v_{i_1\cdots i_r}} dv_{i_1\cdots i_r} \wedge \mu_i + \frac{\partial S^i}{\partial \Sigma^j} d\Sigma^j \wedge \mu_i \\
&\quad + \frac{\partial S^i}{\partial \Sigma} d\Sigma \wedge \mu_i - \Big(\frac{\partial X^j}{\partial x^k} dx^k + \frac{\partial X^j}{\partial u} du + \frac{\partial X^j}{\partial v_k} dv_k\Big) \wedge d\Sigma^i \wedge \mu_{ji} \\
&= \nu\omega + \sum_{r=0}^{m-1} \sigma_{i_1\cdots i_r} \wedge A^{i_1\cdots i_r} + \Omega^i \wedge B_i + \Omega \wedge B
\end{aligned}$$

$$+\sum_{r=0}^{m-1} d\sigma_{i_1\cdots i_r}\wedge C^{i_1\cdots i_r}+d\Omega^i\wedge D_i+d\Omega\wedge D.$$

If we follow a path similar to what we have followed in the case of $N>1$, we easily obtain the expression

$$\begin{aligned}&\Big[\Gamma-\nu\Big(\Sigma+s^i_i+\sum_{r=0}^{m-1}s^{ii_1\cdots i_r}v_{i_1\cdots i_r i}\Big)\Big]\mu+\sigma\wedge(\gamma-A-\nu\, s^i\mu_i) \quad (9.8.16)\\&+\sigma_i\wedge(\gamma^i-A^i-\nu s^{ji}\mu_j dx^i\wedge C)+\sum_{r=2}^{m-1}\sigma_{i_1\cdots i_r}\wedge\big(\gamma^{i_1\cdots i_r}-A^{i_1\cdots i_r}\\&-\nu s^{ii_1\cdots i_r}\mu_i+dx^{i_r}\wedge C^{i_1\cdots i_{r-1}}\big)+\Omega^i\wedge(\lambda_i-B_i-\nu\mu_i)\\&+\Omega\wedge\Big(\frac{\partial S^i}{\partial\Sigma}\mu_i-B\Big)\\&+(\Gamma^{ii_1\cdots i_m}-\nu s^{ii_1\cdots i_m})\,dv_{i_1\cdots i_m}\wedge\mu_i\\&+dv_{i_1\cdots i_m}\wedge dx^{i_m}\wedge C^{i_1\cdots i_{m-1}}=0,\end{aligned}$$

The functions and forms entering into the single equation (9.8.16) are essentially associated with unknown functions X^i, S^i, T and are given by

$$\begin{aligned}\Gamma=&\,T+\frac{\partial S^i}{\partial x^i}+\frac{\partial S^i}{\partial u}v_i+\frac{\partial S^i}{\partial v_j}v_{ji}+\Sigma\Big(\frac{\partial X^i}{\partial x^i}+\frac{\partial X^i}{\partial u}v_i+\frac{\partial X^i}{\partial v_j}v_{ji}\Big)\\&+\sum_{r=2}^{m-1}\frac{\partial S^i}{\partial v_{i_1\cdots i_r}}v^i_{i_1\cdots i_r i}\\&+\Big[\frac{\partial S^i}{\partial\Sigma^j}+\Big(\frac{\partial X^k}{\partial x^k}+\frac{\partial X^k}{\partial u}v_k+\frac{\partial X^k}{\partial v_l}v_{lk}\Big)\delta_j\\&-\Big(\frac{\partial X^i}{\partial x^j}+\frac{\partial X^i}{\partial u}v_j+\frac{\partial X^i}{\partial v_k}v_{kj}\Big)\Big]\Big(s^j_i+\sum_{r=0}^{m-1}s^{ji_1\cdots i_r}v_{i_1\cdots i_r i}\Big)\\&+\frac{\partial S^i}{\partial\Sigma}\Big(t_i+\sum_{r=0}^{m-1}t^{i_1\cdots i_r}v_{i_1\cdots i_r i}\Big)\in\Lambda^0(\mathcal{K}_m),\\\gamma=&\Big[\frac{\partial S^i}{\partial u}+\Sigma\frac{\partial X^i}{\partial u}+\frac{\partial S^i}{\partial\Sigma^j}s^j+\frac{\partial S^i}{\partial\Sigma}t+\Big(\frac{\partial X^j}{\partial x^j}+\frac{\partial X^j}{\partial u}v_j+\frac{\partial X^j}{\partial v_k}v_{kj}\Big)s^i\\&-\Big(\frac{\partial X^i}{\partial x^j}+\frac{\partial X^i}{\partial u}v_j+\frac{\partial X^i}{\partial v_k}v_{kj}\Big)s^j+\frac{\partial X^i}{\partial u}\Big(s^j_j+\sum_{r=0}^{m-1}s^{ji_1\cdots i_r}v_{i_1\cdots i_r j}\Big)\\&-\frac{\partial X^j}{\partial u}\Big(s^i_j+\sum_{r=0}^{m-1}s^{ii_1\cdots i_r}v_{i_1\cdots i_r j}\Big)\Big]\mu_i+\Big(\frac{\partial X^j}{\partial v_k}s^i-\frac{\partial X^j}{\partial u}s^{ik}\Big)\sigma_k\wedge\mu_{ji}\end{aligned}$$

$$
- \frac{\partial X^j}{\partial u}\Big(\sum_{r=2}^{m-1} s^{ii_1\cdots i_r}\sigma_{i_1\cdots i_r}\wedge\mu_{ji}\Big) - \frac{\partial X^j}{\partial u} s^{ii_1\cdots i_m} dv_{i_1\cdots i_m}\wedge\mu_{ji}\in\Lambda^{n-1}(\mathcal{K}_m),
$$

$$
\begin{aligned}
\gamma^j = \Big[& \frac{\partial S^i}{\partial v_j} + \Sigma\,\frac{\partial X^i}{\partial v_j} + \frac{\partial S^i}{\partial \Sigma^k}s^{kj} + \frac{\partial S^i}{\partial \Sigma}\,t^j + \Big(\frac{\partial X^k}{\partial x^k} + \frac{\partial X^k}{\partial u}v_k + \frac{\partial X^k}{\partial v_l}v_{lk}\Big)s^{ij} \\
& - \Big(\frac{\partial X^i}{\partial x^k} + \frac{\partial X^i}{\partial u}v_k + \frac{\partial X^i}{\partial v_l}v_{lk}\Big)s^{kj} + \frac{\partial X^i}{\partial v_j}\Big(s^k_k + \sum_{r=0}^{m-1} s^{ki_1\cdots i_r}v_{i_1\cdots i_r k}\Big) \\
& - \frac{\partial X^k}{\partial v_j}\Big(s^i_k + \sum_{r=0}^{m-1} s^{ii_1\cdots i_r}v_{i_1\cdots i_r k}\Big)\Big]\mu_i - \frac{\partial X^k}{\partial v_j}s^{il}\sigma_l\wedge\mu_{ki} \\
& - \frac{\partial X^k}{\partial v_j}\Big(\sum_{r=2}^{m-1} s^{ii_1\cdots i_r}\sigma_{i_1\cdots i_r}\wedge\mu_{ki}\Big) \\
& - \frac{\partial X^k}{\partial v_j}s^{ii_1\cdots i_m}dv_{i_1\cdots i_m}\wedge\mu_{ki}\in\Lambda^{n-1}(\mathcal{K}_m),
\end{aligned}
$$

$$
\begin{aligned}
\lambda_j = \Big[& \frac{\partial S^i}{\partial \Sigma^j} + \Big(\frac{\partial X^k}{\partial x^k} + \frac{\partial X^k}{\partial u}\,v_k + \frac{\partial X^k}{\partial v_l}v_{lk}\Big)\delta^i_j - \Big(\frac{\partial X^i}{\partial x^j} + \frac{\partial X^i}{\partial u}\,v_j \\
& + \frac{\partial X^i}{\partial v_k}v_{kj}\Big)\Big]\mu_i + \frac{\partial X^i}{\partial u}\sigma\wedge\mu_{ij} + \frac{\partial X^i}{\partial v_k}\sigma_k\wedge\mu_{ij}\in\Lambda^{n-1}(\mathcal{K}_m),
\end{aligned}
$$

$$
\begin{aligned}
\gamma^{i_1\cdots i_r} = \Big[& \frac{\partial S^i}{\partial v_{i_1\cdots i_r}} + \frac{\partial S^i}{\partial \Sigma^j}s^{ji_1\cdots i_r} + \frac{\partial S^i}{\partial \Sigma}t^{i_1\cdots i_r} \\
& + \Big(\frac{\partial X^j}{\partial x^j} + \frac{\partial X^j}{\partial u}\,v_j + \frac{\partial X^j}{\partial v_k}v_{kj}\Big)s^{ii_1\cdots i_r} \\
& - \Big(\frac{\partial X^i}{\partial x^j} + \frac{\partial X^i}{\partial u}\,v_j + \frac{\partial X^i}{\partial v_k}v_{kj}\Big)s^{ji_1\cdots i_r}\Big]\mu_i\in\Lambda^{n-1}(\mathcal{K}_m),\,2\le r\le m-1,
\end{aligned}
$$

$$
\begin{aligned}
\Gamma^{ii_1\cdots i_m} = & \frac{\partial S^i}{\partial v_{i_1\cdots i_m}} + \frac{\partial S^i}{\partial \Sigma^j}s^{ji_1\cdots i_m} + \frac{\partial S^i}{\partial \Sigma}t^{i_1\cdots i_m} \\
& + \Big(\frac{\partial X^j}{\partial x^j} + \frac{\partial X^j}{\partial u}\,v_j + \frac{\partial X^j}{\partial v_k}v_{kj}\Big)s^{ii_1\cdots i_m} \\
& - \Big(\frac{\partial X^i}{\partial x^j} + \frac{\partial X^i}{\partial u}\,v_j + \frac{\partial X^i}{\partial v_k}v_{kj}\Big)s^{ji_1\cdots i_m}\in\Lambda^0(\mathcal{K}_m).
\end{aligned}
$$

To satisfy the equation (9.8.16), it would suffice now to select

$$
\begin{aligned}
& A = \gamma - \nu\, s^i\mu_i, \quad A^i = \gamma^i - \nu s^{ji}\mu_j + dx^i\wedge C, \\
& B_i = \lambda_i - \nu\mu_i, \quad B = \frac{\partial S^i}{\partial \Sigma}\mu_i, \\
& A^{i_1\cdots i_r} = \gamma^{i_1\cdots i_r} - \nu s^{ii_1\cdots i_r}\mu_i + dx^{i_r}\wedge C^{i_1\cdots i_{r-1}},\ 0\le r\le m-1
\end{aligned}
$$

In this case, (9.8.16) reduces to

$$\Big[\Gamma-\nu\Big(\Sigma+s^i_i+\sum_{r=0}^{m-1}s^{ii_1\cdots i_r}v_{i_1\cdots i_ri}\Big)\Big]\mu$$
$$+\,(\Gamma^{ii_1\cdots i_m}-\nu s^{ii_1\cdots i_m})\,dv_{i_1\cdots i_m}\wedge\mu_i$$
$$+\,dv_{i_1\cdots i_m}\wedge dx^{i_m}\wedge C^{i_1\cdots i_{m-1}}=0.$$

If we introduce the forms

$$C^{i_1\cdots i_{m-1}}=C^{i_1\cdots i_{m-1}ij}\mu_{ij}\in\Lambda^{n-2}(\mathcal{K}_m),$$
$$C^{i_1\cdots i_{m-1}ij}=-\,C^{i_1\cdots i_{m-1}ji}\in\Lambda^0(\mathcal{K}_m)$$

we find

$$\begin{aligned}dv_{i_1\cdots i_m}\wedge dx^{i_m}\wedge C^{i_1\cdots i_{m-1}ij}\mu_{ij}&\\ &=-\,dv_{i_1\cdots i_m}\wedge dx^{i_m}\wedge C^{i_1\cdots i_{m-1}ij}\mu_{ji}\\ &=-C^{i_1\cdots i_{m-1}ij}dv_{i_1\cdots i_m}\wedge dx^{i_m}\wedge\mu_{ij}\\ &=-C^{i_1\cdots i_{m-1}ij}dv_{i_1\cdots i_m}\wedge(\delta^{i_m}_j\mu_i-\delta^{i_m}_i\mu_j)\\ &=-\,2C^{i_1\cdots i_{m-1}ii_m}dv_{i_1\cdots i_m}\wedge\mu_i.\end{aligned}$$

We thus obtain

$$\Big[\Gamma-\nu\Big(\Sigma+s^i_i+\sum_{r=0}^{m-1}s^{ii_1\cdots i_r}v_{i_1\cdots i_ri}\Big)\Big]\mu$$
$$+\,(\Gamma^{ii_1\cdots i_m}-\nu s^{ii_1\cdots i_m}-2C^{i_1\cdots i_{m-1}ii_m})dv_{i_1\cdots i_m}\wedge\mu_i=0$$

implying that the coefficients of independent forms μ and $dv_{i_1\cdots i_m}\wedge\mu_i$ must vanish. Obviously the antisymmetric parts of the terms $\Gamma^{ii_1\cdots i_m}-\nu s^{ii_1\cdots i_m}$ with respect to indices i and i_m will determine the coefficients $2C^{i_1\cdots i_{m-1}ii_m}$ that are antisymmetric in those two indices. Thus the symmetric parts of these terms with respect to the same indices should vanish. Hence, the determining equations for isovector components take eventually the forms below

$$\Gamma-\nu\Big(\Sigma+s^i_i+\sum_{r=0}^{m-1}s^{ii_1\cdots i_r}v_{i_1\cdots i_ri}\Big)=0, \tag{9.8.17}$$
$$\Gamma^{ii_1\cdots i_m}+\Gamma^{i_mi_1\cdots i}-\nu(s^{ii_1\cdots i_m}+s^{i_mi_1\cdots i})=0.$$

We cannot extract the determining equations corresponding to the case $m=1$ directly from (9.8.17). The relation

$$\pounds_V\omega=\nu\omega+\sigma\wedge A+d\sigma\wedge B+\Omega^i\wedge C_i+\Omega\wedge C$$

can now be written as

$$\begin{aligned}
&\big[\Gamma - \nu(\Sigma + s^i_i + s^i v_i)\big]\mu + \sigma \wedge (\Gamma^i \mu_i - A - \nu\, s^i \mu_i) \\
&+ \Omega^j \wedge (\Gamma^i_j\, \mu_i - C_j - \nu \delta^i_j\, \mu_i) + \Omega \wedge \Big(\frac{\partial S^i}{\partial \Sigma}\mu_i - C\Big) + \frac{\partial X^j}{\partial u}\Omega^i \wedge \sigma \wedge \mu_{ji} \\
&+ \Big(\frac{\partial X^j}{\partial v_k} s^i - \frac{\partial X^j}{\partial u} s^{ik}\Big)\sigma \wedge dv_k \wedge \mu_{ji} + \frac{\partial X^j}{\partial v_k}\Omega^i \wedge dv_k \wedge \mu_{ji} \\
&+ (\Gamma^{ij} - \nu s^{ij})\, dv_j \wedge \mu_i + dv_i \wedge dx^i \wedge B + \Gamma^{ijkl} dv_l \wedge dv_k \wedge \mu_{ji} = 0
\end{aligned}$$

where we defined

$$\begin{aligned}
\Gamma &= T + \frac{\partial S^i}{\partial x^i} + \Sigma \frac{\partial X^i}{\partial x^i} + \Big(\frac{\partial S^i}{\partial u} + \Sigma \frac{\partial X^i}{\partial u}\Big) v_i + \frac{\partial S^i}{\partial \Sigma}(t_i + t v_i) \\
&\quad + \Big(\frac{\partial S^i}{\partial \Sigma^j} + \frac{\partial X^k}{\partial x^k}\delta^i_j - \frac{\partial X^i}{\partial x^j}\Big)(s^j_i + s^j v_i) + \frac{\partial X^j}{\partial u}(s^i_i v_j - s^i_j v_i), \\
\Gamma^{ij} &= \frac{\partial S^i}{\partial v_j} + \Sigma \frac{\partial X^i}{\partial v_j} + \Big(\frac{\partial S^i}{\partial \Sigma^k} + \frac{\partial X^l}{\partial x^l}\delta^i_k - \frac{\partial X^i}{\partial x^k}\Big) s^{kj} + \frac{\partial S^i}{\partial \Sigma} t^j \\
&\quad \frac{\partial X^k}{\partial v_j}(s^i_k + s^i v_k) + \frac{\partial X^i}{\partial v_j}(s^k_k + s^k\, v_k) + \Big(\frac{\partial X^k}{\partial u} s^{ij} - \frac{\partial X^i}{\partial u} s^{kj}\Big) v_k, \\
\Gamma^{ijkl} &= -\Gamma^{jikl} = -\Gamma^{ijlk} = \frac{1}{4}\Big(s^{il}\frac{\partial X^j}{\partial v_k} - s^{ik}\frac{\partial X^j}{\partial v_l} + s^{jk}\frac{\partial X^i}{\partial v_l} - s^{jl}\frac{\partial X^i}{\partial v_k}\Big) \\
&= \frac{1}{4}\,\delta^{ij}_{mn}\delta^{kl}_{pq} s^{mq}\frac{\partial X^n}{\partial v_p}.
\end{aligned}$$

To combine the form $B \in \Lambda^{n-2}(\mathcal{K}_1)$ with other terms, we take it in the form $B = B^{jk}\mu_{jk} + B^{jklm} dv_m \wedge \mu_{jkl}$. Antisymmetry properties impose the restrictions $B^{jk} = -B^{kj}$, $B^{jklm} = B^{[jkl]m} \in \Lambda^0(\mathcal{K}_1)$. We thereby find

$$-\, dv_i \wedge dx^i \wedge B = 2B^{ij} dv_j \wedge \mu_i + 3B^{ijlk} dv_l \wedge dv_k \wedge \mu_{ji}.$$

On the other hand, by selecting

$$\begin{aligned}
A &= (\Gamma^i - \nu\sigma^i)\mu_i + \Big(\frac{\partial X^j}{\partial v_k} s^i - \frac{\partial X^j}{\partial u} s^{ik}\Big) dv_k \wedge \mu_{ji}, \\
C &= \frac{\partial S^i}{\partial \Sigma}\mu_i, \\
C_i &= (\Gamma^j_i - \nu\delta^j_i)\mu_j + \frac{\partial X^j}{\partial u}\sigma \wedge \mu_{ji} + \frac{\partial X^j}{\partial v_k} dv_k \wedge \mu_{ji},
\end{aligned}$$

we arrive at the relation

$$\big[\Gamma - \nu(\Sigma + s^i_i + s^i v_i)\big]\mu + (\Gamma^{ij} - \nu s^{ij} - 2B^{ij})\,dv_j \wedge \mu_i$$
$$+ (\Gamma^{ijkl} - 3B^{ijkl})dv_l \wedge dv_k \wedge \mu_{ji} = 0$$

which requires that $2B^{ij} = \Gamma^{[ij]} - \nu s^{[ij]}$ and $3B^{ijkl} = \Gamma^{i[jk]l}$. Hence, the determining equations are reduced to

$$\Gamma = \nu(\Sigma + s^i_i + \sigma^i v_i), \quad \Gamma^{ij} + \Gamma^{ji} = \nu(s^{ij} + s^{ji}), \tag{9.8.18}$$
$$\Gamma^{ijkl} + \Gamma^{ikjl} = (\delta^{ij}_{mn}\delta^{kl}_{pq} + \delta^{ik}_{mn}\delta^{jl}_{pq})s^{mq}\frac{\partial X^n}{\partial v_p} = 0.$$

The third set of equations in (9.8.18) govern the dependence of functions $X^i(x^j, u, v_j)$ on variables v_j. In fact, these equations imply that the coefficients of the variables s^{mq} vanish if only the relations

$$(\delta^{ij}_{mn}\delta^{kl}_{pq} + \delta^{ik}_{mn}\delta^{jl}_{pq})\frac{\partial X^n}{\partial v_p} = 0$$

are satisfied. Contractions on indices (i, m) and (l, q) above yield

$$(n-1)^2(\delta^j_n\delta^k_p + \delta^k_n\delta^j_p)\frac{\partial X^n}{\partial v_p} = 0$$

whence we obtain the equations

$$\frac{\partial X^j}{\partial v_k} + \frac{\partial X^k}{\partial v_j} = 0$$

for $n > 1$. The solution of this set of partial differential equations can be written simply as

$$X^i = a^{ij}(x^k, u)v_j + b^i(x^k, u), \;\; a^{ij} = -\,a^{ji}.$$

We then see that a^{ij} must also satisfy the conditions

$$(\delta^{ij}_{mn}\delta^{kl}_{pq} + \delta^{ik}_{mn}\delta^{jl}_{pq})a^{np} = 0$$

that can be expanded into

$$a^{ik}\delta^j_m\delta^l_q + a^{jl}\delta^i_m\delta^k_q - a^{il}\delta^j_m\delta^k_q + a^{ij}\delta^k_m\delta^l_q + a^{kl}\delta^i_m\delta^j_q - a^{il}\delta^k_m\delta^j_q = 0.$$

By contractions on the indices (k, m) and (l, q), we get

$$(n^2 - 1)a^{ij} = 0.$$

Therefore, we have to take $a^{ij} = 0$ for $n > 1$. Consequently, we finally reach to the conclusion $X^i = b^i(x^k, u)$.

For $m = 0$, the case $N = 1$ does not need a special care. Since all Greek indices are equal to 1, the determining equations are directly deduced from (9.8.12) as

$$\Gamma - \nu(\Sigma + s^i_i) = 0, \qquad G^i + \Gamma^i t - \nu s^i = 0.$$

We can easily reproduce the determining equations for the symmetry groups discussed in Sec. 9.4 from the determining equations for the equivalence groups. To this end, it suffices to note that in symmetry transformations $\Sigma^{\alpha i}$ and Σ^α $(N > 1)$, and their derivatives with respect to their arguments can no longer be chosen as independent variables as we have done in equivalence transformations. Therefore, the only surviving isovector components should be X^i and $V^\alpha_{i_1\cdots i_r}$, $0 \le r \le m$. Moreover, we have to keep in mind that the basis vectors $\partial/\partial x^i$ and $\partial/\partial v^\alpha_{i_1\cdots i_r}$, $0 \le r \le m$ appearing in (9.8.3) are calculated by holding $\Sigma^{\alpha i}$ and Σ^α as constants. On taking into notice of the functional forms of $\Sigma^{\alpha i}$ and Σ^α, we immediately conclude that we have to write by using the chain rule

$$\left.\frac{\partial}{\partial x^i}\right|_{\Sigma^{\alpha i},\Sigma^\alpha} = \frac{\partial}{\partial x^i} - \frac{\partial \Sigma^{\alpha j}}{\partial x^i}\frac{\partial}{\partial \Sigma^{\alpha j}} - \frac{\partial \Sigma^\alpha}{\partial x^i}\frac{\partial}{\partial \Sigma^\alpha}$$

$$\left.\frac{\partial}{\partial v^\alpha_{i_1\cdots i_r}}\right|_{\Sigma^{\alpha i},\Sigma^\alpha} = \frac{\partial}{\partial v^\alpha_{i_1\cdots i_r}} - \frac{\partial \Sigma^{\beta j}}{\partial v^\alpha_{i_1\cdots i_r}}\frac{\partial}{\partial \Sigma^{\beta j}} - \frac{\partial \Sigma^\beta}{\partial v^\alpha_{i_1\cdots i_r}}\frac{\partial}{\partial \Sigma^\beta}, \; 0 \le r \le m.$$

Thus an isovector field is now expressible as

$$V = X^i\frac{\partial}{\partial x^i} + \sum_{r=0}^{m} V^\alpha_{i_1\cdots i_r}\frac{\partial}{\partial v^\alpha_{i_1\cdots i_r}} + \left[S^{\alpha i} - V(\Sigma^{\alpha i})\right]\frac{\partial}{\partial \Sigma^{\alpha i}}$$
$$+ \left[T^\alpha - V(\Sigma^\alpha)\right]\frac{\partial}{\partial \Sigma^\alpha}.$$

However, the coefficients of the vectors $\partial/\partial\Sigma^{\beta j}$ and $\partial/\partial\Sigma^\beta$ must vanish. Hence, it is obvious that we are compelled to take $S^{\alpha i} = V(\Sigma^{\alpha i})$ and $T^\alpha = V(\Sigma^\alpha)$. As is well-known, we define

$$V(f) = \frac{\partial f}{\partial x^i}X^i + \sum_{r=0}^{m}\frac{\partial f}{\partial v^\alpha_{i_1\cdots i_r}}V^\alpha_{i_1\cdots i_r}$$

for a function $f \in \Lambda^0(\mathcal{C}_m)$. If we also recall the definitions (9.8.1) and insert all we have found so far into the determining equations (9.8.11), we readily observe that we can recover the determining equations (9.4.11). The case $N = 1$ can be treated in exactly similar fashion. We find $S^i = V(\Sigma^i)$, $T =$

$V(\Sigma)$ and it is straightforward to verify that the equations (9.8.17) give rise to the equations (9.4.17).

We shall now try to determine the general solutions of the determining equations (9.8.11) or (9.8.17).

The case $N > 1$: The explicit form of the equations (9.8.11) can be written as

$$\begin{aligned}T^\alpha &+ \frac{\partial S^{\alpha i}}{\partial x^i} + \Sigma^\beta\Big(\frac{\partial X^i}{\partial x^i}\delta^\alpha_\beta - \nu^\alpha_\beta\Big) + \Big(\frac{\partial S^{\alpha i}}{\partial u^\beta} + \Sigma^\alpha\frac{\partial X^i}{\partial u^\beta}\Big)v^\beta_i \qquad (9.8.19)\\ &+ \sum_{r=1}^{m-1}\frac{\partial S^{\alpha i}}{\partial v^\beta_{i_1\cdots i_r}}v^\beta_{i_1\cdots i_r i} + \Big\{\frac{\partial S^{\alpha i}}{\partial \Sigma^{\beta j}} + \Big[\Big(\frac{\partial X^k}{\partial x^k} + \frac{\partial X^k}{\partial u^\gamma}v^\gamma_k\Big)\delta^i_j\\ &- \Big(\frac{\partial X^i}{\partial x^j} + \frac{\partial X^i}{\partial u^\gamma}v^\gamma_j\Big)\Big]\delta^\alpha_\beta - \nu^\alpha_\beta\delta^i_j\Big\}\Big(s^{\beta j}_i + \sum_{r=0}^{m-1}s^{\beta j i_1\cdots i_r}_\gamma v^\gamma_{i_1\cdots i_r i}\Big)\\ &+ \frac{\partial S^{\alpha i}}{\partial \Sigma^\beta}\Big(t^\beta_i + \sum_{r=0}^{m-1}t^{\beta i_1\cdots i_r}_\gamma v^\gamma_{i_1\cdots i_r i}\Big) = 0\\ &\Lambda^{\alpha i i_1\cdots i_{m-1} i_m}_\beta + \Lambda^{\alpha i_m i_1\cdots i_{m-1} i}_\beta = 0\end{aligned}$$

Here, we have introduced the following functions

$$\begin{aligned}\Gamma^{\alpha i i_1\cdots i_m}_\beta - \nu^\alpha_\gamma s^{\gamma i i_1\cdots i_m}_\beta &= \frac{\partial S^{\alpha i}}{\partial v^\beta_{i_1\cdots i_m}} + \Big\{\frac{\partial S^{\alpha i}}{\partial \Sigma^{\gamma j}} + \Big[\Big(\frac{\partial X^k}{\partial x^k} + \frac{\partial X^k}{\partial u^\delta}v^\delta_k\Big)\delta^i_j\\ - \Big(\frac{\partial X^i}{\partial x^j} + \frac{\partial X^i}{\partial u^\delta}v^\delta_j\Big)\Big]\delta^\alpha_\gamma - \nu^\alpha_\gamma\delta^i_j\Big\}s^{\gamma j i_1\cdots i_m}_\beta &+ \frac{\partial S^{\alpha i}}{\partial \Sigma^\gamma}t^{\gamma i_1\cdots i_m}_\beta = \Lambda^{\alpha i i_1\cdots i_m}_\beta.\end{aligned}$$

If we carefully examine the equations $(9.8.19)_1$, we realise that the coefficients ν^α_β cannot depend on the variables $s^{\alpha i i_1\cdots i_m}_\beta$ and $t^{\alpha i_1\cdots i_m}_\beta$ whereas the equations $(9.8.19)_2$ imply that they are independent of the variables $s^{\alpha i}_j$, t^α_i and $\{s^{\alpha i i_1\cdots i_r}_\beta, t^{\alpha i_1\cdots i_r}_\beta, 0 \le r \le m-1\}$. We thus obtain

$$\nu^\alpha_\beta = \nu^\alpha_\beta\big(x^i, \{v^\gamma_{i_1\cdots i_r}, 0 \le r \le m\}, \Sigma^{\gamma i}, \Sigma^\gamma\big).$$

Therefore, on recalling arguments of X^i, $S^{\alpha i}$ and T^α, we realise at once that $(9.8.19)_1$ makes way for the following relations

$$\frac{\partial S^{\alpha i}}{\partial \Sigma^\beta} = 0, \qquad (9.8.20)$$

$$\nu^\alpha_\beta\delta^i_j = \frac{\partial S^{\alpha i}}{\partial \Sigma^{\beta j}} + \Big[\Big(\frac{\partial X^k}{\partial x^k} + \frac{\partial X^k}{\partial u^\gamma}v^\gamma_k\Big)\delta^i_j - \Big(\frac{\partial X^i}{\partial x^j} + \frac{\partial X^i}{\partial u^\gamma}v^\gamma_j\Big)\Big]\delta^\alpha_\beta,$$

$$T^\alpha + \frac{\partial S^{\alpha i}}{\partial x^i} + \Sigma^\beta\Big(\frac{\partial X^i}{\partial x^i}\delta^\alpha_\beta - \nu^\alpha_\beta\Big) + \Big(\frac{\partial S^{\alpha i}}{\partial u^\beta} + \Sigma^\alpha\frac{\partial X^i}{\partial u^\beta}\Big)v^\beta_i$$
$$+ \sum_{r=1}^{m-1}\frac{\partial S^{\alpha i}}{\partial v^\beta_{i_1\cdots i_r}}v^\beta_{i_1\cdots i_r i} = 0.$$

Let us now define the functions

$$\mathfrak{S}^{\alpha i}_{\beta j} = \frac{\partial S^{\alpha i}}{\partial \Sigma^{\beta j}} - \Big(\frac{\partial X^i}{\partial x^j} + \frac{\partial X^i}{\partial u^\gamma}v^\gamma_j\Big)\delta^\alpha_\beta.$$

If we contract the indices i and j in $(9.8.20)_2$ and take into consideration the equations $(9.8.20)_1$, we reach to the relation

$$\nu^\alpha_\beta - \Big(\frac{\partial X^k}{\partial x^k} + \frac{\partial X^k}{\partial u^\gamma}v^\gamma_k\Big)\delta^\alpha_\beta = \frac{1}{n}\,\mathfrak{S}^{\alpha k}_{\beta k}$$
$$= f^\alpha_\beta(x^i, \{v^\gamma_{i_1\cdots i_r}, 0 \le r \le m\}, \Sigma^{\gamma i}).$$

where f^α_β are arbitrary functions of their arguments. Eventually, $(9.8.20)_2$ would lead to

$$\mathfrak{S}^{\alpha i}_{\beta j} - \frac{1}{n}\,\mathfrak{S}^{\alpha k}_{\beta k}\delta^i_j = 0 \quad \text{or} \quad \frac{\partial S^{\alpha i}}{\partial \Sigma^{\beta j}} = f^\alpha_\beta\delta^i_j + \Big(\frac{\partial X^i}{\partial x^j} + \frac{\partial X^i}{\partial u^\gamma}v^\gamma_j\Big)\delta^\alpha_\beta.$$

Differentiating the above expression with respect to $\Sigma^{\gamma k}$, we get

$$\frac{\partial^2 S^{\alpha i}}{\partial \Sigma^{\beta j}\partial \Sigma^{\gamma k}} = \frac{\partial f^\alpha_\beta}{\partial \Sigma^{\gamma k}}\delta^i_j = \frac{\partial f^\alpha_\gamma}{\partial \Sigma^{\beta j}}\delta^i_k.$$

The extreme right hand side in the foregoing relation arises from the symmetry of second order derivatives with respect to the variables $\Sigma^{\beta j}$ requiring that this expression must be invariant under interchanges (β, γ) and (j, k). A contraction on indices i and j yields

$$n\frac{\partial f^\alpha_\beta}{\partial \Sigma^{\gamma k}} = \frac{\partial f^\alpha_\gamma}{\partial \Sigma^{\beta k}}.$$

Hence, we can write

$$\frac{\partial f^\alpha_\beta}{\partial \Sigma^{\gamma k}}\delta^i_j = n\frac{\partial f^\alpha_\beta}{\partial \Sigma^{\gamma j}}\delta^i_k$$

and on contracting the indices i and k, we arrive at the result

$$(n^2-1)\frac{\partial f^\alpha_\beta}{\partial \Sigma^{\gamma j}} = 0.$$

Since $n > 1$ in partial differential equations, we conclude that the functions f^α_β must be independent of $\Sigma^{\beta j}$, that is, their explicit dependence should be given as follows

$$f^\alpha_\beta = f^\alpha_\beta(x^i, u^\gamma, v^\gamma_{i_1}, v^\gamma_{i_1 i_2}, \ldots, v^\gamma_{i_1\cdots i_m}).$$

Then, we easily obtain

$$S^{\alpha i} = f^\alpha_\beta \Sigma^{\beta i} + \Big(\frac{\partial X^i}{\partial x^j} + \frac{\partial X^i}{\partial u^\beta} v^\beta_j\Big)\Sigma^{\alpha j} \\ + \chi^{\alpha i}(x^j, u^\beta, v^\beta_{i_1}, \ldots, v^\beta_{i_1\cdots i_m}) \tag{9.8.21}$$

where $\chi^{\alpha i}$ are arbitrary functions. With these relations at hand, we find from $(9.8.20)_3$ that

$$T^\alpha = f^\alpha_\beta \Sigma^\beta - \frac{\partial S^{\alpha i}}{\partial x^i} - \sum_{r=0}^{m-1} \frac{\partial S^{\alpha i}}{\partial v^\beta_{i_1\cdots i_r}} v^\beta_{i_1\cdots i_r i} \tag{9.8.22}$$

On the other hand, we can write

$$\Lambda^{\alpha i i_1\cdots i_m}_\beta = \frac{\partial S^{\alpha i}}{\partial v^\beta_{i_1\cdots i_m}} = \frac{\partial f^\alpha_\gamma}{\partial v^\beta_{i_1\cdots i_m}}\Sigma^{\gamma i} + \frac{\partial \sigma^{\alpha i}}{\partial v^\beta_{i_1\cdots i_m}}.$$

where we assume that $m \neq 1$. Then $(9.8.19)_2$ gives

$$\frac{\partial f^\alpha_\gamma}{\partial v^\beta_{i_1\cdots i_m}}\Sigma^{\gamma i} + \frac{\partial f^\alpha_\gamma}{\partial v^\beta_{i_1\cdots i_{m-1} i}}\Sigma^{\gamma i_m} = 0, \quad \frac{\partial \chi^{\alpha i}}{\partial v^\beta_{i_1\cdots i_m}} + \frac{\partial \chi^{\alpha i_m}}{\partial v^\beta_{i_1\cdots i_{m-1} i}} = 0$$

On differentiating the first set of equations with respect to $\Sigma^{\delta k}$, we get

$$\frac{\partial f^\alpha_\gamma}{\partial v^\beta_{i_1\cdots i_m}}\delta^i_k + \frac{\partial f^\alpha_\gamma}{\partial v^\beta_{i_1\cdots i_{m-1} i}}\delta^{i_m}_k = 0 \quad \text{or} \quad (n+1)\frac{\partial f^\alpha_\gamma}{\partial v^\beta_{i_1\cdots i_m}} = 0$$

where we have performed contraction on the indices i and k. Thus, the functions f^α_β possess the following form

$$f^\alpha_\beta = f^\alpha_\beta(x^i, u^\gamma, v^\gamma_{i_1}, v^\gamma_{i_1 i_2}, \ldots, v^\gamma_{i_1\cdots i_{m-1}}).$$

If we recall that the variables $v^\alpha_{i_1\cdots i_m}$ are completely symmetric in their subscripts, the partial differential equations satisfied by the functions $\chi^{\alpha i}$

may be cast into the form

$$\frac{\partial \chi^{\alpha i}}{\partial v^{\beta}_{ji_1\cdots i_{m-1}}} + \frac{\partial \chi^{\alpha j}}{\partial v^{\beta}_{ii_1\cdots i_{m-1}}} = 0. \tag{9.8.23}$$

The number of variables in this system is $N\binom{n+m-1}{m}$ whereas the number of equations is $N^2\frac{n(n+1)}{2}\binom{n+m-2}{m-1}$. Hence, the number of equations is larger by a factor $N\frac{mn(n+1)}{2(n+m-1)}$ than the number of variables. In order to find the solution of (9.8.23), let us start by taking $i = j$ to get

$$\frac{\partial \chi^{\alpha \underline{i}}}{\partial v^{\beta}_{\underline{i}i_1\cdots i_{m-1}}} = 0$$

for all $1 \le i, i_1, \cdots, i_{m-1} \le n$. Let us recall that the summation convention will be suspended on underlined indices. The above equations mean that the functions $\chi^{\alpha i}$ cannot depend on $v^{\beta}_{ii_1\cdots i_{m-1}}$. To simplify the notation, let us introduce the sets $\mathbf{v}_{ii_1\cdots i_{m-1}} = \{v^1_{\underline{i}i_1\cdots i_{m-1}}, \cdots, v^N_{\underline{i}i_1\cdots i_{m-1}}\}$ and $\mathbf{W}_{i_1\cdots i_{m-1}} = \{\mathbf{v}_{1i_1\cdots i_{m-1}}, \mathbf{v}_{2i_1\cdots i_{m-1}}, \cdots, \mathbf{v}_{ni_1\cdots i_{m-1}}\}$. We can thus write

$$\chi^{\alpha i} = f^{\alpha i}(\mathbf{W}_{i_1\cdots i_{m-1}} \backslash \{\mathbf{v}_{\underline{i}i_1\cdots i_{m-1}}\}),\ 1 \le \alpha \le N$$

where the symbol $\backslash$ denotes the set difference. For notational simplicity, we omit the dependence of $f^{\alpha i}$ on the variables $x^i, u^\alpha, v^\alpha_{i_1}, v^\alpha_{i_1 i_2}, \dots, v^\alpha_{i_1\cdots i_{m-1}}$. We now differentiate the equations (9.8.23) with respect to $v^{\gamma}_{jj_1\cdots j_{m-1}}$ with $j \ne i$ to obtain

$$\frac{\partial^2 \chi^{\alpha i}}{\partial v^{\beta}_{\underline{j}i_1\cdots i_{m-1}}\partial v^{\gamma}_{\underline{j}j_1\cdots j_{m-1}}} + \frac{\partial^2 \chi^{\alpha \underline{j}}}{\partial v^{\beta}_{ii_1\cdots i_{m-1}}\partial v^{\gamma}_{\underline{j}j_1\cdots j_{m-1}}} = 0.$$

Because of the relations $\partial \chi^{\alpha \underline{j}}/\partial v^{\gamma}_{\underline{j}j_1\cdots j_{m-1}} = 0$, we find

$$\frac{\partial^2 \chi^{\alpha i}}{\partial v^{\beta}_{\underline{j}i_1\cdots i_{m-1}}\partial v^{\gamma}_{\underline{j}j_1\cdots j_{m-1}}} = 0$$

implying that

$$\frac{\partial \chi^{\alpha i}}{\partial v^{\beta}_{ji_1\cdots i_{m-1}}} = f^{\alpha i j i_1\cdots i_{m-1}}_{\beta}(\mathbf{W}_{i_1\cdots i_{m-1}} \backslash \{\mathbf{v}_{ii_1\cdots i_{m-1}}, \mathbf{v}_{ji_1\cdots i_{m-1}}\}).$$

Obviously, the functions $f^{\alpha i j i_1\cdots i_{m-1}}_{\beta}$ are subject to the restrictions

$$f_{\beta}^{\alpha i j i_1 \cdots i_{m-1}} = -f_{\beta}^{\alpha j i i_1 \cdots i_{m-1}}.$$

On the other hand, because of the symmetry of mixed derivatives, we have

$$\frac{\partial^2 \chi^{\alpha i}}{\partial v^{\beta}_{j i_1 \cdots i_{m-1}} \partial v^{\gamma}_{k j_1 \cdots j_{m-1}}} = \frac{\partial f_{\beta}^{\alpha i j i_1 \cdots i_{m-1}}}{\partial v^{\gamma}_{k j_1 \cdots j_{m-1}}} = \frac{\partial f_{\gamma}^{\alpha i k j_1 \cdots j_{m-1}}}{\partial v^{\beta}_{j i_1 \cdots i_{m-1}}}.$$

Since the functions $f_{\gamma}^{\alpha i k j_1 \cdots j_{m-1}}$ cannot depend on $\mathbf{v}_{k j_1 \cdots j_{m-1}}$, we get at once

$$\frac{\partial^2 f_{\beta}^{\alpha i j i_1 \cdots i_{m-1}}}{\partial v^{\gamma}_{\underline{k} j_1 \cdots j_{m-1}} \partial v^{\delta}_{\underline{k} k_1 \cdots k_{m-1}}} = 0$$

leading to the relations

$$\frac{\partial^2 \chi^{\alpha i}}{\partial v^{\beta}_{j i_1 \cdots i_{m-1}} \partial v^{\gamma}_{k j_1 \cdots j_{m-1}}} = \frac{\partial f_{\beta}^{\alpha i j i_1 \cdots i_{m-1}}}{\partial v^{\gamma}_{k j_1 \cdots j_{m-1}}} =$$
$$f_{\beta\gamma}^{\alpha i j k i_1 \cdots i_{m-1} j_1 \cdots j_{m-1}}(\mathbf{W}_{i_1 \cdots i_{m-1}} \backslash \{\mathbf{v}_{i i_1 \cdots i_{m-1}}, \mathbf{v}_{j i_1 \cdots i_{m-1}}, \mathbf{v}_{k i_1 \cdots i_{m-1}}\}),\ i \neq j \neq k$$

where the functions $f_{\beta\gamma}^{\alpha i j k i_1 \cdots i_{m-1} j_1 \cdots j_{m-1}}$ must satisfy the following symmetry conditions

$$f_{\beta\gamma}^{\alpha i j k i_1 \cdots i_{m-1} j_1 \cdots j_{m-1}} = -f_{\beta\gamma}^{\alpha j i k i_1 \cdots i_{m-1} j_1 \cdots j_{m-1}} = f_{\gamma\beta}^{\alpha i k j i_1 \cdots i_{m-1} j_1 \cdots j_{m-1}}.$$

Continuing this way, we can readily reach to the recurrence relations

$$\frac{\partial^{r+1} \chi^{\alpha i}}{\partial v^{\alpha_1}_{i_1 i_1^{(1)} \cdots i_{m-1}^{(1)}} \cdots \partial v^{\alpha_{r+1}}_{i_{r+1} i_1^{(r+1)} \cdots i_{m-1}^{(r+1)}}} = \frac{\partial f_{\alpha_1 \cdots \alpha_r}^{\alpha i i_1 \cdots i_r i_1^{(1)} \cdots i_{m-1}^{(1)} \cdots i_1^{(r)} \cdots i_{m-1}^{(r)}}}{\partial v^{\alpha_{r+1}}_{i_{r+1} i_1^{(r+1)} \cdots i_{m-1}^{(r+1)}}} =$$
$$f_{\alpha_1 \cdots \alpha_{r+1}}^{\alpha i i_1 \cdots i_{r+1} i_1^{(1)} \cdots i_{m-1}^{(1)} \cdots i_1^{(r+1)} \cdots i_{m-1}^{(r+1)}}(\mathbf{W}_{j_1 \cdots j_{m-1}} \backslash \{\mathbf{v}_{i \underline{j}_1 \cdots j_{m-1}}, \mathbf{v}_{i_1 \underline{j}_1 \cdots j_{m-1}}, \ldots, \mathbf{v}_{i_{r+1} \underline{j}_1 \cdots j_{m-1}}\})$$

whence by taking $r = n - 2$ we draw the conclusion

$$\frac{\partial^{n-1} \chi^{\alpha i}}{\partial v^{\alpha_1}_{i_1 i_1^{(1)} \cdots i_{m-1}^{(1)}} \cdots \partial v^{\alpha_{n-1}}_{i_{n-1} i_1^{(n-1)} \cdots i_{m-1}^{(n-1)}}} = f_{\alpha_1 \cdots \alpha_{n-1}}^{\alpha i i_1 \cdots i_{n-1} i_1^{(1)} \cdots i_{m-1}^{(1)} \cdots i_1^{(n-1)} \cdots i_{m-1}^{(n-1)}}(\emptyset)$$

since $\mathbf{W}_{j_1 \cdots j_{m-1}} \backslash \{\mathbf{v}_{i \underline{j}_1 \cdots j_{m-1}}, \mathbf{v}_{i_1 \underline{j}_1 \cdots j_{m-1}}, \ldots, \mathbf{v}_{i_{n-1} \underline{j}_1 \cdots j_{m-1}}\} = \emptyset$. This is tantamount to say that the functions $f_{\alpha_1 \cdots \alpha_{n-1}}^{\alpha i i_1 \cdots i_{n-1} i_1^{(1)} \cdots i_{m-1}^{(1)} \cdots i_1^{(n-1)} \cdots i_{m-1}^{(n-1)}}$ are independent of the variables $v^{\alpha}_{i_1 i_2 \cdots i_m}$. Thus it becomes rather straightforward to integrate those hierarchical system of partial differential equations for $\chi^{\alpha i}$ in the backward direction starting from the last equations above. We then readily obtain the polynomial expressions

$$
\begin{aligned}
\chi^{\alpha i} &= \sum_{k=1}^{n-1} f_{\alpha_1\cdots\alpha_k}^{\alpha i i_1\cdots i_k j_1^{(1)}\cdots j_{m-1}^{(1)}\cdots j_1^{(k)}\cdots j_{m-1}^{(k)}} v^{\alpha_1}_{i_1 j_1^{(1)}\cdots j_{m-1}^{(1)}} \cdots v^{\alpha_k}_{i_k j_1^{(k)}\cdots j_{m-1}^{(k)}} + f^{\alpha i} \qquad (9.8.24)\\
&= f^{\alpha i} + f_{\alpha_1}^{\alpha i i_1 j_1^{(1)}\cdots j_{m-1}^{(1)}} v^{\alpha_1}_{i_1 j_1^{(1)}\cdots j_{m-1}^{(1)}} \\
&\qquad + f_{\alpha_1\alpha_2}^{\alpha i i_1 i_2 j_1^{(1)}\cdots j_{m-1}^{(1)} j_1^{(2)}\cdots j_{m-1}^{(2)}} v^{\alpha_1}_{i_1 j_1^{(1)}\cdots j_{m-1}^{(1)}} v^{\alpha_2}_{i_2 j_1^{(2)}\cdots j_{m-1}^{(2)}} \\
&\quad + f_{\alpha_1\alpha_2\alpha v^{\alpha_1}_{i_1 j_1^{(1)}\cdots j_{m-1}^{(1)}} \cdots v_{i_{n-1} j_1^{(n-1)}\cdots j_{m-1}^{(n-1)}} 3}^{\alpha i i_1 i_2 i_3 j_1^{(1)}\cdots j_{m-1}^{(1)} j_1^{(2)}\cdots j_{m-1}^{(2)} j_1^{(3)}\cdots j_{m-1}^{(3)}} v^{\alpha_1}_{i_1 j_1^{(1)}\cdots j_{m-1}^{(1)}} v^{\alpha_2}_{i_2 j_1^{(2)}\cdots j_{m-1}^{(2)}} v^{\alpha_3}_{i_3 j_1^{(3)}\cdots j_{m-1}^{(3)}} \\
&\quad + \cdots + f_{\alpha_1\cdots\alpha_{n-1}}^{\alpha i i_1\cdots i_{n-1} j_1^{(1)}\cdots j_{m-1}^{(1)}\cdots j_1^{(n-1)}\cdots j_{m-1}^{(n-1)}} v^{\alpha_1}_{i_1 j_1^{(1)}\cdots j_{m-1}^{(1)}} \cdots v^{\alpha_{n-1}}_{i_{n-1} j_1^{(n-1)}\cdots j_{m-1}^{(n-1)}}
\end{aligned}
$$

The functions $f_{\alpha_1\cdots\alpha_k}^{\alpha i i_1\cdots i_k j_1^{(1)}\cdots j_{m-1}^{(1)}\cdots j_1^{(k)}\cdots j_{m-1}^{(k)}}$ where $1 \le k \le n-1$ and $f^{\alpha i}$ are arbitrary and depend only on the variables $x^i, u^\alpha,\ v^\alpha_{i_1}, v^\alpha_{i_1 i_2}, \dots, v^\alpha_{i_1\cdots i_{m-1}}$. We can easily verify that the functions $f_{\alpha_1\cdots\alpha_r}^{\alpha i i_1\cdots i_r j_1^{(1)}\cdots j_{m-1}^{(1)}\cdots j_1^{(k)}\cdots j_{m-1}^{(k)}}$ where $r = 1, \dots, n-1$ enjoy several symmetry requirements. There are antisymmetry with respect to first two roman superscripts and complete symmetry within the groups of indices, $(i_1, j_1^{(1)}, \cdots, j_{m-1}^{(1)}), \dots, (i_{n-1}, j_1^{(n-1)}, \cdots, j_{m-1}^{(n-1)})$. Furthermore, we immediately observe that block symmetries with respect to the groups of indices $(\alpha_l, i_l, j_1^{(l)}, \cdots, j_{m-1}^{(l)}), \dots, (\alpha_k, i_k, j_1^{(k)}, \cdots, j_{m-1}^{(k)})$ must be obeyed.

We thus see that all components of the isovector fields characterising equivalence transformations of balance equations are determined by means of arbitrary functions $X^i, U^\alpha, f^\alpha_\beta, f^{\alpha i}$ and $f_{\alpha_1\cdots\alpha_k}^{\alpha i i_1\cdots i j_1^{(1)}\cdots j_{m-1}^{(1)}\cdots j_1^{(k)}\cdots j_{m-1}^{(k)}}$ where $1 \le k \le n-1$ depending on certain coordinates of $\mathcal{K}_m$ through the relations (9.8.7), (9.8.21), (9.8.22) and (9.8.24). When $\Sigma^{\alpha i}$ and/or Σ^α are independent of some coordinates, the components of isovector fields corresponding to them must of course vanish. That kind of restrictions removes naturally to some extent the arbitrariness in the determining functions X^i, $U^\alpha, f^\alpha_\beta, \chi^{\alpha i}$.

When $m = 1$, namely, when we take into account second order balance equations, we have to modify slightly the previous analysis. In this case, we have again

$$\Lambda^{\alpha ij}_\beta = \Gamma^{\alpha ij}_\beta - \nu^\alpha_\gamma s^{\gamma ij}_\beta = \frac{\partial S^{\alpha i}}{\partial v^\beta_j}$$

But, this time, (9.8.21) yields

$$\Lambda^{\alpha ij}_{\beta} = \frac{\partial f^{\alpha}_{\gamma}}{\partial v^{\beta}_{j}}\Sigma^{\gamma i} + \frac{\partial X^{i}}{\partial u^{\beta}}\Sigma^{\alpha j} + \frac{\partial \chi^{\alpha i}}{\partial v^{\beta}_{j}}$$

Hence, the solution of the equations $\Lambda^{\alpha ij}_{\beta} + \Lambda^{\alpha ji}_{\beta} = 0$ is found has a distinct structure from above as

$$f^{\alpha}_{\beta}(x^{i}, u^{\gamma}, v^{\gamma}_{j}) = -\frac{\partial X^{i}}{\partial u^{\gamma}}v^{\gamma}_{i}\delta^{\alpha}_{\beta} + g^{\alpha}_{\beta}(x^{i}, u^{\gamma})$$

where g^{α}_{β} are arbitrary functions. The solution of the equations

$$\frac{\partial \chi^{\alpha i}}{\partial v^{\beta}_{j}} + \frac{\partial \chi^{\alpha j}}{\partial v^{\beta}_{i}} = 0$$

can be extracted from the foregoing general solution as follows

$$\chi^{\alpha i} = \sum_{k=1}^{n-1} f^{\alpha i i_{1}\cdots i_{k}}_{\alpha_{1}\cdots\alpha_{k}}(\mathbf{x}, \mathbf{u})v^{\alpha_{1}}_{i_{1}}\cdots v^{\alpha_{k}}_{i_{k}} + f^{\alpha i}(\mathbf{x}, \mathbf{u}).$$

Thus, the relevant isovector components are found as

$$\begin{aligned}
&X^{i} = X^{i}(\mathbf{x}, \mathbf{u}), \;\; U^{\alpha} = U^{\alpha}(\mathbf{x}, \mathbf{u}),\\
&V^{\alpha}_{i} = \frac{\partial U^{\alpha}}{\partial x^{i}} - \frac{\partial X^{j}}{\partial x^{i}}\,v^{\alpha}_{j} + \frac{\partial U^{\alpha}}{\partial u^{\beta}}\,v^{\beta}_{i} - \frac{\partial X^{j}}{\partial u^{\beta}}\,v^{\alpha}_{j}v^{\beta}_{i},\\
&S^{\alpha i} = g^{\alpha}_{\beta}\Sigma^{\beta i} + \left(\frac{\partial X^{i}}{\partial x^{j}} + \frac{\partial X^{i}}{\partial u^{\beta}}v^{\beta}_{j}\right)\Sigma^{\alpha j} - \frac{\partial X^{j}}{\partial u^{\beta}}v^{\beta}_{j}\Sigma^{\alpha i} + \chi^{\alpha i},\\
&T^{\alpha} = g^{\alpha}_{\beta}\Sigma^{\beta} - \frac{\partial X^{i}}{\partial u^{\gamma}}v^{\gamma}_{i}\Sigma^{\alpha} - \frac{\partial g^{\alpha}_{\beta}}{\partial x^{i}}\Sigma^{\beta i} - \frac{\partial g^{\alpha}_{\beta}}{\partial u^{\gamma}}v^{\gamma}_{i}\Sigma^{\beta i}\\
&\qquad - \left(\frac{\partial^{2} X^{i}}{\partial x^{i}\partial x^{j}} + \frac{\partial^{2} X^{i}}{\partial x^{i}\partial u^{\beta}}v^{\beta}_{j}\right)\Sigma^{\alpha j} - \frac{\partial \chi^{\alpha i}}{\partial x^{i}} - \frac{\partial \chi^{\alpha i}}{\partial u^{\beta}}v^{\beta}_{i}.
\end{aligned} \tag{9.8.25}$$

If $m = 0$, that is, when we consider first order balance equations, then we have to search for the solution of the equations (9.8.12). The equations $(9.8.12)_3$ lead to

$$\frac{\partial X^{j}}{\partial u^{\beta}} = 0, \;\; \frac{\partial X^{j}}{\partial \Sigma^{\delta k}} = 0, \;\; \frac{\partial X^{j}}{\partial \Sigma^{\delta}} = 0.$$

We thus get $X^{i} = X^{i}(\mathbf{x})$. The form of the equations $(9.8.12)_1$ reflects the fact that the functions ν^{α}_{β} must be independent of the variables $s^{\alpha i}_{\beta}$. Hence, the relations $(9.8.12)_2$ reduce to the equations

$$\frac{\partial S^{\alpha i}}{\partial u^\beta}+\frac{\partial S^{\alpha i}}{\partial \Sigma^\gamma}t^\gamma_\beta+\Big(\frac{\partial S^{\alpha i}}{\partial \Sigma^{\gamma j}}+\frac{\partial X^k}{\partial x^k}\delta^\alpha_\gamma\delta^i_j-\frac{\partial X^i}{\partial x^j}\delta^\alpha_\gamma-\nu^\alpha_\gamma\delta^i_j\Big)s^{\gamma j}_\beta=0$$

the satisfaction of which requires that

$$\frac{\partial S^{\alpha i}}{\partial u^\beta}=0,\ \frac{\partial S^{\alpha i}}{\partial \Sigma^\beta}=0,\ \nu^\alpha_\gamma\delta^i_j=\frac{\partial S^{\alpha i}}{\partial \Sigma^{\gamma j}}+\Big(\frac{\partial X^k}{\partial x^k}\delta^i_j-\frac{\partial X^i}{\partial x^j}\Big)\delta^\alpha_\gamma.$$

We therefore find $S^{\alpha i}=S^{\alpha i}(x^j,\Sigma^{\beta j})$ and, consequently, $\nu^\alpha_\beta=\nu^\alpha_\beta(x^j,\Sigma^{\gamma j})$. Similar to the approach employed previously, we immediately obtain after some calculations

$$\frac{\partial S^{\alpha i}}{\partial \Sigma^{\beta j}}=f^\alpha_\beta(x^k,\Sigma^{\gamma k})\,\delta^i_j+\frac{\partial X^i}{\partial x^j}\delta^\alpha_\beta$$

from which we deduce that $\partial f^\alpha_\beta/\partial\Sigma^{\gamma j}=0$ for $n>1$ by considering second order derivatives of functions $S^{\alpha i}$ with respect to the variables $\Sigma^{\gamma k}$. This of course means that $f^\alpha_\beta=f^\alpha_\beta(\mathbf{x})$. We then obtain $S^{\alpha i}$ by simply integrating the above equations. If we introduce these expressions into $(9.8.12)_1$, we get the isovector components T^α. The results so obtained are listed below:

$$\begin{aligned}X^i&=X^i(\mathbf{x}),\\ U^\alpha&=U^\alpha(\mathbf{x},\mathbf{u},\Sigma^{\alpha i},\Sigma^\alpha),\\ S^{\alpha i}&=f^\alpha_\beta(\mathbf{x})\Sigma^{\beta i}+\frac{\partial X^i}{\partial x^j}\Sigma^{\alpha j}+g^{\alpha i}(\mathbf{x}),\\ T^\alpha&=f^\alpha_\beta(\mathbf{x})\Sigma^\beta-\frac{\partial f^\alpha_\beta(\mathbf{x})}{\partial x^i}\Sigma^{\beta i}-\frac{\partial^2X^i}{\partial x^i\partial x^j}\Sigma^{\alpha j}-\frac{\partial g^{\alpha i}(\mathbf{x})}{\partial x^i}\end{aligned}\tag{9.8.26}$$

where $X^i,U^\alpha,f^\alpha_\beta$ and $g^{\alpha i}(\mathbf{x})$ are arbitrary functions.

The case $N=1$: In the case of only one dependent variable, we have to look for the solution of the equations (9.8.17). The explicit form of these equations are given below

$$\begin{aligned}&T+\frac{\partial S^i}{\partial x^i}+\frac{\partial S^i}{\partial u}v_i+\frac{\partial S^i}{\partial v_j}v_{ji}+\sum_{r=2}^{m-1}\frac{\partial S^i}{\partial v_{i_1\cdots i_r}}v_{i_1\cdots i_r i}\\ &\quad+\Sigma\Big(\frac{\partial X^i}{\partial x^i}+\frac{\partial X^i}{\partial u}v_i+\frac{\partial X^i}{\partial v_j}v_{ji}-\nu\Big)+\Big[\frac{\partial S^i}{\partial\Sigma^j}+\Big(\frac{\partial X^k}{\partial x^k}+\frac{\partial X^k}{\partial u}v_k\\ &+\frac{\partial X^k}{\partial v_l}v_{lk}-\nu\Big)\delta^i_j-\Big(\frac{\partial X^i}{\partial x^j}+\frac{\partial X^i}{\partial u}v_j+\frac{\partial X^i}{\partial v_k}v_{kj}\Big)\Big]\Big(s^j_i+\sum_{r=0}^{m-1}s^{ji_1\cdots i_r}v_{i_1\cdots i_r i}\Big)\end{aligned}\tag{9.8.27}$$

$$+\frac{\partial S^i}{\partial \Sigma}\Big(t_i+\sum_{r=0}^{m-1}t^{i_1\cdots i_r}v_{i_1\cdots i_r i}\Big)=0$$

$$\Lambda^{ii_1\cdots i_{m-1}i_m}+\Lambda^{i_m i_1\cdots i_{m-1}i}=0$$

where we have defined

$$\begin{aligned}\Lambda^{ii_1\cdots i_m}=&\frac{\partial S^i}{\partial v_{i_1\cdots i_m}}+\Big[\frac{\partial S^i}{\partial \Sigma^j}+\Big(\frac{\partial X^k}{\partial x^k}+\frac{\partial X^k}{\partial u}v_k+\frac{\partial X^k}{\partial v_l}v_{lk}-\nu\Big)\delta^i_j\\&-\Big(\frac{\partial X^i}{\partial x^j}+\frac{\partial X^i}{\partial u}v_j+\frac{\partial X^i}{\partial v_k}v_{kj}\Big)\Big]s^{ji_1\cdots i_m}+\frac{\partial S^i}{\partial \Sigma}t^{i_1\cdots i_m}\\=&\,\Gamma^{ii_1\cdots i_m}-\nu s^{ii_1\cdots i_m}.\end{aligned}$$

When we carefully scrutinise the equations (9.8.27), we realise that ν is independent of the variables s^j_i, t_i, $s^{ii_1\cdots i_m}$ and $t^{i_1\cdots i_m}$ so that one writes

$$\nu=\nu\big(x^i,\{v_{i_1\cdots i_r},0\le r\le m\},\Sigma^i,\Sigma\big).$$

It then follows from $(9.8.27)_1$ that

$$\begin{aligned}&\frac{\partial S^i}{\partial \Sigma}=0, &(9.8.28)\\&\nu\delta^i_j=\frac{\partial S^i}{\partial \Sigma^j}+\Big(\frac{\partial X^k}{\partial x^k}+\frac{\partial X^k}{\partial u}v_k+\frac{\partial X^k}{\partial v_l}v_{lk}\Big)\delta^i_j-\Big(\frac{\partial X^i}{\partial x^j}+\frac{\partial X^i}{\partial u}v_j+\frac{\partial X^i}{\partial v_k}v_{kj}\Big),\\&T+\frac{\partial S^i}{\partial x^i}+\frac{\partial S^i}{\partial u}v_i+\frac{\partial S^i}{\partial v_j}v_{ji}+\sum_{r=2}^{m-1}\frac{\partial S^i}{\partial v_{i_1\cdots i_r}}v_{i_1\cdots i_r i}\\&\qquad+\Sigma\Big(\frac{\partial X^i}{\partial x^i}+\frac{\partial X^i}{\partial u}v_i+\frac{\partial X^i}{\partial v_j}v_{ji}-\nu\Big)=0.\end{aligned}$$

Hence, the functions S^i are independent of Σ. With the definition

$$\mathfrak{S}^i_j=\frac{\partial S^i}{\partial \Sigma^j}-\Big(\frac{\partial X^i}{\partial x^j}+\frac{\partial X^i}{\partial u}v_j+\frac{\partial X^i}{\partial v_k}v_{kj}\Big)$$

we deduce from $(9.8.27)_2$ that

$$\nu-\Big(\frac{\partial X^i}{\partial x^i}+\frac{\partial X^i}{\partial u}v_i+\frac{\partial X^i}{\partial v_j}v_{ji}\Big)=\frac{1}{n}\,\mathfrak{S}^i_i=f$$

where $f(x^i,u,v_{i_1},\dots,v_{i_1\cdots i_m},\Sigma^i)$ is an arbitrary function. Thus $(9.8.28)_2$ takes the form

$$\frac{\partial S^i}{\partial \Sigma^j} = f\delta^i_j + \Big(\frac{\partial X^i}{\partial x^j} + \frac{\partial X^i}{\partial u} v_j + \frac{\partial X^i}{\partial v_k} v_{kj}\Big) \tag{9.8.29}$$

whence we get

$$\frac{\partial^2 S^i}{\partial \Sigma^j \partial \Sigma^m} = \frac{\partial f}{\partial \Sigma^m}\delta^i_j = \frac{\partial f}{\partial \Sigma^j}\delta^i_m$$

and consequently

$$(n-1)\frac{\partial f}{\partial \Sigma^m} = 0.$$

This equation signifies that the function f is independent of Σ^i if $n > 1$. Then, the integration of the simple partial differential equations (9.8.29) leads to the result

$$S^i = f\Sigma^i + \Big(\frac{\partial X^i}{\partial x^j} + \frac{\partial X^i}{\partial u} v_j + \frac{\partial X^i}{\partial v_k} v_{kj}\Big)\Sigma^j + \chi^i \tag{9.8.30}$$

where $\chi^i(x^j, u, v_{i_1}, \dots, v_{i_1\cdots i_m})$ are arbitrary functions. Hence, the functions $\Lambda^{ii_1\cdots i_m}$ can be written in the form

$$\Lambda^{ii_1\cdots i_m} = \frac{\partial S^i}{\partial v_{i_1\cdots i_m}} = \frac{\partial f}{\partial v_{i_1\cdots i_m}}\Sigma^i + \frac{\partial \chi^i}{\partial v_{i_1\cdots i_m}}$$

and the equations $(9.8.27)_2$ lead to the conclusion

$$\frac{\partial f}{\partial v_{i_1\cdots i_m}}\Sigma^i + \frac{\partial f}{\partial v_{i_1\cdots i}}\Sigma^{i_m} + \frac{\partial \chi^i}{\partial v_{i_1\cdots i_m}} + \frac{\partial \chi^{i_m}}{\partial v_{i_1\cdots i}} = 0.$$

Therefrom, we easily arrive at the partial differential equations

$$\frac{\partial f}{\partial v_{ii_1\cdots i_{m-1}}} = 0, \quad \frac{\partial \chi^i}{\partial v_{ji_1\cdots i_{m-1}}} + \frac{\partial \chi^j}{\partial v_{ii_1\cdots i_{m-1}}} = 0.$$

Hence, we have $f = f(x^i, u, v_{i_1}, \dots, v_{i_1\cdots i_{m-1}})$ and the integration of the set of equations for χ^i yield

$$\chi^i = \alpha^{iji_1\cdots i_{m-1}} v_{ji_1\cdots i_{m-1}} + \beta^i, \quad \alpha^{iji_1\cdots i_{m-1}} = -\alpha^{jii_1\cdots i_{m-1}}$$

where the functions $\alpha^{iji_1\cdots i_{m-1}}$ and β^i depend on the variables $x^i, u, v_{i_1}, \dots, v_{i_1\cdots i_{m-1}}$. $\alpha^{iji_1\cdots i_{m-1}}$ are completely symmetric with respect to indices $j, i_1, i_2, \dots, i_{m-1}$. Finally, we obtain from $(9.8.28)_3$ that

$$T = f\Sigma - \frac{\partial S^i}{\partial x^i} - \sum_{r=0}^{m-1} \frac{\partial S^i}{\partial v_{i_1\cdots i_r}} v_{i_1\cdots i_r i}. \tag{9.8.31}$$

Consequently, isovector components characterising equivalence transformations of a single balance equation are determined through arbitrary functions $F, f, \alpha^{iji_1\cdots i_{m-1}}$ and β^i. Naturally, particular structures of the functions Σ^i and Σ may limit arbitrariness on these functions.

In case $m = 1$, we have to modify slightly the analysis above. The determining equations for isovector components are now given by (9.8.18). If we closely examine these equations, we realise at once that the function ν must be independent of the variables t_i and t. This, in turn, implies that we get $\partial S^i/\partial\Sigma = 0$, namely, $S^i = S^i(x^j, u, v_j, \Sigma^j)$. If we eliminate the function ν between the first two equations in (9.8.18), we then obtain

$$(\Gamma^{ij} + \Gamma^{ji})(\Sigma + s_k^k + \sigma^k v_k) = \Gamma(s^{ij} + s^{ji}).$$

The explicit form of the above relations become

$$\begin{aligned}
&\Sigma^2 \mathcal{X}^{ij} + \Sigma\Big[\mathcal{S}^{ij} + (\delta_n^j \mathcal{A}_m^i + \delta_n^i \mathcal{A}_m^j)s^{mn} + \\
&\quad \Big\{2\mathcal{X}^{ij}\delta_l^k - \frac{\partial X^k}{\partial v_j}\delta_l^i - \frac{\partial X^k}{\partial v_i}\delta_l^j\Big\}(s_k^l + s^l v_k)\Big] \\
&\qquad + \mathcal{S}^{ij}(s_k^k + s^k v_k) + (\delta_n^j \mathcal{A}_m^i + \delta_n^i \mathcal{A}_m^j)s^{mn}(s_k^k + s^k v_k) \\
&\qquad + \Big[\mathcal{X}^{ij}\delta_l^k - \frac{\partial X^k}{\partial v_j}\delta_l^i - \frac{\partial X^k}{\partial v_i}\delta_l^j\Big](s_k^l + s^l v_k)(s_m^m + s^m v_m) \\
&\quad = \Big[T + D_k S^k + \Sigma D_k X^k + \mathcal{A}_l^k(s_k^l + s^l v_k)\Big](\delta_m^i \delta_n^j + \delta_n^i \delta_m^j)s^{mn}
\end{aligned}$$

where we defined

$$\begin{aligned}
D_i &= \frac{\partial}{\partial x^i} + v_i\frac{\partial}{\partial u}, \quad \mathcal{X}^{ij} = \frac{\partial X^i}{\partial v_j} + \frac{\partial X^j}{\partial v_i}, \\
\mathcal{S}^{ij} &= \frac{\partial S^i}{\partial v_j} + \frac{\partial S^j}{\partial v_i}, \\
\mathcal{A}_j^i &= \frac{\partial S^i}{\partial \Sigma^j} + \delta_j^i D_k X^k - D_j X^i
\end{aligned}$$

If we recall that

$$X^i = -\frac{\partial F}{\partial v_i}, \ U = F - v_i\frac{\partial F}{\partial v_i}, \ S^i = \mathcal{F}^i(x^j, u, v_j, \Sigma^j)$$

where $F = F(x^i, u, v_i)$, we immediately observe that the following equations must be satisfied when $\Sigma \neq 0$:

$$\begin{aligned} \frac{\partial S^i}{\partial v_j} + \frac{\partial S^j}{\partial v_i} &= 0 \\ 2\Big(\frac{\partial X^i}{\partial v_j} + \frac{\partial X^j}{\partial v_i}\Big)\,\delta^k_l - \frac{\partial X^k}{\partial v_j}\,\delta^i_l - \frac{\partial X^k}{\partial v_i}\,\delta^j_l &= 0 \\ \Big(\frac{\partial X^i}{\partial v_j} + \frac{\partial X^j}{\partial v_i}\Big)\,\delta^k_l - \frac{\partial X^k}{\partial v_j}\,\delta^i_l - \frac{\partial X^k}{\partial v_i}\,\delta^j_l &= 0 \end{aligned} \tag{9.8.32}$$

From the last two equations in (9.8.32), we get

$$\frac{\partial X^i}{\partial v_j} + \frac{\partial X^j}{\partial v_i} = -\,2\frac{\partial^2 F}{\partial v_i \partial v_j} = 0$$

whence we obtain

$$F(\mathbf{x}, u, \mathbf{v}) = \phi^i(\mathbf{x}, u)v_i + \gamma(\mathbf{x}, u)$$

where ϕ^i and γ are arbitrary functions. We thus find $X^i = -\,\phi^i(\mathbf{x}, u)$. In these circumstances, the remaining terms give, after some rather simple manipulations

$$\begin{aligned} &\big(\delta^i_m\delta^j_n + \delta^i_n\delta^j_m\big)\big(T + D_k S^k\big) = \\ &\qquad \Sigma\Big[\delta^j_n\Big(\frac{\partial S^i}{\partial \Sigma^m} - D_m X^i\Big) + \delta^i_n\Big(\frac{\partial S^j}{\partial \Sigma^m} - D_m X^j\Big)\Big], \\ &\big(\delta^i_m\delta^j_n + \delta^i_n\delta^j_m\big)\Big(\frac{\partial S^k}{\partial \Sigma^l} - D_l X^k\Big) = \\ &\qquad \Big[\delta^j_n\Big(\frac{\partial S^i}{\partial \Sigma^m} - D_m X^i\Big) + \delta^i_n\Big(\frac{\partial S^j}{\partial \Sigma^m} - D_m X^j\Big)\Big]\delta^k_l. \end{aligned}$$

On contracting the indices (k, l) and (j, n) in the second set of equations above, we obtain

$$\frac{\partial S^i}{\partial \Sigma^m} - D_m X^i = \frac{1}{n}\Big(\frac{\partial S^k}{\partial \Sigma^k} - D_k X^k\Big)\delta^i_m = \psi(x^j, u, v_j, \Sigma^j)\,\delta^i_m$$

while the relations

$$\frac{\partial^2 S^i}{\partial \Sigma^j \partial \Sigma^k} = \frac{\partial \psi}{\partial \Sigma^k}\delta^i_j = \frac{\partial \psi}{\partial \Sigma^j}\delta^i_k$$

result in $\partial\psi/\partial\Sigma^k = 0$ for $n > 1$. We thus have $\psi = \psi(x^j, u, v_j)$. But, in this case the integration of the equations

$$\frac{\partial S^i}{\partial \Sigma^j} = \psi\delta^i_j + D_j X^i$$

yields simply

$$S^i = \psi\Sigma^i - \Big(\frac{\partial \phi^i}{\partial x^j} + \frac{\partial \phi^i}{\partial u} v_j\Big)\Sigma^j + \chi^i(x^j, u, v_j)$$

where χ^i are arbitrary functions. The equations $(9.8.32)_1$ are now expressed as follows

$$\Big[\Big(\frac{\partial \psi}{\partial v_j} - \frac{\partial \phi^j}{\partial u}\Big)\delta^i_k + \Big(\frac{\partial \psi}{\partial v_i} - \frac{\partial \phi^i}{\partial u}\Big)\delta^j_k\Big]\Sigma^k + \frac{\partial \chi^i}{\partial v_j} + \frac{\partial \chi^j}{\partial v_i} = 0$$

so that we find

$$\Big(\frac{\partial \psi}{\partial v_j} - \frac{\partial \phi^j}{\partial u}\Big)\delta^i_k + \Big(\frac{\partial \psi}{\partial v_i} - \frac{\partial \phi^i}{\partial u}\Big)\delta^j_k = 0,$$
$$\frac{\partial \chi^i}{\partial v_j} + \frac{\partial \chi^j}{\partial v_i} = 0.$$

Contraction on indices (j, k) in the first set of equations above gives

$$\frac{\partial \psi}{\partial v_i} = \frac{\partial \phi^i}{\partial u} \quad \text{and} \quad \psi = \frac{\partial \phi^i}{\partial u} v_i + \omega(\mathbf{x}, u)$$

whereas the solution of the second set is known to be

$$\chi^i = \alpha^{ij}(\mathbf{x}, u)v_j + \beta^i(\mathbf{x}, u), \;\; \alpha^{ij}(\mathbf{x}, u) = -\alpha^{ji}(\mathbf{x}, u).$$

The isovector component T is determined by the relation

$$T = \psi\Sigma - D_i S^i.$$

Hence, the relevant isovector components are given as follows

$$\begin{aligned}
X^i &= -\phi^i(\mathbf{x}, u), \qquad\qquad (9.8.33)\\
U &= \gamma(\mathbf{x}, u),\\
V_i &= \frac{\partial \gamma}{\partial x^i} + v_i\frac{\partial \gamma}{\partial u} + \Big(\frac{\partial \phi^j}{\partial x^i} + v_i\frac{\partial \phi^j}{\partial u}\Big)v_j,\\
S^i &= \Big(\omega + \frac{\partial \phi^j}{\partial u}v_j\Big)\Sigma^i - \Big(\frac{\partial \phi^i}{\partial x^j} + v_j\frac{\partial \phi^i}{\partial u}\Big)\Sigma^j + \alpha^{ij}v_j + \beta^i,\\
T &= \Big(\omega + \frac{\partial \phi^i}{\partial u}v_i\Big)\Sigma - \Big(\frac{\partial \omega}{\partial x^i} + v_i\frac{\partial \omega}{\partial u}\Big)\Sigma^i + \Big(\frac{\partial^2 \phi^i}{\partial x^i \partial x^j} + v_j\frac{\partial^2 \phi^i}{\partial x^i \partial u}\Big)\Sigma^j
\end{aligned}$$

$$-\Big(\frac{\partial\alpha^{ij}}{\partial x^i}+v_i\frac{\partial\alpha^{ij}}{\partial u}\Big)v_j-\frac{\partial\beta^i}{\partial x^i}+v_i\frac{\partial\beta^i}{\partial u}.$$

When $\Sigma=0$, we are compelled to take $T=0$. This imposes additional conditions on the foregoing solution leading to the relations

$$\omega=\frac{\partial\phi^i}{\partial x^i}+c,\quad \frac{\partial\alpha^{ij}}{\partial x^i}=\frac{\partial\beta^j}{\partial u},\quad \frac{\partial\beta^i}{\partial x^i}=0$$

where c is a constant. A detailed discussion of this case is left to the reader as an exercise.

The case $m=0$ presents no difficulty in determining the isovector components which can be found as

$$\begin{aligned}
X^i &= X^i(\mathbf{x}), \\
U &= U(\mathbf{x},\mathbf{u},\Sigma^i,\Sigma), \\
S^i &= f(\mathbf{x})\Sigma^i+\frac{\partial X^i}{\partial x^j}\Sigma^j+g^i(\mathbf{x}), \\
T &= f(\mathbf{x})\Sigma-\frac{\partial f(\mathbf{x})}{\partial x^i}\Sigma^i-\frac{\partial^2 X^i}{\partial x^i\partial x^j}\Sigma^j-\frac{\partial g^i(\mathbf{x})}{\partial x^i}.
\end{aligned} \tag{9.8.34}$$

Any reader who wish to get more detailed information about calculations concerning this section may be referred to the works below[1].

Example 9.8.1. ***Non-linear wave equation.*** Let us consider a second order non-linear partial differential equation

$$[f(x,t,u,u_x,u_t)]_x-u_{tt}+g(x,t,u,u_x,u_t)=0.$$

Since $n=2,\ N=1$ and $m=1$, let us write

---

[1] Şuhubi, E. S., Equivalence Groups for Second Order Balance Equations, *International Journal of Engineering Science*, **37**, 1901-1925, 1999.
Şuhubi, E. S., Explicit Determination of Isovector Fields of Equivalence Groups for Second Order Balance Equations, *International Journal of Engineering Science*, **38**, 715-736, 2000.
Özer, S. and E. S. Şuhubi, E. S, Equivalence Transformations for First Order Balance Equations, *International Journal of Engineering Science*, **42**, 1305-1324, 2004.
Şuhubi, E. S., Equivalence Groups for Balance Equations of Arbitrary Order - Part I, *International Journal of Engineering Science*, **42**, 1729-1751, 2004.
Şuhubi, E. S., Explicit Determination of Isovector Fields of Equivalence Groups for Balance Equations of Arbitrary Order - Part II, *International Journal of Engineering Science*, **43**, 1-15, 2005.

$$x^1 = x,\ x^2 = t,\ v_1 = u_x = p,\ v_2 = u_t = v$$

so that we have

$$\Sigma^1 = f(x,t,u,p,v),\ \Sigma^2 = -v,\ \Sigma = g(x,t,u,p,v).$$

Hence, we have to take into account the submanifold of the manifold $\mathcal{K}_1$ specified by

$$s^2_1 = 0,\quad s^2_2 = 0,\quad s^2 = 0,\quad s^{21} = 0,\quad s^{22} = -1.$$

In the relations (9.8.33), let us denote

$$\phi^1 = \alpha(x,t,u),\quad \phi^2 = \beta(x,t,u),\quad \alpha^{12} = -\alpha^{21} = \lambda(x,t,u)$$
$$\beta^1 = \Psi(x,t,u),\quad \beta^2 = \Phi(x,t,u).$$

As is clearly observed, S^2 is no longer an independent component of the isovector field, but it is equal to $-V_2$. After having resorted to $(9.8.33)_3$ and $(9.8.33)_4$, we find that

$$V_2 = \gamma_t + \alpha_t p + (\gamma_u + \beta_t)v + \alpha_u pv + \beta_u v^2,$$
$$S^2 = -(\omega + \alpha_u p + \beta_u v)v - (\beta_x + \beta_u p)f + (\beta_t + \beta_u v)v - \lambda p + \Phi.$$

Thus the relation $S^2 = -V_2$ leads to

$$\beta_u v^2 + (\alpha_t - \lambda)p + (2\beta_t + \gamma_u - \omega)v + \Phi + \gamma_t - (\beta_x + \beta_u p)f = 0.$$

Whenever f is an arbitrary function, it follows from this equality that the following equations must be satisfied

$$\beta_x + \beta_u p = 0,\ \beta_u = 0,\ \alpha_t - \lambda = 0,$$
$$2\beta_t + \gamma_u - \omega = 0,\qquad \Phi + \gamma_t = 0$$

whence we obtain $\beta_x = 0, \beta_u = 0$ and consequently

$$\beta = \beta(t),\ \lambda = \alpha_t,\ \omega = 2\beta_{,t} + \gamma_u,\ \Phi = -\gamma_t.$$

The components $\overline{S}^2_j$ and S^{2j} of the isovector field must also vanish. But, because of the relation $F^2 = -s^2_j X^j - s^2 U - s^{2j} V_j + S^2 = 0$ these two conditions are satisfied identically. Therefore, the relevant components of the isovector field are determined as follows

$$X^1 = -\alpha(x,t,u),\ X^2 = -\beta(t),$$
$$U = \gamma(x,t,u),$$

$$
\begin{aligned}
V_1 &= \gamma_x + (\alpha_x + \gamma_u)p + \alpha_u p^2,\\
V_2 &= -S^2 = \gamma_t + \alpha_t p + (\beta_t + \gamma_u)v + \alpha_u pv,\\
S^1 &= (2\beta_t - \alpha_x + \gamma_u)f + \alpha_u v^2 + 2\alpha_t v + \Psi,\\
T &= [\alpha_{xx} - \gamma_{xu} + p(\alpha_{xu} - \gamma_{uu})]f + (2\beta_t + \gamma_u + \alpha_u p)g\\
&\quad + v^2(\gamma_{uu} - \alpha_{xu}) + p(\alpha_{tt} - \Psi_u) + v(\beta_{tt} - 2\alpha_{xt} + 2\gamma_{tu}) + \gamma_{tt} - \Psi_x
\end{aligned}
$$

where $\alpha(x,t,u), \beta(t), \gamma(x,t,u)$ and $\Psi(x,t,u)$ are functions which may be chosen arbitrarily.

As a simple example, let us take

$$\alpha = 0,\ \beta = 0,\ \gamma = au^2,\ \Psi = 0$$

so that we obtain

$$
\begin{aligned}
&X^1 = X^2 = 0,\ U = au^2,\ V_1 = 2aup,\ V_2 = 2auv,\\
&S^1 = 2auf,\ T = -2apf + 2aug + 2av^2.
\end{aligned}
$$

In order to determine the equivalence group associated with this isovector, we have to solve the following ordinary differential equations

$$
\begin{aligned}
&\frac{d\overline{x}}{d\epsilon} = 0,\ \frac{d\overline{t}}{d\epsilon} = 0,\ \frac{d\overline{u}}{d\epsilon} = a\overline{u}^2,\ \frac{d\overline{p}}{d\epsilon} = 2a\overline{u}\,\overline{p},\ \frac{d\overline{v}}{d\epsilon} = 2a\overline{u}\,\overline{v}\\
&\frac{d\overline{f}}{d\epsilon} = 2a\overline{u}\overline{f},\ \frac{d\overline{g}}{d\epsilon} = -2a\overline{p}\overline{f} + 2a\overline{u}\overline{g} + 2a\overline{v}^2
\end{aligned}
$$

under the initial conditions $\overline{x}(0) = x, \overline{t}(0) = t, \overline{u}(0) = u, \overline{p}(0) = p,\ \overline{v}(0) = v, \overline{f}(0) = f, \overline{g}(0) = g$. We can then easily reach to the particular equivalence transformation in which independent variables remain unchanged

$$
\begin{aligned}
&\overline{x} = x,\quad \overline{t} = t,\quad \overline{u}(\epsilon) = \frac{u}{1 - \epsilon au},\\
&\overline{p}(\epsilon) = \frac{p}{(1 - \epsilon au)^2},\ \overline{v}(\epsilon) = \frac{v}{(1 - \epsilon au)^2}\\
&\overline{f}(\epsilon) = \frac{f}{(1 - \epsilon au)^2},\ \overline{g}(\epsilon) = \frac{g - \epsilon a(2pf + ug - 2v^2)}{(1 - \epsilon au)^3}.
\end{aligned}
$$

As a simple application to equivalence transformations, let us apply this group of diffeomorphisms to the linear wave equation

$$u_{xx} - u_{tt} = 0$$

where $f = u_x$ and $g = 0$. We know that the general solution of this equation is given by

$$u(x,t) = \phi(x+t) + \psi(x-t).$$

ϕ and ψ are arbitrary functions of their arguments. If we employ the inverse transformation, we can write

$$u = \frac{\overline{u}}{1 + k\,\overline{u}}$$

where $k = \epsilon a$, then the foregoing linear partial differential equation is cast into a family of quasilinear second order equations

$$\overline{u}_{xx} - \overline{u}_{tt} - 2k\frac{\overline{u}_x^2 - \overline{u}_t^2}{1 + k\,\overline{u}} = 0.$$

by this particular equivalence transformation. We can then readily verify by inspection that a solution of that non-linear, second order partial differential equation is indeed given by

$$\overline{u}(x,t) = \frac{\phi(x+t) + \psi(x-t)}{1 - k[\phi(x+t) + \psi(x-t)]}$$

where $\phi(x+t)$ and $\psi(x-t)$ are arbitrary functions.

As a slightly more general case, let us assume that

$$\alpha = 0,\ \beta = 0,\ \gamma = au^2 + bu,\ \Psi = 0.$$

We thus obtain

$$\begin{aligned} &X^1 = X^2 = 0,\ U = au^2 + bu,\ V_1 = (2au+b)p,\ V_2 = (2au+b)v,\\ &S^1 = (2au+b)f,\ T = -2apf + (2au+b)g + 2av^2. \end{aligned}$$

To find the corresponding equivalence transformation we have to integrate the differential equations below

$$\begin{aligned} &\frac{d\overline{x}}{d\epsilon} = 0,\ \frac{d\overline{t}}{d\epsilon} = 0,\ \frac{d\overline{u}}{d\epsilon} = a\overline{u}^2 + b\overline{u},\ \frac{d\overline{p}}{d\epsilon} = (2a\overline{u}+b)\overline{p},\ \frac{d\overline{v}}{d\epsilon} = (2a\overline{u}+b)\overline{v}\\ &\frac{d\overline{f}}{d\epsilon} = (2a\overline{u}+b)\overline{f},\ \frac{d\overline{g}}{d\epsilon} = -2a\overline{p}\overline{f} + (2a\overline{u}+b)\overline{g} + 2a\overline{v}^2 \end{aligned}$$

under the initial conditions $\overline{x}(0) = x, \overline{t}(0) = t, \overline{u}(0) = u, \overline{p}(0) = p, \overline{v}(0) = v, \overline{f}(0) = f, \overline{g}(0) = g$. We then easily find that

$$\overline{x} = x,\quad \overline{t} = t,\quad \overline{u}(\epsilon) = \frac{be^{b\epsilon}u}{b - a(e^{b\epsilon} - 1)u},$$

$$\overline{p}(\epsilon) = \frac{b^2 e^{b\epsilon} p}{[b - a(e^{b\epsilon} - 1)u]^2},$$
$$\overline{v}(\epsilon) = \frac{b^2 e^{b\epsilon} v}{[b - a(e^{b\epsilon} - 1)u]^2}$$
$$\overline{f}(\epsilon) = \frac{b^2 e^{b\epsilon} f}{[b - a(e^{b\epsilon} - 1)u]^2},$$
$$\overline{g}(\epsilon) = \frac{b^2 e^{b\epsilon}[bg - a(e^{b\epsilon} - 1)(2pf + ug - 2v^2)]}{[b - a(e^{b\epsilon} - 1)u]^3}.$$

If we write

$$u = \frac{b\overline{u}}{be^{b\epsilon} + a(e^{b\epsilon} - 1)\,\overline{u}}$$

then the one-dimensional homogeneous wave equation is transformed into

$$\overline{u}_{xx} - \overline{u}_{tt} + \frac{2a(e^{b\epsilon} - 1)(\overline{u}_x^2 - \overline{u}_t^2)}{[be^{b\epsilon} - a(e^{b\epsilon} - 1)\,\overline{u}]} = 0.$$

Therefore a solution for this family of quasilinear second order differential equations is expressible in the form

$$\overline{u}(x, t; \epsilon) = \frac{b[\phi(x + t) + \psi(x - t)]}{be^{b\epsilon} + a(e^{b\epsilon} - 1)[\phi(x + t) + \psi(x - t)]}$$

depending on the parameter ϵ and constants a and b. ■

Example 9.8.2. ***Homogeneous hyperelasticity.*** We have discussed the symmetry transformations of the equations of motion of a homogeneous hyperelastic material in Example 9.4.4. The equations of motion depend heavily on the stress potential $\Sigma = \Sigma(\mathbf{F})$ characterising the physical constitution of the material, thus differing for different types of materials. Hence, they constitute a family of balance equations. We shall now try to determine the equivalence transformations associated with that family. Since, we will be employing the notations introduced earlier, we abstain from repeating them here. The equations of motion corresponding to the case $m = 1,\ n = 4, N = 3$ can be now written

$$\frac{\partial \Sigma_{kK}}{\partial X_K} + \frac{\partial \Sigma_{k4}}{\partial X_4} = 0$$

with $X_4 = t$ and k, K take the values $1, 2, 3$. Then, the coordinates of the manifold $\mathcal{K}_1$ is easily identified as

$$\Sigma_{kK}(\mathbf{X},\mathbf{F}) = \frac{\partial \Sigma}{\partial F_{kK}},\ \Sigma_{k4} = -\,v_k,\ \Sigma^k = 0,\ v_{kK} = F_{kK},\ v_{k4} = v_k,$$
$$s_{kKlL} = \frac{\partial \Sigma_{kK}}{\partial F_{lL}} = \frac{\partial^2 \Sigma}{\partial F_{kK}\partial F_{lL}},\ s_{k4l4} = -\,\frac{\partial v_k}{\partial v_l} = -\,\delta_{kl},\ s_{kKl4} = 0,$$
$$s_{k4lL} = 0,\ s_{kKL} = 0,\ s_{kK4} = 0,\ s_{k4K} = 0,\ s_{k44} = 0.$$

In this circumstance, the coordinate cover of the manifold $\mathcal{K}_1$ should be taken as $\{X_K, t, x_k, F_{kK}, v_k, \Sigma_{kK}, s_{kKlL}\}$. Obviously, the variables s_{kKlL} enjoy the block symmetry $s_{kKlL} = s_{lLkK}$. We denote the isovector field by

$$V = -\,\phi_K \frac{\partial}{\partial X_K} - \psi \frac{\partial}{\partial t} + U_k \frac{\partial}{\partial x_k} + V_{kK} \frac{\partial}{\partial F_{kK}} + V_k \frac{\partial}{\partial v_k} + S_{kK} \frac{\partial}{\partial \Sigma_{kK}} + S_{kKlL} \frac{\partial}{\partial s_{kKlL}}.$$

We may assume without loss of generality that $S_{kKlL} = S_{lLkK}$. Negative signs above are inserted for convenience, We know that the isovector components ϕ_K, ψ, U_k are functions only of the variables X_K, t, x_k. Furthermore, we have to impose the restrictions

$$S_{k4} = -\,V_{k4} = -\,V_k, T_k = 0, S_{kKl4} = 0, S_{k4lL} = 0, \overline{S}_{kKL} = 0$$
$$\overline{S}_{kK4} = 0,\ \overline{S}_{k4K} = 0,\ \overline{S}_{k44} = 0 \tag{9.8.35}$$

on the isovector components. In order not to confuse the functions $F^{\alpha i}$ defined in (9.8.8) with the components of the deformation gradients F_{kK}, we will replace the functions F^{Kk} by G_{kK}. We also have to take $F^{k4} = 0$. We thus get

$$G_{kK} = -\,s_{kKlL}V_{lL} + S_{kK} \tag{9.8.36}$$

and the conditions

$$\overline{S}_{k4K} = 0,\quad \overline{S}_{k44} = 0,\quad S_{k4lL} = 0$$

are satisfied identically. We can then directly deduce from (9.8.25) that

$$V_{kK} = U_{k,K} + \phi_{L,K}F_{kL} + \psi_{,K}v_k + U_{k,l}F_{lK} + \phi_{L,l}F_{kL}F_{lK} + \psi_{,l}v_kF_{lK},$$
$$V_k = \dot{U}_k + \dot{\phi}_K F_{kK} + \dot{\psi}\, v_k + U_{k,l}v_l + \phi_{K,l}F_{kK}v_l + \psi_{,l}v_kv_l,$$
$$\begin{aligned}S_{kK} = {} & g_{kl}\Sigma_{lK} - (\phi_{K,L} + \phi_{K,l}F_{lL})\Sigma_{kL} + (\dot{\phi}_K + \phi_{K,l}v_l)v_k \\ & + (\phi_{L,l}F_{lL} + \psi_{,l}v_l)\Sigma_{kK} + f_{kKLM4lmn}F_{lL}F_{mM}v_n \\ & + f_{kKLMlm}F_{lL}F_{mM} + f_{kKL4lm}F_{lL}v_m \\ & + f_{kKLl}F_{lL} + f_{kK4l}v_l + f_{kK},\end{aligned}$$

$$
\begin{aligned}
S_{k4} = &- g_{kl}v_l - (\psi_{,K} + \psi_{,l}F_{lK})\Sigma_{kK} + \dot{\psi} + \psi_{,l}v_l)v_k \\
&- (\phi_{L,l}F_{lL} + \psi_{,l}v_l)v_k + f_{k4LMNlmn}F_{lL}F_{mM}F_{nN} \\
&+ f_{k4LMlm}F_{lL}F_{mM} + f_{k4Ll}F_{lL} + f_{k4}, \\
T_k = &- S_{kK,K} - \dot{S}_{kK} - S_{kK,l}\,F_{lK} - S_{k4,l}\,v_l.
\end{aligned}
\tag{9.8.37}
$$

g_{kl} and all multi-indexed functions f depend only on the variables $\mathbf{X}$, t and $\mathbf{x}$. The functions f must be so chosen as to obey the symmetry requirements on capital and small indices (Greek and roman superscripts and subscripts in general expressions) for $n = 4$ and $N = 3$. An overdot represent the derivative with respect to the time variable t.

Let us first deal with the relation $(9.8.35)_1$. It follows from (9.8.37) that we have to satisfy the equations

$$
\begin{aligned}
&\psi_{,K} = 0,\ \ \psi_{,l} = 0,\ \ g_{kl} = 2\dot{\psi}\,\delta_{kl} + U_{k,l},\ \ \phi_{M,n}\,\delta_{km} - \phi_{M,m}\,\delta_{kn} = 0, \\
&f_{k4Kl} = -\,\dot{\phi}_K\,\delta_{kl},\ \ f_{k4} = -\,\dot{U}_k,\ \ f_{k4LMNlmn} = 0,\ \ f_{k4LMlm} = 0.
\end{aligned}
$$

The contraction on indices (k, m) in the fourth equation above yields $\phi_{M,n} = 0$. We thus obtain

$$\phi_K = \phi_K(\mathbf{X}, t),\quad \psi = \psi(t).$$

On the other hand, we get from p. 610 that

$$
\begin{aligned}
&\overline{S}_{kKL} = G_{kK,L} + G_{kK,l}F_{lL} = 0, \overline{S}_{kK4} = \dot{G}_K + G_{kK,l}v_l = 0, \\
&S_{kKl4} = \frac{\partial G_{kK}}{\partial v_l} = 0.
\end{aligned}
$$

Hence, the functions G_{kK} must be independent of v_l. We then further deduce from above the relations

$$G_{kK,l} = 0,\ \ G_{kK,L} = 0,\ \ \dot{G}_K = 0.$$

On the other hand, we get

$$V_{kK} = U_{k,K} + \phi_{L,K}F_{kL} + U_{k,l}\,F_{lK}.$$

This implies that the functions V_{kK} do not depend on v_l. When we take the relation (9.8.36) into consideration, we reach to the conclusion

$$
\begin{aligned}
&V_{kK,l} = 0,\ \ V_{kK,L} = 0,\ \ \dot{V}_{kK} = 0, \\
&\frac{\partial S_{kK}}{\partial v_l} = 0,\ \ S_{kK,l} = 0,\ \ S_{kK,L} = 0,\ \ \dot{S}_{kK} = 0.
\end{aligned}
\tag{9.8.38}
$$

The isovector components S_{kK} can now be written as

$$\begin{aligned}S_{kK} = (2\dot{\psi}\,\delta_{kl} + U_{k,l})\Sigma_{lK} - \phi_{K,L}\Sigma_{kL} + \dot{\phi}_K v_k \\ + f_{kKLM4lmn}F_{lL}F_{mM}v_n + f_{kKLMlm}F_{lL}F_{mM} \\ + f_{kKL4lm}F_{lL}v_m + f_{kKLl}F_{lL} + f_{kK4l}v_l + f_{kK}\end{aligned}$$

and $(9.8.38)_4$ leads to

$$\dot{\phi}_K\,\delta_{kp} + f_{kK4p} + f_{kKLM4lmp}\,F_{lL}\,F_{mM} + f_{kKL4lp}\,F_{lL} = 0$$

from which we find that

$$f_{kKLM4lmn} = 0,\;\; f_{kKL4lm} = 0,\;\; f_{kK4l} = -\,\dot{\phi}_K\,\delta_{kl}.$$

However, because of the relation $f_{kK4l} = -\,f_{k4Kl}$ we get $\dot{\phi}_K = 0$, hence we obtain

$$\phi_K = \phi_K(\mathbf{X}).$$

Derivatives of V_{kK} with respect to t, x_m and X_M give, respectively,

$$\dot{U}_{k,K} = 0,\; \dot{U}_{k,l} = 0,\; U_{k,Km} = 0,\; U_{k,lm} = 0,\; U_{k,KM} = 0,\; \phi_{K,LM} = 0.$$

Similar expressions for S_{kK} leads to the equation $\ddot{\psi}(t) = 0$ implying further that all non-zero functions f must be constants. Thus, if we recall the antisymmetry properties, we are able to write

$$\begin{aligned}\psi = a_1 t + a_2, \quad f_{kKLMlm} = e_{KLM}\,e_{lmn}\,c_{kn}, \\ f_{kKLl} = e_{KLM}\,c_{Mkl}, \quad f_{kK} = c_{kK}\end{aligned}$$

where c_{kn}, c_{Mkl} and c_{kK} are constants. e_{KLM}, e_{klm} are, of course, three-dimensional permutation symbols. The solutions of the differential equations satisfied by the functions ϕ_K and U_k are readily obtained as

$$\phi_K = B_{KL}X_L + B_K, \quad U_k = a_{kl}x_l + A_{kK}X_K + A_k(t)$$

where $B_{KL}, B_K, a_{kl}, A_{kK}$ are constants. Finally, we find from $(9.8.37)_5$ that

$$T_k = \dot{V}_k + V_{k,l}v_l = 0.$$

This equations yield $\ddot{U}_k = 0$ or $\ddot{A}_k = 0$ and

$$A_k(t) = \alpha_k t + A_k.$$

Hence, the relevant isovector components take the form

$$\begin{aligned}\phi_K = B_{KL}X_L + B_K,\; \psi = a_1 t + a_2,\; U_k = a_{kl}x_l + A_{kK}X_K + \alpha_k t + A_k, \\ V_{kK} = B_{LK}F_{kL} + a_{kl}F_{lK} + A_{kK}, \qquad V_k = (a_1\delta_{kl} + a_{kl})v_l + \alpha_k,\end{aligned}$$

$$S_{kK} = (2a_1\delta_{kl} + a_{kl})\Sigma_{lK} - B_{KL}\Sigma_{kL} + e_{KLM}\, e_{lmn}\, c_{kn}\, F_{lL}\, F_{mM} + e_{KLM}\, c_{Mkl}\, F_{lL} + c_{kK}.$$

The equivalence transformation is then found by integrating the ordinary differential equations

$$\begin{aligned}
\frac{d\overline{X}_K}{d\epsilon} &= -\,(B_{KL}\overline{X}_L + B_K), \qquad \frac{d\overline{t}}{d\epsilon} = -\,(a_1\overline{t} + a_2),\\
\frac{d\overline{x}_k}{d\epsilon} &= a_{kl}\overline{x}_l + A_{kK}\overline{X}_K + \alpha_k\overline{t} + A_k,\\
\frac{d\overline{F}_{kK}}{d\epsilon} &= B_{LK}\overline{F}_{kL} + a_{kl}\overline{F}_{lK} + A_{kK}, \qquad \frac{d\overline{v}_k}{d\epsilon} = (a_1\delta_{kl} + a_{kl})\overline{v}_l + \alpha_k\\
\frac{d\overline{\Sigma}_{kK}}{d\epsilon} &= (2a_1\delta_{kl} + a_{kl})\overline{\Sigma}_{lK} - B_{KL}\overline{\Sigma}_{kL} + e_{KLM}\, e_{lmn}\, c_{kn}\, \overline{F}_{lL}\, \overline{F}_{mM}\\
&\quad + e_{KLM}\, c_{Mkl}\, \overline{F}_{lL} + c_{kK}.
\end{aligned}$$

under the initial conditions $\overline{X}_K(0) = X_K, \overline{t}(0) = t, \overline{x}_k(0) = x_k, \overline{F}_{kK}(0) = F_{kK}, \overline{v}_k(0) = v_k$ and $\overline{\Sigma}_{kK}(0) = \Sigma_{kK}$. ϵ is the group parameter. That Σ_{kK} are actually dependent on the deformation tensor $\mathbf{C} = \mathbf{F}^{\mathrm{T}}\mathbf{F}$ instead of $\mathbf{F}$ may impose additional restrictions on some constants appearing in the above expressions. Since $\Sigma_{kK} = \partial\Sigma/\partial F_{kK}$ the transformed expressions help us to determine the stress potential $\overline{\Sigma}$. ■

Example 9.8.3. As a last example, we consider the third order nonlinear partial differential equation

$$\big[u_{xx} + \phi(x,t,u)\big]_x + u_t = u_{xxx} + \phi_u u_x + \phi_x + u_t = 0. \quad (9.8.39)$$

We may regard this equation as a kind of generalised Korteweg-de Vries equation. If we take $\phi = u^2/2$, we obtain the known form of the Korteweg-de Vries equation. In this case, it is clear that we have $m = 2$, $N = 1$ and $n = 2$. The manifold $\mathcal{K}_2$ is now generated by taking

$$\begin{aligned}
&x^1 = x,\ x^2 = t,\ v_1 = u_x,\ v_2 = u_t,\ v_{11} = u_{xx},\ v_{12} = v_{21} = u_{xt},\\
&v_{22} = u_{tt},\ \Sigma^1 = v_{11} + \phi,\ \Sigma^2 = u,\ \Sigma = 0,\\
&s^1_1 = \phi_x,\ s^1_2 = \phi_t,\ s^1 = \phi_u,\ s^2 = 1,\ s^{111} = 1.
\end{aligned}$$

Therefore, the coordinate cover of the enlarged manifold $\mathcal{K}_2$ is specified by the following list

$$\{x, t, v_1, v_2, v_{11}, v_{12}, v_{22}, \phi, s^1_1, s^1_2, s^1\}.$$

Hence, an isovector field should be represented by

$$V = X^1\frac{\partial}{\partial x} + X^2\frac{\partial}{\partial t} + U\frac{\partial}{\partial u} + V_1\frac{\partial}{\partial v_1} + V_2\frac{\partial}{\partial v_2} + V_{11}\frac{\partial}{\partial v_{11}} + V_{12}\frac{\partial}{\partial v_{12}}$$
$$+ V_{22}\frac{\partial}{\partial v_{22}} + \Phi\frac{\partial}{\partial \phi} + \overline{S}^1_1\frac{\partial}{\partial s^1_1} + \overline{S}^1_2\frac{\partial}{\partial s^1_2} + S^{111}\frac{\partial}{\partial s^{111}}$$

However, if we note the forms of Σ^1 and Σ^2, we see that we can write

$$\frac{\partial}{\partial u} \to \frac{\partial}{\partial u} + \frac{\partial}{\partial \Sigma^2}\frac{\partial \Sigma^2}{\partial u} = \frac{\partial}{\partial u} + \frac{\partial}{\partial \Sigma^2},$$
$$\frac{\partial}{\partial v_{11}} \to \frac{\partial}{\partial v_{11}} + \frac{\partial}{\partial \Sigma^1}\frac{\partial \Sigma^1}{\partial v_{11}} = \frac{\partial}{\partial v_{11}} + \frac{\partial}{\partial \Sigma^1},$$
$$\frac{\partial}{\partial \phi} \to \frac{\partial}{\partial \Sigma^1}\frac{\partial \Sigma^1}{\partial \phi} = \frac{\partial}{\partial \Sigma^1}$$

so that the isovector field is expressed in the standard form in terms of quantities entering into the balance equation as follows

$$V = X\frac{\partial}{\partial x} + T\frac{\partial}{\partial t} + U\frac{\partial}{\partial u} + V_1\frac{\partial}{\partial v_1} + V_2\frac{\partial}{\partial v_2} + V_{11}\frac{\partial}{\partial v_{11}} + V_{12}\frac{\partial}{\partial v_{12}}$$
$$+ V_{22}\frac{\partial}{\partial v_{22}} + S^1\frac{\partial}{\partial \Sigma^1} + S^2\frac{\partial}{\partial \Sigma^2} + \overline{S}^1_1\frac{\partial}{\partial s^1_1} + \overline{S}^1_2\frac{\partial}{\partial s^1_2} + S^{111}\frac{\partial}{\partial s^{111}}.$$

It is straightforward to notice that

$$S^1 = V_{11} + \Phi,$$
$$S^2 = U.$$

We know that some of the isovector components are determined by a presently arbitrary function $F = F(x, t, u, v_1, v_2)$ through relations given below [*see* (9.3.26)]:

$$X^1 = X = -\frac{\partial F}{\partial v_1},\; X^2 = T = -\frac{\partial F}{\partial v_2},\; U = F - v_1\frac{\partial F}{\partial v_1} - v_2\frac{\partial F}{\partial v_2}$$
$$V_1 = \frac{\partial F}{\partial x} + v_1\frac{\partial F}{\partial u},\quad V_2 = \frac{\partial F}{\partial t} + v_2\frac{\partial F}{\partial u}, \tag{9.8.40}$$
$$V_{11} = \frac{\partial^2 F}{\partial x^2} + 2v_1\frac{\partial^2 F}{\partial x \partial u} + v_1^2\frac{\partial^2 F}{\partial u^2} + v_{11}\Big(2\frac{\partial^2 F}{\partial x \partial v_1} + 2v_1\frac{\partial^2 F}{\partial u \partial v_1} + \frac{\partial F}{\partial u}\Big)$$
$$+ 2v_{12}\Big(\frac{\partial^2 F}{\partial x \partial v_2} + v_1\frac{\partial^2 F}{\partial u \partial v_2}\Big) + v_{11}^2\frac{\partial^2 F}{\partial v_1^2} + 2v_{11}v_{12}\frac{\partial^2 F}{\partial v_1 \partial v_2} + v_{12}^2\frac{\partial^2 F}{\partial v_2^2},$$
$$V_{12} = \frac{\partial^2 F}{\partial x \partial t} + v_1\frac{\partial^2 F}{\partial t \partial u} + v_2\frac{\partial^2 F}{\partial x \partial u} + v_1 v_2\frac{\partial^2 F}{\partial u^2} + v_{11}\Big(\frac{\partial^2 F}{\partial t \partial v_1} + v_2\frac{\partial^2 F}{\partial u \partial v_1}\Big)$$

$$
\begin{aligned}
&+v_{12}\Big(\frac{\partial^2 F}{\partial t\partial v_2}+v_2\frac{\partial^2 F}{\partial u\partial v_2}+\frac{\partial^2 F}{\partial x\partial v_1}+v_1\frac{\partial^2 F}{\partial u\partial v_1}+\frac{\partial F}{\partial u}\Big)+v_{22}\Big(\frac{\partial^2 F}{\partial x\partial v_2}+\\
&\quad v_1\frac{\partial^2 F}{\partial u\partial v_2}\Big)+v_{11}v_{12}\frac{\partial^2 F}{\partial v_1^2}+(v_{11}v_{22}+v_{12}^2)\frac{\partial^2 F}{\partial v_1\partial v_2}+v_{12}v_{22}\frac{\partial^2 F}{\partial v_2^2},\\
V_{22}=&\frac{\partial^2 F}{\partial t^2}+2v_2\frac{\partial^2 F}{\partial t\partial u}+v_2^2\frac{\partial^2 F}{\partial u^2}+2v_{12}\Big(\frac{\partial^2 F}{\partial t\partial v_1}+v_2\frac{\partial^2 F}{\partial u\partial v_1}\Big)+v_{22}\Big(\frac{\partial F}{\partial u}\\
&+2\frac{\partial^2 F}{\partial t\partial v_2}+2v_2\frac{\partial^2 F}{\partial u\partial v_2}\Big)+v_{12}^2\frac{\partial^2 F}{\partial v_1^2}+2v_{12}v_{22}\frac{\partial^2 F}{\partial v_1\partial v_2}+v_{22}^2\frac{\partial^2 F}{\partial v_2^2}.
\end{aligned}
$$

On the other hand, the relations (9.8.8) indicates that we may introduce the functions

$$
\begin{aligned}
F^1 &= -\,s_1^1X-s_2^1T-s^1U-V_{11}+S^1\\
&= -\,s_1^1X-s_2^1T-s^1U+\Phi,\\
F^2 &= -\,U+S^2=0.
\end{aligned}
$$

Moreover, the relations

$$
S^{112}=\frac{\partial F^1}{\partial v_{12}}=\frac{\partial \Phi}{\partial v_{12}}=0,\quad S^{122}=\frac{\partial F^1}{\partial v_{22}}=\frac{\partial \Phi}{\partial v_{22}}=0
$$

imply that the function Φ must be independent of the variables v_{12} and v_{22}. The isovector components that are obtained from the zero function F^2 will naturally become zero. The isovector components S^1 and S^2 follow from the general definitions as

$$
\begin{aligned}
S^1=&-\Big(\frac{\partial^2 F}{\partial x\partial v_1}+\frac{\partial^2 F}{\partial u\partial v_1}v_1+\frac{\partial^2 F}{\partial v_1^2}v_{11}+\frac{\partial^2 F}{\partial v_1\partial v_2}v_{12}\Big)(v_{11}+\phi)\\
&+f(v_{11}+\phi)-\Big(\frac{\partial^2 F}{\partial t\partial v_1}+\frac{\partial^2 F}{\partial u\partial v_1}v_2+\frac{\partial^2 F}{\partial v_1^2}v_{12}+\frac{\partial^2 F}{\partial v_1\partial v_2}v_{22}\Big)u\\
&+\alpha^{121}v_{12}+\alpha^{122}v_{22}+\beta^1, \qquad (9.8.41)\\
S^2=&\,fu-\Big(\frac{\partial^2 F}{\partial x\partial v_2}+\frac{\partial^2 F}{\partial u\partial v_2}v_1+\frac{\partial^2 F}{\partial v_1\partial v_2}v_{11}+\frac{\partial^2 F}{\partial v_2^2}v_{12}\Big)(v_{11}+\phi)\\
&-\Big(\frac{\partial^2 F}{\partial t\partial v_2}+\frac{\partial^2 F}{\partial u\partial v_2}v_{21}+\frac{\partial^2 F}{\partial v_1\partial v_2}v_{12}+\frac{\partial^2 F}{\partial v_2^2}v_{22}\Big)u\\
&-\alpha^{121}v_{11}-\alpha^{122}v_{12}+\beta^2
\end{aligned}
$$

where $f,\alpha^{121}=-\alpha^{211},\alpha^{122}=-\alpha^{212},\beta^1$ and β^2 are arbitrary functions of the variables x,t,u,v_1 and v_2. But we can readily observe that to satisfy the constraint

$$S^2 = U = F - v_1\frac{\partial F}{\partial v_1} - v_2\frac{\partial F}{\partial v_2} \tag{9.8.42}$$

we are required to take

$$\frac{\partial^2 F}{\partial v_1 \partial v_2} = 0, \frac{\partial^2 F}{\partial v_2^2} = 0, \frac{\partial^2 F}{\partial x \partial v_2} + \frac{\partial^2 F}{\partial u \partial v_2} v_1 = 0, \alpha^{121} = \alpha^{122} = 0.$$

These equations lead obviously to

$$F = \alpha(x, t, u, v_1) + \beta(t) v_2.$$

Then (9.8.42) reduces to the equation

$$\beta^2 = (\dot{\beta} - f)u + \alpha - v_1\frac{\partial \alpha}{\partial v_1}.$$

An overdot denotes again the derivative with respect to the variable t. On the other hand, since the expression

$$\begin{aligned}\Phi = &\Big(f - \frac{\partial^2 \alpha}{\partial x \partial v_1} - \frac{\partial^2 \alpha}{\partial u \partial v_1} v_1 - \frac{\partial^2 \alpha}{\partial v_1^2} v_{11}\Big)(v_{11} + \phi) \\ &- \Big(\frac{\partial^2 \alpha}{\partial t \partial v_1} + \frac{\partial^2 \alpha}{\partial u \partial v_1} v_2 + \frac{\partial^2 \alpha}{\partial v_1^2} v_{12}\Big) u + \beta^1 - \frac{\partial^2 \alpha}{\partial x^2} - 2v_1\frac{\partial^2 \alpha}{\partial x \partial u} \\ &- v_1^2\frac{\partial^2 \alpha}{\partial u^2} - \Big(2\frac{\partial^2 \alpha}{\partial x \partial v_1} + 2v_1\frac{\partial^2 \alpha}{\partial u \partial v_1} + \frac{\partial \alpha}{\partial u}\Big) v_{11} - \frac{\partial^2 \alpha}{\partial v_1^2} v_{11}^2\end{aligned}$$

does not depend on v_{12}, we get $\partial^2\alpha/\partial v_1^2 = 0$ the integration of which yields simply

$$\alpha = \lambda(x, t, u)v_1 + \mu(x, t, u).$$

We thus conclude that

$$\beta^2 = \big[\dot{\beta}(t) - f(x, t, u, v_1, v_2)\big]u + \mu(x, t, u), U = \mu(x, t, u). \tag{9.8.43}$$

Finally, the condition $T = 0$ leads us to the equation

$$\frac{\partial S^1}{\partial x} + \frac{\partial S^2}{\partial t} + \frac{\partial S^1}{\partial u} v_1 + \frac{\partial S^2}{\partial u} v_2 + \frac{\partial S^1}{\partial v_1} v_{11} + \frac{\partial S^1}{\partial v_2} v_{12} = 0$$

whose explicit form can be written as

$$-(\lambda_{uu}u + \lambda_u)v_1 v_2 - (\lambda_{tu}u + \lambda_t)v_1 + (\mu_u - \lambda_{ux}u)v_2 - \lambda_{xt}u + \mu_t +$$

$$+ (v_{11} + \phi)\Big[f_x - \lambda_{xx} - \lambda_{ux}v_1 + (f_u - \lambda_{ux} - \lambda_{uu}v_1)v_1 + \Big(\frac{\partial f}{\partial v_1} - \lambda_u\Big)v_{11}$$
$$\frac{\partial f}{\partial v_2}v_{12}\Big] + \frac{\partial \beta^1}{\partial v_1}v_{11} + \Big(\frac{\partial \beta^1}{\partial v_2} - \lambda_u u\Big)v_{12} + \frac{\partial \beta^1}{\partial u}v_1 + \frac{\partial \beta^1}{\partial x} = 0$$

whence we evidently deduce the following equations

$$f_x - \lambda_{xx} - \lambda_{ux}v_1 + (f_u - \lambda_{ux} - \lambda_{uu}v_1)v_1 = 0, \qquad (9.8.44)$$
$$\frac{\partial f}{\partial v_1} - \lambda_u = 0, \ \frac{\partial f}{\partial v_2} = 0, \ \frac{\partial \beta^1}{\partial v_1} = 0, \ \frac{\partial \beta^1}{\partial v_2} - \lambda_u u = 0,$$
$$-(\lambda_u u)_u v_1 v_2 + \big[\beta^1_u - (\lambda_t u)_u\big]v_1 + (\mu_u - \lambda_{ux}u)v_2 + \beta^1_x - \lambda_{xt}u + \mu_t = 0.$$

We first obtain from $(9.8.44)_{2-3}$ that

$$f = \lambda_u v_1 + \varphi(x,t,u).$$

Then $(9.8.44)_1$ gives rise to

$$\varphi_x - \lambda_{xx} + (\varphi_u - \lambda_{ux})v_1 = 0 \quad \text{or} \quad \varphi_x = \lambda_{xx}, \ \ \varphi_u = \lambda_{ux}.$$

These equations determine, in turn, the function φ in the form

$$\varphi = \lambda_x + \psi(t).$$

Once again, $(9.8.44)_{4-5}$ yields easily

$$\beta^1 = \lambda_u u v_2 + b(x,t,u).$$

Thus, the equation $(9.8.44)_6$ is reduced to the form

$$\big[b_u - (\lambda_t u)_u\big]v_1 + \mu_u v_2 + b_x - \lambda_{xt}u - \mu_t = 0$$

whence we find that

$$b_u - (\lambda_t u)_u = 0, \ \ \mu_u = 0, \ \ b_x - \lambda_{xt}u - \mu_t = 0$$

and, respectively,

$$b = \lambda_t u + \nu(x,t), \ \ \mu = \mu(x,t), \ \ \nu_x + \mu_t = 0.$$

Therefore, we obtain $\mu = m_x$ and $\nu = -m_t$ where $m = m(x,t)$ is an arbitrary function. We can thus write

$$\beta^1 = \lambda_u u v_2 + \lambda_t u - m_t, \ \ f = \lambda_u v_1 + \lambda_x + \psi(t)$$

On the other hand, (9.8.43) leads us to the relation

$$\beta^2 = (\dot{\beta} - \lambda_x - \lambda_u v_1)u + m_x$$

where the arbitrary function $\psi(t)$ is absorbed into the function $\beta(t)$ which is arbitrary as well. Consequently, the relevant isovector components that will be used in determining equivalence transformations are obtained as follows

$$\begin{aligned}
X^1 &= -\lambda(x,t,u),\ X^2 = -\beta(t),\ U = m_x(x,t),\\
V_1 &= \lambda_u v_1^2 + \lambda_x v_1 + m_{xx},\\
V_{11} &= \lambda_{uu} v_1^3 + 2\lambda_{ux} v_1^2 + \lambda_{xx} v_1 + (2\lambda_x + 3\lambda_u v_1)v_{11} + m_{xxx},\\
S^1 &= \psi(t)(v_{11} + \phi) - m_t = \psi(t)\Sigma^1 - m_t
\end{aligned}$$

where the functions $\lambda(x,t,u)$, $\beta(t)$, $m(x,t)$ and $\psi(t)$ are arbitrary. The equivalence transformations are then obtained as the solution of the following ordinary differential equations

$$\begin{aligned}
&\frac{d\overline{x}}{d\epsilon} = -\lambda(\overline{x},\overline{t},\overline{u}),\ \frac{d\overline{t}}{d\epsilon} = -\beta(\overline{t}),\ \frac{d\overline{u}}{d\epsilon} = m_{\overline{x}}(\overline{x},\overline{t}),\\
&\frac{d\overline{v}_1}{d\epsilon} = \lambda_{\overline{u}}\overline{v}_1^2 + \lambda_{\overline{x}}\overline{v}_1 + m_{\overline{x}\overline{x}},\\
&\frac{d\overline{v}_{11}}{d\epsilon} = \lambda_{\overline{u}\overline{u}}\overline{v}_1^3 + 2\lambda_{\overline{u}\overline{x}}\overline{v}_1^2 + \lambda_{\overline{x}\overline{x}}v_1 + (2\lambda_{\overline{x}} + 3\lambda_{\overline{u}}\overline{v}_1)\overline{v}_{11} + m_{\overline{x}\overline{x}\overline{x}},\\
&\frac{d\overline{\Sigma}^1}{d\epsilon} = \psi(\overline{t})\overline{\Sigma}^1 - \overline{m}_t.
\end{aligned}$$

under the initial conditions $\overline{x}(0) = x, \overline{t}(0) = t, \overline{u}(0) = u, \overline{v}_1(0) = v_1, \overline{v}_{11}(0) = v_{11}$ and $\overline{\Sigma}^1(0) = \Sigma^1$. ■

IX. EXERCISES

9.1. Discuss the solutions of the equation below:

$$\sum_{i=1}^{n}(u_{,i})^2 - u^2 = 0.$$

9.2. Discuss the solutions of the equation below:

$$uu_x + u_y - 1 = 0$$

9.3. Discuss the solutions of the set of equations below:

$$(u+v)u_x + (u-v)u_y - u^2 = 0, \quad (u+v)v_x + (u-v)v_y - 2v^2 = 0$$

9.4. Find the symmetry groups of *Laplace equation* $u_{xx} + u_{yy} = 0$ and explore its similarity solutions.

9.5. Find the symmetry groups of the wave equation $u_{xx} - u_{yy} = 0$ and explore its similarity solutions.

9.6. Find the symmetry groups of the equation $u_{xx} + u_{yy} = f(u)$ and admissible forms of the function f. Explore its similarity solutions.

9.7. Find the symmetry groups of the equation $u_{xx} - u_{yy} = f(u)$ and admissible forms of the function f. Explore its similarity solutions.

9.8. Find the symmetry groups of non-dimensionalised *Fokker-Planck equation* $u_t = u_{xx} + xu_x + u$ [after Dutch physicist Adriaan Daniël Fokker (1887-1972) and German physicist Max Karl Ernst Ludwig Planck (1858-1947)] encountered in statistical mechanics and explore its similarity solutions.

9.9. Find the symmetry groups of non-dimensionalised *Burgers equation* $u_y + uu_x + u_{xx} = 0$ [after Dutch physicist Johannes Martinus Burgers (1895-1981)] encountered in fluid mechanics and modelling of traffic flow and explore its similarity solutions.

9.10. Find the symmetry groups of the n-dimensional heat conduction equation $u_t = u_{,ii}$ where $\mathbf{x} \in \mathbb{R}^n$.

9.11. Discuss the symmetry groups of the biharmonic equation $\Delta^2 u = u_{,iijj} = 0$ in the manifold $\mathbb{R}^n$.

9.12. Find the symmetry groups of *Helmholtz equation* $u_{xx} + u_{yy} + u_{zz} + \lambda u = 0$ encountered in the propagation of waves in $\mathbb{R}^3$. λ is a constant.

9.13. Let us consider the first order, homogeneous partial differential equation

$$V(u) = v^i(\mathbf{x})u_{,i} = 0, \quad V = v^i(\mathbf{x})\partial_i \in T(\mathbb{R}^n)$$

in $\mathbb{R}^n$. Show that a vector field $U = u^i(\mathbf{x})\,\partial_i \in T(\mathbb{R}^n)$ generates a symmetry group of this differential equation if and only if it satisfies the condition $[V, U] = \lambda(\mathbf{x})\,V$. $\lambda(\mathbf{x})$ is a scalar-valued function.

9.14. Find the symmetry groups of *Euler equations*

$$(\partial v_i/\partial t) + v_j v_{i,j} = \; - p_{,i}, \quad v_{i,i} = 0, \;\; i = 1, 2, 3$$

governing the motion of incompressible fluids where v_i are components of the velocity vector and p is the pressure.

9.15. Determine the symmetry groups of non-dimensional *Navier-Stokes equations*

$$(\partial v_i/\partial t) + v_j v_{i,j} = \; - p_{,i} + (1/R_e)v_{i,jj}, \quad v_{i,i} = 0, \;\; i = 1, 2, 3$$

[after Stokes and French engineer and mathematician Claude Louis Henri Navier (1785-1836)] governing the motion of incompressible viscous fluids. The constant R_e is called *Reynolds number* [after English engineer and mathematician Osborne Reynolds (1842-1912)].

9.16. Determine equivalence transformations of the non-linear, 1-dimensional heat conduction equation

$$\frac{\partial u}{\partial t} = \frac{\partial}{\partial x}\Big(\kappa(u)\frac{\partial u}{\partial x}\Big) + h(x, t, u)$$

CHAPTER X

CALCULUS OF VARIATIONS

10.1. SCOPE OF THE CHAPTER

This chapter is devoted to a brief study of the calculus of variations that deals with determining functions rendering a functional defined as an integral over a function space stationary. However, in contrast to the classical approach, we will be employing exterior analysis throughout the discussion. We shall start with a relatively simple case in Sec. 10.2 by assuming that a functional is specified by an integral of a function over a region of a smooth manifold. This function that is frequently called the Lagrangian function, will be supposed to depend on independent variables, a certain number of dependent variables and their first order partial derivatives. We then examine the conditions under which the functional so formed becomes stationary, that is, its first order variation vanishes. We show in Sec. 10.3 that a function that renders such a functional stationary must satisfy the Euler-Lagrange equations which arise from the vanishing of the Lie derivative of an appropriately prescribed form with respcct to an isovector field of the contact ideal. We discuss in Sec. 10.4 the variational symmetries that leave the functional invariant, the Noetherian vector fields that conduce to such symmetries and conservation laws generated by these symmetries. Finally, Sec. 10.5 is concerned with the case in which the Lagrangian function involves partial derivatives of dependent variables up to any finite order. This is, however, handled by resorting to previous findings.

10.2. STATIONARY FUNCTIONALS

Let us consider an n-dimensional connected region B_n of an n-dimensional smooth manifold. We denote the piecewise smooth boundary of this region by ∂B_n. The local coordinates of the manifold are given by x^i, $1 \leq i \leq n$. Suppose that sufficiently differentiable functions $u^\alpha : B_n \to \mathbb{R}$ defined on B_n constitute the set $\mathcal{B} = \{u^\alpha(x^i),\ 1 \leq \alpha \leq N\}$. Let $L(\mathbf{x}, \mathbf{u})$ be a real-valued function depending on functions u^α and their partial

Exterior Analysis, DOI: 10.1016/B978-0-12-415902-0.50010-4

derivatives of various orders. The mapping $A : \mathcal{B} \to \mathbb{R}$ defined by

$$A(\mathbf{u}) = \int_{B_n} L(\mathbf{x}, \mathbf{u})\, dx^1 dx^2 \cdots dx^n$$

is called a ***functional*** or an ***action integral*** whereas the function L is named as a ***Lagrangian function*** [French mathematician Joseph-Louis Lagrange (1736-1813)]. The source of these terms is the terminology widely used in the analytical mechanics. The ***calculus of variations*** copes with the task of determining the functions **u** that extremise the action integral, namely, either maximise or minimise it. We here deal with this problem as an application of the exterior analysis. We first start with a simpler functional that helps us to comprehend much better our treatment. The general case will be considered later. Thus, we select the Lagrangian function L in the form

$$L(\mathbf{x}, \mathbf{u}) = L(x^i, u^\alpha, v_i^\alpha), \;\; v_i^\alpha = u_{,i}^\alpha = \frac{\partial u^\alpha}{\partial x^i}.$$

We again denote the $(n + N + nN)$-dimensional smooth contact manifold whose coordinate cover is given by $\{x^i, u^\alpha, v_i^\alpha\}$ by $\mathcal{C}_1$ and consider the closed contact ideal $\mathcal{I}_1 = \overline{\mathcal{I}}(\sigma^\alpha, d\sigma^\alpha)$ of the exterior algebra $\Lambda(\mathcal{C}_1)$. We know that the contact forms are given by

$$\sigma^\alpha = du^\alpha - v_i^\alpha dx^i \in \Lambda^1(\mathcal{C}_1)$$

We characterise the set of all *regular mappings* from the region B_n into the manifold $\mathcal{C}_1$ as follows

$$\mathcal{R}(B_n) = \{\phi : B_n \to \mathcal{C}_1 : \phi^* \mathcal{I}_1 = 0, \; \phi^* \mu \neq 0\} \qquad (10.2.1)$$

where $\mu = dx^1 \wedge dx^2 \wedge \cdots \wedge dx^n$ is the volume form. Thus, we must have $\phi^* \sigma^\alpha = 0$ with $1 \leq \alpha \leq N$. According to this definition, regular mappings may be taken as *mappings depending on the same independent variables* x^i *annihilating the ideal* $\mathcal{I}_1$. Hence, if $\phi \in \mathcal{R}(B_n)$, then it is represented by the relations

$$\phi^* x^i = x^i, \;\; \phi^* u^\alpha = \phi^\alpha(\mathbf{x}), \;\; \phi^* v_i^\alpha = \phi_{,i}^\alpha(\mathbf{x}) \qquad (10.2.2)$$

Therefore, the action integral corresponding to a mapping $\phi \in \mathcal{R}(B_n)$ can be expressed as

$$A(\phi) = \int_{B_n} \phi^*(L\mu), \qquad (10.2.3)$$

namely, it becomes the integral of the form $L\mu = L(x^i, u^\alpha, v_i^\alpha)\mu \in \Lambda^n(\mathcal{C}_1)$

on B_n. If we take into account two regular mappings $\phi, \psi \in \mathcal{R}(B_n)$, then the difference of corresponding action integrals are obviously given by

$$A(\phi) - A(\psi) = \int_{B_n} \big[\phi^*(L\mu) - \psi^*(L\mu)\big].$$

Let us note that although the mappings ϕ and ψ are prescribed by different functions $\phi^\alpha(x^i)$ and $\psi^\alpha(x^i)$, their arguments $\{x^i\}$ remain the same. To make the comparison of two action integrals more productive, let us embed the mappings ϕ and ψ into a 1-parameter family of regular mappings in such a way that $\phi = \psi$ for the value $s = 0$ of the parameter. In order to determine such a family systematically, we make use of an isovector field of the contact ideal. If a vector field

$$U = X^i \frac{\partial}{\partial x^i} + U^\alpha \frac{\partial}{\partial u^\alpha} + V_i^\alpha \frac{\partial}{\partial v_i^\alpha}$$

is an isovector field of the ideal $\mathcal{I}_1$, we then know from Sec. 9.3 that its components are given by

$$X^i = X^i(\mathbf{x}, \mathbf{u}), \quad U^\alpha = U^\alpha(\mathbf{x}, \mathbf{u}), \quad V_i^\alpha = D_i^{(0)}(U^\alpha - v_j^\alpha X^j)$$

for $N > 1$. The flow generated by this isovector leaves the ideal $\mathcal{I}_1$ invariant. Here, we have employed the notation

$$D_i^{(0)} = \frac{\partial}{\partial x^i} + v_i^\alpha \frac{\partial}{\partial u^\alpha}$$

that has already been introduced on p. 506. However, since we do not want to transform the independent variables x^i, we have to take $X^i = 0$. Such an isovector field will be called a ***vertical isovector field***. Hence, vertical isovector fields of the ideal $\mathcal{I}_1$ are characterised by

$$V = U^\alpha(\mathbf{x}, \mathbf{u}) \frac{\partial}{\partial u^\alpha} + D_i^{(0)} U^\alpha(\mathbf{x}, \mathbf{u}) \frac{\partial}{\partial v_i^\alpha}.$$

When $N = 1$, the condition $X^i = 0$ requires that $\partial F / \partial v_i = 0$. We thus get $U = F = F(\mathbf{x}, u)$ and the structure of vertical isovectors does not change except we have to take, of course, $\alpha = 1$. Consequently, the flow arisen from a vertical isovector field is obtained as the solution of the set of ordinary differential equations

$$\frac{d\overline{x}^i}{ds} = 0, \frac{d\overline{u}^\alpha}{ds} = U^\alpha(\overline{x}^j, \overline{u}^\beta), \frac{d\overline{v}_i^\alpha}{ds} = \overline{D}_i^{(0)} U^\alpha(\overline{\mathbf{x}}, \overline{\mathbf{u}}) = \frac{\partial U^\alpha}{\partial \overline{x}^i} + \overline{v}_i^\beta \frac{\partial U^\alpha}{\partial \overline{u}^\beta}$$

under the initial conditions $\overline{x}^i(0)=x^i, \overline{u}^\alpha(0)=u^\alpha, \overline{v}_i^\alpha(0)=v_i^\alpha$. It is clear that $\overline{x}^i(s)=x^i$. As is well-known, this flow is specified by the mapping $\psi_V(s)=e^{sV}:\mathcal{C}_1\to\mathcal{C}_1$. As we had mentioned before, the MacLaurin series of this mapping for very small values of the parameter s about $s=0$ may be expressed as

$$\overline{x}^i(s)=x^i,\ \overline{u}^\alpha(s)=u^\alpha+U^\alpha(x^j,u^\beta)s+o(s),$$
$$\overline{v}_i^\alpha(s)=v_i^\alpha+D_i^{(0)}U^\alpha(x^j,u^\beta)s+o(s)$$

Let us now define the mapping $\phi_V(s)=\psi_V(s)\circ\phi=e^{sV}\circ\phi$ where ϕ is a regular mapping. We obviously get $\phi_V(0)=\phi$ and since

$$\phi_V(s)^*\omega=\phi^*(e^{s\pounds_V}\omega)=0$$

for all $\omega\in\mathcal{I}_1$, the 1-parameter mappings $\phi_V(s)$ are also regular. For a regular mapping $\phi(\mathbf{x})$ and small values of the parameter s, the mapping $\phi_V(s)$ can be represented as follows

$$\overline{x}^i(s)=x^i,\ \overline{\phi}^\alpha(s)=\phi^\alpha+U^\alpha(x^j,\phi^\beta)s+o(s),$$
$$\overline{\phi}_{,i}^\alpha(s)=\phi_{,i}^\alpha+\Big(\frac{\partial}{\partial x^i}+\phi_{,i}^\beta\frac{\partial}{\partial\phi^\beta}\Big)U^\alpha(x^j,\phi^\gamma)s+o(s).$$

Let us now define a ***variation operator*** δ_V such that the ***variation*** $\delta_V\phi$ means that

$$\delta_V\phi^\alpha=U^\alpha(x^j,\phi^\beta)=\phi^*\mathbf{i}_V(du^\alpha),$$
$$\delta_V\phi_{,i}^\alpha=D_i^{(0)}U^\alpha(x^j,\phi^\beta)=\phi^*\mathbf{i}_V(dv_i^\alpha).$$

It is immediately seen that $\delta_V\phi_{,i}^\alpha=D_i^{(0)}\delta_V\phi^\alpha$. Therefore, the operators of variation and partial differentiation commute. We can thus write

$$\overline{\phi}^\alpha(s)=\phi^\alpha+(\delta_V\phi^\alpha)s+o(s),\ \overline{\phi}_{,i}^\alpha(s)=\phi_{,i}^\alpha+(D_i^{(0)}\delta_V\phi^\alpha)s+o(s).$$

The action functional $A:\mathcal{R}(B_n)\to\mathbb{R}$ becomes stationary at a regular mapping ϕ if and only if the ***variation of the functional*** *$A(\phi)$ satisfies the condition*

$$\delta_V A(\phi)=\lim_{s\to0}\frac{A(\phi_V(s))-A(\phi)}{s}=0 \tag{10.2.4}$$

for every vertical isovector field V verifying the boundary conditions $\delta_V\phi^\alpha|_{\partial B_n}=0$. In fact, under this circumstances, the functional $A\big(\phi_V(s)\big)$ that may be expressible in the neighbourhood of the mapping ϕ for small values of the parameter s as

$$A(\phi_V(s)) = \int_{B_n} \phi^* \left[e^{s\pounds_V}(L\mu)\right] = A(\phi) + \delta_V A(\phi)\, s + o(s),$$

does not experience a change of first order and it is reduced to the form

$$A(\phi_V(s)) = A(\phi) + o(s).$$

Let $n \geq 1$. If $o(s) \simeq s^{2n}$ then ϕ corresponds to an extremum point while if $o(s) \simeq s^{2n+1}$ then to an inflection point. The restriction $\delta_V \phi^\alpha|_{\partial B_n} = 0$ on the boundary means that admissible functions ϕ^α take on specified values on ∂B_n and all varied functions $\phi^\alpha + (\delta_V \phi^\alpha)s$ assume the same values on the boundary.

10.3. EULER-LAGRANGE EQUATIONS

Let us consider the action functional

$$A(\phi) = \int_{B_n} \phi^*(L\mu)$$

where $L \in \Lambda^0(\mathcal{C}_1)$ corresponding to a regular mapping ϕ. We would like to evaluate the expression (10.2.4) when V is a vertical isovector field of the contact ideal. We easily obtain

$$\begin{aligned}\delta_V A(\phi) &= \lim_{s\to 0} \frac{1}{s} \int_{B_n} \left[\phi_V(s)^*(L\mu) - \phi^*(L\mu)\right] \\ &= \lim_{s\to 0} \int_{B_n} \frac{\phi^*(e^{s\pounds_V} - 1)(L\mu)}{s}\end{aligned}$$

from which we immediately arrive at the result

$$\delta_V A(\phi) = \int_{B_n} \phi^* \pounds_V(L\mu).$$

We shall now try to convert this expression into another one on which we will be able to work more efficiently although it would creates exactly the same effect. We define the ***generalised Cartan n-form*** $J \in \Lambda^n(\mathcal{C}_1)$ [it had been essentially proposed by Cartan as a 1-form within a limited scope. *See* Cartan (1922)] by the relation

$$J = \frac{\partial L}{\partial v_i^\alpha} \sigma^\alpha \wedge \mu_i = \frac{\partial L}{\partial v_i^\alpha}(du^\alpha \wedge \mu_i - v_i^\alpha \mu). \tag{10.3.1}$$

It is clear that $J \in \mathcal{I}_1$ so that we automatically find $\phi^* J = 0$. On the other hand, since V is an isovector of the contact ideal, we have $\pounds_V J \in \mathcal{I}_1$ and we thus obtain $\phi^* \pounds_V J = 0$. Therefore, we are able to introduce the relations

$$A(\phi) = \int_{B_n} \phi^*(L\mu + J), \tag{10.3.2}$$
$$\delta_V A(\phi) = \int_{B_n} \phi^* \pounds_V (L\mu + J).$$

Furthermore, according to the Cartan magic formula employed in evaluating the Lie derivatives, we write

$$\pounds_V (L\mu + J) = \mathbf{i}_V \big(d(L\mu + J)\big) + d\big(\mathbf{i}_V (L\mu + J)\big).$$

If we utilise the Stokes theorem by taking notice of the commutation rule $\phi^* d = d\phi^*$ we arrive at the result

$$\begin{aligned}\delta_V A(\phi) &= \int_{B_n} \big[\phi^*\{\mathbf{i}_V \big(d(L\mu + J)\big)\} + d\{\phi^*\big(\mathbf{i}_V (L\mu + J)\big)\}\big] \\ &= \int_{B_n} \phi^*\{\mathbf{i}_V \big(d(L\mu + J)\big)\} + \int_{\partial B_n} \phi^*\big(\mathbf{i}_V (L\mu + J)\big).\end{aligned}$$

But because $X^i = 0$, we get $\mathbf{i}_V \mu = 0$ and $\mathbf{i}_V \mu_i = 0$ so that we obtain

$$\begin{aligned}\phi^*\big(\mathbf{i}_V (L\mu + J)\big)\big|_{\partial B_n} &= \phi^*\Big(\frac{\partial L}{\partial v_i^\alpha} U^\alpha \mu_i\Big)\Big|_{\partial B_n} \\ &= \phi^*\Big(\frac{\partial L}{\partial v_i^\alpha} \mu_i\Big)\, \delta_V \phi^\alpha\Big|_{\partial B_n} = 0\end{aligned}$$

since $\delta_V \phi^\alpha$ vanishes on the boundary. We thus conclude that

$$\delta_V A(\phi) = \int_{B_n} \phi^*\{\mathbf{i}_V \big(d(L\mu + J)\big)\}. \tag{10.3.3}$$

By making use of the relation $d\sigma^\alpha = -\, dv_i^\alpha \wedge dx^i$, we can now introduce the *Cartan* $(n+1)$*-form* C as follows

$$\begin{aligned}C = d(L\mu + J) = dL \wedge \mu + dJ &= \frac{\partial L}{\partial u^\alpha} du^\alpha \wedge \mu + \frac{\partial L}{\partial v_i^\alpha} dv_i^\alpha \wedge \mu \\ &\quad + d\Big(\frac{\partial L}{\partial v_i^\alpha}\Big) \wedge \sigma^\alpha \wedge \mu_i - \frac{\partial L}{\partial v_i^\alpha} dv_i^\alpha \wedge \mu \\ &= \sigma^\alpha \wedge E_\alpha\end{aligned}$$

where the n-forms

$$E_\alpha = \frac{\partial L}{\partial u^\alpha}\,\mu - d\Big(\frac{\partial L}{\partial v_i^\alpha}\Big) \wedge \mu_i \in \Lambda^n(K_1) \tag{10.3.4}$$

will henceforth be called as ***Euler-Lagrange forms***.

Theorem 10.3.1. *A regular mapping $\phi : B_n \to C_1$ described by the relations* (10.2.2) *renders the action functional A stationary if only if the equations*

$$\phi^* E_\alpha = 0$$

are satisfied in all interior points of the region B_n.

Since $X^i = 0$, we can write

$$\mathbf{i}_V(C) = U^\alpha E_\alpha - \sigma^\alpha \wedge \mathbf{i}_V(E_\alpha)$$

Hence, because of the relations $\phi^*\sigma^\alpha = 0$, we find from (10.3.3) that

$$\delta_V A(\phi) = \int_{B_n} \phi^* U^\alpha\, \phi^* E_\alpha = \int_{B_n} \delta_V \phi^\alpha\, \phi^* E_\alpha.$$

If $\phi^* E_\alpha = 0$, we naturally get $\delta_V A(\phi) = 0$ for all vertical isovectors. Conversely, let us assume that the condition $\delta_V A(\phi) = 0$ holds *for all vertical isovectors*. Then, it can easily be shown that the equations

$$\begin{aligned}\phi^* E_\alpha &= \Big[\frac{\partial L}{\partial \phi^\alpha} - \frac{\partial}{\partial x^i}\Big(\frac{\partial L}{\partial \phi^\alpha_{,i}}\Big)\Big]\mu \\ &= \mathcal{E}_\alpha(L)\mu = 0\end{aligned}$$

are to be satisfied at every interior point of B_n due to the fact that the integrand above is a continuous function (this is known as the fundamental lemma of the calculus of variations). Consequently, the action functional becomes stationary for functions $u^\alpha = \phi^\alpha(\mathbf{x})$ where $\mathbf{x} \in \mathring{B}_n$ if and only if they are determined as solutions of the following N second order quasilinear partial differential equations called ***Euler-Lagrange equations***

$$\begin{aligned}\mathcal{E}_\alpha(L) &= \frac{\partial L}{\partial u^\alpha} - \frac{\partial}{\partial x^i}\Big(\frac{\partial L}{\partial u^\alpha_{,i}}\Big) = \frac{\partial L}{\partial u^\alpha} - \mathcal{D}_i\Big(\frac{\partial L}{\partial u^\alpha_{,i}}\Big) \\ &= \frac{\partial L}{\partial u^\alpha} - \frac{\partial^2 L}{\partial u^\alpha_{,i}\partial x^i} - \frac{\partial^2 L}{\partial u^\alpha_{,i}\partial u^\beta}\frac{\partial u^\beta}{\partial x^i} - \frac{\partial^2 L}{\partial u^\alpha_{,i}\partial u^\beta_{,j}}\frac{\partial^2 u^\beta}{\partial x^i \partial x^j} = 0\end{aligned} \tag{10.3.5}$$

and satisfy the prescribed conditions on the boundary ∂B_n. Here, the operator $\mathcal{D}_i$ is defined as $\mathcal{D}_i = \phi^* D_i^{(1)}$ where

$$D_i^{(1)} = \frac{\partial}{\partial x^i} + v_i^\alpha \frac{\partial}{\partial u^\alpha} + v_{ij}^\alpha \frac{\partial}{\partial v_j^\alpha}. \qquad \square$$

Instead of a vertical isovector V, let us now consider any isovector $U = X^i \partial/\partial x^i + V$ of the contact ideal $\mathcal{I}_1$. Let us suppose that a regular mapping $\phi : B_n \to \mathcal{C}_1$ is satisfying the equations $\phi^* E_\alpha = 0$. Since we have now $X^i \neq 0$, we find that

$$\mathbf{i}_U(L\mu + J) = \Big[\Big(L\delta_j^i - v_j^\alpha \frac{\partial L}{\partial v_i^\alpha}\Big)X^j + \frac{\partial L}{\partial v_i^\alpha} U^\alpha\Big]\mu_i - X^j \frac{\partial L}{\partial v_i^\alpha}\sigma^\alpha \wedge \mu_{ji}.$$

Therefore, $\phi^*\big(\mathbf{i}_U(L\mu + J)\big)$ is no longer zero on ∂B_n. In this case, the variation of the action functional A must be calculated as follows

$$\delta_U A(\phi) = \int_{\partial B_n} \phi^*\big(\mathbf{i}_U(L\mu + J)\big).$$

Some properties of Euler-Lagrange forms can be readily observed:
(i). *Euler-Lagrange forms are balance n-forms.*
In fact, if we write

$$\Sigma_\alpha^i(x^j, u^\beta, v_j^\beta) = -\frac{\partial L}{\partial v_i^\alpha},$$
$$\Sigma_\alpha(x^j, u^\beta, v_j^\beta) = \frac{\partial L}{\partial u^\alpha}$$

we obtain the balance forms

$$E_\alpha = d\Sigma_\alpha^i \wedge \mu_i + \Sigma_\alpha \mu \in \Lambda^n(\mathcal{C}_1)$$

and Euler-Lagrange equations becomes expressible in the form of balance equations

$$\frac{\partial \Sigma_\alpha^i}{\partial x^i} + \Sigma_\alpha = 0.$$

However, the converse statement is generally not true, that is, every balance forms cannot be represented as the Euler-Lagrange forms.

(ii). *The balance n-forms $\omega_\alpha = d\Sigma_\alpha^i \wedge \mu_i + \Sigma_\alpha \mu$ are to be the Euler-Lagrange forms corresponding to a variational principle if and only if the $(n+1)$-form $\Omega \wedge \mu$ is closed. The 1-form Ω is defined by*

$$\Omega = -\Sigma_\alpha^i \, dv_i^\alpha + \Sigma_\alpha \, du^\alpha \in \Lambda^1(\mathcal{C}_1). \tag{10.3.6}$$

If the forms ω_α are Euler-Lagrange forms we can write

$$\Omega \wedge \mu = \Big(\frac{\partial L}{\partial v_i^\alpha} dv_i^\alpha + \frac{\partial L}{\partial u^\alpha} du^\alpha \Big) \wedge \mu = dL \wedge \mu = d(L\mu)$$

because $dx^i \wedge \mu = 0$. Thus the form $\Omega \wedge \mu$ is closed. Conversely, let us assume that the form $\Omega \wedge \mu$ is closed, namely, $d(\Omega \wedge \mu) = 0$. Since we are occupied in the Euclidean manifold, the Poincaré lemma will be valid. Thus, there exists a form $\lambda \in \Lambda^n(\mathcal{C}_1)$ such that $\Omega \wedge \mu = d\lambda$. μ is a simple n-form. As a result, the form $\Omega \wedge \mu$ and thus $d\lambda$ become simple $(n+1)$-forms. Let us write $\lambda = L\mu + \Lambda$ where $L \in \Lambda^0(\mathcal{C}_1)$. The form $\Lambda \in \Lambda^n(\mathcal{C}_1)$ cannot contain the form μ. In this case, we have to write $d\lambda = dL \wedge \mu + d\Lambda = \Omega \wedge \mu$. But, we get $d\Lambda = 0$ because μ is not present in the form Λ. We thus obtain $\Omega \wedge \mu = dL \wedge \mu$ and

$$\Sigma_\alpha^i = -\frac{\partial L}{\partial v_i^\alpha}, \quad \Sigma_\alpha = \frac{\partial L}{\partial u^\alpha}.$$

Hence, the forms ω_α are Euler-Lagrange forms.

Example 10.3.1. Let us consider one dimensional non-linear diffusion equation

$$\frac{\partial^2 u}{\partial x^2} = \kappa(u)\frac{\partial u}{\partial t}, \quad \kappa(u) \neq 0.$$

On denoting $x^1 = x$, $x^2 = t$, $v_1 = u_x$, $v_2 = u_t$, $\mu = dx \wedge dt$, $\mu_1 = dt$ and $\mu_2 = -dx$, the balance form producing this equation can be written as

$$\omega = dv_1 \wedge dt + dK \wedge dx = dv_1 \wedge \mu_1 - dK \wedge \mu_2$$

where we defined $\kappa(u) = K'(u)$. We thus have $\Sigma^1 = v_1$, $\Sigma^2 = -K(u)$, $\Sigma = 0$ and find

$$\Omega = -v_1 dv_1 + K(u)\, dv_2 = -\frac{1}{2}\, d(v_1^2) + K(u)\, dv_2.$$

Therefore, we obtain $d(\Omega \wedge \mu) = d\Omega \wedge \mu = \kappa(u)\, du \wedge dv_2 \wedge dx \wedge dt \neq 0$ which implies that this balance form cannot be derived from a variational principle. ■

We shall now try to show that every balance forms that cannot be generated by a variational principle can be transformed into Euler-Lagrange forms by suitably enlarging that system. We first assume that an arbitrary system of balance forms

$$\omega_\alpha = d\Sigma_\alpha^i \wedge \mu_i + \Sigma_\alpha \mu$$

is given on the contact manifold $\mathcal{C}_1$ whose coordinate cover is $\{x^i, u^\alpha, v_i^\alpha\}$.

In the first step, let us extend the $(n+N+nN)$-dimensional manifold $\mathcal{C}_1$ to an $(n+2N+2nN)$-dimensional manifold $\overline{\mathcal{C}}_1$ by introducing the *adjoint variables* w^α and w_i^α so that its coordinate cover is $\{x^i, u^\alpha, v_i^\alpha, w^\alpha, w_i^\alpha\}$. We then define, respectively, the additional contact and balance forms depending also on adjoint variables

$$\overline{\sigma}^\alpha = dw^\alpha - w_i^\alpha dx^i, \quad \overline{\omega}_\alpha = dW_\alpha^i \wedge \mu_i + W_\alpha \mu, \quad W_\alpha^i, W_\alpha \in \Lambda^0(\overline{\mathcal{C}}_1)$$

where the functions W_α^i and W_α are undetermined as yet. We can thus easily prove the following theorem.

Theorem 10.3.2. *The enlarged balance system* $(\omega_\alpha, \overline{\omega}_\alpha)$ *admits a variational principle if the additional functions* W_α^i *and* W_α *are chosen as follows*

$$\begin{aligned} W_\alpha^i &= -\frac{\partial}{\partial v_i^\alpha}\big(-\Sigma_\beta^j\, w_j^\beta + \Sigma_\beta\, w^\beta\big), \\ W_\alpha &= \frac{\partial}{\partial u^\alpha}\big(-\Sigma_\beta^i\, w_i^\beta + \Sigma_\beta\, w^\beta\big). \end{aligned}$$

Let us introduce a function $L \in \Lambda^0(\overline{\mathcal{C}}_1)$ by the relation

$$L = -\Sigma_\alpha^i\, w_i^\alpha + \Sigma_\alpha\, w^\alpha. \tag{10.3.7}$$

Obviously L depends linearly on the adjoint variables. This definition leads directly to the results

$$\Sigma_\alpha^i = -\frac{\partial L}{\partial w_i^\alpha}, \quad \Sigma_\alpha = \frac{\partial L}{\partial w^\alpha}, \quad W_\alpha^i = -\frac{\partial L}{\partial v_i^\alpha}, \quad W_\alpha = \frac{\partial L}{\partial u^\alpha}.$$

Hence, the form Ω defined above and associated with the enlarged system becomes

$$\begin{aligned} \Omega &= -\Sigma_\alpha^i\, dw_i^\alpha - W_\alpha^i\, dv_i^\alpha + \Sigma_\alpha\, dw^\alpha + W_\alpha\, du^\alpha \\ &= \frac{\partial L}{\partial w_i^\alpha}\, dw_i^\alpha + \frac{\partial L}{\partial v_i^\alpha}\, dv_i^\alpha + \frac{\partial L}{\partial w^\alpha}\, dw^\alpha + \frac{\partial L}{\partial u^\alpha}\, du^\alpha \\ &= dL - \frac{\partial L}{\partial x^i}\, dx^i. \end{aligned}$$

in view of the definitions given above. It then follows that

$$\Omega \wedge \mu = dL \wedge \mu = d(L\mu)$$

which means that the form $\Omega \wedge \mu$ is now closed. Consequently, the balance forms $(\omega_\alpha, \overline{\omega}_\alpha)$ are Euler-Lagrange forms corresponding to a variational principle that can be written as

$$E_\alpha = \frac{\partial L}{\partial w^\alpha}\,\mu - d\Big(\frac{\partial L}{\partial w_i^\alpha}\Big) \wedge \mu_i = d\Sigma_\alpha^i \wedge \mu_i + \Sigma_\alpha \mu,$$

$$\overline{E}_\alpha = \frac{\partial L}{\partial u^\alpha}\,\mu - d\Big(\frac{\partial L}{\partial v_i^\alpha}\Big) \wedge \mu_i = dW_\alpha^i \wedge \mu_i + W_\alpha \mu$$

$$= d\Big(\frac{\partial \Sigma_\beta^j}{\partial v_i^\alpha}\, w_j^\beta - \frac{\partial \Sigma_\beta}{\partial v_i^\alpha} w^\beta\Big) \wedge \mu_i + \Big(\frac{\partial \Sigma_\beta}{\partial u^\alpha} w^\beta - \frac{\partial \Sigma_\beta^i}{\partial u^\alpha} w_i^\beta\Big) \wedge \mu$$

Therefore, if $\phi : B_n \to \overline{\mathcal{C}}_1$ is a regular mapping,, that is, if it satisfies the conditions $\phi^*\sigma^\alpha = 0$, $\phi^*\overline{\sigma}^\alpha = 0$, $\phi^*\mu \neq 0$, then the functions $u^\alpha = \phi^\alpha(\mathbf{x})$, $w^\alpha = \psi^\alpha(\mathbf{x})$ that render the action functional

$$A(\phi) = \int_{B_n} \phi^*\big[L(x^i, u^\alpha, v_i^\alpha, w^\alpha, w_i^\alpha)\mu\big]$$

stationary are found as solutions of differential equations

$$\frac{\partial L}{\partial w^\alpha} - \frac{\partial}{\partial x^i}\Big(\frac{\partial L}{\partial w_{,i}^\alpha}\Big) = 0, \quad \frac{\partial L}{\partial u^\alpha} - \frac{\partial}{\partial x^i}\Big(\frac{\partial L}{\partial u_{,i}^\alpha}\Big) = 0 \qquad (10.3.8)$$

corresponding to $\phi^* E_\alpha = 0$ and $\phi^*\overline{E}_\alpha = 0$, respectively. The Lagrangian function L is given by (10.3.7). Generally non-linear partial differential equations $(10.3.8)_1$ depend only on x^i, u^α and $u_{,i}^\alpha$. They are exactly in the form of balance equations and determine only the functions u^α whereas $(10.3.8)_2$ yield the *adjoint equations*

$$\frac{\partial \Sigma_\beta^j}{\partial u_{,i}^\alpha}\frac{\partial^2 w^\beta}{\partial x^i \partial x^j} + \Big[\mathcal{D}_i\Big(\frac{\partial \Sigma_\beta^j}{\partial u_{,i}^\alpha}\Big) - \frac{\partial \Sigma_\beta^j}{\partial u^\alpha} - \frac{\partial \Sigma_\beta}{\partial u_{,i}^\alpha}\delta_i^j\Big]\frac{\partial w^\beta}{\partial x^j} \qquad (10.3.9)$$

$$+ \Big[\frac{\partial \Sigma_\beta}{\partial u^\alpha} - \mathcal{D}_i\Big(\frac{\partial \Sigma_\beta}{\partial u_{,i}^\alpha}\Big)\Big] w^\beta = 0$$

where the operator $\mathcal{D}_i$ was introduced on *p.* 663. The equations (10.3.9) are linear second order partial differential equations to determine the functions w^α since their coefficients are now, in principle, known functions in terms of $u^\alpha(\mathbf{x})$. □

Example 10.3.2. Let us reconsider the field equations taken into account in Example 10.3.1 on the enlarged manifold $\overline{\mathcal{C}}_1$ this time. With the Lagrangian function

$$L = -v_1 w_1 + K(u) w_2$$

we then obtain

$$W^1 = -\frac{\partial L}{\partial v_1} = w_1,\ W^2 = -\frac{\partial L}{\partial v_2} = 0,\ W = \frac{\partial L}{\partial u} = \kappa(u)w_2.$$

Hence, the adjoint balance form becomes

$$\overline{E} = dw_1 \wedge \mu_1 + \kappa(u)\, w_2\, \mu = dw_1 \wedge dt + \kappa(u)\, w_2\, dx \wedge dt$$

and Euler-Lagrange equations corresponding to the enlarged system are found as

$$\frac{\partial^2 u}{\partial x^2} - \kappa(u)\frac{\partial u}{\partial t} = 0,\ \frac{\partial^2 w}{\partial x^2} + \kappa(u)\frac{\partial w}{\partial t} = 0.$$

If we determine the function u from the first equation, then we see that the adjoint equation turns out to be linear in terms of w. Solutions of these equations stationarises the functional

$$A(u,w) = \int_{B_2}\Big(-\frac{\partial u}{\partial x}\frac{\partial w}{\partial x} + K(u)\frac{\partial w}{\partial t}\Big)dxdt. \quad \blacksquare$$

We shall now try to obtain another relation associated with a regular mapping $\phi \in \mathcal{R}(B_n)$ satisfying the Euler-Lagrange equations. Let us introduce the tensor

$$T^i_j(L) = v^\alpha_j \frac{\partial L}{\partial v^\alpha_i} - L\delta^i_j \tag{10.3.10}$$

that depends linearly on the Lagrangian function L. Inspired by the analytical mechanics, we shall also call (10.3.10) as the ***energy-momentum tensor***. *For any regular mapping ϕ, the following identity holds*:

$$\phi^*(v^\alpha_i E_\alpha + dT^j_i \wedge \mu_j) = -\phi^*\frac{\partial L}{\partial x^i}\mu. \tag{10.3.11}$$

Indeed, routine calculations yield

$$\begin{aligned} v^\alpha_i E_\alpha + dT^j_i \wedge \mu_j &= v^\alpha_i \frac{\partial L}{\partial u^\alpha}\mu - v^\alpha_i d\Big(\frac{\partial L}{\partial v^\alpha_j}\Big) \wedge \mu_j + \frac{\partial L}{\partial v^\alpha_j} dv^\alpha_i \wedge \mu_j \\ &\quad + v^\alpha_i d\Big(\frac{\partial L}{\partial v^\alpha_j}\Big) \wedge \mu_j - dL \wedge \mu_i \\ &= \Big(v^\alpha_i \frac{\partial L}{\partial u^\alpha} - \frac{\partial L}{\partial x^i}\Big)\mu - \frac{\partial L}{\partial u^\alpha} du^\alpha \wedge \mu_i + \frac{\partial L}{\partial v^\alpha_j}(dv^\alpha_i \wedge \mu_j - dv^\alpha_j \wedge \mu_i). \end{aligned}$$

We thus obtain

$$\phi^*(v_i^\alpha E_\alpha + dT_i^j \wedge \mu_j) = \Big(u_{,i}^\alpha \phi^* \frac{\partial L}{\partial u^\alpha} - \phi^* \frac{\partial L}{\partial x^i}\Big)\mu - \phi^* \frac{\partial L}{\partial u^\alpha} u_{,i}^\alpha \mu$$
$$+ \phi^* \frac{\partial L}{\partial v_j^\alpha}(u_{,ij} - u_{,ji})\mu = -\, \phi^* \frac{\partial L}{\partial x^i}\mu.$$

If the mapping ϕ is making the action functional stationary, then we ought to get $\phi^* E_\alpha = 0$. Therefore, the foregoing relations lead in that case to

$$\phi^* dT_i^j \wedge \mu_j = d\phi^* T_i^j \wedge \mu_j = \frac{\partial(\phi^* T_i^j)}{\partial x^j}\mu = -\, \phi^* \frac{\partial L}{\partial x^i}\mu$$

where partial derivatives with respect to the variables x^j are actually designating total derivatives with respect to those variables. We finally arrive at the relations

$$\frac{\partial(\phi^* T_i^j)}{\partial x^j} = \mathcal{D}_j(\phi^* T_i^j) = \frac{\partial T_i^j}{\partial x^j} + u_{,j}^\alpha \frac{\partial T_i^j}{\partial u^\alpha} + u_{,jk}^\alpha \frac{\partial T_i^j}{\partial u_{,k}^\alpha} = -\, \phi^* \frac{\partial L}{\partial x^i}.$$

If the function L does not depend explicitly on the variables x^i, then we get $\partial L/\partial x^i = 0$ so that we reach to the result

$$\frac{\partial(\phi^* T_i^j)}{\partial x^j} = 0.$$

We thus produce n conservation laws [*see* (8.7.3)] associated with this variational principle. However, if the function L is explicitly dependent on the variables x^i, then we plainly conclude that the conserved quantities are $\phi^*(T_i^j + L\delta_i^j)$.

Example 10.3.3. We choose the Lagrangian function as

$$L = \frac{1}{2}\sum_{i=1}^{n} v_i^2$$

where $v_i = \partial u/\partial x^i$ with $1 \le i \le n$. Since we have

$$\frac{\partial L}{\partial x^i} = 0, \quad \frac{\partial L}{\partial u} = 0, \quad \frac{\partial L}{\partial v_i} = v_i,$$

the corresponding Euler-Lagrange equation turns out to be none other than the familiar Laplace equation

$$-\, u_{,ii} = -\, \nabla^2 u = 0.$$

The energy-momentum tensor

$$T_{ij} = v_i v_j - \tfrac{1}{2} v_k v_k \delta_{ij}$$

associated with this case is obviously symmetric and the conservation law takes the form

$$\frac{\partial}{\partial x^j}(u_{,i}u_{,j} - \tfrac{1}{2}u_{,k}u_{,k}\delta_{ij}) = 0.$$

Hence, if a function u satisfies the Laplace equation inside a region B_n, then the relation

$$\int_{\partial B_n} (u_{,i}u_{,j}n_j - \tfrac{1}{2}u_{,k}u_{,k}n_i)dS = 0$$

must hold on the boundary ∂B_n of that region. This relation may also be represented in the form

$$\int_{\partial B_n} \Big(\frac{\partial u}{\partial n}\boldsymbol{\nabla} u - \frac{1}{2}|\boldsymbol{\nabla} u|^2 \mathbf{n}\Big) ds = 0$$

where $\partial u/\partial n$ denotes the normal derivative $u_{,j}n_j$. ■

10.4. NOETHERIAN VECTOR FIELDS

Let us consider a group of transformations acting on the graph space G replacing the variables x^i, u^α by the new variables $\overline{x}^i, \overline{u}^\alpha$. If this group leaves the action integral

$$A(\mathbf{u}) = \int_{B_n} L\Big(x^i, u^\alpha, \frac{\partial u^\alpha}{\partial x^i}\Big) dx^1 dx^2 \cdots dx^n$$

invariant, that is, if it enforces the satisfaction of the numerical equality

$$\int_{B_n} L\Big(x^i, u^\alpha, \frac{\partial u^\alpha}{\partial x^i}\Big) dx^1 \cdots dx^n = \int_{\overline{B}_n} L\Big(\overline{x}^i, \overline{u}^\alpha, \frac{\partial \overline{u}^\alpha}{\partial \overline{x}^i}\Big) d\overline{x}^1 \cdots d\overline{x}^n,$$

then it is called a ***variational symmetry group***. That a variational symmetry anticipating the transformations of independent variables as well gives rise to a conservation law is established by the Noether theorem in its most illustrative form [German mathematician Emmy Amalie Noether (1882-1935)]. This theorem has proven to be a principal agent in revealing important results in many physical areas. In this section, we shall try to discuss the Noether theorem and some of its generalisations by an approach based on exterior analysis.

Let $U \in T(\mathcal{C}_1)$ be an isovector field of the contact ideal $\mathcal{I}_1$ and $\phi : B_n \to \mathcal{C}_1$ be a regular mapping satisfying the condition (10.2.1). We consider the action functional

$$A(\phi) = \int_{B_n} \phi^*(L\mu + J)$$

where the form J is given by (10.3.1). An *isovector field* U of the contact ideal $\mathcal{I}_1$ is a ***Noetherian vector field*** if and only if the condition

$$\pounds_U(L\mu + J) = 0 \tag{10.4.1}$$

is fulfilled. We denote the set of all Noetherian vector fields associated with a given Lagrangian function $L \in \Lambda^0(\mathcal{C}_1)$ by $\mathcal{N}(L, \mathcal{C}_1)$. It is quite clear that *the form $L\mu + J \in \Lambda^n(\mathcal{C}_1)$ will be invariant with respect to a Noetherian vector field.*

An *isovector field* U of the contact ideal $\mathcal{I}_1$ is called a ***Noetherian vector field of the first kind*** if and only if it satisfies the condition

$$\pounds_U(L\mu + J) \in \mathcal{I}_1 \tag{10.4.2}$$

We denote the set of all Noetherian vector fields of the first kind associated with a given Lagrangian function L by $\mathcal{N}_1(L, \mathcal{C}_1)$.

An *isovector field* U of the contact ideal $\mathcal{I}_1$ is called a ***Noetherian vector field of the second kind*** if and only if it satisfies the condition

$$\pounds_U(L\mu + J) - d\Omega(U) \in \mathcal{I}_1 \tag{10.4.3}$$

for a form $\Omega(U) \in \Lambda^{n-1}(\mathcal{C}_1)$ depending on the vector U. We denote the set of all Noetherian vector fields of the second kind associated with a given Lagrangian function L by $\mathcal{N}_2(L, \mathcal{C}_1)$.

Theorem 10.4.1. *The sets $\mathcal{N}(L, \mathcal{C}_1)$, $\mathcal{N}_1(L, \mathcal{C}_1)$ and $\mathcal{N}_2(L, \mathcal{C}_1)$ constitute Lie subalgebras on $\mathbb{R}$ of the Lie algebra of isovectors of the contact ideal $\mathcal{I}_1$. These Lie algebras satisfy the set inclusion relations below*

$$\mathcal{N}(L, \mathcal{C}_1) \subset \mathcal{N}_1(L, \mathcal{C}_1) \subset \mathcal{N}_2(L, \mathcal{C}_1).$$

If $U_1, U_2 \in \mathcal{N}_2(L, \mathcal{C}_1)$, then the following rules for $\Omega(U)$ must be valid

$$\begin{aligned} &d\Omega(a_1U_1 + a_2U_2) - d\big[a_1\Omega(U_1) + a_2\Omega(U_2)\big] \in \mathcal{I}_1, \\ &d\Omega([U_1, U_2]) - d\big[\pounds_{U_1}\Omega(U_2) - \pounds_{U_2}\Omega(U_1)\big] \in \mathcal{I}_1 \end{aligned} \tag{10.4.4}$$

where $a_1, a_2 \in \mathbb{R}$.

The set inclusion relations are immediately observable from the definitions. Indeed, by taking $d\Omega = 0$ in $\mathcal{N}_2(L, \mathcal{C}_1)$ we obtain $\mathcal{N}_1(L, \mathcal{C}_1)$ from which we get, in turn, $\mathcal{N}(L,\mathcal{C}_1)$ by simply considering only the zero n-form in $\mathcal{I}_1$. Because of the inclusion relations, it suffices to show that $\mathcal{N}_2(L,\mathcal{C}_1)$ is a Lie subalgebra.

Let us assume that $U_1, U_2 \in \mathcal{N}_2(L,\mathcal{C}_1)$ and $a_1, a_2 \in \mathbb{R}$. Since we can write

$$£_{a_1U_1+a_2U_2}(L\mu + J) = (a_1£_{U_1} + a_2£_{U_2})(L\mu + J)$$

we obtain

$$£_{a_1U_1+a_2U_2}(L\mu + J) - a_1 d\Omega(U_1) - a_2 d\Omega(U_2) \in \mathcal{I}_1.$$

According to $(10.4.4)_1$, we find that $a_1U_1 + a_2U_2 \in \mathcal{N}_2(L,\mathcal{C}_1)$. Therefore, $\mathcal{N}_2(L,\mathcal{C}_1)$ becomes a linear vector space on $\mathbb{R}$. On the other hand, in view of (5.11.12), we are able to write

$$£_{[U_1,U_2]}(L\mu + J) = (£_{U_1}£_{U_2} - £_{U_2}£_{U_1})(L\mu + J).$$

Since the operators $£_U$ and d commute, we get

$$£_{[U_1,U_2]}(L\mu + J) - d\big[£_{U_1}\Omega(U_2) - £_{U_2}\Omega(U_1)\big] \in \mathcal{I}_1.$$

If we note $(10.4.4)_2$, we are led to $[U_1,U_2] \in \mathcal{N}_2(L,\mathcal{C}_1)$. This proves that, the vector space $\mathcal{N}_2(L,\mathcal{C}_1)$ is a Lie algebra. In a similar fashion, we can easily show that vector spaces $\mathcal{N}_1(L,\mathcal{C}_1)$ and $\mathcal{N}(L,\mathcal{C}_1)$ are also Lie algebras. □

Let U be an isovector field of the contact ideal $\mathcal{I}_1$. The value of the action functional under the group of transformation, or under the flow, generated by this vector on the manifold $\mathcal{C}_1$ may be computed through the relation

$$A(\phi_U(s)) = \int_{B_n} \phi^*\big[e^{s£_U}(L\mu + J)\big].$$

We represent the isovector field by

$$U = X^i\frac{\partial}{\partial x^i} + U^\alpha\frac{\partial}{\partial u^\alpha} + V_i^\alpha\frac{\partial}{\partial v_i^\alpha}.$$

We know that the isovector components are given by

$$\begin{aligned} &X^i = X^i(\mathbf{x},\mathbf{u}), \quad U^\alpha = U^\alpha(\mathbf{x},\mathbf{u}),\\ &V_i^\alpha = D_i^{(0)}(U^\alpha - v_j^\alpha X^j),\ D_i^{(0)} = \frac{\partial}{\partial x^i} + v_i^\alpha\frac{\partial}{\partial u^\alpha} \end{aligned}$$

when $N > 1$. By making use of (10.3.2) we can write

$$A(\phi_U(s)) - A(\phi) = \int_{B_n} \phi^* \pounds_U (L\mu + J) + o(s).$$

whence follows the *condition of infinitesimal invariance* as

$$\phi^* \pounds_U (L\mu + J) = \phi^* \pounds_U (L\mu) + \phi^* \pounds_U J = 0$$

Since U is an isovector of the contact ideal, we have $\pounds_U J \in \mathcal{I}_1$. We thus obtain $\phi^* \pounds_U (L\mu + J) = \phi^* \pounds_U (L\mu)$. On the other hand, the relation

$$\pounds_U (L\mu) = \mathbf{i}_U (dL \wedge \mu) + d\,\mathbf{i}_U (L\mu) = U(L)\mu - X^i dL \wedge \mu_i + d(LX^i \mu_i)$$

yields

$$\pounds_U (L\mu) = U(L)\mu + LdX^i \wedge \mu_i = \big[U(L) + LD_i X^i\big]\mu + L\frac{\partial X^i}{\partial u^\alpha}\sigma^\alpha \wedge \mu_i.$$

Consequently, the *criterion for infinitesimal invariance* takes the form

$$\phi^* \pounds_U (L\mu) = \phi^* \big[U(L) + LD_i X^i\big]\mu = 0 \qquad (10.4.5)$$

where the function $\phi^* U(L)$ is obviously defined as

$$\phi^* U(L) = \phi^* \Big[X^i \frac{\partial L}{\partial x^i} + U^\alpha \frac{\partial L}{\partial u^\alpha} + V_i^\alpha \frac{\partial L}{\partial v_i^\alpha}\Big].$$

However, as is clearly emphasised in the following theorem, the global invariance will also be realised if this criterion occurs.

Let us now attempt to reveal the interrelation between Noetherian vector fields and the invariance of the action functional and conservation laws.

Theorem 10.4.2 (The Noether Theorem). *Let $U \in \mathcal{N}(L, \mathcal{C}_1)$ be a Noetherian vector field. For every regular mapping $\phi \in \mathcal{R}(B_n)$, we recover the global invariance*

$$A(\phi_U(s)) = A(\phi).$$

Furthermore, if the regular mapping ϕ satisfies the Euler-Lagrange equations, then the identity

$$d\phi^* \big(\mathbf{i}_U (L\mu + J)\big) = 0$$

is also fulfilled. This identity may be as well written in the form

$$d\phi^* \mathcal{J} = 0$$

*where the **current form** $\mathcal{J} \in \Lambda^{n-1}(\mathcal{C}_1)$ is defined by*

$$\mathcal{J} = \Big(U^\alpha \frac{\partial L}{\partial v_i^\alpha} - X^j T_j^i\Big)\mu_i$$

Therefore, one has $\phi^* \mathcal{J} = J^i \mu_i$. The functions J^i are given by

$$J^i = \phi^*\Big[U^\alpha \frac{\partial L}{\partial v_i^\alpha} - X^j T_j^i\Big] = \phi^*\Big[(U^\alpha - X^j v_j^\alpha)\frac{\partial L}{\partial v_i^\alpha} + LX^i\Big] \tag{10.4.6}$$

Since $U \in \mathcal{N}(L, \mathcal{C}_1)$, the relation $£_U(L\mu + J) = 0$ is satisfied so that we obtain

$$\begin{aligned} A\big(\phi_U(s)\big) - A(\phi) &= \int_{B_n} \phi^*\big[(e^{s£_U} - I)(L\mu + J)\big] \\ &= \int_{B_n} \phi^*\Big(s£_U + \frac{s^2}{2!}£_U^2 + \cdots\Big)(L\mu + J) = 0 \end{aligned}$$

implying that $A\big(\phi_U(s)\big) = A(\phi)$ for every mapping $\phi \in \mathcal{R}(B_n)$. Hence, the integral curves of a Noetherian vector field generate transformation groups eliciting variational symmetries. On the other hand, $£_U(L\mu + J) = 0$ can be expressed explicitly as

$$\mathbf{i}_U\big(d(L\mu + J)\big) + d\big(\mathbf{i}_U(L\mu + J)\big) = 0$$

so that we can write $d\big(\mathbf{i}_U(L\mu + J)\big) = -\,\mathbf{i}_U\big(d(L\mu + J)\big) = -\,\mathbf{i}_U C$. We thus get

$$d\big(\mathbf{i}_U(L\mu + J)\big) = -\,(U^\alpha - v_i^\alpha X^i)E_\alpha + \sigma^\alpha \wedge \mathbf{i}_U(E_\alpha).$$

Let us now assume that $\phi^* E_\alpha = 0$. Since we also have $\phi^* \sigma^\alpha = 0$, the above expression results in

$$\phi^* d\big(\mathbf{i}_U(L\mu + J)\big) = d\phi^*\big(\mathbf{i}_U(L\mu + J)\big) = 0.$$

If we recall the definition (10.3.10), we understand that we can write

$$\mathbf{i}_U(L\mu + J) = \mathcal{J} - X^j \frac{\partial L}{\partial v_i^\alpha}\sigma^\alpha \wedge \mu_{ji} \tag{10.4.7}$$

and we obtain $\phi^*\big(\mathbf{i}_U(L\mu + J)\big) = \phi^* \mathcal{J}$ whence we deduce that $d\phi^* \mathcal{J} = 0$. This relation gives then rise to the integral conservation law below

$$\int_{B_n} d\phi^* \mathcal{J} = \int_{\partial B_n} \phi^* \mathcal{J} = 0.$$

At interior points of the region B_n, the equation $d\phi^*\mathcal{J} = 0$ produces the conservation law in the differential form

$$\frac{\partial J^i}{\partial x^i} = \frac{\partial}{\partial x^i}\phi^*\Big[(U^\alpha - X^j u^\alpha_{,j})\frac{\partial L}{\partial u^\alpha_{,i}} + LX^i\Big] = 0. \qquad \square$$

Theorem 10.4.3. *Let $U \in \mathcal{N}_1(L, \mathcal{C}_1)$. For every regular mapping $\phi \in \mathcal{R}(B_n)$, the global invariance*

$$A(\phi_U(s)) = A(\phi)$$

is valid. Furthermore, if the regular mapping ϕ satisfies Euler-Lagrange equations, then the following identity

$$d\phi^*\mathcal{J} = 0$$

is fulfilled.

If $U \in \mathcal{N}_1(L, \mathcal{C}_1)$, then the relation $\pounds_U(L\mu + J) \in \mathcal{I}_1$ must be satisfied. Since $\phi^*\mathcal{I}_1 = 0$, we evidently obtain exactly the same result as in Theorem 10.4.2. However, although the results of those two theorems seem to be remarkably alike, we have to point out that they possess actually rather different structures because variational symmetry groups are generated by significantly different Noetherian vector fields in Theorems 10.4.2 and 10.4.3. $\square$

Theorem 10.4.4. *Let $U \in \mathcal{N}_2(L, \mathcal{C}_1)$. For every regular mapping $\phi \in \mathcal{R}(B_n)$ the relation*

$$A(\phi_U(s)) = A(\phi) + \int_{\partial B_n} \phi^* \int_0^s e^{t\pounds_U}\Omega(U)\, dt$$

is valid. Furthermore, if the regular mapping ϕ satisfies Euler-Lagrange equations, then the following identity

$$d\phi^*(\mathcal{J} - \Omega) = 0$$

is fulfilled.

If $U \in \mathcal{N}_2(L, \mathcal{C}_1)$, we have the relation $\pounds_U(L\mu + J) - d\Omega(U) = \alpha$ where $\alpha \in \mathcal{I}_1$. On the other hand, utilising the commutation rule $d \circ \pounds_U = \pounds_U \circ d$ we come up with the expression

$$\begin{aligned}(e^{s\pounds_U} - I)(L\mu + J) &= \int_0^s \frac{d}{dt}\big[e^{t\pounds_U}(L\mu + J)\big]\, dt \\ &= \int_0^s e^{t\pounds_U}\pounds_U(L\mu + J)\, dt\end{aligned}$$

$$
\begin{aligned}
&= \int_0^s e^{t\pounds_U} d\Omega(U)\, dt + \int_0^s e^{t\pounds_U} \alpha\, dt \\
&= d\int_0^s e^{t\pounds_U} \Omega(U)\, dt + \int_0^s e^{t\pounds_U} \alpha\, dt.
\end{aligned}
$$

In the last line, we have made use of the plainly observable rule $e^{t\pounds_U} \circ d = d \circ e^{t\pounds_U}$. Because $\phi^*\alpha = 0$, we thus conclude that

$$
\begin{aligned}
A\big(\phi_U(s)\big) &= A(\phi) + \int_{B_n} \phi^*\big[(e^{s\pounds_U} - I)(L\mu + J)\big] \\
&= A(\phi) + \int_{B_n} d\,\phi^* \int_0^s e^{t\pounds_U} \Omega(U)\, dt \\
&= A(\phi) + \int_{\partial B_n} \phi^* \int_0^s e^{t\pounds_U} \Omega(U)\, dt.
\end{aligned}
$$

Employing this time, the relation (10.4.7) we can reach to the expression

$$
\begin{aligned}
\pounds_U(L\mu + J) &= \mathbf{i}_U(\sigma^\alpha)E_\alpha - \sigma^\alpha \wedge \mathbf{i}_U(E_\alpha) + d\big(\mathbf{i}_U(L\mu + J)\big) \\
&= \mathbf{i}_U(\sigma^\alpha)E_\alpha + d\mathcal{J} - X^j \frac{\partial L}{\partial v_i^\alpha}\, d\sigma^\alpha \wedge \mu_{ji} \\
&\quad + \sigma^\alpha \wedge \Big[d\Big(X^j \frac{\partial L}{\partial v_i^\alpha}\Big) \wedge \mu_{ji} - \mathbf{i}_U(E_\alpha)\Big] \\
&= d\Omega + \alpha
\end{aligned}
$$

so that we can write $d(\mathcal{J} - \Omega) + \mathbf{i}_U(\sigma^\alpha)E_\alpha \in \mathcal{I}_1$. This relation helps us to deduce the result

$$
\phi^* d(\mathcal{J} - \Omega) = d\phi^*(\mathcal{J} - \Omega) = 0
$$

whenever $\phi^* E_\alpha = 0$. Then the conservation law in integral form leads to the boundary integral

$$
\int_{\partial B_n} \phi^*(\mathcal{J} - \Omega) = 0
$$

when we employ the Stokes theorem. If we particularly choose $\Omega = \Omega^i \mu_i$, then we obtain the conservation law

$$
\frac{\partial \phi^*(J^i - \Omega^i)}{\partial x^i} = 0
$$

at interior points of B_n. □

Example 10.4.1. As an application of the Noether theorem, we will obtain once more the conservation laws given by the relations (8.7.7) of the equations of motion of an hyperelastic medium in Sec. 8.7 by means of variational symmetries. We employ the notation in Example 8.7.4 and denote the isovector field by

$$U = \Phi_K \frac{\partial}{\partial X_K} + \Psi \frac{\partial}{\partial t} + \Omega_k \frac{\partial}{\partial x_k} + U_{kK} \frac{\partial}{\partial F_{kK}} + U_k \frac{\partial}{\partial v_k}$$

where $F_{kK} = \partial x_k / \partial X_K$, $v_k = \partial x_k / \partial t$. Introducing the Lagrangian function $L = \Sigma(F_{kK}) - \frac{1}{2}\rho_0 v_k v_k = \Sigma(\mathbf{F}) - \frac{1}{2}\rho_0 |\mathbf{v}|^2$, we may define the action functional by

$$A(\phi) = \int_{B_4} \phi^*(L\mu)$$

where $\mu = dX_1 \wedge dX_2 \wedge dX_3 \wedge dt$ is the volume form. Then the Euler-Lagrange equations (10.3.5) give the equations of motion (8.7.4)

$$-\frac{\partial}{\partial X_K}\Big(\frac{\partial \Sigma}{\partial F_{kK}}\Big) + \rho_0 \frac{\partial v_k}{\partial t} = 0$$

in the absence of body forces. The components U_{kK} and U_k of the isovector field U can be written as

$$U_{kK} = \frac{\partial \mathcal{F}_k}{\partial X_K} + F_{lK} \frac{\partial \mathcal{F}_k}{\partial x_l}, \quad U_k = \frac{\partial \mathcal{F}_k}{\partial t} + v_l \frac{\partial \mathcal{F}_k}{\partial x_l}$$

where $\mathcal{F}_k = \Omega_k - F_{kL}\Phi_L - v_k \Psi$. The isovector field U satisfies the criterion for infinitesimal invariance (10.4.5) provided that

$$\phi^*\Big[U_{kK}\frac{\partial \Sigma}{\partial F_{kK}} - \rho_0 U_k v_k + L\Big(\frac{\partial \Phi_K}{\partial X_K} + F_{kK}\frac{\partial \Phi_K}{\partial x_k} + \frac{\partial \Psi}{\partial t} + v_k \frac{\partial \Psi}{\partial x_k}\Big)\Big] = 0.$$

Keeping in mind this relation, we then obtain from (10.4.6)

$$\begin{aligned} J_K &= \frac{\partial \Sigma}{\partial F_{kK}}\Omega_k + \Big(L\delta_{KL} - F_{kL}\frac{\partial \Sigma}{\partial F_{kK}}\Big)\Phi_L - v_k \frac{\partial \Sigma}{\partial F_{kK}}\Psi \\ J_4 &= -\rho_0 v_k \Omega_k + \rho_0 F_{kL} v_k \Phi_L + (L + \rho_0 v_k v_k)\Psi \end{aligned}$$

and the conservation law takes the form

$$\frac{\partial J_K}{\partial X_K} + \frac{\partial J_4}{\partial t} = 0.$$

To find the general expressions satisfying the criterion of infinitesimal invariance for an arbitrary stress potential Σ is next to impossible. However, we can designate some conservation laws just by inspection corresponding to certain isovector fields satisfying the criterion of infinitesimal invariance trivially.

(i). We take $\Phi_K = 0, \Psi = 0, \Omega_k = a_k =$ *constant.* (10.4.5) is clearly satisfied. Since then

$$J_K = \frac{\partial \Sigma}{\partial F_{kK}} a_k, \quad J_4 = -\rho_0 v_k a_k$$

the conservation law is found to be

$$a_k \left[\frac{\partial}{\partial X_K} \left(\frac{\partial \Sigma}{\partial F_{kK}} \right) - \rho_0 \frac{\partial v_k}{\partial t} \right] = 0.$$

Because the coefficients a_k are arbitrary, we thus get

$$\frac{\partial T_{Kk}}{\partial X_K} - \rho_0 \frac{\partial v_k}{\partial t} = 0$$

[*see* $(8.7.7)_2$].

(ii). We take $\Phi_K = 0, \Psi = -A =$ *constant*, $\Omega_k = 0$. (10.4.5) is satisfied automatically. We then get

$$J_K = \frac{\partial \Sigma}{\partial F_{kK}} v_k A, \quad J_4 = -(L + \rho_0 v_k v_k) A = -\left(\Sigma + \frac{1}{2} \rho_0 v_k v_k \right) A$$

so that the conservation law becomes

$$\frac{\partial}{\partial X_K} (T_{Kk} v_k) - \frac{\partial}{\partial t} \left(\Sigma + \frac{1}{2} \rho_0 v_k v_k \right) = 0$$

[*see* $(8.7.7)_1$].

(iii). We take $\Phi_K = A_K =$ *constant*, $\Psi = 0, \Omega_k = 0$. (10.4.5) is satisfied identically and we have

$$J_K = \left(L \delta_{KL} - F_{kL} \frac{\partial \Sigma}{\partial F_{kK}} \right) A_L, \quad J_4 = \rho_0 F_{kL} v_k A_L.$$

Hence, the corresponding conservation laws are

$$\frac{\partial}{\partial X_K} \left[\left(\Sigma - \frac{1}{2} \rho_0 v_k v_k \right) \delta_{KL} - T_{Kk} x_{k,L} \right] + \rho_0 \frac{\partial}{\partial t} (x_{k,L} v_k) = 0$$

[*see.* $(8.7.7)_4$].

(iv). We take $\Phi_K = 0, \Psi = 0, \Omega_k = e_{klm} a_l x_m$ where $a_l =$ *constant.* In this case we obtain

$$U_{kK} = e_{klm}a_l F_{mK}, \quad U_k = e_{klm}a_l v_m$$

and (10.4.5) cannot satisfied unless

$$a_l\Big(e_{klm}F_{mK}\frac{\partial \Sigma}{\partial F_{kK}} - e_{klm}\rho_0 v_m v_k\Big) = a_l e_{klm}F_{mK}\frac{\partial \Sigma}{\partial F_{kK}} = 0.$$

Due to $(8.7.5)_2$, $F_{mK}T_{Kk}$ is symmetric in indices k and m. It is immediate then to observe that the above condition is satisfied and the current components

$$J_K = e_{klm}\frac{\partial \Sigma}{\partial F_{kK}}x_m a_l, \quad J_4 = -\,e_{klm}\rho_0 v_k x_m a_l$$

yield the conservation laws

$$\frac{\partial}{\partial X_K}(e_{klm}x_m T_{Kk}) - \rho_0\frac{\partial}{\partial t}(e_{klm}x_m v_k) = 0$$

[*see* $(8.7.7)_3$]. It is seen that the above introduced invariances of the action functional characterising the motion of an hyperelastic material give rise to the following conservation laws: *invariance under spatial translations leads to the balance of linear momentum, invariance under time translations leads to the conservation of energy, invariance under spatial rotations leads to the balance of angular momentum and invariance under material translations leads to the J-integral of fracture mechanics.* ■

Since variational symmetries leave action functionals invariant, the transformations of the extremals that are solutions of Euler-Lagrange equations under these groups yield solutions of the same equations. Therefore, a variational symmetry group of an action functional becomes likewise a symmetry group of Euler-Lagrange equations arisen from that functional. However, the converse statement is generally not true. It is possible to observe that although Euler-Lagrange equations may possess some symmetry groups, these groups may not correspond to a variational symmetry.

10.5. VARIATIONAL PROBLEM FOR A GENERAL ACTION FUNCTIONAL

As in Sec. 10.2, we consider a functional A associated with functions $u^\alpha(\mathbf{x})$, where $\alpha = 1, \dots, N$, defined on a region $B_n \subseteq \mathbb{R}$. However, this time we shall assume that the Lagrangian function L generating this functional is also dependent on the partial derivatives of functions u^α up to and including the order m. Hence, such a functional is expressible as

$$A(\mathbf{u}) = \int_{B_n} L(x^i, u^\alpha, u^\alpha_{,i_1}, u^\alpha_{,i_1 i_2}, \dots, u^\alpha_{,i_1 i_2 \cdots i_m})\, dx^1 dx^2 \cdots dx^n.$$

As in (9.2.4), we introduce auxiliary variables $u^\alpha_{,i_1\cdots i_r} = v^\alpha_{i_1\cdots i_r}, 0 \le r \le m$, $1 \le i_1, \cdots, i_r \le n$ to construct the contact manifold $\mathcal{C}_m$ whose coordinate cover is now given by $\{x^i, v^\alpha_{i_1 i_2 \cdots i_r} : 0 \le r \le m\}$. The closed contact ideal of this manifold is

$$\mathcal{I}_m = \overline{\mathcal{I}}(\sigma^\alpha, \sigma^\alpha_{i_1}, \sigma^\alpha_{i_1 i_2}, \dots \sigma^\alpha_{i_1 i_2 \cdots i_{m-1}}; d\sigma^\alpha_{i_1 i_2 \cdots i_{m-1}}).$$

As we have done before, we define

$$\sigma^\alpha_{i_1 i_2 \cdots i_r} = dv^\alpha_{i_1 i_2 \cdots i_r} - v^\alpha_{i_1 i_2 \cdots i_r i} dx^i, 0 \le r \le m-1.$$

We denote the set all *regular mappings* from B_n into the manifold $\mathcal{C}_m$ by

$$\mathcal{R}(B_n) = \{\phi : B_n \to \mathcal{C}_m : \phi^* \mathcal{I}_m = 0,\ \phi^* \mu \neq 0\}.$$

Consequently, we must have $\phi^* \sigma^\alpha_{i_1 i_2 \cdots i_r} = 0, 1 \le \alpha \le N, 0 \le r \le m-1$. The condition $\phi^* d\sigma^\alpha_{i_1 \cdots i_{m-1}} = 0$ is satisfied automatically due to the commutation rule $\phi^* d = d\phi^*$. We know that we can write

$$d\sigma^\alpha_{i_1 i_2 \cdots i_{m-1}} \wedge \mu_{i_m} = -\, dv^\alpha_{i_1 \cdots i_{m-1} i_m} \wedge \mu.$$

We thus obviously get $\phi^* dv^\alpha_{i_1 \cdots i_m} = 0$. According to this definitions the regular mappings are *functions depending on the same independent variables and annihilating the ideal* $\mathcal{I}_m$. Hence, if $\phi \in \mathcal{R}(B_n)$, then it must be represented by the relations

$$x^i = x^i, \quad v^\alpha_{i_1 \cdots i_r} = \phi^\alpha_{,i_1 \cdots i_r}(\mathbf{x}), \quad 0 \le r \le m. \qquad (10.5.1)$$

Therefore, the Lagrangian function

$$L(x^i, u^\alpha, v^\alpha_{i_1}, v^\alpha_{i_1 i_2}, \dots, v^\alpha_{i_1 i_2 \cdots i_m}) \in \Lambda^0(\mathcal{C}_m)$$

is dependent only on the coordinates of the manifold $\mathcal{C}_m$. In order to be able to utilise the exterior analysis that involves only first order derivatives, we shall not take into account the relations between arguments of L as types of derivatives when it was pulled back by ϕ. We, therefore, write $\phi^* L$ as follows

$$\phi^* L = L\big(\mathbf{x}, u^\alpha(\mathbf{x}), v^\alpha_{i_1}(\mathbf{x}), v^\alpha_{i_1 i_2}(\mathbf{x}), \dots, v^\alpha_{i_1 i_2 \cdots i_m}(\mathbf{x})\big).$$

But, with the aim of revealing the existing interrelations between these variables, we shall resort to the method of *Lagrange multipliers* introduced by Lagrange to provide an effective strategy to determine extrema of functions subject to constraints. To further simplify the analysis we now define a form $\mathcal{L} \in \Lambda^n(\mathcal{C}_m)$ by the expression below

$$\begin{aligned}\mathcal{L} &= L\mu + \sum_{r=0}^{m-1} \lambda_\alpha^{i_1\cdots i_r i}(\mathbf{x})\, \sigma^\alpha_{i_1 i_2\cdots i_r} \wedge \mu_i + \sum_{r=0}^{m-1} \frac{\partial L}{\partial v^\alpha_{i_1\cdots i_r i}}\, \sigma^\alpha_{i_1 i_2\cdots i_r} \wedge \mu_i \\ &= \Big[L - \sum_{r=0}^{m-1}\Big(\lambda_\alpha^{i_1\cdots i_r i}(\mathbf{x}) + \frac{\partial L}{\partial v^\alpha_{i_1\cdots i_r i}}\Big) v^\alpha_{i_1 i_2\cdots i_r i}\Big]\mu \\ &\quad + \sum_{r=0}^{m-1}\Big(\lambda_\alpha^{i_1\cdots i_r i}(\mathbf{x}) + \frac{\partial L}{\partial v^\alpha_{i_1\cdots i_r i}}\Big) dv^\alpha_{i_1 i_2\cdots i_r} \wedge \mu_i \end{aligned} \tag{10.5.2}$$

where the Lagrange multipliers $\lambda_\alpha^{i_1\cdots i_r i}(\mathbf{x}), 0 \le r \le m-1$ are presently arbitrary functions. Since (10.5.2) implies that $\mathcal{L} - L\mu \in \mathcal{I}_m$, it is clear that $\phi^*\mathcal{L} = \phi^*(L\mu)$. Hence, there will be no harm in writing the action functional in the form

$$A(\phi) = \int_{B_n} \phi^*\mathcal{L}. \tag{10.5.3}$$

In order to construct the regular mappings, we again make use of the isovector fields of the contact ideal. If $N > 1$, we know that such a vector is expressible as

$$U = X^i(\mathbf{x},\mathbf{u})\frac{\partial}{\partial x^i} + U^\alpha(\mathbf{x},\mathbf{u})\frac{\partial}{\partial u^\alpha} + \sum_{r=1}^{m} V^\alpha_{i_1\cdots i_r}\frac{\partial}{\partial v^\alpha_{i_1\cdots i_r}} \tag{10.5.4}$$

[*see* p. 506]. The functions $V^\alpha_{i_1\cdots i_r}$ can be calculated through the recurrence relations

$$V^\alpha_{i_1\cdots i_r i} = D_i^{(r)}(V^\alpha_{i_1\cdots i_r} - v^\alpha_{i_1\cdots i_r j}X^j). \tag{10.5.5}$$

The operator $D_i^{(r)}$ is defined by the relation (9.3.18) and accounts for the *total derivative* of any function $f\big(\mathbf{x}, u^\alpha(\mathbf{x}), v^\alpha_{i_1}(\mathbf{x}), \ldots, v^\alpha_{i_1\cdots i_r}(\mathbf{x})\big)$ with respect to the variable x^i. Under the mapping ϕ, we get

$$\phi^* D_i^{(r)} f = \frac{\partial \phi^* f}{\partial x^i}.$$

Regular mappings can now be selected as flows generated by isovector fields of the contact manifold. However, since we do not wish to change the independent variables, we have to take $X^i = 0$. As we have mentioned before, this property allows us to convey all findings established in this situation directly to the case $N = 1$. We thus consider the ***vertical isovector fields*** in the form

$$V = U^\alpha(\mathbf{x}, \mathbf{u})\frac{\partial}{\partial u^\alpha} + \sum_{r=1}^{m} V^\alpha_{i_1\cdots i_r}\frac{\partial}{\partial v^\alpha_{i_1\cdots i_r}}. \tag{10.5.6}$$

where we have of course the recurrence relations $V^\alpha_{i_1\cdots i_r i} = D^{(r)}_i V^\alpha_{i_1\cdots i_r}$. The flow created by a vertical isovector field will be obtained, as usual, by integrating the set of ordinary differential equations

$$\frac{d\overline{x}^i}{ds} = 0,$$
$$\frac{d\overline{v}^\alpha_{i_1\cdots i_r}}{ds} = V^\alpha_{i_1\cdots i_r}(\overline{x}^j, \overline{u}^\beta, \overline{v}^\alpha_{i_1}, \ldots, \overline{v}^\alpha_{i_1\cdots i_r}),\ \ 0 \le r \le m$$

under the initial conditions $\overline{x}^i(0) = x^i, \overline{v}^\alpha_{i_1\cdots i_r}(0) = v^\alpha_{i_1\cdots i_r}, 0 \le r \le m$. It is obvious that $\overline{x}^i(s) = x^i$. We know that this flow is represented by the mapping $\psi_V(s) = e^{sV} : \mathcal{C}_m \to \mathcal{C}_m$. Expanding this exponential mapping into a Maclaurin series about $s = 0$ and retain only up to the first order terms, we can write

$$\overline{x}^i(s) = x^i,\ \overline{u}^\alpha(s) = u^\alpha + U^\alpha(x^j, u^\beta)s + o(s),$$
$$\overline{v}^\alpha_{i_1\cdots i_r}(s) = v^\alpha_{i_1\cdots i_r} + V^\alpha_{i_1\cdots i_r}s + o(s), \quad 1 \le r \le m$$

for small values of the parameter s. Let us now define the mapping $\phi_V(s) = \psi_V(s) \circ \phi = e^{sV} \circ \phi$ for a regular mapping ϕ. We have $\phi_V(0) = \phi$ and

$$\phi_V(s)^*\omega = \phi^*(e^{s\pounds_V}\omega) = 0$$

for every form $\omega \in \mathcal{I}_m$. We introduce the variation operator δ_V as follows

$$\delta_V\phi^\alpha = \phi^*U^\alpha; \quad \delta_V\phi^\alpha_{,i_1\cdots i_r} = \phi^*V^\alpha_{i_1\cdots i_r},\ 1 \le r \le m\,.$$

The action functional $A : \mathcal{R}(B_n) \to \mathbb{R}$ becomes stationary at a regular mapping ϕ if and only if the variation of the functional $A(\phi)$ vanishes, that is, if the relation

$$\delta_V A(\phi) = \lim_{s\to 0}\frac{A(\phi_V(s)) - A(\phi)}{s} = 0$$

is satisfied for any vertical isovector field V of the contact ideal satisfying the boundary conditions $\delta_V\phi^\alpha_{,i_1\cdots i_r}\big|_{\partial B_n} = 0,\ 0 \le r \le m-1$. These boundary conditions mean that the values of the functions $\phi^\alpha_{,i_1\cdots i_r}$, $0 \le r \le m-1$ on the boundary ∂B_n are prescribed and for any isovector *field V all varied functions* $\phi^{\alpha}_{,i_1\cdots i_r} + \delta_V \phi^{\alpha}_{,i_1\cdots i_r}$*s*, $0 \le r \le m-1$ *take on the same values on the boundary.*

With the aim of determining the regular mapping ϕ that makes the functional $A(\phi)$ stationary, we have to discuss again the satisfaction of the condition

$$\delta_V A(\phi) = \int_{B_n} \phi^* \pounds_V \mathcal{L} = 0. \tag{10.5.7}$$

Since V is an isovector field, we must have $\pounds_V(\mathcal{L} - L\mu) \in \mathcal{I}_m$ so that we get $\phi^* \pounds_V \mathcal{L} = \phi^* \pounds_V (L\mu)$. If we substitute into the expression (10.5.7) the Lie derivative $\pounds_V \mathcal{L} = \mathbf{i}_V(d\mathcal{L}) + d\mathbf{i}_V(\mathcal{L})$ and recall the commutation rule $d \circ \phi^* = \phi^* \circ d$, then the application of the Stokes theorem casts (10.5.7) into the form

$$\delta_V A(\phi) = \int_{B_n} \phi^* \mathbf{i}_V(d\mathcal{L}) + \int_{\partial B_n} \phi^* \mathbf{i}_V(\mathcal{L}) = 0.$$

On the other hand, because of the relation

$$\mathbf{i}_V(\mathcal{L}) = \sum_{r=0}^{m-1} \Big(\lambda_\alpha^{i_1\cdots i_r i}(\mathbf{x}) + \frac{\partial L}{\partial v^\alpha_{i_1\cdots i_r i}} \Big) V^\alpha_{i_1\cdots i_r} \mu_i = \sum_{r=0}^{m-1} \Lambda_\alpha^{i_1\cdots i_r i} V^\alpha_{i_1\cdots i_r} \mu_i$$

we obtain

$$\phi^* \mathbf{i}_V(\mathcal{L})|_{\partial B_n} = \sum_{r=0}^{m-1} (\phi^* \Lambda_\alpha^{i_1\cdots i_r i}) \delta_V v^\alpha_{i_1\cdots i_r} \mu_i \Big|_{\partial B_n} = 0$$

where we have obviously defined

$$\Lambda_\alpha^{i_1\cdots i_r i} = \lambda_\alpha^{i_1\cdots i_r i}(\mathbf{x}) + \frac{\partial L}{\partial v^\alpha_{i_1\cdots i_r i}}.$$

Next, we shall evaluate the form $d\mathcal{L}$. After some almost trivial manipulations, we end up with the expression

$$\begin{aligned} d\mathcal{L} = \Big(\frac{\partial L}{\partial u^\alpha} - \frac{\partial \lambda^i_\alpha}{\partial x^i} \Big) \sigma^\alpha \wedge \mu - \sum_{r=1}^{m-1} \Big(\frac{\partial \lambda_\alpha^{i_1\cdots i_r i}}{\partial x^i} + \lambda_\alpha^{i_1\cdots i_r} \Big) \sigma^\alpha_{i_1\cdots i_r} \wedge \mu \\ + \sum_{r=0}^{m-1} d\Big(\frac{\partial L}{\partial v^\alpha_{i_1\cdots i_r i}} \Big) \wedge \sigma^\alpha_{i_1\cdots i_r} \wedge \mu_i - \lambda_\alpha^{i_1\cdots i_m} dv^\alpha_{i_1\cdots i_m} \wedge \mu. \end{aligned}$$

But we can write

$$\sum_{r=0}^{m-1} d\Big(\frac{\partial L}{\partial v^\alpha_{i_1\cdots i_r i}}\Big)\wedge\sigma^\alpha_{i_1\cdots i_r}\wedge\mu_i = -\sum_{r=0}^{m-1}\frac{\partial}{\partial x^i}\Big(\frac{\partial L}{\partial v^\alpha_{i_1\cdots i_r i}}\Big)\sigma^\alpha_{i_1\cdots i_r}\wedge\mu$$
$$+\sum_{s=0}^{m}\sum_{r=0}^{m-1}\frac{\partial^2 L}{\partial v^\alpha_{i_1\cdots i_r i}\partial v^\beta_{j_1\cdots j_s}}\,dv^\beta_{j_1\cdots j_s}\wedge\sigma^\alpha_{i_1\cdots i_r}\wedge\mu_i.$$

Thus, introducing this expression into $d\mathcal{L}$ and arranging the terms appropriately we eventually obtain

$$d\mathcal{L} = \Big[\frac{\partial L}{\partial u^\alpha} - \frac{\partial\lambda^i_\alpha}{\partial x^i} - \frac{\partial}{\partial x^i}\Big(\frac{\partial L}{\partial v^\alpha_i}\Big)\Big]\sigma^\alpha\wedge\mu - \sum_{r=1}^{m-1}\Big[\frac{\partial\lambda^{i_1\cdots i_r i}_\alpha}{\partial x^i} + \lambda^{i_1\cdots i_r}_\alpha$$
$$+\frac{\partial}{\partial x^i}\Big(\frac{\partial L}{\partial v^\alpha_{i_1\cdots i_r i}}\Big)\Big]\sigma^\alpha_{i_1\cdots i_r}\wedge\mu - \lambda^{i_1\cdots i_m}_\alpha dv^\alpha_{i_1\cdots i_m}\wedge\mu$$
$$+\sum_{s=0}^{m-1}\sum_{r=0}^{m-1}\frac{\partial^2 L}{\partial v^\alpha_{i_1\cdots i_r i}\partial v^\beta_{j_1\cdots j_s}}\,\sigma^\beta_{j_1\cdots j_s}\wedge\sigma^\alpha_{i_1\cdots i_r}\wedge\mu_i$$
$$+\sum_{r=0}^{m-1}\frac{\partial^2 L}{\partial v^\alpha_{i_1\cdots i_r i}\partial v^\beta_{j_1\cdots j_m}}\,dv^\beta_{j_1\cdots j_m}\wedge\sigma^\alpha_{i_1\cdots i_r}\wedge\mu_i.$$

From the above expression we deduce that

$$\mathbf{i}_V(d\mathcal{L}) = U^\alpha\Big[\frac{\partial L}{\partial u^\alpha} - \frac{\partial\lambda^i_\alpha}{\partial x^i} - \frac{\partial}{\partial x^i}\Big(\frac{\partial L}{\partial v^\alpha_i}\Big)\Big]\mu$$
$$-\sum_{r=1}^{m-1}\Big[\frac{\partial\lambda^{i_1\cdots i_r i}_\alpha}{\partial x^i} + \lambda^{i_1\cdots i_r}_\alpha$$
$$+\frac{\partial}{\partial x^i}\Big(\frac{\partial L}{\partial v^\alpha_{i_1\cdots i_r i}}\Big)\Big]V^\alpha_{i_1\cdots i_r}\mu - \lambda^{i_1\cdots i_m}_\alpha V^\alpha_{i_1\cdots i_m}\mu$$
$$+\sum_{s=0}^{m-1}\sum_{r=0}^{m-1}\frac{\partial^2 L}{\partial v^\alpha_{i_1\cdots i_r i}\partial v^\beta_{j_1\cdots j_s}}\,(V^\beta_{j_1\cdots j_s}\sigma^\alpha_{i_1\cdots i_r} - V^\alpha_{i_1\cdots i_r}\sigma^\beta_{j_1\cdots j_s})\wedge\mu_i$$
$$+\sum_{r=0}^{m-1}\frac{\partial^2 L}{\partial v^\alpha_{i_1\cdots i_r i}\partial v^\beta_{j_1\cdots j_m}}\,(V^\beta_{j_1\cdots j_m}\sigma^\alpha_{i_1\cdots i_r} - V^\alpha_{i_1\cdots i_r}dv^\beta_{j_1\cdots j_m})\wedge\mu_i.$$

However, since $\phi^*\mathcal{I}_m = 0$ we finally obtain

$$\int_{B_n}\phi^*\mathbf{i}_V(d\mathcal{L}) = \int_{B_n}\Big[\phi^* F_\alpha\delta_V u^\alpha - \sum_{r=1}^{m-1}\phi^* F^{i_1\cdots i_r}_\alpha\delta_V v^\alpha_{i_1\cdots i_r} - \lambda^{i_1\cdots i_m}_\alpha\delta_V v^\alpha_{i_1\cdots i_m}\Big]\mu$$

where we have defined

$$F_\alpha = \frac{\partial L}{\partial u^\alpha} - \frac{\partial \lambda_\alpha^i}{\partial x^i} - \frac{\partial}{\partial x^i}\Big(\frac{\partial L}{\partial v_i^\alpha}\Big),$$

$$F_\alpha^{i_1\cdots i_r} = \frac{\partial \lambda_\alpha^{i_1\cdots i_r i}}{\partial x^i} + \lambda_\alpha^{i_1\cdots i_r} + \frac{\partial}{\partial x^i}\Big(\frac{\partial L}{\partial v_{i_1\cdots i_r i}^\alpha}\Big), \quad 1 \le r \le m-1.$$

Since we have assumed that all variations $\delta_V v_{i_1\cdots i_r}^\alpha, 0 \le r \le m$ are independent, the condition

$$\int_{B_n} \phi^* \mathbf{i}_V(d\mathcal{L}) = 0$$

is satisfied if and only if the inverse recurrence relations

$$\phi^* \mathcal{F}_\alpha = \phi^* \Big[\frac{\partial L}{\partial u^\alpha} - \frac{\partial \lambda_\alpha^i}{\partial x^i} - \frac{\partial}{\partial x^i}\Big(\frac{\partial L}{\partial v_i^\alpha}\Big)\Big] = 0, \tag{10.5.8}$$

$$\phi^* \mathcal{F}_\alpha^{i_1\cdots i_r} =$$

$$\phi^* \Big[\frac{\partial \lambda_\alpha^{i_1\cdots i_r i}}{\partial x^i} + \lambda_\alpha^{i_1\cdots i_r} + \frac{\partial}{\partial x^i}\Big(\frac{\partial L}{\partial v_{i_1\cdots i_r i}^\alpha}\Big)\Big] = 0, 1 \le r \le m-1$$

$$\lambda_\alpha^{i_1\cdots i_m} = 0$$

are verified. We shall try to find the solution of these recurrence relations by discarding the operator ϕ^* for notational simplicity. Therefore, $\partial/\partial x^i$ will denote the total derivative with respect to that variable and we should, in fact, employ the substitution $v_{i_1\cdots i_r}^\alpha = u_{,i_1\cdots i_r}^\alpha$. Noting that $\lambda_\alpha^{i_1\cdots i_m} = 0$, we extract from $(10.5.8)_2$, respectively,

$$\lambda_\alpha^{i_1\cdots i_{m-1}} = -\frac{\partial}{\partial x^{i_m}}\Big(\frac{\partial L}{\partial v_{i_1\cdots i_{m-1} i_m}^\alpha}\Big)$$

$$\lambda_\alpha^{i_1\cdots i_{m-2}} = -\frac{\partial \lambda_\alpha^{i_1\cdots i_{m-2} i_{m-1}}}{\partial x^{i_{m-1}}} - \frac{\partial}{\partial x^{i_{m-1}}}\Big(\frac{\partial L}{\partial v_{i_1\cdots i_{m-2} i_{m-1}}^\alpha}\Big)$$

$$= -\frac{\partial}{\partial x^{i_{m-1}}}\Big(\frac{\partial L}{\partial v_{i_1\cdots i_{m-2} i_{m-1}}^\alpha}\Big) + \frac{\partial^2}{\partial x^{i_{m-1}} \partial x^{i_m}}\Big(\frac{\partial L}{\partial v_{i_1\cdots i_{m-2} i_{m-1} i_m}^\alpha}\Big).$$

....

We shall now show by mathematical induction that

$$\lambda_\alpha^{i_1\cdots i_r} = \sum_{s=1}^{m-r} (-1)^s \frac{\partial^s}{\partial x^{i_{r+1}} \cdots \partial x^{i_{r+s}}}\Big(\frac{\partial L}{\partial v_{i_1\cdots i_r i_{r+1}\cdots i_{r+s}}^\alpha}\Big). \tag{10.5.9}$$

Indeed, let us assume that

$$\lambda_\alpha^{i_1\cdots i_r i_{r+1}} = \sum_{s=1}^{m-r-1}(-1)^s \frac{\partial^s}{\partial x^{i_{r+2}}\cdots\partial x^{i_{r+s+1}}}\Big(\frac{\partial L}{\partial v^\alpha_{i_1\cdots i_r i_{r+1} i_{r+2}\cdots i_{r+s+1}}}\Big).$$

Because of the relation

$$\begin{aligned}
-\frac{\partial\lambda_\alpha^{i_1\cdots i_r i_{r+1}}}{\partial x^{i_{r+1}}} \\
&= \sum_{s=1}^{m-r-1}(-1)^{s+1}\frac{\partial^{s+1}}{\partial x^{i_{r+1}}\partial x^{i_{r+2}}\cdots\partial x^{i_{r+s+1}}}\Big(\frac{\partial L}{\partial v^\alpha_{i_1\cdots i_r i_{r+1} i_{r+2}\cdots i_{r+s+1}}}\Big) \\
&= \sum_{t=2}^{m-r}(-1)^t\frac{\partial^t}{\partial x^{i_{r+1}}\cdots\partial x^{i_{r+t}}}\Big(\frac{\partial L}{\partial v^\alpha_{i_1\cdots i_r i_{r+1}\cdots i_{r+t}}}\Big),
\end{aligned}$$

we conclude that

$$\begin{aligned}
\lambda_\alpha^{i_1\cdots i_r} &= -\frac{\partial\lambda_\alpha^{i_1\cdots i_r i_{r+1}}}{\partial x^{i_{r+1}}} - \frac{\partial}{\partial x^{i_{r+1}}}\Big(\frac{\partial L}{\partial v^\alpha_{i_1\cdots i_r i_{r+1}}}\Big) \\
&= \sum_{s=1}^{m-r}(-1)^s\frac{\partial^s}{\partial x^{i_{r+1}}\cdots\partial x^{i_{r+s}}}\Big(\frac{\partial L}{\partial v^\alpha_{i_1\cdots i_r i_{r+1}\cdots i_{r+s}}}\Big), 1 \le r \le m-1.
\end{aligned}$$

It is clear that the terms corresponding to the case $s > m-r$ must be supposed to be nil. Hence, we get

$$\begin{aligned}
\lambda_\alpha^{i_1} &= \sum_{s=1}^{m-1}(-1)^s\frac{\partial^s}{\partial x^{i_2}\cdots\partial x^{i_{1+s}}}\Big(\frac{\partial L}{\partial v^\alpha_{i_1 i_2\cdots i_{1+s}}}\Big) \\
&= \sum_{s=2}^{m}(-1)^{s-1}\frac{\partial^{s-1}}{\partial x^{i_2}\cdots\partial x^{i_s}}\Big(\frac{\partial L}{\partial v^\alpha_{i_1 i_2\cdots i_s}}\Big)
\end{aligned}$$

and the equations $(10.5.8)_1$ take finally the form below in which Lagrange multipliers are eliminated

$$\begin{aligned}
\frac{\partial L}{\partial u^\alpha} - \frac{\partial\lambda_\alpha^{i_1}}{\partial x^{i_1}} - \frac{\partial}{\partial x^{i_1}}\Big(\frac{\partial L}{\partial v^\alpha_{i_1}}\Big) &= \\
\frac{\partial L}{\partial u^\alpha} - \frac{\partial}{\partial x^{i_1}}\Big(\frac{\partial L}{\partial v^\alpha_{i_1}}\Big) &+ \sum_{t=2}^{m}(-1)^t\frac{\partial^t}{\partial x^{i_1}\cdots\partial x^{i_t}}\Big(\frac{\partial L}{\partial v^\alpha_{i_1\cdots i_t}}\Big) \\
&= \frac{\partial L}{\partial u^\alpha} + \sum_{s=1}^{m}(-1)^s\frac{\partial^s}{\partial x^{i_1}\cdots\partial x^{i_s}}\Big(\frac{\partial L}{\partial v^\alpha_{i_1\cdots i_s}}\Big) = 0.
\end{aligned}$$

Consequently, the functions $u^\alpha = \phi^\alpha(\mathbf{x})$ that render the action functional stationary must satisfy the *Euler-Lagrange equations*

$$\mathcal{E}_\alpha(L) = \sum_{s=0}^{m}(-1)^s \frac{\partial^s}{\partial x^{i_1}\cdots\partial x^{i_s}}\Big(\frac{\partial L}{\partial u^\alpha_{,i_1\cdots i_s}}\Big) = 0 \qquad (10.5.10)$$

where $\alpha = 1, \dots, N$. *We have to take heed that the operators $\partial/\partial x^i$ here represent total derivatives with respect to the independent variables x^i*. For instance, if $m = 2$, then the Lagrangian function is $L = L(x^i, u^\alpha, u^\alpha_{,i}, u^\alpha_{,ij})$ and the Euler-Lagrange equations turn out to be the following fourth order quasilinear partial differential equations

$$\frac{\partial L}{\partial u^\alpha} - \frac{\partial}{\partial x^i}\Big(\frac{\partial L}{\partial u^\alpha_{,i}}\Big) + \frac{\partial^2}{\partial x^i \partial x^j}\Big(\frac{\partial L}{\partial u^\alpha_{,ij}}\Big) = \frac{\partial L}{\partial u^\alpha} - \frac{\partial^2 L}{\partial x^i \partial u^\alpha_{,i}} + \frac{\partial^3 L}{\partial x^i \partial x^j \partial u^\alpha_{,ij}}$$
$$+ \Big(2\frac{\partial^3 L}{\partial x^j \partial u^\beta \partial u^\alpha_{,ij}} - \frac{\partial^2 L}{\partial u^\beta \partial u^\alpha_{,i}}\Big)\frac{\partial u^\beta}{\partial x^i} + \Big(\frac{\partial^2 L}{\partial u^\beta \partial u^\alpha_{,ij}} + 2\frac{\partial^3 L}{\partial x^k \partial u^\beta_{,j} \partial u^\alpha_{,ik}}$$
$$- \frac{\partial^2 L}{\partial u^\alpha_{,i} \partial u^\beta_{,j}}\Big)\frac{\partial^2 u^\beta}{\partial x^i \partial x^j} + \Big(\frac{\partial^2 L}{\partial u^\beta_{,k} \partial u^\alpha_{,ij}} + 2\frac{\partial^3 L}{\partial x^l \partial u^\alpha_{,il} \partial u^\beta_{,jk}}$$
$$- \frac{\partial^2 L}{\partial u^\alpha_{,i} \partial u^\beta_{,jk}}\Big)\frac{\partial^3 u^\beta}{\partial x^i \partial x^j \partial x^k} + \frac{\partial^3 L}{\partial u^\beta \partial u^\gamma \partial u^\alpha_{,ij}}\frac{\partial u^\gamma}{\partial x^i}\frac{\partial u^\beta}{\partial x^j}$$
$$+ 2\frac{\partial^3 L}{\partial u^\beta \partial u^\gamma_{,k} \partial u^\alpha_{,ij}}\frac{\partial u^\beta}{\partial x^j}\frac{\partial^2 u^\gamma}{\partial x^i \partial x^k} + 2\frac{\partial^3 L}{\partial u^\beta \partial u^\alpha_{,ij} \partial u^\gamma_{,kl}}\frac{\partial u^\beta}{\partial x^j}\frac{\partial^2 u^\gamma}{\partial x^i \partial x^k \partial x^l}$$
$$+ \frac{\partial^3 L}{\partial u^\beta_{,k} \partial u^\gamma_{,l} \partial u^\alpha_{,ij}}\frac{\partial^2 u^\beta}{\partial x^j \partial x^k}\frac{\partial^2 u^\gamma}{\partial x^i \partial x^l} + 2\frac{\partial^3 L}{\partial u^\beta_{,k} \partial u^\alpha_{,ij} \partial u^\gamma_{,lm}}\frac{\partial^2 u^\beta}{\partial x^j \partial x^k}\frac{\partial^3 u^\gamma}{\partial x^i \partial x^l \partial x^m}$$
$$+ \frac{\partial^3 L}{\partial u^\alpha_{,ij} \partial u^\beta_{,kl} \partial u^\gamma_{,mn}}\frac{\partial^3 u^\beta}{\partial x^j \partial x^k \partial x^l}\frac{\partial^3 u^\gamma}{\partial x^i \partial x^m \partial x^n}$$
$$+ \frac{\partial^2 L}{\partial u^\alpha_{,ij} \partial u^\beta_{,kl}}\frac{\partial^4 u^\beta}{\partial x^i \partial x^j \partial x^k \partial x^l} = 0.$$

As a simple example, let us consider the Lagrangian function

$$L = \frac{1}{2}(u_{xx} + u_{yy})^2$$

where $u = u(x, y)$. Since

$$\frac{\partial L}{\partial u_{xx}} = \frac{\partial L}{\partial u_{yy}} = u_{xx} + u_{yy}$$

The Euler-Lagrange equation yields the *biharmonic equation*

$$(u_{xx}+u_{yy})_{xx}+(u_{xx}+u_{yy})_{yy}=u_{xxxx}+2u_{xxyy}+u_{yyyy}=\nabla^4 u=0.$$

As another example, let us take $L=u_{xx}u_{yy}-u_{xy}^2$. Since

$$\frac{\partial L}{\partial u_{xx}}=u_{yy},\ \frac{\partial L}{\partial u_{yy}}=u_{xx},\ \frac{\partial L}{\partial u_{xy}}=-2u_{xy}$$

the Euler-Lagrange equation is satisfied identically:

$$\mathcal{E}(L)=u_{yyxx}+u_{xxyy}-2u_{xyxy}\equiv 0.$$

In other words, the action functional becomes stationary for every function u obeying the boundary conditions. This Lagrangian function can also be written in the shape

$$L=(u_x u_{yy})_x-(u_x u_{xy})_y$$

so that it is in the form of a divergence. This is, in fact, a general feature of Lagrangian functions rendered by every functions obeying the boundary conditions. In effect, if we can express a Lagrangian functions as

$$L=\frac{\partial P^i}{\partial x^i}$$

where the functions P^i depend on x^j, u^α and derivatives of various orders of u^α, then the action integral may be transformed to an integral over the boundary through the Stokes theorem as follows

$$\int_{\mathcal{B}_n} L\,dV=\int_{\partial\mathcal{B}_n} P^i n_i\,dS.$$

Hence, it becomes dependent only on data describing the behaviours of functions u^α on the boundary. Since we have assumed that the variations on the boundary vanish, this amounts to say that the integral on the left hand side above is not affected by variations inside the region. In consequence, the Euler-Lagrange equations are satisfied identically by all functions meeting the conditions on the boundary. *It can be proven that a divergence form for the Lagrangian function provides a necessary and sufficient condition for the existence of this property.*

The result (10.5.10) can also be obtained from the equations (10.3.5) as a conditional variation problem. The function $L(x^i,u^\alpha,u^\alpha_{,i_1},\dots,u^\alpha_{,i_1\cdots i_m})$ can be written in the form $L(x^i,u^\alpha,v^\alpha_{i_1},\dots,v^\alpha_{i_1\cdots i_m})$ together with the additional conditions $v^\alpha_{i_1\cdots i_r i}=v^\alpha_{i_1\cdots i_r,i}$. In this case, the arguments of L are interrelated. In order to replace these relations in the action functional, we again use the method of Lagrange multipliers and define the function

$$\mathfrak{L} = L + \sum_{r=0}^{m-1} \lambda_\alpha^{i_1\cdots i_r i}(\mathbf{x})(v_{i_1\cdots i_r i}^\alpha - v_{i_1\cdots i_r,i}^\alpha) \tag{10.5.11}$$

where $\lambda_\alpha^{i_1\cdots i_r i}(\mathbf{x})$ are Lagrange multipliers. Therefore, the action functional becomes stationary if the following equations are to be satisfied

$$\frac{\partial \mathfrak{L}}{\partial v_{i_1\cdots i_r}^\alpha} - \frac{\partial}{\partial x^i}\Big(\frac{\partial \mathfrak{L}}{\partial v_{i_1\cdots i_r,i}^\alpha}\Big) = 0,\ 0 \le r \le m.$$

It then follows from (10.5.11) that

$$\frac{\partial L}{\partial v_{i_1\cdots i_r}^\alpha} + \lambda_\alpha^{i_1\cdots i_r} + \frac{\partial \lambda_\alpha^{i_1\cdots i_r i}}{\partial x^i} = 0,\ 0 < r \le m-1. \tag{10.5.12}$$

We obviously find

$$\lambda_\alpha^{i_1\cdots i_m} = -\frac{\partial L}{\partial v_{i_1\cdots i_m}^\alpha}$$

for $r = m$ and

$$\frac{\partial L}{\partial u^\alpha} + \frac{\partial \lambda_\alpha^i}{\partial x^i} = 0$$

for $r = 0$. Utilising these results, one can readily show that the solution of the recurrence relation (10.5.12) is still given by the equations (10.5.10).

The Euler-Lagrange equations (10.5.10) can also be written as balance equations

$$\frac{\partial \Sigma_\alpha^i}{\partial x^i} + \Sigma_\alpha = 0$$

if we introduce the definitions

$$\Sigma_\alpha^i = \sum_{r=1}^{m}(-1)^r \frac{\partial^{r-1}}{\partial x^{i_1}\cdots\partial x^{i_{r-1}}}\Big(\frac{\partial L}{\partial u_{,ii_1\cdots i_{r-1}}^\alpha}\Big),\ \Sigma_\alpha = \frac{\partial L}{\partial u^\alpha}$$

However, balance equations are evidently not expressible in general in the form of the Euler-Lagrange equations.

The variational symmetries that pave the way for the Noether theorem were dealt with in Sec. 10.4. We shall now try to generalise those results for the functional given by (10.5.3). Let us consider an isovector field U of the contact ideal given by (10.5.4) together with (10.5.5). We know that we have in the first order

$$A(\phi_U(s)) = \int_{B_n} \phi^*\left(e^{s\pounds_U}\mathcal{L}\right) = A(\phi) + \delta_U A(\phi)\, s + o(s)$$

so that we can write

$$\delta_U A(\phi) = \int_{B_n} \phi^* \pounds_U \mathcal{L}.$$

Since the contact ideal is invariant under the isovector field U, the criterion of infinitesimal invariance is again recovered from the condition $\phi^*\pounds_U\mathcal{L} = \phi^*\pounds_U(L\mu) = 0$ giving

$$\phi^*\pounds_U(L\mu) = \phi^*\big[U(L) + LD_i X^i\big]\mu = 0$$

as in (10.4.5). However, this time we have

$$\phi^* U(L) = \phi^*\Big[X^i\frac{\partial L}{\partial x^i} + U^\alpha\frac{\partial L}{\partial u^\alpha} + \sum_{r=1}^m V^\alpha_{i_1\cdots i_r}\frac{\partial L}{\partial v^\alpha_{i_1\cdots i_r}}\Big].$$

We shall now attempt to transform the expression $\phi^*\pounds_U\mathcal{L} = 0$ into another form in case the regular mapping ϕ satisfies the Euler-Lagrange equations (10.5.10) or, in other words, the relations (10.5.8) leading to those equations. Because of the relation $\phi^*\pounds_U\mathcal{L} = 0$, we obtain owing to the Cartan formula

$$\phi^* d(\mathbf{i}_U\mathcal{L}) = d\phi^*(\mathbf{i}_U\mathcal{L}) = -\phi^*\mathbf{i}_U(d\mathcal{L}). \tag{10.5.13}$$

Let us now try to calculate this expression explicitly. On making use of the formula for $d\mathcal{L}$ given on p. 684, we first get

$$\begin{aligned}\mathbf{i}_U(d\mathcal{L}) = {} & (U^\alpha - v^\alpha_j X^j)\Big[\Big(\frac{\partial L}{\partial u^\alpha} - \frac{\partial \lambda^i_\alpha}{\partial x^i}\Big)\mu - d\Big(\frac{\partial L}{\partial v^\alpha_i}\Big)\wedge\mu_i\Big] \\ & - \sum_{r=1}^{m-1}(V^\alpha_{i_1\cdots i_r} - v^\alpha_{i_1\cdots i_r j}X^j)\Big[\Big(\frac{\partial \lambda^{i_1\cdots i_r i}_\alpha}{\partial x^i} + \lambda^{i_1\cdots i_r}_\alpha\Big)\mu \\ & + d\Big(\frac{\partial L}{\partial v^\alpha_{i_1\cdots i_r i}}\Big)\wedge\mu_i\Big] \\ & - \lambda^{i_1\cdots i_m}_\alpha V^\alpha_{i_1\cdots i_m}\mu + \Big\{-X^j\Big(\frac{\partial L}{\partial u^\alpha} - \frac{\partial \lambda^i_\alpha}{\partial x^i}\Big)\sigma^\alpha\wedge\mu_j\end{aligned}$$

$$+\sum_{r=1}^{m-1} X^j\Big(\frac{\partial\lambda_\alpha^{i_1\cdots i_r i}}{\partial x^i}+\lambda_\alpha^{i_1\cdots i_r}\Big)\sigma^\alpha_{i_1\cdots i_r}\wedge\mu_j$$

$$+\sum_{r=1}^{m-1} U\Big(\frac{\partial L}{\partial v^\alpha_{i_1\cdots i_r i}}\Big)\sigma^\alpha_{i_1\cdots i_r}\wedge\mu_i$$

$$+\sum_{r=1}^{m-1} X^j\, d\Big(\frac{\partial L}{\partial v^\alpha_{i_1\cdots i_r i}}\Big)\wedge\sigma^\alpha_{i_1\cdots i_r}\wedge\mu_{ji}+X^i\lambda_\alpha^{i_1\cdots i_m}dv^\alpha_{i_1\cdots i_m}\wedge\mu_i\Big\}$$

where the terms within braces belong to the contact ideal $\mathcal{I}_m$. On the other hand, owing to the relations

$$\phi^* d\Big(\frac{\partial L}{\partial v^\alpha_{i_1\cdots i_r i}}\Big)=\frac{\partial}{\partial x^j}\phi^*\Big(\frac{\partial L}{\partial v^\alpha_{i_1\cdots i_r i}}\Big)dx^j$$

and $\phi^*\mathcal{I}_m=0$, we conclude that

$$\phi^*\mathbf{i}_U(d\mathcal{L})=\phi^*(U^\alpha-v^\alpha_j X^j)\,\phi^*\Big[\frac{\partial L}{\partial u^\alpha}-\frac{\partial\lambda^i_\alpha}{\partial x^i}-\frac{\partial}{\partial x^i}\Big(\frac{\partial L}{\partial v^\alpha_i}\Big)\Big]\mu$$

$$-\sum_{r=1}^{m-1}\phi^*(V^\alpha_{i_1\cdots i_r}-v^\alpha_{i_1\cdots i_r j}X^j)\,\phi^*\Big[\frac{\partial\lambda_\alpha^{i_1\cdots i_r i}}{\partial x^i}+\lambda_\alpha^{i_1\cdots i_r}+\frac{\partial}{\partial x^i}\Big(\frac{\partial L}{\partial v^\alpha_{i_1\cdots i_r i}}\Big)\Big]\mu$$

$$-\lambda_\alpha^{i_1\cdots i_m}\,\phi^* V^\alpha_{i_1\cdots i_m}\mu.$$

If the mapping ϕ is a solution of the Euler-Lagrange equations the right hand side in the foregoing expression vanishes in view of (10.5.8) so that we obtain

$$\phi^*\mathbf{i}_U(d\mathcal{L})=0.$$

Thus, it follows from (10.5.13) that

$$d\phi^*(\mathbf{i}_U\mathcal{L})=d\phi^*\mathcal{J}=0$$

where the form $\mathcal{J}\in\Lambda^{n-1}(\mathcal{C}_m)$ is defined by

$$\mathcal{J}=\mathbf{i}_U\mathcal{L}=LX^i\mu_i$$

$$+\sum_{r=0}^{m-1}\Big(\lambda_\alpha^{i_1\cdots i_r i}+\frac{\partial L}{\partial v^\alpha_{i_1\cdots i_r i}}\Big)(V^\alpha_{i_1\cdots i_r}-v^\alpha_{i_1\cdots i_r j}X^j)\mu_i$$

$$-\sum_{r=0}^{m-1}\Big(\lambda_\alpha^{i_1\cdots i_r i}+\frac{\partial L}{\partial v^\alpha_{i_1\cdots i_r i}}\Big)X^j\sigma^\alpha_{i_1\cdots i_r}\wedge\mu_{ji}$$

Hence, we can again write $\phi^*\mathcal{J} = J^i\mu_i$ where the functions J^i are determined as follows

$$\begin{aligned}J^i &= \phi^*\Big[LX^i + \sum_{r=0}^{m-1}\Big(\lambda_\alpha^{i_1\cdots i_r i} + \frac{\partial L}{\partial v^\alpha_{i_1\cdots i_r i}}\Big)(V^\alpha_{i_1\cdots i_r} - v^\alpha_{i_1\cdots i_r j}X^j)\Big]\\ &= \phi^*\Big[LX^i + \sum_{r=0}^{m-1}(V^\alpha_{i_1\cdots i_r} - v^\alpha_{i_1\cdots i_r j}X^j)\Big\{\frac{\partial L}{\partial v^\alpha_{i_1\cdots i_r i}}\\ &\quad + \sum_{s=2}^{m-r}(-1)^{s-1}\frac{\partial^{s-1}}{\partial x^{i_{r+1}}\cdots\partial x^{i_{r+s-1}}}\Big(\frac{\partial L}{\partial v^\alpha_{i_1\cdots i_r i_{r+1}\cdots i_{r+s-1} i}}\Big)\Big\}\Big]\end{aligned} \tag{10.5.14}$$

Accordingly, the relation $d\phi^*\mathcal{J} = 0$ leads again to the conservation law

$$\frac{\partial J^i}{\partial x^i} = 0.$$

It is immediate to observe that these expressions are also valid in the case $N = 1$. But, it is evident that we have then to employ isovector components generated by a function $F(x^i, u, v_i)$.

For $m = 1$, we obtain from (10.5.14)

$$\begin{aligned}J^i &= \phi^*\Big[LX^i + \frac{\partial L}{\partial v^\alpha_i}(U^\alpha - v^\alpha_j X^j)\Big],\\ \phi^* v^\alpha_i &= u^\alpha_{,i}.\end{aligned}$$

This result is exactly the same as the expression (10.4.6) previously obtained. When we choose $m = 2$, we find that

$$\begin{aligned}J^i = \phi^*\Big[&LX^i + \Big\{\frac{\partial L}{\partial v^\alpha_i} - \frac{\partial}{\partial x^k}\Big(\frac{\partial L}{\partial v^\alpha_{ik}}\Big)\Big\}(U^\alpha - v^\alpha_j X^j)\\ &+ \frac{\partial L}{\partial v^\alpha_{ik}}(V^\alpha_k - v^\alpha_{jk}X^j)\Big]\end{aligned}$$

where we have $\phi^* v^\alpha_i = u^\alpha_{,i}$, $\phi^* v^\alpha_{ij} = u^\alpha_{,ij}$ and the functions V^α_i are expressible as

$$V^\alpha_i = \frac{\partial U^\alpha}{\partial x^i} + v^\beta_i\frac{\partial U^\alpha}{\partial u^\beta} - v^\alpha_j\frac{\partial X^j}{\partial x^i} - v^\alpha_j v^\beta_i\frac{\partial X^j}{\partial u^\beta}$$

so that we can write

$$\phi^* V^\alpha_i = \frac{\partial \phi^* U^\alpha}{\partial x^i} - u^\alpha_{,j}\frac{\partial \phi^* X^j}{\partial x^i}$$

X. EXERCISES

10.1. Show that the balance equation

$$\frac{\partial^2 u}{\partial x^2} - \frac{\partial^2 u}{\partial t^2} = c(u)\frac{\partial u}{\partial t}, \quad c(u) \neq 0$$

modelling the propagation of one-dimensional waves in a dissipative medium cannot be derived from a variational principle. Determine the enlarged system that gives rise to a variational principle and find this principle.

10.2. The following non-linear ordinary second order equation

$$u'' + \frac{k}{x}u' - \lambda\, x^\alpha u^\beta = 0$$

is known as the *Emden-Fowler equation* [Swiss astrophysicist and meteorologist Jacob Robert Emden (1862-1940) and English physicist and astronomer Sir Ralph Howard Fowler (1889-1944)]. k, λ, α and β are given real constants.

(a). Show that this equation is derivable from a variational principle (*Hint*: multiply the differential equation by x^k).

(b). Show that this variational principle has a variational symmetry in the form of scaling transformation if the condition $\alpha = [(\beta - 1)k - \beta - 3]/2$ is satisfied.

(c). Utilising that symmetry, show that the Emden-Fowler equation has a solution in the form

$$u(x) = \Big(-\frac{(1-k)^2}{4\lambda}\Big)^{1/(\beta-1)} x^{(1-k)/2}.$$

For instance, when we take $k = 2$, $\alpha = 0$, $\beta = 5$ and $\lambda = -1$ we get $u = 1/\sqrt{2x}$.

10.3. Determine the variational symmetry groups of the variational principle generating the Laplace equation $\Delta u = u_{,ii} = 0$ in $\mathbb{R}^n$ and find the corresponding conservation laws.

10.4. Determine the variational symmetries of a variational principle generating the following differential equations

$$\Delta u^1 + \lambda u^2 = 0, \quad \Delta u^2 + \lambda u^1 = 0$$

in $\mathbb{R}^n$ where λ is a constant. Find the corresponding conservation laws.

10.5. The *Navier equations* governing the motion of a homogeneous and isotropic linearly elastic three-dimensional body are given by

$$(c_1^2 - c_2^2)\frac{\partial^2 u_j}{\partial x_i \partial x_j} + c_2^2\frac{\partial^2 u_i}{\partial x_j \partial x_j} = \frac{\partial^2 u_i}{\partial t^2}$$

where u_i with $i = 1, 2, 3$ denote the Cartesian components of the displacement vector field. The constants c_1 and c_2 are the velocities of propagation of the longitudinal and transversal waves in the medium, respectively. Find a variational principle generating these equations. Determine the variational symmetries and corresponding conservation laws.

10.6. Find the variational principle generating the equation $\Delta^2 u + \lambda u = 0$ in $\mathbb{R}^n$. Determine the variational symmetries and corresponding conservation laws.

10.7. Show that the *BBM* (*Benjamin-Bona-Mahony*) *equation*

$$u_t + u_x + uu_x - u_{xxt} = 0$$

that is encountered in the shallow water theory in hydrodynamics is derivable from a variational principle after the substitution $u = v_x$. Find the variational symmetries and corresponding conservation laws.

CHAPTER XI

SOME PHYSICAL APPLICATIONS

11.1. SCOPE OF THE CHAPTER

This chapter deals with the exploration of some physical applications of exterior differential forms. In the first four successive sections we discuss the analytical mechanics for which exterior forms prove to be a very powerful tool to reveal its various fundamental properties. We first investigate in Sec. 11.2 the behaviour of a dynamical system with m degrees of freedom, whose constraints are holonomic and are not changing with time. We further assume that forces acting on the system are derivable from a time-independent potential. Such a system is depicted by the Lagrangian function. Then, the Lagrange equations are given, and by defining the generalised momenta, the Hamiltonian function and the Hamilton equations are introduced. It is shown that the generalised coordinates and momenta are local coordinates of a $2m$-dimensional symplectic manifold S. A symplectic 2-form then provides an isomorphism between the module of 1-forms on the manifold S and its tangent bundle. This enables us to define Hamiltonian vector fields and to express equations of motions as an exterior equation on S. We then introduce the Poisson bracket of 1-forms on S in Sec. 11.3 and we examine properties of these brackets. We further show that 1-forms constitute a Lie algebra with respect to a product identified as a Poisson bracket. Making use of the relations involving such Poisson brackets, we obtain Poisson brackets of 0-forms, namely, differentiable functions and we show that these functions also constitute a Lie algebra with respect to Poisson brackets. Then the connection between Poisson brackets and equations of motion is established. We deal with canonical transformations in Sec. 11.4 that are characterised as mappings under which the symplectic form remains invariant. It turns out that these transformations leave also the Hamilton equations of motions invariant. Afterwards, we discuss non-conservative mechanics in Sec. 11.5. Dynamical system now occupies a $2m + 1$-dimensional non-symplectic manifold. The Hamilton equations are then reduced again

Exterior Analysis, DOI: 10.1016/B978-0-12-415902-0.50011-6

to an exterior form equation by means of a 1-form involving a time-dependent Hamiltonian function and a 2-form that is its exterior derivative. It is shown that the canonical transformations leave this 2-form invariant. It is then proven that the Hamilton equations remain invariant under canonical transformations. The structural properties of canonical transformations are investigated. We finalise the study of analytical mechanics by exploring the Hamilton-Jacobi theory that help reduce the Hamilton equations to their simplest possible form. Our next topic is the electromagnetic theory studied in Sec. 11.6. The Maxwell equations are expressed as vanishing divergences of two second order antisymmetric tensors on a 4-dimensional manifold and it is found that these equations are equivalent to an exterior system involving two 2-forms. The general solution of these equations is constructed by employing the homotopy operator. When constitutive relations are taken into account, it is shown that this solution leads to the classical solution that are expressed in terms of scalar and vectorial potentials satisfying wave equations. In the final Sec. 11.7, the classical thermodynamics is briefly treated in a rather elementary level. A thermodynamic system whose state is determined by external and internal variables, and the empirical temperature is considered. An isothermal work function is defined assuming that external agents are conservative. By employing the first law of thermodynamics which states that the work done by external effects plus the heat energy input is equal to the rate of change of the internal energy and the physical fact that thermodynamic functions are additive, admissible versions of work and heat energy forms are obtained. Furthermore, the thermodynamic (absolute) temperature is introduced by an appropriate transformation, the existence of the entropy is proven under the conditions of complete integrability of the heat form. Then, the relations between the internal energy, free energy and heat forms are illustrated.

11.2. CONSERVATIVE MECHANICS

Let us consider a dynamical system consisting of several particles and rigid bodies moving in the space $\mathbb{R}^3$. In this space, the position of a particle is determined by at most 3 numbers corresponding to its coordinates implying that a particle has *three degrees of freedom*. On the other hand the position of a rigid body is prescribed by at most 6 numbers (for instance, 3 coordinates of one of its points, frequently of its centroid, and 3 Euler angles prescribing its orientation in the space). Therefore, a rigid body has *six degrees of freedom*. We can thus represent the position of a dynamical system as a point in some space $\mathbb{R}^N$ where $1 \leq N \leq \infty$ and the time evolution of such a system can be depicted by a curve in this space.

However, the system may possess ***constraints*** that restrict its motion so that it has a lower degrees of freedom. For instance, if we restrict the motion of a particle to a plane, then it has only two degrees of freedom. For another example, let us consider a rigid body. Although it has infinitely many particles, due to the fact that the distance between any two particles does not change during the motion, its degrees of freedom become just six and its motion is completely determined by specifying only six functions depending on time. Constraints that can be expressed by functional relations are called the ***holonomic constraints*** while they are known as the ***anholonomic constraints*** if they are prescribed by non-integrable differential forms. Moreover, if their structure is rigid, i.e., it does not change with time they are called the ***scleronomic constraints*** whereas if it varies with time they are named as the ***rheonomic constraints***. We first consider a system with scleronomic holonomic constraints. Let us assume that the system has now m degrees of freedom with constraints. The position of the system, thereby of every member of the system are completely determined by m variables $\mathbf{q} = \{q^1, q^2, \ldots, q^m\}$ called the ***generalised coordinates*** through the relations $x^i = x^i(\mathbf{q}), i = 1, 2, \ldots, m$. If we denote the time by t, the functions $\mathbf{q}(t) = \{q^i(t),\ i = 1, 2, \ldots, m\}$ now describe fully the evolution of the dynamical system. This coordinate transformation produces a differentiable m-dimensional submanifold M of the simple manifold $\mathbb{R}^N$. We call this manifold, which might acquire quite a complicated structure due to this transformation, as the ***configuration manifold***. Hence, the motion of the system is represented by a curve on this manifold. If we can find this curve on M, then we can carry it over the physical space by using appropriate coordinate transformations. This task is, of course, conceptually quite simple, but it may prove to be rather difficult to realise it operationally.

It is evident that the generalised coordinates need not to be determined uniquely. A new set of generalised coordinates $\mathbf{Q} = \{Q^1, Q^2, \ldots, Q^m\}$ for the configuration manifold may be defined by the help of functions

$$Q^i = \mathcal{Q}^i(q^1, q^2, \ldots, q^m),\ \ i = 1, 2, \ldots, m$$

However, in order that the degrees of freedom of the system are preserved, the new coordinates should be functionally independent. Therefore, we have to be sure that the condition

$$\det \boldsymbol{\Omega} = \det\Big[\frac{\partial Q^i}{\partial q^j}\Big] \neq 0$$

must be satisfied. That we are somewhat free in choosing the generalised coordinates suggests the possibility of searching for a particular choice of them to simplify the investigation of the system to a great extent.

Example 11.2.1. Let us consider three particles moving in a plane with masses m_1, m_2, m_3. The masses m_1 and m_2 are connected by a rod of length l_1 whereas m_2 and m_3 are connected by a rod of length l_2. Both rods are assumed to be rigid and massless. Connections are provided by freely rotating joints. This system is, of course, taking place in the manifold $\mathbb{R}^6$ with the coordinate cover $(x_1, y_1, x_2, y_2, x_3, y_3)$. But, if we denote the angles between rods and the horizontal line by θ_1 and θ_2, we can write

$$\begin{aligned} x_2 &= x_1 + l_1 \cos\theta_1, \qquad y_2 = y_1 + l_1 \sin\theta_1, \\ x_3 &= x_1 + l_1\cos\theta_1 + l_2\cos\theta_2, y_3 = y_1 + l_1\sin\theta_1 + l_2\sin\theta_2. \end{aligned}$$

Since the motion of the system is now determined by generalised coordinates $(x_1, y_1, \theta_1, \theta_2)$, it has four degrees of freedom. So the system will evolve with time on a 4-dimensional configuration manifold $M^4 \subset \mathbb{R}^6$ with a coordinate cover $(x_1, y_1, \theta_1, \theta_2)$. This manifold may be defined by the following algebraic equations

$$(x_2 - x_1)^2 + (y_2 - y_1)^2 = l_1^2, \ (x_3 - x_2)^2 + (y_3 - y_2)^2 = l_2^2. \qquad \blacksquare$$

When the constraints are both scleronomic and holonomic, the kinetic energy of the system can be expressed as follows

$$T = \frac{1}{2}\, g_{ij}(q^1, q^2, \dots, q^m)\, \dot{q}^i \dot{q}^j = \frac{1}{2}\, g_{ij}(\mathbf{q})\, \dot{q}^i \dot{q}^j \geq 0. \tag{11.2.1}$$

An overdot denotes as usual the time derivative. $\mathbf{G}(\mathbf{q}) = [g_{ij}(\mathbf{q})]$ must be a symmetric and positive definite $m \times m$ matrix. The functions $\dot{q}^i(t)$ are called the ***generalised velocities***. We know that the generalised velocities at a point $\mathbf{q} \in M$ take place in the tangent space of the manifold M at that point. The ***velocity phase space*** is the tangent bundle $T(M)$ of the configuration manifold M. It is a $2m$-dimensional differentiable manifold whose coordinate cover is $(q^i, \dot{q}^i)$. If the system is ***conservative***, then there exists a scalar-valued ***potential function***

$$V = V(q^1, q^2, \dots, q^m) = V(\mathbf{q})$$

and the gradient $\{\partial V / \partial q^i\}$ of this function with respect to the generalised coordinates determines, somewhat indirectly, the actual forces acting on the physical system. The differentiable function $L : T(M) \to \mathbb{R}$ defined by the relation

$$L(\mathbf{q}, \dot{\mathbf{q}}) = T - V \tag{11.2.2}$$

is called the ***Lagrangian function*** of the system and the dynamical evolution of the system is governed by the following ***Lagrange equations***

$$\frac{d}{dt}\Big(\frac{\partial L}{\partial \dot{q}^i}\Big) - \frac{\partial L}{\partial q^i} = 0, \quad i = 1, 2, \ldots, m. \tag{11.2.3}$$

These are a set of second order ordinary differential equations satisfied by functions $q^i(t)$. One must easily recognise that these equations are none other than Euler-Lagrange equations for functions $q^i(t)$ extremising the *action functional*

$$A(\mathbf{q}) = \int_{t_1}^{t_2} L(\mathbf{q}, \dot{\mathbf{q}})\, dt$$

[*see* (10.3.5)].

Lagrange had obtained the equations (11.2.3) and similar equations corresponding to more general systems in 1760. However, the importance of these equations and, particularly of the approach leading to these equations has been fully understood only after he has published in 1788 *Mécanique Analytique*, which is a groundbreaking and probably one of the most influential books in the history of science. In this work, Lagrange has succeeded to convert the rational mechanics to a branch of mathematical analysis. In contrast to the geometrical approach prevalent at that time, his priding himself on not including even a single figure in his book[1] is a striking statement reflecting his new philosophy to which he had subscribed in treating the rational mechanics.

For scleronomic systems the kinetic energy given by (11.2.1) enables us to equip the configuration manifold M with a metric so that M becomes a complete Riemannian manifold. We define the metric tensor by using the coefficient functions $g_{ij}(\mathbf{q})$ in the expression for the kinetic energy just like in (5.9.1) as follows

$$\boldsymbol{\mathcal{G}} = g_{ij}(\mathbf{q})\, dq^i \otimes dq^j \in \mathfrak{T}(M)^0_2.$$

Therefore, we can introduce an inner product on $T(M)$ by the relation

$$(U, V) = \boldsymbol{\mathcal{G}}(U, V) = g_{ij}u^i v^j, \quad U, V \in T(M).$$

The arc element on the manifold M in the direction of the generalised velocity vector is then given by

$$ds^2 = g_{ij}\, dq^i dq^j = g_{ij}\, \dot{q}^i\, \dot{q}^j\, dt^2 = 2T\, dt^2$$

or $ds = \sqrt{2T}\, dt$. Therefore, in such kind of systems when we insert the

[1]"On ne trouvera point de Figures dans cet Ouvrage."

Lagrangian function

$$L = \frac{1}{2}\, g_{ij}(\mathbf{q})\,\dot{q}^{\,i}\,\dot{q}^{\,j} - V(\mathbf{q})$$

into the equations (11.2.3), we arrive at the following equations of motion on the manifold M

$$\big[g_{ij}(\mathbf{q})\,\dot{q}^{\,j}\big]^{\cdot} - \frac{1}{2}\frac{\partial g_{jk}}{\partial q^i}\,\dot{q}^{\,j}\,\dot{q}^{\,k} + \frac{\partial V}{\partial q^i} = 0.$$

On evaluating the time derivatives above and arranging the resulting terms, we obtain

$$g_{ij}\,\ddot{q}^j + \Big(\frac{\partial g_{ij}}{\partial q^k} - \frac{1}{2}\frac{\partial g_{jk}}{\partial q^i}\Big)\dot{q}^{\,j}\,\dot{q}^{\,k} + \frac{\partial V}{\partial q^i} = 0.$$

Nevertheless, if we notice that only the symmetric part with respect to indices j and k of the expression within parentheses in the above equations would survive, then it is straightforward to see that these set of equations can be cast into the form

$$g_{ij}\,\ddot{q}^j + \frac{1}{2}\Big(\frac{\partial g_{ij}}{\partial q^k} + \frac{\partial g_{ik}}{\partial q^j} - \frac{\partial g_{jk}}{\partial q^i}\Big)\dot{q}^{\,j}\,\dot{q}^{\,k} + \frac{\partial V}{\partial q^i} = 0.$$

If we utilise the relation $g^{jk}g_{ik} = \delta_i^j$ and recall the definition (7.4.5) of the Christoffel symbols of the second kind, we end up with the following set of second order, generally non-linear ordinary differential equations by inverting the coefficient matrix $[g_{ij}]$

$$\ddot{q}^i + \Gamma^i_{jk}(\mathbf{q})\,\dot{q}^{\,j}\,\dot{q}^{\,k} = -\,g^{ij}\frac{\partial V}{\partial q^j}.$$

When we suppose that $V = 0$, these equations reveal the fact that *points representing dynamical systems that are free of forces must move on some geodesics in the configuration manifold* [see (7.2.16)].

The set of second order differential equations (11.2.3) can be transformed into an equivalent but larger set of first order ordinary differential equations by introducing certain auxiliary variables. To this end, we shall select the new variables p_i with $i = 1, 2, \ldots, m$ that will be called the ***generalised momenta*** as follows

$$p_i = \frac{\partial L}{\partial \dot{q}^{\,i}}. \tag{11.2.4}$$

When the condition

$$\det\left[\frac{\partial p_i}{\partial \dot{q}^j}\right] = \det\left[\frac{\partial^2 L}{\partial \dot{q}^i \partial \dot{q}^j}\right] \neq 0$$

is met, then by resorting to the inverse mapping, (11.2.4) yields in principle

$$\dot{q}^i = \mathbf{q}^i(p_1, \dots, p_m, q^1, \dots, q^m). \tag{11.2.5}$$

So long as the quantities $\dot{q}^i$ are given by (11.2.5), the ***Hamiltonian function*** $H = H(\mathbf{p}, \mathbf{q})$ can now be defined by the *Legendre transformation*

$$H(\mathbf{p}, \mathbf{q}) = p_i \dot{q}^i - L(\mathbf{q}, \dot{\mathbf{q}}). \tag{11.2.6}$$

When we evaluate the differential of the function (11.2.6) and employ the equations (11.2.4) and (11.2.3), we conclude that

$$\begin{aligned} dH = \frac{\partial H}{\partial p_i} dp_i + \frac{\partial H}{\partial q^i} dq^i &= \dot{q}^i dp_i + p_i\, d\dot{q}^i - \frac{\partial L}{\partial q^i}\, dq^i - \frac{\partial L}{\partial \dot{q}^i}\, d\dot{q}^i \\ &= \dot{q}^i dp_i - \dot{p}_i\, dq^i \end{aligned}$$

from which we derive the first order ***Hamilton equations***

$$\dot{q}^i = \frac{\partial H}{\partial p_i}, \quad \dot{p}_i = -\frac{\partial H}{\partial q^i}, \quad i = 1, 2, \dots, m. \tag{11.2.7}$$

In order to fully understand the exact nature of generalised momenta p_i, we wish to examine their behaviour under a coordinate transformation $Q^i = \mathcal{Q}^i(q^j)$ in the configuration manifold. To this end, let us define an $m \times m$ matrix $\boldsymbol{\mathcal{Q}}$ by

$$\boldsymbol{\mathcal{Q}} = [\mathfrak{Q}^i_j(\mathbf{q})] = \left[\frac{\partial \mathcal{Q}^i}{\partial q^j}\right].$$

so that the time derivative of the coordinate transformation is expressible as

$$\dot{Q}^i = \frac{\partial \mathcal{Q}^i}{\partial q^j}\,\dot{q}^j = \mathfrak{Q}^i_j\,\dot{q}^j \text{ or } \dot{\mathbf{Q}} = \boldsymbol{\mathfrak{Q}}\dot{\mathbf{q}} \text{ or } \dot{\mathbf{q}} = \boldsymbol{\mathfrak{Q}}^{-1}\dot{\mathbf{Q}}.$$

Making use of these relations we obtain

$$P_i = \frac{\partial L}{\partial \dot{Q}^i} = \frac{\partial L}{\partial \dot{q}^j}\frac{\partial \dot{q}^j}{\partial \dot{Q}^i} = (\boldsymbol{\mathfrak{Q}}^{-1})^j_i\, p_j = \frac{\partial q^j}{\partial Q^i} p_j$$

which means that the elements $\{p_1, p_2, \cdots, p_m\}$ behaves like components of a covariant vector that is a member of the cotangent bundle, in other words, they are the components of a 1-form. Therefore, the coordinate cover of the

$2m$-dimensional differentiable manifold $S = T^*(M)$ is given by $\{q^i, p_i\}$. We name this manifold as the ***momentum phase space***, or in short, merely the ***phase space***. Thus, the vector fields $\{\dot{q}^i, \dot{p}_i\}$ satisfying the Hamilton equations (11.2.6) inhabit the tangent bundle $T(T^*(M))$. On the other hand, the 1-form defined by

$$\theta = p_i\, dq^i = p_1\, dq^1 + \cdots + p_m\, dq^m \in \Lambda^1(S) \tag{11.2.8}$$

and usually known as the ***Liouville form*** is a member of the cotangent bundle $T^*(M)$. From the exterior derivative of the form (11.2.8) we can generate a closed 2-form

$$\begin{aligned} \omega &= -\, d\theta = -\, dp_i \wedge dq^i = dq^i \wedge dp_i \\ &= dq^1 \wedge dp_1 + \cdots + dq^m \wedge dp_m \in \Lambda^2(S). \end{aligned} \tag{11.2.9}$$

Let us denote the coordinate cover of the manifold S by $\{x^a, a = 1, 2, \ldots, 2m\}$. These coordinates will represent the coordinates q^i when we take $a = i$ with $1 \le i \le m$, and the coordinates p_i when we take $a = m + i$ if we do not mind a slight abuse of notation due to the unfamiliar positions of superscripts and subscripts. Hence, the form ω can now be written as follows

$$\omega = \frac{1}{2}\,\omega_{ab}(\mathbf{x})dx^a \wedge dx^b, \quad 1 \le a, b \le 2m.$$

In this case, the coefficients ω_{ab} of the form ω can now be expressed by the $2m \times 2m$ antisymmetric matrix

$$\mathbf{J} = [\,\omega_{ab}] = \begin{bmatrix} \mathbf{0} & \mathbf{I}_m \\ -\mathbf{I}_m & \mathbf{0} \end{bmatrix}. \tag{11.2.10}$$

where $m \times m$ identity matrix is denoted by $\mathbf{I}_m$. Since $\det \mathbf{J} = 1$, then the rank of 2-form ω is maximal, namely, it is $2m$. We shall see a little later that ω is also non-degenerate. Hence, ω is a symplectic form [*see* p. 46]. We shall call this form whose structure has been manifested by (11.2.9) as the ***canonical symplectic form***. The generalised coordinates $\{q^i, p_i\}$ that enable us to write the symplectic form locally in this way are also called ***canonical coordinates***. We refer a manifold S endowed with a symplectic form as a ***symplectic manifold***. Inasmuch as the rank of ω is $2m$, the Darboux class of the form θ is m. Consequently, Theorem 6.6.2 states that we can always find canonical coordinates that make it possible to write the symplectic form locally in the canonical form (11.2.9).

The matrix $\boldsymbol{J}$ is called a ***symplectic matrix*** [*see* Exercise **3.4**]. We can immediately see that this matrix enjoys the following properties

$$\boldsymbol{J}^{\mathrm{T}} = \begin{bmatrix} \mathbf{0} & -\mathbf{I}_m \\ \mathbf{I}_m & \mathbf{0} \end{bmatrix} = -\boldsymbol{J}$$

$$\boldsymbol{J}^2 = \begin{bmatrix} \mathbf{0} & \mathbf{I}_m \\ -\mathbf{I}_m & \mathbf{0} \end{bmatrix}\begin{bmatrix} \mathbf{0} & \mathbf{I}_m \\ -\mathbf{I}_m & \mathbf{0} \end{bmatrix} = \begin{bmatrix} -\mathbf{I}_m & \mathbf{0} \\ \mathbf{0} & -\mathbf{I}_m \end{bmatrix}$$
$$= -\mathbf{I}_{2m}$$

whence we deduce that

$$\boldsymbol{J}^{-1} = -\boldsymbol{J} = \boldsymbol{J}^{\mathrm{T}}.$$

We would like now to introduce a mapping $S_\omega : T(S) \to T^*(S)$ between the tangent and cotangent bundles of a symplectic manifold S which may be equivalently interpreted as a mapping $S_\omega : \mathfrak{V}(S) \to \Lambda^1(S)$ between the tangent module $\mathfrak{V}(S)$ and the module of 1-forms $\Lambda^1(S)$. For each vector field $V \in T(S)$, we define this mapping by employing the symplectic form in the following fashion

$$S_\omega V = \mathbf{i}_V(\omega) \in \Lambda^1(S). \tag{11.2.11}$$

Because of the properties (5.4.7), we immediately see that S_ω *is a linear operator on the module* $\mathfrak{V}(S)$. The value of this 1-form on a vector field $U \in T(S)$ is naturally given by

$$S_\omega V(U) = \mathbf{i}_U\big(\mathbf{i}_V(\omega)\big) = \omega(V, U) \in \Lambda^0(S).$$

If we write

$$V = v^a \frac{\partial}{\partial x^a} = v^i \frac{\partial}{\partial q^i} + \mathfrak{v}_i \frac{\partial}{\partial p_i} \in T(S),$$

then (11.2.11) yields

$$S_\omega V = \mathbf{i}_V(\omega) = \omega_{ab}\, v^a\, dx^b = v^i\, dp_i - \mathfrak{v}_i\, dq^i \in \Lambda^1(S).$$

Let us consider a form $\alpha \in \Lambda^1(S)$ by

$$\alpha = \xi_i dq^i + \eta^i dp_i.$$

It is straightforward to observe immediately that this 1-form is the image $S_\omega V_\alpha$ of the vector

$$V_\alpha = \eta^i \frac{\partial}{\partial q^i} - \xi_i \frac{\partial}{\partial p_i}, \quad \alpha = \mathbf{i}_{V_\alpha}(\omega). \tag{11.2.12}$$

Hence, the operator S_ω will be surjective. On the other hand, if we write

$S_\omega V = 0$ we end up with the expression $\omega_{ab}\, v^a = 0$. Since the matrix $\mathbf{J} = [\omega_{ab}]$ is regular, we find only the trivial solution $v^a = 0$ or $V = 0$, that is, the operator S_ω is injective, and consequently it is bijective. Therefore, the operator S_ω is one of the isomorphisms between tangent and cotangent spaces. Thus, the inverse mapping $S_\omega^{-1} : \Lambda^1(S) \to \mathfrak{V}(S)$ assigns to each 1-form field

$$\alpha = \xi_i dq^i + \eta^i dp_i \in \Lambda^1(S)$$

a unique vector field

$$S_\omega^{-1}\alpha = V_\alpha = \eta^i \frac{\partial}{\partial q^i} - \xi_i \frac{\partial}{\partial p_i} \in T(S).$$

It is evident that one can write $\alpha = S_\omega V_\alpha$. Since the relation $\mathbf{i}_V(\omega) = \omega(V) = 0$ is satisfied if and only if $V = 0$, we gather that the form ω is *non-degenerate.*

Let us next consider the smooth function $H \in \Lambda^0(S)$. A vector field V_H complying with the condition

$$S_\omega V_H = \mathbf{i}_{V_H}(\omega) = dH \in \Lambda^1(S) \qquad (11.2.13)$$

is called a ***Hamiltonian vector field***. Since S_ω is an isomorphism, when a function H is chosen, the Hamiltonian vector field corresponding to this function is *uniquely* determined through the relation $V_H = S_\omega^{-1}(dH)$. If we explicitly write dH as

$$dH = \frac{\partial H}{\partial q^i} dq^i + \frac{\partial H}{\partial p_i} dp_i,$$

then the corresponding Hamiltonian vector field is given by the relation

$$V_H = \frac{\partial H}{\partial p_i}\frac{\partial}{\partial q^i} - \frac{\partial H}{\partial q^i}\frac{\partial}{\partial p_i}. \qquad (11.2.14)$$

It is now obvious that trajectories of such a vector field will have to satisfy the Hamilton equations

$$\dot{q}^i = \frac{\partial H}{\partial p_i}, \quad \dot{p}_i = -\frac{\partial H}{\partial q^i}$$

cited in (11.2.7). Thus, it would be then quite reasonable to state that the equation (11.2.13) is the *symplectic form of the Hamilton equations.*

Example 11.2.1. Suppose that $H = qp^2 - qp + ap$ where a is a constant. In that case, the Hamilton equations become

$$\frac{dq}{dt} = 2qp - q + a,$$
$$\frac{dp}{dt} = -p^2 + p.$$

The integration of these differential equations yields easily

$$p(t) = \frac{1}{1 - e^{-(t-c_1)}},$$
$$q(t) = \left(1 - e^{-(t-c_1)}\right)\left[c_2\, e^t\left(1 - e^{-(t-c_1)}\right) - a\right].$$

c_1 and c_2 are integration constants to be determined through the initial conditions. If we evaluate the given Hamiltonian function on these trajectories, we find that

$$H = q(t)p(t)^2 - q(t)p(t) + ap(t) = a + e^{c_1}c_2 = \mathit{constant}.$$

Clearly, this constant will generally be different on each trajectory in the phase space. ■

11.3. POISSON BRACKET OF 1-FORMS AND SMOOTH FUNCTIONS

(S, ω) is a symplectic manifold. We consider the forms $\alpha, \beta \in \Lambda^1(S)$. The ***Poisson bracket*** *of 1-forms α and β is also a 1 form* $\{\alpha, \beta\} \in \Lambda^1(S)$ defined by the following relation

$$\begin{aligned}\{\alpha, \beta\} &= -\,S_\omega([V_\alpha, V_\beta]) = -\,\mathbf{i}_{[V_\alpha, V_\beta]}(\omega) \\ &= -\,S_\omega([S_\omega^{-1}\alpha, S_\omega^{-1}\beta])\end{aligned} \tag{11.3.1}$$

where the vector fields $V_\alpha = S_\omega^{-1}\alpha$ and $V_\beta = S_\omega^{-1}\beta$ are generated from the forms α and β, respectively, through the isomorphism S_ω. Consequently, on the module $\mathfrak{V}(S)$ the expression

$$S_\omega^{-1}(\{\alpha, \beta\}) = -\,[S_\omega^{-1}\alpha, S_\omega^{-1}\beta] \tag{11.3.2}$$

would be valid. On the other hand, if we recall that $d\omega = 0$ we can write

$$\begin{aligned}\mathbf{i}_{[V_\alpha, V_\beta]}(\omega) &= \mathbf{i}_{\pounds_{V_\alpha} V_\beta}(\omega) = [\pounds_{V_\alpha}, \mathbf{i}_{V_\beta}](\omega) = \pounds_{V_\alpha}\mathbf{i}_{V_\beta}(\omega) - \mathbf{i}_{V_\beta}\pounds_{V_\alpha}(\omega) \\ &= \pounds_{V_\alpha}\beta - \mathbf{i}_{V_\beta}\big(\mathbf{i}_{V_\alpha}(d\omega) + d\mathbf{i}_{V_\alpha}(\omega)\big) \\ &= \pounds_{V_\alpha}\beta - \mathbf{i}_{V_\beta}(d\alpha)\end{aligned}$$

owing to the equality (5.11.7). However, if we take into account the Cartan magic formula $\mathbf{i}_{V_\beta}(d\alpha) = \pounds_{V_\beta}\alpha - d\mathbf{i}_{V_\beta}(\alpha)$ and anticommutativity (5.4.4) of the interior product, the Poisson bracket becomes expressible as

$$\{\alpha, \beta\} = -\,\mathbf{i}_{[V_\alpha, V_\beta]}(\omega) = \pounds_{V_\beta}\alpha - \pounds_{V_\alpha}\beta + d\big(\mathbf{i}_{V_\alpha}\mathbf{i}_{V_\beta}(\omega)\big). \quad (11.3.3)$$

It then follows from (11.3.3) that *the Poisson bracket of two closed* 1*-forms is an exact* 1*-form*. In fact, when $d\alpha = 0$ and $d\beta = 0$, then we have

$$\pounds_{V_\alpha}\beta = d\big(\mathbf{i}_{V_\alpha}\mathbf{i}_{V_\beta}(\omega)\big), \;\; \pounds_{V_\beta}\alpha = -\,d\big(\mathbf{i}_{V_\alpha}\mathbf{i}_{V_\beta}(\omega)\big)$$

and (11.3.3) leads to

$$\begin{aligned}\{\alpha, \beta\} &= -\,d\big(\mathbf{i}_{V_\alpha}\mathbf{i}_{V_\beta}(\omega)\big) = -\,d\big(\omega(V_\beta, V_\alpha)\big) \\ &= d\big(\omega(V_\alpha, V_\beta)\big).\end{aligned} \quad (11.3.4)$$

From the definition of the Poisson bracket and the linearity of the operator S_ω, we see that the following properties are valid:

(i). $\{\alpha, \beta\} = -\,\{\beta, \alpha\}$ (*Antisymmetry*),
(ii). $\{\alpha, b\beta + c\gamma\} = b\{\alpha, \beta\} + c\{\alpha, \gamma\}, \quad b, c \in \mathbb{R}$ (*Linearity*).

Furthermore, the Poisson bracket satisfies the *Jacobi identity*

(iii). $\{\alpha, \{\beta, \gamma\}\} + \{\beta, \{\gamma, \alpha\}\} + \{\gamma, \{\alpha, \beta\}\} = 0.$

In order to see this, it only suffices to notice that one can write

$$\{\alpha, \{\beta, \gamma\}\} = -\,S_\omega([V_\alpha, [V_\beta, V_\gamma]])$$

and the operator S_ω is linear.
(iv). Let us consider a function $f \in \Lambda^0(S)$. The linearity of the operator S_ω and the relation (2.10.19) result in

$$\begin{aligned}\{\alpha, f\beta\} &= -\,S_\omega([V_\alpha, fV_\beta]) \\ &= -\,S_\omega\big(f[V_\alpha, V_\beta] + V_\alpha(f)V_\beta\big) \\ &= -\,fS_\omega([V_\alpha, V_\beta]) - V_\alpha(f)S_\omega V_\beta \\ &= f\{\alpha, \beta\} + V_\alpha(f)\beta.\end{aligned}$$

The properties (i), (ii) and (iii) demonstrate that the module $\Lambda^1(S)$ of 1-forms constitutes a Lie algebra with respect to the Poisson bracket if we rightly interpret the Poisson bracket as the Lie product of 1-forms. Since the Poisson bracket of two closed form is an exact, consequently, a closed form, we realise at once that closed 1-forms is a subalgebra of such a Lie algebra of 1-forms. It then obviously follows from the relation (11.3.1) that $\{\alpha, \beta\} = 0$ whenever $[V_\alpha, V_\beta] = 0$.

Let us next take into account 1-forms $df, dg \in \Lambda^1(S)$ that are exterior derivatives of functions $f, g \in \Lambda^0(S)$. We know that the isomorphism S_ω generates the vectors

$$V_f = S_\omega^{-1} df = \frac{\partial f}{\partial p_i}\frac{\partial}{\partial q^i} - \frac{\partial f}{\partial q^i}\frac{\partial}{\partial p_i},$$
$$V_g = S_\omega^{-1} dg = \frac{\partial g}{\partial p_i}\frac{\partial}{\partial q^i} - \frac{\partial g}{\partial q^i}\frac{\partial}{\partial p_i}.$$

In view of the relation (11.2.13), these vectors are Hamiltonian vector fields associated with functions f and g. Since df and dg are closed forms, then (11.3.4) yields

$$\{df, dg\} = -\, d\big(\mathbf{i}_{V_f}\mathbf{i}_{V_g}(\omega)\big) = d\{f, g\} \tag{11.3.5}$$

where the ***Poisson bracket*** *of the functions f and g* are defined by the following relation

$$\{f, g\} = -\,\mathbf{i}_{V_f}\mathbf{i}_{V_g}(\omega) = \mathbf{i}_{V_g}\mathbf{i}_{V_f}(\omega) \in \Lambda^0(S). \tag{11.3.6}$$

On the other hand, we can easily evaluate that

$$\mathbf{i}_{V_g}(\omega) = \frac{\partial g}{\partial p_i}dp_i + \frac{\partial g}{\partial q^i}dq^i = dg,$$
$$\mathbf{i}_{V_f}(dg) = V_f(g).$$

Therefore, we conclude that the Poisson bracket of two functions f and g is determined by the expression

$$\{f, g\} = -\,V_f(g) = V_g(f) = \frac{\partial f}{\partial q^i}\frac{\partial g}{\partial p_i} - \frac{\partial f}{\partial p_i}\frac{\partial g}{\partial q^i}. \tag{11.3.7}$$

We can thereby deduce from the relations (11.3.2) and (11.3.5) that

$$S_\omega^{-1}(\{df, dg\}) = S_\omega^{-1} d\{f, g\} = -\,[S_\omega^{-1} df, S_\omega^{-1} dg].$$

This simply implies that Hamiltonian vector fields generated by functions f, g and $\{f, g\}$ are connected by the relation

$$V_{\{f,g\}} = -\,[V_f, V_g]. \tag{11.3.8}$$

The equation (11.3.8) amounts to say that if V_f and V_g are Hamiltonian vector fields, then their Lie product $[V_f, V_g]$ is also a Hamiltonian vector field. This, of course, means that Hamiltonian vector fields constitute a Lie subalgebra. We then observe from the expression (11.3.6) that the equality

$$\{f,g\} = -\{g,f\}$$

holds. On the other hand, for three functions $f,g,h\in\Lambda^0(S)$ (11.3.7) leads to the relations

$$\begin{aligned}&\{f,\{g,h\}\}=V_fV_g(h),\ \{g,\{h,f\}\}=V_gV_h(f)=-V_gV_f(h)\\&\{h,\{f,g\}\}=-\{\{f,g\},h\}=V_{\{f,g\}}(h)=-[V_f,V_g](h)\end{aligned}$$

from which we deduce the identity

$$\{f,\{g,h\}\}+\{g,\{h,f\}\}+\{h,\{f,g\}\}=([V_f,V_g]-[V_f,V_g])(h)=0.$$

Hence, Poisson brackets on smooth functions verify the Jacobi identity as well. Accordingly, the module $\Lambda^0(S)$ equipped with the Poisson bracket is a Lie algebra. One readily sees that the Poisson bracket $\{f,g\}$ is a bilinear function on real numbers. Moreover, we find that

$$\begin{aligned}\{f,hg\}&=-V_f(hg)=-V_f(h)g-V_f(g)h\\&=g\{f,h\}+h\{f,g\}.\end{aligned}$$

Let us now consider canonical local coordinates. Then (11.3.7) leads to the relations

$$\{q^k,q^l\}=0,\ \{p_k,p_l\}=0,\ \{q^k,p_l\}=-\{p_l,q^k\}=\delta^k_l$$

and for a function $f\in\Lambda^0(S)$ we obtain

$$\{f,q^i\}=-\{q^i,f\}=-\frac{\partial f}{\partial p_i},\ \{f,p_i\}=-\{p_i,f\}=\frac{\partial f}{\partial q^i}.$$

Hence, the Hamilton equations can now be written in the form

$$\dot q^i=\{q^i,H\},\quad \dot p_i=\{p_i,H\}. \tag{11.3.9}$$

We shall next try to evaluate the change in a function $f\in\Lambda^0(S)$ on the flow $e^{tV_g}:S\to S$ generated by a Hamiltonian vector field V_g on the manifold S. We know that we can write $f(t)=e^{t\pounds_{V_g}}f$ and the necessary and sufficient condition for the function f to remain invariant under this flow is $\pounds_{V_g}f=0$. However, this condition means that

$$\pounds_{V_g}f=V_g(f)=\{f,g\}=0. \tag{11.3.10}$$

Accordingly, if the Poisson bracket, or equivalently the Lie product, of functions f and g vanishes, then the function f has to remain 'constant' on the flow on S generated by the Hamiltonian vector field V_g, i.e., $f(\mathbf{q},\mathbf{p})=f(\mathbf{q}_0,\mathbf{p}_0)$ where $(\mathbf{q},\mathbf{p})=e^{tV_g}(\mathbf{q}_0,\mathbf{p}_0)$ although this constant may take different values on each trajectory. The relation (11.3.10) requires, of course, that the function g will, in turn, remain constant on the flow produced by the Hamiltonian vector field V_f. We already know that a trajectory of the Hamiltonian vector field V_H associated with the Hamiltonian function H determines the evolution of a dynamical system with particular initial conditions on the symplectic manifold S. The time rate of change of a function $f \in \Lambda^0(S)$ during the evolution of the dynamical system can now be calculated by

$$\frac{df}{dt} = \frac{\partial f}{\partial q^i}\dot{q}^i + \frac{\partial f}{\partial p_i}\dot{p}_i = \frac{\partial f}{\partial q^i}\frac{\partial H}{\partial p_i} - \frac{\partial f}{\partial p_i}\frac{\partial H}{\partial q^i} = \{f, H\}.$$

Thus, in case a function $f \in \Lambda^0(S)$ verifies the condition

$$\{f, H\} = 0,$$

it remains constant in association with the evolution of the dynamical system. A relation between generalised coordinates and generalised momenta in the form

$$f(\mathbf{q}(t), \mathbf{p}(t)) = c = constant$$

corresponds to an ***integral of the motion*** and help us to reduce the number of the dependent variables $\mathbf{p}(t)$. Due to the property of the Poisson bracket, we clearly obtain $\{H, H\} = 0$. Hence, the Hamiltonian function H is an integral of the motion. The relation

$$H(\mathbf{p}, \mathbf{q}) = p_i\dot{q}^i - L(\mathbf{q}, \dot{\mathbf{q}}) = constant$$

is known as the ***conservation of energy***. As a matter of fact, when the kinetic energy is prescribed by (11.2.1) we find at once that

$$p_i = \frac{\partial L}{\partial \dot{q}^i} = g_{ij}\,\dot{q}^j \text{ and } p_i\dot{q}^i = g_{ij}\,\dot{q}^i\dot{q}^j = 2T.$$

Thus, the Hamiltonian function

$$H = 2T - T + V = T + V$$

represents now the total energy of the dynamical system that is conserved during the motion of the system.

Finally, we attempt to calculate the Lie derivative of the symplectic form ω with respect to a Hamiltonian vector field V_f. Since $d\omega = 0$, we easily obtain

$$\pounds_{V_f}\,\omega = d\,\mathbf{i}_{V_f}(\omega) = d(df) = d^2 f = 0. \tag{11.3.11}$$

Consequently, the symplectic form ω remains invariant under the flow produced by a Hamiltonian vector field. In other words, under the mapping $\phi_t = e^{tV_f} : S \to S$, we get

$$\omega^* = \phi_t^*\omega = e^{t\pounds_{V_f}}\omega = \omega.$$

The volume form of the symplectic manifold S is of course

$$\mu = dq^1 \wedge dq^2 \wedge \cdots \wedge dq^m \wedge dp_1 \wedge dp_2 \wedge \cdots \wedge dp_m \in \Lambda^{2m}(S).$$

It is quite easy now to prove the following theorem .

Theorem 11.3.1 (The Liouville Theorem). *Let (S,ω) be a $2m$-dimensional symplectic manifold and ϕ_t be the flow of a Hamiltonian vector field. The mapping ϕ_t^* preserves the volume form μ of the symplectic manifold for all t, namely, the invariance condition $\phi_t^*\mu = \mu$ is satisfied.*

Indeed, the volume form $\mu \in \Lambda^{2m}(S)$ is expressible as

$$\mu = \frac{(-1)^{\frac{m(m-1)}{2}}}{m!} \underbrace{\omega \wedge \omega \wedge \cdots \wedge \omega}_{m} = C\omega^m.$$

Nevertheless, on account of the relations (5.7.4) and $\phi_t^*\omega = \omega$ we obtain

$$\phi_t^*\mu = C\phi_t^*\omega \wedge \phi_t^*\omega \wedge \cdots \wedge \phi_t^*\omega = C\omega \wedge \omega \wedge \cdots \wedge \omega = \mu. \qquad \square$$

According to this theorem *the volume of the phase space is conserved under a flow generated by trajectories of a Hamiltonian vector field.* This statement is of course true for every volume elements in the phase space.

Next, let us consider a form $\Omega_k = \omega^k = \underbrace{\omega \wedge \omega \wedge \cdots \wedge \omega}_{k} \in \Lambda^{2k}(S)$, $1 \le k \le m$. For an arbitrary Hamiltonian vector field V_f, we obviously find that $\pounds_{V_f}\Omega_k = 0$. Hence, all the forms $\Omega_k, 1 \le k \le m$ remain invariant under flows generated by Hamiltonian vector fields.

11.4. CANONICAL TRANSFORMATIONS

(S,ω) is a symplectic manifold with canonical coordinates $\{q^i, p_i\}$. If a mapping $\phi : S \to S$ transforming this manifold into itself leaves the symplectic form invariant, that is, if it satisfies the condition

$$\phi^*\omega = \omega, \tag{11.4.1}$$

then it is called a ***canonical*** or ***symplectic transformation***. Accordingly, the flow of every Hamiltonian vector field produces a canonical transformation. Because of (11.4.1), we have $\phi^*\mu = \mu$ so that one obtains $\det\phi = 1$ [*see*

(5.7.11)]. Therefore, a canonical transformation will preserve a volume form and its orientation. Furthermore, it must be locally a diffeomorphism. A canonical transformation ϕ will best be expressed as a local coordinate transformations $(Q^i, P_i) \to (q^i, p_i)$ with $i = 1, \ldots, m$ in the phase space specified by the functions

$$q^i = q^i(Q^1, \ldots, Q^m, P_1, \ldots, P_m), \; p_i = p_i(Q^1, \ldots, Q^m, P_1, \ldots, P_m).$$

Insofar as we have assumed that ϕ is a diffeomorphism, the inverse transformations

$$Q^i = Q^i(q^1, \ldots, q^m, p_1, \ldots, p_m), \; P_i = P_i(q^1, \ldots, q^m, p_1, \ldots, p_m)$$

will exist, at least, locally. Our expectation from such a canonical transformation would be to make the equations of motion acquire a simpler structure. The relation (11.4.1) now takes the form

$$\begin{aligned} \phi^*(dq^i \wedge dp_i) &= d(\phi^* q^i) \wedge d(\phi^* p_i) \\ &= dQ^i \wedge dP_i = dq^i \wedge dp_i. \end{aligned} \tag{11.4.2}$$

Thus, in order that ϕ turns out to be a canonical transformation, the functions Q^i and P_i must satisfy the relations

$$\begin{aligned} &\frac{\partial Q^i}{\partial q^j}\frac{\partial P_i}{\partial q^k}\, dq^j \wedge dq^k + \frac{\partial Q^i}{\partial p_j}\frac{\partial P_i}{\partial p_k}\, dp_j \wedge dp_k \\ &\qquad + \Big(\frac{\partial Q^i}{\partial q^j}\frac{\partial P_i}{\partial p_k} - \frac{\partial Q^i}{\partial p_k}\frac{\partial P_i}{\partial q^j}\Big)\, dq^j \wedge dp_k = dq^i \wedge dp_i. \end{aligned}$$

If we take into account the antisymmetry of the exterior product, the above relation leads to the following equations

$$\frac{\partial Q^i}{\partial q^j}\frac{\partial P_i}{\partial q^k} - \frac{\partial Q^i}{\partial q^k}\frac{\partial P_i}{\partial q^j} = 0, \qquad \frac{\partial Q^i}{\partial p_j}\frac{\partial P_i}{\partial p_k} - \frac{\partial Q^i}{\partial p_k}\frac{\partial P_i}{\partial p_j} = 0$$

$$\frac{\partial Q^i}{\partial q^j}\frac{\partial P_i}{\partial p_k} - \frac{\partial Q^i}{\partial p_k}\frac{\partial P_i}{\partial q^j} = \delta^k_j.$$

To treat this matter in a more general context, let us consider two $2m$-dimensional symplectic manifolds (S_1, ω_1) and (S_2, ω_2). A local diffeomorphism $\phi : S_1 \to S_2$ satisfying the relation

$$\phi^* \omega_2 = \omega_1 \tag{11.4.3}$$

is called a ***canonical*** or ***symplectic mapping***. If ϕ is a symplectic mapping, it will preserve the volume form so we must have the condition $\det \phi = 1$.

Theorem 11.4.1 (The Jacobi Theorem). *Let* (S_1, ω_1) *and* (S_2, ω_2) *be* $2m$*-dimensional symplectic manifolds. A diffeomorphism* $\phi : S_1 \to S_2$ *is a symplectic mapping if and only if Hamiltonian vector fields* $V_f \in T(S_2)$ *and* $V_{\phi^* f} \in T(S_1)$ *are to satisfy the relation*

$$(d\phi)^{-1} V_f = \phi_*^{-1} V_f = V_{\phi^* f} \quad or \quad V_f = \phi_* V_{\phi^* f}. \tag{11.4.4}$$

for all $f \in \Lambda^0(S_2)$.

Let us first demonstrate that the relation

$$S_{\omega_1}^{-1}(\phi^* \alpha) = \phi_*^{-1} S_{\omega_2}^{-1}(\alpha) \tag{11.4.5}$$

is satisfied for all $\alpha \in \Lambda^1(S_2)$ if and only if ϕ is a symplectic mapping. Let $V = S_{\omega_2}^{-1}(\alpha) \in T(S_2)$. Utilising the relation (5.7.7), we find that

$$\phi^* \alpha = \phi^*(\mathbf{i}_V \omega_2) = \mathbf{i}_{\phi_*^{-1} V}\, \phi^* \omega_2.$$

By applying the operator $S_{\omega_1}^{-1}$ to this expression, we obtain

$$S_{\omega_1}^{-1}(\phi^* \alpha) = S_{\omega_1}^{-1} S_{\phi^* \omega_2}(\phi_*^{-1} V) = S_{\omega_1}^{-1} S_{\phi^* \omega_2}\big(\phi_*^{-1} S_{\omega_2}^{-1}(\alpha)\big).$$

Since the condition $\phi^* \omega_2 = \omega_1$ must be obeyed when ϕ is a symplectic mapping, we simply find the identity mapping $S_{\omega_1}^{-1} S_{\phi^* \omega_2} = S_{\omega_1}^{-1} S_{\omega_1} = i_{\mathfrak{V}(S_1)}$ and the relation (11.4.5) follows immediately. Conversely, if we suppose that the relation (11.4.5) is satisfied for all forms $\alpha \in \Lambda^1(S_2)$, we readily observe that the equality $S_{\omega_1}^{-1}(\phi^* \alpha) = S_{\omega_1}^{-1} S_{\phi^* \omega_2}\big(S_{\omega_1}^{-1}(\phi^* \alpha)\big)$ must result in $S_{\omega_1}^{-1} S_{\phi^* \omega_2} = I_{\mathfrak{V}(S_1)}$. This is, of course, realisable if only $\phi^* \omega_2 = \omega_1$, i.e., if ϕ is a symplectic mapping.

Let us now assume that ϕ is a symplectic mapping. In this case, the relation (11.4.5) leads to the result

$$\phi_*^{-1} V_f = \phi_*^{-1} S_{\omega_2}^{-1}(df) = S_{\omega_1}^{-1}(\phi^* df) = S_{\omega_1}^{-1}\big(d(\phi^* f)\big) = V_{\phi^* f}$$

in view of Theorem 5.8.2. Conversely, let us now assume that the equality (11.4.4) is satisfied for all $f \in \Lambda^0(S_2)$. We then successively obtain

$$d(\phi^* f) = \phi^* df = \phi^* S_{\omega_2} V_f = \phi^*(\mathbf{i}_{V_f} \omega_2) = \mathbf{i}_{\phi_*^{-1} V_f}\, \phi^* \omega_2 = \mathbf{i}_{V_{\phi^* f}}\, \phi^* \omega_2.$$

On the other hand, the same expression can also be written in the form

$$d(\phi^* f) = S_{\omega_1} V_{\phi^* f} = \mathbf{i}_{V_{\phi^* f}}\, \omega_1.$$

This implies that we obtain the relation

$$\mathbf{i}_{V_{\phi^* f}}\, \phi^* \omega_2 = \mathbf{i}_{V_{\phi^* f}}\, \omega_1$$

for each function $f \in \Lambda^0(S_2)$ and Hamiltonian vector field $V_{\phi^* f} \in T(S_1)$ from which it follows that

$$\mathbf{i}_{V_{\phi^* f}}(\phi^*\omega_2 - \omega_1) = 0.$$

Since the rank of symplectic forms must be maximal, the above equation is satisfied if and only if $\phi^*\omega_2 = \omega_1$, that is, if the diffeomorphism ϕ is a symplectic mapping.

If a diffeomorphism $\phi : S \to S$ is mapping a symplectic manifold S onto itself, then the above conditions are reduced to the ones such that the condition

$$\phi_*^{-1} V_f = V_{\phi^* f}$$

and consequently,

$$S_\omega^{-1}(\phi^*\alpha) = \phi_*^{-1} S_\omega^{-1}(\alpha)$$

must hold for all functions $f \in \Lambda^0(S)$ and forms $\alpha \in \Lambda^1(S)$. □

We can easily prove the existence of an important property related to symplectic diffeomorphisms and Poisson brackets.

Theorem 11.4.2. *Let (S_1, ω_1) and (S_2, ω_2) be $2m$-dimensional symplectic manifolds. A diffeomorphism $\phi : S_1 \to S_2$ is symplectic if and only if it preserves Poisson brackets, that is, if and only if the relation*

$$\phi^*\{f, g\} = \{\phi^* f, \phi^* g\} \tag{11.4.6}$$

is satisfied for all $f, g \in \Lambda^0(S_2)$.

In view of (11.3.7) and (11.3.10), we can write the Poisson bracket $\{f, g\} \in \Lambda^0(S_2)$ in the form

$$\{f, g\} = V_g(f) = \pounds_{V_g} f.$$

On account of (11.4.4), we have

$$V_g = \phi_* V_{\phi^* g}$$

if and only if ϕ is a symplectic mapping. Hence, making use of (5.11.17) we obtain

$$\phi^*\{f, g\} = \phi^* \pounds_{\phi_* V_{\phi^* g}} f = \pounds_{V_{\phi^* g}}(\phi^* f) = \{\phi^* f, \phi^* g\}.$$

This is tantamount to say that a symplectic mapping provides a homomorphism on $\Lambda^0(S_2)$ with respect to the Lie product defined by the Poisson bracket. □

It is quite straightforward to show that Theorem 11.4.2 will also be in effect for Poisson brackets of 1*-forms*. If we employ the relation (2.10.21), we obtain

$$\begin{aligned}\phi^*\{\alpha,\beta\} &= -\,\phi^*\mathbf{i}_{[V_\alpha,V_\beta]}(\omega_2) = -\,\phi^*\mathbf{i}_{\phi_*\phi_*^{-1}[V_\alpha,V_\beta]}(\omega_2)\\ &= -\,\phi^*\mathbf{i}_{\phi_*[\phi_*^{-1}V_\alpha,\,\phi_*^{-1}V_\beta]}(\omega_2) = -\,\mathbf{i}_{[\phi_*^{-1}V_\alpha,\,\phi_*^{-1}V_\beta]}(\phi^*\omega_2).\end{aligned}$$

On the other hand, the definition $\alpha = \mathbf{i}_{V_\alpha}(\omega_2)$ yields

$$\phi^*\alpha = \phi^*\mathbf{i}_{V_\alpha}(\omega_2) = \mathbf{i}_{\phi_*^{-1}V_\alpha}(\phi^*\omega_2).$$

If ϕ is a symplectic mapping, one must have $\phi^*\omega_2 = \omega_1$ and the relation $\phi^*\alpha = \mathbf{i}_{\phi_*^{-1}V_\alpha}(\omega_1)$ will follow. So we easily obtain

$$\phi^*\{\alpha,\beta\} = -\,\mathbf{i}_{[\phi_*^{-1}V_\alpha,\,\phi_*^{-1}V_\beta]}(\omega_1) = \{\phi^*\alpha,\phi^*\beta\}. \qquad (11.4.7)$$

Conversely, if (11.4.7) is to be satisfied for all forms $\alpha,\beta\in\Lambda^1(S_2)$, then the relation

$$-\,\mathbf{i}_{[\phi_*^{-1}V_\alpha,\,\phi_*^{-1}V_\beta]}(\phi^*\omega_2) = -\,\mathbf{i}_{[\phi_*^{-1}V_\alpha,\,\phi_*^{-1}V_\beta]}(\omega_1)$$

requires that $\phi^*\omega_2 = \omega_1$. □

Let us now consider a canonical mapping $\phi : S \to S$ on a symplectic manifold (S,ω). This mapping is of course represented by transformations between local canonical coordinates. We know that in this situation both sets of canonical coordinates must satisfy the relation (11.4.2). In order to systematically investigate the implication of (11.4.2), let us first define 1-forms θ and Θ as follows

$$\theta = p_i\,dq^i, \quad \Theta = P_i\,dQ^i.$$

The relation (11.4.2) compels us to write

$$\phi^*d\theta = d\Theta \quad \text{or} \quad d(\Theta - \phi^*\theta) = 0.$$

Thus, according to the Poincaré lemma we obtain at least locally

$$\Theta - \phi^*\theta = dF, \quad F\in\Lambda^0(S).$$

where F is an arbitrary function. Hence, whenever $\phi : S \to S$ satisfies locally the expression

$$P_i\,dQ^i - p_i\,dq^i = dF, \qquad (11.4.8)$$

then it becomes a canonical mapping. However, we have also to keep in mind that the function F should be so chosen that the mapping ϕ must be a diffeomorphism. For instance, if we choose a smooth function $F = F(\mathbf{p},\mathbf{q})$

in (11.4.8), the canonical mapping is determined by the equations

$$P_i \frac{\partial Q^i}{\partial q^j} = p_j + \frac{\partial F}{\partial q^j}, \; P_i \frac{\partial Q^i}{\partial p_j} = \frac{\partial F}{\partial p_j}$$

Example 11.4.1. For $m = 1$, we define a mapping $(p, q) \to (P, Q)$ by

$$P = \frac{1}{2}(p^2 + q^2), \quad Q = \arctan \frac{q}{p}.$$

We can then write

$$\begin{aligned} P\,dQ - p\,dq &= \frac{1}{2}(p^2 + q^2) \frac{p\,dq - q\,dp}{p^2\Big(1 + \dfrac{q^2}{p^2}\Big)} - p\,dq \\ &= -\frac{1}{2}(p\,dq + q\,dp) = -\frac{1}{2}\,d(pq) \end{aligned}$$

and find that

$$P\,dQ - p\,dq = dF, \quad F = -\frac{1}{2}\,pq.$$

Hence, this diffeomorphism is a canonical mapping. ■

Theorem 11.4.3. *Let (S, ω) be a $2m$-dimensional symplectic manifold. A canonical mapping $\phi : S \to S$ preserves the form of the Hamilton equations governing the motion of a dynamical system on this manifold.*

Let the canonical mapping $\phi : S \to S$ be prescribed by the coordinate transformation $(Q^i, P_i) \to (q^i, p_i)$. In the local coordinates (q^i, p_i) of the manifold S we shall assume that the Hamilton equations are specified in the symplectic form by

$$\mathbf{i}_{V_H}(\omega) = dH$$

where $H(\mathbf{p}, \mathbf{q})$ is the Hamiltonian function. On applying the pull-back operation on this equation, we obtain

$$\phi^* \mathbf{i}_{V_H}(\omega) = \mathbf{i}_{\phi_*^{-1} V_H}(\phi^* \omega) = \phi^* dH = d(\phi^* H)$$

Since ϕ is a canonical mapping, we can, of course, write $\phi^* \omega = \omega$ so we finally find that

$$\mathbf{i}_{V_{\phi^* H}}(\omega) = d(\phi^* H).$$

Therefore, by defining the function $K(\mathbf{P}, \mathbf{Q}) = \phi^* H = H \circ \phi$, we end up in the following expression

$$\mathbf{i}_{V_K}(\omega) = dK,\ \ K(\mathbf{P},\mathbf{Q}) = H\big(\mathbf{p}(\mathbf{P},\mathbf{Q})\mathbf{q}(\mathbf{P},\mathbf{Q})\big).$$

Thus, the vector field V_K associated with the function K is a Hamiltonian vector field and its trajectories satisfy the Hamilton equations

$$\dot{Q}^i = \frac{\partial K}{\partial P_i},\quad \dot{P}_i = -\frac{\partial K}{\partial Q^i}$$

corresponding to the Hamiltonian function K. □

This result brings to mind to search for an appropriate canonical transformation that simplifies the structure of the function K to a great extent so much so that the Hamilton equations take a much simpler form in the new canonical coordinates. Achievement of such a strategy entails, of course, much facile integration of differential equations. After having obtained the solution corresponding to the simplified system, we need to perform only some algebraic operations concerning canonical coordinates in order to obtain the actual solution associated with the physical system. We shall discuss this approach later in detail in Sec. 11.5 in a more general context.

11.5. NON-CONSERVATIVE MECHANICS

Let us consider a dynamical system of m degrees of freedom. We assume that the constraints between members of the system may be rheonomic, namely, they may be time-dependent. Or some parameters describing the system may be time-dependent. We further suppose that the potential function associated with the system may also be depending on time. In this situation, the kinetic energy of the system and its potential function now take in general the following forms

$$T = \frac{1}{2}g_{ij}(\mathbf{q},t)\,\dot{q}^i\,\dot{q}^j + g_i(\mathbf{q},t)\,\dot{q}^i + g(\mathbf{q},t),\ \ V = V(\mathbf{q},t).$$

Thus, the Lagrangian function becomes explicitly dependent on time:

$$L(\mathbf{q},\dot{\mathbf{q}},\mathrm{t}) = T(\mathbf{q},\dot{\mathbf{q}},\mathrm{t}) - V(\mathbf{q},t). \tag{11.5.1}$$

As a result of this both the Hamiltonian function and thus the generalised momenta become dependent explicitly on time:

$$p_i = \frac{\partial L(\mathbf{q},\dot{\mathbf{q}},t)}{\partial \dot{q}^i}, \tag{11.5.2}$$
$$H(\mathbf{p},\mathbf{q},t) = p_i\dot{q}^i - L(\mathbf{q},\dot{\mathbf{q}},t).$$

Thus, the Hamiltonian function is now a mapping like $H : S\times\mathbb{R}\to\mathbb{R}$

where $(S, \omega = dq^i \wedge dp_i)$ is again a symplectic manifold. The generalised coordinates $\mathbf{q}$ and the generalised momenta $\mathbf{p}$ still satisfy the Hamilton equations (11.2.7). We next like to introduce a $(2m+1)$-dimensional manifold $\mathfrak{S} = S \times \mathbb{R}$ whose coordinate cover is evidently given by $\{\mathbf{p}, \mathbf{q}, t\}$. We then define the following form $\theta_H \in \Lambda^1(\mathfrak{S})$:

$$\theta_H = p_i dq^i - H dt = \theta - H dt. \tag{11.5.3}$$

We further introduce the form $\omega_H \in \Lambda^2(\mathfrak{S})$ by

$$\begin{aligned}\omega_H &= -\,d\theta_H = dq^i \wedge dp_i + dH \wedge dt \\ &= \omega + dH \wedge dt.\end{aligned} \tag{11.5.4}$$

Insofar as $d\omega_H = -\,d^2\theta_H = 0$, the 2-form ω_H, too, is closed. Due to the fact that the dimension of the manifold $\mathfrak{S}$ is $2m+1$, the rank of the form ω_H would be at most $2m$. On the other hand, if one takes $t =$ *constant*, one finds $\omega_H = \omega$ so that the rank of ω_H cannot be less than $2m$. Consequently, the rank of the form ω_H is $2m$, that is, it is maximal. But $\mathfrak{S}$ is no longer a symplectic manifold because its dimension is an odd number. Nevertheless, although $(\mathfrak{S}, \omega_H)$ is not a symplectic manifold, it is straightforward to see that its restriction on a submanifold $t =$ *constant* is a symplectic manifold that is diffeomorphic to the manifold (S, ω).

We now define a vector field $\mathcal{V}_H \in T(\mathfrak{S})$ depending on a Hamiltonian function H in the following way

$$\mathcal{V}_H = \frac{\partial}{\partial t} + V_H \tag{11.5.5}$$

where the *Hamiltonian vector field* generated by the Hamiltonian function $H(\mathbf{p}, \mathbf{q}, t)$ is again given by

$$V_H(\mathbf{p}, \mathbf{q}, t) = \frac{\partial H}{\partial p_i}\frac{\partial}{\partial q^i} - \frac{\partial H}{\partial q^i}\frac{\partial}{\partial p_i}.$$

Since $V_H(H) = 0$, it follows from (11.5.5) that

$$\mathcal{V}_H(H) = \frac{\partial H}{\partial t}. \tag{11.5.6}$$

Because we obviously have $\mathbf{i}_{\mathcal{V}_H}(\omega) = \mathbf{i}_{V_H}(\omega)$, we conclude that

$$\begin{aligned}\mathbf{i}_{\mathcal{V}_H}(\omega_H) &= \mathbf{i}_{\mathcal{V}_H}(\omega) + \mathbf{i}_{\mathcal{V}_H}(dH \wedge dt) = \mathbf{i}_{V_H}(\omega) + \mathcal{V}_H(H)dt - dH \\ &= \frac{\partial H}{\partial q^i}dq^i + \frac{\partial H}{\partial p_i}dp_i + \frac{\partial H}{\partial t}dt - dH = 0.\end{aligned}$$

As a matter of fact, we can readily show that *each vector field* $\mathcal{V} \in T(\mathfrak{S})$ *satisfying the relation* $\mathbf{i}_{\mathcal{V}}(\omega_H) = 0$ *can only be a multiple of the vector field* $\mathcal{V}_H$ *by a scalar function.* Indeed, let us consider an arbitrary vector

$$\mathcal{V} = \xi^i \frac{\partial}{\partial q^i} + \eta_i \frac{\partial}{\partial p_i} + \tau \frac{\partial}{\partial t}$$

where $\xi^i, \eta_i, \tau \in \Lambda^0(\mathfrak{S})$. The condition

$$\begin{aligned}\mathbf{i}_{\mathcal{V}}(\omega_H) &= \xi^i \, dp_i - \eta_i \, dq^i + V(H) \, dt - \tau \, dH \\ &= \Big(\xi^i - \tau \frac{\partial H}{\partial p_i}\Big) dp_i - \Big(\eta_i + \tau \frac{\partial H}{\partial q^i}\Big) dq^i + \Big(\xi^i \frac{\partial H}{\partial q^i} + \eta_i \frac{\partial H}{\partial p_i}\Big) dt = 0\end{aligned}$$

now requires that the components of the vector $\mathcal{V}$ must satisfy

$$\xi^i = \tau \frac{\partial H}{\partial p_i}, \quad \eta_i = -\tau \frac{\partial H}{\partial q^i}, \quad \xi^i \frac{\partial H}{\partial q^i} + \eta_i \frac{\partial H}{\partial p_i} = 0.$$

It is obvious that the last expression vanishes identically. Thus, the desired vector field is obtained as follows

$$\mathcal{V} = \tau \Big(\frac{\partial H}{\partial p_i} \frac{\partial}{\partial q^i} - \frac{\partial H}{\partial q^i} \frac{\partial}{\partial p_i} + \frac{\partial}{\partial t}\Big) = \tau \, \mathcal{V}_H$$

where $\tau \in \Lambda^0(\mathfrak{S})$ is an arbitrary function. It is easily observed that trajectories of such a vector field determine the time evolution of the dynamical system under various initial conditions. If we denote the parameter of a trajectory by s, we can write

$$\frac{dq^i}{ds} = \tau \frac{\partial H}{\partial p_i}, \quad \frac{dp_i}{ds} = -\tau \frac{\partial H}{\partial q^i}, \quad \frac{dt}{ds} = \tau.$$

However, once we eliminate the parameter s, we again arrive at the usual Hamilton equations

$$\frac{dq^i}{dt} = \frac{\partial H}{\partial p_i}, \quad \frac{dp_i}{dt} = -\frac{\partial H}{\partial q^i}.$$

This result reveals then the possibility of determining the vector field $\mathcal{V}_H$ uniquely by imposing the conditions

$$\mathbf{i}_{\mathcal{V}}(\omega_H) = 0, \quad \mathbf{i}_{\mathcal{V}}(dt) = \tau = 1. \qquad (11.5.7)$$

Therefore, the equations (11.5.7) can now be regarded as equivalent to the Hamilton equations associated with a time dependent Hamiltonian function $H(\mathbf{p}, \mathbf{q}, t)$. On account of the satisfaction of the relation $\mathbf{i}_{\mathcal{V}}(\omega_H) = 0$ by a non-zero vector $\mathcal{V}$, we realise that the form ω_H happens to be *a degenerate 2-form*. This conclusion should, of course, be expected.

We would like now to evaluate the Lie derivative of the form ω_H with respect to the vector field $\mathcal{V}_H$. Because of the relations $d\omega_H = 0$ and $\mathbf{i}_{\mathcal{V}_H}(\omega_H) = 0$, we find that

$$\pounds_{\mathcal{V}_H}\omega_H = d\mathbf{i}_{\mathcal{V}_H}(\omega_H) + \mathbf{i}_{\mathcal{V}_H}(d\omega_H) = 0.$$

Hence, the form ω_H will remain invariant under the flow on the manifold $\mathfrak{S}$ which is brought into being by the vector field $\mathcal{V}_H$. This result will naturally imply that the forms $\Omega_k = \omega_H^k = \underbrace{\omega_H \wedge \omega_H \wedge \cdots \wedge \omega_H}_{k}, 1 \le k \le m$ remain invariant as well under the same flow.

Inasmuch as the vector field $\mathcal{V}_g$ corresponding to a function $g \in \Lambda^0(\mathfrak{S})$ has been given by

$$\mathcal{V}_g = \frac{\partial g}{\partial p_i}\frac{\partial}{\partial q^i} - \frac{\partial g}{\partial q^i}\frac{\partial}{\partial p_i} + \frac{\partial}{\partial t},$$

we then obtain

$$\begin{aligned}\pounds_{\mathcal{V}_g}(f) = \mathcal{V}_g(f) &= \frac{\partial g}{\partial p_i}\frac{\partial f}{\partial q^i} - \frac{\partial g}{\partial q^i}\frac{\partial f}{\partial p_i} + \frac{\partial f}{\partial t} \\ &= \frac{\partial f}{\partial t} + \{f, g\}\end{aligned}$$

for a function $f \in \Lambda^0(\mathfrak{S})$. This result means that the necessary and sufficient condition in order that a given function f remains invariant, or in other words, constant under the flow generated by a vector field $\mathcal{V}_g$ is the satisfaction of the following equation

$$\frac{\partial f}{\partial t} + \{f, g\} = 0. \tag{11.5.8}$$

In this case, when we consider the motion of a dynamical system described by the flow produced by a vector field associated with a given Hamiltonian function H, a function $f \in \Lambda^0(\mathfrak{S})$ verifying the equation

$$\pounds_{\mathcal{V}_H}(f) = \frac{\partial f}{\partial t} + \{f, H\} = 0 \tag{11.5.9}$$

must satisfy the relation

$$f\big(\mathbf{q}(t), \mathbf{p}(t), t\big) = f\big(\mathbf{q}_0(t_0), \mathbf{p}_0(t_0), t_0\big) = \textit{constant}.$$

Therefore, it corresponds to an integral of the motion of the system. On the other hand, when H is time dependent, we of course obtain

$$\frac{\partial H}{\partial t} + \{H, H\} = \frac{\partial H}{\partial t} \neq 0.$$

Hence, such a Hamiltonian function is no longer an integral of the motion. In other words, the conservation of energy loses its validity in such systems.

Let us now specify a diffeomorphism $\phi : \mathfrak{S}_1 \to \mathfrak{S}_2$ by the following transformations between local coordinates (Q^i, P_i, T) and (q^i, p_i, t):

$$q^i = q^i(\mathbf{Q}, \mathbf{P}, T), \quad p_i = p_i(\mathbf{Q}, \mathbf{P}, T), \quad t = T \qquad (11.5.10)$$

Suppose that $H = H(\mathbf{q}, \mathbf{p}, t) \in \Lambda^0(\mathfrak{S}_2)$, $K = K(\mathbf{Q}, \mathbf{P}, t) \in \Lambda^0(\mathfrak{S}_1)$. If the relation

$$\phi^*\big(dq^i \wedge dp_i + dH(\mathbf{q}, \mathbf{p}, t) \wedge dt\big) = dQ^i \wedge dP_i + dK(\mathbf{Q}, \mathbf{P}, t) \wedge dt$$

is satisfied, then we say that ϕ is a ***canonical transformation***. It is clear in this case that the mapping ϕ^{-1} is also a canonical transformation. The short version of the foregoing expression can, of course, be written as follows

$$\phi^* \omega_H = \omega_K. \qquad (11.5.11)$$

Since $\omega_1 = dQ^i \wedge dP_i$, $\omega_2 = dq^i \wedge dp_i$ and $\phi^* t = t$, we readily deduce from (11.5.11) that the relation

$$\phi^* \omega_2 = \omega_1 + d\mathcal{F} \wedge dt$$

must be satisfied if ϕ is a canonical transformation. Here, $\mathcal{F}$ is an arbitrary smooth function defined by

$$\mathcal{F} = K - \phi^* H \in \Lambda^0(\mathfrak{S}_1).$$

Thus, we can write $K = \phi^* H + \mathcal{F}$. We can immediately realise that *every canonical transformation preserves the form of the Hamilton equations*. A vector field $\mathcal{V}_H \in T(\mathfrak{S}_2)$ on the manifold $\mathfrak{S}_2$ satisfying the conditions $\mathbf{i}_{\mathcal{V}_H}(\omega_H) = 0$ and $\mathbf{i}_{\mathcal{V}_H}(dt) = 1$ gives rise to the Hamilton equations on $\mathfrak{S}_2$:

$$\frac{dq^i}{dt} = \frac{\partial H}{\partial p_i}, \quad \frac{dp_i}{dt} = -\frac{\partial H}{\partial q^i}.$$

On the other hand, we know that a vector field $\mathcal{V} \in T(\mathfrak{S}_1)$ satisfying the conditions $\mathbf{i}_{\mathcal{V}}(\omega_K) = 0$, $\mathbf{i}_{\mathcal{V}}(dt) = 1$ is a uniquely determined vector field $\mathcal{V}_K$ generated by a Hamiltonian vector field. *Trajectories of this vector field will also satisfy the Hamilton equations on* $\mathfrak{S}_1$:

$$\frac{dQ^i}{dt} = \frac{\partial K}{\partial P_i}, \quad \frac{dP_i}{dt} = -\frac{\partial K}{\partial Q^i}. \tag{11.5.12}$$

The connection between the vectors $\mathcal{V}_H$ and $\mathcal{V}_K$ associated with a canonical transformation can easily be found. Consider a vector field $\mathcal{V}_H \in T(\mathfrak{S}_2)$ satisfying the conditions $\mathbf{i}_{\mathcal{V}_H}(\omega_H) = 0$ and $\mathbf{i}_{\mathcal{V}_H}(dt) = 1$. If $\phi : \mathfrak{S}_1 \to \mathfrak{S}_2$ is a canonical transformation, then the pull-back operator $\phi^* : \Lambda(\mathfrak{S}_2) \to \Lambda(\mathfrak{S}_1)$ yields the equalities

$$0 = \phi^* \mathbf{i}_{\mathcal{V}_H}(\omega_H) = \mathbf{i}_{\phi_*^{-1}\mathcal{V}_H}(\phi^*\omega_H) = \mathbf{i}_{\phi_*^{-1}\mathcal{V}_H}(\omega_K),$$
$$1 = \phi^* \mathbf{i}_{\mathcal{V}_H}(dt) = \mathbf{i}_{\phi_*^{-1}\mathcal{V}_H}(\phi^* dt) = \mathbf{i}_{\phi_*^{-1}\mathcal{V}_H}(d\phi^* t) = \mathbf{i}_{\phi_*^{-1}\mathcal{V}_H}(dt).$$

When we are given the form ω_K, we know that a vector field $\mathcal{V}_K \in T(\mathcal{S}_1)$ satisfying the conditions $\mathbf{i}_{\mathcal{V}_K}(\omega_K) = 0$ and $\mathbf{i}_{\mathcal{V}_K}(dt) = 1$ will be determined uniquely. Hence, the above relations show unequivocally that the connection between $\mathcal{V}_H$ and $\mathcal{V}_K$ is provided by

$$\mathcal{V}_K = \phi_*^{-1}\mathcal{V}_H \quad \text{or} \quad \mathcal{V}_H = \phi_* \mathcal{V}_K.$$

Since ϕ is a diffeomorphism, $\phi_* : T(\mathfrak{S}_1) \to T(\mathfrak{S}_2)$ is an isomorphism.

In order to illuminate the local structure of canonical transformations $\phi : \mathfrak{S} \to \mathfrak{S}$ that maps the manifold $\mathfrak{S}$ onto itself and to disclose unambiguously the interrelation between Hamiltonian functions H and K, we can make use of the expression (11.5.4). From the relation

$$-\phi^* d\theta_H = -d\phi^*\theta_H = -d\theta_K$$

we find that the equation $d(\phi^*\theta_H - \theta_K) = 0$ has to be satisfied. Thus, according to the Poincaré lemma canonical transformations must obey at least locally to the condition $\phi^*\theta_H - \theta_K = dF$ where $F \in \Lambda^0(\mathfrak{S})$ is an arbitrary function. This expression can be written explicitly as

$$\phi^*(p_i dq^i - H dt) = P_i dQ^i - K dt + dF \tag{11.5.13}$$

The function F is called a ***generating function*** because it is instrumental in designating a canonical transformation. In order to specify a transformation between old and new canonical coordinates, the function F must depend on $4m + 1$ variables $\mathbf{q}, \mathbf{p}, \mathbf{Q}, \mathbf{P}, t$. But, owing to equations (11.5.10), we are allowed to choose only $2m + 1$ independent variables. Therefore, we can consider only four different alternatives characterising a canonical transformation between old and new coordinates that are listed below:

$$\{\mathbf{q}, \mathbf{Q}, t\}, \ \{\mathbf{q}, \mathbf{P}, t\}, \ \{\mathbf{p}, \mathbf{Q}, t\}, \ \{\mathbf{p}, \mathbf{P}, t\}.$$

We shall now discuss these choices separately.

(i). Let us chose $F = F(\mathbf{q}, \mathbf{Q}, t)$. When we insert this function into (11.5.13) and arrange the resulting terms, we obtain

$$\Big(p_i - \frac{\partial F}{\partial q^i}\Big)dq^i - \Big(P_i + \frac{\partial F}{\partial Q^i}\Big)dQ^i + \Big(K - H - \frac{\partial F}{\partial t}\Big)dt = 0.$$

Hence, the canonical transformation is specified by the equations

$$p_i = \frac{\partial F}{\partial q^i}, \quad P_i = -\frac{\partial F}{\partial Q^i}, \quad K = H + \frac{\partial F}{\partial t} \tag{11.5.14}$$

In case $\det\Big(\dfrac{\partial^2 F}{\partial Q^i \partial q^j}\Big) \neq 0$, then $(11.5.14)_2$ yields $q^i = q^i(\mathbf{Q}, \mathbf{P}, t)$ through the inverse function theorem. On introducing these relations into equations $(11.5.14)_1$ we are led to $p_i = p_i(\mathbf{Q}, \mathbf{P}, t)$. Substituting functions so obtained into the functions $H(\mathbf{q}, \mathbf{p}, t)$ and $F(\mathbf{q}, \mathbf{Q}, t)$ we can determine the Hamiltonian function $K(\mathbf{Q}, \mathbf{P}, t)$.

For instance, let us choose $F = a_{ij}(t)\, q^i Q^j$, $\det[a_{ij}] \neq 0$. Then, the canonical transformation becomes

$$p_i = a_{ij}(t)\, Q^j,\ P_i = -a_{ji}(t)\, q^j \text{ or } q^i = -b^{ij}(t) P_j,\ \mathbf{B} = (\mathbf{A}^{\mathrm{T}})^{-1}.$$

If we take $\mathbf{A} = \mathbf{I}$, this transformation merely interchanges the generalised coordinates and generalised momenta.

(ii). Let us choose $F = -P_i Q^i + F_1(\mathbf{q}, \mathbf{P}, t)$. (11.5.13) gives then

$$\Big(p_i - \frac{\partial F_1}{\partial q^i}\Big)dq^i + \Big(Q^i - \frac{\partial F_1}{\partial P_i}\Big)dQ^i + \Big(K - H - \frac{\partial F_1}{\partial t}\Big)dt = 0$$

from which it follows that

$$p_i = \frac{\partial F_1}{\partial q^i}, \quad Q^i = \frac{\partial F_1}{\partial P_i}, \quad K = H + \frac{\partial F_1}{\partial t}. \tag{11.5.15}$$

If $\det\Big(\dfrac{\partial^2 F_1}{\partial P_i \partial q^j}\Big) \neq 0$, then $(11.5.15)_2$ yields the relation $q^i = q^i(\mathbf{Q}, \mathbf{P}, t)$ and $(11.5.15)_1$ provides $p_i = p_i(\mathbf{Q}, \mathbf{P}, t)$. Finally, the transformed Hamiltonian function $K(\mathbf{Q}, \mathbf{P}, t)$ follows from $(11.5.15)_3$.

For instance, let us choose $F_1 = a_i^j(t)\, q^i P_j$, $\det[a_i^j] \neq 0$. The canonical transformation becomes

$$p_i = a_i^j(t)\, P_j,\ Q^i = a_j^i(t)\, q^j \quad \text{or} \quad q^i = b_j^i(t) Q^j,\ \mathbf{B} = (\mathbf{A}^{\mathrm{T}})^{-1}.$$

If we take $\mathbf{A} = \mathbf{I}$, this transformation does not change at all the canonical variables.

(iii). Let us chose $F = p_i q^i + F_2(\mathbf{p}, \mathbf{Q}, t)$. Then, (11.5.13) results in

$$-\Big(q^i + \frac{\partial F_2}{\partial p_i}\Big)dp_i - \Big(P_i + \frac{\partial F_2}{\partial Q^i}\Big)dQ^i + \Big(K - H - \frac{\partial F_2}{\partial t}\Big)dt = 0$$

and the canonical transformation is prescribed by the equations

$$q^i = -\frac{\partial F_2}{\partial p_i}, \quad P_i = -\frac{\partial F_2}{\partial Q^i}, \quad K = H + \frac{\partial F_2}{\partial t}. \qquad (11.5.16)$$

If $\det\Big(\dfrac{\partial^2 F_2}{\partial Q^i \partial p_j}\Big) \neq 0$, the expression $(11.5.16)_2$ determines the functions $p_i = p_i(\mathbf{Q}, \mathbf{P}, t)$, and $(11.5.16)_1$ yields the functions $q^i = q^i(\mathbf{Q}, \mathbf{P}, t)$. By employing these expressions, we deduce the transformed Hamiltonian function from $(11.5.16)_3$.

For instance, if we choose $F_2 = a^i_j(t)\, p_i Q^j$, $\det[a^i_j] \neq 0$, then the canonical transformation is found to be

$$q^i = -\, a^i_j(t)\, Q^j,\ P_i = -\, a^j_i(t)\, p_j \quad \text{or} \quad p_i = -\, b^j_i(t) P_j,\ \mathbf{B} = (\mathbf{A}^\mathrm{T})^{-1}.$$

If we take $\mathbf{A} = \mathbf{I}$, then this transformation changes only the signs of the canonical variables.

(iv). Let us choose $F = p_i q^i - P_i Q^i + F_3(\mathbf{p}, \mathbf{P}, t)$. Then, it follows from (11.5.13) that

$$-\Big(q^i + \frac{\partial F_3}{\partial p_i}\Big)dp_i + \Big(Q^i - \frac{\partial F_3}{\partial P_i}\Big)dP_i + \Big(K - H - \frac{\partial F_3}{\partial t}\Big)dt = 0.$$

Consequently, the canonical transformation is specified by the equations

$$q^i = -\frac{\partial F_3}{\partial p_i}, \quad Q^i = \frac{\partial F_3}{\partial P_i}, \quad K = H + \frac{\partial F_3}{\partial t}. \qquad (11.5.17)$$

Indeed, if $\det\Big(\dfrac{\partial^2 F_3}{\partial P_i \partial p_j}\Big) \neq 0$, then $(11.5.17)_2$ leads to $p_i = p_i(\mathbf{Q}, \mathbf{P}, t)$ and $(11.5.17)_1$ gives $q^i = q^i(\mathbf{Q}, \mathbf{P}, t)$. We obtain the transformed Hamiltonian function $K(\mathbf{Q}, \mathbf{P}, t)$ from the equation $(11.5.17)_3$.

For instance, if we choose $F_3 = a^{ij}(t)\, p_i P_j$, $\det[a^{ij}] \neq 0$, then the canonical transformation becomes

$$q^i = -\, a^{ij}(t)\, P_j,\ Q^i = a^{ji}(t)\, p_j \quad \text{or} \quad p_i = b_{ij}(t) Q^j,\ \mathbf{B} = (\mathbf{A}^\mathrm{T})^{-1}.$$

If we take $\mathbf{A} = \mathbf{I}$, this transformation interchanges the generalised coordinates and generalised momenta with a change of sign in one set of variables.

The equations (11.5.14-17) specify also the canonical transformations in conservative mechanics. However, in this situation the functions F, F_1, F_2, F_3 are either independent of time or may only be certain particular functions of time. The transformation

$$K(\mathbf{Q}, \mathbf{P}) = H\big(\mathbf{q}(\mathbf{Q}, \mathbf{P}), \mathbf{p}(\mathbf{Q}, \mathbf{P})\big)$$

gives then the Hamiltonian function.

After having established the structure of the canonical transformations, Jacobi thought quite an ingenious idea for that time which seems to be rather natural to us now and he had asked this question: whether is it possible to determine a canonical transformation in such a manner that the transformed Hamiltonian function K turns out to be a constant that can be taken zero without loss of generality? If we can make such a choice leading to

$$K(\mathbf{Q}, \mathbf{P}, t) = 0,$$

then the corresponding Hamilton equations (11.5.12) take their simplest possible form

$$\frac{dQ^i}{dt} = 0, \quad \frac{dP_i}{dt} = 0.$$

Thus, $2m$ integrals of motion are simply obtained as follows

$$Q^i(\mathbf{q}, \mathbf{p}, t) = a^i = constant, \;\; P_i(\mathbf{q}, \mathbf{p}, t) = b_i = constant$$

in terms of new canonical variables. On the other hand, such a canonical transformation can be prescribed by selecting the generating function F as to satisfy the equation

$$\frac{\partial F(\mathbf{q}, \mathbf{Q}, t)}{\partial t} + H(\mathbf{q}, \mathbf{p}, t) = 0.$$

In this case, however, the coordinates Q^i are constant so that the function F depends only on variables $q^1, q^2, \ldots, q^m, t$ and the constants $Q^1 = a^1$, $Q^2 = a^2, \ldots, Q^m = a^m$. The generalised momenta p_i are then given by $(11.5.14)_1$. Consequently, the function F must satisfy the ***Hamilton-Jacobi differential equation***

$$\frac{\partial F(\mathbf{q}, \mathbf{a}, t)}{\partial t} + H(\mathbf{q}, \frac{\partial F}{\partial \mathbf{q}}, \mathbf{a}, t) = 0. \tag{11.5.18}$$

where $\mathbf{a} = (a^1, a^2, \ldots, a^m)$.

Although this equation is always attributed today to two mathematicians W. R. Hamilton and C. G. J. Jacobi, it has been actually published first by Hamilton in 1834. Here, we have obviously employed the abbreviated notations

$$F(\mathbf{q},t) = F(q^1,q^2,\ldots,q^m,t)$$
$$H(\mathbf{q},\frac{\partial F}{\partial \mathbf{q}},t) = H(q^1,q^2,\ldots,q^m,\frac{\partial F}{\partial q^1},\frac{\partial F}{\partial q^2},\ldots,\frac{\partial F}{\partial q^m},t).$$

The Hamilton-Jacobi equation is a first order, generally non-linear, partial differential equation with $m+1$ independent variables q^i,t. Therefore, the function F depends on $m+1$ integration constants $a^1,a^2,\ldots,a^m,a^{m+1}$. But, it is evident that the function $F+a^{m+1}$ satisfies likewise the equation (11.5.18). Since transformation equations involve only some derivatives of F, this constant will have no effect in this approach. Hence, it can be discarded. Thus, by using the representation $\mathbf{a}=\{a^1,a^2,\ldots,a^m\}$, the function F that is the solution of the equation (11.5.18) may be expressible in the form

$$F = F(\mathbf{q},\mathbf{a},t)$$

where we obviously have $Q^i = a^i$. Hence, the following equations are deduced from the relations (11.5.14)

$$p_i = \frac{\partial F}{\partial q^i} = p_i(\mathbf{q},\mathbf{a},t), \tag{11.5.19}$$
$$P_i = -\frac{\partial F}{\partial Q^i} = -\frac{\partial F}{\partial a^i} = P_i(\mathbf{q},\mathbf{a},t) = b_i.$$

The initial conditions of the dynamical system corresponding to generalised positions and velocities may be given in the following way

$$\mathbf{q}(t_0) = \mathbf{q}_0 = \textit{constant}, \quad \mathbf{p}(t_0) = \mathbf{p}_0 = \textit{constant}.$$

Since we have assumed that $\det\Big(\partial^2 F/\partial a^i \partial q^j\Big) \neq 0$, insertion of the initial conditions into $(11.5.19)_1$ leads to the determination of the constants $\mathbf{a}$ that are arbitrary at the outset as

$$\mathbf{a} = \mathbf{a}(\mathbf{q}_0,\mathbf{p}_0,t_0).$$

Substituting the constants a_i so obtained together with initial conditions into $(11.5.19)_2$, we determine the constants b^i:

$$\mathbf{b} = \mathbf{P}\big(\mathbf{q}_0,\mathbf{a}(\mathbf{q}_0,\mathbf{p}_0,t_0),t_0\big).$$

Recalling again the condition $\det\left(\partial^2 F/\partial a^i \partial q^j\right) \neq 0$, we can in principle construct the inverse function from $(11.5.19)_2$, to arrive eventually at the following equations

$$\mathbf{q} = \mathbf{q}(t; \mathbf{a}, \mathbf{b})$$

that describe the evolution of the system with time. It is thus evident that in case we can determine the function $F(\mathbf{q}, t)$, sometimes called the ***Hamilton principal function***, satisfying the first order partial differential equation (11.5.18), then the expressions describing the motion of the system are found by almost algebraic manipulations. Therefore, this method seems, at first glance, to be a much more effective approach to determine the motion of a system than trying to solve directly the Hamilton equations. In reality, it is highly unlikely to be able solve directly the Hamilton-Jacobi equation, except in very few cases. The standard technique of characteristics to solve this non-linear partial differential equation requires again to obtain the solution of the Hamilton equations [*see* Example 9.2.3]. Therefore, it does not bring about a fresh approach. Nonetheless the discussion of the Hamilton-Jacobi equation may provide rather significant qualitative information about the behaviour of a dynamical system.

In order to comprehend better the meaning of the function F, let us calculate its derivative with respect to time along the trajectory of the system. When (11.5.14) and (11.5.18) are taken into account, it is easily found that

$$\frac{dF}{dt} = \frac{\partial F}{\partial t} + \frac{\partial F}{\partial q^i}\frac{dq^i}{dt} = p_i \dot{q}^i - H = L.$$

Hence, this time rate of change of F yields the Lagrangian function.

If the Hamiltonian function H does not explicitly depend on time the equation (11.5.18) leads of course to the result

$$\frac{\partial^2 F(\mathbf{q}, \mathbf{Q}, t)}{\partial t^2} = 0$$

since $\partial H/\partial t = 0$. A simple integration then gives

$$F(\mathbf{q}, \mathbf{Q}, t) = -E(\mathbf{q}, \mathbf{Q})t + W(\mathbf{q}, \mathbf{Q}), \quad \mathbf{Q} = \mathbf{a}.$$

Consequently, the generalised momenta become

$$p_i = \frac{\partial F}{\partial q^i} = -\frac{\partial E}{\partial q^i}t + \frac{\partial W}{\partial q^i}.$$

Hence, H will be explicitly independent of time t if only $\partial E/\partial q^i = 0$. According to the definition of the function F, we have to take $\mathbf{Q} = \mathbf{a} =$ *constant*. Hence, we must write $E(\mathbf{Q}) = a^m =$ *constant*. Thus, F will now be expressible in the form

$$F(\mathbf{q}, \mathbf{a}, t) = -Et + W(q^1, \ldots, q^m; a^1, \ldots, a^{m-1}, E) \quad (11.5.20)$$

whence we can deduce the following relations

$$\begin{aligned} p_i &= \frac{\partial W}{\partial q^i}, & i &= 1, 2, \ldots, m \quad (11.5.21) \\ b_i &= -\frac{\partial W}{\partial a^i}, & i &= 1, 2, \ldots, m-1 \\ b_m &= -\frac{\partial F}{\partial E} = t - \frac{\partial W}{\partial E}. \end{aligned}$$

In this case, the Hamilton-Jacobi equation reduces to the non-linear partial differential equation

$$H(\mathbf{q}, \frac{\partial W}{\partial \mathbf{q}}) = E \quad (11.5.22)$$

that helps determine the function W that is called sometimes the ***Hamilton characteristic function***.

Example 11.5.1. ***Harmonic Oscillator***. Let us denote by q the coordinate of the 1-dimensional configuration manifold associated with the rectilinear harmonic motion of a particle with mass m. Then, the kinetic and potential energies are prescribed in the following manner

$$T = \frac{1}{2}m\dot{q}^2, \quad V = \frac{1}{2}kq^2.$$

Introducing the definition $\omega^2 = k/m$, we see that the Lagrangian function and the generalised momentum that is equal to the ordinary momentum in this case are given by

$$\begin{aligned} L &= \frac{m}{2}(\dot{q}^2 - \omega^2 q^2), \\ p &= \frac{\partial L}{\partial \dot{q}} = m\dot{q}. \end{aligned}$$

Hence, the Hamiltonian function takes the form

$$H = \frac{1}{2m}(p^2 + m^2\omega^2 q^2) = E.$$

and the equation (11.5.20) becomes

$$\left(\frac{dW}{dq}\right)^2 + m^2\omega^2 q^2 = 2mE.$$

When we choose the $+$ sign in front of the square root, we obtain the differential equation

$$\frac{dW}{dq} = m\omega\sqrt{\frac{2E}{m\omega^2} - q^2}$$

whose solution is easily found as follows

$$\begin{aligned} W(q;E) &= m\omega\int\sqrt{\frac{2E}{m\omega^2} - q^2}\,dq + a \\ &= \frac{m\omega}{2}\left[q\sqrt{\frac{2E}{m\omega^2} - q^2} + \frac{2E}{m\omega^2}\arctan\frac{q}{\sqrt{\dfrac{2E}{m\omega^2} - q^2}}\right] + a. \end{aligned}$$

Therefore, $(11.5.21)_3$ yields

$$b = -\frac{\partial F}{\partial E} = t - \frac{\partial W}{\partial E} = t - \frac{1}{\omega}\arctan\frac{q}{\sqrt{\dfrac{2E}{m\omega^2} - q^2}}$$

and we finally obtain by using inverse trigonometric functions

$$q = \sqrt{\frac{2E}{m\omega^2}}\sin\omega(t-b).$$

The constants E and b are to be determined from the initial conditions. ■

Example 11.5.2. ***Central-Force Motion***. Let us denote the polar coordinates of the 2-dimensional configuration manifold that is associated with the central motion of a particle with mass m by $q^1 = r, q^2 = \theta$. Then, its Lagrangian function can be written as follows

$$L = \frac{1}{2}m(\dot{r}^2 + r^2\dot{\theta}^2) - V(r).$$

Hence, the generalised momenta become

$$p_1 = p_r = \frac{\partial L}{\partial \dot{r}} = m\dot{r}, \; p_2 = p_\theta = \frac{\partial L}{\partial \dot{\theta}} = mr^2\dot{\theta}$$

and the Hamiltonian function is prescribed by the expression

$$H = \frac{1}{2}m(\dot{r}^2 + r^2\dot{\theta}^2) + V(r) = \frac{1}{2m}\Big(p_r^2 + \frac{p_\theta^2}{r^2}\Big) + V(r).$$

Equation (11.5.22) now takes the form

$$\frac{1}{2m}\Big[\Big(\frac{\partial W}{\partial r}\Big)^2 + \frac{1}{r^2}\Big(\frac{\partial W}{\partial \theta}\Big)^2\Big] + V(r) = E$$

with $W = W(r, \theta; a^1, E)$. Since the function H does not depend on θ explicitly, the Hamilton equations yield $\dot{p}_\theta = 0$, and $p_\theta = a^1 = constant$. Thus, we are allowed to write from (11.5.21)

$$p_r = \frac{\partial W}{\partial r}, \quad p_\theta = \frac{\partial W}{\partial \theta} = a^1,$$
$$b_1 = -\frac{\partial W}{\partial a^1}, \quad b_2 = t - \frac{\partial W}{\partial E}.$$

It then follows from the second equation above

$$W(r, \theta; a^1, E) = a^1\theta + w(r; a^1, E).$$

If we take into consideration the $+$ sign in front of the square root, we observe that the function w must satisfy the following differential equation

$$\frac{dw}{dr} = \sqrt{2m[E - V(r)] - \Big(\frac{a^1}{r}\Big)^2}$$

whose solution is easily obtainable in the form

$$w(r; a^1, E) = \int_{r_0}^{r} \sqrt{2m[E - V(s)] - \Big(\frac{a^1}{s}\Big)^2}\, ds$$

where r_0 is yet an arbitrary constant. Upon introducing this relation into $(11.5.21)_{2\text{-}3}$, we find that

$$b_1 = -\theta + \int_{r_0}^{r} \frac{a^1 ds}{s^2\sqrt{2m[E - V(s)] - (a^1/s)^2}},$$
$$b_2 = t - \int_{r_0}^{r} \frac{m\, ds}{\sqrt{2m[E - V(s)] - (a^1/s)^2}}.$$

If the initial conditions are such that $r = r_0$ and $\theta = \theta_0$ for $t = t_0$, we get

$$b_1 = -\theta_0, \quad b_2 = t_0.$$

Hence, in terms of the parameter r, the equations describing the motion of the particle are expressible as

$$\theta - \theta_0 = \int_{r_0}^{r} \frac{a^1 ds}{s\sqrt{2ms^2\big[E - V(s)\big] - (a^1)^2}},$$

$$t - t_0 = \int_{r_0}^{r} \frac{ms\, ds}{\sqrt{2ms^2\big[E - V(s)\big] - (a^1)^2}}.$$

We have to take $V(r) = k/r$ in order to discuss perhaps the most important application of the central-force motion. When $k < 0$, this potential corresponds to the Newton law of gravitational attraction [the English mathematician and physicist Sir Isaac Newton (1643-1727)]. If the constant k may be taken either negative or positive, this potential represents the Coulomb law describing the force between point electric charges that can be attractive or repulsive [the French engineer and physicist Charles Augustin de Coulomb (1736-1806)]. When $V(r) = k/r$, we readily find that

$$\theta - \theta_0 = \int_{r_0}^{r} \frac{a^1 ds}{s\sqrt{2mE\, s^2 - 2mks - (a^1)^2}},$$

$$t - t_0 = \int_{r_0}^{r} \frac{ms\, ds}{\sqrt{2mE\, s^2 - 2mks - (a^1)^2}}.$$

If we make the substitution $s = 1/\sigma$ in the first equation above and rename the constant a^1 by l, we obtain

$$\theta - \theta_0 = -\int_{1/r_0}^{1/r} \frac{d\sigma}{\sqrt{\dfrac{2mE}{l^2} - \dfrac{2mk}{l^2}\sigma - \sigma^2}}$$

$$= -\arccos \left. \frac{\dfrac{\sigma l^2}{mk} - 1}{\sqrt{1 + \dfrac{2El^2}{mk^2}}} \right|_{1/r_0}^{1/r}.$$

Let us define an angle θ_1 by $\theta_1 = \arccos\left[\left(\dfrac{l^2}{mkr_0} - 1\right)\Big/\sqrt{1 + \dfrac{2El^2}{mk^2}}\right]$. Then, we finally reach to the conclusion

$$\frac{1}{r} = \frac{mk}{l^2}\left[1 + \sqrt{1 + \frac{2El^2}{mk^2}}\cos\left(\theta - \theta_0 - \theta_1\right)\right].$$

Introducing the definition $p = l^2/mk$ and $e = \sqrt{1 + (2El^2/mk^2)}$ we arrive at the standard equation describing conics in polar coordinates

$$r = \frac{p}{1 + e\cos(\theta - \theta')}, \quad \theta' = \theta_0 + \theta_1.$$

On defining a function

$$f(\sigma) = \sqrt{\frac{2mE}{l^2} - \frac{2mk}{l^2}\sigma - \sigma^2}$$

we determine the function $t = t(r)$ in a similar way as follows

$$t(r) - t_0 = t - t_0 = -\frac{m}{l}\int_{1/r_0}^{1/r} \frac{d\sigma}{\sigma^2 f(\sigma)} = T(\sigma)\big|_{1/r_0}^{1/r}$$

where the function $T(\sigma)$ is given by

$$T(\sigma) = \frac{2E^{1/2}lf(\sigma) + (2m)^{1/2}k\sigma\log\dfrac{2mE - km\sigma + (2mE)^{1/2}lf(\sigma)}{\sigma}}{4E^{3/2}}$$

The function $T(\sigma)$ determines the time taken by the particle on its trajectory traversing from the radial distance r_0 to the radial distance r. ■

11.6. ELECTROMAGNETISM

Let us consider the 4-dimensional manifold $\mathbb{R}^4$. Its coordinates will be denoted by $x^\mu, \mu = 1, 2, 3, 4$. $x^i, i = 1, 2, 3$ correspond to spatial coordinates while $x^4 = t$ denotes the time coordinate. Electromagnetic fields on this manifold, representing a material medium or the vacuum, are governed by the ***Maxwell equations*** [the English mathematician and physicist James Clerk Maxwell (1831-1879)] that are given, in rationalised M.K.S. units, by

$$\nabla \times \mathbf{E} + \frac{\partial \mathbf{B}}{\partial t} = \mathbf{0}, \quad \nabla \cdot \mathbf{B} = 0 \qquad (11.6.1)$$
$$\nabla \times \mathbf{H} - \frac{\partial \mathbf{D}}{\partial t} = \mathbf{J}, \quad \nabla \cdot \mathbf{D} = \rho$$

where the vectors **E**, **H**, **B** and **D** specify, respectively, *the electric field, the magnetic field, the magnetic induction* and *the electric displacement field.* The vector **J** is *the free electric current density* whereas the scalar ρ is *the free electric charge density*. The divergence of the equation $(11.6.1)_3$ yields an equation corresponding to *the conservation of electric charge*

$$\frac{\partial \rho}{\partial t} + \boldsymbol{\nabla} \cdot \mathbf{J} = 0. \tag{11.6.2}$$

Actually, the vector fields **E**, **H**, **B**, **D** and **J** cannot be utterly independent of one another. There are relations among them called *constitutive equations* reflecting the physical properties of the medium. The simplest physically meaningful relations of that kind can be given by

$$\mathbf{D} = \epsilon \mathbf{E}, \quad \mathbf{B} = \mu \mathbf{H}, \quad \mathbf{J} = \sigma \mathbf{E}$$

where the three physical constants ϵ, μ, σ are known, respectively, as *the dielectric* and *the magnetic permittivities* and *the electric conduction coefficient*. The values of these constants in the *vacuum* are

$$\epsilon = \epsilon_0 = 8.854187817620 \times 10^{-12}\, \text{F/m (Farad/metre)}$$
$$\mu = \mu_0 = 4\pi \times 10^{-7}\, \text{N/A}^2\ (\text{Newton/ampere}^2)$$

and $\sigma = 0$. These constants satisfy the relation $c = 1/\sqrt{\epsilon_0 \mu_0}$ where the physical constant c is the speed of light in the vacuum. The most recent value of c is 299792.458 km/sec.

The equation $(11.6.1)_1$ is known as the ***Faraday induction law*** [the autodidact English physicist and chemist Michael Faraday (1791-1867)] that shows that a mechanical energy causing a magnetic induction in a region to change with time can be converted to the electrical energy. The equation $(11.6.1)_2$ is the ***Gauss law*** implying that magnetic charges (monopoles) do not exist in nature in the realm of the classical physics. The equation $(11.6.1)_3$ is a somewhat modified version of the ***Ampère law*** expressing the fact that electric currents create magnetic fields. Equation $(11.6.1)_4$, when written in the form $\boldsymbol{\nabla} \cdot \mathbf{E} = \rho/\epsilon$, is originally obtained from the ***Coulomb law*** that specifies the repulsive or attractive force between two electric point charge as the expression $q_1 q_2/\epsilon r^2$ by exactly following the path leading to the Gauss law. In the original version of the Ampère law the term $\partial \mathbf{D}/\partial t$ which will be called later *the displacement current* does not exist. However, the governing equations at that form are not consistent because they violate the equation (11.6.2) associated with the conservation of charge that can also be derived independently. The genius of Maxwell has caused the creation of a consistent theory of electromagnetism. He has cleverly introduced a displacement vector **D** to recover the conservation of charge. It has been realised, however, that only in particular, but practically very important, cases this vector could be identified as $\epsilon\mathbf{E}$. This theory of electromagnetism was perhaps the greatest scientific achievement in the

19th century and has paved the way for incredible technological developments in the 20th century.

Let us now express equations (11.6.1) in terms of their components in Cartesian coordinates by paying attention to the dictates of the summation convention as follows

$$e^{ijk}E_{k,j}+\frac{\partial B^i}{\partial t}=0,\ B^i_{,i}=0;\ e^{ijk}H_{k,j}-\frac{\partial D^i}{\partial t}=J^i,\ D^i_{,i}=\rho.$$

Next, we wish to introduce the 4×4 *antisymmetric matrices* $\mathbb{F}=[F^{\mu\nu}]$ and $\mathbb{H}=[H^{\mu\nu}]$ by the following entries

$$\begin{aligned} F^{ij}&=-F^{ji}=e^{jik}E_k, & F^{4i}&=-F^{i4}=B^i,\\ H^{ij}&=-H^{ji}=e^{jik}H_k, & H^{4i}&=-H^{i4}=-D^i. \end{aligned} \tag{11.6.3}$$

In matrix notation, we can, of course, write

$$\mathbb{F}=\begin{bmatrix} 0 & -E_3 & E_2 & -B^1\\ E_3 & 0 & -E_1 & -B^2\\ -E_2 & E_1 & 0 & -B^3\\ B^1 & B^2 & B^3 & 0\end{bmatrix},$$

$$\mathbb{H}=\begin{bmatrix} 0 & -H_3 & H_2 & D^1\\ H_3 & 0 & -H_1 & D^2\\ -H_2 & H_1 & 0 & D^3\\ -D^1 & -D^2 & -D^3 & 0\end{bmatrix}.$$

Hence, the Maxwell equations are now expressible in the form

$$\begin{aligned} &\frac{\partial F^{ji}}{\partial x^j}+\frac{\partial F^{4i}}{\partial x^4}=0, && \frac{\partial F^{i4}}{\partial x^i}=0;\\ &\frac{\partial H^{ji}}{\partial x^j}+\frac{\partial H^{4i}}{\partial x^4}=J^i, && \frac{\partial H^{i4}}{\partial x^i}=\rho. \end{aligned}$$

Let us now define a 4-vector $\{J^\mu\}=\{J^i,J^4=\rho\}$ and note that $F^{44}=0$ and $H^{44}=0$. Then, it is straightforward to see that the Maxwell equations can be written concisely as follows

$$\frac{\partial F^{\mu\nu}}{\partial x^\mu}=0,\quad \frac{\partial H^{\mu\nu}}{\partial x^\mu}=J^\nu. \tag{11.6.4}$$

Let $\mu=dx^1\wedge dx^2\wedge dx^3\wedge dx^4$ be the volume form in the manifold $\mathbb{R}^4$. By using the familiar bases $\mu_{\mu\nu}$ induced by this volume form, we can introduce the two 2-forms $\mathcal{F}\in\Lambda^2(\mathbb{R}^4)$ and $\mathcal{H}\in\Lambda^2(\mathbb{R}^4)$ by employing the antisymmetric coefficients $F^{\mu\nu}$ and $H^{\mu\nu}$ through

$$\mathcal{F} = \frac{1}{2} F^{\mu\nu} \mu_{\mu\nu},$$
$$\mathcal{H} = \frac{1}{2} H^{\mu\nu} \mu_{\mu\nu}$$

where $\mu_{\mu\nu} = \mathbf{i}_{\partial\mu} \circ \mathbf{i}_{\partial_\nu}(\mu)$ and a vector field $\mathcal{J} \in T(\mathbb{R}^4)$ by

$$\mathcal{J} = J^i \frac{\partial}{\partial x^i} + \rho \frac{\partial}{\partial x^4} = J^\mu \frac{\partial}{\partial x^\mu}. \tag{11.6.5}$$

As is easily seen, we can now write

$$\begin{aligned} d\mathcal{F} &= \frac{1}{2} F^{\mu\nu}_{,\gamma} dx^\gamma \wedge \mu_{\mu\nu} \\ &= \frac{1}{2} F^{\mu\nu}_{,\gamma} (\delta^\gamma_\mu \mu_\nu - \delta^\gamma_\nu \mu_\mu) = F^{\mu\nu}_{,\mu} \mu_\nu. \end{aligned}$$

Hence, the equations $(11.6.4)_1$ are equivalent to the exterior equation

$$d\mathcal{F} = 0. \tag{11.6.6}$$

On the other hand, because of the relation

$$\mathbf{i}_\mathcal{J}(\mu) = J^\nu \mu_\nu \in \Lambda^3(\mathbb{R}^4)$$

the equations $(11.6.4)_2$ become equivalent to the exterior equation

$$d\mathcal{H} = \mathbf{i}_\mathcal{J}(\mu). \tag{11.6.7}$$

Let us now express the form $\mathcal{F}$ with respect to the natural basis dx^μ as follows:

$$\mathcal{F} = \frac{1}{2} F^{\mu\nu} \mu_{\mu\nu} = \frac{1}{2} \mathcal{F}_{\alpha\beta}\, dx^\alpha \wedge dx^\beta.$$

On the other hand, for $m = 4$ and $k = 2$ the relation (5.5.10) yields the following expression

$$\mu_{\mu\nu} = \frac{1}{2}\, e_{\nu\mu\alpha\beta}\, dx^\alpha \wedge dx^\beta.$$

Therefore, we find that

$$\begin{aligned} \mathcal{F}_{\alpha\beta} &= -\mathcal{F}_{\beta\alpha} = \frac{1}{2}\, e_{\nu\mu\alpha\beta} F^{\mu\nu} \\ &= -\frac{1}{2}\, e_{\alpha\beta\mu\nu}\, F^{\mu\nu}. \end{aligned}$$

Consequently, we obtain

$$\mathcal{F}_{ij} = -\, e_{ijk4}F^{k4} = e_{ijk4}B^k = e_{ijk}B^k,$$
$$\mathcal{F}_{i4} = -\,\frac{1}{2}\, e_{i4jk}F^{jk} = \frac{1}{2}\, e_{ijk}e^{jkl}E_l = \delta_i^l E_l = E_i$$

from which we deduce that

$$\mathcal{F}_{12} = -\, F^{34} = B^3,\ \mathcal{F}_{13} = F^{24} = -\, B^2,\ \mathcal{F}_{14} = -\, F^{23} = E_1,$$
$$\mathcal{F}_{23} = -\, F^{14} = B^1,\ \ \mathcal{F}_{24} = F^{13} = E_2,\ \ \mathcal{F}_{34} = -\, F^{12} = E_3.$$

Thus, the antisymmetric matrix $\boldsymbol{\mathcal{F}} = [\mathcal{F}_{\alpha\beta}]$ is given by

$$\boldsymbol{\mathcal{F}} = \begin{bmatrix} 0 & B^3 & -\,B^2 & E_1 \\ -\,B^3 & 0 & B^1 & E_2 \\ B^2 & -\,B^1 & 0 & E_3 \\ -\,E_1 & -\,E_2 & -\,E_3 & 0 \end{bmatrix}. \tag{11.6.8}$$

Similarly, the form $\mathcal{H}$ can be rewritten as

$$\mathcal{H} = \frac{1}{2}\mathcal{H}_{\alpha\beta}\, dx^\alpha \wedge dx^\beta,\ \ \mathcal{H}_{\alpha\beta} = -\,\frac{1}{2}\, e_{\alpha\beta\mu\nu}\, H^{\mu\nu}$$

and one finds that

$$\mathcal{H}_{ij} = -\, e_{ijk}D^k,\ \ \mathcal{H}_{i4} = H_i.$$

Hence, the antisymmetric matrix $\boldsymbol{\mathcal{H}} = [\mathcal{H}_{\alpha\beta}]$ is given by

$$\boldsymbol{\mathcal{H}} = \begin{bmatrix} 0 & -\,D^3 & D^2 & H_1 \\ D^3 & 0 & -\,D^1 & H_2 \\ -\,D^2 & D^1 & 0 & H_3 \\ -\,H_1 & -\,H_2 & -\,H_3 & 0 \end{bmatrix}. \tag{11.6.9}$$

With these representations, Equation (11.6.6) leads to

$$d\mathcal{F} = \frac{1}{2}\mathcal{F}_{\alpha\beta,\gamma}\, dx^\gamma \wedge dx^\alpha \wedge dx^\beta$$
$$= \frac{1}{3!}\mathcal{F}_{\gamma\alpha\beta}\, dx^\gamma \wedge dx^\alpha \wedge dx^\beta = 0$$

and we arrive at the equations [*see p.* 265]

$$\mathcal{F}_{\gamma\alpha\beta} = \frac{\partial \mathcal{F}_{\alpha\beta}}{\partial x^\gamma} + \frac{\partial \mathcal{F}_{\beta\gamma}}{\partial x^\alpha} + \frac{\partial \mathcal{F}_{\gamma\alpha}}{\partial x^\beta} = 0 \tag{11.6.10}$$

that are counterparts of Equations $(11.6.4)_1$. Similarly, on considering the definition (5.5.8), Equation (11.6.7) leads to

$$
\begin{aligned}
d\mathcal{H} &= \frac{1}{3!}\mathcal{H}_{\gamma\alpha\beta}\, dx^\gamma \wedge dx^\alpha \wedge dx^\beta \\
&= \frac{1}{3!} e_{\nu\gamma\alpha\beta} J^\nu dx^\gamma \wedge dx^\alpha \wedge dx^\beta
\end{aligned}
$$

from which follow the equations

$$
\mathcal{H}_{\gamma\alpha\beta} = \frac{\partial \mathcal{H}_{\alpha\beta}}{\partial x^\gamma} + \frac{\partial \mathcal{H}_{\beta\gamma}}{\partial x^\alpha} + \frac{\partial \mathcal{H}_{\gamma\alpha}}{\partial x^\beta} = e_{\nu\gamma\alpha\beta} J^\nu \qquad (11.6.11)
$$

that are counterparts of Equations $(11.6.4)_2$.

It is immediately seen that the equation (11.6.7) elicits the condition

$$
d\mathbf{i}_{\mathcal{J}}(\mu) = J^\nu_{\ ,\nu}\, \mu = 0
$$

that is none other than the equation (11.6.2) for the conservation of charge:

$$
J^\nu_{\ ,\nu} = J^i_{\ ,i} + \frac{\partial \rho}{\partial t} = 0.
$$

The exterior equations (11.6.6) and (11.6.7) are obviously coordinate free versions of Maxwell equations. We shall now try to establish a general solution of these equations. Since the manifold $\mathbb{R}^4$ is star-shaped with respect to each of its points, the use the homotopy operator with the centre $\mathbf{x}_0 = \mathbf{0}$ enables us to represent the forms $\mathcal{F}$ and $\mathcal{H}$ as follows

$$
\begin{aligned}
&\mathcal{F} = dH(\mathcal{F}) + H(d\mathcal{F}) = dH(\mathcal{F}) \\
&\mathcal{H} = dH(\mathcal{H}) + H(d\mathcal{H}) = dH(\mathcal{H}) + H\big(\mathbf{i}_{\mathcal{J}}(\mu)\big).
\end{aligned}
$$

Let us now introduce 1-forms

$$
H(\mathcal{F}) = \Phi \in \Lambda^1(\mathbb{R}^4), \quad H(\mathcal{H}) = \Psi \in \Lambda^1(\mathbb{R}^4).
$$

In accordance with (6.3.1), we obtain for $k = 3$ the following 2-form

$$
H\big(\mathbf{i}_{\mathcal{J}}(\mu)\big) = x^\mu \int_0^1 \mathbf{i}_{\partial_\mu}\big(J^\nu(s\mathbf{x})\mu_\nu\big) s^2 ds = x^\mu \Big(\int_0^1 s^2 J^\nu(s\mathbf{x})\, ds\Big)\mu_{\mu\nu}
$$

If we define a linear operator $A : \Lambda^0(\mathbb{R}^4) \to \Lambda^0(\mathbb{R}^4)$ by the rule

$$
A(f)(\mathbf{x}) = \int_0^1 s^2 f(s\mathbf{x})\, ds, \quad f \in \Lambda^0(\mathbb{R}^4)
$$

we find at once that

$$H(\mathbf{i}_{\mathcal{J}}(\mu)) = x^\mu A(J^\nu)\,\mu_{\mu\nu} = x^{[\mu} A(J^{\nu]})\,\mu_{\mu\nu}.$$

Therefore, the general structure of exterior forms determining electromagnetic fields takes the shape

$$\begin{aligned}\mathcal{F} &= d\Phi, \\ \mathcal{H} &= d\Psi + x^{[\mu} A(J^{\nu]})\,\mu_{\mu\nu}.\end{aligned} \tag{11.6.12}$$

It is clear that the forms Φ and Ψ cannot be prescribed uniquely unless we impose some restrictions. As a matter of fact, for arbitrary smooth functions f, g, the exterior forms $\Phi + df$ and $\Psi + dg$ satisfy perfectly again the equations (11.6.12).

Let us now consider 1-forms $\Phi = \Phi_\mu(\mathbf{x})\,dx^\mu$ and $\Psi = \Psi_\mu(\mathbf{x})\,dx^\mu$. If we explicitly write the equation $(11.6.12)_1$, we get

$$\frac{1}{2}\mathcal{F}_{\mu\nu}\,dx^\mu \wedge dx^\nu = \Phi_{\mu,\nu}\,dx^\nu \wedge dx^\mu = -\,\Phi_{[\mu,\nu]}\,dx^\mu \wedge dx^\nu$$

from which we find that

$$\mathcal{F}_{\mu\nu} = -\,2\Phi_{[\mu,\nu]} = \Phi_{\nu,\mu} - \Phi_{\mu,\nu}.$$

We thus obtain

$$\mathcal{F}_{ij} = e_{ijk}B^k = \Phi_{j,i} - \Phi_{i,j}.$$

From this relation, we can easily deduce that

$$e^{ijl}e_{ijk}B^k = 2B^l = e^{ijl}(\Phi_{j,i} - \Phi_{i,j}) = 2e^{lij}\Phi_{j,i}$$

and finally $B^i = e^{ijk}\Phi_{k,j}$. Let us now regard three functions (Φ_1, Φ_2, Φ_3) defined on the manifold $\mathbb{R}^4$ as the components of a 3-vector field $\boldsymbol{\Phi}(\mathbf{x}, t)$. Thus, we come to the conclusion that we can write

$$\mathbf{B} = \boldsymbol{\nabla} \times \boldsymbol{\Phi}$$

by using the familiar curl operator. On the other hand, the remaining relations yield

$$\mathcal{F}_{i4} = E_i = \Phi_{4,i} - \Phi_{i,4}$$

that can be expressed obviously as

$$\mathbf{E} = \boldsymbol{\nabla}\phi - \frac{\partial \boldsymbol{\Phi}}{\partial t}$$

when we introduce the scalar function $\Phi_4 = \phi(\mathbf{x}, t)$. If we repeat the same operations for $(11.6.12)_2$, then we easily obtain

$$\frac{1}{2}\mathcal{H}_{\mu\nu}\,dx^\mu \wedge dx^\nu = \Psi_{[\nu,\mu]}\,dx^\mu \wedge dx^\nu + \frac{1}{2}x^{[\alpha}A(J^{\beta]})\,e_{\beta\alpha\mu\nu}\,dx^\mu \wedge dx^\nu$$

and hence we find that

$$\mathcal{H}_{\mu\nu} = \Psi_{\nu,\mu} - \Psi_{\mu,\nu} - e_{\mu\nu\alpha\beta}x^\alpha A(J^\beta).$$

Consequently, we reach to the relations

$$\mathcal{H}_{ij} = -\,e_{ijk}D^k = \Psi_{j,i} - \Psi_{i,j} - e_{ijk}x^k A(J^4) + e_{ijk}x^4 A(J^k).$$

If we note that $e^{ijl}e_{ijk} = 2\delta^l_k$, the components D^i are then determined by the relations

$$D^i = -\,e^{ijk}\Psi_{k,j} - x^i A(\rho) + tA(J^i).$$

Therefore, the representation of the vector field $\mathbf{D}$ becomes

$$\mathbf{D} = -\,\boldsymbol{\nabla} \times \boldsymbol{\Psi} - A(\rho)\,\mathbf{x} + tA(\mathbf{J})$$

where we define the 3-vector $\boldsymbol{\Psi}(\mathbf{x}, t) = (\Psi_1, \Psi_2, \Psi_3)$. Finally the remaining expressions lead to the relations

$$\mathcal{H}_{i4} = H_i = \Psi_{4,i} - \Psi_{i,4} - e_{i4jk}x^j A(J^k)$$

from which we infer that $H_i = \psi_{,i} - \Psi_{i,4} - e_{ijk}x^j A(J^k)$ or

$$\mathbf{H} = \boldsymbol{\nabla}\psi - \frac{\partial \boldsymbol{\Psi}}{\partial t} - \mathbf{x} \times A(\mathbf{J})$$

where we have introduced the scalar function $\Psi_4 = \psi(\mathbf{x}, t)$. If we collect all the results that have been obtained so far, we then express the general solution of the Maxwell equations in the following form

$$\mathbf{B} = \boldsymbol{\nabla} \times \boldsymbol{\Phi}, \quad \mathbf{E} = \boldsymbol{\nabla}\phi - \frac{\partial \boldsymbol{\Phi}}{\partial t} \tag{11.6.13}$$
$$\mathbf{D} = -\,\boldsymbol{\nabla} \times \boldsymbol{\Psi} - A(\rho)\,\mathbf{x} + tA(\mathbf{J}), \quad \mathbf{H} = \boldsymbol{\nabla}\psi - \frac{\partial \boldsymbol{\Psi}}{\partial t} - \mathbf{x} \times A(\mathbf{J})$$

in terms of arbitrary vector fields $\boldsymbol{\Phi}$ and $\boldsymbol{\Psi}$, and arbitrary scalar fields ϕ and ψ depending on independent variables $(\mathbf{x}, t)$. As we have mentioned above these fields cannot be prescribed uniquely. Indeed, when $\lambda(\mathbf{x}, t)$ and $\Lambda(\mathbf{x}, t)$ are arbitrary scalar functions, we immediately observe that the relations (11.6.13) remains unchanged if we replace the vector-valued functions $\boldsymbol{\Phi}$ and $\boldsymbol{\Psi}$, and scalar-valued functions ϕ and ψ by

$$\boldsymbol{\Phi}+\boldsymbol{\nabla}\lambda,\ \phi+\frac{\partial\lambda}{\partial t};\ \ \boldsymbol{\Psi}+\boldsymbol{\nabla}\Lambda,\ \psi+\frac{\partial\Lambda}{\partial t}.$$

As we have mentioned before, in a physical medium the field vectors will not be all independent and they will be interconnected by some constitutive relations. Naturally, these relations affect the structure of the equations (11.6.1) to a great extent. As an example, let us choose the constitutive relations $\mathbf{D}=\epsilon\mathbf{E}$, $\mathbf{B}=\mu\mathbf{H}$. Although these relations are quite simple, they are considerably important as far as practical applications are concerned. When we insert these relations into the Maxwell equations, they become

$$\boldsymbol{\nabla}\times\mathbf{E}+\frac{\partial\mathbf{B}}{\partial t}=\mathbf{0},\quad \boldsymbol{\nabla}\cdot\mathbf{B}=0, \tag{11.6.14}$$

$$\boldsymbol{\nabla}\times\mathbf{B}-\frac{1}{c^2}\frac{\partial\mathbf{E}}{\partial t}=\mu\mathbf{J},\quad \boldsymbol{\nabla}\cdot\mathbf{E}=\frac{\rho}{\epsilon}$$

where we define $c=1/\sqrt{\epsilon\mu}$. c is a constant of the dimension of velocity. In terms of the components these equations take the form

$$e^{ijk}E_{k,j}+\frac{\partial B^i}{\partial t}=0,\ B^i_{,i}=0;\ e^{ijk}B_{k,j}-\frac{1}{c^2}\frac{\partial E^i}{\partial t}=\mu J^i,\ E^i_{,i}=\frac{\rho}{\epsilon}.$$

We see now that the same field vectors $\mathbf{E}$ and $\mathbf{B}$ appear in both group of equations. However, the positions of upper and lower indices, that were employed to comply with the summation convention, evoke covariant and contravariant components of vectors. To explore this possibility, we shall try to equip the manifold $\mathbb{R}^4$ with an indefinite metric to make it an incomplete Riemannian manifold. We shall now introduce the indefinite ***Lorentz metric*** by the relation [*see* Exercise **7.8**]

$$g_{\lambda\mu}dx^\lambda dx^\mu=(dx^1)^2+(dx^2)^2+(dx^3)^2-c^2(dx^4)^2 \tag{11.6.15}$$

where $x^4=t$. Hence, the metric tensor and its inverse are given by the following matrices, respectively

$$[g_{\lambda\mu}]=\begin{bmatrix}1&0&0&0\\0&1&0&0\\0&0&1&0\\0&0&0&-c^2\end{bmatrix},\quad [g^{\lambda\mu}]=\begin{bmatrix}1&0&0&0\\0&1&0&0\\0&0&1&0\\0&0&0&-\dfrac{1}{c^2}\end{bmatrix}$$

and we have $g=|\det[g_{\lambda\mu}]|=c^2$. Now, by using this metric we may determine covariant components of an antisymmetric tensor whose contravariant components are given by $F^{\mu\nu}$ as follows

$$F_{\alpha\beta} = -F_{\beta\alpha} = g_{\alpha\mu}g_{\beta\nu}F^{\mu\nu}.$$

Thus, we easily find that

$$\begin{aligned} F_{ij} &= g_{i\mu}g_{j\nu}F^{\mu\nu} = e_{ijk}E^k, \\ F_{4i} &= g_{4\mu}g_{i\nu}F^{\mu\nu} = -c^2 B_i \end{aligned}$$

Next, we define a new antisymmetric tensor by its contravariant components through the relation

$$\widetilde{F}^{\alpha\beta} = \frac{1}{2c}\,\epsilon^{\alpha\beta\mu\nu}F_{\mu\nu} = \frac{1}{2c^2}\,e^{\alpha\beta\mu\nu}F_{\mu\nu}$$

whence we readily deduce that

$$\begin{aligned} \widetilde{F}^{ij} &= \frac{1}{2c^2}\,e^{ij\mu\nu}F_{\mu\nu} = \frac{1}{2c^2}\,e^{ijk4}F_{k4} = e^{ijk}B_k \\ \widetilde{F}^{4j} &= \frac{1}{2c^2}\,e^{4j\mu\nu}F_{\mu\nu} = \frac{1}{2c^2}\,e^{4jkl}F_{kl} = -\frac{1}{2c^2}\,e^{jkl}F_{kl} \\ &= -\frac{1}{2c^2}\,e^{jkl}e_{lkm}E^m = \frac{1}{c^2}\,\delta^j_m E^m = \frac{1}{c^2}E^j. \end{aligned} \tag{11.6.16}$$

The components of the divergence of the tensor $\widetilde{F}^{\mu\nu}$, which is a 4-vector, are clearly given now by the following expressions

$$\begin{aligned} \frac{\partial \widetilde{F}^{\mu i}}{\partial x^\mu} &= \frac{\partial \widetilde{F}^{ji}}{\partial x^j} + \frac{\partial \widetilde{F}^{4i}}{\partial x^4} \\ &= e^{jik}\frac{\partial B_k}{\partial x^j} + \frac{1}{c^2}\frac{\partial E^i}{\partial t} \\ &= -\left(e^{ijk}\frac{\partial B_k}{\partial x^j} - \frac{1}{c^2}\frac{\partial E^i}{\partial t}\right) \\ \frac{\partial \widetilde{F}^{\mu 4}}{\partial x^\mu} &= \frac{\partial \widetilde{F}^{i4}}{\partial x^i} = -\frac{1}{c^2}\frac{\partial E^i}{\partial x^i}. \end{aligned}$$

Consequently, the admitted forms of the constitutive relations transform the Maxwell equations into the form

$$\frac{\partial F^{\alpha\beta}}{\partial x^\alpha} = 0, \quad \frac{\partial \widetilde{F}^{\alpha\beta}}{\partial x^\alpha} = -\mu J^\beta. \tag{11.6.17}$$

The existence of the constitutive relations make it possible to write the field equations (11.6.13) in much more interesting and meaningful forms. In order to simplify the necessary manipulations, we prefer to take $\mathbf{J} = \mathbf{0}$ and

$\rho = 0$. In that case, it is clear that the resolving functions must satisfy the relations

$$-\boldsymbol{\nabla}\times\boldsymbol{\Psi} = \epsilon\Big(\boldsymbol{\nabla}\phi - \frac{\partial\boldsymbol{\Phi}}{\partial t}\Big), \tag{11.6.18}$$
$$\boldsymbol{\nabla}\times\boldsymbol{\Phi} = \mu\Big(\boldsymbol{\nabla}\psi - \frac{\partial\boldsymbol{\Psi}}{\partial t}\Big).$$

When we evaluate the divergences of these two equations, we are led, respectively, to the equations

$$\nabla^2\phi - \frac{\partial}{\partial t}(\boldsymbol{\nabla}\cdot\boldsymbol{\Phi}) = 0, \tag{11.6.19}$$
$$\nabla^2\psi - \frac{\partial}{\partial t}(\boldsymbol{\nabla}\cdot\boldsymbol{\Psi}) = 0$$

where ∇^2 denotes the Laplace operator in the configuration manifold. Let us next evaluate the curls of the equations (11.6.18):

$$\boldsymbol{\nabla}\times\boldsymbol{\nabla}\times\boldsymbol{\Psi} = \epsilon\frac{\partial}{\partial t}(\boldsymbol{\nabla}\times\boldsymbol{\Phi}),$$
$$\boldsymbol{\nabla}\times\boldsymbol{\nabla}\times\boldsymbol{\Phi} = -\mu\frac{\partial}{\partial t}(\boldsymbol{\nabla}\times\boldsymbol{\Psi}).$$

On recalling the well-known and easily verifiable following vectorial identity involving curl, gradient and divergence operators

$$\boldsymbol{\nabla}\times\boldsymbol{\nabla}\times\boldsymbol{\Psi} = \boldsymbol{\nabla}(\boldsymbol{\nabla}\cdot\boldsymbol{\Psi}) - \boldsymbol{\nabla}^2\boldsymbol{\Psi},$$

where $\boldsymbol{\nabla}^2$ denotes the Laplace operator on 3-vector fields, utilising the equations (11.6.18) and introducing again the constant $c^2 = 1/\epsilon\mu$, we easily find that

$$\boldsymbol{\nabla}^2\boldsymbol{\Psi} - \frac{1}{c^2}\frac{\partial^2\boldsymbol{\Psi}}{\partial t^2} = \boldsymbol{\nabla}f,$$
$$\boldsymbol{\nabla}^2\boldsymbol{\Phi} - \frac{1}{c^2}\frac{\partial^2\boldsymbol{\Phi}}{\partial t^2} = \boldsymbol{\nabla}g$$

where the functions f and g are given by

$$f = \boldsymbol{\nabla}\cdot\boldsymbol{\Psi} - \frac{1}{c^2}\frac{\partial\psi}{\partial t},$$
$$g = \boldsymbol{\nabla}\cdot\boldsymbol{\Phi} - \frac{1}{c^2}\frac{\partial\phi}{\partial t}.$$

Differentiating these functions with respect to time and making use of the equations (11.6.19), we obtain

$$\nabla^2\psi - \frac{1}{c^2}\frac{\partial^2\psi}{\partial t^2} = \frac{\partial f}{\partial t},$$
$$\nabla^2\phi - \frac{1}{c^2}\frac{\partial^2\phi}{\partial t^2} = \frac{\partial g}{\partial t}.$$

We would like now to remove the arbitrariness in the selection of functions $(\boldsymbol{\Psi}, \psi)$ and $(\boldsymbol{\Phi}, \phi)$ by imposing that the ***Lorenz conditions*** $f = 0$ and $g = 0$ [after Danish physicist and mathematician Ludvig Valentin Lorenz (1829-1891)] should be satisfied. To this end, it suffices to choose

$$\boldsymbol{\nabla}\cdot\boldsymbol{\Psi} - \frac{1}{c^2}\frac{\partial\psi}{\partial t} = 0,$$
$$\boldsymbol{\nabla}\cdot\boldsymbol{\Phi} - \frac{1}{c^2}\frac{\partial\phi}{\partial t} = 0.$$

In this case, we can easily verify that all field quantities can be prescribed by considering only one pair of functions, say, for instance $(\boldsymbol{\Phi}, \phi)$. Indeed, (11.6.13) now takes the form

$$\mathbf{B} = \boldsymbol{\nabla}\times\boldsymbol{\Phi}, \qquad \mathbf{E} = \boldsymbol{\nabla}\phi - \frac{\partial\boldsymbol{\Phi}}{\partial t},$$
$$\mathbf{D} = \epsilon\Big(\boldsymbol{\nabla}\phi - \frac{\partial\boldsymbol{\Phi}}{\partial t}\Big), \quad \mathbf{H} = \frac{1}{\mu}\boldsymbol{\nabla}\times\boldsymbol{\Phi}.$$

These functions have to satisfy the scalar-valued and vector-valued wave equations

$$\nabla^2\phi - \frac{1}{c^2}\frac{\partial^2\phi}{\partial t^2} = 0,$$
$$\boldsymbol{\nabla}^2\boldsymbol{\Phi} - \frac{1}{c^2}\frac{\partial^2\boldsymbol{\Phi}}{\partial t^2} = \mathbf{0}.$$

ϕ and $\boldsymbol{\Phi}$ are called, respectively, *scalar* and *vector potentials* of electromagnetic fields. The number c denotes naturally the velocity of propagation of electromagnetic waves in such a medium.

11.7. THERMODYNAMICS

Let us consider a thermodynamic system $\mathbb{T}$ occupying a finite region in $\mathbb{R}^3$. The variables that describe the behaviour of the system will be called as the ***state variables***. We distinguish three different sets of substate variables:

the external variables $\{q^i, i = 1, 2, \dots, n\}$, *the internal variables* $\{\nu^\alpha, \alpha = 1, 2, \dots, N\}$ and *the empirical temperature* or solely *the temperature* T. Thus, we may assume that the system $\mathbb{T}$ is incorporated in an $(n + N + 1)$-dimensional Euclidean manifold. The ***external agencies*** acting upon the system $\mathbb{T}$ is called the set of ***external forces*** $\{F_i(\mathbf{q}, \boldsymbol{\nu}, T), i = 1, 2, \dots, n\}$. The ***infinitesimal work*** that will occur during some infinitesimal changes in the external variables will be denoted by the 1-form

$$W = F_i(\mathbf{q}, \boldsymbol{\nu}, T)\, dq^i. \tag{11.7.1}$$

We adopt the convention that the signs of forces F_i are positive when the forces do work on the system $\mathbb{T}$ whereas are negative if the system $\mathbb{T}$ do work on the external agencies. We say that the external forces are *conservative* if they satisfy the relations

$$\frac{\partial F_i}{\partial q^j} = \frac{\partial F_j}{\partial q^i}.$$

In the present context, we shall consider this particular situation. In this case when we restrict the form W to the submanifold given by $\boldsymbol{\nu} = constant$, $T = constant$, we obtain

$$\begin{aligned} dW|_{\boldsymbol{\nu},T} &= F_{i,j}\, dq^j \wedge dq^i \\ &= \frac{1}{2}(F_{i,j} - F_{j,i}) dq^j \wedge dq^i = 0 \end{aligned}$$

where we employed the notation $F_{i,j} = \dfrac{\partial F_i}{\partial q^j}$. Hence the form $W|_{\boldsymbol{\nu},T}$ is closed and it is exact since the manifold is star-shaped so that we can write

$$W|_{\boldsymbol{\nu},T} = d\Psi|_{\boldsymbol{\nu},T}.$$

The function $\Psi(\mathbf{q}, \boldsymbol{\nu}, T)$ which we shall call *isothermal work function* can be evaluated as follows by the homotopy operator

$$\Psi(\mathbf{q}, \boldsymbol{\nu}, T) = \Psi_0(\boldsymbol{\nu}, T) + \int_0^1 (q^i - q_0^i)\, F_i\big[\mathbf{q}_0 + t(\mathbf{q} - \mathbf{q}_0), \boldsymbol{\nu}, T\big]\, dt \tag{11.7.2}$$

on resorting to Theorem 6.3.1(i) and the relation (6.3.1). $\mathbf{q}_0$ is an arbitrary point. From this relation, we immediately deduce that

$$F_i = \frac{\partial \Psi}{\partial q^i}. \tag{11.7.3}$$

Thus, the 1-form $d\Psi$ is now expressible as

$$\begin{aligned} d\Psi &= \frac{\partial \Psi}{\partial q^i}\,dq^i + \frac{\partial \Psi}{\partial \nu^\alpha}\,d\nu^\alpha + \frac{\partial \Psi}{\partial T}\,dT \\ &= F_i\,dq^i + N_\alpha d\nu^\alpha - N\,dT \end{aligned}$$

where we have defined

$$N_\alpha(\mathbf{q},\boldsymbol{\nu},T) = \frac{\partial \Psi}{\partial \nu^\alpha}, \quad N(\mathbf{q},\boldsymbol{\nu},T) = -\,\frac{\partial \Psi}{\partial T} \tag{11.7.4}$$

The trivial condition $d^2\Psi = 0$ then leads at once to the celebrated ***Maxwell reciprocity relations***

$$\begin{aligned} &\frac{\partial F_i}{\partial q^j} = \frac{\partial F_j}{\partial q^i}, \quad \frac{\partial F_i}{\partial \nu^\alpha} = \frac{\partial N_\alpha}{\partial q^i}, \quad \frac{\partial F_i}{\partial T} = -\,\frac{\partial N}{\partial q^i}, \\ &\frac{\partial N_\alpha}{\partial \nu^\beta} = \frac{\partial N_\beta}{\partial \nu^\alpha}, \quad \frac{\partial N_\alpha}{\partial T} = \frac{\partial N}{\partial \nu^\alpha}. \end{aligned}$$

Then, the work form (11.7.1) can be written in the following manner

$$W = d\Psi + N\,dT - N_\alpha d\nu^\alpha \tag{11.7.5}$$

whence we obtain

$$dW = dN \wedge dT - dN_\alpha \wedge d\nu^\alpha.$$

The expression (11.7.5) implies that if all components N_α do not vanish, then the Darboux class of the form W is at least 4. If the external forces are dependent on some of the internal variables, then the Maxwell reciprocity relations clearly show that at least some of the coefficients N_α should not vanish as a consequence of this property.

If we denote the heat input to the system $\mathbb{T}$ by the *heat* 1*-form* Q, then *the first law of thermodynamics* states that in quasistatic situations in which the time change of the system is so slow that its kinetic energy can be neglected, one can write

$$dE = W + Q$$

where $E(\mathbf{q},\boldsymbol{\nu},T)$ is called the ***internal energy function***. Therefore, when we make use of (11.7.5), we obtain

$$Q = d(E - \Psi) - N\,dT + N_\alpha d\nu^\alpha$$

and consequently

$$dQ = -\,dN \wedge dT + dN_\alpha \wedge d\nu^\alpha = -\,dW.$$

In his case, the Darboux class of the form Q is at least 4, that is, in view of Theorem 6.6.3 it does not possess the inaccessibility property. Hence, the form Q is in general not completely integrable and Theorem 5.13.4 requires that $Q \wedge dQ \neq 0$. On the other hand, we naturally find

$$dQ|_{\boldsymbol{\nu}} = -\,dN \wedge dT$$

on the submanifold $\boldsymbol{\nu} = constant$. Experimental information show that we cannot get our system occupying a state on the submanifold $\boldsymbol{\nu} = constant$ to reach to another state on the same submanifold without exchanging heat. In other words, two states on the submanifold $\boldsymbol{\nu} = constant$ cannot be connected by an *adiabatic path* in this manifold. This means that $dQ|_{\boldsymbol{\nu}} \neq 0$. Let us now discuss the condition under which the form $Q|_{\boldsymbol{\nu}}$ becomes completely integrable. The relation

$$Q|_{\boldsymbol{\nu}} \wedge dQ|_{\boldsymbol{\nu}} = 0$$

yields

$$d(E - \Psi) \wedge dN \wedge dT = 0$$

This, in turn, indicates that $E - \Psi$ is functionally dependent on variables N and T. Thus, we can write

$$E(\mathbf{q}, \boldsymbol{\nu}, T) = \Psi(\mathbf{q}, \boldsymbol{\nu}, T) + f\big(N(\mathbf{q}, \boldsymbol{\nu}, T), \boldsymbol{\nu}, T\big). \qquad (11.7.6)$$

So we find that

$$Q = df(N, \boldsymbol{\nu}, T) - N\,dT + N_\alpha d\nu^\alpha. \qquad (11.7.7)$$

Let us now consider two thermodynamic systems $\mathbb{T}_1$ and $\mathbb{T}_2$ having exactly the same temperature and internal variables. In this case, we can obviously write

$$N_1(\mathbf{q}_1, \boldsymbol{\nu}, T) = -\frac{\partial \Psi_1}{\partial T}, \quad N_2(\mathbf{q}_2, \boldsymbol{\nu}, T) = -\frac{\partial \Psi_2}{\partial T}.$$

If we let these two systems to interact, our physical experience tells us that the common isothermal work function should be expressed in the following fashion

$$\Psi_{12}(\mathbf{q}_1, \mathbf{q}_2, \boldsymbol{\nu}, T) = \Psi_1(\mathbf{q}_1, \boldsymbol{\nu}, T) + \Psi_2(\mathbf{q}_2, \boldsymbol{\nu}, T) + \psi(\mathbf{q}_1, \mathbf{q}_2, \boldsymbol{\nu}).$$

This relation implies that isothermal work function must be a *semi-additive* function (in thermodynamics, it is frequently taken $\psi = 0$. In such a case, Ψ will become a strictly additive function). This is tantamount that interaction forces in the composite system are independent of the temperature. It is now clear that we can write

$$N_{12} = -\frac{\partial \Psi_{12}}{\partial T} = -\frac{\partial \Psi_1}{\partial T} - \frac{\partial \Psi_2}{\partial T} = N_1 + N_2.$$

The composition rule of the internal energy will follow from the physical assumption that the function $E - \Psi$ is strictly additive and it is found as

$$E_{12}(\mathbf{q}_1, \mathbf{q}_2, \boldsymbol{\nu}, T) = E_1(\mathbf{q}_1, \boldsymbol{\nu}, T) + E_2(\mathbf{q}_2, \boldsymbol{\nu}, T) + \psi(\mathbf{q}_1, \mathbf{q}_2, \boldsymbol{\nu}).$$

Then, (11.7.6) provides the functional relation

$$\begin{aligned} f(N_{12}, \boldsymbol{\nu}, T) &= f(N_1 + N_2, \boldsymbol{\nu}, T) \\ &= f(N_1, \boldsymbol{\nu}, T) + f(N_2, \boldsymbol{\nu}, T) \end{aligned} \tag{11.7.8}$$

that must be held by the function f. With the definition $u = N_1 + N_2$, this relation leads to

$$\frac{\partial f}{\partial N_1} = \frac{\partial f}{\partial N_2} = \frac{\partial f}{\partial u}.$$

Since N_1 and N_2 are independent variables, we finally obtain

$$f(N, \boldsymbol{\nu}, T) = g(\boldsymbol{\nu}, T)N + g_1(\boldsymbol{\nu}, T).$$

However, (11.7.8) now implies that $g_1(\boldsymbol{\nu}, T) = 0$ and we thus reach to the conclusion

$$f(N, \boldsymbol{\nu}, T) = g(\boldsymbol{\nu}, T)N.$$

Consequently, we can write

$$\begin{aligned} E(\mathbf{q}, \boldsymbol{\nu}, T) &= \Psi(\mathbf{q}, \boldsymbol{\nu}, T) + g(\boldsymbol{\nu}, T)\, N(\mathbf{q}, \boldsymbol{\nu}, T) \\ Q(\mathbf{q}, \boldsymbol{\nu}, T) &= d\big[g(\boldsymbol{\nu}, T)\, N(\mathbf{q}, \boldsymbol{\nu}, T)\big] - N(\mathbf{q}, \boldsymbol{\nu}, T)\, dT + N_\alpha d\nu^\alpha. \end{aligned} \tag{11.7.9}$$

We shall now try to reduce these functional relations for E and Q into simpler forms. To this end, we want to introduce a ***thermodynamic temperature*** depending on the empirical temperature and internal variables by the following expression

$$\theta(\boldsymbol{\nu}, T) = \theta_0(\boldsymbol{\nu}) \exp\Big[\int_{T_0}^{T} \frac{d\tau}{g(\boldsymbol{\nu}, \tau)}\Big]. \tag{11.7.10}$$

Obviously, θ will satisfy the relation

$$\frac{\partial \theta}{\partial T} = \frac{\theta}{g}.$$

On the other hand, we can extract, in principle, from (11.7.10) the inverse function $T = T(\boldsymbol{\nu}, \theta)$ connecting the empirical temperature to the thermodynamic temperature. Let us now introduce the functions

$$\begin{aligned} \psi(\mathbf{q}, \boldsymbol{\nu}, \theta) &= \Psi\big(\mathbf{q}, \boldsymbol{\nu}, T(\boldsymbol{\nu}, \theta)\big), \\ e(\mathbf{q}, \boldsymbol{\nu}, \theta) &= E\big(\mathbf{q}, \boldsymbol{\nu}, T(\boldsymbol{\nu}, \theta)\big) \end{aligned}$$

and the quantity

$$\eta(\mathbf{q}, \boldsymbol{\nu}, \theta) = -\frac{\partial \psi}{\partial \theta}. \tag{11.7.11}$$

From the chain rule of differentiation, we obtain

$$N = -\frac{\partial \Psi}{\partial T} = -\frac{\partial \psi}{\partial \theta}\frac{\partial \theta}{\partial T} = \frac{\theta \eta}{g}$$

that yields the relation $gN = \theta\eta$. Then, the equation $(11.7.9)_2$ leads us to the expression

$$\begin{aligned} Q &= d(\theta\eta) - N dT + N_\alpha d\nu^\alpha \\ &= \theta\, d\eta + \eta\, d\theta - N dT + N_\alpha d\nu^\alpha \\ &= \theta\, d\eta + \eta\Big(\frac{\partial \theta}{\partial T}\, dT + \frac{\partial \theta}{\partial \nu^\alpha}\, d\nu^\alpha\Big) - \frac{\theta\eta}{g}\, dT + N_\alpha d\nu^\alpha. \end{aligned}$$

Let us now consider the expression

$$n_\alpha = \eta\,\frac{\partial \theta}{\partial \nu^\alpha} + N_\alpha.$$

From the chain rule again, we can write

$$N_\alpha = \frac{\partial \psi}{\partial \nu^\alpha} + \frac{\partial \psi}{\partial \theta}\frac{\partial \theta}{\partial \nu^\alpha} = \frac{\partial \psi}{\partial \nu^\alpha} - \eta\,\frac{\partial \theta}{\partial \nu^\alpha}$$

so that we find

$$n_\alpha(\mathbf{q}, \boldsymbol{\nu}, \theta) = \frac{\partial \psi}{\partial \nu^\alpha}. \tag{11.7.12}$$

Thus the equations (11.7.9) take now the forms

$$\begin{aligned} e(\mathbf{q}, \boldsymbol{\nu}, \theta) &= \psi(\mathbf{q}, \boldsymbol{\nu}, \theta) + \theta\, \eta(\mathbf{q}, \boldsymbol{\nu}, \theta) \\ &= \psi - \theta\frac{\partial \psi}{\partial \theta}, \\ Q &= \theta\, d\eta + n_\alpha\, d\nu^\alpha. \end{aligned} \tag{11.7.13}$$

We call the function η as the ***entropy*** and the function ψ as the ***Helmholtz free energy function*** [after German mathematician, physicist and medical doctor Hermann Ludwig Ferdinand von Helmholtz (1821-1894)]. When we take $\boldsymbol{\nu} = constant$, it follows from $(11.7.13)_2$ that

$$Q|_{\boldsymbol{\nu}} = \theta \, d\eta$$

Hence, the heat 1-form Q is completely integrable in this case as it should be expected.

If we choose $\theta_0 > 0$ in (11.7.10), then the condition $\theta > 0$ is satisfied. We can then also write $\inf \theta = 0$. The temperature verifying this condition will be called the ***absolute temperature***. Moreover $(11.7.13)_1$ gives

$$c_{\mathbf{q},\boldsymbol{\nu}} = \frac{\partial e}{\partial \theta} = -\theta \frac{\partial^2 \psi}{\partial \theta^2}.$$

The quantity $c_{\mathbf{q},\boldsymbol{\nu}}$ is known as the ***specific heat*** under the restrictions $\mathbf{q} = constant$ and $\boldsymbol{\nu} = constant$. We know that $c_{\mathbf{q},\boldsymbol{\nu}}$ is positive in real materials. This implies that the following inequality must be satisfied

$$\frac{\partial^2 \psi}{\partial \theta^2} < 0$$

due to the fact that $\theta > 0$.

XI. EXERCISES

11.1. Let (S, ω) be a symplectic manifold. Show that every vector field $V \in T(S)$ satisfying the condition $\pounds_V \omega = 0$ is a Hamiltonian vector field.

11.2. The Hamiltonian function of the *Toda lattice* [after Japanese physicist Morikazu Toda (1917-2010)] involving three particles, which we encounter in the solid state physics, is given by the following function on the symplectic manifold $S = T^*(\mathbb{R}^3)$

$$H(\mathbf{q}, \mathbf{p}) = \frac{1}{2}(p_1^2 + p_2^2 + p_3^2) + e^{q^1 - q^2} + e^{q^2 - q^3} + e^{q^3 - q^1}.$$

(a) Write the Hamilton equations governing the motion of the system. (b) In order that the function $f \in \Lambda^0(S)$ be an integral of the motion, we know that the condition $\{f, H\} = 0$ should be satisfied. Searching for the integrals of the partial differential equation so obtained for the function f, show that the following functions are integrals of the motion

$$f_1(\mathbf{q}, \mathbf{p}) = H,$$
$$f_2(\mathbf{q}, \mathbf{p}) = p_1 + p_2 + p_3$$

$$f_3(\mathbf{q},\mathbf{p}) = \frac{1}{3}(p_1^3 + p_2^3 + p_3^3) + p_1\big(e^{q^1-q^2} + e^{q^3-q^1}\big) + p_2\big(e^{q^2-q^3} + e^{q^1-q^2}\big) + p_3\big(e^{q^3-q^1} + e^{q^2-q^3}\big).$$

11.3. Show that the function

$$g(\mathbf{q},\mathbf{p}) = p_1p_2p_3 - p_1e^{q^2-q^3} - p_2e^{q^3-q^1} - p_3e^{q^1-q^2}$$

is also an integral of the motion for the Toda lattice.

11.4. Is the transformation

$$Q = \log\,(q^{-1}\sin p), \quad P = q\cot p$$

canonical?

11.5. Determine the structure of the function $f(q^1,\dots,q^n)$ so that the mapping

$$Q^i = f(q^1,\dots,q^n)\sin p_i,$$
$$P_i = f(q^1,\dots,q^n)\cos p_i,\ \ i = 1,\dots,n$$

becomes a canonical transformation in the phase space.

11.6. Find the structure of the constant matrix $\mathbf{A} = [a_{ij}]$ so that the mapping

$$Q^i = q^i, \quad P_i = p_i + a_{ij}\,q^j$$

becomes a canonical transformation in the phase space. Determine the generating function $F_1(\mathbf{q},\mathbf{P})$ corresponding to this case.

11.7. We can build a symplectic structure in the non-conservative mechanics by introducing an *energy variable* E. Let us denote the coordinate cover in the manifold $\mathbb{R}^{2n+2}$ by $(\mathbf{q},\mathbf{p},E,t)$. We then define a symplectic form as follows

$$\omega = dq^i \wedge dp_i + dE \wedge dt.$$

Let $H(\mathbf{q},\mathbf{p},t)$ be the Hamiltonian function of the system. We next consider a function

$$P(\mathbf{q},\mathbf{p},E,t) = H(\mathbf{q},\mathbf{p},t) - E.$$

Show that along the integral curves of a Hamiltonian vector field $V_P \in T(\mathbb{R}^{2n+2})$ defined by the relation $\mathbf{i}_{V_P}\omega = dP$ and associated with the function P, the following equations are satisfied

$$\frac{dq^i}{dt} = \frac{\partial H}{\partial p_i}, \quad \frac{dp_i}{dt} = -\frac{\partial H}{\partial q^i}, \quad \frac{dE}{dt} = \frac{\partial H}{\partial t}.$$

11.8. Let the functions f and g be two integrals of the motion in a non-conservative system so that they satisfy the following equations

$$\frac{\partial f}{\partial t} + \{f, H\} = 0,$$

$$\frac{\partial g}{\partial t} + \{g, H\} = 0.$$

Show that the Poisson bracket $\{f, g\}$ is also an integral of the motion (this result is known as the ***Poisson theorem***).

11.9. Utilising the Poisson theorem prove that if the function $f(\mathbf{q}, \mathbf{p}, t)$ is an integral of the motion in a conservative system $\big(H = H(\mathbf{q}, \mathbf{p})\big)$, then all derivatives $\dfrac{\partial^n f}{\partial t^n}$, $n = 1, 2, \ldots$ are also integrals of the motion.

11.10. The energy balance in a perfect fluid can be expressed in the form

$$de = \theta\, d\eta - p\, d\nu$$

where p is the pressure, ν is the specific volume. Find the relations to which the exterior derivatives of the forms de and $\dfrac{de}{\theta}$ give rise under the following assumptions:

$$\begin{aligned}
&(a)\ e = e(\theta, \nu), \quad p = p(\theta, \nu), \\
&(b)\ \theta = \theta(\eta, \nu), \quad p = p(\eta, \nu), \\
&(c)\ \eta = \eta(\theta, \nu), \quad p = p(\theta, \nu).
\end{aligned}$$

REFERENCES

This list of references is not intended to provide an exhaustive bibliography. All the books listed here are utilised in various degrees in preparing this work.

Abraham, R. and J. E. Marsden, *Foundations of Mechanics*, The Benjamin/ Cummins Publ. Co.., Inc., Reading, Massachusetts, 1978.

Abraham, R., J. E. Marsden and T. Ratiu, *Manifolds, Tensor Analysis, and Applications*, 2. Ed., Springer-Verlag, New York, 1988.

Akyıldız, Y., Expository Lectures on Topology, Geometry and Gauge Theories, *Quantum Theory, Groups, Fields and Particles*, pp. 179-234, Edited by A. O. Barut, D. Reidel Publ. Co., Dordrecht, 1983.

Anderson, R. L. and N. H. Ibragimov, *Lie-Bäcklund Transformations in Applications*, SIAM, Philadelphia, 1979.

Arnold, V. I., *Mathematical Methods of Classical Mechanics*, Springer-Verlag New York, Inc., 1978.

Bishop, R. L. and S. I. Goldberg, *Tensor Analysis on Manifolds*, Dover Publications, Inc., New York, 1980.

Bluman, G. W. and J. D. Cole, *Similarity Methods for Differential Equations*, Springer-Verlag, 1974.

Bluman, G. W. and S. Kumei, *Symmetries and Differential Equations*, Springer-Verlag, 1989.

Bredon, G. E., *Topology and Geometry*, Springer-Verlag, Berlin, 1993.

Bryant, R. L., S. S. Chern, R. B. Gardner, H. L. Goldschmidt and P. A. Griffiths, *Exterior Differential Systems*, Springer-Verlag, New York, 1991.

Cartan, É., *Leçon sur les Invariants Integraux*, Hermann, Paris, 1922.

Cartan, É., *Les Systèmes Différentiels Extérieur et Leurs Applications Géometriques*, Hermann, Paris, 1945.

Cartan, H., *Differential Forms*, Hermann, Paris, 1970.

Chern, Shiing-Shen, W. H. Chen and K. S. Lam, *Lectures on Differential Geometry*, **I**, World Scientific Publishing Company, Singapore, 1999.

Choquet-Bruhat, Y., C. DeWitt-Morette and M. Dillard-Bleick, *Analysis, Manifolds and Physics. Part I: Basics*, North-Holland, Amsterdam, 1982.

Coddington, E. A. and N. Levinson, *Theory of Ordinary Differential Equations*, McGraw-Hill Book Company, Inc., New York, 1955.

Cronin, J., *Differential Equations. Introduction and Qualitative Theory*, Marcel Dekker, Inc., New York, 1980.

Darling, R. W. R., *Differential Forms and Connections*, Cambridge University Press, 1994.

de Leon, M. and P. R. Rodrigues, *Methods of Differential Geometry in Analytical Mechanics*, North-Holland, Amsterdam, 1989.

de Rham, G., *Variété Différentiables, Formes, Courants, Formes Harmoniques*, Hermann, Paris, 1955.

do Carmo, M. P., *Differential Forms and Applications*, Springer-Verlag, Berlin, 1991.

do Carmo, M. P., *Riemannian Geometry*, Birkhäuser Boston, 1992.

Edelen, D. G. B., *Isovector methods for equations of balance*, Sijthoff & Noordhoff, Alphen aan den Rijn, Holland, 1980

Edelen, D. G. B., *Applied Exterior Calculus*, John Wiley & Sons, Inc., New York, 1985.

Edelen, D. G. B. and Jian-hua Wang, *Transformation Methods for Nonlinear Partial Differential Equations*, World Scientific, Singapore, 1992.

Edelen, D. G. B., *The College Station Lectures on Thermodynamics*, Texas A&M University, 1993.

Flanders, H., *Differential Forms with Applications to the Physical Sciences*, Dover Publications, Inc., New York, 1989.

Frankel, Th., *The Geometry of Physics: An Introduction*, Cambridge University Press, Cambridge, 1997.

Funk, P., *Variationrechnung und ihre Anwendung in Physik und Technik*, 2. Ed., Springer-Verlag, Berlin, 1970.

Gilmore, R., *Lie Groups, Lie Algebras and Some of Their Applications*, Dover Publications, Inc., New York, 2005.

Gorbatsevich, V. V., A. L. Onishchik and E. B. Vinberg, *Foundations of Lie Theory and Lie Transformation Groups*, Springer-Verlag, Berlin, 1997.

Greub, W. H., *Multilinear Algebra*, Springer-Verlag, New York, 1978.

Greub, W. H., *Linear Algebra*, Springer-Verlag, New York, 1981.

Guggenheimer, H. W., *Differential Geometry*, Dover Publications, Inc., New York, 1977.

Guillemin, V. and A. Pollack, *Differential Topology*, AMS Chelsea Publishing, Providence, Rhode Island, 2010. Originally Published: Prentice-Hall, Englewood Cliffs, N. J., 1974.

Hafkenscheid, P., *De Rham Cohomology of Smooth Manifolds*, Bachelorthesis, VU University, Amsterdam

Hall, B. C., *Lie Groups, Lie Algebras, and Representation: An Elementary Introduction*, Springer-Verlag, New York, 2003.

Heinbockel, J. H., *Introduction to Tensor Calculus and Continuum Mechanics*, Trafford Publishing, Victoria, British Columbia, Canada, 2001.

Hodge, V. W. D., *Theory and Applications of Harmonic Integrals*, 2. Ed., Cambridge University Press, Cambridge, 1952.

Holm, D. D., T. Schmah and C. Stoica, *Geometric Mechanics and Symmetry - From Finite to Infinite Dimensions*, Oxford University Press, 2009.

Hsiang, Wu Yi, *Lectures on Lie Groups*, World Scientific Publishing Company, Singapore, 2000.

Ibragimov, N. H., *Transformation Groups Applied to Mathematical Physics*, D. Reidel Publishing Company, Dordrecht, 1985.

Ibragimov, N. H. (Editor), *CRC Handbook of Lie Group Analysis of Differential Equations.* **I.** Symmetries, Exact Solutions and Conservation Laws, 1994. **II.** Applications in Engineering and Physical Sciences, 1995. **III.** New Trends in Theoretical Developments and Computational Methods, 1996, CRC Press, Boca Raton.

Kobayashi. Sh. and K. Nomizu, *Foundations of Differential Geometry*, **I**, John Wiley & Sons, New York, 1963.

Krasil'shchik, I. S., V. V. Lychagin and A. M. Vinogradov, *Geometry of Jet Spaces and Nonlinear Partial Differential Equations*, Gordon and Breach Science Publishers, New York, 1986.

Lee, J. M., *Introduction to Smooth Manifolds*, Springer-Verlag, New York, 2003.

Madsen, I. H. and J. Tornehave, *From Calculus to Cohomology. De Rham Cohomology and Characteristic Classes*, Cambridge University Press, 1997.

Marmo, G., E. J. Saletan, A. Simoni and B. Vitale, *Dynamical Systems. A Differential Geometric Approach to Symmetry and Reduction*, John Wiley & Sons, New York, 1985.

Marsden, J. E. and T. S. Ratiu, *Introduction to Mechanics and Symmetry*, 2. Ed., Springer-Verlag, New York, 1999.

Munkres, J. R., *Topology*, 2. Ed., Prentice-Hall International, London, 2000.

Noll, W., *Finite-Dimensional Spaces*-I. *Algebra, Geometry, and Analysis,* Martinus Nijhoff Publishers, Dordrecht, 1987.

Olver, P. J., *Applications of Lie Groups to Differential Equations*, Springer-Verlag, New York, 1986.

Olver, P. J., *Equivalence, Invariants, and Symmetry*, Cambridge University Press, Cambridge, 1995.

Ovsiannikov, L. V., *Group Analysis of Differential Equations*, Academic Press, New York, 1982 (Translation of 1962 Russian edition).

Schouten, J. A., *Ricci-Calculus - An Introduction to Tensor Analysis and its Geometrical Applications*, 2. Ed., Springer-Verlag, Berlin, 1954.

Schouten, J. A., *Tensor Analysis for Physicists*, 2. Ed., Oxford University Press, London, 1954

Schutz, B. F., *Geometrical Methods of Mathematical Physics*, Cambridge University Press, Cambridge, 1980.

Singer, I. M. and J. A. Thorpe, *Lecture Notes on Elementary Topology and Geometry*, Scott, Foresman and Company, Dallas, Texas, 1967.

Ślebodziński, W., *Exterior Forms and Their Applications*, PWN - Polish Scientific Publishers, Warszawa, 1970.

Spivak, M., *Calculus on Manifolds*, W. A. Benjamin, Inc., New York, 1965.

Spivak, M., *A Comprehensive Introduction to Differential Geometry,* **I - V**, 2. Ed., Publish or Perish, Inc., Houston, Texas, 1979.

Stamm, E., Introduction to Differential Topology, *Proceedings of the Thirteenth Biennial Seminar of the Canadian Mathematical Congress on Differential Topology; Differential Geometry and Applications,* Vol. 1, s. 281-327, Edited by J. R. Vanstone, Canadian Mathematical Society, Montreal, 1972.

Stepani, H. (Editor M. MacCallum), *Differential Equations. Their Solution using Symmetries*, Cambridge University Press, Cambridge, 1989.

Şuhubi, E. S., *Functional Analysis*, Kluwer Academic Publishers, London, 2003.

Talpaert, Y., *Differential Geometry with Applications to Mechanics and Physics*, Marcel Dekker, Inc., New York-Basel, 2001.

Thomas, G. H., *Introductory Lectures on Fibre Bundles and Topology for Physicists*, Rivista del Nuovo Cimento, Series 3, Vol. 3, No. 4, Bologna, 1980.

Warner, F. W., *Foundations of Differentiable Manifolds and Lie Groups*, Scott, Foresman and Company, Glenview, Illinois, 1971.

Westenholz, C. von, *Differential Forms in Mathematical Physics*, North-Holland, Amsterdam, 1981.

Zorich, V. A., *Mathematical Analysis* **II**, Springer-Verlag, Berlin, 2004.

INDEX OF SYMBOLS

$\mathring{A}$ 54
$\bar{A}$ 54
$\mathcal{A}^k(U)$ 336
$\mathcal{A}(U)$ 338
$\mathcal{I}_g$ 194
Ad_g 195
Ad 195
$Aff(n, \mathbb{R})$ 205
$B_r(p)$ 64
$B_r[p]$ 65
$B_k(M)$ 466
$\mathbb{C}$ 2
$C^r(M)$ 94
$C^\infty(M)$ 94
$C(M)$ 94
$C(r)$ 95
$C[r]$ 95
$\mathcal{C}^k(M)$ 267
$\mathcal{C}(M)$ 268
c_k 421
$C_k(M)$ 421
$C_k^*(M)$ 423
$\mathring{C}_k(M)$ 466
∂A 55
d 260
δ 281
Δ 283
∂ 421
$\mathfrak{d}$ 423
δ_j^i 12
$\delta_{j_1 j_2 \cdots j_k}^{i_1 i_2 \cdots i_k}$ 27
D_r 308
∇_j 367, 370, 371, 378
$\boldsymbol{\nabla}$ 368, 370, 371, 378
∇_U 373. 374
$D_i^{(k)}$ 506
δ_V 660
$e_{i_1 i_2 \cdots i_n}, e^{i_1 i_2 \cdots i_n}$ 31
e^{tV} 139
$\mathcal{E}^k(M)$ 267
$\mathcal{E}(M)$ 268
$\epsilon_{i_1 \cdots i_m}$ 276
$\epsilon^{i_1 \cdots i_m}$ 277
$\epsilon(\omega)$ 342
$\mathbb{F}$ 2
$\mathcal{F}(X, \mathbb{F})$ 4
ϕ^* 98, 253
$\phi_*, d\phi$ 120
$\phi_t(p)$ 135
ϕ_V 536
$gl(n, \mathbb{R})$ 78
$GL(n, \mathbb{R})$ 79
$(G, *)$ 176
$\mathfrak{g}$ 183
G_p 197
g_{ij} 269
Γ_{jk}^i 367
γ_{ki}^j 378, 389
γ_{ijk} 389
Γ_{ljk} 389
$H^n(V^\bullet)$ 17
$\mathbb{H}^n$ 90
H 331
H^n 402
$H^k(M)$ 460
$H(M)$ 461
$H_k(M)$ 466
$\mathcal{H}_k(M)$ 467
$\mathrm{Im}(A)$ 10
ι 176
$\mathbf{i}$ 226
$\mathbf{i}_V$ 226
$\mathrm{Ker}(A)$ 10
$\mathbb{K}^2$ 84
$K(\omega)$ 342
$\pounds_V$ 144
L_g 180
$\Lambda^k(U)$ 35
$\Lambda(U)$ 35
$\Lambda^k(M)$ 221
$\Lambda(M)$ 222
$(M, \mathfrak{M})$ 53
$\mathfrak{M}_R$ 61
$\mu_{i_k i_{k-1} \cdots i_1}$ 238
$n(A)$ 10
$\mathcal{N}(A)$ 10
$O(n)$ 177
$\mathcal{O}_{p_0}$ 198
$\mathbb{R}$ 2
$r(A)$ 10
$\mathcal{R}(A)$ 10
$\mathbb{RP}^n$ 87
R_g 181
R_{jkl}^i 375, 381
supp (f) 62
$\mathbb{S}^n$ 80
σ 176
$\bar{\sigma}$ 176
$SL(n, \mathbb{R})$ 177
$SO(n)$ 177
$SU(n)$ 203
$Sp(n, \mathbb{R})$ 201
$\mathcal{S}(\mathcal{I})$ 247
s_k 416
s_{k-1}^i 417
σ_k 420
$\mathbb{T}^2$ 82
$\mathbb{T}^n$ 84
$T_p(M)$ 117
$T(M)$ 127
$T_p^*(M)$ 208
$T^*(M)$ 209
$\mathrm{T}_p(M)_l^k$ 210
$\mathfrak{T}_p(M)_l^k$ 210
$\mathfrak{T}(M)_l^k$ 211
T_{ij}^k 375, 380
$U \otimes V,\ u \otimes v$ 20
(U, φ) 71

$U(n)$	202
$\mathfrak{V}(M)$	127
$\oplus$	6
$\otimes$	20
$\langle \cdot, \cdot \rangle$	13
$\wedge$	30
$\Vert \cdot \Vert$	66
$(\cdot, \cdot)$	68, 477
$[\cdot, \cdot]$	146
$\sqcup$	461
$\{ \cdot, \cdot \}$	705, 707

NAME INDEX

Abel, N. H. **2**
Ampère, A. M. 577

Bäcklund, A. V. **510**
Beltrami, E. 284
Betti,E. 461
Bianchi, L. 387
Borel, F. É. J. É. 70
Boy, W. 89
Bryant, R. L. 89
Burgers, J. M. 656

Carathéodory, C. **350**
Cartan, É. J. 219
Cauchy, A.-L. 65
Cayley, A. 204
Christoffel, E. B. 368
Clifford, W. K. 51
Coulomb, Ch. A. de 730

Dantzig, D. van **143**
Darboux, J. G. 342

Ehresmann, Ch. **490**
Einstein, A. 13, 208
Euler, L. 461

Faraday, M. **732**
Fokker, A. D. 656
Frobenius, F. G. 156

Galilei, G. **563**
Gauss, J. C. F. 445
Gegenbauer, L. B. 452
Gordon,W. 597
Gram, J. P. 273
Grassmann, H. G. 1, 219
Green, G. 445

Hamel, G. K. W. 8
Hamilton, W. R. 204, 495, 725
Hausdorff, F. 55
Heine, H. E. 70
Helmholtz, H. L. F. von 748
Hermite, Ch. 203
Hodge,W. V. D. 243

Jacobi, C. G. J. **102, 495, 725**

Killing, W. K. **199**
Klein, F. Ch. 84
Korteweg, D. J. 546
Koszul, J.-L. 376
Kronecker, L. 12
Kusner, R. B. 89

Lagrange, J. L. **658**
Laplace, P.-S. 273
Landau, E. G. H. 113
Legendre, A.-M. 600
Leibniz, G. W. von 116
Levi-Civita, T. 31, 208
Lie, M. S. 143,175
Liouville, J. 479
Listing, J. B. 131
Lorentz, H. A. 402
Lorenz, L. V. 742

Maclaurin, C. **113**
Maurer, L. 319
Maxwell, J. C. 731, 732
Mayer, W. 470
Monge, G. 577
Möbius, F. 131

Navier, C. L. H. **656**
Newton, I. 730
Nijenhuis, A. 218
Noether, E. A. 670

Ostrogradski, M. V. **445**
Ovsiannikov, L. V. 604

Pauli, W. E. **203**
Pfaff, J. F. 308
Planck, M. K. E. L. 656
Poincaré, H. 219
Poisson, S. D. 324

Reynolds, O. **656**
Rham, G. de 283
Ricci-Curbastro, G. 208
Riemann, G. F. B. 51
Rodrigues, B. O. 202

Schmidt, E. **273**
Schouten, J. A. 49
Schur, I. 457
Schwarz, K. H. A. 69
Schwarzschild, K. 392
Ślebodziński, W. 143
Stokes, G. G. 429

Taylor, B. **137**
Thomson, W. (Lord Kelvin) 429
Toda, M. 748

Vietoris, L. **470**
Vries, G. de 546

Weyl, H. K. H. **51, 365**
Whitehead, J. H. C. 329
Whitney, H. 51, 75, 123
Wronski, J.-M. H. 47

Zorn, M. **6**

SUBJECT INDEX

Absolute temperature 748
Accessibility 350
Action functional 660, 663, 667, 671, 677, 681, 686, 689, 699
Action integral 658-670, 688
Adjoint operator 478
Affine space 78, 185
Algebraic basis 8-10
Algebraic dual (*see* dual space)
Alternating k-linear functional 24, 28, 213, 214, 220
Alternating multilinear functional 24
Alternation 26, 27, 387
Ampère's law 732
Analytical manifold 75, 80, 82, 87, 89, 123
Antiderivation 227
Antiexact form 336-340, 359, 360
Arc 63, 274, 284, 388, 441, 446, 699
Arc-connected 63
Atlas, 72, 74, 77, 80, 84, 89, 94, 99, 105, 127, 129, 169, 262, 328, 366, 407, 427, 437, 463, 488
 compatible 74, 94
 equivalent 74, 75
Attitude matrix 358

Bäcklund transformation 537
Balance forms 524, 530, 579, 597, 664-666, 668
Balance ideal 524, 529, 534, 536, 565, 594-597, 605, 612, 616, 622
 isovector fields 521, 524, 536, 594, 606, 612, 634
Barycentric coordinates 417
Bases induced by the volume form 237, 242, 257, 265, 269, 278-281, 299, 442
Benjamin-Bona-Mahony equation 694
Betti number 461, 465, 482, 483,
Bianchi identities, 387
 contracted 392
Bilinear form, 13, 48, 49, 468, 476, 482, 483
Bilinear functional (*see* bilineer form)
Black hole 396
Boundary layer 552, 568
Boundary of a set 55
Boundary operator 421, 422, 429, 466, 467

Calculus of variations **658, 663**
Canonical coordinates 191, 702, 710, 714, 716, 721
Canonical projection 62, 87, 569
Canonical symplectic form 702
Canonical system 585, 588, 591, 595, 597, 598, 601
Canonical transformation 710, 711, 716, 720-724, 749
Carathéodory's theorem 350
Cartan annihilators 585, 589, 592
Cartan connection 377-380
 moving frame 365, 377, 384, 389
Cartan form 661, 662
Cartan's lemma 226
Cartan magic formula 293, 294, 332, 512, 571, 662
Cartan's theorem 307, 309, 491
Cartesian product 4, 19, 29, 48, 59, 66, 77, 130, 169, 176, 178, 209, 415, 419, 484
Cauchy-Bunyakowski-Schwarz inequality 68, 69, 476
Cauchy sequence 65, 68
Central motion 728
Chain complex 423, 466
Chain rule 118, 124, 144, 211, 255, 366, 367, 370, 523, 628, 747
Chains 416, 418, 421, 423, 427, 429, 433, 434, 466
Characteristic curve 493, 496
Characteristics 164, 310, 448, 496, 551, 565, 569, 598, 603, 726
Characteristic strip 493
Characteristic submanifold 312
Characteristic subspace 247-250, 253, 305-309, 312-315, 499, 588
Characteristic vector field,
 of an exterior form 246, 247, 252
 of an ideal 247, 249, 252, 305, 308, 311, 492, 496, 499, 538, 574, 588, 592
Chart, 71-75, 82-65, 68-71, 82-89, 101-103, 253, 321, 407, 435, 474
 coordinates 71, 73, 98, 102, 104

coordinate functions 71, 78, 99, 110, 116, 125, 133, 321
coordinate cover 66, 71, 119, 454, 490, 511, 604, 616, 647, 665, 680, 701, 717
local coordinates 71, 98, 116, 136, 159, 183, 209, 253, 290, 343, 407, 423, 488, 569, 657, 708, 720
Christoffel symbols,
first kind 389
second kind 368, 377, 389, 391, 700
Closed ball 65-67, 70, 72
Closed form 267, 269, 298, 334, 343, 347, 349, 434, 438, 460-465, 473, 481, 706, 707
Closed ideal 286, 288, 302, 305, 307, 312, 314, 491, 500, 510, 571, 573, 587, 590, 593, 595, 598
Closed set 53-55, 60, 62, 65, 79, 179, 463
Closure of a set 54, 55, 57, 72
Coboundary operator 423, 467, 471
Cochain complex 14, 269, 423, 460, 467, 470, 471
Cochain homomorphism 14, 15, 17
Cocycle 423, 467
Co-differential 282, 286, 398, 399, 477, 480
Cohomology class 17-19, 460, 467, 481-483
Cohomology group 17, 19, 460, 465-467, 469, 470, 481, 482, 484
Commutator 146, 149, 157, 160, 187, 381, 383, 396
Compact manifold 135, 466, 476, 478, 479
Compactness 56, 57, 72
Compact topological space 56, 70
Complete ideal 306, 308
Complete vector field 135, 171
Configuration manifold 697-701, 727, 741
Conformal operator 477
Congruence 139, 142, 145-148, 151, 155, 158, 190, 290, 592
Conjugate linear 68
Connected manifold 78, 191, 203, 407, 461, 465, 466
Connected spaces 63, 64. 71, 179, 571
Connection 1-forms 384
Conservation laws 448, 451- 453, 458-459, 510, 669,, 673, 677-679
Conservation of energy 459, 679, 709, 720
Conservative mechanics 696, 724
Constraints, 697-698, 716
anholonomic 697
holonomic 697, 698
rheonomic 697, 716
scleronomic 697-699
Contact forms 490, 507, 523, 524, 581, 605, 658, 666
Contact ideal, 491, 500, 506, 521, 524, 528, 533, 538, 553, 564, 577, 579, 581, 584, 589, 593, 606, 611, 616, 619, 621, 658, 661, 664, 671-673, 680-682, 689-691
isovector fields 500, 506, 510, 606, 611, 612, 621, 681
Contact manifold 490, 492, 500, 520, 559, 569, 586, 658, 665, 680, 681
Continuous mappings 51-64, 70-72, 76, 79, 97-100, 102, 127, 135, 335, 604, 663
Contractible manifold 328, 334, 461
Contraction 24, 212, 237, 275, 374, 391, 498, 503, 505, 627, 630, 631, 641, 648
Contravariant vectors 118, 209, 210, 270, 366, 444, 739
Convex set 8, 9, 48, 328, 329, 416, 472, 474
Coordinate basis 119
Cotangent bundle 119, 209, 250, 366, 370, 388, 701-703
Coulomb's law 730, 732
Covariant derivative 279, 366, 368-372, 374, 375, 377-380, 385, 388, 391, 397, 399
Covariant vector 209, 368, 369, 374, 701
Covector 209, 214
Current form 673
Curl 264, 265, 323, 446, 737, 741
Curvature operator 376
Curvature 2-forms 384
Curvature tensor 375, 377, 380, 381, 387, 390, 394, 396, 398, 403
Cycle 422, 423, 434, 466-468, 475, 482

Darboux class 342, 344, 345, 347-350, 702 744, 745
Darboux' theorems 344, 349
Dense set 55, 56, 109
de Rham algebra 461, 462
de Rham cohomology group 460, 466, 473
de Rham homomorphism 469, 472
de Rham manifold 472-475
de Rham's theorem 469, 474, 475
Derivation 116, 152
Diffeomorphic manifolds 98, 101
Diffeomorphisms 98, 101, 122, 124, 126, 133, 138, 142, 158, 178-181, 187, 191, 194, 255-258, 317, 355, 453, 462, 506, 510, 536, 557, 644, 711-715
Differentiable curve 110, 133, 414
Differentiable manifolds 51, 70, 74, 77, 90,

93, 95, 97, 100, 102, 104, 108, 110, 112, 118, 120, 123, 127, 129, 131, 133, 142, 153, 167, 253, 327, 334, 349, 369, 406, 415, 423, 429, 435, 438, 459, 698, 702
Differentiable mapping,71, 93, 98, 110, 120, 123-125, 131, 152, 197, 266, 274
its rank 102-110, 200, 422, 573
Differentiable singular homology group 466
Differentiable singular simplex 420-422, 424, 430, 472
Differentiable structure 74, 77, 90, 93,101, 105, 127, 209, 211, 460
Differential ideal (*see* closed ideal)
Differential of a mapping 120, 124, 126, 129, 131, 133, 150, 152, 155, 163, 181, 188, 191, 195, 199, 253, 256, 259, 422
Diffusion 449, 538, 665
Direct sum 6, 35, 36, 215, 222, 339, 461
Disconnected spaces 63, 83, 86, 177, 461
Disjoint sets 5, 55, 57, 61, 63, 112, 127, 143, 153, 177, 460, 467, 472, 474
Disjunct subspaces 6
Distributions, 153, 155, 163, 164, 166, 259, 589, 596
completely integrable 156
integral submanifold 156
involutive 154, 156, 158-161, 163, 166-168, 192, 259, 297, 307, 309, 312, 589
Divergence 265, 282, 447, 688, 731, 740
Divergence formula 445
Divergence theorem 459, 523
Duality pairing 13
Dual space 28, 29, 32, 125, 208, 272, 317-319, 389, 423, 467, 468, 472, 475

Eiconal equation 494
Einstein tensor 392
Elasticities 454, 558
Electromagnetic fields, 731, 737, 742
scalar potential 742
vector potential 742
Embedding 105, 108, 123, 155, 156, 329, 463, 481
Emden-Fowler equation 761
Energy-momentum tensor 392, 393, 668
Entropy 747
Equivalence class 5, 17-19, 61, 74, 87, 112, 114, 117, 120, 406, 411, 460, 467-468, 473, 475, 481-483
Equivalence group 604, 612, 628, 644
Equivalence relation 5, 61, 74, 87, 112, 114, 406, 460, 466
Equivalence transformations 604, 606, 616, 628, 634, 639, 644-646, 650, 655
Euclidean space 64, 66, 70-72, 78, 112, 123 329, 340, 369, 414, 416, 419, 424, 489
Euler-Lagrange equations 663, 664, 668, 669, 673, 675, 677, 687-691, 699
Euler-Lagrange forms 663-666
Euler-Poincaré characteristic 461, 465, 483
Exact forms 267, 269, 298, 334, 434, 460, 462, 464, 465, 481
Exact sequences 14-18, 470, 471
Exponential mapping 139, 147, 190, 682
Exponential operator 137, 138, 151
Exterior algebra 1, 33, 35, 222, 226, 236, 244, 247, 255, 260, 268, 278, 280, 286, 288, 290, 293, 298, 301, 304, 306, 312-317, 331, 339, 446, 460, 477, 507, 524, 570, 587, 605, 658
Exterior derivative 261-267, 269, 278, 281, 285-287, 289, 293, 299, 304, 309, 319, 339, 352, 354-358, 360, 386, 398, 429, 436, 443, 446, 477, 486, 489, 511, 573, 605, 702, 707
Exterior differential equations230, 352, 355, 360-362, 387
Exterior differential form 71, 214, 220-222, 398, 429, 489, 500, 604
Exterior equations 258, 306-308, 313, 315, 350, 352, 734, 736
Exterior form 29, 33, 37, 46, 214
Exterior normal 435, 444, 445, 459
Exterior product 29, 33-36, 43-45, 213-215, 222, 224, 226
Exterior system, 248, 252, 306-309, 311
completely integrable 308, 309, 311, 312, 314, 350, 590, 596, 745, 748
Externally oriented submanifold 411, 413

Faraday's induction law 732
Fibre bundle 127, 129, 507
base 127, 129, 130
fibres 127, 129, 130, 587
First countable space 54, 65
Five lemma 15, 474
Flow 133, 135, 139, 141-143, 147, 190, 290, 297, 301, 315, 397, 439, 451, 518, 563, 568-570, 659, 672, 681, 708-710, 719
Fokker-Planck equation 656
Foliation 160, 590, 593, 594, 596, 598
Forms on a Lie group, 317
Maurer-Cartan forms 319
right-invariant forms 318

left-invariant forms 317-319, 326
Frobenius' theorem,
for distributions 156, 160, 192, 259
for exterior forms 309, 312, 362
Fundamental ideal 492, 521, 524, 538, 563, 565, 572-575
Fundamental system of neighbourhoods 54,
countable 65

Gauss' law 732
General linear group 79, 129, 177, 192, 196
Generalised characteristics 569, 575
Generalised coordinates 496, 697, 698, 702, 709, 717, 722, 724
Generalised Kronecker delta 27, 28, 45, 50, 213, 214, 217, 236
Generalised momenta 496, 700, 701, 709, 716, 717, 722, 724, 726, 728
Generalised velocities 698
Generating function 452, 721, 724
Generators 190, 245, 250, 301
Geodesic 373, 374, 401, 700
Global cross section 130
Gordon equation 597, 600
Graded algebra 35, 223, 335, 338, 339, 477
Gradient 343, 345, 347, 355, 366, 369, 453, 557, 647, 698
Gram-Schmidt orthormalisation 273
Graph space 489, 493, 505, 509, 510, 536, 537, 550, 551, 584, 670
Grassmann algebra 1, 215
Green-Gauss-Ostrogradski formula 445
Group, 2, 61, 130, 138, 176-181, 195, 198
Abelian group 3, 17, 195, 197, 320, 460, 466
Group-invariant solution 563, 575
Group of diffeomorphisms 139, 506, 507, 536, 644
Group of local diffeomorphisms 139

Hamiltonian function 496, 701, 705, 709, 715-724, 726-728, 748, 749
Hamiltonian vector fields 704, 707-710, 712, 713, 716, 717, 720, 748, 749
Hamilton-Cayley's theorem 204, 363
Hamilton-Jacobi equation 495, 725-727
Hamilton principal function 726
Harmonic forms 284, 476, 478, 479, 481
Harmonic oscillator 727
Hausdorff space 55-57, 60, 65, 67, 71, 168
Heat conduction 449, 542, 545,565, 656
Heine-Borel's theorem 70
Helmholtz free energy function 747
Hodge decomposition 479-481
Hodge dual 243, 269, 227, 279, 285, 476, 483
Hodge star operator 280, 477
Homeomorphic spaces 61, 71, 78-80, 82, 90, 101, 123, 328-330, 334, 340, 342, 349, 356, 463
Homeomorphism 61, 71-73, 75, 77, 80, 82, 86, 89, 91, 98, 101, 105, 108-110, 114, 128, 168, 183, 257, 328, 422, 436, 437
Homologous cycles 466
Homology class 466
Homotopy operator 267, 331, 333-336, 339, 340, 352, 353, 358, 459, 736
change of centre 340
Horizontal ideal 588-590, 594-596, 599, 601
isovector fields 590, 592, 595
Horizontal module 588, 589
Horizontal one-form 588
Hyperelasticity 453, 557, 604, 646, 677, 679
Hyperspherical coordinates 412, 462, 464

Ideals,
of exterior algebra 244-247, 250, 252, 286
its closure 286, 288, 302, 314, 491, 500, 606
Identity mapping 78, 100, 124, 130, 155, 181, 194, 197, 332, 712
Immersed manifold 109, 110
Immersion 104, 105, 107, 109, 110, 122
Inaccessibility 359, 745
Infinitesimal generators 139, 191, 522, 529
Initial data submanifold 571-575, 579
Integral of the motion 709, 720, 724, 748-750
Integration of exterior forms 414, 419, 423, 439
Interior normal 435
Interior of a set 54, 91
Interior point 54, 55, 57, 59, 92, 334, 387, 663, 675, 676
Interior product 226, 229, 232, 234, 238, 242, 247, 249, 251, 256, 296, 324, 586, 706
Internal energy function 744, 746
Invariant,
form field 298
function 141
vector field 151
Isentropic gas flow 451, 510
Isometry 204, 403

Isomorphic spaces 6, 10, 49, 78, 114, 117, 125, 180, 183-185, 190, 320, 467, 469, 472, 475, 481, 483
Isomorphism 6, 7, 10, 12, 15-18, 48, 78, 114, 119, 122, 124, 126, 129, 180, 183, 185, 187-189, 191, 195, 200, 255, 271, 462, 467-469, 472-475, 481, 483, 704, 707, 721
Isothermal work function 743, 745
Isovector fields, 301, 302, 304, 305, 315, 316, 500, 589
 vertical 659-661, 663, 664, 682

Jacobi identity 148, 151, 154, 320, 324, 403, 706, 708
 generalised 218
Jacobi's theorem 712
Jacobian 166, 257, 277, 310, 407, 433
 matrix 102-105, 121, 122, 410
Jet bundle 490

Killing equations 397
Killing vector field 199, 200, 205, 397, 398
Klein bottle 84, 85, 87, 88, 484
k-linear functional 19-22, 24, 25, 77
Korteweg-de Vries equation 546, 550, 566, 567, 579, 650
Kronecker delta 12, 27, 498

Lagrange equations 698
Lagrange multiplier 680, 681, 689
Lagrangian function 698, 700, 716, 726-728
Laplace equation 479, 655, 669, 670, 693
Laplace-Beltrami operator 284
Laplace-de Rham operator 283, 398, 400, 401, 478
Leaf 160, 590, 593, 594, 596, 597, 599
Legendre elliptic integral 600
Legendre transformation 701
Leibniz rule 116, 291, 300, 369-371, 379, 382, 383
Levi-Civita connections 389, 392, 396, 398
Levi-Civita symbols 31, 236, 239, 243, 277
Levi-Civita tensors 277, 278
Lie algebra149, 155, 180, 183-188, 191, 194 -197, 199, 297, 301, 305, 314, 318-320, 506, 510, 536, 544, 592, 671, 706, 708
Lie bracket 146, 147, 149, 152, 154, 183
Lie derivative,
 of form fields 290-293, 295, 298, 303, 307, 332, 397, 501, 521, 530, 610, 614, 662, 683, 709, 719
 of tensor fields 215, 300, 397
 of vector fields 143-146, 148, 151, 155
Lie group, 176, 178-187, 189, 191, 194-199, 201, 317-320, 506, 510, 518, 539, 604
 adjoint representation 196
 exponential mapping 190, 191
 homomorphism 187-189, 196, 318
 inner automorphism 194
 isomorphism 187-191, 192, 195, 196
 left-invariant vector field 181-1183, 185-190, 195, 196
 left translation 180, 185, 186, 188, 317
 one-parameter subgroup 189, 190, 192, 196, 200
 right-invariant vector field 185, 190, 199
 right translation 180, 181, 185, 317
 structure constants 184-186, 189, 192, 297, 302, 319, 320
 structure tensor 184
 subgroup, 129, 178, 179, 188, 192, 194, 197, 198, 200, 507
  normal 179, 198
 transformation group, 197, 302, 500, 519
  acting effectively 197, 200, 205
  acting freely 197, 198, 205
  acting transitively 197, 198, 205
  isotropy group 198, 206
  Killing vector field 200, 206
  orbit 198, 206
Lie product 146, 148, 154, 158, 187, 195, 199, 302, 305, 308, 377, 587, 592, 706-708, 713
Linear combination 7-9, 25, 31, 37, 39, 126, 224, 226, 233, 240, 317, 347, 348, 417, 420, 424
Linear (affine) connection 367-369, 374, 377, 389
Linear (affine) connection coefficients 367-369, 370, 373, 378, 381, 387-389, 398
Linearly dependent 7, 8, 25, 69, 226, 345
Linear functional 10-12, 23, 28, 125, 208, 214, 423, 467, 468, 475, 483
Linear hull 7, 8
Linearly independent 7-10, 12, 23, 31-33, 35-39,41, 43, 78, 118, 122, 153-158, 160, 163, 167, 184, 186, 221, 223-226, 231-235, 238, 240, 242, 246-250, 252, 259, 273, 288, 297, 302, 307-309, 311, 317, 343, 377, 387, 393, 409, 411, 416, 458, 460, 490, 502, 510, 517, 519, 527, 536, 544, 551, 557, 573, 584, 586
Linear operator, 9-11, 14, 16-18, 39, 76, 115-

117, 120, 122, 124, 129, 131, 143, 152, 163, 167, 181, 188, 191, 195, 199, 255, 260, 280, 331, 336, 414, 421, 477, 481, 483, 570, 703, 736
nullity 10
null space 9
range 10
rank 10
regular 9
Linear vector spaces, 2-14, 16, 19-25, 30-32, 35-37, 42, 46, 66-68, 70, 78, 94, 112-115, 117, 124-127, 129, 132, 143, 149, 155, 182-187, 190, 195, 200, 208-210, 217, 221, 267, 269, 272, 296, 301, 317, 418, 421, 423, 460, 466-469, 475-477, 480, 483, 498, 672
basis (set, vectors) 8-14, 21-25, 28, 32, 35-39, 41
change of basis 13, 23
dimension 8-10
dual space (algebraic dual) 8
Hamel basis 8, 9
quotient space 5, 6, 15-17, 48, 460, 466
subspaces, 4-7, 9
complementary 6, 7, 10
direct sum 6
intersection 6
sum 6
Liouville form 702
Liouville's theorems 479, 710
Local cross section 130
Locally compact space 57, 70-72, 95, 469, 470, 474
Locally contractible manifold 328, 329, 349
Lorentz metric 402, 739
Lorenz conditions 742

Maclaurin series 25, 135, 150, 440, 660, 682
Manifold 51, 70
Manifold with boundary, 90, 93, 429, 435, 478
its boundary 90-93
its interior 90-92
Matrix, 11-13, 37, 39, 78, 103, 122, 129, 177, 196
antisymmetric 47
Hermitean 203
Jacobian 102
orthogonal 177
polar decomposition 203
rank 37, 39, 102, 122, 157, 225, 232, 439
symmetric 47
symplectic 702
unitary 202
Maurer-Cartan forms 319
Maxwell equations 731, 733, 736, 738-740
Maxwell reciprocity relations 744
Mayer-Vietoris sequence 469-471, 473
Mayer-Vietoris' theorem 469
Metric 64
Metric space, 64-66, 70, 71
complete 65, 68, 70, 75
Metric tensor 269-271, 273-276, 280, 285, 389, 392, 397, 399, 441, 476, 699, 739
Metric topology 64, 67, 70
Möbius band 130-132, 409, 413, 416
Module, 2, 19, 127, 149, 212, 214, 221, 223, 228, 238, 258, 267, 278, 288, 301, 336, 373, 406, 460, 471, 507, 585, 592, 618, 703, 705, 708
horizontal 588, 589
Momentum phase space 702
Monge-Ampère equation 577
Multilinear functional 19, 20, 23, 24, 210

Natural basis 119, 126, 146, 159, 208, 221, 223, 236, 243, 280, 372, 377, 734
Navier equations 693
Navier-Stokes equations 552, 656
Neighbourhood,
open 54, 57, 59, 62, 72, 90-92, 108, 134, 178, 183, 191, 334, 420, 430
compact 57, 72
Nilpotent operator 261
Noether's theorem 670, 673, 676, 689
Noetherian vector fields 671, 673-675
first kind 671
second kind 671
Non-conservative mechanics 716, 749
Non-linear wave equation 580, 642
Norm 66-70, 75, 76, 78, 123, 272, 273, 476

Open ball 64, 65, 67, 70, 72, 328, 329
Open cover, 56-60, 62, 71, 95-97, 408, 470
locally finite, 57, 59, 62, 408
Open mapping 61
Open set 56-63, 65, 70, 72, 74, 80, 82, 84, 86, 90, 93, 95, 97, 104-107, 110, 114, 118, 127, 130, 134, 178, 183, 261, 328-330, 334, 336, 340, 342, 349, 352, 355, 361, 387, 408, 410, 425, 427, 463, 465, 472, 474, 488, 523, 572
Open submanifold 77-79, 107, 130, 135,

179, 186
Orbit 133, 198
Orientable manifold 275, 406, 409, 484, 486
Orientation, 406, 407, 409, 410, 415, 417, 421, 695, 711
negative 407
positive 407, 417, 435, 437, 444, 463
Oriented manifold 406-408, 411, 418, 466, 469, 474, 479-484
Orthogonal group 177, 194
Orthogonal vectors 272
Orthonormal basis 264, 272, 273, 389
Orthonormal vectors 272, 273

Paracompact spaces 57, 59, 65, 70, 407, 408, 428, 438, 470
Parallel vector field 374
Partition 5, 61, 74, 112, 406, 414, 460
Partition of unity 62, 63, 95, 97, 408, 427-429, 437, 470, 471
Path 63, 64, 350-352, 745
Path-connected 63, 64
Pfaff system 308
Phase space 702, 705, 710, 711, 749
Poincaré duality 483, 484
Poincaré form 482
Poincaré lemma 320, 334, 339, 341, 349, 353, 463, 465, 472, 665, 714, 721
Poisson bracket,
of functions 324, 707-709, 713, 750
of one-forms 705-707, 714
Poisson's theorem 750
Potential function 698, 716
Product manifold 77, 82, 84, 129, 169, 176, 197, 205, 489
Projective space 87, 101, 410, 411
Prolongations 507, 509, 510, 542, 544, 550, 583, 584
Pseudo-Riemannian manifold 272, 402
Pseudotensor 277
Pull-back operator 98, 100, 254, 256, 306, 318, 424, 430, 461, 470, 570, 572, 715, 721

Quadratic form 41, 42, 44, 45
Quotient rule 212, 275, 375
Quotient set 5, 61, 112, 460
Quotient space 5, 15-17, 48, 62, 87, 460, 466

Rank of an exterior form 27-46, 246, 342, 345, 347, 702, 713, 717
Reciprocal basis 13, 21, 24, 28, 37, 125, 208, 224, 234, 248, 250, 254, 271, 273, 318, 320, 377, 389
Recursive form 354, 355
Refinement, 57
locally finite 57, 59
Regular mapping 484, 491, 520, 524, 536, 564, 658-661, 663, 667, 671, 673, 675, 680-683, 690
Relatively compact 57, 58, 96
Resolvent mapping 259, 315, 325
Reynolds number 656
Ricci tensor 391, 394
Riemann integral 414, 415, 424
Riemannian connection 389
Riemannian manifold, 243 269, 272-276, 279, 323, 325, 388, 392, 398, 441, 476, 479-484
complete 272, 275, 402, 441, 476, 699
incomplete 272, 739
Riemann-Christoffel tensor 390

Scalar 2
Scalar (inner) product 68-70, 202,264, 272, 409, 411, 476-478, 699
Schouten's theorem 49
Schouten-Nijenhuis bracket 218
Schwarzschild metric 392, 396
Schwarzschild radius 396
Second countable space 56, 59, 71, 75, 95, 123, 469, 474
Self-adjoint operator 478
Sesquilinear 68
Shallow water theory 451, 694
Short exact sequence 15, 17, 470, 471
Similarity solution 563-567, 575, 655, 656
Simple form 36, 39, 44, 224, 232, 235, 345, 347, 405
Simplex 417-419, 20
its oriented boundary 417-419, 421
Simply connected 84, 185, 191, 484, 486
Singular chains 421-424, 466
Singular cohomology class 467
Singular cohomology group 467
Singular homology group 466, 467
Smooth functions 102, 113, 126, 134, 138, 152, 156, 158, 164, 182, 221, 245, 253, 301, 321, 489, 507, 510, 522, 526, 530, 534, 571, 583, 585, 589, 591, 594, 606, 618, 708, 737
Smooth manifold 75, 78, 123, 133-135, 176, 178, 183, 185, 197-200, 208, 211, 220,

255, 260, 266, 269, 300, 317, 336, 342, 366, 374, 420, 422, 425, 427, 435, 461, 469, 472, 474, 488-490, 657
Smooth mapping 99, 102-104, 122, 130, 133, 135, 176, 180, 253, 255, 258, 266, 300, 335, 414, 425, 438, 461, 464, 469, 489
Special linear group 177
Special orthogonal group 177
Specific heat 451, 748
Sphere 79-82, 101, 108, 161, 168, 329, 409, 462, 465, 466
Spherical coordinates 79, 284, 409
Stable submodule 298, 301
Standard simplex 419-422, 424-426, 430, 431, 433
Star-shaped region 328, 331, 334, 340, 349
State variables 742, 745
Stereographic projection 81, 170
Stokes formula 446
Stokes' theorem, 429, 433, 435, 441, 443, 445-447, 462, 468, 475, 478, 482, 662, 676, 683, 688
on chains 429
on manifolds with boundary 435
Stress potential 562, 604, 646, 650, 677
Structural group 129-131
Submanifold 105-110, 130, 135, 155, 160, 167, 177-179, 186, 192, 258, 260, 306-308, 313, 350, 409, 411, 414, 422-425, 427, 438, 441-443, 446-449, 451, 455, 462, 490, 493, 596, 571, 579, 587, 592, 594, 596, 601, 648, 697, 717, 743, 745
Submersion 103, 104, 106, 107, 122
Support 62, 97, 408, 415, 425, 428, 391
Symmetry group 536, 550, 556, 563,679
Symplectic form 46, 702-704, 709, 715, 749
Symplectic manifold 702, 703, 705, 709, 710, 713-715, 717
Symplectic mapping (canonical mapping) 711-714
Symplectic matrix 201, 702

Tangent bundle, 127-130, 132, 156, 163, 209, 215, 232, 246, 250, 259, 272, 372, 377, 407, 409, 411, 413, 435, 439, 494, 507, 509, 536, 550, 585, 596, 619, 698, 702
Tangent space 112, 114, 117-119, 122, 125-127, 143, 145, 150, 153, 155, 181, 184-186, 190, 208, 213, 259, 270, 318, 369, 407, 698
Tangent vectors 112-116, 118, 124, 189-191, 200, 274, 373, 446
Taylor series 137, 144, 152, 291
Temperature 538, 742, 745, 746, 748
Tensor, 21, 210
contravariant 23, 214, 279, 740
covariant 22, 24, 29, 213, 220, 269, 388, 442
density 277
mixed 23, 24, 210, 215, 275
Tensor bundle 211, 213-215
Tensor product 20, 21, 23, 24, 29, 49, 210, 212, 215, 300, 317, 371, 372, 382
Theory of Hodge-de Rham 406, 476
Thermodynamics 696, 742, 745
first law 744
Thermodynamic temperature 746
Toda lattice 748, 749
Topological manifold 71, 72, 74, 75, 90
Topological space, 53-57, 59-64, 70, 71, 87, 90, 93, 95, 128, 179, 466
separable 55, 56, 71, 75, 123
Topology, 53, 55-57, 59-65, 67, 77, 105, 127, 268, 429, 459-461
basis 41
product 42, 51
relative 53, 56, 57, 60, 63, 90, 91, 93, 105-107, 110
Torsion operator 376
Torsion tensor 375, 380, 385, 390, 396, 401
Torsion two-forms 384
Torus 82, 84, 105, 109, 178, 329
Trajectory 133, 705, 709, 718, 726, 731
Tranversality condition 573, 579

Unimodular group 177
Unitary operator 477

Variation of a functional 660, 664, 682
Variation operator 660, 682
Variational principle 664-666, 663, 693
Variations 685, 688
Variational symmetry 670, 674, 675, 677, 679, 689, 693, 694
Vectors 2, 6-10, 12-14, 17, 20-25, 28, 30-40, 66, 69, 71, 76-78, 81, 94, 112-120, 122, 142, 146, 149, 154, 157, 162-165, 184-186, 191, 209, 223, 229, 232, 248, 250-254, 270-274, 296, 302, 309, 368, 370, 376-378, 382, 389, 416, 439, 444, 592, 707, 721, 731, 739
Vector fields, 119, 126, 129, 132, 136, 139,

141-146, 149-151, 155, 181, 190-192, 196, 199, 215, 226, 229, 233, 247, 249, 259, 265, 270, 283, 290, 293, 295, 297, 300-304, 366, 368, 372, 374, 376, 380, 397, 410, 413, 439, 492, 495, 500, 509, 536, 565, 569, 569-571, 584, 589, 592, 606, 659, 703, 716, 718-721, 734, 737
integral curve 133-135, 138, 141, 155, 161, 189, 200
Velocity phase space 698
Vertical ideal 587, 589
Volume form 223, 236, 240, 257, 265, 275, 278, 280, 285, 299, 406-409, 411-414, 435, 441, 447, 449, 462, 465, 476, 520, 573, 658, 677, 710, 733

Wave equation 486, 644, 646, 656, 742
Whitehead manifold 329
Whitney's theorems 51, 75, 123
Work 696, 743, 744
Wronskian 47

Zigzag lemma 17
Zorn lemma 6, 8

Printed and bound by CPI Group (UK) Ltd, Croydon, CR0 4YY
08/05/2025
01864880-0003